INTRODUCTION TO ALKALOIDS

INTRODUCTION TO ALKALOIDS

A Biogenetic Approach

GEOFFREY A. CORDELL
University of Illinois

Library of Congress Cataloging in Publication Data:

Cordell, Geoffrey A   1946–
   Introduction to alkaloids.
   "A Wiley-Interscience publication."
   Includes bibliographical references and index.
   1. Alkaloids.   I. Title.
QD421.C67   547.7'2   80-39651
ISBN 0-471-03478-9

Printed in the United States of America

10 9 8 7 6 5 4 3 2 1

"It must not be assumed that atoms of every sort can be linked in every variety of combination"

Titus Lucretius Carus (100-55 B.C.)
On the Nature of the Universe

"What is now proved was once only imagined"

William Blake
The Marriage of Heaven and Hell

PREFACE

This monograph arose at the suggestion of the publisher after I had submitted a chapter on alkaloids for the Kirk-Othmer *Encyclopedia of Chemical Technology*. It was realized that there was no current book developing the history, chemistry, biosynthesis, and utility of this group of important compounds for the advanced graduate student or chemist unfamiliar with the area.

The critical series discussing the chemistry of alkaloids is that edited by the late R. H. F. Manske and simply called *The Alkaloids, Chemistry and Physiology*. At the time of writing, this series is at Volume 17. Two drawbacks of this series are that much of the history of alkaloids was presented in the early volumes and that for the beginning reader the coverage is very detailed.

A second series, entitled *The Alkaloids,* published by The Chemical Society of London as part of the Specialist Periodical Reports Series, covers most groups of alkaloids on a yearly basis. Again, however, the reader is generally assumed to have an exceptional background knowledge of alkaloids in order to understand the new developments.

I felt that a book was needed from which a person could become aware of the alkaloids as a group without needing to delve into the more sophisticated reviews, monographs, and series. This book then is primarily intended for the advanced graduate chemistry, biochemistry, or pharmacy student. Chemists unfamiliar with alkaloids may also find it useful, and alkaloid chemists should find the book interesting for the diverse alkaloid groups discussed.

No attempt has been made to cite *all* the references pertaining to a given group of compounds; rather key references are given at the end of each chapter or major section to important research papers, reviews, book chapters, and monographs.

My intent has been to cover the alkaloids from a number of perspectives, including their taxonomic distribution, structure elucidation, chemistry, synthesis, biosynthesis, and pharmacology. In a book of this size however numerous compromises were necessary, and clearly some of my peers will be upset that their favorite group of compounds may have been treated with undue brevity. For this I apologize.

The title of the book also merits some explanation. When I was an undergraduate student at the University of Manchester I was privileged to receive my first lectures on natural products from the then Professor of Organic Chemistry, Arthur J. Birch. These lectures were my first exposure to biosynthesis and biogenesis. Subsequently when I was a graduate student at Manchester, I heard a lecture by Birch on the brevianamides, a particularly interesting group of indole alkaloids. One phrase used by Birch in the lecture has stuck in my mind; paraphrased, it is: "At this time, *biogenetic reasoning* suggested the structure X, which was established by. . . . " It took some years for the significance of this concept to percolate. When contemplated carefully it is essentially that nature will not produce what it cannot produce. Since it is apparent that the molecules of key biosynthetic importance are well known, any proposal for a new structure should be biosynthetically reasonable.

This is not to say that structures are not deduced which defy initial biogenetic reasoning (e.g. securinine, Chapter 8). Nor does it rule out the possibility that initial biogenetic ideas may not be correct (e.g. monoterpenoid indole alkaloids); surprises continue at an alarming rate. Rather, it is to say that nature has few fundamental precursors; be aware of the structural limitations of these, and structure elucidation will be greatly simplified.

This book has its foundations in these ideas: namely that the hundreds of diverse alkaloid skeleta can be drawn together not on the basis of their structural similarity but rather on the basis of their biogenesis or biosynthesis. Such an approach has not been attempted previously.

There is one additional concept which may be gleaned from the comment of Birch: Can the reaction in a biosynthetic pathway be mimicked or preferably duplicated in the laboratory? For years the concept of synthesis for a complex natural produce centered on the skillful, often exquisitely delicate, manipulation of polyfunctionalized intermediates. In the past 10 years a revolution in synthesis has occurred, the concept of biomimetic synthesis. This is the elaboration of a supposed or proven biosynthetic intermediate to a final natural product, preferably under mild conditions. A classic area of these two approaches is in the morphinandienone series. Consider for example the early Gates synthesis of morphine and the biogenetically modeled synthesis of morphine developed more recently (Chapter 8).

In this monograph emphasis is placed on the biogenetically modeled synthesis of alkaloids, although the important nonbiogenetic approaches are also described.

In the period of publishing this book, three exceptional alkaloid chemists have died; R. B. Woodward, R. H. F. Manske, and H. Schmid. This book is dedicated to their scientific endeavors.

GEOFFREY A. CORDELL

University of Illinois
Chicago, Illinois
May 1981

CONTENTS

chapter 1

INTRODUCTION

The alkaloids comprise an array of structure types, biosynthetic pathways, and pharmacologic activities unmatched by any other group of natural products. How did this group of compounds reach such prominence, and why are they of such continuing interest? One aim of this book is to answer these questions by example.

This introductory chapter describes some important aspects of the history of alkaloids, an approach to their classification, some information concerning the occurrence of alkaloids, and examples of the techniques used to isolate, purify, and characterize alkaloids.

1.1 HISTORY

The history of alkaloids is almost as old as civilization. Humankind has used drugs containing alkaloids in potions, medicines, teas, poultices, and poisons for 4000 years. Yet no attempts were made to isolate any of the therapeutically active ingredients from the crude drugs until the early nineteenth century.

The Austrian apothecary Storck is generally attributed with the reintroduction of many plant drugs into medical practice. Because of their potent yet variable toxicities, drugs had fared poorly at a time when life and the sustainment of it was being carefully scrutinized. Some of the drugs reintroduced at this time are of classical importance and include aconite, colchicum, stramonuim, henbane, and belladonna.

The first crude drug to be investigated chemically was opium, the dried latex of the poppy *Papaver somniferum*. Opium had been used for centuries in popular medicine, and both its analgesic and narcotic properties were well known. In 1803, Derosne isolated a semipure alkaloid from opium and named it narcotine. Further examination of opium by Serturner in 1805 led to the isolation of morphine, and it was Serturner who first discovered the basic character of morphine.

Twelve years passed before the lead of Serturner was pursued. What followed, in the years 1817–1820 in the laboratory of Pelletier and Caventou

1

at the Faculty of Pharmacy in Paris, continues to amaze alkaloid chemists. Not since that time has one laboratory isolated so many active principles of pharmaceutical importance. Among the alkaloids obtained in this brief period were strychnine, emetine, brucine, piperine, caffeine, quinine, cinchonine, and colchicine. These alkaloids are the cornerstones of all that has transpired in alkaloid chemistry in the past 160 years.

In 1826, Pelletier and Caventou also obtained coniine, an alkaloid of considerable historical significance. Not only is it the alkaloid responsible for the death of Socrates from a draught of poison hemlock, but because of its simple molecular structure, it was the first alkaloid to be characterized (1870) and the first to be synthesized (1886). This is not to imply that investigators had forgotten alkaloids in the intervening years—far from it, for by 1884 at least 25 alkaloids had been obtained from *Cinchona* bark alone.

The molecular complexity of the majority of these alkaloids precluded their structure elucidation during the nineteenth century or even the early twentieth century. The case of strychnine is a good example. First obtained in 1819 by Pelletier and Caventou, it took nearly 140 years of extremely arduous, very frustrating chemical investigation before the structure was finally determined in 1946 by Robinson and co-workers.

It should be noted that while organic chemistry grew immensely in stature during this period, to become the sophisticated science that it is today, efforts in natural products chemistry were growing at a similar pace. Indeed many reactions that are now classics in organic chemistry were first discovered from the careful study of natural product degradations.

But let us look at the pace of alkaloid research in the past years. By 1939 nearly 300 alkaloids had been isolated and about 200 of these had at least reasonably well defined structures. In the first edition of Mankse's *Alkaloids* series, published beginning in 1950, more than 1000 alkaloids are noted.

With the introduction of preparative chromatographic techniques and sophisticated spectroscopic instrumentation, the number of known alkaloids has risen dramatically. A review to the middle of 1973 counted 4959 alkaloids, of which 3293 had known structures. By late 1978, the number stood at nearly 4000, structurally defined alkaloids.

1.2   OCCURRENCE

The major source of alkaloids in the past has been the flowering plants, the angiosperms. In recent years however there have been increasingly numerous examples of the occurrence of alkaloids in animals, insects, marine organisms, microorganisms, and the lower plants. Some examples of this very diverse occurrence of alkaloids are the isolation of muscopyridine (1) from the musk deer; castoramine (2) from the Canadian beaver; the pyrrole derivative (3), a sex pheromone of several insects; saxitoxin (4), the neurotoxic constituent of the red tide *Gonyaulax catenella*; pyocyanine (5) from the

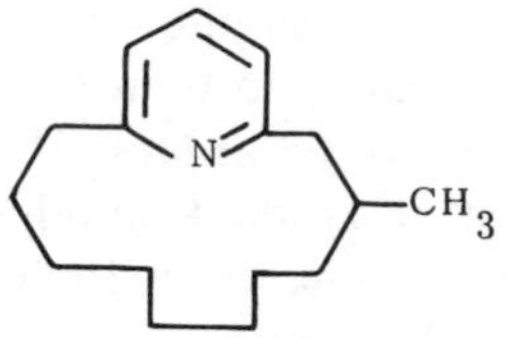

1 muscopyridine

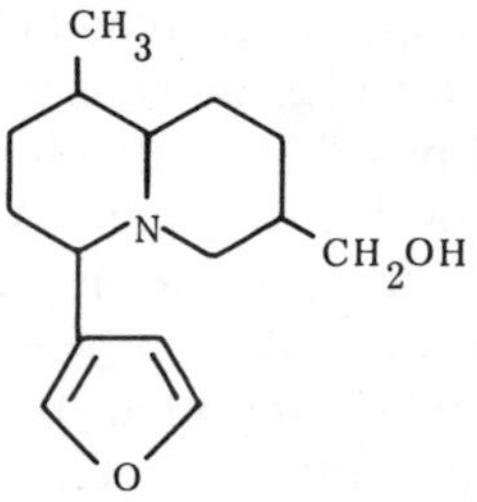

2 castoramine

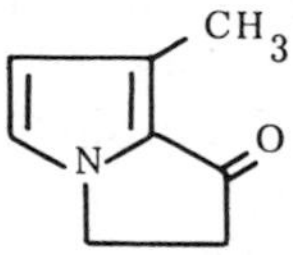

3

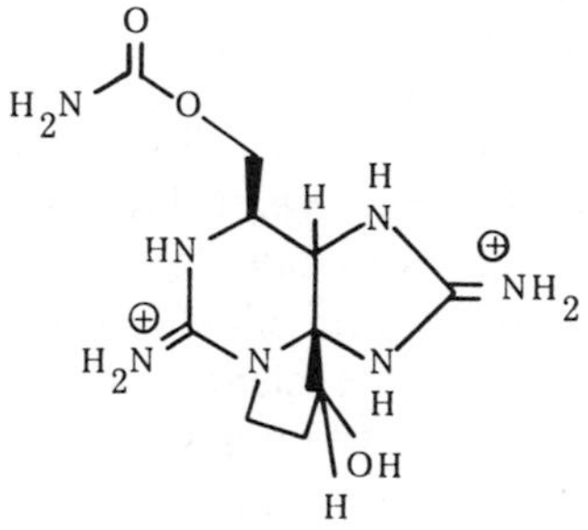

4 saxitoxin

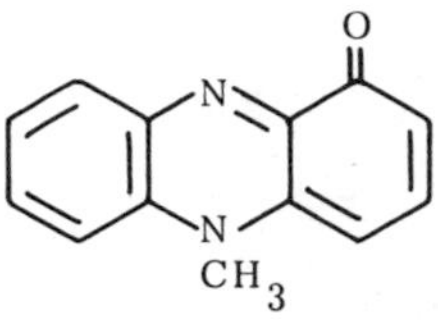

5 pyocyanine

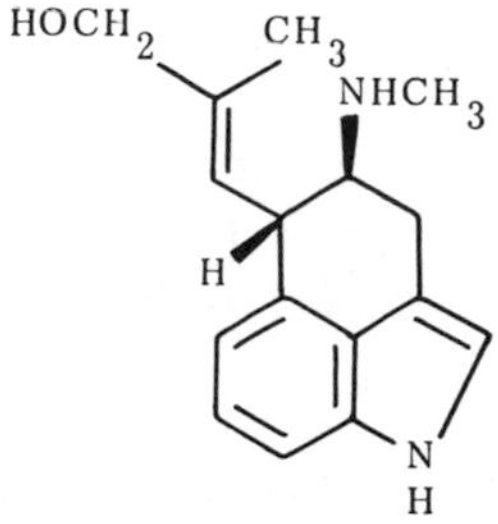

6 chanoclavine-I

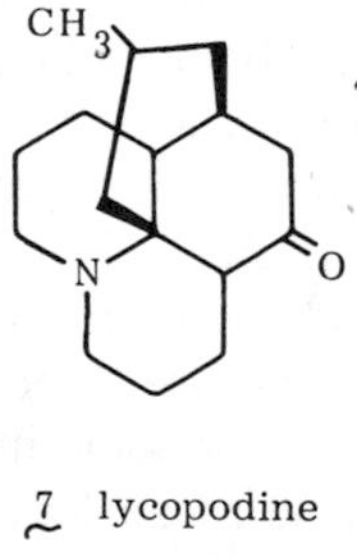

7 lycopodine

bacterium *Pseudomonas aeruginosa*; chanoclavine-I (**6**) from the ergot fungus, *Claviceps purpurea*; and lycopodine (**7**) from the genus of club mosses, *Lycopodium*.

Because the alkaloids as a class of compounds have been found predominantly in the flowering plants, scientists interested in the systematic organization of plants have long been interested in the alkaloids.

Certain groups (chemical classes) of alkaloids are associated with particular families or genera of plants. However, only quite recently has it proved possible to attach any taxonomic significance to their distribution.

In the higher plant system of Engler, there are 60 orders. Of these, 34 contain alkaloid-bearing species. Forty percent of all plant families contain at least one alkaloid-bearing plant. However, of the over 10,000 genera, alkaloids are reported in only 8.7%, and this distribution is very uneven. The most important alkaloid-containing families are the Liliaceae, Amaryllidaceae, Compositae, Ranunculaceae, Menispermaceae, Lauraceae, Papaveraceae, Leguminosae, Rutaceae, Loganiaceae, Apocynaceae, Solanaceae, and Rubiaceae. The Papaveraceae is an unusual family in that all the species of all the genera so far studied contain alkaloids. In most plant families that contain alkaloids, some genera contain alkaloids whereas others do not. A given genus will often yield the same or structurally related alkaloids, and even several different genera within a family may contain the same alkaloid. For example hyoscyamine (**8**) has been obtained from seven different genera of the plant family Solanaceae. In addition, simple alkaloids such as nicotine frequently occur in botanically unrelated plants. On the other hand the more complex alkaloids, such as vindoline (**9**) and morphine (**10**), are often limited to one species or genus of plant.

8 hyoscyamine   9 vindoline

Within a given alkaloid-containing plant, the alkaloids may be highly localized (concentrated) in a particular plant part. For example reserpine is concentrated in (and therefore isolated from) the roots of *Rauvolfia* spp.; quinine occurs in the bark, but not the leaves, of *Cinchona ledgeriana*; and morphine occurs in the latex of *Papaver somniferum*. Indeed certain parts of a plant may be devoid of alkaloids but another plant part may be quite rich in alkaloids. This does not necessarily mean that the alkaloids are formed in that plant part. For example alkaloids in *Datura* and *Nicotiana* species are produced in the roots but are translocated quite rapidly to the leaves. An additional complexity is that plants in the same genus may even produce the same alkaloid in different plant parts.

Even when the alkaloids do occur in a particular plant part, the range of concentration not only of the total alkaloids, but also of the alkaloid of pharmacologic significance, may vary enormously. For example, reserpine (**11**) may occur in concentrations of up to 1% in the roots of *Rauvolfia serpentina*, but vincristine (**12**) from the leaves of *Catharanthus roseus* is ob-

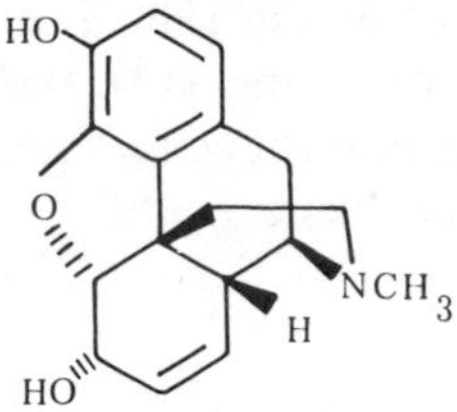

10   morphine

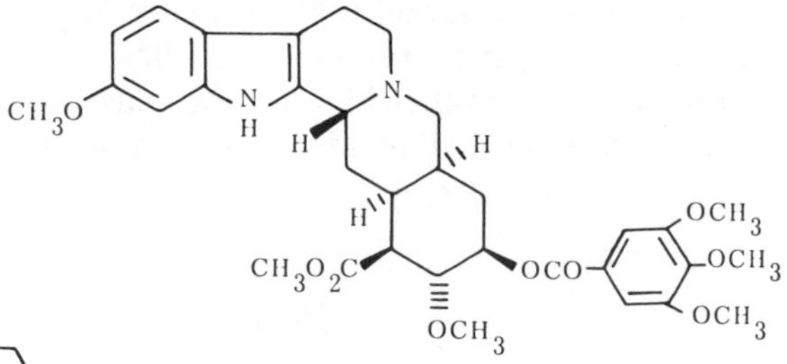

11   reserpine

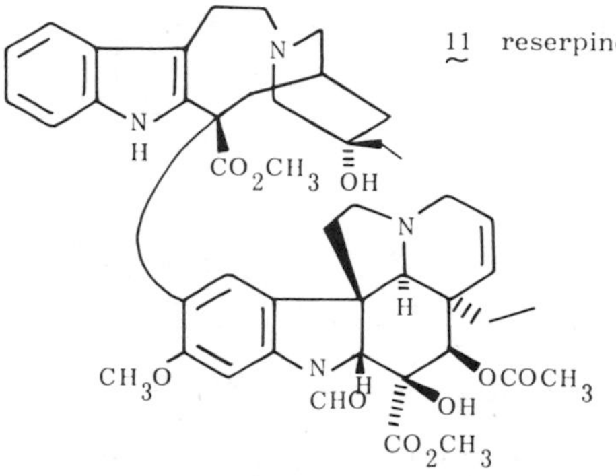

12   vincristine

tained in only $4 \times 10^{-6}\%$ yield. As might be imagined, the problems involved in the industrial production of an alkaloid occurring in such small quantities are substantial.

Many seeds of alkaloid-containing plants (e.g. *Nicotiana* sp., *Papaver somniferum*, and *Catharanthus roseus*) contain little or no alkaloid material. Yet very soon (within 12 hr) a range of alkaloids may be detectable in the germinated seedlings.

1.3   CLASSIFICATION

In the section dealing with the history of alkaloids, it was made clear that for the alkaloids as a group of compounds, no single definition is all-embracing. In place of this, numerous attempts have been made to provide a system of classification into which most alkaloids can be placed. The most widely accepted classification system, due to Hegnauer, groups the alkaloids as (*a*) true alkaloids, (*b*) protoalkaloids, and (*c*) pseudoalkaloids. Let us discuss this system of classification and point out some of the exceptions.

1.3.1   True Alkaloids

The true alkaloids are toxic; they show a wide range of physiological activity; they are almost invariably basic; they normally contain nitrogen in a heterocyclic ring; they are derived from amino acids; they are of limited taxonomic distribution, and they normally occur in the plant as the salt of an organic acid. Some exceptions to these "rules" are colchicine (**13**) and aristolochic acid (**14**), which are not basic and have no heterocyclic ring, and the quaternary alkaloids, which are acidic rather than basic.

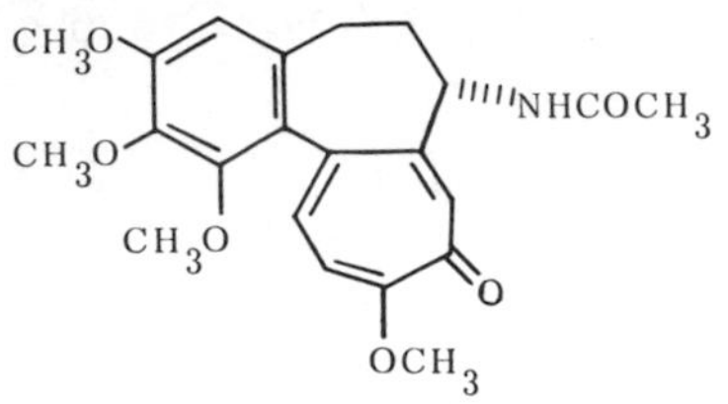

13   colchicine 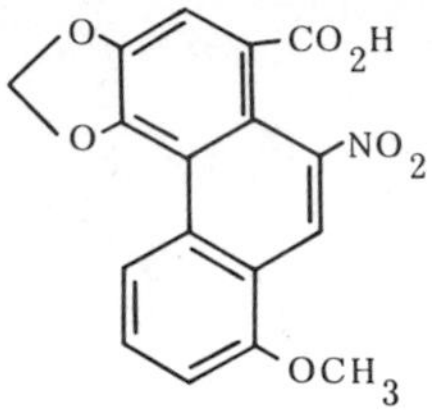   14   aristolochic acid-II

1.3.2 Protoalkaloids

The protoalkaloids are relatively simple amines in which the amino acid nitrogen is not in a heterocyclic ring. They are biosynthesized from amino acids and are basic. The term "biological amines" is often used for this group of compounds. Examples are mescaline (**15**), ephedrine (**16**), and *N,N*-dimethyltryptamine (**17**).

1.3.3 Pseudoalkaloids

The pseudoalkaloids are not derived from an amino acid precursor. They are usually basic. There are two important series of alkaloids in this class, the steroidal alkaloids [e.g., conessine (**18**)] and the purines [e.g. caffeine (**19**)].

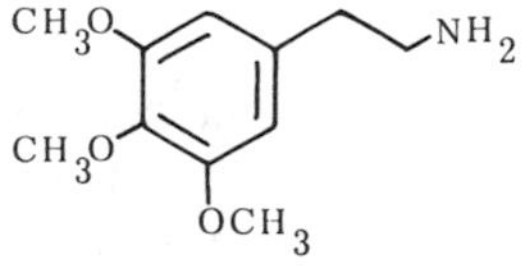

15   mescaline 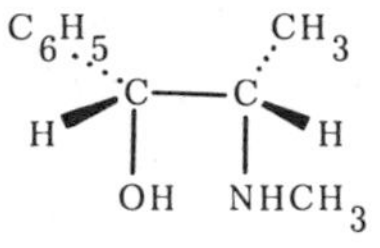   16   ephedrine 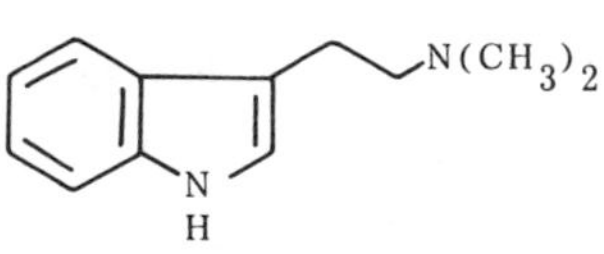   17   N,N-dimethyltryptamine

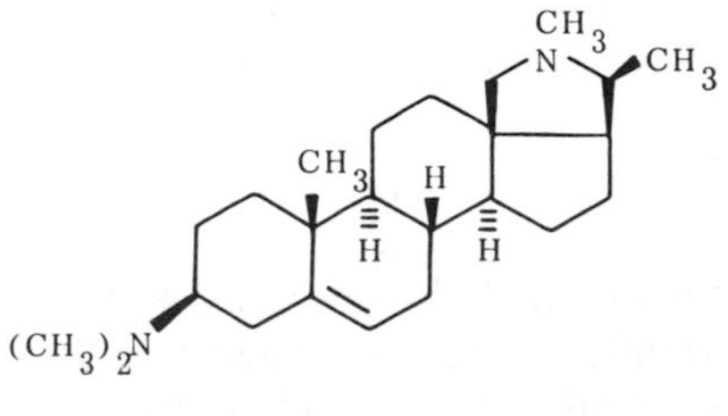

18   conessine 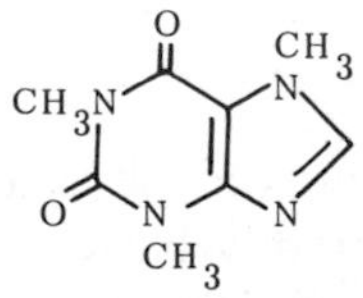   19   caffeine

1.4 NOMENCLATURE

With so many different alkaloid types, a single unifying nomenclature is of course not possible. Even within a given group of alkaloids, it is not often that there is a consistent system of nomenclature or numbering. One example is in the indole alkaloids, where many different skeleta are found. Most workers in this field use a numbering system based on the biogenesis for all the alkaloids, but unfortunately *Chemical Abstracts* has a highly confusing numbering system for each individual skeleton.

The only common characteristic of alkaloid names is that they terminate in "-ine." Beyond this, alkaloids, like other natural products, are given so-called "trivial" (i.e. nonsystematic) names. They may derive from genus names (e.g. atropine from *Atropa belladonna*); from the species name (e.g. cocaine from *Erythroxylon coca*); from a common name for the drug (e.g. ergotamine from ergot); from the physiological action of the compound (e.g. emetine, an emetic); or from the name of a famous alkaloid chemist (e.g. pelletierine).

Where appropriate, the numbering system for the various alkaloid groups will be given as each of the groups is discussed.

1.5 PHYSICAL AND CHEMICAL PROPERTIES

1.5.1 Physical Properties

Most of the alkaloids that have been isolated are crystalline solids with a defined melting point or decomposition range. A few alkaloids are amorphous gums, and some, such as nicotine (**20**) and coniine (**21**), are liquids.

Most alkaloids are colorless, but some of the complex, highly aromatic species are colored [e.g berberine (**22**) is yellow and betanin (**23**) is red].

The solubility of the alkaloids and their salts is of considerable significance in the pharmaceutical industry, both in the extraction of the alkaloid from the plant or fungus, and in the formulation of the final pharmaceutical preparation. In general, the free base of the alkaloid is soluble only in an organic solvent, although some of pseudo- and protoalkaloids are substantially soluble in water. The salts of alkaloids and the quaternary alkaloids are normally highly water soluble.

20 nicotine   21 coniine   22 berberine

$$\underset{\sim}{23} \quad \text{betanin}$$

1.5.2 Chemical Properties

The most distinct chemical property of most alkaloids is that they are basic. This property is of course dependent on the availability of the lone pair of electrons on nitrogen. If the functional groups adjacent to nitrogen are electron releasing, for example an alkyl group, the availability of the electrons on nitrogen is increased and the compound is more basic. Thus triethylamine (**24**) is more basic than diethylamine (**25**), which in turn is more basic than ethylamine (**26**). Alternatively, if the adjacent functional group is electron

$$(C_2H_5)_3N \qquad\qquad (C_2H_5)_2NH \qquad\qquad C_2H_5NH_2$$

$$\underset{\sim}{24} \qquad\qquad\qquad \underset{\sim}{25} \qquad\qquad\qquad \underset{\sim}{26}$$

withdrawing (e.g. a carbonyl group), the availability of the lone pair is decreased and the effect is to make the alkaloid neutral or even slightly acidic. A typical example is the amide group of compounds.

It will become evident that the alkaloids contain a diversity of nitrogen heterocyclic systems, and therefore it is pertinent to discuss the basicity of some of the most common heterocyclic systems found in alkaloids.

The pyridine nucleus (**27**) contains 6π electrons within the heterocyclic ring. The lone pair on nitrogen is therefore available and pyridine is basic. The carbon–nitrogen double bonds reduce this basicity somewhat however, and pyridine is less basic than its saturated analogue piperidine (**28**). Quinoline (**29**) and isoquinoline (**30**) are of similar basicity as pyridine.

Turning attention to the corresponding five-membered ring systems, pyrrole (**31**) is only fully aromatic ($4\pi + 2$ electrons) when the lone pair of

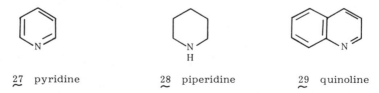

$$\underset{\sim}{27} \quad \text{pyridine} \qquad\qquad \underset{\sim}{28} \quad \text{piperidine} \qquad\qquad \underset{\sim}{29} \quad \text{quinoline}$$

electrons in nitrogen is involved in the aromaticity. Thus pyrrole (**31**) and its benzenoid analogue indole (**32**) are not basic. They are in fact acidic because formation of the anion increases the availability of electrons on nitrogen for aromaticity. However, pyrrolidine (**33**), for which there is no question of aromaticity, is, like piperidine (**28**), quite strongly basic.

30 isoquinoline   31 pyrrole   32 indole

33 pyrrolidine

The basicity of the alkaloids makes them extremely susceptible to decomposition, particularly by heat and light in the presence of oxygen. The product of such a reaction is often an N-oxide, and alkaloid N-oxides are becoming of increasing biosynthetic importance. More will be said about these compounds in subsequent chapters.

The decomposition of alkaloids during or after isolation can be a very serious problem if storage is likely to be for prolonged periods. Salt formation with an organic (tartaric, citric) or inorganic (hydrochloric or sulfuric) acid often prevents decomposition. It is therefore common to see alkaloids available commercially in their salt form.

1.6 DETECTION, ISOLATION, PURIFICATION, AND STRUCTURE ELUCIDATION

It is rare that alkaloids are found by chance. Alkaloid chemists normally search for alkaloids quite deliberately. Indeed often one group of alkaloids will be sought to the exclusion of others.

How is it possible to be able to evaluate a plant specifically or alkaloids, and how are these natural products isolated and characterized?

1.6.1 Detection

Alkaloids continue to provide new structures and interesting pharmacologic activities. It has therefore been necessary to develop simple methods for the detection of alkaloids in plant materials. Preferably these techniques should be (a) rapid and utilize minimal sample and equipment, (b) reproducible, and (c) sensitive.

Two methods are probably the most reliable for the screening of potential alkaloid-containing plants. The Wall procedure involves the extraction of approximately 20 g of dried plant material with 80% refluxing ethanol. After cooling and filtering, the residue is washed with 80% ethanol and the combined filtrates evaporated. This residue is taken up in water, filtered, acidified with 1% hydrochloric acid, and the alkaloids precipitated either with Mayer's reagent or with silicotungstic acid. If either test is positive, a confirmatory test is made in which the acid solution is basified, the alkaloids are extracted into organic solvent, and then the alkaloids are backextracted into aqueous acid. If this acid solution yields a precipitate with either reagent, the plant contains alkaloids. The basified aqueous phase should also be examined for the presence of quaternary alkaloids.

The Kiang–Douglas procedure is somewhat different in that the alkaloidal salts present in the plant (normally citrates, tartrates, or lactates) are first converted to free bases by moistening the dried plant material with dilute aqueous ammonia. The drug is then extracted with chloroform, the extract concentrated, and the alkaloids removed as their hydrochlorides by the addition of 2 N hydrochloric acid. The filtered aqueous solution is then screened for alkaloids by the addition of Mayer's, Dragendorff's, or Bouchardat's reagent. An approximate estimate of the potential alkaloid content can be obtained by using standard dilute solutions of a typical alkaloid such as brucine.

One disadvantage of the second procedure is that quaternary ammonium compounds, which are not converted to their free bases by the addition of ammonia, remain in the plant and are not detected. Similarly in the standard Wall procedure quaternary alkaloids appear as a "false-positive" since they are not extracted into organic solvent in the acid–base partition.

As indicated above, these methods have their limitations, but overall they are probably the best presently available. Neither method is suitable for field work.

Mention has been made of several precipitating reagents used in screening for alkaloids. The reagents are often based on the ability of alkaloids to combine with high-atomic-weight metals such as mercury, bismuth, tungsten, or iodine. Thus Mayer's reagent, undoubtedly the most popular, contains potassium iodide and mercuric chloride, and Dragendorff's reagent contains bismuth nitrate and potassium iodide in aqueous nitric acid. Bouchardat's reagent is similar to Wagner's reagent and contains potassium iodide and iodine, and reacts by halogenation. The silicotungstic acid reagent contains a complex of silicon dioxide and tungsten trioxide. Those various reagents display wide differences in sensitivity for dissimilar alkaloid groups. In spite of its popularity, Mayer's formulation is considerably less sensitive than Wagner's or Dragendorff's reagent.

Unfortunately, several other types of compounds may also give precipitates with these heavy metal reagents. Examples are proteins, coumarins, α-pyrones, hydroxyflavones, and tannins. Such reactions are termed "false-positive," for on attempted confirmation by the back-extraction technique precipitation is not usually successful.

Chromatography on a suitable absorbent (see later) is the normal method for the isolation of pure alkaloids from crude mixtures. As with other natural products the column fractions are monitored by thin-layer chromatography (TLC). What reagents are used for the detection of alkaloids chromatographically?

One quite general reagent is Dragendorff's reagent, which in the form of a spray produces orange-colored spots for alkaloidal materials. Note however that some unsaturated systems, particularly coumarins and α-pyrones, may also produce orange-colored spots with this reagent. Other general reagents but ones that are less widely used include phosphomolybdic acid, iodoplatinate, iodine vapor, and antimony (III) chloride.

Most alkaloids with these reagents react without distinction as to the class of alkaloid. If specific alkaloid types are being sought or detected, a number of reagents are available. Ehrlich's reagent (acidified p-dimethylaminobenzaldehyde) gives a quite characteristic blue or gray-green color with the ergot alkaloids (see Chapter 9). Acidified (sulfuric or phosphoric acid) ceric ammonium sulfate (CAS) reagent yields different, quite distinctive colors with many indole alkaloids. The colors are dependent on the ultraviolet chromophore of the alkaloid and are therefore of quite considerable structural significance. Some examples are illustrated in Figure 1.

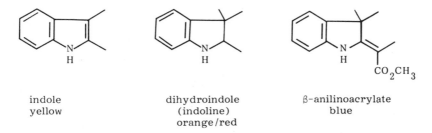

indole	dihydroindole	β-anilinoacrylate
yellow	(indoline)	blue
	orange/red	

Figure 1 Typical colors produced with ceric ammonium sulfate (CAS) reagent.

A mixture of ferric chloride and perchloric acid has found use for the detection of *Rauvolfia* alkaloids. The *Cinchona* alkaloids give an intense blue fluorescence in ultraviolet (UV) light after treatment with formic acid, and phenylalkylamines may be visualized with ninhydrin. Steroidal glycoalkaloids are often detected with a vanillin–phosphoric acid spray.

The Oberlin–Zeisel reagent, a 1–5% solution of ferric chloride in 0.5 N hydrochloric acid, is particularly sensitive to the tropolone nucleus of colchicine alkaloids, and amounts as small as 1 μg have been detected. The preparation of these various reagents is summarized in the Appendix.

1.6.2 Isolation

The basic character of many alkaloids is used to advantage in their isolation. Thus the alkaloids are taken into an acidic aqueous solution (normally hydrochloric, citric, or tartaric acids are used) and the neutral and acidic components of the original mixture are removed by solvent extraction. After the aqueous solution has been basified, the alkaloids are obtained by extraction into an appropriate solvent. This is a somewhat simplified description of a process that is considerably more complex in practice.

Extraction

A typical extraction procedure for alkaloids is shown in Figure 2.

Plant materials, particularly seed and leaf materials, often contain quite substantial quantities of very nonpolar fats and waxes. Because these compounds frequently cause problems due to emulsions when they are subjected to partition, they are often removed from the plant material as an initial step by percolation of the plant material with petroleum ether.

Most alkaloids are not very soluble in petroleum ether. But this extract should always be checked for the presence of alkaloids using one of the alkaloid-precipitating reagents described previously. If some of the required alkaloids are soluble in petroleum ether, the plant material may be pretreated (moistened) with aqueous acid to bind the alkaloids as their salts. This procedure has been used in the extraction of ergotamine (**34**) from the ergot fungus, *Claviceps purpurea*.

34 ergotamine

After defatting, several procedural choices are available. The plant material may be extracted with water, with ethanol or methanol, with aqueous alcohol mixtures, or with acidified aqueous alcohol solutions. Most alkaloids occur in plants as organic salts, and these salts are normally soluble in 95% ethanol. Pigments, sugars, and other organic secondary constituents are almost completely removed with alcohol, but many of the more complex

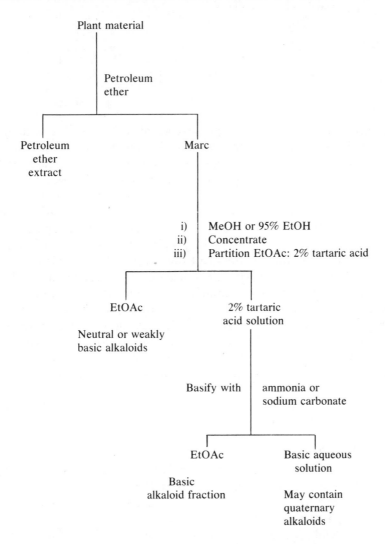

Figure 2   Typical extraction of a plant material for alkaloids.

organic and inorganic salts are only partially removed. This usually reduces the problems of precipitation and emulsification in the next step.

The alcohol solution is evaporated to a thick syrup and the residue partitioned between an aqueous acid solution and an organic solvent. Emulsions or precipitates are frequently observed at this stage. After repeated extraction with the organic solvent, the aqueous phase is made basic with sodium carbonate or ammonia. In some instances ammonia may give rise to new alkaloids not present in the original plant. A classic example is the very facile conversion of the iridoid sweroside (**35**) to the monoterpene pyridine alkaloid gentianine (**36**).

35   sweroside   36   gentianine

The basic aqueous solution is then extracted with a suitable organic solvent, normally chloroform or ethyl acetate. The alkaloid-containing solution is dried with an agent such as sodium sulfate (magnesium sulfate tends to be sufficiently acidic to bind strongly basic alkaloids), filtered, and evaporated *in vacuo* to afford the crude alkaloid residue. The basic aqueous solution may contain quaternary alkaloids and is normally tested with an alkaloid-precipitating reagent. The quaternary alkaloids can be separated from the other water-soluble components by precipitation as the Reineckate salt, followed by filtration and treatment of the precipitated complex with acetone:water (1:1). The quaternary alkaloids are now contained in the filtrate, which after cleanup with silver sulfate and an equivalent of barium chloride solution can be lyophilized to give the crude quaternary alkaloid chlorides.

A second quite general method for alkaloid extraction involves treatment of the plant material with ammonia to convert the alkaloid salts to free bases, which can then be extracted with an appropriate organic solvent. In this case the alkaloids obtained are mixed with the neutral and acidic components and must be separated by acid–base workup. Any quaternary alkaloids present in the plant would not be removed by this technique, but they may be obtained by extraction with alcohol.

Selective Extraction

For the most part the procedures just described are aimed at extracting all the alkaloids contained in a plant as a group to be separated subsequently. Normally this results in a very complex mixture which can complicate further purification steps.

A more careful procedure has been developed by Svoboda, formerly of Eli Lilly, and has been used for both investigational and commercial work on several plants in the family Apocynaceae. The background to the procedure is based on the concept that not all the tartrates of alkaloids are insoluble in organic solvent.

The defatted plant material is treated with 2% tartaric acid solution and extracted with benzene to remove the weak bases. The plant material is then made alkaline with ammonia, and the stronger bases extracted with organic solvent (benzene, chloroform, or ethyl acetate). The alkalinized drug is finally extracted with alcohol to give phenolic and quaternary alkaloids.

1.6.3 Purification of the Alkaloid Extract

With the crude, complex alkaloid extract in hand, the next step is separation into the individual components. There are a number of conventional methods available, and the choice of appropriate method or combination of methods will depend entirely on the particular alkaloid mixture.

Direct Crystallization

Although this is potentially the simplest procedure, it is rarely successful for the isolation of a pure alkaloid unless one alkaloid is present as a very important constituent or if it is particularly insoluble. It is a very useful technique however when particular purification has been effected by chromatography or some other technique. Some solvent combinations that are frequently used for crystallizing alkaloids include methanol, aqueous ethanol, methanol–chloroform, methanol–ether, methanol–acetone, and ethanol–acetone.

Steam Distillation

The high molecular weight of most alkaloids precludes steam distillation as a technique for purification. Some exceptions are the quite simple, low-molecular-weight alkaloids coniine (**21**), nicotine (**20**), and sparteine (**37**).

$\underset{\sim}{37}$ sparteine

Gradient pH Technique

This technique was devised by Svoboda for the isolation of the antileukemic alkaloids of *Catharanthus roseus*. It relies on the fact that the structurally diverse indole alkaloids present in the plant have quite different basicities. The crude alkaloid mixture is dissolved in 2% tartaric acid solution and extracted with benzene or ethyl acetate. This first fraction will contain the neutral and/or very weakly basic alkaloids. The pH of the aqueous solution is then raised by 0.5-pH increments to pH 9.0, extracting at each pH with the organic solvent. The subtle changes in pH permit the gradual separation of weakly basic from medium basic and strongly basic alkaloids. The very strongly basic alkaloids are extracted last. This concept is represented in Figure 3.

1.6.4   Chromatography of Alkaloids

The vast increase in the number of alkaloids isolated and characterized in the past 20 or so years can be traced directly to the introduction of chromatography as a purification technique. Prior to carrying out any preparative-scale separation of alkaloid mixtures, it is normal to develop thin-layer chromatographic systems. Such separation systems may be on silica gel,

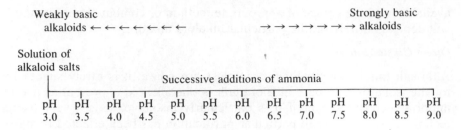

Figure 3   pH gradient separation of alkaloids.

alumina, cellulose powder, or kieselguhr. This order generally reflects the activity of the stationary phase and of the use to which these phases are put. The most popular phase currently employed is Silica gel G containing a fluorescent indicator, manufactured by E. Merck, Darmstadt. Not only is this phase commonly used for thin- and preparative-layer chromatography, but it is also used for preparative-scale chromatography.

Table 1 indicates some useful thin-layer chromatographic systems for the separation of alkaloids of various types.

1.6.5   Structure Elucidation

It was discussed previously that although a number of alkaloids were described in the nineteenth and early twentieth centuries, their structures took many years to deduce or were left unsolved. This was the era of classical organic chemistry when structures were obtained by degradation and the demonstration that certain functional groups were present in the degradation product. Separation techniques were crude (mainly crystallization) and it is amazing that so many pure alkaloids were isolated and their structures obtained.

At present, if the isolated alkaloid is known, it may take only minutes to prove identity. If the alkaloid is new, a concerted effort may result in a tentative structure within days and be proven within a week. Particularly difficult structures may take up to 6 months, but these instances are becoming increasingly rare.

The reasons for this dramatic change in the efficacy of structure elucidation have been the introduction of physical techniques developed to determine and define both major and minor structural fragments. The days of degradation to characterize every individual carbon atom are over.

A second factor should also be included at this point since it is central to the theme of the book. In the author's opinion, if one is to be interested in the structure elucidation of natural products, it is vital that one be cognizant of the biosynthesis or biogenesis of this group of compounds. This concept is developed more fully in chapter 2, but it is a truism that nature does not produce what it cannot produce. Certain structures are allowable

Table 1 Selected Thin-Layer Chromatographic Systems for Alkaloids

Alkaloid Type		Solid Phase	Eluent
I.	General	Silica gel G	Chloroform/acetone/diethylamine (5:4:1)
			Chloroform/triethylamine (9:1)
			Chloroform/cyclohexane/diethylamine (4:5:1)
			Cyclohexane/diethylamine (9:1)
			Benzene/ethyl acetate/diethylamine (7:2:1)
			Chloroform/methanol (19:1)
		Alumina G	Chloroform
			Cyclohexane/chloroform (3:7) + 0.5% diethylamine
II.	Phenylalkylamines	Silica gel G	Isopropanol/5% NH_4OH
			Acetone/methanol/acetic acid (5:4:1)
III.	Simple amines	Silica gel G	Ethanol/25% NH_4OH (8:2)
			n-Butanol/acetic acid/water (4:1:5)
IV.	Pyridine–pyrrolidine	Silica gel G	Chloroform/methanol (3:1)
			Chloroform/methanol/NH_4OH (60:10:1)
			Chloroform/methanol/acetic acid (60:10:1)
			Benzene/chloroform/methanol (7:3:2)
		Acidic alumina	Chloroform/ethanol (19:1)
V.	Pyrrolizidine	Silica gel G	Benzene/ethyl acetate/diethylamine (70:20:10)
VI.	Piperidine–quinolizidine	Silica gel G	Ethyl methyl ketone/methanol/7.5% NH_4OH (3:1.5:0.5)
			Chloroform: methanol (4:1)
			Chloroform/cyclohexane/diethylamine (7:3:1)
			Chloroform/absolute ethanol/25% NH_4OH (9:1:1)
			Cyclohexane/diethylamine (7:3)

Table 1 (*Continued*)

Alkaloid Type		Solid Phase	Eluent
VII.	Quinoline	Silica gel G	Isopropanol/benzene/diethylamine (2:4:1)
VIII.	Isoquinoline	Silica gel G	Benzene/ether/diethylamine (20:12:5)
			Benzene/methanol (9:1)
			Xylene/butanone/methanol/diethylamine (40:40:6:2)
			Chloroform/ethanol/ethyl acetate/acetone (6:2:1:1)
			Hexane/cyclohexane/cyclohexanol (1:1:1) + 5% diethylamine
			Cyclohexane/diethylamine (9:1)
IX.	Bisbenzylisoquinoline	Silica gel G (0.1 N NaOH)	Chloroform/ethyl acetate/methanol (2:2:1)
X.	Amaryllidaceae	Silica gel G	Ethyl acetate/chloroform/methanol (2:2:1)
XI.	*Colchicum*	Silica gel G	Chloroform/acetone/diethylamine (7:2:1)
			Benzene/ethyl acetate/diethylamine (5:4:1)
			Ethyl acetate/ethanol (4:1)
XII.	Ipecac	Silica gel G	Chloroform/methanol (85:15)
XIII.	*Lobelia*	Silica gel G	Cyclohexane/chloroform/diethylamine (5:4:1)
			Chloroform/methanol (3:1)
		Acidified alumina	Chloroform/ethanol (19:1)
XIV.	Tropane	Silica gel G	Chloroform/acetone/diethylamine (5:4:1)
			Chloroform/diethylamine (9:1)
			Butanone/methanol/7.5% ammonium hydroxide (6:3:1)
		Alumina G	Cyclohexane/chloroform (3:7) + 0.05% diethylamine

		Adsorbent	Solvent system
XV.	Indole		
	A. Harmala	Alumina G	Acetone/ethanol (85:15)
	B. Ergot	Silica gel G	Ethyl acetate/*N, N*-dimethylformamide/ethanol (13:1.9:0.1)
			Chloroform/methanol (95:5)
	C. Terpenoid	Alumina G	Chloroform/ethanol/water (3:1:1)
		Silica gel G	*n*-Butanol/acetic acid/water (10:10:1)
			Benzene/triethylamine (9:1)
			Benzene/ethylacetate/ethanol (2:2:1)
			Chloroform/methanol (19:1)
XVI.	*Cinchona*	Silica gel G	Benzene/methanol (4:1)
			Chloroform/ethanol (9:1)
			Chloroform/methanol/diethylamine (8:2:1)
			Hexane/carbon tetrachloride/diethylamine (5:4:1)
XVII.	Furanoquinoline	Alumina G	Chloroform/methanol (49:1)
		Silica gel G	Cyclohexane/chloroform/diethylamine (6:3:1)
XVIII.	Steroid	Silica gel G	Chloroform/methanol (19:1)
			Benzene saturated with ammonia
			Ether saturated with ammonia
XIX.	Steroid (glycosidic)	Silica gel G	Ethyl acetate/pyridine/water (3:1:3)
			Cyclohexane/ethyl acetate (1:1)
			Chloroform/methanol (3:2)

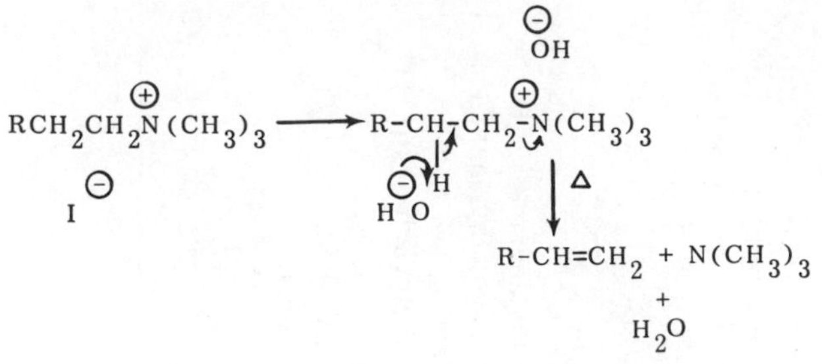

$$RCH_2CH_2\overset{\oplus}{N}(CH_3)_3 \longrightarrow R\text{-}CH\text{-}CH_2\text{-}\overset{\oplus}{N}(CH_3)_3$$

I

$$R\text{-}CH=CH_2 + N(CH_3)_3$$
$$+$$
$$H_2O$$

Scheme 1   Hofmann elimination reaction.

because of their biosynthetic probability; others can be totally disregarded because they are highly improbable or impossible on biosynthetic grounds. This so-called "biogenetic reasoning" approach cuts down on the number of structures that are considered in a given case and greatly facilitates the structure-elucidation process.

In order to appreciate the modern techniques that are available, examples of the classical approach to alkaloid structure elucidation will be given.

If we recall the classification of alkaloids discussed previously, most alkaloids contain nitrogen in a heterocyclic ring, and the Hofmann exhaustive methylation and degradation procedure was formerly crucial in determining the nature of the heterocyclic nitrogen. The efficacy of this sequence depends on the ability of a quaternary ammonium hydroxide, having a hy-

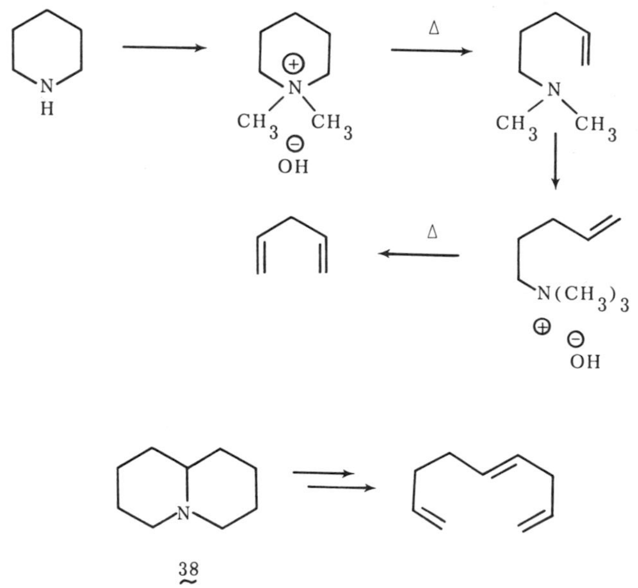

38

Scheme 2   Hofmann degradation sequence.

drogen β to the nitrogen, to eliminate water on heating to give an olefin with cleavage of the bond between nitrogen and the α-carbon (Scheme 1). The quaternary hydroxide is produced from the iodide by silver oxide in the presence of water or more efficiently by an ion exchange resin.

If the nitrogen is part of one heterocyclic ring [e.g. piperidine (28)], the process must be repeated twice before the nitrogen is completely eliminated (Scheme 2). A quinolizidine nucleus (38) in which the nitrogen is at the junction of two rings requires three elimination steps. From the number of steps required to eliminate nitrogen, it is therefore possible to define the nature of the nitrogen atom. The Hofmann degradation applies only to reduced systems. A modification involves cleavage of the quaternary ammonium halide with sodium amalgam (Emde degradation).

A third degradative reaction, developed by von Braun, uses cyanogen bromide on a tertiary amine. The product of reduction with lithium aluminum hydride affords in amine (Scheme 3).

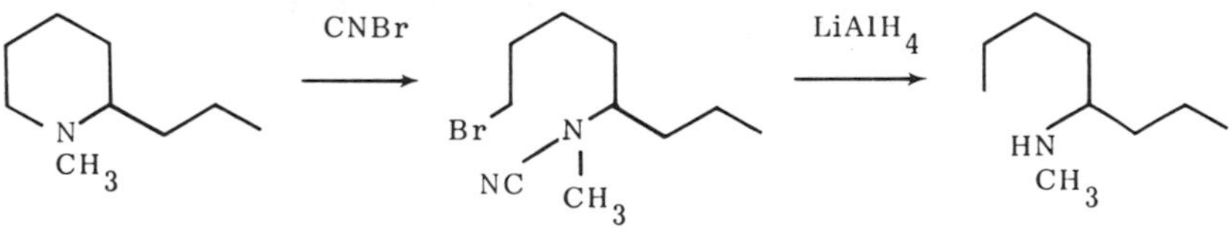

Scheme 3 Von Braun degradation.

In succeeding chapters structural relationships and spectral properties will be an important theme.

1.7 PHARMACOLOGY—THE BASIC CONCEPTS

There are two important information-processing systems in humans: the nervous system and the endocrine system. For the most part the principal action of alkaloids is on the nervous system.

The nervous system transmits preceived environmental changes rapidly to a specific processing center. The fundamental unit involved in this process is the neuron, with the information being transferred across a minute gap $(2 \times 10^{-5}$ mm) called a synapse.

Although within a neuron information is passed electrically, *between* neurons information is passed chemically by compounds known as neurotransmitters. The four neurotransmitters generally recognized are acetylcholine, noradrenaline, dopamine, and serotonin. The reactive site for these chemicals on the adjacent neuron is called a receptor.

Neurons that release acetylcholine are termed cholinergic and those that release norandrenaline, dopamine, and serotonin are called adrenergic, dopaminergic, and serotonergic respectively.

1.7.1 The Nervous System

The nervous system is comprised of two parts, the *central nervous system* (CNS), which is the major integrating system of the body, and the *autonomic nervous system* (ANS), which connects the spinal cord at each vertebra. The systems are not totally independent, since part of the ANS is in the CNS and information from the CNS travels *via* the ANS to muscles and glands. The CNS is primarily concerned with integrating information from both internal and external sources, and it also processes appropriate responses to muscles and glands. The ANS on the other hand is primarily a motor system and is concerned with visceral functions and balance. The ANS, whose control center is the hypothalamus, is classically divided into two separate systems, the sympathetic and parasympathetic. The effects of these on a given function or organ are usually opposite (Table 2).

Table 2 Typical Sympathetic and Parasympathetic Effects on Various Bodily Functions

Bodily Function	Parasympathetic Reaction	Sympathetic Reaction
Heart rate	Decrease	Increase
Breathing rate	Slow and deep	Fast and shallow
Sweat glands	No effect	Increase secretion
Salivary glands	?	Increase secretion
Skin blood vessels	Dilation	Constriction
Stomach and intestinal glands	Activation	Inhibition
Pupil of eye	Constriction	Dilation

Drug Action

A drug may mimic, facilitate, or antagonize a normal process. It may also increase, decrease, or intermittently disrupt the normal activity of the cell.

Curare (Chapters 8 and 9) played an important role in developing what is a fundamental concept in both drug action and drug design, namely the specificity of action of a given drug. Bernard in 1856 was using curare to study the interaction between the nervous system and the skeletal muscles. As a result of this work, it was demonstrated that curare acts as a specific blocker of the skeletal neuromuscular junction by occupying the acetylcholine receptor.

Acetylcholine is liberated from motor nerve terminals and transmits the nerve impulse over a synaptic gap to an end-plate region on skeletal muscle effector cells. Acetylcholine also acts as a neurotransmitter at autonomic ganglia and at certain cells in the spinal cord. A similar mechanism of transmission occurs at certain smooth muscle motor terminals.

The formation of acetylcholine in motor nerve terminals is carried out by the enzyme choline acetyl transferase and requires sodium ions, glucose, and oxygen. Blocking this synthesis causes transmission to fail.

Clearly a drug that affects acetylcholine levels in any way will elicit important clinical pharmacologic responses.

Cholinergic Drugs   Cholinergic drugs elicit responses that mimic stimulation of cholinergic nerves by any one of three basic mechanisms:

1. Direct action of effector cells supplied by cholinergic nerves. This is the mechanism of action of choline and related synthetics.
2. Indirect action involving the inhibition of cholinesterase. This preserves or elevates acetylcholine levels for continued stimulation of effector cells postsynaptically. A typical alkaloid that has this action is physostigmine (Chapter 9).
3. Direct stimulation of cholinergic effector cells. There are several alkaloids which display this mode of action. Among these, pilocarpine and arecoline have a selective action on smooth muscle; muscarine acts on the smooth muscle, heart, and glands; and nicotine acts on the ganglia and skeletal muscle. These specific actions led to the naming of the receptor sites after the alkaloid of action. Thus there are "muscarinic" and "nicotinic" receptors.

Cholinergic Blocking Agents   In opposition to the cholinergic drugs are the cholinergic blocking agents and three groups of compounds are generally recognized.

1. THE ATROPINELIKE COMPOUNDS   Atropine and several related natural and synthetic compounds are capable of occupying the postsynaptic receptor site. By doing so they prevent the normal neurotransmitter from acting. Depending on the chemical structure, action may be on both the CNS and the peripheral nervous system, or only on the latter.
2. THE CURARE AND CURARE-MIMETIC DRUGS   These compounds act by antagonizing the nicotinic action of acetylcholine at the motor end plates of skeletal muscles.
3. GANGLIONIC BLOCKING AGENTS   These compounds depress autonomic ganglia by depolarizing the end plate. Typical compounds displaying this activity are the methonium compounds. These are bisquaternary salts where cationic centers are separated by a chain of 10 to 12 carbon atoms. Nicotine in the second phase of its action is another example.

Adrenergic Drugs   Adrenergic drugs elicit responses that resemble stimulation of the adrenergic nerves (i.e. release of noradrenaline). These drugs differ markedly in the intensity and specificity of their response. This has become the basis for using the correct drug for a particular disease state. Epinephrine and ephedrine are typical examples, but there are also numerous synthetic drugs that elicit related reactions.

Adrenergic Blocking Agents   Adrenergic blocking agents prevent the stimulation of sympathetic nerves by antagonizing the effects of drugs such as epinephrine. Their usefulness is somewhat limited because they frequently have other pronounced effects, such as histaminelike, ganglionic blocking, or cholinergic actions. The ergot alkaloids, such as ergonovine, methergine, ergotamine, and dihydroergotamine, are examples of such compounds.

Of necessity this has been a very brief introduction to what is in reality a vast area. As the various alkaloid groups are discussed however, the reader should be considering not only the site of action and the pharmacologic response, but also the mechanism of action.

LITERATURE

Annual Reports of the Chemical Society, London.

Battersby, A. R., *Quart. Rev.* **15**, 259 (1961).

Bentley, K. W., *The Alkaloids*, Interscience, New York, 1957, 1965.

Bentley, K. W., *The Isoquinoline Alkaloids*, Pergamon, Oxford, 1965, p. 264.

Boit, H. G., *Ergebnisse der Alkaloide—Chemie bis 1960*, Akademie-Verlag, Berlin, 1961.

Bové, F. J., *The Story of Ergot*, S. Karger, Basel, 1970, p. 297.

Budzikiewicz, H., C. Djerassi, and D. H. Williams, *Structure Elucidation of Natural Products by Mass Spectrometry*, Vol. 1, *Alkaloids*, Holden-Day, San Francisco, 1964.

Closs, G. L., in H. S. Ashmore (Ed.), *Encyclopedia Brittanica*, Benton, Chicago, 1963, p. 637.

Farnsworth, N. R., *J. Pharm. Sci.* **55**, 225 (1966).

Glasby, J. S., *Encyclopedia of the Alkaloids*, Vols. 1 and 2, Plenum, New York, 1975; Vol. 3, 1978.

Goodman, L. S., and A. Gilman (Eds.), *The Pharmacological Basis of Therpaeutics*, 5th ed., Macmillan, New York, 1975, p. 1704.

Guggenheim, M., *Die Biogener Amine*, 4th ed., Karger Verlag, Basel, 1951.

Hegnauer, R., in T. Swain (Ed.), *Chemical Plant Taxonomy*, Academic, New York, 1963, p. 389.

Hegnauer, R., in T. Swain (Ed.), *Comparative Phytochemistry*, Academic, New York, 1966, p. 211.

Hendrickson, J. B., *The Molecules of Nature*, W. A. Benajamin, New York, 1965, p. 143.

Henry, T. A., *The Plant Alkaloids*, 4th ed., Blakeston, Philadelphia, 1949.

Hesse, M., *Alkaloidchemie*, Thieme, Stuttgart, Germany, 1978, pp. 231.

Hesse, M. *Indolalkaloide in Tabellen*, Springer-Verlag, Berlin, 1964, 1968.

Hesse, M. and H. O. Bernhard, in H. Budzikiewicz (Ed.), *Progress in Mass Spectrometry*, Volume 3, *Alkaloids*, Verlag Chemie, Weinheim, 1975, p. 372.

Hirata, Y., T. Kametani, and M. Ihara, in F. Korte (Ed.). *Methodicum Chimicum*, Volume 11, Part 3, Academic, New York, 1978, p. 101.

Hughes, D. W., and K. Genest, in L. P. Miller (Ed.), *Phytochemistry*, Vol. 3, Van Nostrand Reinhold, New York, 1973, p. 118.

Ito, S., in K. Nakanishi, T. Goto, S. Ito, S. Natori, and S. Nozoe (Eds.), *Natural Products Chemistry*, Vol. 2, Academic, New York, 1975, p. 255.

Kametani, T., *The Chemistry of the Isoquinoline Alkaloids*, Hirokawa, Tokyo, 1969, p. 265.

Leete, E., in P. Bernfeld (Ed.), *Biogenesis of Natural Compounds*, Macmillan, New York, 1963, p. 739.

Leete, E., *Ann. Rev. Plant. Physiol.* **18**, 179 (1967).

Manske, R. H. F. (Ed.), *The Alkaloids, Chemistry and Physiology*, Vols. 1–16, Academic New York, 1950–1977.

Manske, R. H. F. and R. Rodrigo, (Eds.), *The Alkaloids, Chemistry and Physiology*, Volume 17, Academic, New York, 1979.

Mothes, K., and H. R. Schutte, *Angew. Chem. Int. Ed. Engl.* **2**, 341, 441 (1963).

Neuss, K., *Physical Data of Indole and Dihydro Indole Alkaloids*, rev. ed., Eli Lilly and Co., Indianapolis, Ind., 1966, 1968.

Pelletier, S. W. (Ed.), *Chemistry of the Alkaloids*, Van Nostrand Reinhold, New York, 1970.

Pelletier, S. W., in G. L. Clark and G. G. Harvey (Eds.), *The Encyclopedia of Chemistry*, Van Nostrand Reinhold, New York, 1966, p. 43.

Raffauf, R. F., *A Handbook of Alkaloids and Alkaloid-Containing Plants*, Wiley–Interscience, New York, 1970.

Robinson, R., *The Structural Relations of Natural Products*, Clarendon, Oxford, 1955.

Robinson, T., *The Biochemistry of Alkaloids*, Springer-Verlag, New York, 1968, p. 149.

Rodd, E. H. (Ed.), *Chemistry of Carbon Compounds*, Vol. IV-C, Elsevier, Amsterdam, 1960, pp. 1799–2130.

Santavy, F., in E. Stahl (Ed.), *Thin-Layer Chromatography*, Springer-Verlag, New York, 1969, p. 421.

Shamma, M. and D. M. Hindenlang, *Carbon-13 NMR Shift Assignments of Amines and Alkaloids*, Plenum, New York, 1979, p. 303.

Sim, S. K., *Medicinal Plant Alkaloids*, Univ. Toronto, 1965, p. 181.

Specialist Periodical Reports, Alkaloids, Vols. 1–9, The Chemical Society, London, 1971–1979.

Spenser, I. D., in M. Florkin and E. H. Stotz (Eds.), *Comprehensive Biochemistry*, Vol. 20, Elsevier, Amsterdam, 1968, p. 231.

Stanek, J., *Alkaloidy*, Ceskoslovenske' Akademie Ved, Prague, 1957.

Swan, G. A., *An Introduction to the Alkaloids*, Wiley, New York, 1967.

Wiesner, K. (Ed.), *MTP International Review of Science, Organic Chemistry*, Ser. 1, Vol. 9, *Alkaloids*, Butterworths, London, 1973, p. 346.

Wiesner, K. (Ed.), *MTP International Review of Science, Organic Chemistry* Ser. 2, Vol. 9, *Alkaloids*, Butterworths, London, 1975, p. 318.

Willaman, J. J., and H. L. Li, *Alkaloid-Bearing Plants and Their Contained Alkaloids*, 1957–1968, *Lloydia* **33**, Suppl. 3A (1970).

Willaman, J. J., and B. G. Schubert, *Alkaloid-Bearing Plants and Their Contained Alkaloids*, Tech. Bull. No. 1234. Agricultural Research Service, U.S. Department of Agriculture, 1961.

Winterstein, E. W., and G. Trier, *Die Alkaloide*, 2nd ed., Verlag Gebruder, Bornträger, Berlin, 1931.

chapter 2

BIOSYNTHESIS AND BIOGENESIS

2.1 INTRODUCTION

In the Preface mention was made of the rationale for the title. It is now appropriate to begin to discuss the significance of biogenesis and biosynthesis as applied to alkaloids. To do this we need to look at the fundamental building blocks of nature which are involved in the formation of the thousands of known alkaloids. Surprisingly there are very few molecules involved. Indeed it is a function of the range of enzymes involved at the intermediate stages, rather than a multiplicity of precursors in the initial stages, which accounts for the numerous alkaloids that are known.

The question of why plants, animals, and fungi produce alkaloids (or any other secondary metabolite) is often posed. For the most part it remains unanswered. Although some compounds are known to be insect attractants, insect repellants, antifungal agents, or toxins, these are minor uses in comparison with the multitude of products. What is certain is that alkaloids are not the end products of metabolism as was once thought. Indeed in several instances it has been found that they are merely intermediates in a highly dynamic metabolic process. They cannot be seriously regarded as an important metabolic reserve however, for they form only a small part of the total nitrogen content. Further discussion of this fascinating topic is beyond the intended scope of this section. What we are concerned with at this time are the molecules which together will build the alkaloids, and some of the enzymes that are important in these processes.

What then are the key alkaloid-producing precursors? For the most part they are a number of quite simple amino acids (see the definition of alkaloids and protoalkaloids in Chapter 1). Typical of these are ornithine (**1**) and its next higher homologue lysine (**2**), phenylalanine (**3**) and its *p*-hydroxyderivative tyrosine (**4**), tryptophan (**5**), histidine (**6**), and anthranilic acid (**7**). In combination with simple acetate or terpenoid units, aromatic hydroxyla-

1 ornithine   2 lysine   3 phenylalanine, R=H
                                               4 tyrosine, R=OH

5 tryptophan   6 histidine   7 anthranilic acid

tions, and the occasional O- and N-methylation with methionine, the majority of alkaloids are produced. The remaining alkaloids are derived either
from the purine nucleus or by the addition of ammonia or an alkylamine to
any one of several terpenoid nuclei. In essence, the rest of the book is a
discussion of these processes, the products that result, and the chemical and
physiological properties of these compounds.

The preceding paragraph is obviously an oversimplification of an extremely complex process, and by way of illustration it is worth considering
that details of the formation of even the most simple of alkaloids are known
only very incompletely. One major reason for this is the lack of enzymologic
work devoted to secondary metabolism, alkaloids in particular. Only now
are plant enzymes beginning to be isolated which will take two complex
molecules and join them stereospecifically or which will carry out a complex
series of reactions to give a unique product. Undoubtedly this will be an
enormously fruitful area of research in the future.

Previous studies with the enzymes of certain microorganisms have en-
abled a few to be identified, purified, and characterized. As expected there
are some enzymes of primary metabolism which show little substrate specificity and which therefore can be used in secondary metabolic process. On
the other hand there are myriad enzymes that are wholly or partly substrate
specific. In other words, unless the substrate(s) is correct, no reaction occurs
and no product is formed.

There is also a third type of reaction which we should expect, namely
those which proceed "spontaneously." These reactions are of quite limited
scope and proceed without the benefit of induced asymmetry.

Let us consider first some of the important enzymes involved in the
formation of alkaloids.

2.2 ENZYMES OF ALKALOID FORMATION

2.2.1 Thiokinases

The thioesters of coenzyme A, for example acetyl coenzyme A (**8**), are vital intermediates in carboxylic acid metabolism, for the resulting products are more reactive than the normal carboxyl group. The carbonyl carbon in particular is considerably more susceptible to nucleophilic substitution under these conditions. Some examples of the reactivity of the carbonyl group of acetyl CoA (**8**) are indicated in Figure 1. Acetoacetyl CoA (**9**) is an important

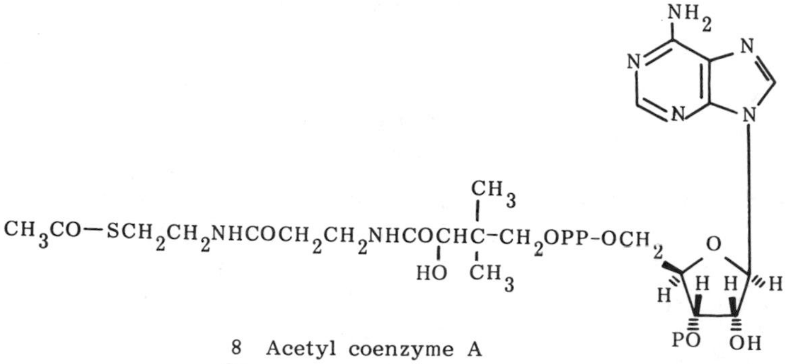

$$CH_3CO\!-\!SCH_2CH_2NHCOCH_2CH_2NHCOCHC\!-\!CH_2OPP\!-\!OCH_2$$

8   Acetyl coenzyme A

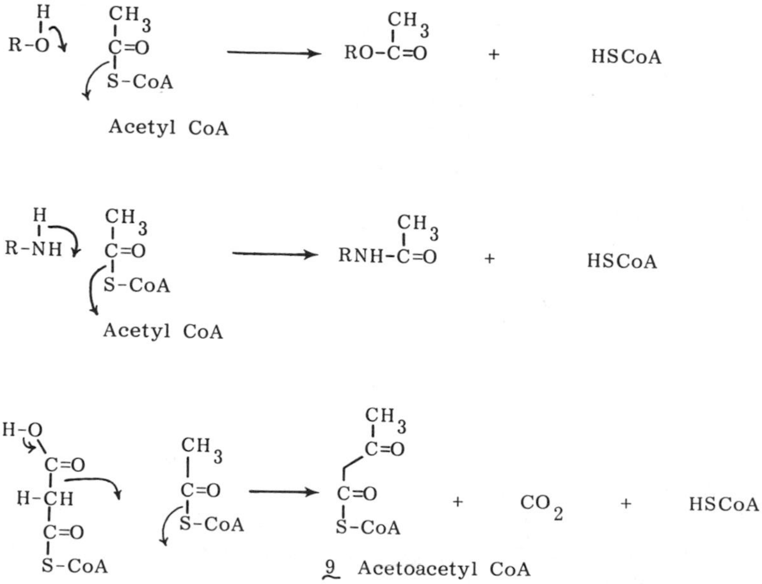

Figure 1   Acetyl coenzyme A in alkaloid biosynthesis.

intermediate in the formation of the tropane alkaloids (Section 3.3), several alkaloids contain an *N*-acetyl group (e.g. colchicine, Section 8.35) and acetylation of secondary hydroxyl groups is quite common (e.g. vindoline, Section 9.23.3).

2.2.2 Mixed-Function Oxygenases

One of the most important groups of enzymes responsible for the many diverse alkaloid products are the mixed function oxygenases. Functionally they are extremely complex, although the overall reactions which they carry out are apparently quite simple. Some examples follow.

1. The conversion of an amine to its N-oxide. The N-oxides of alkaloids are a sorely neglected group of compounds. They may well prove to be of critical importance in the metabolism and transport of alkaloids and may not be the result of unwanted side reactions as has been supposed.

2. Substitution of a hydrogen atom on an aliphatic carbon atom by a hydroxy group is a stereospecific process involving only the appropriate C—H bond. Many examples of this process will be described as the various alkaloid groups are discussed. For now it is sufficient to realize that this is a stereospecific process proceeding in almost all the cases studied with retention of configuration at the site of substitution.

3. When an isolated double bond is attacked by a mixed function oxidase, any of several products may result. The initial reaction product is an epoxide **10** which is introduced stereospecifically. Several alternatives are possible at this point, the most important of which are the protonation–deprotonation sequence leading to either enol form of a ketone, and the addition of water to give a diol **11**.

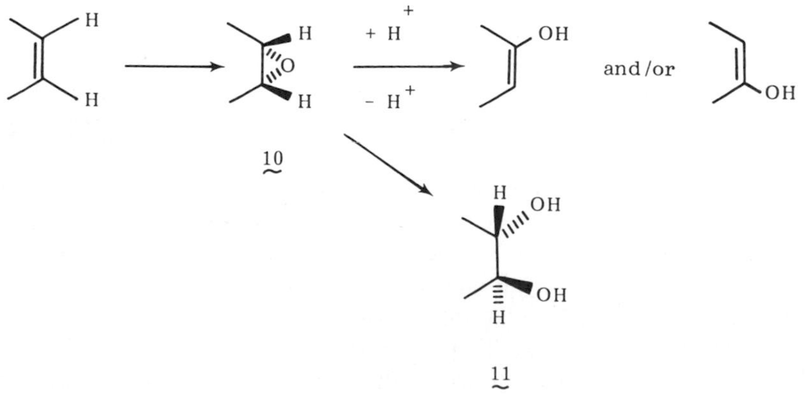

The pathway leading to either of the two ketones has an important place in the formation of hydroxylated aromatic compounds, for in this case the enol forms are stable phenolic groups. More interestingly, there is a stereo-specific hydride shift involved in this process. This shift, the so-called NIH shift, was first discovered by Witkop and co-workers at the National Institutes of Health. One system that carries out this reaction is phenylalanine hydroxylase, the enzyme responsible for the transformation of phenylalanine (3) to tyrosine (4), and this system will be discussed subsequently.

When an aromatic system forms an epoxide, the compound is called an arene oxide. Many such arene oxides are now known to be intermediates in the metabolism, particularly by liver monoxygenases, of aromatic hydro-carbons to phenols or diols.

In vitro, two reactions are important in the chemistry of arene oxides: the rearrangement to phenols and the valence-bond tautomerism to oxepins. The latter reaction is typified by an equilibrium for benzene oxide (12).

One alkaloid that contains this oxepin ring system is aranotin (13), a metabolite of *Arachniotus aureus*. The biosynthesis of this and related compounds poses a number of interesting problems and will be discussed in Chapter 8.

At this point we are concerned with the rearrangement of an arene oxide to a phenol, a reaction which occurs with an intramolecular migration and retention of substituents. The kinetics indicate that two reactions may occur depending on the pH. At physiological pH a spontaneous reaction is involved in which the zwitterion 14 is an intermediate.

This zwitterion has two distinct alternative pathways to follow. One involves direct loss of a proton and rearrangement to a phenolate anion which is then reprotonated. The second route involves the 1,2-shift of a hydride (NIH shift) to give a ketone **15,** which may then lose either hydrogen on rearrangement to the phenol. If we being with a labeled benzenoid substrate, two possible phenols may be produced, one retaining the label (e.g. **16**), the other losing it completely (e.g. **17**).

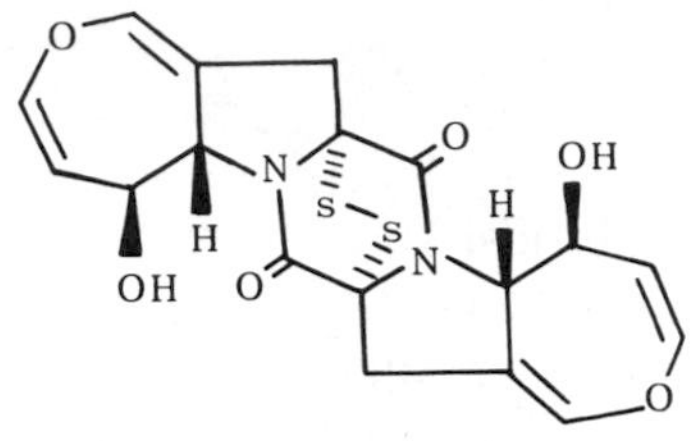

13   aranotin

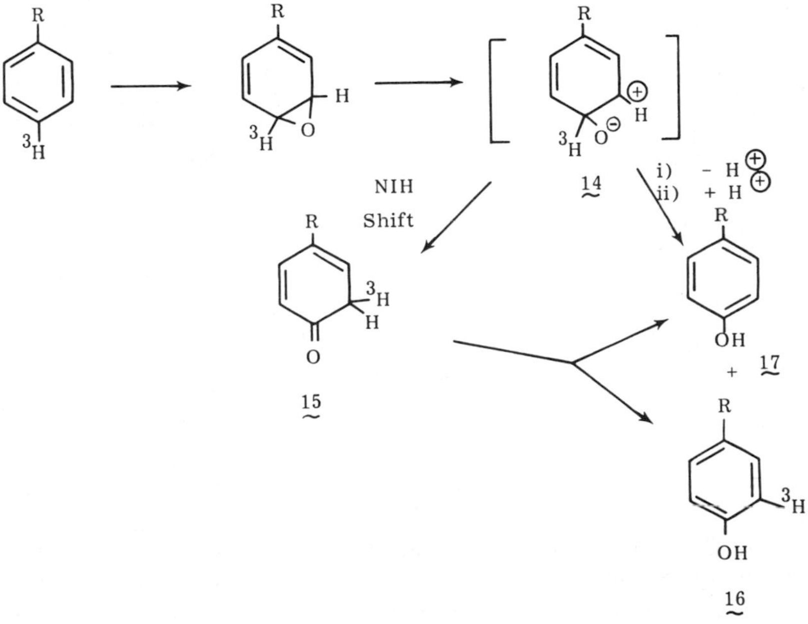

In vivo, the NIH-shift mechanism predominates and retention of isotope occurs. In almost all of the cases studied to date the percent of isotope retention is independent of the source of the enzyme. For example the conversion of [4′-³H]phenylalanine (**18**) to tyrosine (**4**) proceeds with greater than 90% retention of label regardless of whether the enzyme is derived from a liver microsomal system *Pseudomonas* or *Penicillium* sp. In addition, the expected isotope effect operates and tritium is retained to a greater extent than is deuterium.

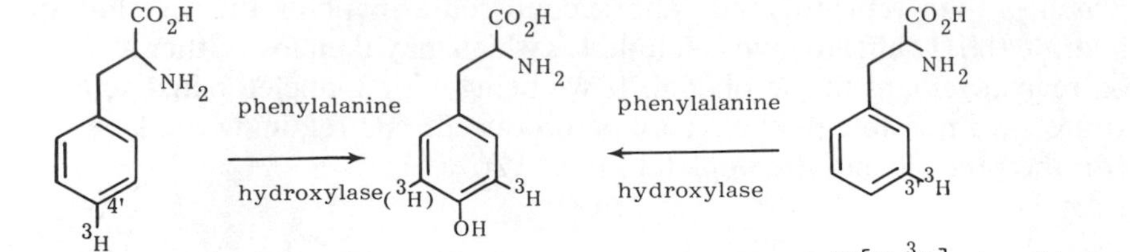

18 [4'-³H]-phenylalanine   4   22 [3'-³H]-phenylalanine

In the case where a symmetrical substrate is used, two positions, the 3- and the 5-, effectively become labeled to an equal extent. For an unsymmetrical system the situation is potentially more complex, although in practice it appears that there is greater selectivity in the migration. Thus tryptophan hydroxylase transforms [5-³H]tryptophan (19) to 5-hydroxytryptophan (20) with better that 85% retention of ³H at C-4. The initial arene oxide must therefore have selectively formed between C-4 and C-5 (i.e. 21) and not between C-5 and C-6.

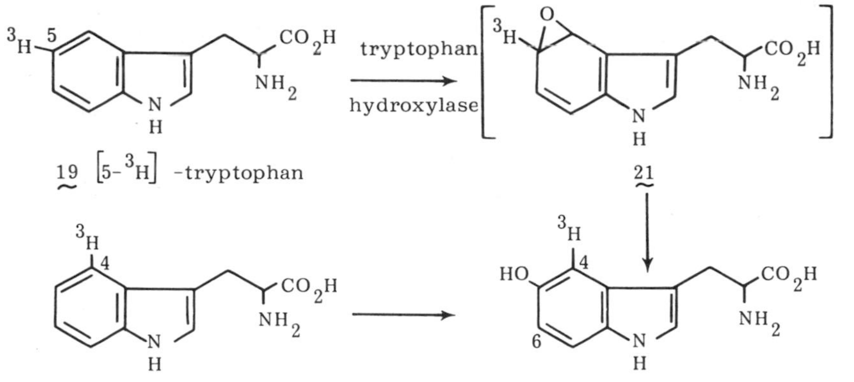

19 [5-³H] -tryptophan   21

23 [4-³H] -tryptophan   20   5-hydroxytryptophan

There is another point to be considered here: namely that in the conversion of [3'-³H]phenylalanine (22) to tyrosine (4), or [4-³H]tryptophan (23) to 5-hydroxytryptophan (20), the retention of tritium is exactly the same as in the cases where migration of the tritium occurred.

Migration of hydride is also observed in the case where hydroxylation occurs *ortho* to an alkyl group. In the formation of *o*-coumaric acid (24) from cinnamic acid (25), migration of the 2-hydrogen occurs with greater that 85% retention at C-3 in the product.

Many alkaloids contain two adjacent oxygen functions and the introduction of the second oxygen atom has also been studied. For example in the conversion of tyrosine (labeled equivalently at C-3' and C-5' with ³H; see above) by tyrosine hydroxylase to dopa (26), half of the original tritium

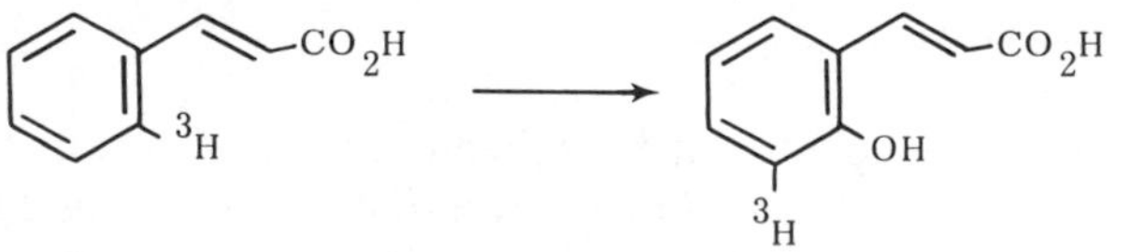

25 cinnamic acid, trans-   24 o-coumaric acid, trans-

activity is lost. One might expect that a second migration of tritium would occur to give a product essentially labeled at C-2′ and C-5′. This is not the case; the intermediate cation 27 is not stabilized by migration of the hydride (i.e. 28) but rather by tautomerization to the hydroxy dienone 29, which then undergoes further tautomerization to the dihydroxy compound.

In the biosynthesis of mescaline (30), an hallucinogenic principle of the peyote cactus, from [3′,5′-³H₂]tyrosine (4), how much tritium should be retained? As expected, none was found in the isolated product.

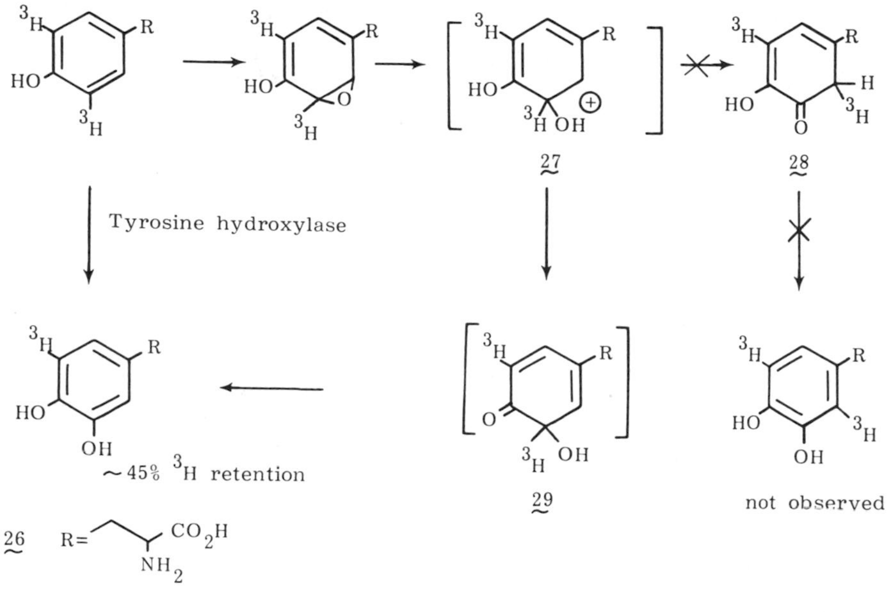

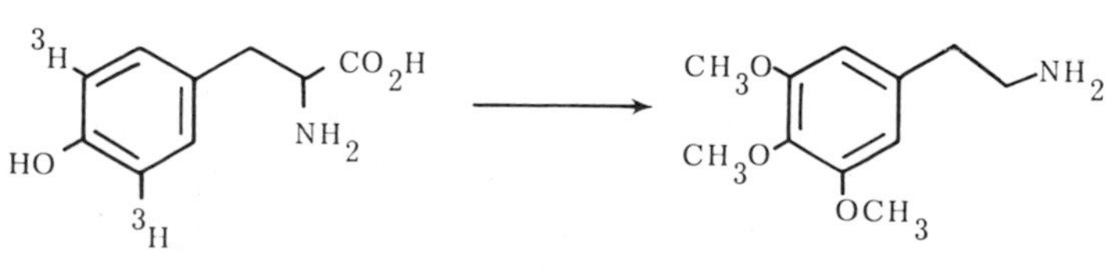

Tyrosine hydroxylase

27

28

~ 45% ³H retention

29

not observed

26   R=

30   mescaline

no ³H retained

2.2.3   Methyl Transferases

From the few alkaloid structures that have been presented so far it will be apparent that there are a number of one-carbon units introduced. The most common examples are the introduction of O-methyl and N-methyl groups.

All the evidence accumulated indicates that the methyl groups attached to oxygen and nitrogen in alkaloids are derived from the S-methyl group of methionine (**31**), by means of a transmethylation reaction mediated by a methyl transferase.

Methionine itself is not sufficiently reactive, and activation of the thiomethyl group is achieved by the formation of an intermediate sulfinium cation. In the cases studied in depth to date, methionine reacts with adenosine triphosphate (ATP, **32**) to form S-adenosylmethionine (**33**). The S-methyl group is now susceptible to nucleophilic attack by either N or O with transfer of the methyl group and formation of S-adenosylhomocysteine (**34**) (Scheme 1).

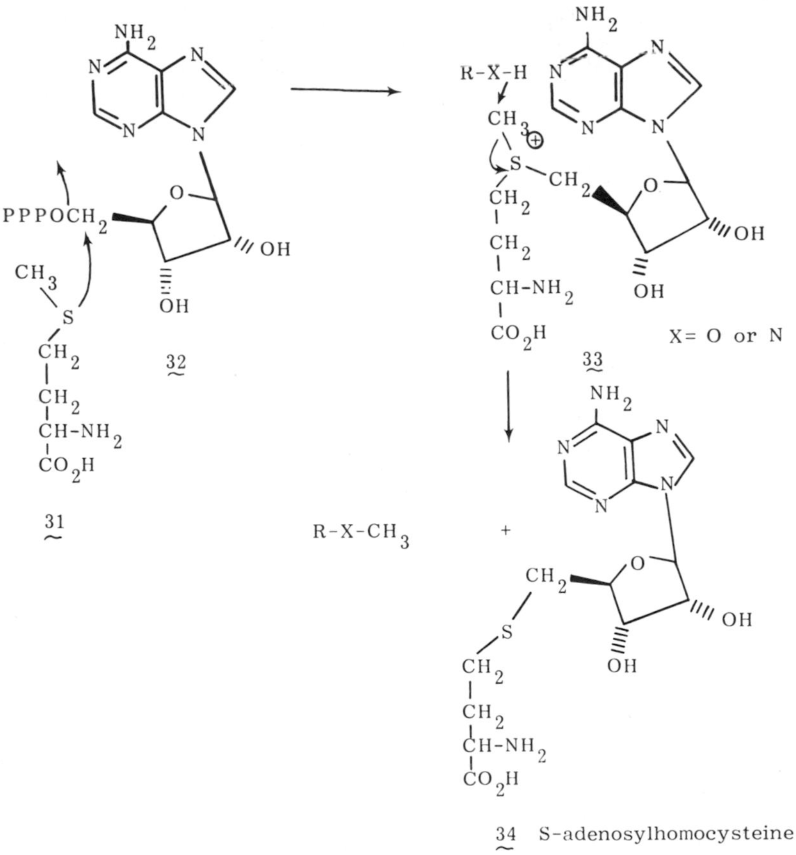

Scheme 1   Methylation of —OH and —NH groups by methionine.

This is not an oxidation–reduction process and all the hydrogens present on the methyl group are retained in the methylated product.

2.2.4 Amino Acid Decarboxylases and Transaminases

Relatively few alkaloids contain the carboxylic acid group derived from the parent α-amino acid if the nitrogen atom is retained. In other cases an amino acid may be utilized as substrate but the nitrogen atom not retained. The enzymes that carry out these processes are respectively amino acid decarboxylases and amino acid transaminases. The coenzyme for both of these is pyridoxal-5′-phosphate (35), and it is the aldehyde of this compound that reacts with the amine group of the amino acid to give a Schiff base (36). Two reactions are now possible for this intermediate. Loss of the α-proton gives a 1-carboxy imine such as 37, or loss of carbon dioxide gives an imine 38. In each of the subsequent reaction pathways, the driving force is the reformation of the aromatic pyridine nucleus as shown in Scheme 2. Hydrolysis of the resulting aromatic intermediates then affords an α-keto acid (39) or. an amine (40). Only in the latter case is the pyridoxal-5′-phosphate regenerated.

2.3 TECHNIQUES OF BIOSYNTHESIS

In the past 25 years enormous progress has been made in elucidating the pathways of alkaloid biosynthesis. Without such information it would not be possible to organize this monograph along biogenetic lines. Two terms, biosynthesis and biogenesis, are commonly used in discussing the formation of the products of secondary metabolism. It is important at this early stage that a clear distinction be made between these two concepts. *Biosynthesis* is the experimental study of the formation of secondary metabolites. *Biogenesis* is the hypothetical speculation on the precursor–product relationships in a biosynthetic pathway. It is based mainly on the visual dissection of a molecule into recognizable precursor fragments.

Biogenesis in the past 70 years has proved to be a very interesting and fruitful area of organic chemistry. Although not all of these speculations have proved to be correct, a triple purpose has been served: to enhance the concept of natural product structure elucidation along biogenetic lines, to encourage the development of experimental techniques to deduce the origin of the many complex skeleta, and to stimulate total synthesis along biogenetic lines.

There are two reactions that are of crucial importance in the biosynthesis of alkaloids. The first, the so-called Mannich reaction, is the reaction of an amine with an aldehyde or ketone. The intermediate Schiff base may then be (*a*) reduced or (*b*) condense with a nucleophilic carbon. These reactions are shown in Scheme 3.

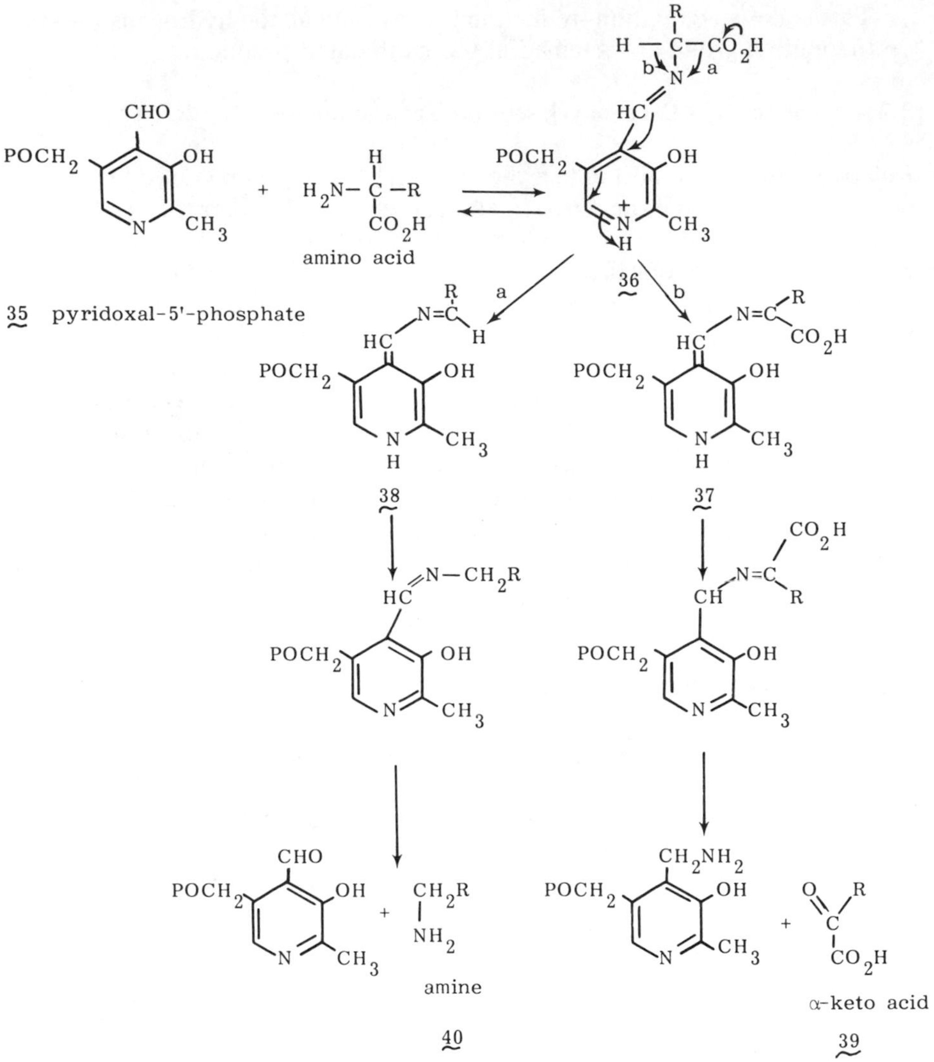

Scheme 2 The decarboxylation and transamination of amino acids.

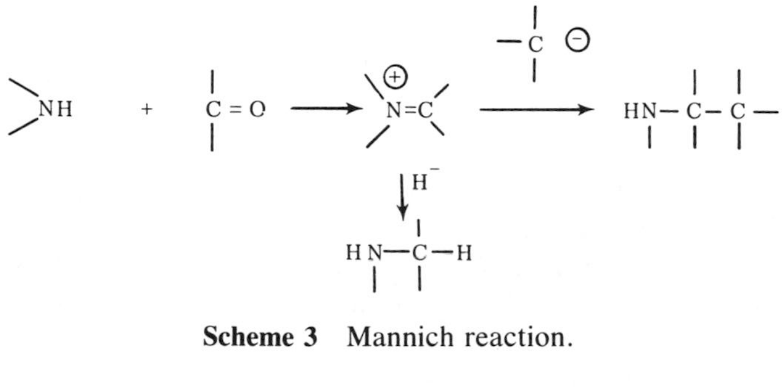

Scheme 3 Mannich reaction.

The second important reaction is the oxidative coupling of phenolic radicals. In this reaction a new carbon–carbon bond is formed by the joining of two radicals *ortho* or *para* to phenolic groups. An example is shown in Scheme 4.

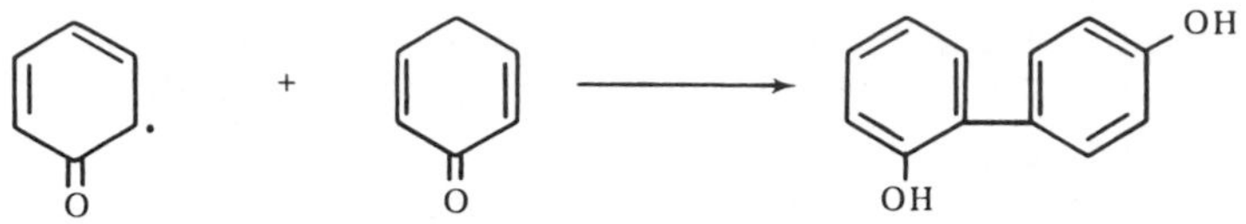

Scheme 4 Phenolic coupling.

As we shall see in subsequent chapters, either or both of these reactions have been used extensively in the so-called "biogenetic-type" synthesis of alkaloids of many classes.

2.3.1 Experimental Approaches

The study of alkaloid biosynthesis is fraught with problems both in technique and in logic. These problems have been discussed in detail elsewhere (see Literature) and will be the subject of only limited discussion here.

The general experimental approach has been to feed the projected isotopically labeled precursor to the intact organism (plant or fungus). After a suitable period of time the organism is worked up for the purified product(s) and the isotope content measured and preferably located. Most of these experiments have been carried out with the radioactive isotopes of carbon and hydrogen, ^{14}C and ^{3}H. Several stable isotopes, including deuterium, nitrogen-15, and carbon-13 have also been used. The latter isotope has recently been of considerable interest to workers in the field of biosynthesis, because the method permits the observation of individual carbon atoms. The use of ^{13}C nuclear magnetic resonance (NMR) in the study of biosynthetic pathways in higher plants has so far been somewhat limited, because of generally low levels of incorporation. A separate section will deal with the use of ^{13}C NMR in biosynthesis.

Studies of the biosynthesis of secondary metabolites using isotopic tracers fall into two quite general areas:

1. Investigation of the overall pathway
2. Investigation of the stereospecificity of intermediate reaction steps

To date, the former studies have predominated since the identity of at least some of the intermediates in a biosynthetic scheme must be established before any detailed consideration can be given to the mechanism of the interconversions. In addition, the labeling requirements for the substrate are much less stringent for elucidating the overall pathway than for the investigation of stereospecific processes.

Because of the structural complexity and variety of the products isolated from many intact organisms, precursors and intermediates can frequently be suggested based on a rational view of the necessary chemical steps and may be identified with confidence. A preferred approach however would be identification of the intermediates using tracers followed by purification of the enzymes that mediate the individual steps. By certain manipulations it may even be possible to block late steps in a pathway and thereby build up intermediate products.

In 1953, Adelberg illuminated a very important qualification to all biosynthetic studies. The problem relates to the possibility that a given compound may be interconvertible with a metabolite not on the direct pathway. An example is the reversible equilibrium between C and X in the sequence

$$A \longrightarrow B \longrightarrow C \longrightarrow D \longrightarrow products$$
$$\Updownarrow$$
$$X$$

Such interconversions are probably the exception rather than the rule, but it is something that we should be wary of. If the C to X to C process is rapid so that the pool of X is negligible, the problem is especially difficult. If the C to X to C process does not occur, sequential analysis may be useful technique.

Sequential analysis is a technique whereby each or many of the intermediates on a biosynthetic route are isolated or detected as their concentration increases and decreases with time. It will be evident that slow growth and slow metabolism are important for the success of this technique. Because metabolism is very rapid in microorganisms, it is rare to see this technique used successfully in this field. There are several examples where this technique has been used in the study of alkaloid biosynthesis in plants.

One simple situation which has been studied by Fairbairn and co-workers is the conversion γ-coniceine (41) to coniine (42) in *Conium maculatum* L. (Umbelliferae). Using paper chromatography it was found that the concentration of γ-coniceine (41) was at a maximum 1 week before that of coniine (42). This was in agreement with the idea that 41 was a precursor of 42.

A second example occurs in the indole alkaloid field, where Scott has studied the sequential formation of alkaloids in *Catharanthus roseus*. This series of experiments is discussed in more detail in Chapter 9.

41 γ-coniceine   42 coniine

2.3.2   Tracer Methods

The basic principles of radioisotope methodology are covered in several available textbooks and these are included in the reading list.

One point must be made clear at the very beginning of any discussion concerning biosynthetic experimentation. It is axiomatic that the labeled compound reaches the site of synthesis of the metabolite under study at a time when the enzyme systems mediating the synthesis are present and active.

Time is a crucial factor in biosynthetic studies, particularly in plants. Although many compounds, particularly the cell nutrients, are produced throughout the complete life cycle of the organism, it is erroneous to consider that this situation also obtains for secondary metabolites such as alkaloids. The carefully timed release of hormones in female mammals is one general example. In the alkaloid field mention can be made of the biosynthesis of scopolamine (43) in the roots of *Datura stramonium* (Solanaceae) at the preflowering stage, but not after fruiting.

Even if the organism (plant or microbe) is known to be producing alkaloids at the time of feeding and even if a known precursor, such as acetate or an amino acid is used, poor or no incorporation may result. There are two additional factors that should be taken into account, permeability barriers and transport. It is well known for example that the metabolically active phosphate esters penetrate cells walls poorly, if at all, and there is evidence that several other simple precursors may not reach the cell interior.

When a potential alkaloid precursor is fed to a growing plant, a very common procedure is to administer the compound in aqueous solution *via* a series of wicks passed through the stem of the plant. The assumption here is an obvious one; that the precursor will reach the site of alkaloid synthesis, be that the roots, the leaves, or the flowers, before it is completely metabolized. Transport or translocation of the precursor is always a problem if the site of alkaloid synthesis is not known, as is usually the case. It is probably one of the fundamental reasons why the levels of specific incorporation levels of precursors into alkaloids in plants is so low. From all the available data, it appears that nicotine (44) is synthesized in the roots of *Nicotiana*

43   scopolamine

44   nicotine

tabacum; it would therefore be erroneous to feed potential nicotine precursors *via* the stems on leaves and expect meaningful results.

There are, needless to say, numerous problems with microbial systems. It might be supposed that production of the secondary metabolite(s) would parallel the growth of the organism. Such a situation is rare. It is far more common that the secondary metabolites are produced near the end or after the active cellular growth phase. Consequently administering the labeled precursor at a time prior to active production of the metabolite may lead to poor incorporation at best, or at least extensive randomization of the label.

Two criteria, dilution value and degree of incorporation, are used to assess the effectiveness of a precursor.

The dilution value is defined as the ratio of the specific activity of the precursor (A_1) to the specific activity of the product (A_2), each being expressed on a molar basis. A low dilution value would therefore indicate a better precursor. One advantage of reporting dilution values is that the product need not be recovered quantitatively. A distinct disadvantage is that because pool sizes of intermediates and products may vary from one plant to another, a closer intermediate may afford a lower incorporation value (higher dilution). An additional disadvantage involves the problems surrounding a comparison of the efficiency with which a given precursor is incorporated into different compounds in the same organism.

Some of these disadvantages may be overcome by consideration of a second criterion, degree of incorporation, defined as the ratio $A_2M_1/A_1M_2 \times 100$, where A_1 and A_2 are defined as before, M_1 is the molar quantity of administered precursor, and M_2 is the molar quantity of recovered product. The term M_2 implies quantitative recovery of chemically and radiochemically pure material. The degree of incorporation is therefore a minimum value.

Even though incorporation into a product may be observed, yet another criterion should be applied to precursor assessment: namely the degree of randomization incurred during the conversion of a precursor to a product. The degree of randomization will be a function of the precursor (how general a metabolite it is) and of time. For example, because acetate is involved in the tricarboxylic acid cycle, $[2\text{-}^{14}C]$acetate will eventually give a mixture of acetates labeled at the 1 and 2 positions.

There is yet another cause of randomization which has already been observed in the discussion of the NIH shift. That is if an intermediate is symmetrical, labeling may effectively be randomized between two or more positions. One example occurs in the biosynthesis of nicotine (Chapter 3).

Implicit in the previous discussion was that only one site in the molecule was labeled. A more sophisticated approach to the study of precursor relationships involves the introduction of isotopic label into more than one atom of a molecule, so-called double-labeling studies.

There is an important theoretical point here. It is only very rarely necessary to doubly label the *same* molecule. Important exceptions occur in the use of ^{13}C-labeled precursors in which *adjacent* carbon atoms in the *same* molecule must be labeled for the needed coupling to be observed. In working with ^{14}C and ^{3}H it is necessary only to label one molecule with one label, a second molecule with another label, and mix the two molecules. In other words to obtain [4R-^{3}H,2-^{14}C]mevalonate, [4R-^{3}H]mevalonate (**45**), and [2-^{14}C]mevalonate (**46**), each of known specific activity, are physically

45 [4R-^{3}H] – mevalonic   **46** [2-^{14}C] –
                acid   mevalonic acid

mixed in known proportions. The advantages of double- or multiple-labeling studies are twofold:

1. Detection of *in vivo* cleavage.
2. Possibility of studying stereospecific reactions in the biosynthetic scheme.

Consider the incorporation of a unit X–Y into a complex alkaloid. We can label each of X and Y and show that each is incorporated. We can also put a label (*) separately into each part of the precursor X–Y (i.e. use X*–Y and X–Y*) and show that each is incorporated. However, if we mix X*–Y and X–Y* so that the ratio of specific activities is *n*, any unaccountable deviation from *n* in the product would indicate that cleavage of X–Y (degradation) had occurred at some point. Numerous examples will be discussed in the various chapters of this type of experiment.

An extension of the experiment discussed above is often useful in distinguishing mechanistic pathways or indicating the stereospecificity of a particular reaction. In the latter case it is normally necessary to begin with the precursor labeled sterospecifically. At this point it is worthwhile to recollect the terms R and S which serve to designate the configuration of a particular group at a given carbon atom.

The substituents around a given center are arranged in order of decreasing atomic number of the atom attached to the center. If two or more of these atoms have the same atomic number, the directly attached substituents are considered.

The substituents having been rank-ordered, observation is made down the bond joining the smallest group (often H) to the center, with the smallest group toward the observer and the remaining three away. A clockwise array of decreasing atomic number is designated S and an anticlockwise array is

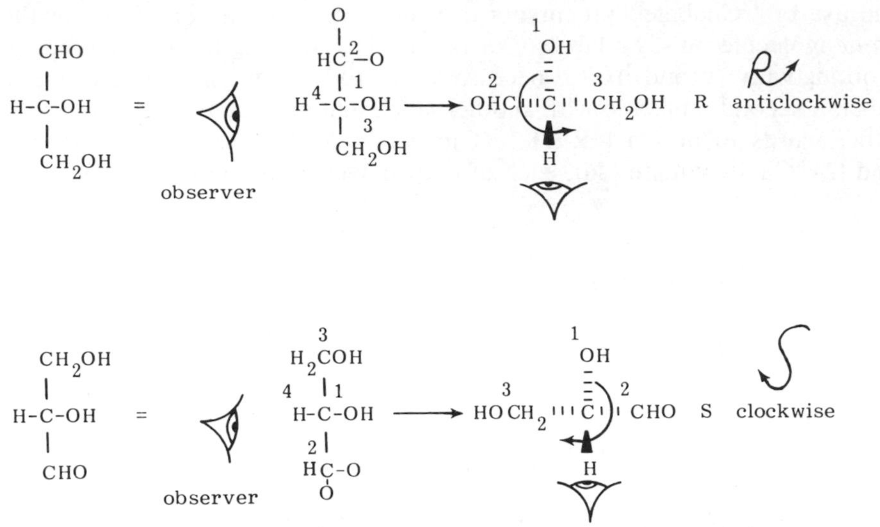

Figure 2 The R and S configurations of glyceraldehyde.

designated R. An application to the two glyceraldehydes is shown in Figure 2.

In biosynthetic studies it is often necessary to distinguish between two hydrogens on a methylene group. Of particular interest in terpenoid biosynthesis is the stereochemistry of the isomerization of isopentenyl pyrophosphate (**47**) to dimethylallyl pyrophosphate (**48**). Biologically, the two

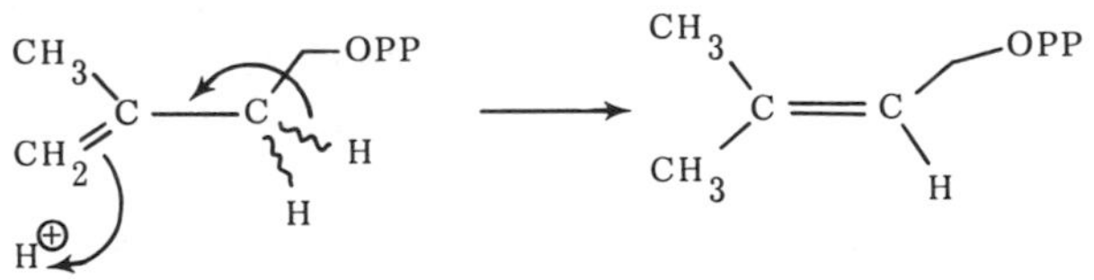

47 isopentenyl   48 dimethylallyl
~   pyrophosphate (IPP)   ~   pyrophosphate (DMAPP)

hydrogens at the carbon labeled 4 (from carbon 4 of mevalonate) are different. The reader would do well to assign the R and S configurations to the hydrogens on the carbon indicated of isopentenyl pyrophosphate.

Besides showing that an intermediate is important in the biosynthetic scheme by demonstrating that it is a precursor, are there any other techniques which can be used to show that a particular compound X is an intermediate?

Initial evidence may come from the isolation of X from the organism.

The condition here is that further metabolism of X is relatively slow. If X is metabolized rapidly, then at no time will an isolable pool of X build up.

The technique of *trapping* is designed to overcome this problem. Supposing that we are examining the sequence $Q \rightarrow X \rightarrow Y$. The experiment is begun by feeding labeled Q (a known precursor of Y) and at the same time nonlabeled X is added. The pool of X is therefore increased and the nonlabeled X is metabolized to Y. At the same time labeled X is produced from labeled Q. Therefore on a temporary basis at least, a pool of labeled X is available for isolation. Labeled X is then shown to be incorporated into Y.

There are many problems associated with this technique: (*a*) although a positive result may indicate that X is an intermediate in the formation of Y, it does *not* indicate that this is the only route between Q and Y; (*b*) the increased pool size of X may shut down the conversion of $Q \rightarrow X$ so that the pool of X never becomes labeled; and (*c*) the increased pool size of X may give rise to numerous products other than Y.

2.3.3 Stable Isotopes

The discussion to this point has centered on the use of radioactive isotopes in biosynthetic work, and there is no doubt that ^3H and ^{14}C have provided information of certain biosynthetic pathways in exquisite detail.

There is however a whole other side of biosynthetic work involving the use of precursors labeled with nonradioactive (so-called stable) isotopes. It is a valid question to ask why one would want to use stable isotopes in biosynthetic studies. Fundamentally there are probably three reasons:

1. No radioactive isotope is available for that particular element.
2. A stable isotope is required or is preferable for a certain type of experiment.
3. The corresponding radioactively labeled compound is less readily available.

Situation 1 commonly occurs in any work dealing with oxygen and nitrogen where the lowest decaying radioactive isotopes (^{15}O and ^{13}N) are very short lived ($t_{1/2}$ 2 and 10 min respectively). Subsequent examples throughout this monograph will illustrate the second and third cases.

The impression should not be left however that stable isotope work is a simple proposition; it is not. In fact there are many disadvantages and relatively few advantages to working with such compounds:

ADVANTAGES

1. The detection methods normally permit both quantitation and specificity of the labeled site directly.
2. The health and contamination problems of radioactive isotopes are substantially reduced.

DISADVANTAGES

1. The methods of detection are relatively insensitive.
2. The detection equipment is usually sophisticated (mass or NMR spectrometer).
3. The number of commercially available precursors is limited.
4. Reasonable quantities of the precursors are needed and these are expensive.
5. Enrichment of the precursor with label should be at least 90%.

The dilution factor normally encountered in biosynthetic studies is a particular problem. Initially it means that more substantial quantities of a precursor must be used and this may in itself alter the metabolism of the system. In addition, these requirements indicate that the specific incorporation should be relatively high (>1%), and for the most part this has prevented extensive use of stable isotopes in the study of secondary metabolites in plants.

The two most common techniques used in the detection of stable isotopes are mass spectrometry and NMR spectroscopy. The former technique is used for the detection of ^{18}O and ^{15}N, and the latter for ^{13}C and ^{2}H.

Mass spectrometry is a very useful technique for biosynthetic studies because only very small quantities of labeled product are required. A particularly good example of the use of ^{15}N in biosynthetic studies are the experiments by Floss and co-workers on the biosynthesis of pyrrolnitrin (**49**) using ^{15}N-labeled tryptophan derivatives. The fundamentals of the ^{15}N experiments are discussed here, but the other details of pyrrolnitrin biosynthesis are presented in Chapter 9.

Pyrrolnitrin (**49**) had previously been shown to be derived from tryptophan (**5**), but a question arose as to the origin of the two nitrogen atoms. The problem was solved by separately feeding the two tryptophans in which the nitrogen atoms were labeled with ^{15}N. Before doing the necessary experiments however it was vital to understand some aspects of the mass spectrum of pyrrolnitrin (**49**). Normally the molecular ion is observed at m/e 256 with a major fragment ion at m/e 229 derived by loss of HCN from the pyrrole ring.

When the amino acid nitrogen was labeled, the resulting pyrrolnitrin (**49**) displayed the isotope satellites associated with the molecular ion but not with the M^+–HCN fragment. The amino acid nitrogen is therefore the one present in the pyrrole nitrogen. In the experiment where the indole ring nitrogen is labeled with ^{15}N both the molecular ion *and* the M^+–HCN ion showed isotopic enrichment. The indole ring nitrogen therefore becomes the nitro-group nitrogen of pyrrolnitrin (**49**), as shown in Figure 3.

When deuterium labeling is used, detection of the site of labeling may be made with either proton NMR spectroscopy or mass spectrometry. The following example relies on the use of the NMR technique and concerns the

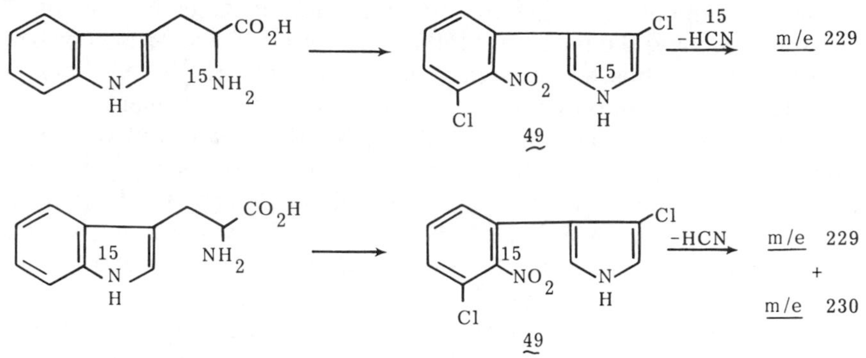

Figure 3 Biosynthetic origin of the nitrogen atoms of pyrrolnitrin (**49**).

stereospecificity of the phenylalanine ammonia lyase (PAL) reaction as deduced by Ife and Haslam. When the two phenylalanines where deuterium is stereospecifically located at the 3R and 3S positions were used as precursors, in one case deuterium was lost and in the other it was retained. NMR spectroscopy was used in this example to establish the configuration of the deuteriums in the starting materials and the location of the deuterium in the products.

Typically all-proton cinnamic acid methyl ester shows two doublets at 6.45 and 7.75 ppm due to the C-2 and C-3 protons respectively. This pattern was observed when [3S-^{2}H]L-phenylalanine (**50**) was used as a precursor, but the methyl cinnamate from [3R-^{2}H]L-phenylalanine (**51**) showed only a broad singlet at 6.45 ppm. In this instance the deuterium is stereospecifically retained. Together these results demonstrate that phenylalanine ammonia lyase proceeds by way of a *trans* elimination of ammonia.

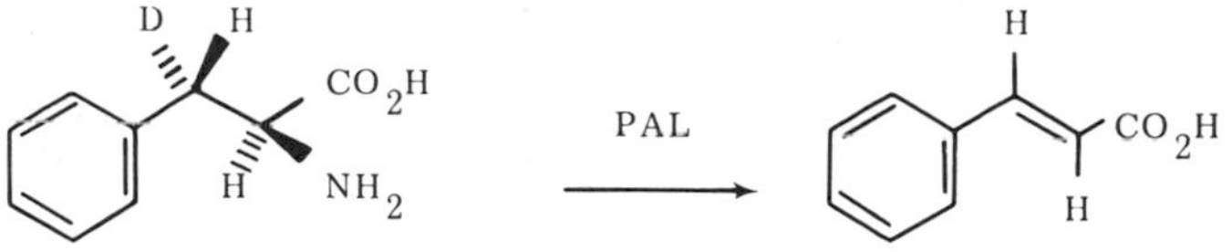

50 [3S - ^{2}H] - L-phenylalanine

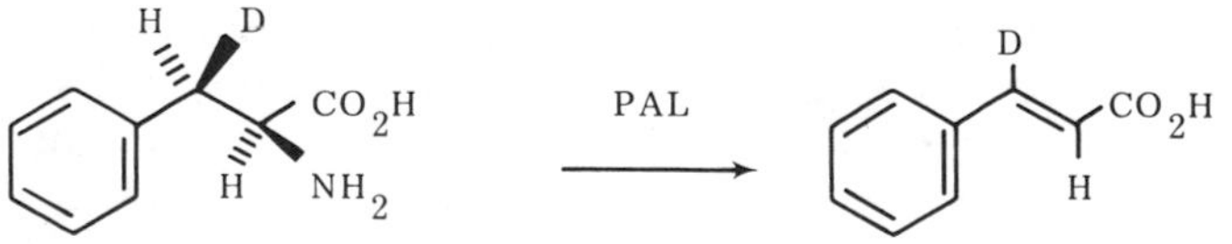

51 [3R - ^{2}H] - L-phenylalanine

This is not the place for an intimate discussion of the techniques of ^{13}C NMR spectroscopy; the books of Stothers and of Levy and Nelson provide excellent introductions to this area. Some comments on ^{13}C NMR as the technique applies to biosynthetic work are appropriate however.

The problems of the accumulation of ^{13}C NMR data are well known and center mainly on the low (1.1%) natural abundance of the isotope. Similarly well known are the broad (250 ppm) chemical shift range and the fact that since the spin of carbon-13 is the same as that of the proton, the spin multiplicities of carbon spectra are the same as those in proton spectra.

The technological developments which have made ^{13}C NMR a fundamental necessity in both biosynthetic and structure elicidation work are the availability of high-stability, high-field spectrometers and the development of Fourier transform spectroscopy. The combination of these developments has made obtaining the ^{13}C NMR spectrum of 20 mg of a compound of molecular weight 300 relatively routine.

The proton-noise decoupled spectrum collapses all the multiplets due to ^{13}C-H coupling to a single line. In addition, the low probability of two ^{13}C atoms being adjacent, and therefore coupling, assures that the resulting line will be a singlet unless the sample is particularly concentrated. Unfortunately although ^{13}C-^{13}C coupling constants would be of considerable value in structure-elucidation studies, they are rarely observed. However in biosynthetic studies these couplings are observed in certain experiments and this information may be extremely useful.

Any biosynthetic work involving ^{13}C NMR spectroscopy requires a number of criteria to be met before the experiment can (should) be performed. One of the most critical factors is the level of incorporation of the corresponding radioactive isotope. Although lower levels can be examined, the practical lower limit of specific incorporation is about 0.1%.

A second important factor is that all the carbon resonances in the product under investigation should be unambiguously assigned. With all previously determined data available it is possible that the compound under study will be just a simple variant of a well-established series. However, in other instances model compounds must be used in order to establish the correct assignments.

Other factors to be considered include the reproducibility of the intensities in the proton-noise decoupled spectrum and the availability of an efficient method of synthesis for the labeled precursor.

When all these problems have been resolved the use of ^{13}C NMR in a biosynthetic study becomes an enormously powerful technique. For potentially it can provide the degree of incorporation and the precise location of the site of label from a single NMR spectrum, all without chemical degradation.

A number of studies of alkaloid biosynthesis have been made using ^{13}C NMR and these are discussed in the appropriate chapter. At this juncture it is instructive to discuss the incorporation of singly and doubly labeled ^{13}C precursors into a simple compound.

Consider a simple theoretical example such as hexanoic acid (52), which is derived from three acetate units. When [1-^{13}C]acetate is used as a precursor, labeling is expected at carbons 1, 3, and 5 and from [2-^{13}C]acetate

$$\overset{5}{CH_3}-CH_2\overset{3}{CH_2}-CH_2\overset{1}{CH_2}CO_2H$$

52 hexanoic acid

at carbons 2, 4, and 6. Each precursor will therefore give rise to the specific enrichment of the appropriate carbons in comparison with the natural abundance spectrum.

What is the expected result if an acetate precursor is used in which both the 1- and 2-carbons are 90% enriched with ^{13}C within the same precursor? This is *not* the same as the physical mixing of [1-^{13}C]- and [2-^{13}C]acetates. This is a compound that has ^{13}C labeling on the *adjacent* carbon atoms in the same molecule.

Feeding experiments with compounds such as [1,2-$^{13}C_2$]acetate are very useful because they establish whether the bond between carbon 1 and carbon 2 remains intact in the biosynthetic process. If this bond is not broken, there is a high probability of ^{13}C being at adjacent carbons and this will lead to a situation in which coupling can occur. If the bond is broken during the biosynthetic process, the probability of a molecule having two adjacent ^{13}C atoms is now low again, and no coupling is to be expected. In this instance only the enriched singlets comparable to a simple single-labeling experiments will be observed. When the doubly labeled precursor is incorporated intact, the two adjacent carbon atoms carrying the label will appear as "triplets" in the product. The immediate question is, why a "triplet"? The two outside lines represent the doublet from the ^{13}C-^{13}C coupling of the adjacent atoms present from the original doubly labeled acetate. The center line is a singlet which occurs at the normal chemical shift for that carbon. Its magnitude has contributions from three sources: (*a*) a natural abundance component from unlabeled product, (*b*) a labeled component from acetate which was originally only labeled at one site, and (*c*) a labeled component from doubly labeled acetate which has been recycled through the citric acid cycle.

It is the doublet components in this system which are of interest. First, they establish that an intact acetate unit is incorporated into the compound under study. Second, the other "triplet" signal in the spectrum indicates the chemical shift of the second ^{13}C which is coupling. If more than two triplets are observed, the ^{13}C-^{13}C coupling constants can be used to establish which carbons are coupling.

But why in a molecule such a hexanoic acid is there no coupling of ^{13}C-labeled carbons at C-2 and C-3? In other words why is not the C-2 resonance considerably more complex? The reason is again one of probability, namely that the probability of two isotopically enriched carbon atoms being adjacent in the labeled product is very low, unless the level of incorporation is exceptionally high.

Returning to our original problem of hexanoic acid, it should be apparent that according to the polyacetate theory, three intact units of acetate are involved in the biosynthesis, and consequently three sets of ^{13}C-^{13}C coupling constants will be observed.

One aspect of this work which deserves special mention is that these experiments thoroughly confirm the earlier observations (see Section 2.4) that in the formation of malonate from acetate, the same unit of carbon dioxide is first added and then removed. If this were not so there would be a 50% loss of ^{13}C label from the C-1 of acetate during the biosynthetic process. This would selectively increase the isotopic enrichment in carbons labeled by C-2 of acetate.

For two particularly elegant uses of ^{13}C NMR in biosynthesis the reader is referred to Section 3.3.6, where the biosynthesis of the tropic acid protion of atropine is discussed, and to Section 9.3.2, where the biosynthesis of pyrrolnitrin is presented.

2.3.4   Administration of the Precursors

Most of the alkaloids under discussion occur in plants, so we should examine methods for the administration of precursors to plants.

Plants absorb nutrients in two ways: (*a*) in the form of carbon dioxide *via* their photosynthetic organs or (*b*) in aqueous solution through their roots. The roots would therefore appear to be a very logical place to administer a precursor. In these cases the plants are cultivated hydroponically to avoid microbial contamination.

An alternative technique is to remove the roots and place the cut ends of the stem in the nutrient solution. Uptake is improved by cutting the stems under water and exposing the plant to bright light and a good air supply. If it is necessary to retain the roots to permit long-term growth, wick feeding, as discussed previously in this chapter, can be used. Direct injection into a hallow leaf or stem has also been utilized, and when small amounts of plant material will suffice leaf discs can be floated on the substrate solution.

With all the problems and difficulties involved in the study of the biosynthesis of alkaloids it is somewhat surprising that so much is known about how alkaloids are produced from simple precursors. As we shall see this information is still incomplete and was gained with considerable effort.

The subsequent chapters are organized on the basis of the known or probable biosynthetic pathways of the various alkaloid groups, beginning with the simplest precursors. The aim of each chapter is to discuss the occurrence, isolation, biosynthesis, synthesis, and pharmacologic activity of the alkaloid groups.

Complete coverage of such a vast field is an impossible task in a book of this size. However, it is hoped that the principal alkaloids of interest can be covered in sufficient detail to whet the appetite of the interested reader.

2.4 BIOSYNTHESIS OF ALKALOID PRECURSORS

The preceding section discussed some of the techniques of biosynthesis and the problems involved. It is now time to turn our attention to the precursors of the alkaloids. Since this book is organized on the basis of certain precursor relationships to the alkaloids, it would seem appropriate to discuss the way in which nature produces these precursors. An in-depth treatment of this topic is beyond the scope of this book; for that the reader is referred to the major texts on biochemistry.

It was mentioned previously that the amino acids ornithine (1), lysine (2), anthranilic acid (7), phenylalanine (3)/tyrosine (4), tryptophan (5), and histidine (6) are important in the formation of alkaloids. The biosynthesis of the terpenes is also discussed in this chapter.

2.4.1 Biosynthesis of Ornithine

Ornithine (1) is one member of a highly reversible sequence of amino acids having five carbon atoms, including glutamic acid (53) and proline (54). All of these compounds are derived from α-ketoglutaric acid (55).

Although glutamic acid (53) is the primary metabolite from which the other two amino acids are derived, the steps involved in these processes differ in various organisms.

Figure 4 indicates the biosynthetic relationships between these compounds. The clear distinction should be noted between the origin of the nitrogen atom in Δ^1-pyrroline-2-carboxylic acid (56) and Δ^1-pyrroline-5-carboxylic acid (57). In the first case it is derived from the δ-amino group of glutamic acid (53) and in the second instance from the α-amino group.

2.4.2 Biosynthesis of Lysine

The next higher homologue to the ornithine series of amino acids is the lysine–pipecolic acid group, having six carbon atoms.

The biosynthesis of lysine (2) is somewhat more complex than that of ornithine (1), for two quite distinct pathways have been recognized. In bacteria, green algae, some fungi, and some higher plants, aspartic acid (58) and pyruvic acid (59) are involved and α,ε-diaminopimelic acid (60) is a key intermediate. In other algae, fungi, and certain plants lysine (2) is derived

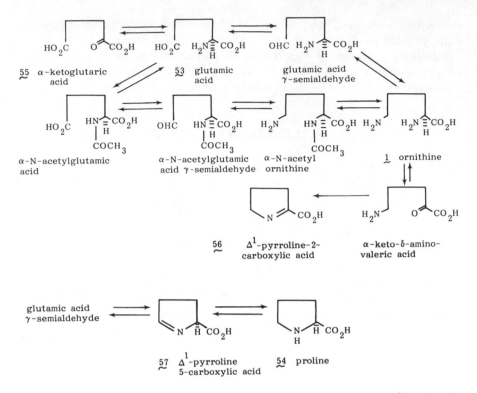

Figure 4 Biosynthetic interrelationships of amino acids in the glutamic acid–proline–ornithine group.

from acetate and α-ketoglutaric acid (**55**) *via* α-aminoadipic acid (**61**). These two schemes are outlined in Schemes 5 and 6 respectively.

The pathways are distinguished by using [4-^{14}C]aspartic acid as a precursor. Such an intermediate affords lysine *specifically* labeled at the 4-position only from the diaminopimelic acid pathway.

2.4.3 Biosynthesis of Phenylalanine and Tyrosine

In general terms the biosynthesis of the C_6-C_3 unit of phenylalanine (**3**) and tyrosine (**4**) follows the same metabolic route in all the organisms thus far investigated.

Shikimic acid (**62**) is the key precursor in this sequence and the first step is the formation of 5-enolpyruvylshikimate-3-phosphate (**63**). This reaction involves the transfer of an unchanged phosphoenolpyruvate to shikimate-3-phosphate (**64**). Sprinson and co-workers have suggested that the mechanism involves the formation of an intermediate methyl group as shown in Scheme 7.

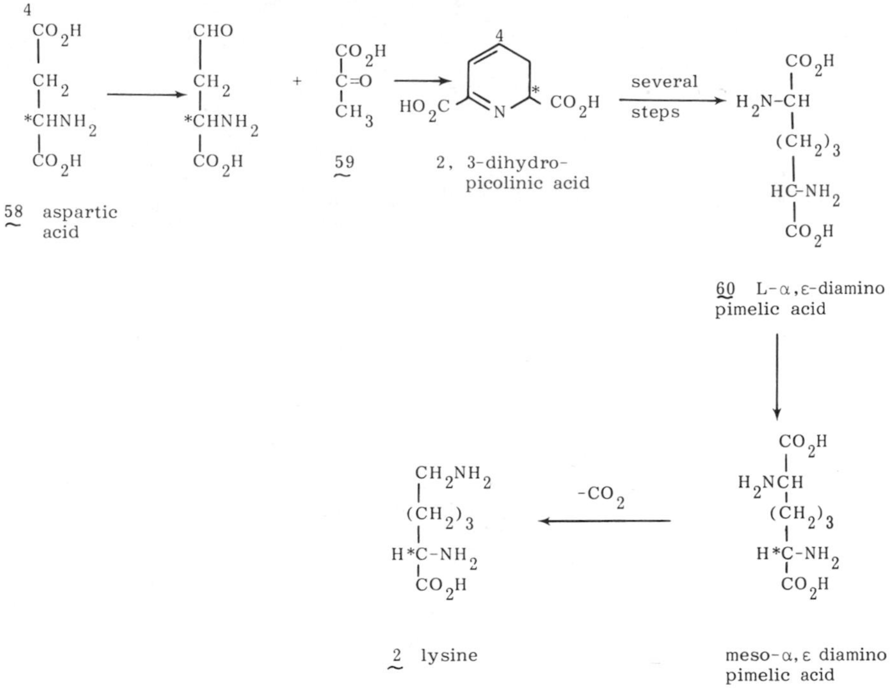

Scheme 5   The diaminopimelic acid pathway for the formation of lysine (**2**).

Trans-elimination of phosphoric acid transforms 5-enolpyruvylshiki-mate-3-phosphate (**63**) into chorismic acid (**65**) with specific loss of the 6R-proton. A two-stage mechanism has been postulated to explain this stereochemical result.

At this point a number of synthetic routes to essential metabolites evolve. Two of these routes, the formation of phenylalanine (**3**)/tyrosine (**4**), and the formation of anthranilic acid (**7**) and tryptophan (**5**), are pertinent to the present discussion.

Claisen rearrangement of chorismate (**65**) by chorismate mutase affords yet another famous acid, prephenic acid (**66**), which now has the three-carbon unit at the 1-position. It is at the stage of prephenic acid (**66**) that the metabolic routes to phenylalanine (**3**) and tyrosine (**4**) diverge, for **3** is not converted into **4** in bacteria and microorganisms. Phenylalanine hydroxylase from mammalian liver transforms **3** to **4** under aerobic conditions.

In the subsequent discussions of specific alkaloid groups there are many examples where tyrosine (**4**) is a specific precursor, but phenylalanine (**3**) is not, even though phenylalanine (**3**) can be a precursor of tyrosine (**4**) in higher plants. Clearly compartmentalization of enzyme activity is crucial here.

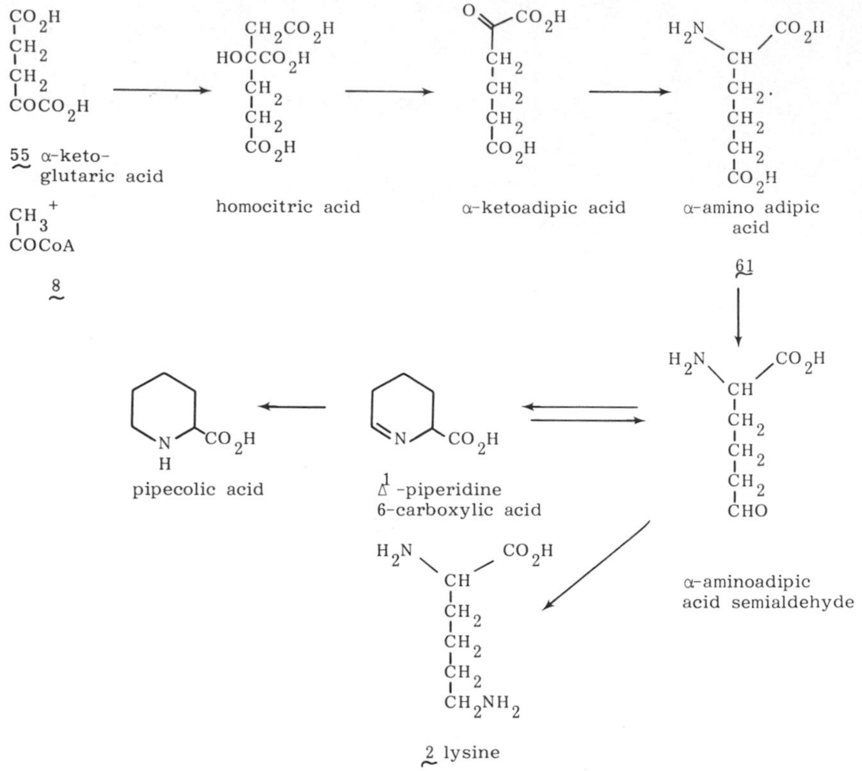

Scheme 6 The α-aminoadipic acid pathway for the formation of lysine.

The enzymes involved in the steps leading to phenylpyruvic acid (**67**) and *p*-hydroxyphenylpyruvic acid (**68**) are prephenate dehydratase and prephenate dehydrogenase respectively. A transaminase then stereospecifically introduces the amino group.

Two enzymes carrying out these steps and having multiple activity have been obtained from several bacteria. One of these, the so-called P-complex, contains chorismate mutase and prephenate dehydratase which could not be separated by diethylaminoethyl (DEAE) cellulose chromatography. The second complex, the T-protein, was purified by gel filtration to give a constant ratio of two enzyme activities (chorismate mutase and prephenate dehydrogenase). In yeast however, two chorismate mutases were found, each readily separable from the associated prephenate dehydratase and prephenate dehydrogenase activities.

The schemes just outlined have been deduced mainly from data obtained by studying intermediates and enzyme systems in bacteria and microorganisms.

Tracer studies have confirmed that the formation of shikimic acid (**62**) in plants follows the same route as that in microorganisms, namely a condensation of D-erythrose-4-phosphate (69) and phosphoenolpyruvate (70) (Scheme 8). Several of the enzymes involved in these transformations have been obtained from plants. Shikimic acid (62) is also a precursor of aromatic amino acids in plants, but the enzymes involved in these processes in plants have been less widely studied.

2.4.4 Biosynthesis of Anthranilic Acid

We have seen that chorismic acid (65) is a very important intermediate on the biosynthetic route from shikimic acid (62) to the aromatic amino acids phenylalanine (3) and tyrosine (4). It is also important in the formation of

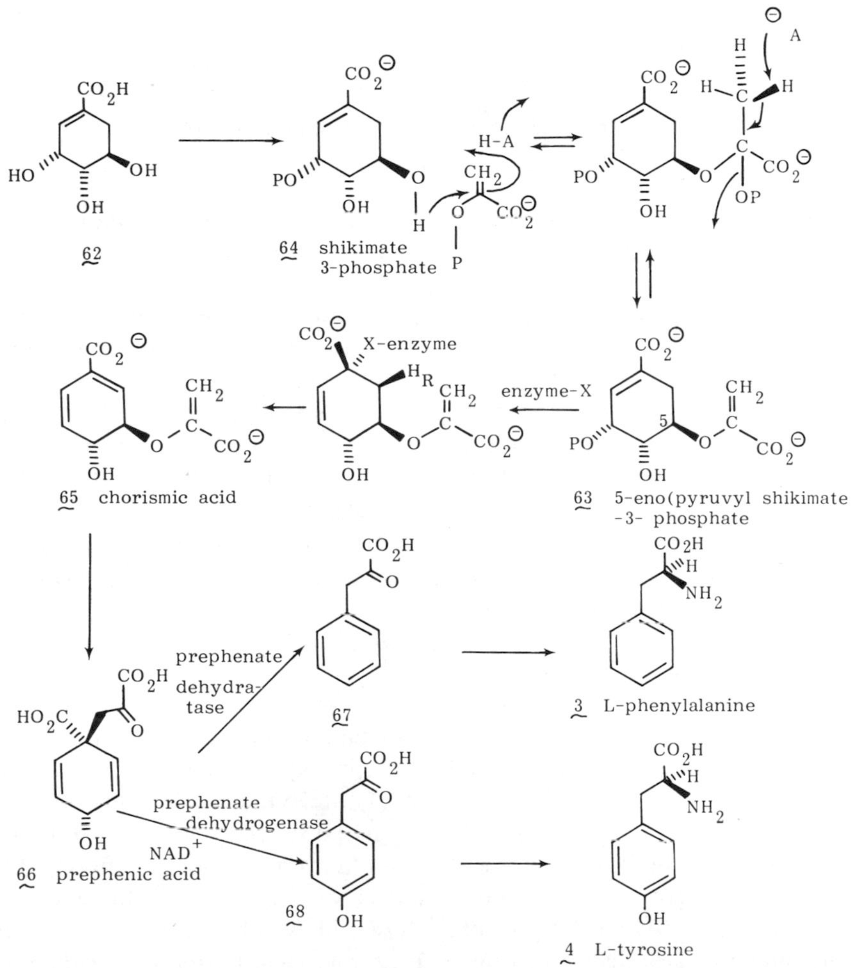

Scheme 7 Formation of L-phenylalanine (3) and tyrosine (4) from shikimic acid (62).

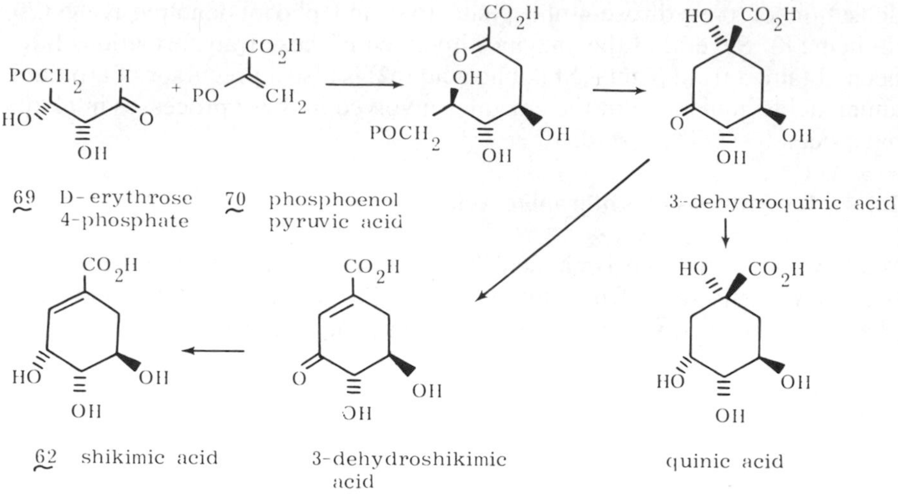

Scheme 8   The biosynthesis of shikimic acid (**62**).

two other aromatic amino acids, anthranilic acid (**7**) and tryptophan (**5**), and indeed anthranilic acid (**7**) is a precursor of tryptophan (**5**).

The first steps in this complex reaction sequence involve the enzyme anthranilate synthetase, which catalyzes the formation of anthranilate from L-glutamine (**71**) and chorismic acid (**65**). The amine nitrogen of glutamine (**71**) is transferred during this process, although ammonia can act as an alternative (in some instance sole) source. No intermediates have been detected in this pathway, which is thought to proceed through the cyclic intermediate **72**. The enolpyruvate group has been detected as pyruvic acid from this sequence. To date, this scheme (Scheme 9) and the enzyme system involved have only been studied in bacteria. It is generally assumed that the same pathway also operates in fungi and higher plants, although evidence for this is quite preliminary.

2.4.5   Biosynthesis of Tryptophan

There is only one route to the important amino acid tryptophan (**5**), but, depending on the organism, intermediate steps may be observed. In bacteria, it was found that both anthranilic acid (**7**) and indole (**73**) could replace tryptophan (**5**) in L-tryptophan-requiring mutants. Some of these mutants could selectively be blocked to accumulate anthranilic acid or indole. The carboxylic acid group of anthranilic acid is lost in the early stages of formation of indole glycerol-3-phosphate (**74**), in which C-2 and C-3 of the indole nucleus are derived from C-1 and C-2 and 5-phosphoribosyl-1-pyrophosphate (**75**) by an Amadori-type rearrangement.

Tryptophan synthetase obtained from *Escherichia coli* is comprised of

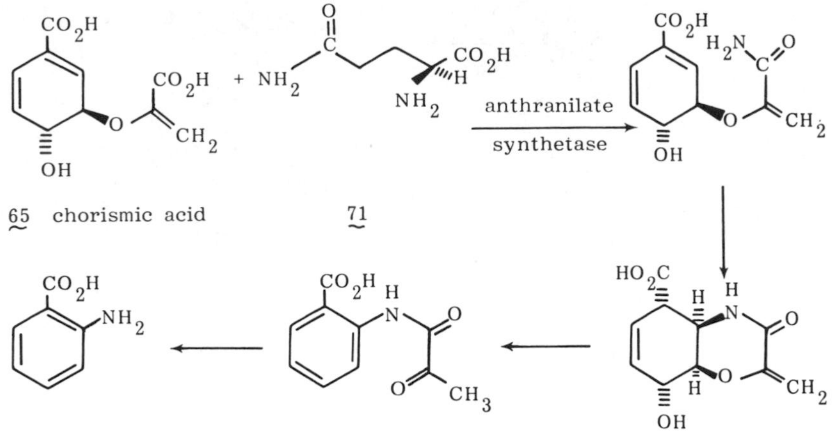

65 chorismic acid   71

7 anthranilic
acid   72

Scheme 9 Formation of anthranilic acid (7) from chorismic acid (65).

two subunits. One of these, the α subunit, releases D-glyceraldehyde (26) from 74 to afford indole (73). The second, β$_2$, subunit catalyzes the condensation of indole (73) with L-serine (77), for which pyridoxal 5-phosphate is a cofactor, to afford L-tryptophan (5) (Scheme 10). When the α and β$_2$ subunits are mixed however to give the complete L-tryptophan synthetase complex, free indole was not observed at any point of analysis, and the reaction was considerably faster than the sum of each of the individual reactions using the α and β$_2$ subunits.

The L-tryptophan synthetases of fungal origin appear to be undissociable protein species which convert indole glycerol-3-phosphate (74) to L-tryptophan (5) directly. The synthesis of tryptophan in plants is thought to follow the same route, and tryptophan synthetase that catalyzes the reaction between serine (77) and indole (73) has been obtained from *Nicotiana tabacum*.

A second route to anthranilic acid (7) operates in mammalian tissues and some microorganisms, where tryptophan (5) may be catabolized to produce 7 *via* formylkynurenin (78) and kynurenin (79) (Scheme 11). Nicotinic acid is also an end product of this route, but this is not of significance in higher plants. There are several alkaloids derived from anthranilic acid (7) by this route, and these are discussed in Chapter 7.

2.4.6 Biosynthesis of Histidine

The biosynthesis of L-histidine (6) is very closely connected with purine metabolism. Condensation of 5-phosphoribosyl-1-pyrophosphate (75) with adenosine tryphosphate (ATP) (32) affords an intermediate quaternary species which is cleaved and undergoes an Amadori rearrangement. The inter-

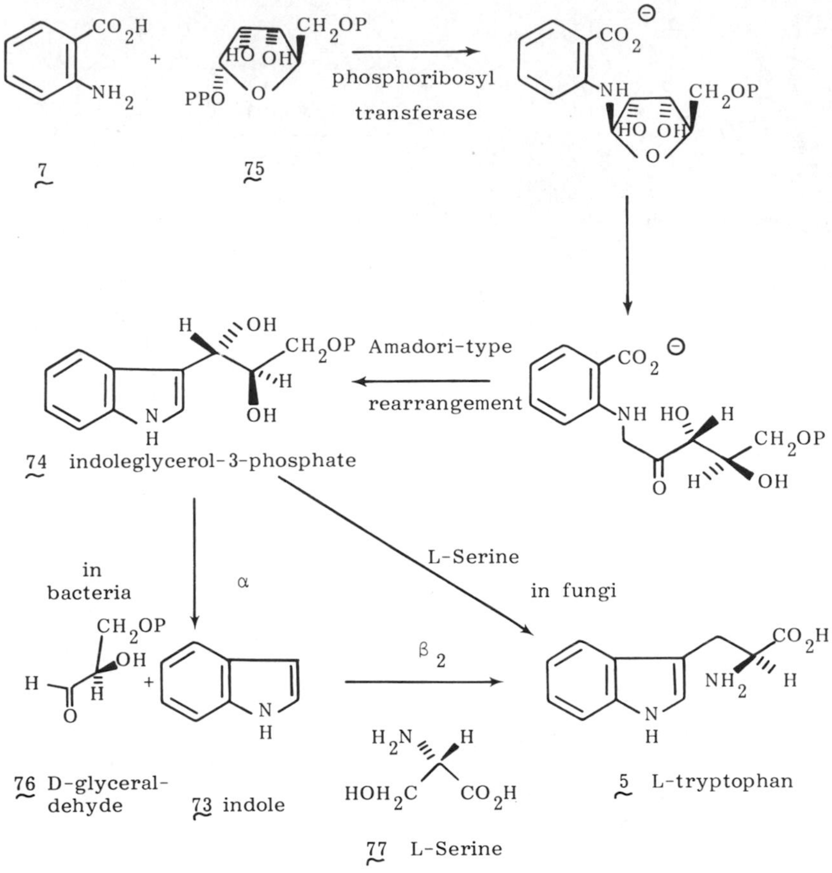

Scheme 10 Formation of L-tryptophan (**5**).

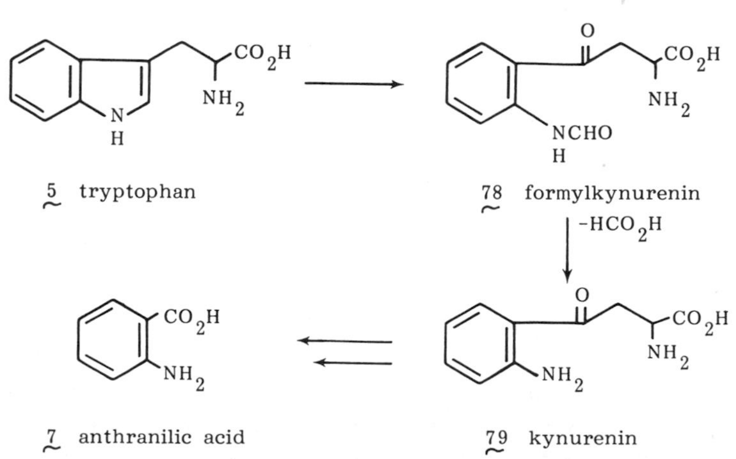

Scheme 11 Catabolism of tryptophan (**5**) to anthranilic acid (**7**).

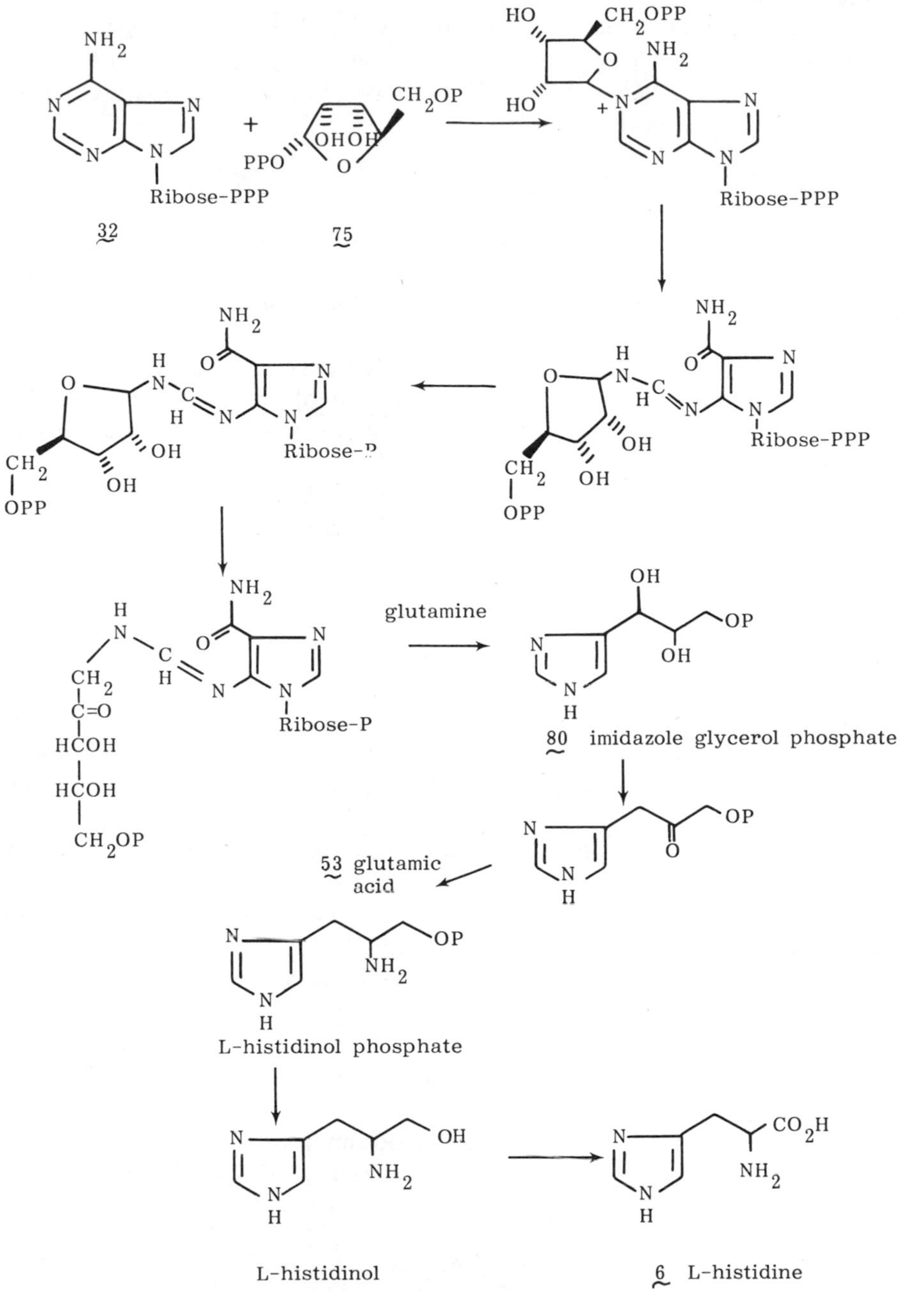

Scheme 12 Formation of histidine (**6**).

80 imidazole glycerol phosphate

53 glutamic acid

L-histidinol phosphate

L-histidinol

6 L-histidine

32

75

glutamine

Ribose-PPP

Ribose-P

mediate subsequently reacts with glutamine (**71**) to give D-erythroimidazole glycerol phosphate (**80**).

The conversion of **80** to L-histidine (**6**) has been well investigated in microorganisms and has been shown to follow the sequence outlined in Scheme 12. Once again it is thought that an identical route operates in the formation of L-histidine in plants.

2.4.7 The Polyacetate Pathway

There exists in nature an extremely diverse group of natural products based on the linear combination of two-carbon units. These compounds range from the fatty acids through the numerous aromatic compounds derived by cyclization of the chains of two-carbon units.

Whereas most ideas concerning the biosynthesis of natural products developed during the 1950s and 1960s, it was actually in the last century that the first evidence for the importance of acetate in the formation of secondary metabolites was recognized.

Collie and Myers in 1893 were investigating the structure of dehydroacetic acid (**81**), and found that treatment with strong sodium hydroxide gave the aromatic compound orcinol (**82**). The structure deduced for dehydroacetic acid on the basis of this rearrangement was incorrect, but the stimulus was there. For in the same year, Collie indicated that such condensations might be models for the formation of certain organic natural products. Fourteen years later Collie wrote of the "manner in which the group —CH₂—CO— can be made to yield . . . a very large number of interesting compounds: the chief point of interest being that these compounds belong to groups largely represented in plants." Collie called these compounds *polyketides*.

Thirty years later experiments with yeast suggested that acetate might be involved in steroid biosynthesis. By the early 1950s tracer studies had established that acetate was crucial to primary metabolism and that the active form of acetate was acetyl coenzyme A (**8**).

It was not until 1953 that the original ideas of Collie, as applied to natural aromatic compounds, were further considered. Birch and Donovan found that the location of the oxygen atoms (the oxygenation pattern) of many aromatic compounds could be explained in terms of cyclization of $CH_3(COCH_2)_n CO_2H$ units.

It was also Birch who, in 1955, provided the first experimental evidence in support of these ideas when 6-methylsalicylic acid (**83**) was shown to be

81 dehydroacetic acid   82 orcinol   83 6-methyl
                                                 salicylic acid

derived from [1-^{14}C]acetate according to the polyacetate hypothesis. Subsequent work in many systems and on many different aromatic compounds has provided extensive confirmation of the polyacetate theory.

Although the term polyacetate theory is used, the concept is actually a much wider one for there are several so-called starter units (i.e. groups which begin the process of adding acetate units). Some examples are propionyl, benzoyl, and cinnamoyl.

In practice there are relatively few alkaloids derived solely from acetate, if alkaloids containing a terpene unit are excluded. However, they do form an interesting, rapidly expanding group and are discussed separately in Chapter 6.

2.4.8 Biosynthesis of Terpenoids—Introduction

The third major pathway in the biosynthesis of alkaloids is concerned with those alkaloids derived not from amino acids but from terpenoid precursors. In addition there are over 1000 indole alkaloids derived from tryptophan and a monoterpene unit, so that this discussion is highly relevant as a prelude to a discussion of this important group of alkaloids.

The terpenes are a unique, highly diverse group of compounds. They are structurally diverse, yet a thread of commonality concerning their biosynthetic origin allows many apparently unrelated compounds to be viewed in unifying perspective. Some illustrative examples will be presented shortly.

The terpenes have held a special interest to organic chemists for nearly 100 years. Indeed, the names of many great organic chemists are associated with this area of research; they include Perkin, Baeyer, Wieland, Meerwein, Karrer, Butenandt, Ruzicka, Doisy, Sir Robert Robinson, Reichstein, Diels, Alder, Bloch, Lynen, Sir Derek Barton, Sir John Cornforth, Prelog, and of course Woodward. The name of Woodward, like that of Sir Robert Robinson, will come up time and again in this book.

The composition of the main terpene groups increases from a five-carbon unit by five-carbon units:

C_5	Hemiterpene	C_{20}	Diterpene
C_{10}	Monoterpene	C_{25}	Sesterterpene
C_{15}	Sesquiterpene	C_{30}	Triterpenes

There are examples of alkaloids in each of these groups except the sesterterpenes.

The five-carbon unit has the 2-methylbutane structure and is colloquially known as the isoprene unit (84). It is not uncommon to find the terpenes referred to as "isoprenoids."

A characteristic of many terpenes is that their skeleta can be dissected in terms of five-carbon units having the isoprene skeleton. When that is possible the compound is said to "follow" the isoprene rule. But how did these compounds come to be regarded and grouped in this way?

In all fairness it cannot be said to have started with Faraday, who in 1826 deduced that rubber was a polyunsaturated polymer of pentadiene having the molecular formula $(C_5H_8)_n$. Carotene was subsequently shown to be a $(C_5)_n$ compound, and Wallach determined that some volatile terpene hydrocarbons had the molecular formula $C_{10}H_{16}$ whereas others had a formula $C_{15}H_{24}$. Several other workers were also busy investigating the nature of rubber and found that pyrolysis afforded isoprene (**85**), which could be polymerized to rubber.

<u>84</u>   isoprene unit   85   isoprene

However it was Wallach who in 1887 wrote: "Such a structure for isoprene . . . allows a polymerization to terpenes, sesquiterpenes, etc., to appear reasonable. . . ." Wallach then went on to delineate some examples of his ideas. Although the structures were not correct, the idea was there. As with many good ideas it remained dormant for a long period even as more and more structures were being elucidated. Ruzicka over a period of many years is responsible for the development of the isoprene rule, which could assist in the elucidation of new structures and correct previously suggested structures.

In *the* classic paper of the isoprene theory Eschenmoser *et al.* finally demonstrated how the biogenetic isoprene rule could explain the formation of so many diverse compounds *not* on the basis of their structure but rather on their biogenesis. Just prior to this paper, Woodward and Bloch indicated how cyclization of isoprene units might lead to the steroids.

2.4.9   Formation and Rearrangement of Terpenoid Units

One of the most fascinating areas of biosynthesis in the past 25 or so years has been the study of how the many thousands of compounds referred to as terpenoids are produced naturally. The story is a long and fascinating one and only a simple outline of these reactions will be presented.

Overall the steps we are looking at are shown in Scheme 13. Three units of acetate are converted into mevalonic acid (**86**), which is oxidatively decarboxylated to isopentenyl pyrophosphate (IPP) (**47**). This compound rearranges to dimethylallyl pyrophosphate (DMAPP) (**48**), which combines with another IPP unit to produce geranyl pyrophosphate, the crucial compound

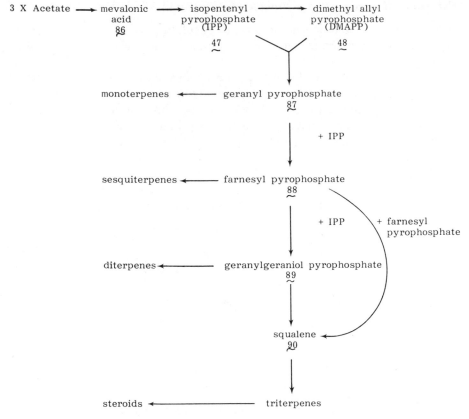

Scheme 13 The overall scheme of terpene biosynthesis.

in terpenoid biosynthesis. Oxidation with new C—C bond formations affords a wide array of monoterpenes. Addition of another IPP unit to geranyl pyrophosphate (**87**) affords farnesyl pyrophosphate (**88**), a sesquiterpene that gives rise to several alkaloid types.

Farnesyl pyrophosphate (**88**) may also add another IPP unit to give the precursor of the diterpenes, geranylgeraniol pyrophosphate (**89**). Alternatively, two farnesyl units may join together to afford squalene (**90**), the precursor of the triterpenes and ultimately the steroids.

With this background in mind let us focus in on some of intimate details of these processes. Of particular interest are:

1. The formation of geranyl pyrophosphate.
2. The rearrangement of geranyl pyrophosphate to loganin.
3. The formation of squalene from farnesyl pyrophosphate.
4. The rearrangement of squalene to the triterpenes.
5. The sequence of reactions leading from the triterpenes to the steroids.

Formation of Geranyl Pyrophosphate (87)

Although much could be said about the speculations of how a C_5 unit could be formed and dimerized, we will begin this discussion with a "vitamin," vitamin B_{13}.

In 1948 it was observed that the concentrates of dried distillers' solubles had a growth-promoting effect in chicks and rats and the product was named vitamin B_{13}. Folkers, then at Merck, Sharpe and Dohme, and his group studied the growth of a bacterium with this product. They found that in an acetate-devoid medium, the response was similar to that of acetate: it was an "acetate replacing factor." Knowing the importance of acetate in biosynthesis, anything which can replace it must be equally or more important.

Fractionation of this "factor" afforded a colorless oil characterized as 3,5-dihydroxy-3-methylpentanoic acid (**86**), which was given the trivial name

86   3R-mevalonic
     acid (MVA)

91   cholesterol

mevalonic acid (MVA). The compound was optically active and in many subsequent studies it has been found, without exception, that only the R-isomer of mevalonic acid is important in terpenoid biosynthesis.

At virtually the same time Tamura was studying how sake was spoiled by bacteria, for there was evidence of a bacterial growth promotor being involved. This activity was traced to "hiochic acid" isolated from *Asperigillus oryzae* as the quinine salt. The compound was found to be identical to mevalonic acid (MVA).

In virtually the same laboratory at Merck as the MVA work was being done, another group was studying the formation of the steroid cholesterol (**91**). Racemic [2-^{14}C]mevalonate (**46**) was incubated with rat liver homogenate to give a 43% incorporation into the crude sterol fraction. Since only one enantiomer is utilized, the incorporation is actually 86%. The five-carbon precursor of the steroids had been found! Numerous experiments followed and completely verified these early findings. In particular by labeling the carboxylic acid of mevalonic acid and finding no label in cholesterol (**91**) the five-carbons of MVA which give rise to the isopentene unit were determined. It then became critical to determine the mode of formation of MVA and of its conversion to the isoprene units **47** and **48**.

Rudney and his associates, over a period of several years, have shown that the early stages in this sequence involve the condensation of two acetyl CoA units to acetoacetyl CoA. A third molecule of acetyl CoA then condenses with the ketone of acetoacetyl CoA, with loss of CoA to afford hydroxymethylglutarate CoA (HMG CoA) (92) as shown in Scheme 14. The enzyme responsible for this process, β-ketoacylthiolase, has been obtained from bakers' yeast.

In the next steps HMG is reduced to MVA *via* an aldehyde, mevaldic acid, which is probably not a true intermediate, but rather remains bound as a hemithioacetal 93 prior to its reduction to mevalonic acid (86).

Work in several laboratories, particularly those of Lynen, Bloch, and

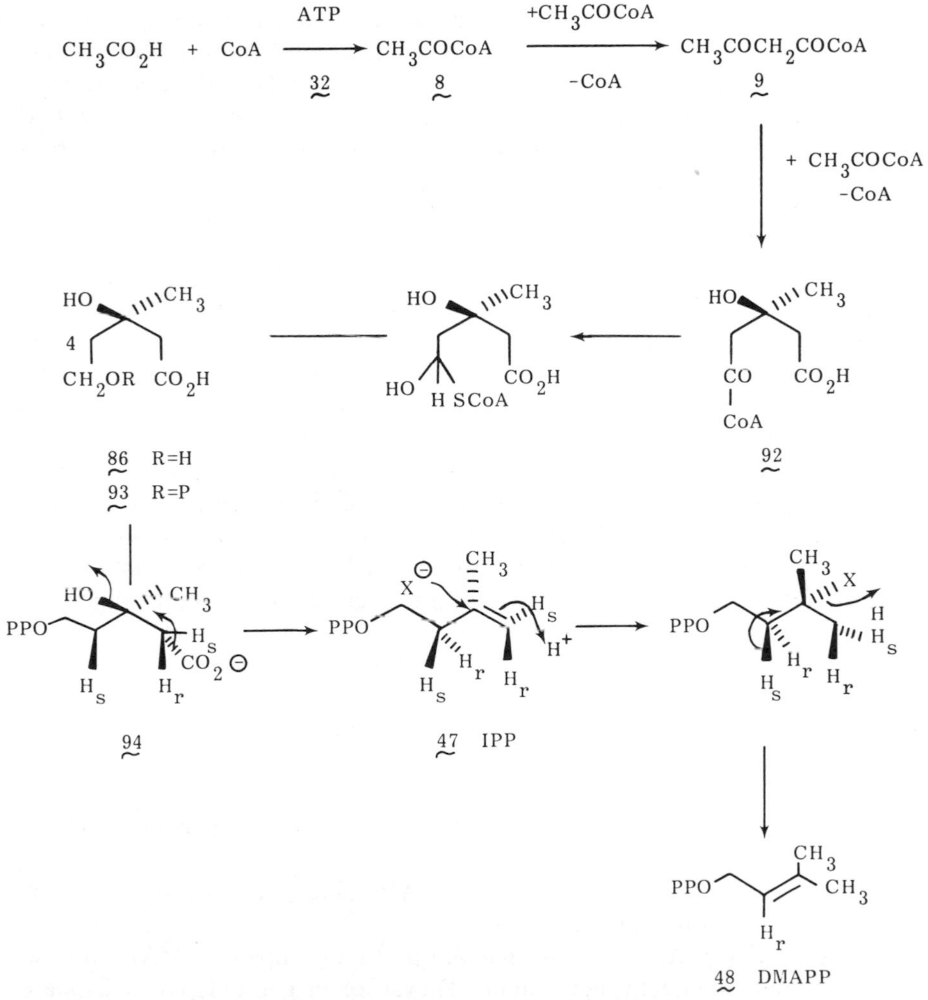

Scheme 14 The formation of isopentenyl pyrophosphate, IPP (47), and dimethylallyl pyrophosphate, DMAPP (48).

Popjak, defined the next stages in the sequence, the conversion of MVA to the true five-carbon intermediates. The stereochemistry of these processes has been deduced mainly by Cornforth and Popjak and co-workers.

Mevalonic acid (**86**) is first phosphorylated to mevalonic acid 5-phosphate (**93**) in the presence of ATP. Involvement of a second ATP molecule produces mevalonic acid 5-pyrophosphate (**94**), and yet another ATP molecule is involved in the elimination of CO_2 and water to afford isopentenyl pyrophosphate (**47**). Popjak and Cornforth determined in 1966 that this elimination follows a *trans* stereochemistry.

Lynen's group was the first to demonstrate that a second C_5 unit was involved when an isomerase was obtained from yeast which produced dimethylallyl pyrophosphate (**48**) from IPP (**47**).

This isomerization, which is of fundamental biosynthetic importance, is enzyme-mediated and involves stereospecific loss of the *pro*-4S hydrogen and stereoselective addition of hydrogen to the *re* side of the double bond. It has been studied in detail by Cornforth and co-workers. In further work by many groups, the stereochemical nature of this isomerization has been verified in plant, fungal and microsomal systems.

There is only one known exception to the stereochemistry of this isomerization, the formation of rubber in the tree *Hevea brasiliensis*. In this case, where a *cis* rather than a *trans* double bond results in the product, it is the *pro*-4R hydrogen which is lost. The formation of IPP and DMAPP is summarized in Scheme 14.

The next steps involve chain extension of dimethylallyl pyrophosphate (DMAPP) (**48**) by isopentenyl pyrophosphate (IPP) (**47**), whose mechanism was first illustrated by Lynen and demonstrated by Cornforth.

The enzymes mediating these reactions, prenyl transferases, have been isolated from several animal, microbial, and plant sources. Once again, in the cases studied so far the mechanism involves loss of the *pro*-4S proton in the IPP unit as the double bond effects nucleophilic displacement of the pyrophosphate group of DMAPP. The product geranyl pyrophosphate (**87**)

may be regarded as a C-4 alkylated DMAPP and is thus set up admirably for further reaction with IPP.

It is very important to note that the methyl groups of DMAPP are not initially biosynthetically equivalent. Therefore in geranyl pyrophosphate, C-10 is specifically derived from C-2 of mevalonic acid (**86**) and C-8 and C-9 from C-6 of MVA. In subsequent interconversions C-8 and C-10 of geraniol *may* become equivalent. This is a point of some interest and appears to be based to a large extent on the individual system.

Rearrangement of Geranyl Pyrophosphate (87) to Loganin (95)

Although the biosynthesis of indole alkaloids, particularly those derived from tryptophan and a monoterpene unit, will be discussed in detail in the appropriate chapter (Chapter 9), it can be disclosed at this time that a key compound in providing the C_{10} unit is loganin (95). In addition, loganin is

95 loganin

87

a precursor of a number of monoterpenoid alkaloids (Chapter 12). It is therefore pertinent to discuss the formation of loganin (95) from geranyl pyrophosphate (87) at this point.

No carbon atoms are gained or lost in this interconversion: we begin with a C_{10} unit and end with one. Only one C—C bond is formed (between C-2 and C-6 geraniol) in the process, but several oxidations are performed at various stages.

Somewhat surprisingly the route between geraniol (96) and loganin (95) is not well understood. It appears that the early steps in the sequence involve (*a*) a *cis-trans* isomerization of the 2,3-double bond of geraniol (96) to give nerol (97), in which the hydrogen at C-2 of 96 is retained in 97, and (*b*) hydroxylation of nerol (97) at C-10 to give 10-hydroxynerol (98). There is evidence to suggest that at this point further oxidation of C-8 and C-10 occurs to give a trialdehyde such as 99 in which C-8 and C-10 have become equivalent (by tautomerization). Probably ring closure occurs at this point to give the monocyclic trialdehyde (100), which exists as the cyclized hemiacetal 101.

The next known intermediate is deoxyloganin (102), and it is not difficult to imagine the steps from 101 to 102. Glycosylation, possibly of the hemi-

CH₃ structures row (geraniol, nerol, 10-hydroxynerol series):

96 geraniol 97 nerol 98 10-hydroxy-nerol 99

95 loganin 102 deoxyloganin 101 100

acetal **101**, undoubtedly aids transport. Hydroxylation at C-7 of deoxylo-ganin (**102**) occurs stereospecifically to give loganin (**95**).

Formation of Squalene (90) from Farnesyl Pyrophosphate (88)

We have seen in previous sections how DMAPP (**48**), geranyl pyrophos-phate (**87**), and farnesyl pyrophosphate (**88**) can be extended by five-carbon atoms with IPP (**47**). Because carbon 1 of DMAPP etc. joins with carbon 4 of IPP, this mechanism is often referred to as *head-to-tail* chain extension.

There is another very important mechanism, which results in *tail-to-tail* joining of units. *The* classic example of such bond formation is involved in the biosynthesis of squalene (**90**) from farnesyl pyrophosphate (**88**).

For many years, squalene (**90**) remained somewhat of a curiosity: a linear C_{30} compound of extremely limited distribution, shark liver oil. Discovered in 1916, the structure of squalene was confirmed by synthesis in 1931, but it was many years before its importance in biosynthesis was realized. It is now known to be the key precursor of triterpenoids and steroids. There are no known exceptions.

In 1958, Lynen and co-workers found that when yeast enzyme prepa-rations were incubated in the presence of NADPH, squalene was produced. Five years earlier, Langdon and Bloch had shown that [¹⁴C]squalene was a precursor of cholesterol (**91**). Cholesterol, it will be noted, is a C_{27} com-pound, and this was the first evidence that it was derived from an isoprenoid precursor.

But how is squalene derived from farnesyl pyrophosphate? This appar-ently simple question has been the subject of considerable controversy be-tween several eminent groups.

When [1,1-²H₂]farnesyl pyrophosphate (**88**) was used as precursor of squalene (**90**) in liver microsomal enzyme systems, three of the deuterium atoms were retained in squalene. The remaining hydrogen on the central unit was found by labeling experiments to be derived from NADPH. Clearly the mechanism of squalene biosynthesis is not a simple one.

It was Rilling who, in 1966, isolated an intermediate between farnesyl pyrophosphate (**88**) and squalene (**90**). This compound is now known as pre-squalene pyrophosphate.

Several structures were suggested for this intermediate but the structure **103** is now regarded as being correct. A number of syntheses have been reported and the efficient conversion of synthetic **103** to squalene (**90**) by a yeast system in the presence of NADPH has further substantiated the validity of this structure.

One mechanism for the formation of presqualene pyrophosphate (**103**) is shown in Scheme 15. Nucleophilic attack at C-3 of one farnesyl pyro-phosphate (**88**) results in displacement of the pyrophosphate of the second unit of **88**, and this reaction proceeds with inversion at C-1 of the second

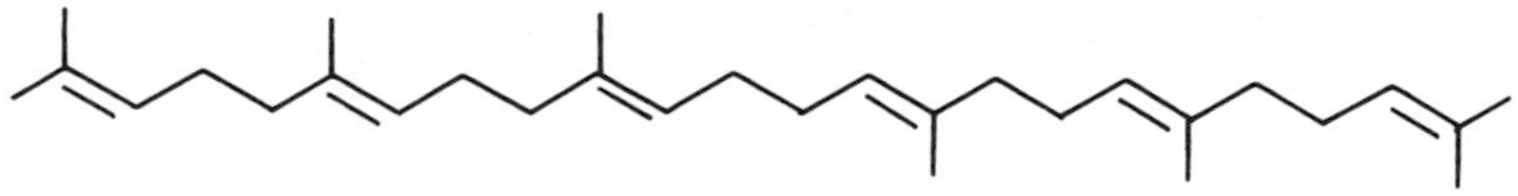

90   squalene

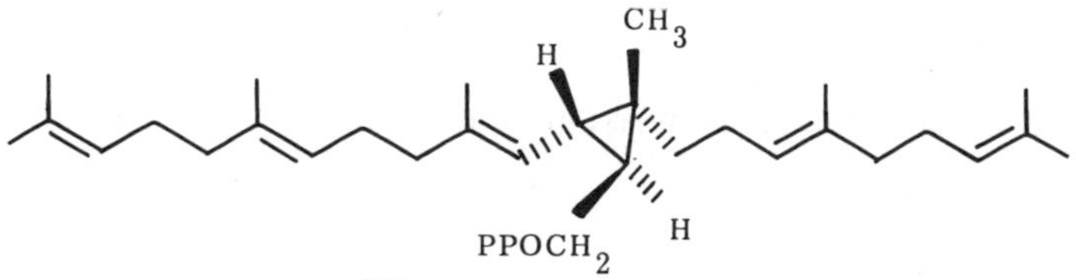

103   presqualene pyrophosphate

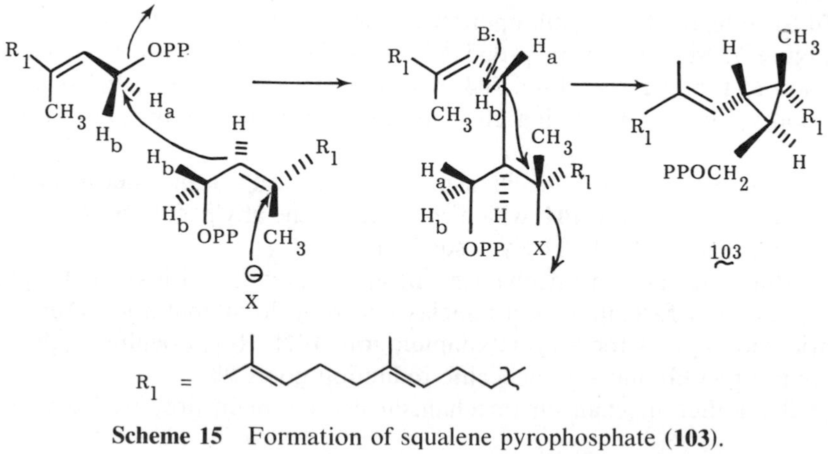

Scheme 15   Formation of squalene pyrophosphate (**103**).

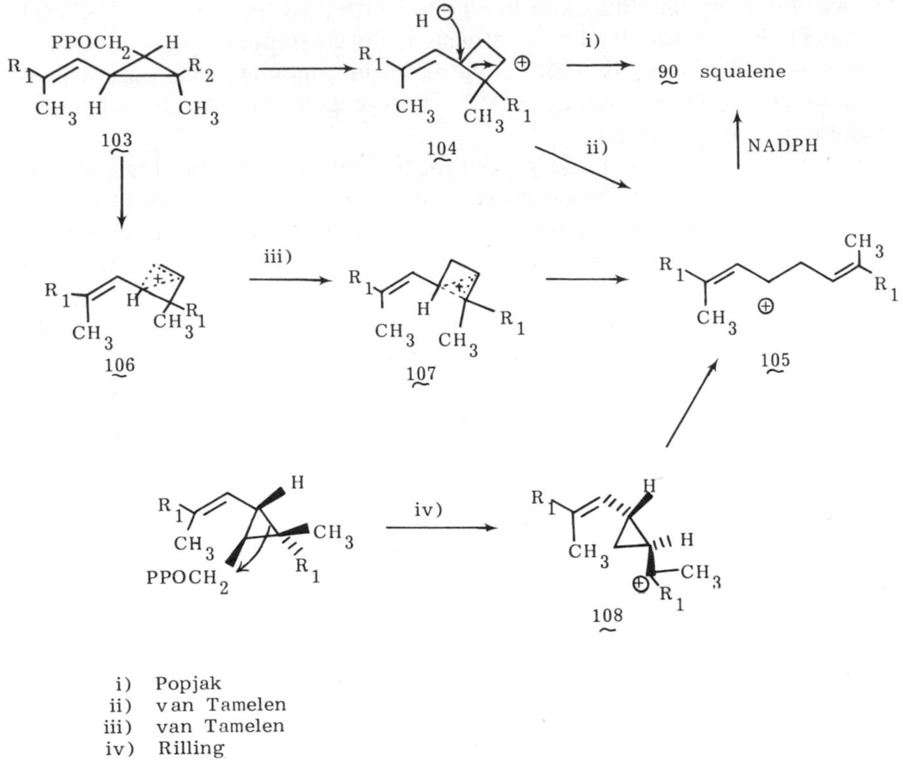

i) Popjak
ii) van Tamelen
iii) van Tamelen
iv) Rilling

Figure 5 Mechanisms for the conversion of presqualene pyrophosphate (**103**) to squalene (**90**).

unit. The second step in the formation of **103** is envisaged as *trans*-elimination of H_b and the original nucleophile.

The mechanism of reductive cleavage of presqualene pyrophosphate (**103**) to squalene (**90**) is still unknown. Some of the possibilities are shown in Figure 5. Mechanism i), proposed by Popjak's group, involves ring expansion to the cyclobutyl ion **104**; hydride attack (from NADPH) α to the cyclobutyl ring leads to ring cleavage and formation of the *trans* double bond.

An alternative route (mechanism ii) also envisages the intermediacy of a cyclobutyl ion cation (**104**) which collapses to the allylic cation **105**. Stereospecific attack by NADPH gives squalene.

A third mechanism (mechanism iii) anticipates initial loss of the pyrophosphate and formation of a nonclassical bicyclobutanium ion (**106**). Rehybridization gives the bicyclobutanium ion (**107**). Ring opening to the carbonium ion (**105**) and stereospecific reduction gives **90**.

Yet another mechanism (mechanism iv) has been proposed which involves a cyclopropylcarbinyl–cyclopropylcarbinyl rearrangement to **108**, which then rearranges to the allylic cation **105**, and is reduced as before.

Yeast extracts have afforded a *trans*-farnesylpyrophosphate synthetase which in the absence of a reducing nucleotide accumulates presqualene pyrophosphate. A squalene synthetase has also been obtained from tobacco cell cultures.

There are undoubtedly some aspects of this work which remain to be clarified, but the route farnesyl pyrophosphate (**88**) → presqualene pyrophosphate (**103**) → (**90**) is probably the only route to squalene in nature.

Rearrangement of Squalene to the Triterpenes

Squalene, as indicated previously, is the C_{30} compound which is a precursor of all triterpenoids and steroids. The steps from squalene to the steroids in mammals and plants have been and continue to be the subject of numerous studies. A detailed discussion of these results is beyond the scope of this book and the interested reader is referred to the most recent treatment of this subject by Nes and McKean for additional information. Here we will discuss only the quite basic details of this complex scheme.

In 1934, Robinson indicated how a particular conformation of squalene could lead to lanosterol. When the structure of lanosterol was finally proven in 1952, the unique talents of Woodward and Bloch combined to suggest a cyclization of squalene which differed significantly from Robinson's in that the gem-dimethyl group at C-4 is derived directly from the terminal carbons of squalene and not by subsequent alkylation. The schemes of Robinson and of Woodward and Bloch are shown in Figure 6, indicating the positions expected to be labeled by the methyl group of acetate. As a result of the two labeling patterns the same carbom atom may come from a different source. The carbon atoms of interest in this respect, and thus those which could serve to distinguish between the two schemes, are C-7, C-8, C-12, and C-13 of cholesterol. Experiments with labeled acetate established that C-13 was derived from the methyl group of acetate, indicating that the Woodward–Bloch hypothesis, in which lanosterol (**109**) is an intermediate, was correct. All subsequent experiments with labeled acetate, mevalonate, and farnesol have substantiated this pattern of folding for the formation of the triterpene nucleus.

A very important difference between diterpenes and triterpenes should be mentioned at this point. Whereas the latter almost always have an oxygen functionality at C-3, such a group is quite rare in the diterpenes.

Bloch determined that the oxygen atom of this hydroxy group was derived from molecular oxygen, and for some years it was felt that $^+$OH, produced by reduction of molecular oxygen by a reduced pyridine nucleotide (e.g. NADPH), was responsible for the initiation of cyclization. The groups of Corey, Clayton, and Van Tamelen however established that the processes of oxidation and cyclization are separate steps involving squalene-2,3-oxide

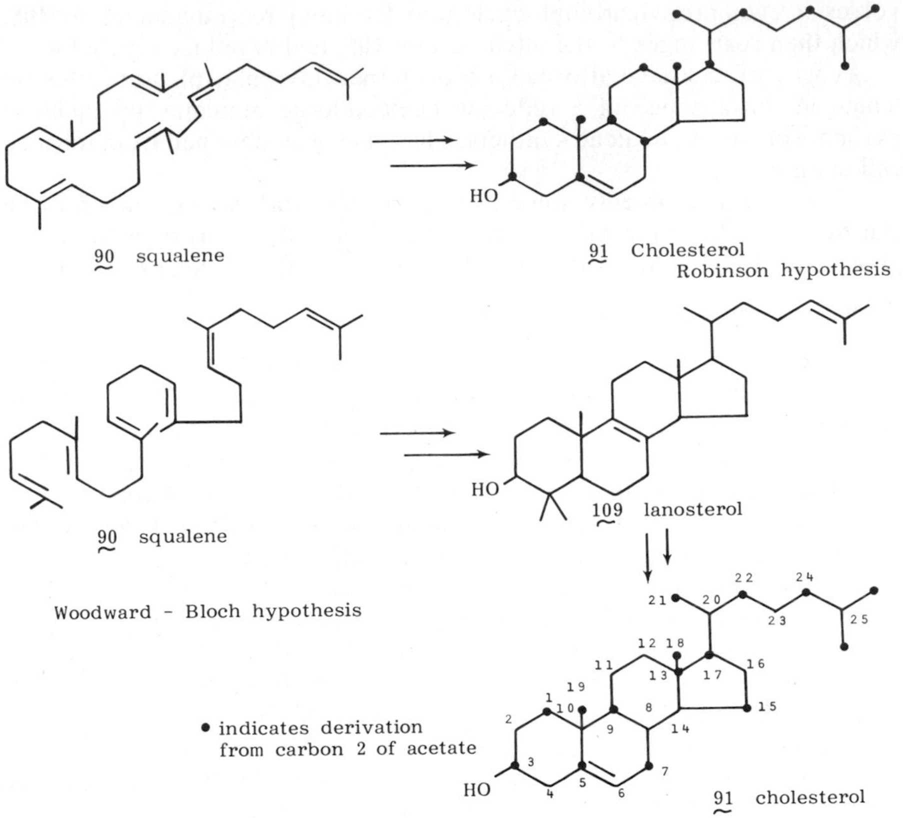

Figure 6 Mechanisms for the conversion of squalene (**90**) to cholesterol (**91**).

(**110**) as an intermediate. Some of this evidence includes: (*a*) when 2,3-imino squalene is used as an inhibitor of the cyclization, **110** accumulates; (*b*) squalene-2,3-oxide (**110**) was converted by rat liver homogenates to lanosterol (**109**) and other products; and (*c*) the oxygen atom of **110** is the same oxygen as appears at C-3 in lanosterol (**109**). Additionally it has been found that the

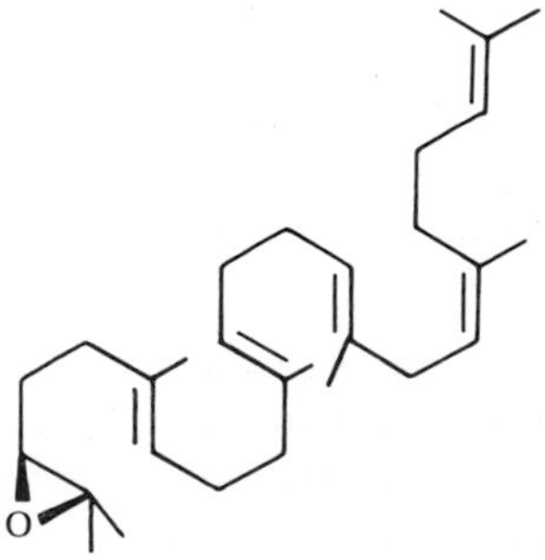

110   squalene oxide

S-isomer of **110** is the sole precursor of lanosterol (**109**) in yeast and pig liver and of β-amyrin, lupeol, and cycloartenol (**111**) in *Pisum sativum* (pea) seedlings.

The cyclizations of squalene oxide (**110**) and the subsequent rearrangements are a very complex subject and the reader is referred to Nes and McKean for an extensive discussion.

One clear point should be made at this early stage, namely that in mammalian systems and in most fungi, lanosterol (**109**) is *the* intermediate to the steroids. In photosynthetic plants however, it is cycloartenol (**111**) which is *the* precursor of steroids (Scheme 16).

At this point the student should attempt to define the six isoprene units present in the nuclei of cycloartenol (**111**) and lanosterol (**109**). No matter what end you start from, neither compound obeys the biogenetic isoprene rule! The methyl group at C-13 should be at C-8. Evidently at a very early stage a rearrangement has taken place for, as shown in Scheme 17, the initial product of cyclization of squalene oxide (**110**) should be the cation **112**, the so-called protosteroid cation. Also note that compared to the typical stereochemistry of cycloartenol (**111**), several of the ring junctions have their stereochemistry inverted.

It is at the point of the protosteroid cation (**112**) where rearrangement takes place on one hand to lanosterol (**109**) and on the other to cycloartenol

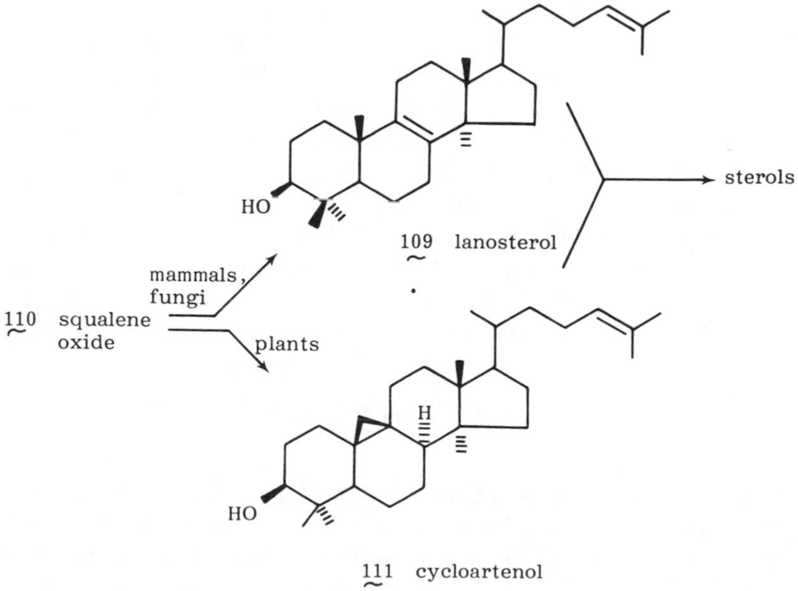

Scheme 16

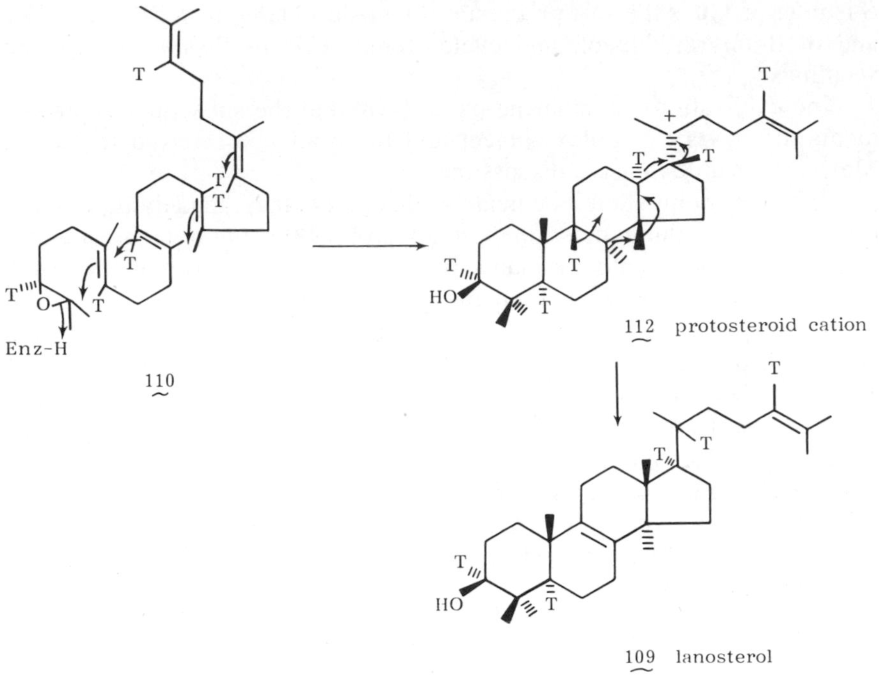

Scheme 17

(111). The mechanism of the rearrangement to lanosterol **(109)** is shown in Scheme 17. Initiation of what is apparently a concerted process is by enzyme removal of a proton at C-9. This begins a series of suprafacial shifts in which a new double bond is introduced at $\Delta^{8,9}$, the C-8 methyl migrates to C-14, the C-14 methyl to C-13, and hydrogens at C-13 and C-17 migrate to C-17 and C-20 respectively. When [4R-^{3}H]mevalonic acid **(45)** is used as a precursor of lanosterol **(109)**, only five of the possible six tritiums are incorporated as expected in this scheme.

The rearrangement to cycloartenol **(111)** is thought to proceed not by proton loss from C-9 but rather by hydride shift from C-9 to C-8 with stereospecific complexation of an enzyme at C-9 to give an intermediate such as **113**. Loss of a proton from the methyl group at C-10 with displacement of the enzyme gives cycloartenol **(111)** (Scheme 18).

The Sequence of Reactions Leading from the Triterpenes to the Steroidal Alkaloids

There are four important skeleta in the steroidal alkaloids: (*a*) the buxenone type (e.g. **114**), which contains the cyclopropane ring of **111** but which has lost the six side-chain carbon atoms; (*b*) the *Solanum* type, in which all the side-chain carbons remain but where three methyl groups (two at C-4, one at C-14) have been lost (e.g. **115**); (*c*) the Cevine type, where ring C is

112
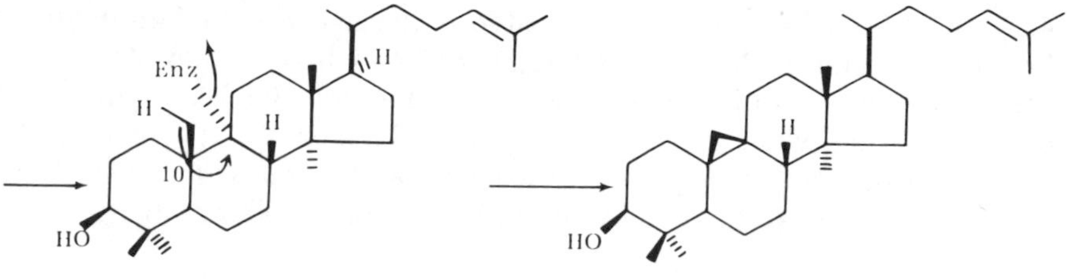

113 111 cycloartenol

Scheme 18

missing a carbon atom and ring C has gained one, and three methyl groups have been lost (e.g. **116**); and (*d*) the *Pachysandra* or *Holarrhena* type, in which both the six carbons from the side chain and the three methyl groups are missing (e.g. **117** and **118** respectively).

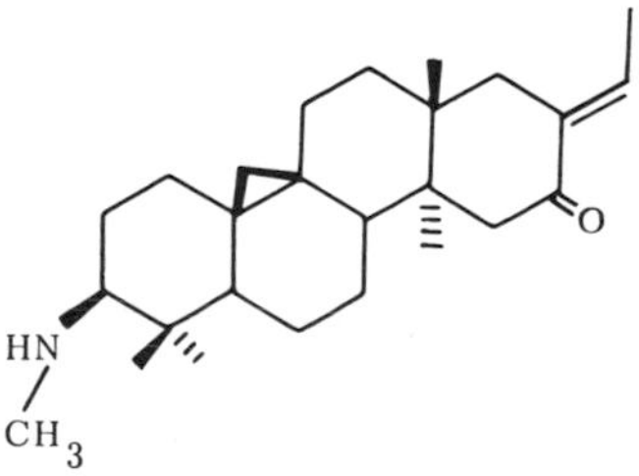

114 Buxenone

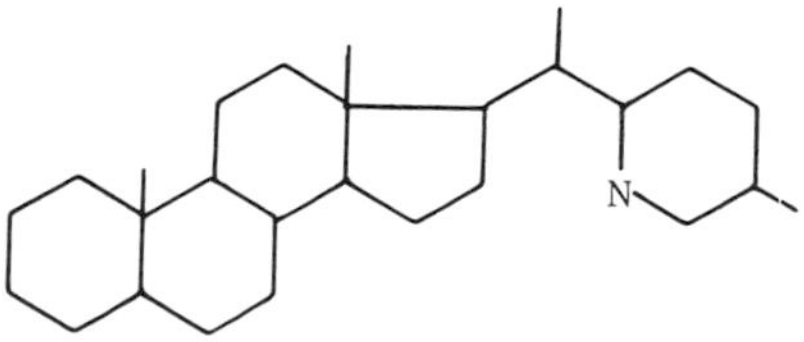

115 Solanum - type

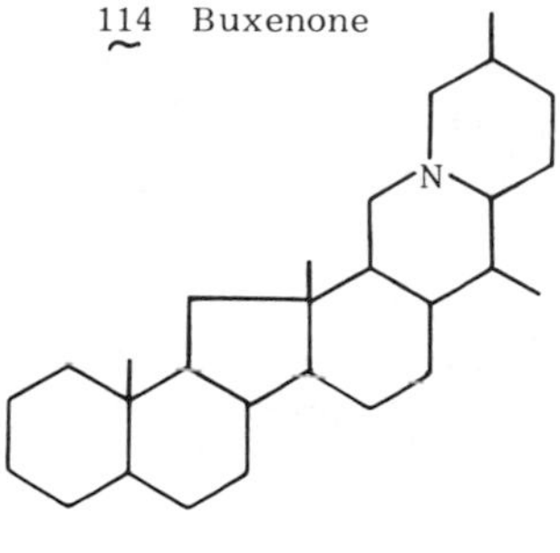

116 Cevine type

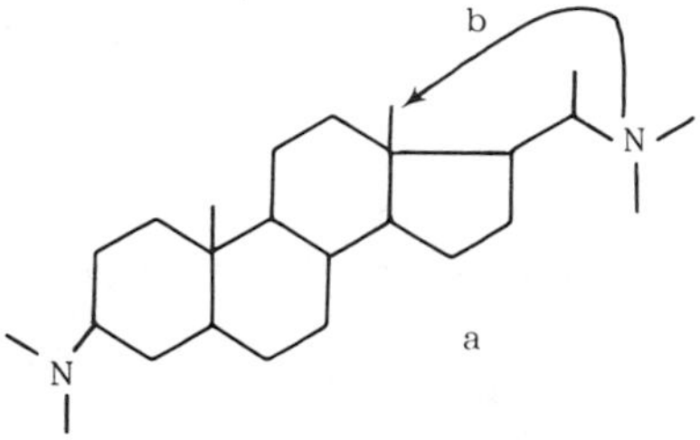

117 a Pachysandra

118 b Holarrhena

In the remaining brief discussion we will concentrate on these transformations.

Removal of the C-4 and C-14 Methyl Groups The sequence of reactions leading to removal of the two C-4 methyl groups involves an enzyme complex (methyl sterol oxidase) which by a series of sequential oxidations forms the

4α-carboxylic acid (**119**). The next step is the decarboxylation of the carboxylic acid and oxidation of the adjacent 3β-hydroxy group. At this point the 4β-methyl group is epimerized to a 4α-methyl, the 3-keto group is reduced, and the whole process is repeated (Scheme 19). These steps have been verified time and again in both plants and animals. In fungi, there is evidence that the 4β-methyl group may be removed preferentially.

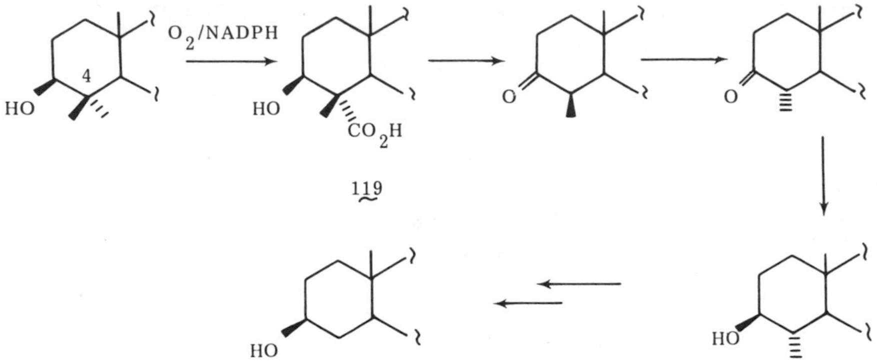

Scheme 19   Removal of the two C-4 methyl groups in steroid biosynthesis.

The third methyl group eliminated (at C-14) is probably removed before the two C-4 methyl groups, although this is by no means established in all cases. However, it is known that removal of this group occurs by oxidation of the C-14 methyl group to formic acid. As a prelude to this sequence of reactions, isomerization of the $\Delta^{8,9}$-bond (**120**) to a $\Delta^{7,8}$-bond (**121**) occurs by elimination of the 7β-hydrogen (rat liver, plants) or the 7α-hydrogen (fungi). The steps leading to loss of the 14α-methyl group are shown in Scheme 20.

Formation of Cholesterol from Lanosterol and Cycloartenol   The biosynthesis of steroids apparently returns to a unifying compound, cholesterol (**91**), in both plants and animals. We have seen how lanosterol (**109**) loses three methyl groups, and in a series of further processes isomerization of this $\Delta^{7,8}$-bond to $\Delta^{5,6}$ and reduction of the $\Delta^{24,25}$-bond must also occur.

Present evidence suggest that the reduction actually precedes loss of the methyl group; such a process would place lathosterol (**122**) as a key intermediate in mammalian and possibly plant systems.

In the isomerization of $\Delta^{7,8}$ to $\Delta^{5,6}$ there is a *cis*-elimination of the 5α- and 6α-hydrogens to afford the conjugated diene **123**. Selective reduction of the $\Delta^{7,8}$-double bond then affords cholesterol (**91**) (Scheme 21).

Removal of the Side Chain at C-20 and Formation of Progesterone   For many years after the first isolation from sow ovaries in 1934, the hormone progesterone (**124**) was regarded as a mammalian product. Now it is regarded

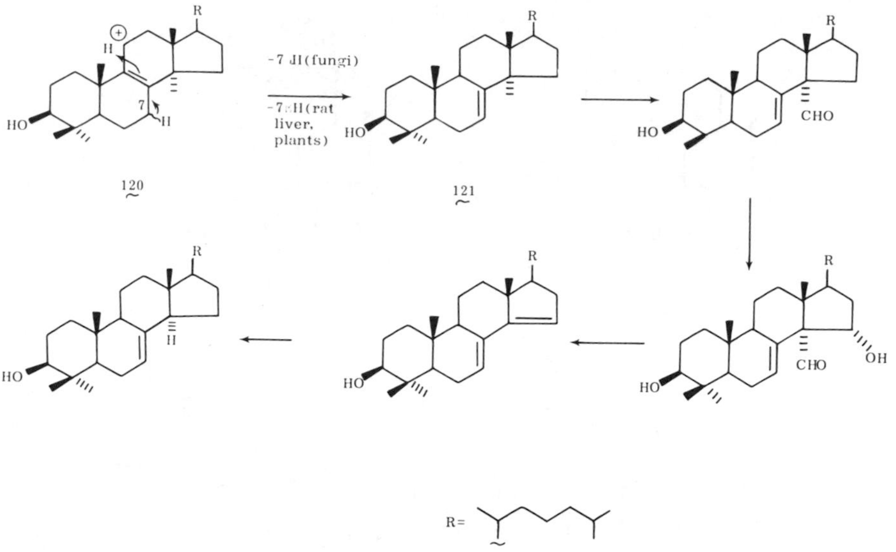

120 121

Scheme 20 Removal of the 14α-methyl group in steroid biosynthesis.

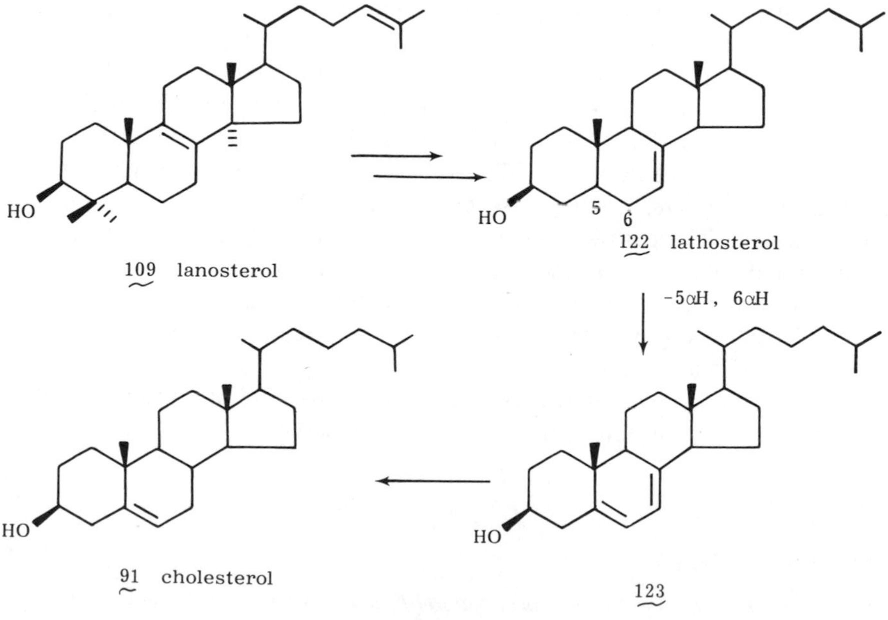

109 lanosterol

122 lathosterol

91 cholesterol

123

Scheme 21

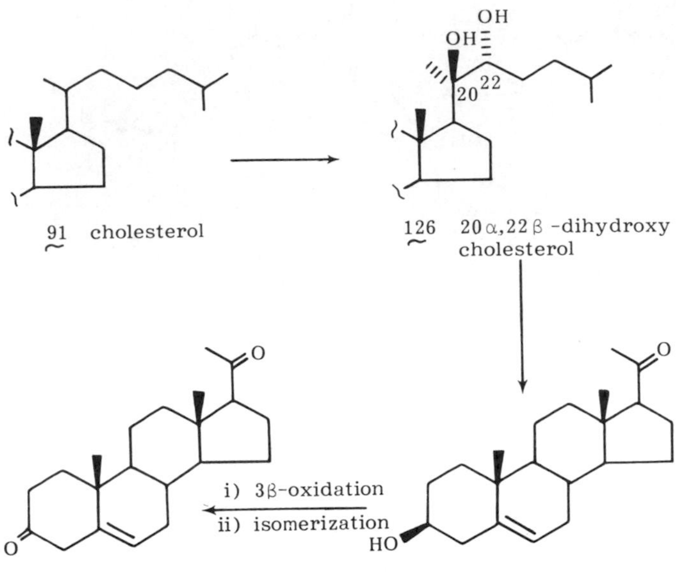

91 cholesterol

126 20α,22β-dihydroxy cholesterol

124 progesterone

125 pregnenolone

Scheme 22

not only as critical in the formation of the hormones of the adrenal cortex, testes, and ovaries, but also of the cardenolides and some steroidal alkaloids in plants. The key intermediate in the formation of progesterone (**124**) from cholesterol (**91**) is pregnenolone (**125**) in which the side chain at C-20 has been oxidatively removed.

The oxidative loss of the side chain proceeds by a mixed-function oxidase system which in the presence of oxygen and reduced pyridine nucleotide introduces a 20α- and a 22β-hydroxy group and then oxidatively cleaves the carbon–carbon bond between C-20 and C-22. Both additional oxygens in the intermediate 20α,22β-dihydroxycholesterol (**126**) are derived from molecular oxygen. The pathway has been established using adrenal glands and it is assumed (perhaps erroneously) that similar intermediates are involved in plant systems. Subsequent steps to progesterone (**124**) involve oxidation of the 3β-hydroxy group to a ketone and isomerization of the $\Delta^{5,6}$-double bond to $\Delta^{4,5}$ (Scheme 22).

With this background in the formation of the alkaloid precursors, we can now begin our discussion of the alkaloids themselves.

LITERATURE

Biosynthesis: General

Robinson, R., *The Structural Relations of Natural Products*, Clarendon, Oxford, 1955.

Fieser, L. F., and M. Fieser, *Steroids*, Reinhold, New York, 1959.

Richards, J. H., and J. B. Hendrickson, *The Biosynthesis of Steroids, Terpenes and Aceto-genins*, W. A. Benjamin, New York, 1964.

Clayton, R. B., Biosynthesis of sterols, steroids and terpenoids, *Quart. Rev.* **19**, 168 (1965); **19**, 201 (1965).

Bu'Lock, J. D., *The Biosynthesis of Natural Products*, McGraw-Hill, London, 1965.

Hendrickson, J. B., *The Molecules of Nature*, W. A. Benjamin, New York, 1965.

Bernfeld, P. (Ed.), *The Biogenesis of Natural Compounds*, 2nd ed., Pergamon, Oxford, 1967.

Grisebach, H., *Biosynthetic Patterns in Microorganisms and Higher Plants*, Wiley, New York, 1967.

Bentley, R., and I. M. Campbell, in M. Florkin and E. H. Stotz (Eds.), *Comprehensive Biochemistry*, Vol. 20, Elsevier, New York, 1969, p. 415.

Geissman, T. A., and D. H. G. Crout, *Organic Chemistry of Secondary Plant Metabolism*, Freeman, Cooper, San Francisco, 1969.

Turner, W. B., *Fungal Metabolites*, Academic, New York, 1971.

Luckner, M., *Secondary Metabolism in Plants and Animals*, Academic, New York, 1972.

Tedder, J. M., A. Nechvatal, A. W. Murray, and J. Carnduff, *Basic Organic Chemistry*, Part 4, Wiley, New York, 1972.

Chemical Society Specialist Periodical Reports, Biosynthesis, Vols. 1–3, T. A. Geissman (Ed.), Chemical Society, London, 1972–1975.

Chemical Society Specialist Periodical Reports, Biosynthesis, Vols. 4, 5, J. D. Bu'Lock, (Ed.), Chemical Society, London, 1976–1977.

Nes, W. R., and M. L. McKean, *Biochemistry of Steroids and Other Isopentenoids*, University Park Press, Baltimore, Md., 1977.

Techniques in Biosynthesis

Comar, C. L., *Radioisotopes in Biology Agriculture*, McGraw-Hill, New York, 1955.

Kamen, M. D., *Isotopic Tracers in Biology*, Academic, New York, 1957.

Broda, E., *Radioactive Isotopes in Biochemistry*, Elsevier, Amsterdam, 1960.

Catch, J. R., *Carbon-14 Compounds*, Butterworths, London, 1961.

Birks, J. B., *The Theory and Practice of Scintillation Counting*, Pergamon, Oxford, 1964.

Wolf, G., *Isotopes in Biology*, Academic, New York, 1964.

Evans, E. A., *Tritium and Its Compounds*, Van Nostrand, London, 1966.

Chase, G. D., and J. L. Rabinowitz, *Principles of Radioisotope Methodology*, Burgess, Minneapolis, Minn., 1967.

Simon, H., and H. G. Floss, *Bestimmung der Isotopenverteilung in markierten Verbindungen*, Springer-Verlag, Berlin, 1967.

Brown, S. A., *Biosynthesis* **1**, 1 (1972).

Stable Isotopes

Tanabe, M., *Biosynthesis* **1**, 241 (1971).

Stothers, J. B., *Carbon-13 NMR Spectroscopy*, Academic, New York, 1972.

Levy, G. C., and G. L. Nelson, *Carbon-13 Nuclear Magnetic Resonance for Organic Chemists*, Wiley–Interscience, New York, 1972.

Levy, G. C., R. L. Lichter and G. L. Nelson, *Carbon-13 Nuclear Magnetic Resonance Spectroscopy*, 2nd ed., Wiley–Interscience, New York, 1980.

Grutzner, J. B., *Lloydia* **35,** 375 (1972).

Floss, H. G., *Lloydia* **35,** 399 (1972).

Tanabe, M., *Biosynthesis* **2,** 247 (1973).

Tanabe, M., *Biosynthesis* **4,** 204 (1976).

NIH Shift

Daly, J. W., D. M. Jerina, and B. Witkop, *Experientia* **28,** 1129 (1972).

Alkaloid Biosynthesis: General

Battersby, A. R., *Quart. Rev.* **15,** 259 (1961).

Leete, E., *Ann. Rev. Plant. Physiol.* **18,** 179 (1967).

Leete, E., in *Advan. Enzymol. Related Areas Mol. Biol.* **32,** 373 (1969).

Mothes, K., and H. R. Schutte (Eds.), *Biosynthese der Alkaloide*, VEB Deutsche Verlag der Wissenschaften, Berlin, 1969.

Spenser, I. D., in M. Florkin and E. H. Stotz (Eds.), *Comprehensive Biochemistry*, Vol. 20, Elsevier, New York, 1969, p. 231.

Spenser, I. D., in S. W. Pelletier (Ed.), *Chemistry of the Alkaloids*, Van Nostrand Reinhold, New York, 1970, p. 669.

Haslam, E., *The Shikimate Pathway*, Butterworths, London, 1974.

Polyacetate Pathway: General

Richards, R. W., in W. D. Ollis (Eds.), *Recent Developments in the Chemistry of Natural Phenolic Compounds*, Pergamon Press, Oxford, 1961.

Birch, A. J., *Proc. Chem. Soc.* 3 (1962).

Bu'Lock, J. D., *The Biosynthesis of Natural Products*, McGraw-Hill, London, 1965.

Birch, A. J., *Science* **156,** 202 (1967).

Geissman, T. A., in P. Bernfeld (Ed.), *Biogenesis of Natural Compounds*, 2nd ed., Pergamon, Oxford, 1967.

Shibata, S., *Chem. Brit.* **3,** 110 (1967).

Isoprene Units

Ruzicka, L., *Proc. Chem. Soc.* 341 (1959).

Cornforth, J. W., *Chem. Brit.* **4,** 102 (1968).

Clayton, R. B., *Quart. Rev.* **19,** 168, 201 (1968).

Wallach, O., *Justus Liebigs Ann. Chem.* **239,** 34 (1887).

Eschenmoser, A., L. Ruzicka, O. Jeger, and D. Arigoni, *Helv. Chim. Acta* **38,** 1890 (1955).

Woodward, R. B., and K. Bloch, *J. Amer. Chem. Soc.* **75,** 2023 (1953).

Discovery and Early Studies with MVA

Folkers, K., C. H. Shunk, B. O. Linn, F. M. Robinson, P. E. Wittreich, J. W. Huff, J. L. Gilfillian, and H. R. Skeggs, in G. E. W. Wolstenholme and M. O'Connor (Eds.), *Biosynthesis of Terpenes and Sterols*, Little, Brown, Boston, 1958, p. 20.

Tamura, G., *J. Gen. Microbiol.* **2,** 431 (1956).

Tavormina, R. A., M. H. Gibbs, and J. W. Huff, *J. Amer. Chem. Soc.* **78,** 4498 (1956).

Rudney, H., and J. J. Ferguson, *J. Biol. Chem.* **234,** 1076 (1959).

Formation of IPP, DMAPP and Geranyl Pyrophosphate

Popjak, G., and J. W. Cornforth, *Biochem. J.* **101,** 553 (1966).

Agranoff, B. W., H. Eggerer, U. Henning, and F. Lynen, *J. Biol. Chem.* **235,** 326 (1960).

Cornforth, J. W., R. H. Cornforth, C. Donninger, and G. Popjak. *Proc. Roy. Soc. Ser. B* **163,** 492 (1966).

Cornforth, J. W., R. H. Cornforth, G. Popjak, and L. Vengoyan, *J. Biol. Chem.* **241,** 3970 (1966).

Clifford, K., J. W. Cornforth, R. Mallaby, and G. T. Phillips. *Chem. Commun.* 1599 (1971).

Squalene Biosynthesis

Lynen, F., H. Eggerer, U. Hemming, and I. Kessel, *Angew. Chem.* **70,** 738 (1958).

Langdon, R. G., and K. Bloch, *J. Chem.* **200,** 135 (1953).

Rilling, H. C., *J. Biol. Chem.* **241,** 3233 (1966).

Goodwin, T. W. (Ed.), *Aspects of Terpenoid Chemistry and Biochemistry*, Academic, London, 1971.

Rees, H. H., and T. W. Goodwin, *Biosynthesis* **1,** 59 (1972).

Sterol Biosynthesis

Goad, L. J., in T. W. Goodwin (Ed.), *Natural Substances Formed Biologically from Mevalonic Acid*, Biochemical Society Symposium No. 29, Academic, London, 1970, p. 45.

Nes, W. R., and M. L. McKean, *Biochemistry of Steroids and Other Isopentenoids*, University Park Press, Baltimore, Md., 1977.

chapter 3

ALKALOIDS DERIVED FROM ORNITHINE

Chapter 2 discussed how ornithine (**1**) is produced biosynthetically; now let us turn our attention to the utility of ornithine in the formation of alkaloids.

There are three important alkaloid groups derived from ornithine; the tropane alkaloids, the nicotine group, and the pyrrolizidine alkaloids, and their skeleta are shown in Figure 1. In addition, there are a number of simple alkaloids derived more directly from ornithine.

3.1 SIMPLE PYRROLIDINE ALKALOIDS

The simplest pyrrolidine alkaloid is pyrrolidine (**2**) itself, a minor alkaloid of tobacco, *Nicotiana tabacum* L., which is also found in the wild carrot, *Daucus carota* L. It is a typical strongly basic (pK_a 11.3), secondary amine.

N-Methylpyrrolidine (**3**) is also found in tobacco as well as deadly night-shade (*Atropa belladonna* L.); being a tertiary amine it is more basic (pK_a 10.4) than **2**.

Stachydrine (**4**) occurs in alfalfa (*Medicago sativa* L.) as the (−) form, but as the racemate in betony (*Stachys officinale* Franch.). On melting a methyl transfer occurs to give methyl hygrinate (**5**).

Several alkaloids are derived by combination of a cyclized ornithine and an acetoacetate or polyacetate unit. Two of these, hygrine (**6**) and cusco-hygrine (**7**),* occur in the Peruvian coca shrub (*Erythroxylon truxillense* Rusby) and were originally isolated by Liebermann in 1889. Hygrine is a colorless liquid which gives derivatives both for the tertiary amine (e.g. picrate) and ketone (e. g. oxime) functionalites. Chromic acid oxidation of hygrine (**6**) afforded hygrinic acid (**8**), which could be related to L-(−)-proline (**9**), and thus the absolute configuration of hygrine could be determined. Sorm has reported a synthesis of racemic hygrine which begins with elec-

---

* This alkaloid is sometimes referred to as cuskhygrine.

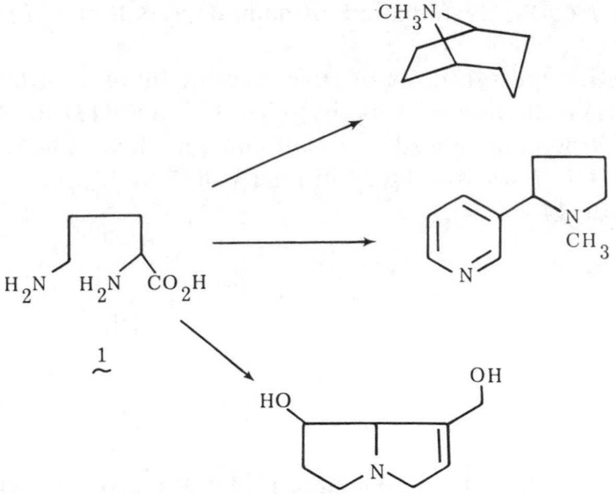

Figure 1 Principal alkaloid types derived from ornithine (**1**).

trophilic substitution of *N*-methylpyrrole (**10**) by diazoacetone in the presence of copper. Hydrogenation with a platinum catalyst followed by oxidation of the secondary alcohol **11** gave **6**.

Cuscohygrine (**7**), another liquid alkaloid, also occurs in deadly nightshade (*Atropa belladonna*), thornapple (*Datura stramonium* L.), and several

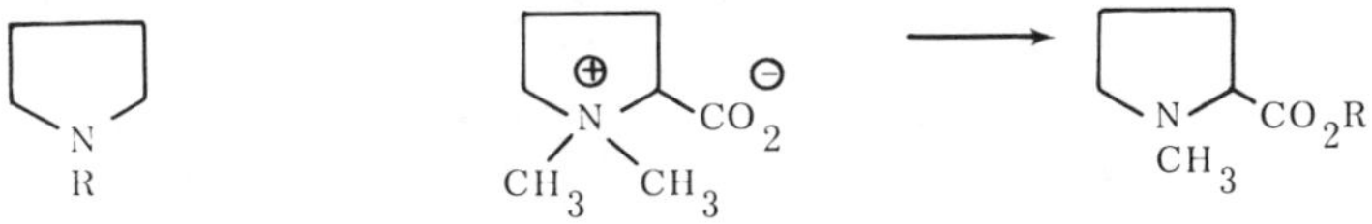

2 R=H
3 R=CH$_3$

4 stachydrine

5 R=CH$_3$
8 R=H

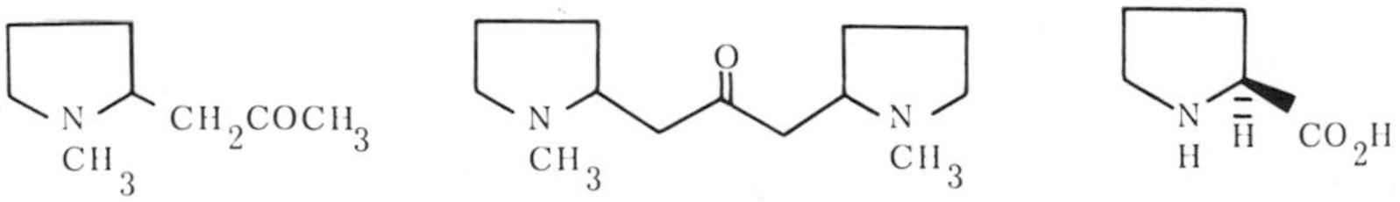

6 hygrine

7 cuscohygrine

9 L-proline

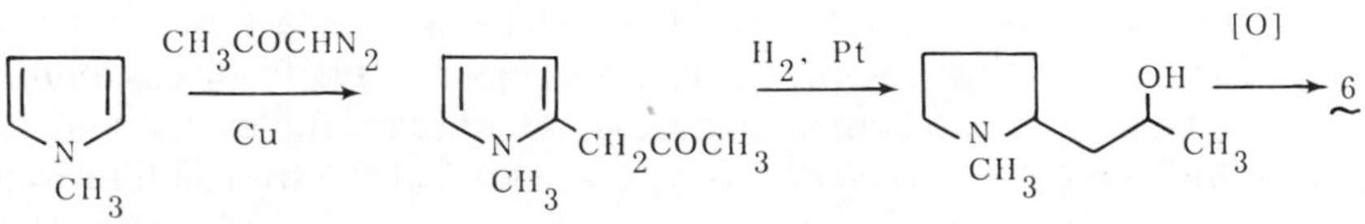

10

11

other *Datura* species. Systematic Hofmann degradation of **7** gave undecan-6-ol.

A biogenetic-type synthesis of cuscohygrine involves lithium aluminum hydride (LAH) reduction of 1-methylpyrrolid-2-one (**12**) to afford a carbinolamine **13**. When condensed with half an equivalent of acetone dicarboxylic acid at pH 7, **7** was produced in nearly 40% yield, presumably *via* the amino aldehyde **14**.

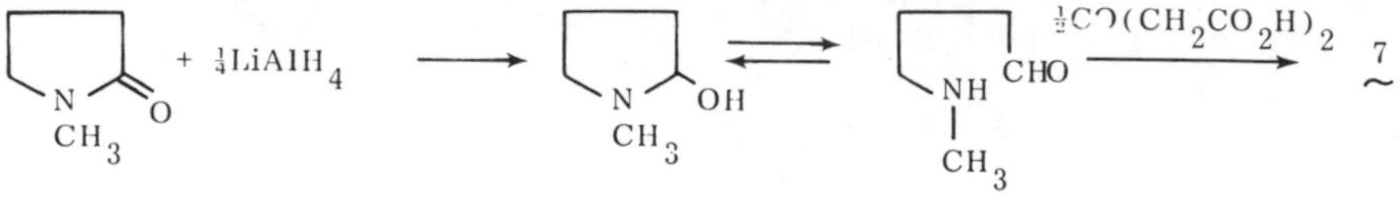

In nature the anion of acetoacetate (**15**) is the probable intermediate in an initial attack on the *N*-methylpyrrolinium cation (**16**). This intermediate can undergo double decarboxylation to hygrine (**6**), or decarboxylation, carbanion formation, and nucleophilic attack on a second *N*-methylpyrrolinium species to give cuscohygrine (**7**) (Scheme 1).

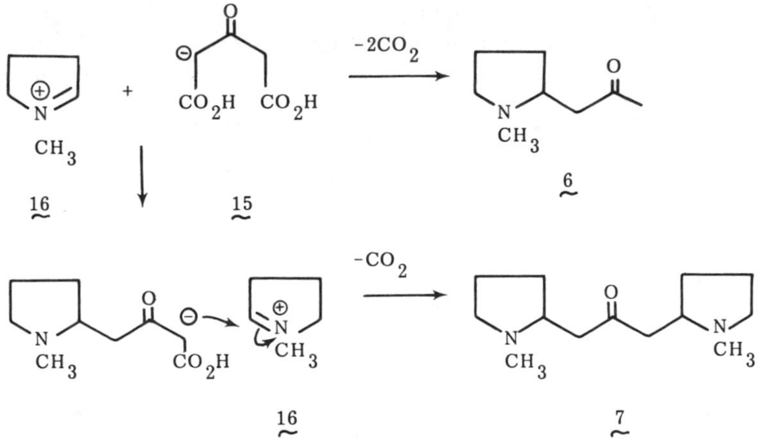

Scheme 1

There are several alkaloids known which, like nicotine, contain a pyrrolidine ring attached to an aromatic system.

One such compound is ficine (**17**), a flavonoid alkaloid from *Ficus pantoniana*. Ficine has been synthesized from **18** using standard procedures.

A second interesting compound is codonopsine (**19**) from *Codonopsis clematidea* C. B. Clarke (Campanulaceae), and other examples include brevicolline (**20**) from *Carex brevicollis* (Cyperaceae) and the pyrrolidinoisoquinoline macrostomine (**21**) from *Papaver macrostomum* Boiss. (Papaveraceae).

It should not be assumed that in each case a derivation of the pyrrolidine

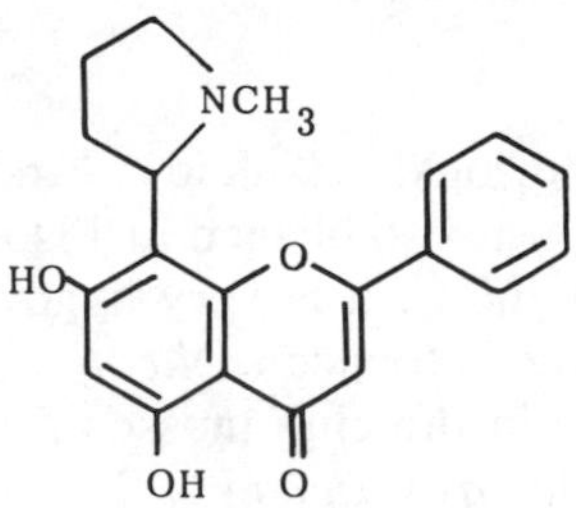

17 ficine

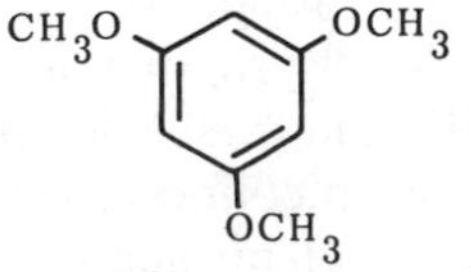

18

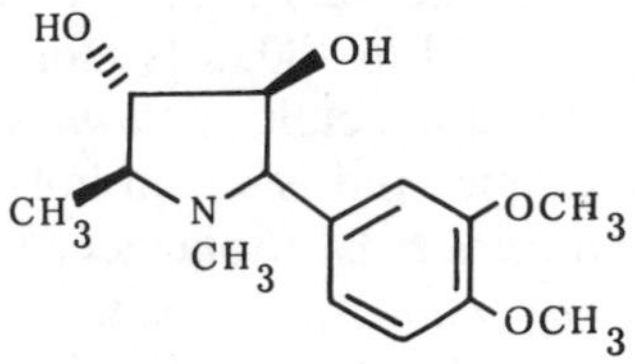

19

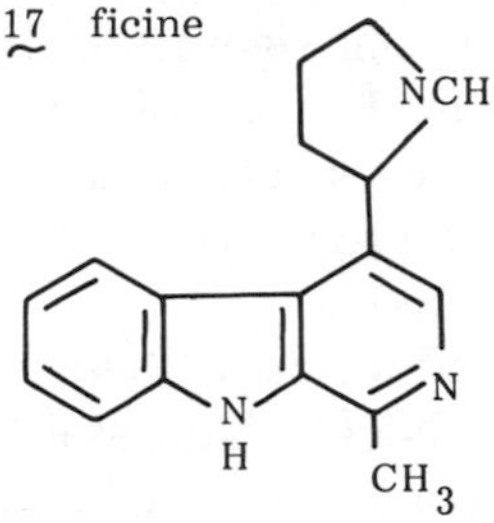

20

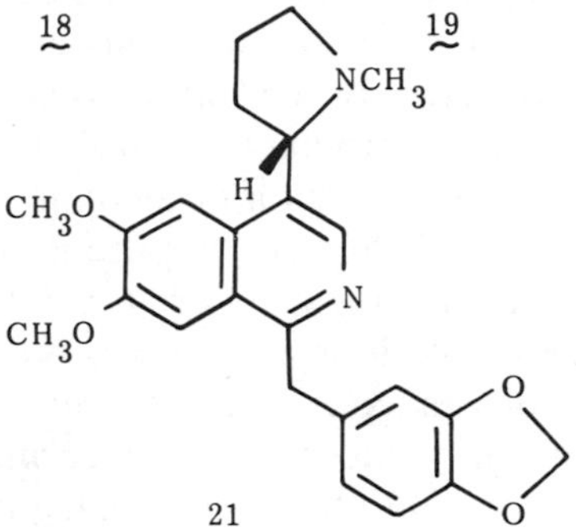

21

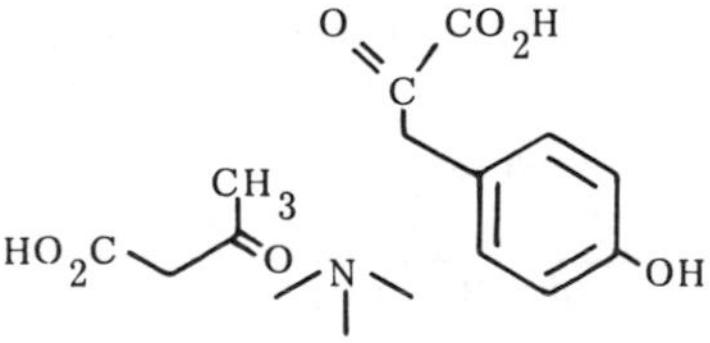

22

ring from ornithine is implied. Indeed for codonopsine a biogenesis from phenyl pyruvic acid and acetoacetate (i.e. 22) would seem more likely.

LITERATURE

Simple Pyrrolidine Alkaloids

Marion, L., *Alkaloids NY* **1,** 91 (1950).

Marion, L., *Alkaloids NY* **6,** 31 (1960).

Snieckus, V. A., *Alkaloids, London* **1,** 48 (1971).

Snieckus, V. A., *Alkaloids, London* **2,** 33 (1972).

Snieckus, V. A., *Alkaloids, London* **3,** 43 (1973).

Snieckus, V. A., *Alkaloids, London* **4,** 50 (1974).

Snieckus, V. A., *Alkaloids, London* **5,** 56 (1975).

Pinder, A. R., *Alkaloids, London* **6,** 54 (1976).

Pinder, A. R., *Alkaloids, London* **7,** 35 (1977).

Pinder, A. R., *Alkaloids, London* **8,** 37 (1978).

3.2 NICOTINE

Nicotine (**23**) is one of the oldest (and most notorious!) alkaloids. First isolated in 1809, its molecular formula ($C_{10}H_{14}N_2$) was established in 1843. Being miscible with water in all proportions below 60°C, it is very hygroscopic, and is steam volatile. The best known source is tobacco, *Nicotiana tabacum* L. (Solanaceae), but nicotine also occurs in the club mosses *Lycopodium* sp., as well as higher plants such as *Ascelepias syriaca* L., *Eclipta alba* (L.) Hassk, *Erythroxylon coca* Lam., and *Withania somnifera* Dunal.

Tobacco itself is of enormous commercial and social significance, and annual production of the leaf of *Nicotiana tabacum* is in excess of 5 million tons. Part of this production is used for cigarettes, pipe tobacco, and cigars, and part for the isolation of nicotine. Because of the problems of toxicity, plants of low alkaloid content are preferred for human use.

Tobacco was originally brought into Europe from Florida in about 1560. The story is told how a plant was presented to Jean Nicot, Ambassador of the King of France to Portugal. Subsequently the plant developed a reputation for curing ulcers and ringworm, and later many other diseases. It is thought (ironically) that the smoking of tobacco may have started as a way of curing diseases of the cardiovascular or respiratory system.

3.2.1 Chemistry and Spectral Properties

Nicotine, the principal alkaloid of tobacco, is dibasic, having two quite different tertiary nitrogens: the strongly basic pyrrolidine nitrogen (pK_a 8.2) and the weakly basic pyridine nitrogen (pK_a 3.4). Two methiodides are possible, and degradation of one of these, nicotine isomethiodide (**24**), offers some important structure information on further reaction. Oxidation with permanganate gave trigonelline (**25**), indicating the pyridine nucleus to be present. Oxidation with alkali potassium ferricyanide gave the 2-pyridone **26**, which was further oxidized by chromic acid to $(-)$-hygrinic acid (**8**). The 2-position of *N*-methylpyrrolidine is therefore joined to the 3-position of pyridine, and nicotine has the structure **23**.

There are many methods available for the quantitative analysis of nicotine in tobacco, several of them automated. Spectrometrically its concentration can be measured by the difference in absorbance at 259 nm between the neutral and acidified alkaloid fraction obtained by steam distillation. Gas chromatography has proved highly successful in the analysis of tobacco alkaloid mixtures. Three systems of particular use are 10% DC 550, 10% Versamid 900, and 20% Carbowax 20 M.

There have been several syntheses of nicotine. The first was that of Pictet in 1904, which involved pyrolysis of the pyrrole **27** to the isomeric pyridine pyrrole **28**, which by an extensive series of reactions gave nicotine (**23**). An alternative synthesis begins with a Claisen-type condensation of 1-methylpyrrolidone (**12**) and ethyl nicotinate (**29**). Acid hydrolysis of the product cleaved the amide and permitted decarboxylation of the β-keto acid to give the amino ketone **30**. Reduction of the ketone to an alcohol and formation of the secondary iodide with hydrogen iodide followed by treatment with base gave racemic nicotine (**23**).

Leete and his associates have more recently developed two new syntheses of nicotine. One of these is a quite general synthesis and involves the carbanion **31** as an intermediate. This anion has been used for the synthesis of several nicotine analogues, but in the synthesis of nicotine itself condensation is made with acrylonitrile to give **32**. Hydrolysis and catalytic hydrogenation gives myosmine (**33**) by internal cyclization of an intermediate keto amine. Sodium borohydride reduction of the imine followed by N-methylation gave **23**.

The second route developed by Leete is more biomimetic in character and involves the mixing of glutaraldehyde (**34**), ammonia and *N*-methyl-Δ^1-pyrrolinium acetate (**35**) at pH 10.3 for 24 hr to give nicotine (**23**) in 21% yield. The reaction is considered to proceed through the dienamine, 1,4-dihydropyridine (**36**). These syntheses are summarized in Figure 2.

Nicotine (**23**) shows λ_{max} 262 nm (log ϵ 3.46) in ethanol with little change on the addition of acid. Myosmine (**33**) on the other hand shows two maxima at 234 and 266 nm (log ϵ 4.05 and 3.58).

The conformation of chiral nicotine-type alkaloids in aqueous solution can be determined by circular dichroism using the sector rules for the $\pi \rightarrow \pi^*$ transition in monosubstituted benzenes. In this way negative Cotton effects were observed for the pyridine $\pi \rightarrow \pi^*$ band of (S)-$(-)$-nicotine (**37**), (S)-$(-)$-anabasine (**38**), and (S)-$(-)$-cotinine (**39**). These data also supported values of the torsional angle τ, as indicated in **37**, of about 120° which had been obtained from Hückel MO calculations and X-ray analysis.

The ^{1}H NMR spectrum of nicotine has recently been examined in detail using selectively deuterated analogues. This has permitted the first unam-

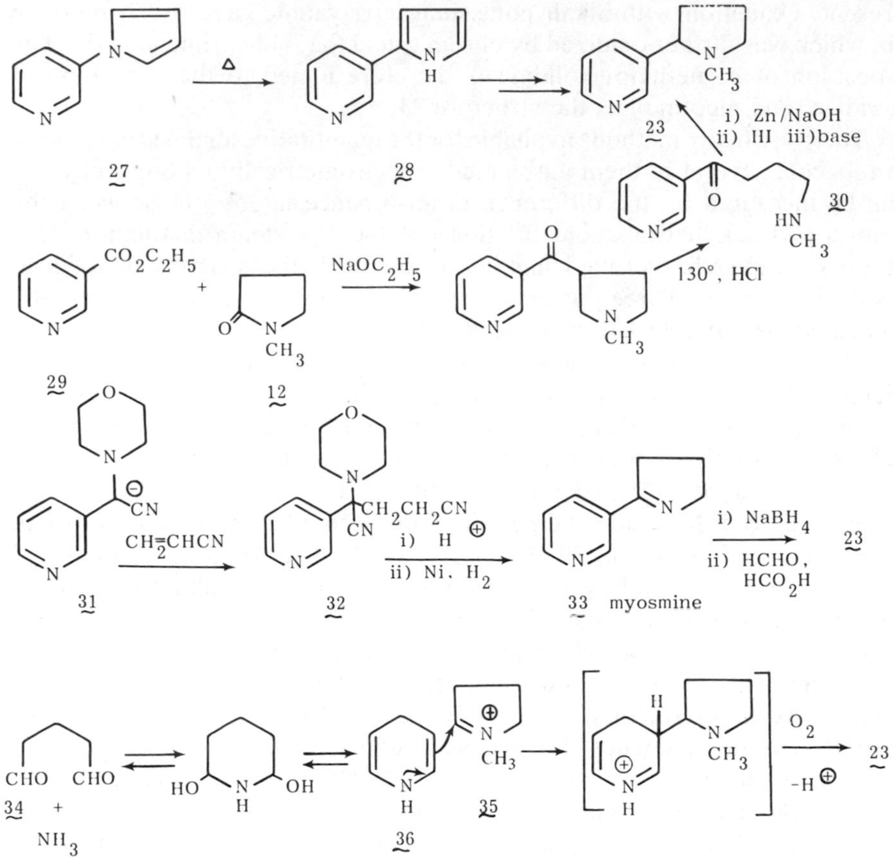

Figure 2 Syntheses of nicotine (**23**).

biguous assignment of the pyrrolidine ring protons (Table 1). The data establish an envelope conformation for the five-membered ring, where the methyl and pyridine moieties are equatorial. The pyridine ring was found to be oriented perpendicular to the pyrrolidine ring, as evidenced by the nuclear Overhauser effect observed for the H-2' resonances on the irradiation of H-2 and H-4.

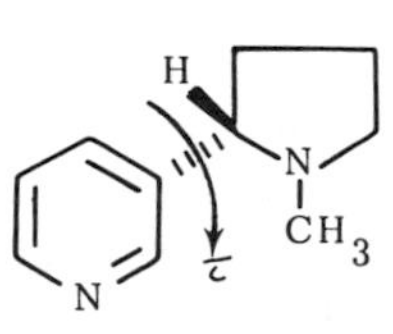

37   nicotine
showing torsional
angle

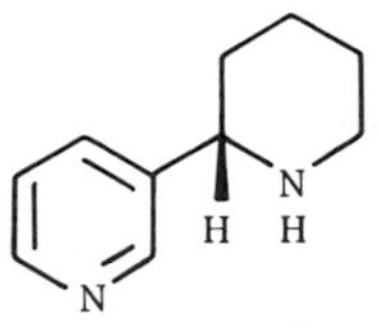

38   (S)-(-)-anabasine

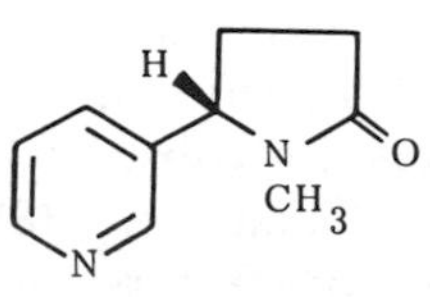

39   cotinine

Table 1 The ^{1}H Proton Assignments of Nicotine[a]

		Coupling Constants (Hz)			
Proton	Chemical Shift	2	4	5	6
2	8.540	−0.51	−0.45	0.37	<0.05
4	7.683	—	2.27	0.89	0.0
5	7.218	—	—	7.86	1.70
6	8.475	—	—	—	4.79

		Coupling Constants (Hz)					
		$3'\alpha$	$3'\beta$	$4'\alpha$	$4'\beta$	$5'\alpha$	$5'\beta$
$2'$	3.084	8.99	7.96	[b]	[b]	0.50	[b]
$3'\alpha$	1.719	—	−12.75	10.92	5.68	[b]	[b]
$3'\beta$	2.189	—	—	5.43	9.67	[b]	[b]
$4'\alpha$	1.952	—	—	—	—	7.93	9.68
$4'\beta$	1.806	—	—	12.50	—	2.19	8.30
$5'\alpha$	3.248	—	—	—	—	—	−9.17
$5'\beta$	2.316	—	—	—	—	—	—

Source: Data from Pitner *et al., J. Amer. Chem Soc.* **100**, 246 (1978).
[a] Duplicate coupling constants are not indicated (i.e. $J_{3'\alpha,3'\beta}$ is not repeated as $J_{3'\beta,3'\alpha}$).
[b] Not measurable.

In trifluoroacetic acid the *N*-methyl group of nicotine gives two signals at 3.13 and 2.82 ppm, attributed to the *trans* and *cis* isomers respectively. The ^{13}C assignments of nicotine as the free base are shown in **40**.

$$\underset{\sim}{40}$$

3.2.2 Biosynthesis

In 1912, Trier suggested that nicotine was derived by a combination of proline (**9**) and nicotinic acid (**41**). When feeding experiments with [2-^{14}C]- and [5-^{14}C]ornithine began it was quickly shown that in each case carbons 2 and 5 of the pyrrolidine ring of nicotine were equally labeled. This led to the suggestion that the pyrrolidine ring is formed *via* the symmetrical intermediate putrescine (42). N-Methylation, oxidation of the amine to an aldehyde, followed by cyclization gives the *N*-methyl-Δ^1-pyrrolinium cation (16). The enzymes that carry out these steps have been obtained from *N. tabacum* roots.

In 1960, Dawson showed that nicotinic acid (41) was an efficient precursor of the pyridine ring of nicotine. There was one anomalous result however, for although [5-^{3}H]nicotinic acid (43) gave 14% incorporation, [6-^{3}H]nicotinic acid (44) gave only 1.1% incorporation into 37. When these

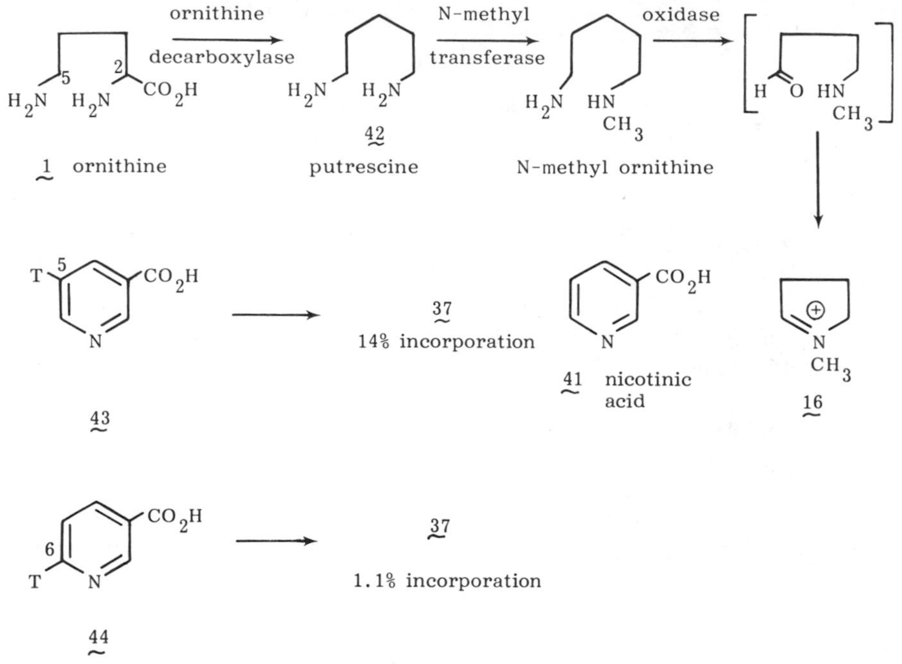

results were followed up by Leete years later, it was found that loss of tritium from position 6 occurred in the process of condensing a reduced nicotinic acid (45) and the pyrrolinium cation 16. The mechanism proposed for this sequence is shown in Scheme 2. Considerable steric control operates in this reaction in order to permit the stereospecific, concerted loss of the hydrogen originally at C-6 in nicotinic acid.

Unfortunately the formation of the pyrrolidine ring is not as simple as the scheme given above. When [δ-^{15}N, 2-^{14}C]- and [α-^{15}N, 2-^{14}C]ornithines were used as precursors, the δ-N was found to be incorporated and the ^{14}C/^{15}N ratio was twice that in the starting material. Putrescine (42) is therefore eliminated as a free intermediate, *if* it is produced by direct decarboxylation of ornithine.

A pathway to explain these results is shown in Scheme 3. Crucial in this suggestion is the concept that ornithine rapidly loses the α-nitrogen by transamination. The key step is N-methylation of putrescine where, depending

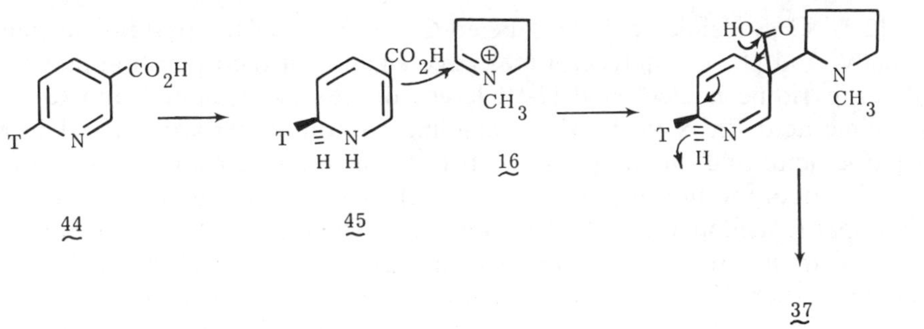

Scheme 2

on the site of the N-methylation, in subsequent steps either the labeled or unlabeled nitrogen is lost. Thus in the final product, half of the nitrogen label is lost and all the [^{14}C] activity retained.

When the origin of the pyridine nucleus was first investigated the only route to this nucleus known was that *via* tryptophan, the fungal route. When labeled tryptophan was used as a precursor it was not incorporated into nicotine. Similar results were also obtained for other pyridine alkaloids. There must therefore be a previously unknown pathway to nicotinic acid involved in plants.

The nature of the early intermediates on the pathway came from work with bacterial systems. Using a variety of precursors, [1-^{14}C]acetic acid, [1,4-^{14}C$_2$]aspartic acid, and [2-^{14}C]- and [5-^{14}C]glutamic acid (**46**) were found to be precursors of nicotinic acid.

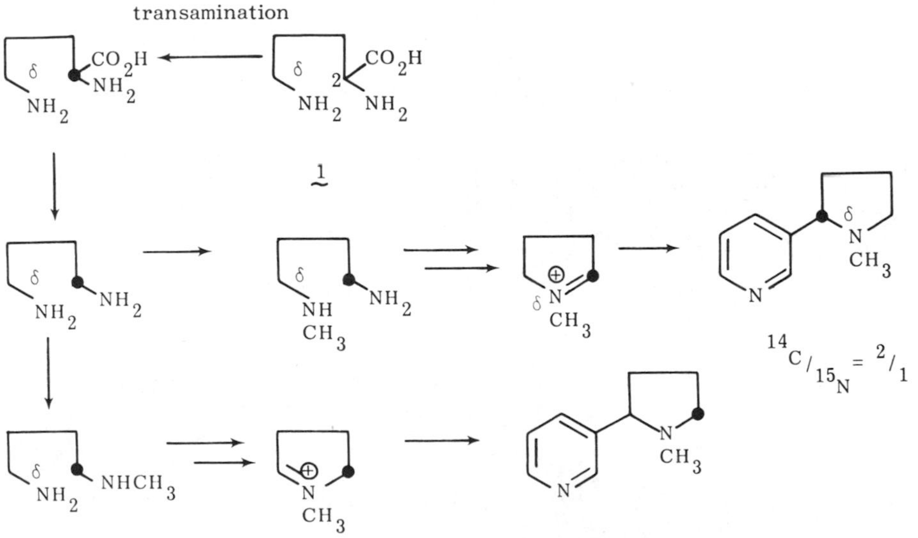

Scheme 3 Source of the pyrrolidine ring nitrogen in nicotine (**37**).

[2,3-$^{14}C_2$]Succinic acid (**47**) labeled C-2 and C-3 of the pyridine nucleus equally, and [1,3-$^{14}C_2$]glycerol (**48**) gave significant incorporation into C-6 of the pyridine nucleus and [2-^{14}C]glycerol was incorporated into C-5 of nicotinic acid. In bacteria, the remaining carbon atoms are derived from aspartic acid, and this may also be true for nicotine in plants, where preferential incorporation into C-2 and C-3 of nicotine was found for both [3-^{14}C]aspartic acid and [2,3-$^{14}C_2$]succinic acid. The route is clearly more complicated than direct incorporation of aspartic acid, which should lead to specific rather than the partially random incorporation observed.

The mechanism by which this randomization between C-2 and C-3 of aspartic acid occurs and the mechanism by which glycerol and aspartic acid (or some derivative thereof) combine remains unresolved.

The first major intermediate is quinolinic acid (**49**) and this has been shown to be a precursor of nicotine (**37**). Even this step is not a direct decarboxylation, but rather proceeds *via* nicotinic acid mononucleotide (**50**) by condensation with 5-phosphoribosyl-1-pyrophosphate and decarboxylation. Removal of the ribose phosphate may then occur by a simple or a complex process (Scheme 4).

3.2.3 Catabolism

Nicotine is not the end product of metabolism in *Nicotiana* sp. In 1945, Dawson showed that nicotine was demethylated to nornicotine (**51**), and

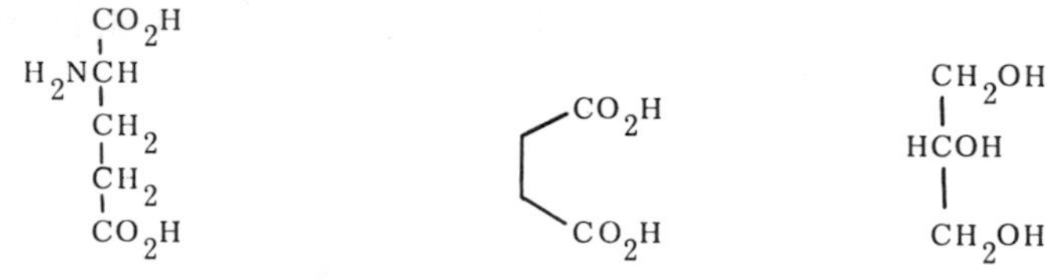

46 L-glutamic acid 47 succinic acid 48 glycerol

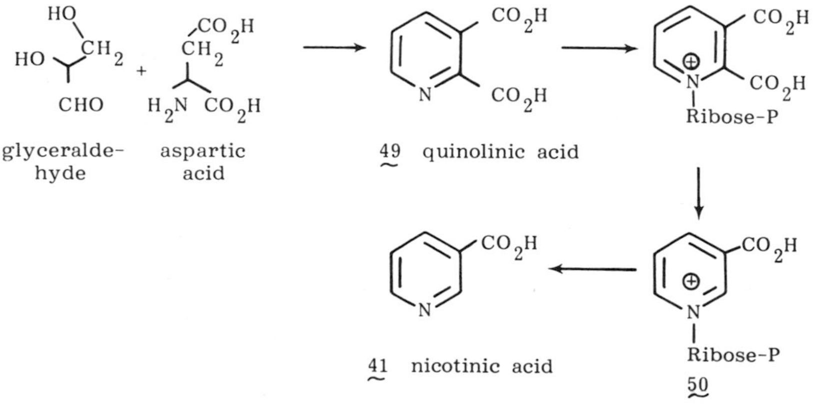

Scheme 4 Biosynthesis of nicotinic acid (**41**).

later workers showed that some racemization occurs in this nonreversible process. It has been suggested that the process of demethylation and racemization proceeds by way of the N-oxide, but definitive evidence is lacking. Myosmine (33) is also a significant metabolite.

3-Acetylpyridine (52) is one of the products of the degradation of nicotine (37) in microorganisms.

The metabolism of nicotine in mammals involves oxidation of the pyrrolidine ring to give the weakly active cotinine (39), the 3β-hydroxy derivative (53) of which has been obtained from the urine of smokers. A synthesis of 53 actually proceeds through the 3α-isomer 54, which is then epimerized by an SN_2 sequence.

37   (−)−nicotine   51   nornicotine   33   myosmine

52   3−acetyl−   39   cotinine, R=H
     pyridine   53   R=β − OH
                           54   R=α − OH

In the hepatic supernatants of several small mammals, nicotine was converted to the diastereomeric N-oxides, but no demethylation occurred.

3.2.4   Pharmacology

Nicotine is a toxic alkaloid and a dose of 40 mg is said to be fatal to humans (cf. strychnine, Section 9.21). It is fatal to many forms of animal life. Small doses of nicotine stimulate respiratory processes, but larger doses cause inhibition in all sympathetic and parasympathetic ganglia. Death results from respiratory paralysis.

Nicotine imitates the effects of muscarine (ganglion stimulation of cholinergic nerves) and of catecholamines (stimulation of sympathetic ganglia). It also causes vomiting, antidiuresis, and an increase in bowel motor activity.

The average cigarette may contain 10–14 mg of nicotine and a cigar about 70 mg, and much attention has focused on the relationship among lung cancer, smoking, and nicotine. However, nicotine has pronounced effects on the cardiovascular system, where peripheral vasoconstriction, atrial tachycardia, and an increase in both systolic and diastolic blood pressures are observed. It is worth noting that 50% of all smokers die of heart disease and 20% of lung cancer. Nicotine is metabolized quite rapidly by the body and is not accumulated to any great extent.

The carcinogenicity of tobacco is probably not due to nicotine but rather to a far more potent carcinogen. Tobacco chewing, as well as cigarette, pipe, and cigar smoking, are strongly correlated with cancer of the oral cavity, so that smoking of tobacco is not necessary to induce cancer. Recently N-nitrosonornicotine (55) was obtained from cigarettes, cigars, and

55

chewing tobacco at levels in the range 2–90 ppm. The carcinogenicity of N-nitroso derivatives is well known and parts per billion concentrations of N-nitrosamines are considered hazardous to health.

We have seen that nornicotine (51) is a product of bacteriological degradation of nicotine (37) during curing. Nitrosamine formation probably occurs at this stage either from natural nitrite or from the biological reduction of nitrate (added as fertilizer) to nitrite.

Nicotine injected into pregnant mice either subcutaneously or intraperitoneally caused fetal deaths, decreased litter size, and produced skeletal defects. Nicotine sulfate injected into fertilized chicken eggs had a pronounced teratogenic effect on the hatched chicks. Similar results concerning the teratogenicity of tobacco have been observed for swine, but it is not known with certainty that nicotine is responsible for these effects.

Other Alkaloids of Tobacco

In addition to nicotine (37), nornicotine (55), and myosmine (33) mentioned previously, several other alkaloids occur in tobacco. The alkaloid content and distribution in tobacco at various times in the life of the plant and during curing are the subject of continued study. We have seen previously that nicotine is produced almost exclusively in the roots of tobacco. Recently it was shown that the *stem* callus tissue of *N. tabacum* did not produce alkaloids until root development took place.

Some of the other alkaloids which occur in tobacco include nicotyrine (56) and several N-acylated derivatives (57–59).

Nicotyrine (56) was prepared as an intermediate in the original synthesis of nicotine long before it was detected in tobacco. Reduction with zinc and hydrochloric acid affords dihydronicotyrine, which is a more potent insecticide than nicotine (37). Dihydronicotyrine has the structure 60, as expected for the reduction of a pyrrole, based on NMR evidence (two olefinic protons at 5.75 and 6.06 ppm).

An interesting new synthesis of nicotyrine (56) involves the photolytic reaction of 3-iodopyridine (61) and N-methylpyrrole (10). The precise mechanism has not been established.

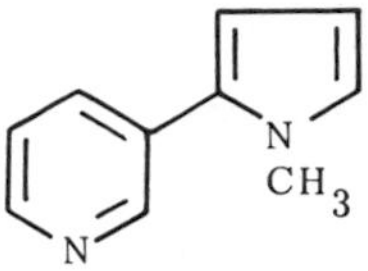

56　nicotyrine

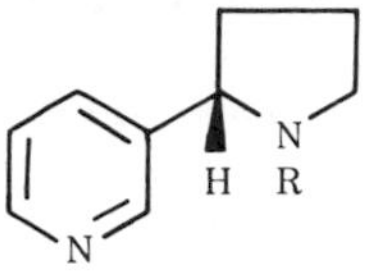

57　R=CHO
58　R=COCH$_3$
59　R=COn-C$_5$H$_{11}$

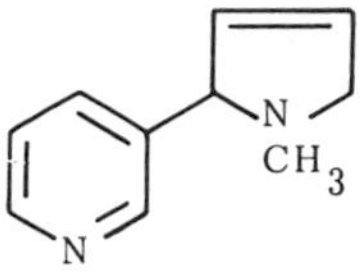

60

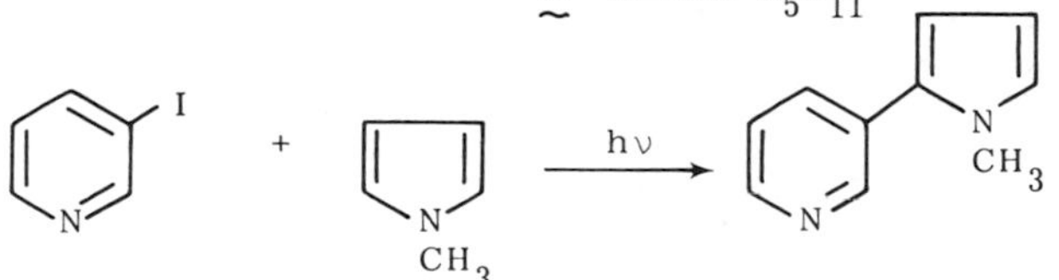

61

10

56

The other major series of alkaloids of N. tabacum is a group based on the union of a pyridine and a piperidine. Typical of these is anabasine (38), which is discussed in Chapter 4. Related to anabasine are the alkaloids 3',2-bipyridyl (62) and nicotelline (63).

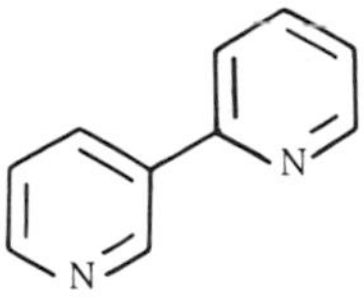

62　3',2 – bipyridyl

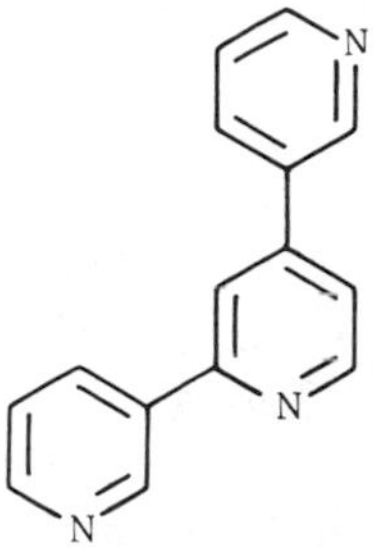

63　nicotelline

LITERATURE

Nicotine

Enzell, C. R., I. Wahlberg, and A. J. Aasen, *Fortschr. Org. Chem. Naturs.* **34**, 1 (1977).

Biosynthesis

Dawson, R. F., D. R. Christman, R. C. Anderson, M. L. Solt, A. F. D'Adamo, and U. Weiss, *J. Amer. Chem. Soc.* **78,** 2645 (1956).

Christman, D. R., and R. F. Dawson, *Biochemistry* **2,** 182 (1963).

Leete, E., E. G. Gros, and T. J. Gilbertson, *Tetrahedron Lett.* 587 (1964).

Leete, E., and Y.-Y. Liu, *Phytochemistry* **12,** 593 (1973).

Mizusaki, S., Y. Yanabe, M. Nognchi, and E. Tamaki, *Plant Cell. Physiol.* **14,** 103 (1973) and references therein.

Leete, E., in N. Fina (Ed.), *Proc. First Philip Morris Sci. Symp.*, New York, 1973, p. 92.

Metabolism

Dawson, R. F., *J. Amer. Chem. Soc.* **67,** 503 (1945).

Leete, E., and M. R. Chedekel, *Phytochemistry* **13,** 1853 (1974).

Frankenburg, W. G., and A. A. Vaitekunas, *Arch. Biochem. Biophys.* **58,** 509 (1955).

McKennis, H., in U. S. von Euler (Ed.), *Tobacco Alkaloids and Related Compounds*, Pergamon, Oxford, 1965, p. 53.

Toxicity

Gerrod, J. W., and P. Jenner, *Essays Toxicol.* **6,** 35 (1975).

3.3   TROPANE ALKALOIDS

Plants in the family Solanaceae produce a variety of alkaloids (see also Chapter 11), some of considerable therapeutic importance. One such group of alkaloids are those possessing the tropane (**64**) nucleus. These alkaloids also occur in the Erythroxylaceae and Convolvulaceae, but most attention has focused on their occurrence in such solanaceous plants as henbane (*Hyoscyamus niger* L.), the thorn apple (*Datura stramonium* L.), and deadly nightshade (*Atropa belladona* L.). These plants are some of the most poisonous known, and three of the attractive berries can cause infant mortality.

The tropane alkaloids are esters of an organic acid and a tropan-3-ol having either the α- *or* β-configuration. Some representative tropane alkaloids are shown in Table 2.

3.3.1   Atropine and Scopolamine—Isolation and Detection

Atropine (**65**), the optically inactive form of *l*-hyoscyamine, was first isolated by Mein in 1833 from *Atropa belladonna*. In most of the plants from which it has been isolated it is not a natural product, but is rather the result of a facile isomerization of *l*-hyoscyamine (**66**). This racemization occurs under a variety of very mild conditions.

The uses of *Atropa belladonna* were first recorded in the early sixteenth century, but the leaves were not introduced into the *London Pharmacopoea* until 1809. Alkaloid content of the leaves varies between 0.15 and 0.6%.

Table 2   Representative Tropane Alkaloids

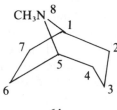

64

| | Substituent | | |
Compound	C-3	C-6	C-7
Atropine (**65**)	α(±)-*O*-tropyl	H	H
(−)-Hyoscyamine (**66**)	α(−)-*O*-tropyl	H	H
Meteloidine (**67**)	α-*O*-tigloyl	β-OH	β-OH
Scopolamine (**68**)	α-*O*-tropyl	β-oxido	
Hyoscine (**69**)	α-*O*-tropyl	β-oxido	

l-Hyoscyamine is the most common tropane alkaloid. First isolated by Geiger in 1833 from henbane, it has subsequently been isolated from *Datura* sp., *Duboisia* sp., *Mandragora* sp., and *Scopolia* sp. The major current source of *l*-hyoscyamine is *Hyoscyamus muticus*, L., indigenous to Egypt, but now cultivated in southern California. Alternative sources are Jimson

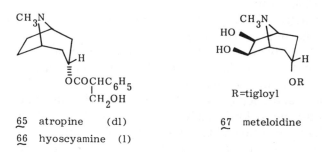

65   atropine   (dl)
66   hyoscyamine   (l)

67   meteloidine

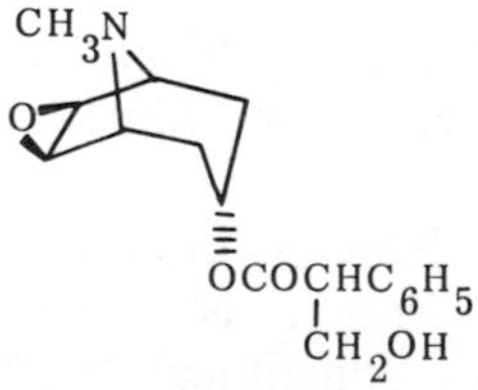

68   scopolamine   (dl)
69   hyoscine   (l)

weed, *Datura stramonium*, and *Duboisia myoporides* R. Br., a plant native to Australia.

Hyoscyamine (**66**) is rarely isolated commercially, but rather is totally racemized during the isolation process to atropine (**65**). Atropine sulfate is prepared by the careful addition of methanolic sulfuric acid to a cold methanolic solution of atropine. Final purification may be made by extracting at pH 7.8–8.2. The salts of atropine and *l*-hyoscyamine are affected by light, moisture, and heat.

Scopolamine was first obtained by Ladenburg in 1881 from *Hyoscyamus muticus* and subsequently from *Scopolia atropides* Berch. and Presl. It is the major alkaloid in the leaves of *Datura metel* L., *D. meteloides* L., and *D. fastuosa* var. *alba*. As with *l*-hyoscyamine (**66**), racemization occurs easily with alkali, and *Duboisia leichardtii* F. Muell. is a commercal source of the racemate.

The most sensitive test for atropine, *l*-hyoscyamine, and *l*-scopolamine is their mydriatic reaction in the pupil of the eyes of young dogs, cats, and rabbits. A neutral alcohol-free aqeous solution of the sulfate salt is dropped into the conjunctival sac of the eye. As little as 0.4 μg is said to be effective. The Vitali reaction is important in determining tropane alkaloids on a small scale. The test involves treating 0.1 mg of the alkaloid with a drop of fuming nitric acid, evaporating to dryness at 100°C, and adding a drop of freshly prepared ethanolic potassium hydroxide. A bright purple color develops, which fades slowly to dark red.

Potassium iodoplatinate is a good TLC-detecting reagent, affording blue or violet spots. Bromothymol is also effective and gives dark blue spots on a light blue background.

A colorimetric assay, specific to the esters of the tropane alkaloids rather than their inactive hydrolysis products, involves formation of the ferric hydroxamate derivative and measurement of the absorbance at 540 nm.

An alternative procedure is somewhat longer but does give a stable end product. The method involves nitration of the alkaloid, Zn/HCl reduction, diazotization, and coupling with *N*-naphthylethylenediamine dihydrochloride. The red-violet color is analyzed quantitatively at 550 nm.

There have been several gas chromatographic systems used with success to separate and quantitate the various tropane alkaloids. Among these are 20% SE-30, 10% XE-60, and 5% Versamid on Chromasorb-W, and 3% OV-17 on Gas Chrom Q.

Several new types of tropane alkaloid have been isolated in recent years from the family Proteaceae, previously not known to contain tropane alkaloids.

Bellendine (**70**) from *Bellendena montana* Boiss. was the first alkaloid to be obtained from the Proteaceae and is a pyranotropane derivative. 2-Methylbellendine (**71**) is a constituent of *Darlingia ferruginea* J. F. Bailey, a plant that also gave the novel tropane deriative ferrugine (**72**). *Knightia deplanchei* Vieill. ex Brongn. et Gris. has afforded a new group of 2-benzyl tropane alkaloids (e.g. **73**) whose stereochemistry was determined by ^{13}C NMR.

Physalis peruviana L. (Solanaceae) has afforded the alkaloid physoperuvine (**74**) in which the C-2-to-N-8 linkage of tropinone is broken. It is not known if this alkaloid co-occurs with normal tropane alkaloids.

70 bellendine, R=H
71 2-methylbellendine, R=CH$_3$

72 ferrugine

73

74 physoperuvine

An interesting example of the distribution of alkaloids within a plant concerns the Australian plant *Duboisia hopwoodii* F. Muell. The leaves of this plant are a narcotic when chewed by the Australian aborigines, who also use the leaves to poison water holes in emu hunting. Chromatographically, nicotine and nornicotine were detected in the leaves, but the roots, as well as containing the nicotine derivatives, also contained hyoscyamine (major alkaloid) and hyoscine. The tropane alkaloids are restricted to the roots of the plant!

3.3.2 Chemistry of the Tropane Nucleus

Hydrolysis of atropine (**65**) affords tropine (**75**), having the molecular formula $C_8H_{15}NO$ and ($\pm$)-tropic acid; *l*-hyoscyamine (**66**) affords **75** and ($-$)-tropic acid (**76**). The configuration of the latter was demonstrated by correlation with alanine (Figure 3).

Curtius degradation of ($+$)-α-methylphenylacetic acid (**77**) gave ($-$)-α-phenylethylamine (**78**), which on potassium permangamate oxidation gave L-($+$)-alanine (**79**). The absolute configuration of **77** is therefore S. Correlation with ($-$)-tropic acid (**76**) was carried out by catalytic hydrogenolysis of ($-$)-β-chlorohydratropic acid (**80**).

Tropine is optically inactive, strongly basic (pK_a 10.5), and contains an *N*-methyl group. Reaction with hydriodic acid and zinc–hydrochloric acid reduction gave tropane (**64**), which with zinc dust was degraded to 2-ethyl pyridine (**81**).

The seven-membered carbocyclic system was deduced by dehydration

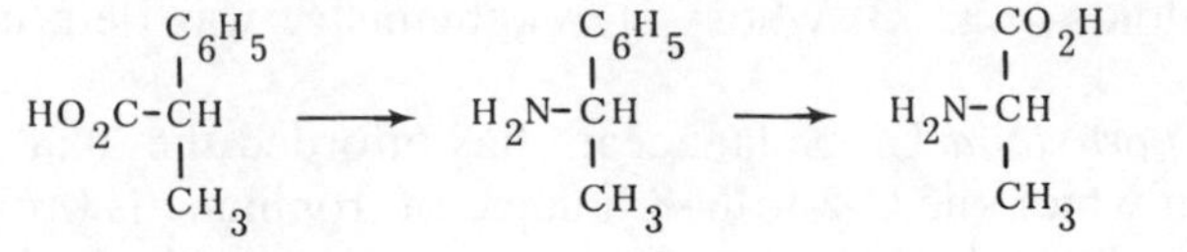

$$HO_2C-\underset{\underset{CH_3}{|}}{\overset{\overset{C_6H_5}{|}}{C}}H \longrightarrow H_2N-\underset{\underset{CH_3}{|}}{\overset{\overset{C_6H_5}{|}}{C}}H \longrightarrow H_2N-\underset{\underset{CH_3}{|}}{\overset{\overset{CO_2H}{|}}{C}}H$$

77 (+)-methyl   78 (−)-phenyl   79 (+)-alanine
~ phenylacetic   ~ ethylamine   ~
   acid

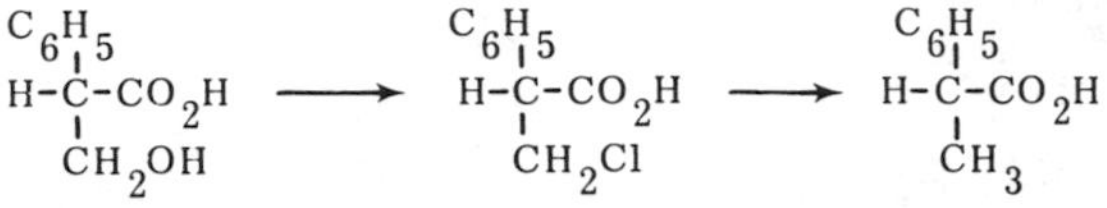

$$H-\underset{\underset{CH_2OH}{|}}{\overset{\overset{C_6H_5}{|}}{C}}-CO_2H \longrightarrow H-\underset{\underset{CH_2Cl}{|}}{\overset{\overset{C_6H_5}{|}}{C}}-CO_2H \longrightarrow H-\underset{\underset{CH_3}{|}}{\overset{\overset{C_6H_5}{|}}{C}}-CO_2H$$

76 (−)-tropic   80   (−)-methyl
~ acid   ~   phenylacetic
                                                 acid

Figure 3   Correlation of (−)-tropic acid (**76**) with (+)-alanine (**79**).

of tropine (**75**) followed by exhaustive Hofmann degradation to tropylidene
(**82**).

Chromic acid oxidation of **75** in sulfuric acid gave *N*-methylsuccinimide
(**83**), good evidence for the presence of a pyrrolidine ring in **75**. This work
was carried out mainly by Wilstätter, and it was he who assigned the structure **75** to tropine.

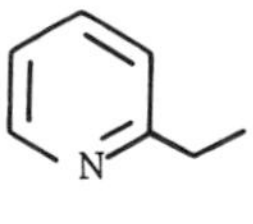

81
~

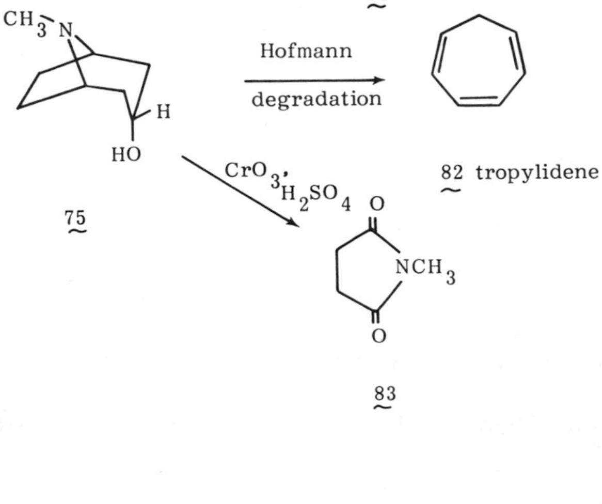

Hofmann
───────────→
degradation

82 tropylidene
~

CrO_3, H_2SO_4

75
~

83
~

It was Wilstätter too who first synthesized tropine, even if by a laborious route, but the most dramatic synthesis is that due to Robinson. In this approach, which is biomimetic in character, succindialdehyde, methylamine, and acetone dicarboxylic acid are mixed at pH 5.6 for 30 min to give tropinone (**84**) in one step! This procedure has been widely used over the

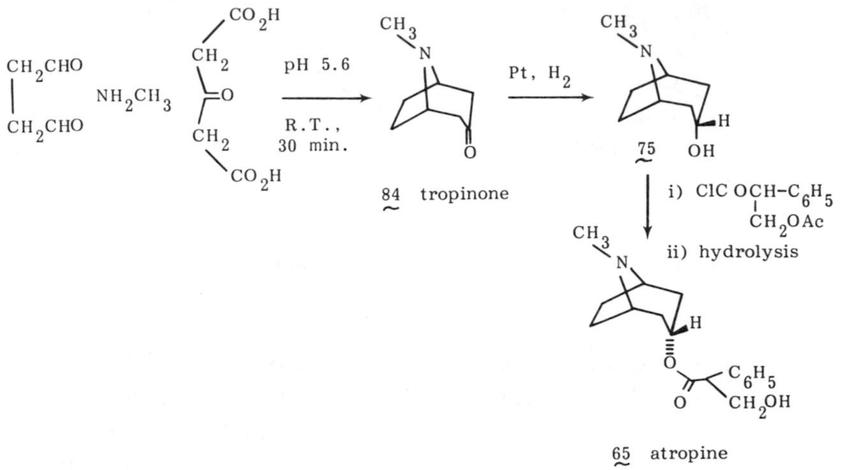

years for the synthesis of numerous substituted tropinones. Completion of a synthesis of atropine was achieved by Wilstätter in 1922, by hydrogenation of **84** to tropine (**75**) followed by reesterification with acetyltropyl chloride and mild hydrolysis of the acetyl group.

The C-3 configuration of tropine was deduced when *N*-acetyl-nor-pseudotropine (**85**), as the hydrochloride, was heated to 160°C. *O*-Acetyl-nor-pseudotropine (**86**) was the product, the rearrangement occurring through

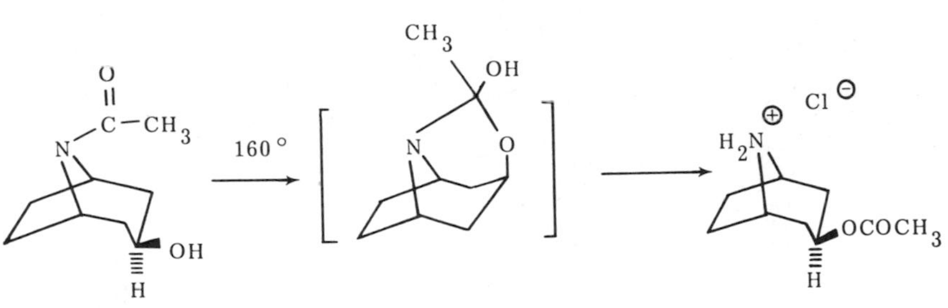

an intermediate oxazine. Such a compound can only be produced when the piperidine ring adopts a boat configuration, and the *N*-acyl and hydroxyl groups are proximate. The C-3 configuration of the hydroxyl group in **85** is therefore β.

Further chemical evidence for this configurational assignment was obtained from the rates of hydrolysis of the epimeric benzoate esters of tropine (75) and pseudotropine (87). As expected for an axial ester (chair configuration), tropine benzoate (88) is hydrolyzed at a slower rate than pseudotropine benzoate (89).

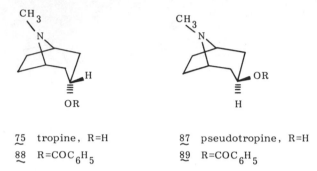

75 tropine, R=H   87 pseudotropine, R=H
88 R=COC$_6$H$_5$ 89 R=COC$_6$H$_5$

Scopolamine (68) is a somewhat weaker base (pK_a 7.6) than hyoscyamine (66). Alkaline hydrolysis of scopolamine afforded tropic acid (76) and oscine having the molecular formula $C_8H_{13}NO_2$, which can be resolved by D-(+)-tartaric acid. A different isomeric product, scopine, results when scopolamine is hydrolyzed with buffered pancreatic lipase. This product cannot be resolved by D-(+)-tartaric acid, but both products can be acetylated to give monoacetate derivatives. Scopine is easily converted to oscine under mild basic conditions. These data were interpreted in terms of structures 90 and 91 for scopine and oscine respectively. Stereochemically, the rearrangement of 90 to 91 can occur only by rearside nucleophilic attack of the 3α-

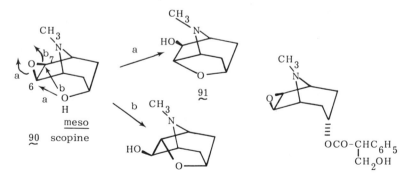

68 scopolamine

hydroxy group on the 6,7-epoxide in the β-configuration. Evidence to support this comes from the inability of pseudoscopine to undergo rearrangement. Scopolamine therefore has the structure 68.

Substantial effort has been directed toward determining the ground state conformation of the piperidine ring and of the N-methyl group.

X-ray crystallographic examination of tropine hydrobromide (75) indi-cated the piperidine ring to be in the chair form with an equatorial N-methyl group. The crystal structure of the base pseudotropine (87) also indicated an equatorial N-methyl group, but with quite considerable distortion in the piperidine ring.

The configuration of the respective groups in N-alkylated quaternary salts has also been investigated by both X-ray and NMR techniques. Tropine ethiodide for example has an equatorial N-ethyl group, but the methiodide of N-ethyl nortropine (92) has an equatorial N-methyl group. Apparently the quaternizing group consistently enters by equatorial attack. The precise shape of the tropane nucleus determines the orientation of these quaterni-zation reactions. Pseudotropine for example has both the piperidine and pyrrolidine nuclei deformed, as shown in 93. It is thought that compression of the N-substituent with the axial 2β and 4β-hydrogens (equatorial attack) is less than its compression between the C-6 and C-7 hydrogens (axial at-tack). The N-substituent therefore preferentially adopts an axial configu-ration under quaternizing conditions.

Similarly, oxidation of tropine with hydrogen peroxide leads to two N-oxides. The major N-oxide (65% yield) is the equatorial N-oxide 94, whereas the minor compound (2–8% yield) is the axial isomer 95.

The NMR spectrum of pseudotropine hydrochloride (87) shows two sig-nals for the N-methyl group, indicating the presence of two conformational isomers in solution and the tropanes 96 and 97 were found to exist as a 9 : 1 mixture of the equatorial : axial N-methyl conformations by NMR analysis.

There are several routes available to remove the N-methyl group of tropane alkaloids. Chromium trioxide in pyridine is one method, but alter-natives include treatment with ethyl chloroformate or trichloroethylchlo-roformate and zinc–acetic acid reduction, or photolysis in benzene solution saturated with oxygen to give the *N*-formyl derivative, which can be hydrolyzed to the secondary amine.

3.3.3 Cocaine

The use of coca leaves in South America as a masticatory has been known since early times. Explorers to South America found in the early sixteenth century that natives chewed the leaves, mixed with lime, to travel great distances without fatigue and with little food. Indeed coca chewing is still part of the native culture in both the highlands and lowlands of the north-western Amazon Valley in Columbia and Peru. Coca chewing does not lead to addiction.

Coca is the dried leaves of the shrubs *Erythroxylon coca*, known as Bolivian or Huanaco coca, or *Erythoxylon truxillense* (Peruvian or Truxillo coca), each in the family Erythroxylaceae. The plants are indigenous to Peru, Bolivia, and Colombia, but crops are also harvested in Java and Ceylon. Annual requirements for the leaves are in excess of 10,000 tons.

The yield of total alkaloids is the range 0.7–1.5%, but since there are two other alkaloid types present of no commercial significance, the more important data are for the ester alkaloids of the tropane type. Huanaco coca contains 0.5–1% of the ester alkaloids, principally cocaine. The Truxillo coca has the same alkaloid content and may contain up to 75% of cocaine.

Cocaine was first obtained by Wöhler in 1862, and the anesthetic properties rediscovered by the Austrian physician Koller in 1882. It is a member of the group of compounds called ecgonines, which in general are weaker bases than the corresponding tropanes (e.g. pK_a cocaine 8.7).

Acid hydrolysis of cocaine gives benzoic acid, methanol, and a carboxylic acid, $C_9H_{15}NO_3$, named ecgonine. The skeleton of the latter was determined by chromic acid oxidation to ecgoninone (**98**) and decarboxylation of the β-keto acid to tropinone (**84**). Ecgonine should therefore have the gross structure **99** and cocaine the structure **100**. Treatment of (−)-ecgonine (**99**) with mild alkali gave (+)-pseudoecgonine (**101**), in which it was thought that C-3 epimerization had occurred. With the improved interpretation of the stereochemical aspects of the tropinones, subsequent work indicated that this was probably a C-2 epimerization. Note how the ecgonine nucleus is now optically active due to the C-2-carboxylic acid.

The stereochemical relationships of the basic nitrogen and the C-3 hydroxyl group were determined as described previously for tropan-3β-ol. Thus the ethyl ester of *N*-acetyl-nor-pseudoecgonine (**102**) gave the ethyl ester of *O*-acetyl-nor-pseudoecgonine (**103**) on treatment with acid. The reverse reaction occurs under basic conditions. Once again it is necessary for the piperidine ring to assume a boat conformation during this reaction. The C-3 hydroxyl group in ecgonine is therefore β.

Intermolecular rearrangements, or lack of them, are also important in determining the relative stereochemistries at C-2 and C-3. Curtius degra-

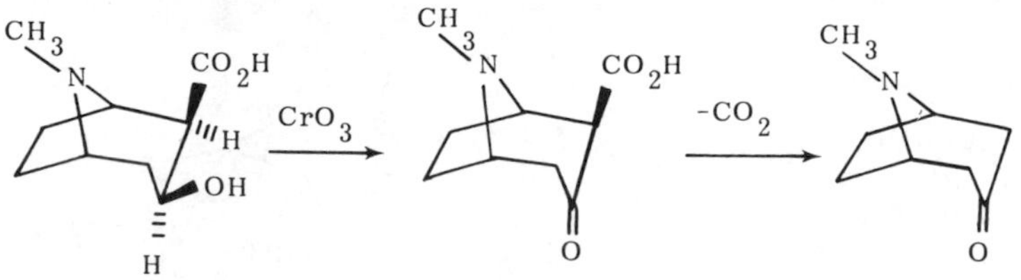

99 ecgonine   98 ecgoninone   84 tropinone

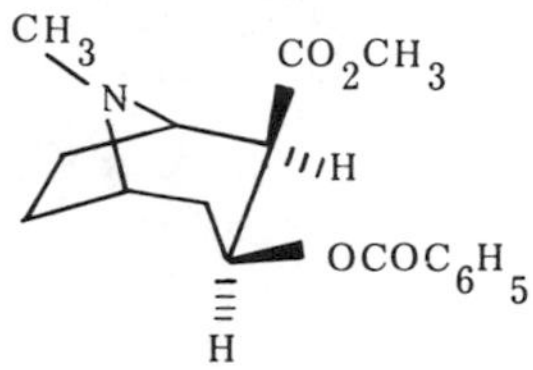

100   cocaine 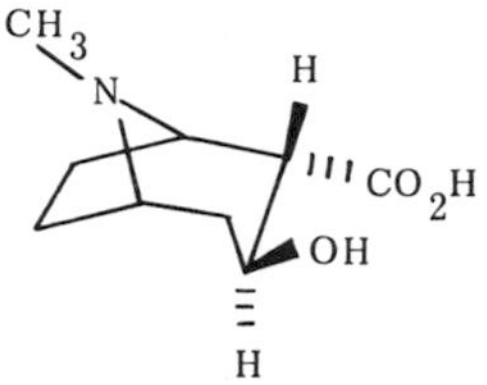   101   pseudoecgonine

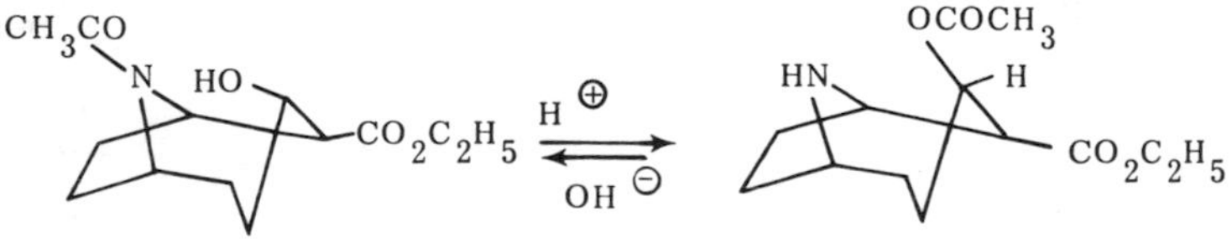

102   103

dation of *O*-benzoylecgonine (**104**) and *O*-benzoyl-pseudoecgonine (**105**) gave epimeric 2-benzamido-tropan-3-ols **106** and **107**, in which the stereochemical integrity of the C-2 position is retained. When **106** and **107** were treated with acid only the tropanol **106** underwent N → O acyl migration to give the benzoate ester of 2-aminotropan-3β-ol (**108**). In this isomer the hydroxyl and amino groups must therefore be *cis* and ecgonine has the structure **99**, where both the C-2 and C-3 substituents are β.

Confirming evidence for this assignment was obtained by lithium aluminum hydride reduction of cocaine, which gave a diol **109**. The ready formation of a benzylidene derivative (**110**) from this compound but not from pseudococaine (**111**) also demonstrates the *cis* relationship of the C-2 and C-3 substituents in **100**.

The absolute configuration of (−)-cocaine was determined by correlation with L-(+)-glutamic acid (**46**). Chromic acid oxidation of ecgonine (**99**) eventually gives ecgoninic acid (**112**). Internal cyclization of L-(+)-glutamic acid (**46**) gave L-(+)-pyroglutamic acid (**113**), which by predictable series of reactions *via* **114** was converted to ecgoninic acid (**112**), identical with that obtained from cocaine. The absolute configuration of cocaine is therefore the one depicted in **100**.

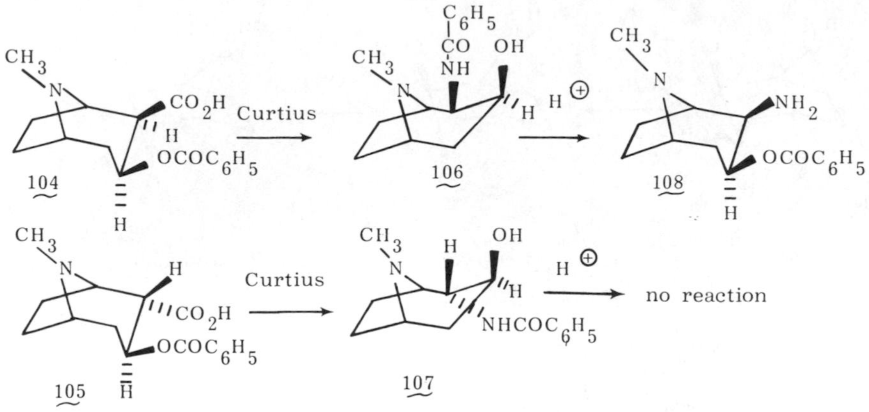
104
Curtius
106
$H^{\oplus}$
108
105
Curtius
107
$H^{\oplus}$
no reaction

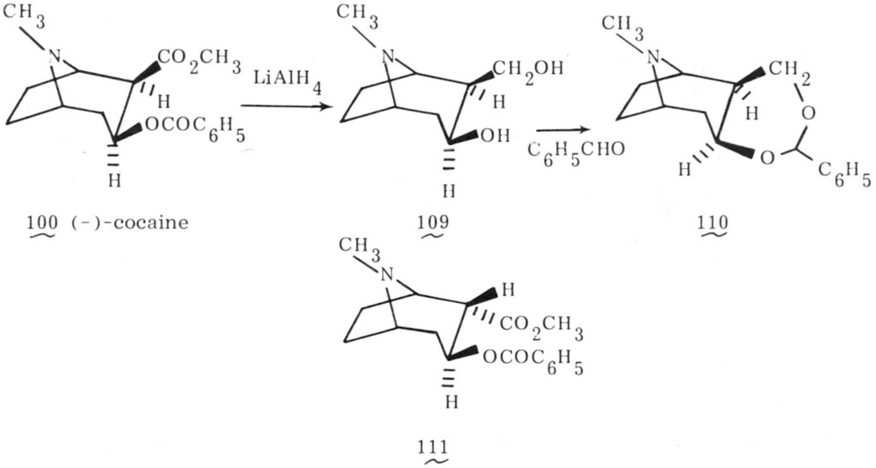
LiAlH4
C6H5CHO
100 (-)-cocaine
109
110
111

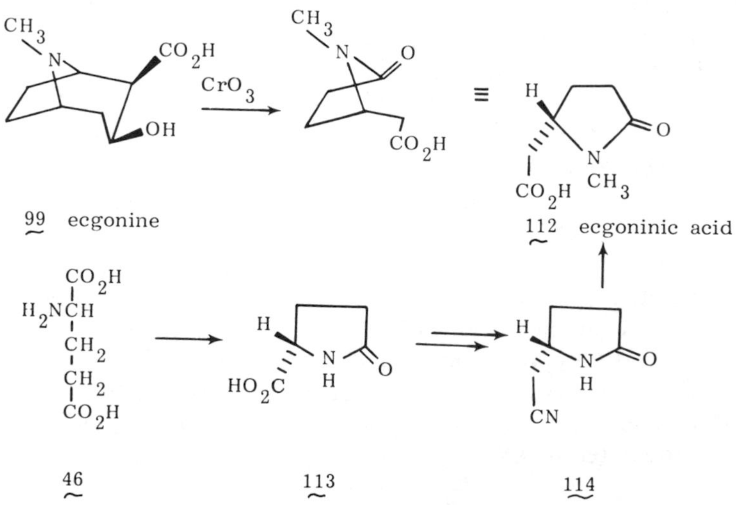
CrO3
99 ecgonine
112 ecgoninic acid
46
113
114

3.3.4 Synthesis

The Robinson biomimetic type of synthesis of the tropinone nucleus as originally described proceeds in poor yield. But Schöpf and Lehmann, who investigated the reaction extensively, found that a dilute, buffered solution of succindialdehyde (**115**), methylamine hydrochloride, and acetone dicarboxylic acid at 25°C and pH 5 gave tropinone in 83% yield.

The greatest problem with this synthesis is the preparation of succindialdehyde, which used to be prepared from pyrrole, but which is now obtained from the commercially available 2,5-dimethoxytetrahydrofuran (**116**).

An alternative synthesis of the nucleus is due to Raphael and co-workers and involves the diyne diester **117**, which is condensed with methylamine to give the pyrrolidine diester **118**. Catalytic hydrogenation gives **119**, which on Dieckmann cyclization, hydrolysis, and decarboxylation yields tropinone (**84**). Ecgoninone (**98**) was synthesized by Wilstätter using Robinson's route, using the monomethyl ester of acetone dicarboxylic acid (**120**). Reduction of the keto group with sodium amalgam gave a mixture of isomers **121** which could be separated by fractional crystallization of their benzoates. Hydrolysis of the more soluble benzoate, resolution with tartaric acid, O-methylation, and benzoylation then gave (−)-cocaine (**100**).

116 115

117 118 119 84

115 120 98 121

100 (−)-cocaine

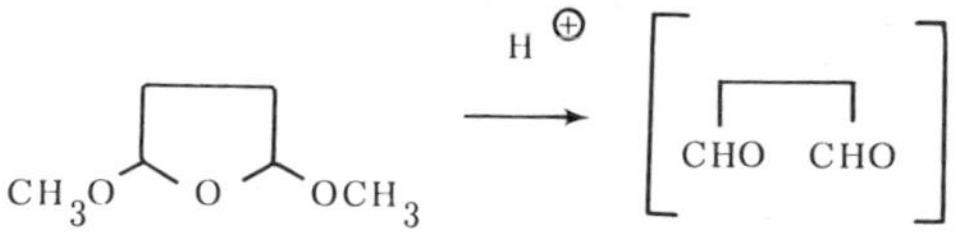

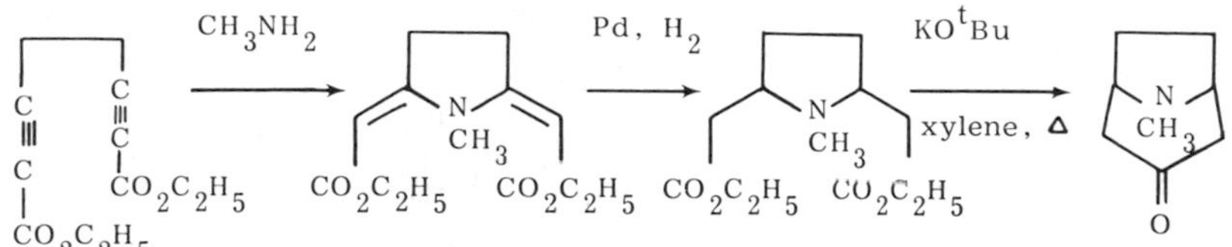

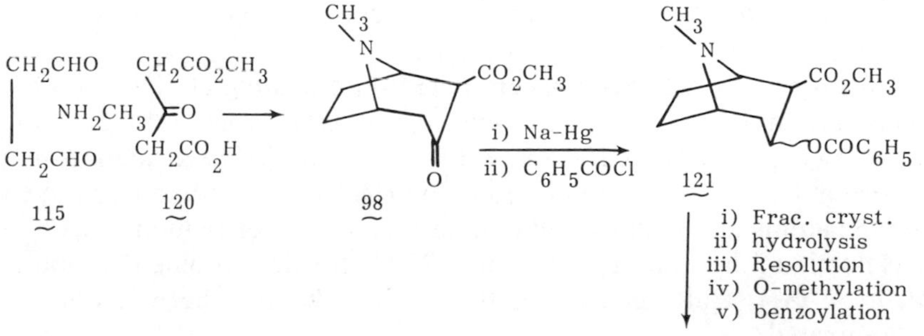

The structure of scopolamine (68) was also confirmed by total synthesis. Methoxysuccindialdehyde (122) is condensed with methylamine and acetone dicarboxylic acid to give 6-methoxytropinone (123), which is reduced to 6-methoxytropan-3α-ol (124) and de-O-methylated with aqueous hydrobromic acid to give tropane-3α,6β-diol (125). Internal cyclization with p-toluenesulfonic acid anhydride followed by treatment with acetyl bromide gave 6β-bromo-3α-acetoxytropane (126). Piperidine was used to eliminate hydrogen bromide and the ester exchanged in a two-step process to give 6,7-dehydrohyoscyamine (127). Epoxidation of the hydrobromide with 30% hydrogen peroxide in the presence of tungstic acid gave (−)-scopolamine hydrobromide (68).

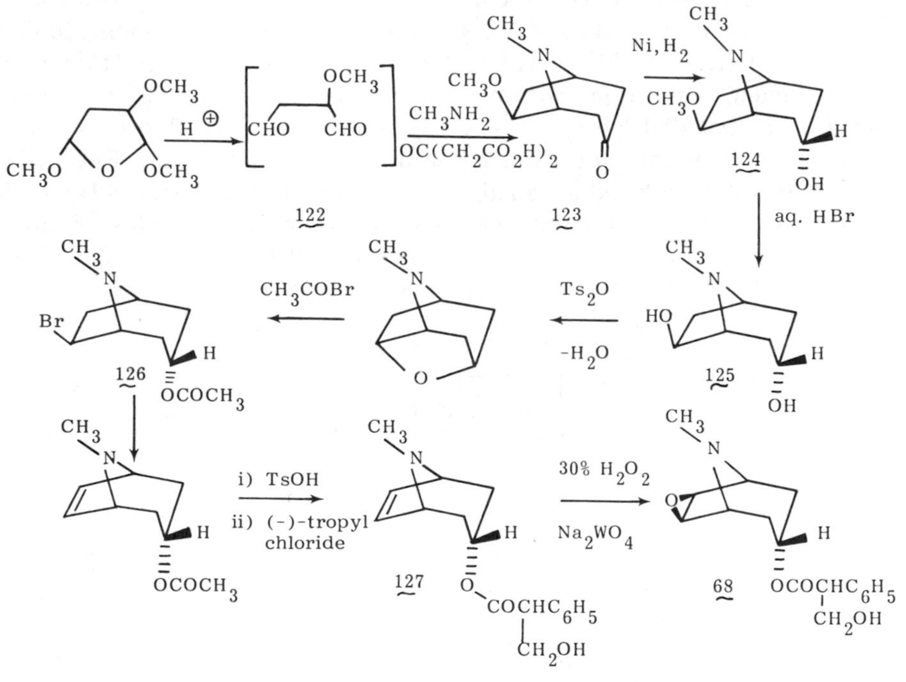

A more modern approach to the synthesis of the tropane skeleton was reported in 1974. The key step is the cyclo addition of a three-carbon unit at the 2 and 5 positions of a pyrrole; a reaction carried out in the presence of transition metal carbonyl and heat.

Thus α,α,α′,α′-tetrabromoacetone (128) on heating at 50°C with diironennea carbonyl [$Fe_2(CO)_9$] and 1-carbomethoxypyrrole (129) gave the dibromotropenone adducts 130 and 131 in a 2:1 ratio. These adducts may be reduced in two ways. Catalytic reduction (H_2,Pd/C) in ethanol removed the bromine atoms and reduced the double bond. Further reduction with diisobutylaluminum hydride (DIBAL) at −78°C afforded tropine (75) and pseudotropine (87). Quite unusually, tropine (75) having the α-alcohol group predominated.

Zinc–copper reduction of the adducts in methanol gave the tropenone (132). This compound can be used for the synthesis of both scopine (90) and meteloidine (67) as shown (Scheme 5).

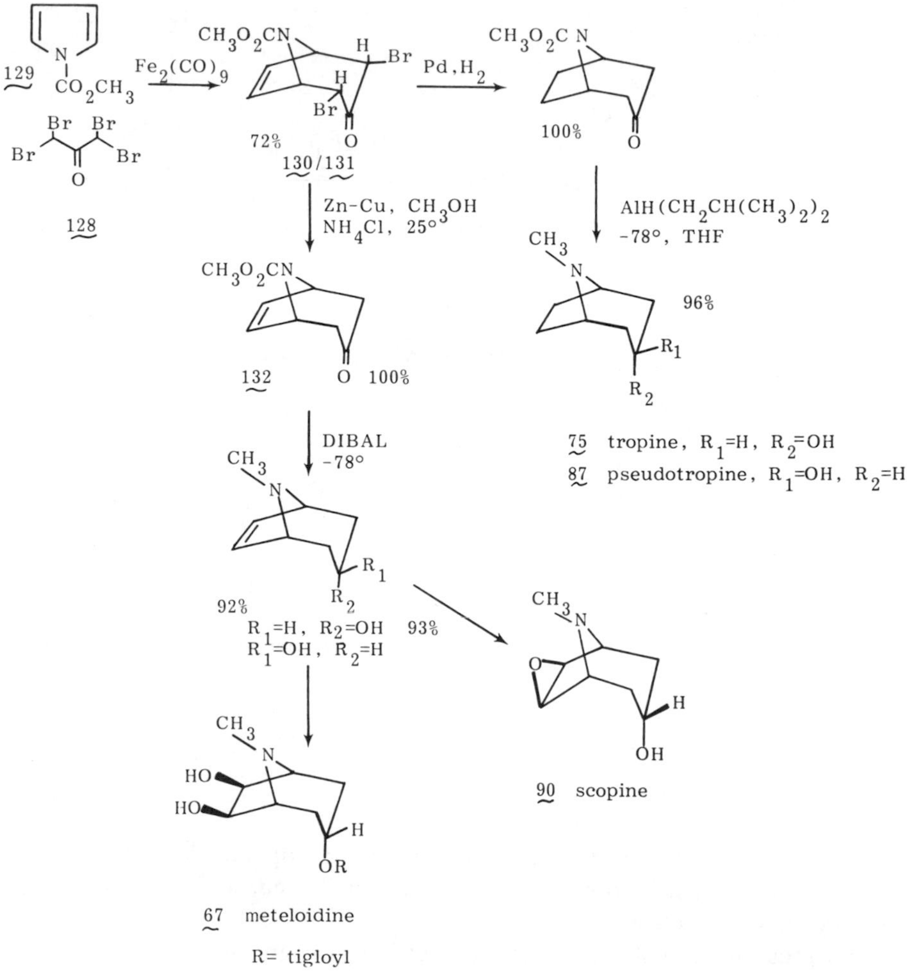

Scheme 5

Another novel synthesis of pseudotropine (87) involves a nitrone-induced oxidative cyclization. 4-Nitrobut-1-ene (133) undergoes Michael addition with acrolein in sodium methoxide–methanol to give the nitro acetal 134, which on treatment with zinc in aqueous ammonium chloride gave the nitrone olefin 135 (Scheme 6). This compound cyclized on heating to give the isoxazolidine 136.

A quite different synthesis of this nucleus involves the direct addition of methylamine to cyclohepta-2,6-dienone (137), where tropinone (84) is produced in 64% yield.

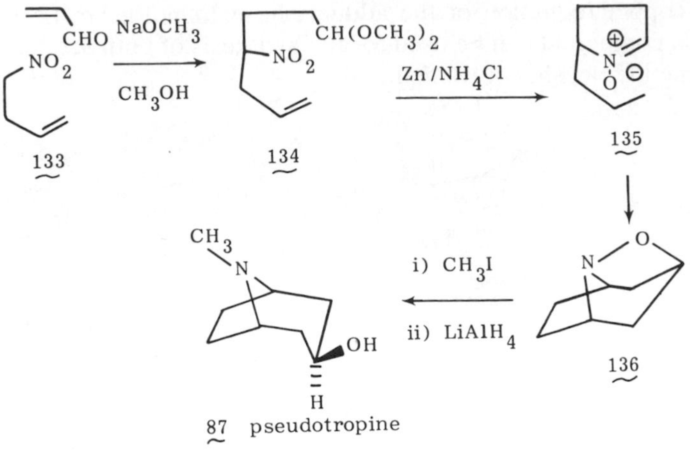

Scheme 6

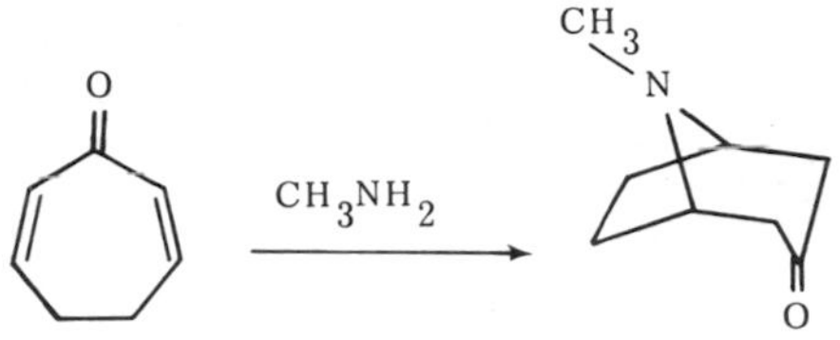

3.3.5   Spectral Properties

The infrared (IR) and ultraviolet (UV) spectra of tropane alkaloids are quite predictable and useful for the determination of standard functional groups that occur in this class.

Of greater interest are the proton, carbon-13, and mass spectra of the tropanes. The proton magnetic resonance spectra of some tropane alkaloids are shown in Table 3. As expected the spectra in the region of the H-2, H-4, H-6, and H-7 protons are extremely complex and not subject to easy interpretation.

The ^{13}C NMR spectra of the symmetrical tropane alkaloids are classically simple (Table 4), and this technique is undoubtedly the most efficient means of structural analysis of this group.

The axial-equatorial possibilities for the N-methyl group in the tropane nucleus have been elegantly verified by ^{13}C NMR. Some data for the axial-equatorial conformers of tropine (**75**) and pseudotropine (**87**) at $-70°C$ in $CHCl_3 : CH_3OH$ (2 : 1) are shown in Table 5.

The mass spectra of the tropanes have been investigated with the aid of

Table 3 Proton Chemical Shifts of Representative Tropane Alkaloids (ppm from TMS)

Compound	H-1/H-5	H-6/H-7	H-2/H-4	H-3	NCH$_3$
Tropine (75)	3.0	2.05	1.6–1.9	3.90	2.16
Pseudotropine (87)	2.07	1.5–1.8	1.5–1.8	3.70	2.22
Atropine (65)	2.93	1.2–2.05	1.2–2.05	4.97	2.15
Scopine (90)	3.12	3.6	1.25–2.33	3.90	2.47
Scopolamine (68)	2.98	2.84–3.37	1.1–2.3	4.98	2.4
Cocaine (100)	3.55	1.58–2.18	2.95–2.6	5.28	2.22

Source: Data from G. Fodor, in S. W. Pelletier (Ed.), *Chemistry of the Alkaloids,* Van Nostrand Reinhold, New York, 1970, p. 445.

Table 4 Carbon Chemical Shifts of Representative Tropane Alkaloids (ppm from TMS)

Alkaloid	C-/C-5	C-2/C-4	C-3	C-6/C-7	NCH$_3$
Nortropane	54.7	32.9	17.2	29.0	—
Tropane (64)	61.2	29.9	15.9	25.6	40.4
Tropinone (84)	60.2	47.1	207.8	27.3	37.8
Tropine (75)	59.8	39.1	63.6	25.7	40.0
Pseudotropine (87)	60.1	38.3	62.7	26.7	39.2
Atropine (65)	59.3	35.7	67.6	24.8	39.7
Scopolamine (68)	58.2	31.7	66.6	55.9	43.4

Source: Data from E. Wenkert *et al., Acc. Chem. Res.* **7,** 46 (1974).

deuterated derivatives. In the case of tropanone (**84**), three bond-breaking routes have been deduced.

One major route is 1-7 fission to give the radical ion **140**, which can either lose ethylene and CO to give *m/e* 83 (**141**), or can fragment by hydrogen radical transfer from C-4 and loss of an ethyl group to afford an ion *m/e* 110 (**142**). The other major route is 1-2 fission followed by loss of ketone to give the ion *m/e* 96 (**143**). In the spectrum of tropan-3α-ol (**75**), loss of hydroxyl radical occurs preferentially to give the bridged piperidinium ion (**144**) (Scheme 7).

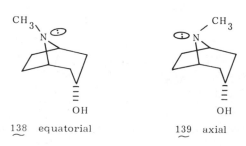

138 equatorial   139 axial

Table 5　^{13}C NMR Spectra of the Axial-Equatorial Nitrogen Inversion in the Tropane Series

Compound	C-1/C-5	C-2/C-4	C-3	C-6/C-7	NCH$_3$
Tropine					
Equatorial-CH$_3$ (**138**)	60.74	39.50	63.41	25.34	41.13
Axial-CH$_3$ (**139**)	56.36	29.87^a	63.41	28.54^a	32.64
Pseudotropine					
Equatorial-CH$_3$	61.85	41.58	63.08	26.25	40.80
Axial-CH$_3$	58.16	32.61	63.08	29.17	32.82

Source: Data from H.-J. Schneider and L. Storm, *Angew. Chem.* **88,** 574 (1976).
^{a} Signals tentatively assigned.

3.3.6　Biosynthesis

The tropane alkaloids such as atropine (**65**) are ester derivatives of an alcohol, normally a tropan-3α-ol and an acid such as tropic acid. In developing the biosynthesis of tropane alkaloids therefore, consideration must be given to the biosynthesis of each unit. The biosynthesis of the smaller,

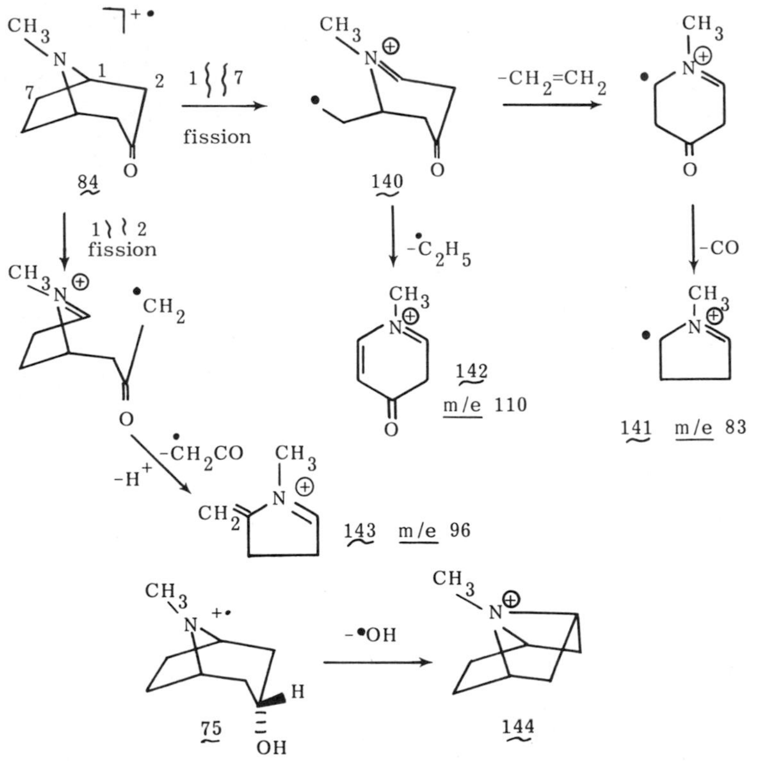

Scheme 7　Mass spectral fragmentation of tropinone (**84**) and tropine (**75**).

nonalkaloidal unit has proved to be more problematic than that of the alkaloid moiety.

Formation of Tropan-3α-ol (75)

The tropane alkaloids seem always to have been at the forefront of biosynthetic studies in plants. Thus it was in 1954 that [2-^{14}C]ornithine (1) was shown to be a precursor of hyoscyamine (66) in *D. stramonium*. Degradation *via* N-methyl succinimide (83) to benzoic acid (145) indicated that either one or both of the bridge carbon atoms was labeled. This point, whether a symmetrical or unsymmetrical intermediate was involved, remained unresolved for many years, but was eventually solved in an exceptionally ingenious way.

Tropine is a symmetrical compound, but being *meso*, the two bridge carbon atoms are of opposite chirality. Thus if these two are optically resolved at some point in a nonsymmetrical compound, the extent of radioactive label in each will determine the nature of the intermediate.

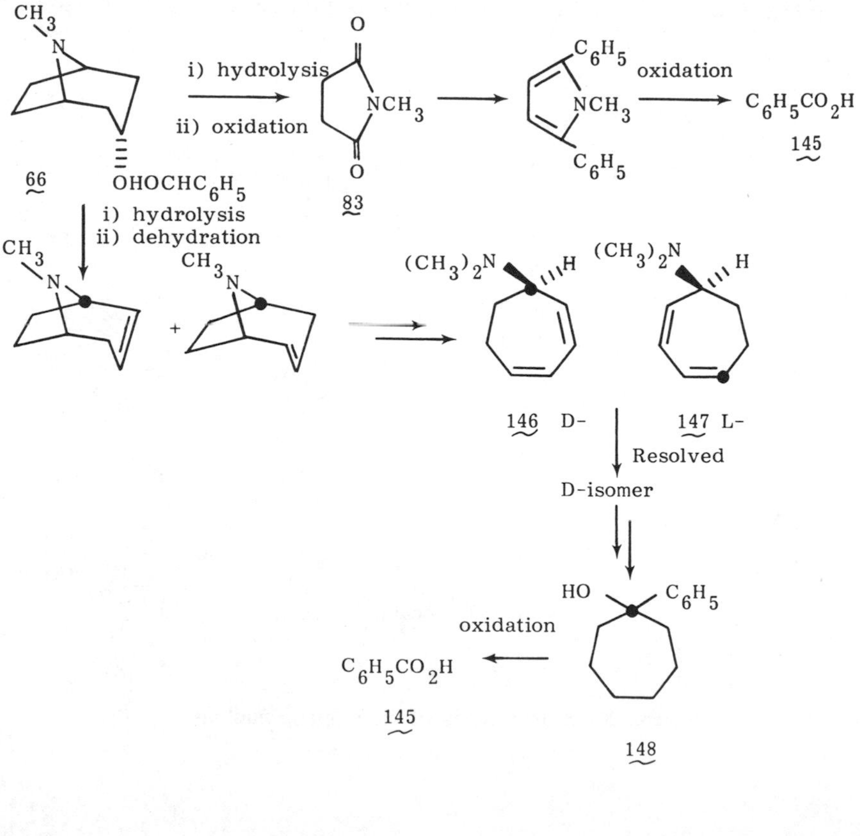

Dehydration and ring cleavage of tropine derived from [2-^{14}C]ornithine gave the enantiomeric tropidines (146) and (147), only one of which has a label at the carbon-bearing nitrogen, if incorporation is asymmetric. These compounds were resolved and D-isomer converted to 4-phenylcycloheptanol (148). Degradation indicated the labeled benzoic acid (145) to have a specific activity identical to that of the original alkaloid. This result not only establishes the unsymmetrical nature of an intermediate but also indicates that only L-ornithine is utilized by the plant and that the fragment which gives C-2, C-3, and C-4 enters with inversion of configuration (see later).

As with the biosynthesis of the pyrrolidine ring of nicotine (23), the next question then became which of the two nitrogen atoms of ornithine is incorporated into the tropane nucleus.

δ-N-Methyl[³H]ornithine (149) served as a good precursor with the majority of the activity confined to the N-methyl groups, the corresponding α-N-methyl[³H]ornithine (150) was poorly incorporated (Scheme 8). Note however that no [¹⁴C]-labeled compound was used to check intact incorporation. When the nitrogen atoms of ornithine were labeled with ¹⁵N it was found that each nitrogen was almost equally incorporated, implying a symmetrical intermediate. It will be recalled that a similar result was obtained in the nicotine series. Spenser has explained this phenomenon in terms of a massive overload of the natural pools of ornithine which by degradation and reincorporation leads to an ornithine pool in which each nitrogen atom is heavily labeled with ¹⁵N, no matter the precursor. Except for the N-methyl group, which is methionine- or formate-derived, the remaining carbon atoms of the tropane nucleus are from acetate. Thus [1-¹⁴C]acetate specifically labeled C-3, but [2-¹⁴C]acetate did not. These data are in agreement with nucleophilic attack by the anion of acetoacetate (15) at C-2 of the pyrrolidinium ion (16) to give 151. If decarboxylation occurs at this point hygrine (152) is produced.

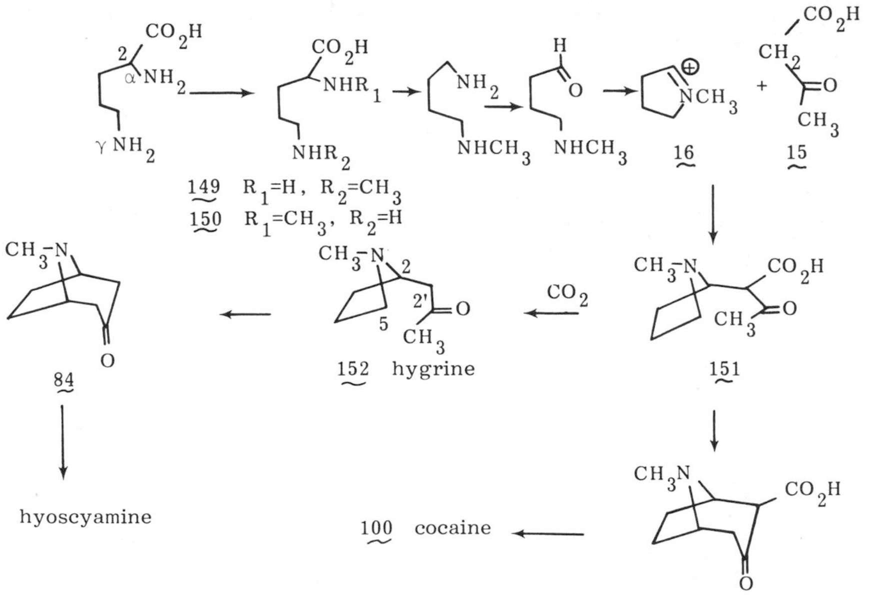

Scheme 8 Biosynthesis of the tropane nucleus.

When [N-methyl-^{14}C, 2'-^{14}C]hygrine (**152**) was used as a precursor of hyoscyamine (**66**) in *D. stramonium*, a 2.1% specific, intact incorporation was observed. Oxidation between the N and C-5 apparently affords tropinone (**84**) from hygrine (**152**).

A more subtle point concerning the involvement of hygrine has been established by McGaw and Wooley. When D(+)- and L(−)-hygrines (**152**) and (**153**) labeled at the 2'-position with ^{14}C were used as precursors, only

152   D (+)-hygrine   153   L(−)-hygrine

the D(+)-isomer **152** was incorporated into hyoscyamine (**66**) in *D. innoxia*. The L(−)-isomer was not an effective precursor. Both compounds were however precursors of cuscohygrine (**7**).

In cocaine (**100**) biosynthesis decarboxylation of the intermediate **151** does not occur. Scheme 8 shows the biosynthetic pathway for the tropane nucleus in these compounds.

Formation of Tropic Acid (76)

Tropic acid (**76**) occurs frequently as an esterifying acid in tropane alkaloids, but nowhere else in nature.

Considerable effort has been expended in trying to deduce the biosynthesis of this simple molecule, and it is only just recently that the problem has been resolved with any degree of certainty.

In 1960 it was shown that [1,3-^{14}C$_2$]phenylalanine (**154**) gave rise to tropic acid (**76**) having essentially all the activity at C-2. Further work established that the intact phenylalanine unit was incorporated, suggesting that some type of molecular rearrangement was occurring within phenylalanine.

A somewhat misleading result was obtained when [1-^{14}C]phenylacetic acid (**155**) was found to be a precursor of tropic acid giving specific labeling at C-3. It was suggested on this basis that phenylpyruvic acid was an intermediate which underwent oxidative decarboxylation to give phenylacetic acid (**155**). Recarboxylation and reduction then affords tropic acid (**76**) (Scheme 9). Subsequent work in 1975 by Leete and co-workers indicated that phenylacetic acid was not an intermediate.

The problem of the formation of tropic acid (**76**) was clarified somewhat by Leete and co-workers using [1,3-^{13}C$_2$]phenylalanine (**156**), and at this point it is worthwhile to consider the theory behind this elegant experiment.

If the rearrangement of the side chain of phenylalanine is intramolecular, then the resulting tropic acid (**157**) should have two adjacent ^{13}C atoms. These, in the proton-noise decoupled spectrum, should afford satellite peaks

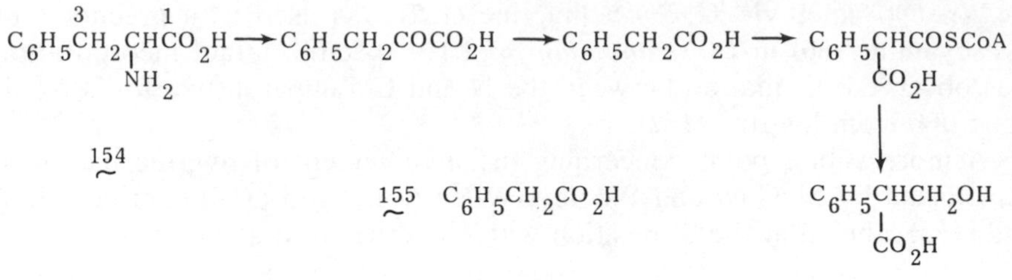

$$C_6H_5CH_2\overset{3}{C}HCO_2H \longrightarrow C_6H_5CH_2COCO_2H \longrightarrow C_6H_5CH_2CO_2H \longrightarrow C_6H_5CHCOSCoA$$

with NH$_2$ below C-2 of first structure, CO$_2$H below last structure.

154

155 C$_6$H$_5$CH$_2$CO$_2$H

$$C_6H_5CHCH_2OH$$
$$CO_2H$$

Scheme 9

due to ^{13}C-^{13}C coupling. An intermolecular rearrangement would lead to no such satellite peaks.

In practice, a 0.32% incorporation into hyoscyamine (**66**) was observed from [1,3-^{13}C$_2$]phenylalanine, but more important, satellite peaks ($J = 56$ Hz) were observed for the C-1 and C-2 carbons of the tropic acid moiety

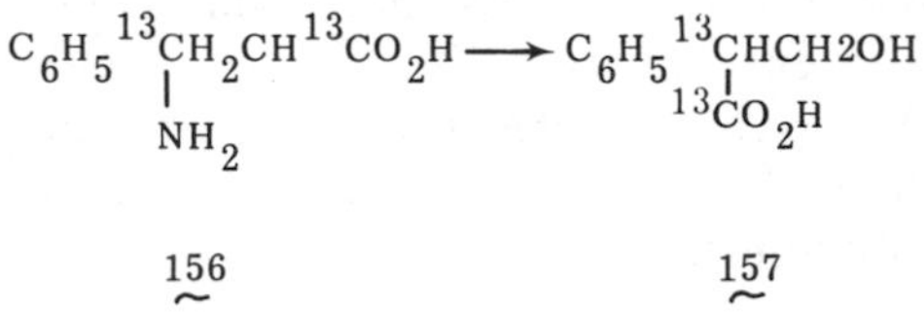

$$C_6H_5\,^{13}CH_2CH^{13}CO_2H \longrightarrow C_6H_5\,^{13}CHCH2OH$$
$$NH_2 \qquad\qquad\qquad ^{13}CO_2H$$

156   157

of both scopolamine (**68**) and hyoscyamine (**66**), thereby establishing the intermolecular nature of the rearrangement, a rearrangement whose mechanism remains to be determined *in vivo*.

The mechanism of the rearrangement from L-phenylalanine to (S)-(−)-tropic acid has been investigated *in vitro*. The nitrous acid deamination of L-phenylalanine ester in acetic acid gives the (R)-(+)-tropic acid ester acetate. However, nitrous acid deamination of L-phenylalanine (**154**) itself in trifluoroacetic acid gives mainly the trifluoroacetate of (S)-(−)-tropic acid (**158**). It is thought that this reaction proceeds *via* the α-lactone **159**, which undergoes aryl migration with retention of configuration prior to attack by a nucleophile.

The biosynthesis of some of the more complex tropane esters has also been studied, with interesting results. Thus tropine (**75**) is a precursor of

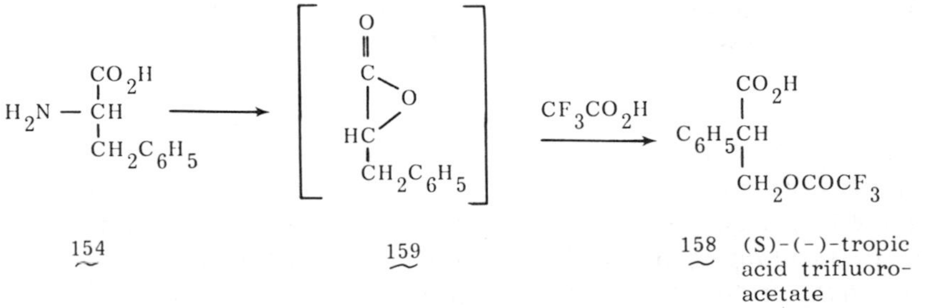

$$H_2N-\underset{CH_2C_6H_5}{\overset{CO_2H}{\underset{|}{\overset{|}{CH}}}} \longrightarrow \left[\ \underset{CH_2C_6H_5}{\overset{O}{\underset{|}{\overset{\|}{C}}}} \underset{}{\overset{}{O}} \right] \xrightarrow{CF_3CO_2H} C_6H_5\underset{CH_2OCOCF_3}{\overset{CO_2H}{\underset{|}{\overset{|}{CH}}}}$$

154   159   158   (S)-(−)-tropic
                                                              acid trifluoro-
                                                              acetate

meteloidine (67), and so is 3α-tigloyloxytropane (160). In addition, hyoscyamine (66) was a precursor of scopolamine (68). Clearly oxidation at the 6 and 7 positions occurs at a late stage in the pathway.

Leete has recently investigated the stereochemistry of these oxidations using [N-methyl-^{14}C, 6β,7β-^{3}H$_2$]tropine (75) as a precursor. As expected hyoscyamine (66) showed no loss of activity, but both scopolamine (68) and meteloidine (67) were essentially devoid of ^{3}H activity (Scheme 10). Thus oxygenation of the tropane nucleus in both cases occurs with retention of configuration. The precise mechanism remains to be determined.

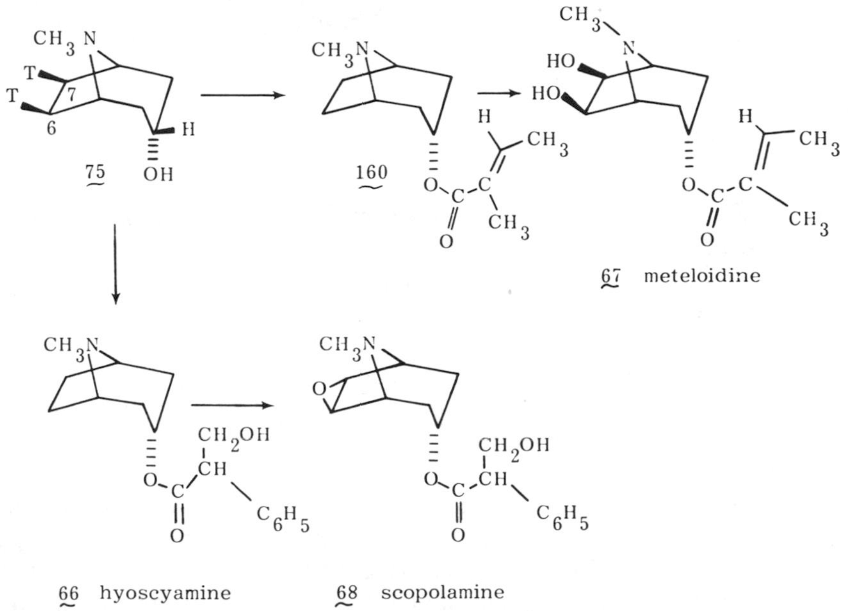

Scheme 10

The alkaloid content of *A. belladonna* and related plants varies considerably during the growing season and in different plant parts. In addition many environmental and climatic factors influence total alkaloid content and individual alkaloid production. For example, in *D. metel* maximum alkaloid content occurs at the flowering and fruiting stage, and at elevations greater than 2000 m.

Alkaloid production is even dynamic within a day. A study of the scopolamine (68) and hyoscyamine (66) content of *Scopolia tangutica* revealed that alkaloid content was maximal at noon and 8 P.M. and minimal at midnight and 4 P.M.

3.3.7 Pharmacology

The pharmacology of the tropane alkaloids is typified by atropine (65). Probably the most important action of atropine is antagonism to muscarinic receptors (parasympathetic inhibition). These receptors are responsible for slowing of the heart, constriction of the pupil of the eye, vasodilation, and stimulation of secretions. Typical compounds which exhibit these effects are acetylcholine, pilocarpine (161), physostigmine (162), and arecoline (163). Atropine (65) is therefore an acetylcholine antagonist.

161 pilocarpine   162 physostigmine   163 arecoline

Clinically, the atropinic drugs cause mydriasis and cycloplegia and can be used locally to induce mydriasis for a thorough eye examination. Atropine is used to reduce salivary and bronchial secretions by smooth muscle relaxation of bronchi.

The principal effect of atropine on the heart is to alter rate. At low doses the rate is slowed without a change in blood pressure or cardiac output. Large doses cause an increase heart rate. Atropine administered may be used in the initial treatment of a myocardial infarction or high-grade atroventricular block.

The tropane alkaloids are also used in peptic ulcer therapy, because they decrease gastric secretion and total acid content in moderate doses. Tablets containing sodium bicarbonate deteriorate on storage and bismuth(II) salicylate, calcium silicate, or magnesium silicate are preferred as excipients.

Because motility of the gastrointestinal tract is decreased, the belladonna alkaloids have been used in combination in the treatment of severe infectious diarrhea in infants.

Clinical doses of atropine cause mild excitation of the central nervous system. At steadily increasing doses central excitation is increased, but then central depression follows. These effects are used in products to prevent motion sickness (scopolamine), and in preanesthetic products.

Atropine is also useful in cases of poisoning. Natives in New Caledonia have used a mixture of tropane alkaloids from *Duboisia myoporoides* to treat ciguatera poisoning. In large doses atropine may also be used in the treatment of anticholinesterase poisoning by organophosphorus insecticides, and of the muscarinic effects due to *Amanita muscaria* ingestion.

Cocaine is too toxic to be used as an anesthetic by injection, but the hydrochloride is used in 0.1–1% aqueous solution as a topical anesthetic in ophthalmology and in 10–20% solutions in face, nose, and throat work. Cocaine is a potent central nervous system stimulant and adrenergic blocking agent. Addiction and some tolerance arise from continued use as a central stimulant; withdrawal symptoms are not serious.

The importance of cocaine lies in the fact that it served as a model for a tremendous synthetic effort designed to produce an anesthetic of increased stability and reduced toxicity.

LITERATURE

Reviews

Holmes, H. L., *Alkaloids NY* **1**, 27 (1950).

Fodor, G., *Alkaloids NY* **6**, 145 (1960).

Fodor, G., *Alkaloids NY* **9**, 269 (1967).

Fodor, G., *Alkaloids NY* **13**, 351 (1971).

Fodor, G,, *Progr. Phytochem.* **1**, 491 (1968).

Fodor, G., in S. W. Pelletier (Ed.), *Chemistry of the Alkaloids*, Van Nostrand Reinhold, New York, 1970, p. 431.

Saxton, J. E., *Alkaloids, London* **1**, 55 (1971).

Saxton, J. E., *Alkaloids, London* **2**, 54 (1972).

Saxton, J. E., *Alkaloids, London* **3**, 64 (1973).

Saxton, J. E., *Alkaloids, London* **4**, 78 (1974).

Saxton, J. E., *Alkaloids, London* **5**, 69 (1975).

Saxton, J. E., *Alkaloids, London* **6**, 65 (1976).

Fodor, G., *Alkaloids, London* **7**, 47 (1977).

Synthesis

Robinson, R., *J. Chem. Soc.* **111**, 762 (1917).

Schöpf, C., and G. Lehmann, *Ann.* **518**, 1 (1935).

Parker, W., R. A. Raphael, and D. I. Wilkinson, *J. Chem. Soc.* 2453 (1959).

Wilstatter, R., O. Wolfes, and O. Mader, *Ann.* **434**, 11 (1932).

Fodor, G., S. Kiss, and J. Rakocsi, *Chim. Ind.* **90**, 225 (1963).

Noyori, R., Y. Baba, and Y. Haykawa, *J. Amer. Chem. Soc.* **96**, 3336 (1974).

Tufariello, J. J., and E. J. Trybulski, *Chem. Commun.* 720 (1973).

Bottini, A. T., and J. Gal, *J. Org. Chem.* **36**, 1718 (1971).

Biosynthesis: Tropanone Moiety

Leete, E., L. Marion, and I. D. Spenser, *Can. J. Chem.* **32**, 1116 (1954).

Leete, E., *J. Amer. Chem. Soc.* **84**, 55 (1962).

Baralle, F. E., and E. G. Gros, *Chem. Commun.* 721 (1969).

O'Donovan, D. G., and M. F. Keogh, *J. Chem. Soc., Ser. C* 223 (1969).

McGaw, B. A., and J. G. Wooley, *Phytochemistry* **17**, 259 (1978).

Biosynthesis: Tropic Acid Moiety

Leete, E., *J. Amer. Chem. Soc.* **82**, 612 (1960).

Louden, M. L., and E., Leete, *J. Amer. Chem. Soc.* **84**, 1510, 4507 (1962).

Underhill, E. W., and H. W., Youngken, *J. Pharm. Sci.* **51**, 121 (1962).

Leete, E., N., Kowanko, and R. A. Newmark, *J. Amer. Chem. Soc.* **97**, 6826 (1975).

More Complex Tropane Esters

Leete, E., *Phytochemistry* **11**, 1713 (1972).

Leete, E., and D. H. Lucast, *Phytochemistry* **14**, 2199 (1975).

Leete, E., and D. H. Lucast, *Tetrahedron Lett.* 3401 (1976).

3.4 PYRROLIZIDINE ALKALOIDS

3.4.1 Occurrence

The pyrrolizidine alkaloids contain one nitrogen at the bridge of two five-membered rings (**164**), a system also known as 1-azabicyclo[0,3,3]octane. The alkaloids were thought to have a quite limited generic distribution in three plant families, *Senecio* and *Petasites* in the Compositae; *Heliotropium, Trichodesma, Echium*, and *Trachelanthus* in the Boraginaceae; and *Crotalaria* in the Leguminosae. It is now realized that these alkaloids also occur in at least five other plant families.

The genus *Crotalaria* is restricted to tropical and subtropical regions, but *Senecio* and *Heliotropium* are of worldwide distribution.

Poisoning of cattle and horses by *Senecio* species has been suspected for nearly 200 years. It invariably takes the form of cirrhosis of the liver and in different countries various names have been used. Thus in Germany it is known as Schweinberger disease, in Canada as Pictou disease, in New Zealand as Winton disease, and elsewhere as "walking disease," "horse staggers," and "Missouri river bottom disease." The reason for the poisoning was traced to plants of the *Senecio* genus in 1950.

Although a number of simple pyrrolizidine alkaloids are known, those of greatest interest are esters, which on alkaline hydrolysis afford an amino alcohol (the necine portion) and a carboxylic acid (the necic acid). Many of the alkaloids contain a 1,2-double bond and are thus allylic esters that can be cleaved by hydrogenolysis. Several of the alkaloids occur naturally as quaternary N-oxides from which the parent base can be obtained by reduction with zinc dust.

Some of the simple alkaloids known include 1-methylenepyrrolizidine (**165**) from *Crotalaria* species, laburnine (**166**) from laburnum, festucine (**167**) from fescue hay, and sex pheromone **168** from the hair-pencils of the tropical butterfly, *Lycorea ceres ceres*.

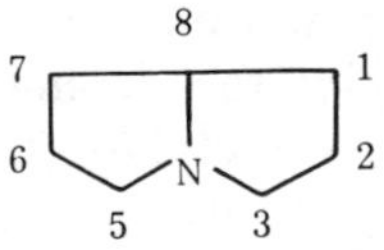

164

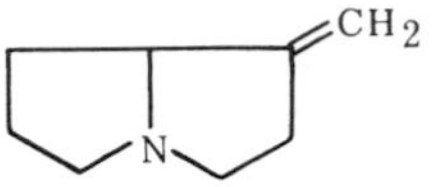

165

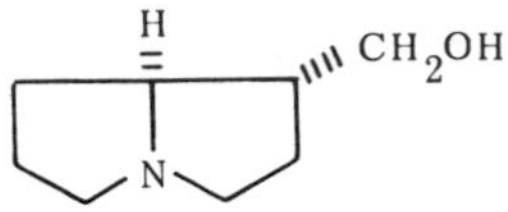

166

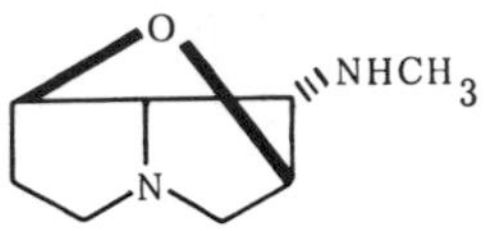

167

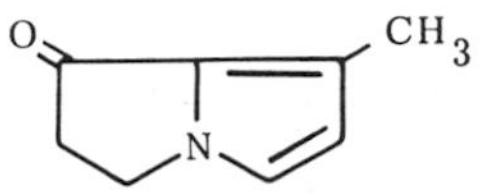

168

It is not the intention to discuss the ester groups of the alkaloids to any great extent but rather to concentrate on the necine portion. The principal necine alcohols found in the pyrrolizidine alkaloids are shown in Figure 3. They are supinidine (**169**), heliotridine (**170**), retronecine (**171**), and platynecine (**172**). Of these, the most important is retronecine (**171**).

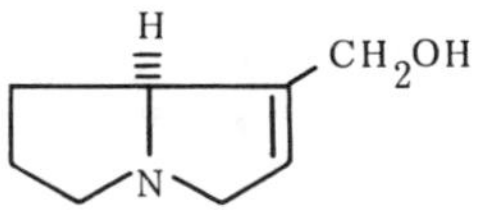

169   supinidine

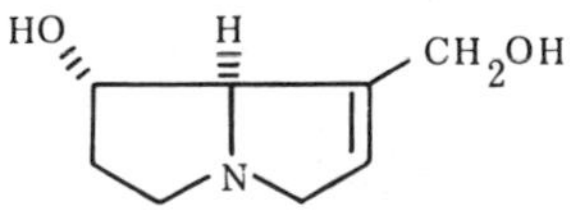

170   heliotridine

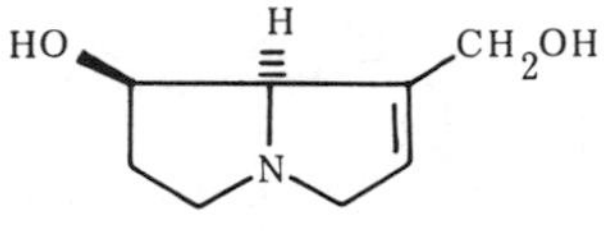

171   retronecine

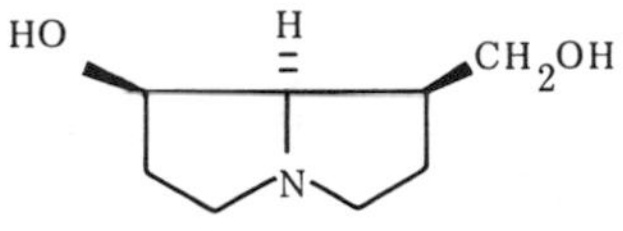

172   platynecine

The structure of retronecine (**171**) was originally determined by a classic degradation sequence. Catalytic hydrogenation gave **173**, indicating an allylic alcohol in **171**. The alcohol group in **173** was removed by treatment with thionyl chloride and hydrogenolysis of the chloride.

A series of Hofmann degradations established that the nitrogen was common to both rings, and that the chloride hydrogenolysis product had the structure 1-methylpyrrolizidine (**174**).

Retronecine (**171**) has two asymmetric centers and the stereochemistry was deduced by hydrogenation to give the alcohol **175**, which on tosylation afforded the internal cyclization product **176**. This product can only be produced if the C-8 and C-9 hydrogens are *cis*.

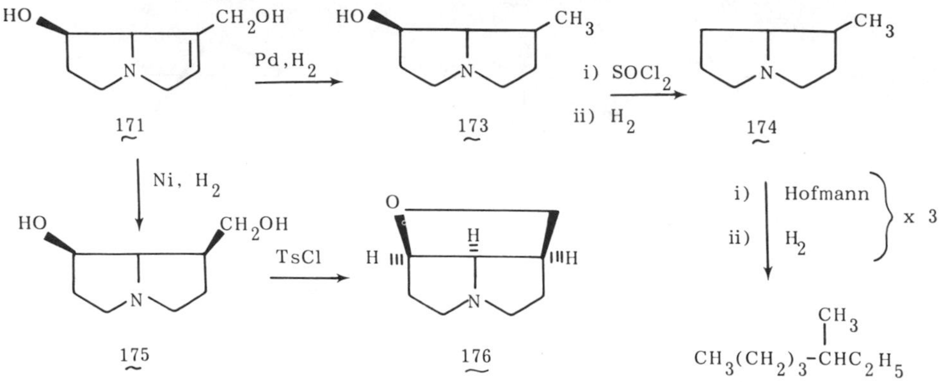

3.4.2   Isolation and Detection

There are several problems associated with the detection and isolation of pyrrolizidine alkaloids. They include: (*a*) pyrrolizidine alkaloids are notorious for failing to give precipitation reactions with the commonly used reagents; (*b*) although some alkaloids are stable to the drying process of the plant material, losses of up to 80% have been observed with some *Crotolaria* species; (*c*) many of the N-oxides are not organic-solvent soluble; and (*d*) because strong bases (e.g. NaOH) may be necessary to completely free the strongly basic alkaloids, extraction with chloroform can lead to products from the insertion of dichlorocarbene (e.g. **177**).

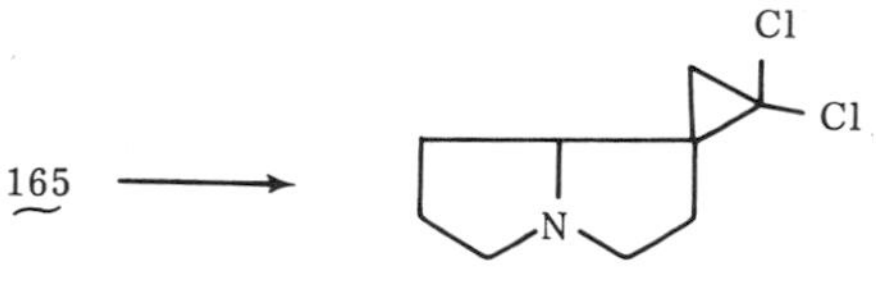

The N-oxides pose some interesting problems; the low-molecular-weight N-oxides are water soluble, whereas the high-molecular-weight diester N-oxides may be completely chloroform soluble. Overall the best isolation method for this group is an ion exchange technique.

The hepatotoxicity and varied distribution of pyrrolizidine alkaloids led to a need for a specific reaction which could be used as a screening test. One test that has been developed involves an oxidation sequence which affords a pyrrole derivative. This compound can be detected by the use of a modified Ehrlich's reagent. Unless accurate spectrophotometric measurements are made the color test will not distinguish between an α-unsubstituted indole nucleus. The test is specific for pyrrolizidine alkaloids having a 1,2-double bond. The procedure is to form the N-oxide of the base and then heat with acetic anhydride to carry out a Polonovskii reaction. The conjugated iminium then undergoes rearrangement to an *N*-alkyl pyrrole as shown in Figure 4. Modified Ehrlich's reagent is *N*,*N*-dimethylaminobenzaldehyde in the presence of the Lewis acid boron trifluoride, which acts to catalyze the electrophilic substitution. The red color ($\lambda_{max} \sim$ 565 nm) is the formation of the tautomeric species.

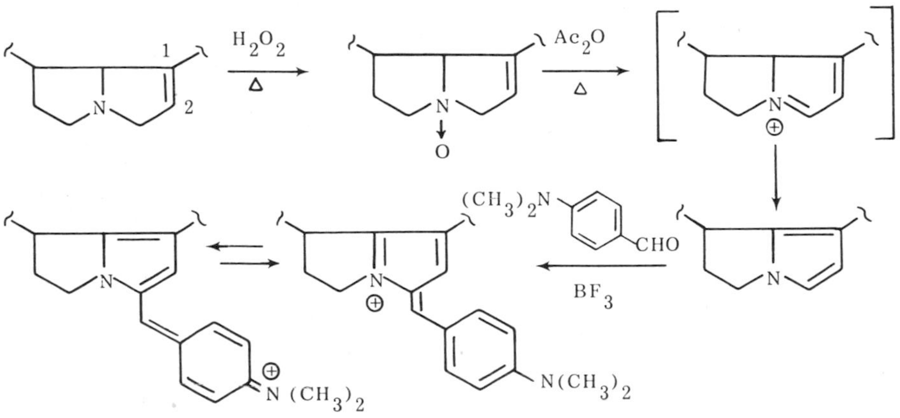

Figure 4 Detection of $\Delta^{1,2}$-pyrrolizidine alkaloids.

3.4.3 Synthesis of the Necine Nucleus

The most facile synthesis of the necine nucleus begins with mild hydrolysis of the symmetrical amino diacetal **178** to the dialdehyde **179**. On standing at pH 7 for several days, the aminoaldehyde **180** was produced, and sodium borohydride reduction gave the necine **181**. As we shall see this scheme closely follows the biogenetic route to this nucleus.

A more difficult goal is retronecine (**171**) with a C-8 hydroxyl group and 1,2-double bond. Michael addition of diethyl fumarate to the urethane ester **182** followed by Dieckmann cyclization gave the keto urethane ester **183** after hydrolysis. Sodium borohydride reduction afforded an alcohol which cyclized to the lactone **184**. Hydrolysis and N-alkylation with ethyl bromoacetate followed by cyclization yielded the hydroxy β-ketoester **185**. Reduction and elimination gave α,β-unsaturated ester **186**, and LAH reduction afforded **171** (Scheme 11).

A different strategy to the synthesis of pyrrolizidines involves osmium tetroxide–sodium metaperiodate cleavage of an aminocarbethoxycycloheptenone (**187**). The intermediate **188** is reduced with sodium cyanoborohy-

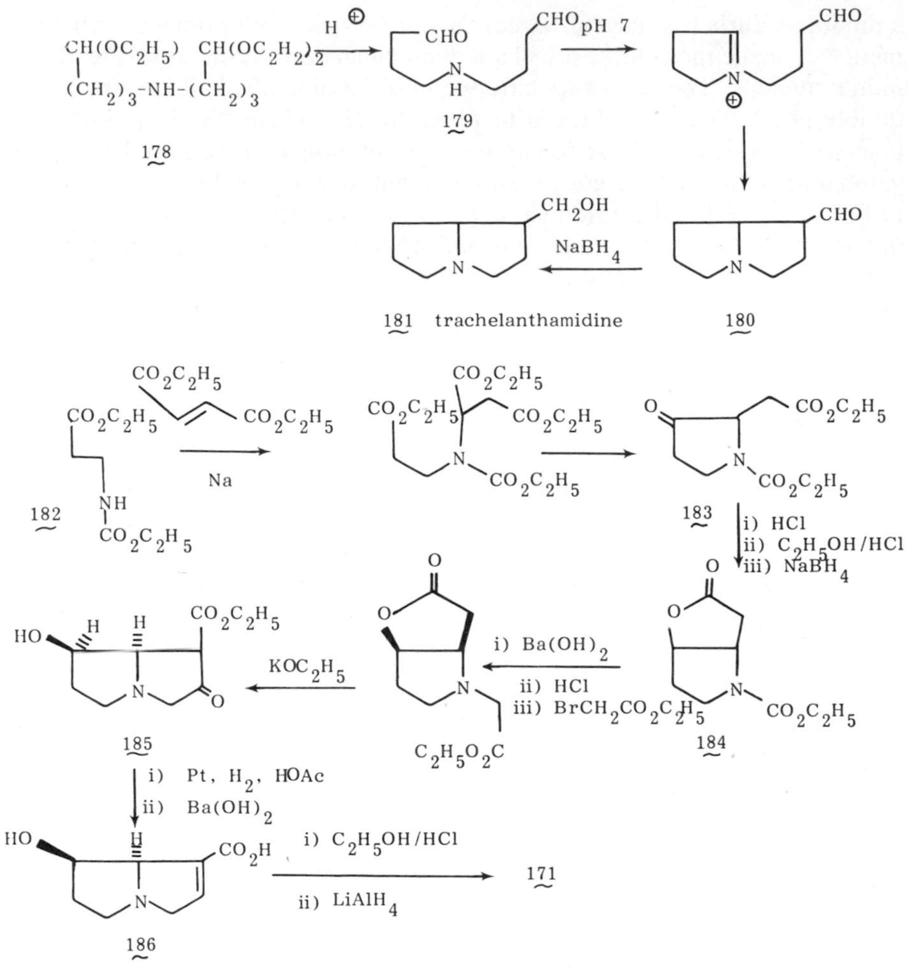

CH(OC$_2$H$_5$) CH(OC$_2$H$_2$)$_2$
(CH$_2$)$_3$-NH-(CH$_2$)$_3$

178

→

CHO
N
H

179

CHO pH 7
→

CHO
N$^{\oplus}$

↓

CH$_2$OH
N

181 trachelanthamidine

← NaBH$_4$

CHO
N

180

CO$_2$C$_2$H$_5$
CO$_2$C$_2$H$_5$ ⟍⟍ CO$_2$C$_2$H$_5$

NH
CO$_2$C$_2$H$_5$

182

Na
→

CO$_2$C$_2$H$_5$
CO$_2$C$_2$H$_5$ CO$_2$C$_2$H$_5$
N
CO$_2$C$_2$H$_5$

→

O
CO$_2$C$_2$H$_5$
N
CO$_2$C$_2$H$_5$

183
i) HCl
ii) C$_2$H$_5$OH/HCl
iii) NaBH$_4$

HO ⟍ H H CO$_2$C$_2$H$_5$
N
O

185

← KOC$_2$H$_5$

O
O
N
C$_2$H$_5$O$_2$C

i) Ba(OH)$_2$
ii) HCl
iii) BrCH$_2$CO$_2$C$_2$H$_5$
→

O
O
N
CO$_2$C$_2$H$_5$

184

185
i) Pt, H$_2$, HOAc
ii) Ba(OH)$_2$
↓

HO ⟍ H
N
CO$_2$H

186

i) C$_2$H$_5$OH/HCl
ii) LiAlH$_4$
→ **171**

Scheme 11

dride to give a mixture of simple pyrrolizidines (**189**). So far this synthesis has not been extended to compounds in the 7-hydroxy series.

A simpler route to the synthesis of the nucleus involves condensation of N-formyl-L proline (**190**) with ethyl propiolate to give the pyrrole **191**. This compound may be reduced catalytically to racemic ethyl isoretronecanolate (**192**).

Yet another synthetic approach to the pyrrolizidine nucleus involves cyclopropylimine rearrangement of an imine such as **193**, in refluxing xylene in the presence of NH$_4$Cl. Deprotection of the acetal gives the cyclized product **194**, which can be desulfurized with Raney nickel to isoretronecanol (**195**).

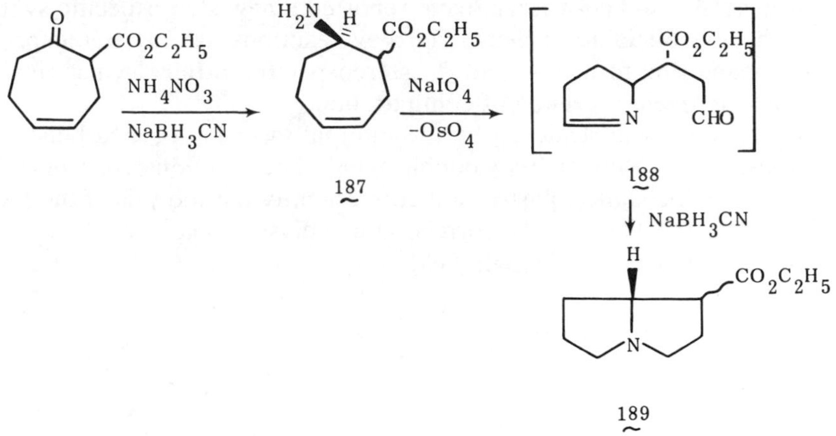
CO2C2H5
NH4NO3
NaBH3CN
H2N
H
CO2C2H5
NaIO4
-OsO4
187
CO2C2H5
N
CHO
188
NaBH3CN
H
CO2C2H5
N
189

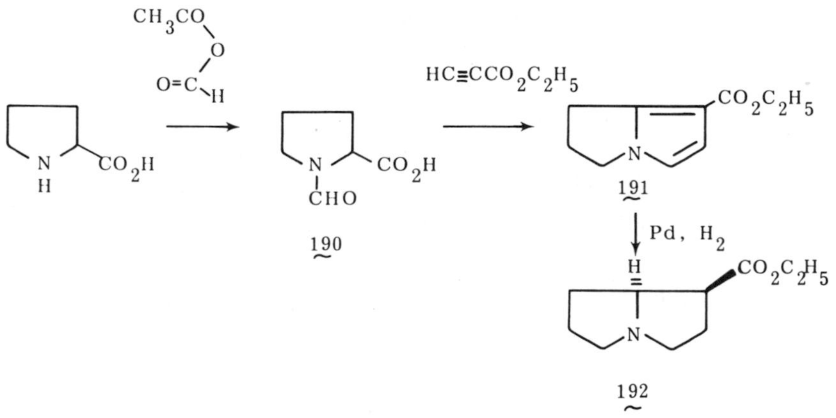
CH3CO
O
O=C
H
N
H
CO2H
N
CHO
CO2H
190
HC≡CCO2C2H5
CO2C2H5
N
191
Pd, H2
H
CO2C2H5
N
192

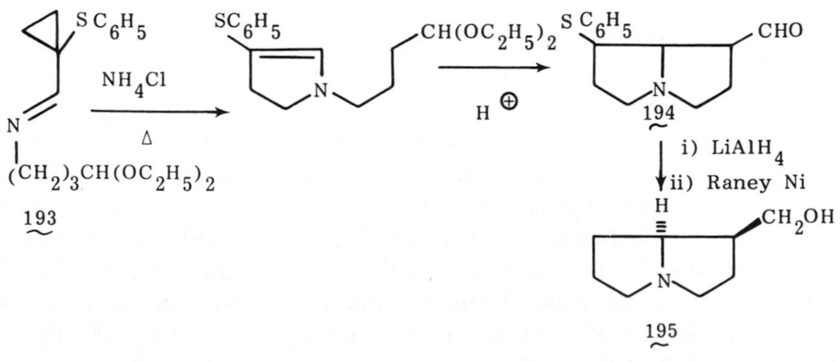
SC6H5
NH4Cl
Δ
N
(CH2)3CH(OC2H5)2
193
SC6H5
N
CH(OC2H5)2
H ⊕
SC6H5
CHO
N
194
i) LiAlH4
ii) Raney Ni
H
CH2OH
N
195

Danishefsky and co-workers have reported a new stereospecific synthesis of the pyrrolizidine nucleus. The key reactions are (a) stereospecific cyclopropane ring formation and (b) stereospecific intramolecular ring fission with complete inversion of configuration.

Objective (a) was achieved by the internal insertion of a carbene from a diazoester into a cis or trans double bond. The reaction proceeded with complete stereochemical purity, and consequently thermolysis of the E and Z olefins 196 and 197 gave the corresponding diastereomeric cyclopropanes 198 and 199 respectively in high yield.

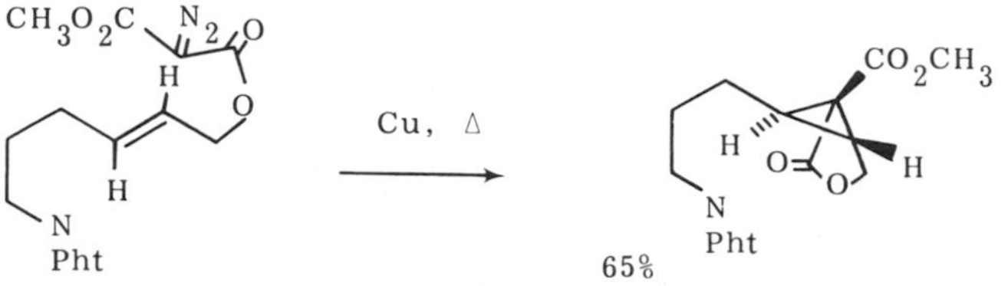

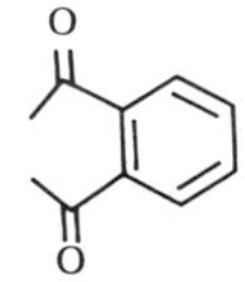

Objective (b), the key step in the preservation of the two asymmetric centers transferred from the cyclopropane to the pyrrolizidines, was induced by refluxing 196 with hydrazine in methanol. The intermediates were not characterized, but rather the crude product (probably 200) was hydrolyzed with aqueous acid, under which conditions decarboxylation of the β-dicarbonyl took place. The product 201 could be cyclized with base to the lactam alcohol 202. The mechanism of this reaction is regarded as occurring as shown in Scheme 12. Quite analogously 199 afforded the corresponding C-1 epimeric lactam alcohol 203. Lithium aluminum hydride reduction of 202 and 203 proceeded smoothly to yield isoretroneccanol (195) and its C-1 epimer trachelanthamidine (181). In neither case was any of the C-1 isomer pro-

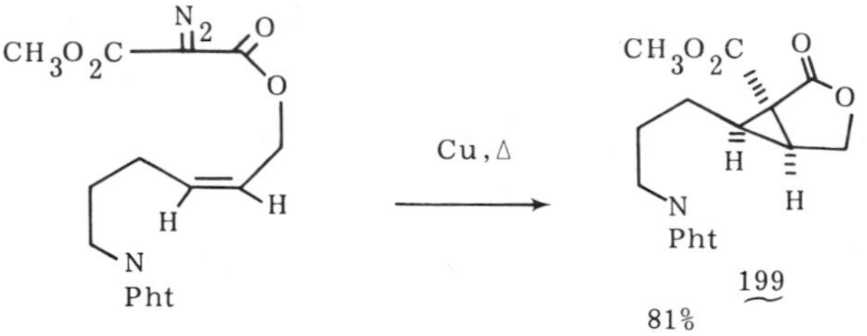

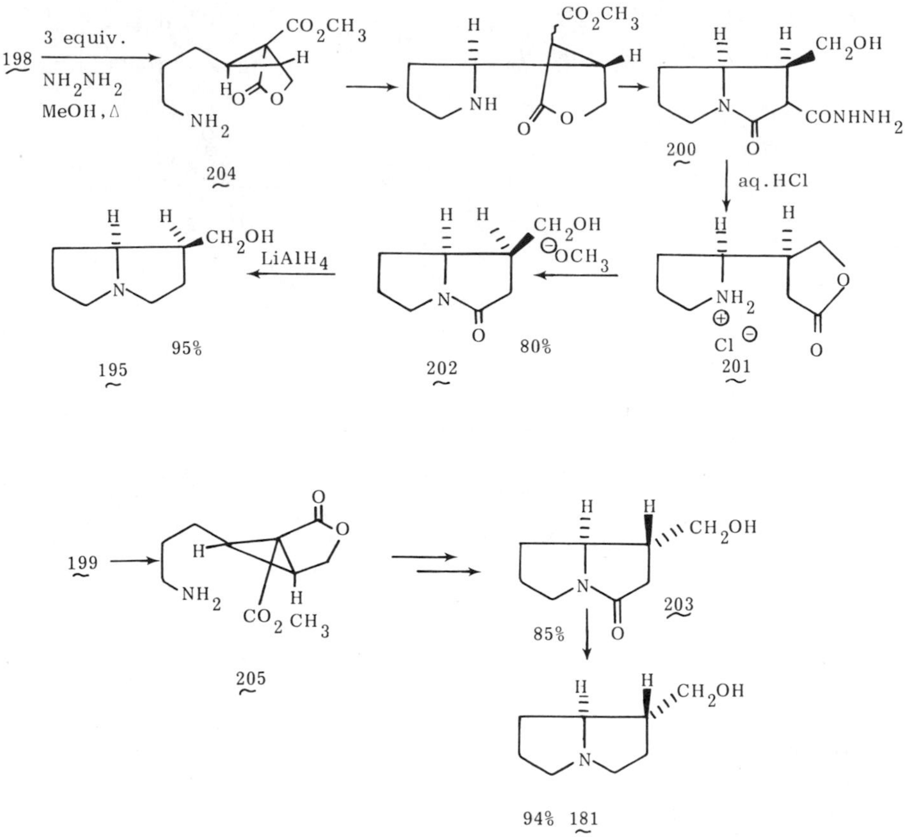

Scheme 12 Danishefsky and co-workers synthesis of (+)-isoretronecanol (**195**) and (±)-trachelanthamidine (**181**).

duced, thereby confirming the stereospecific inversion which occurs in the intramolecular opening of the cyclopropane ring of both **204** and **205**.

3.4.4 The Necic Acid Portion

The necic acids are mono- or dicarboxylic acids in the range $C_3–C_7$, which may contain additional double bonds and/or hydroxy groups. The necine portion contains two acylatable groups, and combinations of these and the necic acids amount to over 100 isolated alkaloids. Some are monoesters, some diesters, and others cyclic diesters. Representative examples of these alkaloids are shown in Figure 5.

The acids may be obtained from the alkaloids by alkaline hydrolysis or hydrogenolysis. Each technique has its problems, the first because of decomposition of some of the less stable acid moieties, the second because of reductions that may occur within the acid moiety leading to additional

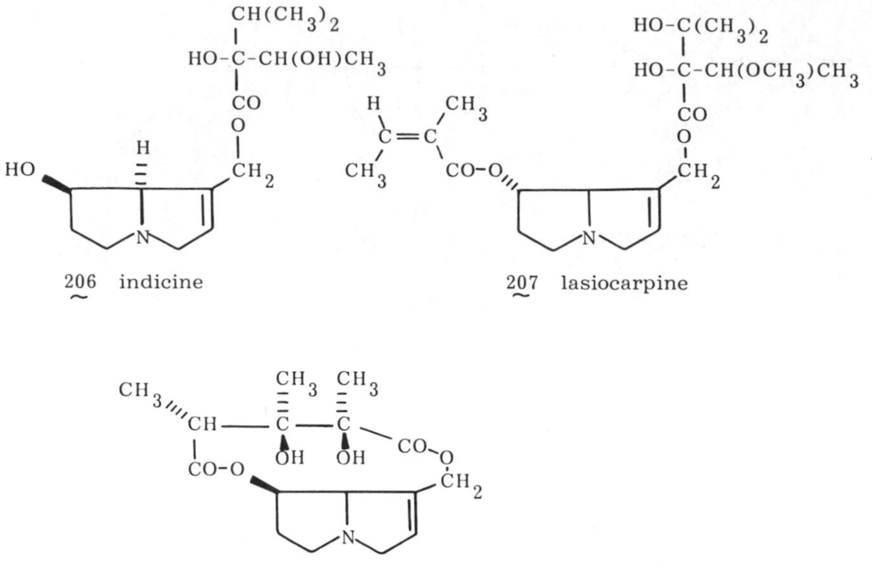

206 indicine   207 lasiocarpine

208 monocrotaline

Figure 5   Representative pyrrolizidine esters.

stereoisomers. In a diester alkaloid having a 1,2-double bond, mild hydro-genolysis will only cleave the allylic ester moiety.

The ester portion of monocrotaline (**208**) has three centers of asymmetry (i.e. eight possible stereoisomers). Hydrogenolysis of monocrotaline (**208**) gives an acid, monocrotalic acid, which since it has the formula $C_8H_{12}O_5$ must be lactonized. Lithium aluminum hydride reduction of monocrotalic acid methyl ester gave a tetraol **209**, which was cleaved by periodate to give the products indicated. Monocrotalic acid therefore has the gross structure **210**.

It was 20 years before the absolute configuration of **210** was deduced. Lithium aluminum hydride reduction of monocrotaline (**208**) gave the tetraol **209**, which on periodate oxidation and hypobromite oxidation gave 3-hydro-xy-2-methyl propanoic acid (**211**). The hydroxyl of this acid was shown to have the S-configuration. Carbon-4 in methyl monocrotalate therefore has the R-configuration.

The absolute configuration at C-2 was determined by degradation of methyl monocrotalate by successive dehydration, ozonolysis, and selective hydrolysis to the ester **212**. Sodium borotritiide reduction followed by hy-drolysis afforded in labeled form the *threo* isomers of 2,3-dihydroxy-2-methyl butanoic acid (**213**). The 2R,3S-acid was synthesized from tiglic acid followed by resolution of the brucine salt. When the tritiated and separately resolved samples were mixed and the brucine salts recrystallized to constant activity, all of the activity remained with the 2R-isomers. Hence C-2 of methylmono-crotolate has the R-configuration.

Phosphorylchloride in pyridine has a rigid *trans*-orientation requirement for the elimination of water. Since this reaction proceeded slowly for methyl monocrotalate, it was deduced that the relative stereochemistries at C-3 and C-4 were *cis*. Methyl monocrotalate therefore has the absolute configuration shown in **214**, and monocrotaline the structure **208**.

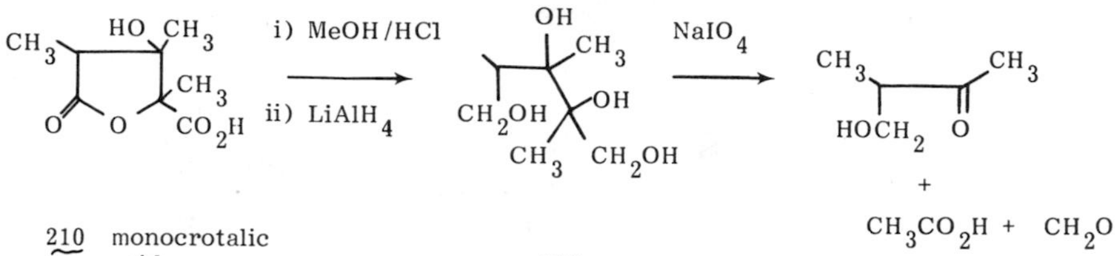

210 monocrotalic acid

209

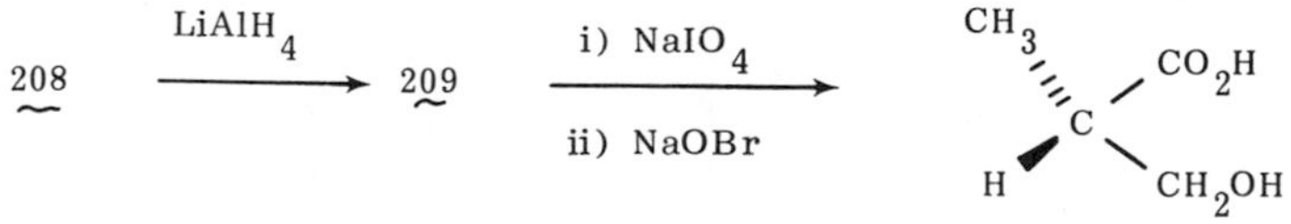

$CH_3CO_2H + CH_2O$

211

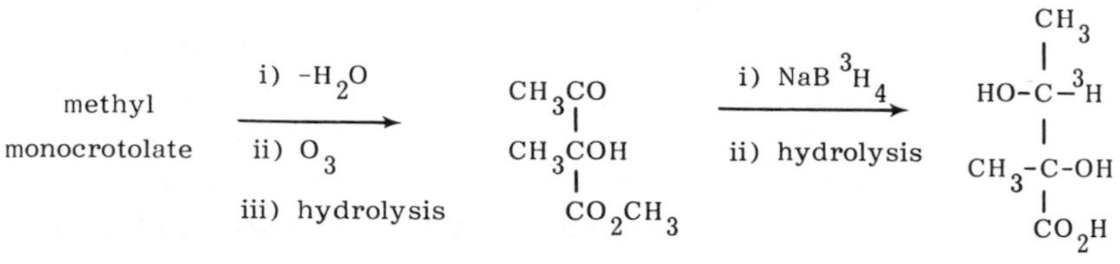

212

213

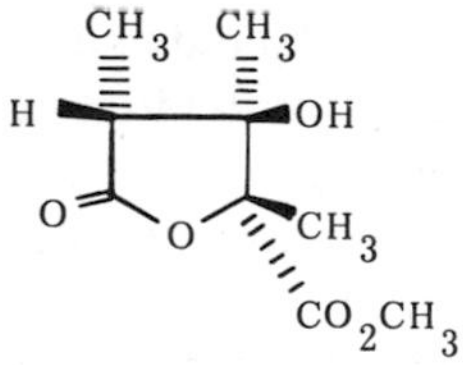

214

3.4.5 Spectral Properties

UV spectroscopy has been of little value in determining the structure of these alkaloids, except to show the presence of *cis* and *trans* α,β-unsaturated ester systems.

The pyrrolizidine alkaloids are an interesting group as far as IR spectroscopy is concerned because of the varied basicity and variety of carbonyl frequencies. Some examples of the latter include γ-lactones, 1770–1790 cm^{-1}; δ-lactones, 1740–1750 cm^{-1}; saturated esters, 1735–1745 cm^{-1}; and α,β-unsaturated esters, 1715–1725 cm^{-1}. In addition there may be substantial differences in the carbonyl region for spectra obtained in different media. Thus senecionine (215), which has one saturated ester and one α,β-unsat-

215   senecionine   216   heliotrine

urated ester, shows a single carbonyl frequency (at 1722 cm^{-1}) in carbon tetrachloride solution, but two distinct frequencies (at 1715 and 1735 cm^{-1}) in a Nujol mull. For most of these alkaloids the Nujol mull gives the better defined carbonyl region.

Other potential sources of problems are the hydroxyl groups, which not only reduce the carbonyl frequencies when hydrogen bonding can occur, but may also not be observed because of intramolecular hydrogen bonding with a strongly basic nitrogen. This phenomenon is reduced considerably in solution spectra and it is therefore advisable to take IR spectra of pyrrolizidine alkaloids as both a solution and a mull in order to gain the maximum information.

The NMR spectra of most pyrrolizidine alkaloids are extremely complicated, and even the simple alkaloids such as retronecine (171) pose considerable problems in assigning individual resonances, because of magnetic nonequivalence of methylene protons.

The chemical shifts and the coupling constants for retronecine (171) in pyridine are shown in Table 6. The β-protons at C-3 and C-5 are shielded by the lone pair of electrons on nitrogen, and therefore absorb at a higher field than the α-protons. But the main cause of the complexity is the extensive cross-ring coupling which occurs. Culvenor and Woods have made a very valuable study of pyrrolizidine alkaloid types and the data for some representative alkaloids are shown in Table 7.

As expected, the chemical shifts of H-7 and H-9 protons are important in assigning the position of any esterifying group. Compare retronecine (171), heliotrine (216), and lasiocarpine (207) and note how esterification deshields

Table 6 Chemical Shifts and Coupling Constants of Retronecine (**171**) in d_5-Pyridine

Chemical Proton	Shift, δ (ppm)	Coupling Constants (Hz)			
H-2	5.91	$J_{2,3\alpha}$	1.6	$J_{5\alpha,5\beta}$	−8.6
H-3α	4.04	$J_{2,3\beta}$	1.6	$J_{5\alpha,6\alpha}$	7.2
H-3β	3.54	$J_{2,8\alpha}$	−1.6	$J_{5\alpha,6\beta}$	1.2
H-5α	3.32	$J_{2,9}$	−1.6	$J_{5\beta,6\alpha}$	12.0
H-5β	2.96	$J_{3\alpha,3\beta}$	−15.9	$J_{5\beta,6\beta}$	6.0
H-6α	1.96	$J_{3\beta,8\alpha}$	3.5	$J_{6\alpha,6\beta}$	−12.8
H-6β	2.09	$J_{3\beta,8\alpha}$	5.4	$J_{6\alpha,7\alpha}$	3.6
H-7α	4.58	$J_{3\alpha,9}$	1.6	$J_{6\beta,7\alpha}$	1.3
H-8α	4.47	$J_{3\beta,9}$	1.6	$J_{7\alpha,8\alpha}$	3.9
H-9	4.58				

the acyl methine (C-7) or methylene (C-9) protons. In the latter case, the C-9 methylene protons often become magnetically nonequivalent.

The fragmentation of a simple pyrrolizidine nucleus follows an expected pattern as illustrated in Scheme 13 for retronecine (**171**). The initial process is loss of the C-6 and C-7 atoms to give a stable allylic radical/iminium ion m/e 111 (**217**) which may subsequently lose either the hydroxy or the hydroxy methyl to afford the pyridinium ions at m/e 94 (**218**) and 80 (**219**).

Esters at the 9-position which are allylic (e.g. **220**) are cleaved through the C$_9$—O bond in three different ways to afford the ions m/e 137, 138, and 139. The m/e 138 ion (**221**) arises through simple allylic fission and the ion at m/e 137 (**222**) by a double six-membered ring rearrangement. A more complex process is needed to explain the m/e 139 ion (**223**) involving hydrogen transfer from the side chain ester unit.

Table 7 Chemical Shift Data for the Ring Protons of Representative Pyrrolizidine Alkaloids (in CDCl$_3$)

Compound	Chemical Shifts						
	H-2	H-3	H-5	H-6	H-7	H-8	H-9
Retronecine (**171**)	5.7	3.4, 3.8	2.75, 3.2	1.95	4.3	4.2	4.25
Heliotridine (**170**)	5.58	3.3, 3.9	2.75, 3.3	1.9	4.1	4.0	4.32
Platynecine (**172**)	1.9	2.3, 3.2	2.3, 3.2	1.9	4.21	3.2	3.92
Heliotrine (**216**)	5.77	3.2, 3.8	2.6, 3.3	2.05	4.1	4.0	4.73, 5.10
Lasiocarpine (**207**)	5.85	3.4, 4.0	[a]	1.9	5.18	4.13	4.95
Monocrotaline (**208**)	5.03	[a]	[a]	2.13	5.08		4.71, 4.87

Source: Data from C. C. J. Culvenor and W. G. Woods, *Aust. J. Chem.* **18**, 1625 (1965).

[a] Not observed because of the complexity of the protons from the ester groups.

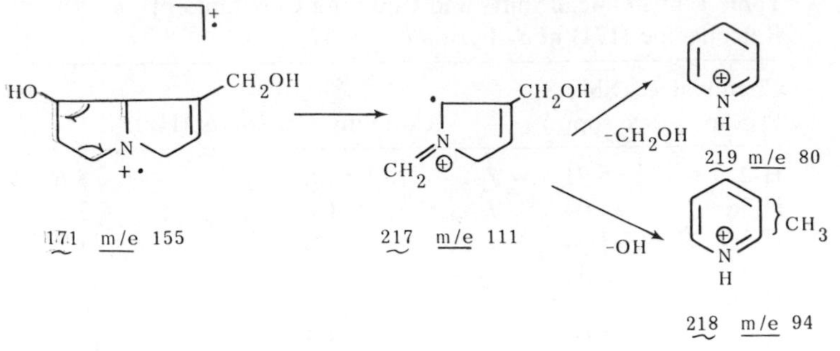

Scheme 13

The secondary esters at C-7 may undergo fission of the oxygen–carbonyl bond or elimination of RCO_2H.

In the macrocyclic diesters [e.g. senecionine (**215**)] it can often be difficult to decide which carboxyl is esterfied by which alcohol. An α-hydroxy ester may undergo the rearrangement shown in Scheme 14, with loss of CO_2. The intermediate ion **224** is not very stable and undergoes subsequent fragmentation as shown. Production of such ions is only possible if the hydroxy ester is attached to the allylic primary alcohol site.

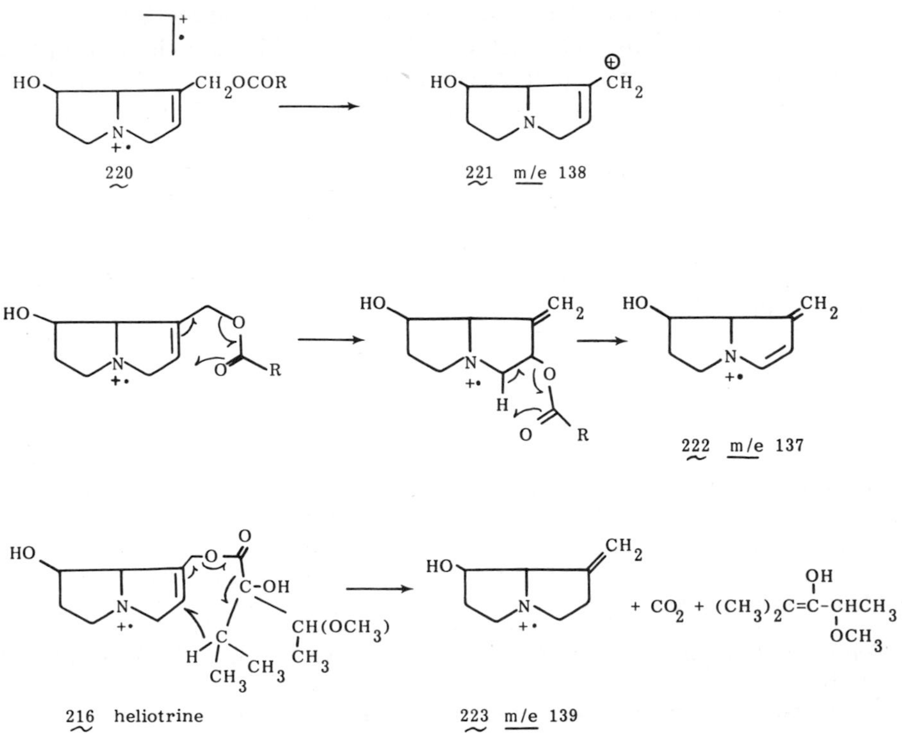

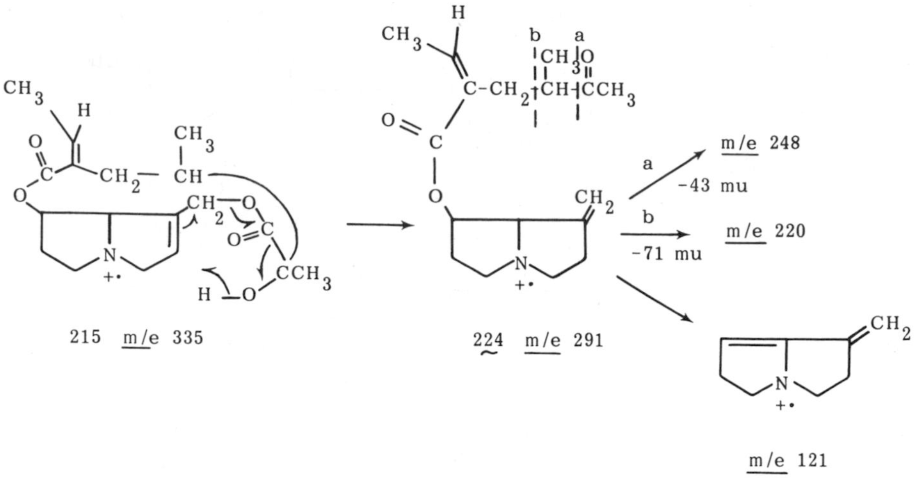

Scheme 14   Mass spectral fragmentation of senecionine (215).

3.4.6   Biosynthesis

There are two aspects to the biosynthesis of the pyrrolizidine alkaloids: (*a*) the formation of the necine portion and (*b*) the formation of the necic acid group. The biosynthesis of the latter is better known than that of the former.

Initial experiments indicated that ornithine was a precursor of the retro-necine portion of several alkaloids in *Crotalaria* and *Senecio* species, with essentially no activity in the acid portion. Degradation of retronecine (171) after administration of [2-^{14}C]ornithine (1) to *Senecio isatideus* indicated activity at C-1' (26%), and at C-7 or more probably at C-8 (71%). Because the C-1' and C-8 or C-7 activities are different, the two uniting fragments are also different. In addition since C-3 and C-5 are not labeled, each fragment enters the nucleus by way of a nonsymmetrical intermediate. This should be contrasted with the biosynthesis of lupinine in Chapter 4.

In a separate series of experiments with *S. douglasii*, incorporations of [2-^{14}C]- or [5-^{14}C]ornithine gave retronecine (171) labeled in identical fashion, in which C-1' of 181/195 contained 25% of the activity. These two contradictory results are in need of clarification.

If symmetrical intermediates are involved, the biosynthetic scheme shown in Figure 6 may be operating in which the symmetrical aminodialdehyde 225 is important. Mannich condensation and reduction can then afford the alcohol 181/195. The point of introduction and stereospecificity of the C-7 hydroxy group and of any 1,2-double bond in the scheme is unknown.

It was shown quite early on that neither acetate nor mevalonate were precursors of the necic acid units [e.g. senecic acid (226) and seneciphyllic acid (227)]. In *S. douglasii*, senecic acid (226) was found to be derived specifically from two isoleucine units. [2-^{14}C]Isoleucine (228) afforded se-

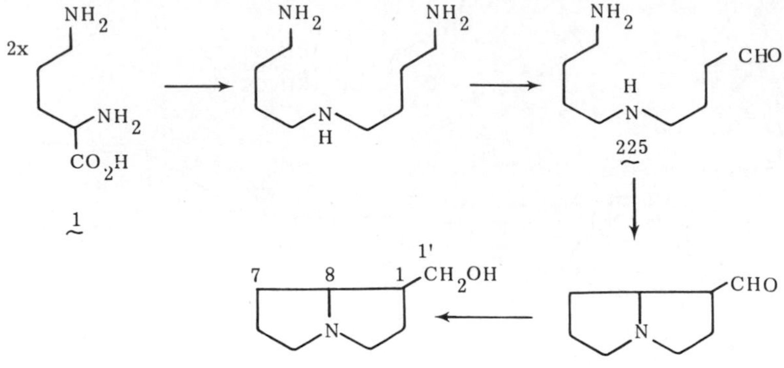

Figure 6

necic acid (**226**) labeled at C-1 and C-10, and [5-^{14}C]isoleucine (**228**) gave **226** labeled equally at C-7 and C-9. The mechanism of joining the two isoleucine units is not known, but the mode of this linkage is shown in Figure 7. Subsequent work determined that of the possible isomers of isoleucine, only L-isoleucine was a precursor.

Seneciphyllic acid (**227**) has posed a more interesting problem. Isoleucine was a precursor, but degradation of the labeled **227** established that only *one* unit was incorporated. Additional experiments with labeled methionine and formate indicated the specific derivation (~25% of total activity) of C-8 from a one-carbon unit. The remaining carbon atoms are at present of uncertain origin.

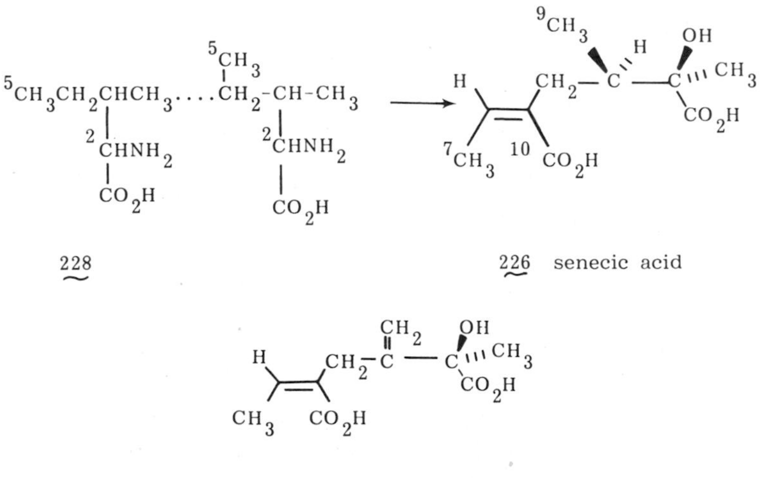

228   226   senecic acid

227   seneciphyllic acid

Figure 7   Formation of senecic acid (**226**) from isoleucine.

The distribution of pyrrolizidine alkaloids appears to be widening with the recent isolation of nitropolizonamine (**229**), a component of the defensive secretion of the millipede *Polyzonium rosalbum*.

Two plausible biogenetic routes involve either a pyrrolizidine derivative **230** with a exomethylene group which can cyclize with IPP; or alternatively an internal cyclization of a monoterpene such as **231**, followed by condensation of the resulting imine with a three-carbon unit. The intermediate imine **232** co-occurs with **229**.

229   nitropolizonamine   230

231   232

3.4.7 Pharmacology

The so-called pyrrolizidine alkaloidosis is characterized by cirrhosis of the liver. Some examples of the genera of plants giving this effect, the animals involved, and the country where the poisoning occurred are shown in Table 8. The death rate is dependent on the feeding time, the age of the animals, and which particular species is involved.

The metastatic spread of liver tumors by pyrrolizidine alkaloids was first observed in 1963. With alkaloids such as lasiocarpine, seneciphylline, monocrotaline, and heliotrine, one dose is sufficient to produce chronic liver lesions and the N-oxide compounds are almost as toxic. The diesters and cyclic esters are more toxic than the monoesters [e.g. heliotrine (**216**)], and branching in the chain is also an important contributing factor to toxicity.

The liver is not the only site of action, for monocrotaline (**208**) also causes the development of chronic lung lesions and bronchopneumonia has been induced in horses from *Trichodesma* sp.

The median LD_{50} values for some pyrrolizidine alkaloids in male rats at 72 hr are given in Table 9. Twice-weekly doses of 0.1 LD_{50} causes chronic

Table 8　Pyrrolizidine Alkaloid-Containing Plants
Causing Poisoning in Farm Animals

Genus	Animal	Country
Senecio	Cattle, horses	Canada, New Zealand, South Africa, England, United States
	Sheep	New Zealand
Crotalaria	Cattle, horses	United States
	Pigs	
	Sheep	Rhodesia
Heliotropium	Sheep	Russia, Australia
	Cattle (rare)	Australia
Echium	Sheep, pigs	Australia
Trichodesma	Horses, cattle	Russia

liver damage and death in 6 months. A 40–50-fold increase in copper concentration in the liver has been observed at death.

Not all pyrrolizidine alkaloids are hepatotoxic; platyphylline (**233**), and heliotridine (**170**) are just two examples that do not have this activity. However, ester alkaloids derived from the amino alcohols retronecine (**171**), heliotridine (**170**), and supinidine (**169**) are hepatotoxic.

From the available data the necessary features for hepatotoxicity appear to be a 1,2-double bond and esterification of the allylic primary hydroxyl group by a branched-chain acid.

Subsequently Culvenor and associates have carried out more detailed investigations of both the chronic hepato- and pneumotoxicities of over 60

Table 9　Median Lethal Dose of
Pyrrolizidine Alkaloids for Male Rats
(i.p.[a] Injection, 72-hr Survivors)

Alkaloid	MLD_{50} (mg/kg)
Heliotrine (**216**)	300
Lasiocarpine (**207**)	72
Lasiocarpine-*N*-oxide	547
Platyphylline (**233**)	252
Monocrotaline (**208**)	175
Seneciphylline (**234**)	77
Senecionine (**215**)	85

[a] i.p., intraperitoneal.

233 platyphylline, R=CH$_3$,H

234 seneciphylline, R=CH$_2$

pyrrolizidine alkaloids. The diesters were found to be four times as toxic as the monoesters, with the heliotridine esters being more toxic than the retronecine esters. N-Methylation of monocrotaline (208) and senecionine (215) gave nonhepatotoxic compounds.

The structure requirements for pneumotoxicity were found to be the same as for hepatotoxicity, suggesting that possibly similar metabolites are involved.

The metabolites of the hepatotoxic pyrrolizidine alkaloids are now quite well established; they are pyrroles formed by dehydrogenation between N-4 and C-8. First observed in the urine of rats, these compounds have now been shown to be produced in both rat liver microsomal and human embryonic liver systems.

Dehydromonocrotaline (235) administered intravenously (i.v.) close to the target organ caused lung and liver lesions in rats similar to those induced by the parent alkaloid. The different ester groups appear to have little or no effect on the toxicity.

The mechanism of formation of the pyrrole metabolites is not certain. At one time it was thought that the N-oxides were important intermediates which could undergo a Polonovskii-type reaction as discussed previously. However, little of the pyrrole derivatives are produced when the N-oxides are metabolized, and there is some evidence that they are actually reduced to the parent base in the gut.

A clue to the possible first step comes from the observation that alkaloids such as otosenine (236) are also hepatotoxic. In the rat liver microsomal system N-demethylation occurs to give a product thought to be dehydroretronecine (237). A plausible scheme, which may also generalize to retronecine (171) metabolism, is shown in Scheme 15. Even simple pyrrole derivatives such as 238 are potent hepatotoxins.

Liver and lung lesions are not the only pharmacologic results after administration of pyrrolizidine alkaloids.

The spasmolytic action of several pyrrolizidine alkaloids in rabbit intestine has shown that unsaturation in both the acid and nucleus are important for activity.

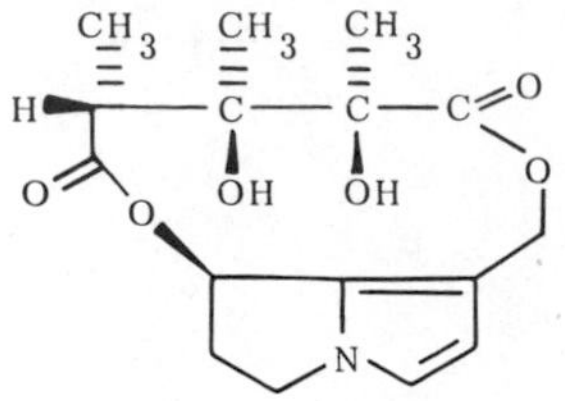

235 dehydromonocrotaline

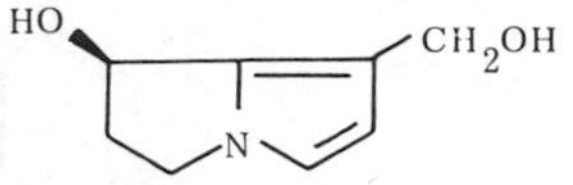

237 dehydroretronecine

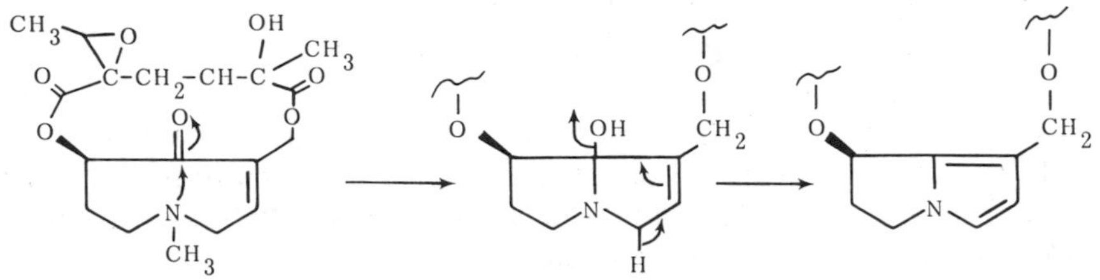

236 otosenine

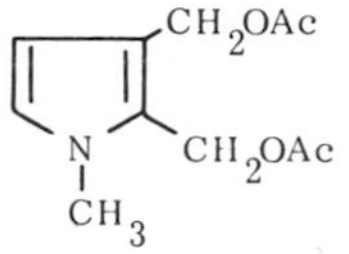

238

Scheme 15

Quaternization of simple pyrrolizidines with a dihalo alkane gives salts which exhibit neuromuscular blocking activity *in vitro* (e.g. **239**). Other derivatives have hypotensive or local anesthetic activity.

It is well known that many classes of carcinogenic or cocarcinogenic compounds exhibit antitumor activity at lower doses or with slight structure modification. One example of importance in the pyrrolizidines is indicine-*N*-oxide (**240**) from *Heliotropium indicum*. This compound is active against a number of tumor systems in mice and is being evaluated clinically in humans.

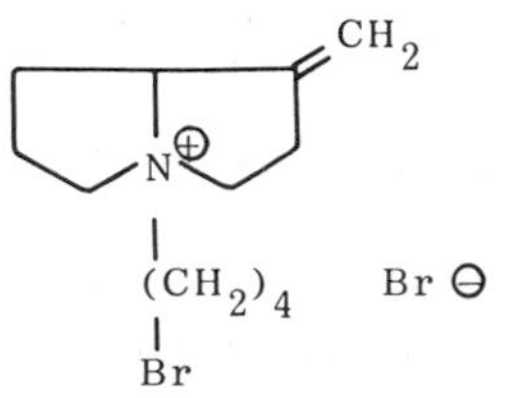

239

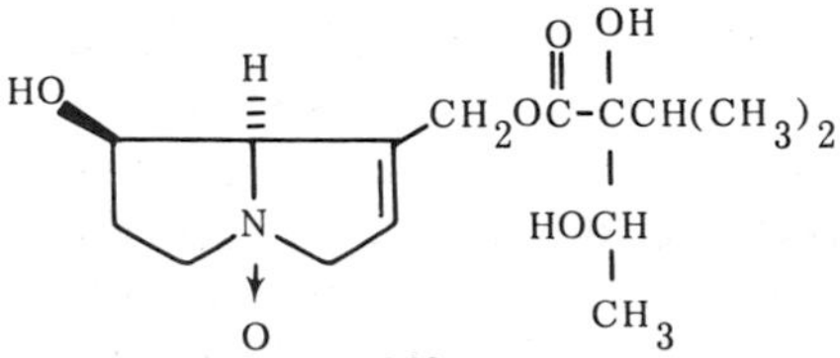

240

LITERATURE

Reviews

Leonard, N. J., *Alkaloids NY* **1,** 107 (1950).

Warren, F. L., *Fortschr. Chem. Org. Naturs.* **12,** 198 (1955).

Leonard, N. J., *Alkaloids NY* **6,** 37 (1960).

Warren, F. L., *Fortschr. Chem. Org. Naturs.* **24,** 329 (1966).

Bull, L. B., C. C. J. Culvenor, and A. T. Dick, *The Pyrrolizidine Alkaloids*, North-Holland, Amsterdam, 1968.

Warren, F. L., *Alkaloids NY* **12,** 245 (1970).

Saxton, J. E., *Alkaloids, London* **1,** 59 (1971).

Saxton, J. E., *Alkaloids, London* **2,** 59 (1972).

Saxton, J. E., *Alkaloids, London* **3,** 76 (1973).

Saxton, J. E., *Alkaloids, London* **4,** 84 (1974).

Klasek, A., and O. Weinbergova, *Rec. Dev. Chem. Nat. Carbon Compd.* **6,** 35 (1975).

Saxton, J. E., *Alkaloids, London* **5,** 77 (1975).

Crout, D. H. G., *Alkaloids, London* **6,** 72 (1976).

Crout, D. H. G., *Alkaloids, London* **7,** 54 (1977).

Robins, D. J., *Alkaloids, London* **8,** 47 (1978).

Synthesis

Borch, R. F., and B. C. Ho, *J. Org. Chem.* **42,** 1225 (1977).

Pizzorno, M. T., and S. M. Albonico, *J. Org. Chem.* **39,** 731 (1974).

Danishefsky, S., R. McKee, and R. K. Singh, *J. Amer. Chem. Soc.* **99,** 4783 (1977).

chapter 4

ALKALOIDS DERIVED FROM LYSINE

Lysine (**1**) is the next member in the homologous series of amino acids from ornithine. And it affords in many instances alkaloids similar to those derived from ornithine but containing one extra methylene group in the heterocyclic ring. Thus the simplest alkaloids are derived from the saturated hetereocyclic nucleus piperidine (**2**). The absence of unsaturation also means that few of these alkaloids will display meaningful UV spectra.

Some of the lysine-derived alkaloids which are quite analogous to those derived from ornithine are pelletierine (**3**), anaferine (**4**), pseudopelletierine (**5**), anabasine (**6**), and lupinine (**7**). However, there are several groups of alkaloids derived from lysine which have no comparison in the ornithine-derived series; examples include lobeline (**8**), sparteine (**9**), matrine (**10**), lythrine (**11**), and lycopodine (**12**).

This is not to say that all alkaloids containing a piperidine nucleus are derived from lysine; far from it. Arecoline (**13**) is probably derived from nicotinic acid (**14**). Coniine (**15**) is acetate-derived, and probably so are pinidine (**16**) and carpaine (**17**). Skytanthine (**18**) is a monoterpene alkaloid, nupharidine (**19**) is a sesquiterpene alkaloid, solasodine (**20**) is a steroidal alkaloid, and secodine (**21**) is a monoterpene indole alkaloid. These alkaloids are consequently discussed elsewhere in this monograph.

4.1 PELLETIERINE AND RELATED ALKALOIDS

Few alkaloids are named after chemists, but the bark and root of the pomegranate tree (*Punica granatum* L.) contains a number of quite simple alkaloids which were named in honor of the French alkaloid chemist Pelletier (see Chapter 1).

Isopelletierine (**22**) forms an *N*-methyl derivative (**23**) which gives *N*-methyl coniine (**24**) on Wolff–Kischner reduction and *N*-methyl pipecolic acid (**25**) on chromic acid oxidation. The structure **22** was confirmed by

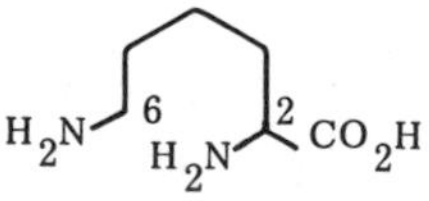

 H_2N 6 H_2N 2 CO_2H

1 lysine

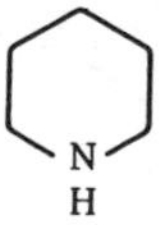

2 piperidine

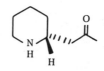

3 (+)-pelletierine

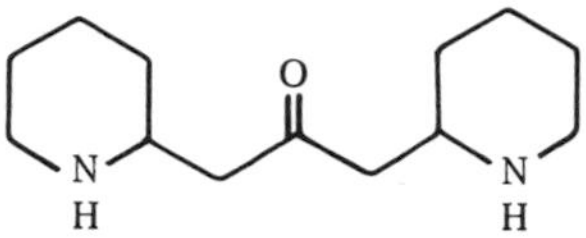

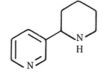

4 anaferine

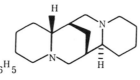

5 pseudopelletierine   **6** anabasine

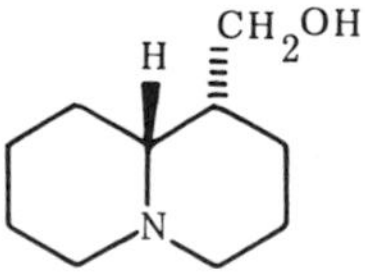

CH_2OH

7 lupinine

OH

C_6H_5 N CH_3 O C_6H_5

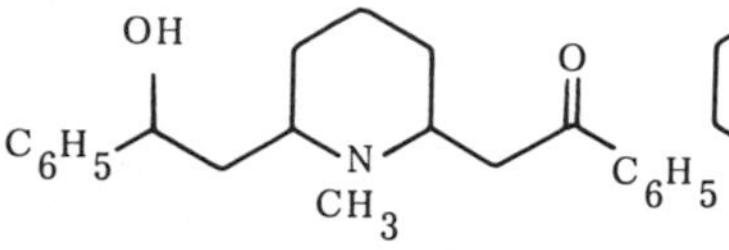

8 lobeline

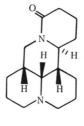

9 sparteine

10 matrine

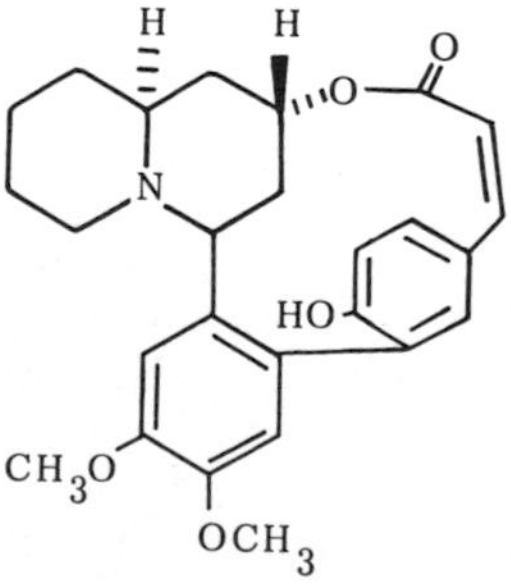

CH_3O HO OCH_3

11 lythrine

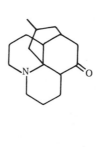

12 lycopodine

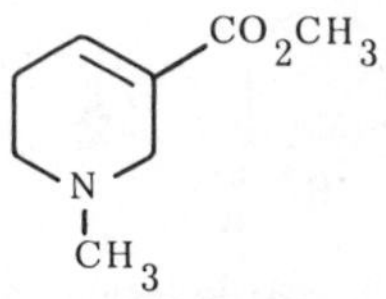

13 arecoline

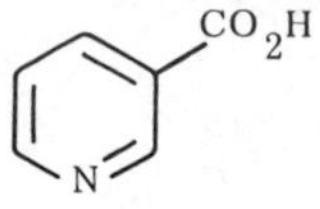

14 nicotinic acid

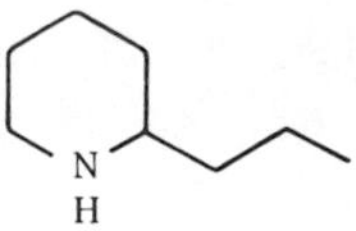

15 coniine

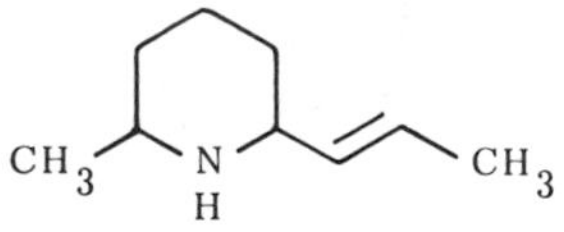

16 pinidine

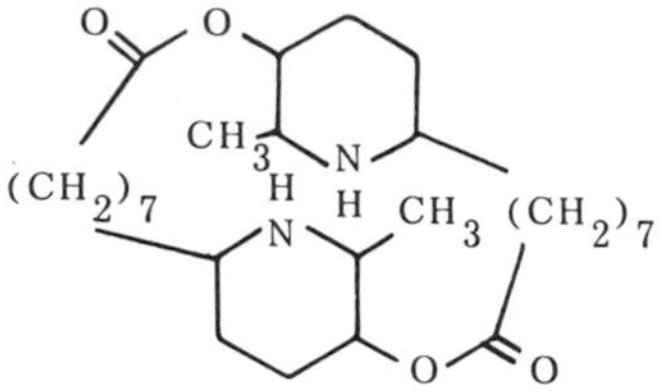

17 carpaine

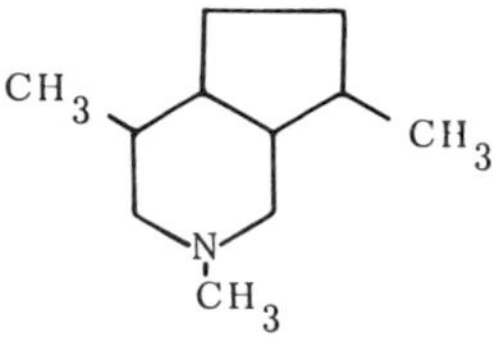

18 β-skytanthine

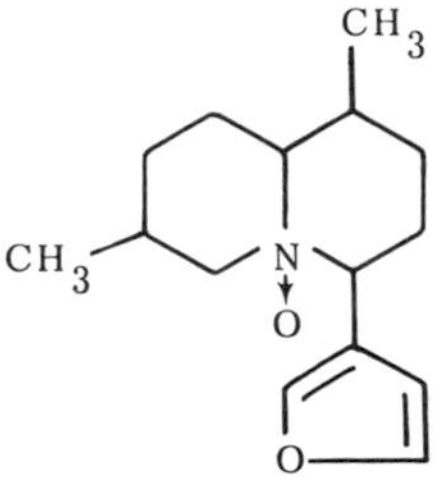

19 nupharidine

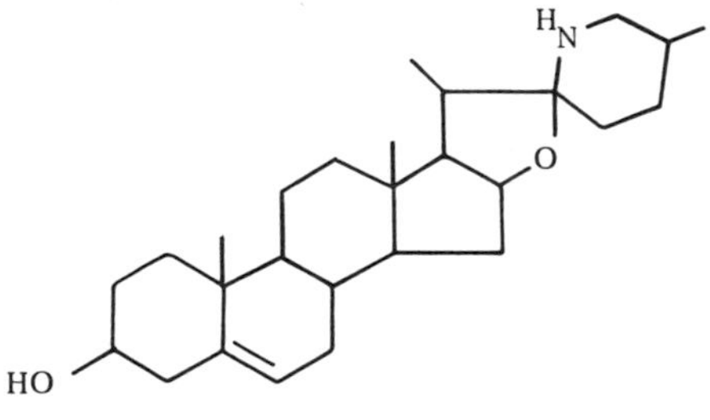

20 solasodine

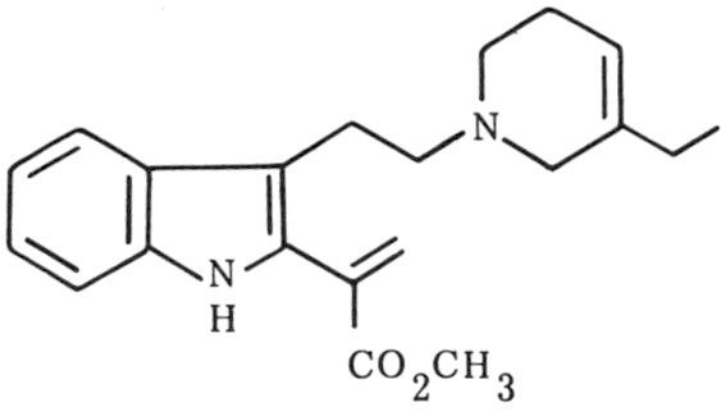

21 secodine

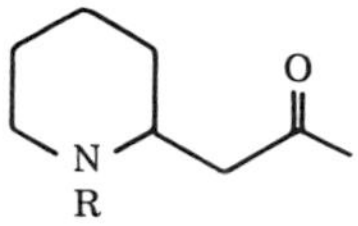

22 isopelletierine, R=H

23 R=CH$_3$

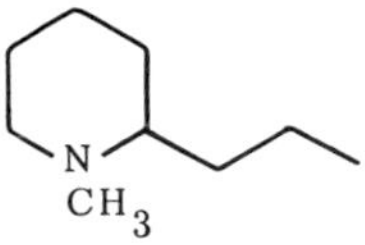

24 N-methylconiine

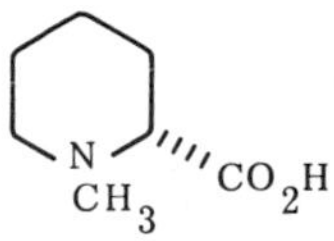

25 N-methyl pipecolic acid

140

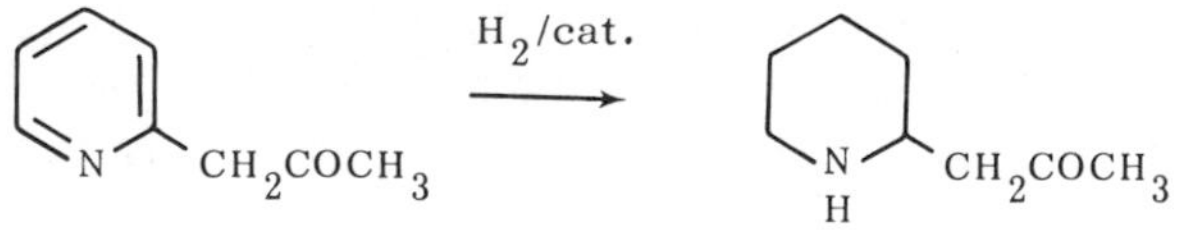

26   22

synthesis from the lithium salt of 2-methyl pyridine. Acylation with acetic anhydride gave the ketone **26**, which on hydrogenation afforded isopelletierine (**22**). Like hygrine, isopelletierine racemized easily with base, but careful isolation affords the R-(−)-isomer.

Pelletierine caused numerous problems before the structure was finally (in 1955) deduced to be **3**, the (−)-isomer of isopelletierine. The problems arose because the oxime of pelletierine was dehydrated to what was considered to be a nitrile; pelletierine was therefore thought to be aldehyde **27**. Now this dehydration product is regarded as the cyclized Beckmann rearrangement product **28**.

(±)-Pelletierine (**22**) has been prepared from piperidine (**2**) by N-chlorination with N-chlorosuccinimide followed by dehydrochlorination to 1,2-dehydropiperidine (**29**). Condensation with ethyl acetoacetate then gave the racemate **22**.

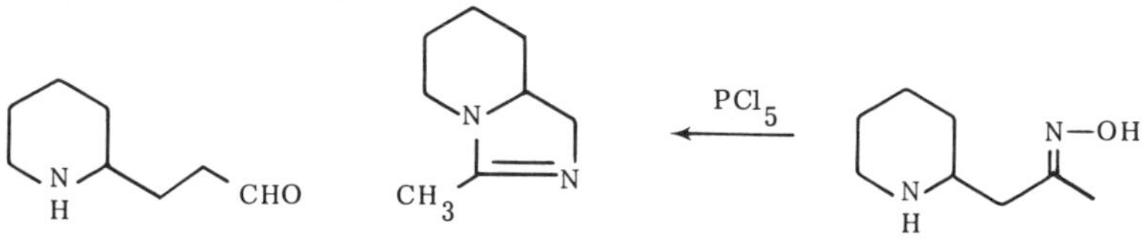

27   28   pelletierine
                                                  oxime

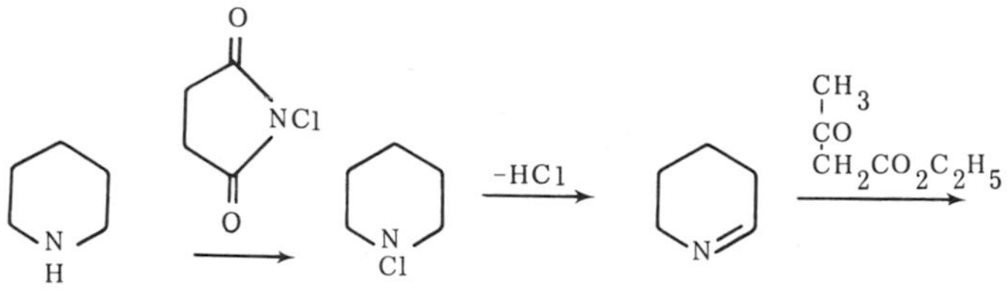

                          29

The optical rotatory dispersion (ORD) and circular dichroism (CD) spectra of several simple piperidine alkaloids have been studied and compounds such as (−)-pelletierine (**3**), which show conformational changes in solution, display reversed Cotton effects depending on the solvent. Thus (−)-pelletierine shows $[\theta]_{max}$ − 127 at 285 nm in 95% ethanol and $[\theta]_{max}$ + 140 at 292 nm in water, indicating a reversal of the $n \rightarrow \pi^*$ Cotton effect. It was suggested that this change might occur by solvation of the equatorial electron pair of nitrogen, causing the carbonyl oxygen to form a *cis*-fused pseudoring structure such as **30**, which would have a positive Cotton effect.

Pelletierine is highly toxic to tapeworms and has been used as an anthelmintic.

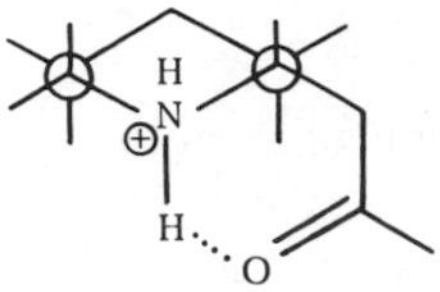

30

Another major alkaloid of pomegranate is the tertiary amine pseudopelletierine (**5**) having the formula $C_9H_{15}NO$. Two successive Hofmann eliminations gave 1,5-cyclooctadiene (**31**), and since the original compound was optically inactive and formed a dibenzylidene derivative, the nitrogen must be present in a symmetrical array. Dehydrogenation of pseudopelletierine gave 2-propyl pyridine (**32**). The only structure compatible with this information is **5**, which makes pseudopelletierine the next higher homologue of tropinone.

Synthetically pseudopelletierine can be prepared by modification of the Robinson route using glutaraldehyde, methylamine, and acetone dicarboxylic acid (Scheme 1).

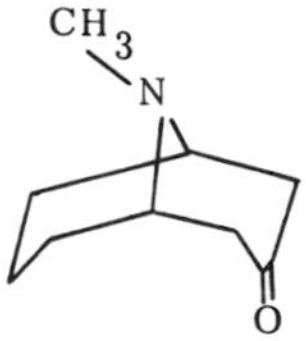

5   pseudopelletierine    31   cis, trans-cyclo   32
                                     -octadiene

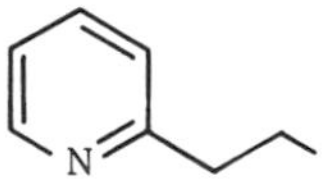

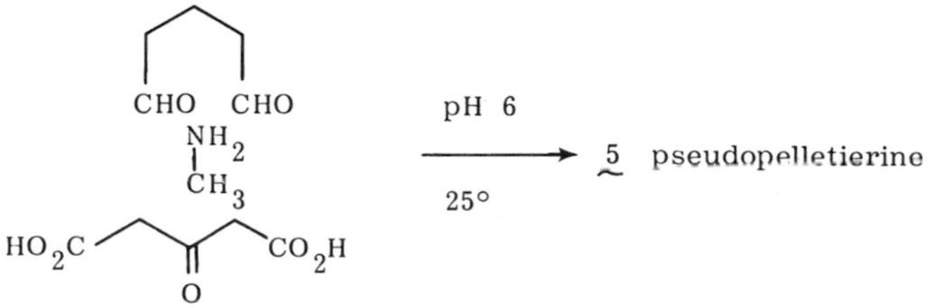

Scheme 1

The original work of Wilstätter on this problem opened up two major areas of organic chemistry, the chemistry of medium-size rings containing double bonds and the valence tautomerism of cyclooctatetraene.

Myrtine is a quinolizidine alkaloid from *Vaccinium myrtillus* Cham. & Schlecht. (Ericaceae) which was shown by spectroscopy and synthesis to have the structure **33**. Mannich condensation of R-(−)-pelletierine (**3**) with acetaldehyde in acetic acid gave (−)-myrtine (**33**) of slightly greater optical purity than the isolated product. Biosynthetically, myrtine is probably derived along similar lines from pyruvic acid and R-(−)-pelletierine (**3**) followed by decarboxylation.

33 myrtine   3 R-(-)-pelletierine

4.2 ANABASINE AND RELATED ALKALOIDS

Although nicotine (**34**) is the best known alkaloid of tobacco, anabasine (**6**) is the major alkaloid, as it is in *Nicotiana glauca* R. Grah. and *Anabasis aphylla* L. (Chenopodiaceae) and *Haloxylon persicum* Bunge (Chenopodiaceae). Oxidation yields nicotinic acid (**14**) and dehydrogenation 3′, 2-bipyridyl (**35**).

34 nicotine   35   36 anatabine

l- and *dl*-Anabasine (**6**) can be separated by benzoylation, formation of the picrate derivatives, and separation of these on alumina.

Anabasine can also be purified by formation of a complex with aqueous cobalt(II) sulfate solution. The precipitate is recrystallized from water, and the free base liberated with alkali.

Anabasine, like lobeline (**8**), has antismoking and respiratory muscle stimulatory action, and like nicotine it exhibits insecticidal properties. Anabasine has also been used as a metal anticorrosive.

Anatabine (**36**) cooccurs with anabasine (**6**) in *N. glutinosa*. However, because it is not derived from lysine it is discussed further in Chapter 5.

Several N-acylated derivatives of anabasine (**6**) are known.

Astrocasia phyllanthoides Rob & Mills (Euphorbiaceae) produces a number of interesting alkaloids, including astrophylline (**37**) and astrocasine (**38**), which are *N-cis*-cinnamoyl derivatives of 3(5)-[2'(R)-piperidyl]piperidine.

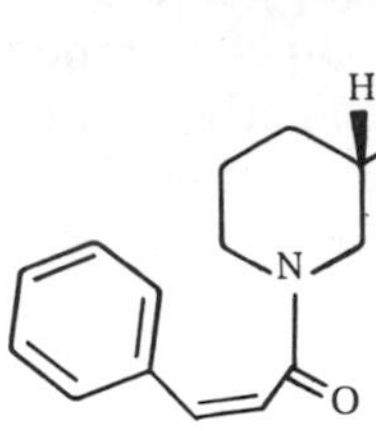

37   astrophylline 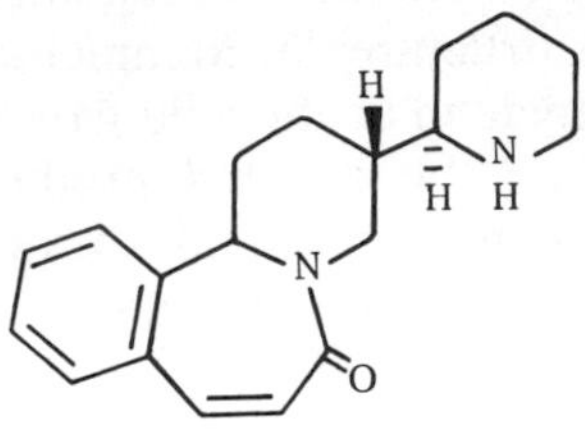   38   astrocasine

Other examples include adenocarpine (**39**) (orensine?) and isoorensine (**40**). But a more interesting alkaloid is santiaguine (**41**) from *Adenocarpus foliosus* (Leguminosae) and the closely related (−)-hoveine (**42**) from *Hovea longipes* Benth. (Leguminosae), an alkaloid that exhibits marked hypotensive activity.

The biosynthesis of santiaguine (**41**) is quite an interesting one. Both [2-

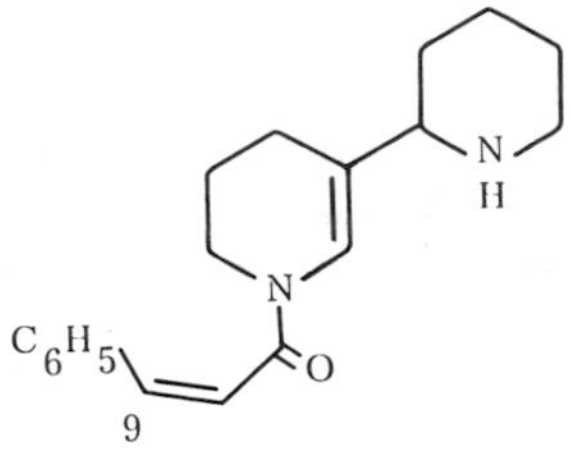

39   adenocarpine 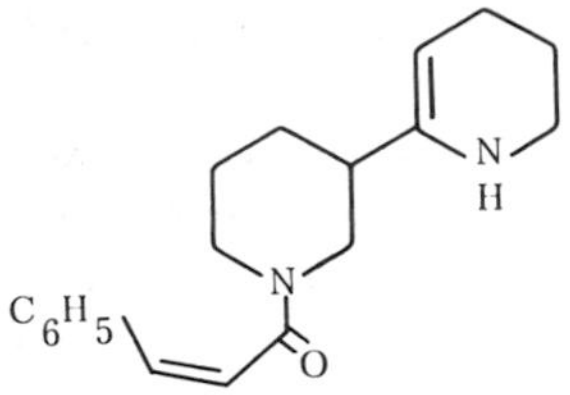   40   isoorensine

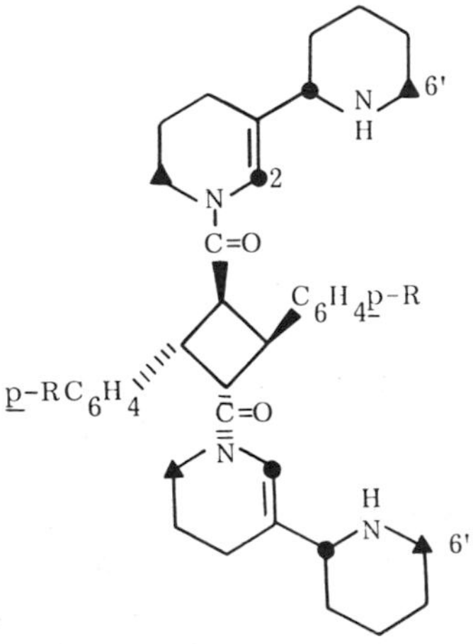

●   label from $\left[2\text{-}^{14}C\right]$-lysine

▲   label from $\left[6\text{-}^{14}C\right]$-lysine

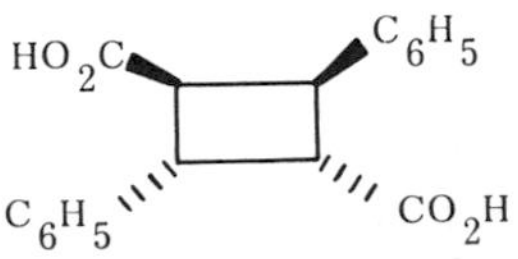

41   santiaguine,   R=H

42   hoveine,   R=OH

43   α-truxillic acid

[14]C]- and [6-^{14}C]lysine (1) were specifically incorporated without the involvement of a symmetrical intermediate, and as expected the central portion was derived from phenylalanine and cinnamic acid. The next steps are somewhat confused, for α-truxillic acid (43) and [9-^{14}C]adenocarpine (39) were precursors of 41. The level of incorporation of 39 was higher in the dark than in the light, and this was interpreted as being evidence that the cyclobutane ring of 41 is formed enzymatically and not photochemically.

4.3   THE *SEDUM* ALKALOIDS

The stonecrop family [e.g., *Sedum acre* (Crassulaceae)] contains a number of simple piperidine alkaloids closely related to those of hemlock, pomegranate, and *Lobelia inflata* L. Oxidation of each gave the same ketone, which was levorotatory, and further oxidation gave (−)-*N*-methylpipecolic acid (25) and benzoic acid. Sedamine (44) and allosedamine (45) are therefore isomers at the hydroxy group and the sterochemistries were deduced by the Hofmann degradation of allosedamine to 46, followed by oxidation to acid 47 of known stereochemistry.

    A dimeric alkaloid, lobinaline (48), the principal alkaloid of *Lobelia cardinalis*, appears to be derived biosynthetically from two sedamine-like units, followed by reduction and N-methylation as shown in 49.

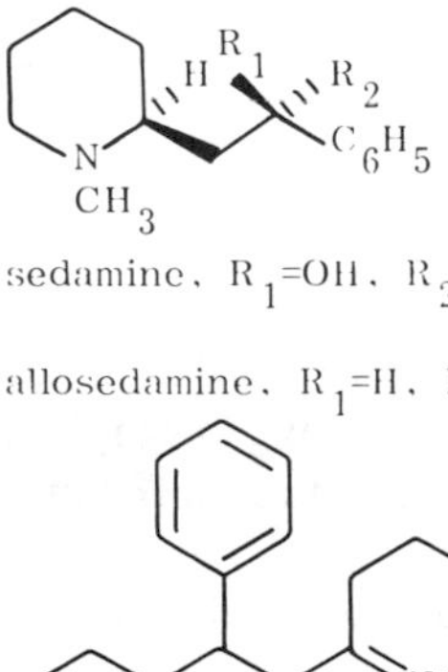

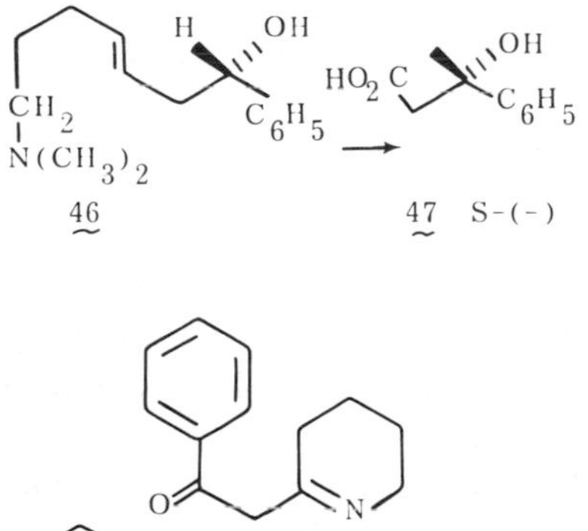

44   sedamine, R$_1$=OH, R$_2$=H

45   allosedamine, R$_1$=H, R$_2$=OH

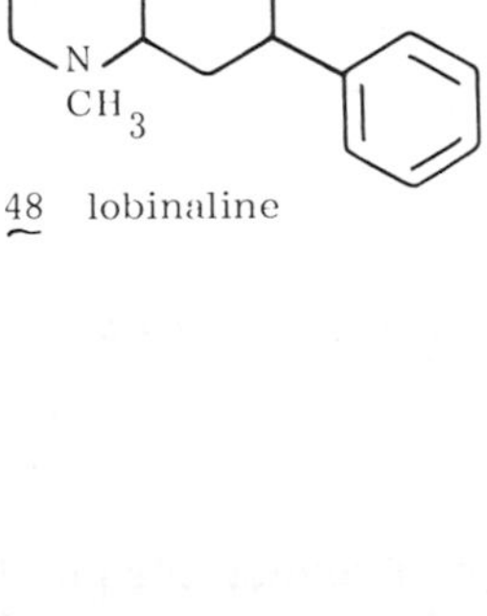

46

47   S-(−)

48   lobinaline

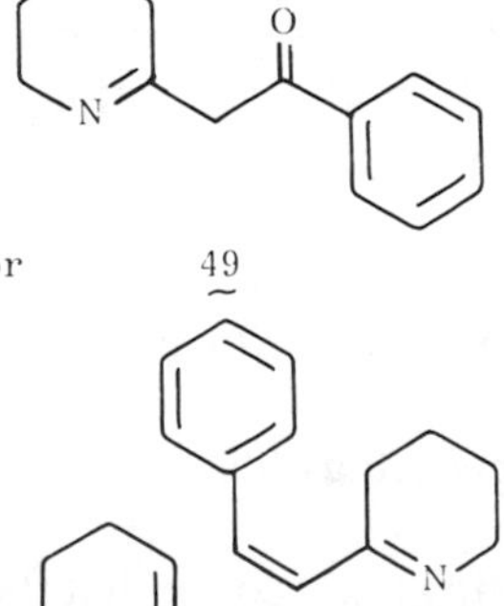

or   49

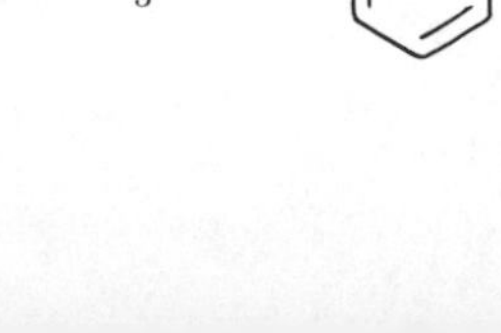

4.4   THE ALKALOIDS OF *LOBELIA INFLATA*

Lobelia inflata L., known as Indian tobacco, is native to the eastern and southeastern United States. The principal alkaloid of the leaves and tops is lobeline (**8**), and several other related alkaloids co-occur. Lobeline is also the major alkaloid of several other *Lobelia* species, and is now being efficiently isolated by the use of ion exchange resins.

Lobeline, $C_{22}H_{27}NO_2$, is an optically active tertiary amine containing both keto and secondary hydroxyl groups. Reduction of the keto group gives a diol, lobelanidine (**50**), which is optically inactive and is therefore *meso*. Double Hofmann elimination of lobelanine (**51**) gave the diene **52**, which could be hydrogenated to 1,7-dibenzoylheptane (**53**).

$$\underset{\sim}{8}\quad \text{lobeline } R_1\text{=H, OH, } R_2\text{=O}$$
$$\underset{\sim}{50}\quad \text{lobelanidine } R_1\text{=}R_2\text{=H, OH}$$
$$\underset{\sim}{51}\quad \text{lobelanine } R_1\text{=}R_2\text{=O}$$

Many different syntheses of lobeline have been reported, but the most efficient is a Robinson-type biomimetic synthesis using methylamine, glutaraldehyde, and benzoylacetic acid at pH 4 to produce lobelanine (**51**) in 90% yield.

The expectorant properties of *Lobelia inflata* were first noted in 1785 and the drug was widely used for that purpose. Lobeline is quite similar to nicotine in its pharmacologic action, although it is less potent. The sulfate salt, which stimulates the respiratory system and induces coughing, is used in antismoking tablets and lozenges.

4.5   MISCELLANEOUS SIMPLE PIPERIDINE ALKALOIDS

4.5.1   The Pepper Alkaloids

It may be surprising to discover that the pungent, irritant principles of pepper are alkaloids. Black pepper is the unripe fruit of *Piper nigrum* L. and contains as its major (6–11%) constitutent piperine (**54**). Being an amide piperine is

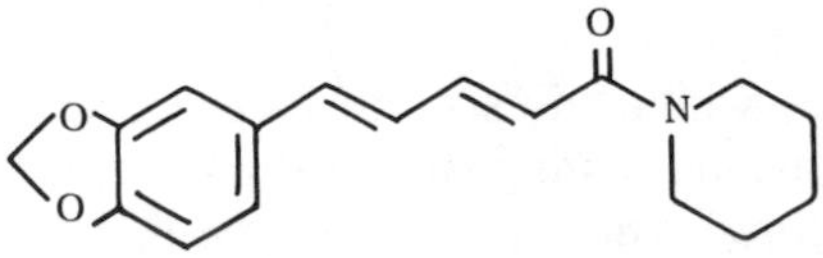

54 piperine

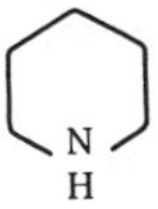

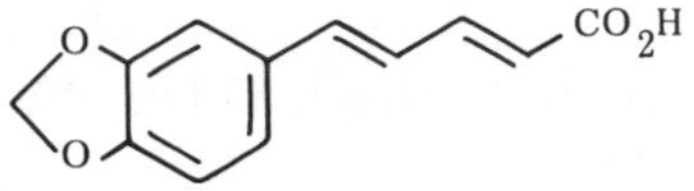

55

not basic, but it does undergo hydrolysis to piperidine (2) and piperic acid **55**. The stereochemistry of piperine is *trans,trans* [λ_{max} 345 nm (log ϵ 4.47)] and the proton and ^{13}C NMR data are shown in **56** and **57** respectively.

Piperine is at least partly responsible for the "hot" taste of pepper.

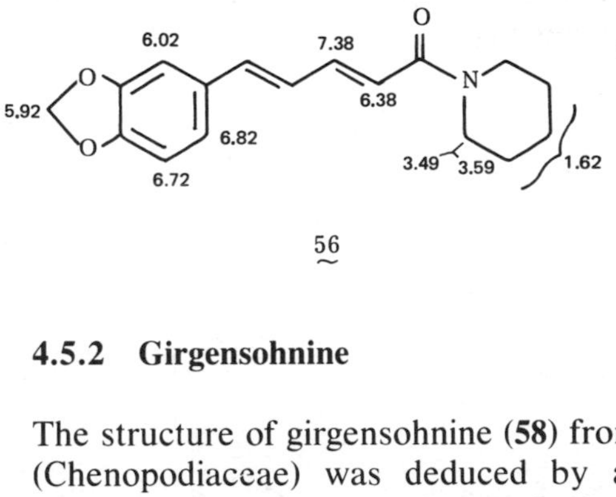

56

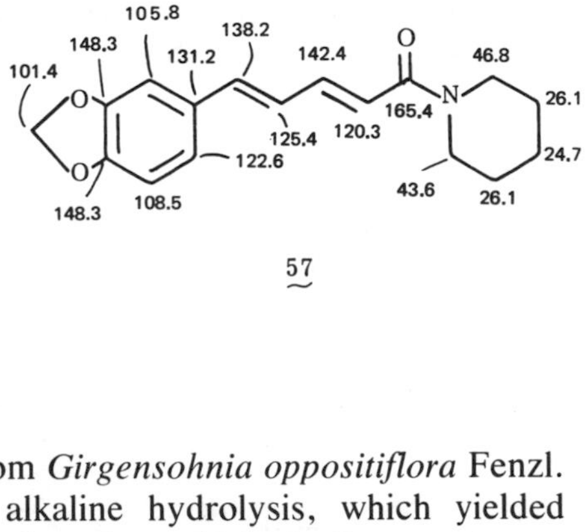

57

4.5.2 Girgensohnine

The structure of girgensohnine (**58**) from *Girgensohnia oppositiflora* Fenzl. (Chenopodiaceae) was deduced by alkaline hydrolysis, which yielded piperidine, hydrogen cyanide, and *p*-hydroxybenzaldehyde. In a precise reversal of this process, condensation of *p*-hydroxybenzaldehyde cyanohydrin (**59**) with piperidine (2) afforded **58**. Biogenetically one would expect a derivation from tyrosine and lysine, but this remains to be established.

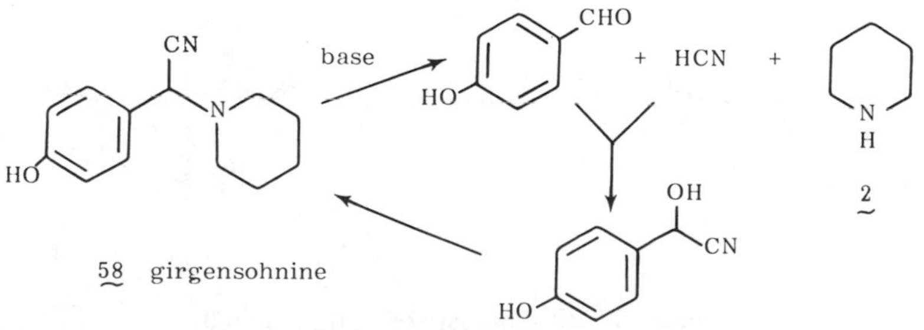

58 girgensohnine

59

4.5.3 Slaframine

The fungus *Rhizoctonia leguminicola*, which infests the forage crop, red clover (*Trifolium pratense*), produces the interesting alkaloid slaframine (**60**). The compound is the "salivation factor," which is responsible for the excessive salivation and refusal to feed by cattle, horses, and sheep who graze or feed on the hays. Other effects noted in farm animals include diarrhea and anorexia.

The pathway deduced for the biosynthesis of slaframine (**60**) is shown in Scheme 2. Unlike most alkaloids derived from lysine, *all* the carbon atoms are retained in the product. Thus both [1-^{14}C]- and [6-^{14}C]lysines labeled slaframine (**60**), as did various ring-labeled pipecolic acids (**61**).

Further experiments established **62**, **63**, and **64** to be intermediates, and consequently the overall route is envisaged to be that shown. The origin of the remaining two carbons is obscure, and the sequence and stereochemistry of any of the final steps remains unknown.

The fungus also produces the novel 1-pyrindine alkaloid **65**. [1-^{14}C]- and [6-^{14}C]Lysines and tritiated pipecolic acid were precursors, but the origin of the additional two carbon atoms is unknown.

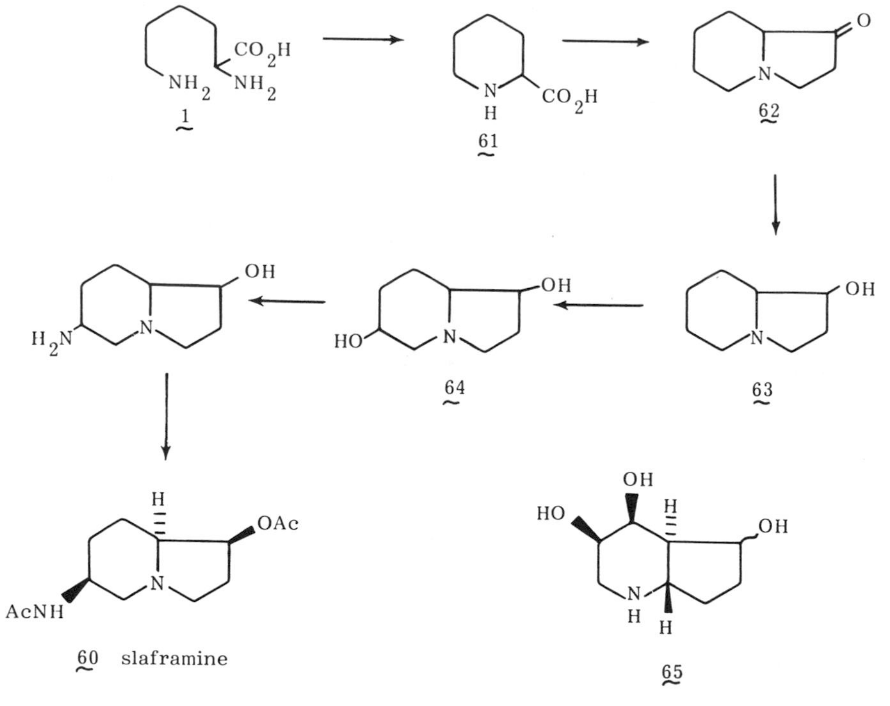

Scheme 2 Biosynthesis of slaframine (**60**).

LITERATURE

Broquist, H. P., and J. J. Snyder, in S. Kadis, A. Ciegler, and S. J. Ajl (Eds.), *Microbial Toxins*, Vol. 7, Academic, New York, 1971, p. 319.

Guengerich, F. P., J. J. Snyder, and H. P. Broquist, *Biochemistry* **12**, 4264 (1973).

Guengerich, F. P., and H. P. Broquist, *Biochemistry* **12**, 4270 (1973).

4.5.4 Cryptopleurine

Cryptopleurine (**66**) was originally obtained from *Cryptocarya pleurosperma* Wh. and Fr. (Lauraceae) but has since been obtained from *Boehmeria cylindrica* (L.) Sw. (Urticaceae).

The alkaloid shows both cytotoxic and antiviral activities and is discussed in detail in Chapter 8.

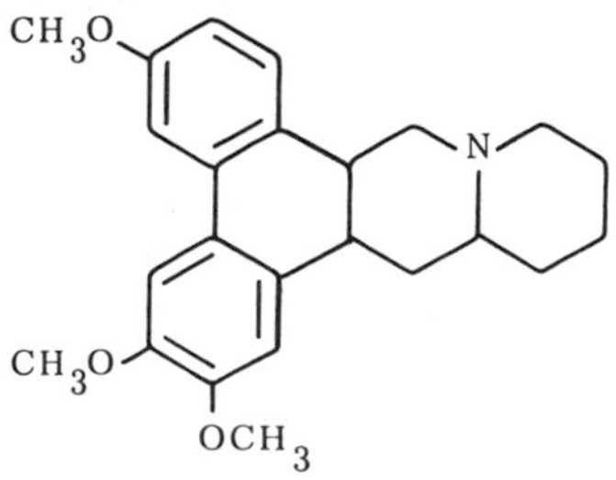

66 cryptopleurine

LITERATURE

Reviews

Ayer, W. A., and T. E. Habgood, *Alkaloids NY* **11**, 459 (1968).

Hill, R. K., in S. W. Pelletier (Ed.), *Chemistry of the Alkaloids*, Van Nostrand Reinhold, New York, 1970, p. 385.

Gross, D., *Fortschr. Chem. Org. Naturs.* **28**, 109 (1970).

Gross, D., *Fortschr. Chem. Org. Naturs.* **29**, 1 (1971).

Snieckus, V. A., *Alkaloids, London* **1**, 48 (1971).

Snieckus, V. A., *Alkaloids, London* **2**, 33 (1972).

Snieckus, V. A., *Alkaloids, London* **3**, 43 (1973).

Snieckus, V. A., *Alkaloids, London* **4**, 50 (1974).

Snieckus, V. A., *Alkaloids, London* **5**, 56 (1975).

Pinder, A. R., *Alkaloids, London* **6**, 54 (1976).

Pinder, A. R., *Alkaloids, London* **7**, 37 (1977).

Pinder, A. R., *Alkaloids, London* **8**, 37 (1978).

4.6   LYTHRACEAE ALKALOIDS

Plants in the Lythraceae family are widely distributed in the tropics and subtropics of South America, and the genera *Heimia* and *Lythrum* are particularly interesting sources of alkaloids. Very little was known about these alkaloids until the early 1960s when Ferris reported the isolation of several alkaloids from *Decodon verticillatus* (L.) Ell.

Several features of these alkaloids are of interest, including a quinolizidine nucleus, a biphenyl moiety, sometimes a *cis*-cinnamoyl ester, and a macrocyclic lactone. Because of their complexity, many of the structures were determined by X-ray analysis. There are two main groups, depending on the stereochemistry of the quinolizidine nucleus (*cis* or *trans*), and typical representatives are lythrine (**11**) and cryogenine (**67**), which differ only in the C-10 stereochemistry and which cooccur in *Heimia salicifolia*, Link and Otto. Some compounds with a dihydrocinnamoyl moiety have also been isolated [e.g. decodine (**68**)].

11   lythrine, 10α–H
67   cryogenine, 10β–H

68   decodine

A second major group of compounds have been obtained from *D. verticillatus*, which contain a biphenyl ether linkage; an example in this series is decaline (**69**).

Yet another series of alkaloids contains no quinolizidine unit, but instead has a piperidine ring as the nucleus; such a compound is lythranine (**70**) from *Lythrum anceps* Makino.

The structure elucidation of lythranine is quite instructive, for although it was completed in 1971, a considerable amount of degradative work was needed to firmly establish even the gross structural features.

Lythranine has the molecular formula $C_{28}H_{37}NO_5$, and from spectroscopic data and simple chemical reactions, evidence was obtained for secondary and phenolic hydroxy groups, a secondary acetoxy group, a secondary amine, an aromatic methoxy group, and six aromatic hydrogens.

Alkaline hydrolysis of lythranine (**70**) gave lythranidine (**71**), which on treatment with diazomethane gave *O*-methyllythranidine (**72**). Oxidation with potassium permanganate and esterification gave the known biphenyl derivative **73**. Dehydrogenation of lythranine (**70**) with palladium–charcoal followed by permanganate oxidation and esterification afforded methyl pyridine 2,6-dicarboxylate (**74**), thereby establishing the piperidine ring and the 2,6-substitution in **70**.

The presence of a macrocyclic ring was demonstrated by Hofmann degradation of O-methyllythranidine (**72**). Catalytic hydrogenation gave a diol **75**, which on oxidation of the two secondary alcohols afforded a symmetrical diketone. The presence of azelaic acid, $(CH_2)_7(CO_2H)_2$, in the mixture of acids after oxidation of the diketone established the positions of the original hydroxyl groups as indicated.

When O-methyllythranidine was treated with ethyl orthoformate and toluene-p-sulfonic acid an aminoacetal **76** was produced, and this in itself

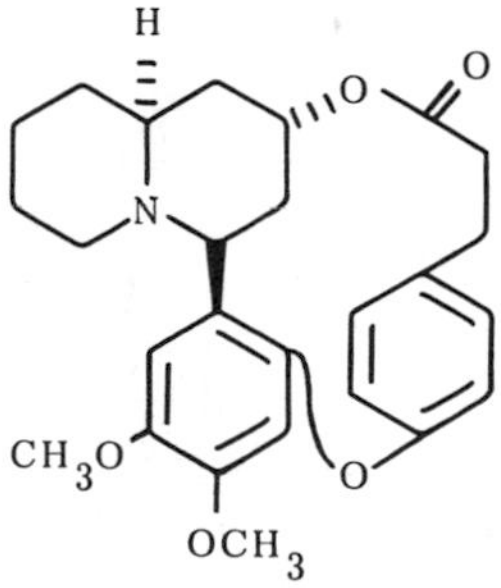

69 decaline

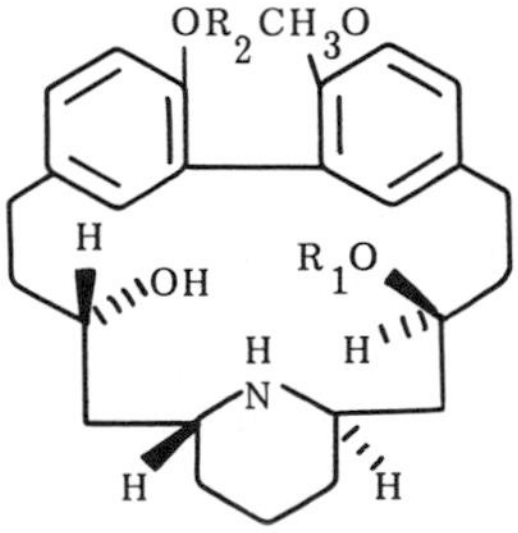

73

70 lythranine, R_1=CH$_3$CO, R_2=H

71 lythranidine, R_1=H, R_2=H

72 O-methyl lythranidine, R_1=H, R_2=CH$_3$

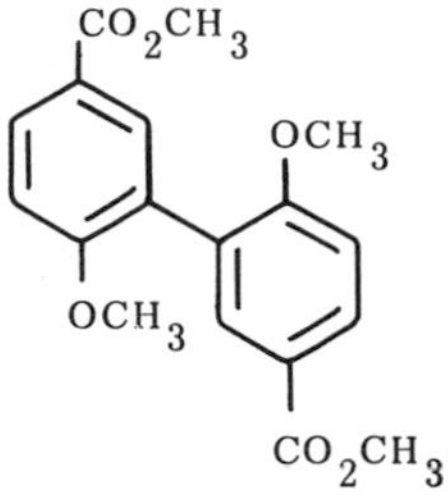

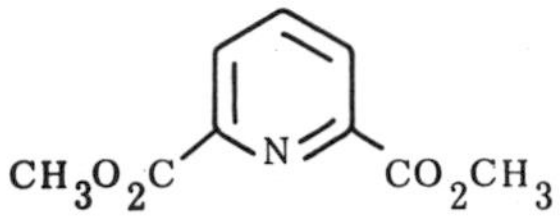

74

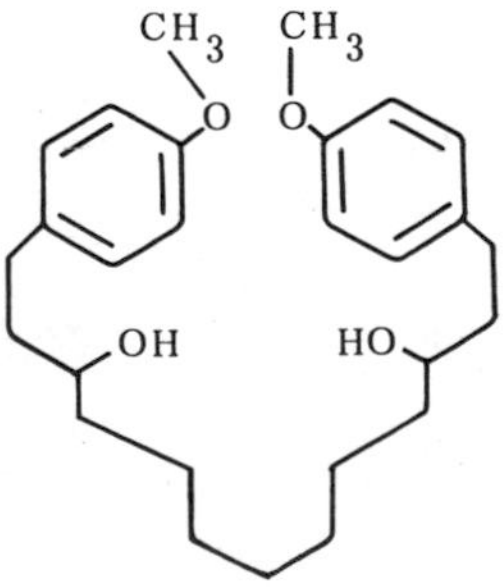

75

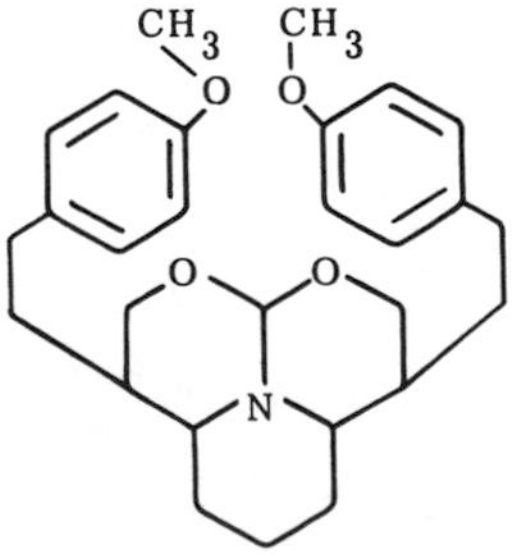

76

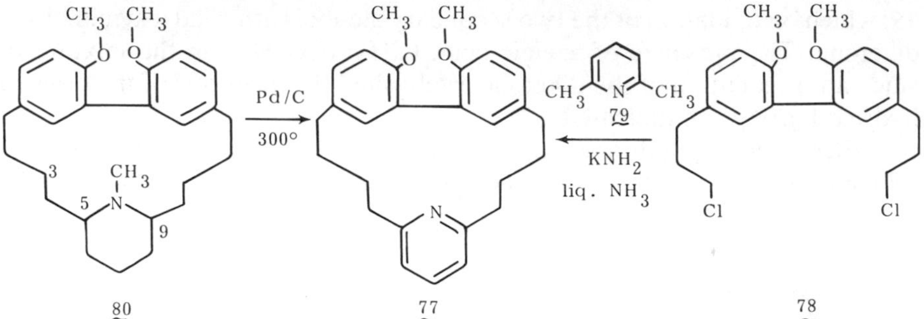

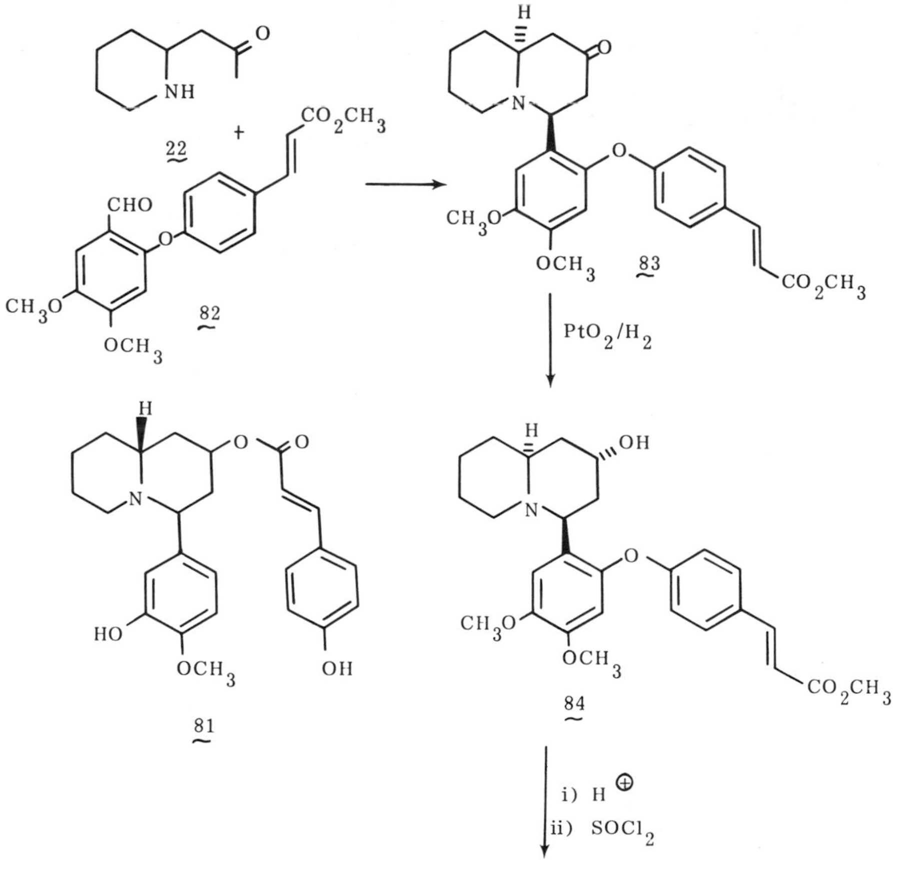

is evidence for the gross structure **72** for *O*-methyllythranidine. This skeleton was confirmed by synthesis of **77** from the dichloride **78**, by condensation with 2,6-dimethylpyridine (**79**) in the presence of potassium amide in liquid ammonia. The same compound, **77**, was also available by palladium–charcoal dehydrogenation of bisdeoxy-*O,N*-dimethyllythranidine (**80**). Catalytic hydrogenation of 2,6-disubstituted pyridines gives a *cis* product and reduction of **77** followed by N-methylation gave a product not identical to **80**, which must therefore have a 5,9-*trans* stereochemistry.

The diol **75** was optically inactive and should therefore have *trans* related hydroxy groups. In this way some of the stereochemical problems could be resolved, leaving only the relative locations of the hydroxy groups (phenolic and secondary) and the relationships of H-3 to H-5 and H-9 to H-11 to be determined. Although this problem was solved by X-ray analysis, perhaps you can think of a chemical way in which this could have been achieved. In fact the alcoholic and phenolic groups are on the same side of the molecule and lythranine has the structure **70**.

In determining the absolute configuration, a positive Cotton effect at 232 nm was observed for the π-π* transition. From previous work this indicates the configuration for the diphenyl system to be as shown.

There have been several synthetic approaches reported for the various Lythraceae alkaloids, but although derivatives such as **81** could be obtained with relative ease, oxidative coupling has not been very successful. One compound that has been synthesized is decaline (**69**), and in each of the two available routes the lactone ring was produced as the last step. One sequence involved as a key step the condensation of isopelletierine (**22**) with the diphenyl ether aldehyde (**82**) to give the *trans*-fused quinolizidine **83**, having the C-4 substituent equatorial. Catalytic reduction (PtO_2/H_2) gave a 4:1 mixture of axial:equatorial alcohols which could be separated, hydrolyzed, and the axial isomer **84** cyclized to racemic decalinc (**69**).

LITERATURE

Fujita, E., and K. Fuji, in K. Wiesner (Ed.), *International Review of Science, Organic Chemistry*, Ser. 2, Vol. 9 *Alkaloids*, Butterworths, London, 1976, p. 119.

Wrobel, J. T., and W. B. Golebiewski, *Tetrahedron Lett.* 4293 (1973).

4.7 QUINOLIZIDINE ALKALOIDS

Numerous alkaloids contain the quinolizidine nucleus (**85**). Many of them however belong to the isoquinoline and indole groups, and these are discussed in Chapters 8 and 9 respectively. Two other important alkaloid groups containing the quinolizidine nucleus are the lupin and the lycopodium alkaloids. The latter are discussed in a subsequent section of this chapter. Initially we will discuss the simple quinolizidine alkaloids, which are also sometimes known as the lupin alkaloids.

These compounds occur in broom (*Cytisus scoparius* Lind.), laburnum (*Laburnum anagyroides* Medic.), and lupins (*Lupinus* sp.) and are widespread in certain tribes of the subfamily Lotoideae of the Leguminosae.

The alkaloids are readily organized in terms of the complexity of the ring system, beginning with the simple bicyclic alkaloids such as lupinine (7) and proceeding through to the tricyclic alkaloids, such as angustifoline (86), and to the various tetracyclic alkaloids, such as (−)-sparteine (9) and matrine (10). The structural variant cytisine (87) will then be discussed, followed by a quite interesting group of alkaloids from *Ormosia* species.

4.7.1 Bicyclic Quinolizidine Alkaloids

Lupinine, the simplest quinolizidine alkaloid, was isolated from the yellow lupin (*L. luteus* L.) in 1835. The molecular formula was deduced in 1902 and a structure suggested in 1928. It is a strong tertiary base, forms a number of crystalline salts, and by its easy oxidation with chromic acid to a carboxylic acid contains a primary alcohol. The skeleton and location of the hydroxymethyl group were obtained by degradation *via* 88 to 3-methylpyridine-2-carboxylic acid (89). Lupinine therefore has the gross structure 7.

Several syntheses of lupinine have been reported since the initial efforts

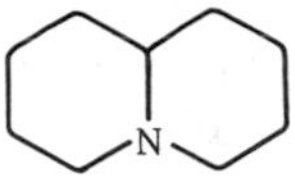

85

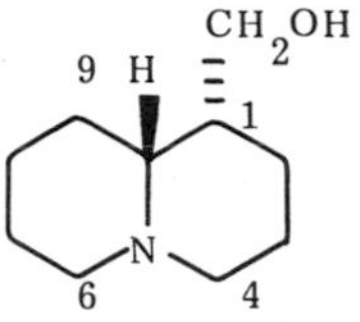

7 (−)-lupinine

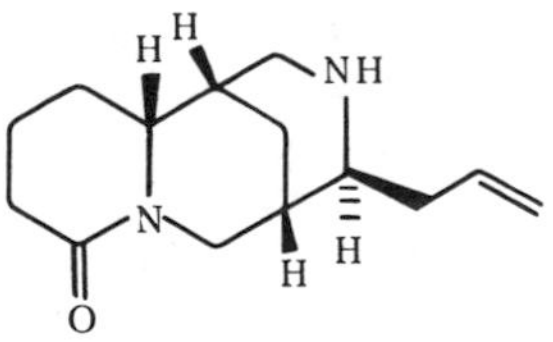

86 angustifoline

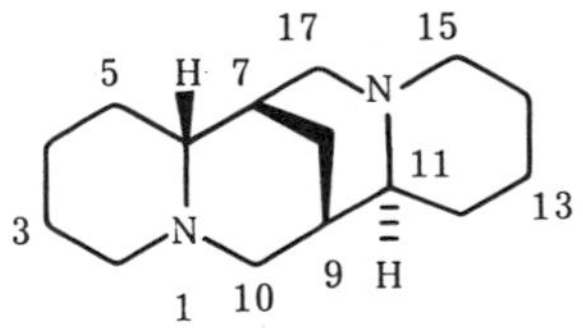

9 (−)-sparteine

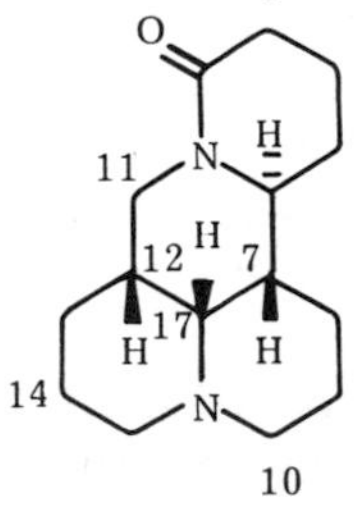

10 matrine

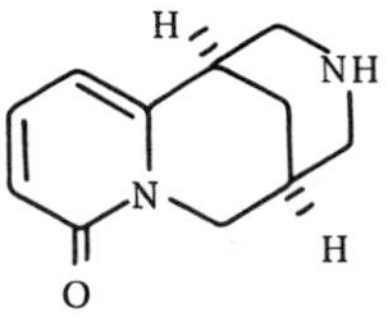

87 (−)-cytisine

of Clemo *et al.* in 1937. A "biomimetic" synthesis of epilupinine (**90**) was described by van Tamelen and Foltz, who prepared the diol **91** from ethyl *N*-benzyliminodivalerate (**92**) as shown. Oxidation of **91** with periodic acid at pH 5 and room temperature gave an intermediate dialdehyde **93** which underwent double internal cyclization. Workup with addition of a hydride reduction agent gave epilupinine (**90**) (Scheme 3).

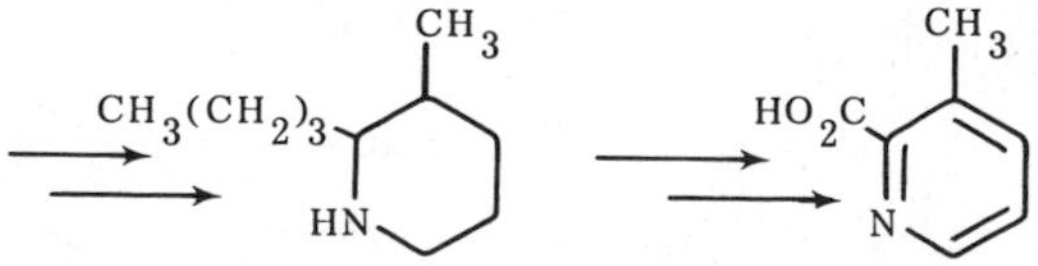

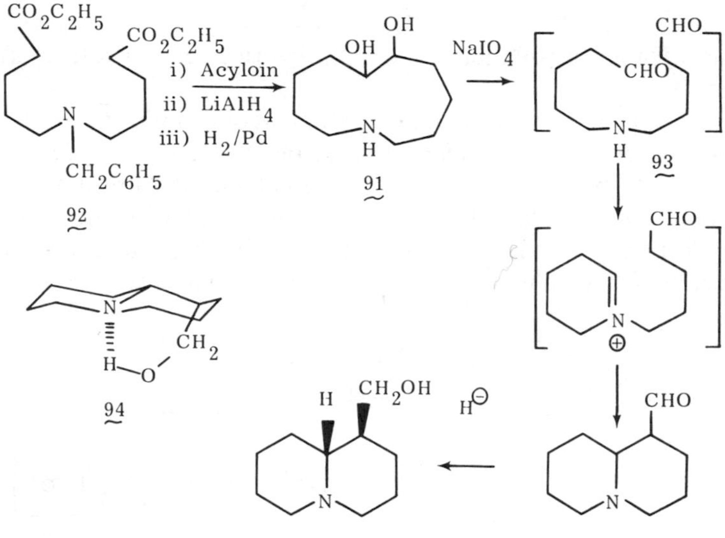

90   epilupinine

Scheme 3   Synthesis of epilupinine (**90**).

When (−)-lupinine (**7**) is heated with sodium in dry benzene, (+)-epi-lupinine (**90**) is produced in which the hydroxymethyl group is now in the thermodynamically more stable equatorial orientation. A preferred method involves photochemical isomerization using acetophenone as sensitizer. Lupinine and epilupinine are readily distinguished by their IR spectra. Thus lupinine, having an axial hydroxymethyl group, displays a broad hydroxyl band centered at 3270 cm^{-1} due to **94**, whereas epilupinine shows only a sharp band at 3580 cm^{-1} since no internal hydrogen bonding can occur.

An alternative synthesis of racemic epilupinine (**90**) involves the Michael addition of acrylonitrile to 2-cyanomethylenepiperidine (**95**) to give 1-cyano-4-oxo-1,10-dehydroquinolizidine (**96**), followed by reduction to **97**. Hydrol-

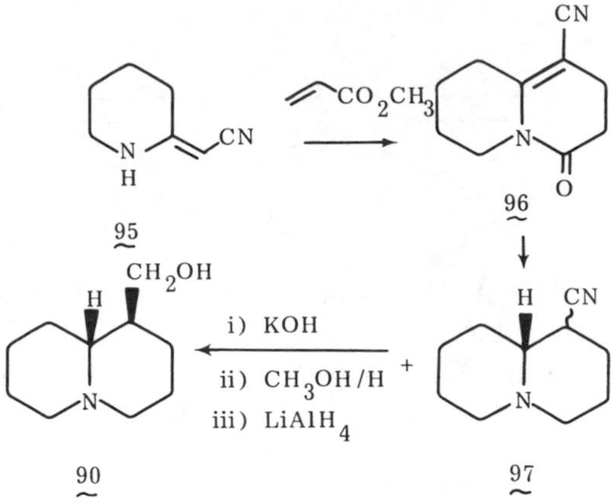

Scheme 4

ysis, esterification, and lithium aluminum hydride reduction then affords **90** (Scheme 4).

A novel approach to the lupinine system involving nitrone cyclo addition has also been described. Condensation of the N-oxide of 1,2-dehydropiperidine (**98**) with the ester **99** gave the internal salt **100** presumably *via* the intermediate isoxazolidine **101**. Reduction with zinc and acetic acid followed by dehydration gave the α,β-unsaturated ester **102**, which on successive catalytic and lithium aluminum hydride reduction gave racemic lupinine (**7**).

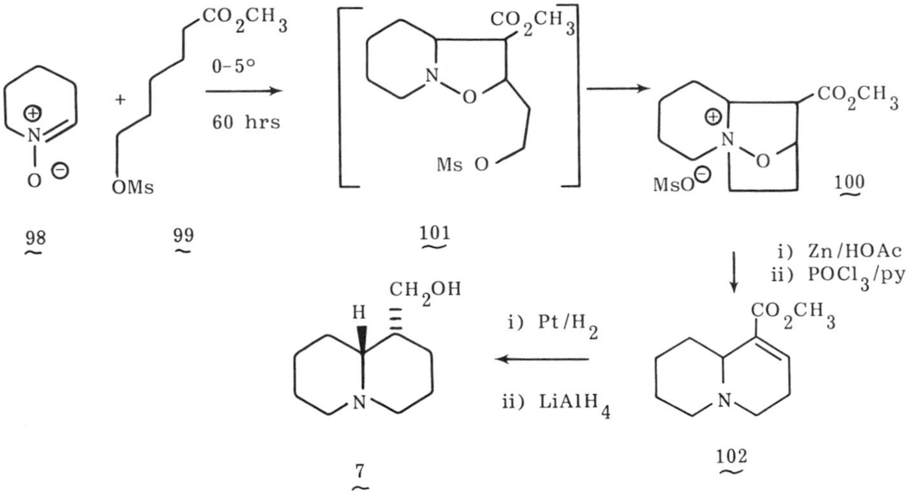

4.7.2 Tetracyclic Quinolizidine Alkaloids

Sparteine (**9**) is one of a group of closely related tetracyclic quinolizidine alkaloids, and some other members are lupanine (**103**) and anagyrine (**104**). The latter alkaloid was actually crucial in determining the skeleton of this group, which was deduced by degradation. Selective reduction can afford both (−)-lupanine (103) and (+)-sparteine (105).

Anagyrine (104) cooccurs with cytisine (87) in gorse (*Ulex europaeus* L.) and has a number of similar properties. Each form only mono salts since the second nitrogen is neutral, and each gives a red color with ferric chloride. Anagyrine (104) has been synthesized by condensing 2-(α-pyridyl)allylmalonic acid (106) with 1,2-dehydropiperidine (29) in hot ethanol. The product 107 is thought to arise in two steps *via* 108. A standard sequence then produces the quaternary bromide (109), which is converted to the pyridone anagyrine (104) with alkaline potassium ferricyanide (Scheme 5).

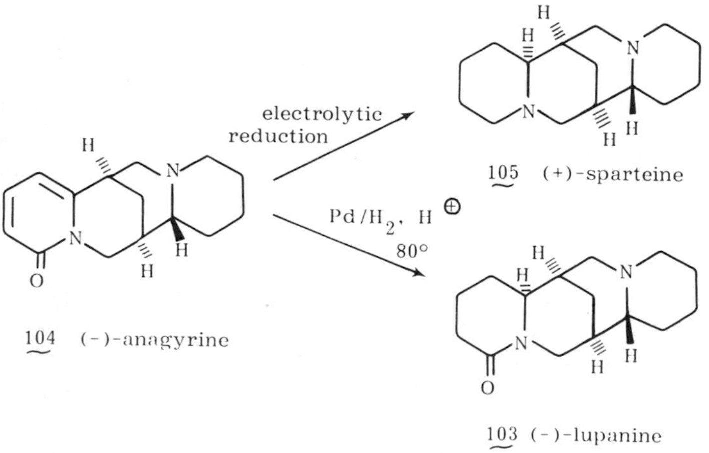

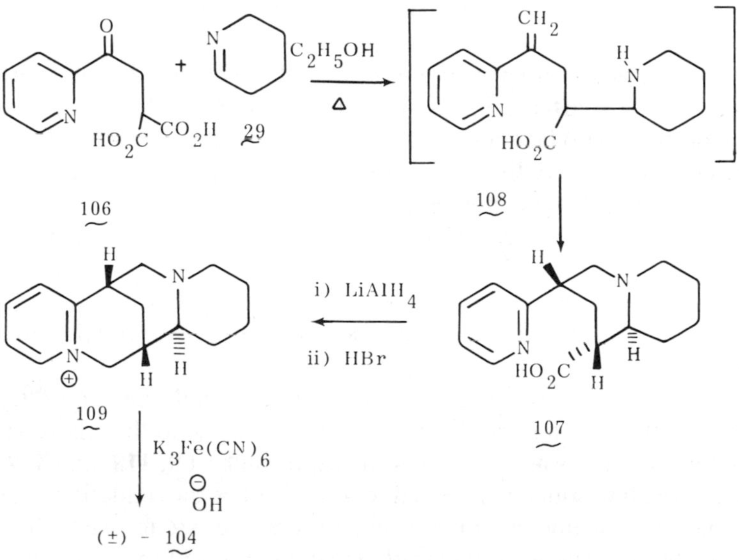

Scheme 5   Synthesis of anagyrine (104).

An alternative synthesis of anagyrine (**104**) begins with the condensation of methyl-2-pyridylacetate (**110**) with triethyl orthoformate and acetic anhydride to afford the quinolizinone (**111**). Hydrogenation and equilibration gave the diequatorial lactam acid **112**, which could be reduced and brominated to the quaternary bromide **109** (Scheme 6).

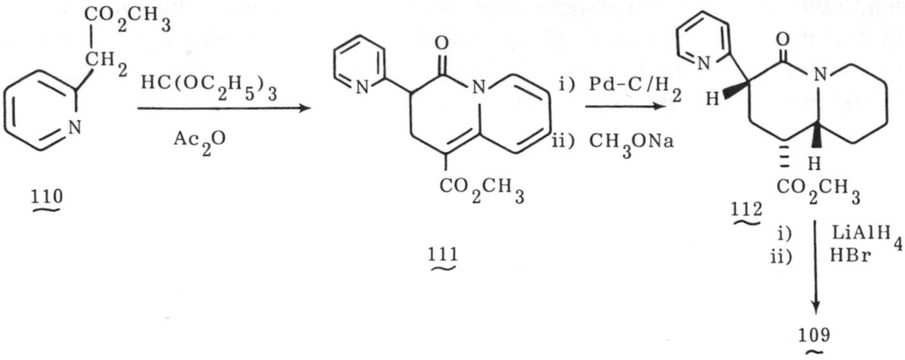

Scheme 6

Sparteine is a common alkaloid of the lupin family occurring in *Cytisus, Lupinus, Sarothamnus*, and *Spartium* species in both (+)- and (−)-forms. First isolated in 1851, it is colorless oil whose structure was first proposed in 1933. Commercially the alkaloid may be obtained from the common broom, *C. scoparius*.

Each nitrogen of sparteine is basic and two stereoisomerically different monomethiodides can be obtained. Complete Hofmann degradation (six steps), reducing catalytically at each step, eventually gave 6,8-dimethyltridecane (**113**).

Several syntheses of sparteine or close relatives have been reported and the symmetry of sparteine poses some interesting problems. For example, when methyl-2-pyridylacetate (**110**) was condensed with formaldehyde, the product **114** could be hydrogenated over copper chromite directly to a mixture of sparteine stereoisomers **115**.

An alternative approach involves the mercuric acetate oxidation of the symmetrical dipiperidino ketone **116** to give 8-oxosparteine (**117**) directly. The ketone **116** is produced by the condensation of piperidine, acetone, and formaldehyde. Wolff–Kishner reduction of **117** gave *dl*-sparteine (**115**) (Scheme 7).

With four asymmetric centers, sparteine potentially has 16 optically active forms, but the C-8 methylene bridge must be *cis* so that there are three racemic forms, one form of which is shown in each of **9, 118**, and **119**. These racemates are all produced in the reduction of **114** or the oxidative cyclization of **116**, and at least one example of each racemate occurs naturally.

A new stereospecific synthesis of sparteine devised by Bohlmann begins

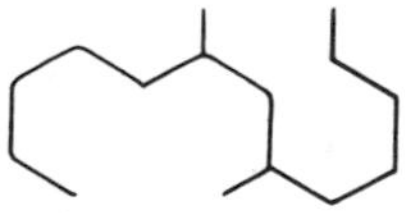

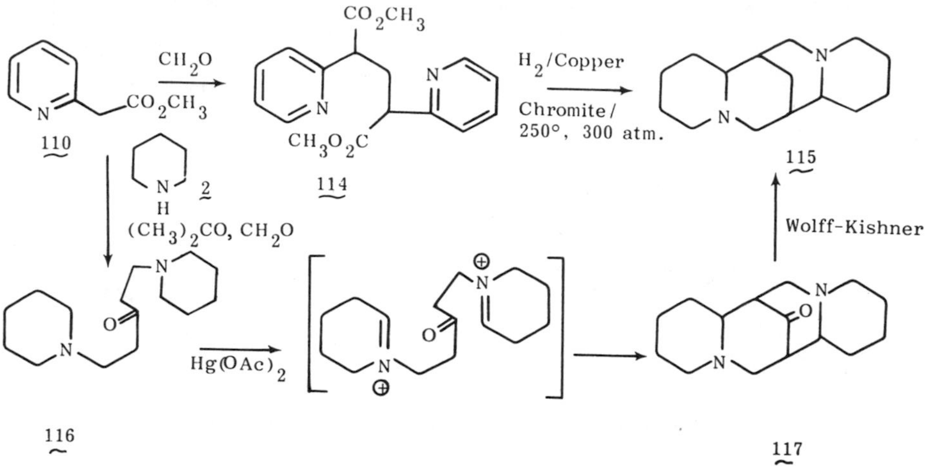

Scheme 7 Synthesis of sparteine (**115**).

with the bromolactam **120**, which can be condensed with piperidone to the dilactam **121**. Reduction with diisobutyl aluminum hydride gave the enamine imine **122**, which cyclized spontaneously to the imminium ion **123**. Reduction with sodium borohydride yielded racemic sparteine (**124**) (Scheme 8).

There is some confusion over the structure of β-isosparteine, which was originally assigned the A/B-*cis*-C,D-*cis* structure **119** because no Bohlmann bands were observed in the IR spectrum, and the proton NMR spectrum was in apparent agreement with these data. It is now reported that β-isosparteine does show Bohlmann bands and the revised stereochemistry **125** has been suggested. The obvious problems with the NMR spectral data have not been resolved.

Microbiological oxidation of (−)-sparteine (**9**) by the basidiomycete *Trametes gibbosa* gave 17-hydroxysparteine (**126**) of uncertain stereochemistry.

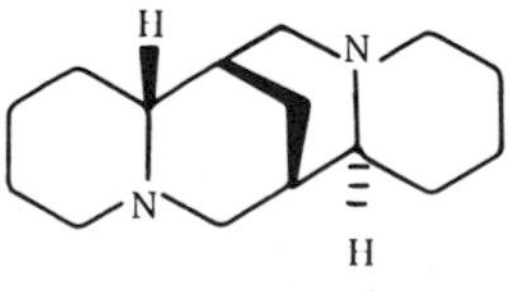

9 (−)-sparteine 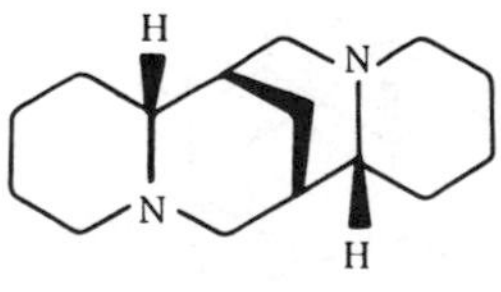   118 α-isosparteine 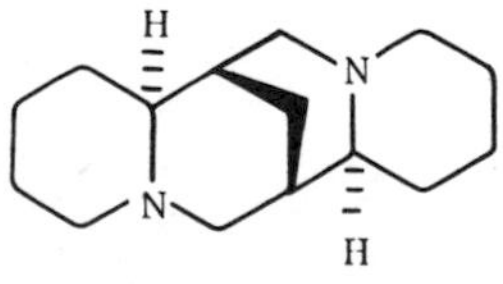   119 β-isosparteine

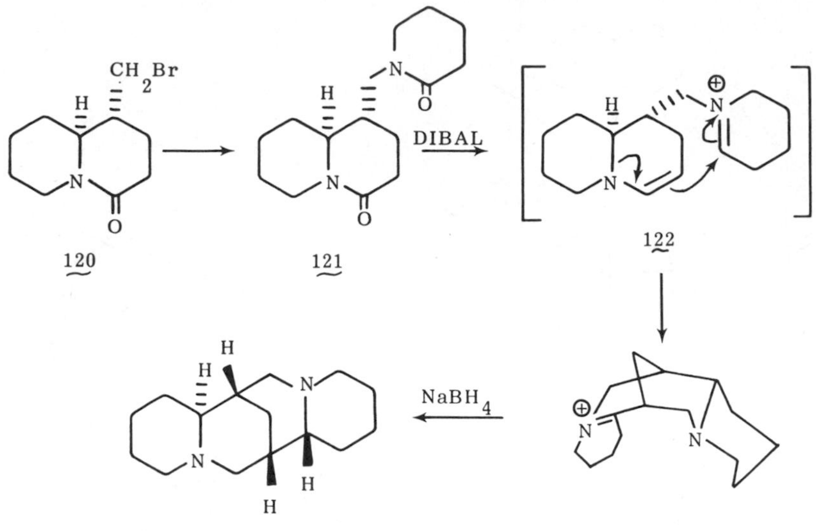

124 (±)-sparteine 123

Scheme 8 Bohlmann synthesis of (±)-sparteine.

Sparteine is an exceptionally good bidentate ligand for magnesium, and with magnesium dialkyls gives complexes such as **127**. This has been suggested as an explanation for the partial asymmetric synthesis of allenes in the presence of (−)-sparteine (**9**).

Sparteine is also capable of inducing asymmetry in the Reformatsky reaction. For example the product **128** from the reaction of benzaldehyde, zinc, sparteine, and ethyl bromoacetate was obtained in 95 ± 3% optical

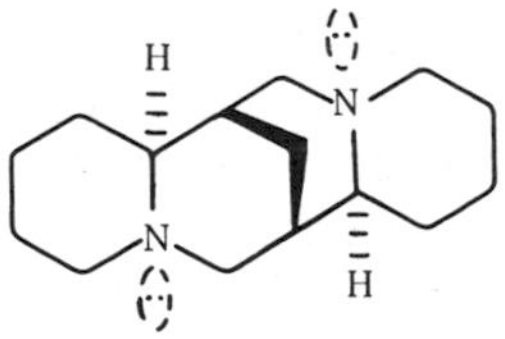

old 119

(+)-β-isosparteine

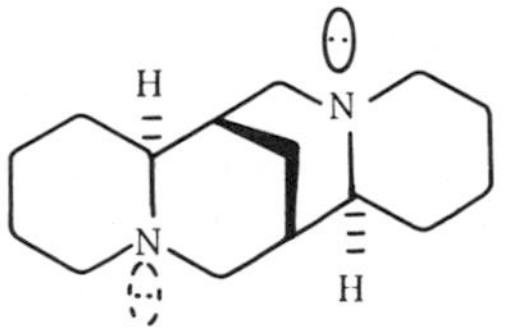

new 125

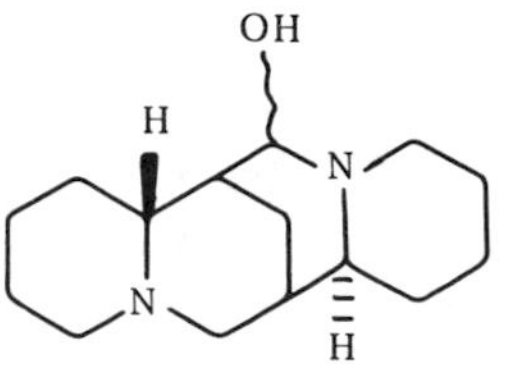

126 17-hydroxysparteine

127

129   128

yield. This has been explained in terms of a complex **129** in which the zinc enolate of the bromoester is coordinated with sparteine and benzaldehyde. The preferred transition state places the bulky aryl group away from the methylene group at C-15, and an S-configuration results in the product.

Both Dioscorides and Pliny comment on the action of spartium as a diuretic, and this was the principal use for centuries. Subsequently, sparteine sulfate was used for a period in the treatment of certain cardiac problems, a pharmacologic action not realized until the pure compound became available. The sulfate salt is also used as an oxytocic and acts by stimulation of hypotonic uterine contractions.

In general pharmacologic terms sparteine resembles coniine but is far less toxic. It acts to paralyze motor nerve endings and sympathetic ganglia. Some sparteine derivatives substantially increase the refractory period before fibrillation of a guinea pig atrium preparation, and the adenylate derivative is used in Europe in the treatment of cardiac insufficiencies.

A related isomeric tetracyclic quinolizidine alkaloid is matrine (**10**), which occurs in several *Sophora* species. The main structural elements were deduced by degradative experiments, but more definitive structural evidence came from a synthesis of matridine (**130**), a reduction product of matrine (**10**).

Although in previous discussions we have encountered 1,2-dehydropiperidine (**29**), and implied its ready use as a monomer, in actuality it exists as the trimer **131**. The monomer is available in aqueous solution only as part of an equilibrium process. Preparation of **131** may be carried out by dehydrohalogenation of N-chloropiperidine (**132**), but the product is a mixture of **131** together with several other trimeric species. Schöpf found that heating

133, another form of the trimer, with ammonium chloride gave a further isomer **134** in which the fragile N—C—N linkage could be reductively cleaved to give an amine **135**. Nitrosation and bromination with HBr gave the tetracyclic quinolizidine **130**, which could be resolved and shown to be identical to the reduction product of matrine, matridine (Scheme 9).

Any synthesis of matrine (**10**) must also allow for allomatrine (**136**), which has the *dl*-chair equatorial possibility, and is the more thermodynamically stable isomer.

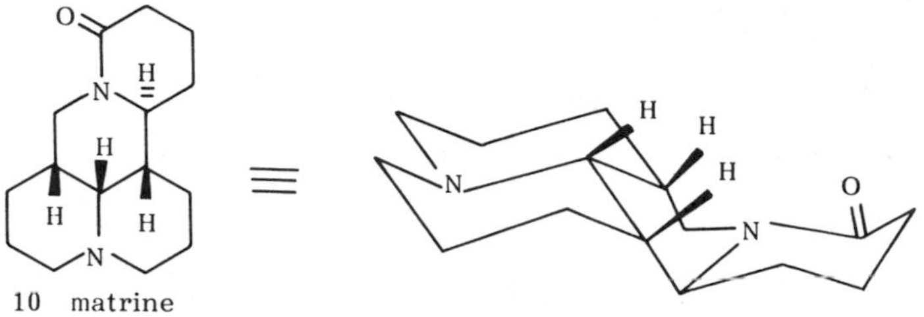

10 matrine

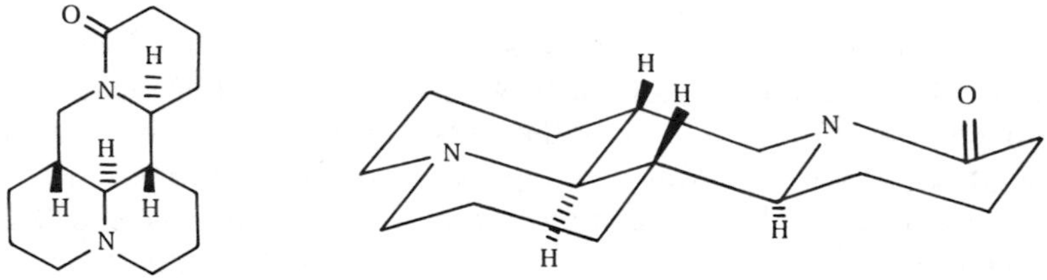

136 allomatrine

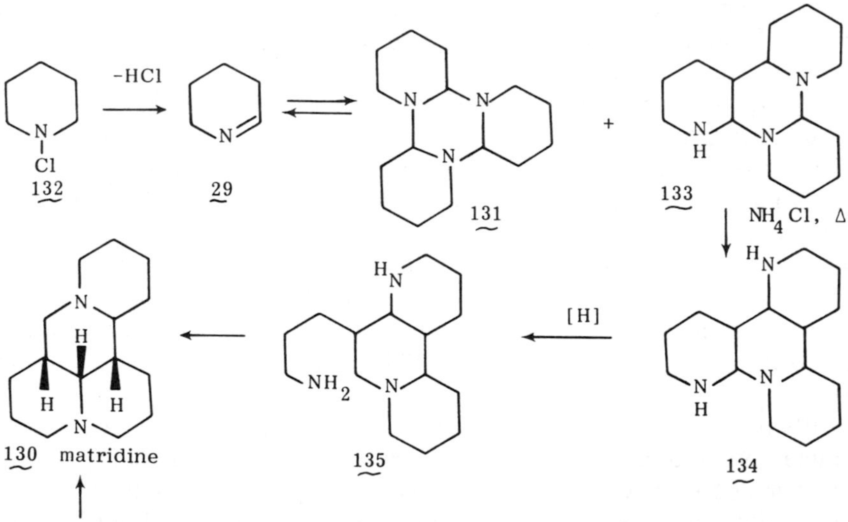

Scheme 9 Synthesis of matridine (**130**).

It is therefore somewhat surprising to find that reduction of the dinitrile **137** gave matridine (**130**), and not the more stable isomer. In a similar way, condensation of 2,6-dioxoquinolizidine (**138**) with acrylonitrile gave **139**, which on catalytic hydrogenation gave *dl*-matrine (**10**).

137   130

138   139   10 dl-matrine

4.7.3 Cytisine

The laburnum tree *Cytisus laburnum* L. is a common ornamental throughout Europe, yet all parts of the tree are poisonous due to the presence of the interesting tricyclic alkaloid cytisine (**87**). The alkaloid also occurs in *Anagyris, Baptisia, Genista, Sophora*, and *Thermopsis* sp., often with sparteine-type derivatives. Commercially, cytisine may be obtained from *Thermopsis dolichocarpa* V. Nikitin. Cytisine is a crystalline (m.p. 155°C) secondary amine (pK_a 8.2) which forms a monoacetyl derivative and contains an α-pyridone ring. Evidence for the structure came from a systematic Hofmann degradation and was subsequently confirmed by a synthesis closely akin to that used for anagyrine (**104**). Michael addition of diethyl malonate to the vinyl pyridine **140** gave **141**, which after hydrolysis to the diacid was condensed with benzylamine and formaldehyde to afford **142**. Lithium aluminum hydride reduction followed by bromination gave the pyridinium salt **143**. Alkaline ferricyanide introduced the pyridone, and the benzyl group was removed by acid to give *dl*-cytisine (**87**) (Scheme 10). As with sparteine, the bridge can only have a *cis* stereochemistry, so that there is only one racemate of cytisine.

Cytisine is highly toxic to both humans and livestock, and numerous poisonings have been reported. Pharmacologic responses include nausea, convulsions, and death by respiratory failure.

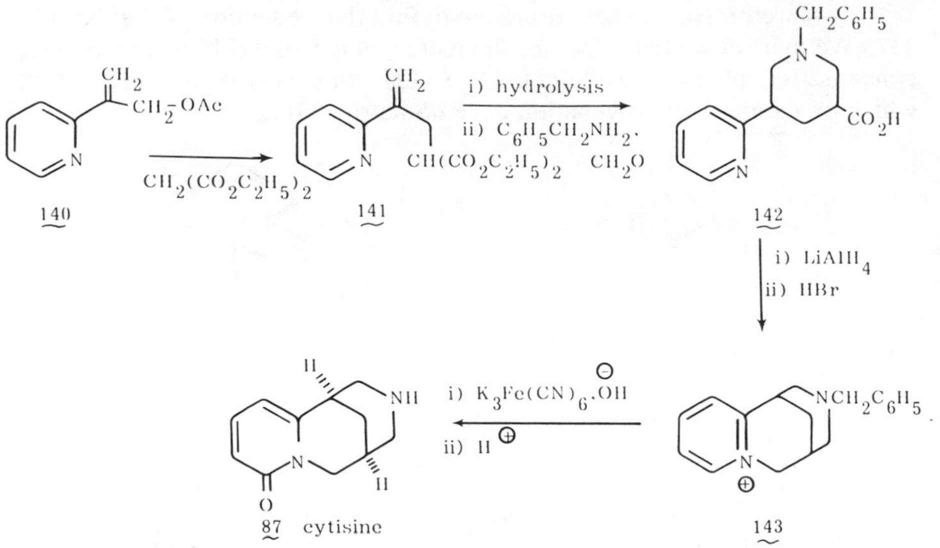

Scheme 10   Synthesis of cytisine (**87**).

Grazing by pregnant cows on certain members of the lupine genus has been shown to lead to congenital deformities known as the "crooked calf disease."

In 1960 it was reported that cytisine (**87**) was teratogenic in rabbits, but analysis of plants known to be responsible for the defects in calves indicated that anagyrine (**104**) was closely associated with the teratogenicity.

In recent years several interesting quinolizidine alkaloids have been isolated, for example cineverine (**144**) from *Genista cinera* L., tinctorine (**145**) from *Genista tinctoria* L., lamproboline (**146**) from *Lamprolobium fruti-*

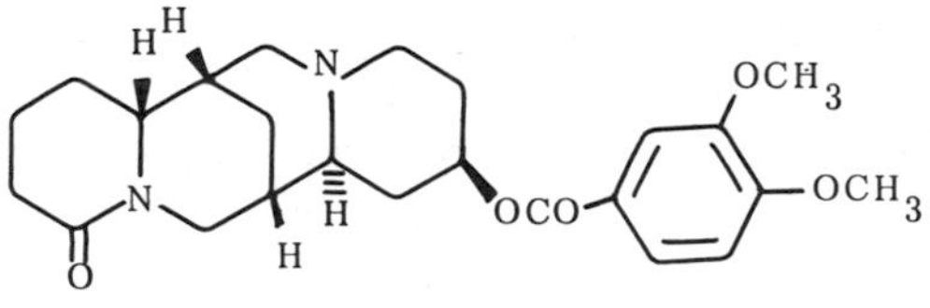

144   cineverine 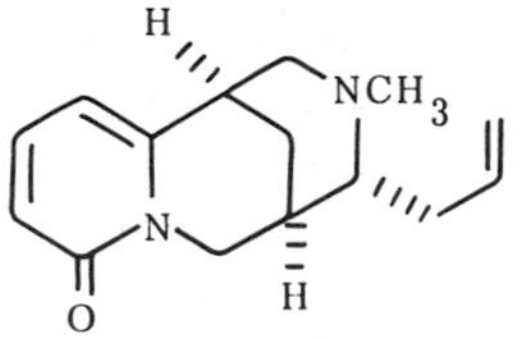   145   (−)-tinctorine

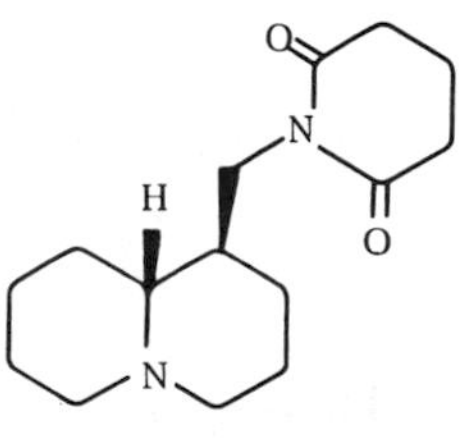

146   lamproboline

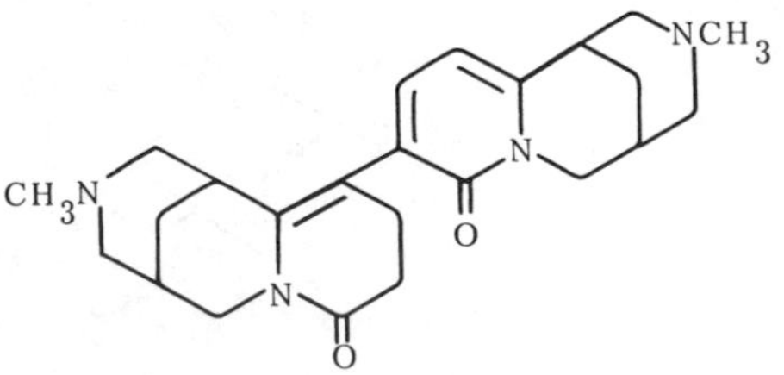

147 dimethamine

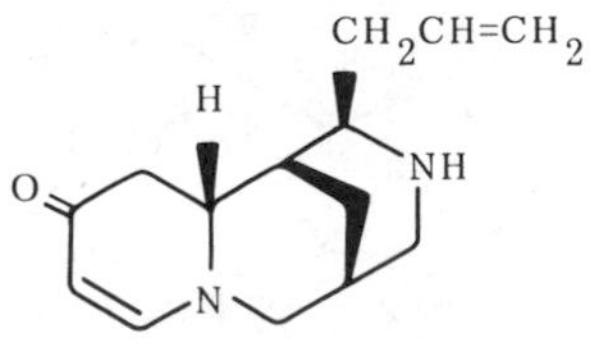

148 dehydroalbine

cosum Benth., the dimeric alkaloid dimethamine (**147**) from *Thermopsis alternariflora*, and dehydroalbine (**148**) from *Lupinus albus* L.

4.7.4 Spectral Properties

In general terms the proton NMR spectra of the quinolizidine alkaloids are difficult to interpret. However, deuterium-labeling studies have established that the chemical shifts of protons α to the quinolizidine nitrogen are highly dependent on the stereochemistry. Thus the equatorial protons are deshielded to 2.80 ppm and the axial protons shielded to ~2.0 ppm as shown in **149**.

When an amide group is present, the α-equatorial proton is even more deshielded, as shown for the quinolizidinone **150**.

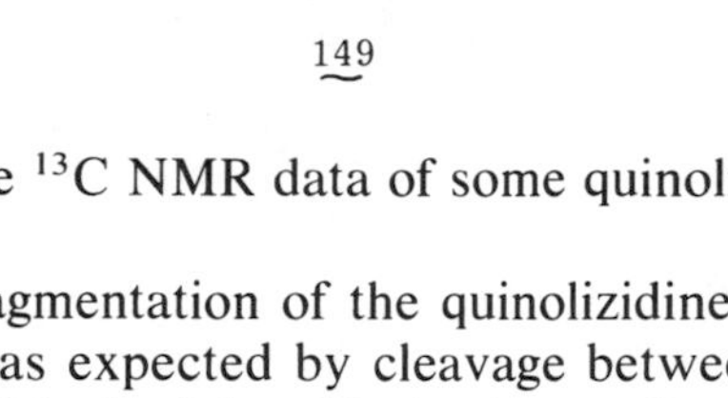

149

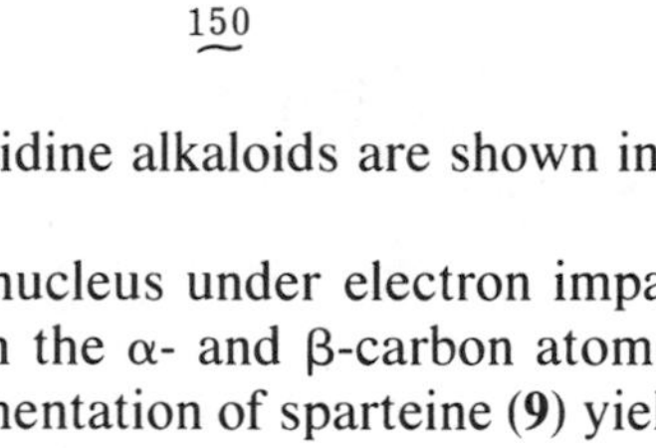

150

The ^{13}C NMR data of some quinolizidine alkaloids are shown in Figure 1.

Fragmentation of the quinolizidine nucleus under electron impact proceeds as expected by cleavage between the α- and β-carbon atoms as indicated for lupinine (**7**). Analogous fragmentation of sparteine (**9**) yields ions at *m/e* 98 and 136 (Scheme 11).

4.7.5 *Ormosia* Alkaloids

The genus *Ormosia*, comprising approximately 100 species, is native to the tropical Americas and has proved to be an interesting source of alkaloids. The structures are stereochemically complex and demanding. Most of these alkaloids contain 20 carbon atoms and three nitrogens and often cooccur

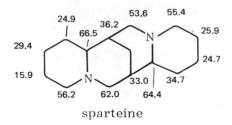

sparteine

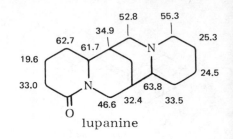

lupanine

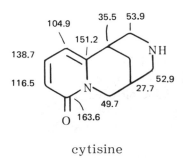

cytisine

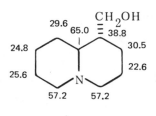

lupinine

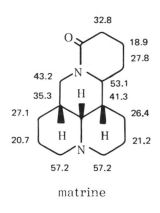

matrine

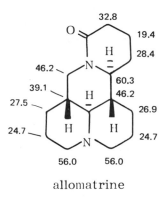

allomatrine

Figure 1 ^{13}C NMR spectral data of some representative quinolizidine alkaloids. [From F. Bohlmann and R. Zeisberg, *Chem. Ber.* **108,** 1043 (1975).]

with other quinolizidine alkaloids. The skeleton and numbering system is typified by (+)-ormosanine (**151**). Common modifications are a 16, 17-double bond and an N—C-22 bond [e.g. ormojanine (**152**)]. These alkaloids have also been obtained from *Piptanthus, Podopetalum,* and *Templetonia* species in the Leguminosae.

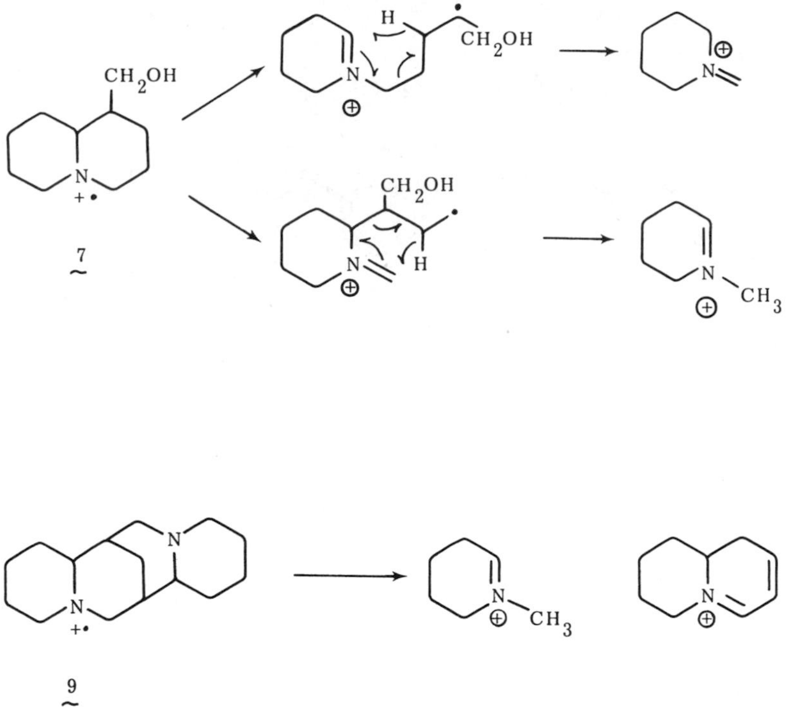

Scheme 11 Mass spectral fragmentation of lupinine (**7**) and sparteine (**9**).

The synthesis of ormosanine (**151**) employs **153** as a key intermediate, a compound whose formation was discussed previously. Alkylation with ethyl acrylate and sodium hydride afforded the ester **154**, which has a defined C-6 stereochemistry proven by subsequent conversions. Acid hydrolysis and alkylation of the ketone with ethyl acrylate and sodium hydride gave **155**. A bromination–dehydrobromination sequence gave an unsaturated ketone which could be condensed with ammonia to the keto dilactam **156**. Huang–Minlon reduction of the keto group results in epimerization at C-11 to give **157** (Wolff–Kishner reduction did not epimerize C-11). Catalytic

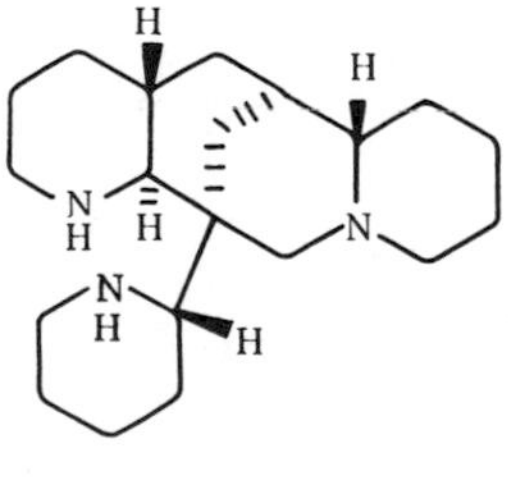

151   ormosanine 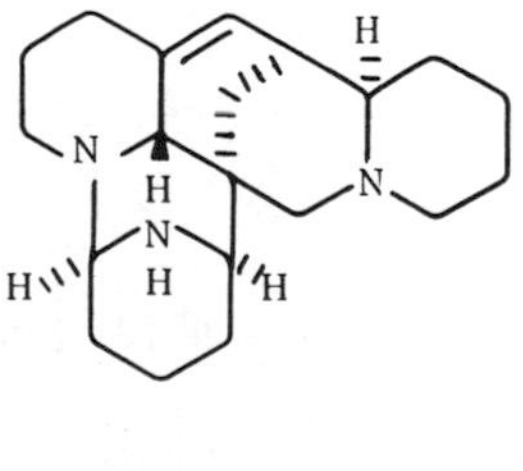   152   ormojanine

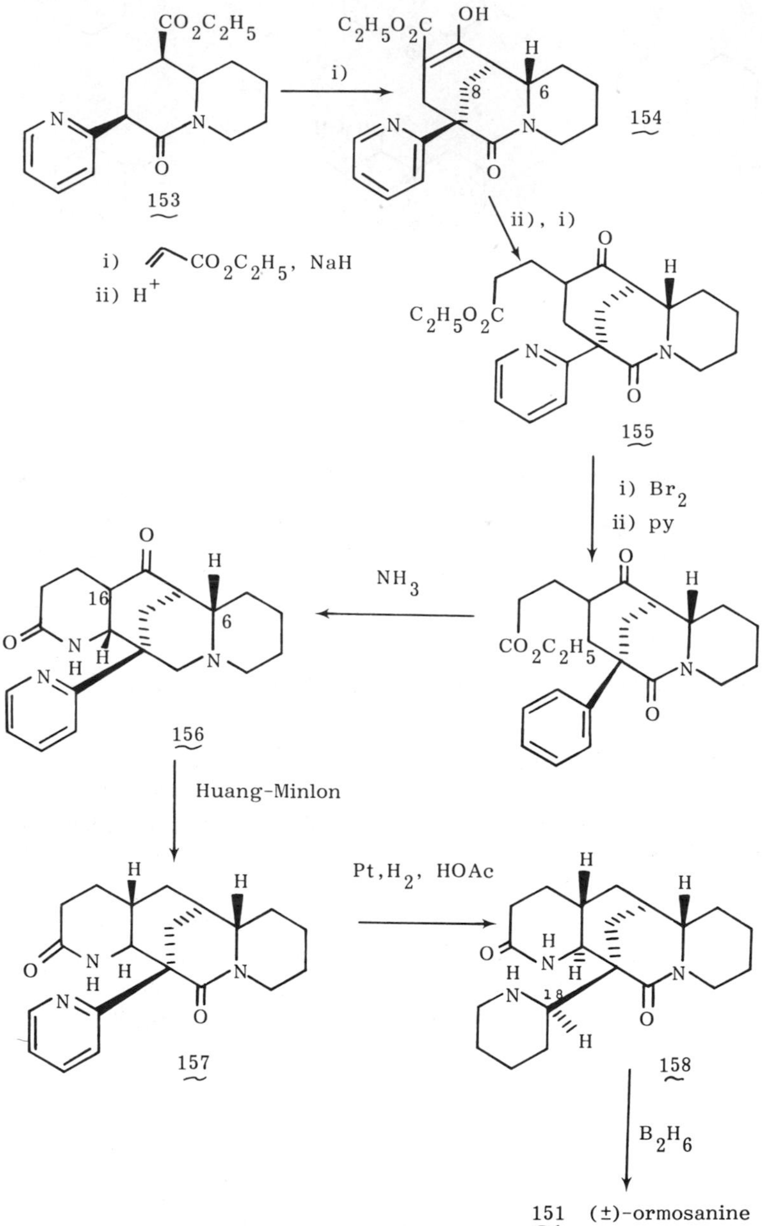

Scheme 12 Synthesis of (±)-ormosanine (**151**).

reduction in acetic acid gave mainly the dilactam **158**, and diborane reduction afforded racemic ormosanine (**151**) (Scheme 12).

The formation of the *Ormosia* alkaloids has not been investigated, but it does pose some interesting biogenetic problems, particularly since these alkaloids do co-occur with the lupin alkaloids. One hypothesis suggests an intermediate such as **159** derived from four lysine molecules. A biogenetic clue comes possibly from the isolation of pohakuline (**160**) from *Sophora chrysophylla* Seem. (Leguminosae).

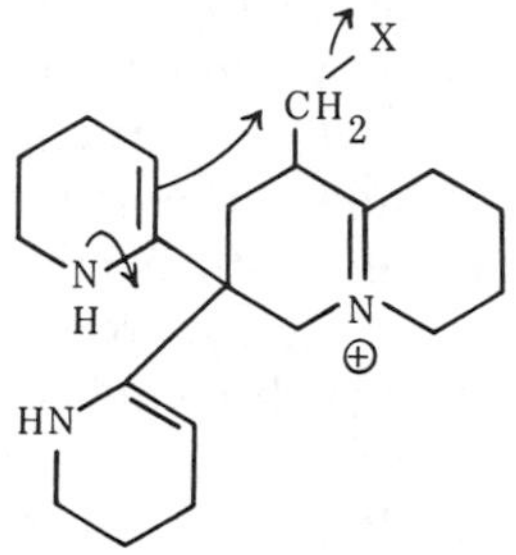

159

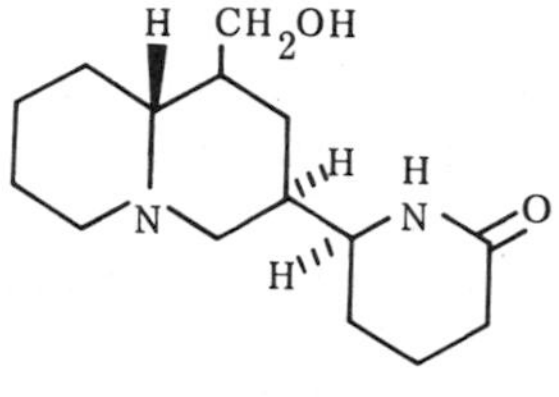

160   pohakuline

LITERATURE

Reviews: General

Saxton, J. E., *Alkaloids, London* **1**, 86 (1971).

Saxton, J. E., *Alkaloids, London* **2**, 79 (1972).

Saxton, J. E., *Alkaloids, London* **3**, 95 (1973).

Saxton, J. E., *Alkaloids, London* **4**, 104 (1974).

Saxton, J. E., *Alkaloids, London* **5**, 93 (1975).

Grundon, M. F., *Alkaloids, London* **6**, 90 (1976).

Grundon, M. F., *Alkaloids, London* **7**, 69 (1977).

Grundon, M. F., *Alkaloids, London* **8**, 66 (1978).

Lupin Alkaloids

Leonard, N. J., *Alkaloids NY* **3**, 119 (1953).

Leonard, N. J., *Alkaloids NY* **7**, 253 (1960).

Bohlmann, F., and D. Schumann, *Alkaloids NY* **9**, 175 (1967).

Ormosia *Alkaloids*

Valenta, Z., and H. J. Liu, in K. Wiesner (Ed.), *International Review of Science, Organic Chemistry*, Ser. 2, Vol. 9, *Alkaloids*, Butterworths, London 1976, p. 1.

4.8 *LYCOPODIUM* ALKALOIDS

The vascular cryptogams or club mosses are among the very lowest forms of plant life. Surprisingly though the club mosses, in particular the genus *Lycopodium*, are quite a prolific producer of alkaloids, and approximately 100 are now known.

4.8.1 Lycopodine and Annotinine

Bodeker in 1881 reported the first isolation of an alkaloid (probably lycopodine) from *L. complanatum*. Fifty-seven years passed before serious phytochemical work began, and it took another 19 years before a structure, that of annotinine, was deduced for any of these alkaloids. The structure of lycopodine (**12**) was deduced in 1960, and numerous phytochemical, synthetic, and biosynthetic studies have followed. Two skeleta are of importance. One of these is the hydrojulolidine system, which includes several structure types [e.g. lycopodine (**12**) and annotinine (**161**) and the majority of the alkaloids]; the second is a group of alkaloids having two nitrogen atoms, and cernuine (**162**) and lycodine (**163**) are good examples.

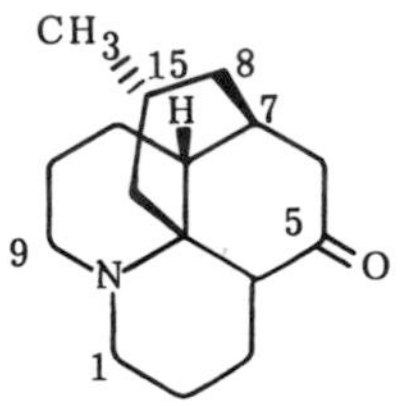

161   annotinine 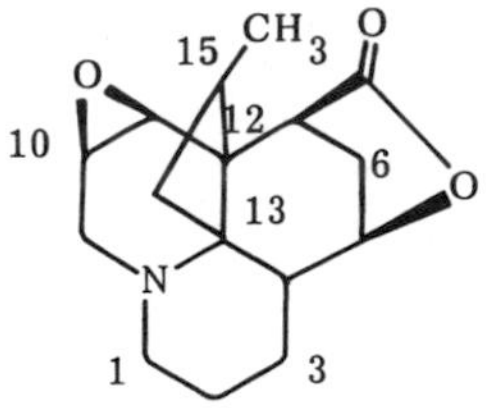   12   lycopodine

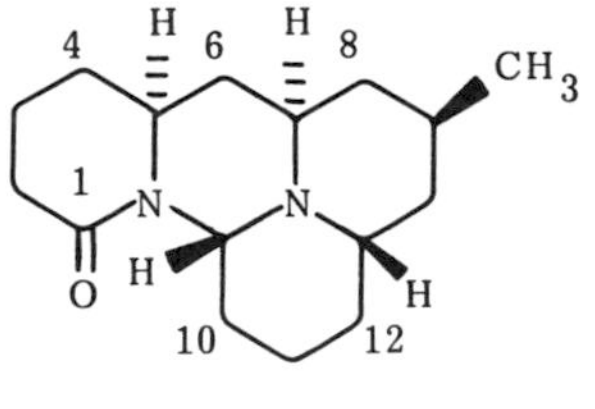

162   cernuine 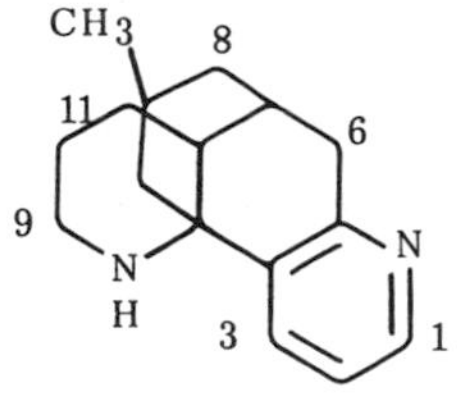   163   lycodine

Annotinine (**161**), the major alkaloid of *L. annotinum*, was first obtained in 1947 by Manske and Marion, who deduced the presence of an ether linkage and a lactone.

The nature of the parent ring system was deduced by careful dehydrogenation studies. Crucial in this work was the isolation and structure proof of **164**, which contains 13 of the 16 carbons. Considering the original molecular formula, and the evidence of a single methyl group, a part structure

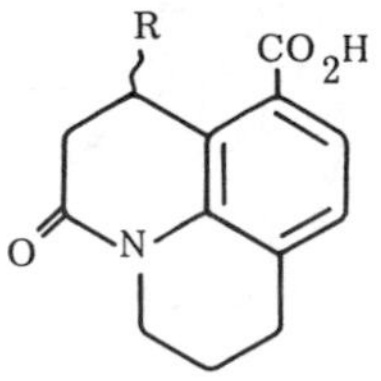

164   R=H
166   R=CH$_3$

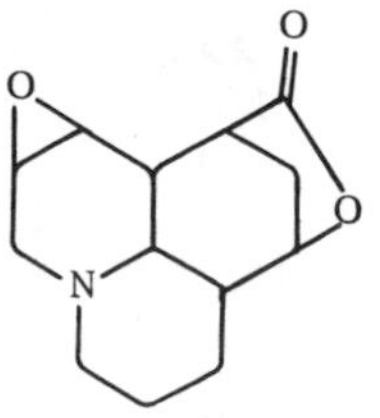

165
+   CH$_3$CHCH$_2$–
or (CH$_3$)$_2$C$<$

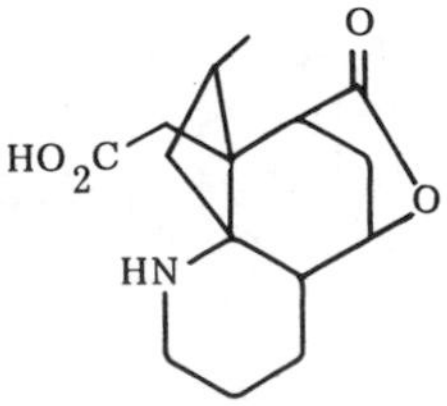

167

165 was proposed in which the three carbons act as a bridge between C-12 and C-13.

The structure, including most of the stereochemistry (except C-15), was finally assembled by Wiesner, who analyzed the structure and stereochemistry of 166, a product of the dehydrogenation of 167. Only the bridge shown can account for the formation of this structure. X-ray analysis confirmed the structure and stereochemistry of annotinine. Lycopodine (12) is the most abundant of the *Lycopodium* alkaloids and an interesting rearrangement takes place when lycopodine methiodide (168) is subjected to Hofmann degradation. Proton removal takes place not from C-8 or C-12 but rather from C-4, to give an anti-Bredt intermediate 169. The end product 170 may now be rationalized in terms of a transannular hydride shift from C-9 to C-13 followed by bond formation between C-9 and C-4 (Scheme 13).

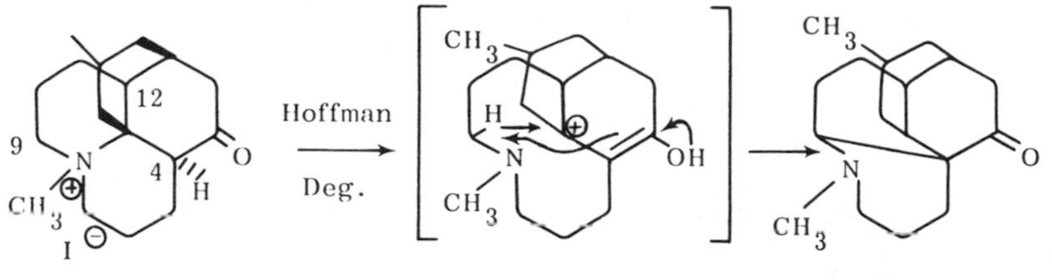

168   169   170

Scheme 13

The discrete structural and stereochemical differences between certain members of this group of compounds have led to some interesting interconversions. One example is the transformation of lycopodine (12) to annofoline (171) (Scheme 14). The strategy here was to remove the reactive basic nitrogen by conversion to an amide at C-9 with potassium permanganate; sodium borohydride reduction then gave the alcohol 172 stereospecifically. Oxidation with lead tetraacetate afforded the ether 173, which could be solvolyzed with boron trifluoride in acetic anhydride to the un-

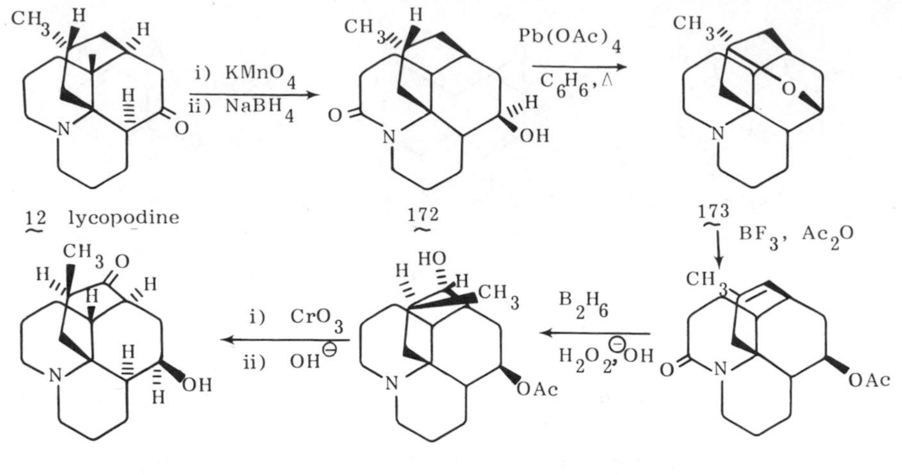

Scheme 14 Synthesis of annofoline (**171**).

saturated acetate **174**. Diborane/alkaline hydrogen peroxide carries out both reduction of the unwanted lactam and anti-Markovnikov hydration stereospecifically to give the alcohol **175**, which can be readily oxidized and hydrolyzed to annofoline (**171**).

A second example is the transformation of α-obscurine (**176**), an alkaloid in the lycodine series, to dihydrolycopodine (**177**), along what may well be biogenetic lines (see the discussion of biosynthesis) (Scheme 15). Demethylation of **176** followed by hydrolysis gave the amino keto acid **78**, which

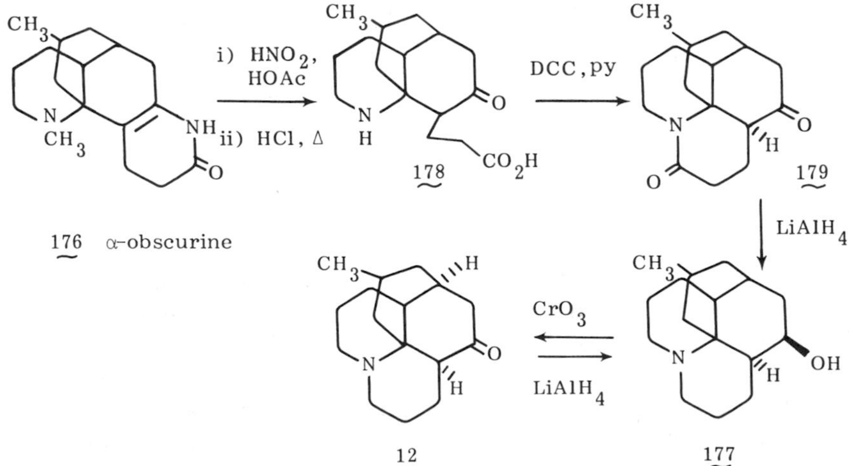

Scheme 15

could be cyclized to the amido ketone **179**. Lithium aluminum hydride reduction of this compound or lycopodine (**12**) gave the same amino alcohol, dihydrolycopodine (**177**).

There has been quite considerable success in the synthesis of these alkaloids in recent years, and both annotinine and lycopodine have been synthesized by quite individual approaches.

For example in Wiesner's synthesis of annotinine (**161**), the two-carbon bridge is introduced by photochemical addition of allene to the vinylogous amide. The many remaining steps are too complex for detailed discussion here, but some of the key intermediates are shown in Scheme 16.

Stork and Ayer and their respective co-workers completed syntheses of lycopodine (**12**) at the same time. Again the details are discussed in detail elsewhere, but several features of the synthetic strategy are significant. Conjugated addition of *m*-methoxybenzyl Grignard to the methyl cyclohexenone **180** at high dilution, in the presence of cuprous chloride, gave the expected enolate **181** in which the bulky benzyl group is equatorial and the methyl group axial. The next critical step is alkylation, which could be carried out with allyl bromide directly on **181** using HMPA as the solvent. Ketalization followed by an oxidative hydroboration sequence gave the keto acid **182** and condensation with ammonia afforded **183**. Internal cyclization to give **184** proceeded in 55% yield with phosphoric acid–formic acid. Potentially, two epimeric iminium species would be formed, but only one of these has the benzyl group axial in an all-chair conformation, and therefore

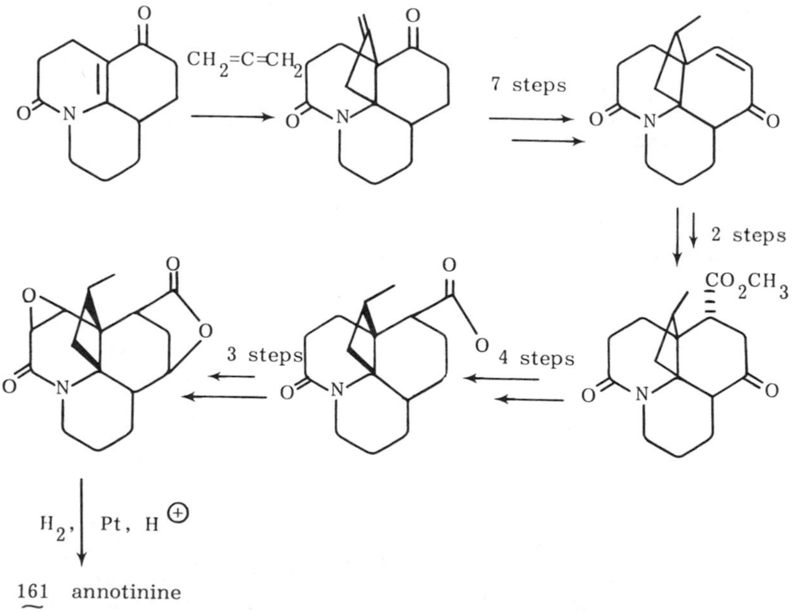

Scheme 16 Wiesner synthesis of annotinine (**161**).

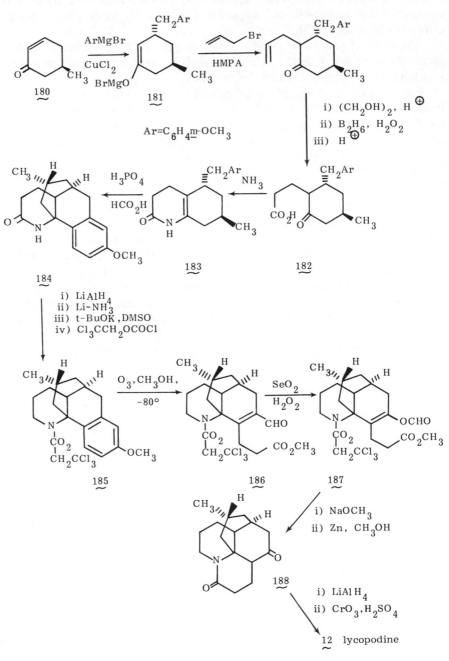

Scheme 17 Synthesis of lycopodine (**12**.

Stork predicted, correctly, that this reaction could be stereospecific. Lithium aluminum hydride reduction followed by Birch reduction and N-protection with an easily removable group afforded **185**. Ozonolysis yielded the aldehyde ester **186**, which would be converted to the formate ester **187** on reaction with selenium dioxide/hydrogen peroxide. Treatment with base and removal of the N-protecting group with zinc in methanol resulted in spontaneous cyclization to afford the amido ketone **188**, which is easily transformed into **12** (Scheme 17). A third synthesis of lycopodine (**12**) has recently been reported by Kim and co-workers.

The skeleta of many indole alkaloids, particularly those in the *Aspidosperma* and Eburna series, contain the hydrojulolidine skeleton or a closely related form (see Chapter 9), and many of the stereochemical problems involved overlap.

For example it is important to be able to distinguish between the alternative stereochemical possibilities **189**, **190**, and **191**. Bohlmann and Wenkert were independently successful in solving this problem.

The infrared spectra of **189** and **190** display quite characteristic absorption

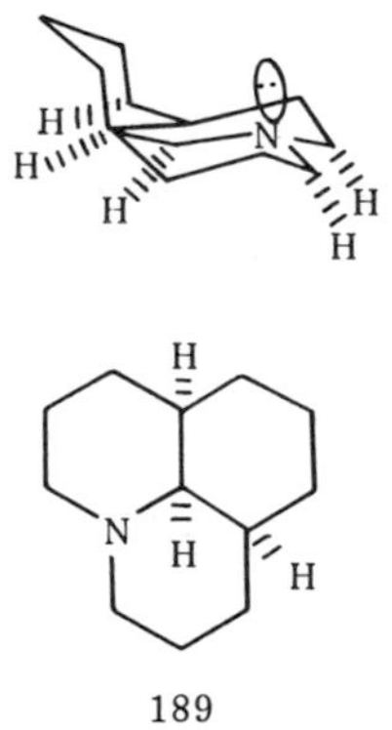

189

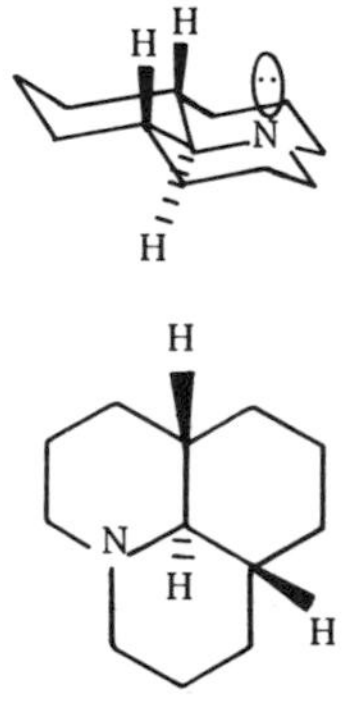

190

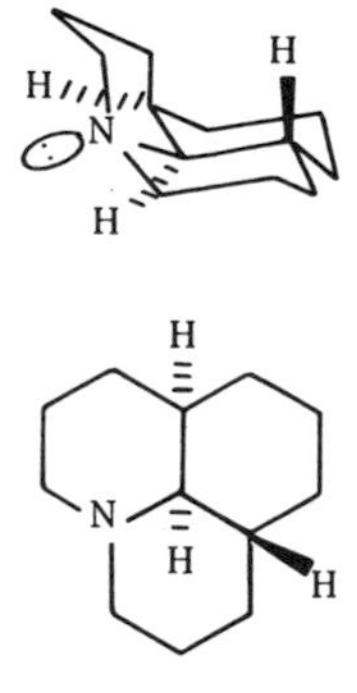

191

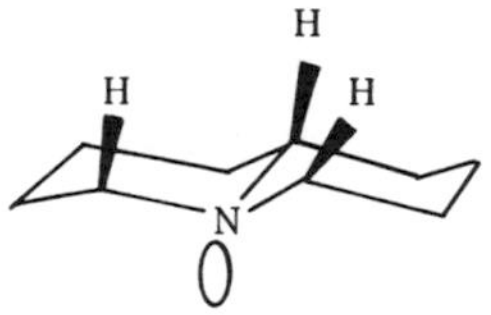

193 trans

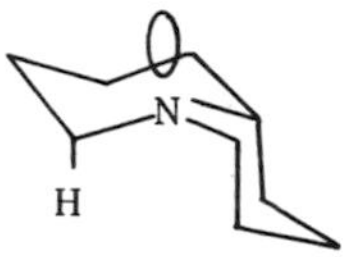

192 cis

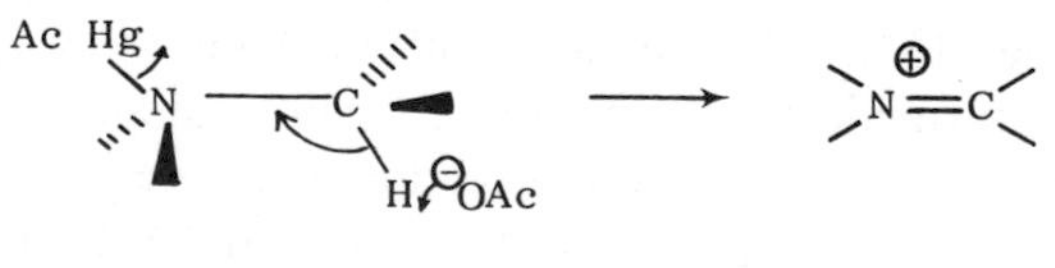

194

in the region 2700–2800 cm^{-1} (Bohlmann bands). The requirements for this absorption are that there are two (or more) protons adjacent, anti- and coplanar with the lone pair of electrons on nitrogen. The examples show the situation for *cis*- and *trans*-quinolizidine **192** and **193**, where the former has only one proton suitably disposed but the *trans* isomer has three.

A secondary criterion is the rate of oxidation with mercuric acetate. This reaction has a requirement for an antiperiplanar arrangement for the proton lost and the departing mercurous acetate in the transition state (i.e. **194**). When applied to the hydrojulolidine isomers, **191** did not react at an appreciable rate and must therefore be the *cis,trans* isomer having only one proton suitable disposed (axial at C-9). Isomers **189** and **190** each have three available protons, but **189** reacts five times faster than **190**. This has been explained in terms of relief of steric strain in the transition state of **189**.

N-Methylation also distinguishes between **189** and **190**; for steric reasons the former is not N-methylated by methyl iodide in ether, indicating that it is the *cis,cis* isomer.

The ^{13}C NMR data of lycopodine and obscurine are shown in Figure 2.

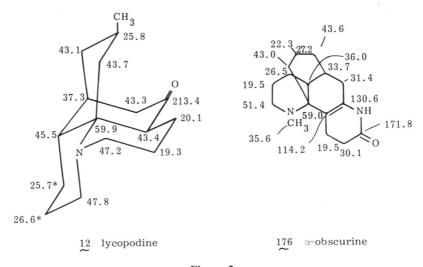

Figure 2

4.8.2   Cernuine

Lycopodium cernuum L. produces a quite different array of alkaloids, one of which is cernuine (**162**). The alkaloid differs from other *Lycopodium* alkaloids in having both nitrogens at bridge heads, and consideration of chemical and spectroscopic evidence led to the structure and stereochemistry shown. The synthesis of the transformation product **195** confirmed both the skeleton and the stereochemical assignments.

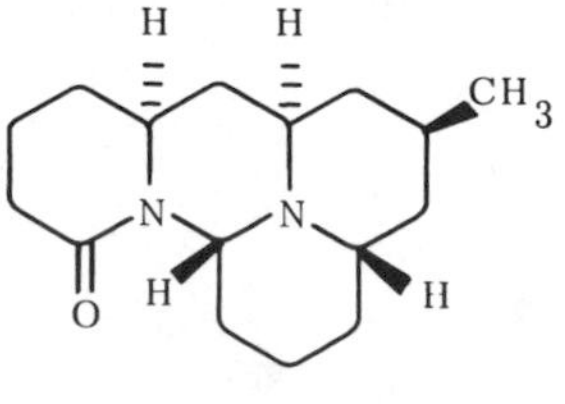

162 cernuine 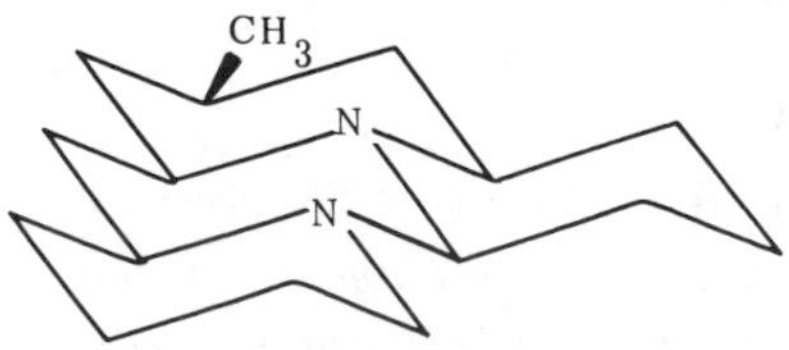   195

4.8.3 Luciduline

Luciduline (**196**) is a neutral alkaloid occurring in *L. lucidulum* which was recently synthesized in a novel fashion by Oppolzer and Petrzilka. Diels–Alder condensation of butadiene and 5α-methyl cyclohexenone (**197**) gave the decalone **198** as a mixture of *cis* and *trans* isomers. Formation of the oxime **199** in the presence of mild base gave only the *cis* derivative, which was reduced with sodium cyanoborohydride to the hydroxylamine **200**. The key step is then the treatment of **200** with paraformaldehyde in the presence of molecular sieve to afford the isoxazolidine **201**. The intermediate in this step is probably a nitrone **202**, and for reasons that are not clear, this nitrone undergoes regioselective addition to the double bond. Methylation and lithium aluminum hydride reduction gave the alcohol **203**, which could be easily oxidized to luciduline (**196**) (Scheme 18).

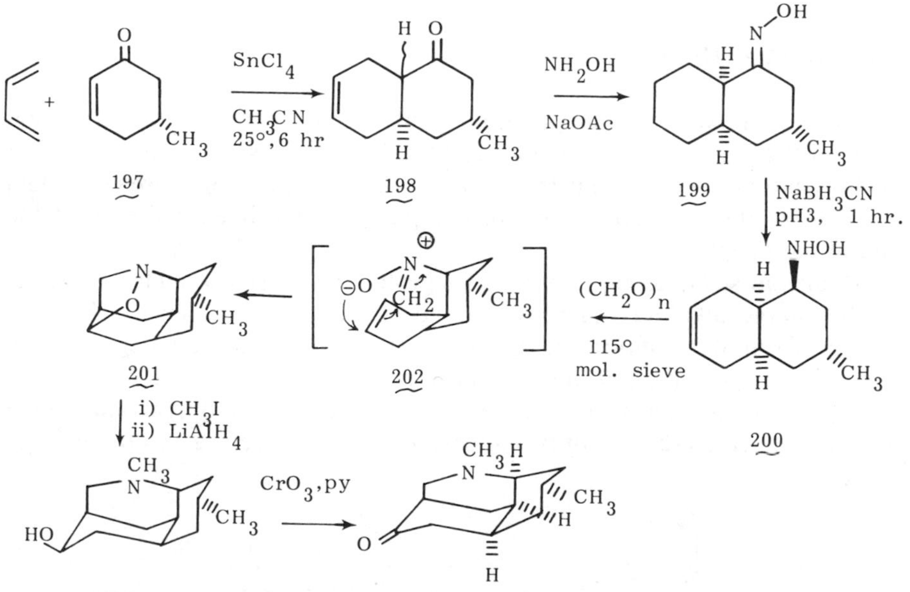

Scheme 18 Synthesis of luciduline (**196**).

LITERATURE

Reviews

Manske, R. H. F., *Alkaloids NY* **5**, 295 (1955).

Wiesner, K. F., *Fortschr. Chem. Org. Naturs.* **20**, 271 (1962).

MacLean, D. B., *Alkaloids NY* **10**, 305 (1968).

MacLean, D. B., in S. W. Pelletier (Ed.), *Chemistry of the Alkaloids,* Van Nostrand Reinhold, New York, 1970, p. 469.

Ayer, W. A., in K. F. Wiesner (Ed.), *MTP International Review of Science, Organic Chemistry,* Ser. 1, Vol. 9, *Alkaloids,* University Park Press, Baltimore, Md., 1973, p. 1.

MacLean, D. B., *Alkaloids NY* **14**, 347 (1975).

Synthesis

Stevens, R. V., in J. ApSimon (Ed.), *The Total Synthesis of Natural Products,* Vol. 3, Wiley, New York, 1977, p. 439.

Wiesner, K., L. Poon, I. Jirokovsky, and M. Fishman, *Can. J. Chem.* **47**, 433 (1969).

Stork, G., R. A. Kretchmer, and R. H. Schlessinger, *J. Amer. Chem. Soc.* **90**, 1647 (1968).

Ayer, W. A., W. R. Bowman, T. C. Joseph, and P. Smith, *J. Amer. Chem. Soc.* **90**, 1648 (1968).

Kim, S. W., Y. Bando, and Z. Horii, *Tetrahedron Lett.* 2293 (1978).

Luciduline

Oppolzer, W., and M. Petrzilka, *J. Amer. Chem. Soc.* **98**, 6722 (1976).

4.9 BIOSYNTHESIS

Lysine is a precursor of the piperidine ring of many types of alkaloids, and some of the principal representatives have been discussed in the early part of this chapter. Because this is proving to be quite a complex area, the data for most of the separate groups of compounds have been gathered at the end of the chapter.

Robinson's classic paper in 1917 suggested that the C_5N nucleus of some of the simple alkaloids such as coniine (**15**) and isopelletierine (**22**) was derived from the amino acid lysine. As mentioned at the beginning of this chapter, not all piperidine rings are derived from lysine. Even coniine (**15**), which for so many years appeared to be merely the dihydroderivative of isopelletierine (**22**), is not lysine-derived and is therefore discussed elsewhere (Chapter 6).

15 coniine   22 isopelletierine

Just as in the biosynthesis of alkaloids derived from ornithine (Chapter 3), it became important to determine whether a symmetrical or a nonsymmetrical intermediate is involved, so it has been for the lysine-derived alkaloids. Several groups of alkaloids do show incorporation *via* a nonsymmetrical intermediate. In other instances it has been categorically shown that lysine is incorporated *via* a symmetrical intermediate.

From the very straightforward experiments using ^{14}C-labeled precursors of the simple piperidine alkaloids, we will move on to discuss some of the more complex stereochemical issues of the utilization of lysine. Finally there is a whole host of alkaloids for which interesting biosynthetic data are only just beginning to accumulate.

4.9.1 Simple Piperidine Alkaloids

[6-^{14}C]Lysine (**1**) was specifically incorporated into *N*-methylpelletierine (**204**) in *Sedum sarmentosum* and anabasine (**6**) in *Nicotiana glauca*, giving rise to a label only at the 6-position. As a corollary, L-[4,5-^{3}H$_2$]lysine was incorporated into **204** without loss of ^{3}H, and [6-^{14}C, 2-^{3}H] lysine without change in the ^{3}H/^{14}C ratio. The more advanced intermediate [6-^{14}C]1,2-dehydropiperidine (**29**) was incorporated into anabasine (**6**) without randomization.

With the derivation from lysine established, attention was turned to the mode of incorporation of lysine (i.e. which of the two nitrogen atoms is retained). Using both [ε-^{15}N, 2-^{14}C]- and [α-^{15}N, 2-^{14}C]lysine, it was shown that the piperidine nitrogen of anabasine (**6**) was derived from the ε-nitrogen. A similar conclusion was reached for **204** using [6-^{3}H, 6-^{14}C]lysine, from which no label was lost.

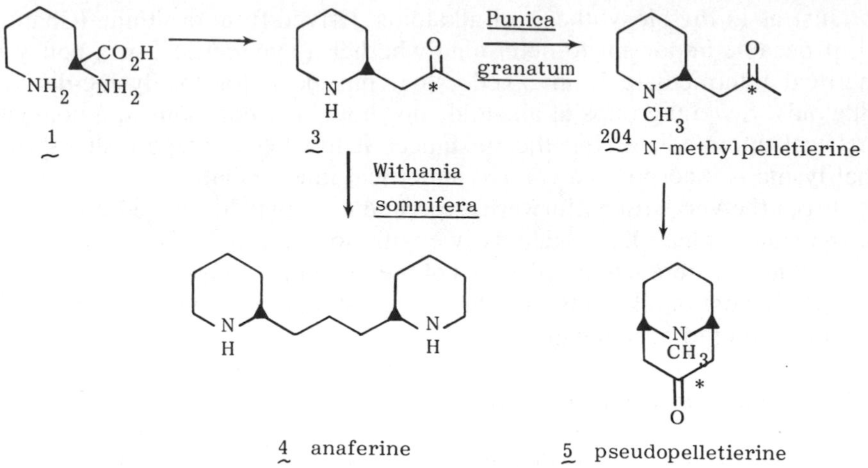

Scheme 19   Biosynthesis of simple piperidine alkaloids.

The side chain of several of the simple alkaloids is acetate-derived, for [1-^{14}C]acetate specifically labeled the carbonyl carbon of N-methylpelletierine (**204**), pseudopelletierine (**5**), and anaferine (**4**). These data suggest that like hygrine, the corresponding compound derived from ornithine, **204** is derived by condensation of **29** with the anion of acetoacetate (**205**) followed by decarboxylation. In addition, N-methylpelletierine (**204**) was a precursor of pseudopelletierine (**5**), and pelletierine (**3**) was a precursor of anaferine (**4**). These results are well accommodated by a pathway such as that in Scheme 19.

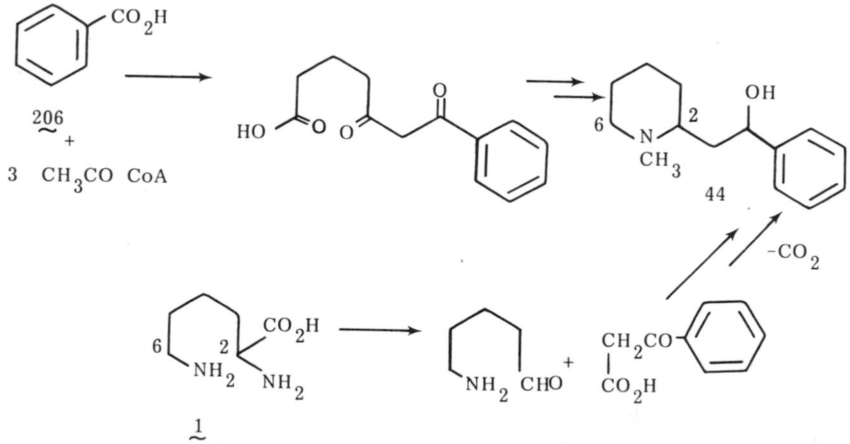

Scheme 20   Biogenesis of sedamine (**44**).

4.9.2 *Sedum* and *Lobelia* Alkaloids

The *Sedum* and *Lobelia* alkaloids pose an interesting biosynthetic problem because, as shown in Scheme 20, they could potentially be derived either by a polyacetate pathway from benzoic acid (206) or from lysine and phenylalanine.

Feeding experiments clearly indicated that the piperidine ring of sedamine (44) is derived from lysine *via* a nonsymmetrical intermediate, since both [2-^{14}C]- and [6-^{14}C]lysine gave specific incorporation into C-2 and C-6 of 44 respectively. Pipecolic acid is not an intermediate because [2-^{3}H, 6-^{14}C]lysine was incorporated with no loss of label into 44.

The side chain of sedamine is specifically derived from phenylalanine *via* cinnamic acid, although details of the union of the two coupling units of 44 are not known.

In previous sections we discussed how lobinaline (48) could possibly be a modified dimer of sedamine (44). However, when [6-^{14}C]lysine was used as a precursor, only 25% of the total activity was found at C-2; therefore at least one of the two lysine-derived units is derived from a symmetrical intermediate. Both [3-^{14}C]phenylalanine and [3-^{14}C]cinnamic acid were specifically incorporated into lobinaline (48).

61

48 lobinaline   ● from $\left[6-^{14}C\right]$-lysine

▲ from $\left[3-^{14}C\right]$-cinnamic acid

When [2-^{14}C]lysine was used as a precursor of lobeline (8), C-2 was found to have 50% of the activity. If the remaining activity is at C-6, then incorporation into lobeline is *via* a symmetrical intermediate. Such a scheme would appear to be in contrast to the data obtained for sedamine (44), considering that 207 is probably an intermediate. However, the reason for the localization of activity at C-2 and C-6 may be due to the intermediacy of a symmetrical compound such as lobelanine (51), which was shown to be incorporated into 8. Additional work indicated that cadaverine (208) was probably not an intermediate, but that 1,2-dehydropiperidine (29) was well incorporated.

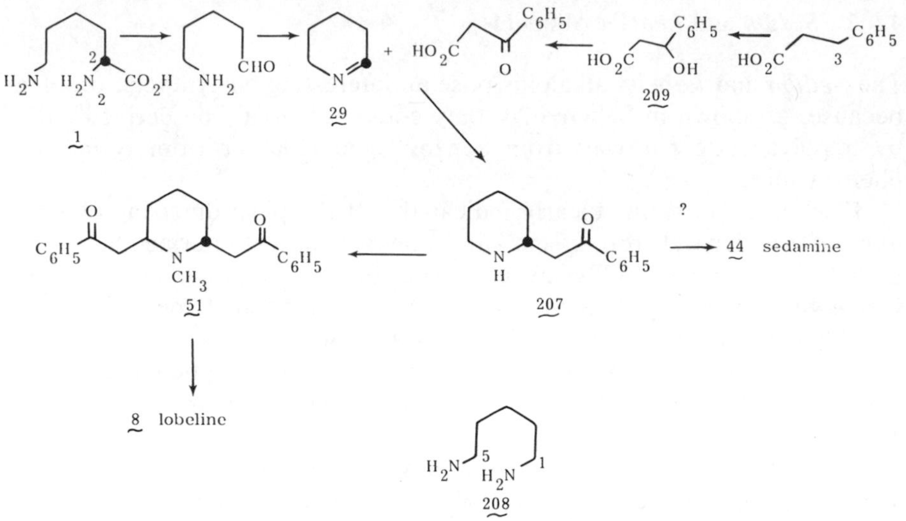

Scheme 21   Biosynthesis of lobeline (8).

Phenylalanine, cinnamic acid, and 3-hydroxy-3-phenylpropanoic acid (209) were also good precursors of lobeline (8). Hence the present scheme proposed for the formation of lobeline (8) is that shown in Scheme 21.

4.9.3   Pipecolic Acid

Pipecolic acid (61) is the homologue of proline (210) and is a product of lysine metabolism in plants, animals, and microorganisms. In all of the organisms so far examined it is derived by loss of the α-nitrogen of lysine, and ε-amino-α-keto caproic acid (211) is therefore probably a key intermediate (Scheme 22).

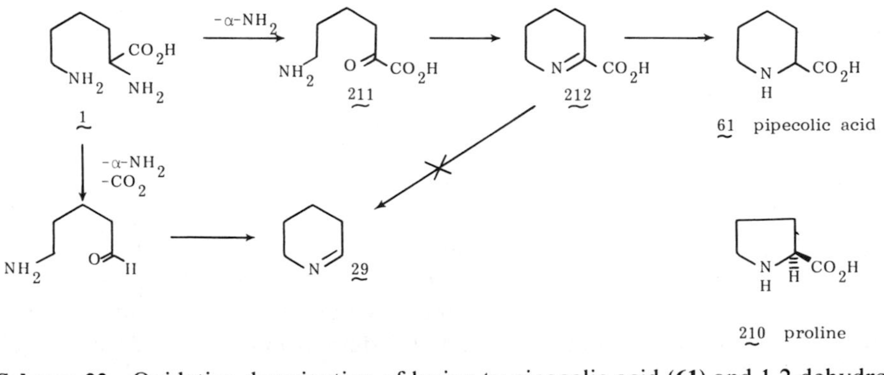

Scheme 22   Oxidative deamination of lysine to pipecolic acid (61) and 1,2-dehydropiperidine (29).

From the previously discussed biosynthesis of sedamine (**44**) however, it would appear that this is not the biosynthetic route to the simple alkaloids. Rather, oxidation and decarboxylation occur *prior* to cyclization, implying that 1,2-dehydropiperidine (**29**) is a true intermediate with essentially no decarboxylation of **212** taking place.

4.9.4 Lythraceae Alkaloids

Biogenetic analysis of the Lythraceae alkaloids indicates that probably the 2-hydroxy-4-phenyl quinolizidine nucleus is produced first. This unit is then acylated with a cinnamic acid (or phenyl propionic acid) residue and the phenyl rings oxidatively coupled. Such a scheme would make **213**, or a very close relative, a key intermediate. But even this intermediate can be produced by at least three different routes as shown (Scheme 23). From two of these one would anticipate lysine to be a precursor of part of the quinolizidine nucleus, and but the third (**214**) would involve only acetate as a precursor. Such a distinction was easily made when Spenser found that both [2-^{14}C]- and [6-^{14}C]lysines were incorporated into decodine (**68**) in *D. verticillatus*, and degradation indicated that positions 6 and 9 were labeled equally. Lysine is therefore incorporated *via* a symmetrical intermediate.

Feeding experiments with [1,3-^{14}C$_2$]phenylalanine established that one

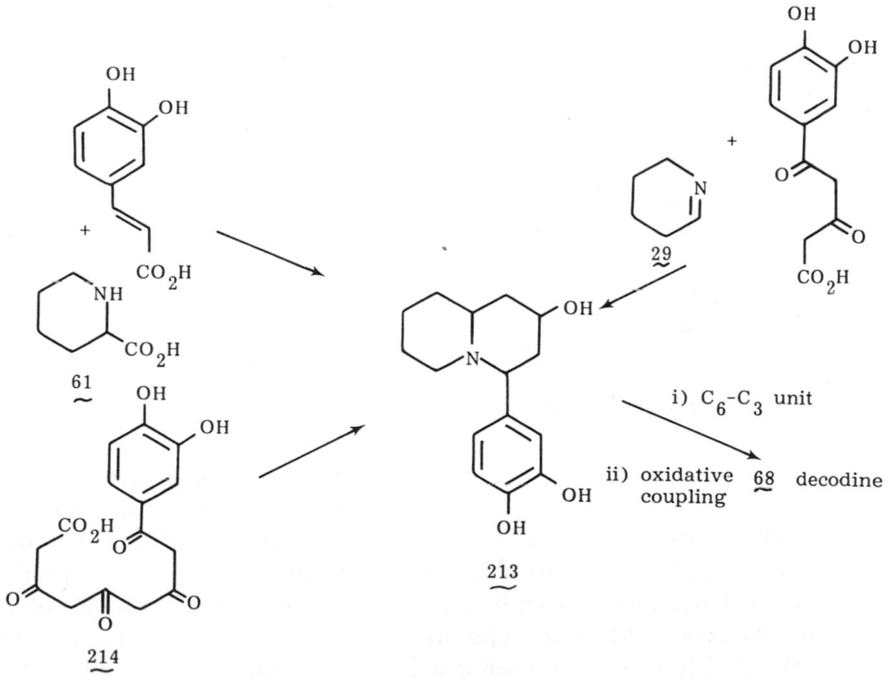

Scheme 23 Biogenesis of Lythraceae alkaloids.

phenylalanine was incorporated intact, but that only C-1 was derived from the second unit, which is therefore a C_6-C_1 unit. At the present time the origin of the remaining three carbon atoms is not firmly established, but there are clearly two main alternatives: one in which the acetoacetate combines with the C_6-C_1 unit prior to uniting with 1,2-dehydropiperidine (**29**); or alternatively union of the C_6-C_1 unit with isopelletierine (**22**) followed by condensation with acetoacetate. In either of these sequences the quinolizidine **213** is a probable intermediate.

In support of this overall scheme, two interesting metabolites **215** and **216** have been obtained by Schwarting and co-workers from *Heimia salicifolia*. Indeed the alcohol **215** was synthesized by condensation of isopelletierine (**22**) and isovanillin (**217**), followed by borohydride reduction. On this basis oxidative coupling of the two aromatic units may well be the final step in the biosynthetic sequence.

68   decodine

215   R=H
216   R=CO-CH=CH—⟨aromatic⟩—OH

22

217

4.9.5   Quinolizidine Alkaloids

In the earlier discussion of the lupin and related alkaloids a diverse array of structure types was presented for these alkaloids; some examples included lupinine (**7**), sparteine (**9**), cytisine (**87**), lupanine (**103**), matrine (**10**), pohakuline (**160**), and ormosanine (**151**). All these structures can be interpreted in terms of a biosynthesis from lysine units as shown in Scheme 24. (Stereochemical considerations are ignored at this point.) Some of the experiments that lead to this conclusion are described below. Almost all of this work was carried out by Schütte and his co-workers.

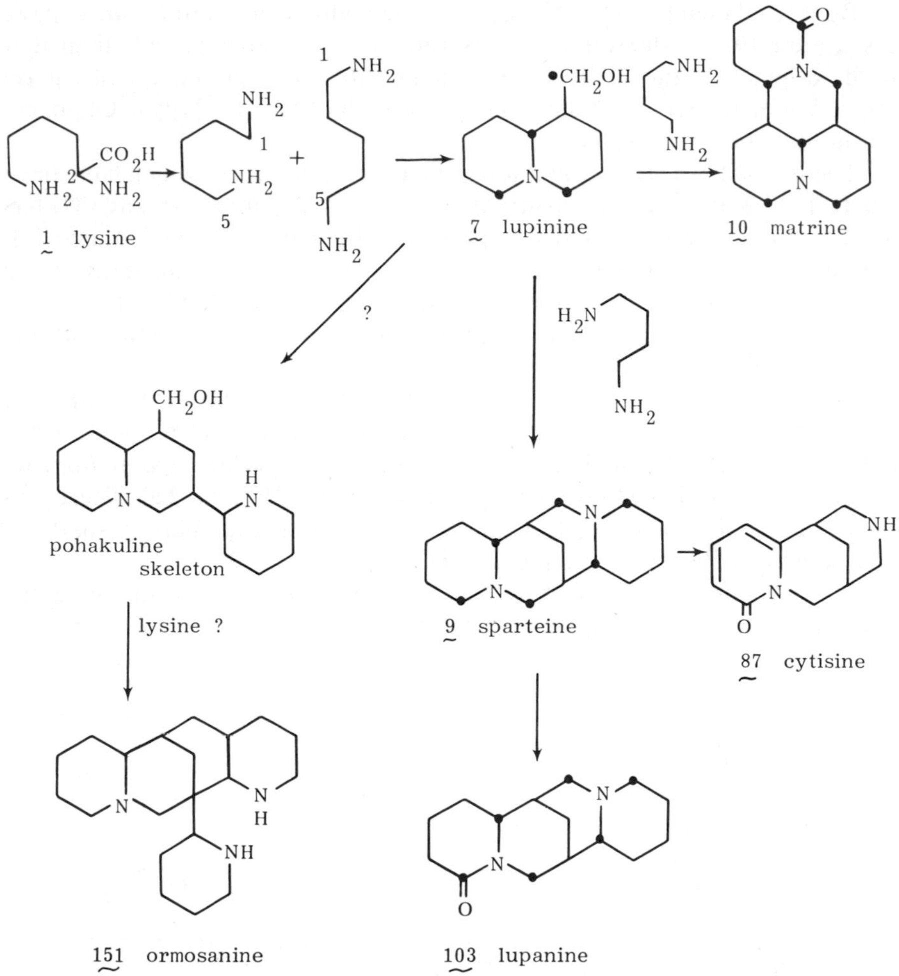

Scheme 24 Biogenetic interrelationships of lupin and related alkaloids.

As a result of numerous experiments using [2-^{14}C]lysine (**1**) and [1,5--^{14}C$_2$]cadaverine (**208**) into lupinine (**7**) and sparteine (**9**) in *L. luteus*, lupanine (**103**) in *L. angustifolius*, and matrine (**10**) in two *Sophora* sp., it is established beyond doubt that activity appears at the starred carbon atoms in these alkaloids. In essence the results indicate that unlike the nicotine and *Sedum* alkaloids, lysine enters the lupin alkaloids *via* a symmetrical intermediate. Since [2-^{14}C]1,2-dehydropiperidine (**29**) labeled matrine (**10**) only at positions 2, 10, and 11, it would appear that cadaverine is the symmetrical intermediate. The observation of diamine oxidase activity in lupin seedlings may therefore be of some significance. Again there is a frequently overlooked point to be made here: namely that the three precursor fragments are labeling each of the units in sparteine (**9**) equally.

Doubly labeled [2-^{14}C, α-^{15}N]lysine when administered to *L. luteus* gave a sparteine (**9**) in which the ^{14}C/^{15}N ratio was six times greater than that in the original. Is this result in agreement with the intermediacy of cadaverine? Unfortunately a similar experiment with [2-^{14}C, ε-^{15}N]lysine appears not to have been attempted.

There are also a number of interesting interrelationships which have been elucidated. A structural curiosity in terms of its skeleton, cytisine (**87**) has been found to be derived from sparteine (**9**), and not from lupinine (**7**), plus a C$_1$-N unit. Thus both [2-^{14}C]lysine and [1,5-^{14}C$_2$]cadaverine were incorporated into cytisine (**87**) in *Cytisus laburnum*, and 20% of the total activity was found at C-13. In addition, labeled lupinine (**7**) and sparteine (**9**) were each incorporated.

Further experiments by Nowacki and Waller have shown that lupanine (**103**) can serve as a quite effective precursor of alkaloids such as thermopsine (**218**), cytisine (**87**), *N*-methyl cytisine (**219**), and baptifoline (**220**) in *Baptisia leucopheya*, and that **218** is also a precursor of **87**, **219**, and **220**. There was no evidence of any reversal of the scheme to lupanine (**103**) or sparteine (**9**). Some possible relationships are shown in Scheme 25.

It was mentioned above that three units of lysine (cadaverine) label the various parts of sparteine (**9**) almost equally. This significant point was

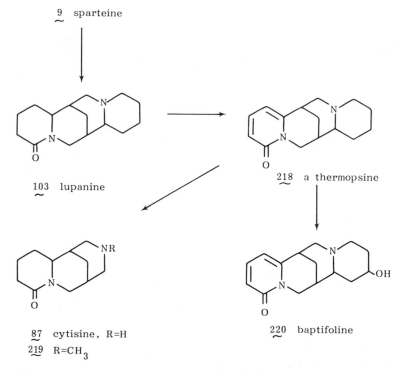

Scheme 25   Possible precursor relationships in *Baptisia leucopheya*.

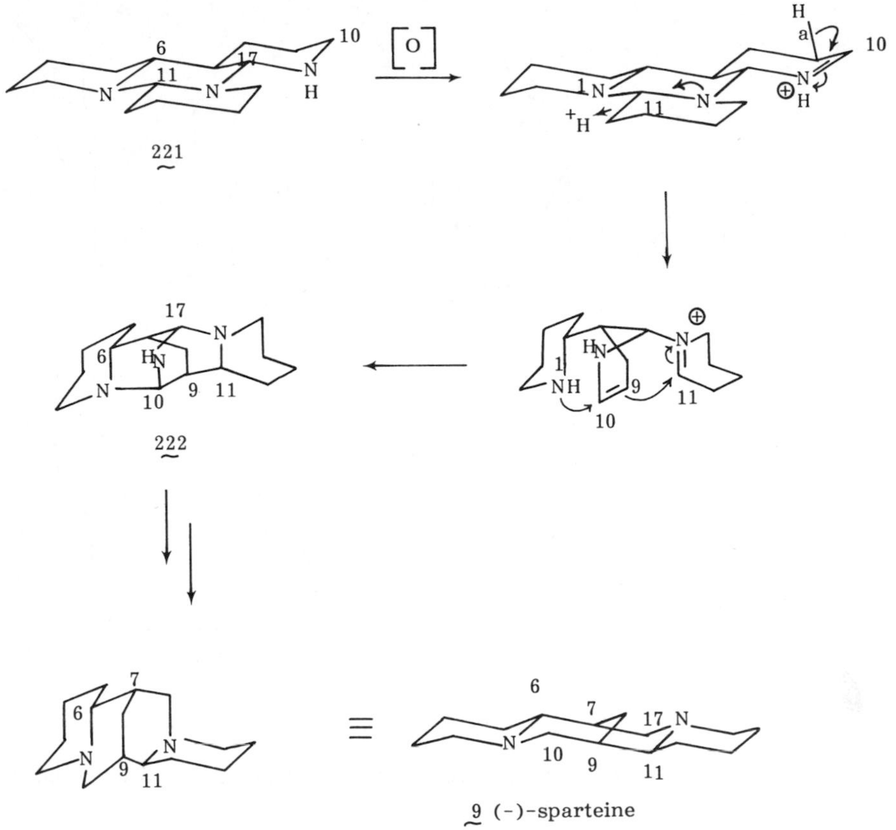

Scheme 26 Biogenesis of (−)-sparteine (**9**) from a trimer of 1,2-dehydropiperidine.

overlooked for many years, but has recently been reinvestigated by Gole-biewski and Spenser.

Key to the new concept for the formation of the lupin alkaloids is the knowledge that 1,2-dehydropiperidine (**29**) does not exist in the free state but as a mixture of trimers, the most thermodynamically favored of which is all-*trans*-isotripiperideine (**221**). The situation is complicated by the stereoisomerism of sparteine and its derivatives, but let us consider the case of (−)-sparteine (**9**), which has the 6R:7S:9S:11S configuration. The trimer **221** has the stereochemistries 6β, 7α, 11β, and 17β and it is suggested that the C$_{15}$-lupine alkaloids can be derived from this trimer in four steps with a rearranged trimer **222** as the key intermediate. The proposed formation of (−)-sparteine (**9**) is shown in Scheme 26, and it should be noted that the stereochemistry of **9** at C-6, C-7, and C-9 has been predetermined by the original stereochemistry of **221**. The configuration at 11 is determined by *re* or *si* attack on the iminium precursor of **222**, which leads to either (−)-sparteine or (−)-α-isosparteine having the 11R-configuration.

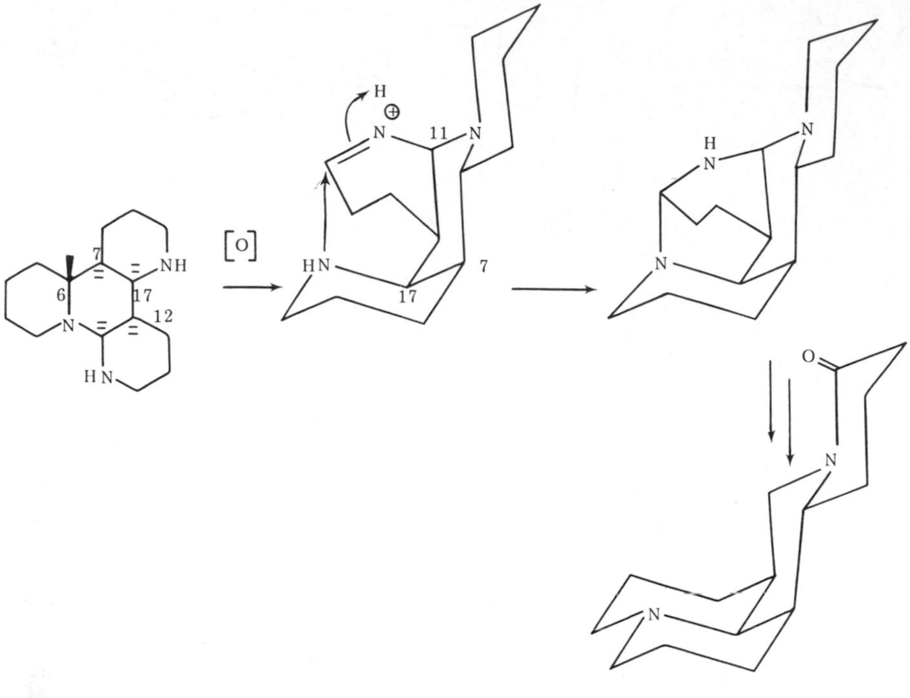

10   matrine

Scheme 27   Biogenesis of matrine (**10**) from a trimer of 1,2-dehydropiperidine.

In experiments with [2-^{14}C]- and [6-^{14}C]1,2-dehydropiperidine (**29**) it was demonstrated that in agreement with this theory, the former precursor labeled C-11 and C-17 (C-6 was not determined) and the latter precursor labeled C-2 and C-15 (C-10 was not determined). What would be the expected result if lupinine (**7**) and **29** were direct precursors of **9**?

This concept has been extended to cover alkaloids in the matrine series as shown in Scheme 27. Note that in each scheme removal of the imino group involves double reductive cleavage of a quite fragile N—C—N network. The *Ormosia* alkaloids could then be derived from a beginning tetramer such as **223**.

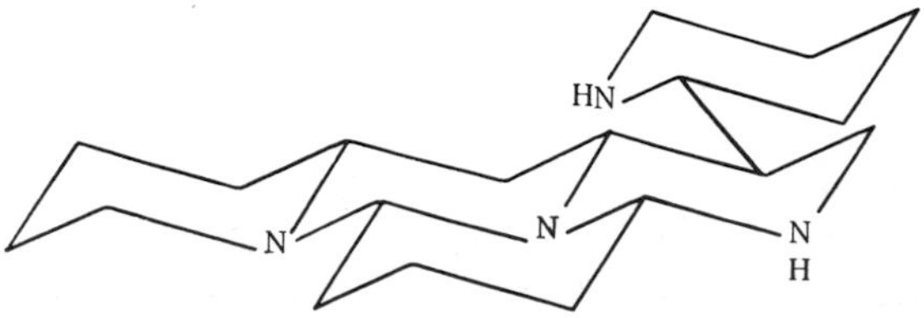

223

There is still some controversy over the formation of lupanine (**103**) from sparteine (**9**). Both positive and negative data have been obtained and the matter cannot be easily resolved.

4.9.6 *Lycopodium* Alkaloids

The diverse skeleta of the *Lycopodium* alkaloids pose some interesting bio-genetic problems, and the biosynthetic data that are available have made this even more intriguing.

Early hypotheses developed by Conroy envisaged a formation from C_8 units derived from acetate for the lycopodine, lycodine and cernuine skeleta as shown in **224**, **225**, and **226**. Alternative pathways to the latter two structures involve the joining of two pelletierine (**3**) units as indicated in **227** and **228**. Alkaloids such as lycopodine would be derived by fission and recyclization of the intermediate **229**.

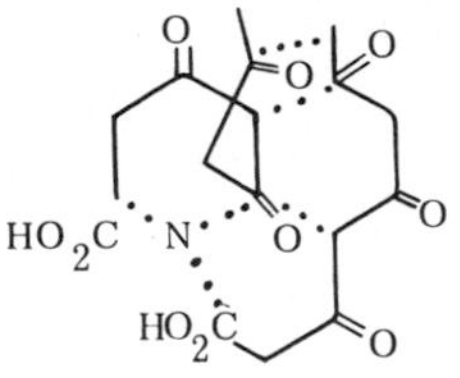

224 lycopodine skeleton 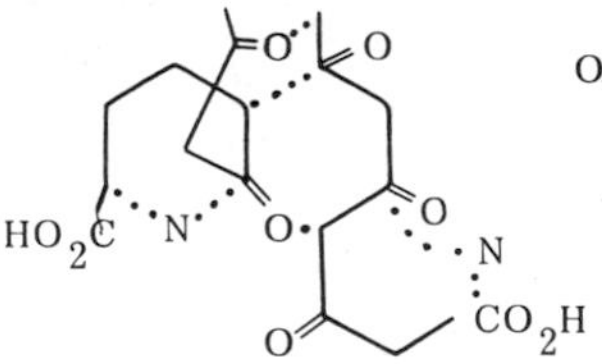   225 lycodine skeleton 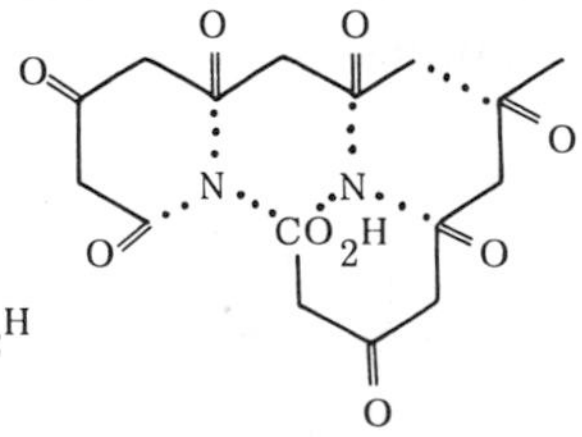   226 cernuine skeleton

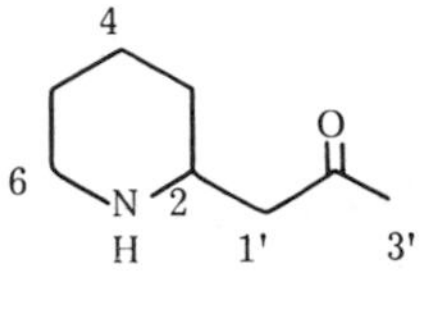

3 pelletierine 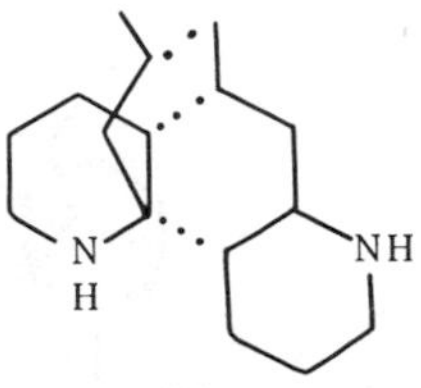   227 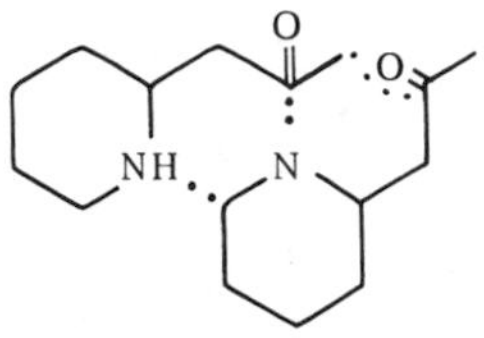   228

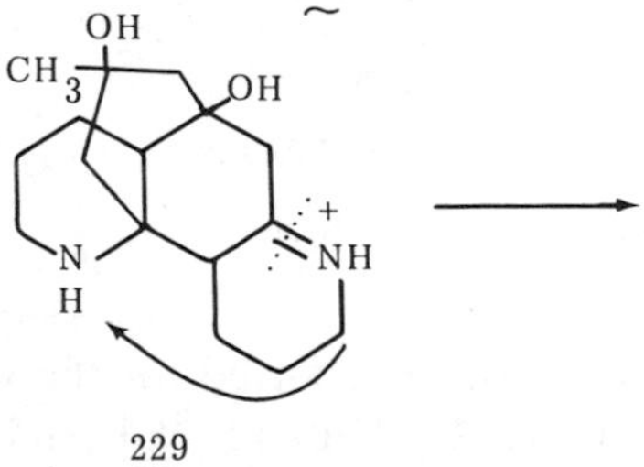

229

12 lycopodine

Either pathway would lead to incorporation from acetate, but according to the acetate hypothesis, [1-^{14}C]acetate would label carbons 1, 3, 5, 7, 9, 11, 13, and 15, each with 12.5% of the total activity, whereas the pelletierine hypothesis would lead to labeling only at C-7 and C-15.

Experimentally, [1-^{14}C]acetate labeled C-15 to the extent of 47%, and [2-^{14}C]acetate labeled C-16 to the level of 21% of the total activity. These results negate the acetate hypothesis for the formation of the whole skeleton.

When [2-^{14}C]- and [6-^{14}C]lysine were used as precursors, degradation indicated that from either precursor 25% of the label was at C-5 and 25% at C-9. The lysine is therefore incorporated by way of a symmetrical intermediate as shown in **230**. [6-^{14}C]1,2-Dehydropiperidine (**29**) afforded lycopodine (**12**) labeled at C-1 and C-9, as in **231**, indicating that no equilibration of activity occurs at this stage. In addition, the concept that cleavage of a lycodine-type intermediate **229** occurs is therefore well substantiated. Yet another point of interest is that the two lysine or 1,2-dehydropiperidine moieties are equally incorporated into each half of the molecule.

A problem arose however when pelletierine (**3**) was used as a precursor. Quite unexpectedly specific incorporation was observed only into that portion of the molecule indicated in **232**. Thus although the second unit is both lysine- and acetate-derived, the precise nature of the next intermediate is unknown.

230

• labeling from $\left[2\text{-}^{14}C\right]$- or $\left[6\text{-}^{14}C\right]$- lysine

34

231

labeling from $\left[6\text{-}^{14}C\right]$- 1,2-dehydropelletierine (**34**)

232

labelling from $\left[4\text{-}^{3}H, 6, 2'\text{-}^{14}C\right]$- pelletierine

Essentially similar data have been reported for the incorporation of labeled precursors into cernuine (**162**). Thus [2-^{14}C]- and [6-^{14}C]lysine gave labeling of cernuine as shown in **233**, and [6-^{14}C]1,2-dehydropiperidine (**29**) gave equivalent labeling only at C-1 and C-9 (i.e. **234**). But [4-^{3}H, 6,2'-^{14}C$_2$]pelletierine was incorporated into only one half of the molecule as shown in **235**, with no extraneous ^{14}C or ^{3}H label at other positions! A further series of experiments established the obligatory nature of pel-

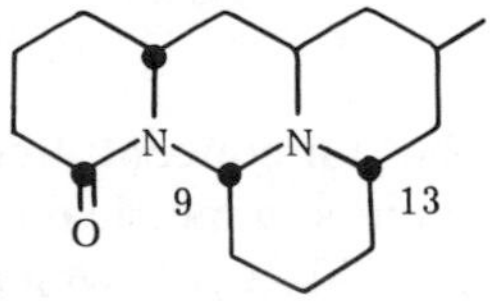

233
labeling from $[2-^{14}C]$-
or $[6-^{14}C]$ - lysine

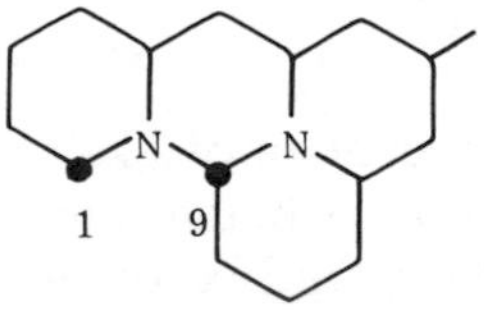

234
labeling from $[6-^{14}C]$-1,2-
dehydropiperidine

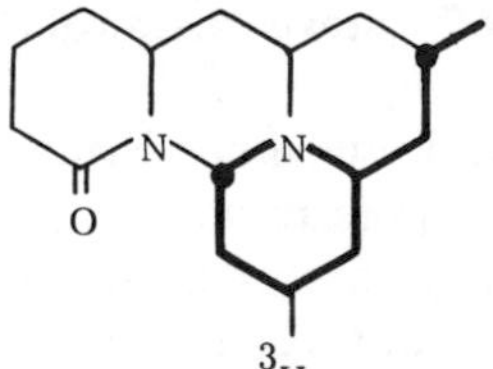

235
labeling from $[4-^3H,$
$6,2'-^{14}C_2]$- pelletierine

letierine as an intermediate. For example when $[2-^{14}C]1,2$-dehydropiperidine (**29**) was used as precursor in the presence of inactive pelletierine, the efficiency of incorporation is markedly reduced and active pelletierine is isolated. In addition 88% of the activity in lycopodine from this experiment was at C-5, implying that the inactive excess pelletierine had been incorporated into the C-9–C-16 unit as observed previously. If pelletierine had not been an intermediate, the $[2-^{14}C]1,2$-dehydropiperidine would still have labeled C-5 and C-13 equally.

Spenser has suggested a scheme whereby the route to lycopodine diverges at an intermediate such as the coenzyme derivative of 4-(2-piperidyl)acetoacetic acid (**236**), and that the conversion of **236** to pelletierine (**3**) is not reversible (Scheme 28).

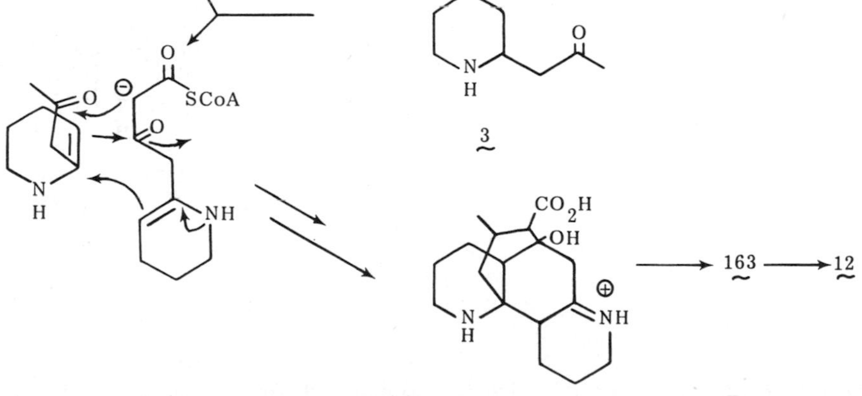

Scheme 28

4.9.7 Stereochemical Aspects

The discussion of lysine as a precursor of alkaloids has thus far concentrated on simple precursor relationships whether these resulted in symmetrical or nonsymmetrical labeling. There are two other aspects of this work which have not yet been touched upon. One of these concerns a possible comparison with the pathway for the pyrrolidine alkaloids where compounds containing N-methyl groups are repeatedly invoked as intermediates. All the data thus far available indicate that this is not the case for the utilization of lysine. Thus the asymmetric incorporation is due to an asymmetric intermediate containing the 6-amino group, and 6-aminohexanal has been proposed.

However cadaverine (208), a symmetrical compound, is an efficient precursor of anabasine (101), N-methylpelletierine (204), pseudopelletierine (5), and sedamine (44). Indeed, cadaverine is a normal plant constituent and is

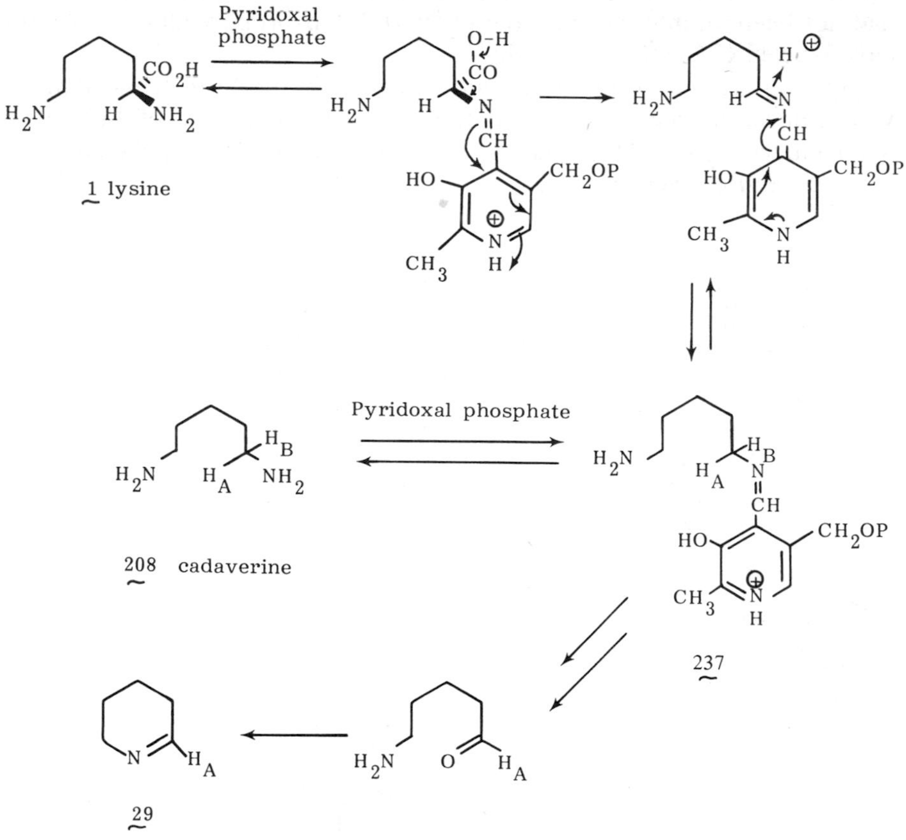

Scheme 29 Biogenesis of 1,2-dehydropiperidine (**29**) from lysine and cadaverine (**208**).

produced from lysine. Thus cadaverine must be considered on any biosynthetic scheme.

Spenser and co-workers have begun to follow this idea to its logical conclusion: namely how can a symmetrical intermediate be involved in a pathway that gives an unsymmetrically labeled product? A scheme accounting for such data is illustrated (Scheme 29), and several important points are evident: (*a*) many of the steps are reversible; (*b*) pyridoxal phosphate is a crucial requirement; (*c*) a one-proton (stereospecific?) loss from the intermediate **237** is postulated; and (*d*) exogenous cadaverine may enter the pathway to give the intermediate **237** directly. Is there any evidence for this scheme? Both lysine decarboxylase and diamine oxidase require pyridoxal phosphate as a cofactor, and when cadaverine chirally labeled at the 1-position with tritium was used as a precursor of N-methylpelletierine (**204**) and sedamine (**44**), only one of these introduced label at C-2. The loss of tritium is explained by invoking the process (*d*) above to be stereospecific. Notably though, **204** and **44** have the opposite stereochemistry at C-2; therefore attack of the side chain at C-2 of 1,2-dehydropiperidine (**29**) can clearly occur on either of the *re* and *si* faces.

204   (−)−N−methylpelletierine   44   (−)−sedamine

The crucial point however is that for those groups of alkaloids which incorporate lysine in a symmetrical fashion, we can postulate a rapid equilibrium between the bound cadaverine and unbound cadaverine. In the cases where incorporation is asymmetrical, we would postulate only the bound cadaverine being carried through to 1,2-dehydropiperidine (**29**), with no equilibration with unbound cadaverine.

Yet another interesting result is that of the fate of a DL-amino acid substrate. Spenser and co-workers found that in *Nicotiana glauca*, L-lysine was utilized mainly for anabasine biosynthesis and D-lysine for pipecolic acid formation! These data would imply that both decarboxylation and oxidative deamination of the α-nitrogen are enzymically controlled processes.

LITERATURE

General: Reviews

Gupta, R. N., *Lloydia* **31**, 218 (1968).

Spenser, I. D., *Compr. Biochem.* **20**, 231 (1968).

Leete, E., *Biosynthesis* **2**, 106 (1974).

Leete, E., *Biosynthesis* **3**, 113 (1975).

Herbert, R. B., *Alkaloids, London* **1**, 1 (1971).

Herbert, R. B., *Alkaloids, London* **2**, 1 (1972).

Herbert, R. B., *Alkaloids, London* **3**, 1 (1973).

Herbert, R. B., *Alkaloids, London* **4**, 1 (1974).

Herbert, R. B., *Alkaloids, London* **5**, 1 (1975).

Herbert, R. B., *Alkaloids, London* **6**, 1 (1976).

Herbert, R. B., *Alkaloids, London* **7**, 2 (1977).

Herbert, R. B., *Alkaloids, London* **8**, 1 (1978).

Simple Piperidine Alkaloids

Leete, E., *Acc. Chem. Res.* **4**, 100 (1971).

Leete, E., E. G. Gros, and T. J. Gilbertson, *J. Amer. Chem. Soc.* **86**, 3907 (1964).

Hanson, J. R., and B. Achilladelis, *Tetrahedron Lett.* 1295 (1967).

Keogh, M. F., and D. G. O'Donovan, *J. Chem. Soc., Ser. C* 1792 (1970).

Sedum *and* Lobelia

Gupta, R. N., and I. D. Spenser, *Can. J. Chem.* **45**, 2375 (1967).

Gupta, R. N., and I. D. Spenser, *Phytochemistry* **9**, 2329 (1970).

O'Donovan, D. G., D. J. Long, E. Forde, and P. Geary, *J. Chem. Soc. Perkin Trans. I* 415 (1975).

Pipecolic Acid

Gupta, R. N., and I. D. Spenser, *J. Biol. Chem.* **244**, 88 (1969).

Lythraceae Alkaloids

Koo, S. H., R. N. Gupta, I. D. Spenser, and J. T. Wrobel, *Chem. Commun.* 396 (1970).

Koo, S. H., F. Comer, and I. D. Spenser, *Chem. Commun.* 897 (1970).

Hörhammer, R. B., A. E. Schwarting, and J. M. Edwards, *J. Org. Chem.* **40**, 656 (1975).

Rother, A., and A. E. Schwarting, *Experientia* **30**, 222 (1974).

Lupin Alkaloids

Schütte, H. R., and H. Hindorf, *Z. Naturforsch.* **19B**, 855 (1964).

Schütte, H. R., H. Hindorf, K. Mothes, and G. Hubner, *Ann.* **680**, 93 (1964).

Schütte, H. R., and H. Hindorf, *Ann.* **685**, 187 (1965).

Shibata, S., and Y. Sankawa, *Chem. Ind. (London)* 1161 (1963).

Schütte, H. R., and J. Lehfeldt, *J. Prakt. Chem.* **24**, 143 (1964).

Nowacki, E. K., and G. R. Waller, *Phytochemistry* **14**, 161 (1975).

Golebiewski, W. M., and I. D. Spenser, *J. Amer. Chem. Soc.* **98**, 6726 (1976).

Lycopodium Alkaloids

Conroy, H., *Tetrahedron Lett.* 34 (1960).

Castillo, M., R. N. Gupta, D. B. MacLean, and I. D. Spenser, *Can. J. Chem.* **48**, 1893 (1970).

Castillo, M., R. N. Gupta, Y. K. Ho, D. B. MacLean, and I. D. Spenser, *Can. J. Chem.* **48**, 2911 (1970).

Gupta, R. N., Y. K. Ho, D. B. MacLean, and I. D. Spenser, *Chem. Commun.* 409 (1970).

Ho, Y. K., R. N. Gupta, D. B. MacLean, and I. D. Spenser, *Can. J. Chem.* **49**, 3352 (1971).

Braekman, J.-C., R. N. Gupta, D. B. MacLean, and I. D. Spenser, *Can. J. Chem.* **50**, 2591 (1972).

Stereochemical Aspects

Leistner, E., and I. D. Spenser, *J. Amer. Chem. Soc.* **95**, 4715 (1973).

Leistner, E., R. N. Gupta, and I. D. Spenser, *J. Amer. Chem. Soc.* **95**, 4040 (1973).

chapter 5

ALKALOIDS DERIVED FROM NICOTINIC ACID

The biosynthesis of alkaloids from a nonessential amino acid such as nicotinic acid (**1**) could be regarded by purists as not pertinent to the discussion.

At a time when only nicotine was regarded as being derived from nicotinic acid this might have been a valid point. In recent years however, a number of developments have led to the conclusion that nicotinic acid should be regarded as an important alkaloid precursor.

There are now at least five alkaloid groups derived from nicotinic acid; they include arecoline (**2**), ricinine (**3**), anatabine (**4**), dioscorine (**5**), and nicotine (**6**). The latter alkaloid has been discussed in Chapter 3; the remaining alkaloids are discussed here.

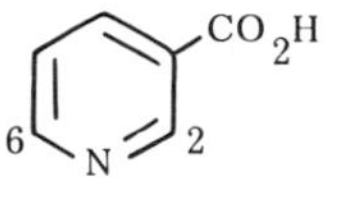

1 nicotinic acid 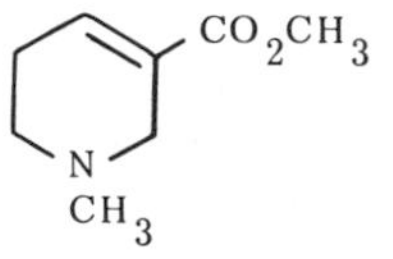   2 arecoline 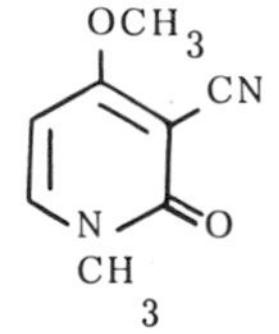   3 ricinine

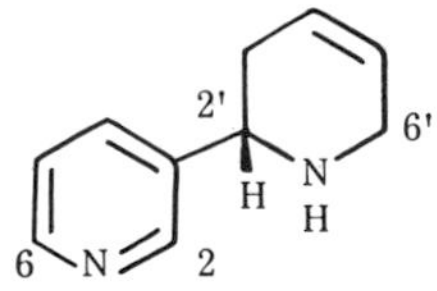

4 anatabine 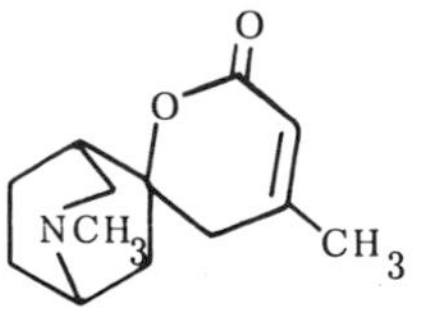   5 dioscorine 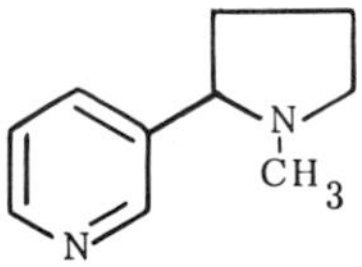   6 nicotine

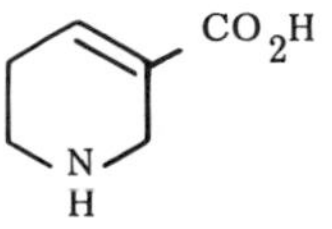

7 guvacine

5.1 ARECOLINE

Areca or betel nut chewing has been practiced by natives in various tropical and subtropical countries since early times. The fresh nuts are mixed with lime and the leaves of *Piper betle* L. (Piperaceae) and the mixture is used as an addictive euphoretic. Note the use of lime to liberate the alkaloid base from any salts. The nut is the fruit of the palm *Areca catechu* L. (Palmae), which is cultivated in India and much of Southeast Asia. Several piperidine alkaloids, including guvacine (7), co-occur, but the principal (up to 0.8%) and most important compound is arecoline (2). The biosynthesis is not known with certainty but probably involves nicotinic acid (1).

Pharmacologically, arecoline has a pronounced central stimulant action, although large doses may cause depression and paralysis. It is a classic para-sympathomimetic acting also as a sialogogue and diaphoretic.

In China the drug has been used as a vermifuge since the sixth century, and the hydrobromide of arecoline has been used as an anthelmintic and diaphoretic. Excessive chewing of the nut is associated with a high incidence of oral cancer, but this may not be due to arecoline.

5.2 RICININE

Ricinine (3) is the principal (sole?) alkaloid obtained from the seeds of the castor bean plant, *Ricinus communis* L. (Euphorbiaceae), native to India. The plant is now cultivated in South America, several countries in Africa, and Italy.

First isolated in 1864, it is only mildly toxic, the principal toxin being ricin of unknown structure. Hydrolysis of ricinine with sodium hydroxide gives 8, which on heating with concentrated hydrochloric acid affords 4-hydroxy-1-methyl-2-pyridone (9).

A quite classic synthesis of ricinine (3) takes advantage of some of the

peculiarities of the chemistry of the pyridine nucleus. For example, although pyridine itself is resistant to electrophilic substitution, the N-oxide is quite susceptible to substitution at the 4-position. In addition, groups at the 2- and 4-positions are readily displaced by nucleophiles.

Thus 3-methylpyridine (10) is converted to the N-oxide and nitrated to 11. After oxidation of the methyl group to a carboxylic acid, the nitro group displaced by methoxide and the acid converted to the corresponding amide 12. Chlorination of this amide gave 2,4-dichloro-3-cyanopyridine (13), in which the two chloro groups could be replaced by methoxide and the 2-methoxy group isomerized with methyl iodide to give ricinine (3).

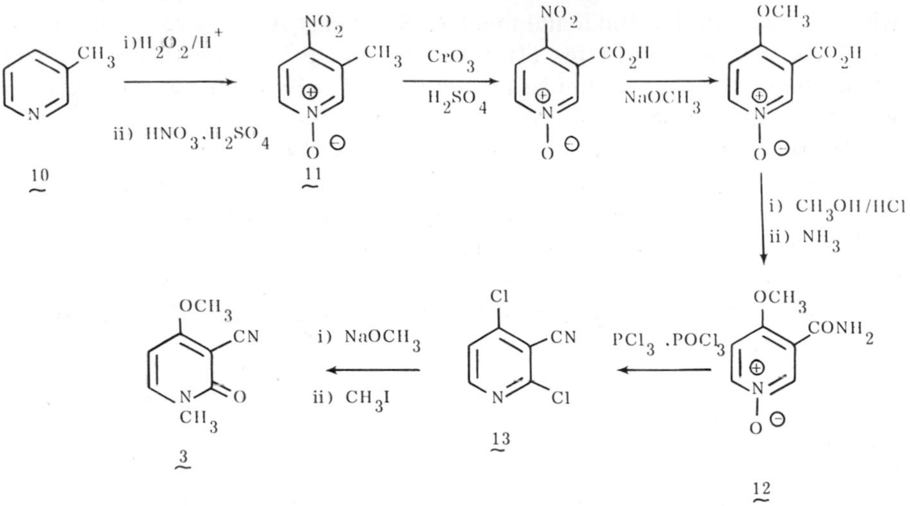

5.2.1 Biosynthesis

In Chapter 3 it was discussed how nicotinic acid (1) is an important intermediate in the biosynthesis of both nicotine (6) and anabasine (14). Numerous experiments have demonstrated that ricinine (3) is produced by a similar route. Thus [4-^{14}C]aspartic acid (15) gave specific incorporation into the nitrile of 3 and [2,3-^{14}C$_2$]succinic acid (16) and [3-^{14}C]aspartic acid (15) gave randomization of label at C-2 and C-3 of 3.

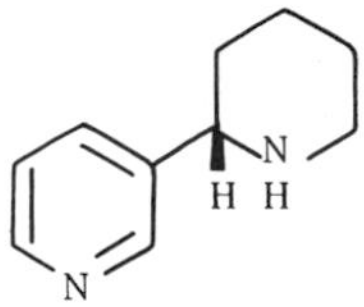

14   anabasine

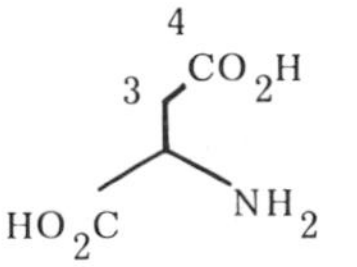

15   aspartic acid

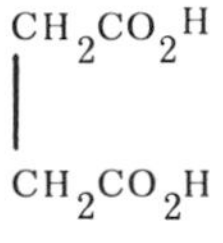

16

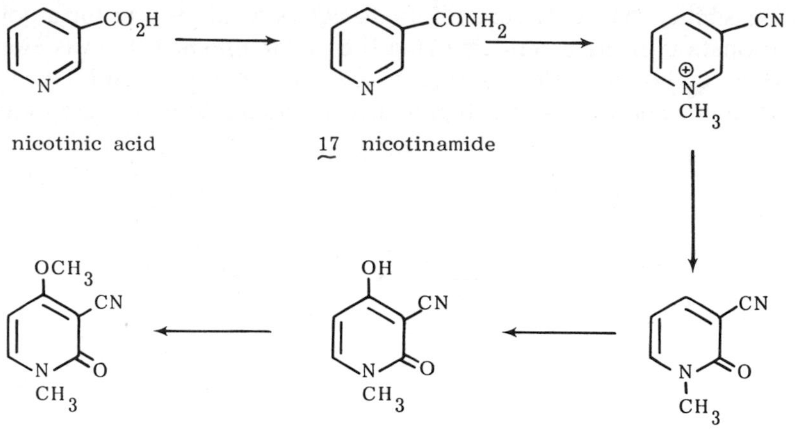

1 nicotinic acid   17 nicotinamide

3 ricinine   18 N–methyl–3–
                                                              cyano–2–pyridone

Scheme 1 Biogenesis of ricinine (3) from nicotinic acid (1).

Nicotinic acid (**1**) and nicotinamide (**17**), with carboxyl group and the nitrogen of the amide labeled, are incorporated intact into ricinine (**3**). Methionine is a precursor of both the *N*-methyl and *O*-methyl groups and no marked difference between the levels of incorporation was observed.

The further steps from nicotinamide (**17**) are not well worked out, but the quite reasonable (0.4%) incorporation of *N*-methyl-3-cyano-2-pyridone (**18**) suggests the route shown in Scheme 1. The remaining intermediates have still to be demonstrated.

5.3 ANATABINE

The alkaloid anatabine (**4**) may be regarded as 4′,5′-dehydroanabasine. Anatabine and anabasine (**14**) co-occur in *N. glutinosa*, where the former predominates. Not surprisingly lysine was considered as a precursor.

However, although anabasine was specifically labeled, neither [2-^{14}C] lysine (**19**) nor [2-^{14}C]4-hydroxylysine (**20**) were precursors of **4**.

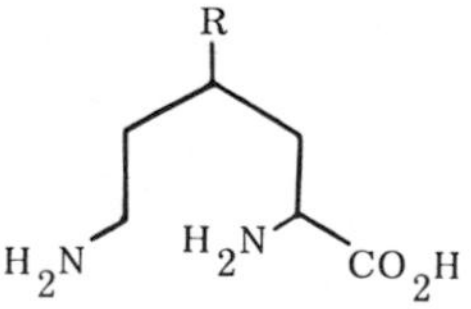

19   R=H
20   R=OH

The possibility that nicotinic acid might act as a starter unit for a poly-acetate formation of the piperideine ring [i.e. as in anibine (21)] was elimi-nated when carboxyl-labeled nicotinic acid was not incorporated. But [6-^{14}C]nicotinic acid (1) was specifically incorporated into *both* the pyridine

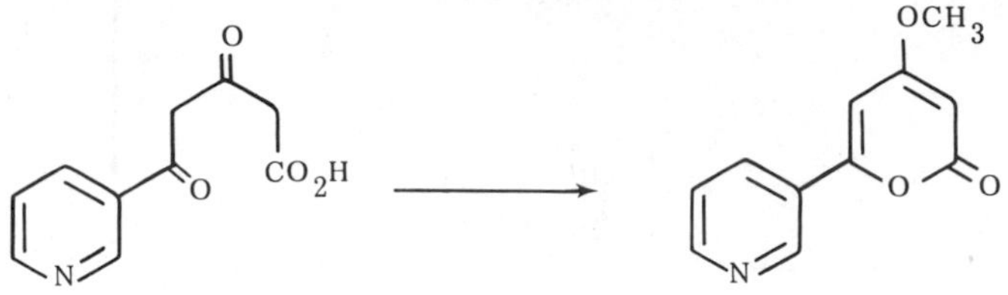

21   anibine

and piperdeine nuclei as shown. Similarly, [2-^{14}C]nicotinic acid (1) gave anatabine (4) specifically labeled at C-2 and C-2′. Anabasine (14) was labeled only in the pyridine ring.

The most curious aspect of the labeling of anatabine (4) was that the specific activity at C-2 and C-2′ (and C-6 and C-6′) were essentially equal, suggesting that possibly dimerization of a single intermediate was involved. Leete has suggested that this intermediate is 2,5-dihydropyridine (22), which can dimerize by the overall route shown in Scheme 2, although the details remain to be elucidated.

5.4   DIOSCORINE

Dioscorine (5) is another alkaloid which poses a biogenetic problem; for clearly, like coniine or isopelletierine, there are two plausible routes, one from acetate alone (e.g. 23), and the second by a combination of lysine and acetate (e.g. 24 or 25). The second route derived from lysine would also place pelletierine as a key intermediate.

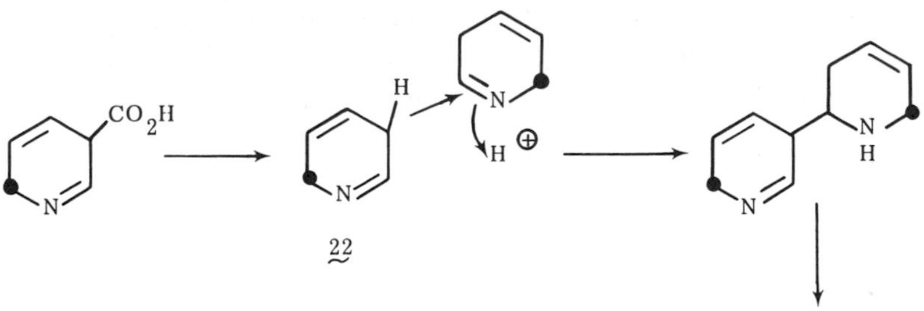

22

4   anatabine

Scheme 2   Biogenesis of anatabine (4).

When [1-^{14}C]acetate was used as a precursor of dioscorine in *Dioscorea hispida* Dennst., 30% of the total label appeared at C-5, suggesting that only four acetate units were involved, the implication clearly being that lysine provided the piperidine ring carbons. Unfortunately neither [2-^{14}C]lysine (**19**) nor [6-^{14}C]1,2-dehydropiperidine (**26**) labeled dioscorine, and this was thought to be due to the precursor not reaching the site of synthesis. [6-^{14}C]1,2-Dehydropiperidine was subsequently incorporated, but degradation indicated that the incorporation was randomized rather than specific.

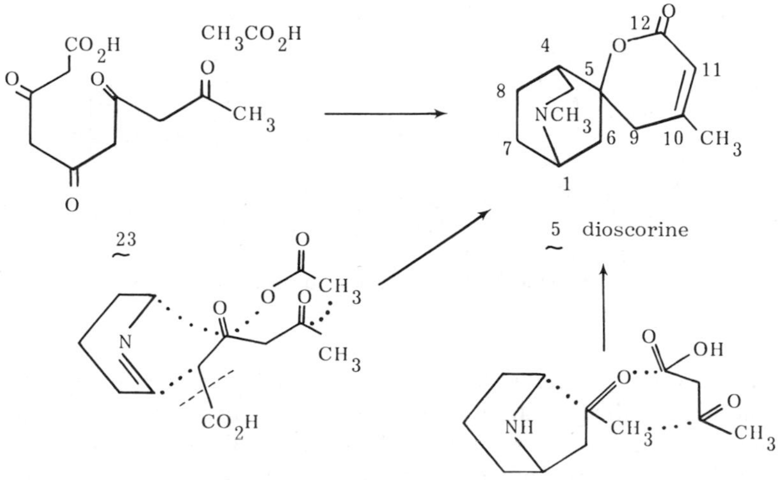

23

24

5   dioscorine

25   pelletierine

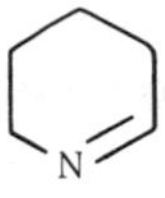

26

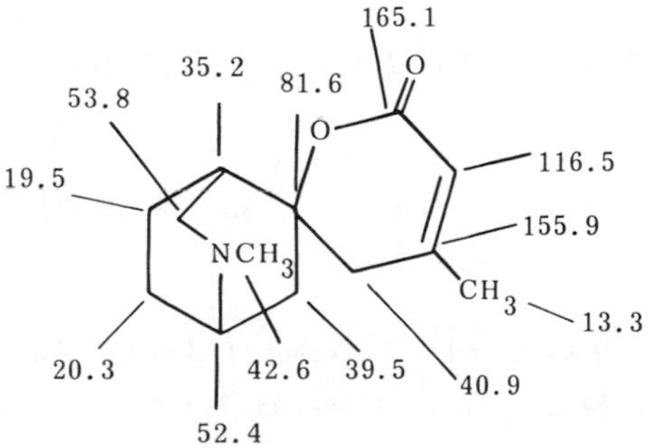

27

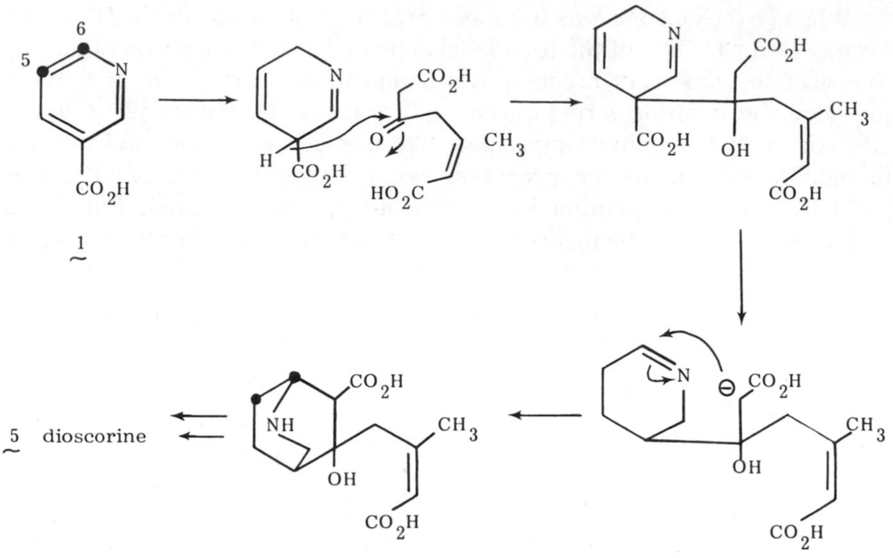

Scheme 3   Biogenesis of dioscorine (5).

Leete has reinvestigated the biosynthesis of dioscorine (5) and found that nicotinic acid is the precursor of the piperidine ring! Thus [2-^{14}C]nicotinic acid (1) was specifically incorporated into the quinuclidine nucleus to the extent of 1.9%. A more demanding experiment utilized [5,6-^{14}C, ^{13}C$_2$]nicotinic acid (1) and gave a 2.9% absolute incorporation, but more important, afforded dioscorine specifically labeled at C-7 and C-1 by adjacent ^{13}C labels. The ^{13}C assignments of dioscorine are shown in 27.

A new scheme has therefore been suggested for dioscorine (Scheme 3) and the initial steps bear a similarity to nicotine biosynthesis.

LITERATURE

Arecoline

Coutts, R. T., and J. R. Scott, *Can. J. Pharm. Sci.* **6**, 78 (1971).

Antabine

Leete, E., and S. A. Slattery, *J. Amer. Chem. Soc.* **98**, 6326 (1976).

Ricinine

Yang, K. S., R. K. Gholson, and G. R. Waller, *J. Amer. Chem. Soc.* **87**, 484 (1965).

Johns, S. R., and L. Marion, *Can. J. Chem.* **44**, 23 (1966).

Juby, P. F., and L. Marion, *Can. J. Chem.* **41**, 117 (1963).

Gross, D., in K. Mothes (Ed.), *Biochem. Physiol. Alkaloide, Int. Symp. 4th. 1969*, Akademie-Verlag, Berlin, 1972, p. 197.

Dioscorine

Leete, E., and A. R. Pinder, *Phytochemistry* **11,** 3219 (1972).

Leete, E., *J. Amer. Chem. Soc.* **99,** 648 (1977).

Leete, E., *Phytochemistry* **16,** 1705 (1977).

chapter 6

ALKALOIDS DERIVED FROM A POLYACETATE PRECURSOR

It has not been fashionable in the past to consider nitrogen-containing compounds derived from a polyacetate precursor as alkaloids. The insertion of nitrogen into a preformed terpenoid unit (Chapter 11) was barely acceptable to purists, hence the classification pseudoalkaloids.

Increasingly though there have been reports concerning the isolation and chemistry of compounds derived from a polyacetate chain, and incorporating a nitrogen atom as a key component of the ring system. In the author's opinion these alkaloids are now of such importance that they should be regarded as a distinct group of compounds within the pseudoalkaloid class.

These compounds range from simple monocyclic species such as coniine to some of the most complex nonaromatic alkaloids known. Four interesting new groups of alkaloids which are included here are the ladybird alkaloids, having the 9b-azaphenalene skeleton, the naphthalene–isoquinoline alkaloids, the *Galbulimima* alkaloids, and the cytochalasins.

6.1 SHIHUNINE

The Chinese orchid *Dendrobium lolohense* has afforded the interesting alkaloid shihunine (**1**). Permanganate oxidation gave phthalic acid **2** and hydrogenolysis gave the amino acid **3**.

The structure of shihunine (**1**) was confirmed by synthesis from *o*-carboxyphenyl cyclopropyl ketone (**4**). Condensation with methylamine gave **5**, which on treatment with aqueous HBr afforded shihunine (**1**).

Biosythetically however, the pyrrolidine ring is not derived from ornithine, but rather from insertion of nitrogen into a preformed carbon skeleton. *o*-Succinyl benzoic acid (**6**) was shown to be this key precursor (13%

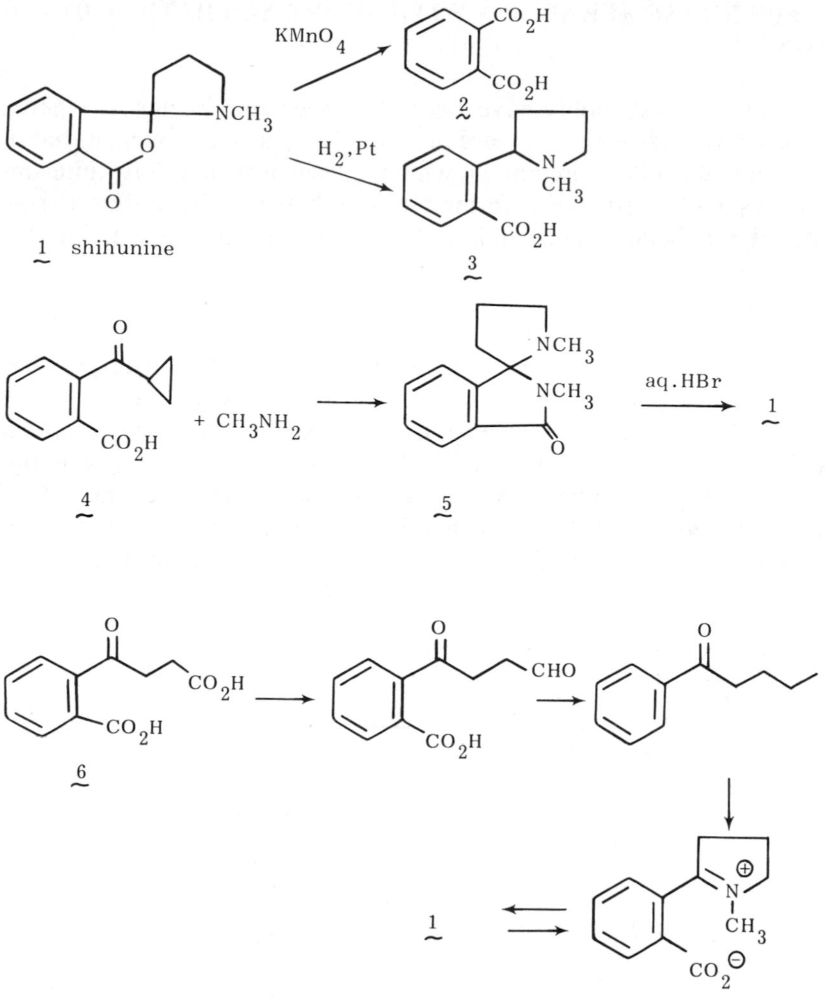

Scheme 1 Biogenesis of shihunine (**1**).

incorporation), and a scheme outlining a possible route is shown in Scheme 1.

LITERATURE

Elander, M., L. Gawell, and K. Lander, *Acta Chem. Scand.* **25**, 721 (1971).

Breuer, E., and S. Zbaida, *Tetrahedron* **31**, 499 (1975).

Leete, E., and G. B. Bodem, *Chem. Commun.* 522 (1973).

6.2 PIPERIDINE ALKALOIDS WITH SHORT ALIPHATIC SIDE CHAINS

Many piperidine alkaloids have been discussed in Chapter 4, where the alkaloids derived from lysine were presented. It was mentioned at that point that several alkaloids were known which, although structurally quite similar to the lysine-derived, were formed biosynthetically by different routes. Among these alkaloids are pinidine (**7**), coniine (**8**), and nigrifactine (**9**).

6.2.1 Pinidine

Pinus species (Pinaceae) are noted for the presence of their terpinoid constituents; alkaloid constituents are quite rare, and when they are present they are quite simple. A typical example is pinidine (**7**) from *Pinus sabiniana* Dougl., whose structure was deduced by hydrogenation/dehydrogenation to 6-methyl-2-propyl pyridine (**10**), and ozonolysis to give acetaldehyde. The relative configuration was established by conversion to 2,6-dimethyl-*N*-benzylpiperidine (**11**), which since it was optically inactive must have the *cis*

7 pinidine   8 (+)-coniine   9 nigrifactine

10

11

stereochemistry. The *cis* and *trans* isomers can also be distinguished by the multiplicity of the benzyl methylene protons. The dissymmetry of the *trans* isomer results in two doublets for these protons because of magnetic nonequivalence, but they appear as a singlet in the *cis* isomer.

The susceptibility of 2-alkyl pyridines to anion formation has been used in a recent synthesis of pinidine (**7**). 2,6-Dimethylpyridine (**12**) is converted to the anion, which is then condensed with acetaldehyde to give the alcohol **13**. Catalytic reduction and dehydration gave (±)-pinidine (**14**), which could be resolved with the optically active 6,6'-dinitrodiphenic acids.

Pinidine (**7**) was shown to be specifically derived from acetate and it was thought that possibly decanoic acid (**15**), 3,7-dioxodecanoic acid (**16**),

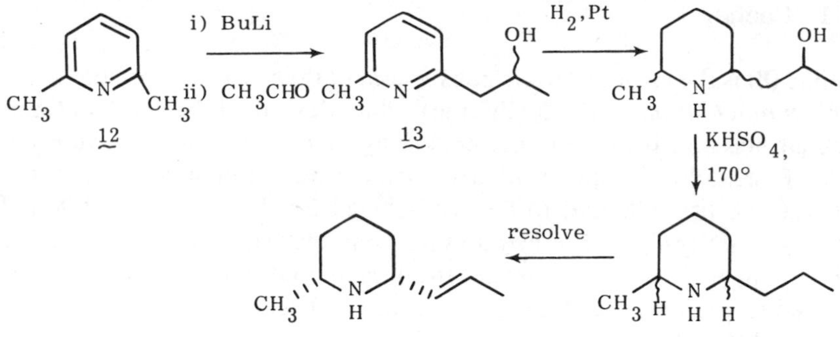

7 (−)-pinidine   **14**

or 2,6-nonadione (**17**) were intermediates. However, negligible incorporation of the precursors was observed.

The explanation of this dilema was uncovered when it was found that the "starter" acetate unit in the chain was at carbons 7 and 2, in essence at the opposite end than had been previously considered likely (compare coniine). The failure of decanoic acid (**15**) to be incorporated was suggested

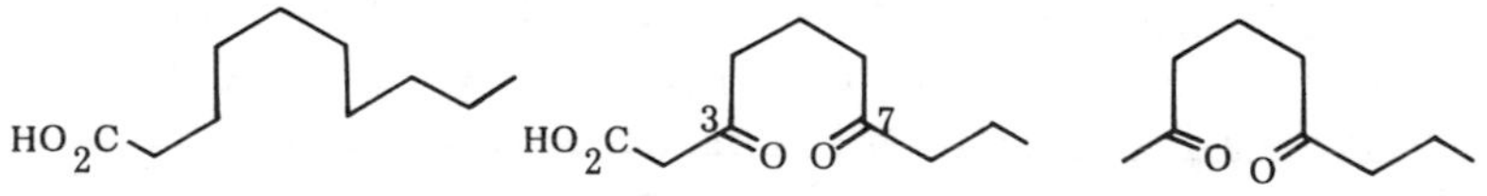

15 decanoic acid   **16** 3,7-dioxodecanoic   **17** 2,6-nonadione
                                     acid

to be due to the importance of the presence of a double bond in the fatty acid precursor.

Some even simpler alkaloids, including **18** and **19**, have been obtained from *Nanophyton erinaceum* (Chenopodiaceae); note that these are *trans* isomers.

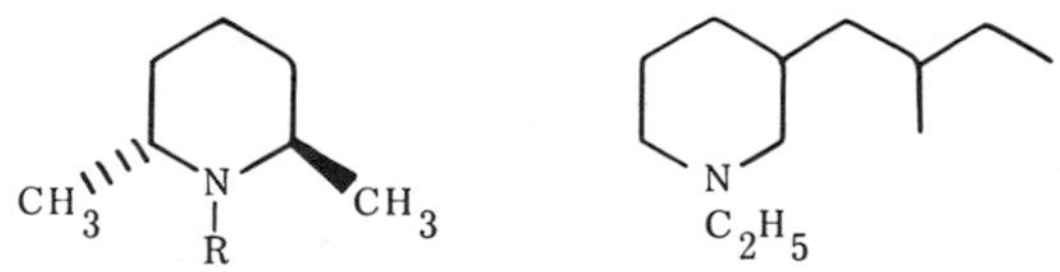

18 R=H   **20** stenusine

19 R=CH$_3$

A novel application of an alkaloid by the host organism is that of stenusine (**20**) by the staphylinid *Stenus comma*. The compound is excreted by the defence gland and acts as a water-spreading agent, enabling the animal to move rapidly over water.

6.2.2   Coniine

Coniine (**8**) is the principal toxic constituent of the notorious poison hemlock, *Conium maculatum* L. (Umbelliferae). Plato described the death of Socrates in about 400 B.C. from hemlock poisoning, and a draught was widely used by the Greeks to execute criminals. Coniine was originally isolated in 1827 and was the first alkaloid to be synthesized (by Ladenburg in 1886). It is strongly basic (pK_a 11.25), has a quite characteristic odor, and is optically active ($[\alpha]_D$ + 15.7°). Dehydrogenation gave conyrine (**21**), which could be oxidized to pyridine-2-carboxylic acid (**22**) and reduced to (±)-coniine (**23**).

Ladenburg's synthesis is of historical, rather than practical significance and involves the condensation of 2-methylpyridine (**24**) with acetaldehyde at 250°C. The resulting 2-propenylpyridine (**25**) was reduced by sodium in alcohol to (±)-coniine (**23**), which could be resolved as the tartrate salt. The overall yield was very low and alkaloid synthesis has come a long way since these early days only 90 years ago.

Hemlock also contains a number of other simple piperidine alkaloids, including *N*-methylconiine (**26**), γ-coniceine (**27**), conhydrine (**28**), and pseudoconhydrine (**29**). The structures and stereochemistries of these were determined by degradation. For example, Hofmann degradation of pseudoconhydrine (**29**) gave an alkene **30**, which on further Hofmann degradation afforded an epoxide **31** in a manner typical for an α-amino alcohol.

The biosynthesis of coniine (**8**) has been mainly deduced by the excellent work of Leete over many years. Initial experiments with [2-^{14}C]lysine (**32**),

21   conyrine

22

24

25

23   (±)-coniine

26

27   γ-coniceine

28   conhydrine

29 pseudoconhydrine   30   31

[1,5-^{14}C$_2$]cadaverine (**33**), and [6-^{14}C]1,2-dehydropiperidine (**34**) established that these were not intermediates or precursors of coniine, but when [1-^{14}C]acetate and [2-^{14}C]acetate were used specific incorporation was observed as indicated. The late stages of the biosynthetic scheme were investigated with [1-^{14}C]γ-coniceine (**27**) and excellent incorporation into both coniine (**8**) and N-methylconiine (**26**) was observed. The γ-coniceine (**27**) to coniine (**8**) step is reversible however.

8 (+)-coniine   • Label from [1-^{14}C]-acetate

32   33   34

The intermediate steps were elucidated with the aid of labeled fatty acid derivatives. [1-^{14}C]-, [7-^{14}C]-, and [8-^{14}C]Octanoic acids (**35**) were well incorporated into coniine, which became labeled at positions 6, 2', and 3' respectively. Even more significant however was the incorporation of [6-^{14}C]5-oxooctanoic acid (**36**) and [6-^{14}C]5-oxooctanal (**37**) which are clearly the immediate precursors of γ-coniceine (**27**). Whether octanoic acid (**35**) is a true intermediate remains to be firmly established. It can certainly be oxidized to **36** in *C. maculatum*, but this may be an aberrant pathway (Scheme 2).

The enzyme that converts γ-coniceine (**27**) to coniine (**8**), γ-coniceine reductase, is NADPH dependent and delivers a hydride from the *pro*-S side of the nucleoside.

Both coniine and γ-coniceine are extremely toxic and induce paralysis of the motor nerve endings. After an initial acceleration, respiration is slowed, and death results due to paralysis of the respiratory system.

The pitcher plant, *Sarracenia flava*, L., is insectivorous and seems to use coniine as a paralyzing agent.

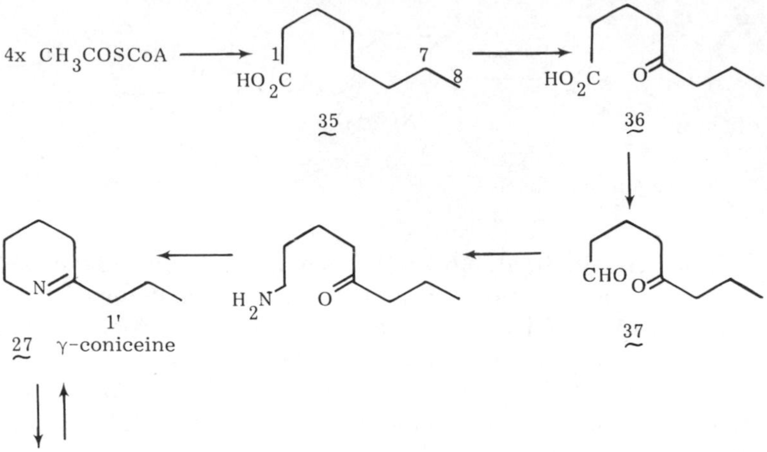

Scheme 2   Biosynthesis of coniine (**2**).

It was only recently that *Conium maculatum* has been shown to have teratogenic activity in both calves and swine. Further work established that coniine (**8**) fed to pregnant cows during days 55–75 of gestation resulted in limb and spinal defects similar to those from the whole plant.

6.2.3   Nigrifactine

Nigrifactine (**9**) from several *Streptomyces* strains is an extremely unstable piperidine alkaloid which has been synthesized by two quite interesting routes. One of these involves the unsaturated chloro ketone **38**, which was converted to **9** on treatment with sodium azide followed by reaction with

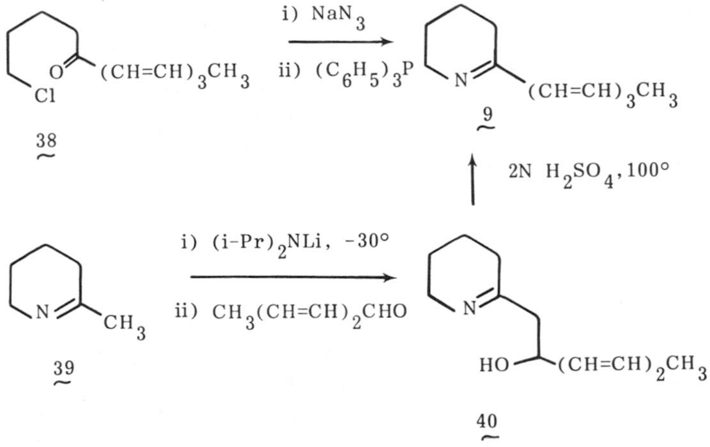

Scheme 3

triphenylphosphine. The mechanism of this reaction is not known with certainty, but may involve a phosphorous imine intermediate.

The second synthesis elaborates the side chain after formation of the heterocyclic ring. Thus 2-methyl-1,2-dehydropiperidine (39) is converted to the corresponding anion, which is condensed with hexa-2,4-dienal at −70°C to give the alcohol 40. Surprisingly, this alcohol was dehydrogenated at 100°C with 2 N sulfuric acid to give 9 (Scheme 3).

LITERATURE

Pinidine

Tallent, W. H., V. L. Stromberg, and E. C. Horning, *J. Amer. Chem. Soc.* **77**, 6361 (1955).

Hill, R. K., and T. H. Chan, *Tetrahedron* **21**, 2015 (1965).

Schildknecht, H., D. Krauss, J. Connert, H. Essenbreis, and N. Orfanides, *Angew. Chem. Int. Ed.* **14**, 427 (1975).

Leete, E., and R. A. Carver, *J. Org. Chem.* **40**, 2151 (1975).

Leete, E., J. C. Lechleiter, and R. A. Carver, *Tetrahedron Lett.* 3779 (1975).

Coniine

Giesecke, L., *Arch. Pharm. (Weinheim)* **20**, 97 (1827).

Ladenburg, A., *Chem. Ber.* **19**, 439 (1886).

Leete, E., *Acc. Chem. Res.* **4**, 100 (1971).

Nigrifactine

Terashima, T., Y. Kuroda, and Y. Kaneko, *Tetrahedron Lett.* 2535 (1969).

Pailer, M., and E. Haslinger, *Monatsh. Chem.* **101**, 508 (1970).

Gschwend, H. W., *Tetrahedron Lett.* 2711 (1970).

6.3 PIPERIDINE ALKALOIDS WITH LONG ALIPHATIC SIDE CHAINS

Although their biosynthesis remains unknown, there are an increasing number of alkaloids containing a piperidine ring and a long alkyl side chain at the 6-position, with occasionally a short alkyl side chain at the 2-position. Biogenetically, these alkaloids would appear to be derived from a polyacetate intermediate in a manner similar to coniine; hence they are included here rather than in Chapter 4.

6.3.1 Carpaine

The oldest of these alkaloids, and in some respects the most complex, is carpaine, a cardioactive principle from the leaves of the papaya tree *Carica papaya* L. (Caricaceae). The alkaloid is a lactone which can be readily hydrolyzed to carpamic acid. Hofmann exhaustive methylation procedures gave a 12-hydroxymyristic acid (straight-chain C_{14} acid) from carpamic acid and palladium dehydrogenation of carpaine gave pyridine 2,6-dicarboxylic acid (41). Taken together these results indicated a structure 42 for carpamic acid, and for many years it was thought that carpaine was simply the internal lactone 43. Mass spectrometric evaluation of carpaine however indicated that the molecular weight was twice that predicted and that carpaine actually had the structure 44, containing a 26-membered ring. When heated with thionyl chloride at high dilution, carpaine (44) can be obtained in low yield from carpamic acid (42).

A new synthetic route to carpamic acid (42) involves the ketone 45, which is converted to the keto ester 46. Ozonolysis, condensation with nitroethane, and reductive cyclization gives 42. A similar route has been used for the synthesis of pseudoconhydrine (29).

Carpaine induces bradycardia, depresses the central nervous system, and is a potent amoebicide.

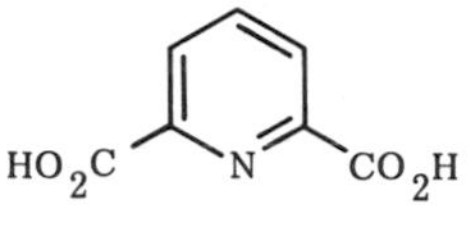

41

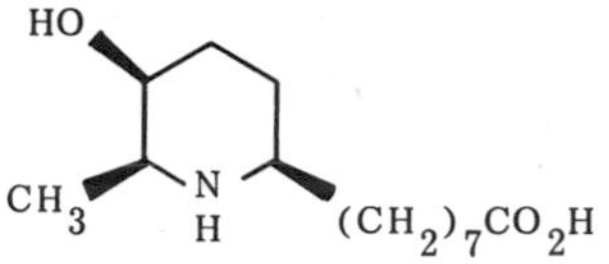

42 carpamic acid

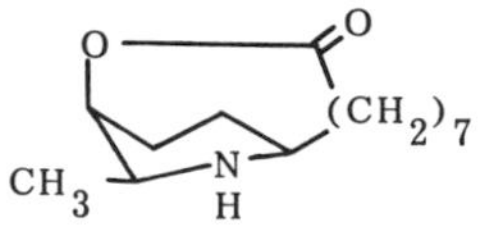

43 carpaine
(incorrect)

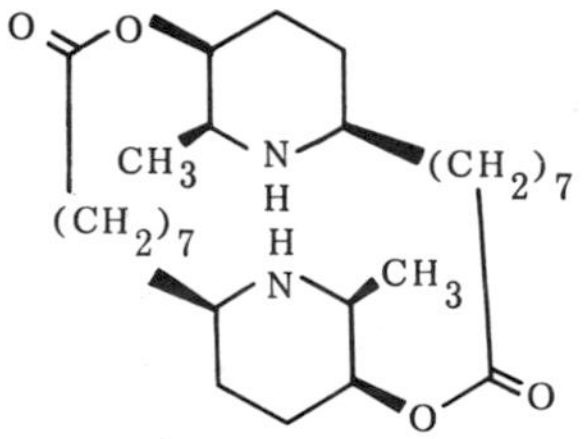

44 carpaine (correct)

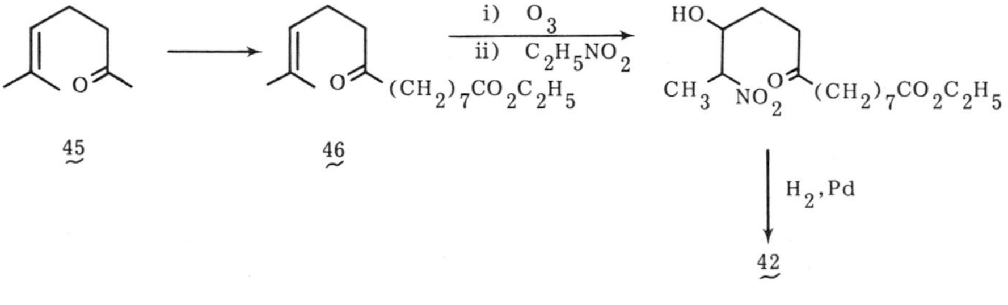

45

46

42

6.3.2 Cassine and Related Alkaloids

A closely related alkaloid, cassine (**47**), has been obtained from the tropical shrub *Cassia excelsa* Shrad (Leguminosae). The NMR spectrum showed the presence of a secondary methyl group at 1.02 ppm and a methyl ketone (singlet at 2.05 ppm). Palladium dehydrogenation gave a 3-hydroxypyridine **48**, having an aromatic methyl group (singlet at 2.5 ppm) and two *ortho*-related protons. Two structures are therefore possible for the parent alkaloid, depending on the substitution of the alkyl side chain relative to the hydroxy group. A distinction between these structures was made in a novel way, and

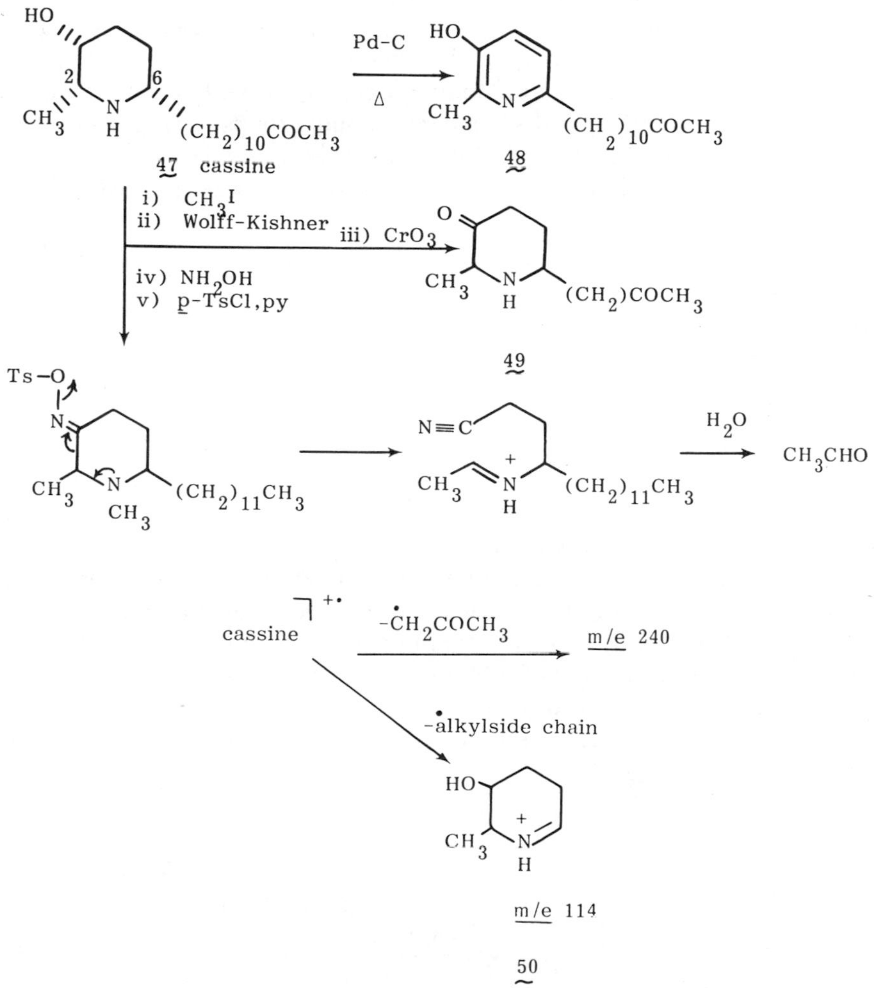

Scheme 4

relies on a known fragmentation of the oximes of α-amino ketones. Wolff–Kishner reduction of N-methyl cassine followed by oxidation of the secondary alcohol gave a ketone **49**. When the oxime of this ketone was treated with p-toluenesulfonyl chloride and pyridine, acetaldehyde was produced (Scheme 4), thereby locating the methyl group at C-2 on the piperidine ring.

The structure of cassine (**47**) was confirmed by synthesis from carpamic acid (**42**) using standard procedures. The sign of the optical rotation of the products indicated that **42** and **47** are in enantiomeric series.

The two most intense ions in the mass spectrum of cassine (**47**) occur at m/e 114 (**50**), cleavage of the alkyl chain adjacent to nitrogen, and at m/e 240, loss of the acetonyl group ($\dot{C}H_2COCH_3$).

The leaves of *Bathiorhamnus cryptophorus* (Rhamnaceae) have also afforded some alkaloids of this type. The major one was cryptophorine (**51**), which contained four conjugated double bonds from the UV spectrum, and also had N-methyl, ethyl, and secondary alcohol groups. Using a europium shift reagent, crytophorine was found to have a —CH(CH₃)CH(OH)— group. Dehydrogenation afforded a 3-hydroxypyridine **52**, which was

51 cryptophorine   52

2,3,6-trisubstituted. The all-*cis* ring stereochemistry was deduced by NMR analysis of the parent compound.

Prosopis africana Taub. (Leguminosae) has afforded several piperidine alkaloids of this general type, a typical example being prosopinine (**53**). The stereochemical relationship of the functional groups on the piperidine ring was deduced by examination of NMR spectrum of O,O'-benzylidene deoxoprosopinine (**54**). The protons at C-2 and C-3 appeared as doublets of doublets of doublets at 2.88 ppm (J = 10, 10 and 5 Hz) and 3.30 ppm (J

53 prosopinine

54

= 10, 10, and 5 Hz) respectively. These data suggested a *trans* relationship for the 2-hydroxymethyl and 3-hydroxy groups. The equatorial orientation of H-6 was assigned on the basis of the half-width of the proton, and hence the structure of prosopinine is **53**.

6.3.3 Fire Ant Venoms

These alkaloids are not however limited in their distribution to the plant kingdom, for the fire ant *Solenopsis saevissima* produces the alkaloids **55**, where $n = 10, 12,$ and 14, and **56**, where $n = 3$ and 5. The small amounts

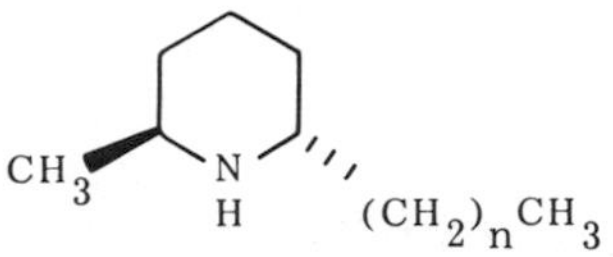

55

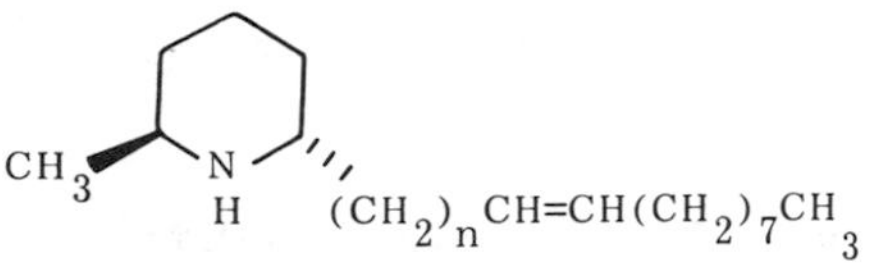

56

of these materials available has led to some interesting analytic identification techniques.

Further analysis of fire ant venoms by gas chromatography–mass spectrometry has revealed the presence of several additional derivatives having both the *cis* and *trans* stereochemistries. Some of these compounds appear to have a metabolic (and social?) significance, for the C_{11}-side chain piperidines (*cis* and *trans*) were predominant in the workers of *S. germinata* and *S. xyloni*, but the C_{13}- and C_{15}-side chain derivatives were dominant in the workers of *S. richteri* and *S. invicta*. The venoms of queens of all species contained only the C_{11}-side chain derivatives.

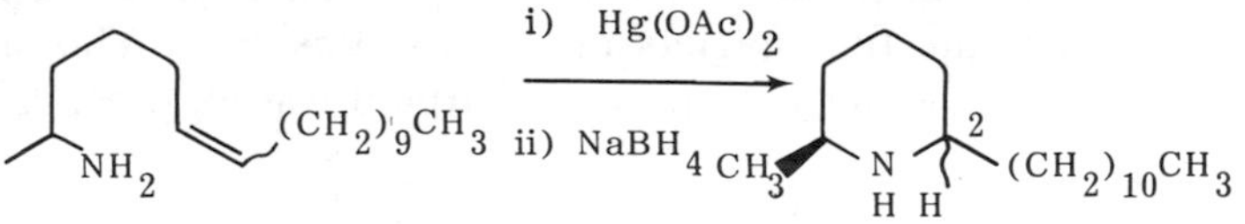

57 solenopsine A, 2 β-H

58 isosolenopsine A, 2 α-H

A new synthesis of solenopsine A (**57**) involves an interesting internal cyclization, mediated by acetate in methanol. Reduction of the mercury–amine link by sodium borohydride gave a mixture of solenopsine A (**57**) and isosolenopsine A (**58**).

LITERATURE

Carpaine

Rapoport, H., H. D. Baldridge, Jr., and E. J. Volcheck, Jr., *J. Amer. Chem. Soc.* **75**, 5290 (1953).

Spiteller-Friedmann, M., and G. Spiteller, *Monatsh, Chem.* **95**, 1234 (1964).

Brown, E., and A. Bourgaouin, *Chem. Lett.* 109 (1974).

Cassine and Related Compounds

Highet, R. J., and P. F. Highet, *J. Org. Chem.* **31**, 1275 (1966).

Khoung-Huu, Q., G. Ratle, X. Monseur, and R. Goutarel, *Bull, Soc. Chem. Belges* **81**, 425 (1972).

Bruneton, J., and A. Cavé, *Tetrahedron Lett.* 739 (1975).

Fire Ant Venoms

MacConnell, J. G., M. S. Blum, and H. M. Fales, *Tetrahedron* **27**, 1129 (1971).

Brand, J. M., M. S. Blum, and H. H. Ross, *Insect Biochem.* **3**, 45 (1973).

Moriyama, Y., D. Doan-Huynh, C. Monneret, and Q. Khoung-Huu, *Tetrahedron Lett.* 825 (1977).

6.4　9B-AZAPHENALENE ALKALOIDS (THE LADYBIRD ALKALOIDS)

Ladybirds and some beetles belong to the family Coccinellidae, and since the first studies were reported in 1971 they have proved to be an interesting new source of alkaloid structures.

Coccinellin (**59**) is an N-oxide and appears to be a defensive agent for the common ladybird *Coccinella septempunctata* L. Both coccinellin and the reduced form precoccinellin (**60**) are optically inactive, contain no olefinic protons, but do show a methyl doublet for a CH₃CH-group. Precoccinellin contains 13 carbons, but the ^{13}C NMR spectrum gives three singlets for one carbon each and five singlets of two carbons each. This unusual symmetry has been interpreted in terms of the structure **60** for precoccinellin.

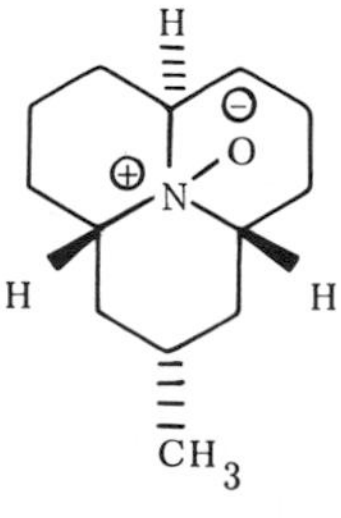

59　coccinellin 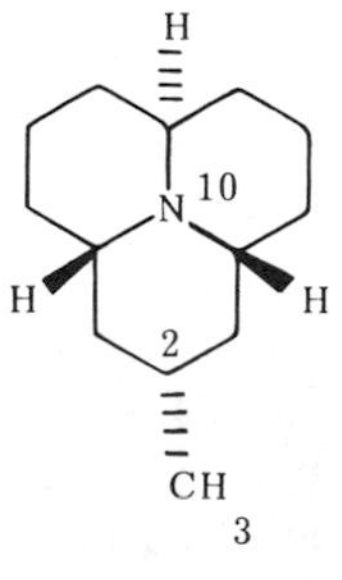　　　　　　　60　precoccinellin

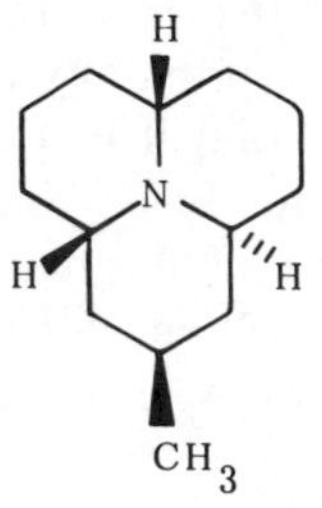

61 hippodamine 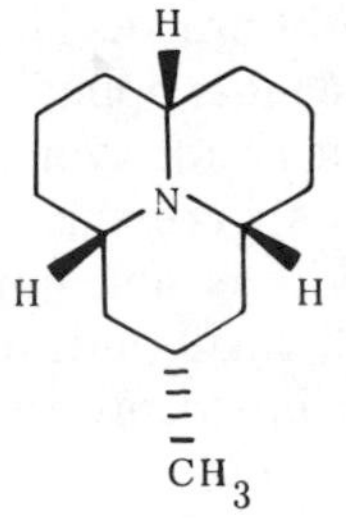   62 myrrhine

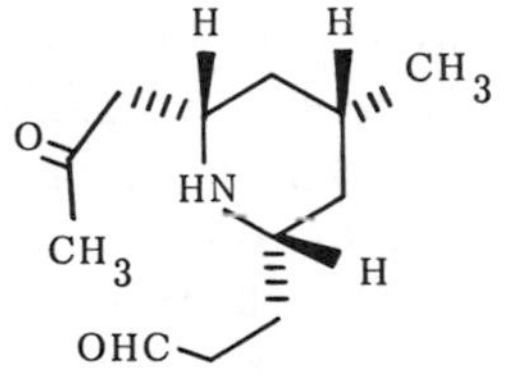

i) C₆H₅Li → i) C_6H_5Li

ii)

iii) (CH₂OH)₂, p-TsOH

Na, isoamyl alcohol

p-TsOH

H⊕ ← [63]

63

64

65 BrCH₂CH₂CH(OCH₃)₂

66

67

68

69

70

i) (CH₂SH)₂, p-TsOH
ii) Raney Ni

62 myrrhine

62 myrrhine
+
60 precoccinellin

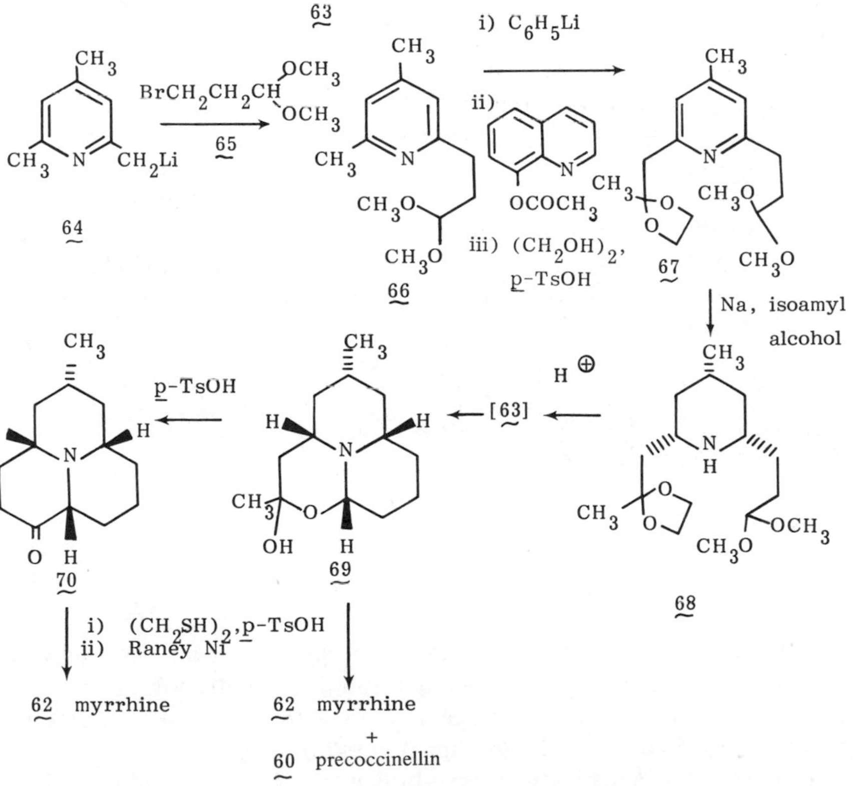

Scheme 5

217

One isomer of precoccinellin is hippodamine (61) from *Hippodamia convergens*. Like 60 it is optically inactive, but the complexity of the ^{13}C NMR indicates that it is not symmetrical, suggesting that it is a racemate of 60. These structures were substantiated by X-ray analysis. The apparent lack of optical activity is not, however, due to the racemic nature of the compound, but to a small optical rotation which is solvent dependent.

The *trans,trans,trans* isomer, myrrhine (62), was later isolated from *Myrra octodecimguttata*.

Ayer and co-workers have recently reported the synthesis of several of these interesting structures. The synthetic strategy was to aim for the keto aldehyde 63, which could undergo a double internal cyclization to either precoccinellin (60) and/or myrrhine.

Alkylation of the monolithium derivative of 2,4,6-trimethylpyridine (64) with the dimethyl acetal of β-bromopropionaldehyde (65) gave 66. Monolithiation with phenyllithium and acylation with 8-acetoxyquinoline afforded a methyl ketone which could be protected as the ethylene ketal 67. Reduction with sodium in isoamyl alcohol gave a mixture of steroisomers containing, as expected, mainly the all-*cis* piperidine 68. Hydrolysis with 5% aqueous hydrochloric acid yielded the hemiketal (69), which was rearranged with *p*-toluenesulfonic acid to the unstable ketone 70. Thioketalization and Raney nickel reduction afforded racemic myrrhine (62).

When the hemiketal 69 was rearranged with pyrrolidine–acetic acid, a mixture of two ketones resulted which could be thioketalized and reduced to a mixture of myrrhine (62) and precoccinellin (60) (Scheme 5).

Distribution of these azaphenalene alkaloids is not limited to the insect kingdom; for in 1971 Johns and co-workers isolated the alkaloids porantherine (71) from the woody shrub *Poranthera corymbosa* Brogn. (Euphorbiaceae) together with several close relatives. Corey and Balanson have reported a computer-assisted retrosynthetic analysis of porantherine (71) from 72 in 10 steps.

71   porantherine   72

Although it has not been used for a total synthesis, a promising general approach to the skeleton involves oxidation of perhydroboraphenalene (73) with a modified von Rudloff reagent to give the triketone 74. Reductive amination afforded the perhydrophenalene (75).

A preliminary investigation of the biosynthesis of coccinellin established by Kuhn–Roth oxidation that carbons 2 and 10 were derived from acetate.

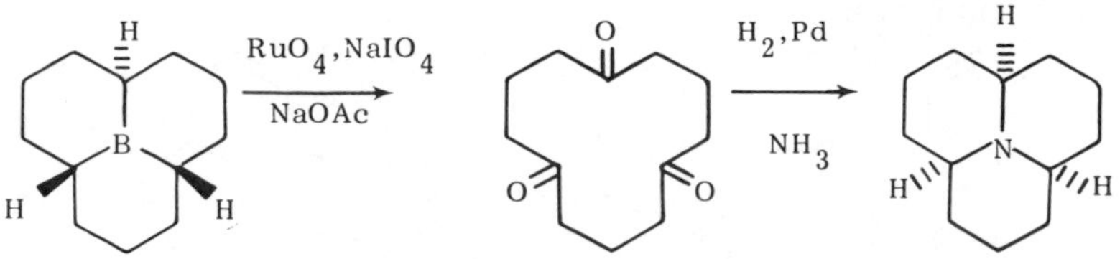

73   74   75

LITERATURE

Isolation

Tursch, B., D. Daloze, M. Dupone, J. M. Pasteels, and M. C. Tricot, *Experientia* **27**, 1380 (1971).

Tursch, B., D. Daloze, J. M. Pasteels, A. Cravador, J. C. Braekman, C. Hootele, and D. Zimmerman, *Bull. Soc. Chim. Belges* **81**, 649 (1972).

Tursch, B., D. Daloze, J. Braekman, C. Hootele, and J. M. Pasteels, *Tetrahedron* **31**, 1545 (1975).

Denne, W. A., S. R. Johns, J. A. Lamberton, and A. McL. Mattieson, *Tetrahedron Lett.* 3107 (1971).

X-ray Studies

Karlsson, R., and D. Losman, *Chem. Commun.* 626 (1972).

Tursch, B., D. Daloze, J. C. Braekman, C. Hootele, A. Cravador, D. Losman, and R. Karlsson, *Tetrahedron Lett.* 409 (1974).

Synthesis

Ayer, W. A., R. Dawe, R. A. Eisner, and K. Furuicki, *Can. J. Chem.* **54**, 473 (1976).

Corey, E. J., and R. D. Balanson, *J. Amer. Chem. Soc.* **96**, 6516 (1974).

Mueller, R. H., and R. M. DiPardo, *Chem. Commun.* 565 (1975).

6.5 NAPHTHALENE–ISOQUINOLINE ALKALOIDS

One of the novel groups of alkaloids to be discovered in the past 10 years is the naphthalene–isoquinoline group. These compounds have been isolated mainly from the plant family Ancistrocladaceae, particularly the genus *Ancistrocladus*, although there is one reported isolation from the Dioncho-phyllaceae (*Triphyophyllum peltatum*).

Three skeletal types have been isolated, and are based on the point of linkage of the 1-oxynaphthalene to the isoquinoline unit. Respectively, the linkages are 5-4′ [e.g., ancistrocladine (**76**)], 7-2′ [e.g. ancistrocladidine (**77**)], and 7-4′ [e.g., ancistrocladisine (**78**)].

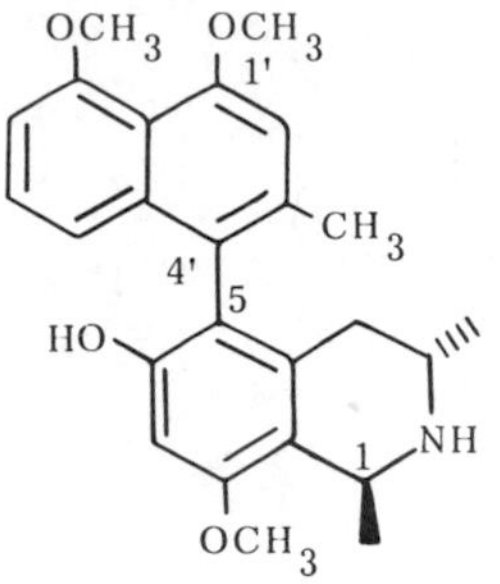

76   ancistrocladine 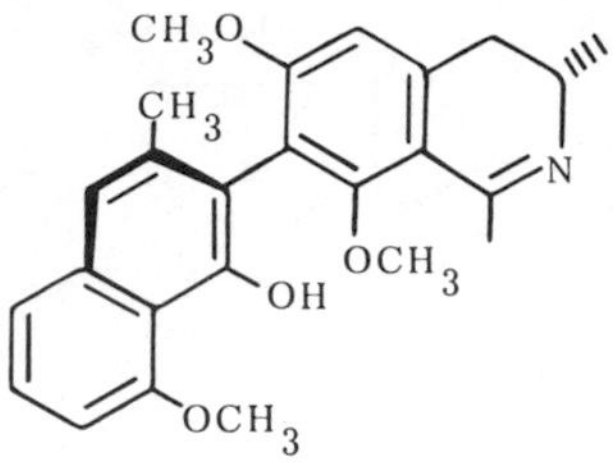   **77**   ancistrocladidine

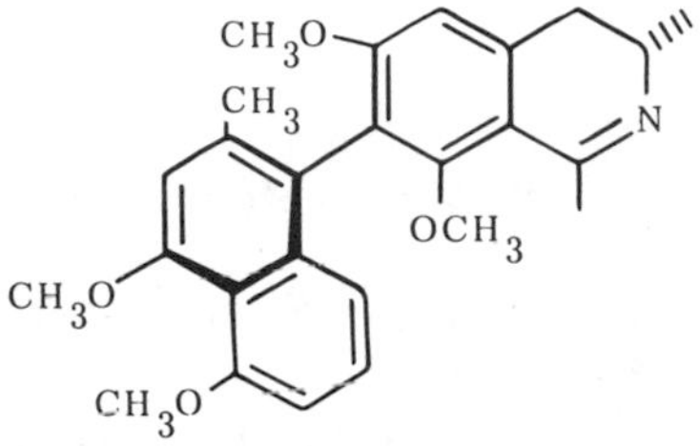

78   ancistrocladisine

These alkaloids are of interest for two reasons. First, they are a novel skeletal type whose biogenesis is probably by the acetate pathway. Second, they exhibit chirality not only due to diastereoisomerism at the methyl groups, but also in the biaryl linkage due to restricted rotation. The first alkaloid to be structurally determined was ancistrocladine (**76**), from the roots of *Ancistrocladus heyneanus* Wall.

The diacetate derivative indicated the presence of OH and NH groups in the parent compound, and the NMR spectrum of the parent indicated two secondary methyl, three aromatic *O*-methyl, and one aromatic methyl group. Oxidation with potassium permanaganate gave a naphthalene carboxylate derivative after treatment with diazomethane.

The nature of the linkage and the substitution on the isoquinoline ring system were determined by Hofmann degradation of the *O,N,N*-trimethyl derivative to **79**, followed by subsequent oxidations to give **80**. The place-

76 Ancistrocladine ⟶

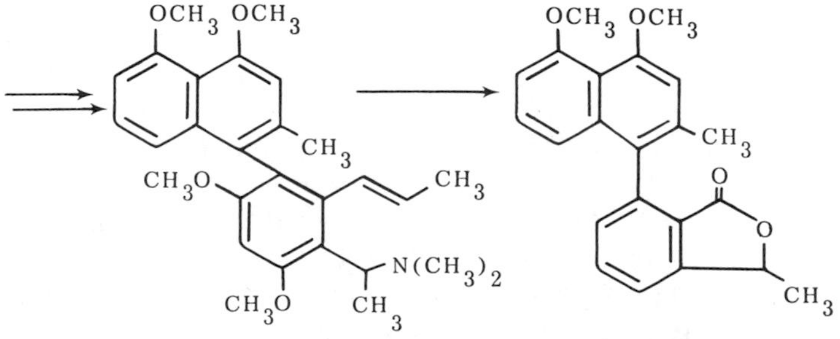

             79   80

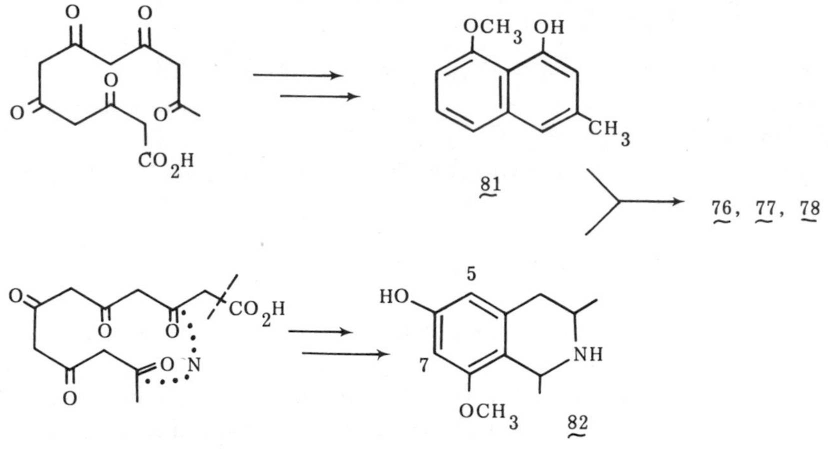

Scheme 6 Biogenesis of naphthalene–isoquinoline alkaloids in plants.

ment of the hydroxyl at C-6 was made on the basis of NMR evidence, and the location of the methoxyl at C-7 or C-8 by the observation of a Claisen rearrangement on an *O*-allyl derivative.

No synthetic or biosynthetic data have been reported for these alkaloids, but one could envisage a biogenesis by oxidative coupling of two polyacetate units, one a naphthalene **81** and the other a 6,8-dioxygenated tetrahydro-isoquinoline **82** (Scheme 6).

One of the simple monomeric isoquinoline alkaloids which appears to be closely related to the naphthalene isoquinoline alkaloids is siamine (**83**), an isoquinolone from *Cassia siamea* Lam. (Leguminosae). Biogenetically this compound could be derived as shown in **84**. The ^{13}C NMR data for siamine are shown in **85**.

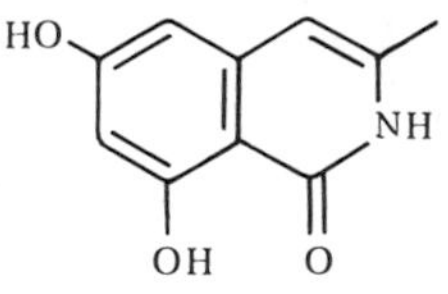

83   siamine

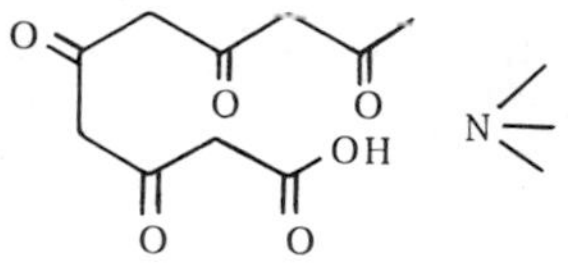

84

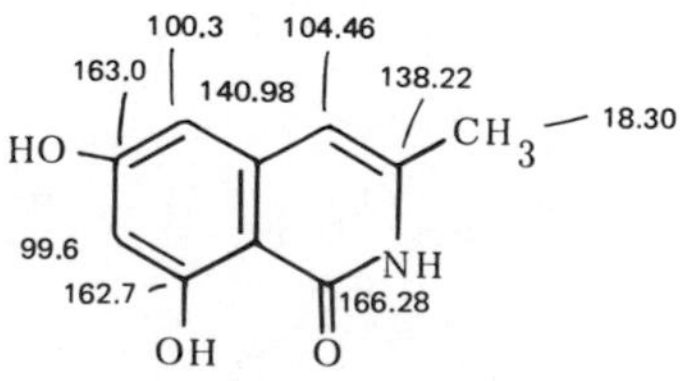

85

LITERATURE

Govindachari, T. R., and P. C. Parthasarathy, *Tetrahedron* **27**, 1013 (1971).

Govindachari, T. R., P. C. Parthasarathy, T. G. Rajagopalan, H. V. Desai, K. S. Ramachandran, and E. Lee, *Ind. J. Chem.* **13**, 641 (1975).

Bruneton, J., A. Bouquet, A. Fournet, and A. Cavé, *Phytochemistry* **15**, 217 (1976).

6.6   *ELAEOCARPUS* ALKALOIDS

The genus *Elaeocarpus* in the family Elaeocarpaceae is comprised of over 200 tropical species and these have begun to yield some very interesting alkaloids. There are three basic structure types and these can be represented by elaeocarpine (**86**), elaeokanine C (**87**), and elaeocarpidine (**88**).

Whereas the first and last compounds cooccur, elaeokanine C (**87**) and related alkaloids are limited in distribution to *E. kaniensis* Schtlr.

The structure of elaeocarpine was determined by X-ray crystallographic analysis and distinguished from isoelaeocarpine (**89**) by NMR. In **86** the C-7 proton appears as a complex multiplet at 4.15 ppm; in **89** this proton appears as a narrow "quartet" at 4.64 ppm, thereby confirming the equa-

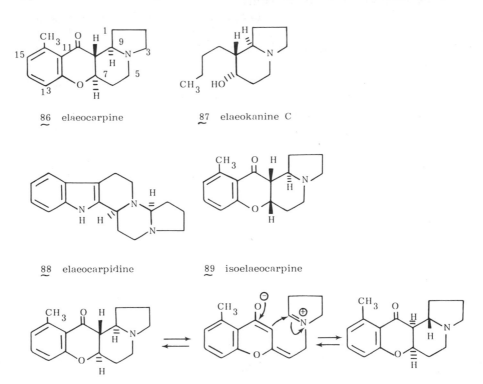

86   elaeocarpine

87   elaeokanine C

88   elaeocarpidine

89   isoelaeocarpine

Scheme 7   Interconversion of elaeocarpine (**86**) and isoelaeocarpine (**89**).

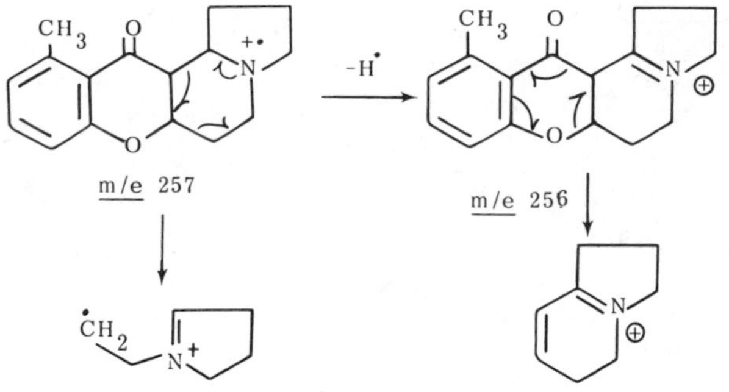

Scheme 8 Mass spectral fragmentation of elaeocarpine (**86**).

torial configuration of this proton in **89**. The isomers may be interconverted by base to a 1:1 equilibrium mixture. Although the mechanism for this interconversion is not known with certainty, it may involve inversion at both C-8 and C-9 (Scheme 7).

The principal ions in the mass spectrum are derived by two alternative retro Diels–Alder fragmentations (Scheme 8).

A relatively simple synthesis of **86** and **89** has been described by Onaka (Scheme 9) and involves the condensation of 6-methylsalicaldehyde with the dienamine **90**. Oxidation with chromium trioxide in pyridine gave an intermediate ketone which cyclized to **86** and **89**.

Elaeocarpidine (**88**) is the only indole alkaloid obtained from *Elaeocarpus* and may easily be prepared by lithium aluminum hydride reduction of the lactam **91** in 1:1 pyrrolidine–tetrahydrofuran at 0°C.

Nothing is known about the biosynthesis of any of the *Elaeocarpus* alkaloids. Biogenetically, elaeocarpine (**86**) is suggested to be formed from

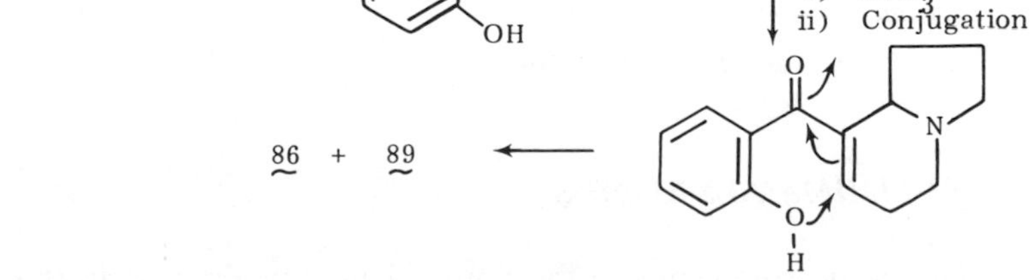

Scheme 9 Synthesis of elaeocarpine (**86**) and isoelaeocarpine (**89**).

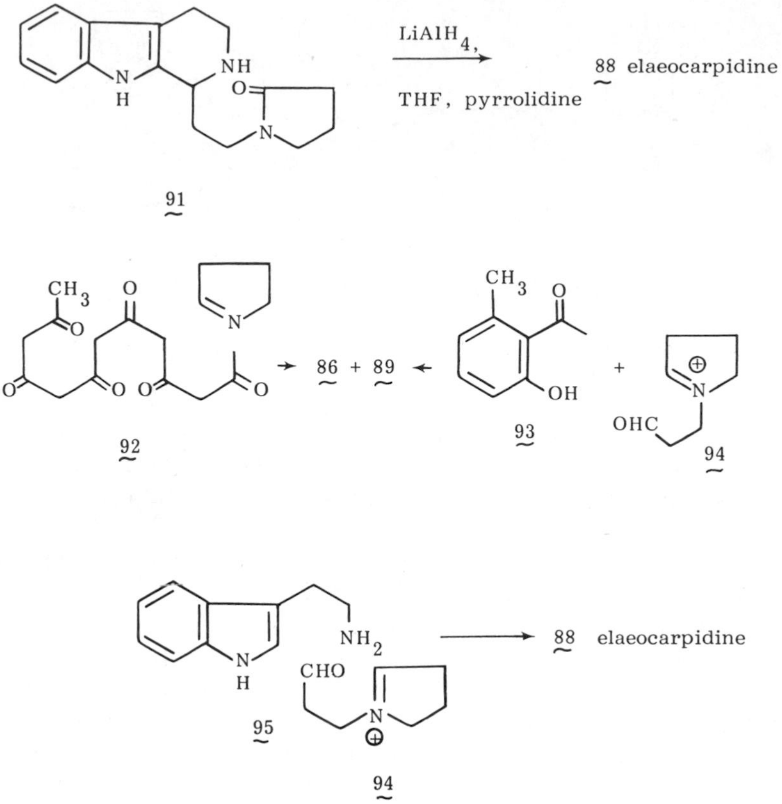

Scheme 10 Biogenesis of *Elaeocarpus* alkaloids.

polyacetate unit and ornithine as shown in **92**. An alternative sequence may involve an acetophenone and the N-alkylated pyrrolidinium species **94**. This latter species is suggested to condense with tryptamine (**95**) in the formation of **88**, and these biogenetic ideas are shown in Scheme 10.

LITERATURE

Johns, S. R., and J. A. Lamberton, *Alkaloids NY* **14**, 325 (1973).

Onaka, T., *Tetrahedron Lett.* 4395 (1971).

Gribble, G. W., *J. Org. Chem.* **35**, 1944 (1970).

6.7 *GALBULIMIMA* ALKALOIDS

Galbulimima is the sole genus in the family Himantandraceae in the order Magnoliales and is indigenous to New Guinea and North Queensland. The genus probably only contains one species, *G. belgraveana* (F. Muell.) Sprague, although this is an area of some confusion.

The leaves and wood contain little basic material, but the bark has yielded over 25 alkaloids, which are quite unlike any other alkaloids in the plant or animal kindgoms. Alkaloid content of the bark is highly variable (trace to 0.5%), but when present the most prolific alkaloids were found to be himbacine (**96**) and himandridine (**97**).

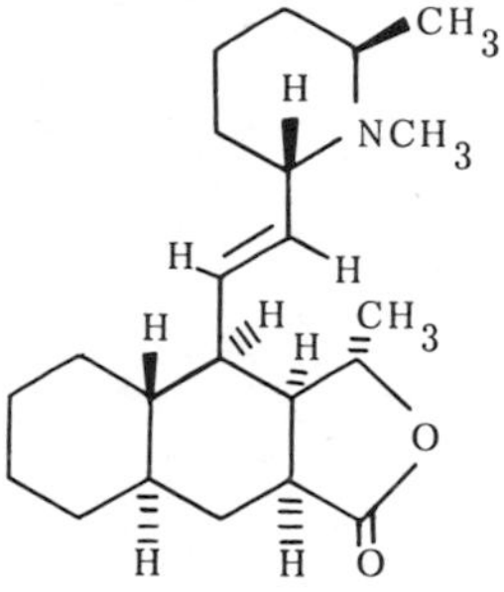

96   himbacine

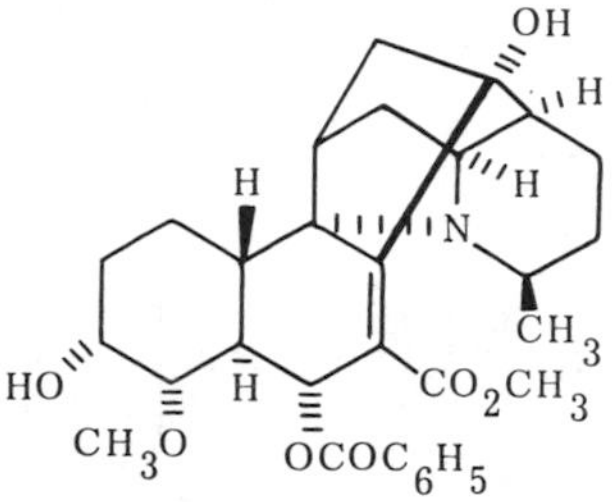

97   himandridine

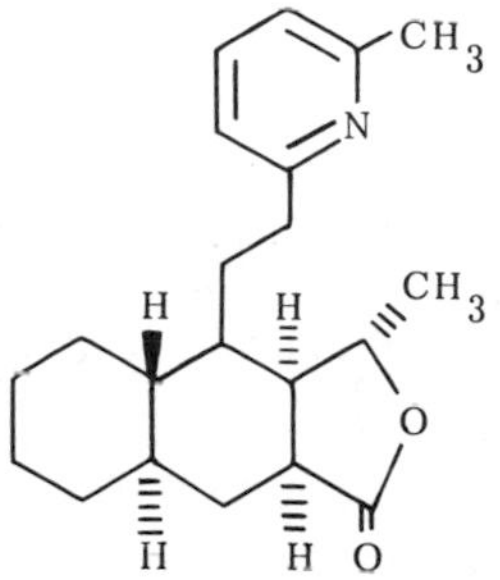

98

The structure of himbacine was established by Ritchie and co-workers in 1961 by degradative techniques, and in a sense this marked the end of an era where degradation played such an important role in structure elucidation. Evidence for the *trans*-double bond and the lactone were obtained from the IR spectrum and simple reductive transformations. Dehydrogenation with palladium–charcoal gave a 2,6-disubstituted pyridine **98**, but under more vigorous conditions with selenium at 350°C, 2,6-dimethylpyridine (**12**), 2-ethyl-6-methylpyridine (**99**), and 2-ethyl-3-methylnaphthalene (**100**) were produced.

Surprisingly no simple cleavage occurred on treatment with ozone or osmium tetroxide, but potassium permanganate did afford the lactone acid

101. Degradation of this compound proceeded through several steps to the keto lactone **102**, and from there by base elimination to acetaldehyde and the tricarboxylic acid **103**. These and many other data were put together in the form of a gross structure without stereochemistry, and the latter was subsequently deduced by X-ray crystallography to be as shown in **96**.

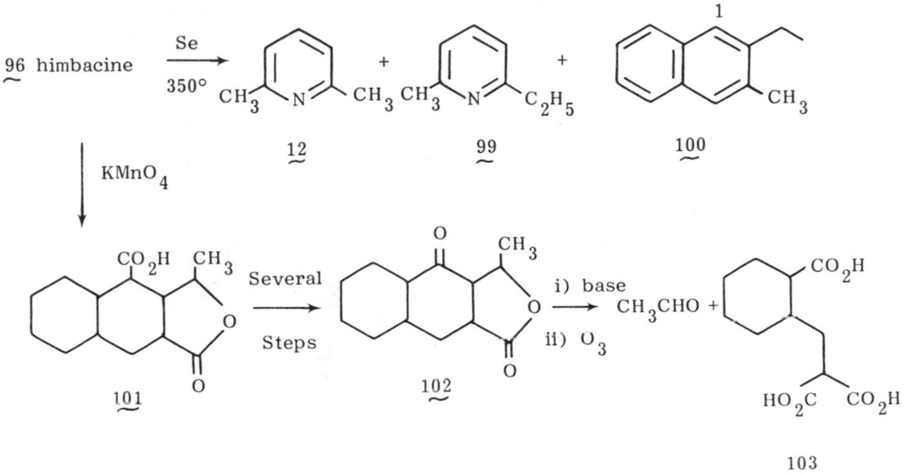

Pharmacologically, himbacine shows antispasmodic activity and little toxicity, and several derivatives and synthetic modifications have been made to potentiate this activity without success.

The structures of the ester alkaloids, of which himandridine (**97**) is the principal representative, have been deduced by relationship to himbosine (**104**), whose structure was determined by X-ray analysis.

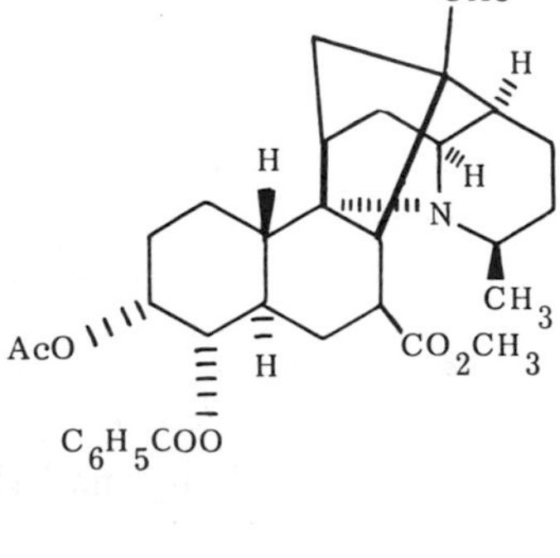

104   himbosine

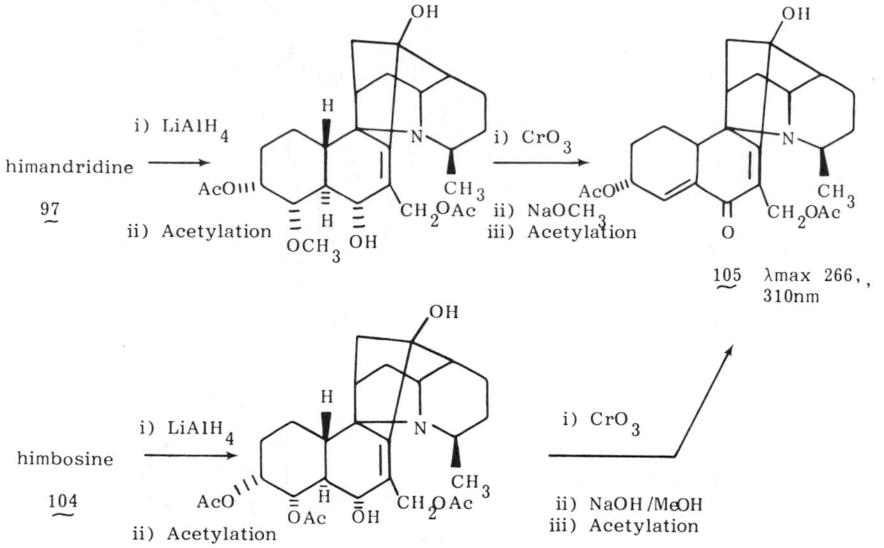

Scheme 11 Chemical correlation of himandridine (**97**) and himbosine (**104**).

Himandridine has also been subjected to numerous degradative reactions in order to establish the functional groups and to perform a chemical correlation with himbosine through the dienone diacetate (**105**) as shown in Scheme 11.

A third type of *Galbulimima* alkaloid is represented by himbadine (**106**),

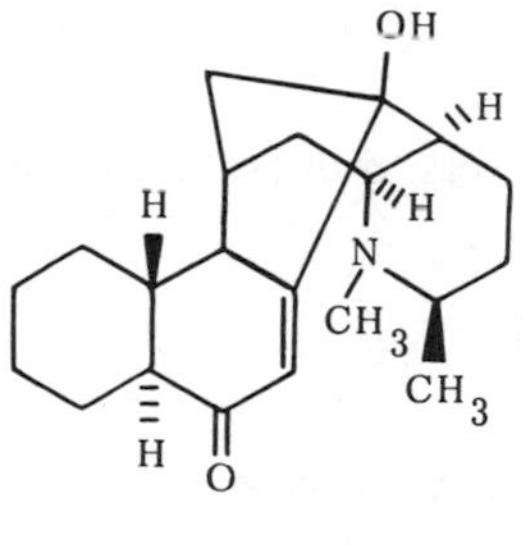

106   himbadine

which is closely related to himandridine (**97**), but lacks some of the oxygen substituents of the latter compound.

Nothing is known concerning the biosynthesis of these alkaloids, but the Australian group have speculated on their biogenesis (Scheme 12).

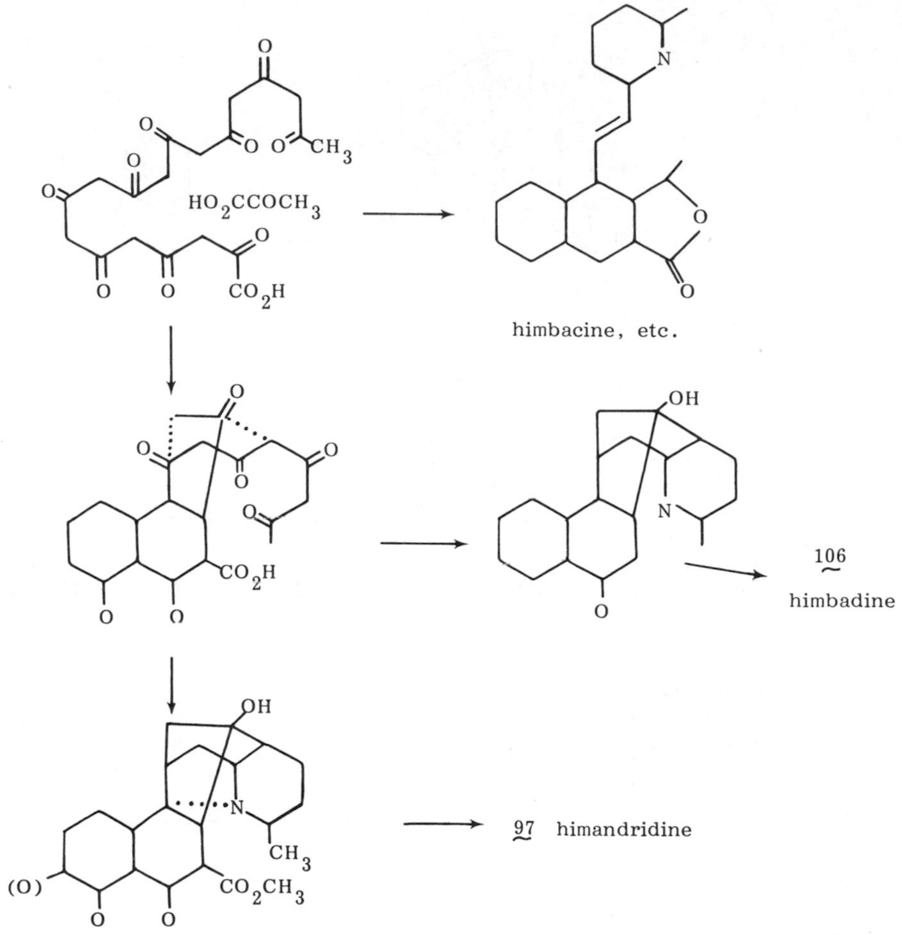

Scheme 12 Ritchie and co-workers' proposal for the biogenesis of the *Galbulimima* alkaloids.

LITERATURE

Ritchie, E., and W. C. Taylor, *Alkaloids NY* **9**, 529 (1967).

Ritchie, E., and W. C. Taylor, *Alkaloids NY* **13**, 227 (1971).

6.8 CYTOCHALASANS

One of the major new groups of compounds to be obtained in the past 10 or so years are the cytochalasans, and studies of their biosynthesis conducted mainly by Tamm and co-workers have established the polyacetate derivation of these interesting metabolites.

The cytochalasans are best known for their dramatic and varied effects on animal cells, the most notable of which is their ability to extrude the nuclei of cells. At somewhat lower doses (ca. 1 μg/ml) the cytochalasans act by blocking the formation of contractile microfilament structures, thereby preventing the final stage of cytoplasmic division.

These compounds also display several other interesting actions, including contraction of smooth muscle, thyroid secretion, growth hormone release, interference with sugar uptake, platelet aggregation, and inhibition of cytoplasmic or protoplasmic streaming.

6.8.1 Structure Elucidation and Chemistry

The cytochalasans were obtained independently by Tamm's group in Basel and by Turners' group at ICI Pharmaceuticals in England. The Swiss group obtained two compounds, phomin and dehydrophomin, from a microorganism *Phoma*, strain S 298, and the English group obtained cytochalasins A and B from the mold *Helminthosporium dematioideum*. The name cytochalasan is derived from the Greek, and means "cell relaxers." By direct comparison phomin was shown to be identical with cytochalasin B, and dehydrophomin with cyctochalasin A.

As interest in these compounds grew, additional structures were added and the nomenclature became even more confusing. The cytochalasins B (**107**) and D (**108**) were obtained from *Metarrhizium anisopliae*, and at about

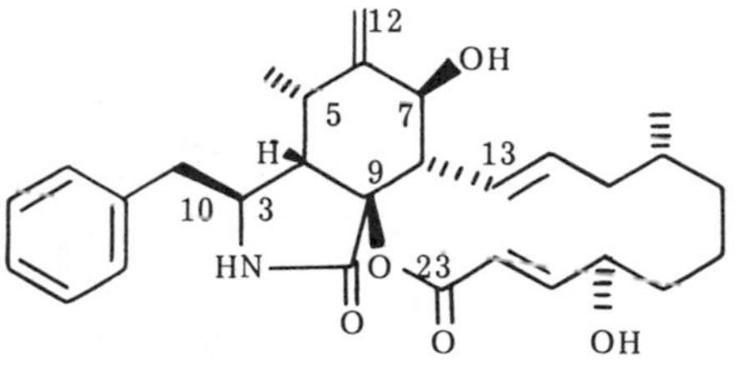

107 cytochalasin B (phomin)
      a 24-oxa-[14] cytochalasan

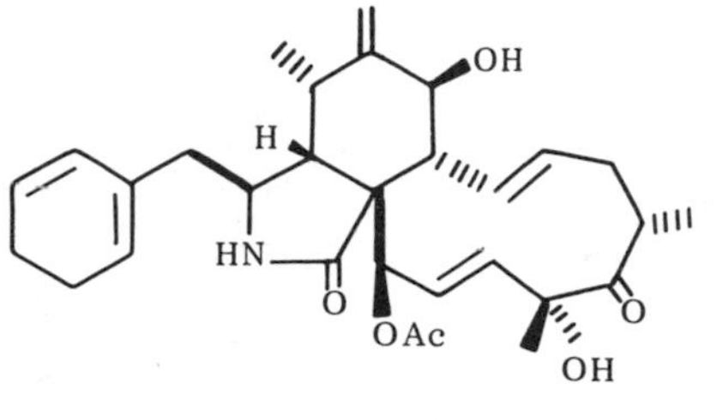

108 cytochalasin D
      an [11] cytochalasan

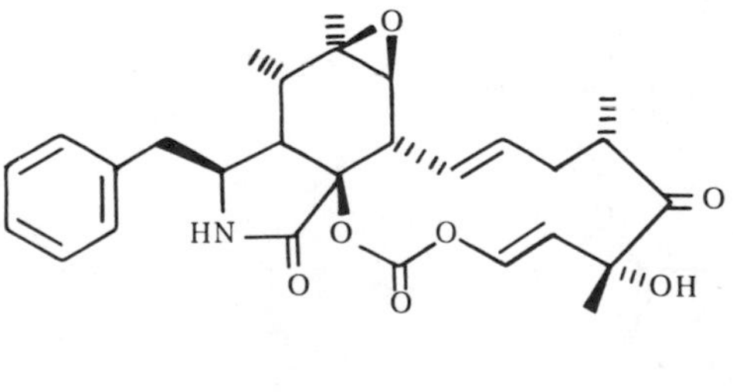

109 cytochalasin E

a 21,23-dioxa-[13] cytochalasan

the same time a new antibiotic, zygosporin A, was obtained by Japanese workers from *Zygosporium masonii*. This compound was found to be identical to cytochalasin D.

The third fundamental type of cytochalasan is represented by cytochalasin E (**109**) obtained from *Rosellinia necatrix*, which has a cyclic carbonate moiety.

The cytochalasans have a number of structural features in common, including (*a*) an oxoisoindole ring of consistent substituent stereochemistry, (*b*) a macrocrylic ring fused to the isoindole at C-8 and C-9, and (*c*) a benzyl or tryptophyl substituent at C-3. Substituents on the central core typically are a methyl group at C-5, a methyl or exomethylene group at C-6, an epoxide at C-6 - C-7 or a hydroxyl group at C-6, and a methyl group at C-16. The most variable feature is the size and functionality of the macrocyclic ring. An additional methyl group is sometimes present in this ring together with at lease one carbonyl group and often an additional hydroxyl group and a double bond. The macrocyclic ring may be a lactone, an alicycle, or a cyclic carbonate and may obtain either 11, 13, or 14 atoms.

There are therefore three fundamental nuclei for the 3-benzyl cytochalasans, (*a*) the 24-oxa[14]cytochalasans, (e.g. **107**), (*b*) the [11]cytochalasans (e.g. **108**), and (*c*) the 21,23-dioxa[13]cytochalasans (e.g. **109**). The numbering system is shown on **107**.

Recently a new group of compounds has been obtained having this overall

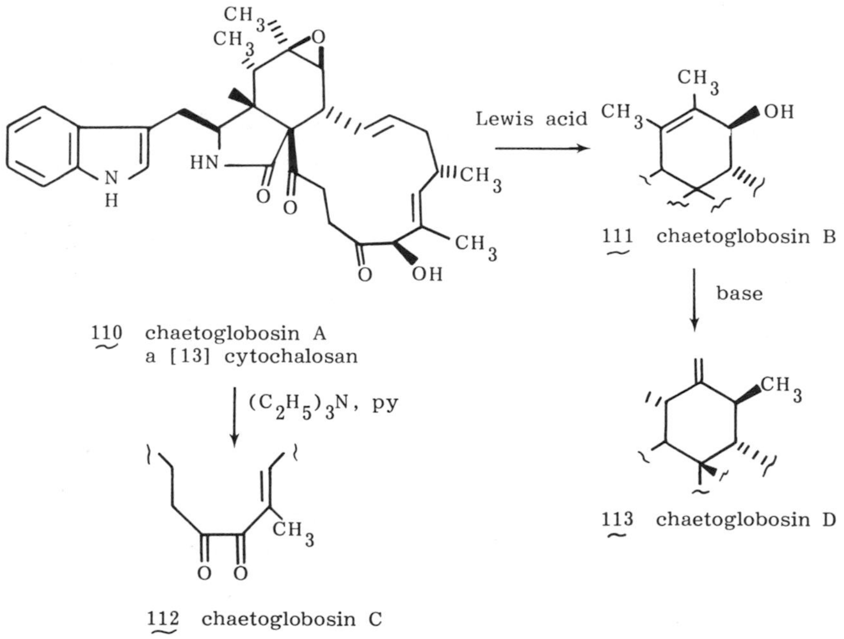

110   chaetoglobosin A
      a [13] cytochalosan

111   chaetoglobosin B

112   chaetoglobosin C

113   chaetoglobosin D

Scheme 13 Interconversion of the chaetoglobosins.

molecular array; these are the chaetoglobosins which are [13]cytochalasan derivatives substituted at the 3-position not by benzyl but by indolyl methyl.

Chaetomium globosum Kuntze ex Fries is a source of the cytotoxic chaetoglobosins A–F. The gross structures of chaetoglobosin A (110) and chaetoglobosin B (111) were confirmed by X-ray analysis. Chaetoglobosin C (112) has been obtained from *Penicillium aurantio-virens* Bunge, which infests weevil-damaged pecans.

Chaetoglobosin A (110) is converted to chaetoglobosin B (111) by a Lewis acid, and the latter is converted to chaetoglobosin D (113) by treatment with base. A series of keto-enol tautomerizations accounts for the conversion of chaetoglobosin A (110) to chaetoglobosin C (112) with triethylamine in pyridine (Scheme 13).

Ozonolytic cleavage of cytochalasin B (107) with reductive (NaBH₄) workup gave a mixture of two lactams 114 and 115 and the triol 116. The latter could be degraded to (3R)-3-methyl pimelic acid (117), whose absolute configuration was known from degradation of (+)-pulegone (118). The lac-

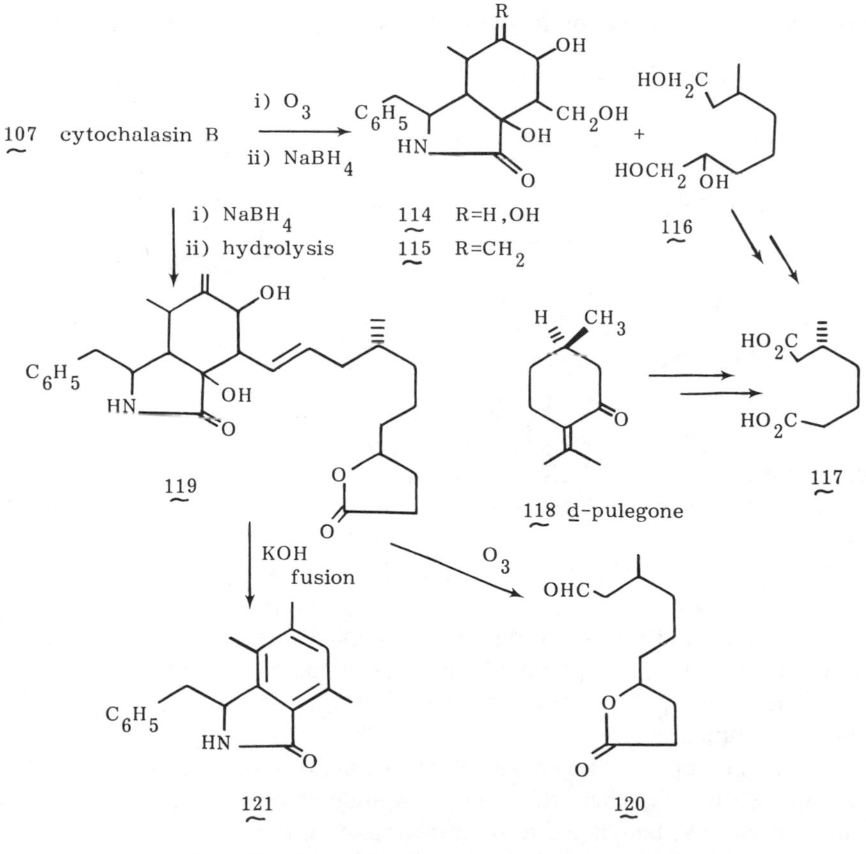

Scheme 14 Degradation of cytochalasin B (107).

tams were examined mainly by NMR spectroscopy, although periodate cleavage of the lactam **114**, did indicate the original position of the exomethylene group.

More vigorous degradations were performed by the English group, who had more material at their disposal, and from 21,22-dihydrocytochalasin B the lactone amide **119** was obtained, which could be ozonized to the aldehyde lactone, **120**. Alkali fusion of the lactone amide **119** gave the substituted phthalimide **121** (Scheme 14).

6.8.2 Biosynthesis

The biosynthesis of the cytochalasins has been studied principally by Tamm and co-workers, and over the past years has involved the use of both radioactive and stable isotopes. The phenylalanine unit was incorporated intact into cytochalasin D (**108**), including both the carboxyl and nitrogen atoms. Methionine provided three of the carbons external to the ring system, whereas [2-^{14}C]acetate labeled only the C-11 methyl group. Several of the ring carbons were also labeled by [2-^{14}C]acetate and the pattern of labeling (**122**) suggested a derivation from a polyacetate precursor. This concept was

substantiated when [1,2-^{13}C$_2$]acetate was used as a precursor and showed that the acetate units were coupled head to tail in a typical polyketide manner.

Both L-phenylalanine and its racemate were equally good precursors, even though cytochalasin D has the S-configuration at C-3 corresponding to the L-amino acid configuration. Vederas and Tamm determined that rapid equilibration of D- and L-phenylalanines occurred by way of phenylpyruvic acid. The major pathway of incorporation was shown to be by way of L-phenylalanine.

A scheme for the biogenesis of cytochalasans has been proposed by Tamm (Scheme 15). The initial step is thought to be formation of an amide linkage between phenylalanine or tryptophan and an octa- or nonaacetate. Condensation, elimination of water, and reduction affords the tricyclic systems **123** and **124**. It is proposed that these compounds are crucial inter-

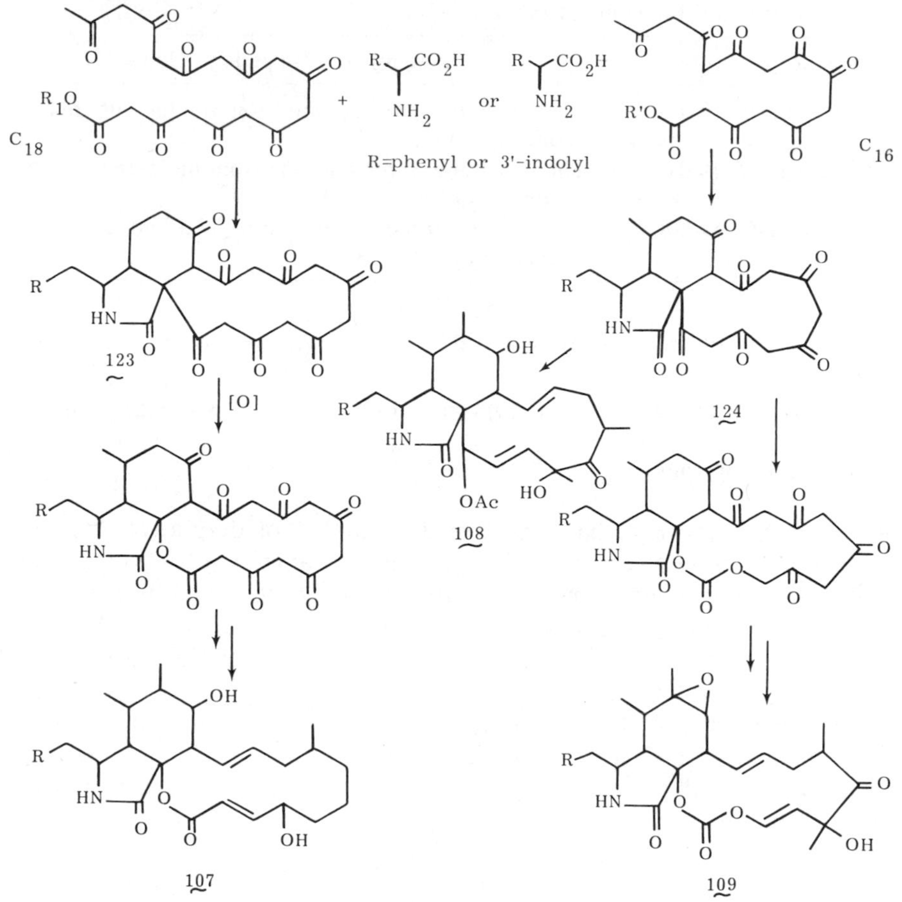

Scheme 15 Biogenesis of the cytochalasans.

mediates on which the major structural modifications such as methylation, reduction, a Baeyer–Villiger-type oxidation to the lactone system of cyto-chalasin B (**107**), and so on, take place. A second Baeyer–Villiger oxidation would afford the carbonate system of cytochalasin E (**109**).

Substantiation of the first Baeyer–Villiger-type oxidation from a feeding experiment with labeled cytochalasin A (dehydrophomin) (**125**) in *Phoma*

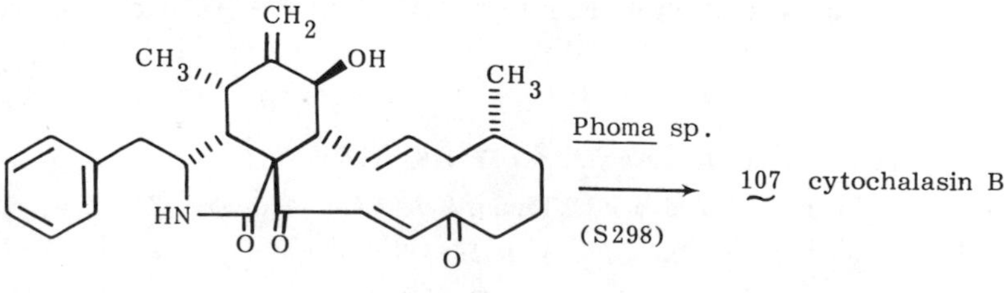

Table 1 Structural Elements of the Main Cytochalasans

Compound	Biosynthetic–Biogenetic Units
Cytochalasin B (**107**)	Phenylalanine, two C_1 units from methionine, nine acetate units
Cytochalasin D (**108**)	Phenylalanine, three C_1 units from methionine, nine acetate units
Chaetoglobosin A (**110**)	Tryptophan, three C_1 units from methionine, nine acetate units

sp. (S298) established an efficient (8%) interconversion to cytochalasin B (**107**).

The structural elements of the main cytochalasans are shown in Table 1.

6.8.3 Pharmacology

Minato and co-workers have examined a number of derivatives of cyto-chalasin D in order to establish the structural parameters for cytotoxic and antitumor activity. Surprisingly, only the benzyl group at C-3, the hydroxyl at C-7, and the macrocyclic ring were found to be essential. Modifications of the ring, such as additional hydroxyl, carbonyl, or olefinic groups, had only a marginal effect on the cytotoxicity.

Clearly then it is the central core of the cytochalasins with its fixed stereochemistries and the sheer bulk of the macrocycle which are important for activity. It will be interesting to see developments in this area as more specific binding studies are made.

LITERATURE

Review

Binder, M., and C. Tamm, *Angew. Chem. Int. Ed.* **12**, 370 (1973).

Nomenclature

Binder, M., C. Tamm, W. B. Turner, and H. Minato, *J. Chem. Soc. Perkin Trans. I* 1146 (1973).

Chaetoglobosins

Springer, J. P., J. Clardy, J. M. Wells, R. J. Cole, J. W. Dirksey, R. D. Macfarlane, and D. F. Torgerson, *Tetrahedron Lett.* 1355 (1976).

Biosynthesis

Lebet, C. R., and C. Tamm, *Helv. Chim. Acta* **57**, 1785 (1974).

Vederas, J. C., W. Graf, L. David, and C. Tamm, *Helv. Chim. Acta* **58**, 1886 (1975).

Vederas, J. C., and C. Tamm, *Helv. Chim. Acta* **59**, 558 (1976).

Robert, J. L., and C. Tamm, *Helv. Chim. Acta* **58**, 2501 (1975).

Pharmacology

Carter, S. B., *Nature* **213**, 261 (1967).

Carter, S. B., *Endeavor* **113**, 77 (1972).

Minato, H., T. Katayama, M. Matsumoto, K. Katagari, S. Matsuura, N. Sunagawa, K. Hori, M. Harada, and M. Takeaki, *Chem. Pharm. Bull.* **21**, 2268 (1973).

chapter 7

ALKALOIDS DERIVED FROM ANTHRANILIC ACID

The biosynthesis of the aromatic amino acid anthranilic acid (**1**) was discussed in Chapter 2. It is a compound that affords a wide range of alkaloids derived from both plant and fungal sources, some of considerable importance.

The plant family Rutaceae is a particularly rich source of alkaloids derived from anthranilic acid, and examples of some of the more important skeleta are shown in Figure 1. Like the previous chapters this one is arranged approximately according to the increasing molecular complexity of the compounds.

7.1 SIMPLE DERIVATIVES

Many volatile oils contain the methyl ester of anthranilic acid, and its *N*-methyl derivative is a constituent of citrus oil.

The trimethylated derivative of 3-hydroxyanthranilic acid is damascenine (**2**), the only detectable alkaloid in *Nigella damescena* L. (Ranunculaceae) seeds, which are also the main site of synthesis. Biosynthetically the alkaloid is derived from anthranilic acid (**1**) (chorismate-derived), which is then hydroxylated and methylated. A component of anthranilic acid oxidase is capable of catalyzing the formation of 3-hydroxyanthranilic acid (**3**) from **1**.

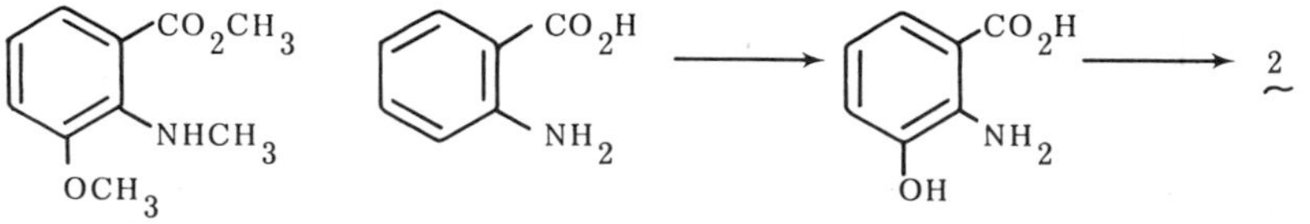

| 2 damascenine | 1 | 3 |

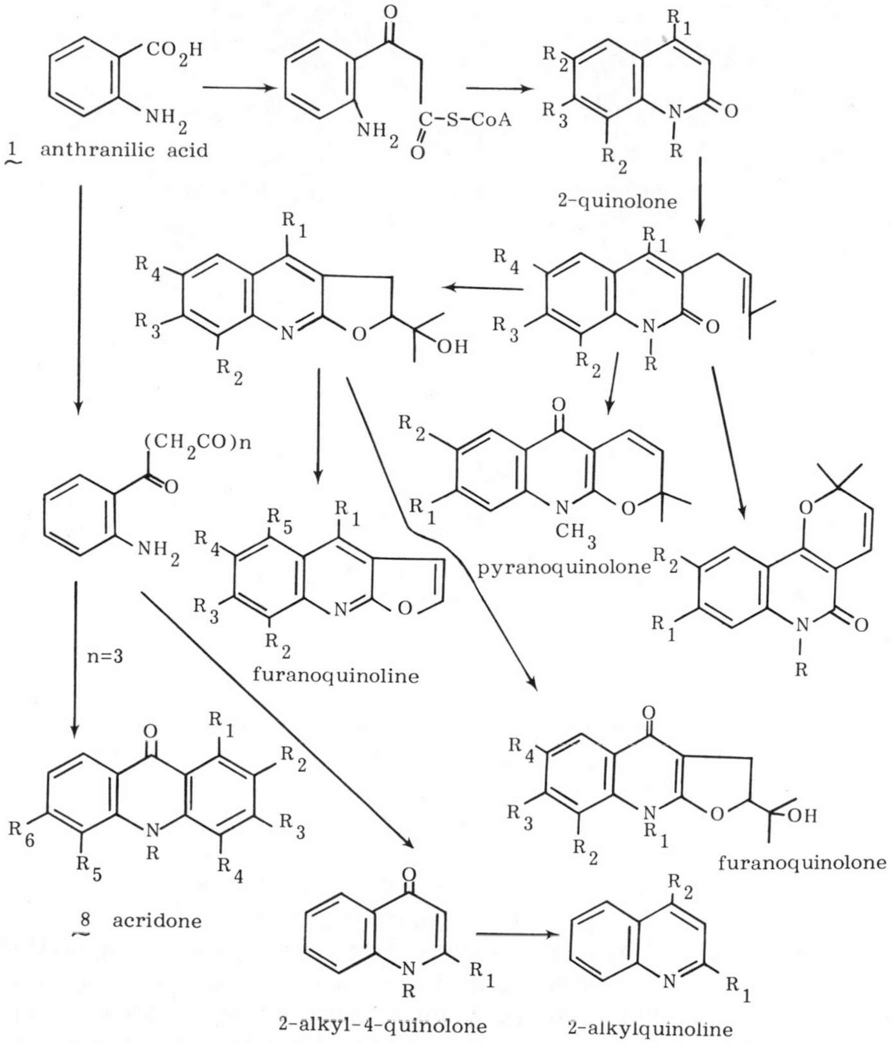

Figure 1 Alkaloids of the Rutaceae derived from anthranilic acid (**1**).

7.2 SIMPLE QUINOLINE DERIVATIVES

A host of alkaloids derived from anthranilic acid are based on the quinoline nucleus (**4**). Some of these are quite simple [e.g. echinopsine (**5**)], others have furan or pyran rings attached to the pyridine ring [e.g. dictamnine (**6**) and flindersine (**7**)], and yet others have a benzene ring attached, and are the acridone (**8**) alkaloids. We will return to the latter two groups subsequently; initially we will concern ourselves with the simple derivatives.

Echinopsine (**5**), from several *Echinops* species in the Compositae, forms

4 quinoline

5 echinopsine

6 dictamnine

7 flindersine

5 echinopsine $\xrightarrow[\text{PCl}_5, 150°]{\text{POCl}_3}$

9

pale yellow crystals that give an intense color with ferric chloride solution. In a reaction typical of quinolines, reaction of echinopsine (**5**) with 1 mole of phosphorus pentachloride in phosphorus oxychloride at 150°C affords 4-chloroquinoline (**9**). Being a vinylogous amide, echinopsine is a very weak base.

The most thoroughly studied simple quinoline alkaloids are those of *Galipea* and *Cusparia* species in the family Rutaceae. The two most important alkaloids are the closely related cusparine (**10**) and galipine (**11**). In a reaction typical of a 4-methoxyquinoline (see also the furoquinoline alkaloid section), heating with methyl iodide causes overall isomerization to an *N*-methyl-4-quinolone [e.g. the conversion of galipine (**11**) to isogalipine (**12**)].

Oxidation of galipine with potassium permanganate gave veratric acid (**13**) and 4-methoxyquinoline 2-carboxylic acid (**14**), and this accounts for all the carbon atoms. The proposed structure was confirmed by Späth, who condensed veratraldehyde (**15**) with 4-methoxy-2-methylquinoline (**16**) in the presence of zinc chloride. Such condensations are typical of the reactivity of a methyl group at the 2 (or 4)-position of a pyridine nucleus. Catalytic reduction of the intermediate styrene **17** gave galipine (**11**).

4-Quinolones having a long alkyl side chain at the 2-position have been obtained from several fungi; one recently isolated example is **18** from *Pseudomonas aeruginosa*. It was not until 1975 that these compounds were obtained from plants when the roots of *Ruta graveolens* L. were found to contain a mixture of the 2-alkyl quinolones (**19**), where $n = 10\text{–}13$.

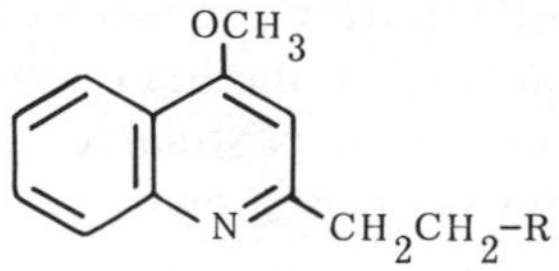

10 cusparine, R =

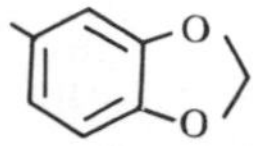

11 galipine, R =

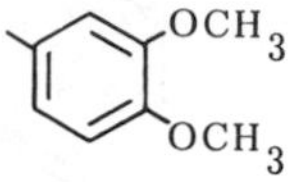

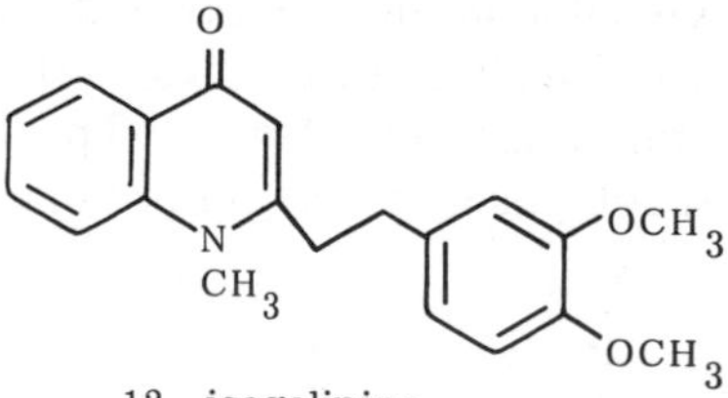

12 isogalipine

11 galipine $\longrightarrow$ 14 + 13

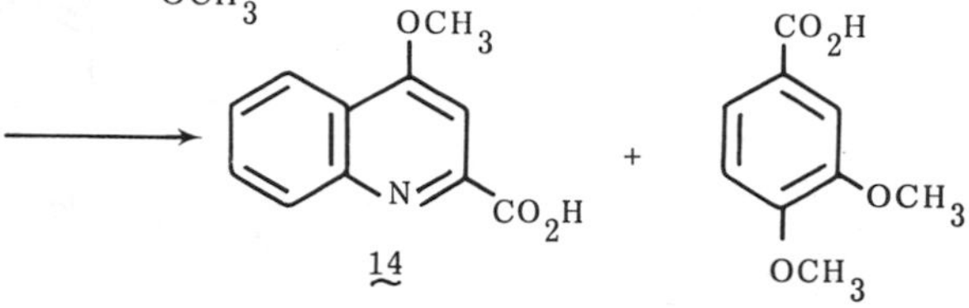

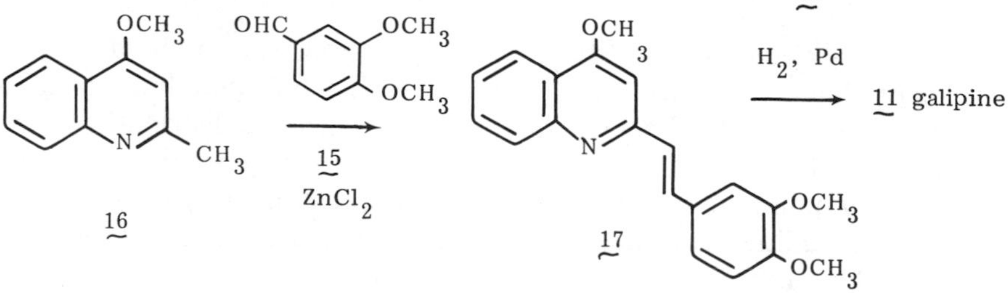

16 + 15 ZnCl$_2$ $\longrightarrow$ 17 $\xrightarrow{\text{H}_2, \text{ Pd}}$ 11 galipine

Japonine is a quinoline alkaloid obtained from the rutaceous plant *Orixa japonica* Thunb. Mainly on the basic spectroscopic data japonine was ascribed the structure **20**. This structure is unusual because it contains a 3-methoxyquinolone nucleus, and was therefore confirmed by synthesis. A second 3-oxygenated derivative is **21** isolated from *Echinops vitro*.

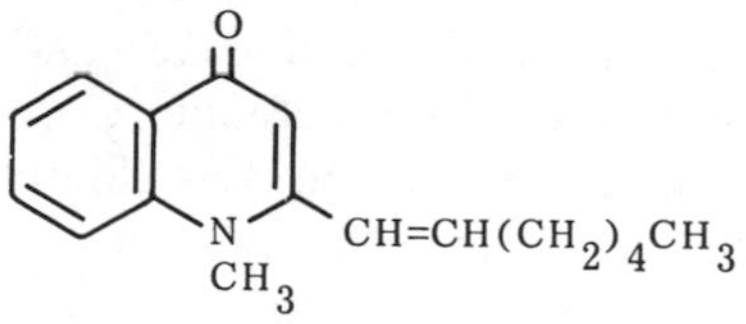

18

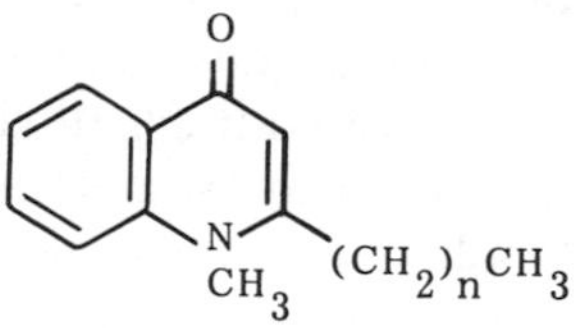

19

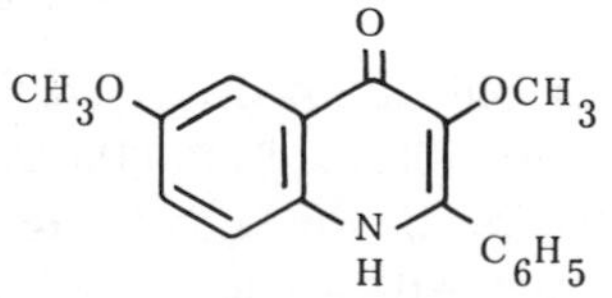

20 japonine

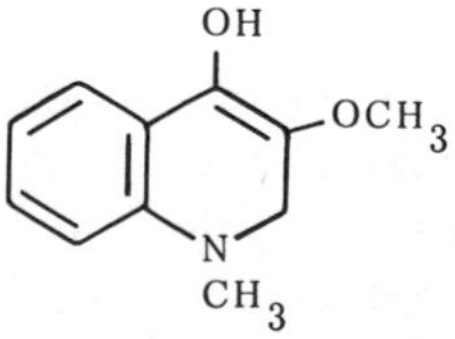

21

The novel alkaloid melochinone (**22**) has been isolated from the roots of *Melochia tomentosa* L. (Sterculiaceae), a plant implicated in the high incidence of oral cancer in Curacao. The biogenesis has been postulated to involve cinnamoylacetic acid, anthranilic acid, and pyruvate as shown in **23**.

22   melochinone

23

7.3 FUROQUINOLINE ALKALOIDS

Some of the alkaloids of the Rutaceae derived from anthranilic acid have been encountered previously in this chapter. The furoquinoline alkaloids are another structurally simple group of alkaloids whose distribution is essentially limited to one plant family, the Rutaceae. They are derivatives of the furo[2,3-*b*]quinoline system(**24**), and the simplest member is dictamnine (**6**). Common additional groups are oxygen functions at C-6, C-7, and C-8 which may be methyl or isopentenyl ethers. A hydroxyisopropyl group is common in the dihydrofuroquinolines and the significance of this will be illustrated in the discussion of the biosynthesis of this group.

7.3.1 Chemistry

Dictamnine (**6**), first isolated by Thoms in 1923, is found quite widely in the Rutaceae. Being a very weak base it does not form a derivative with methyl iodide but rather undergoes isomerization to isodictamnine (**25**), and a similar reaction is observed with dimethyl sulfate or diazomethane. Crucial information about the linear structure of dictamnine came by oxidative degradation with potassium permanganate to the acid **26**, dictamnic acid. The structure of this acid was confirmed by Dieckmann cyclization to the dihydroxyquinoline **27** followed by methylation and hydrolysis. Note that only the 4-hydroxy group of **27** is methylated with diazomethane, suggesting that the 2-hydroxy group exists in the amide form. We will return to this point subsequently.

Hydrogenation of dictamnine (**6**) over palladium affords 2,3-dihydrodictamnine (**28**) and the UV spectrum now resembles 2,4-dimethoxyquinoline (**29**). A platinum oxide catalyst causes fission of the dihydrofuran ring to give 3-ethyl-4-methoxy-2-quinolone (**30**), and with Raney nickel a mixture of the two reduced products is formed.

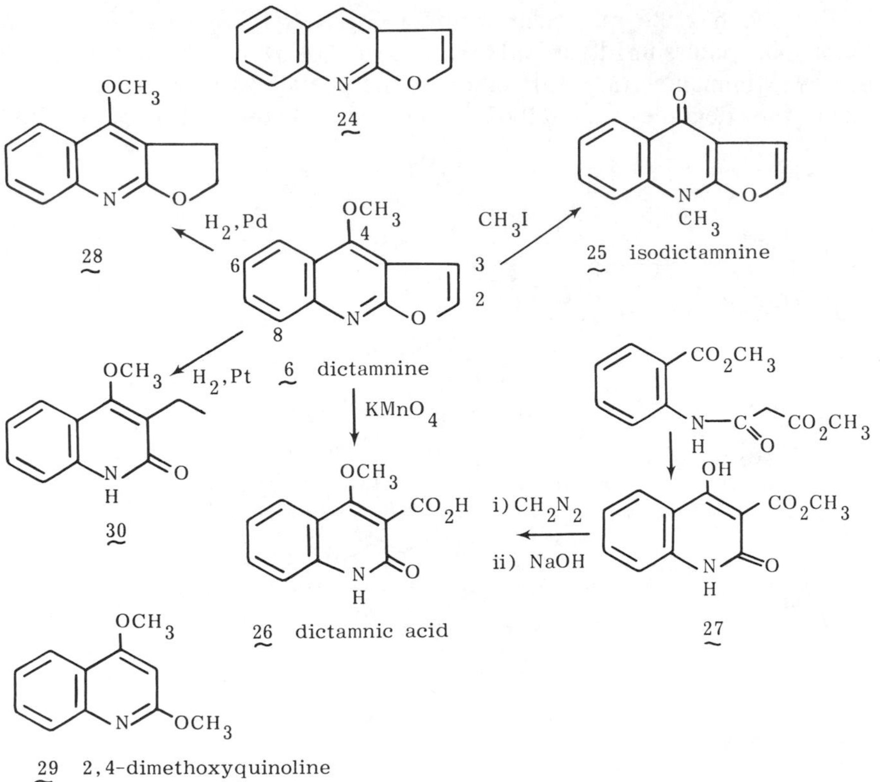

28

24

25 isodictamnine

H$_2$,Pd

OCH$_3$

6 dictamnine

CH$_3$I

H$_2$,Pt

30

KMnO$_4$

26 dictamnic acid

i) CH$_2$N$_2$

ii) NaOH

27

29 2,4-dimethoxyquinoline

Skimmianine (**31**) is the most widespread furoquinoline alkaloid and closely resembles dictamnine in its chemistry. For example, hydrogenolysis of skimmianine gave 3-ethyl-4,7,8-trimethoxy-2-quinolone (**32**), thereby confirming the linear structure of **31**. As we will see, the linear and angular structures can be readily distinguished by spectral means.

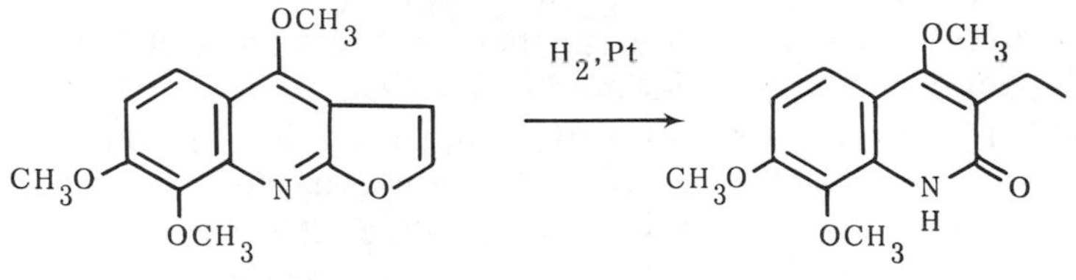

H$_2$,Pt

31 skimmianine

32

So far we have dealt only with methyl ether derivatives of the furoquinoline alkaloids, but as noted previously hydroxy, methylenedioxy, and both C- and O-isoprenylated derivatives are also common.

A typical representative of this type is evoxine (**33**). Although the structure can now be determined quite easily by its NMR spectrum, it is instructive to consider first a chemical approach.

Evoxine has the molecular formula $C_{18}H_{21}NO_6$, and hydrolysis with methanolic potassium hydroxide gave a phenol **34** which on methylation affords skimmianine (**31**). Ethylation of the phenol **34** gave a derivative **35** which was degraded to 7-ethoxy-8-methoxy-4-hydroxy-2-quinolone (**36**),

whose structure was proved by synthesis. The arrangement of the remaining five carbons and two oxygens in the ether side chain was determined by treatment with periodic acid which resulted in the formation of acetone. Evoxine therefore has the structure **33**.

As noted elsewhere a highly characteristic reaction of this group of compounds is their easy isomerization to *N*-methyl-4-quinolones, which in the case of dictamnine (**6**) occurs in methyl iodide at 80°C. The methiodide salts also undergo this isomerization, so that a melting point of the methiodide is really that of the *N*-methyl-4-quinolone.

Another chemical transformation of interest was encountered by Rapoport and Holden when studying the structures of the isomeric alkaloids (−)-balfourodine (**37**) and (+)-isobalfourodine (**38**) from *Balfourodendron riedelianum*. Acetylation of either **37** or **38** gave the same acetate, **39**. Since each alkaloid has only one asymmetric center, an inversion of configuration occurs during the ring expansion.

Dictamnine (**6**) also undergoes a rearrangement on warming in polyphosphoric acid to give an angular derivative **40**, once again indicating the stability of the 4-alkoxy-2-quinolone *vs.* the 2-alkoxy-4-quinolone system. The mechanism is thought to proceed through the aldehyde **41**. The ease of formation of the angular system from the linear system is a particular problem when synthesis of these alkaloids is contemplated.

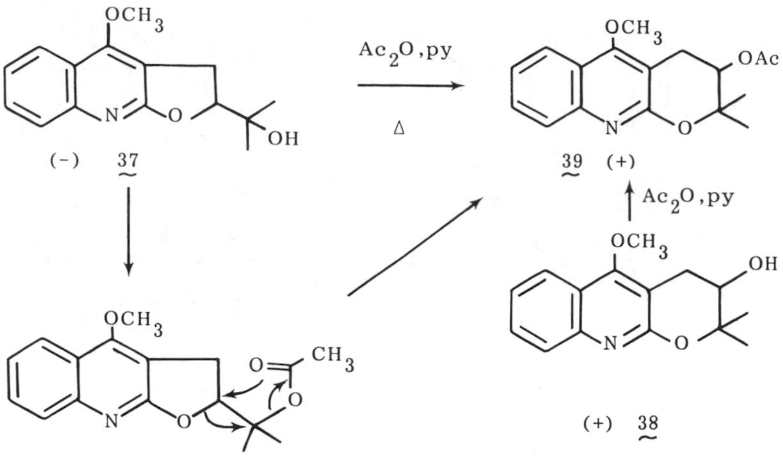

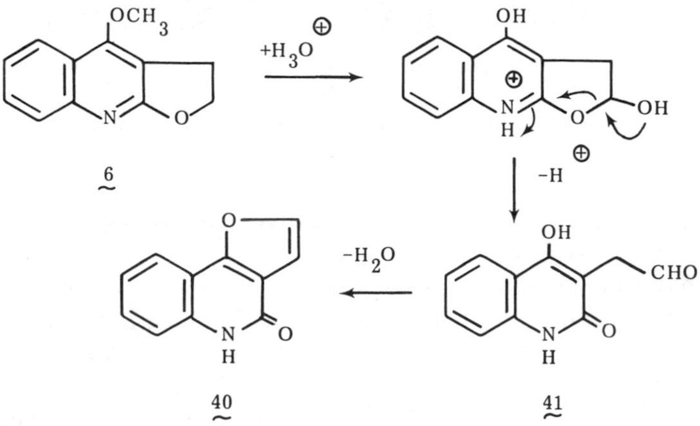

7.3.2 Synthesis

Most of the isolated furoquinoline alkaloids have been synthesized, and a number of quite general routes are available. Mitscher and co-workers have provided a particularly interesting review of the synthetic efforts in this area.

From the previous discussions it should be apparent that the synthesis of linear furanoquinolines poses a number of problems and Tuppy and Böhm were the first to develop a reliable synthesis of these isomers. This route (Scheme 1) involves as the keystep condensation of aniline with an isotetronic acid **42** to give the linear product **43**. Problems of specificity of cyclization only concern substitution in the aromatic ring if a *meta*-substituted aniline is used.

The linear product is reacted with diazomethane to give a mixture of *N*- and *O*-methyl isomers **44** and **45** of which only the latter is normally synthetically useful. Conversion of the β-keto ether to a furan is accom-

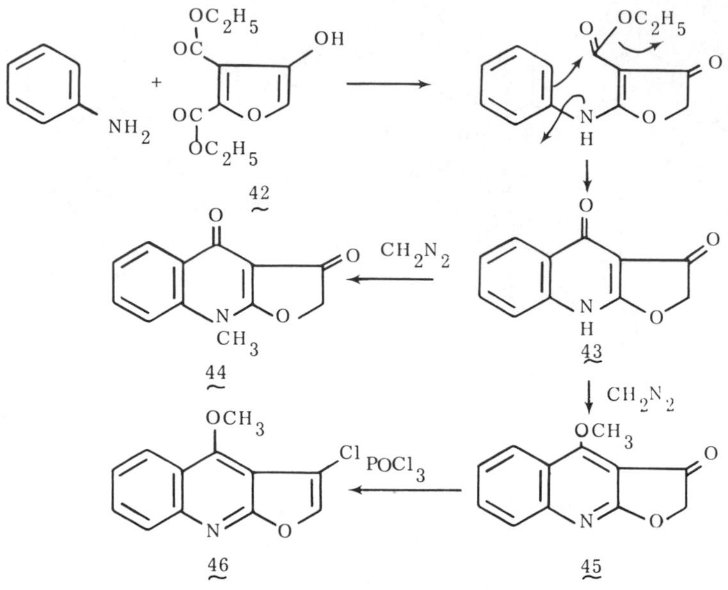

Scheme 1 Tuppy–Böhm synthesis of dictamnine (**6**).

plished by conversion to the chloride **46** with phosphorus oxychloride and hydrogenolysis to afford dictamnine (**6**).

We have already noted how 2,4-dihydroxyquinolines preferentially form an ether at C_4. This knowledge has been used by Grundon as an effective way to prepare the linear isomers, since it essentially blocks cyclization to C_4. Thus condensation of *o*-methoxyaniline (**47**) with the malonate ester **48** gives the 4-hydroxyquinoline **49**, which can be methylated to give **50**. Under acidic conditions, epoxidation of the dimethylallyl side chain leads to internal cyclization and formation of the dihydrofuroquinoline **51**. Unfortunately, this process also leads to substantial quantities of the pyrano analogue (e.g. **52**). When **51** is treated with lead tetraacetate the side chain is lost oxidatively to afford the furoquinoline nucleus, as in **53** (Scheme 2). The overall yield is a respectable 34%.

A more recent synthetic method which offers distinct advantages over the previous routes has been reported by Narasimhan. The strategy involves the tendency of 2,4-dimethoxyquinolines to form a covalent lithium bond at C_3. Reaction of 2,4-dimethoxyquinoline (**29**) with butyl lithium followed by reaction of **54** with ethylene oxide gave the alcohol **55**. Hydrolysis, cyclization with polyphosphoric acid, and dehydrogenation (bromination–dehydrobromination) then afforded dictamnine (**6**).

An alternative route involves 3-formylation of the lithio derivative **54**. Wittig reaction of the aldehyde **56** and selective hydrolysis then leads to the homologous aldehyde **57**, which on acid cyclization gives **6** (Scheme 3).

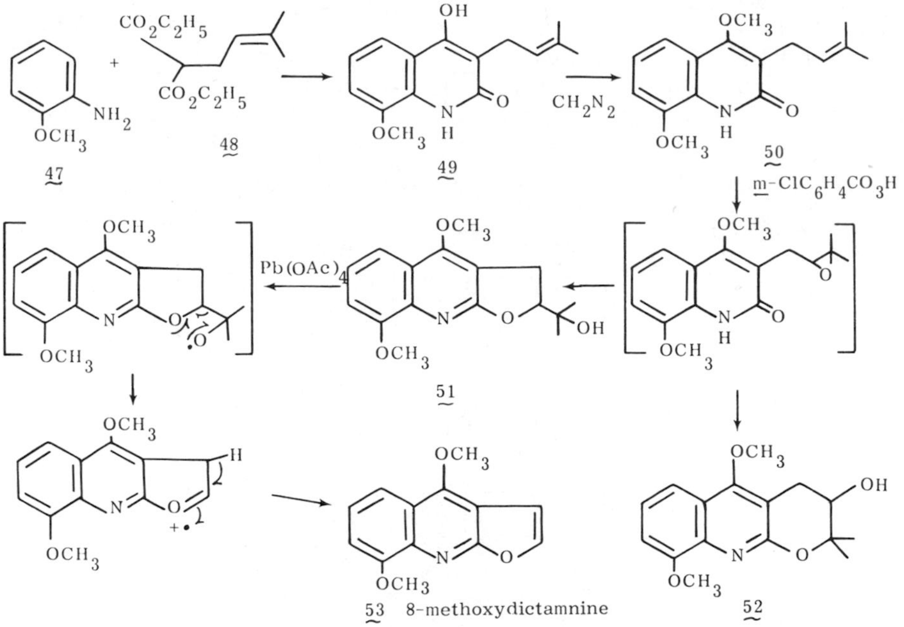

Scheme 2   Synthesis of 8-methoxydictamnine (**53**).

The remaining syntheses involve anthranilic acid derivatives as intermediates, and have distinct advantages for the synthesis of 6- and 7-substituted furoquinolines. Thus condensation of 2-nitro-5-methoxybenzoyl chloride (**58**) with α-acetyl butyrolactone (**59**) using magnesium ethoxide as a catalyst gave **60** and treatment with diazomethane gave the enol ether **61**.

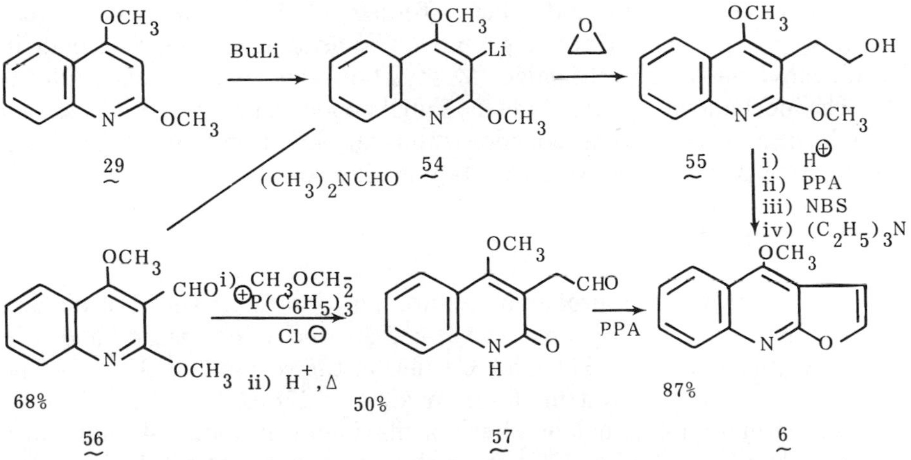

Scheme 3

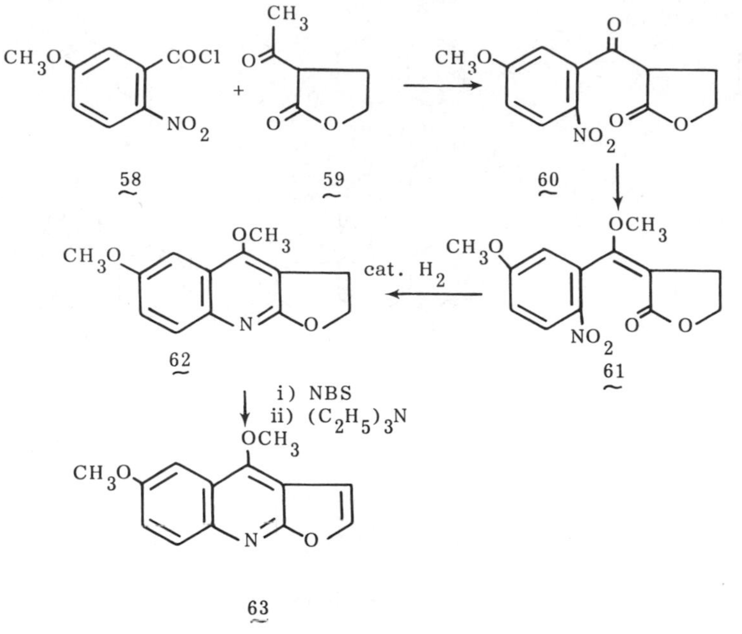

Scheme 4

Reduction of the nitro group catalytically and cyclization afforded the di-hydrofuranoquinoline **62**. Standard bromination–dehydrobromination steps then afforded 6-methoxy dictamnine (**63**) (Scheme 4).

A somewhat different approach to the prenylated quinolone **64** has been described by Mitscher and co-workers. Reaction of **65** with the anion of diethyl malonate gave the 3-carboethoxy-4-hydroxyquinolone **66**. Alkylation with 3,3-dimethylallylbromide then afforded **67**, from which the carbo-methoxy group can be selectively removed by heating with cupric acetate in hexamethyl phosphorus triamide. Methylation with diazomethane affords **64**. The mechanism suggested for the initial condensation reaction is shown in Scheme 5. This reaction sequence is thus far only useful for the formation of the compounds in the *N*-methyl-4-quinolone series.

7.3.3   Spectral Properties

The UV spectra of the furoquinolines are quite characteristic. For dictam-nine (**6**) an intense band is observed at 235 nm and a very broad band with fine structure (λ_{max} 308, 312, and 328 nm) in the region 290–335 nm. The *N*-methyl-2-quinolone system shows maxima at 228, 260, 267 and 320 nm and the *N*-methyl-4-quinolone shows a maximum at about 240 nm and a complex band centered at 320 nm. In the latter case a hypsochromic shift is observed in the spectra of these alkaloids in acid.

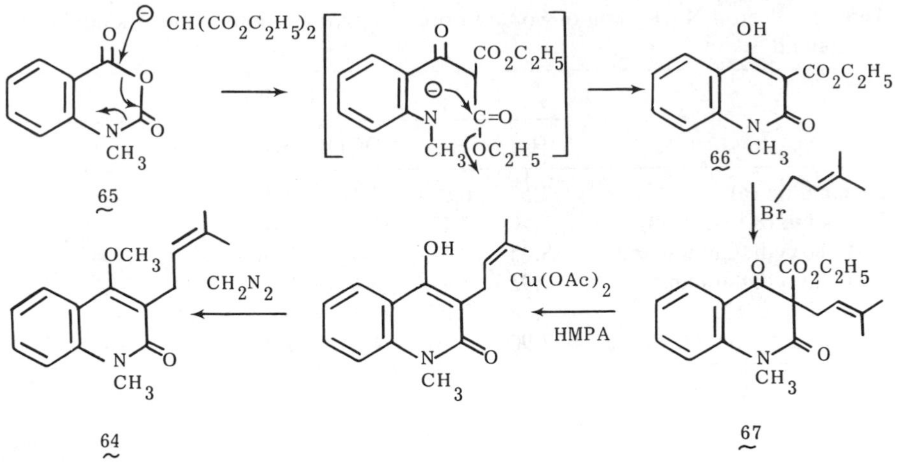

Scheme 5

The IR spectra of furoquinolines do not display many characteristic features; a band in the region 1090–1110 cm^{-1} is prominent but not assigned to a particular vibration. Some attention has been given to the carbonyl frequencies of 2- and 4-quinolones. Although the frequency is not a parameter, the considerably higher intensity of the carbonyl band of 2-quinolones is significant.

Very little is known about the mass spectra of the furoquinoline alkaloids. The molecular ion is usually the base peak, and losses of methyl (M$^+$-15) followed by carbon monoxide (M$^+$-43) are prevalent. In a compound containing an 8-methoxy group, M$^+$-1 and M$^+$-29 species may be also quite intense.

NMR spectroscopy is the key to the structure elucidation of most furoquinoline alkaloids. The furan protons appear as a pair of doublets (J = 3.0 Hz), the C-2 proton appearing in the region 7.50–7.60 ppm and the C-3 proton in the region 6.90–7.10 ppm. The methoxy groups are normally observed in two regions at 4.0–4.2 ppm for an aromatic methoxy group and at ~4.40 ppm for the 4-methoxy group.

A highly significant proton is the one at C-5, for 5-substituted derivatives are extremely rare. Since this signal is often the most downfield (7.5–8.1 ppm) its multiplicity is a key to the substitution on the benzenoid nucleus. The chemical shift of the C-8 proton is also significant, for this proton also occurs in the region 7.8–8.0 ppm. In some instances it can be quite difficult to distinguish these signals. Some examples of the data for simple furoquinolines are shown in Table 1. Note how a methoxy group shields the adjacent aromatic proton by about 0.6 ppm.

The ^{13}C NMR assignments for skimmianine are shown in **68** and for flindersine in **69**.

Table 1 Proton NMR Data of Some Representative Furoquinoline Alkaloids Compound

	Proton						
	H-2	H-3	4-OCH$_3$	H-5	H-6	H-7	H-8
Dictamnine (**6**)	7.53	6.95	4.32	8.11	—	—	7.88
Robustine (8-hydroxy)	7.50	7.13	4.13	—	—	—	—
6-Methoxydictamnine (**63**)	8.05	—	—	7.50	—	7.34	7.90
7-Methoxydictamnine	7.95	—	—	8.10	7.1	—	7.25
Skimmianine (**31**)	7.58	7.07	4.42	8.05	7.25	—	—
6, 7-Dimethoxydictamnine	7.90	7.40	4.40	7.44	—	—	7.30

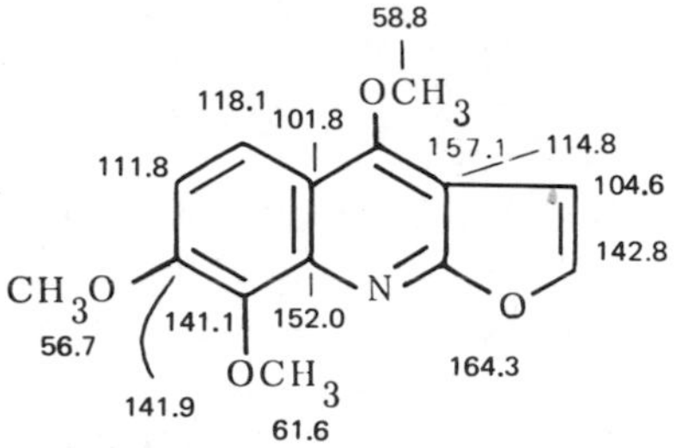

68

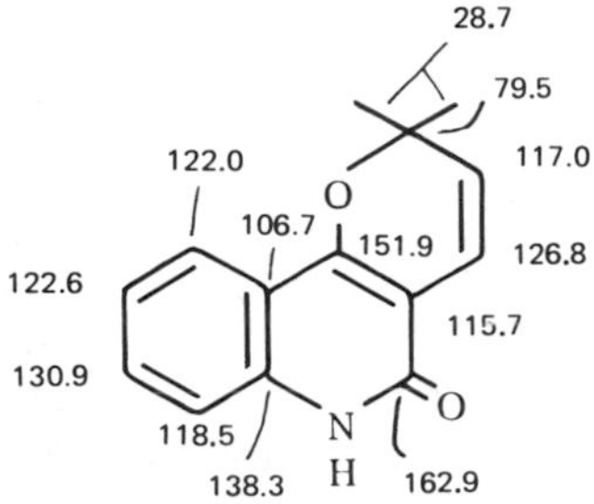

69

7.3.4 Biosynthesis

The first hypothesis for the formation furoquinoline alkaloids was advanced by Wenkert in 1959, who suggested that this nucleus was derived from anthranilic acid and erythrose. This stimulated numerous *in vivo* experiments which have now defined in reasonable detail the biosynthetic pathway.

The principal alkaloids which have been investigated are skimmianine (**31**) in *Skimmia japonica* Lindl. and dictamnine (**6**) in *Dictamnus albus* L.

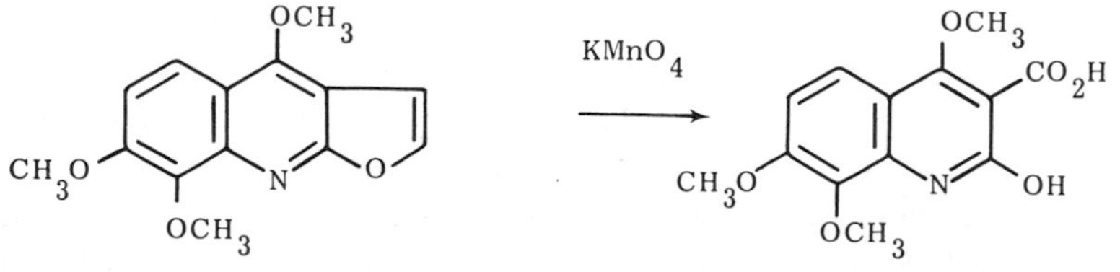

31 skimmianine

70 skimmianic acid

Degradation of **31** after feeding [ring-^{3}H]anthranilic acid (**1**) gave skimmianic acid (**70**) retaining all of the activity. When [1-^{14}C]- and [2-^{14}C]acetates were used as precursors of **31**, labeling was specifically at C-10 and C-11 respectively as shown. Similar results were obtained for dictamnine (**6**).

With the origin of the quinoline nucleus deduced if not defined, the next problem became the origin of the C-2 and C-3 carbon atoms. The furocoumarins are quite closely related to the furoquinolines and these groups frequently co-occur in the Rutaceae. It was therefore of some relevance when in 1966 Floss and Mothes showed that C-2 of the furocoumarin nucleus (e.g. (**71**) was specifically derived from [4-^{14}C]mevalonate. The significance of the occurrence of both the furoquinolines and the furocoumarins with compounds such as marmesin (**72**) and platydesmine (**73**) having an isopropyl side chain at C-2 now became clear. Biogenetically, we can envisage introduction of an isoprenyl unit at C-3 of a quinoline-2,4-diol, cyclization to give platydesmine (**73**), followed by oxidative loss of the isopropyl side chain and hydroxylation/O-methylation to skimmianine (**31**).

31 skimmianine

32

● label from [1-^{14}C]-acetate
▲ label from [2-^{14}C]-acetate

71 furanocoumarin nucleus

72 marmesin

73 platydesmine

This hypothesis was examined in a series of experiments over the next few years. 2,4-Dihydroxyquinoline (**74**) was demonstrated by Luckner to be an intermediate and Collins and Grundon found that 2,4-dihydroxy-3-dimethylallylquinoline (**75**), labeled at carbon 1 of the side chain, was an excellent precursor of platydesmine (**73**) and dictamnine (**6**) in *S. japonica*. Additional experiments utilizing [4-^{14}C]- and [5-^{14}C]mevalonates demonstrated that C-2 and C-3 of **6** respectively originate from an isoprene unit.

A further significant result was the excellent (18.8%) incorporation of
($\pm$)-[3-^{14}C]platydesmine (73) into dictamnine (6). Since it is probable that
only one enantiomer of 73 is utilized biosynthetically, the level of incorporation should be calculated even higher. An isotope-trapping experiment
established that 73 is an intermediate in the biosynthetic scheme. The isopropyl group is therefore lost *after* cyclization to the 2-hydroxyisopropyl
dihydrofuran moiety and not before.

The timing of the introduction of the various methoxyl groups then became of interest. Methylation of the 4-hydroxy group seems to take place
early in the sequence because 2-hydroxy-4-methoxyquinoline (76) and
2,4-dihydroxyquinoline (74) were equally good precursors of dictamnine (6).
Similarly 75 and its 4-O-methyl derivative (77) were just as effective. Hydroxylation of the benzenoid nucleus is established to occur as a late stage
in the formation of skimmianine (31). For using doubly labeled dictamnine
(6) a 2.1% specific, intact incorporation into skimmianine (31) was observed.
How the two oxygen atoms at C-7 and C-8 are incorporated is not known;
it could be in a stepwise fashion involving either 7-hydroxy or 8-hydroxy-
dictamnines (78) and (79), or a dioxygenase could be involved which would
introduce the two functional groups at the same time.

At this point we can therefore write Scheme 6 as a good working hypothesis for the biosynthesis of dictamnine (6) and skimmianine (31).

There are two points remaining to be discussed. In the biosynthetic
scheme shown, there are two intermediates that are postulated but which
have not yet been used in any biosynthetic experiment, namely *N*-acetyl
anthranilic acid (80) and the epoxide (81). In addition, an intermediate 82
is presented for the oxidative removal of the hydroxyisopropyl group as
acetone. There is some preliminary evidence for this.

For example, when [1'-^{3}H$_2$, 1'-^{14}C]3-dimethylallyl-2-hydroxy-4-meth-
oxyquinoline (77) was used half of the tritium activity was lost on conversion
to skimmianine (31). Therefore a 3-keto dihydrofuran intermediate can be
eliminated and since a direct hydride abstraction at C-3 seems unlikely, the
3-hydroxy derivative 82 is an obvious candidate for intermediary.

Biosynthetic hydroxylations are usually highly stereospecific, and in addition the oxidative elimination of acetone is also likely to be stereospecific,
involving groups *trans* to each other. But the stereochemistry of the 2-
hydroxyisopropyl side chain is predetermined by the stereochemistry of the
epoxide 81. Since only (+)-platydesmine (73) is known, it would be both
interesting and instructive for the student to predict the results of feeding
experiments using [5R-^{3}H, 5-^{14}C]- and [5S-^{3}H, 5-^{14}C]mevalonic acid into
dictamnine (6), experiments which at the time of writing had not been
reported.

The question of linear *vs*. angular structure for the furoquinoline (and
furocoumarins) was a perplexing initial problem and is still the first structure
question to be answered. To date there have been no well-characterized
furoquinolines isolated which have the angular structure. This again suggests

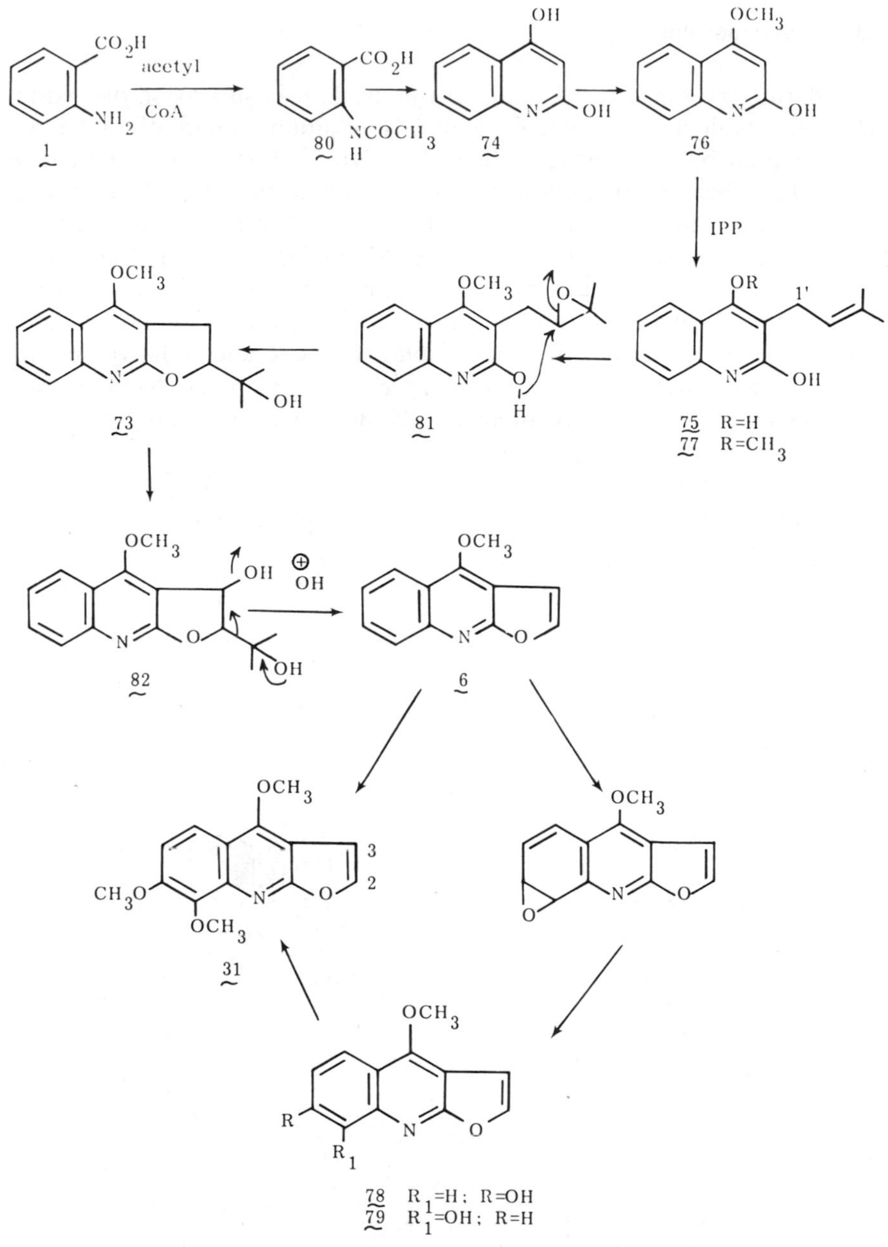

Scheme 6 Biosynthesis of furoquinoline alkaloids.

that 4-O-methylation is an early biosynthetic step, thereby eliminating cyclization to the 4-position.

7.3.5 Pharmacology

Considering the range of alkaloids in the class, it is surprising that so few have been evaluated pharmacologically. In addition, apparently none have been isolated by monitoring biological activity. Most of the pharmalogic studies have been carried out in Russia and details are difficult to obtain.

Dictamnine (**6**) acts to cause smooth muscle contraction, contracts the uterus, and increases heart muscle tone. Skimmianine (**31**) and dubinidine (**83**) have hypothermic, sedative, and antimicrobial activities and evoxine (**33**) and pteleatinium chloride (**84**) also show antimicrobial activity. Haplophyllidine (**85**) and perforine (**86**) display ataractic and sedative activity, and bucharaine (**87**) acts to suppress aggressive tendencies. None of these activities has yet resulted in commercially significant products.

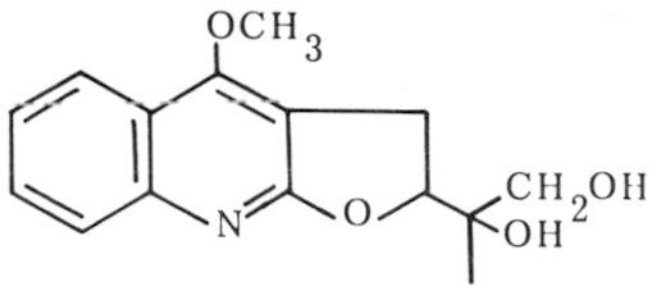

83   dubinidine

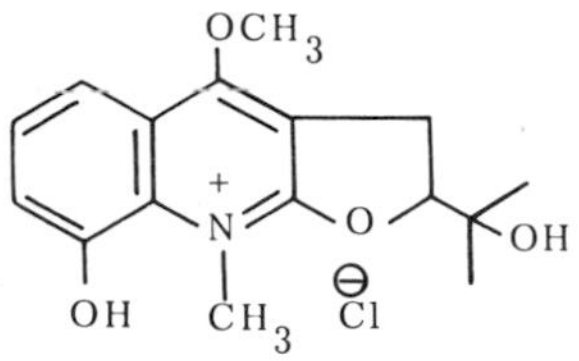

84   pteleatinium chloride

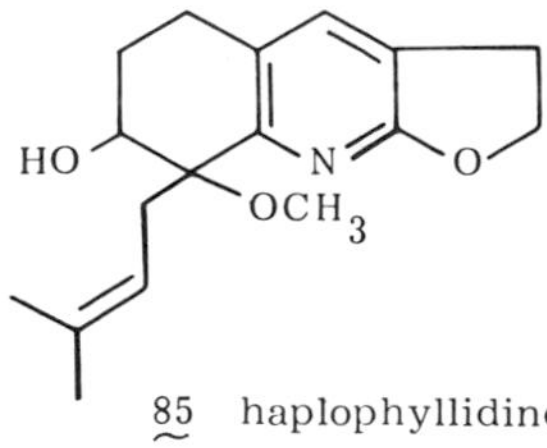

85   haplophyllidine

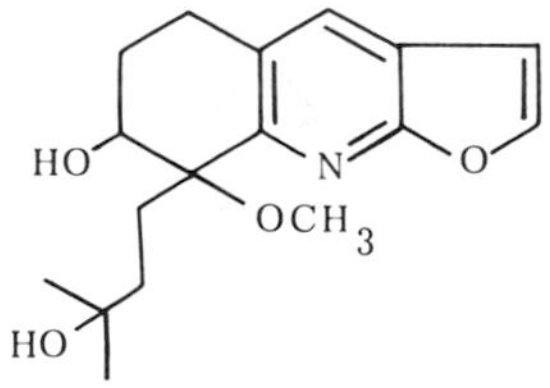

86   perforine

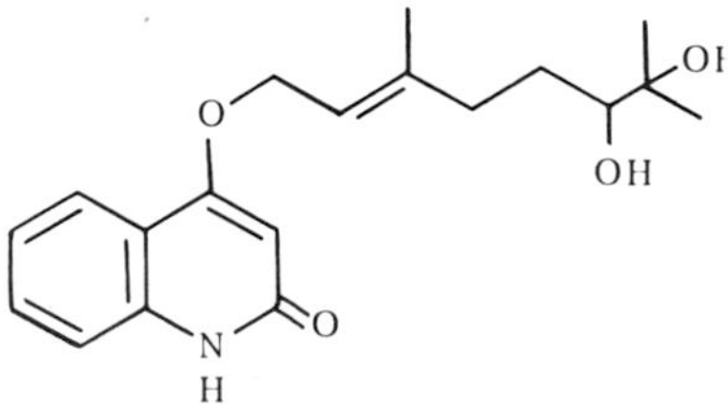

87   bucharaine

Literature

Reviews

Openshaw, H. T., *Alkaloids NY* **3**, 69 (1955).

Price, J. R., *Fortschr. Chem. Org. Naturs.* **13**, 302 (1956).

Openshaw, H. T., *Alkaloids NY* **7**, 233 (1960).

Openshaw, H. T., *Alkaloids NY* **9**, 226 (1967).

Pakrashi, S. C., and J. Bhattacharyya, *J. Sci. Ind. Res. (India)* **24**, 226 (1965).

Synthesis

Mitscher, L. A., T. Suzuki, G. Clark, and M. S. Bathala, *Heterocycles* **5**, 565 (1976).

Tuppy, H., and F. Böhm, *Montatsh. Chem.* **87**, 774 (1956).

Collins, J. F., G. A. Gray, M. F. Grundon, D. M. Harrison, and C. G. Spyropoulos, *J. Chem. Soc. Perkin Trans. I* 94 (1973).

Narasimhan, N. S. and R. S. Mali, *Tetrahedron* **30**, 4153 (1974).

Kuwayama, Y., *Chem. Pharm. Bull.* **9**, 719 (1961).

Biosynthesis

Wenkert, E., *Experientia* **15**, 165 (1959).

Matsuo, M., M. Yamasaki, and Y. Kasida, *Biochem. Biophys. Res. Commun.* **23**, 679 (1966).

Monković, J., I. D. Spenser, and A. O. Plunkett, *Can. J. Chem.* **46**, 1935 (1967).

Cobet, M., and M. Luckner, *Eur. J. Biochem.* **4**, 76 (1968).

Colonna, A., and E. G. Gros, *Phytochemistry* **10**, 1515 (1971).

Collins, J. F., W. J. Donnelly, M. F. Grundon, and K. J. James, *J. Chem. Soc. Perkin Trans. I* 2177 (1974).

Grundon, M. F., D. M. Harrison, and C. G. Spyropoulos, *J. Chem. Soc. Perkin Trans. I* 2181 (1974).

Grundon, M. F., D. M. Harrison, and C. G. Spyropoulos, *J. Chem. Soc. Perkin Trans. I* 302 (1975).

7.4 QUINAZOLINE ALKALOIDS

The quinazoline nucleus occurs in relatively few alkaloids, yet despite its size this group of alkaloids is quite important. Unlike the acridine alkaloids, the quinazolines are not limited in distribution to the Rutaceae, but also occur in the Zygophyllaceae, Saxifragaceae, Malvaceae, and the Acanthaceae. Indeed their taxonomic distribution is quite interesting, for quinazolines occur with harman alkaloids (Chapter 9) in the Zygophyllaceae (*Peganum harmala* L.), but with furoquinolines and acridones in the Rutaceae.

Some examples of this structure type include vasicine (**88**), febrifugine (**89**), rutaecarpine (**90**), and arborine (**91**).

Quinazoline alkaloids are not limited in their distribution to the plant kingdom. The defensive agents of the millipede *Glomeris marginata* are glomerine (**92**) and homoglomerine (**93**), and the bacteria *Pseudomonas aeruginosa* produces the alkaloid **94**.

One of the simple quinazolines obtained from a fungus is chrysogine (**95**)

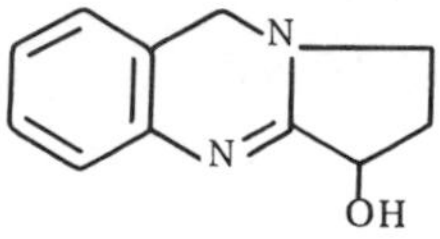

88 vasicine

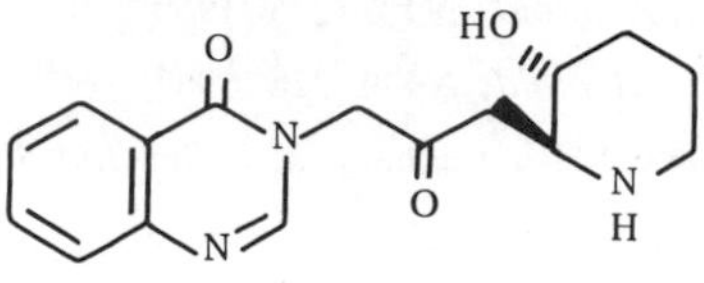

89 febrifugine

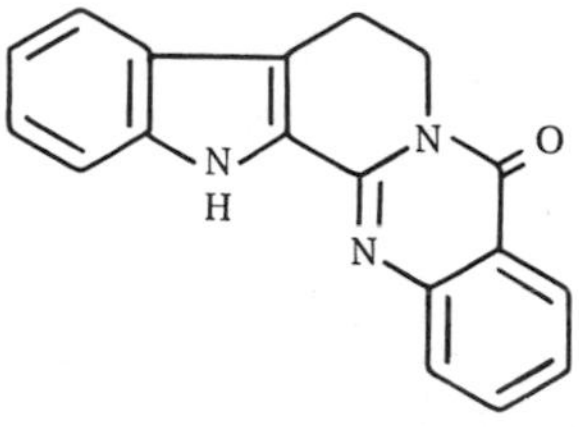

90 rutaecarpine

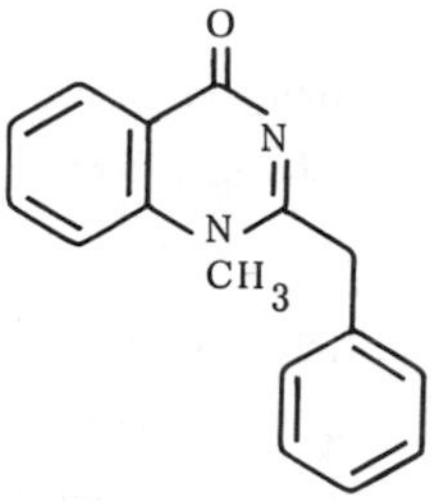

91 arborine

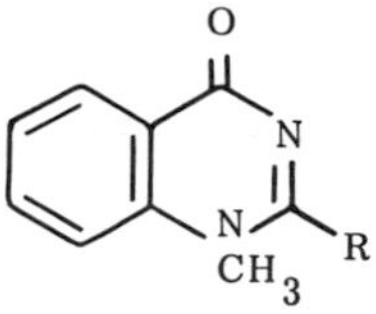

92 glomerine, R=CH₃

93 homoglomerine, R=CH₂CH₃

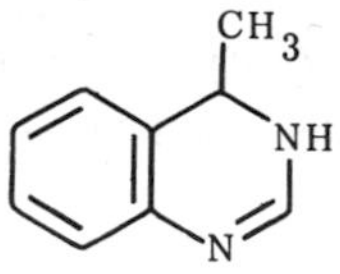

94

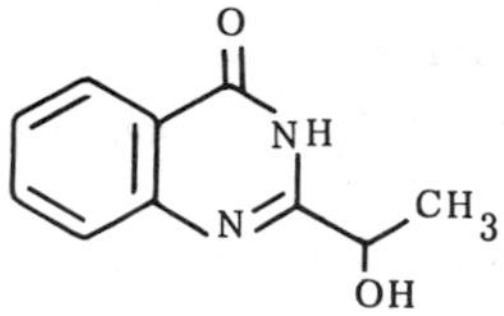

95 chrysogine

isolated from *Pennicillium chrysogenum*, and its yield was increased by the addition of anthranilic acid to the medium.

7.4.1 Vasicine and Related Compounds

Vasicine was obtained from *Adhatoda vasica* Nees by Hooper in 1888. The alkaloid is also known as peganine, having been obtained from alkaloid fractions of *Peganum harmala* L.

Isolation of vasicine usually results in racemization of the laevorotatory form, and since the laevorotatory form melts at the same temperature (211–212°C) as the racemate, it appears that racemization also occurs prior to melting. Vasicine shows a λ_max at 290 nm, is a monoacidic tertiary base, forms a monoacetate but shows no *N*- or *O*-methyl, phenolic, or ketonic groups. It is resistant to catalytic hydrogenation, but the imino double bond can be reduced with sodium amalgam or sodium borohydride. Oxidation of **88** with acidic permanganate gave quinazol-4-one (**96**) and alkaline permanganate, an acid characterized as **97**.

88 96 97

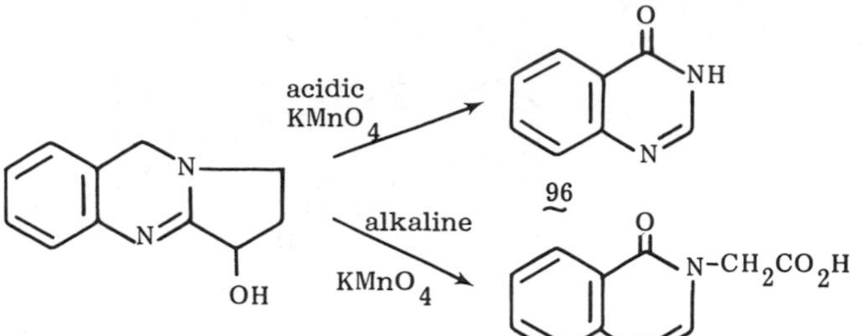

The position of the hydroxy group was the subject of controversy because it was expected that alkaline oxidation would give a dicarboxylic acid, and this problem was finally resolved by synthesis. α-Hydroxybutyrolactone (**99**) was condensed with *o*-aminobenzylamine (**98**) at 200°C to give (±)-vasicine (**88**).

A biomimetic synthesis of deoxyvasicine (**100**) was developed in 1936 by Schöpf and Oechler by reaction of *o*-aminobenzaldehyde (**101**) and the diethyl acetal of γ-aminobutyraldehyde at pH 5 followed by catalytic reduction. A similar reaction with α-hydroxy-γ-aminobenzaldehyde (**102**) gave vasicine (**88**) in 39% yield.

98 99 88

101 102 R=H, or OH

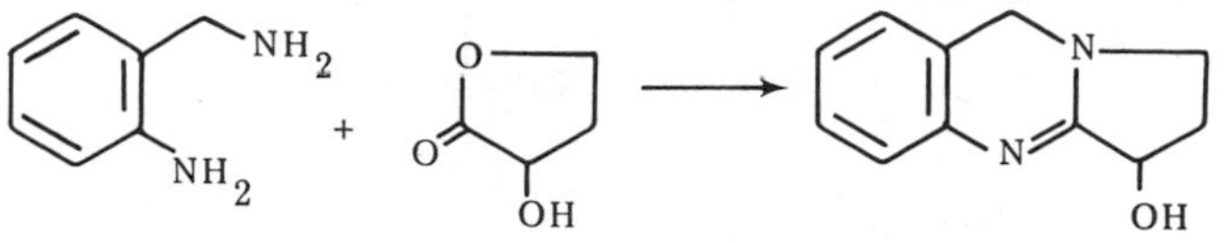

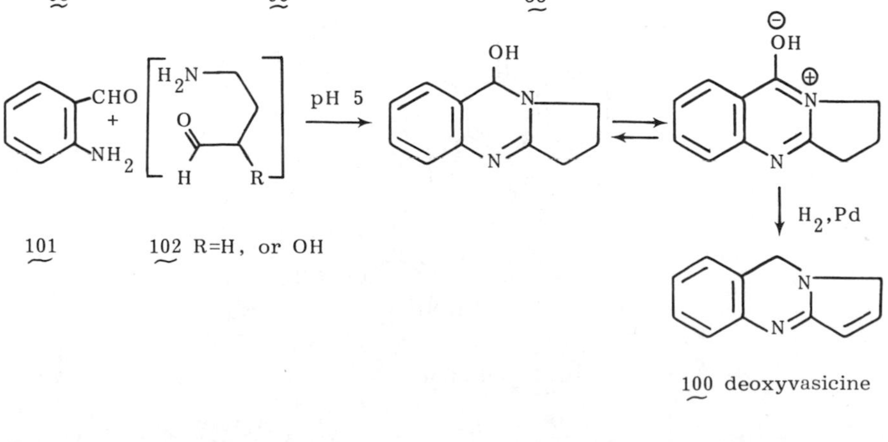

100 deoxyvasicine

A simple synthesis of (±)-vasicine (**88**) was recently reported by Moehrle and Gundlach. The route involves condensation of 3-hydroxypyrrolidine (**103**) and *o*-nitrobenzyl chloride (**104**) to give **105**, which after reduction of the nitro group and oxidation with mercuric acetate–EDTA afforded vasicine (**88**) regiospecifically.

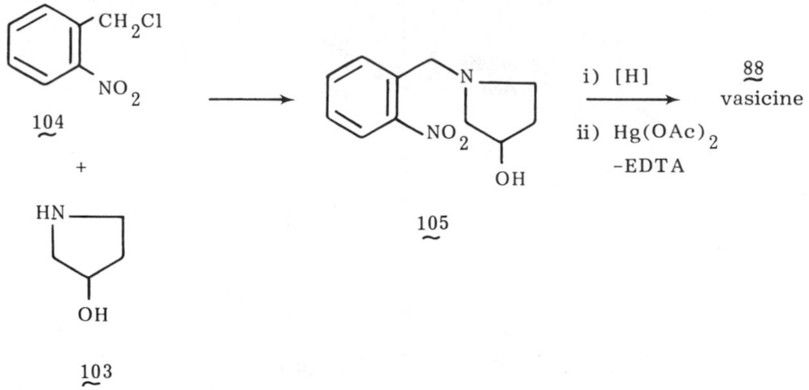

Dimeric quinazoline alkaloids are rare, but dipegine (**106**) has recently been obtained from *Peganum harmala*.

Several alkaloids have been isolated which are apparently derived from two anthranilic acid residues. One such example is anisotine (**107**) obtained from *Adhatoda vasica* Nees and *Arisotes sessiliflorus*. It is thought that vasicine (**88**) may be a biosynthetic intermediate.

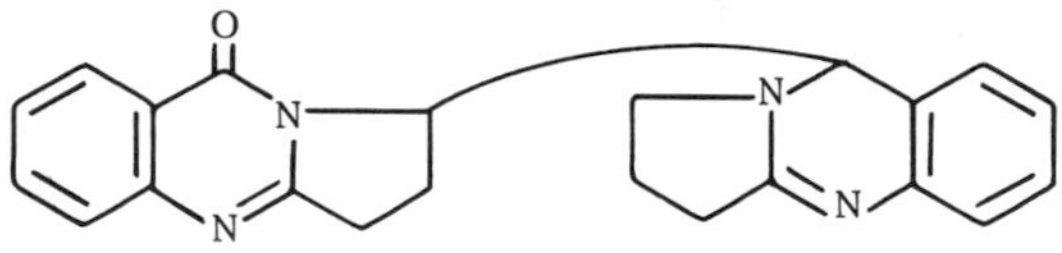

106   dipegine

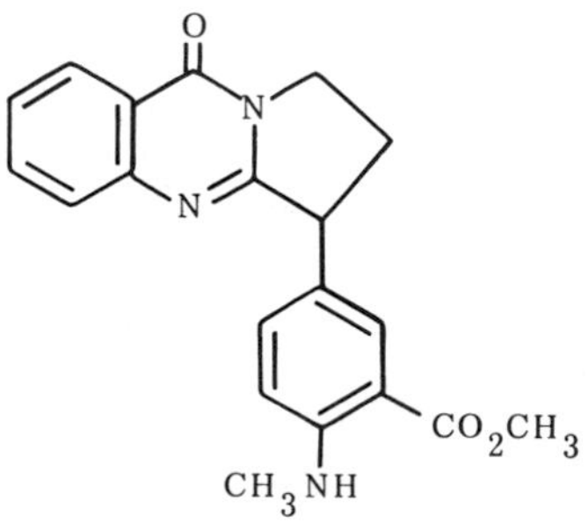

107   anisotine

7.4.2 Febrifugine

The Chinese drug Ch'ang Shan, *Dichroa febriguga* Lorv. (Saxifragaceae), has long been used as a febrifuge, and when aqueous extracts became available antimalarial activity was established and eventually traced to an alkaloid. A similar active compound was also obtained from a horticultural variety of *Hydrangea*.

Extensive NMR spectral data indicated that febrifugine had the *trans* stereochemistry based on syntheses of the *cis* and *trans* isomers **89** and **108** and of comparison of the NMR data of these isomers with those of the piperidones **109** and **110**. In particular the C-2′ proton appeared as a quartet

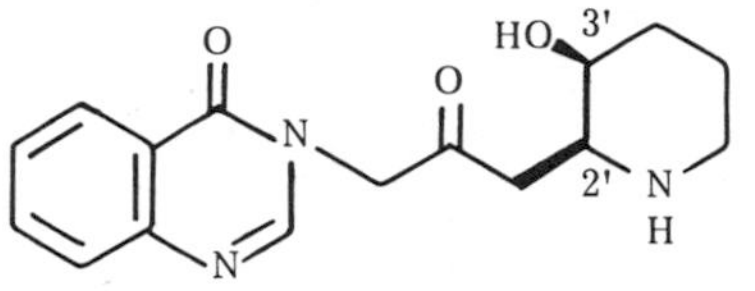

108   original structure for febrifugine

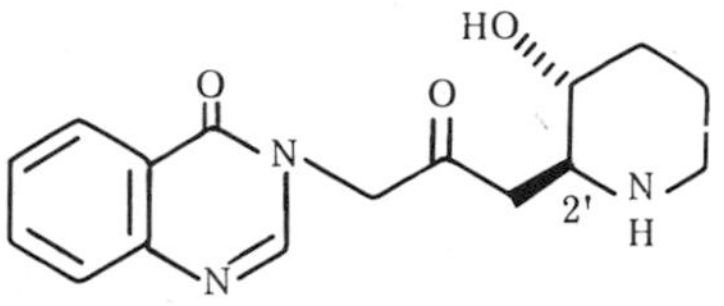

89   febrifugine (revised)

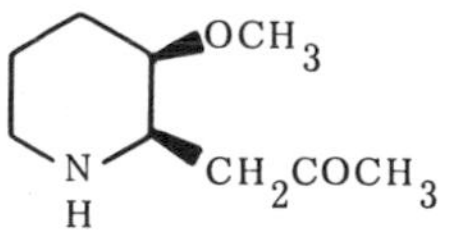

109

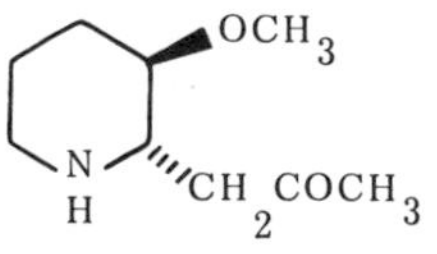

110

(J = 7 Hz) at 3.98 ppm in the *O*-acetate of febrifugine, and irradiation at the frequency of the side-chain methylene collapsed this to a doublet (J = 7 Hz), which must therefore represent the coupling between the C-2′H and the C-3′H.

Previous workers had established the absolute configuration at C-2′ of febrifugine, so that the structure **89** also represents the absolute configuration.

7.4.3 *Evodia* Alkaloids

Rutaecarpine (**90**) and evodiamine (**111**) are constituents of the Chinese drug *Evodia rutaecarpa* Hook. f. and Thou. (Rutaceae), and they also occur in the broad genus *Zanthoxylum*.

Chemistry

These alkaloids are highly susceptible to hydrolytic reactions as shown in Scheme 7. In alcoholic alkali evodiamine (**111**) affords 3,4-dihydronorharman (**112**) and *N*-methyl anthranilic acid (**113**), whereas acidic conditions lead to an intermediate hydroxyquinazolone **114** which can be cleaved in

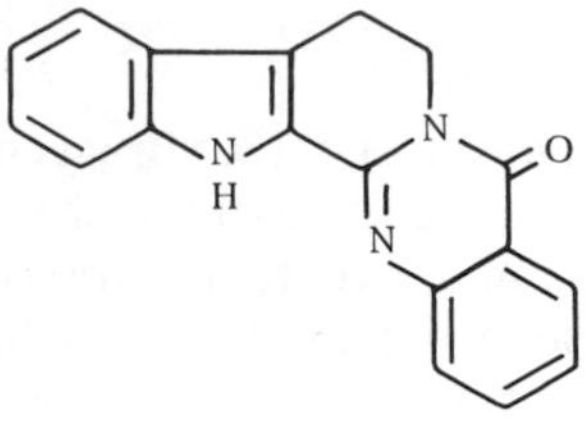

90   rutaecarpine 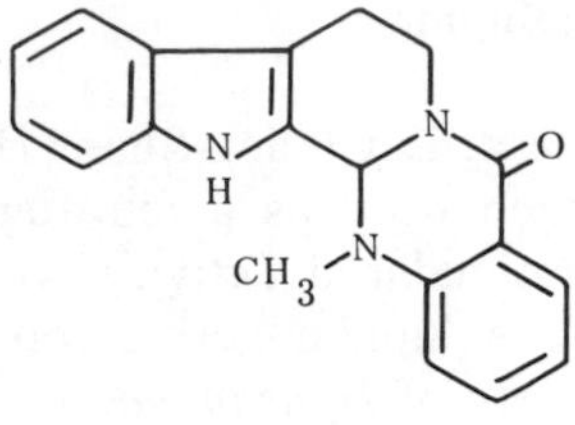   111   evodiamine

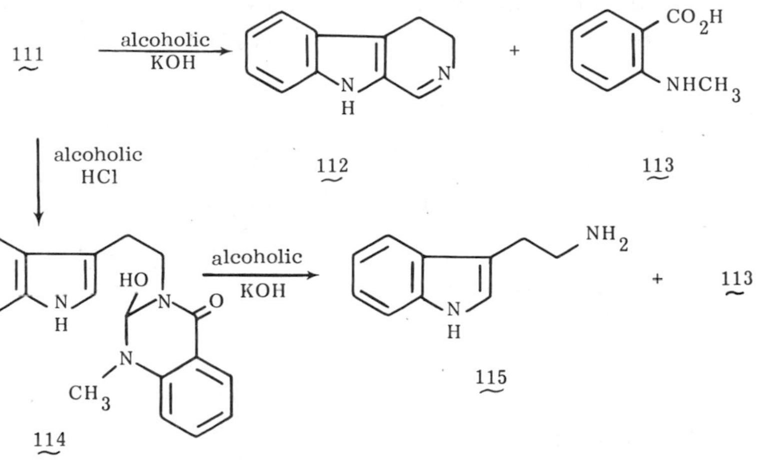

base to tryptamine (115) and 113. When evodiamine hydrochloride is heated, rutaecarpine (90) and methyl chloride are produced.

Several syntheses of rutaecarpine (90) and evodiamine (111) have been reported. For example condensation of 3-keto-3,4,5,6-tetrahydro-β-carboline (116) with methyl anthranilate (117) or of o-aminobenzaldehyde (101) with dihydronorharman (112) at pH 5 followed by oxidation with potassium ferricyanide yield 90.

An interesting synthesis of evodiamine (111) was reported by Asakina and Ohta in 1928. Treatment of N-methyl anthranilic acid (113) with ethyl chloroformate gave the quinoxalone 118, which on reaction with tryptamine gave 119. The cyclization to evodiamine (111) was completed by condensation with ethyl orthoformate.

A more recent synthesis developed by Kametani and co-workers involves the condensation of the sulfur analogue of 118 with 3,4-dihydro-β-carboline (112) at room temperature to give rutaecarpine (90) in 80% yield. Evodiamine (111) was synthesized in a similar manner beginning with 113. It has been suggested that this reaction may proceed via the iminoketene 120 in a concerted cyclo addition process. However, it is more likely that the reaction involves stepwise addition, as shown in the postulated formation of deoxy-vasicinone (**121**) from **118** and the imino ether **122** (Scheme 8). The imino ether is not necessary, for the yields are actually improved (to 93%) when the corresponding lactam is used. This in itself is good evidence for a non-concerted stepwise process. In this scheme an intermediate such as **123** would be envisaged as a key intermediate.

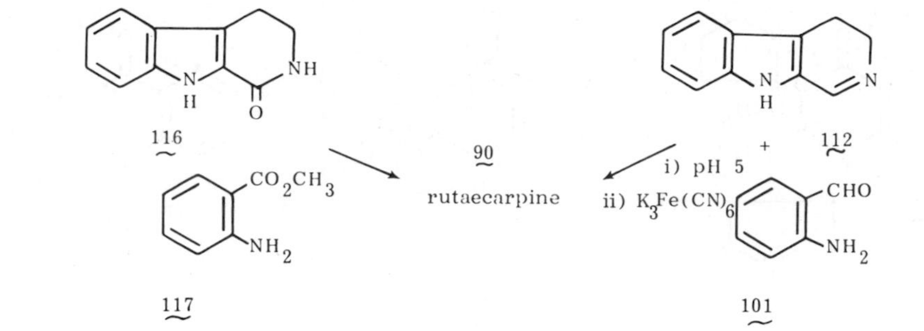

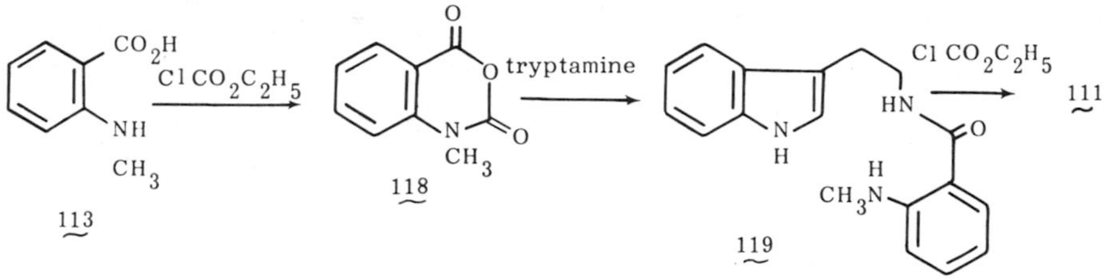

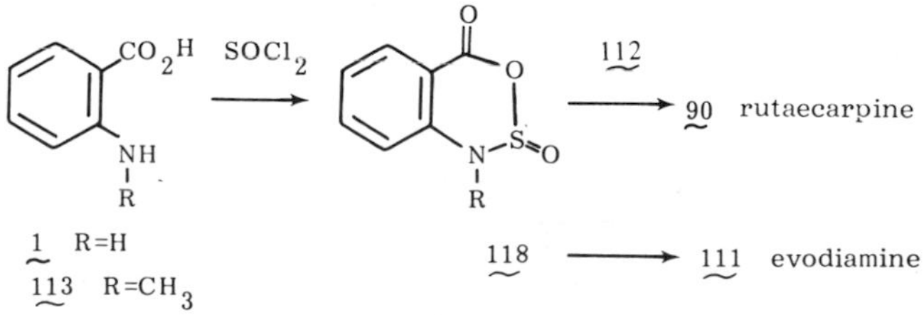

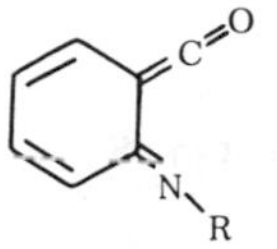

Several syntheses of quinazoline alkaloids have been carried out in this manner, including arborine (**91**) from phenylacetamide (**124**), and rutaecarpine (**90**) from N-formyltryptamine (**125**).

A general route to 4(1H)-quinazolines involves condensation of an aminobenzamide and an aldehyde. Arborine (**91**), the major alkaloid of *Glycosmis arborea* (Roxb.) DC., was synthesized by this route.

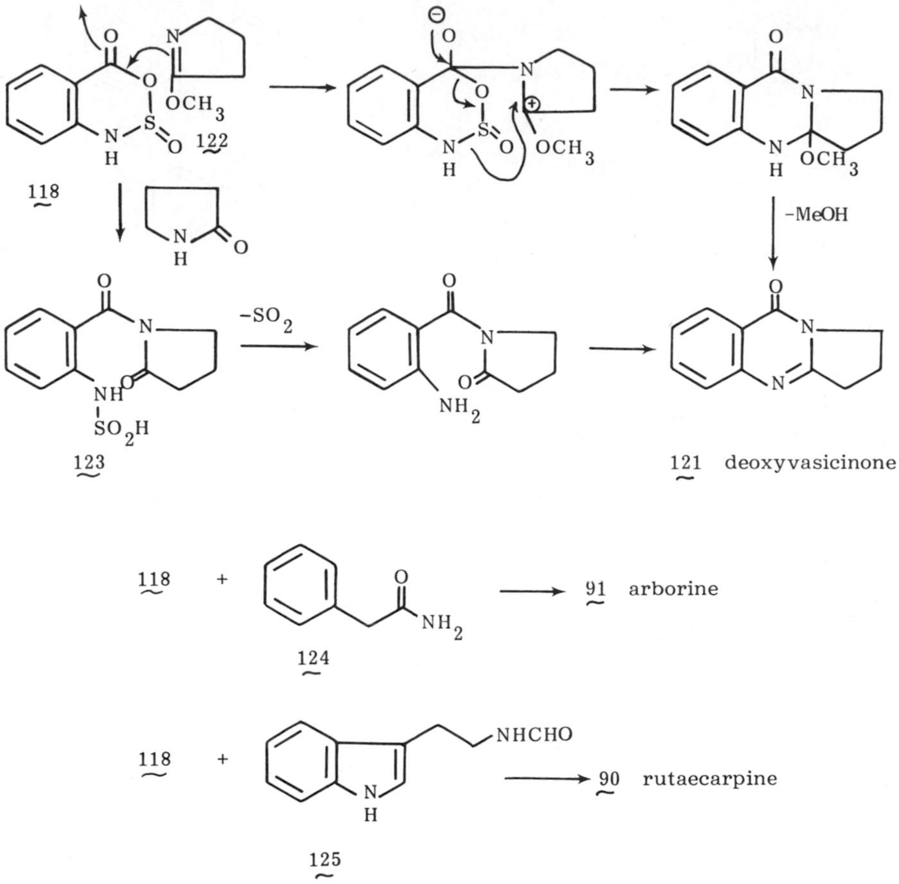

Scheme 8

Biosynthesis

The interesting biosynthetic precursors of many quinazoline alkaloids have been quite well investigated. Robinson first suggested that vasicine (**88**) could arise from anthranilic acid (**1**) and proline (**126**), and arborine (**91**) from N-methylanthranilic acid (**113**), ammonia, and phenylacetic acid.

The biosynthesis of arborine (**91**) has been investigated in *Glycosmis*

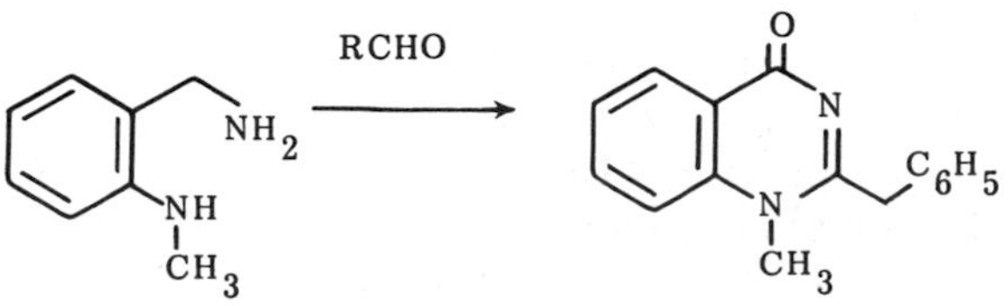

arborea, where both anthranilic acid (**1**) and phenylalanine were regiospecific precursors. The specific origin of the nonanthranilic acid–derived nitrogen has not been established.

The formation of vasicine (**88**), potentially a simple problem, has proved to be quite complex. Gröger and Mothes established the specific incorporation of anthranilic acid (**1**) into vasicine (**88**). The nonanthranilic acid derived part is where problems have arisen. Initially it was thought that deoxyvasicine (**100**) might be an intermediate, but labeling experiments gave a low random incorporation into vasicine (**88**), suggesting degradation of the added precursor. Similar results were obtained with labeled proline (**126**), 4-hydroxyproline (**127**), ornithine (**128**), and putrescine (**129**), evidence that these compounds are used in the general metabolic process rather than specifically for alkaloid production. γ-Hydroxyglutamic acid (**130**) seemed

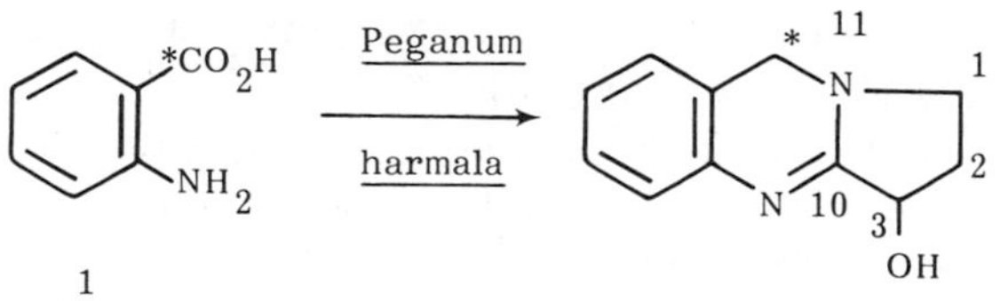

1

88   vasicine

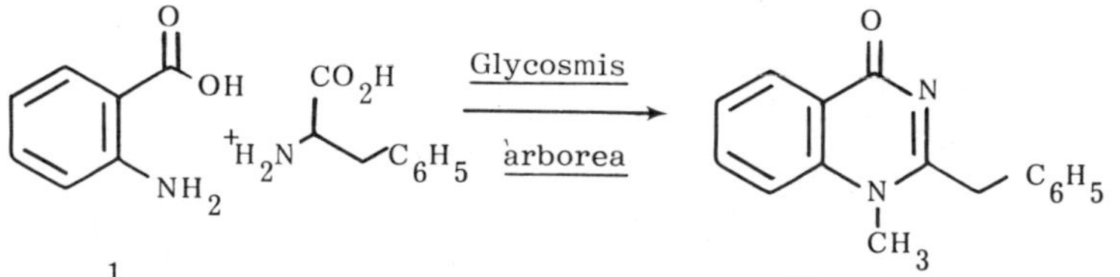

1

91   arborine

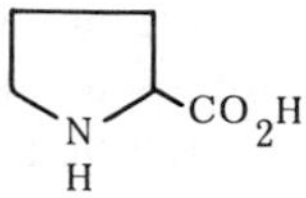

126   proline

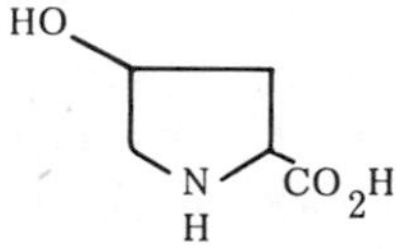

127

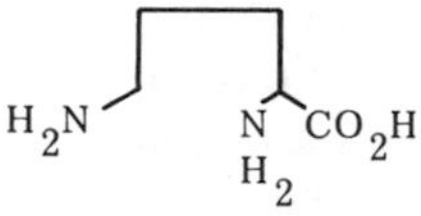

128

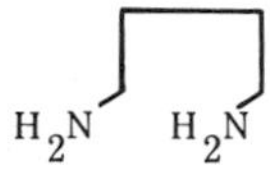

129

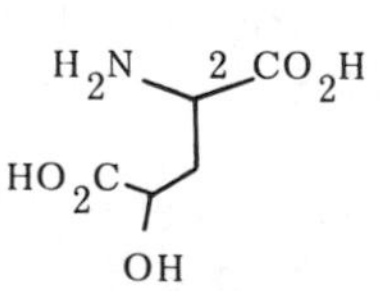

130   γ-hydroxy
glutamic acid

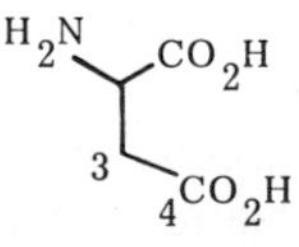

131   aspartic
acid

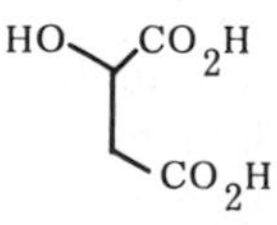

132   malic
acid

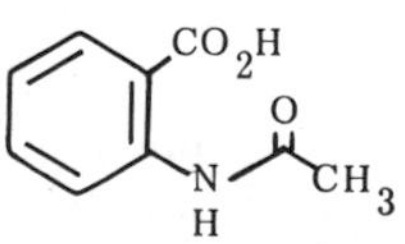

80   N–acetyl
anthranilic
acid

to be a promising precursor since it cooccurred with vasicine in *Lunaria* species, and as indicated previously α-hydroxy-γ-aminobutyraldehyde (**102**) reacts readily *in vitro* to give vasicine (**88**). A specific incorporation of [2-^{14}C]γ-hydroxyglutamic acid (**130**) should give vasicine (**88**) labeled at C-1, but degradation of the labeled alkaloid indicated essentially complete randomization of label. [3-^{14}C]Aspartic acid (**131**) gave 80% of the radioactivity at C-1 and C-2 of vasicine (**88**), of which 50% was at C-2. [4-^{14}C]Aspartic acid (**131**) gave random labeling with a nonspecific concentration of activity at C-3. Malic acid (**132**), specifically labeled at C-3, gave vasicine (**88**) labeled at C-2 (46% total activity) and C-1 (26%). It is not easy to rationalize these results, but it seems, unexpectedly, that a five-carbon precursor is not incorporated intact.

In 1971 the biosynthesis of vasicine (**88**) was reinvestigated by Liljegren. [2-^{14}C]Ornithine (**128**) gave **88** labeled at C-1 (43%) and C-10 (42%), suggesting decarboxylation to give putrescine (**129**), but [1,4-^{14}C$_2$]putrescine gave random labeling at C-1, C-3, and C-10!

Gröger and co-workers have more recently established that *N*-acetylanthranilic acid (**80**) is a specific precursor of both the anthranilic acid and C-3 and C-10 carbon atoms of **88**. The origin of the C-1, C-2, and N-11 atoms is neither known nor speculated on. The synthesis of *Evodia* alkaloids such as rutaecarpine (**90**) and evodiamine (**111**) from a dihydronorharman (**112**)

and *O*-aminobenzaldehyde (**101**) has been discussed previously, and clearly this is an attractive biogenetic process. Feeding experiments have not yet established that this is the biosynthetic route. [3-^{14}C]Tryptophan was specifically incorporated into the tryptamine moiety and [U-^{3}H]anthranilic acid into the quinazoline unit of both **90** and **111**.

The remaining one carbon atom of **90** was specifically derived from [^{14}C]methionine, a precursor that labeled both C-3 and the *N*-methyl group of **111**.

Additional experiments have suggested that **90** is not a precursor of **111**, but that the *N*-methyl group is introduced at an early stage of the biosynthesis possibly *via* *N*-methylanthranilic acid (**113**).

In summary, these data establish the precursors we might expect for **90** and **111**, but there is no available information on the importance of any dihydronorharman-type intermediates in the biosynthetic scheme.

The formation of the quinazoline nucleus in bacteria differs from that in plants because the biosynthetic origin of the anthranilic acid portion is dif-

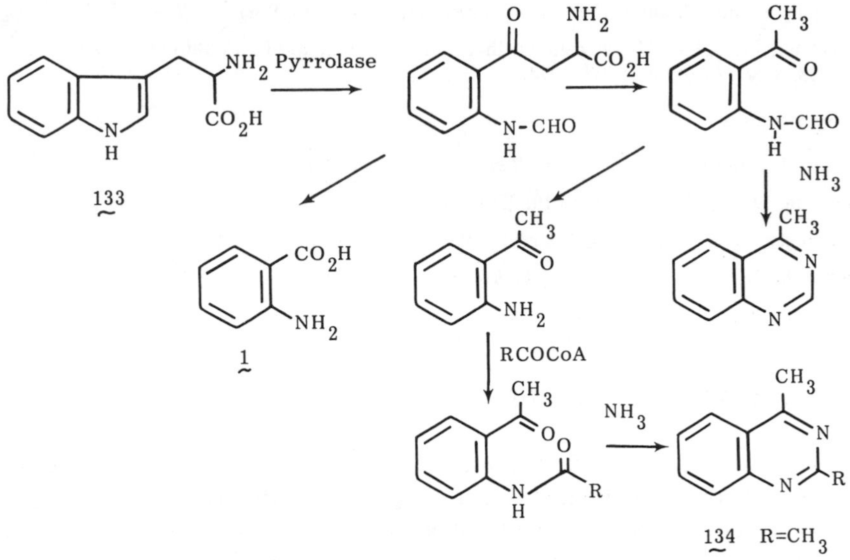

Scheme 9

ferent. In plants, anthranilic acid (**1**) is derived by the shikimate pathway, but in bacteria it is derived from tryptophan (**133**). In *Pseudomonas* three pathways of tryptophan degradation are known and one of these is responsible for the formation of 4-methylquinazoline (**134**) in *P. aeruginosa* (Scheme 9) and other related 4-methyl-2-alkyl quinazolines.

LITERATURE

Reviews

Openshaw, H. T., *Alkaloids NY* **3**, 101 (1953).

Openshaw, H. T., *Alkaloids NY* **7**, 247 (1960).

Gröger, D., *Lloydia* **32**, 221 (1969).

Johne, S., and D. Gröger, *Pharmazie* **25**, 22 (1970).

Synthesis: Vasicine

Späth, E., and N. Platzer, *Chem. Ber.* **69**, 255 (1976).

Leonard, N. J., and M. J. Martell, *Tetrahedron Lett.* 44 (1960).

Moehrle, H., and C. M. Seidel, *Arch. Pharm.* **309**, 503 (1976).

Moehrle, H., and P. Gundlach, *Tetrahedron Lett.* 3249 (1970).

Synthesis: Rutaecarpine/Evodiamine

Asakina, Y., R. H. F. Manske, and R. Robinson, *J. Chem. Soc.* 1708 (1927).

Schopf, C., and H. Stener, *Ann.* **558**, 124 (1947).

Asakina, Y., and T. Ohta, *Ber. Dtsch. Chem. Ges.* **61,** 319 (1928).

Kametani, T., T. Higa, C. V. Loc, M. Ihara, M. Koizumi, and K. Fukutmoto, *J. Amer. Chem. Soc.* **98,** 6186 (1976); **99,** 2307 (1977).

Biosynthesis

Gröger, D., and K. Mothes, *Arch. Pharm.* **293,** 1049 (1960).

Liljegren, D. J., *Phytochemistry* **10,** 2261 (1971).

Waiblinger, K., S. John, and D. Gröger, *Phytochemistry* **11,** 2263 (1972).

Yamazaki, M., and A. Ikuta, *Tetrahedron Lett.* 3221 (1966).

Yamazaki, M., A. Ikuta, T. Mari, and T. Kawana, *Tetrahedron Lett.* 3317 (1967).

7.5 ACRIDINE ALKALOIDS

Alkaloids that are formally derived from acridine (**135**) are very weak bases whose distribution is limited to certain genera of the Rutaceae.

Some representative acridone alkaloids are shown in Figure 2. There are now nearly 40 compounds in this series, and both biologically and chemically, the best known of these is acronycine (**138**).

7.5.1 Chemistry

Several properties of this group of alkaloids are quite characteristic, among these are ready O-demethylation in acid and low solubility in organic solvent

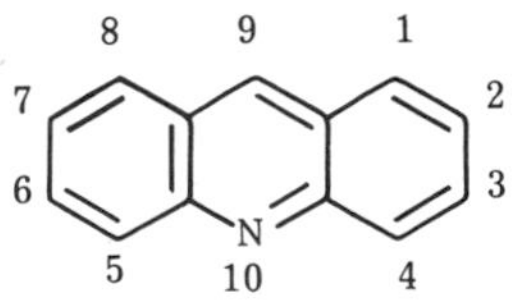

135

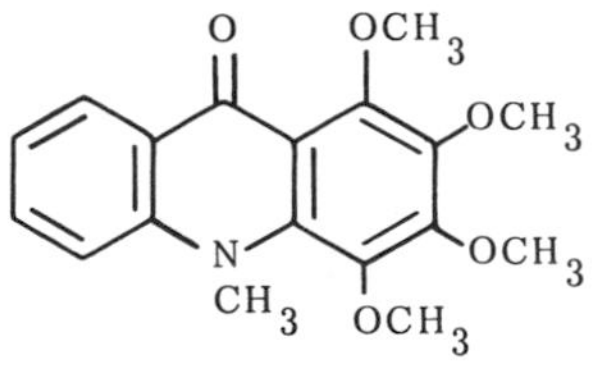

136 melicopicine

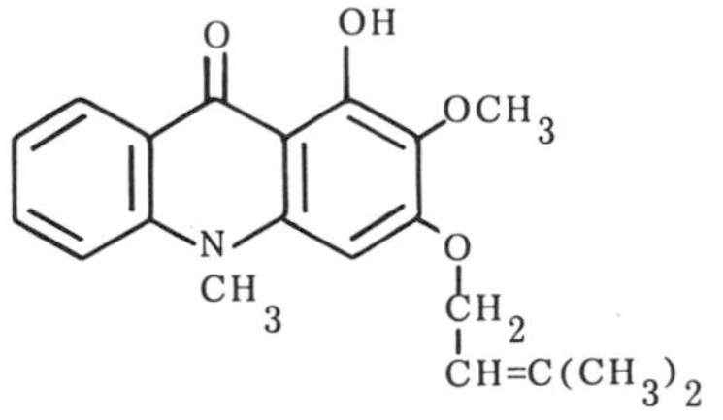

137 evoprenine

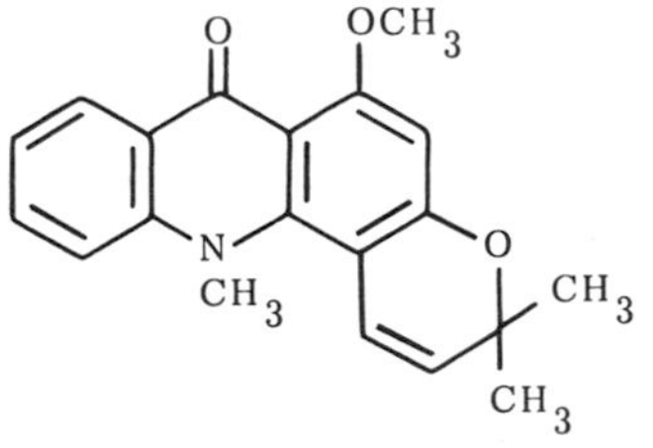

138 acronycine

Figure 2 Representative acridone alkaloids.

due to the near planarity of the ring system. Ring A oxygenation in this series is quite rare.

Melicopicine (136) is a weak base and does not form a picrate or a perchlorate. When refluxed with ethanolic hydrogen chloride the methoxyl group adjacent to the carbonyl group specifically undergoes O-demethylation to give normelicopicine (139). The orange color, insolubility in alkali, and absence of hydroxyl absorption in the infrared indicate that the compound exists in the strongly hydrogen-bonded form shown.

One interesting and unusual reaction of melicopine (140) involves the reaction with methanolic alkali, a reaction which splits the methylene–dioxy group and affords 141. Such reactions are known to take place by nucleopholic attack of the alkoxide anion *para* to an electron-withdrawing group.

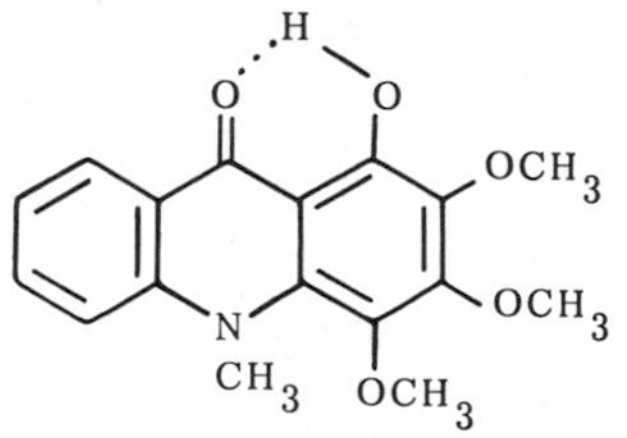

139 normelicopicine

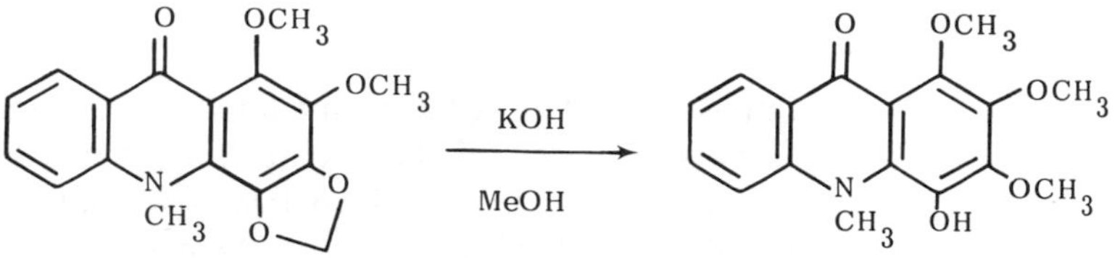

140 141

7.5.2 Synthesis

Melicopicine (136) has been synthesized from tetramethoxyiodobenzene (142) and anthranilic acid (1) by Ritchie *et al.* The secondary amine 143 produced from 1 and 142 is cyclized with phosphorus oxychloride to the chloroacridine 144. This chloro group at C-9 is readily displaced by nucleophiles, and treatment with sodium methoxide gave the pentamethoxyacridine 145. On heating in a sealed tube with methyl iodide, N-methylation and O-demethylation occurs to give melicopicine (136).

A biogenetic-type synthesis of the acridine nucleus involves the condensation of *o*-aminobenzaldehyde (101) and phloroglucinol, a reaction that proceeds to give 2,4-dihydroxyacridine (146) in 90% at pH 8.

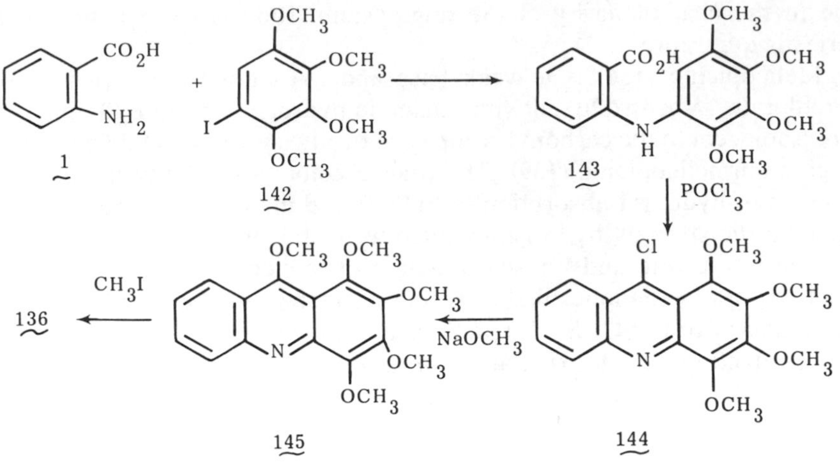

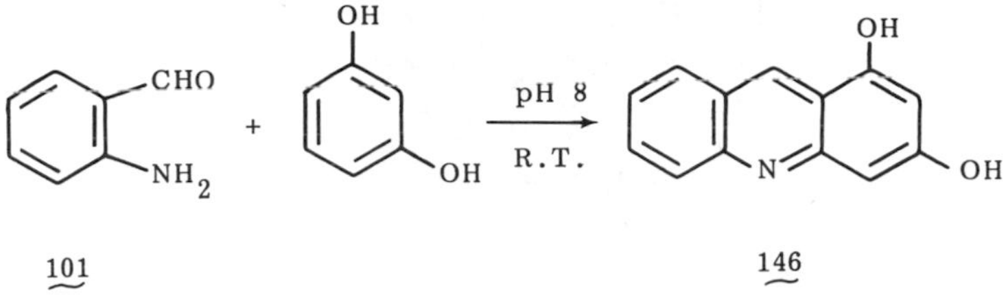

7.5.3 Biosynthesis

Few details of the biosynthesis of other acridone alkaloids have been clarified. For example, the acridone alkaloid melicopicine (136) is not derived from tryptophan or mevalonic acid, but is derived from [5-³H]anthranilic acid (1), [5-³H]N-methylanthranilic acid (113), [1-¹⁴C]acetate, [-³H]4-hydroxyquinol-2-one (74), and [-³H]4-hydroxy-N-methyl-quinol-2-one (147). Since these studies in 1969, very little additional information has been obtained.

At this time therefore the formation of the acridones appears to parallel the furoquinoline alkaloids to the point at which 147 is an intermediate. The most logical step from this compound would be condensation with two molecules of acetyl coenzyme A to give 148, which could then be further elaborated.

Since biosynthetic dehydroxylation of aromatic compounds is an almost unknown process, the formation of compounds such as N-methylacridone (149) and acridone (8) remains somewhat of a curiosity.

7.5.4 Acronycine

The scrub ash, *Acronychia baueri* Schott. (Rutaceae), as well as yielding acridone alkaloids such as melicopine (140) is best known as being a source

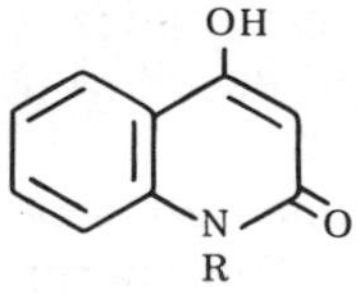

74 R=H
147 R=CH$_3$

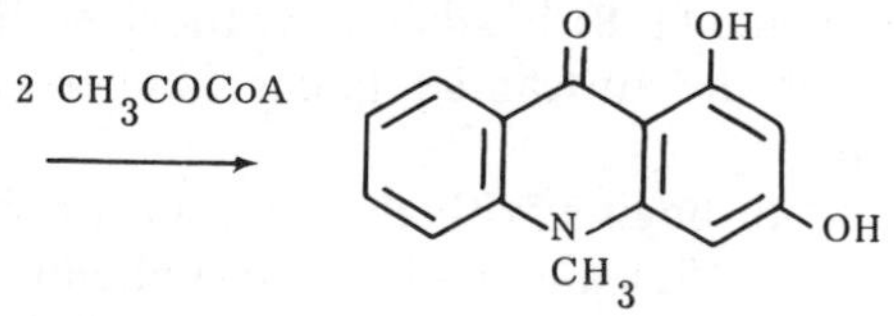

148

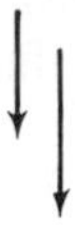

136 or 138

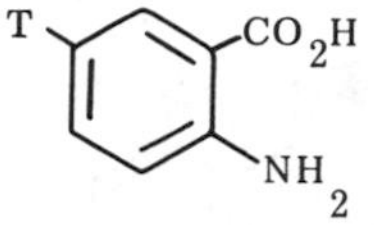

1

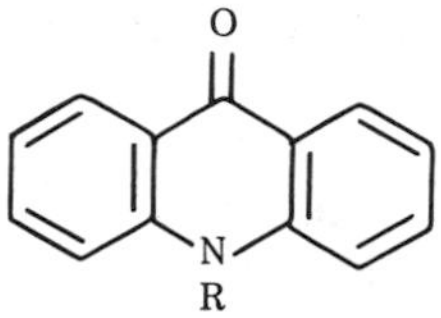

8 R=H
149 R=CH$_3$

of the potent anticancer alkaloid acronycine (**138**). Initial spectroscopic and chemical studies could not distinguish between **138** and the corresponding linear isomer **150**. The structure was eventually deduced by double-resonance NMR studies and confirmed by X-ray crystallographic examination.

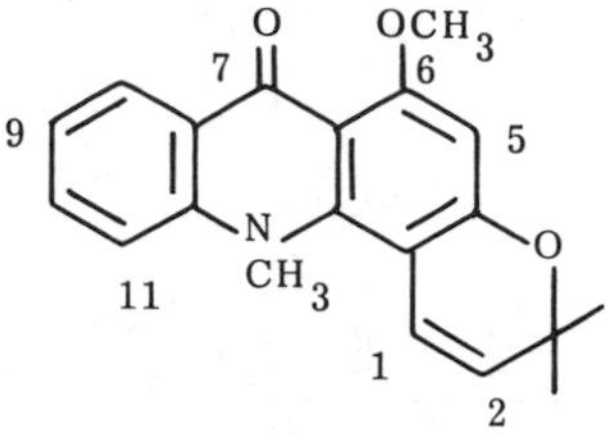

138 acronycine

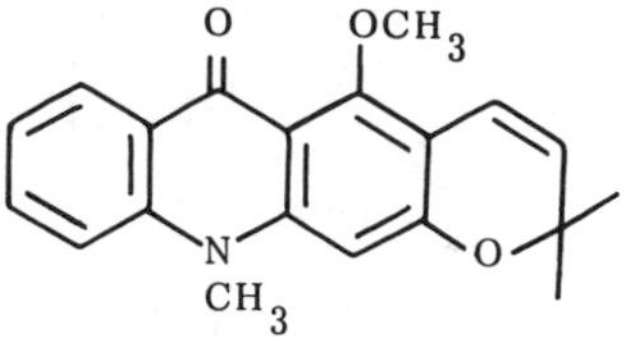

150

In the proton NMR spectrum the N—CH$_3$ and H$_a$ of the chromene ring are sufficiently close for a Nuclear overhauser effect (NOE) to be observed on irradiation at the frequency of the N—CH$_3$ group (i.e. **151**). The *ortho* relationship between the OCH$_3$ and H-2 was established similarly.

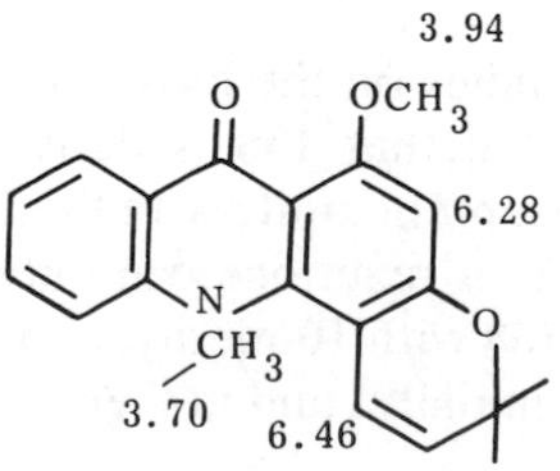

151

Acronycine (**138**) has been synthesized by several routes, but typically mixtures of the unwanted linear isomers are produced which must be separated.

One route begins with the Friedel–Crafts condensation of *o*-nitrobenzoylchloride (**152**) and *m*-dimethoxyphenol (**153**). The intermediate nitrobenzophenone (**154**) was reacted with 3-chloro-3-methylbut-1-yne to give the dimethoxy chromene **155**, which was reduced with zinc dust to the amine **156**.

Treatment of **156** with sodium hydride in dimethyl sulfoxide for 6 days gave a mixture of **157** and norisoacronycine (**158**) (Scheme 10). The weakness inherent in any synthesis of **138** involving a symmetrical intermediate such as **154** should be clear to the reader at this point.

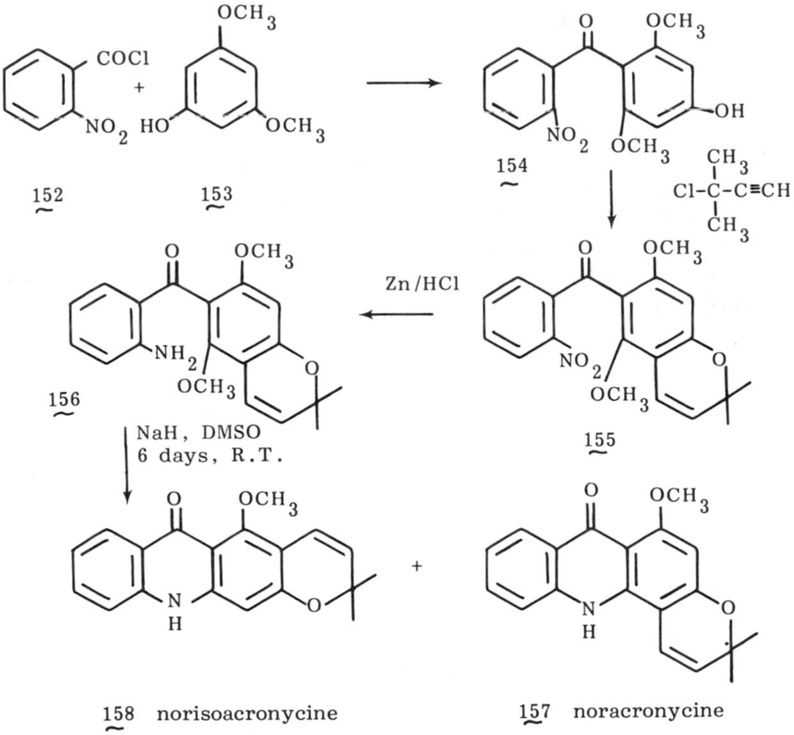

Scheme 10   Synthesis of noracronycine (**157**).

There have been no studies on the biosynthesis of acronycine (**138**), but biogenetically it is expected that **138** is derived from anthranilic acid, a polyacetate unit, and an isoprene unit, with the methyl groups derived from methionine. The sequence of reactions can only be surmised at this point, but the co-occurrence of **138** with 10-methyl acridone alkaloids suggests that thus nucleus is produced initially, and is then isoprenylated and oxidatively cyclized.

Acronycine (**138**) is of interest because it possesses the broadest spectrum of antitumor activity of any alkaloids isolated to date, being effective against 13 of 19 experimental tumor systems. Some cures were also noted.

Studies of the metabolism of acronycine (**138**) have indicated that hydroxylation occurs on the aromatic nucleus at C-9 or C-11 and/or in the side chain. Hydroxylation at the 9-position also occurs on incubation with *Cunninghamella echinulata*.

One important feature of *in vivo* experimental work with acronycine (**40**) is of interest because of its potential applicability to other *in vivo* studies with alkaloids. For the first time in the testing of highly insoluble alkaloids, some efforts have been made to improve the solubility in the administering solvent. Two preparations of interest are an acronycine–polyvinyl pyrrolidone coprecipitate, which is 15 times more soluble than the parent base, and acetyl acronycine perchlorate, which is 1000 times more soluble than **138**.

LITERATURE

Reviews

Price, J. R., *Fortschr. Chem. Org. Naturs.* **13**, 302 (1956).

Acheson, R. M., *Acridines*, Interscience, New York, 1956.

Manske, R. H. F., *Alkaloids NY* **12**, 478 (1970).

Johne, S., and D. Gröger, *Pharmazie* **27**, 195 (1972).

Waterman, P., *Biochem. Syst. Ecol.* **3**, 149 (1975).

Biosynthesis

Prager, R. H., and H. M. Thredgold, *Aust J. Chem.* **22**, 2627 (1969).

Acronycine Synthesis

Adams, J., P. Gupta, and J. R. Lewis, *Chem. Ind.* 109 (1976).

7.6   1,4-BENZOXAZIN-3-ONE ALKALOIDS

A neglected yet significant group of alkaloids are the 1,4-benzoxazin-3-ones having the basic skeleton **159**.

These compounds were first obtained in 1959 from species in the Gramineae, such as maize (*Zea mays*), wheat (*Triticum cereale*), and rye (*Secale cereale*). Since then several additional members of this group have been isolated from maize.

2,4-Dihydroxy-7-methoxy-1,4-benzoxazin-3-one (**160**), known as DIMBOA, is a feeding deterrent for the European corn borer, inhibits spore germination of the pathogenic fungus *Helminthosporium turcicum*, and is positively correlated with fungal resistance of cereal grasses.

159   160

These compounds exist in the plant mainly as glucosides, but cell injury apparently releases a glucosidase which catalyzes hydrolysis to the 2-hydroxy derivatives.

The biosynthesis of these compounds was first studied by Reimann and Byerrum, who found that the aromatic nucleus of DIMBOA was derived from quinic acid and the methoxy carbon from methionine. The remaining two carbons of the oxazine ring are derived from carbons 1 and 2 of ribose. Acetate, glycine, and glycolate were not incorporated into the oxazine ring.

Subsequent work by Tipton and co-workers has shown that the aromatic amine involved is anthranilic acid (1) and that further metabolism to 3-hydroxyanthranilic acid or *o*-aminophenol is not important. It was suggested that *N*-(5'-phosphoribosyl)anthranilate might be the next intermediate, but neither this nor any of the subsequent steps have been investigated further.

LITERATURE

Reimann, J. E., and R. H. Byerrum, *Biochemistry* **3**, 847 (1964).

Tipton, C. L., M. C. Wang, F. H.-C. Tsao, C.-C. Lin Lu, and R. R. Husted, *Phytochemistry* **12**, 347 (1973).

7.7 BENZODIAZEPINE ALKALOIDS

The structurally interesting antibiotics cyclopenin (**161**) and viridicatin (**162**) are obtained from a related group of *Penicillia* species, notably *P. viridicatum* and *P. cyclopium*. The co-occurrence of these two metabolites becomes more significant when it is realized that cyclopenin (**161**) undergoes an efficient acid-catalyzed rearrangement to viridicatin (**162**). The reaction also proceeds thermally, and the intermediate **163** has been proposed to explain this rearrangement.

161   cyclopenin   163   162   viridicatin

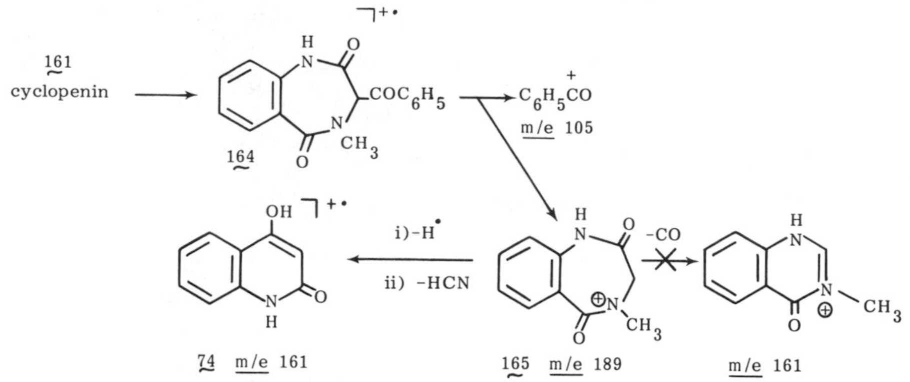

Scheme 11

The mass spectrum of cyclopenin (**161**) has been studied with deuterated derivatives. The compound displays two major fragmentation pathways as shown in Schemes 11 and 12. The first scheme involves rearrangement to the benzophenone **164**, loss of m/e 105 to **165**, followed by successive losses of H and HCN to **74**. It had earlier been thought that loss of CO from **165** was responsible for the ion at m/e 161, but accurate mass measurements proved this to be incorrect.

The second fragmentation pathway (Scheme 12) involves methyl isocyanate loss to **166**. Because this is the thermal loss that occurs in the formation of viridicatin (**162**), it was considered that the ion at m/e 237 in the mass spectrum of **161** was due to **162**. However, the most intense peak at m/e 119, formulated as **167**, is not present in the spectrum of **162**.

Cyclopenin (**161**) has been synthesized stereospecifically from the hippuric acid derivative **168**. Condensation of **168** with benzaldehyde, followed by esterification, gave the methyl ester **169**. When the nitro group was catalytically reduced and the product cyclized in refluxing xylene, the

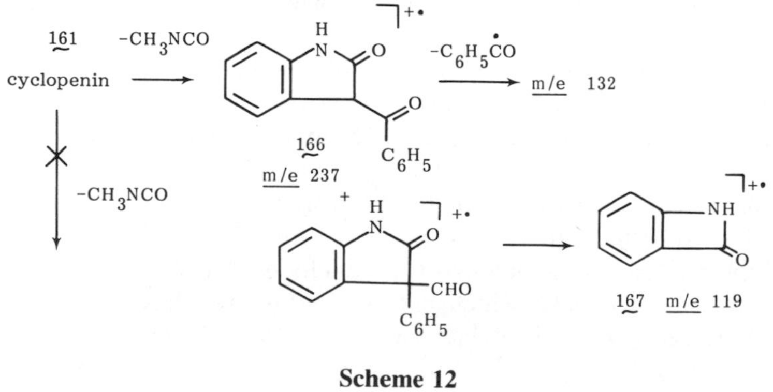

Scheme 12

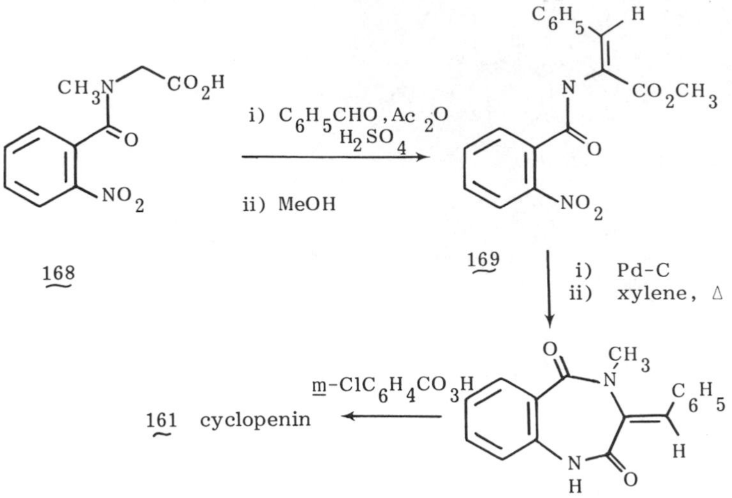

benzodiazepine (170) resulted, which on epoxidation gave cyclopenin (161). Clearly such a sequence also constitutes a total synthesis of viridicatin (162).

Biosynthetically, we can imagine the formation of cyclopenin (161) as occurring by way of a condensation between phenylalanine and anthranilic acid with subsequent, relatively minor, transformations. The conversion of 161 to viridicatin (162) can also be envisaged since it has been demonstrated *in vitro*. Experimental work has substantiated these ideas and has revealed some interesting intermediate stages.

Luckner and Mothes found that the phenyl ring and C-2, C-3, and C-4 of viridactin (162) were derived from phenylalanine, and that the benzene ring of the quinoline system was derived from anthranilic acid (1), but with loss of the carboxyl group at an intermediate stage.

The relationship of cyclopenin (161) and viridicatin (162) *in vivo* was substantiated when an enzyme, cyclopenase, capable of carrying out this transformation was obtained. It is not known if there are intermediates in this step.

One of the early intermediates after the condensation of phenylalanine and anthranilic acid is cyclopeptine (171). This compound was a precursor of cyclopenin (161) and cyclopenol (172) in *P. cyclopium* by way of the dehydroderivative 173. Cyclopeptine (171) and 173 are interconvertable *in vivo*, and the enzyme cyclopeptine dehydrogenase is at maximum activity during alkaloid formation.

Cyclopenin (161) is a precursor of cyclopenol (172) and the enzyme cyclopenin *m*-hydroxylase which catalyzes this step has been identified. This enzyme does not hydroxylate viridactin (162) (Scheme 13).

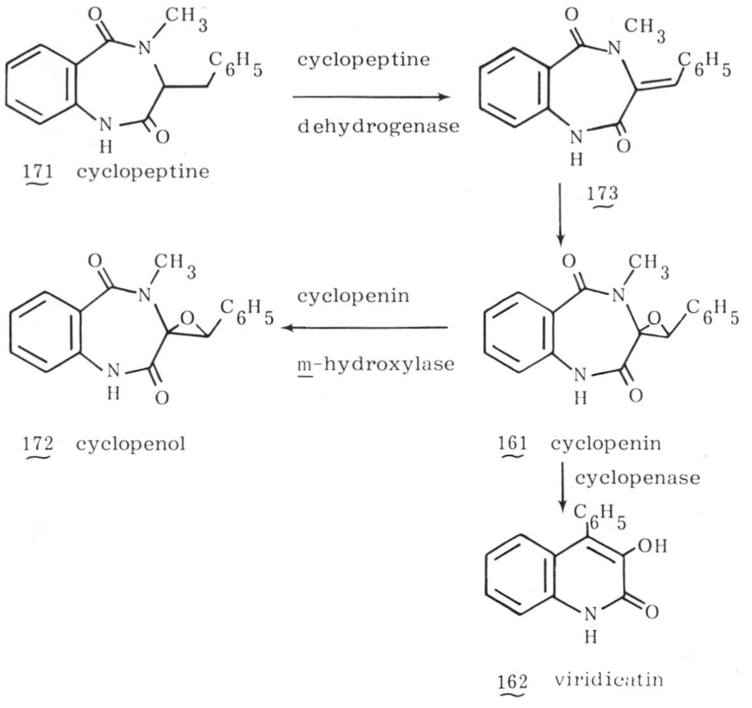

Scheme 13 Biosynthesis of cyclopenin (**161**) and viridicatin (**162**).

LITERATURE

Biosynthesis

Luckner, M., and K. Mothes, *Tetrahedron Lett.* 1035 (1962).

Luckner, M., *Eur. J. Biochem.* **2**, 74 (1967).

Aboutabl, E., and M. Luckner, *Phytochemistry* **14**, 2573 (1975).

Richter, I., and M. Luckner, *Phytochemistry* **15**, 67 (1976).

7.8 CRYPTOLEPINE

The roots of *Cryptolepis triangularis* N.E. Br. (Asclepiadaceae) native to the Belgian Congo were first examined chemically in 1929 by Clinquart, who obtained cryptolepine. Over 20 years later Raymond-Hamet and co-workers obtained the same alkaloid from a Nigerian *Cryptolepis* species *C. sanguin-olenta* (Lindl.) Schlechter and found that depending on the solvent, the solution may vary in color from violet to red, even though it forms violet needles. Some of its salts are however yellow. The structure was deduced

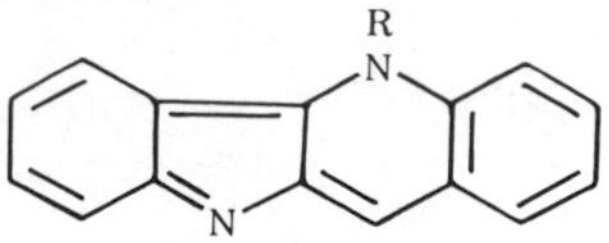

174   cryptolepine, R=CH$_3$ 176   113
175   quindoline, R=H

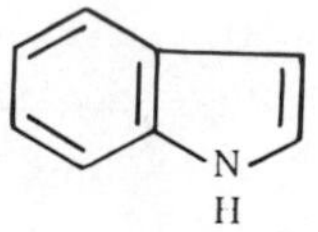

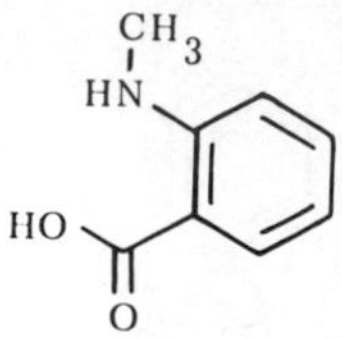

to be **174** by degradation and it was only then that it was realized that the skeleton had been synthesized in the very early 1900s, 20 years before its first isolation. Quindoline (**175**) itself has recently been isolated from *C. sanguinolenta*.

No biosynthetic studies of this skeleton have been reported, but biogenetically one might imagine a derivation from indole (**176**) and *N*-methyl anthranilic acid (**113**) for cryptolepine (**174**).

In dogs cryptolepine causes hypothermia and a pronounced fall in blood pressure.

chapter 8

ALKALOIDS DERIVED FROM PHENYLALANINE AND TYROSINE

The alkaloids derived from the amino acids phenylalanine (**1**) and tyrosine (**2**) are of an extremely diverse nature. They range from quite simple amines such as mescaline (**3**), through simple tetrahydroisoquinolines such as pellotine (**4**) to complex monomeric alkaloids such as morphine (**5**) and dimers of the tetrandrine (**6**) type. There are also examples where the tyrosine-derived portion of the molecule is barely discernible [e.g. betanidin (**7**), aranotin (**8**), and securinine (**9**)]

With so many varied structure types it is not surprising to find that this group of alkaloids covers the gamut of pharmacologic responses, and many of these are of considerable therapeutic significance (Table 1).

One of the special features of this group of compounds is their distribution, which stretches from the bacterial and fungal world into marine organisms, plants, lower animals, and all the way to humans.

Some of the very simple amines such as 3,4-dihydroxyphenylethylamine (**10**), also known as dopamine, appear to be ubiquitous and clearly are of immense biological importance.

The chapter is organized approximately according to molecular complexity and evolving biogenesis.

8.1 SIMPLE TYRAMINE DERIVATIVES

Hydroxylation of L-phenylalanine to L-tyrosine is an important process in higher organisms and, as discussed in Chapter 2, proceeds with involvement of an NIH shift. Most plants however do not use this route as the major source of tyrosine, and in many species no conversion takes place. However a phenylalanine hydroxylase has been obtained from spinach, and it is clear from feeding experiments with peyote, barley, and rye grass that the phenylalanine–tyrosine conversion must occur in these plants as well.

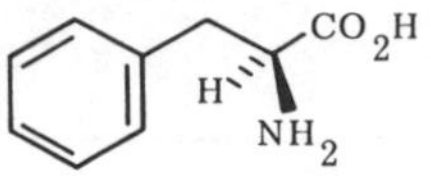

1 phenylalanine 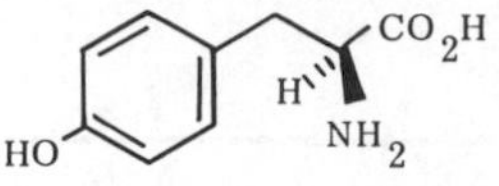   2 tyrosine 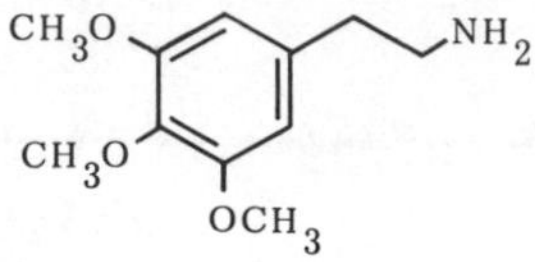   3 mescaline

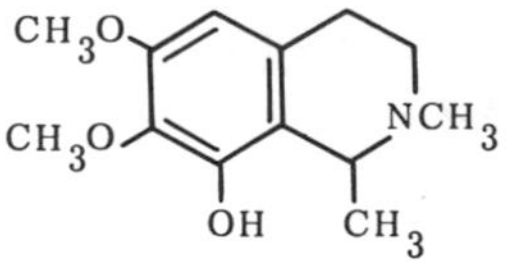

4 pellotine

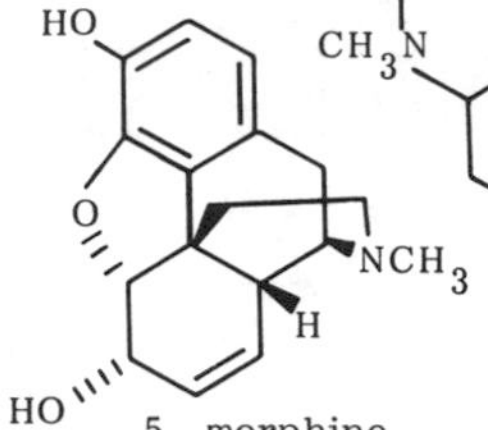

5 morphine 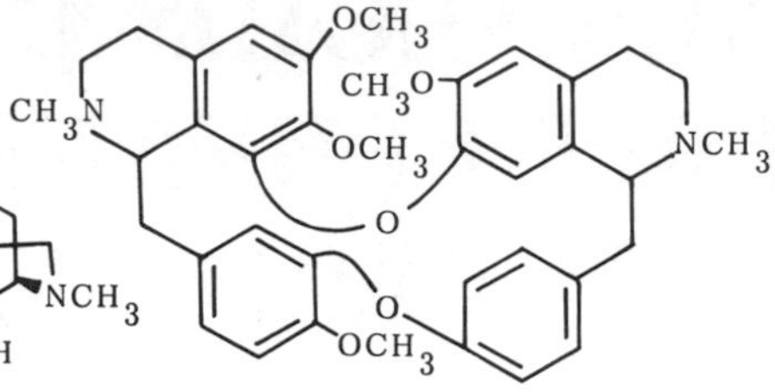   6 tetrandrine

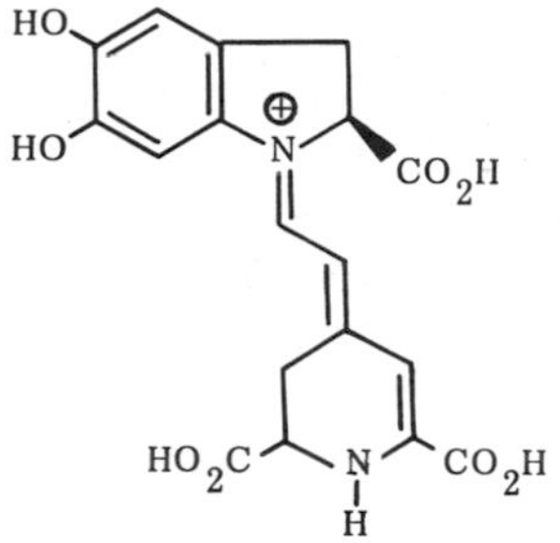

7 betanidin 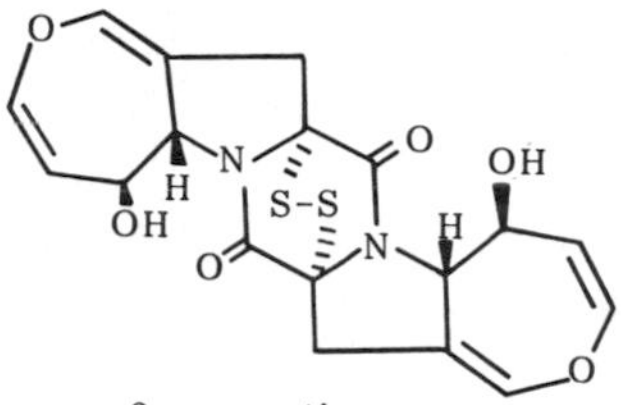   8 aranotin 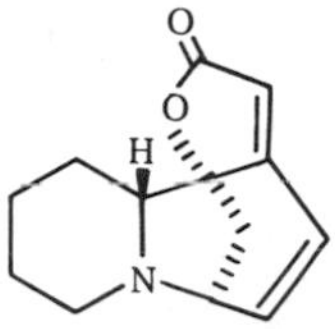   9 securinine

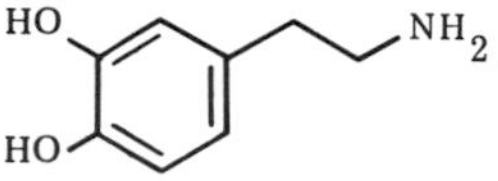

10 dopamine

An important disease state in humans, phenylketonuria, results from the absence of L-phenylalanine hydroxylase activity. It causes severe mental retardation, and children are particularly susceptible.

Hordenine (**11**), the *N,N*-dimethyl derivative of tyramine (**12**) and a constituent of barley, *Hordeum vulgare* L. (Gramineae), is widely distributed, being present in cacti, fungi, and marine algae, and *N,N,N*-trimethyltyramine (**13**), candicine, is also well known.

The β-hydroxyphenethylamines have several biologically important rep-

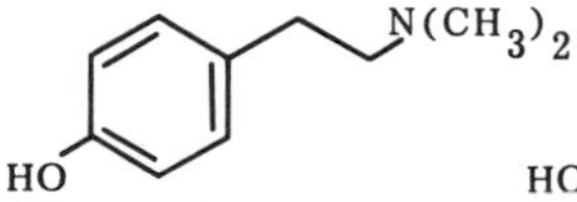

11 hordenine 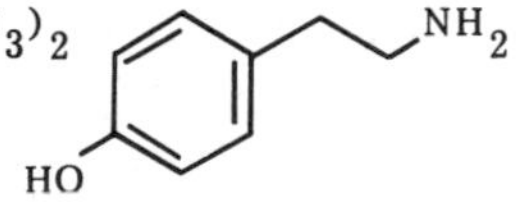   12 tyramine 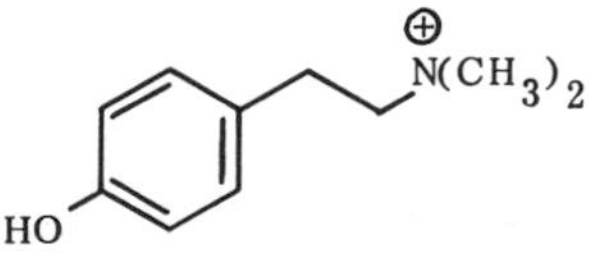   13 candicine

Table 1 Some Phenylalanine-Derived Alkaloids of Pharmacologic Significance

Alkaloid	Source	Pharmacologic Action
Mescaline	*Lophopora williamsii*	Hallucinogen
Ephedrine	*Ephedra sinica*	Sympathomimetic
Papaverine	*Papaver somniferum* and synthesis	Vasodilator
Tubocurarine	*Chondrodendron tomentosum*	Skeletal muscle relaxant
Tetrandrine	*Cyclea peltata*	Antitumor
Glaziovine	*Octoea glaziovii*	Antidepressant
Thalicarpine	*Thalictrum dasycarpum*	Antitumor
Codeine	*Papaver somniferum*	Analgesic, antitussive
Morphine	*Papaver somniferum*	Narcotic analgesic
Fagaronine	*Fagara zanthoxyloides*	Antitumor
Colchicine	*Colchicum autumnale*	Suppression of gout
Harringtonine	*Cephalotaxus harringtonia*	Antitumor
Emetine	*Cephaelis ipecacuanha*	Antiamebic

resentatives, including epinephrine (adrenaline) (**14**) and norepinephrine (noradrenaline) (**15**). Adrenaline does not appear to occur in plants, but noradrenaline has been obtained from bananas. These compounds are both widely distributed in the animal kingdom and are typically found in the neurons, where noradrenaline is a neurotransmitter released from neuron terminals.

The steps involved in the formation of adrenaline (**14**) in mammals are shown in Scheme 1. All the enzymes have been isolated and quite considerable detail is known about their specificity. The first enzyme, L-tyrosine

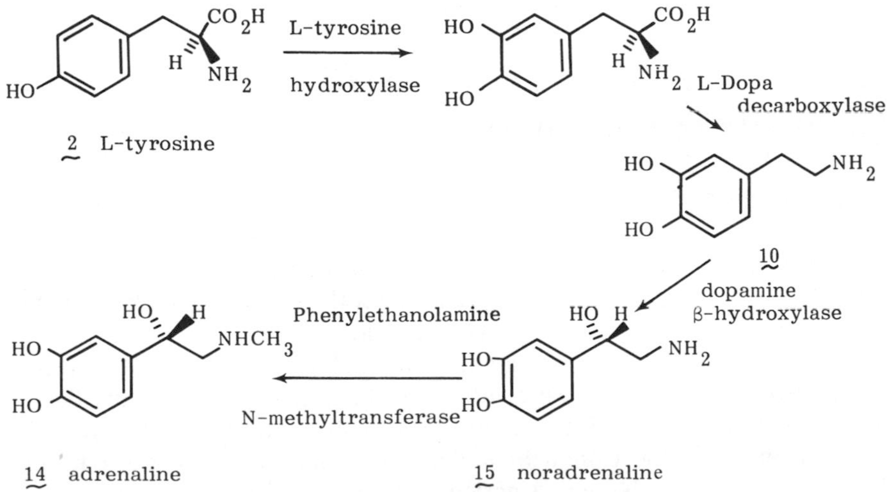

Scheme 1 Formation of adrenaline (**14**).

hydroxylase, is the rate-controlling enzyme in catecholamine metabolism, and no NIH-shift occurs when the hydroxy group is introduced.

The enzyme is not capable of hydroxylating D-tyrosine or tyramine. L-Dopa decarboxylase is widely distributed in mammalian tissues but is not very specific and can decarboxylate other L-amino acids. Similarly dopamine β-hydroxylase, which is also lacking in specificity, has cupric ion as a requirement.

The chiral shift reagent tris[t-butylhydroxymethylene-d-camphorato] europium(III) has been used to determine the enantiomeric purity of chiral β-phenethylamines. Quite substantial shift differences were observed, and for example the methine resonances of (R)- and (S)-amphetamines were separated by 0.7 ppm in a 0.15-mole/liter carbon tetrachloride solution of the complex.

LITERATURE

Reviews

Snieckus, V. A., *Alkaloids, London* **1**, 103 (1971).

Bernhard, H. O., and V. A. Snieckus, *Alkaloids, London* **2**, 96 (1972).

Shamma, M., *Alkaloids, London* **3**, 116 (1973).

Snieckus, V. A., *Alkaloids, London* **4**, 128 (1974).

Turner, A. B., in S. Coffey (Ed.), Rodd's *Chemistry of Carbon Compounds*, 2nd ed., III E, Elsevier, Amsterdam, 1974, p. 79.

Bernhard, H. O., and V. A. Snieckus, *Alkaloids, London* **5**, 111 (1975).

McCorkindale, N. J., *Alkaloids, London* **6**, 110 (1976).

McCorkindale, N. J., *Alkaloids, London* **7**, 92 (1977).

Bentley, K. W., *Alkaloids, London* **8**, 87 (1978).

8.2 PEYOTE AND MESCALINE

Peyote is one of the very old hallucinogenic drugs of the New World. The plant *Lophophora williamsii* (Lemaire) Coulter is a member of the family Cactaceae and is wide ranging in the Chihuahuan Desert of Texas and Mexico. "Mescal buttons" are sun-dried slices of the plant and are brewed with a tea or chewed while drinking other beverages.·

The history of peyote as a panacea and hallucinogen probably predates the Christian era. In the seventeenth and eighteenth centuries its use became widespread among Indian tribes and eventually led at the end of the nineteenth century to the establishment in Oklahoma of a "Peyote Church." This institution, the Native American Church, chartered in 1918, has the only legitimate right to the use of peyote, since the drug is claimed to be part of a religious ceremony. All other uses of peyote are federally controlled.

8.2.1 Chemistry

Peyote was first investigated chemically in 1888 by Lewin, who succeeded in obtaining anhalonine (16). Heffter is credited with the first isolation of

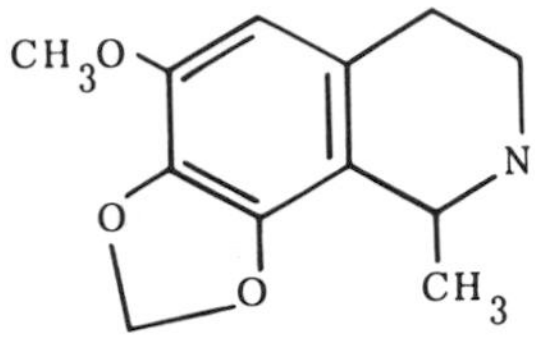

16

mescaline (3) in 1896, and through personal experience demonstrated its hallucinogenic activity. The classic work on peyote and many plants in the Cactaceae is that of Späth and co-workers in the years 1919–1939, and to date over 50 alkaloidal constituents have been obtained. Cacti in the genus *Trichocereus* are an additional source of mescaline.

The first synthesis of mescaline was reported by Späth in 1919 and begins with 3,4,5-trimethoxybenzaldehyde (17). With recent modifications, such as using ammonium acetate in acetic acid for the condensation step with nitromethane, and lithium aluminum hydride for the reduction step, very high yields can be obtained. Some alternative routes to mescaline include using the corresponding nitrile (18), amide (19), or phthalimide (20) derivative.

Several reagents have been used for the qualitative detection or identification of peyote alkaloids. For thin-layer chromatography dansyl chloride is suitable for nonphenolic products and tetrazotized benzidine for phenolic compounds.

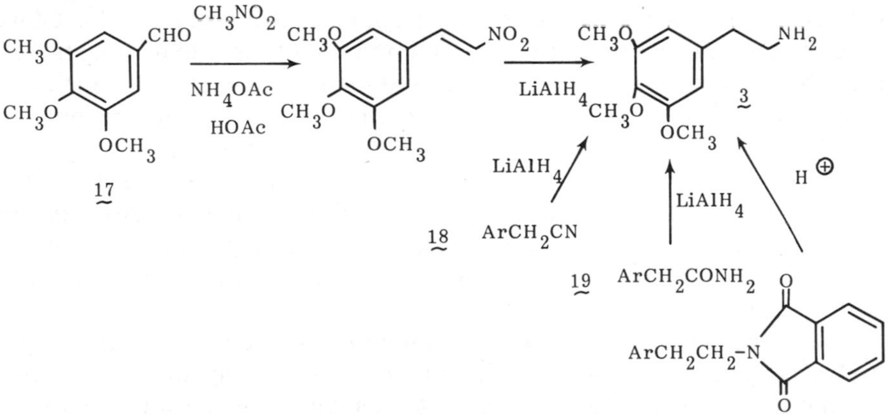

Quantitatively, some of the colorimetric methods which have been used include reactions with *p*-nitrophenyldiazonium chloride, picric acid, and bromcresol purple.

As expected, mass spectral fragmentation of mescaline and mescaline derivatives occurs principally between the side-chain carbon atoms, although transfer of a proton from the nitrogen also apparently takes place to some extent (Scheme 2).

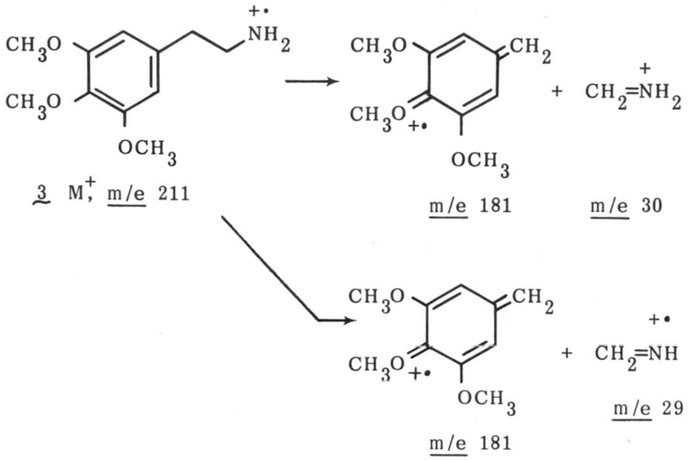

Scheme 2   Mass spectral fragmentation of mescaline (**3**).

8.2.2   Biosynthesis

Reti in 1950 was the first to suggest the biosynthesis of mescaline (**3**) from tyrosine (**2**) through dopa (**21**), and all subsequent investigations have supported this concept.

Mescaline is one of the few alkaloids whose biosynthesis has been studied with essentially all the reasonable precursors, and the results of feeding some of these are shown in Table 2. The purpose of these experiments was to establish the biosynthetic sequence, and the results suggest the pathway in Scheme 3. Note how the feeding experiments with **22** and **23** indicate initial 3- rather than 4-methylation, and how the relative incorporations of **24** and **25** together with the excellent incorporation of **25** indicate that a 5-hydroxyl group is introduced and methylated *before* the 4-hydroxy group is methylated. The good incorporation of **27** is not supported by the incorporation of **25**, and may indicate conversion to **22** prior to incorporation into mescaline (**3**).

The processes involved in the formation of mescaline are quite simple and basically involve aromatic hydroxylation and O-methylation. It is therefore not too surprising to find that an O-methyltransferase was readily obtained from peyote. The enzyme was found to be quite specific taking dop-

Table 2 Biosynthesis of Mescaline from Labeled Precursors

DL-Phenylalanine (**1**)	0.03
DL-Tyrosine (**2**)	0.32
Tyramine (**12**)	1.53
DL-Dopa (**21**)	1.58
Dopamine (**10**)	1.90
4-Hydroxy-3-methoxyphenethylamine (**22**)	1.45
3-Hydroxy-4-methoxyphenethylamine (**23**)	0.10
3, 4-Dimethoxyphenethylamine (**27**)	3.75
3, 4, 5-Trihydroxyphenethylamine (**28**)	0.72
4, 5-Dihydroxy-3-methoxyphenethylamine (**24**)	2.49
3, 4-Dimethoxy-5-hydroxyphenethylamine (**25**)	0.21
3, 5-Dimethoxy-4-hydroxyphenethylamine (**26**)	15.43

Source: Data from A. G. Paul, *Lloydia* **36**, 36 (1973).

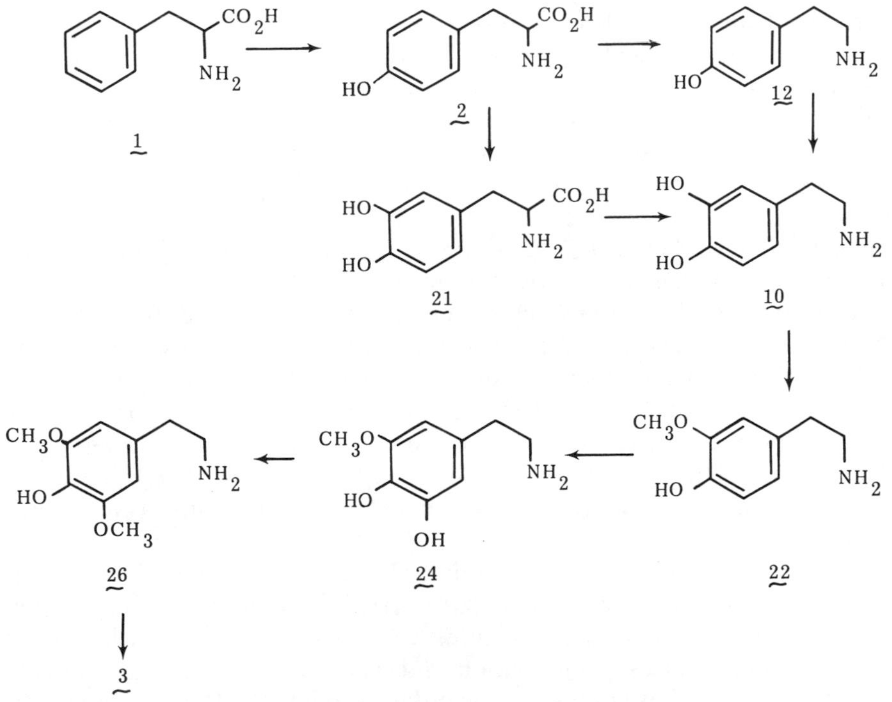

Scheme 3 Biosynthesis of mescaline (**3**).

amine through to **22** but not **23**, and methylating both remaining phenolic groups of **24**, but not methylating **28**. Some of the biosynthetic data concerning other peyote alkaloids is discussed subsequently.

23   R=H   25   28

27   R=CH$_3$

8.2.3 Pharmacology

In moderate doses (20–60 mg/kg) mescaline produces a drop in blood pressure, bradycardia, and respiratory depression. Lower doses inhibit the pressor effect of epinephrine and produce hyperthemia and uterine contractions. The LD$_{50}$'s of mescaline in the rat are about 370 mg/kg i.p. and 155 mg/kg i.v.

In laboratory animals mescaline produces catatonia, and in mice it markedly disturbs socialization, producing an increased excitement, aggressiveness, defensive hostility, and a scratching response. Mescaline also gave aggressive tendencies in mice by intracerebral injection. It potentiates and prolongs the analgesic effect of morphine.

Of more interest are the human clinical responses to mescaline, particularly those of a phantasticant nature. The first reports on the hallucinatory properties were made by Dr. Francisco Hernandes, the court Physician to Philip II of Spain between 1570 and 1575, who reported: "Those who eat or chew it see visions either frightful or laughable. . . ."

Havelock Ellis described his experience with peyote as being mainly sensory, but leaving the intellect unimpaired. Indeed in an article published in 1897, Ellis indicates that "a large part of its charm lies in the halo of beauty it casts around the simplest and commonest things. . . . For a healthy person to be once or twice admitted to the rites of mescal is . . . an influence of no mean value." Other users however have indicated that "The experience . . . was worth one such headache and indigestion, but was not worth a second."

Mescaline is readily absorbed orally, although a latency period for sensory change of 60 min or more is characteristic. The normal dose is 5 mg/kg (i.e. ~300–350 mg for a typical adult), and hallucinations may last up to 6 hr. This dose causes pupil dilation and increases the heart rate, effects that are similar to those of lysergic acid diethylamide (LSD) and psilocybin.

The initial physical reactions are nausea and vomiting after about 30 min, but these usually subside before psychic effects begin. Daily use of mescaline produces tolerance and a cross tolerance to LSD. Habituation and addiction do not result from repeated use and peyote should therefore not be regarded as a narcotic drug.

The mode of action of mescaline is not known in any detail, although there is some evidence that mescaline *per se* is not the true psychomimetic. No psychoactive metabolites have ever been obtained.

There have been several studies made of the importance of the various functional groups present in mescaline and of the effect of additional groups introduced into the aromatic nucleus. N-Methylation or 4-demethylation causes complete loss of psychopharmacologic activity, and removal of the 5-methoxy group causes a 50% loss in activity. If the 5-methoxy group is removed and placed at the 2-position, the activity is increased. α-Methylation of the side chain, to give derivatives such as **29**, also increases potency. At this point the similarity of this compound to elemicin (**30**), the probable hallucinogenic principle of nutmeg, *Myristica fragrans*, is worth noting. The most active mescaline derivative so far prepared appears to be **31**, which is reported to be at least 300 times more active than mescaline itself.

29   30   elemicin   31

LITERATURE

Reviews

Anonymous, *Bull. Narc.* **11**, 16 (1959).

La Barre, W., *The Peyote Cult*, Shoe String, Hamden, Conn., 1964, 260 pp.

Mardesosian, A. D., *Amer. J. Pharm.* **138**, 204 (1966).

Kapadia, G. J., and M. B. E. Fayez, *J. Pharm. Sci.* **59**, 1699 (1970).

Ray, O. S., *Drugs, Society and Human Behavior*, C. V. Mosey, St. Louis, Mo., 1972, pp. 216–220.

Brawley, P., and J. C. Duffield, *Pharm. Rev.* **24**, 31 (1972).

McLaughlin, J. L., *Lloydia* **36**, 1 (1973).

Kapadia, G. J., and M. B. E. Fayez, *Lloydia* **36**, 9 (1973).

Shulgin, A. T., *Lloydia* **36**, 46 (1973).

8.3 EPHEDRINE

"Ma Huang" has been known to Chinese physicians for over 5000 years, and the principal alkaloid of importance, ephedrine (32), was first isolated in 1887 by Nagai. The preferred natural source of ephedrine is still "Ma Huang," which is a mixture of the aboveground parts of *Ephedra equisetina* Bunge, *E. sinica* Stapf., and *E. distachya* L. in the Gnetaceae. The yield of crude alkaloids is in the range 0.5–2.0%, and depending on the species, up to 90% of this may be ephedrine. Little of the ephedrine of U.S. commerce is currently obtained from natural sources.

8.3.1 Chemistry

The structure of ephedrine was first suggested by Ladenburg and Olschägel in 1880 and confirmed by Schmidt and Bumming 20 years later. Subsequent problems revolved around the stereochemical differences between (−)-ephedrine (32) and its isomer (+)-pseudoephedrine (33).

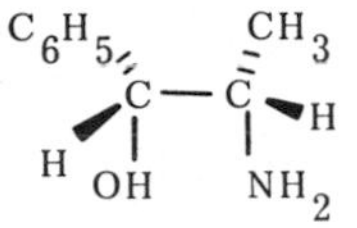

32 (-)-ephedrine 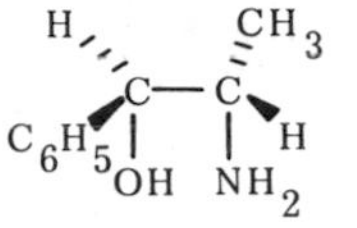   33 (+)-pseudoephedrine 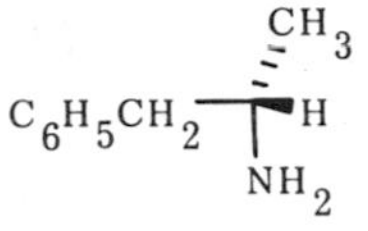   34

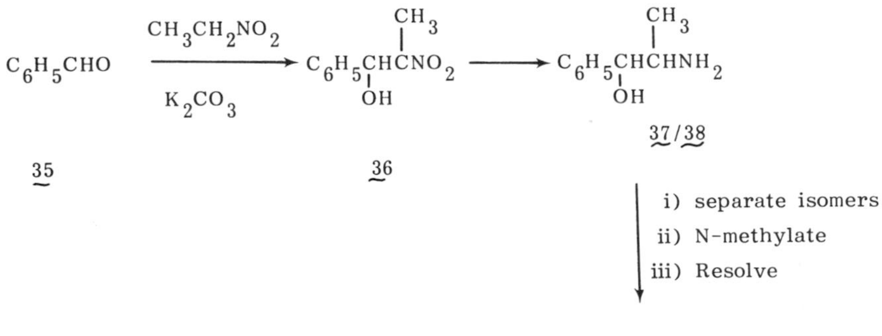

$$C_6H_5CHO \xrightarrow[K_2CO_3]{CH_3CH_2NO_2} C_6H_5\underset{OH}{CH}\overset{CH_3}{C}NO_2 \longrightarrow C_6H_5\underset{OH}{CH}\overset{CH_3}{C}HNH_2$$

35   36   37/38

i) separate isomers
ii) N-methylate
iii) Resolve

(-)-ephedrine

32

These isomers are reversibly interconvertible and by reduction give (+)-desoxyephedrine (34). Ephedrine (32) and pseudoephedrine (33) therefore differ only with respect to the configuration of the carbinol function. Chemical degradation yielded the absolute configuration of the two centers, and this was confirmed by X-ray analysis of (−)-ephedrine hydrochloride, indicating that (−)-ephedrine has the structure **32** and (+)-pseudoephedrine the structure **33**.

There have been numerous syntheses of ephedrine, and two of these are of particular interest. Condensation of benzaldehyde (**35**) with nitroethane in the presence of base (K_2CO_3) afforded a mixture of diastereoisomers **36**. Reduction of the nitro group gave a separable mixture of norephedrine (**37**) and norpseudoephedrine (**38**), and methylation afforded racemic ephedrine (**32**), which could be resolved.

Although the synthesis described above is suitable for commercial use, the most important commercial synthesis of ephedrine is a biomimetic one. In this way (−)-ephedrine can be produced directly. Benzaldehyde (**35**), when added to an actively fermenting (yeast) glucose or sucrose solution, is reduced to benzyl alcohol and L-1-phenyl-2-oxo-1-propanol (**39**). The latter compound can be condensed simultaneously or consecutively with methylamine. Reduction with activated aluminum in moist ether then yields (−)-ephedrine (**32**). In some instances the propanol derivative can be obtained in up to 76% yield.

The Cotton effects of ephedrine (**32**) and some of its derivatives have been studied, and based on the sector rules for the benzene chromophore, the *erythro-* and *cis*-diastereomers (e.g. ephedrine) prefer a conformation in which the oxygen atom is close to the plane of the phenyl group. In the *threo-* and *trans*-isomers (e.g. pseudoephedrine) however the oxygen atom

adopts a configuration such that it is perpendicular to the plane of the phenyl group.

8.3.2 Biosynthesis

The biosynthesis of ephedrine has proved to be somewhat of a surprise, and much of the progress in this area has been achieved by Shibata and co-workers. It is now clear that although the aromatic nucleus of phenylalanine

(1) is incorporated, only one of the side-chain carbons survives the biosynthetic process. This somewhat amazing result was demonstrated by comparison of the results of using [2-^{14}C]- and [3-^{14}C]phenylalanine (1) as precursors. Whereas the former was not incorporated, the latter labeled the benzoic acid produced on oxidation of labeled ephedrine. Thus a C_6-C_1 unit is implied, and indeed both benzoic acid (40) and benzaldehyde were effective precursors of 32.

The origin of the remaining two-carbon unit, the nitrogen atom and the N-methyl group, remains obscure. Serine, alanine, glycine, propionic acid, aspartic acid glucose, and formate all labeled the side chain and N-methyl group, although some differential labeling was observed. For example in the ephedrine labeled by [2-^{14}C]propionic acid, 88% of the activity was at the N-methyl group. From labeled formate however, 51% of the activity was in the N-methyl group and 31% in the C-2–C-3 unit.

No intermediates in the biosynthetic scheme are known, and only a very rudimentary biogenesis can be proposed for the formation of ephedrine (32) at this time (Scheme 4).

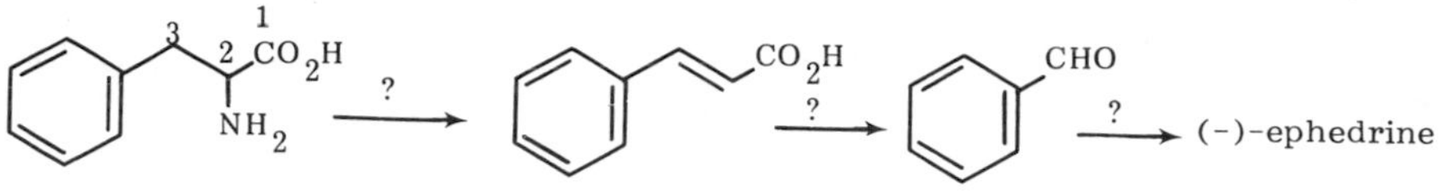

Scheme 4   Biogenesis of ephedrine (32).

8.3.3   Pharmacology

Ephedrine has been and continues to be an important therapeutic agent. The mydriatic effect (pupil dilation) was first noted by Muira in 1887, and it was an important agent for this purpose until Chen and Schmidt demonstrated its adrenalinelike (adrenergic) activity. Such physiological actions are now realized to be quite characteristic of sympathomimetic amines, along with increasing secretions and increasing arterial blood pressure. A 0.5–1.0% solution is used as a nasal decongestant, and a 1–3% solution in oil is available for topical use.

LITERATURE

Reviews

Reti, L., *Alkaloids NY* **3**, 339 (1953).

Chen, K. K., *Alkaloids NY* **5**, 229 (1955).

Biosynthesis

Yamasaki, K., T. Tamaki, S. Uzawa, U. Sankawa, and S. Shibata, *Phytochemistry* **12**, 2877 (1973).

8.4   KHAT

Khat is one of the classic plants in the folklore of Arabia and East Africa, particularly the Yemen and Ethiopia. The drug is derived from the leaves of the shrub *Catha edulis* Forsk. (Celastraceae), and its stimulating effects have probably been known for at least 3000 years. Although it is known to be used in the form of an infusion, the preferred method of ingestion is to chew the fresh leaves and alleviate the bitter taste with sweetened water. Repeated use is said to lead to intellectual deterioration, gastrointestinal problems, and impotency. There is apparently some psychological dependence but no withdrawal symptoms.

   Khat has been the subject of numerous chemical investigations, and several alkaloidal constituents have been claimed. The most important of these is generally agreed to be *d*-norpseudoephedrine (**38**), but some macrolide sesquiterpene alkaloids such as **41** have recently been obtained.

38   d-norpseudoephedrine

41   cathedulin-2

   The social effects of Khat in many parts of the Afro-Arab world cannot be underestimated. Khat replaces alcohol and like a fine wine is said to have its vintage crops. Legislated attempts to prohibit use have for the most part failed, and since it is estimated that 70–80% of the adult population in some Arab countries are regular Khat chewers, perhaps this is not too surprising.

8.5   CINNAMIC ACID AMIDE DERIVATIVES

One of the fundamental reactions in the further elaboration of phenylalanine/ tyrosine is oxidative deamination to cinnamic acid derivatives. The mechanism of this process has been discussed in Chapter 2, and there are many examples of the biosynthetic progress of these cinnamic acid derivatives. Indeed the whole fields of coumarins and flavonoids are based on elaboration of cinnamic or coumaric acids.

8.5.1   Simple Derivatives

Cinnamic acid (**42**) as the corresponding coenzyme A derivative is clearly a highly active acylating agent for amines, comparable even to acetyl CoA. For there are numerous derivatives formed by the cinnamoylation of various amines, both aliphatic and aromatic.

Fagaramide (**43**), from the root of *Fagara zanthoxyloides* (Rutaceae), is a simple member of this series, and herclavine (**44**) from *Zanthoxylum* sp. and subaphylline (**45**) from *Salsola subaphylla* are other examples. These amides are all derived from a *trans*-cinnamic acid; examples such as astrophylline (**46**) which are derived from a *cis*-cinnamic acid (**47**) were discussed in Chapter 4.

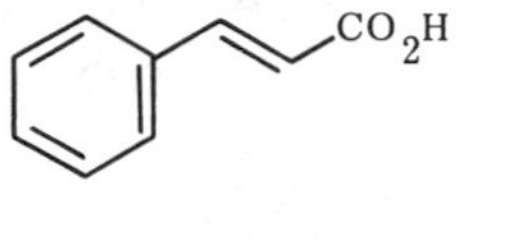

42   trans-cinnamic
        acid

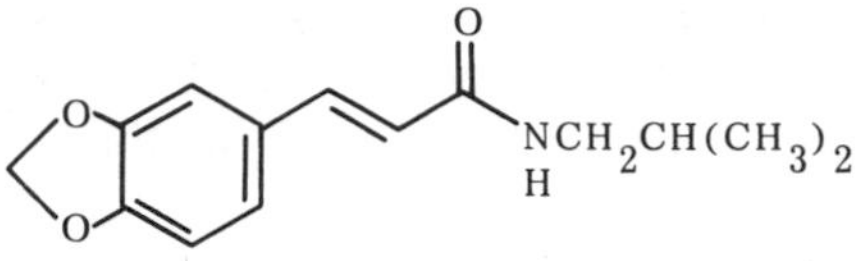

43   fagaramide

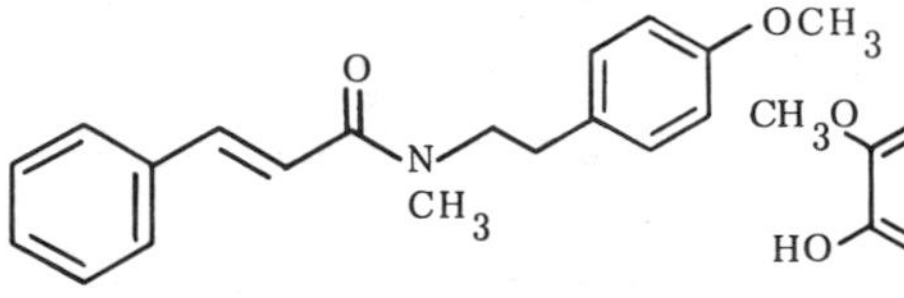

44   herclavine

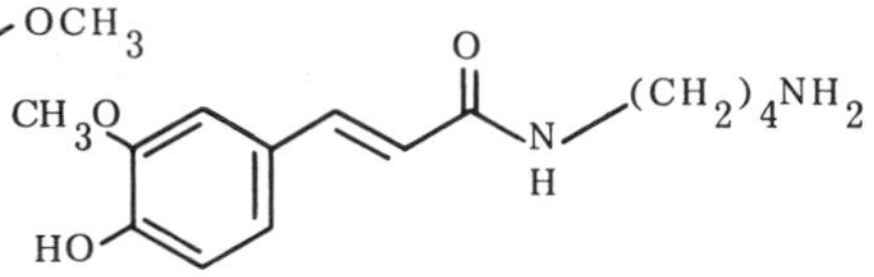

45   subaphylline

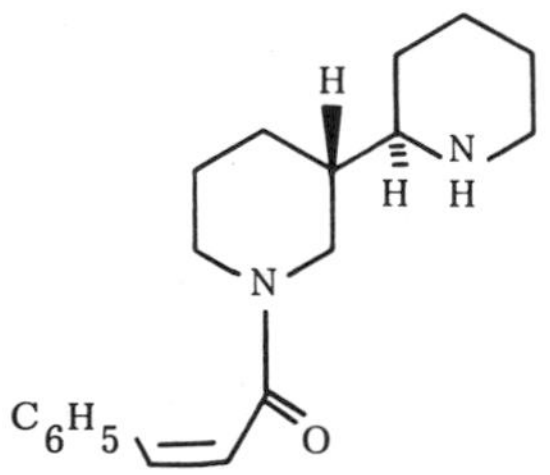

46   astrophylline

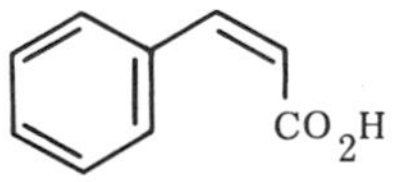

47   cis-cinnamic
        acid

Xanthocillin (**48**) is a yellow pigment of the fungus *Penicillium notatum* and contains the biogenetically interesting isonitrile group. Achenbach and Grisebach found that [2-^{14}C]tyrosine was specifically incorporated as indicated, but [1-^{14}C]tyrosine was not incorporated, indicating that the carboxyl group was lost during biosynthesis. The origin of the isonitrile function remains unknown.

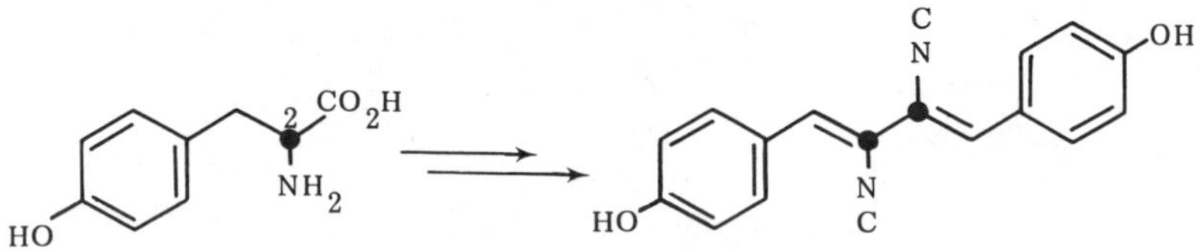

2   48   xanthocillin

8.5.2   Oxazole Alkaloids

A number of oxazole alkaloids have been obtained from various sources, and two examples are halfordinol (49) from *Halfordia schleroxyla* F. Muell. (Rutaceae) and annuloline (50) from *Lolium perenne* L. (Gramineae). They are included here because they appear to be produced by the internal cyclization of cinnamide derivatives.

A facile, one-step synthesis of halfordinol (49) involves the condensation of *p*-hydroxymandelonitrile (51) with pyridine-3-aldehyde (52) in the presence of thionyl chloride. This reaction, which proceeds in about 16% yield, is thought to proceed through the imino chloride 53.

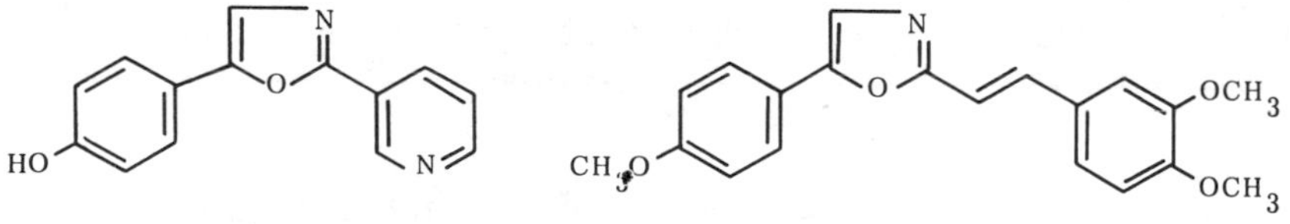

49   halfordinol   50   annuloline

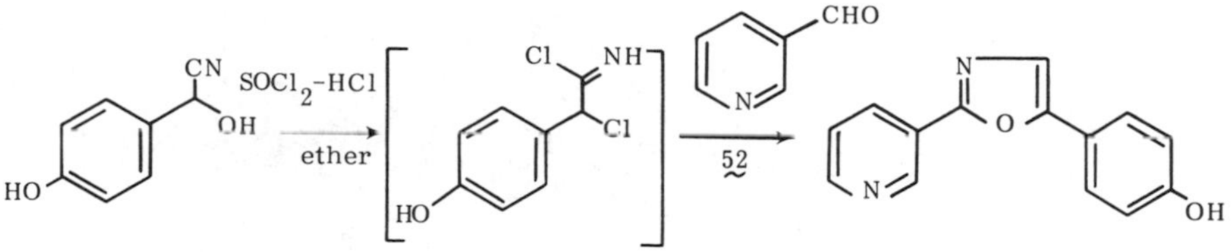

51   53   49

Annuloline (50) has been synthesized biomimetically by phosphorus oxychloride–mediated cyclization of the cinnamide derivative 54. An alternative route involves condensation of veratraldehyde (55) with oxazole 56.

[3-^{14}C]Phenylalanine (1) and [3-^{14}C]tyrosine (2) were each specifically incorporated into C-5 and C-2′ of 50 although not equally, and this is clearly related to available pool sizes, for in subsequent steps the two halves of the molecules are built up discretely. Thus [3-^{14}C]dopa (21) is incorporated only in C-2′ and [2-^{14}C]tyramine (12) only into C-5. These results suggest a quite

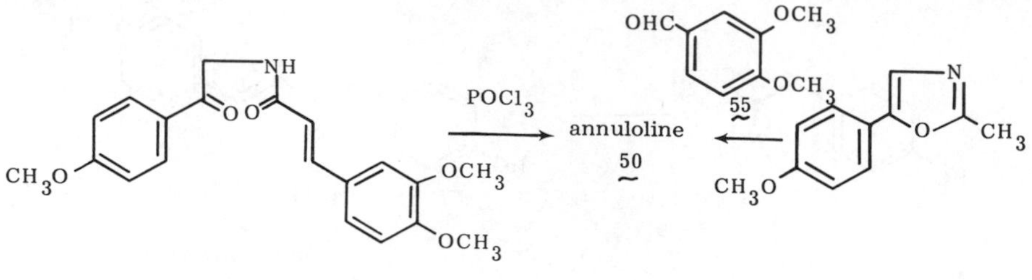

54   55   56   50

simple biosynthetic scheme (Scheme 5) and the incorporation of [2-^{14}C] *p*-methoxyphenethylamine (**57**) supports this. However, the specific incorporation of [3-^{14}C]cinnamic acid (**42**) into C-2′ is more difficult to rationalize by this scheme. [2-^{14}C]Coumaric acid (**58**) and [2-^{14}C]caffeic acid (**59**) were incorporated but no degradations were carried out. Therefore besides C-5, the origin of the remaining atoms is unknown.

8.5.3 Withasomnine

Withania somnifera Dunal (Solanaceae) is indigenous to western India and the leaves are a source of the highly cytotoxic withanolides. The roots have yielded several common tropane alkaloids, but the most interesting alkaloid is withasomnine (**60**), a rare pyrazole derivative.

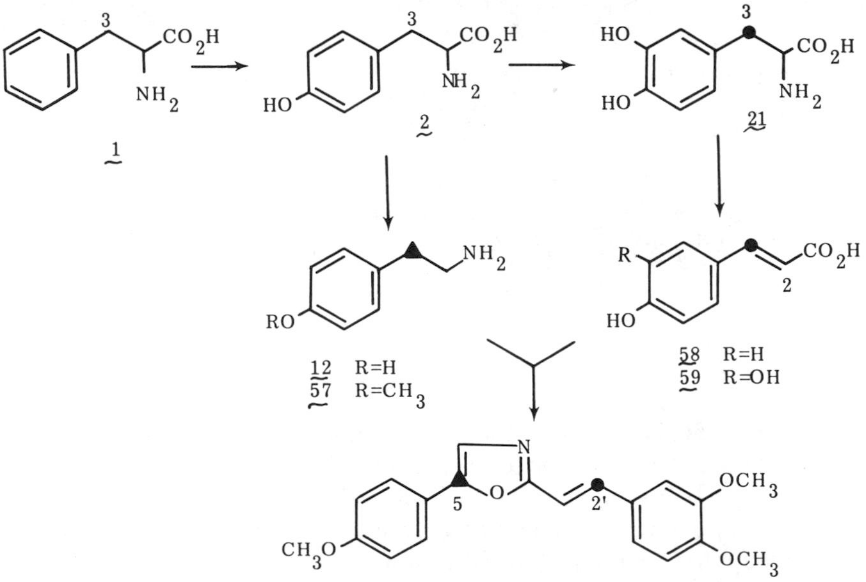

Scheme 5

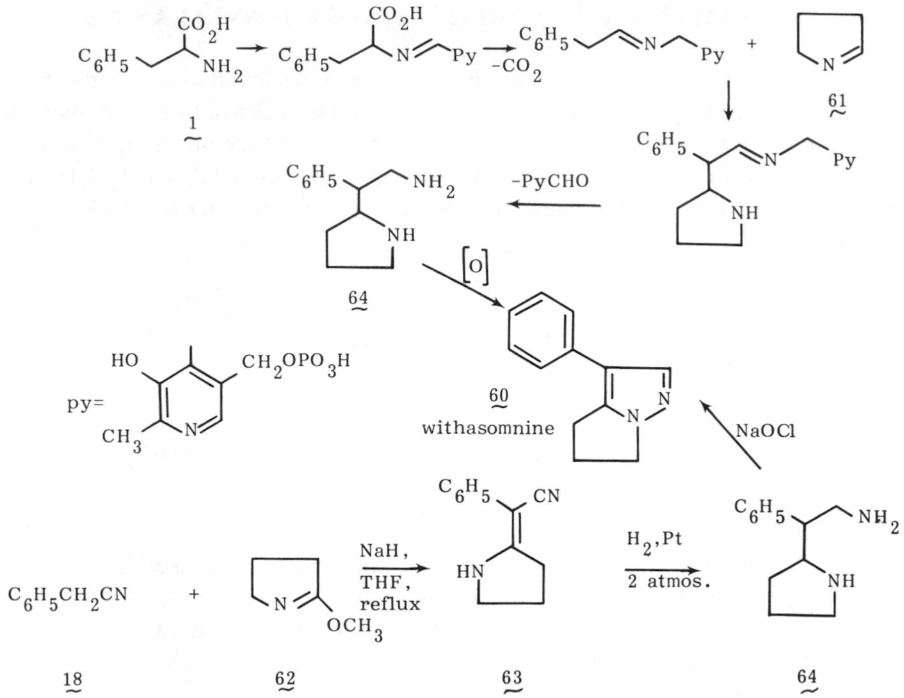

Scheme 6 Biogenesis and synthesis of withamsonine (**60**).

Onaka suggested that **60** was derived from 1,2-dehydropyrrolidine (**61**) and phenylalanine (**1**), and modeled a synthesis along these lines. Condensation of O-methyl butyrolactim (**62**) with phenylacetonitrile (**18**) in the presence of sodium hydride gave **63**, which was catalytically reduced under pressure to the amine **64**. Oxidation of the crude reaction product with sodium hypochlorite gave **60** in low yield (Scheme 6).

LITERATURE

Oxazole Alkaloids

Crow, W. D., and J. H. Hodgkin, *Tetrahedron Lett.* (2), 85 (1963).

Jeffreys, J. A. D., *J. Chem. Soc., Sec C* 1091 (1970).

Karimoto, R. S., B. Axelrod, J. Wolinsky, and E. D. Schall, *Tetrahedron Lett.* (3), 83 (1962).

Onaka, T., *Tetrahedron Lett.* 4393 (1971).

O'Donovan, D. G., and H. Horan, *J. Chem. Soc., Sec. C* 331 (1971).

Withasomnine

Schroter, H.-B., D. Neumann, A. R. Katritsky, and F. W. Swinbourne, *Tetrahedron Lett.* **22**, 2895 (1966).

Onaka, T., *Tetrahedron Lett.* 5711 (1968).

8.6   DIKETOPIPERAZINES DERIVED FROM PHENYLALANINE

It has become clear in recent years that amino acids as well as being involved in protein synthesis, and to a lesser extent alkaloid formation, are also, in the case of fungi, involved in secondary metabolite formation on a biomolecular basis. For example when two amino acids $R—CH(NH_2)CO_2H$ and $R_1—CH(NH_2)CO_2H$ condense, the product is a diketopiperazine **65**. Many

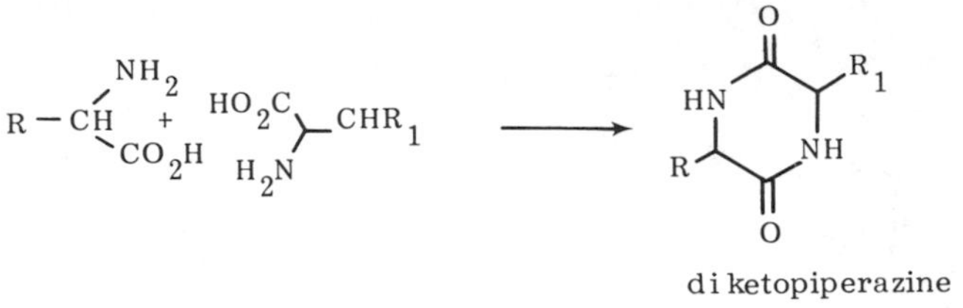

diketopiperazine

$\underset{\sim}{65}$

such compounds are now known and examples will be discussed in both this and the next chapter. Perhaps the most surprising feature of these metabolites is that many of these compounds contain D-amino acid components and that these units are derived from the corresponding L-amino acid with retention of the α-amino nitrogen.

Needless to say the involvement of D-amino acids has been a subject of substantial interest and Vining and Wright have reviewed this area. This racemization process apparently occurs with the amino acid covalently linked to the peptide synthetase in the form of a thioester **66** prior to the peptide bond being formed at the amino group. The mechanism of racemization is unclear but may involve proton transfer to give a planar carbanion intermediate **67**, since it is well established that tritium at C-2 of the amino acid is exchanged rapidly during racemization. An alternative mechanism involves the formation of the thioester enolate **68** (Scheme 7).

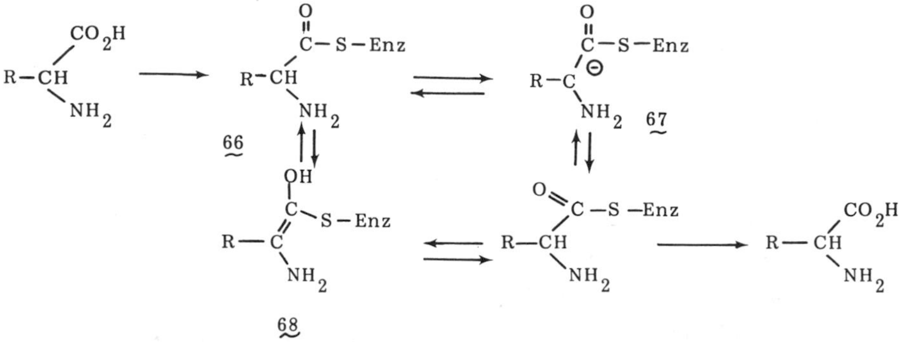

Scheme 7

The question is then raised as to the nature of the true precursors of the secondary metabolites, and this seems to depend very much on the individual system.

In some cases it is established that both forms of the amino acid may act as precursors, with each competitively inhibiting the other. Most cell-free systems that catalyze the formation of the peptides containing D-amino acid residues utilize the D-epimers as substrates. But in culture systems, the D-isomers are normally less effective as precursors and may even inhibit the biosynthesis of the product of which they are a component. Clearly this is an area of some importance, for it has significant implications for all alkaloid biosynthesis when racemic modifications of amino acids are widely used in studying precursor relationships.

Numerous studies with cell-free systems have established that protein, rather than nucleic acid, templates determine the structure of these products. In particular, ribonuclease and deoxyribonuclease do not affect their biosynthesis and neither do classic protein synthesis inhibitors such as chloramphenicol or puromycin. In fact in some instances formation of the peptide products is stimulated.

Some of the enzymes catalyzing peptide synthesis have been considerably purified, and there is accumulating evidence that the assembly process involves thioester-bound amino acids. It has been suggested that this process be referred to as a "protein thiotemplate mechanism."

With this rather extensive introduction to this area, it is time to look at some of the chemical classes involved in their structures. Some compounds are simply condensed forms of two amino acids. This may be the same amino acid as in L-phenylalanine anhydride (69) from *Penicillium nigricans*, different amino acids, as in L-prolyl-L-valyl anhydride (70) from *Aspergillus ochraceus*, or two amino acids with subsequent modifications such as gliotoxin (71) from *Trichoderma viride*. Others may involve two or more amino acids, with other carbon sources such as mevalonate also providing major structural units to the molecule; some examples here are mycelianamide (72) from *Penicillium griseofulvum*, echinulin (73) from several *Aspergillus* species, and brevianamide E (74) from *Penicillium brevicompactum*, and these latter examples are discussed in Chapter 9.

Diketopiperazines may be detected chromatographically by spraying with sodium azide in iodine solution followed by starch solution. The metabolites appear as colorless spots on a blue background. An alternative method involves spraying with neutral aqueous silver nitrate solution to give silver sulfide, which appears as a dark spot.

8.6.1 Gliotoxin and Aranotin

Gliotoxin (71) was first obtained in 1936 from the fungus *Trichoderma viride* and subsequently from *Aspergillus fumigatus* and *Penicillium terlikowskii*.

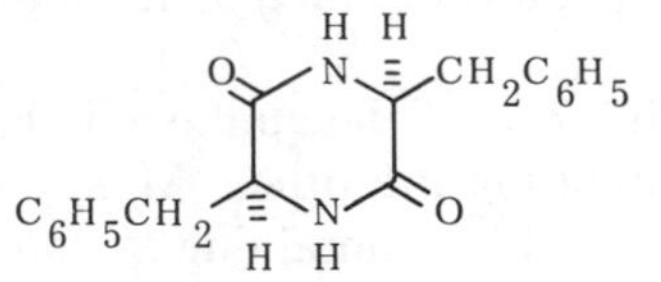

69 L-phenylalanine

anahydride

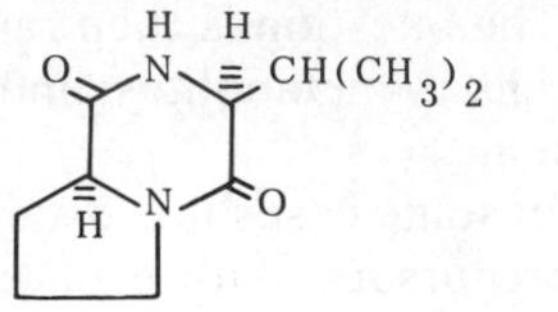

70 L-prolyl-L-valyl

anahydride

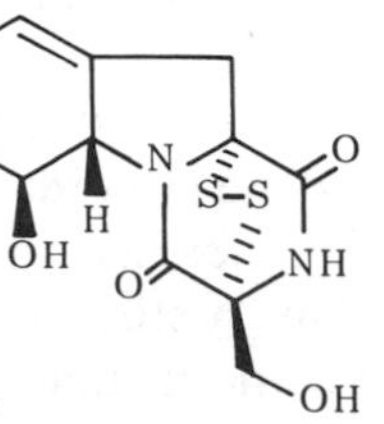

71

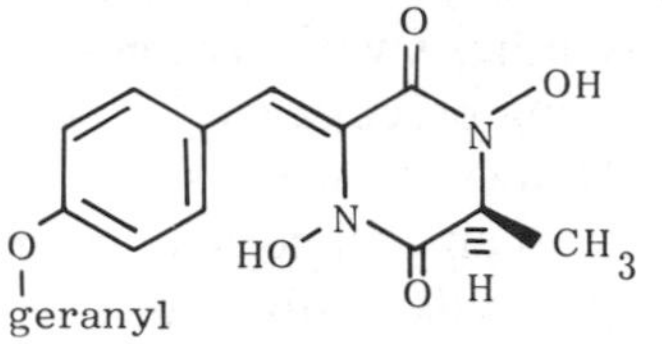

72 mycelianamide

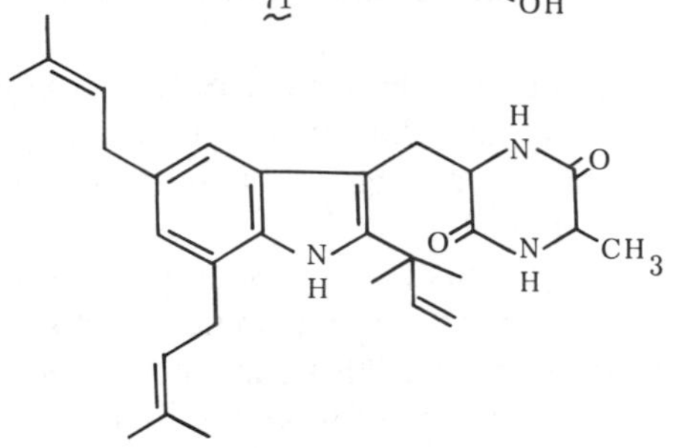

73 echinulin

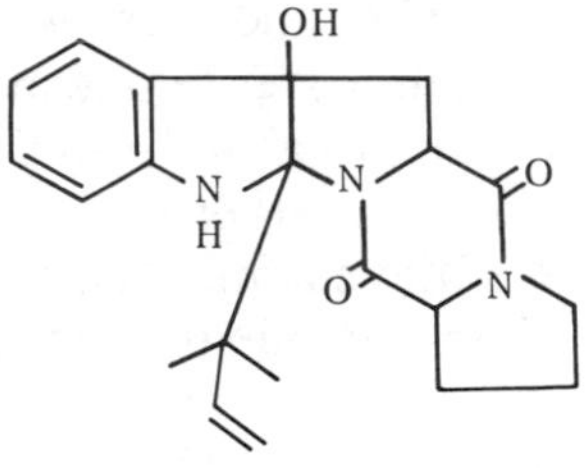

74 brevianamide E

It is a member of a growing list of fungal products containing a diketo-piperazine ring with a disulfide bridge.

Some of the other compounds in this general class include aranotin (8), sporidesmin A (75), and the dimer verticillin A (76). The latter two compounds are discussed in Chapter 9.

Chemistry

The structure of gliotoxin was proposed after extensive chemical degradation, and was subsequently verified by X-ray crystallographic examination. The system exhibits some very interesting chemical reactivity, and a few examples are shown in Scheme 8. Hydrogen iodide–red phosphorus or aluminum amalgam removes the disulfife bridge, in one case reductively, in the other with overall oxidation to give a 1,2-diacylindole. Quinones such as *o*-chloranil effect dehydrogenation of the A ring and the disulfide bridge may now be removed by photolysis.

Facile cleavage of the disulfide bridge is related to the torsional strain of the system, for X-ray analysis established that the C—S—S—C system

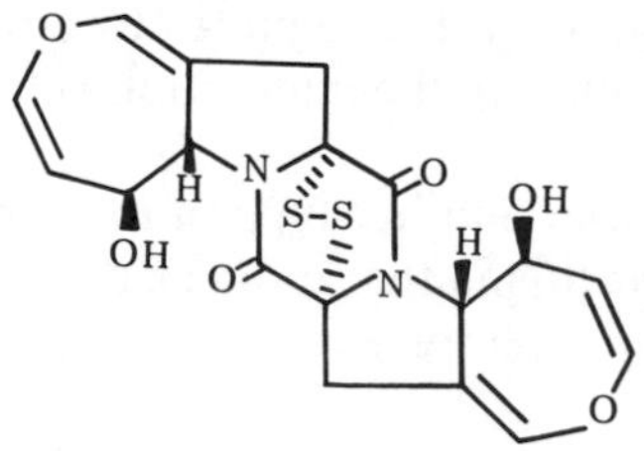

8 aranotin

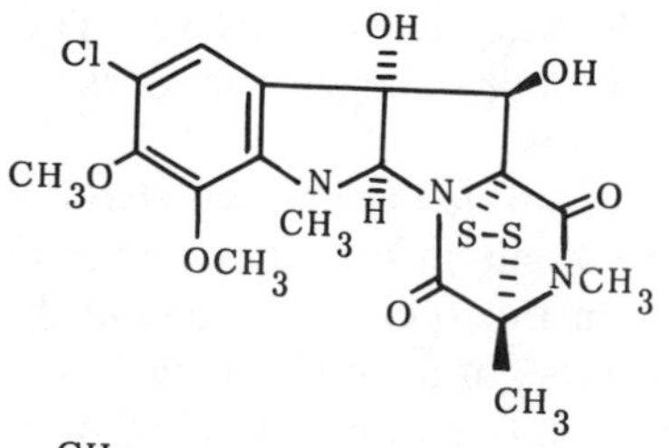

75 sporidesmin A

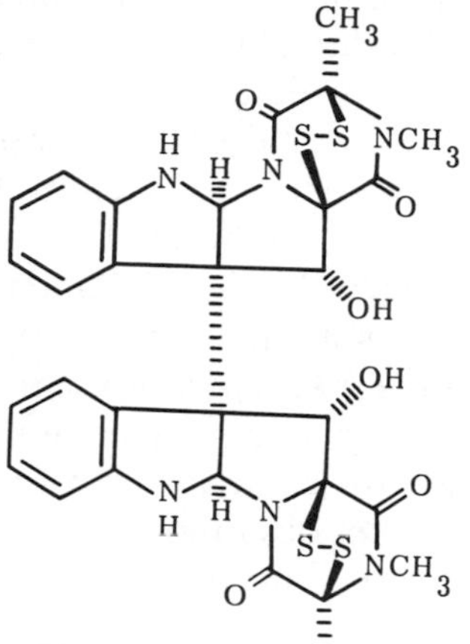

76 verticillin A

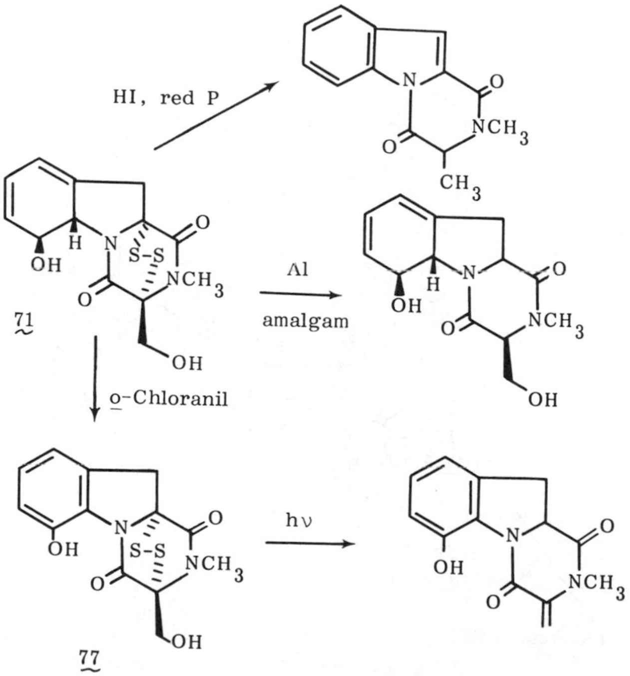

Scheme 8 Chemistry of gliotoxin (71).

has a dihedral angle of only 12°. The sulfur atoms are also quite close to the carbonyl groups and one wonders as to the nature of the interaction between these groups.

The disulfide bridge may also be involved in another type of interaction with the diene chromophore. One reason for this is that circular dichroism studies on the skew angle of the diene chromophore were of the opposite sign to those of a simple diene.

The disulfide function of the bridge epidithiodioxopiperzines can be protected by condensation of a 1,4-dithiol with anisaldehyde to give a thioacetal. The disulfide is regenerated by oxidation with *m*-chloroperbenzoic acid followed by treatment with a Lewis acid such as boron trifluoride etherate or with perchloric acid. This procedure was used by Kishi and coworkers as the final steps in an elegant total synthesis of gliotoxin (**71**) (Scheme 9). A key step is the solvent-dependent Michael addition to give the alcohol **77**.

Aranotin (**8**) is the principal antiviral component of the mold *Arachniotus auseus* (Eidam) Schroeter and the structure was confirmed by X-ray analysis of the acetate derivative, which is also a natural product. The absolute configuration was confirmed by circular dichroism studies on the parent compound and the Raney nickel desulfurization product **78**. Proton NMR data for acetyl aranotin are shown in **79**.

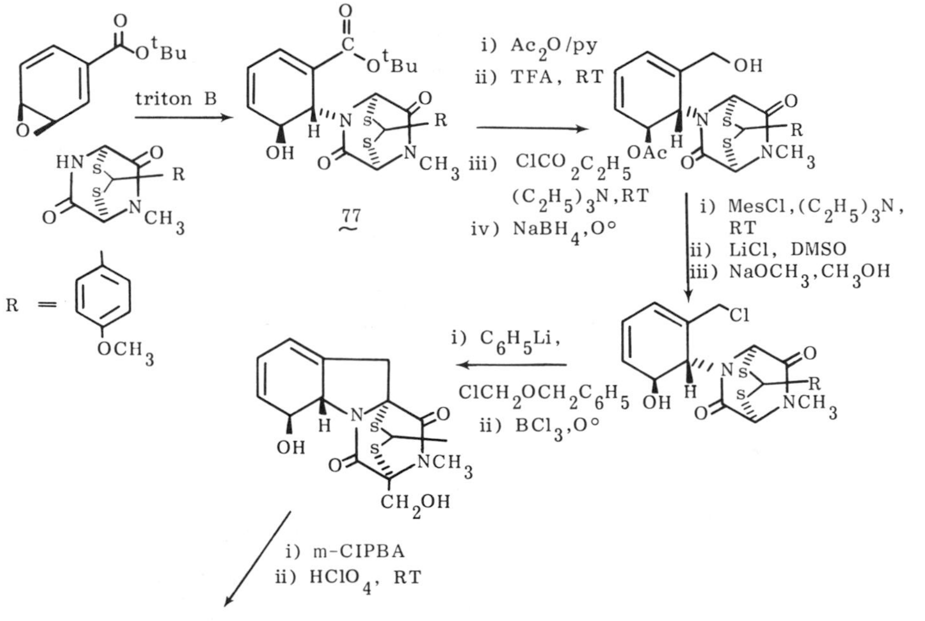

Scheme 9　Kishi synthesis of gliotoxin (**71**).

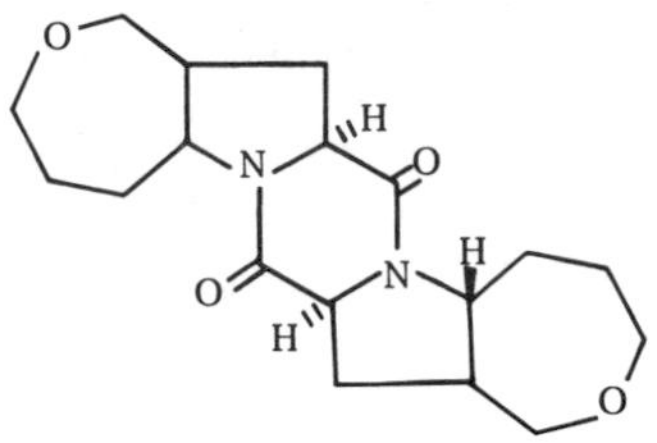

6.63

4.12, 2.70

6.32

4.63

4.70 Ac O

OAc

78

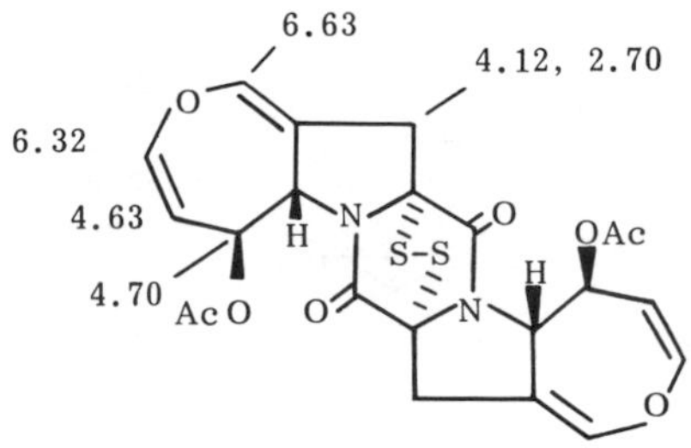

79

Biosynthesis

Early work demonstrated that gliotoxin (71) was derived from phenylalanine (1), serine (80), methionine, and a sulfur donor as shown in 81. At one time it was thought that *m*-tyrosine (82) was a key intermediate, but subsequent work by the groups of Bu'Lock and Kirby has categorically eliminated this possibility. Thus [ar-^{2}H$_5$]phenylalanine (1) was incorporated without loss of aromatic hydrogen atoms, suggesting that the dihydroindole moiety of gliotoxin (71) might be produced by intramolecular attack of the amino acid nitrogen of L-phenylalanine on an epoxide in the aromatic ring. Additional evidence came from the administration of [1-^{14}C, 3',5'-^{3}H$_2$]phenylalanine, where no tritium was lost on conversion to gliotoxin as shown in 83.

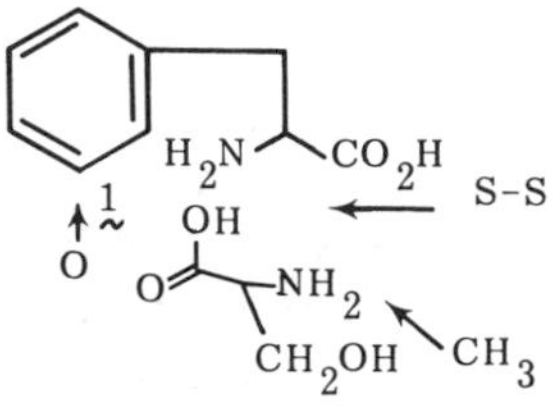

81

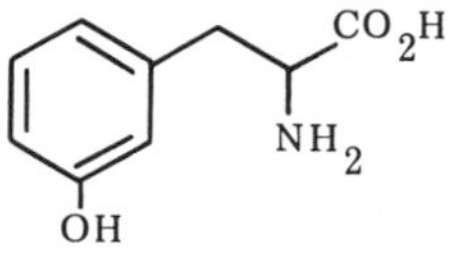

82

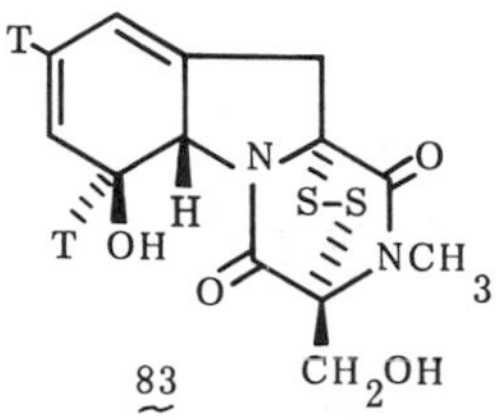

83

labeling from

$\left[1\text{-}^{14}C, 3', 5'\text{-}^3H_2\right]$ –phenylalanine

The first step in the biosynthetic scheme is apparently condensation of L-phenylalanine with L-serine to afford *cyclo*-L-phenylalanyl-L-seryl (84), which was a precursor of 71 in *T. viride*. There is no information as to the timing of introduction of the disulfide bridge, but Scheme 10 has been suggested.

At first sight the biosynthesis of aranotin (8) appears to present a difficult problem, namely the formation of the seven-membered oxepin ring system.

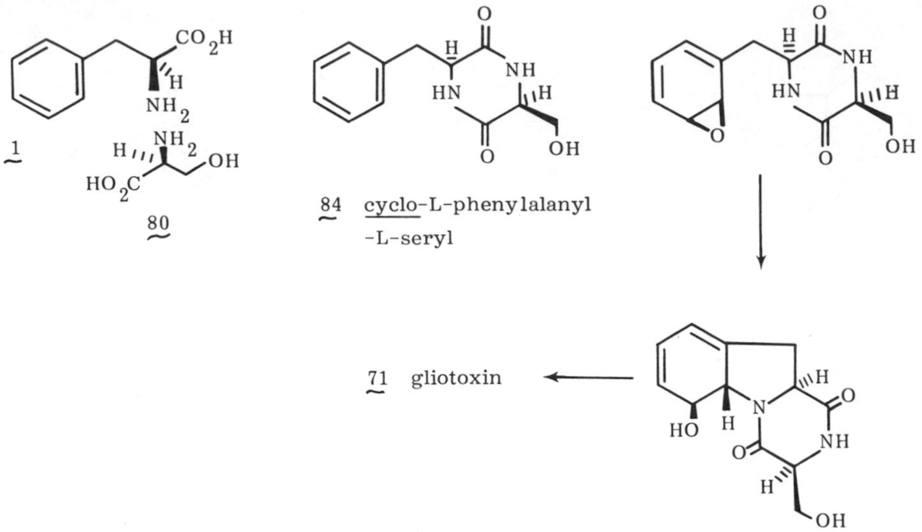

Scheme 10   Biogenesis of gliotoxin (**71**).

The situation is resolved by consideration of the alternative reaction pathway of an arene oxide **85** other than phenol formation, namely valence bond tautomerization to an oxepin **86**. Thus the diepoxide intermediate **87** from L-phenylalanine anhydride (**69**) could undergo valence bond tautomerization to the dioxepin **88** followed by further epoxide formation and nucleophilic attack by the amino acid nitrogen as shown in Scheme 11.

There are some other possibilities which should be considered however, and one of these involves the intermediacy of a 3′,4′-arene oxide such as **89** in gliotoxin biosynthesis. Bis-epoxidation of **69** to the arene dioxide **90** and attack by the amide nitrogens leads to the aranotin skeleton directly (Scheme 12). No evidence is available which would distinguish these alternative pathways.

Pharmacology

Gliotoxin is bacteriostatic against a wide range of gram-positive organisms *in vitro* but has no effect *in vivo*. It also inhibits viral RNA synthesis at about 20 μg/ml, and is cytotoxic to HeLa cells at 0.5 μg/ml.

8.6.2   Mycelianamide

Penicillium griseofulvum and related fungi produce mycelianamide (**72**) which can be envisaged as being derived from tyrosine (**2**), alanine (**91**), and geraniol (**92**). Birch and co-workers established that the geranyl ether group was derived from mevalonate, but more interestingly MacDonald and Slater established that both D-(**93**) and L-[1-^{14}C]tyrosine (**2**) specifically labeled the carbonyl groups. However, neither *cyclo*-L-alanyl-L-[^{14}C]tyrosyl (**94**) nor *cyclo*-L-alanyl-D-[^{14}C]tyrosyl (**95**) were incorporated. The significance of this is not clear, but it may mean that dehydrogenation occurs prior to condensation with alanine. It was subsequently established that the *pro*-S-proton of tyrosine is lost in the dehydrogenation step, but whether this is a *cis*-elimination from an L-amino acid or *cyclo*-peptide as in neoechinulin biosynthesis, or a *trans* elimination from a D-amino acid or *cyclo*-peptide is not known.

8.6.3 Verruculotoxin

The fungus *Penicillium verruculosum* Peyronel obtained from green peanuts produces a metabolite, verruculotoxin, which is toxic (LD$_{50}$ 20 mg/kg) to

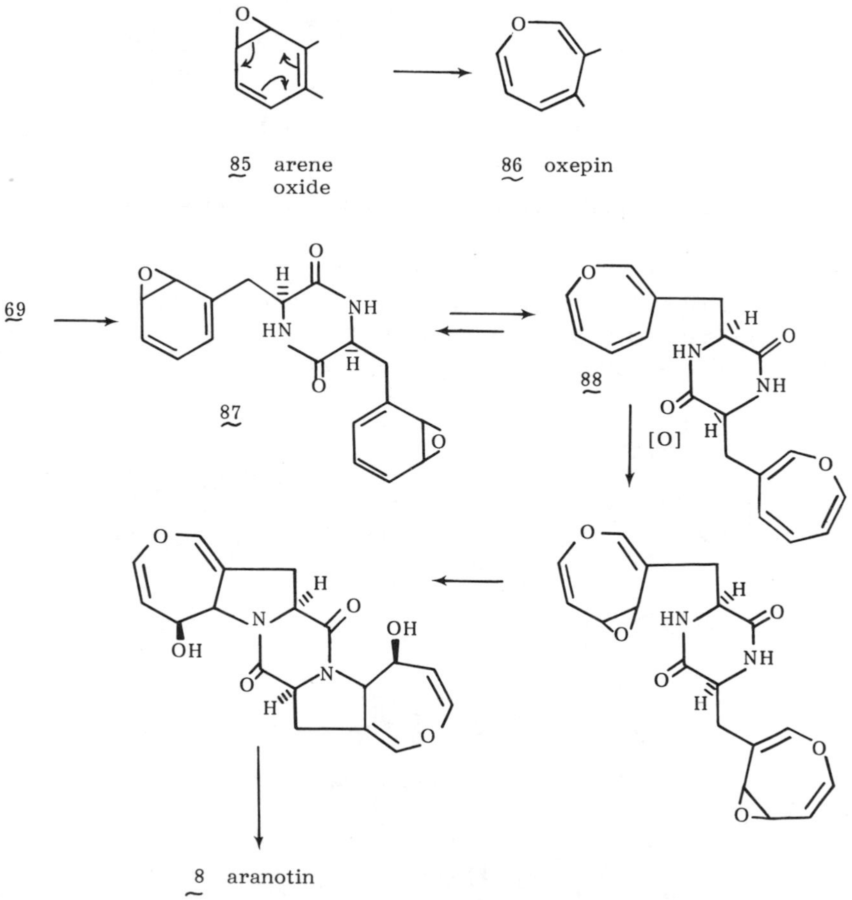

Scheme 11 Biogenesis of aranotin (**8**).

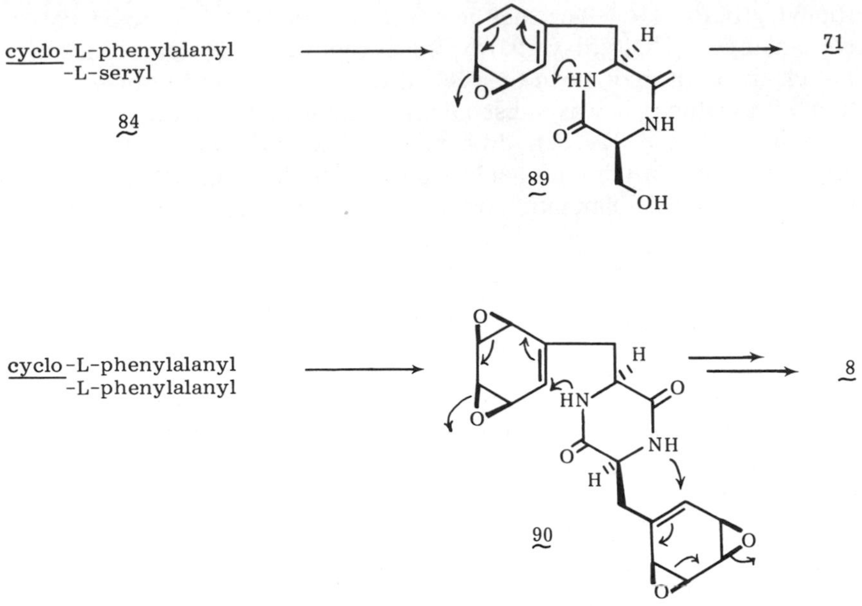

cyclo-L-phenylalanyl
-L-seryl

84

89

71

cyclo-L-phenylalanyl
-L-phenylalanyl

90

8

Scheme 12 Alternative biogenesis of gliotoxin (**71**) and aranotin (**8**).

1-day-old chicks. Toxicity was characterized by ataxia, prostration, and loss of muscle control.

 The structure of verruculotoxin (**96**), deduced by X-ray analysis, can formally be derived from phenylalanine (**1**) and pipecolic acid (**97**).

 A synthesis of verruculotoxin (**96**) along loosely defined biogenetic lines involved the condensation of L-phenylalinol (**98**) with methyl picolinate (**99**)

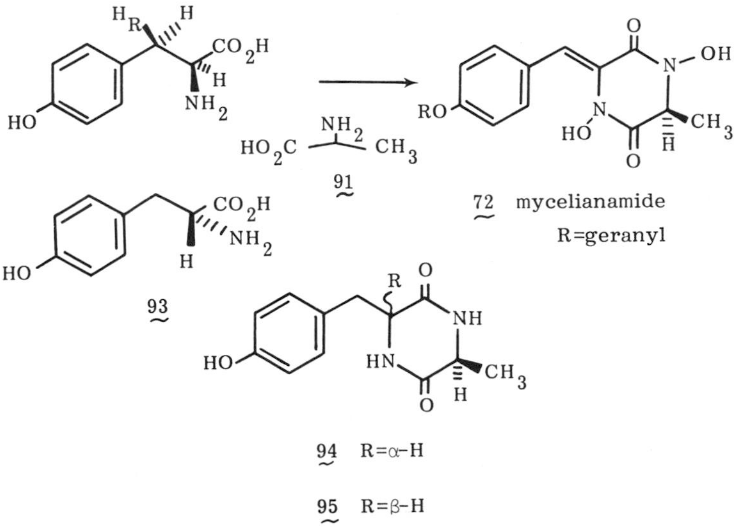

91

72 mycelianamide
  R=geranyl

93

94 R=α-H

95 R=β-H

at 150°C for 1 hr to afford the amide **100** in almost quantitative yield. Treatment with thionyl chloride gave the chloride **101**, which could be converted to a quaternary salt on refluxing in *N,N*-dimethylformamide. Catalytic hydrogenation under pressure afforded a mixture of epimers, in which verruculotoxin (**96**) predominated.

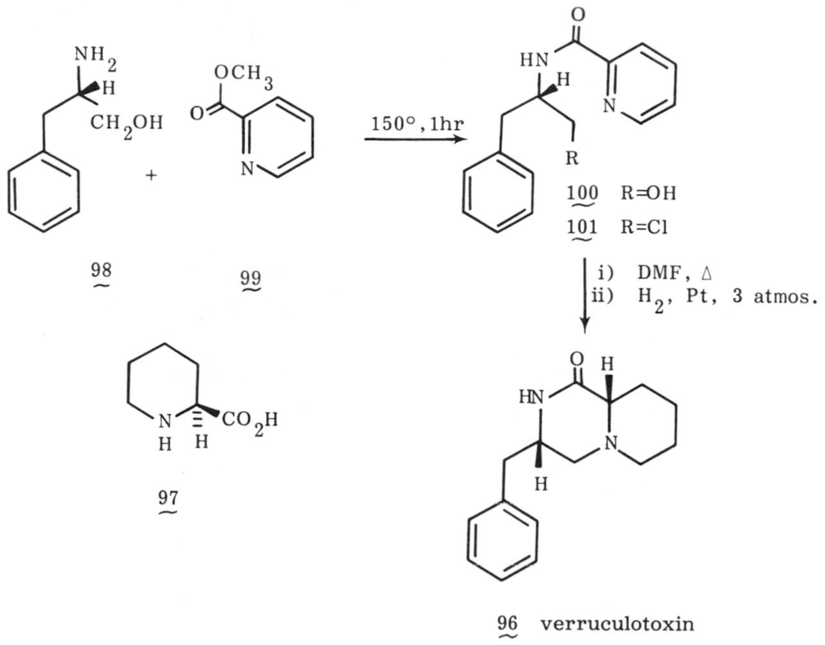

96   verruculotoxin

8.6.4   Anthramycin and Related Compounds

Several *Streptomyces* species produce anthramycin (**102**) and related antibiotics. Tryptophan (**103**) and tyrosine (**2**) were precursors of anthramycin (**102**) in *S. refuineus*, but only the benzene ring carbons of tryptophan were incorporated into the 4-methyl-3-hydroxyanthranilic acid (**104**) component, the aromatic methyl group being derived from methionine.

L-[U-^{14}C]Tyrosine, L-[1-^{14}C]tyrosine (**2**), and L-[1-^{14}C]3',4'-dihydroxyphenylalanine (**21**) were all efficiently incorporated into anthramycin (**102**), but experiments with L-[3- and 5-^{3}H]tyrosine indicated that seven of the nine carbon atoms of tyrosine are used in anthramycin biosynthesis. In addition half of the tritium label was lost from these tyrosines, suggesting that two of the aromatic ring carbons are lost in the biosynthesis.

Biogenetically anthramycin can be broken down as shown, indicating a derivation from 4-methyl-3-hydroxyanthranilic acid (**104**) and the dihydropyrrole **105**. It is the formation of the latter unit which is of interest in these compounds. Before discussing this further, mention should be made of some other closely related compounds.

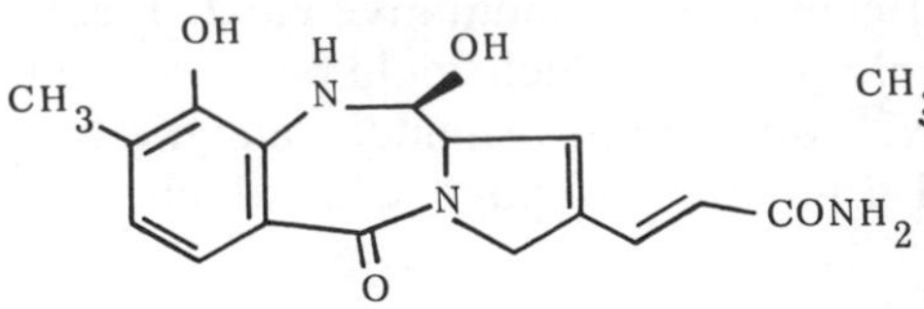

102   anthramycin

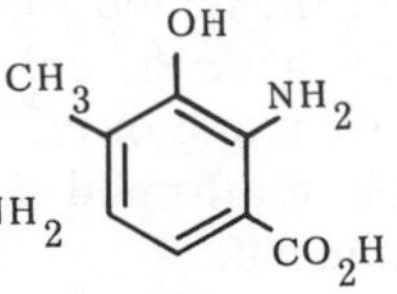

104

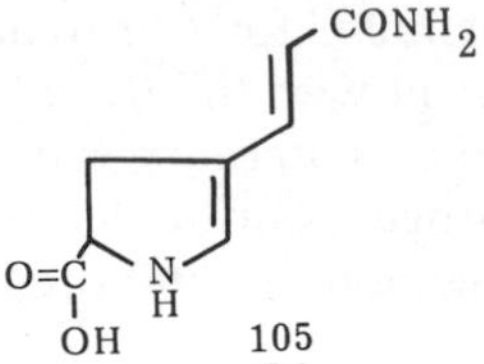

105

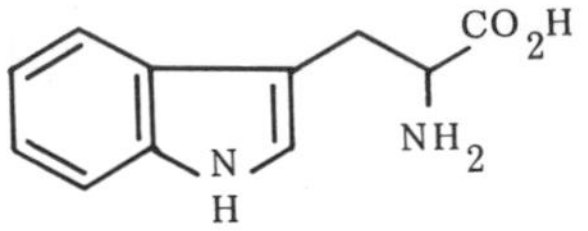

103   tryptophan

Streptomyces achromogenes var. *tomaymyceticus* produces an antibiotic 11-demethyltomaymycin (**106**), and *Streptosporangium sibiricum* has afforded the antitumor antibiotic sibiromycin (**107**). Tryptophan (**103**), tyrosine (**2**), and dopa (**21**) were precursors of **106**. The anthranilate portion was labeled only by tryptophan and methionine, whereas the pyrrolidine unit originated from tyrosine but not methionine. The ethylidene unit of **106** is therefore derived from two of the carbon atoms of tyrosine.

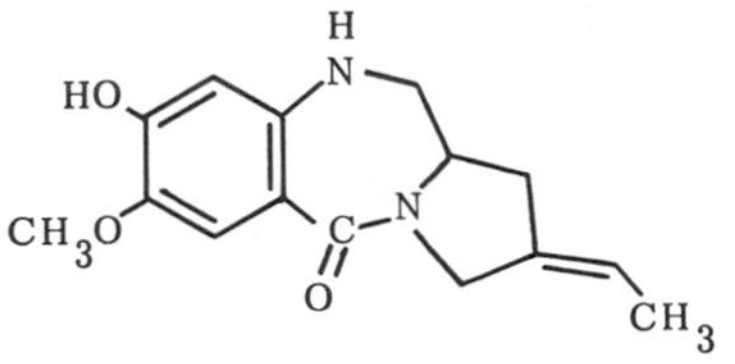

106   11–demethyltomaymycin   107   sibiromycin

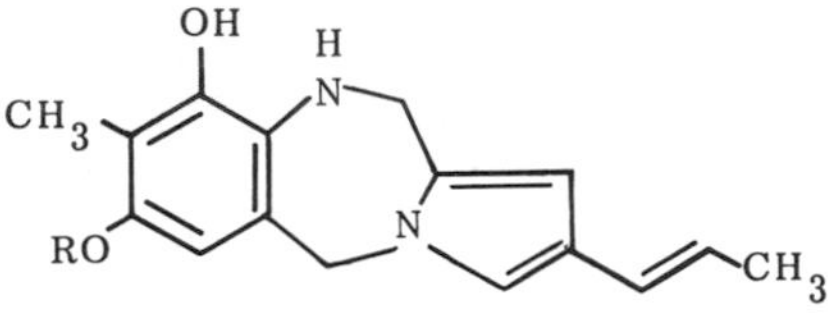

These data suggest the involvement of 3-ethylidene proline (**108**) as a key intermediate derived from tyrosine. Hurley and co-workers have studied the formation of such an intermediate using L-[U-^{14}C]-, [1′-^{14}C]-, [2′,3′-^{3}H$_2$]-, and [3-^{3}H]- or [5-^{3}H]tyrosines. When either of the latter compounds was used, half of the tritium activity was lost and Scheme 13 was proposed for the formation of **102** and **106**.

Although the metabolism of tryptophan to the anthranilate precursor has been studied, the results are inconclusive, although it was apparently established that 5-hydroxytryptophan was not a precursor.

However, in the formation sibiromycin (**107**), [5-^{3}H]tryptophan (**103**) was incorporated without loss of tritium (91% retention). This suggests a specific NIH shift to C-4 rather than migration to C-4 *and* C-6. Alternatively, C-6 methylation could occur prior to 5-hydroxylation and thus migration of tritium to C-6 is prevented and occurs only to C-4.

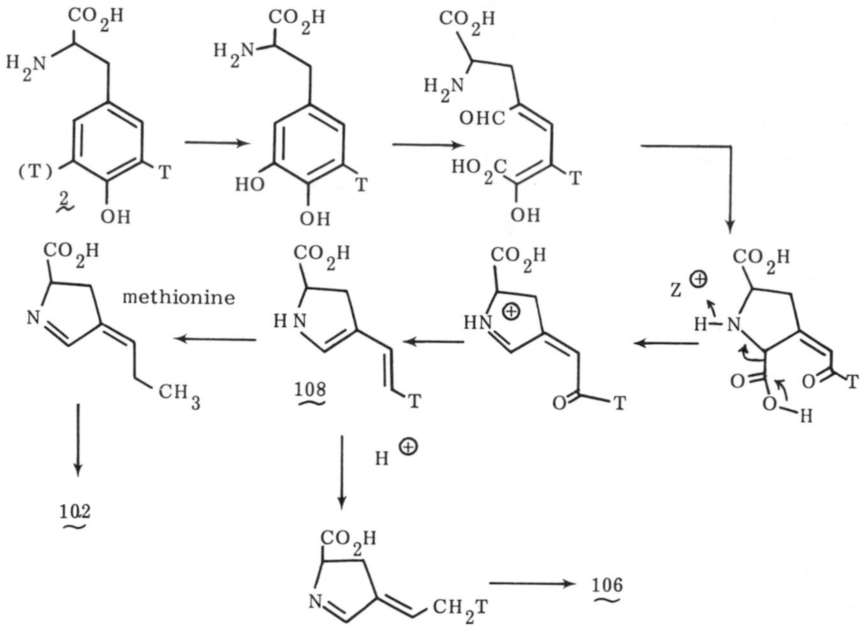

Scheme 13 Formation of the modified proline moiety of anthramycin (**102**) and 11-demethyltomaymycin (**106**).

Sibiromycin (**107**) is closely related structurally to anthramycin (**102**), each having a three-carbon side chain on the proline moiety. Nothing is known of the origin of the additional carbon in **107**, but this has been studied in the case of **102**. When [methyl-^{14}C, methyl-^{3}H]methionine was used as a precursor all the tritium label was located at C-14 in the methyl-3-hydroxyanthranilate moiety. The ^{14}C activity however was equally distributed between C-14 and C-15, the amide carbonyl carbon. These results were substantiated when [methyl-^{13}C]methionine specifically enriched C-14 and C-15. L-[1'-^{13}C]Tyrosine specifically enriched C-11 in anthramycin (**102**). The mechanism of the introduction of this additional carbon atom from methionine is unknown.

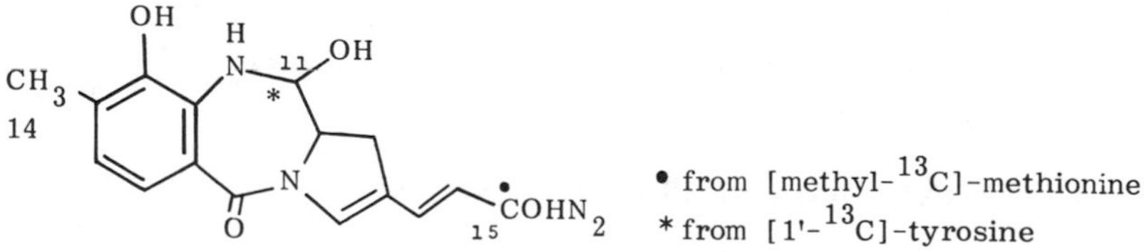

• from [methyl-^{13}C]-methionine

* from [1'-^{13}C]-tyrosine

109

LITERATURE

Reviews

Taylor, W. B., *Fungal Metabolites,* Academic, New York, 1971, p. 320.

Sammes, P. G., *Fortschr. Org. Chem. Naturs.* **32,** 51 (1975).

Daly, J. W., D. M. Jerina, and B. Witkop, *Experientia* **28,** 1129 (1972).

Vining, L. C., and J. L. C. Wright, *Biosynthesis* **5,** 240 (1977).

Gliotoxin

Bell, M. R., J. R. Johnson, B. S. Wildi, and R. B. Woodward, *J. Amer. Chem. Soc.* **80,** 1001 (1958).

Synthesis

Kishi, Y., T. Fukuyama, and S. Nakatsuka, *J. Amer. Chem. Soc.* **95,** 6492 (1973).

Fukuyama, T., and Y. Kishi, *J. Amer. Chem. Soc.* **98,** 6724 (1976).

Biosynthesis

Bu'Lock, J. D., and A. P. Ryles, *Chem. Commun.* 1404 (1970).

Johns, H., and G. W. Kirby, *Chem. Commun.* 163 (1971).

Bu'Lock, J. D., and C. Leigh, *Chem. Commun.* 628 (1975).

Mycelianamide

Birch, A. J., R. A. Massey-Westropp, and R. W. Rickards, *J. Chem. Soc.* 3717 (1956).

Birch, A. J., M. Kocor, N. Sheppard, and J. Winter, *J. Chem. Soc.* 1502 (1962).

MacDonald, J. C., and G. P. Slater, *Can. J. Biochem.* **53,** 475 (1975).

Kirby, G. W., and S. Narayanaswani, *J. Chem. Soc. Perkin Trans. I* 1564 (1976).

Verruculotoxin

Cole, R. J., J. W. Kirksey, and E. Morgan-Jones, *Toxicol. Appl. Pharmacol.* **31,** 465 (1975).

MacMillan, J. G., J. P. Springer, J. Clardy, R. J. Cole, and J. W. Kirksey, *J. Amer. Chem. Soc.* **98,** 246 (1976).

Anthramycins

Hurley, L. H., C. Gairola, and M. J. Zmijewski, *Chem. Commun.* 120 (1975).

Hurley, L. H., C. Gairola, and N. V. Das, *Biochemistry* **15,** 3760 (1976).

Hurley, L. H., M. J. Zmijewski, and C.-J. Chang, *J. Amer. Chem. Soc.* **97,** 4372 (1975).

Hurley, L. H., N. Das, C. Gairola, and M. Zmijewski, *Tetrahedron Lett.* 1419 (1976).

8.7   SECURININE AND RELATED COMPOUNDS

The Euphorbiaceae produces an array of alkaloids which defy precise taxonomic classification. The most important of these groups have been obtained from the genus *Securinega*.

8.7.1 Chemistry and Spectral Properties

Securinine (**9**), the major alkaloid of the group, was first isolated from *Securinega suffruticosa* (Pall.) Rehd. by a Russian group in 1956, and the structure elucidated by Japanese workers in 1962. Several alkaloids have been isolated having this skeleton, including a lower homologue structure type, norsecurinine (**110**), in which ring A has one less carbon atom. Unlike the other securinine-type compounds, it is extremely unstable.

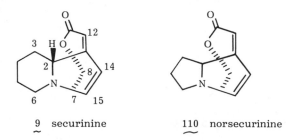

9 securinine   110 norsecurinine

Securinine shows λ_{max} 256 nm in EtOH, and carbonyl absorption in the IR at 1840 and 1760 cm^{-1}, indicating an $\alpha,\beta,\gamma,\delta$-unsaturated lactone. This grouping was substantiated by the proton NMR spectrum, in which a singlet at 5.54 ppm was assigned to H-12 and quartets at 6.67 and 6.42 ppm were attributed to H-15 and H-14 respectively as shown in **111**.

The absolute configuration at C-2 was determined by degradation *via* **112** to (+)-*N*-benzoylpipecolic acid (**113**) having the R-configuration.

The mass fragmentation pathway of securinine (**9**) is shown in Scheme 14. Pathway a with a stable allylic radical leads to effective splitting of the

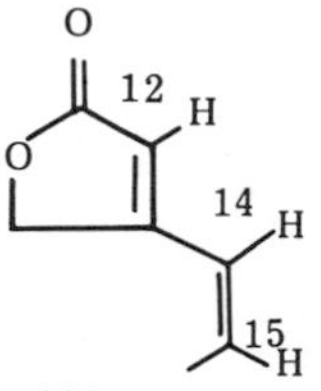

111

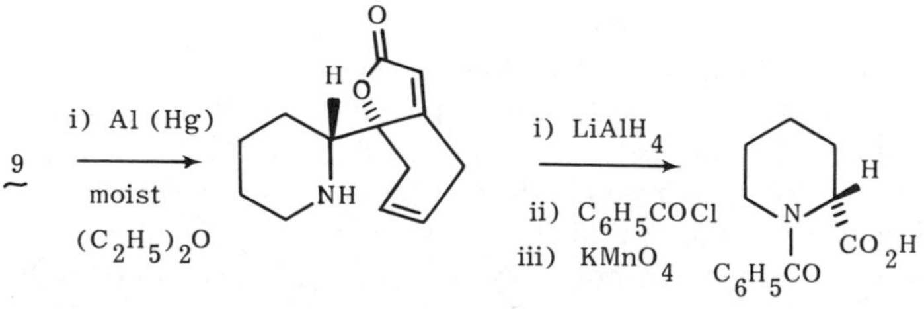

9

i) Al (Hg)

moist

$(C_2H_5)_2O$

112

i) LiAlH$_4$

ii) C$_6$H$_5$COCl

iii) KMnO$_4$

C_6H_5CO CO$_2$H

113

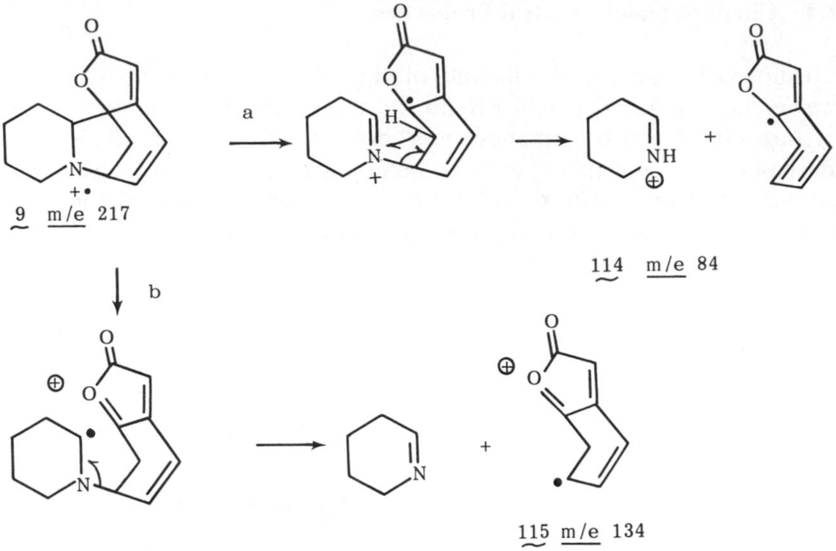

Scheme 14   Mass spectral fragmentation of securinine (**9**).

molecule into two ions, one of which, m/e 84 (**114**), is the base peak. Pathway b proceeds without hydrogen radical transfer to the ion at m/e 134 (**115**).

NMR double resonance studies went a long way to establishing certain aspects of the stereochemistry of securinine and the assignments are given in Table 3 for securinine (**9**) and allosecurinine (**116**). The latter is diaster-

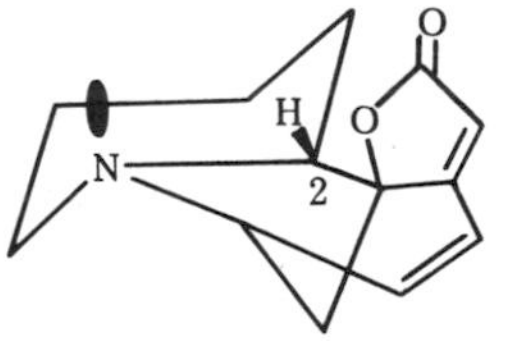

9   securinine 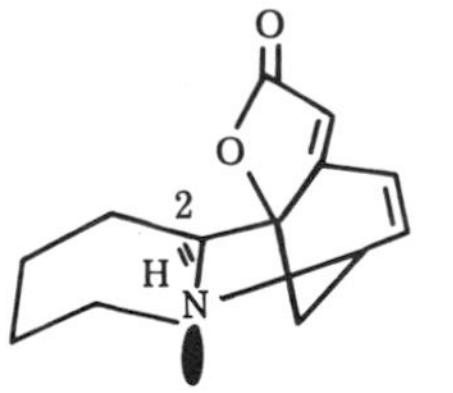   116   allosecurinine

eomeric at C-2, and the substantial chemical shift difference between H-2 in each of the isomers should be noted.

The UV spectra of securinine-type compounds has been a complex problem since the demonstration that the λ_{max} are highly dependent on the solvent. For example, these alkaloids show a quite strong n → π* transition absorption at >300 nm, but this band shifts by 38 nm for allosecurinine when the solvent is changed from carbon tetrachloride (342 nm) to ethanol (304 nm). It disappears when acid is added. The shift is somewhat smaller (7 nm) for virosecurinine (332 → 325 nm).

Table 3 Proton NMR Spectral Data of Securinine and Allosecurinine

	Securinine (9)			Allosecurinine (116)	
Assignment	δ(ppm)	Coupling Constant (Hz)	Assignment	δ(ppm)	Coupling Constant (Hz)
H-3, H-4, H-5	1.0–1.8	—	H-3, H-4, H-5	1.0–1.84	
H-8β	1.77	9.5	H-8β	1.90	9.2
H-2	~2.05	9.5, 4	H-8α	2.64	9.2, 4.5
H-6	~2.4	11, 6.5	H-6, H-6'	2.88	
H-8α	~2.5	9.5, 4.5	H-2	3.67	4.8, 11
H-6	2.99	—			
H-7	3.84		H-7	3.90	5.5, 4.5, 1.1+
H-12	5.54		H-12	5.74	
H-14, H-15	6.30–6.75	9.35+	H-14, H-15	6.54–7.0	9.3+

Single crystal X-ray analysis of allosecurinine (**116**) indicated a *cis*-A/B ring conformation and the presence of a 1,3-transannular interaction between the nitrogen lone pari and the conjugated lactone.

Perhaps the most amazing aspect of these alkaloids is that while securinine (**9**) and allosecurinine (**116**) cooccur in *S. suffruticosa*, their optical antipodes, virosecurinine (**117**) and viroallosecurinine (**118**), cooccur in *S. virosa* Pax. and Hoffm.!

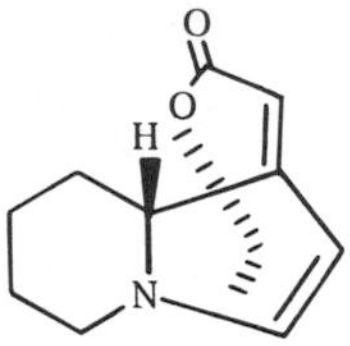

  9   securinine 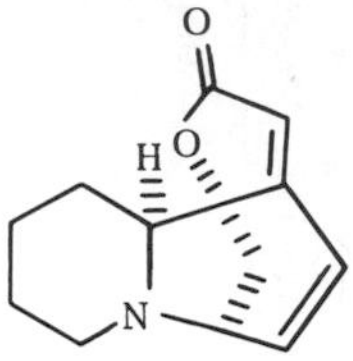   116   allosecurinine

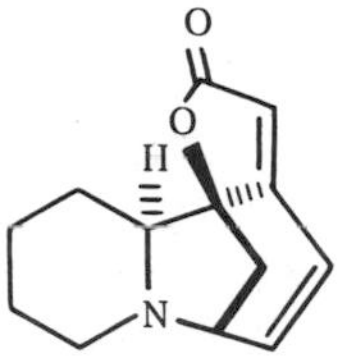

117   virosecurinine 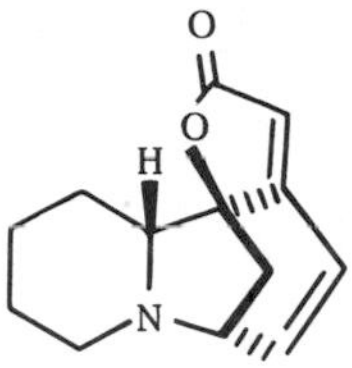   118   viroallosecurinine

8.7.2   Synthesis

The synthesis of securinine poses only one stereochemical problem, the relative relationship between C-2 and C-9. The C-7 stereochemistry follows automatically. In the synthesis developed by Horii and co-workers (Scheme 15), the relative stereochemistry is introduced early in the sequence, the conjugated lactone is constructed, and finally the N to C-7 linkage is made. The overall yield is very low.

8.7.3   Biosynthesis

The novel structure of securinine clearly represents a challenging biogenetic problem, a problem which now has been considerably resolved by the work of three groups.

Parry was the first to demonstrate that 8 of the 13 carbons were derived from tyrosine when [2-^{14}C]tyrosine (**2**) was specifically incorporated into C-11 of securinine (**9**).

More extensive work by Sankawa and co-workers confirmed these initial findings and extended the feeding experiments to show that the molecule was essentially derived from tyrosine (**102**) and lysine (**119**). Acetate, phenyl-

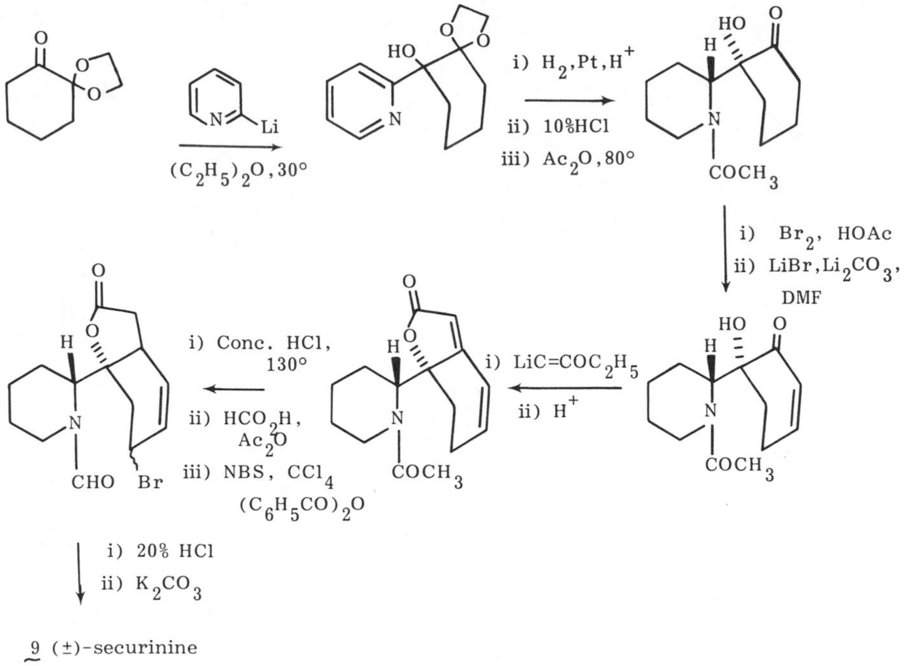

Scheme 15 Synthesis of (±)-securinine (**9**).

alanine (**1**), and tyramine (**12**) were poorly incorporated. Cadaverine (**120**) was also incorporated, labeling positions 2 and 6 as expected. This does not prove that cadaverine is an intermediate, merely that it can be utilized, as discussed in Chapter 4.

Subsequently Spenser and co-workers demonstrated the mode of incorporation of lysine to be *via* an asymmetric intermediate. Thus both [2-^{14}C]lysine (**119**) and [2-^{14}C]1,2-dehydropiperideine (**121**) gave securinine (**9**) labeled at C-2 almost (>90%) exclusively. As discussed in Chapter 4, involvement of cadaverine is therefore in the form of a bound intermediate.

By feeding [6-^{3}H, 6-^{14}C]lysine and demonstrating specific intact incorporation Spenser and co-workers were also able to demonstrate that the nitrogen atom of securinine is derived from the ε-nitrogen of lysine.

Thus the overall route suggested for securinine biosynthesis is that shown in Scheme 16.

A proton originally at C-3 tyrosine (C-12 in securinine) is lost in the biosynthetic process and Parry has investigated the stereochemistry of this loss. When [3RS-^{3}H, 3-^{14}C]tyrosine was used as a precursor, 57% of ^{3}H was lost. That this proton loss was in fact stereospecific was shown by separately feeding 3R and 3S tritiated tyrosines. Essentially all of the label from 3S-tritiated tyrosine was lost and consequently it is the *pro*-3S hydrogen that is lost on conversion to securinine (**9**).

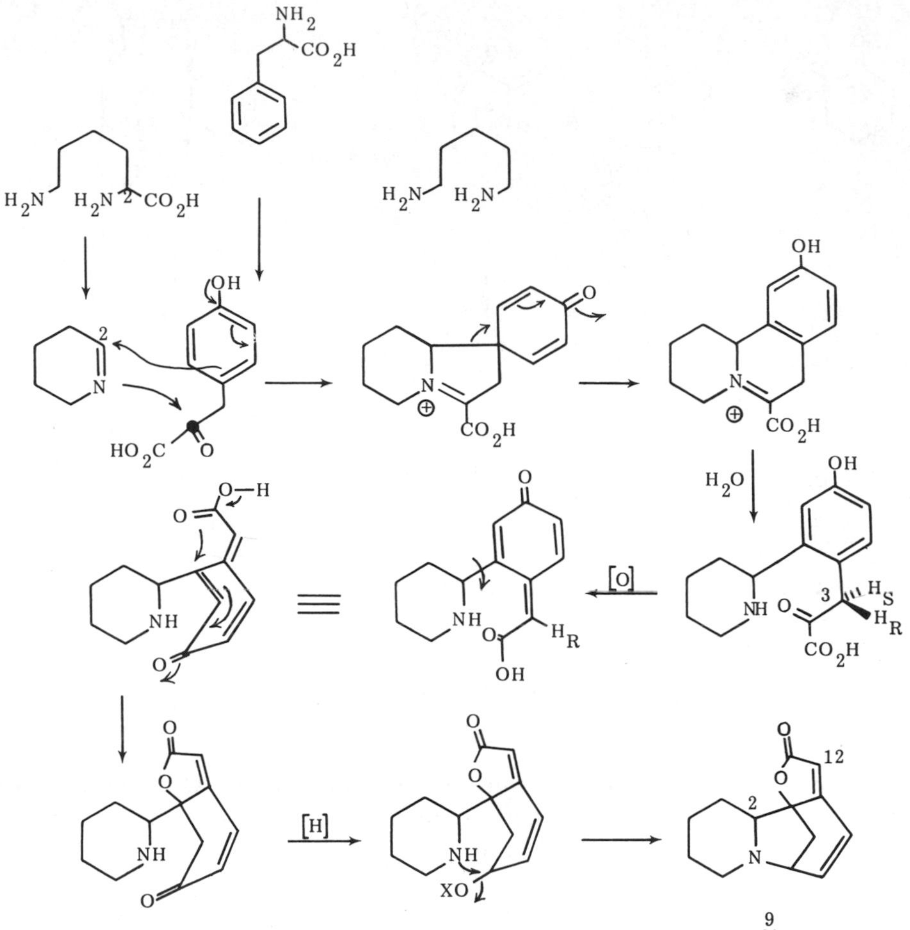

Scheme 16   Biogenesis of securinine (**9**).

8.7.4   Pharmacology

Securinine nitrate is a central nervous system stimulant similar to, but less toxic than, strychnine. The same compound also stimulates respiration, raises blood pressure, and increases cardiac output. It is reportedly useful in the treatment of paralysis following infectious disease and for physical disorders. Allosecurinine is less toxic than securinine.

LITERATURE

Review

Snieckus, V., *Alkaloids NY* **14**, 425 (1973).

Structure

Saito, S., K. Kodera, N. Shigematsu, A. Ide, N. Sugimoto, Z. Horii, M. Hanaoka, Y. Yamawaki, and Y. Tamura, *Tetrahedron* **19**, 2085 (1963).

Horii, Z., M. Ikeda, Y. Yamawaki, Y. Tamura, S. Saito, and K. Kodera, *Chem. Pharm. Bull.* **11**, 817 (1963).

Synthesis

Horii, Z., M. Hanaoka, Y. Yamawaki, Y. Tamura, S. Saito, N. Shigematsu, K. Kodera, H. Yoshikawa, Y. Sato, H. Nakai, and N. Sugmoto, *Tetrahedron* **23**, 1165 (1967).

Spectral Data

Parello, J., *Bull. Soc. Chim. Fr.* 1117 (1968).

Audier, H.-E., and J. Parello, *Bull. Soc. Chim. Fr.* 1552 (1968).

Biosynthesis

Parry, R. J., *Tetrahedron Lett.* 307 (1974).

Sankawa, U., K. Yamasaki, and Y. Ebizuka, *Tetrahedron Lett.* 1867 (1974).

Parry, R. J., *Chem. Commun.* 144 (1975).

Golebiewski, W. M., P. Horsewood, and I. D. Spenser, *Chem. Commun.* 217 (1976).

Sankawa, U., Y. Ebizuka, and K. Yamasaki, *Phytochemistry* **16**, 561 (1977).

8.8 MELANIN

One of the most widely distributed pigments in plants and animals is the dark-colored pigment melanin. A product of the enzymic oxidation of tyrosine, melanin is responsible for both hair and skin pigmentation in humans. It is a weakly acidic, irregular polymer probably derived from highly oxidized indole units as shown in Figure 1. The carboxy group remaining at position 2 of the indole nucleus is thought to be the source of the acidity. The indole nucleus is probably derived not from tryptophan but rather from cylodopa (**122**) and **123**, as indicated. Two linkages are apparent, the 3,7-linked (**124**) and the 4,7-linked (**125**).

8.9 BETALAINS

The betalains are a major group of colored alkaloid pigments whose distribution is limited to 10 families in the order Centrospermae. The most important colored plant pigments are the anthocyanins, but the betalains are of interest because their distribution is mutually exclusive with that of the anthocyanins.

The compounds may occur in the flowers, fruits, leaves, or tubers of a plant and vary in color between yellow and red-violet. As a group they are

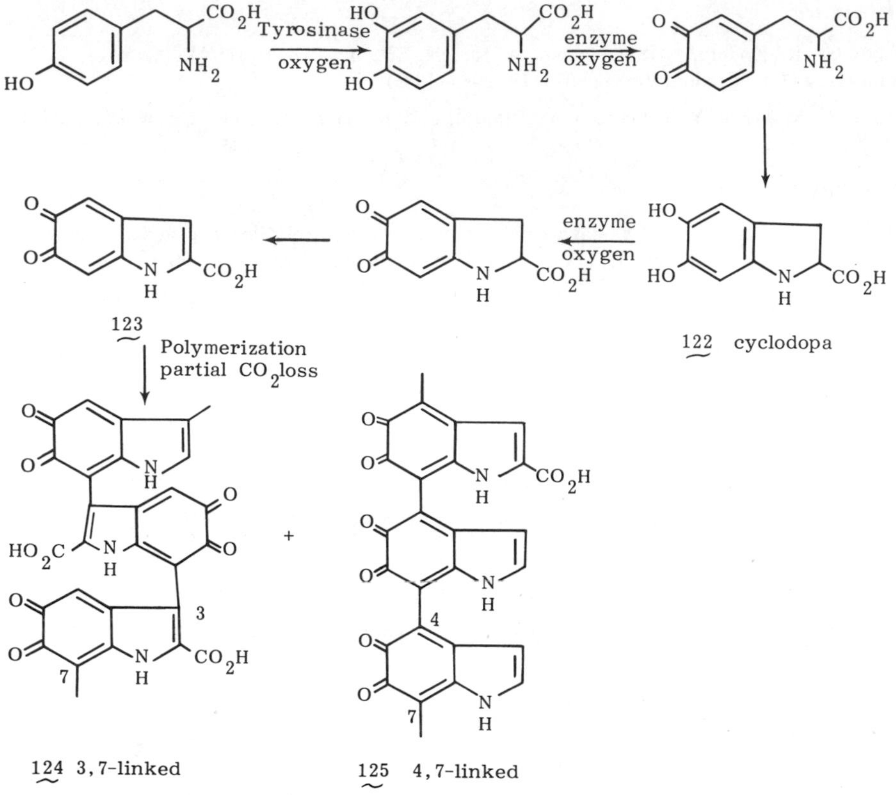

Figure 1 Formation of melanin.

zwitterionic and water-soluble. Examples of plants that contain betalains are cacti, pokeberry, and the red beet.

Early work was foiled by the problems of crystallizing the betacyanins and significant progress was not made until the mid-1950s, when the first structures were deduced. If we consider some of the work carried out on betanidin (**7**) the chemical problems will become evident.

Alkaline degradation of betanidin gave three identifiable products, 5,6-dihydroxy-2,3-dihydroindole 2-carboxylic acid (cyclodopa) (**122**), 4-methyl pyridine 2,6-dicarboxylic acid (**126**), and formic acid. The question was how to put these units together.

Betanidin shows λ_{max} 544 nm, ϵ 51,000, a deep red color, which on treatment with methanolic diazomethane gave a yellow (λ_{max} 403 nm) product, 5,6-di-O-methyl neobetanidin trimethyl ester (**127**). The NMR spectrum of this compound indicated the presence of five methoxy groups. Betanidin gave a trimethyl ester which was oxidized to a diacetate **128** on acetylation in the presence of air. The oxygens in betanidin are therefore three carboxyl groups and two phenolic groups. The NMR spectrum of **128** indicated two phenolic acetate groups, three methoxy groups, two *trans*-related olefinic protons (one of which was highly deshielded to 8.67 ppm), and two equivalent protons (8.10 ppm) at the β-position (C-14 and C-17) of a pyridine ring. These and additional data suggested that betanidin had the structure 7. Betanin (**129**) is an *O*-β-D-glucoside of betanidin and by simple reaction was shown to be attached at the 5-position.

The yellow betalains, also known as betaxanthins, are closely related to the betacyanins. They display UV absorption in the region 475–485 nm and a typical compound in this series is indicaxanthin (**130**) from the cactus *Opuntia ficusindica* Mill. Alkaline degradation of **130** gave L-proline (**131**) and 4-methyl pyridine 2,6-dicarboxylic acid (**126**).

The most interesting aspect of the chemistry of these compounds is their facile interconversion. For example, betanin (**129**) on treatment with excess proline in dilute ammonium hydroxide gave indicaxanthin (**130**), which when treated with excess L-cyclodopa (**122**) gave betanidin (**7**). These reactions were suggested to proceed *via* betalamic acid (**132**).

In 1971 Mabry and co-workers isolated betalamic acid (**132**) as an acid and base labile product from the petals of *Portulacca grandiflora* and found that this compound could be easily converted to betanidin.

Two groups, those of Büchi and Dreiding, expended much effort in attempting to synthesize betalamic acid dimethyl ester (**133**), which because of its instability and polyfunctional nature poses some interesting strategic problems.

7　betanidin R=H
129　betanin R=glucose

122

126

127　R=CH$_3$
128　R=COCH$_3$

130　indicaxanthin

131

132　betalamic acid R=H
133　R=CH$_3$

The synthesis developed by Dreiding and co-workers starts with chelidamic acid (**134**) and proceeds through to the piperidone diester **135**. Reaction with a modified Wadsworth–Emmons reagent containing a semicarbazone group gave **136**, which was oxidized with dicyclohexylcarbodiimide at room temperature to give the protected betalamic acid derivative **137**. Condensation with the methyl ester of L-cyclodopa then afforded the trimethyl ester of betanidin (**138**).

134 135 136 137 138

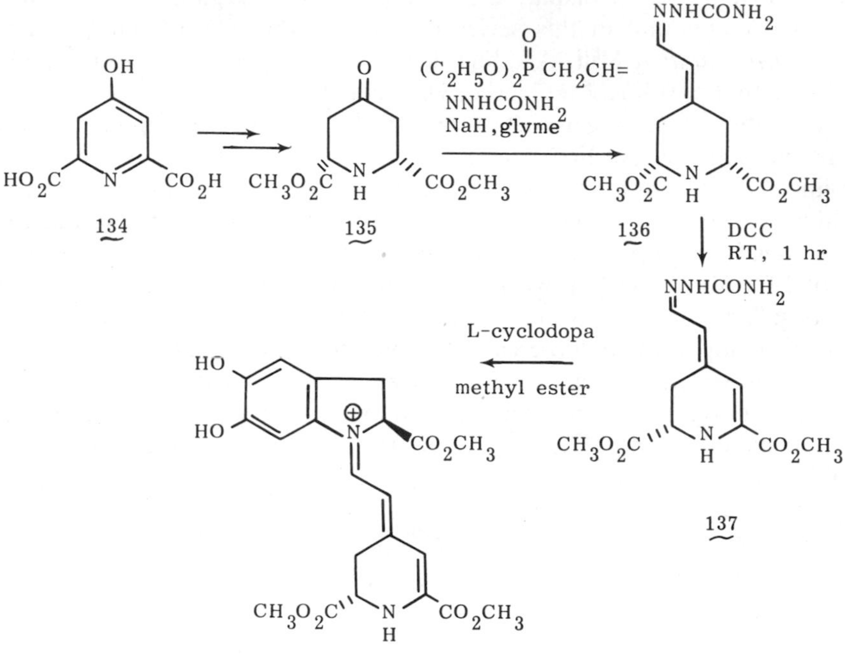

Büchi and co-workers have disclosed syntheses of both the cyclodopa and betalamic acid moieties of betanidin. Oxidation of dopa methyl ester hydrochloride (**139**) with potassium iodate in a two-phase system followed by acetylation gave an iodo diacetate **140**. This compound could be hydrogenated to afford di-*O*-acetyl cyclodopa methyl ester (**141**) in 40% overall yield.

N-Benzylnorteloidinone (**142**), readily prepared by the Robinson–Schöpf synthesis, gave a mixture of orthoesters **143**, by reaction with methyl or-

139 140 141

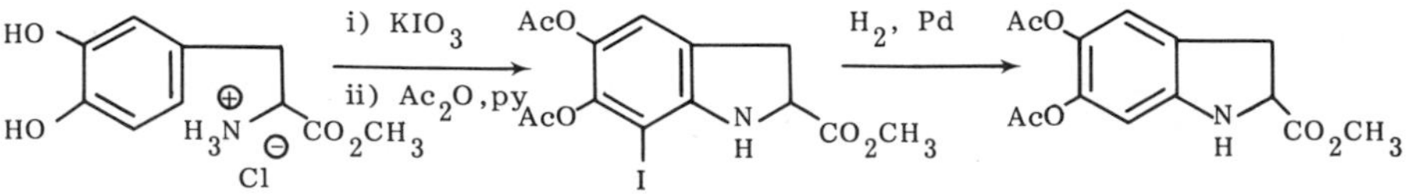

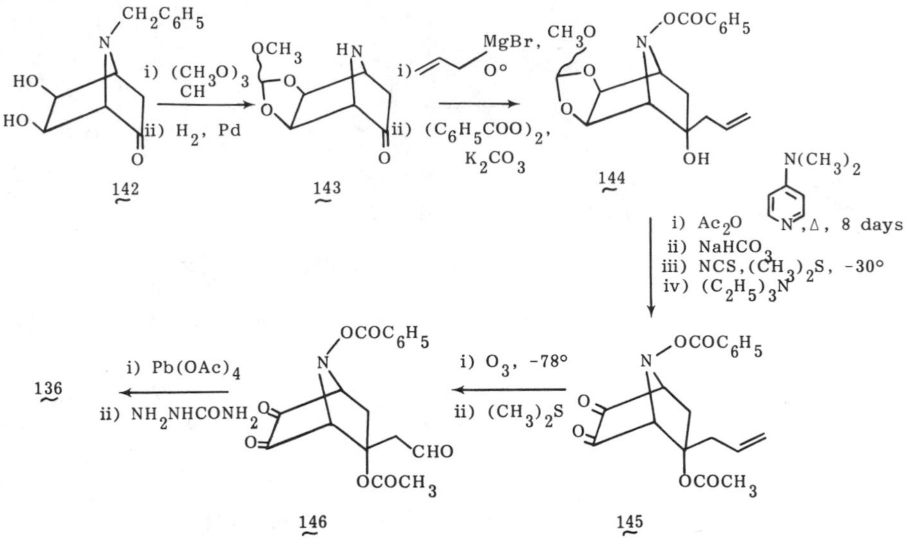

Scheme 17 Büchi synthesis of betalamic acid dimethyl ester semicarbazone (**136**).

thoformate followed by catalytic hydrogenation. A propenyl side chain was introduced at C-3 by reaction with allyl magnesium bromide and the secondary amine protected as the benzolylated hydroxylamine by reaction with dibenzoyl peroxide in DMF, in the presence of base, to afford **144**. Subsequent reactions yielded the diketone **145**, which could be ozonized to the diketoaldehyde **146** and then cleaved with lead tetraacetate in benzene–methanol to give betalamic acid dimethyl ester characterized as the semicarbazone derivative **136** (Scheme 17).

Biogenetically, betanidin (**7**) poses an interesting problem. The formation of cyclodopa (**122**) from dopa (**21**) is readily envisaged, although experimentally it has proved difficult to establish. The more challenging problem is that of betalamic acid (**132**), and it was not until the investigations of Mabry and Dreiding that much light was shed on this problem. Somewhat surprisingly this unit too is produced from dopa by a ring cleavage reaction and recyclization.

When [1-^{14}C]dopa was used as a precursor of betanidin in *Opuntia decumbens* it was shown by degradation that the α-aminocarboxyl group of the betalamic acid unit is derived from the 1-carboxyl group of dopa. Two routes to betalamic acid (**132**) are still possible (Scheme 18). One in which cleavage occurs between the two hydroxyl groups or a second involving cleavage of the bond adjacent to the catechol unit.

A distinction between these two possibilities was made by using [3′,5′-^{3}H$_2$]L-tyrosine (**2**). No NIH shift is expected, so that [5′-^{3}H]L-dopa (**21**) is the initial product and cleavage by pathway a would lead to loss of this tritium, whereas pathway b would leave a label at the position of the

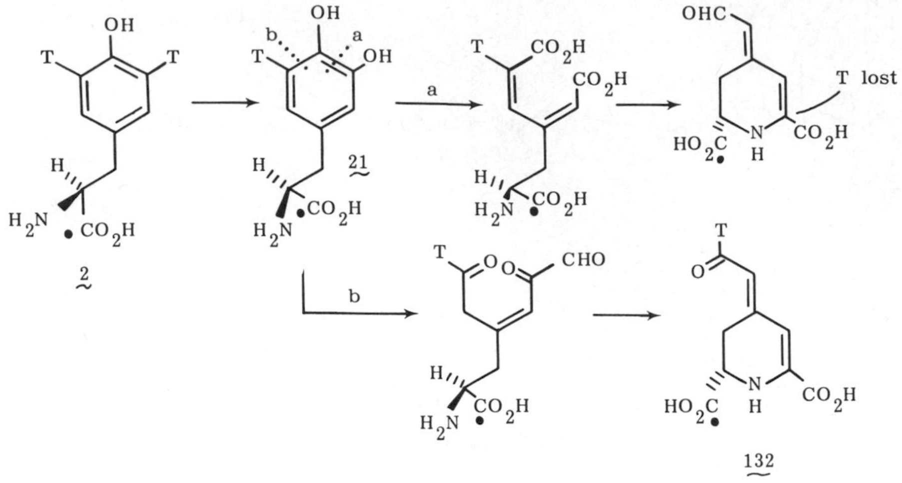

Scheme 18   Biosynthesis of betalamic acid (**132**).

aldehyde proton. In experiments with both betanidin (**7**) and indicaxanthin (**130**) in *Opuntia* species the labeled tyrosine gave products showing a loss of half of the tritium activity, in agreement with pathway b.

Subsequently, [1-^{14}C]- and [2-^{14}C]dopa (**21**) were shown to be specific precursors of betalamic acid (**132**) itself in *Portulacca grandiflora*, but whether betalamic acid is a true biosynthetic intermediate which condenses with either cyclodopa or other amino acids is at present unknown.

LITERATURE

Reviews

Dreiding, A. S. in W. D. Ollis (Ed.), *Recent Developments in the Chemistry of Natural Phenolic Compounds*, Pergamon, Oxford, 1961.

Mabry, T. J., in S. W. Pelletier (Ed.), *Chemistry of the Alkaloids*, Van Nostrand Reinhold, New York, 1970, p. 367.

Piatelli, M., in T. W. Goodwin (Ed.), *Chemistry and Biochemistry of Plant Pigments*, Vol. 1, Academic, London, 1976.

Synthesis

Hermann, K., and A. S. Dreiding, *Helv. Chim. Acta* **58**, 1805 (1975).

Büchi, G., and T. Kamikawa, *J. Org. Chem.* **42**, 4153 (1977).

Büchi, G., H. Fliri, and R. Shapiro, *J. Org. Chem.* **42**, 2192 (1977).

Biosynthesis

Haslam, E., in *The Shikimate Pathway*, Butterworths, London, 1974, p. 279.

Wyler, H., T. J. Mabry, and A. S. Drieding, *Helv. Chim. Acta* **46**, 1745 (1963).

Miller, H. E., H. Rosler, A. Wohlpart, H. Wyler, M. E. Wilcox, H. Frohofer, T. J. Mabry, and A. S. Drieding, *Helv. Chim. Acta* **51**, 1470 (1968).

Fischer, N. H., and A. S. Dreiding, *Helv. Chim. Acta* **55**, 649 (1972).

Impellizzeri, G., and M. Piatelli, *Phytochemistry* **11**, 2499 (1972).

8.10   TETRAHYDROISOQUINOLINE ALKALOIDS

Probably the largest of any of the groups of alkaloids is the one based on the tetrahydroisoquinoline nucleus (**147**). This nucleus occurs in a vast array of structure types and all are derived from phenylalanine or tyrosine. These alkaloids constitute most of the discussion in the remainder of this chapter, such is their importance.

Many of these alkaloids contain this nucleus in a highly modified form from which it may not be immediately recognizable. Thus although the nucleus is quite clear in reticuline (**148**), it is perhaps less obvious in hasubanonine (**149**).

148   reticuline   149   hasubanonine

This is a vast area of alkaloid chemistry and even though Shamma has written a superb treatise on the isoquinoline alkaloids, some major groups were omitted. Thus any discussion in a book of this size must be quite selective and far from complete.

The organization of the subsequent sections is based on the overall biosynthetic pathway from the simplest tetrahydroisoquinoline derivatives to the more highly rearranged species.

8.10.1   Synthesis

There are several routes potentially available for the formation of the tetrahydroisoquinoline nucleus, and because of the importance of this nucleus in so many alkaloids, it is worthwhile at this point to briefly discuss some of these approaches.

Bischler–Napieralski Cyclization

In this synthesis the *N*-acyl derivative of an oxygenated phenethylamine is cyclized with phosphorus pentoxide or phosphorus oxychloride to give a

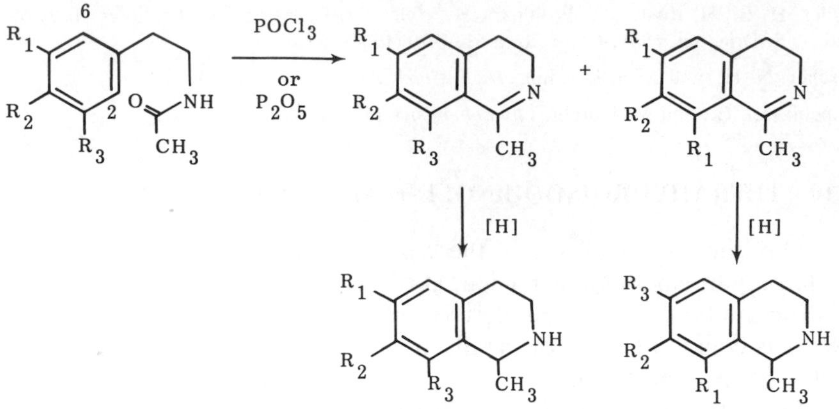

Scheme 19   Bischler–Napieralski synthesis of tetrahydroisoquinolines.

3,4-dihydroisoquinoline which can be reduced (now normally sodium boro-hydride) to give a tetrahydroisoquinoline. If the 3- and 5-groups are different as is often the case, two products are formed in the initial cyclization by condensation at either the 2- or 6-position (Scheme 19).

Pictet–Spengler Cyclization

The Pictet–Spengler cyclization is the reaction between an oxygenated phen-ethylamine and an aldehyde under acidic conditions to give a tetrahydroiso-quinoline directly. Thus Schöpf and Bayerle found that reaction of dopa-mine (10) with acetaldehyde at pH 5 at room temperature for 5 days gave O-norsalsoline (150) in 83% yield. When a trioxygenated phenethylamine is used, two substituted products are formed if the 3- and 5-groups are different.

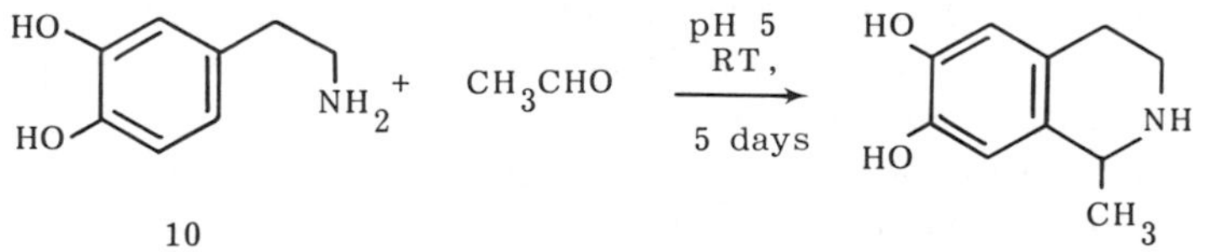

Pomeranz–Fritsch Cyclization

The Pomeranz–Fritsch approach has the advantage that because of the nature of the starting material, cyclization is directed to give only one product.

In the simplest case the reaction sequence is to condense an aromatic aldehyde with aminoacetaldehyde diethyl acetal followed by acid-catalyzed cyclization to give an isoquinoline. Catalytic reduction then affords a te-trahydroisoquinoline. When an acetophenone is used in place of an aromatic aldehyde, yields are invariably lower.

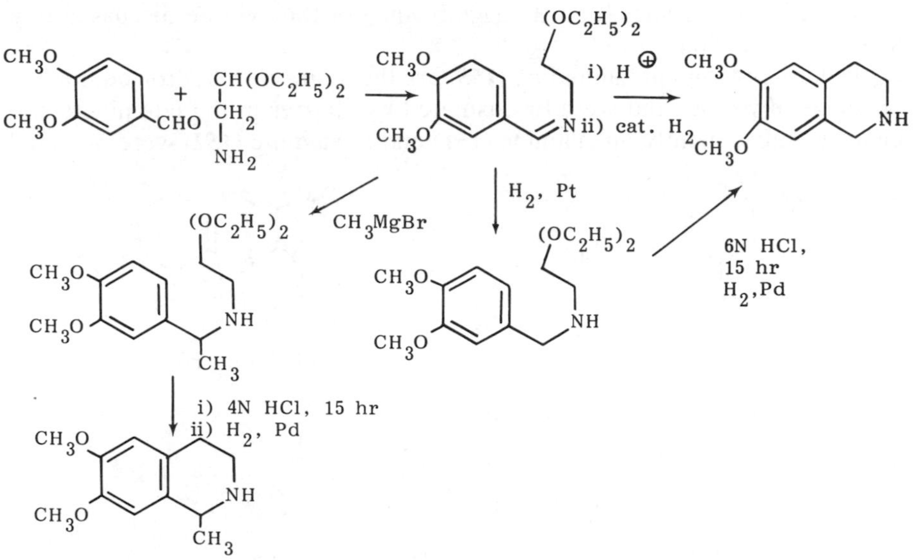

Scheme 20   Modifications of the Pomeranz–Fritsch cyclization.

Bobbitt and co-workers have described some very interesting modifications of this synthetic approach. For example the intermediate imine may be reduced *in situ* and the acetal then cyclized and reduced (Scheme 20). The imine can also be substituted at C-1 by reaction with a Grignard reagent and the sequence completed as before.

LITERATURE

Synthesis

Bobbitt, J. M., D. N. Roy, A. Marchand, and C. W. Allen, *J. Org. Chem.* **32**, 2225 (1967).

Bobbitt, J. M., A. S. Steinfeld, K. H. Weisgraber, and S. Dutta, *J. Org. Chem.* **34**, 2478 (1969).

Bobbitt, J. M., and C. P. Dutta, *J. Org. Chem.* **34**, 2001 (1969).

Takido, M., K. L. Khanna, and A. G. Paul, *J. Pharm. Sci.* **59**, 271 (1970).

8.10.2   Simple Tetrahydroisoquinolines

Chemistry

The peyote cactus, *Lophophora williamsii*, as well as producing phenethylamine derivatives, also produces a wide range of simple tetrahydroisoquinoline derivatives, typical of these are anhalamine (**151**), anhalinine (**152**), anhalidine (**153**), and lophophorine (**154**). Simple alkaloids such as these also occur in other members of the Cactaceae, as well as the families Leguminosae and Papaveraceae. More recently these compounds have been detected in human tissues, and the significance of this will be discussed in a subsequent section.

Pellotine (**4**) and anhalonidine (**155**) are the major tetrahydroisoquinoline alkaloids of peyote and were first isolated by Heffter in the late nineteenth century; subsequently anhalamine (**151**) and anhalinine (**152**) were obtained

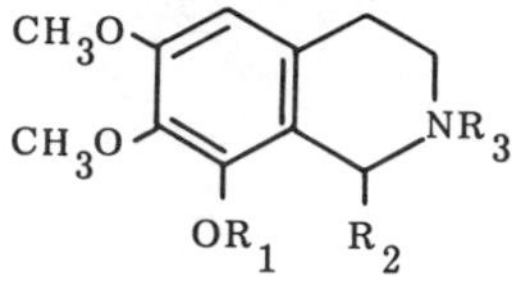

		R₁	R₂	R₃
151	anhalamine	H	H	H
152	anhalinine	CH₃	H	H
153	anhalidine	H	H	CH₃
155	anhalonidine	H	CH₃	H
4	pellotine	H	CH₃	CH₃

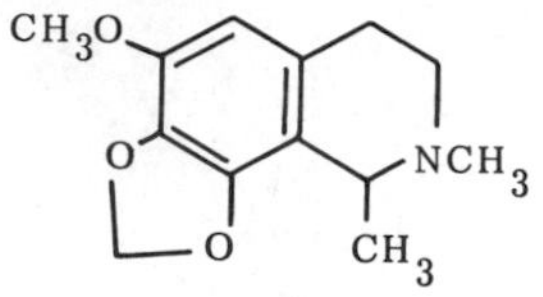

154 lophophorine

and as isolation and detection techniques improved numerous other alkaloids were discovered. When the neutral fraction of peyote was examined for alkaloids by Kapadia a whole new range of derivatives was found, including mescaline succinimide (**156**) and mescalotam (**157**).

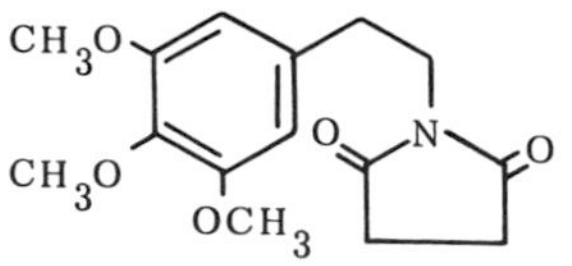

156

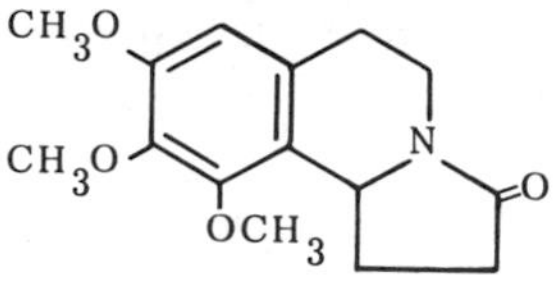

157

A facile synthesis of anhalamine (**151**) and anhalinine (**152**) developed by Bobbitt and co-workers is shown in Scheme 21 and follows the route discussed previously. Substitution takes place regiospecifically at the position *ortho* (rather than *para*) to the phenolic group.

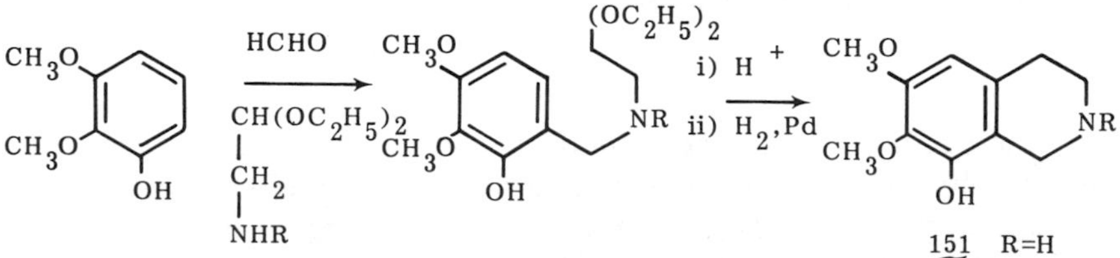

151 R=H

152 R=CH₃

Scheme 21 Bobbitt synthesis of anhalamine (**151**) and anhalinine (**152**).

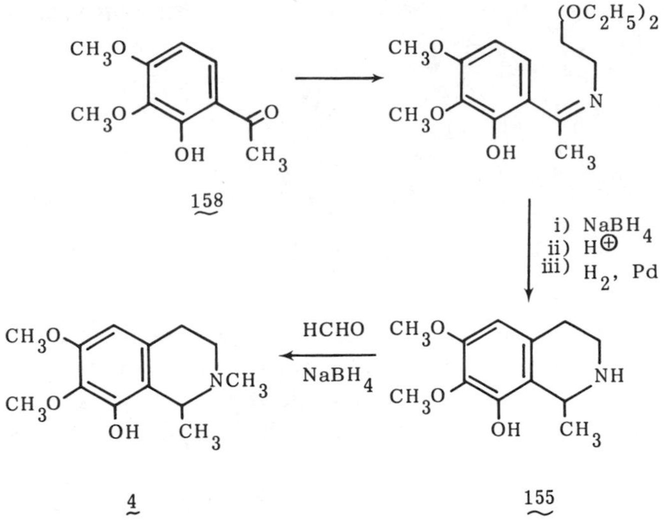

Scheme 22 Paul synthesis of pellotine (**153**).

Anhalonidine (**155**) and pellotine (**4**) were recently synthesized by Paul and co-workers using a modification of the Pomeranz–Fritsch synthesis beginning with the acetophenone derivative **158** (Scheme 22).

Two alkaloids which are closely related structurally to the tetrahydroisoquinolines of peyote are (+)-salsoline (**159**) and (−)-salsolidine (**160**) from

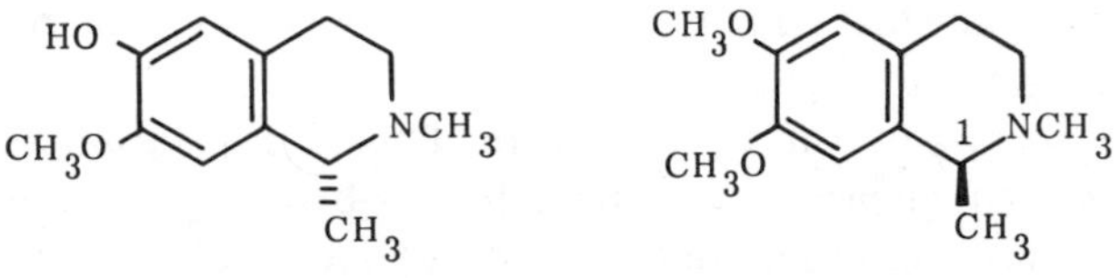

159 (+)-salsoline 160 (-)-salsolidine

Salsola arbuscula Pall. in the Chenopodiaceae and *Desmodium tiliaefolium* (Leguminosae). Although these two bases are closely related, they have the opposite configurations at C-1. In the case of the latter isolation these alkaloids cooccur not only with phenethylamines, but also with tryptamines.

The absolute stereochemistry of (−)-salsolidine (**160**) was determined by chemical correlation with L-alanine (**91**). Ozonolysis of *N*-formylsalsolidine (**161**) followed by peracetic acid treatment and acid hydrolysis gave the diacid **162**. The same product was produced by alkylation of L-alanine (**91**) with acrylonitrile and acid hydrolysis of the intermediate nitrile acid (Scheme 23).

An interesting asymmetric synthesis of the (R)-(+)- and (S)-(−)-salsolidines has been reported by Kametani from homoveratraldehyde and the

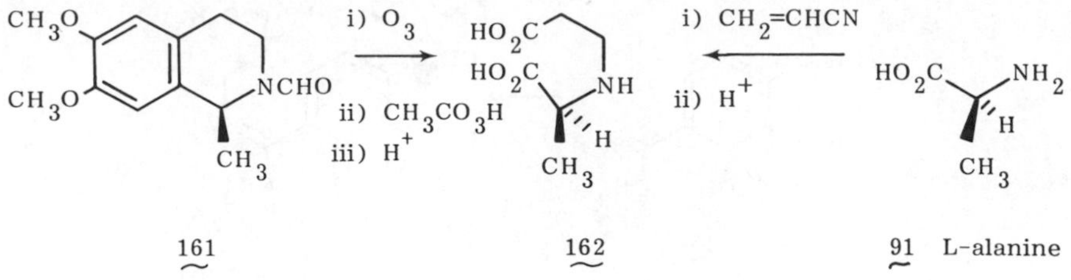

i) O_3
ii) CH_3CO_3H
iii) H^+
i) CH_2=CHCN
ii) H^+
161
162
91 L-alanine

Scheme 23

appropriate (R)-(+)- or (S)-(−)-α-methyl benzylamines. The product was converted to the corresponding dihydroisoquinolinium iodide (e.g. **163**). Asymmetry is induced in the borohydride reduction step due to the bulky phenyl group. Hydrogenolysis of **164** affords the optically active salsolidine.

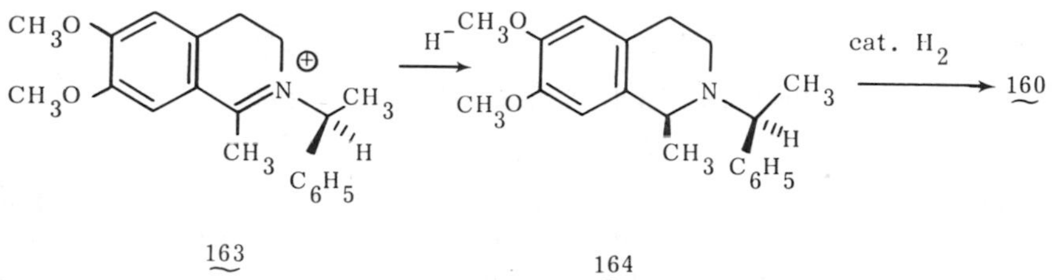

H^-
cat. H_2
160
163
164

Spectral Properties

The UV spectral properties of the simple tetrahydroisoquinolines are well established from substituted benzenoid derivatives and some representative examples are shown in Figure 2.

The NMR spectra of a number of simple tetrahydroisoquinoline chlorides are summarized in Figure 3, and the effects observed are quite typical of this and other series of compounds. From a base value of 7.20 ppm for the protons on tetrahydroisoquinoline itself, introduction of any oxygen substituent shields all the aromatic protons. Shielding by a methoxyl substituent is maximal at the *ortho* position (~0.4–0.5 ppm), followed by the *para* position (~0.05–0.15 ppm). The shielding effects of an hydroxyl group are normally 0.1–0.17 ppm greater than those of a methoxy group.

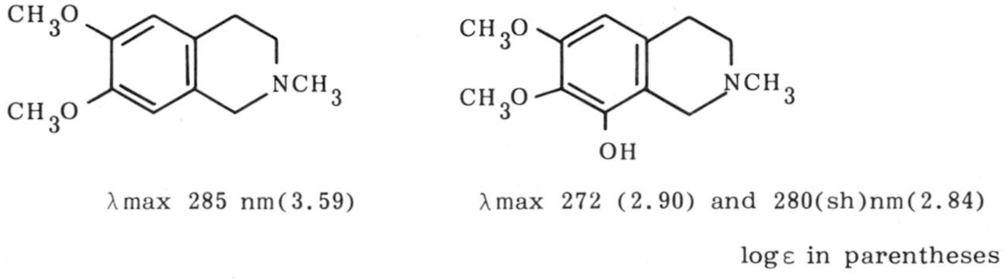

λmax 285 nm(3.59)
λmax 272 (2.90) and 280(sh)nm(2.84)
logε in parentheses

Figure 2

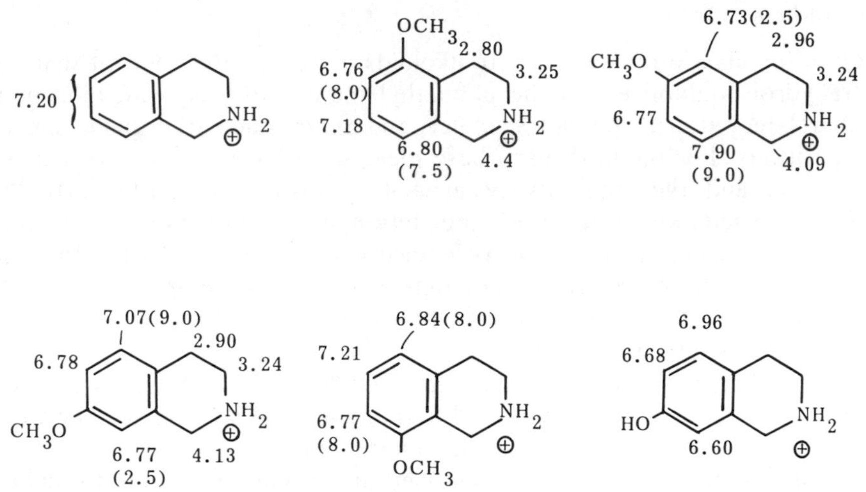

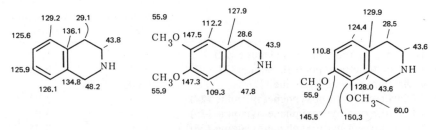

δ ppm in d_6–DMSO
coupling constants in
parentheses

Figure 3

MacLean and co-workers have described the ^{13}C NMR spectra of some of the simple isoquinoline alkaloids and selected examples are shown in Figure 4. The two carbon atoms adjacent to nitrogen are normally easily distinguished because the α-phenyl group further deshields C-1. Substitution at C-8 causes a marked (4.6 ppm) shielding of C-1 due to steric effects. Many isoquinoline alkaloids contain *ortho*-methoxy groups and those having substituents at 6 and 7 give C-5 and C-8 resonances which are often very close in chemical shift (see many additional examples elsewhere in this chapter). Invariably however it appears that C-8 resonates farther downfield. The assignments were made by applying the shift parameters (C-1, +20.8; C-2, −16.9; C-3, −7.6 ppm) to the spectrum of the parent compound tetrahydroisoquinoline (**147**) on oxygenation.

Figure 4

Biosynthesis

Reti in his classic review of cacti alkaloids expanded the concept that the tetrahydroisoquinoline and the phenethylamine alkaloids were biogenetically related, after Pictet and Spengler, and Späth, had earlier made similar propositions. For the most part these ideas were based on *in vitro* experimentation and the similarity of aromatic substitution, particularly the 6,7,8-trioxygenation pattern of the tetrahydroisoquinolines of peyote. Formaldehyde and acetaldehyde were regarded as progenitors of the one- and two-carbon units respectively, although as a result of *in vitro* experiments by Hahn and co-workers in the mid-1930s, the possible involvement of α-keto acids was also proposed.

Battersby and co-workers were the first to show that tyrosine was a precursor of anhalonidine (155) and lophophorine (154), and further work indicated that dopamine (10) was also an intermediate.

A similar range of intermediates used in studying mescaline formation was also used in evaluating the pathway of tetrahydroisoquinoline biosynthesis in peyote. Some of the results are shown in Table 4. In particular, they show the lack of involvement of 3,4-dimethoxyphenethylamine (27), and 3,5-dimethoxy-4-hydroxyphenethylamine (26) in the pathway. As a result of these experiments a pathway (Scheme 24) was proposed quite similar to that for mescaline biosynthesis in which there is a defined sequence of methylation and hydroxylation reactions. The reversible nature of the anhalonidine–pellotine transformation, a quite surprising N-demethylation, was demonstrated by Battersby and co-workers.

The remaining problem is the origin of the C-1 unit [in anhalidine (153)] and the C-1 and C-9 unit [in pellotine (4)]. Methionine was not a precursor, and acetate and formate were not directly incorporated. Leete and Braunstein were the first to provide experimental data in support of Hahn's idea that α-keto acids were involved. Thus [3-^{14}C]pyruvate specifically labeled C-9 of anhalonidine (155) in one experiment, although it was randomized between C-1 and C-9 in another study. The nature of the origin of the C-1

Table 4 Precursors of Tetrahydroisquinoline Alkaloids of *L. williamsii*

| | Percent Recovery of Radioactivity | |
| | Anhalonidine | Anhalamine |
Precursor	(155)	(151)
4-Hydroxy-3-methoxyphenethylamine (22)	2.20	1.72
3-Hydroxy-4-methoxyphenethylamine (23)	0.74	0.90
3, 4-Dimethoxyphenethylamine (27)	0.002	—
4, 5-Dihydroxy-3-methoxyphenethylamine (24)	3.23	3.90
3, 4-Dimethoxy-5-hydroxyphenethylamine (25)	6.24	5.96
3, 5-Dimethoxy-4-hydroxyphenethylamine (26)	0.16	0.05

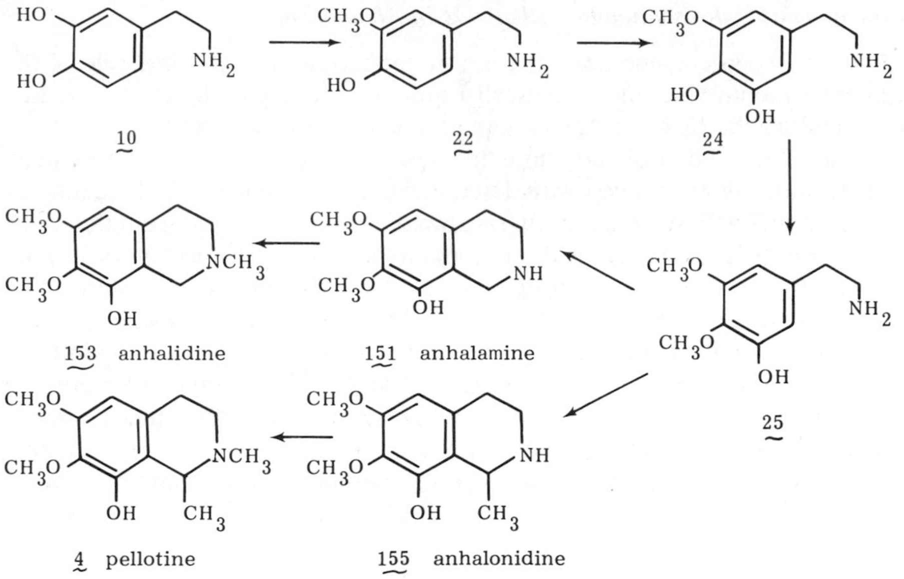

Scheme 24 Biogenesis of the tetrahydroisoquinoline alkaloids of *L. williamsii*.

and C-9 is not known for certain, but there is some further evidence concerning intermediates between 3,4-dimethoxy-5-hydroxyphenethylamine (**25**) and anhalamine (**151**) and anhalonidine (**155**).

Kapadia and co-workers were the first to seriously look at the amino acid fraction of peyote. Two of the amino acids obtained were peyoruvic and peyoxylic acids, **165** and **166** respectively, and they were synthesized by condensation of **25** with pyruvic acid and glyoxylic acid. But more important, each of these was well (~6.0%) incorporated into the respective basic tetrahydroisoquinolines.

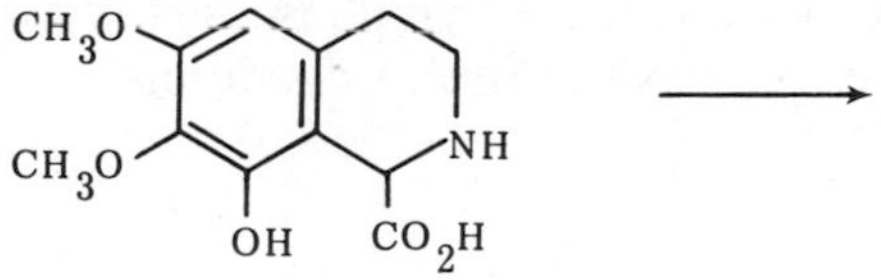

165 peyoxylic acid → **151** anhalamine

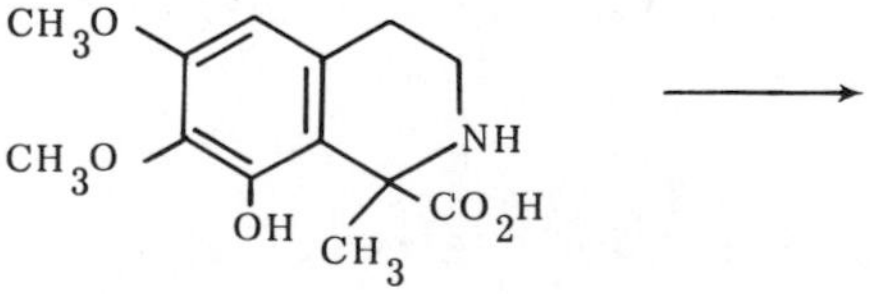

166 peyoruvic acid → **155** anhalonidine

Role of Tetrahydroisoquinoline Alkaloids in Alcoholism

There has been considerable discussion concerning the possible role of te-trahydroisoquinoline and/or benzylisoquinoline alkaloids in alcohol addic-tion, and all the facets of this argument cannot be discussed here.

One of the initial observations involved the finding that rat brain stem homogenates on incubation with dopamine and ethanol gave small quantities of salsolinol (167). More recently salsolinol has been isolated from the urine of alcoholics. In either case the key reactions are the formation of acetal-dehyde by an alcohol dehydrogenase and subsequent condensation with dopamine. Thus alcohol would react to deplete dopamine levels. However, the higher acetaldehyde levels in the brain inhibit the normal dehydrogen-ation of 3,4-dihydroxyphenylacetaldehyde (168). This compound therefore accumulates and is available for condensation with dopamine to give N-norlaudanosoline (169), and indeed this product accumulates when mam-malian liver tissues are incubated with dopamine or in patients on L-dopa therapy.

167   salsolinol   168

169   N-norlaudanosoline

The importance of 169 in the biosynthesis of morphine (5) in plants is well established. The implication, yet to be experimentally verified, is that morphine-like compounds are produced in mammalian tissue and are re-sponsible for alcohol addiction. Clearly these hypotheses are of great interest and should be examined under the most stringent conditions.

LITERATURE

Reviews

Shamma, M., The Isoquinoline Alkaloids, Academic, New York, 1972, p. 2.

Shamma, M. and J. L. Moniot, Isoquinoline Alkaloids Research, 1972–1977, Plenum, New York, 1978, p. 1.

Salsoline-Salsolidine

Battersby, A. R., and T. P. Edwards, J. Chem. Soc. 1214 (1960).

Okawara, T., and T. Kametani, Heterocycles 2, 571 (1974).

Bernath, G., J. Kobor, K. Koczka, L. Radics, and M. Katjar, Tetrahedron Lett. 225 (1968).

Spectral Data

Brossi, A., F. Schenker, and W. Leimgruber, *Helv. Chim. Acta* **47**, 2098 (1964).

Schenker, F., R. A. Schmidt, T. Williams, and A. Brossi, *J. Het. Chem.* 665 (1971).

Hughes, B. W., H. L. Holland, and D. B. MacLean, *Can. J. Chem.* **54**, 2252 (1976).

Biosynthesis

Reti, L., *Fortschr. Chem. Org. Naturs.* **6**, 242 (1950).

Paul, A. G., *Lloydia* **36**, 36 (1973).

Battersby, A. R., R. Binks, and R. Huxtable, *Tetrahedron Lett.* 563 (1967).

Battersby, A. R., R. Binks, and R. Huxtable, *Tetrahedron Lett.* 611 (1968).

Leete, E., and J. D. Braunstein, *Tetrahedron Lett.* 451 (1969).

Kapadia, G. J., G. Subba Rao, E. Leete, M. B. E. Fayez, Y. Vaishnav, and H. M. Fales, *J. Amer. Chem. Soc.* **92**, 6943 (1970).

Lophocerine and Pilocereine

Lophocerine has been obtained from *Lophocereus schottii* (Engelm.) Britt. et Rose (Cactaceae) along with the trimer pilocereine (**171**).

Pilocereine caused some interesting structure problems initially. Potassium (or sodium) in liquid ammonia is capable of cleaving aryl ethers preferentially over alkyl aryl ethers and is widely used in determining the points of attachment of many types of dimeric isoquinoline alkaloids. When this technique was applied to O-ethyl pilocereine four amines were produced

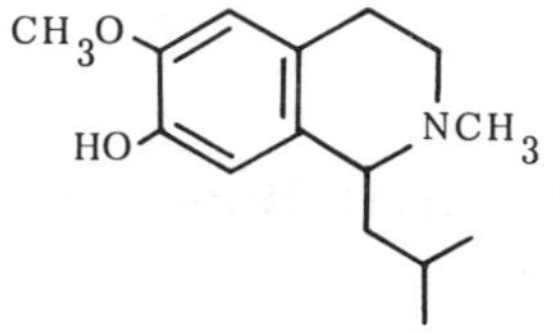

170   lophocerine

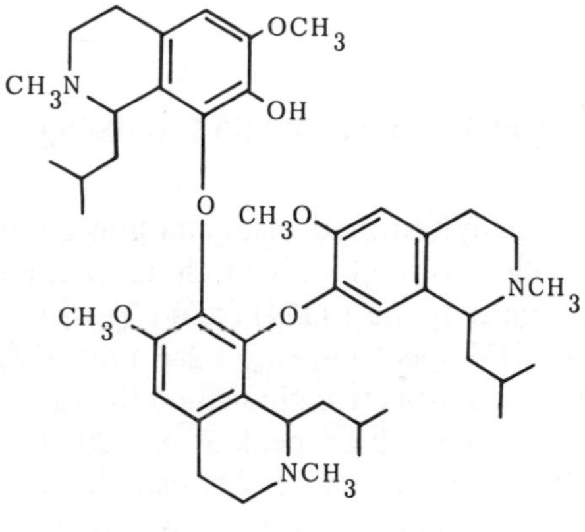

171   pilocereine

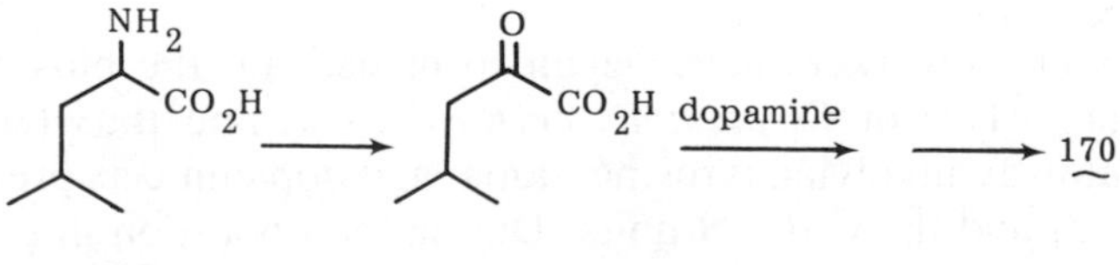

172   173

and from these it was thought that pilocereine was merely a dimer of lophocerine. Mass spectrometry however showed that the correct molecular weight was 757 (i.e. that pilocereine was a trimer).

Pilocereine itself has not been synthesized in pure form, but a mixture of isomers has been obtained by oxidative coupling of lophocerine (170) with potassium ferricyanide.

In spite of its interesting structure lophocerine (170) has not been well investigated from a biosynthetic point of view. Tyrosine and methionine labeled the positions expected, but the precise origin of the five-carbon unit remains unknown since both leucine (172) and mevalonate were incorporated. Bearing in mind the formation of the peyote tetrahydroisoquinolines from keto acids, this unit could well be derived from leucine (172) via the keto acid 173. Doubly labeled lophocerine (170) was incorporated intact into pilocereine (171).

LITERATURE

Lophocerine and Pilocereine

Djerassi, C., T. Nakano, and J. M. Bobbitt, *Tetrahedron* 2, 58 (1958).

Djerassi, C., H. W. Brewer, C. Clarke, and L. J. Durham, *J. Amer. Chem. Soc.* 84, 3210 (1962).

O'Donovan, D. G., and H. Horan, *J. Chem. Soc., Sec. C* 2791 (1968).

O'Donovan, D. G., and H. Horan, *J. Chem. Soc., Sec. C* 1737 (1969).

8.11 1-PHENYLTETRAHYDROISOQUINOLINE ALKALOIDS

The 1-phenyltetrahydroisoquinoline alkaloids are thus far limited in their distribution to the family Orchidaceae, and only three examples are known [e.g. cryptostyline I (174) from *Cryptostylis fulva* Schltr.].

The UV spectrum (λ_{max} 286 nm) is typical of a tetrahydroisoquinoline and the mass spectrum indicated the facile loss of a methylene–dioxyphenyl radical to give a base peak at m/e 206 (175).

The structure was readily established by synthesis (Scheme 25) and the racemic product resolved. In practice, the optical rotation of the resolved (+)-isomer was higher than that of the natural (+)-isomer, indicating the latter to be a partial racemate. The proton NMR spectral values for cryptostyline I are shown.

Agurell and co-workers have begun to investigate the biosynthesis of cryptostyline I (174) in *Cryptostylis erythroglossa*, and thus far the data support a pathway involving tyrosine, dopa, and dopamine as precursors of both the C_6-C_1 and the C_6-C_2-N units. During incorporation into the upper

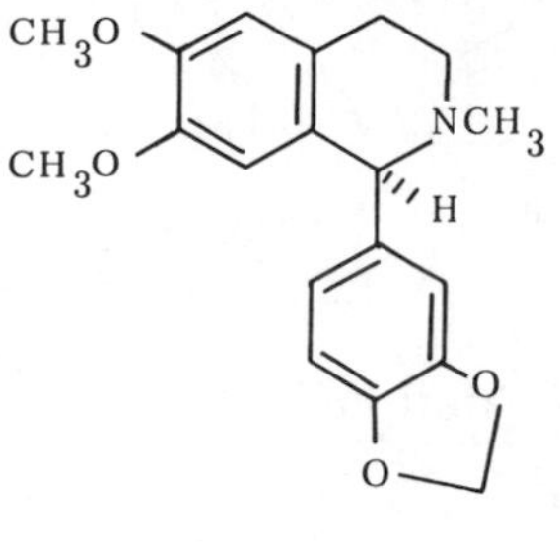

174 (+)-cryptostyline I

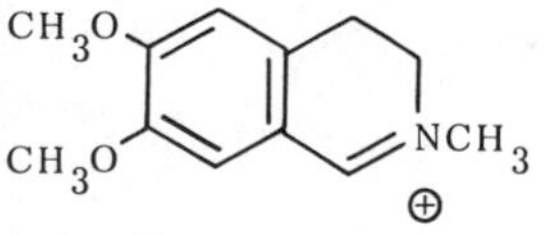

175 m/e 206

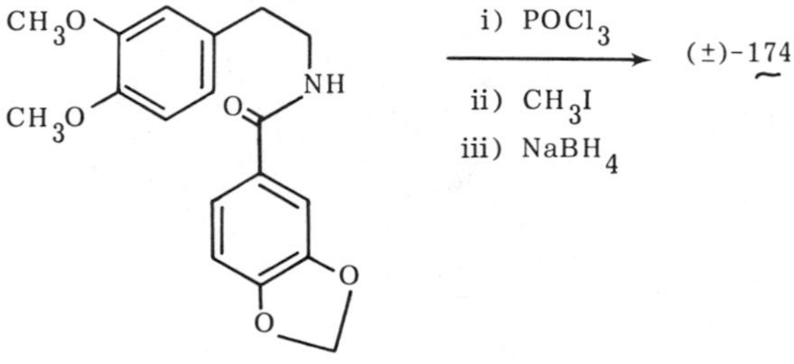

i) POCl$_3$

ii) CH$_3$I

iii) NaBH$_4$

(±)-174

Scheme 25

part of the molecule, dopamine is converted to 3-hydroxy-4-methoxyphen-ethylamine (23).

[2-^{14}C]Dopamine was a good precursor of 174 and 35% of the activity was found at C-4. The remainder of the activity was assumed to be at C-1.

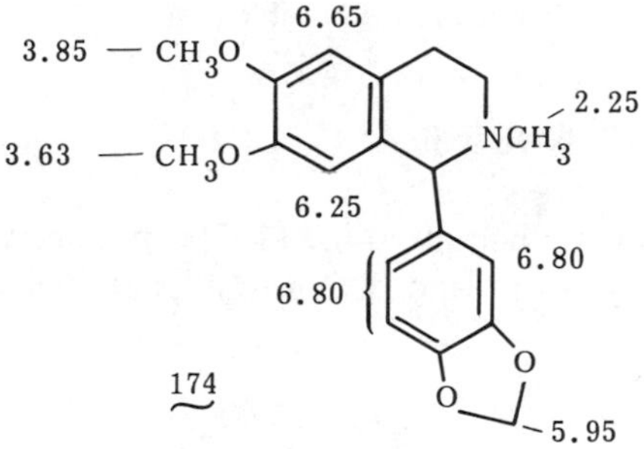

Vanillin (176), but not isovanillin (177), was also a precursor, and as expected the derivative 178 was not incorporated into 174. 1,2-Dehydrocryptostyline (179) was a very effective precursor (15% incorporation) of 174. Proto-catechualdehyde (71), which is known to be a precursor of the C$_6$-C$_1$ unit of the Amaryllidaceae alkaloids (Section 8.37.4), has yet to be tested as a precursor.

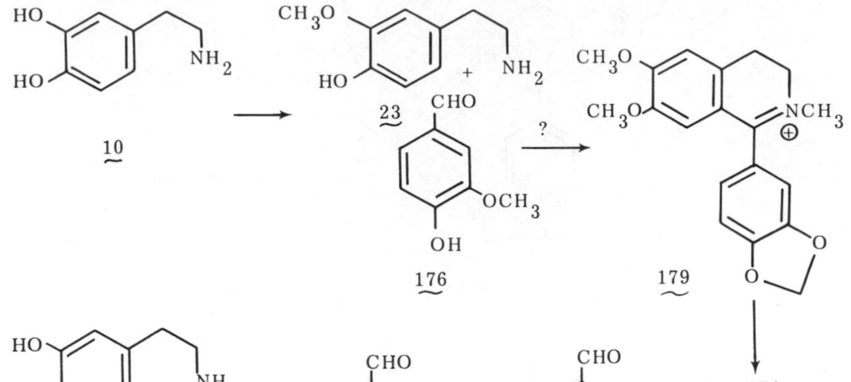

10

176

179

174

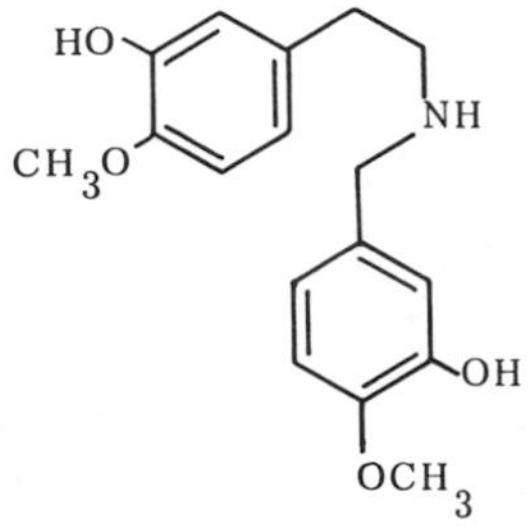

178

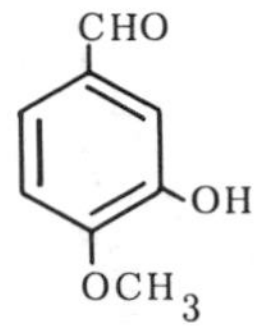

177

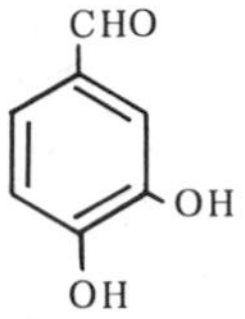

180

LITERATURE

Shamma, M., *The Isoquinoline Alkaloids*, Academic, New York, 1972, p. 490.

Shamma, M. and J. L. Moniot, *Isoquinoline Alkaloid Research, 1972–77*, Plenum, New York, 1978, p. 381.

Agurell, S., I. Granelli, K. Leander, B. Luning, and J. Rosenblom, *Acta Chem. Scand. B* **28**, 239 (1974).

Agurell, S., I. Granelli, K. Leander, and J. Rosenblom, *Acta Chem. Scand. B* **28**, 1175 (1974).

8.12 BENZYLISOQUINOLINE ALKALOIDS

The alkaloids of the benzylisoquinoline (**181**) type, derived in a formal sense from phenylalanine/tyrosine and a β-phenylacetaldehyde, are a vast and

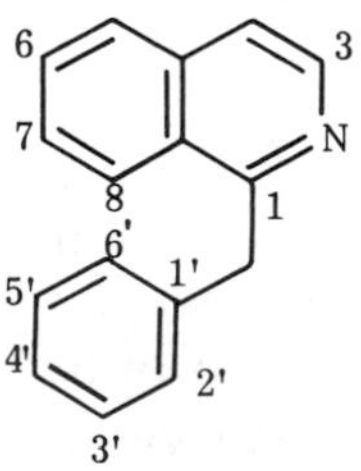

181 benzylisoquinoline nucleus

complex group of alkaloids. There are a quite amazing array of structure types, including simple benzylisoquinoline, bisbenzylisoquinoline dimers, proaporphines, aporphines, aporphine–benzylisoquinoline dimers, oxo-aporphines, protoberberines, benzophenanthridines, protopines, phthalide-isoquinolines, and hasubanane derivatives. Some biogenetic relationships of these alkaloid groups are shown in Scheme 26. In addition, there are a number

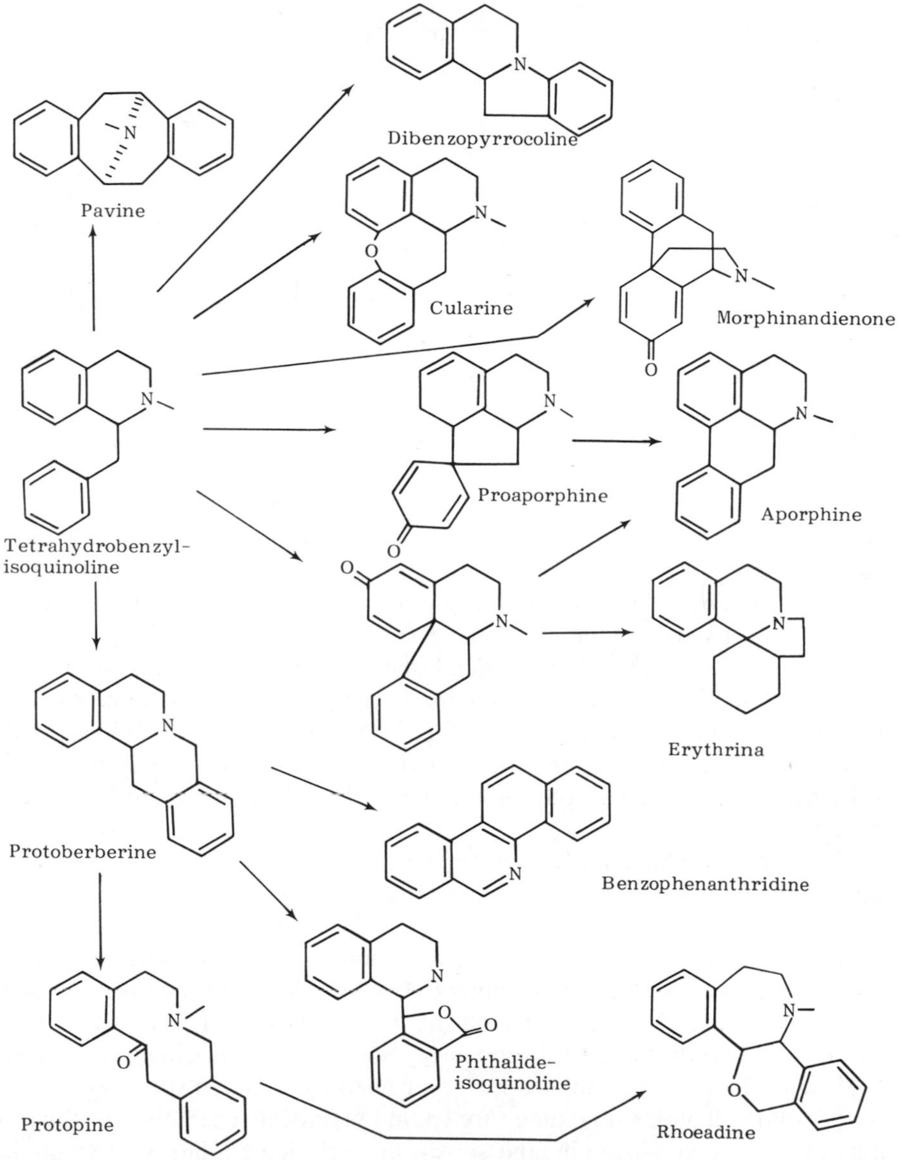

Scheme 26 Biogenetic relationships of the major alkaloid groups derived from a tetrahydrobenzylisoquinoline precursor.

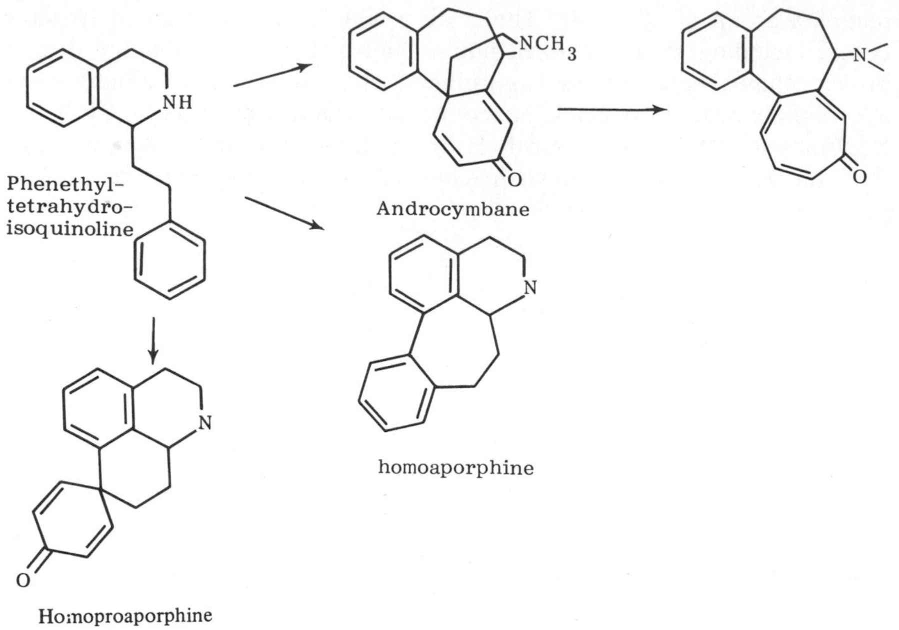

Phenethyl-
tetrahydro-
isoquinoline

Androcymbane

homoaporphine

Homoproaporphine

Scheme 27   Biogenetic relationships of the phenethyltetrahydroisoquinoline alkaloids.

of alkaloid skeleta derived from a γ-phenyl propionaldehyde precursor of the so-called phenethylisoquinolines. Some alkaloid types here include the homoaprophines, the homoproaprophines, and the *Colchicum* alkaloids. These relationships are shown in Scheme 27.

As they play such a central metabolic role, it is not surprising to find that these alkaloids have been obtained from over 13 plant families. The simple nature of the nucleus and the limited substitution possibilities mean that there are really very few benzylisoquinolines known, only about 35, although in many cases both enantiomers of a given structure have been obtained.

There are fundamentally two types of benzylisoquinoline, a 1,2,3,4-tetrahydro type such as (+)-reticuline (**182**), or a completely aromatic type such as papaverine (**183**). The typical oxygenation pattern of the benzylisoquinolines is shown in these compounds, although there are some compounds such as magnocurarine (**184**) which has only one oxygen substituent in ring C. The most unusual substitution is shown by petaline (**185**), and this will be discussed subsequently. Somewhat surprisingly no C-4 hydroxylated benzylisoquinolines have yet been isolated. In the whole of the isoquinoline alkaloids it is quite rare to find glycoside derivatives; a few are known in the benzylisoquinoline series, and latericine (**186**) is an example. It is not clear at present whether failure to isolate these derivatives is due to their absence in the plant or to the method of working up the plant material for alkaloids, which could possibly destroy any glycoside present.

Laudanosoline (**187**) was first isolated in 1871 from opium. Numerous reagents have been used for the O-demethylation of laudanosine (**188**), including 48% hydrobromic acid and aluminum chloride. Under mild conditions each of the four monophenolic isomers has been obtained. Although considerable information is often obtained from the NMR spectra of these

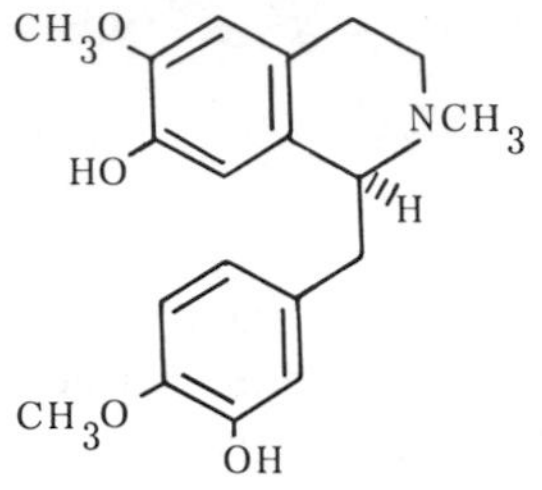

CH$_3$O

HO

NCH$_3$

///H

CH$_3$O

OH

182 (+)-reticuline

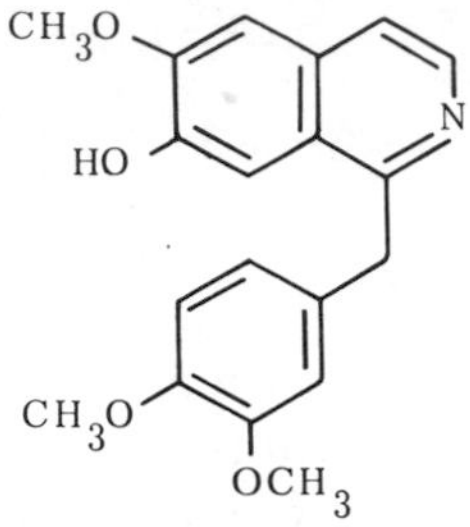

CH$_3$O

HO

N

CH$_3$O

OCH$_3$

183 papaverine

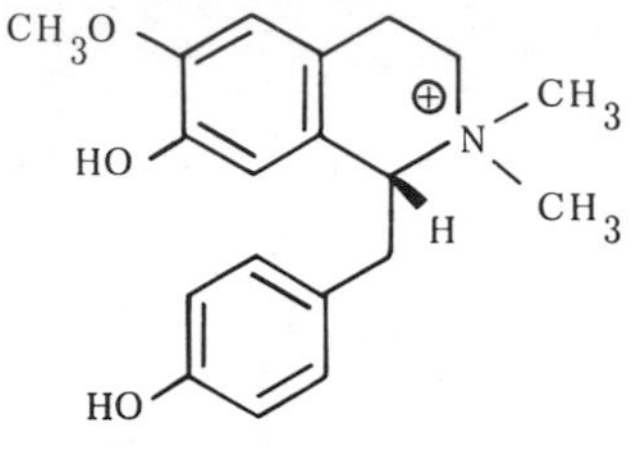

CH$_3$O

HO

⊕N

H

CH$_3$

CH$_3$

HO

184 (−)-magnocurarine

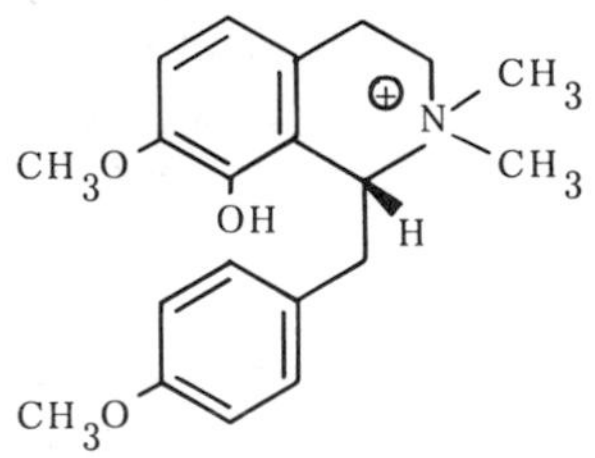

CH$_3$O

CH$_3$O

OH

⊕N

H

CH$_3$

CH$_3$

D-xylose

CH$_3$O

185 (−)-petaline

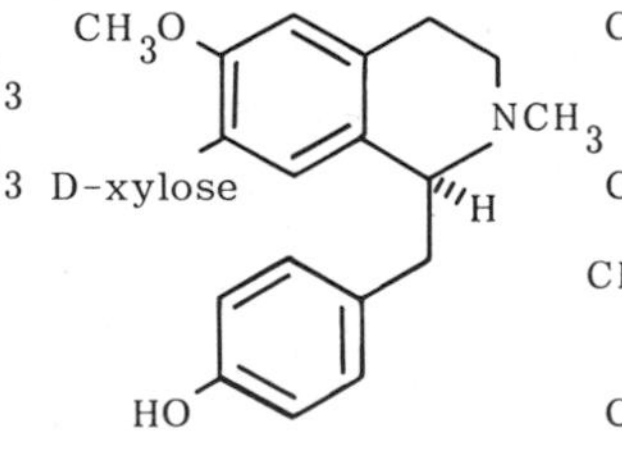

CH$_3$O

NCH$_3$

///H

HO

186 (+)-latericine

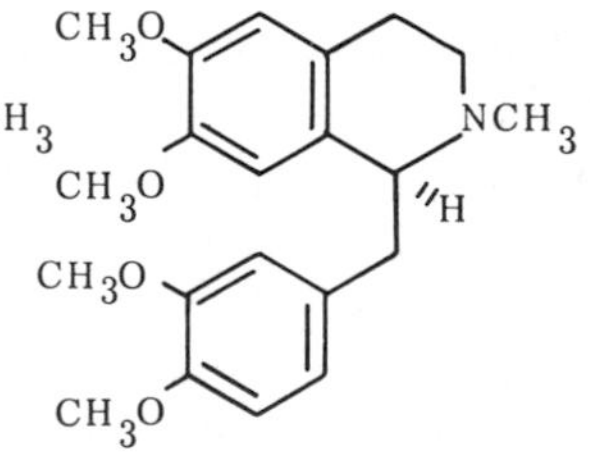

CH$_3$O

CH$_3$O

NCH$_3$

///H

CH$_3$O

CH$_3$O

188 (+)-laudanosine

HO

HO

NCH$_3$

///H

HO

HO

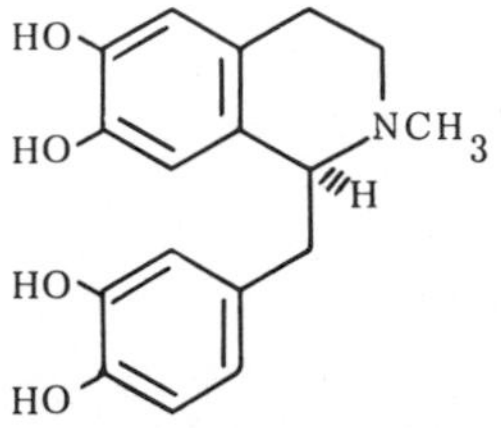

187 (+)-laudanosoline

alkaloids, a new quaternary aromatic alkaloid may be reduced to the tetrahydro species in order to clarify the aromatic region of the spectrum and facilitate determination of the substitution pattern.

Petaline (**185**) was obtained in 1964 from *Leontice leontopetalum* L. (Berberidaceae), a plant native to Lebanon. It is noted for two interesting features: (*a*) the substitution pattern of the A-ring and (*b*) the facile Hofmann degradation which occurs on passage through a basic ion exchange column. It is suggested that this easy reaction is due to an internal abstraction of the β-proton by the phenolate anion (Scheme 28). The only other major group of compounds having this substitution pattern are the cularine alkaloids.

Papaverine (**183**) occurs in *P. somniferum* L. to the extent of 0.8–1.0% and was first obtained by Merck in 1848. The free base is insoluble in water

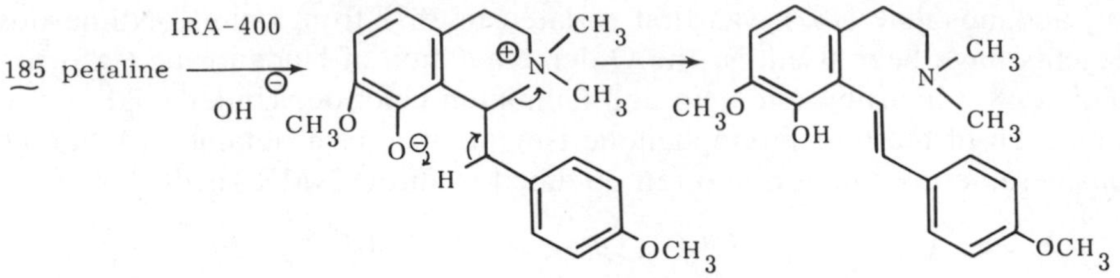

Scheme 28

and only sparingly soluble in ethanol. The quantity of papaverine available from natural sources is insufficient to meet demand and several synthetic routes have been described; these will be discussed subsequently.

The methylenedioxy group is a common feature not only of benzyliso-quinoline alkaloids but of isoquinoline alkaloids in general. A new method for the *selective* removal of this group in the presence of methoxyl groups involves treatment with boron trifluoride followed by etherification with 5-chloro-1-phenyl-1H-tetrazole and hydrogenolysis. Thus the overall reaction sequence converts a methylenedioxy group to a catechol (e.g. **189** to **190**).

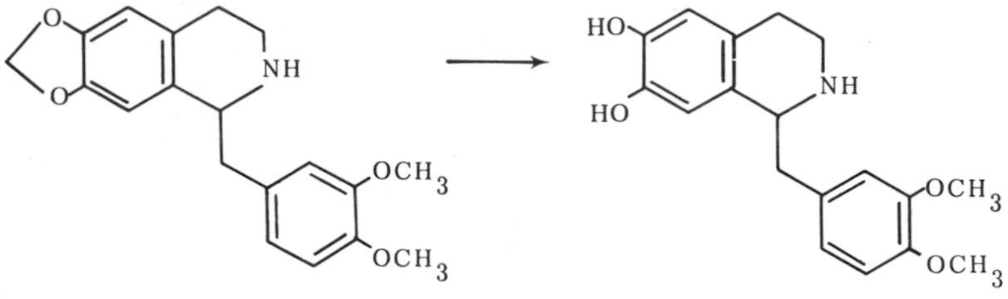

8.12.1 Chemistry

Benzylisoquinolines undergo a variety of interesting reactions, the most important of which both synthetically and biosynthetically is phenolic oxidative coupling, and this subject has been reviewed on several occasions.

It has been assumed that such coupling reactions involve the intermediacy of radicals, which may be visualized as having the electron on oxygen (ArO·) or on carbon (·ArO). Fundamentally two types of process can occur. One in which a phenolate radical couples with one aryl radical, (equation 1) with formation of an ether linkage. The second where two aryl radicals join together to form a new carbon–carbon bond (equation 2). In each case aromatization occurs to yield the final product.

$$ArOH \xrightarrow{-H^+} ArO^- \xrightarrow{-e} ArO· \rightarrow ·ArO$$

$$ArO· + ·ArO \rightarrow [ArOArO] \xrightarrow{aromatize} ArOArOH \qquad (1)$$

$$ArO· + ·ArO \rightarrow [OArArO] \xrightarrow{aromatize} HOArArOH \qquad (2)$$

In a simple system such as catechol monomethyl ether (**191**) we see that there is another problem which arises: namely in equations 1 and 2 does substitution occur *ortho* or *para* to the original phenol group? In other words there are now two available sites for coupling as shown in the radicals **193** and **194**. The various coupling possibilities for these radicals are shown in Figure 5.

In the isoquinoline alkaloids there are examples of all these processes. But this is not all; what if aromatization cannot take place because in the

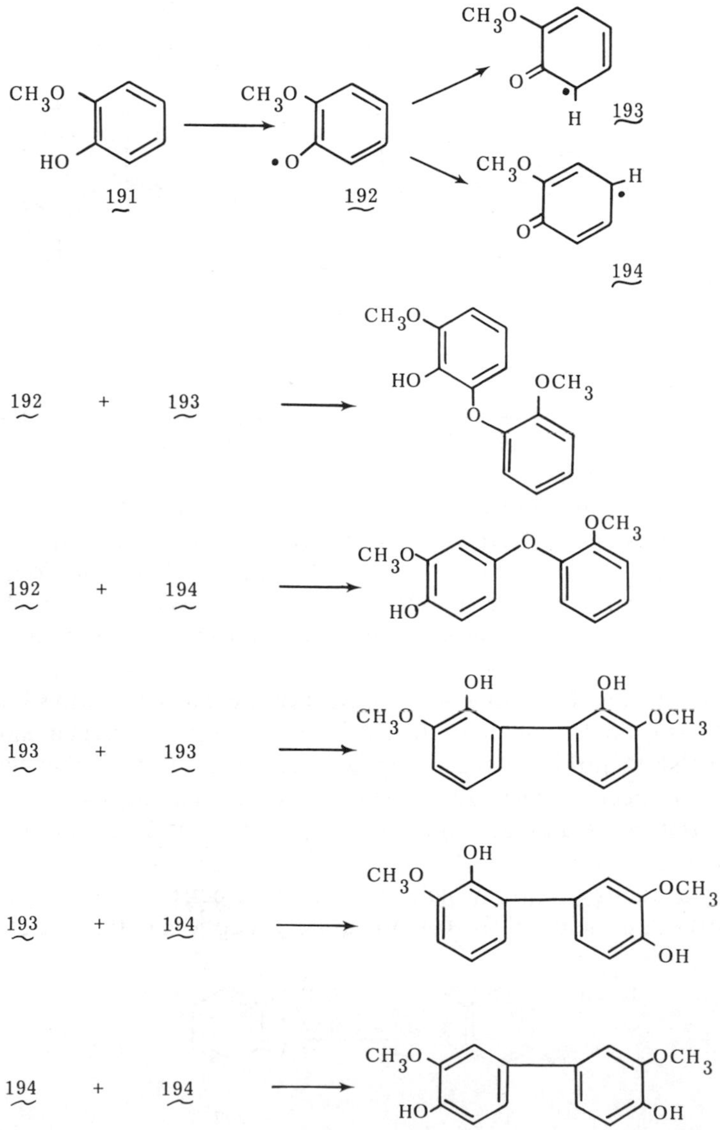

Figure 5

original compound there was no hydrogen *ortho* or *para* to the phonolic group? Consider for example the compound **195**, where R is an alkyl chain.

The coupling reactions of **196** pose no problems, but if **197** couples with **196** the reaction stops at the dienone stage of **198**. Several groups of important alkaloids are derived by this overall scheme. Collectively they are known as the morphinandienones and are discussed in Section 8.24.

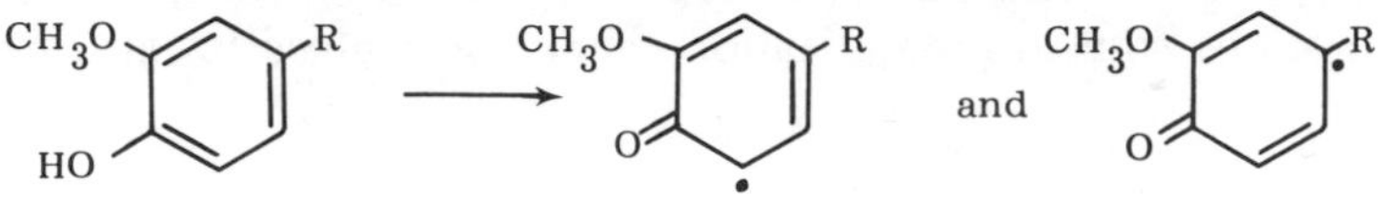

Before going on to show examples of these processes in the benzyliso-quinoline alkaloids it should be mentioned that there are other mechanisms which will also satisfactorily explain these processes.

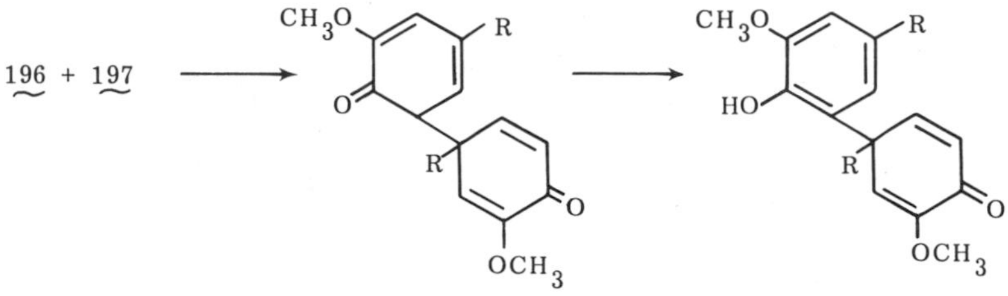

Evidence for radical processes, particularly *in vitro* when one-electron-oxidizing agents such as ferricyanide are used, is quite substantial. However, Dyke has brought forth again the possibility that aryloxonium ions rather than aryl radicals could be involved in these processes. Electrochemical studies have clearly demonstrated that two electrons can be removed from a phenol.

One problem with the total involvement of radicals has always been that a radical ArO should exist with the available electron on oxygen and therefore give rise to carbon–oxygen coupled products. On the other hand, an aryloxonium species such as **199** is expected to exist mainly in the carbonium ion form **200**. Such an electron-deficient species would be expected to be highly electrophilic.

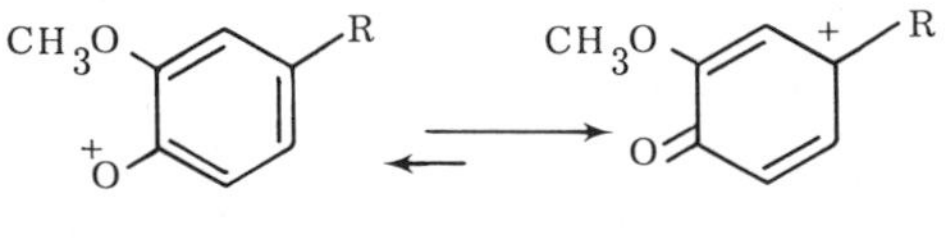

Most reactions *in vitro* which involve one-electron oxidation proceed very poorly and yields of 5–10% are quite exceptional. In contrast when two-electron-oxidizing agents such as thallium trifluoroacetate or vanadium oxychloride are used yields are in the 60–80% range.

Cytochrome P-450 is the enzyme system thought to be involved in this stage of alkaloid biosynthesis. It is highly complex, and may well be capable of "storing" successive one-electron oxidations, either as O_2 or as hydroperoxide.

Dyke has made some interesting speculations to account for some of the isolated natural products in these many classes of isoquinoline alkaloids which do not seem to fit into the "normal" biosynthetic pathway. A classic case is the biosynthesis of mescaline (3), where as discussed previously 4-hydroxy-3-methoxyphenethylamine (22) is a precursor, but not the isomer 3-hydroxy-4-methylphenethylamine (23). Dyke suggests that this is because only the 4-hydroxy isomer can produce an intermediate aryloxonium species 201 capable of nucleophilic attack at the 3-position.

Some of the interesting substitutions of alkaloids where oxygenation has occurred at the 5- or 8-position can be rationalized in a similar way depending on the proximity of a 6- or 7-phenolic group respectively as shown in Scheme 29. There is accumulating evidence that such hydroxylations are late bio-

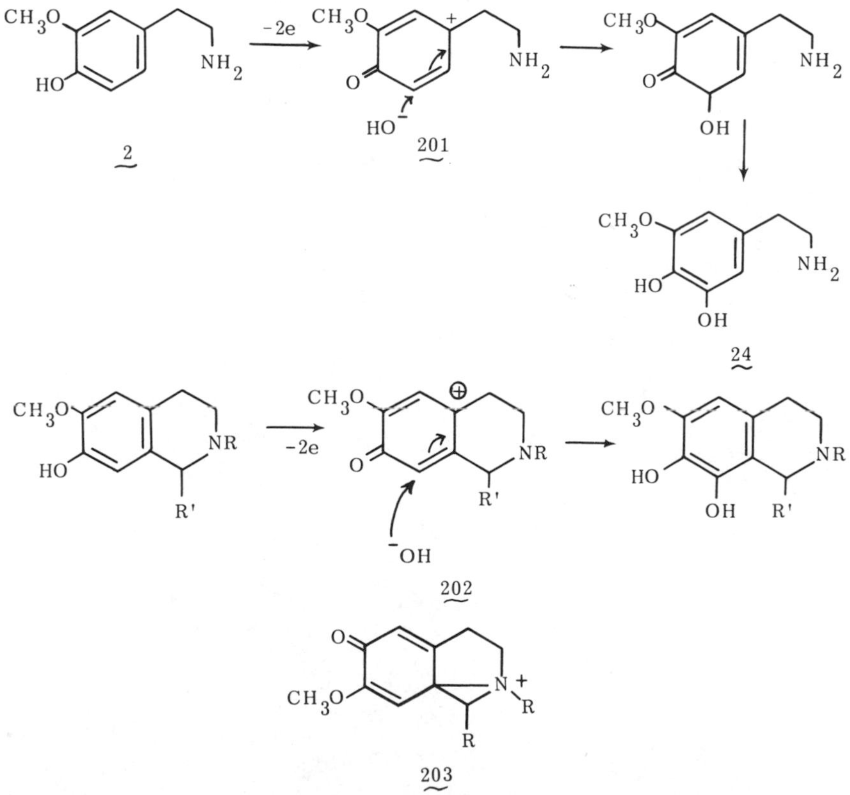

Scheme 29

synthetic stages. Is it possible that intermediates such as **201** and **202** are stabilized by anchimeric assistance from the basic nitrogen (e.g. **203**)? At present there are no answers to this and the many other related questions in this whole mechanistic aspect of *in vitro* phenolic oxidative coupling. For such a fundamental process it is very poorly understood.

Oxidation

Oxidation of a 1,2,3,4-tetrahydrobenzylisoquinoline such as laudanosine (**188**) proceeds effectively with mercuric acetate and EDTA to give a mixture of the *N*-methylpapaverine salt (**204**) and 3,4-dihydropapaverine (**205**). Permanganate oxidation of benzylisoquinolines may lead to cleavage of the benzyl moiety (e.g. **206** to **207**).

Dichromate, permanganate, or selenium dioxide converts papaverine (**183**) to papaveraldine (**208**).

188 204 205

206 207

183 208

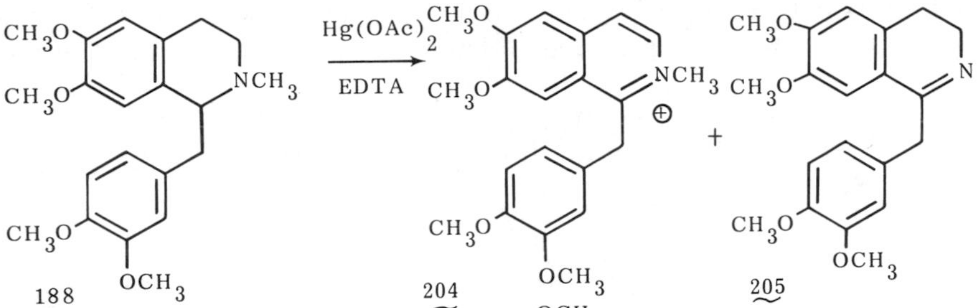

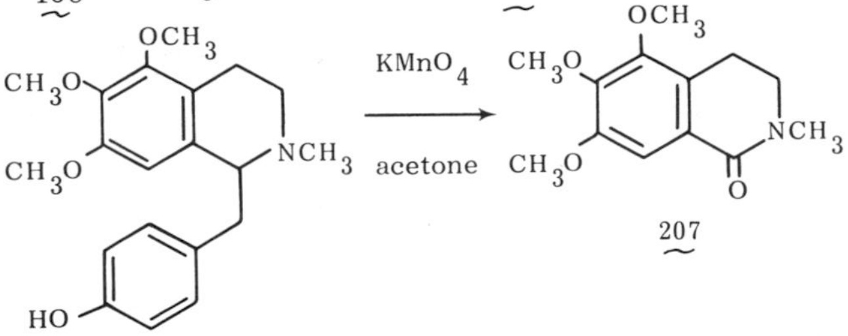

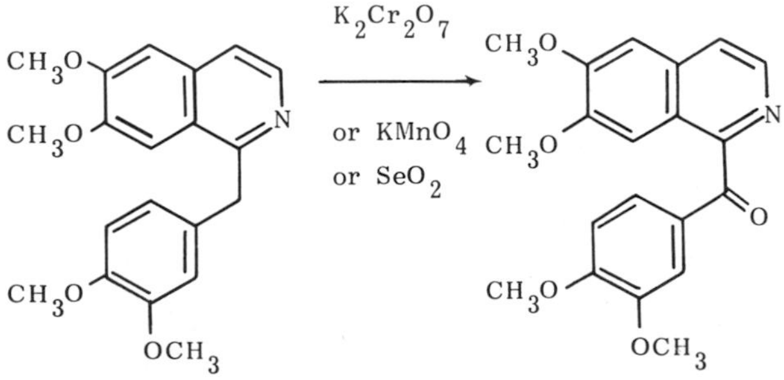

Reduction

Catalytic reduction of papaverine (**183**) afforded *N*-norlaudanosine (**209**) and sodium borohydride reduction of *N*-methyl papaverine (**204**) gave laudanosine (**188**). With other reagents more complex products result. One reaction of interest is the lithium aluminum hydride reduction of *N*-methyl papaverine (**204**) to give the 1,2-dihydroisoquinoline (**210**).

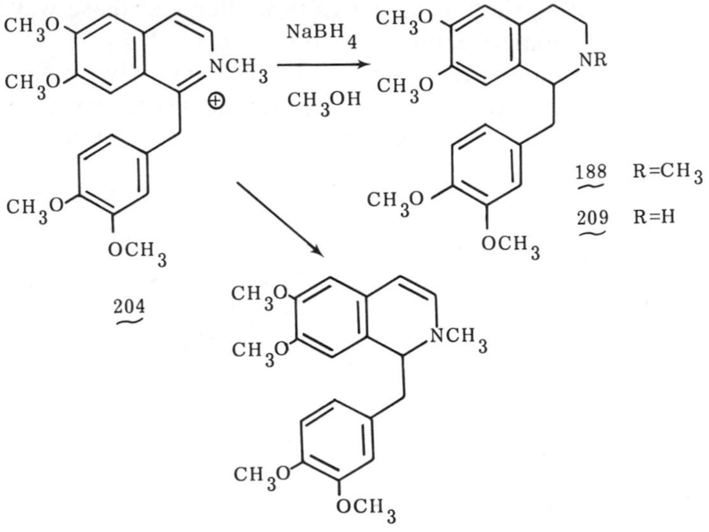

188   R=CH₃

209   R=H

204

210

Treatment of papaverine (**183**) in acetic acid–hydrochloric acid with formaldehyde followed by heating with formic acid–formamide yields the tetrahydroprotoberberine norcoralydine (**211**).

One of the intriguing reactions of optically active tetrahydrobenzylisoquinolines is that they are racemized by Adam's catalyst in ethanol at room temperature. The presence of a phenolic group inhibits the reaction to some extent, and there are instances where a catechol could not be racemized.

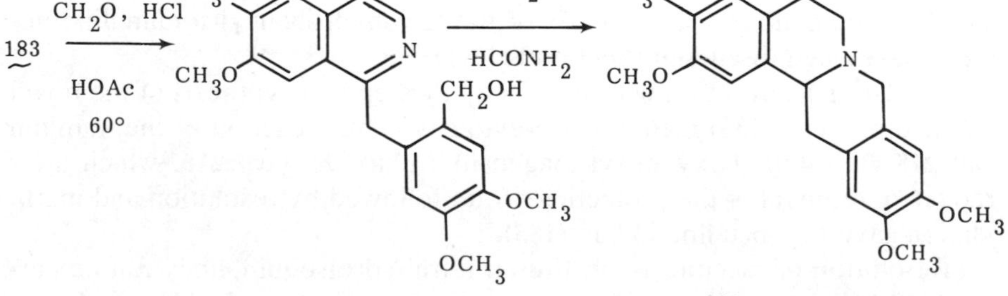

211   norcoralydine

8.12.2   Absolute Configuration

The absolute configuration of the tetrahydroisoquinolines was determined
in 1956 by Corrodi and Hardegger, when (−)-N-norlaudanosine (**209**) was
ozonized and oxidized to N-β-carboxyethyl-L-aspartic acid (**212**).

The best correlation with other derivatives is made by the sign of the
Cotton effect between 270 and 290 nm and between 240 and 225 nm. Compounds in the D-series show negative Cotton effects at these wavelengths,
whereas the L-enantiomers exhibit positive effects.

However, even though the compound may have only one asymmetric
center, optical rotation per se is not an appropriate discerning measurement,
as the situation with D-(−)-armepavine (**213**) and L-(−)-N-norarmepavine
(**214**) illustrates.

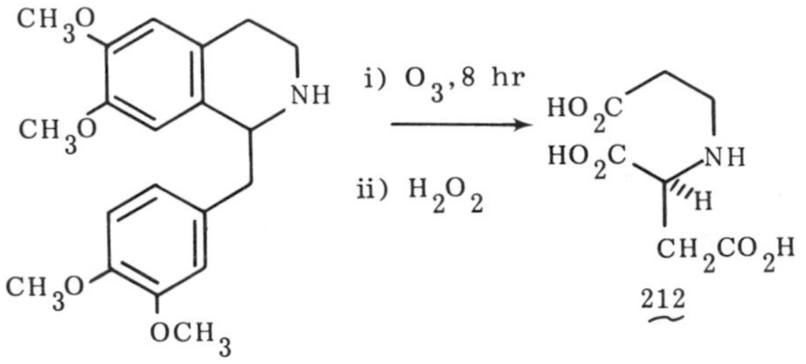

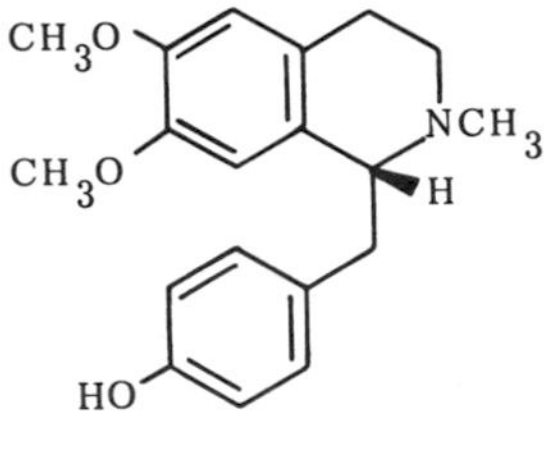

213   D-(-)-armepavine 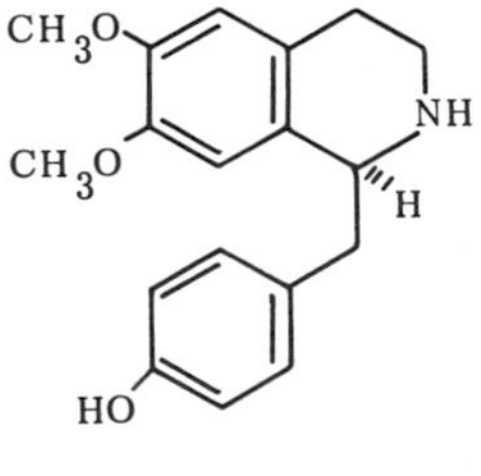   214   L-(-)-N-nor-
                                      armepavine

8.12.3   Synthesis

The fundamentals of the synthesis of the tetrahydrobenzylisoquinoline nucleus have been discussed (Section 8.10.1).

The Grignard reaction was used as a key step in a synthesis of the novel
alkaloid petaline (**185**) mentioned previously. Thus reaction of the iminium
salt **215** with 4-methoxybenzyl magnesium chloride gave **216**, which after
reductive removal of the protecting group followed by resolution and methylation gave (−)-petaline iodide (**185**).

Resolution of racemic 1-substituted tetrahydroisoquinolines can be carried out with several reagents, including (+)-tartaric acid, dibenzoyl-(+)-
tartaric acid, (+)-camphor-10-sulfonic acid, and (−)-quinic acid.

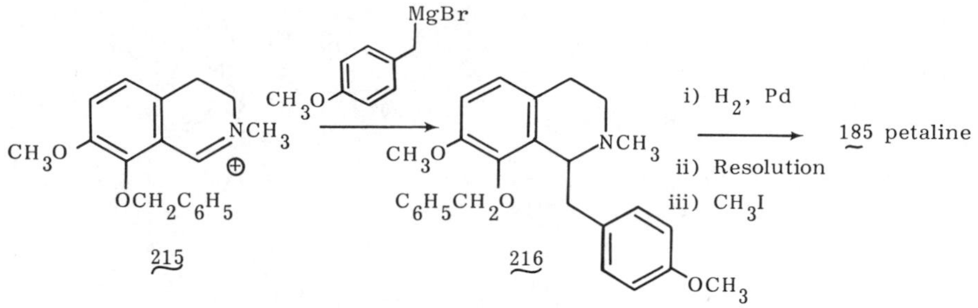

The Bischler–Napieralski cyclization is the most widely used method for the synthesis of the benzylisoquinoline system. Fundamentally, the reaction involves phosphorus oxychloride cyclization of an amide to give a 3,4-dihydrobenzylisoquinoline which can be reduced. Teitel and Brossi have considerably shortened the standard reaction sequence by using acetonitrile in the dehydroation step and catalytic reduction in the formation of the phenethylamine. A typical procedure is shown for the synthesis of coclaurine (**217**) (Scheme 30).

One of the methods of choice for the synthesis of papaverine (**183**) is a modification of the Bischler–Napieralski approach. Under basic conditions the nitrostyrene **218** adds methanol to give **219**. Reduction and acylation affords the amide **220**, which on cyclization with phosphorous pentoxide gives papaverine (**183**) directly (Scheme 31).

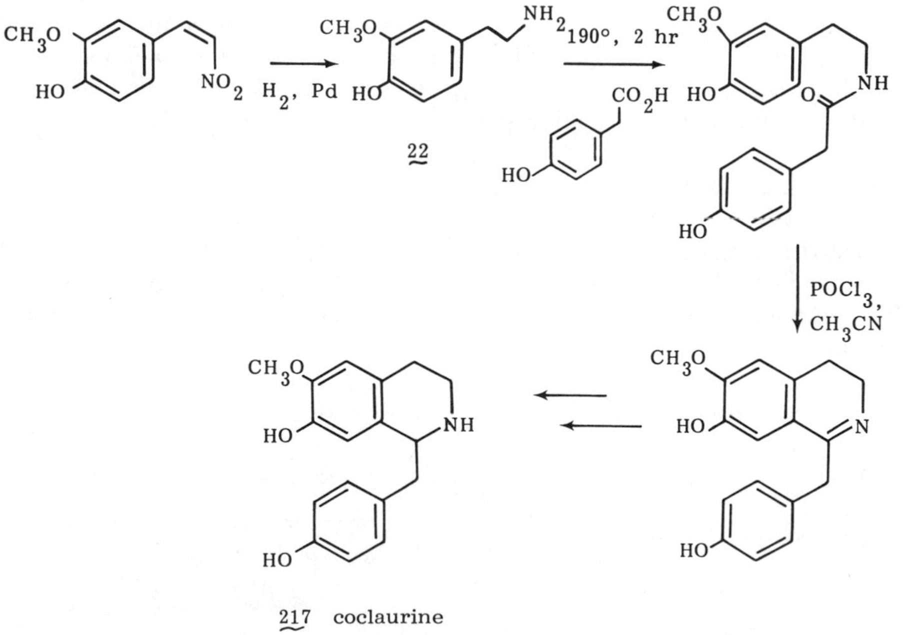

217   coclaurine

Scheme 30   Synthesis of coclaurine (**217**).

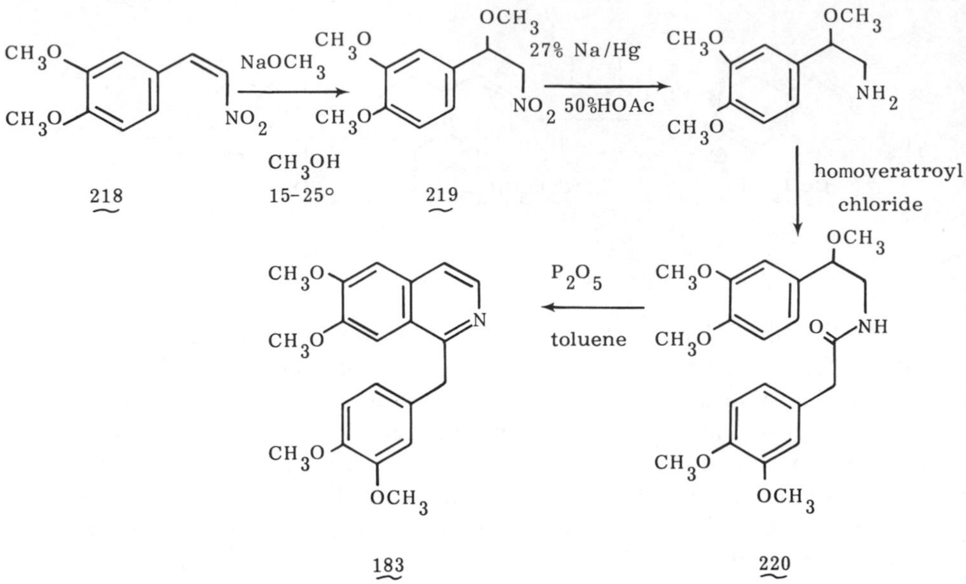

Scheme 31 Brossi–Teitel synthesis of papaverine (**183**).

The Pictet–Spengler condensation is a close approximation to the biosynthesis of this skeleton and was first carried out by Späth and Berger in 1930 as shown in Scheme 32 to give *N*-norlaudanosine (**209**). The reaction is particularly effective when a free phenolic group is *para* to the site of cyclization. An efficient procedure for the synthesis of papaverine (**183**) involves the reaction of the Wittig reagent **221** with 6,7-dimethoxy-1-chloroisoquinoline (**222**) to give **183** directly (Scheme 33).

8.12.4 Spectral Properties

Tetrahydrobenzylisoquinolines show a maximum between 280 and 285 nm which is little affected by additional aromatic substitution. Compounds having methylenedioxy groups show an increased and more intense absorption maximum (e.g. **223**).

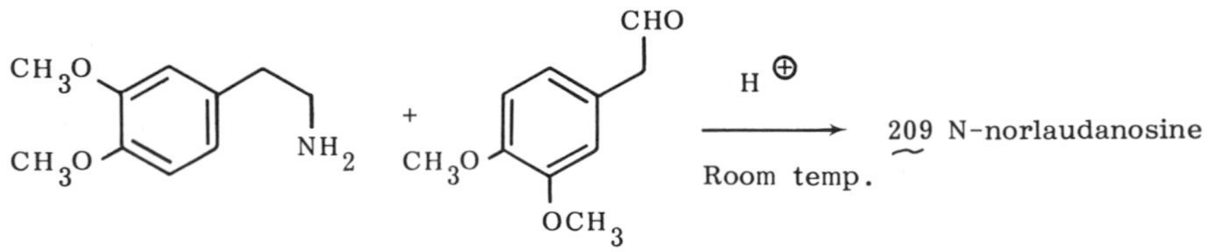

Scheme 32

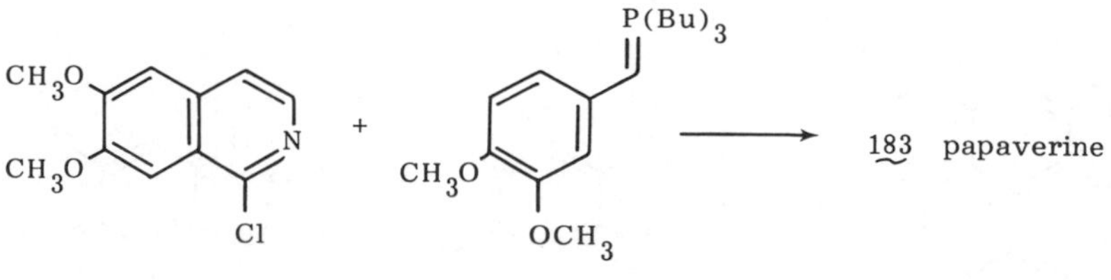

Scheme 33 Taylor–Martin synthesis of papaverine (**183**).

Papaverine shows λ_{max} at 239, 279, 314, and 317 (4.83, 3.86, 3.60, and 3.67), shifted in acid to 250, 280, and 311 nm (4.69, 3.80, and 3.82). The *N*-methyl salts have similar spectra.

The NMR spectra of tetrahydrobenzylisoquinolines show a number of interesting features due to the one asymmetric center. The main effect is analogous to a benzene-induced solvent shift for the C-ring lies over the region of carbons 7 and 8. This results in a shielding effect for both the C-8 proton and any C-7 methoxyl group [e.g. *O*-methyl armepavine (**224**)]. When the C-7 substituent is hydroxyl no shielding of the C-8 proton occurs.

An interesting observation made by Shamma *et al.* concerns the effect of the chemical shift of methoxyl groups on by changing solvents from deuteriochloroform to d_6-DMSO. The solvent shifts are in the order $C_8—CH_3O > C_7—CH_3O > C_4—CH_3O$.

Typical data for the benzylisoquinoline system are shown for 1-(4-methoxybenzyl)-6,7-dimethoxyisoquinoline (**225**).

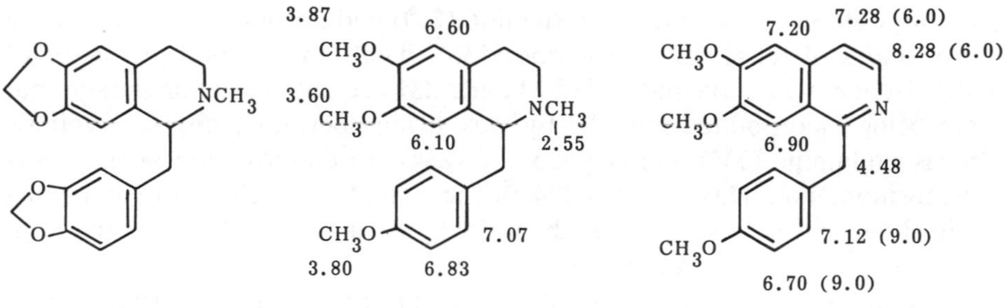

λmax 290nm (4.1)

223 224 O-methylarmepavine 225

In an electron beam the major fragmentation pathway, as expected, is benzylic cleavage; indeed in some instances this process is so facile that the molecular ion is very weak. The charge normally resides on nitrogen so that observed ions are principally due to the A and B rings, (e.g. **226** and **227**).

226   m/e 178   227   m/e 163

8.12.5 Biosynthesis

In subsequent sections there will be considerable discussion of the formation and fate of several important benzylisoquinolines, including reticuline, norlaudanosoline, and orientaline. At this point we will discuss only the formation of papaverine (183) in *P. somniferum*, and of reticuline (182).

In 1910, Winterstein and Trier suggested that the benzylisoquinolines were derived from two units of dopa, one of which was transformed into 3,4-dihydroxyphenylacetaldehyde (168) and the second into dopamine (10). Pictet–Spengler condensation would then afford norlaudanosoline (228).

[2-^{14}C]Tyrosine specifically labeled C-1 and C-3 of papaverine (183) in *P. somniferum* as shown, and further studies established that norlaudanosoline (228) was also a precursor. The question then arises: Is there a defined sequence of O-methylation and aromatization or can both of these processes occur at the same time?

It is only quite recently that the late stages in papaverine (183) biosynthesis were investigated using partially methylated benzyltetrahydroisoquinolines. All four isomers (229–232) were incorporated into papaverine without loss of label, but only nor-reticuline (229) and nororientaline (230) were precursors of tetrahydropapaverine (233), itself a very good precursor of 183. Hence it was concluded that 232 and 231 are not normal precursors but are being incorporated into 183 by way of an aberrant pathway involving norisocodamine (235) and isopacodine (234). In confirmation of this, norprotosinomenine (231) labeled 234 to the extent of 12.2%. In experiments with (−)- and (+)-isomers of 229 and 233 it was found that only the (−)-isomers were precursors of 183.

Complete O-methylation is not a necessary prelude to dehydrogenation, for 236 was well (15.6%) incorporated into palaudine (237).

The main route to papaverine (183) in *P. somniferum* is therefore from (−)-norlaudanosoline (228) via (−)-nor-reticuline (229) or (−)-nororientaline (230) to (−)-norlaudanidine (236) or (−)-norcodamine (238) to (−)-tetrahydropapaverine (233) and thence 183 (Scheme 34).

Kapil and co-workers reached similar conclusions concerning papaverine biosynthesis. (±)-Norlaudanosine (209), (−)-nor-reticuline (229), (±)-nor-

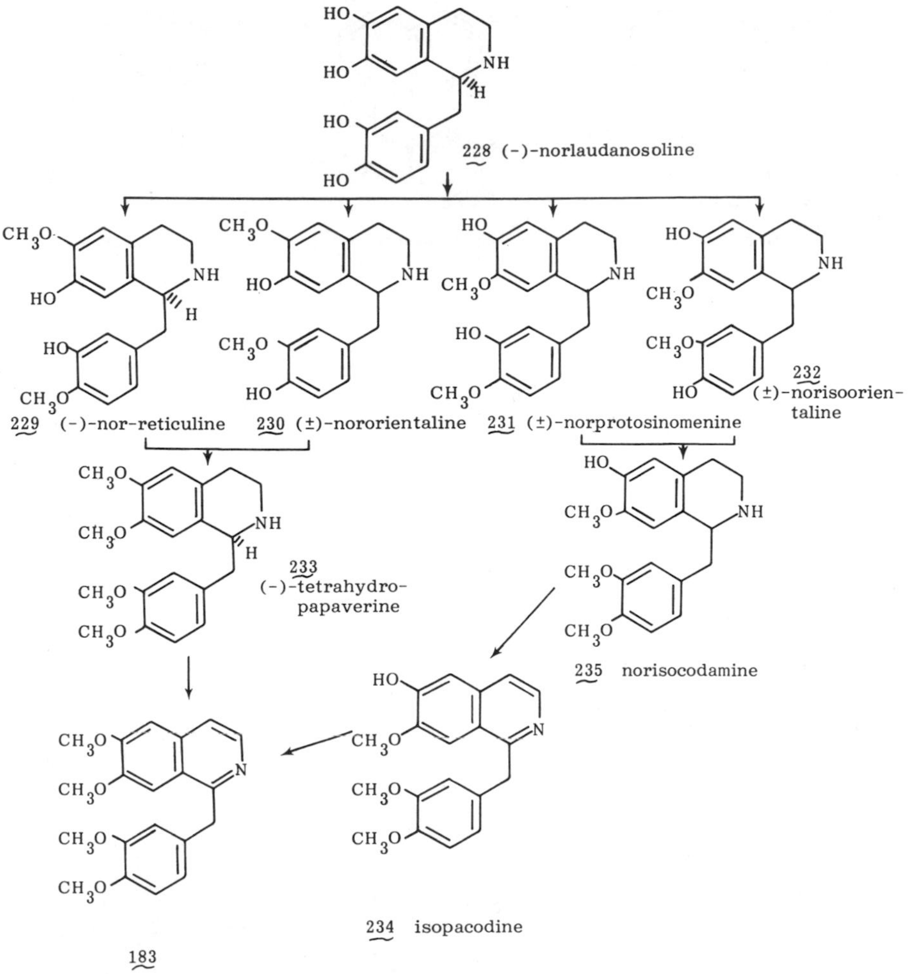

228 (−)-norlaudanosoline

229 (−)-nor-reticuline   230 (±)-nororientaline   231 (±)-norprotosinomenine   232 (±)-norisoorientaline

233 (−)-tetrahydro-papaverine

235 norisocodamine

234 isopacodine

183

Scheme 34 Biosynthesis of papaverine (183) in *P. somniferum*.

laudanidine (236), and (±)-norcodamine (238) were incorporated into 183. (+)-Nor-reticuline (239) was not incorporated. The tenfold difference in incorporation between 237 and 238 suggested that the more important major route involves 237 rather than 238.

Reticuline is regarded as the key intermediate in the biosynthesis of the many alkaloids based on the benzylisoquinoline nucleus, and as a result of the study of these alkaloids much has been learned of the biosynthesis of reticuline. Work by several groups has established that crucial in the overall route is the involvement of norlaudanosoline l-carboxylic acid (240) in the scheme. [1-^{14}C]Dopa (21) was not a precursor of norlaudanosoline (228) but

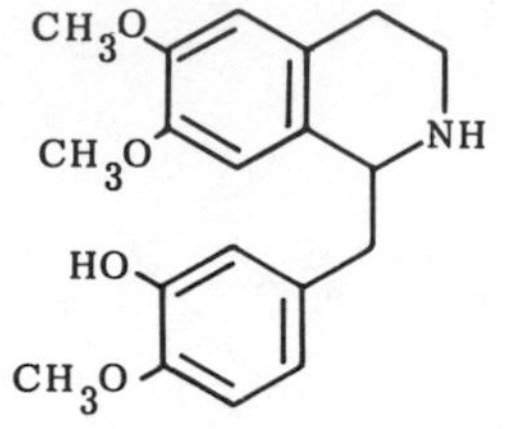

236 (±)-norlaudanidine

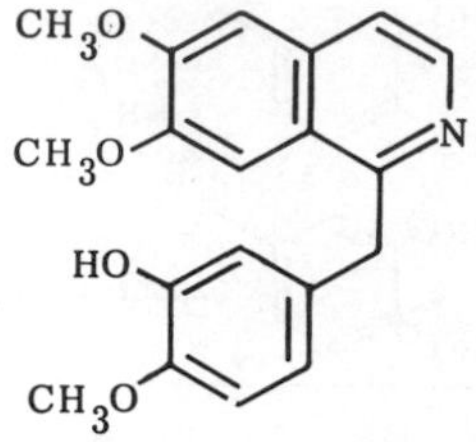

237 palaudine

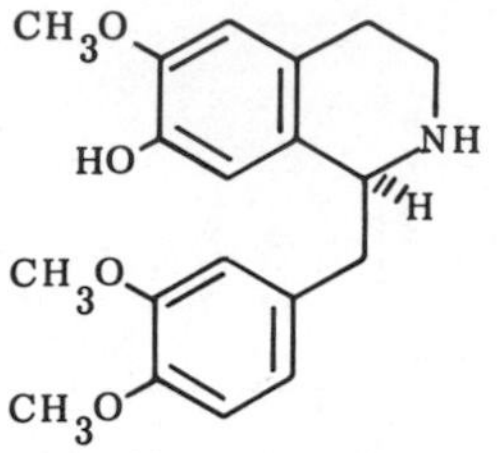

238 norcodamine

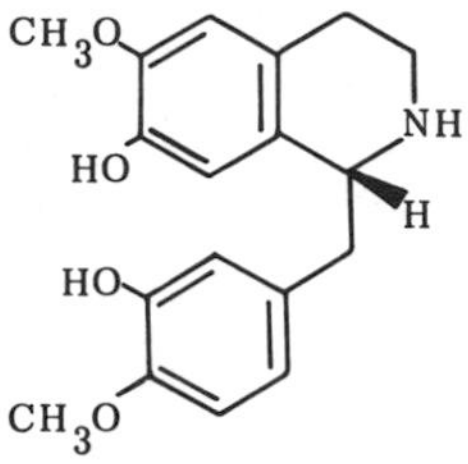

239 (+)-nor-reticuline

was incorporated into **240**, indicating that this compound is decarboxylated on incorporation. The two units which are envisaged to join are therefore dopamine (**10**) and 3,4-dihydroxyphenylpyruvic acid (**241**) (Scheme 35).

Kapil and co-workers have separately examined the formation of reticuline (**182**) in *Litsea glutinosa*. Radioactive tyrosine was a precursor of

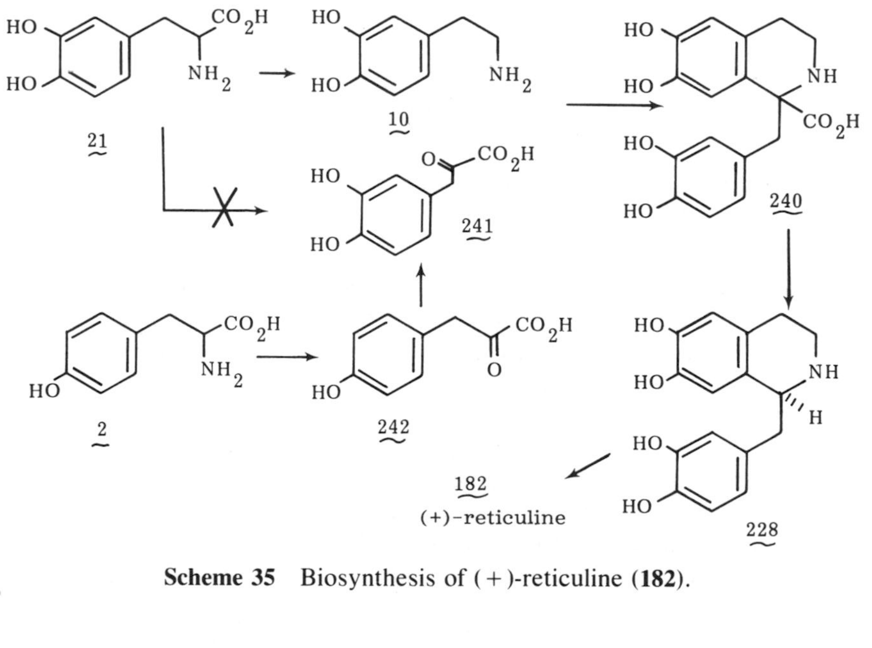

Scheme 35 Biosynthesis of (+)-reticuline (**182**).

both C_6-C_2 units, but [2-^{14}C]dopa labeled only C-3 of reticuline (182). These results agree with data on the biosynthesis of morphine in *Papaver orientale* and on aporphine biosynthesis in *Dicentra eximia*. They imply that dopa is not a precursor of 3,4-dihydroxyphenylpyruvic acid, itself the precursor of the lower C_6-C_2 unit. Rather, it is considered that 4-hydroxyphenylpyruvic acid (242), produced by oxidative deamination of tyrosine, is subsequently hydroxylated to afford 241.

Examination of the relative levels of incorporation of various O- and N-methyl benzylisoquinolines into reticuline (182) has implicated that O-methylation precedes N-methylation.

8.12.6 Pharmacology

The most important benzylisoquinoline alkaloid from a pharmacologic point of view is papaverine (183). In 1914, Pal demonstrated *in vivo* that papaverine decreases the tonus of the smooth muscle and this has stimulated numerous studies of both a synthetic and pharmacologic nature.

In humans, papaverine is completely absorbed in the gastrointestinal tract and is metabolized in the liver to 4'-norpapaverine, which is excreted as the glucuronate.

Papaverine increases coronary artery flow and causes dilation. In angina pectoris papaverine has a beneficial effect but does not alleviate the pain. The hydrochloride has been used in the treatment of vasospasms accompanying pulmonary embolism and cerebrovascular thrombosis. The glyoxylate salt has spasmolytic and vasodilating properties. Papaverine exerts a strong spasmolytic action on the normal and the pregnant uterus.

Numerous studies have confirmed the peripheral vasodilating effects of papaverine, and the 3-methyl analogue 243, known as dioxyline, is used as a coronary and peripheral vasodilator. The corresponding methylenedioxy analogue 244 is a smooth muscle relaxant.

Sulfonamides potentiate the effects of papaverine, and papaverine increases the inhibition of oxygen uptake by sulfonamides in the liver.

Papaverine is neither narcotic nor addictive, and side effects include drowsiness and constipation.

Laudanosine (188) reduces intraocular pressure in rabbits when administered i.v.

Laudanine (245) had effects similar to strychnine in frogs; in low doses it caused convulsions and larger doses caused paralysis.

Demethylcoclaurine (246) was identified as the main cardiac principle of aconite root, *Aconitum japonicum* Thumb. The (−)-isomer was found to be far more potent as a cardiac stimulant than the (+)-isomer on the frog heart. It is however a highly labile compound, as might be expected.

The glycoside veronamine (247) produces systemic hypotension and bradycardia.

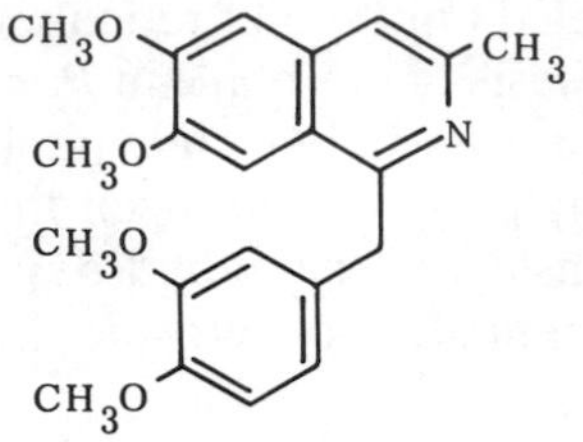

243   dioxyline

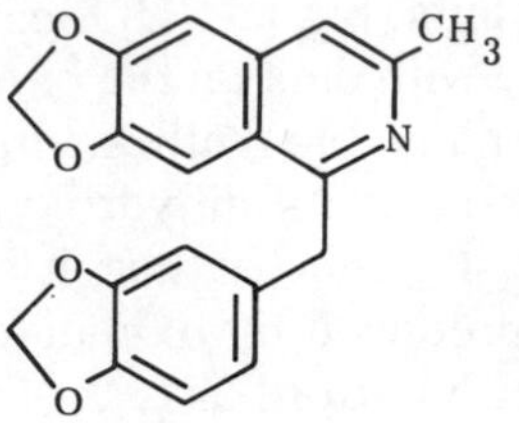

244

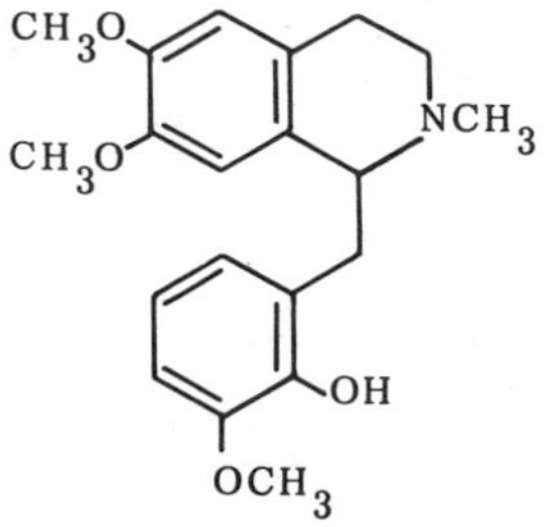

245   laudanine

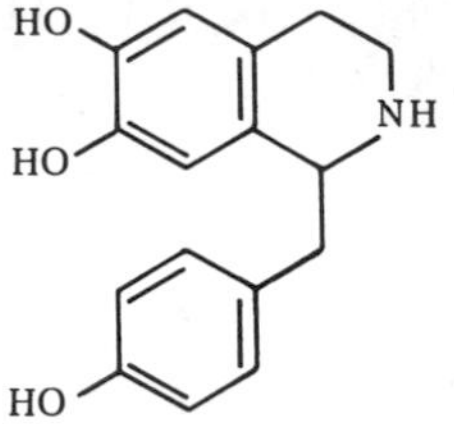

246   demethylcoclaurine

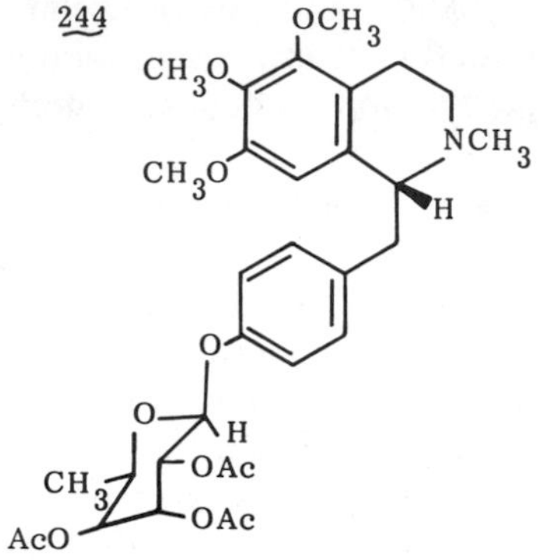

247   (−)-Veronamine

LITERATURE

Reviews

Burger, A., *Alkaloids NY* **4**, 29 (1954).

Deulofeu, V., J. Comin, and M. J. Vernengo, *Alkaloids NY* **10**, 402 (1068).

Santavý, F., *Alkaloids NY* **12**, 333 (1970).

Kametani, T., and K. Fukumoto, in K. F. Weisner (Ed.), *MTP International Review of Science, Organic Chemistry*, Ser. 1, Vol. 9, *Alkaloids*, Butterworths, London, 1973, p. 181.

Oxidative Phenol Coupling

Barton, D. H. R., and T. Cohen, *Festschrift A. Stoll*, Birkhauser, Basel, 1957, p. 117.

Battersby, A. R., *Proc. Chem. Soc.* 189 (1963).

Barton, D. H. R., *Proc. Chem. Soc.* 239 (1963).

Scott, A. I., *Quart. Rev.* **19**, 1 (1965).

Taylor, W. I., and A. R. Battersby (Eds.), *Oxidative Coupling of Phenols*, Marcel Dekker, New York, 1967.

Kametani, T., and K. Fukumoto, *Synthesis* 657 (1972).

Kametani, T., and K. Fukumoto, *Heterocycles* **1**, 129 (1977).

Dyke, S. F., *Heterocycles* **6**, 1441 (1977).

Synthesis

Kametani, T., and K. Fukumoto, in K. F. Weisner (Ed.), *MTP International Review of Science, Organic Chemistry*, Ser. 1, Vol. 9, *Alkaloids*, Butterworths, London, 1973, p. 181.

Grethe, G., M. Uskoković, and A. Brossi, *J. Org. Chem.* **33**, 3500 (1968).

Teitel, S., and A. Brossi, *J. Het. Chem.* **5**, 825 (1968).

Späth, E., and F. Berger, *Chem. Ber.* **63**, 2098 (1930).

Brossi, A., and S. Teitel, *Helv. Chim. Acta* **49**, 1757 (1966).

Taylor, E. C., and S. F. Martin, *J. Amer. Chem. Soc.* **94**, 2874 (1972).

Biosynthesis

Winterstein, E., and E. Trier, *Die Alkaloide*, Borntrager, Berlin, 1910, p. 307.

Battersby, A. R., and B. J. T. Harper, *J. Chem. Soc.* 3526 (1962).

Battersby, A. R., R. Binks, R. J. Francis, D. J. McCaldin, and H. Ramuz, *J. Chem. Soc.* 3600 (1964).

Brochmann-Hanssen, E., C. Chen, C. R. Chen, H. Chang, A. Y. Leung, and K. McMurtey, *J. Chem. Soc. Perkin Trans. I* 1531 (1975).

Uprety, H., D. S. Bhakuni, and R. S. Kapil, *Phytochemistry* **14**, 1535 (1975).

Reticuline Biosynthesis

Wilson, M. L., and C. J. Coscia, *J. Amer. Chem. Soc.* **97**, 431 (1975).

Battersby, A. R., R. C. F. Jones, and R. Kazlauskas, *Tetrahedron Lett.* 1873 (1975).

Tewari, S., D. S. Bhakuni, and R. S. Kapil, *Chem. Commun.* 554 (1975).

Pharmacology

Burger, A., *Alkaloids NY*, **4**, 29 (1954).

Pleininger, V., *Alkaloids NY*, **15**, 209 (1975).

Shamma, M., *The Isoquinoline Alkaloids*, Academic, New York, 1972, p. 77.

8.13 BISBENZYLISOQUINOLINE ALKALOIDS

Over 140 benzylisoquinoline alkaloids are known and this group has been reviewed on a regular basis. Shamma and Moniot have classified this group into 26 structure types and further mention of this classification will be made shortly.

As with the benzylisoquinoline group, aromatic substituents may be hydroxyl, methoxyl, or methylenedioxy. Linkages may be of the diphenyl or diphenyl ether type and a given compound may contain one or more of each of these linkages. A typical bisbenzylisoquinoline contains two asymmetric centers, but sometimes one or both nitrogen atoms are involved in imino groups.

Isolation of these compounds has been reported from 10 plant families, of which the Menispermaceae and Ranunculaceae are the most important sources.

The need for a systematic classification of the bisbenzylisoquinolines was recognized and the first rational attempt at this was made by Shamma in his monograph. Subsequently an extended system was proposed. This revised system is worthy of special mention because of its simplicity yet applicability.

Fundamentally, there are two sets of numbers corresponding to the sites of oxygenation, separated by a hyphen. The more highly oxygenated half of the dimer constitutes the left half of the dimer and is listed first. An asterisk (*) or other symbol (+ or †) on the upper right of a number indicates the diarylether terminals. Numbers in parentheses are used to indicate aryl–aryl bonds. Square brackets are used to indicate the terminals of a methylene dioxy bridge. Representative examples of these structure types are shown in Table 5.

8.13.1 Structure Determination

Crucial in the structure determination of a bisbenzylisoquinoline alkaloid is the result of sodium–liquid ammonia cleavage of the diaryl ether linkages. The use of this reaction is based on the work of Sartoretto and Sowa in 1937, and although the percentages may not now be accurate, the basic data are summarized in Figure 6.

When an *o*-methoxyl group is present, reductive cleavage occurs *ortho* to this methoxyl group, but of greater significance is the knowledge that cleavage of the bond between the methoxylated ring and the oxygen of the

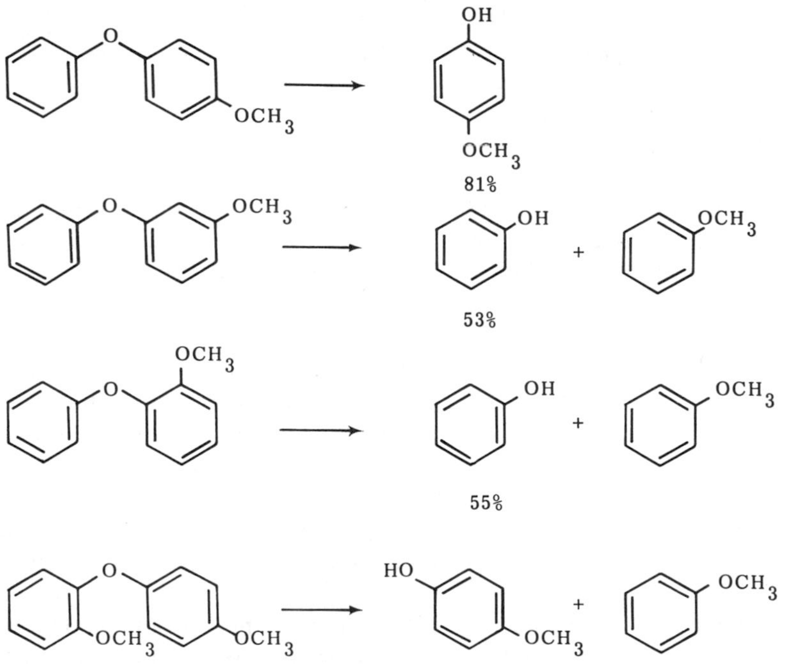

Figure 6

diphenyl ether occurs in the order:

$$o\text{-}CH_3O\text{---} > m\text{-}CH_3O\text{---} > p\text{-}CH_3O\text{---}$$

Tomita and co-workers were the first to use this reaction on a bisben-zylisoquinoline. Classically, when a phenolic group is present it is ethylated rather than methylated in order to define its position in the originial product. In most cases the mixtures obtained in these reactions are extremely complex, and interpretation of the products rarely gives a unique structure. An example occurs in the sodium–liquid ammonia cleavage of thalidasine to afford (+)-O-methylarmepavine (**224**) and a diphenolic dimethoxybenzyl-isoquinoline **248** which on methylation gave **249**. On this basis two structures **250** and **251** can be written for thalidasine. Only the further isolation of (+)-armepavine (**252**) and **253** as minor products of the reduction serve to distinguish the two possible structures. How can the structure **254** be eliminated?

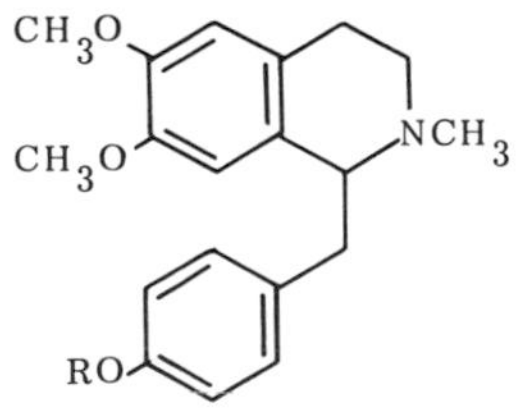

224   O-methylarmepavine
      R=CH₃
252   armepavine, R=H

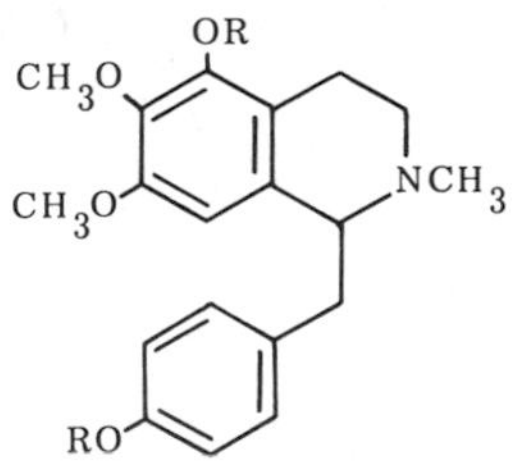

248   R=H
249   R=CH₃

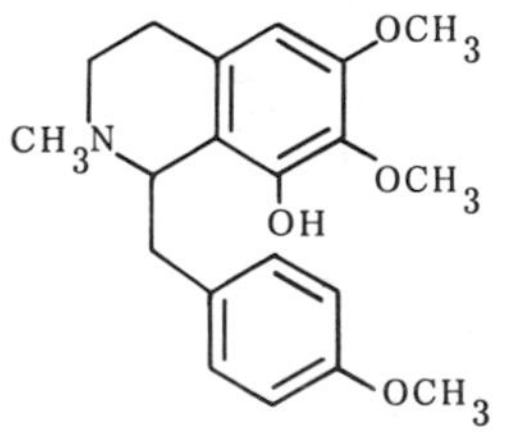

253

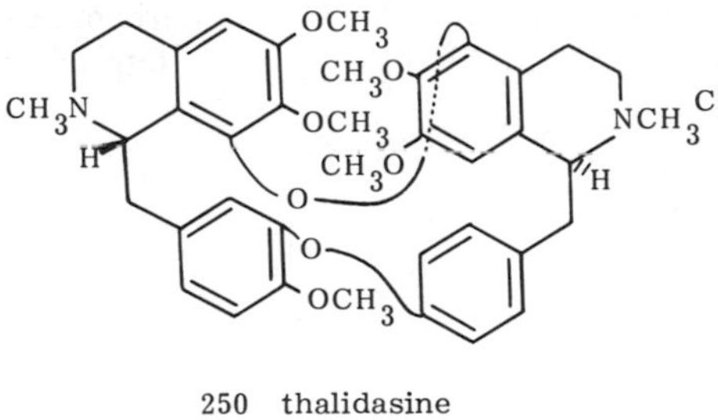

250   thalidasine

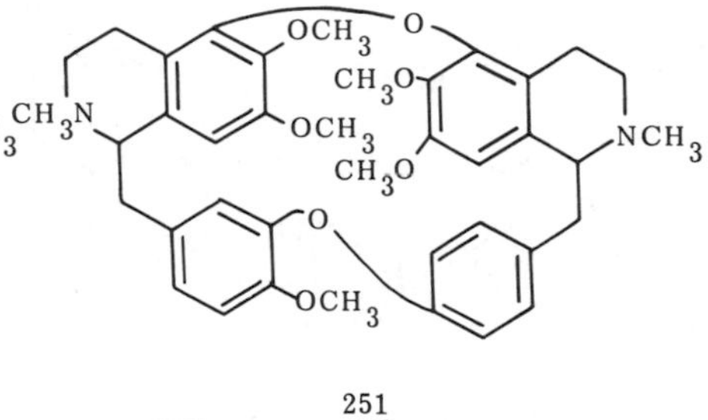

251

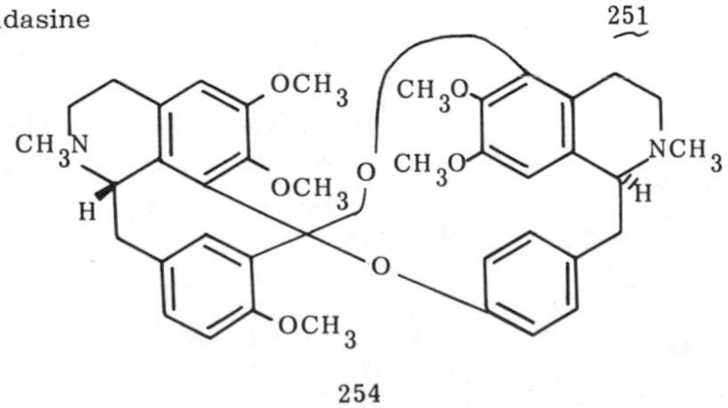

254

Table 5 Representative Examples of Bisbenzylisoquinoline Alkaloid Skeleta[a]

Skeleton	Classification	Alkaloid
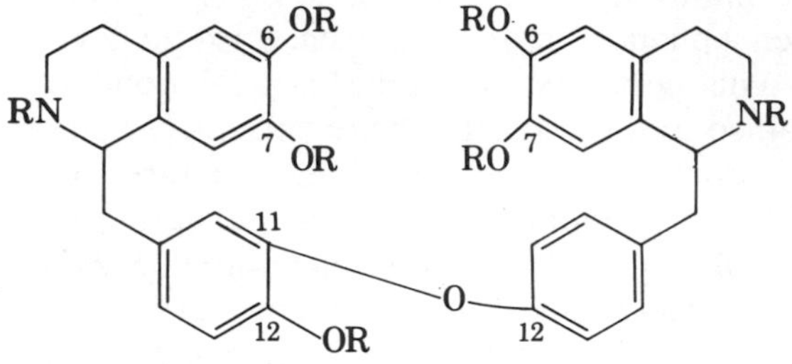	6, 7, 11*, 12-6, 7, 12*	Thalibrine
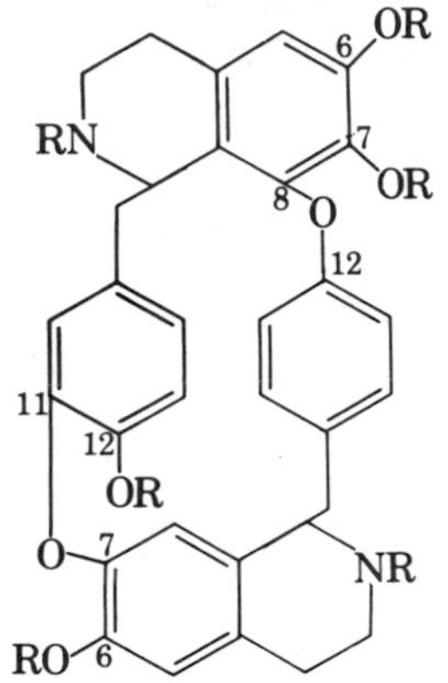	6, 7, 8*, 11†, 12-6, 7†, 12*	Tubocurarine
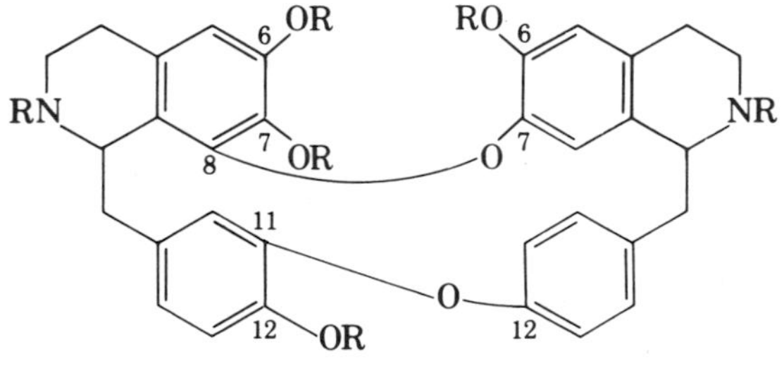	6, 7, 8*, 11†, 12-6, 7*, 12†	Berbamine, tetrandrine
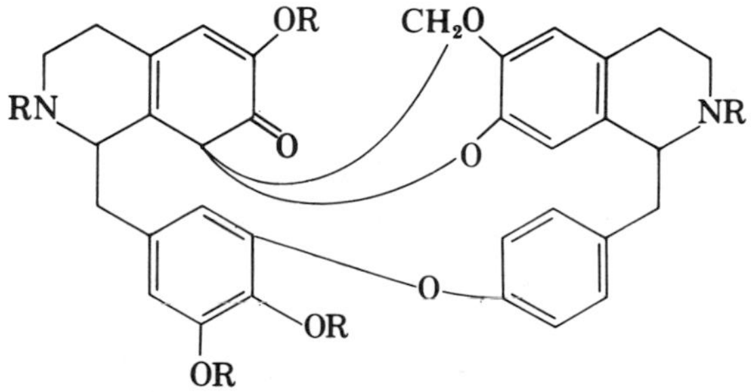	6, 7, 8*, 11†, 12, 13-6, 7*, 12†[8-6]	Repanduline

352

Table 5 (Continued)

Skeleton	Classification	Alkaloid
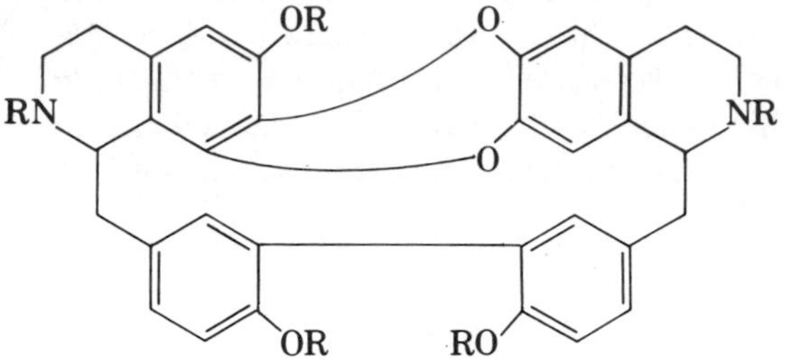	6, 7*, 8†, 12-6*, 7†, 12[11-11]	Tiliacorine
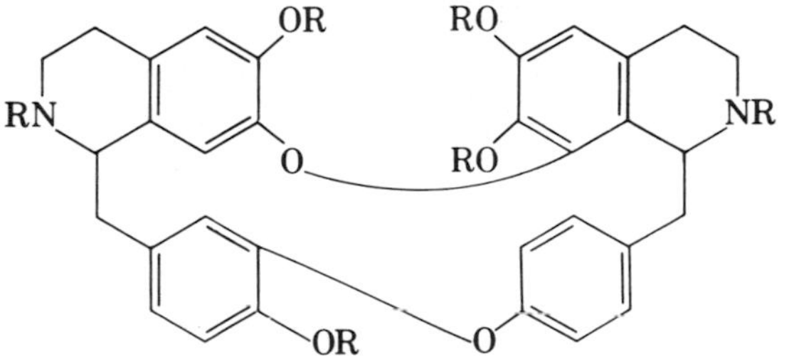	6, 7*, 11†, 12-6, 7, 8*, 12†	Repandine
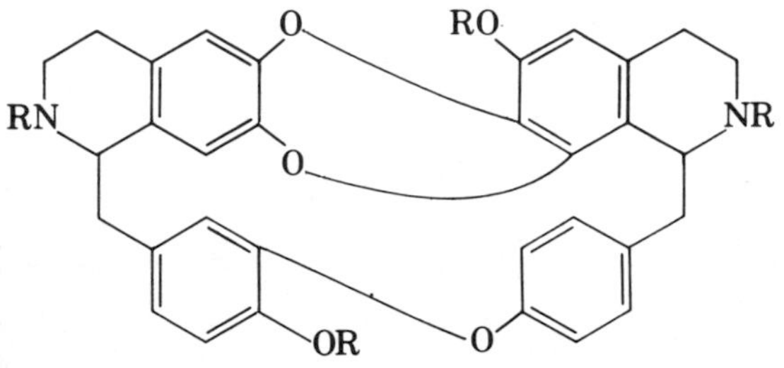	6*, 7†, 11†, 12-6, 7*, 8†, 12†	Trilobine
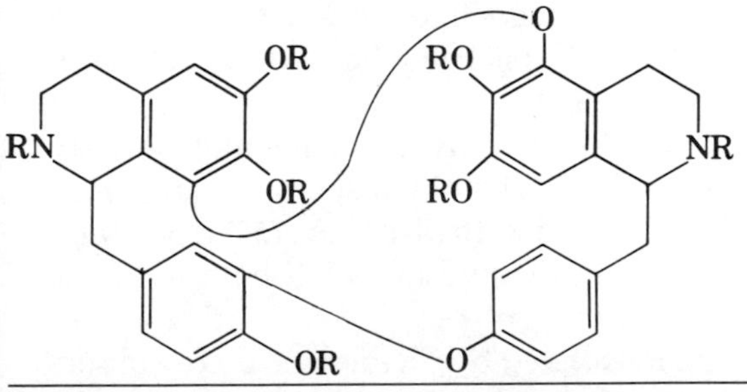	6, 7, 8*, 11†, 12-5*, 6, 7, 12†	Thalidasine

Source: M. Shamma and J. L. Moniot, *Heterocycles* **4,** 1817 (1976).

[a] Numbers indicate sites of oxygenation; − indicates separation of the two units; * indicates first diaryl ether linkage; † indicates second diaryl ether linkage; parentheses indicate a diaryl linkage; square brackets indicate an aryl-to-methyl ether linkage.

Many of the bisbenzylisoquinolines having a diaryl ether linkage can be interrelated or degraded to benzylisoquinolines. For a compound such as tiliacorine (255) which has a biphenyl linkage, sodium–liquid ammonia will not effect reductive cleavage and consequently structure elucidation is more challenging. Tiliacorine was considered to have either structure 255 or 256, and a distinction between the two could be made by potassium permanganate oxidation of tiliacorine acetate which specifically gave a C-1′-cleaved lactam acetate and which could be hydrolyzed to the phenolic aldehyde lactam 257.

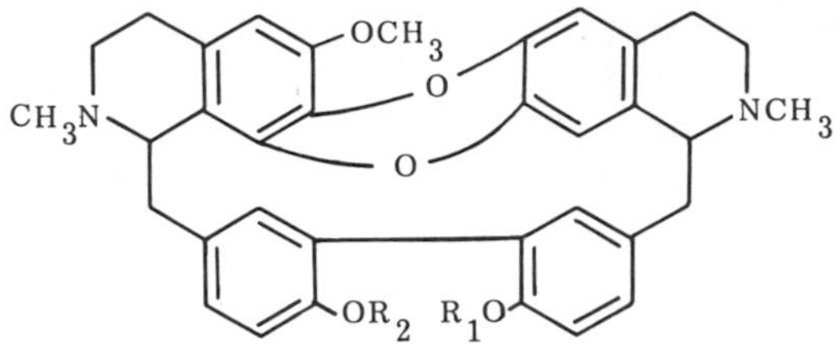

255   Tiliacorine R$_1$=H, R$_2$=CH$_3$
256   R$_1$=CH$_3$, R$_2$=H

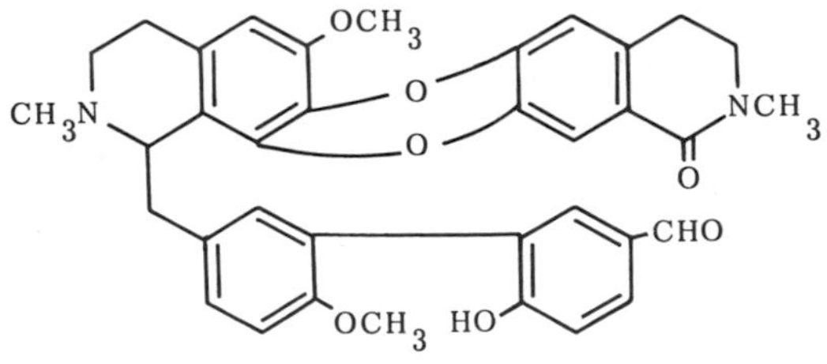

257

The *para* relationship of the phenol and the aldehyde was determined by a substantial bathochromic shift in the UV spectrum on addition of base. Tiliacorine therefore has the structure 255.

Some of the more recently isolated bisbenzylisoquinoline alkaloids are tiliamosine (258) from *Tiliacora racemosa* Colebr. (Menispermaceae), which has a tetraoxygenated isoquinoline nucleus, and thalistyline (259) from *Thalictrum longistylum* DC., a hypotensive derivative in which both aromatic nuclei are trioxygenated.

Another new bisbenzylisoquinoline isolated by Shamma and co-workers actually has one of the benzylisoquinoline units cleaved between the C-1 carbon and the α-carbon atom. This alkaloid is baluchistanamine (260) from *Berberis baluchistanica* Ahrendt (Berberidaceae).

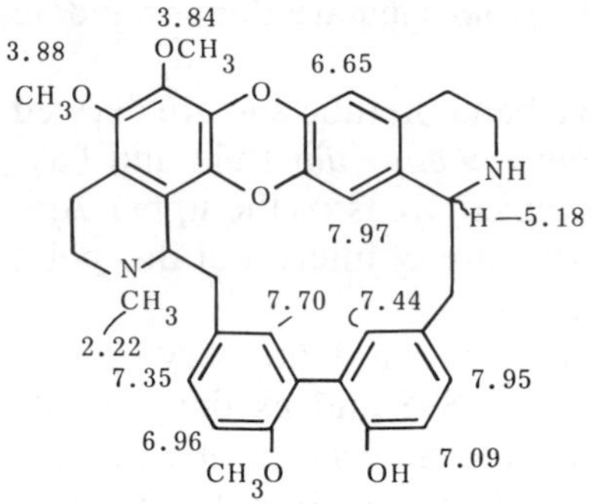

258 tiliamosine

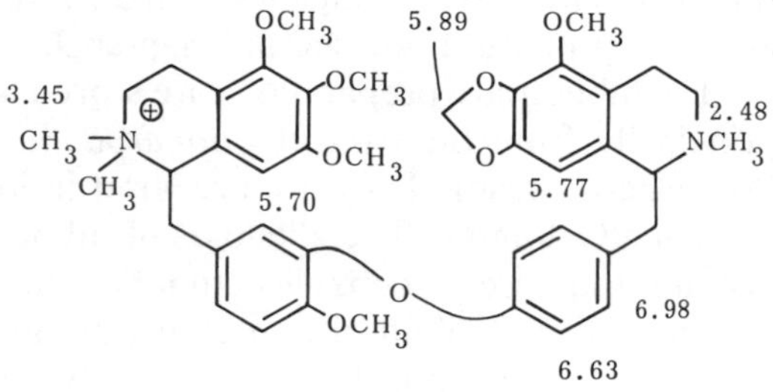

259 thalistyline

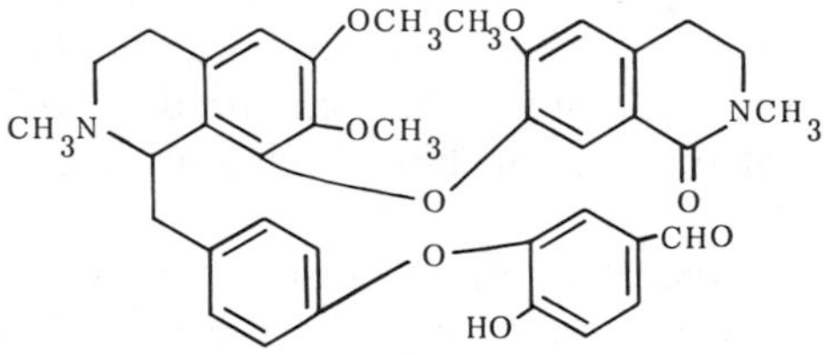

260 baluchistanamine

8.13.2 Curare

De Orbo Novo, a book translated into English in 1555, contains the first printed references to the South American arrow poisons used by the Indians of the Orinoco basin. An experience of an expeditionary force sent from Columbus's fleet told how the Indians "so fiercely assaying oure men with theire venomous arrowes that they slewe of them fortie and seven . . . for that poyson is of such force, that albeit the wounds were not great, yet they dyed thereof immediately."

The Frenchman de la Condamine was the first scientist to venture into the Amazon and in 1745 he collected many interesting plants, including curare. Furthermore, he experimented on the effects of curare in animals and these early results were subsequently elaborated by Herrisant, Brocklesby, and Fontana. In 1800, Bonpland, one of the great botanical explorers, was the first scientist to witness the preparation of curare—this in Esmeralda, Venezuela, described by von Humboldt as "the most celebrated spot on the Orinoco" for making curare. This curare was derived from a *Strychnos* species, subsequently identified as the liana *Strychnos toxifera* Schomb. It is also in the Calabash curare of the much-feared Macusi Indians of British Guiana (now Guyana). These are not forgotten arts, for the natives still use blow darts and arrows dipped in a curare to kill birds and monkeys for food. The active principles in the *Strychnos* species are dimeric indole alkaloids and are discussed in Chapter 9.

The tubocurare preparations are stored in a bamboo tube and are derived principally from the fruit of *Chondrodendrom tomentosum* Ruiz and Pav. (Menispermaceae). They are prepared in localized areas of the upper Amazon and Colombia. The alkaloids of tubocurare are of interest at this point because they are bisbenzylisoquinoline alkaloids.

The major alkaloid of importance is tubocurarine, a quaternary species first isolated in amorphous form by Boehm in 1895 and as the chloride subsequently by King. Commercially the alkaloid may be purified as the picrate derivative and then subjected to anion exchange to afford the chloride salt.

The structure for tubocurarine was determined by classical methods and until quite recently was thought to be **261**. Subsequent work has shown this to be incorrect and the structure is now regarded as **262**. This work also solved the problems of the relationship of tubocurarine to chondocurarine and chondocurine.

The proton NMR spectrum of (+)-tubocurarine (**262**) shows three *N*-methyl groups and two aromatic methoxyl groups. On basification one of the *N*-methyl groups shifts upfield, indicating that it is a tertiary *N*-methyl. (+)-Tubocurarine could therefore have two possible structures depending on the position of the tertiary methyl. This point was readily settled by sodium–liquid ammonia reduction on *O,O*-dimethyl tubocurarine. The two products obtained were **263** and **264**, indicating that the basic nitrogen is the lower half of the molecule. The proposed structure was confirmed by X-ray crystallographic analysis.

Treatment of tubocurarine with sodium thiophenolate under reflux with methyl ethyl ketone gave (+)-chondocurine, which is therefore the ditertiary base **265**. Chondocurarine (**261**) is the quaternary *N*-methyl derivative of tubocurarine.

261 (+)-tubocurarine chloride
     (old structure) now
     chondocurarine

262 (+)-tubocurarine
     chloride, revised
     structure

265 chondocurine

	R_1	R_2
262	CH_3	H
265	H	H

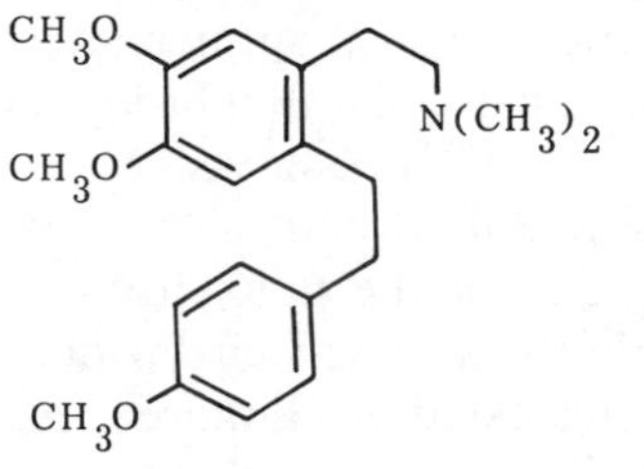

CH$_3$O — N(CH$_3$)$_2$
CH$_3$O

CH$_3$O

263

OH

CH$_3$N — H — OH — OCH$_3$

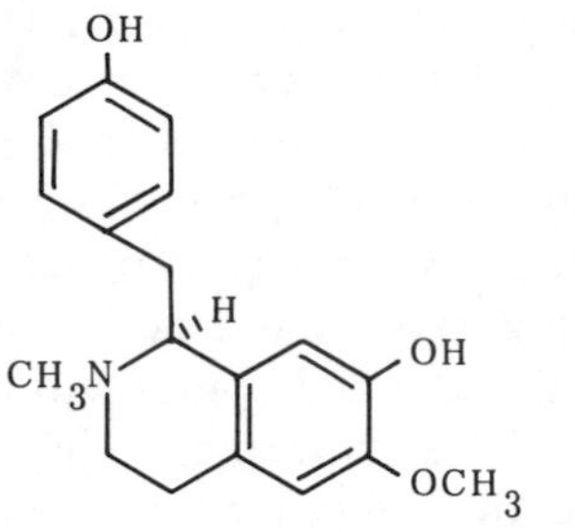

264

8.13.3 Tetrandrine

The bisbenzylisoquinoline alkaloid tetrandrine (266) has been of recent interest because it possesses significant activity against the Walker carcinosarcoma system, and the *d*-isomer was found to be almost as active as the racemate.

Tetrandrine was first isolated from the Japanese plant *Stephania tetrandra* (Menispermaceae) by Kondo and Yano in 1928.

In 1931, Santos reported the isolation of phaeanthine, an alkaloid having antitubercular activity, from the bark of *Phaeanthus ebracteotatus* (Presl) Merr. It was subsequently shown that phaeanthine (267) was the optical antipode of tetrandrine.

The third compound in this series is isotetrandrine (268), in which the two asymmetric centers are antipodal. The three compounds are distinguished by certain characteristics of their proton NMR spectra.

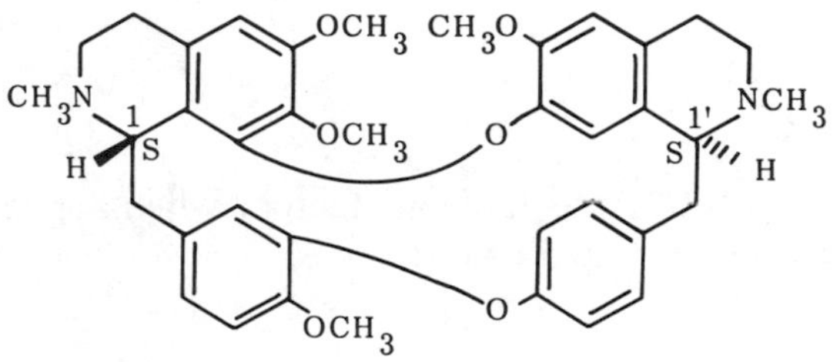

CH$_3$N 1 OCH$_3$ CH$_3$O NCH$_3$
H S OCH$_3$ O S H
1'
OCH$_3$ O

266   (+)-tetrandrine

267   (−)-phaeanthine   1-R, 1'-R

268   (+)-isotetrandrine   1-R, 1'-S

d-Tetrandrine was isolated from the roots of *Cyclea peltata* Diels (Menispermaceae) by Kupchan and co-workers in 1961 as the principal antitumor constituent. It has been the subject of extensive preclinical toxicological studies. In addition to anticancer activity, *d*-tetrandrine (266) has potent hypotensive and vasodilatory effects in several animal species. Other effects observed after intravenous administration include antipyresis, anti-inflammation, and histamine liberation from mast cells. Nephrotoxicity and hepatoxicity in monkeys and dogs have also been observed.

A recent X-ray structure determination of tetrandrine has revealed some interesting molecular characteristics. One of the benzylisoquinoline units adopts an extended conformation while the second benzylisoquinoline unit is folded. This conformation has been suggested by Gilmore *et al.* to account for the quite different chemical reactivities of the two N-methyl groups. One of the N-methyl groups (on N-2) occupies a pseudoaxial position with the lone pair equatorial, but the second N-methyl group (on N-2') is pseudoequatorial with the lone pair axial. Under these circumstances it could be shown that the N-2 methyl group is approximately 100 times more congested than the N-2' methyl group.

Rosazza and co-workers have found that although there is no regiospecificity on chemical N-demethylation, *Cunninghamella blakesleeana* specifically removes the N-2 methyl group of tetrandrine. On the other hand several other organisms specifically gave the N-2' demethyl derivative.

A new method for the degradation of bisbenzylisoquinolines involves treatment with ceric ammonium nitrate (CAN) in acetate buffer solution for brief periods. The crude product is treated with sodium borohydride and the fragments isolated. For example tetrandrine (**266**) on reaction with 8 moles of CAN, followed by reduction, gave the diamine **269** and the diol **270** in

269   270

excellent yield. The reaction was less satisfactory when applied to phenolic and methylenedioxy containing alkaloids.

8.13.4 Synthesis

The synthesis of bisbenzylisoquinolines relies for the most part on the Ullman reaction. In its simplest form this reaction proceeds to give a diaryl ether from a bromobenzene and phenol, in the presence of copper, potassium carbonate, and potassium iodide, in pyridine as a solvent, at 155–160°C (Scheme 36).

A bisbenzylisoquinoline havine two aryl ether alkaloids poses some strategic problems and one approach is shown in the Inubushi synthesis of (+)-isotetrandrine (**268**). In this case the aryl ether linkages must be formed sequentially and the second isoquinoline nucleus formed subsequently.

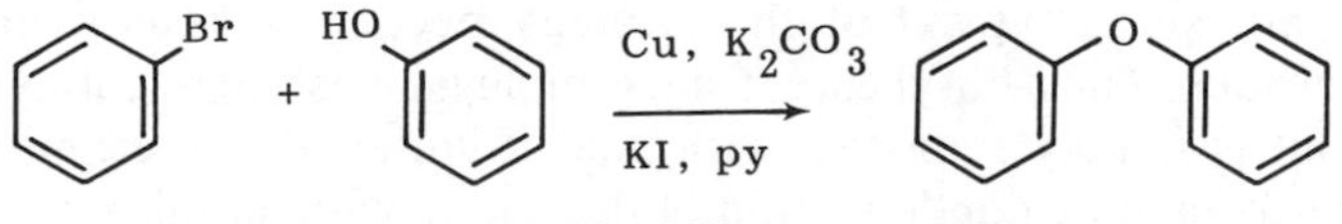

Scheme 36 The Ullmann reaction.

(−)-*O*-Benzyl-8-bromolaudanidine (**271**) is the first key intermediate and was produced by unexceptional methods. This compound is subjected to Ullmann condensation with the N-protected derivative **272**. After debenzylation the phenolic group of **273** is condensed with methyl *p*-bromophenylacetate to produce the second aryl ether linkage as shown in **274**. The remaining steps involve formation of the amide **275** followed by Bischler–Napieralski cyclization, sodium borohydride reduction, and N-methylation. The product is a mixture of (+)-isotetrandrine (**268**) and (−)phaeanthine (**267**).

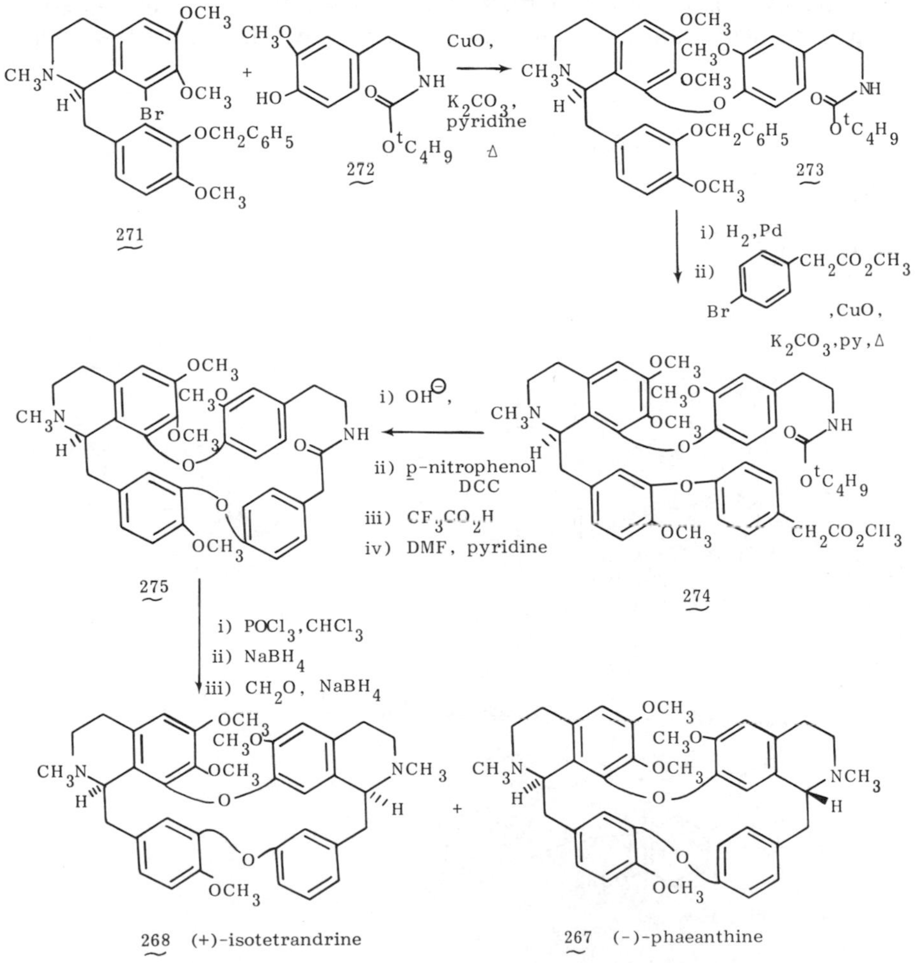

This synthesis is typical of the strategy needed to selectively protect O- and N-groups. The *t*-butyl ester for example is resistant to catalytic hydrogenation but is readily removed with acid. Synthesis of the amide linkage of **275** is performed by condensation of the amine with a *p*-nitrophenol ester.

The next level of complexity is a bisbenzylisoquinoline with three diphenyl ether linkages. The Tomita–Ueda synthesis of racemic **276** is an example. If we consider only the top of the molecule it can be seen that the substitution in each ring is the same. Consequently it was envisaged that the diamine **277** would be a suitable precursor of this unit. This compound was produced by a double Ullmann condensation followed by a series of standard reactions. At high dilution this diamine was condensed with the diacid chloride **278** to give the diamide **279**, which could be subjected to Bischler–Napieralski condensation followed by standard reactions to give **276**.

The most interesting development has been the use by Cava of pentafluorophenyl copper in dry pyridine for the Ullmann reaction. Using this reagent condensation of (+)-6'-bromolaudanosine (**280**) with (+)-armepa-

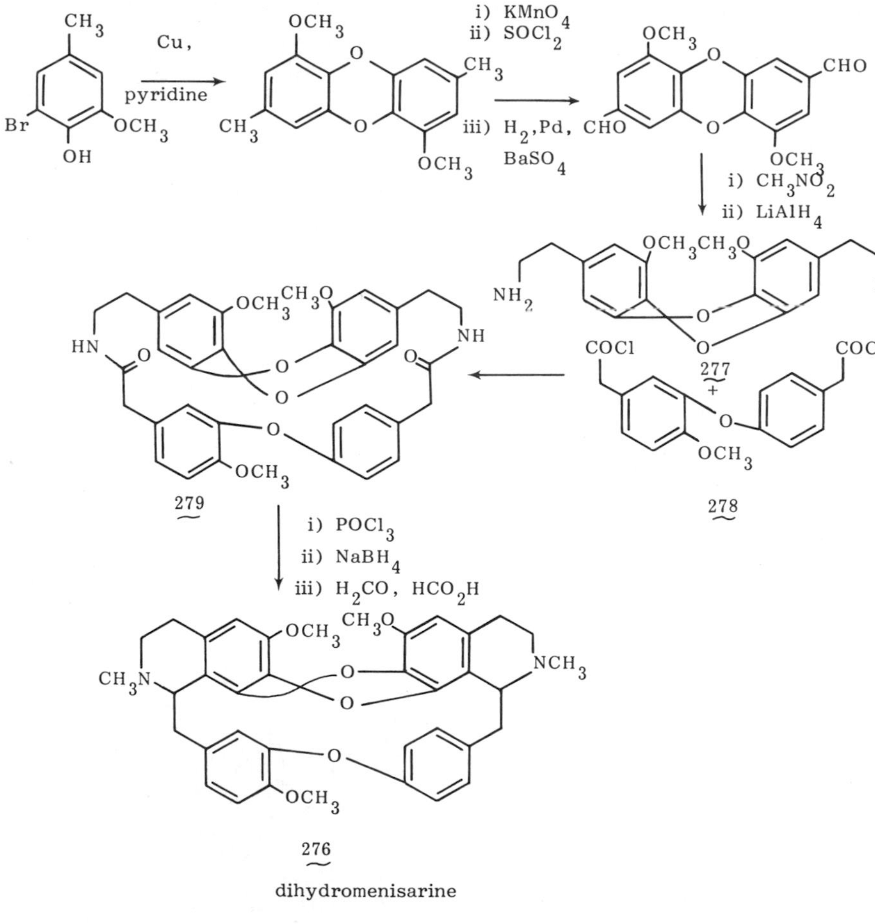

276

dihydromenisarine

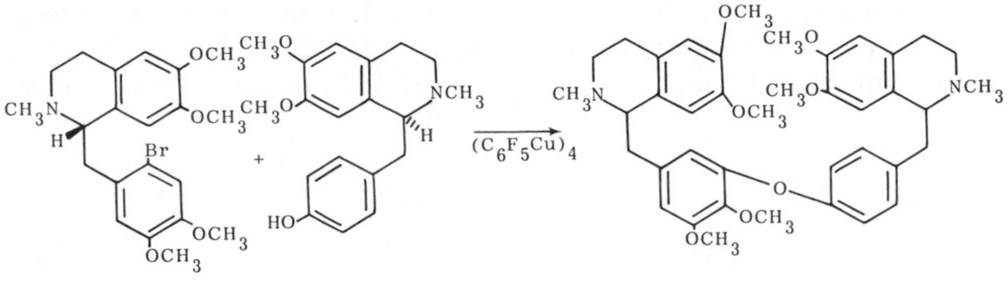

280 213 (+)-armepavine 281

Scheme 37

vine (**213**) gave **281** in 53% yield (Scheme 37). Typically the yields in an Ullmann reaction are in the 10–20% range. To date, this reagent has not been used in the synthesis of any of the more complex bisbenzylisoquinolines.

These are nonbiogenetically modeled synthetic procedures. But attempts to produce bisbenzylisoquinolines by phenol oxidation processes have been rare, possibly because of the danger of forming any one of several other skeleta (see e.g. Section 8.15).

The first successful oxidative coupling to give a bisbenzylisoquinoline was the reaction, due to Franck and Blaschke, of magnocurarine iodide (**184**) at pH 10 with potassium ferricyanide to afford **282** in 18% yield.

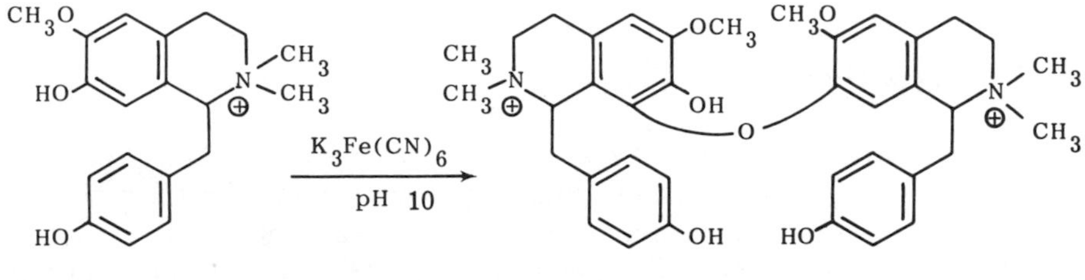

184 282

A biphenyl dimer has been produced by the treatment of the methiodide of armepavine (**213**) with 0.1 *N* sodium carbonate solution and potassium ferricyanide to give the dimer **283** in 15% yield.

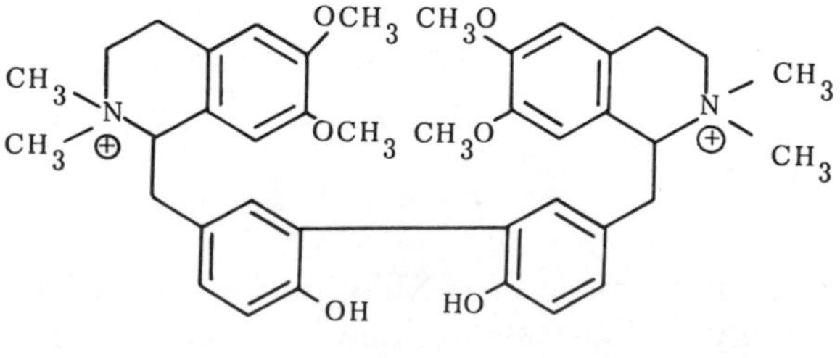

283

Bobbitt and Hallcher have shown however that electrolytic oxidation of the sodium salt of *N*-carbethoxy-*N*-norarmepavine (**284**) afforded a mixture of two dimers, one a biphenyl derivative **285** and the other a bisaryl ether **286**.

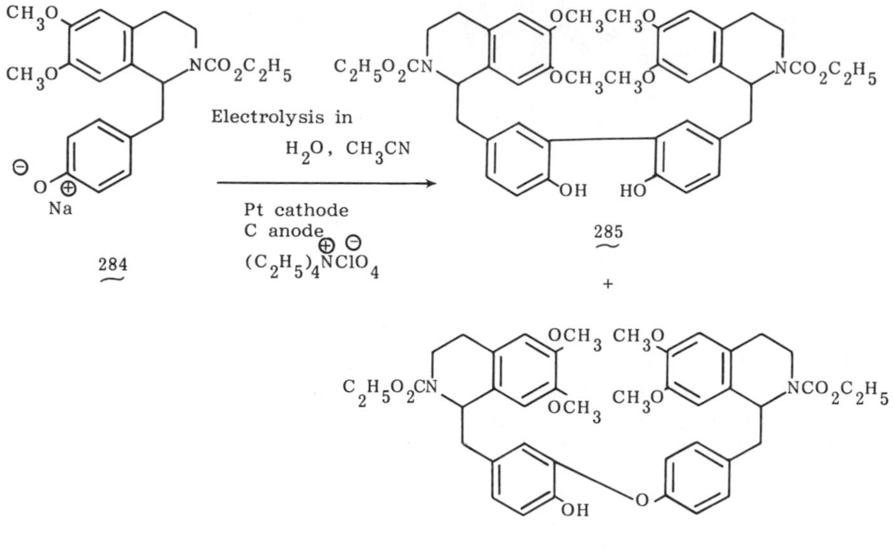

8.13.5   Spectral Properties

The UV spectra of the bisbenzylisoquinoline alkaloids are typical of those of a benzylisoquinoline with a maximum at about 282 nm. For the most part the spectra do not distinguish among the various subgroups.

In sharp contrast, the mass spectra of these alkaloids are quite useful. The most favored fragmentation process is one in which cleavage at the C-1 and C-1' benzylic bonds occurs to give an aryl ether species. Subsequent losses are of small fragments. A typical example is the alkaloid tetrandrine (**266**), which is thought to fragment as shown in Scheme 38. Weak fragment ions are also observed corresponding to $M^+ - 137$ and $M^+ - 191$.

Alkaloids in the tubocurarine series display a similar double-benzylic fragmentation process but the result is quite different, giving rise to the two halves of the structure as shown for chondrofoline (**287**).

8.13.6   Biosynthesis

Few studies have been made of the biosynthesis of bisbenzylisoquinoline alkaloids. [*N*-methyl-^{14}C]*N*-Methylcoclaurine (**288**) was shown to be a pre-

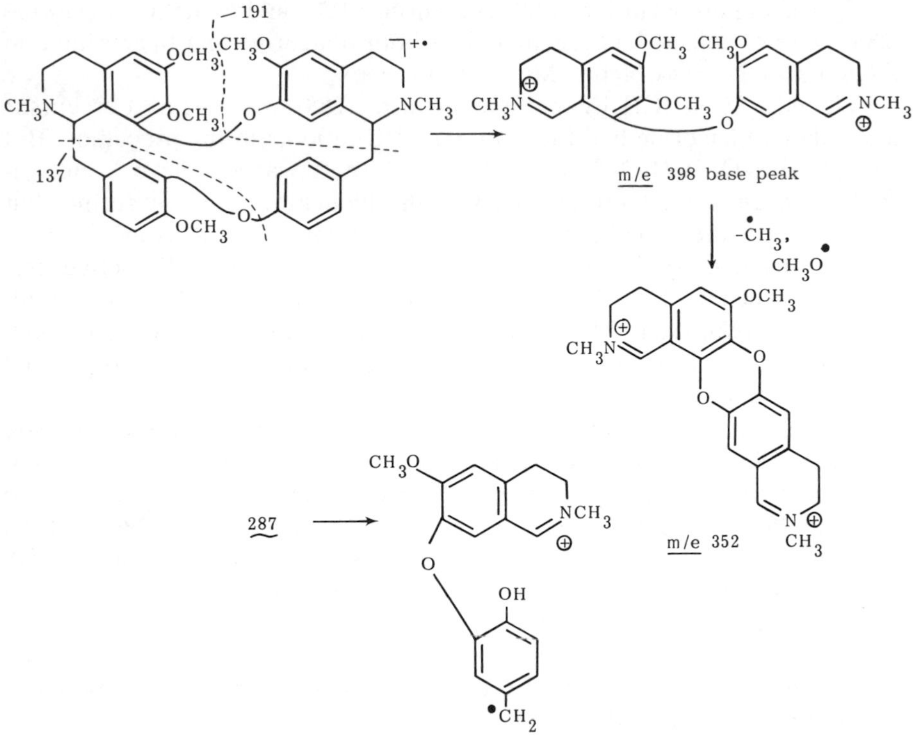

Scheme 38

cursor of (+)-epistephanine (**289**) in *Stephania japonica* Miers., and when doubly labeled **288** was used it was found to be incorporated into only one half of the molecule.

The absolute configuration of tiliacorine could not be deduced by the usual liquid ammonia reduction procedure because of the presence of the diaryl linkage. Instead this point was recently deduced from biosynthetic experimentation.

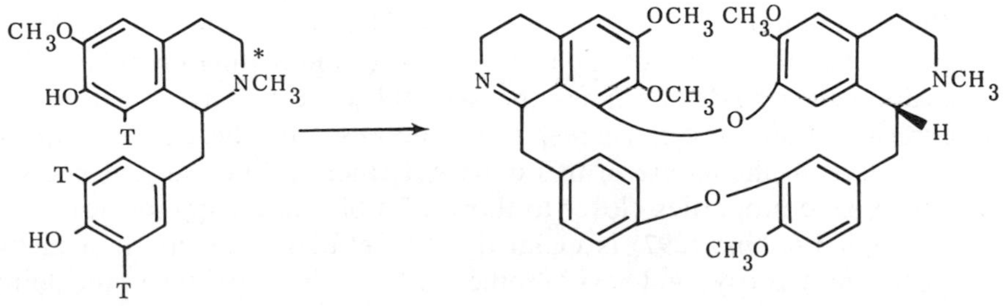

288 N-methylcoclaurine

289 epistephanine

Tyrosine, norcoclaurine (290), coclaurine (217), and N-methylcoclaurine (288) were established as precursors of both tiliacorine and tiliacorinine in *Tiliacora racemosa* Colebr. (Menispermaceae).

When [1-^{3}H,N-^{14}CH$_3$]N-methylcoclaurine (288) was used as a precursor the isolated tiliacorine had the same ^{3}H:^{14}C ratio as in the precursor. But with [1-^{3}H, 6-O-^{14}CH$_3$]288, half of the ^{14}C activity was lost. Two N-methylcoclaurine units are therefore used in the biosynthesis of tiliacorine, but one unit loses the 6-methoxy group at some point in the pathway.

The absolute configuration was determined using optically active precursors. When (−)-(R)-N-methylcoclaurine (291) and (+)-(S)-N-methylcoclaurine (292), each labeled at 3′, 5′, and 8 with ^{3}H were used, both compounds were incorporated into tiliacorine, but only 292 was incorporated into tiliacorinine.

Tiliacorinine therefore has the 1-S, 1′-S configurations and the complete structure 293. For tiliacorine a more subtle determination was required. Alkaline permanganate oxidation of tiliacorine methiodide gave the two products 294 and 295. From the (−)-(R)-291, 294 was active and 295 inactive, but from (+)-(S)-292, 295 was inactive and 294 active. These data establish that C-1 in tiliacorine has the S-configuration and C-1′ the R-configuration, and that the complete structure is represented by 296.

The biogenesis of the various bisbenzylisoquinoline alkaloid types is a fascinating area which has been little discussed. The important compounds tetrandrine (266) and tubocurarine (262) are worthy molecules for biosynthetic evaluation. In addition, compounds having three diphenyl ether linkages pose even more stimulating academic problems.

8.13.7   Pharmacology

There are two bisbenzylisoquinolines of pharmacologic significance, tubocurarine chloride and tetrandrine, and there has already been discussion concerning the antitumor activity of tetrandrine.

Injection of tubocurarine rapidly blocks neuromuscular action, causing respiratory failure and death. A paralyzing dose is in the range of 3–7 mg. The compound has been used in small doses to enhance the action of anesthetics because it causes paralysis of the abdominal muscles without stopping the natural movement of the intestines. The drug is inactive orally.

The di-O-methyl ether of tubocurarine is an interesting compound because although it has three times the skeletal muscle relaxant activity of tubocurarine, it does not cause respiratory depression. There is some question though as to the nature of this dimethyl ether and in many respects it appears to have properties closer to those of a bisquaternary species.

(+)-Cepharanthine (297) is claimed to be effective against human tuberculosis and leprosy, and several other bisbenzylisoquinolines, including thalidasine (250), display good *in vitro* antitubercular activity.

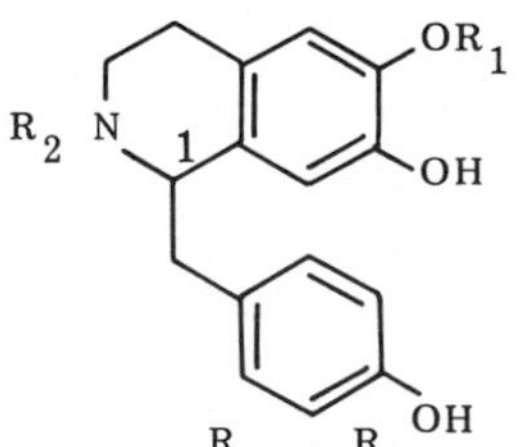

	R_1	R_2
290	H	H
217	CH_3	H
288	CH_3	CH_3

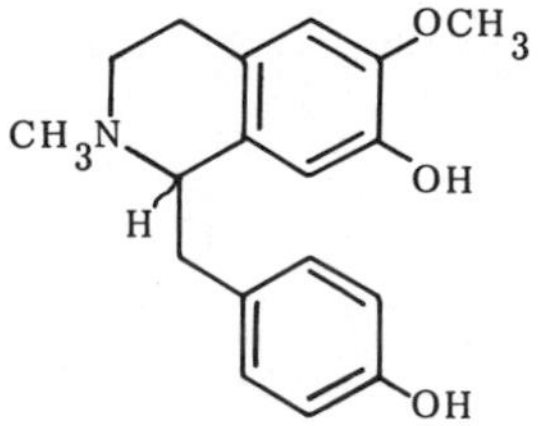

291 R=αH (−)-R-

292 R=βH (+)-S-

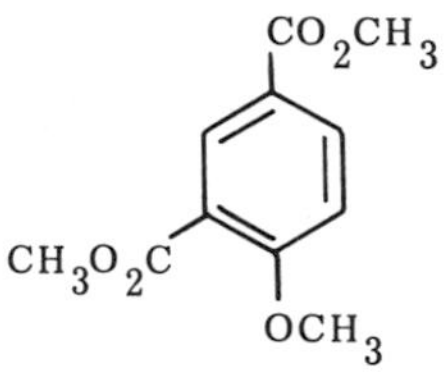

294

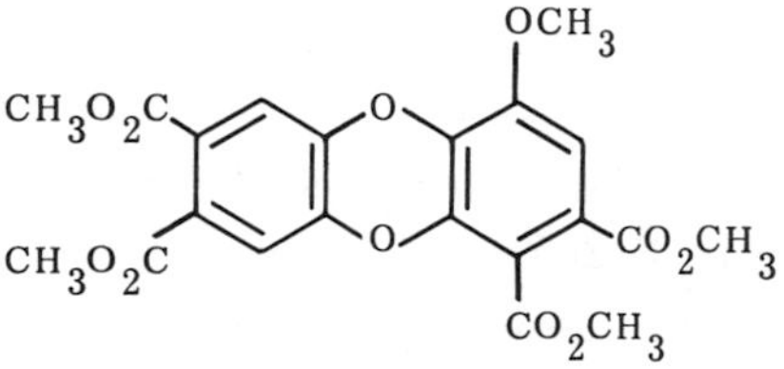

295

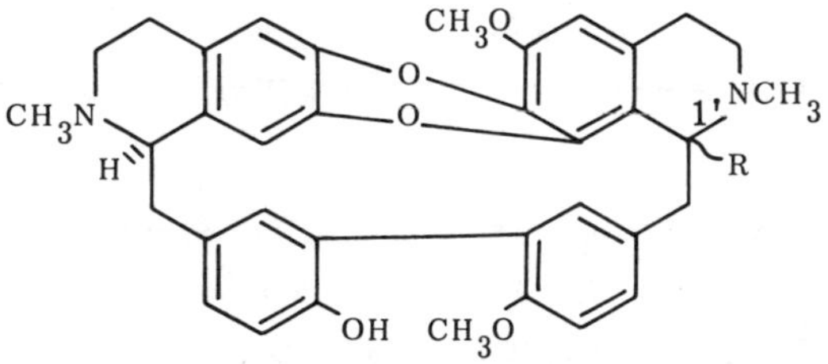

293 tiliacorinine R= β-H

296 tiliacorine R= α-H

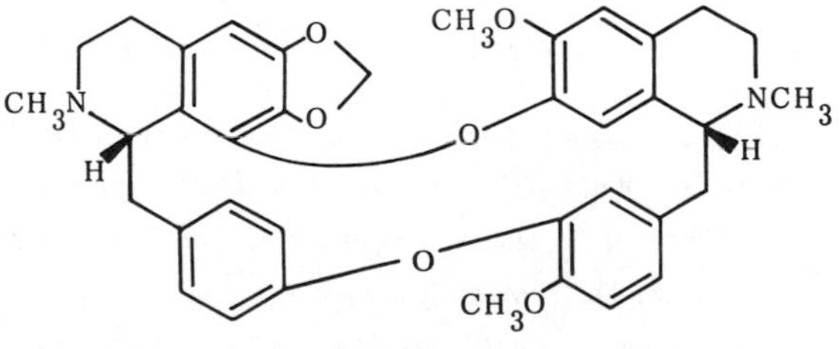

297 (+)-cepharanthine

LITERATURE

Reviews

Kulka, M., *Alkaloids NY* **4,** 199 (1954).

Kulka, M., *Alkaloids NY* **7,** 439 (1960).

Grundon, M. F., *Progr. Org. Chem.* **6,** 38 (1964).

Curcumelli-Rodostamo, M., and M. Kulka, *Alkaloids NY* **9,** 133 (1967).

Thornber, C. W., *Phytochemistry* **9,** 157 (1970).

Curcumelli-Rodastamo, M., *Alkaloids NY* **13,** 304 (1971).

Shamma, M., *The Isoquinoline Alkaloids*, Academic, New York, 1972, p. 115.

Kametani, T., *The Chemistry of the Isoquinoline Alkaloids*, Vol. 2, Sendai Institute of Heterocyclic Chemistry, Sendai, Japan, 1974, p. 107.

Shamma, M., and J. L. Moniot, *Heterocycles* **4,** 1817 (1976).

Cava, M. P., K. T. Buck, and K. L. Stuart, *Alkaloids NY* **16,** 250 (1977).

Shamma, M., and V. St. Georgiev, *Alkaloids NY* **16,** 319 (1977).

Shamma, M. and J. L. Moniot, *Isoquinoline Alkaloid Research, 1972–1977,* Plenum, New York, 1978, p. 71.

Curare

Bryn Thomas, K., *Curare, Its History and Usage*, Pitman Medical, London, 1963.

Everett, A. J., L. A. Lowe, and S. Wilkinson, *Chem. Commun.* 1020 (1970).

Tetrandrine

Kondo, H., and K. J. Yano, *J. Pharm. Soc. Jap.* **48,** 107 (1928).

Santos, A. C., *Rev. Fil. Med. Far.* **22,** 243 (1931).

Santos, A. C., *Acta Manilana Ser. A* 12 (1974).

Kupchan, S. M., N. Yokoyama, and B. S. Thyagarajan, *J. Pharm. Sci.* **50,** 164 (1961).

Kupchan, S. M., A. J. Liepa, R. L. Baxter, and H. P. J. Hintz, *J. Org. Chem.* **38,** 1846 (1973).

Gralla, E. J., G. L. Coleman, and A. M. Jones, *Cancer Chemotherapy Rept., Part 3* **5,** 79 (1974).

Gilmore, G. J., R. F. Bryan, and S. M. Kupchan, *J. Amer. Chem. Soc.* **98,** 1947 (1976).

Davis, P. J., D. R. Wiese, and J. P. Rosazza, *Lloydia* **40,** 240 (1977).

Bick, I. R. C., J. B. Brenner, M. P. Cava, and P. Wiriyachitra, *Aust. J. Chem.* **31,** 321 (1978).

Synthesis

Kametani, T., S. Takano, K. Masuko, and F. Saraki, *Chem. Pharm. Bull.* **14,** 67 (1966).

Inubushi, Y., Y. Masaki, S. Matsumoto, and F. Takami, *J. Chem. Soc., Sec. C* 1547 (1969).

Tomita, M., S. Ueda, and A. Teraoka, *Tetrahedron Lett.* 635 (1962).

Inubushi, Y., Y. Ito, Y. Masaki, and T. Ibuka, *Tetrahedron Lett.* 2857 (1976).

Anjaneyulu, B., T. R. Govindachari, and N. Viswanathan, *Tetrahedron* **27,** 439 (1971).

Cava, M. P., and A. Afzali, *J. Org. Chem.* **40,** 1553 (1975).

Franck, B., and G. Blaschke, *Ann.* **668**, 145 (1963).

Choudhury, A. M., I. G. C. Coults, A. K. Durbin, K. Schofield, and D. J. Humphreys, *J. Chem. Soc., Sec. C* 2070 (1969).

Bobbitt, J. M., and R. C. Hallcher, *Chem. Commun.* 543 (1971).

Biosynthesis

Barton, D. H. R., G. W. Kirby, and A. Wiechers, *Chem. Commun.* 266 (1966).

Bhakuni, D. S., A. M. Singh, S. Jain, and R. S. Kapil, *Chem. Commun.* 266 (1978).

8.14 CULARINE AND RELATED ALKALOIDS

Cularine (**298**) and its close relatives are limited in distribution to the genera *Dicentra* and *Corydalis* in the family Fumariaceae. The structure was deduced in 1950 by a degradative series of reactions, one of which was cleavage of the aryl ether linkage with sodium–liquid ammonia to yield the benzylisoquinoline **299**. This compound in turn was degraded to 4-methoxyphthalic acid (**300**) and asaronic acid (**301**).

The proton NMR spectral data of cularine are shown in **302**, and the

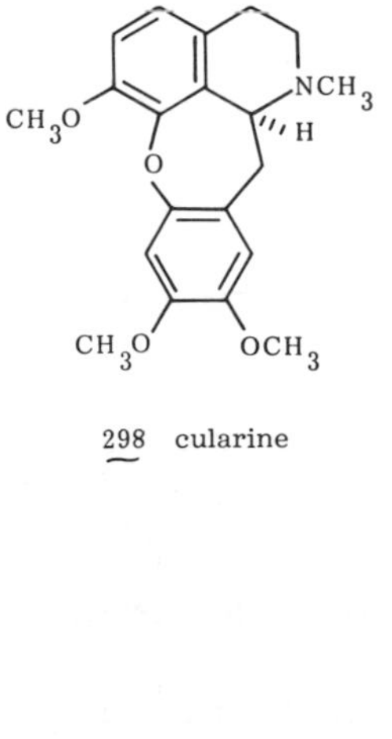

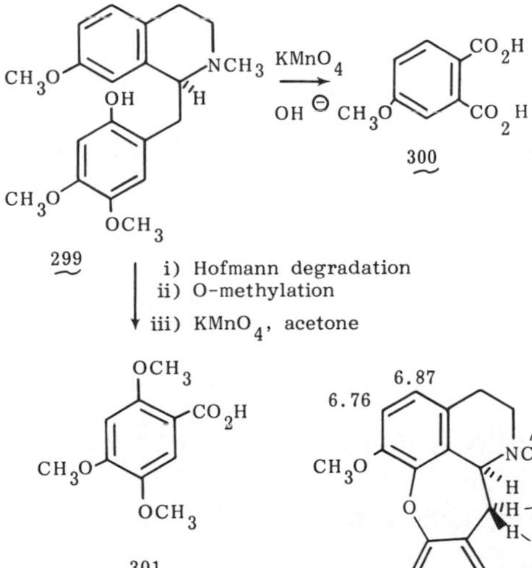

coupling of the C-1 proton with the benzylic methylene protons indicates a dihedral angle of about 180° (J = 12 Hz) between H-1 and the β-proton, and of about 60° (J = 4 Hz) between H-1 and the α-proton. These data suggest a conformation for cularine in which the molecule is essentially bowed. The downfield shift of the C-1 proton is typical of cularine alkaloids.

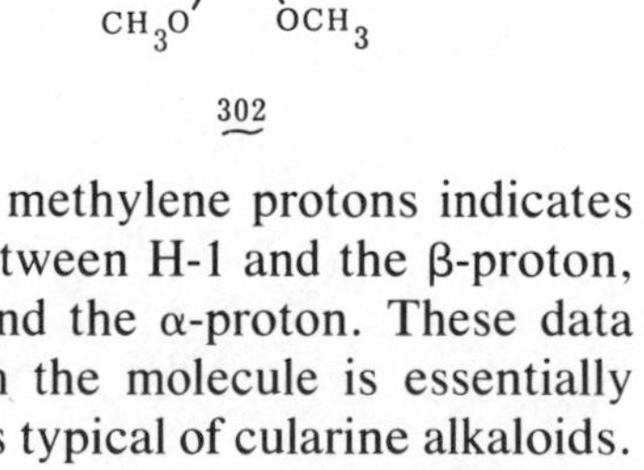

There have been several syntheses of cularine (**298**) and related compounds utilizing two general approaches for the formation of the biphenyl
ether linkage. One route is typified by the Ullmann reaction of the aryl bromide
internally with the phenol, on heating with copper(II) oxide in the
presence of base, to give **303**.

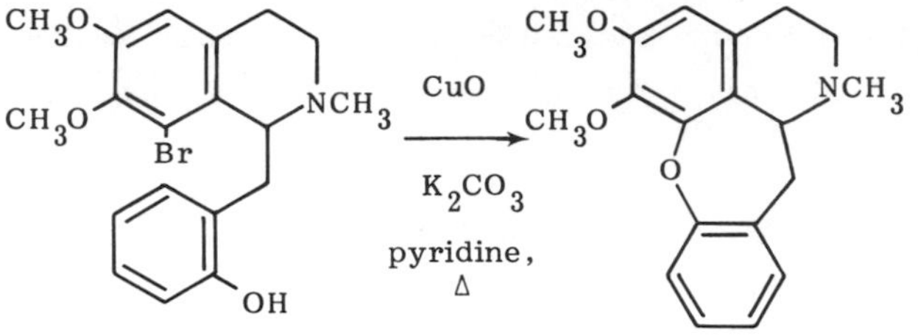

303

The second general approach has been a biogenetic one involving phenolic oxidation of the diphenol **304**. Two aryl ether products are possible
depending on the *ortho* or *para* nature of the coupling reaction. In practice
each isomer was produced in the ratio 2:1 as shown. Methylation of the
minor product gave cularine (**298**).

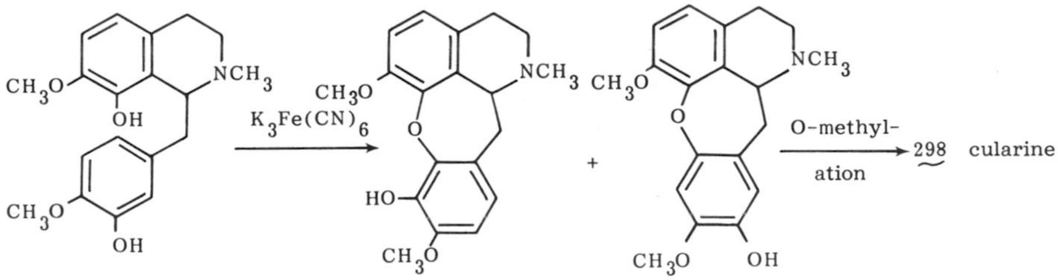

304

5%   2.5%

There have been no biosynthetic studies concerning the formation of the
cularine skeleton.

Cancentrine (**305**) is a yellow alkaloid, first isolated by Manske in 1932

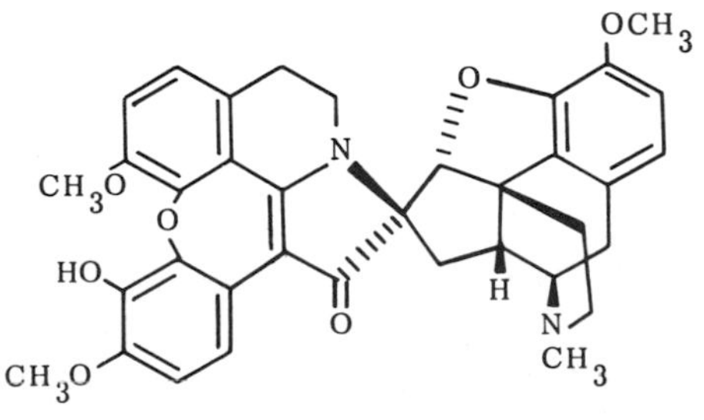

305   cancentrine

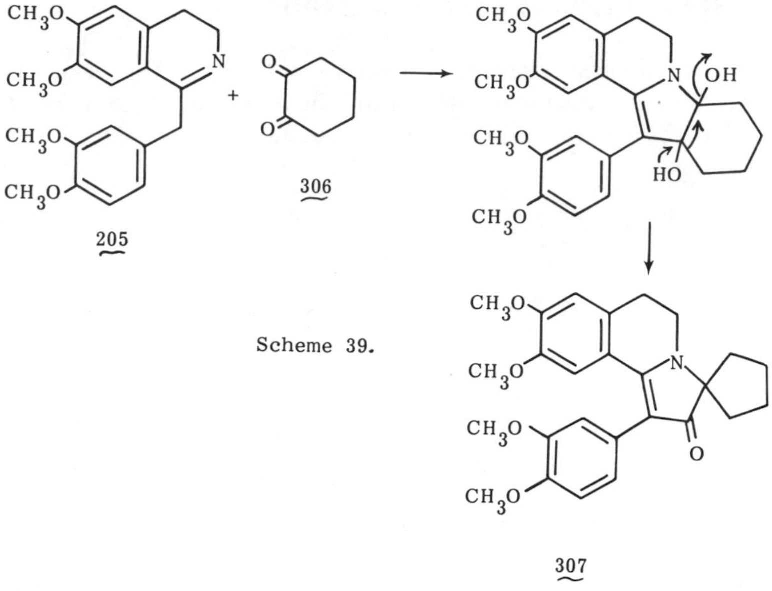

Scheme 39.

Scheme 39

from *Dicentra canadensis* (Goldie) Wald (Fumariaceae), whose structure was elucidated some 38 years later. It was the first member of the morphine–cularine dimers to be obtained and the spiro linkage between the two monomer units poses some interesting biogenetic problems.

Ruchirawat and Somchitman have provided an interesting model for the formation of the spiro center involving condensation of 3,4-dihydropapaverine (205) and 1,2-cyclohexanedione (306) in the presence of triton B. The product 307, obtained in 62% yield, can be envisaged as occurring by the mechanism shown in Scheme 39.

LITERATURE

Cularine

Manske, R. H. F., *J. Amer. Chem. Soc.* **72,** 55 (1950).

Iida, H., T. Kikuchi, K. Sakurai, and T. Watanabe, *J. Pharm. Soc. Japan* **89,** 645 (1969).

Kametani, T., K. Fukumoto, and M. Fujihara, *Chem. Commun.* 352 (1971).

Jackson, A. H., and G. W. Stewart, *Chem. Commun.* 149 (1971).

Cancentrine

Manske, R. H. F., *Can. J. Res., Sec. B* **7,** 258 (1932).

Clark, G. R., R. H. F. Manske, G. J. Palenik, R. Rodrigo, D. B. MacLean, L. Bacynskyj, D. E. F. Gracey, and J. K. Saunders, *J. Amer. Chem. Soc.* **92,** 4998 (1970).

Ruchirawat, S., and V. Somchitman, *Tetrahedron Lett.* 4159 (1977).

8.15 DIBENZOPYRROCOLINE ALKALOIDS

The only two natural dibenzopyrrocolines known are ($-$)-cryptaustoline (**308**) and ($-$)-cryptowoline (**309**) from the bark of *Cryptocaria boweii* (Hook.) Druce. (Lauraceae), and these structures were deduced by degradative studies.

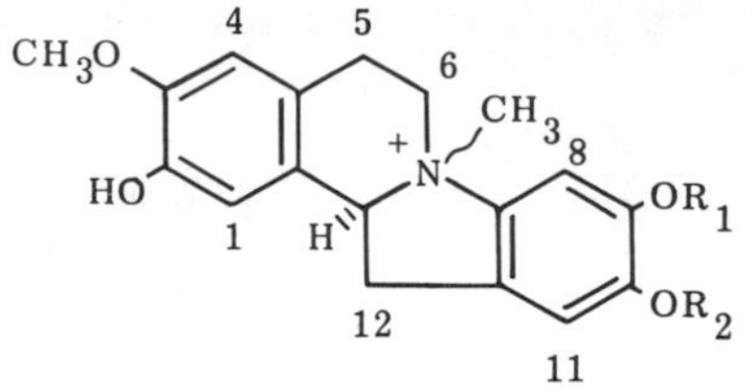

<u>308</u> ($-$)-cryptaustoline $R_1 = R_2 = CH_3$

<u>309</u> ($-$)-cryptowoline R_1, $R_2 = -CH_2-$

The first synthetic efforts predate the isolation from a natural source and inadvertently evolved as Robinson and Schöpf began to study the biogenetic chemistry of laudanosoline (**187**). Quite independently they found that oxidation of **187** with any of several oxidizing agents gave not the expected product, but a compound **310**, having the dibenzopyrrocoline skeleton. This route was used by Hughes and co-workers to synthesize ($-$)-*O*-methyl-cryptaustoline iodide (**311**) from ($+$)-laudanosoline (**187**), thereby deducing the stereochemistry at C-1 in the natural series.

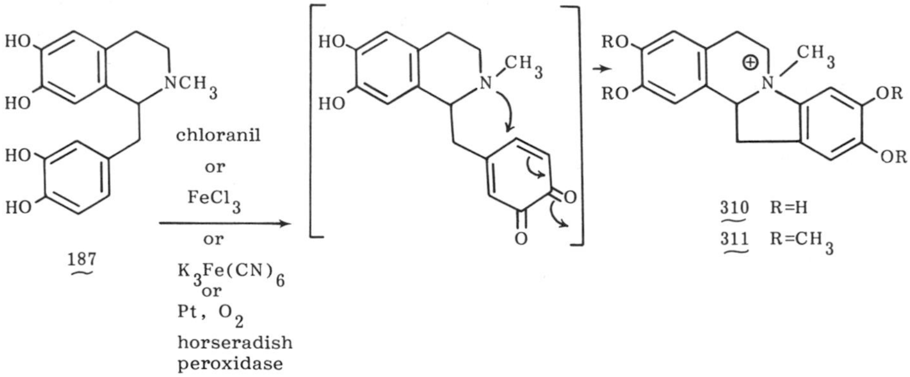

<u>310</u> R=H
<u>311</u> R=CH$_3$

A benzyne intermediate has also proved to be an effective synthetic precursor, as shown in the direct cyclization of 5′-bromocodamine (**312**).

A further synthesis of cryptaustoline (**308**) involves the photocyclization of an enamine precursor. Irradiation of an ethereal solution of the enamine

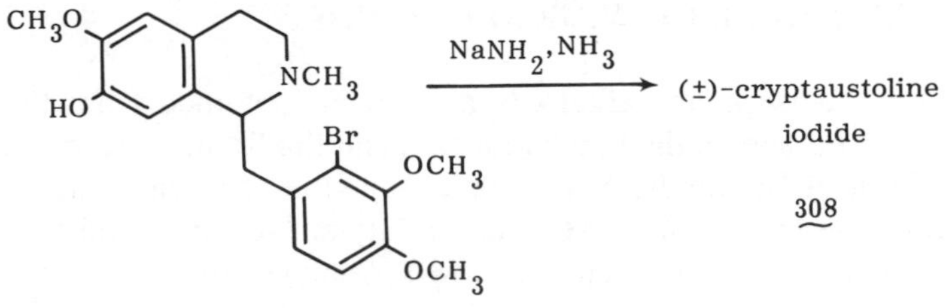

312

313, using the isopropyl ether as a protecting group, gave a mixture of the isomeric dehydrodibenzopyrrocolines **314** and **315** in 12 and 11% yields respectively. After removal of the protecting group with 48% HBr in acetic acid and Sn–HCl reduction, quaternization with methyl iodide afforded **308** (Scheme 40).

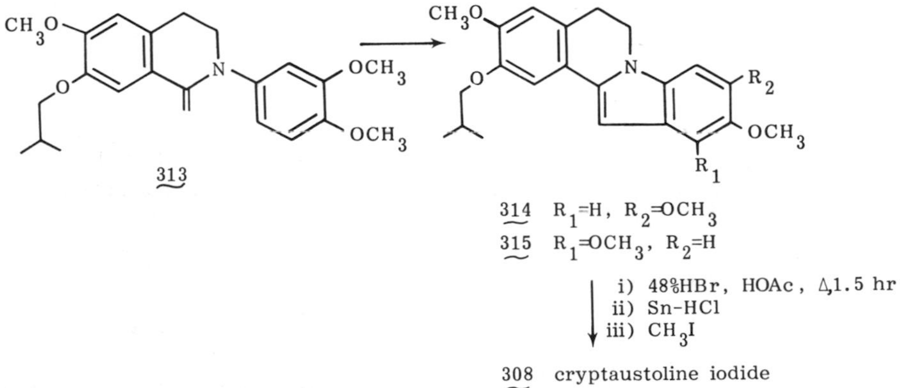

314 R_1=H, R_2=OCH$_3$
315 R_1=OCH$_3$, R_2=H

i) 48%HBr, HOAc, Δ,1.5 hr
ii) Sn–HCl
iii) CH$_3$I

308 cryptaustoline iodide

Scheme 40 Ninomiya synthesis of cryptaustoline iodide (**308**).

LITERATURE

Shamma, M., *The Isoquinoline Alkaloids*, Academic, New York, 1972, p. 169.

Shamma, M. and J. L. Moniot, *Isoquinoline Alkaloid Research, 1972–1977*, Plenum, New York, 1978, p. 113.

Ewing, J., G. K. Hughes, E. Ritchie, and W. C. Taylor, *Nature (London)* **169**, 618 (1952).

Robinson, R., and S. Sugasawa, *J. Chem. Soc.* 789 (1932).

Schöpf, C., and K. Thierfelder, *Ann.* **497**, 22 (1932).

Hughes, G. K., E. Ritchie, and W. C. Taylor, *Aust. J. Chem.* **6**, 315 (1953).

Kametani, T., K. Fukumoto, and T. Nakano, *J. Het. Chem.* **9**, 1363 (1972).

Ninomiya, I., J. Uasui, and T. Kiguchi, *Heterocycles* **6**, 1855 (1977).

8.16   PAVINE AND ISOPAVINE ALKALOIDS

The pavine and isopavine alkaloids have been obtained from three plant families, the Lauraceae, the Papaveraceae, and the Ranunculaceae, and they number about 30 alkaloids. Some typical pavine structures are shown by (−)-argemonine (**316**) and (−)-*O*-methylplatycerine (**317**), and a representative isopavine is (−)-amurensine (**318**). Analysis of these structures in

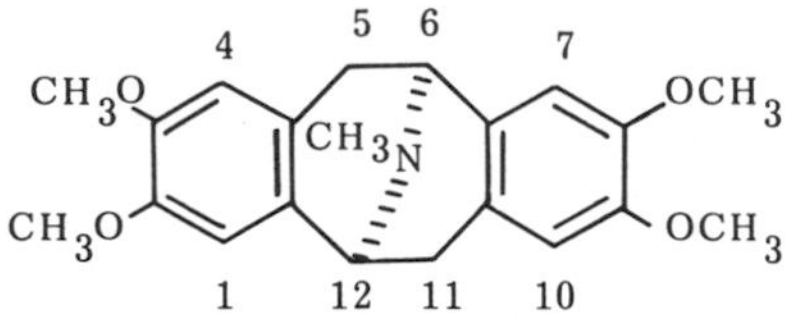

316   (-)-argemonine

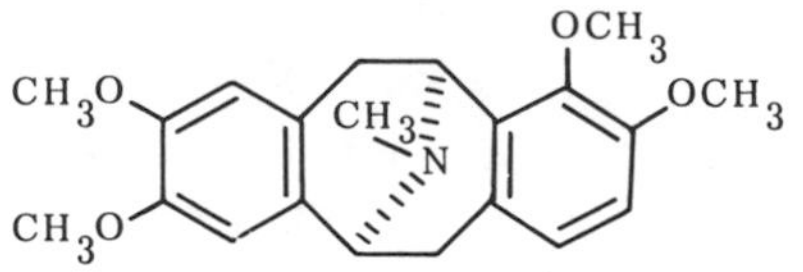

317   (-)-O-methylplatycerine

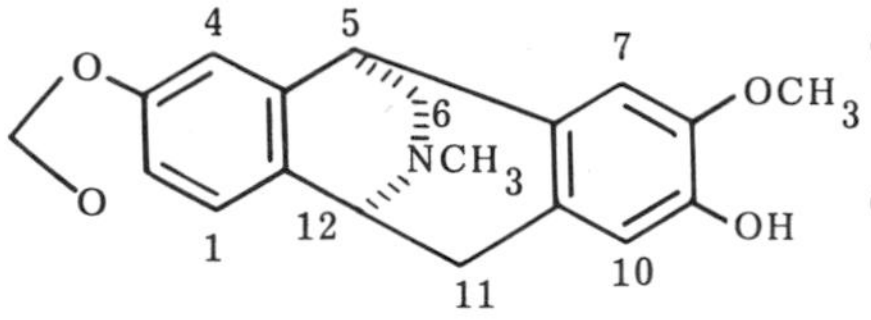

318   (-)-amurensine

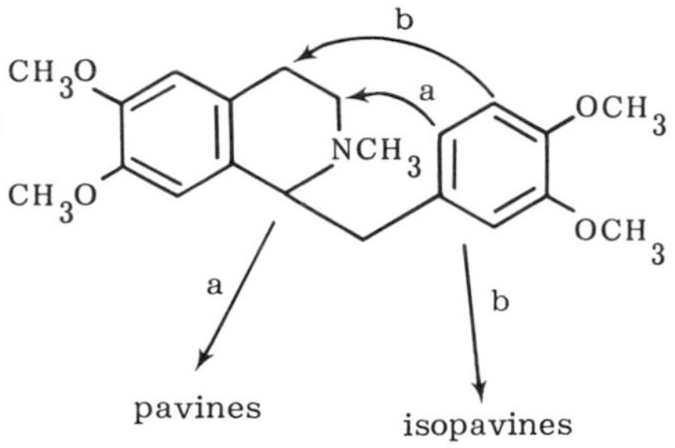

pavines

isopavines

comparison with the tetrahydrobenzylisoquinoline nucleus indicates that they may be formally derived from this nucleus by the joining of bonds a and b.

8.16.1   Chemistry

The pavine skeleton is one of those rare situations in alkaloid chemistry where synthesis of the skeleton predated discovery of a natural source. At the turn of the century Goldsmeidt found that reduction of papaverine (**183**) with tin and hydrochloric acid afforded two products, the expected tetrahydropapaverine (**233**) and a secondary base analyzing as a dihydropapaverine, but which was given the name pavine. The structure of this compound remained unknown until the 1950s when both Schöpf and Battersby deduced its structure to be **319** by degradation.

The isopavine skeleton was also produced synthetically before any natural compounds having this skeleton were known. When the aminoacetal **320** was treated with concentrated sulfuric acid, instead of obtaining pa-

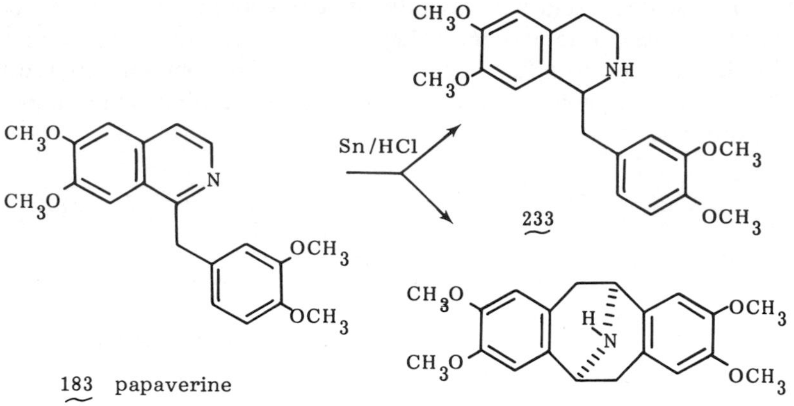

183   papaverine

233

319   pavine

paverine by Pomeranz–Fritsch cyclization, isopavine (**321**) was produced. To date, however, neither isopavine (**321**) nor pavine (**319**) have been isolated from natural sources.

(−)-Argemonine (**316**) was first isolated in 1902 from *Argemone mexicana* L. and has subsequently been found in several *Argemone* specics of

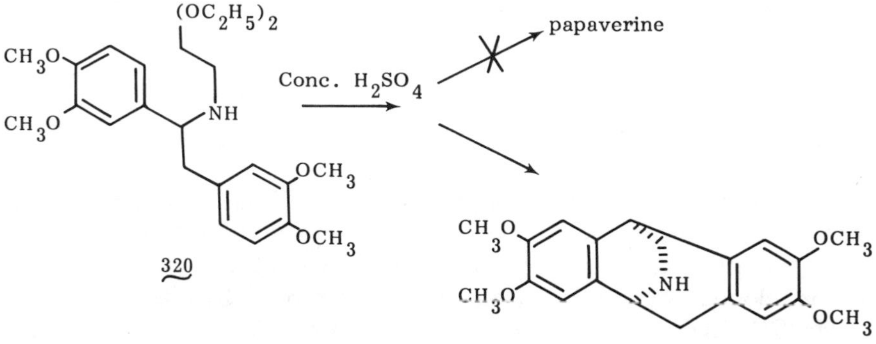

320

321   isopavine

the Papaveraceae, as well as *Leontice smirnowii* and *Thalictrum strictum*. The structure was deduced in 1963 by Stermitz and co-workers to be *N*-methylpavine.

An interesting structure proof by synthesis was required in order to characterize (−)-norargemonine, which frequently cooccurs with argemonine (**316**). The symmetry of the system means that only two structures are possible for this compound, **322** and **323**. It was reasoned by Stermitz that these two possibilities could be distinguished by synthesis from the two nor-*N*-methylpapaverines **324** and **325**. Reduction of each of these products with

tin and hydrochloric acid gave mainly the corresponding tetrahydropapav-
erines, but some norargemonine (**322**) was produced from the isomer **324**,
thereby defining the structure. The pavine derivative obtained by reduction
of **325** was isonorargemonine (**323**), which subsequently was obtained from
natural sources.

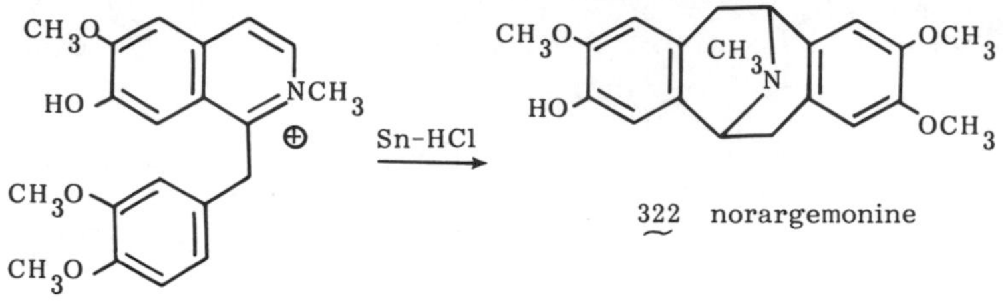

324

322   norargemonine

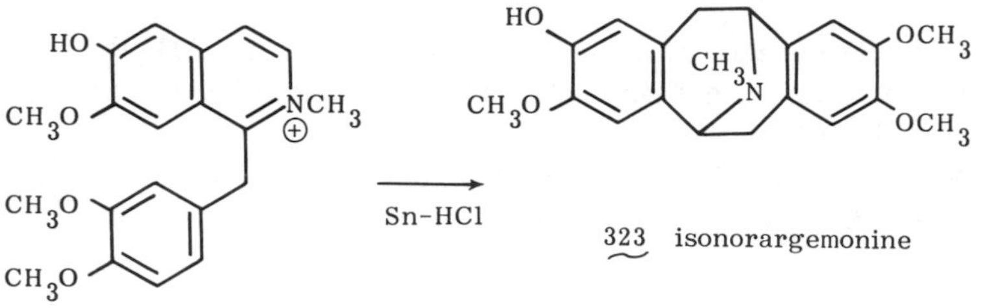

325

323   isonorargemonine

8.16.2   Absolute Configuration

To date, all of the pavines and isopavines isolated have a negative optical
rotation. This corresponds to the absolute configuration shown in **316** for
(−)-argemonine, and has been confirmed by degradation of *N*-benzyllarge-
monine chloride to an aspartic acid derivative of known absolute configuration.

The CD bands of (−)-argemonine also serve to define the absolute configuration based on the exciton theory of the optical rotatory power of
dimeric systems. The absolute configuration of the isopavines has been
determined by the two pairs of split CD curves at 285 and 240 nm. Both of
these are negative, indicating that the chirality of the transition is negative
(left-handed) as shown for (−)-amurensine in **326**. This has been confirmed
by synthesis of optically active pavinane and isopavinane alkaloids from the
same optically active precursor.

8.16.3   Spectral Properties

The UV spectra of the pavine alkaloids are not particularly characteristic,
showing a λ_{max} between 287 and 295 nm. Isopavine alkaloids display quite

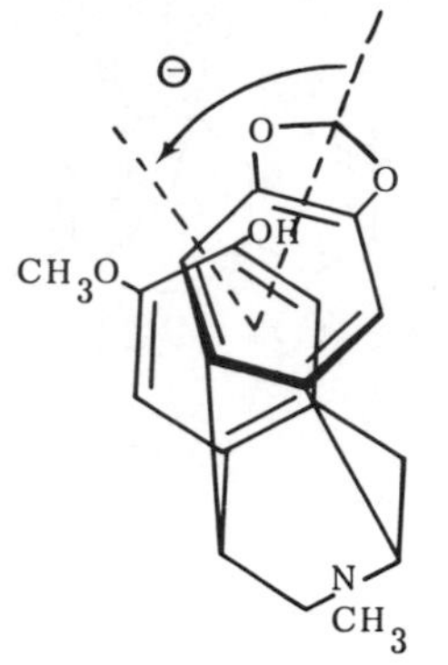

326   (−)−amurensine

similar spectra, although it is considered that a shoulder at 250 nm may be diagnostic for the skeleton.

There is a characteristic difference however between the mass spectra of these compounds. For example the isopavine alkaloid amurensinine (**327**) shows a molecular ion at *m/e* 339, a strong M⁺ − 43 ion and a base peak at *m/e* 188. These ions have been rationalized as shown in Scheme 41.

The proton NMR spectral data of argemonine are shown in **328**, and the deshielding of the C-5 and C-11 protons *cis* to the nitrogen should be noted.

The structure elucidation of munitagine posed some interesting problems in distinguishing between the two possibilities **329** and **330**. Changing the NMR solvent from CDCl₃ to d₆-DMSO, and then adding base, shifted both

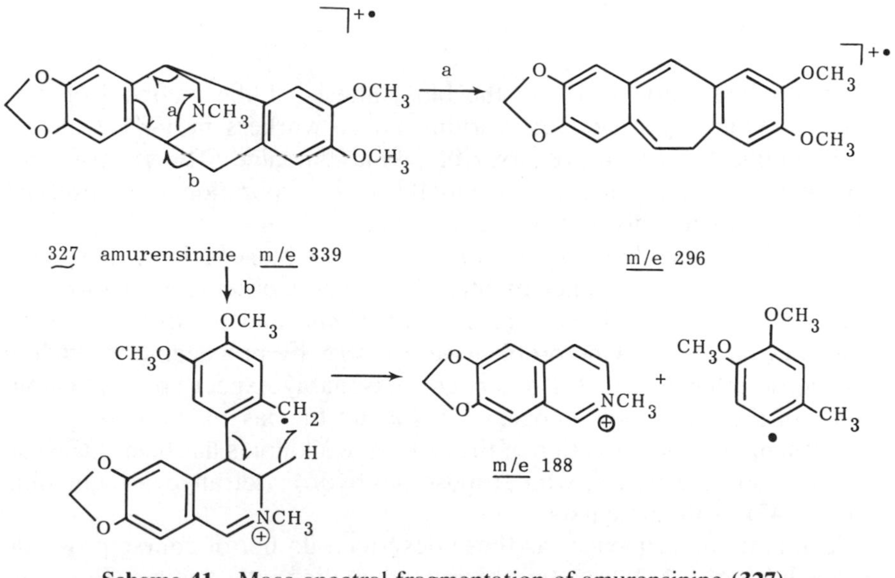

327   amurensinine   *m/e* 339   *m/e* 296

m/e 188

Scheme 41   Mass spectral fragmentation of amurensinine (**327**).

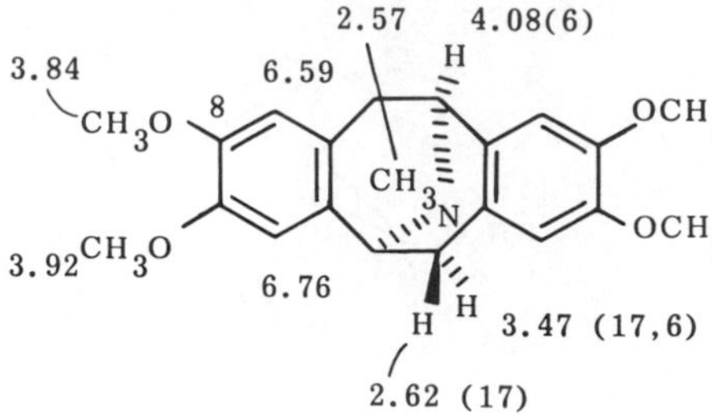

328   argemonine

the C-1 and C-10 protons upfield. The magnitude (0.7 ppm) of the shift for the C-10 proton demonstrated that this proton was *para* to a phenolic group. Two Ar-CH-N protons were observed at 3.93 and 4.38 ppm, each being a doublet, $J = 6$ Hz. The marked deshielding of one of these protons indicates that H-6 is adjacent to a phenolic group and therefore 329 is the correct structure for munitagine.

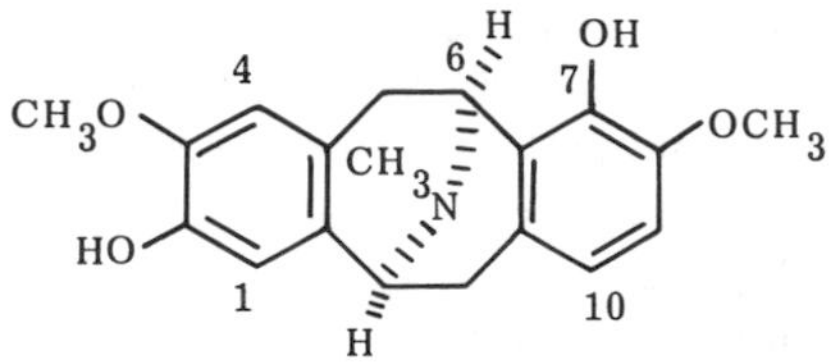

329 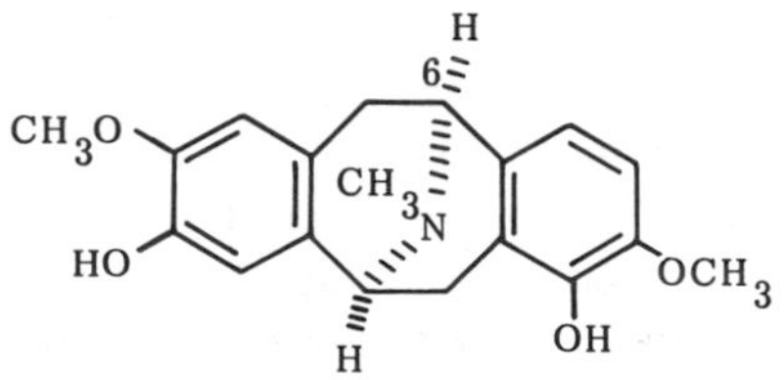   330

8.16.4   Biosynthesis

There has been little work on the biosynthesis of the pavine–isopavine alkaloids. One experiment by Barton and co-workers in 1965 evaluated (+)-reticuline (182) as a precursor of (−)-argemonine (316) in *Argemone mexicana* L, (Papaveraceae), Essentially no incorporation was observed, but the reasons for this were not established.

Several schemes have been proposed for the biogenesis of these alkaloids. One of these, developed by Stermitz and co-workers for the formation of the principal alkaloids of *Argemone mexicana* and *A. hispida* Gray, is shown in Scheme 42. One problem here is that by proceeding through an asymmetric intermediate 331 one might reasonably expect both optical antipodes of these alkaloids to occur. This is not the case.

A scheme for the formation of the isopavine alkaloids has been suggested by Dyke and co-workers, who propose a 4-hydroxytetrahydroisoquinoline (Scheme 43) as the precursor.

Such reaction sequences as those described do not of course prove the intermediacy of a 4-hydroxytetrahydroisoquinoline intermediate. But as we

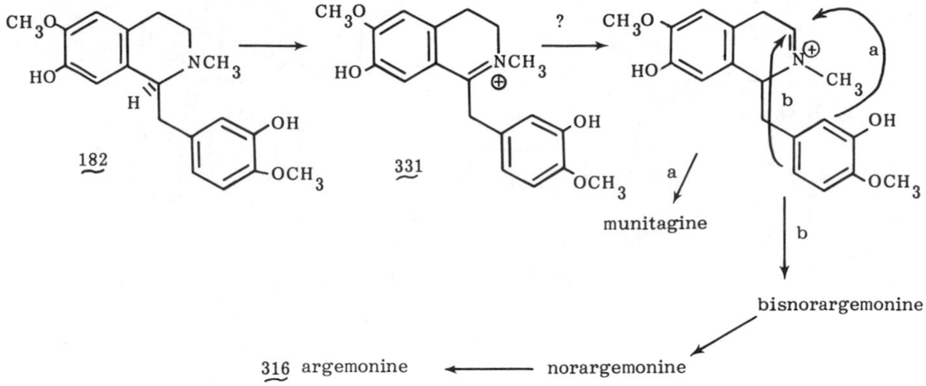

Scheme 42

shall see in subsequent sections there are several other cases where such an intermediate explains quite elegantly the occurrence of a particular compound type.

A biogenetic-type synthesis along these lines was observed when the acetal **332** was treated with acid, for both an isopavine (**333**) and a pavine (**324**) derivative were produced (Scheme 44). In 1971, Dyke and Ellis reported the first hydration of a 1,2-dihydrobenzylisoquinoline (**210**) to a 4-hydroxy-1,2,3,4-tetrahydrobenzylisoquinoline (**335**) and showed that such

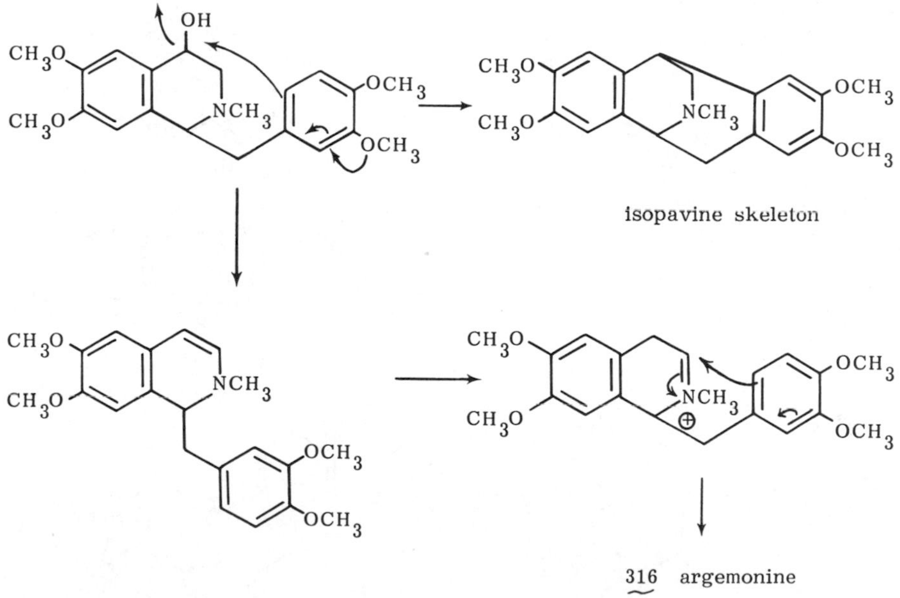

Scheme 43 Dyke biogenesis for the pavine–isopavine alkaloids.

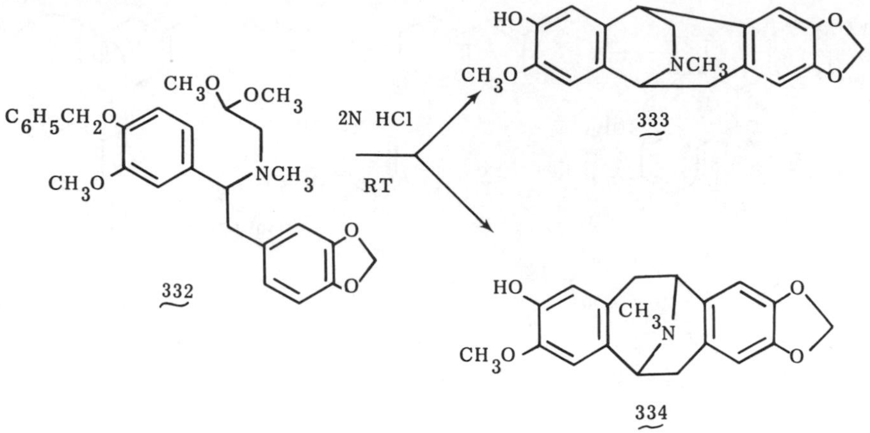

Scheme 44

a compound could be converted to the isopavine 336 by acid treatment. These 4-hydroxylated products can also be prepared by the lead tetraacetate oxidation of the appropriate tetrahydroisoquinoline.

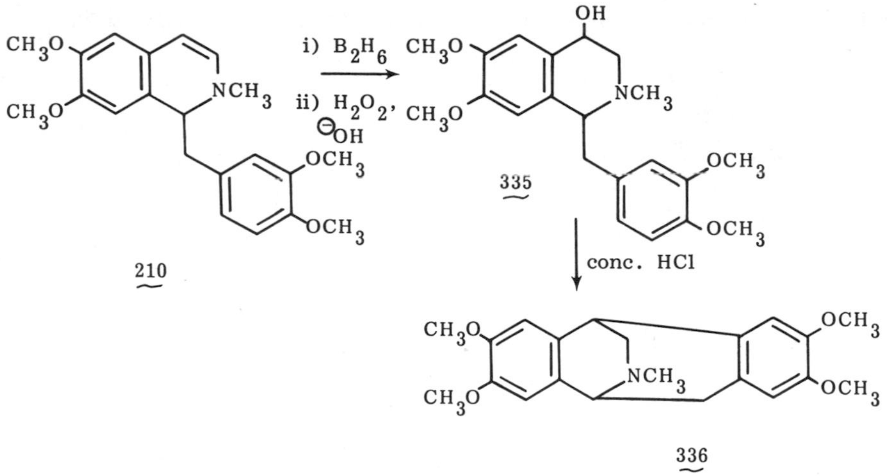

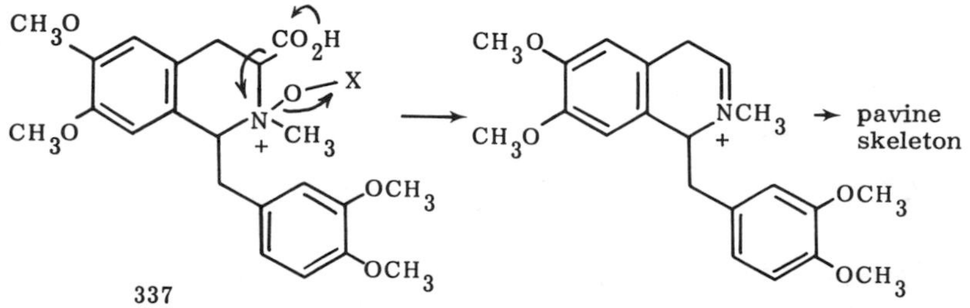

A hypothesis for the pavine skeleton involves generation of an iminium species from an amino acid N-oxide precursor such as 337. Oxidative decarboxylation reactions such as this can be carried out *in vitro* and have been postulated *in vivo* in other systems, (e.g. ajmaline biosynthesis, Section 9.19).

LITERATURE

Reviews

Shamma, M. *The Isoquinoline Alkaloids*, Academic, New York, 1972, p. 96.

Shamma, M. and J. L. Moniot, *Isoquinoline Alkaloid Research, 1972–1977*, Plenum, New York, 1978, p. 61.

Santavý, F., *Alkaloids, NY* **12**, 370 (1970).

Kametani, T., and K. Fukumoto, *Heterocycles* **3**, 931 (1975).

Chemistry and Synthesis

Stermitz, F. R., and J. N. Seiber, *Tetrahedron Lett.* 1177 (1966).

Coomes, R. M., J. R. Falck, D. K. Williams, and F. R. Stermitz, *J. Org. Chem.* **38**, 3701 (1973).

Dyke, S. F., and A. C. Ellis, *Tetrahedron* **27**, 3803 (1971).

Dyke, S. F., R. G. Kinsman, P. Warren, and A. W. C. White, *Tetrahedron* **34**, 241 (1978).

Biosynthesis and Biogenesis

Barton, D. H. R., R. H. Heese, and G. W. Kirby, *J. Chem. Soc.* 6379 (1965).

Brown, D. W., S. F. Dyke, G. Hardy, and M. Sainsbury, *Tetrahedron Lett.* 1515 (1964).

Shamma, M., J. L. Moniot, W. K. Chan, and K. Nakanishi, *Tetrahedron Lett.* 3425 (1971).

Dyke, S. F., *Heterocycles* **6**, 1441 (1977).

8.17 PROAPORPHINE ALKALOIDS

There are over 30 proaporphine alkaloids distributed in the plant families Euphorbiaceae, Lauraceae, Menispermaceae, Monimiaceae, Nymphaceae, and Papaveraceae. Fundamentally they are of two structure types, those with a dienone system, such as (+)-pronuciferine (**338**) and (+)-glaziovine (**339**), and those in which the dienone system has been completely or partially reduced, such as (+)-linearisine (**340**) and (−)-oreoline (**341**). Compounds

338 (+)-pronuciferine R=CH$_3$

339 (+)-glaziovine R=H

340 (+)-linearisine

341 (−)-oreoline

having either stereochemistry at C-6 are known and the two sides of the cyclohexadienone system are not equivalent because of this asymmetry.

(+)-Pronuciferine (338) has been obtained from several *Papaver* species, and is one of the most widely distributed proaporphine alkaloids. Other reported occurrences include *Croton linearis* Jacq. (Euphorbiaceae), *Stephania glabra* Miers (Menispermaceae), and *Nelumbo nucifera* Gaertn. (Nymphaceae).

8.17.1   Chemistry

The structure of pronuciferine was deduced by Bernauer and co-workers in 1963, and some of the typical reactions of a proaporphine are shown in Scheme 45. Of particular importance are the dienone–phenol rearrangement to afford the aporphine 342, and the sodium–liquid ammonia cleavage to give the tetrahydrobenzylisoquinoline (−)-armepavine (343). Sodium borohydride reduction affords a mixture of dienols which on treatment with mineral acid undergo a dienol–benzene rearrangement to give the aporphine (−)-nuciferine (344).

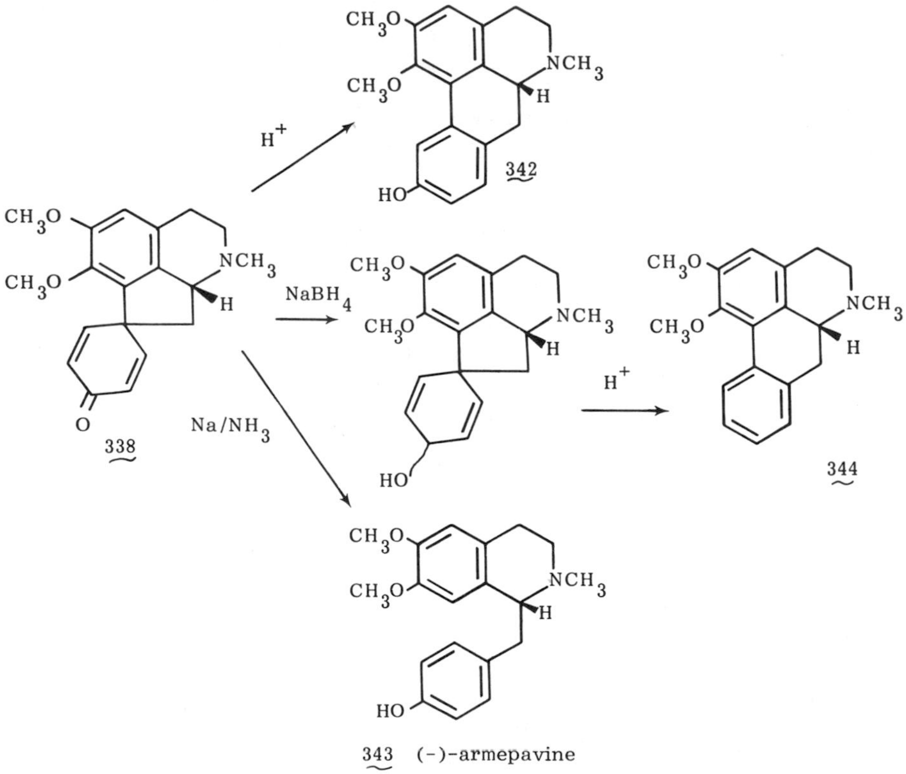

Scheme 45   Chemistry of pronuciferine (338).

The structure elucidation of a reduced proaporphine is usually made by correlation with a dienone proaporphine. Thus (+)-linearisine (340) could be correlated with (+)-crotonosine (345) through enantiomeric 346 and 347, which differ only in configuration at C-6a.

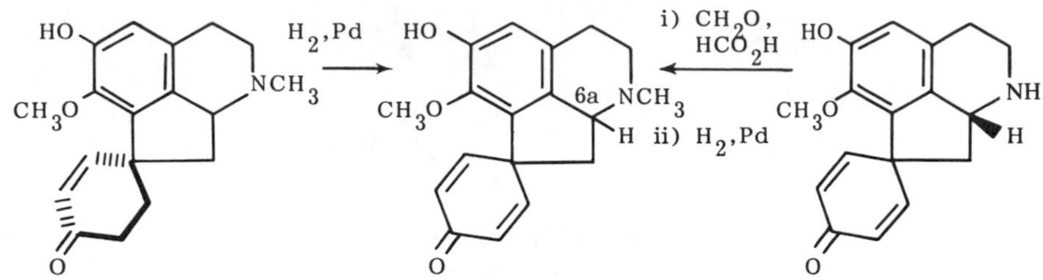

340 (+)-linearisine

346 (−)-dihydrolinearisine
     H-6 α

347 (+)-N-methyltetrahydro-
     crotonosine H-6β

345 (+)-crotonosine

8.17.2   Synthesis

Several successful synthetic approaches to the proaporphine alkaloids have been developed. These fall into the two broad categories, the nonbiogenetic and the biogenetic.

An early synthesis developed by Bernauer involves condensation of the aldehyde 348 with methyl vinyl ketone to give racemic amuronine (349). Dehydrogenation with DDQ under acidic conditions afforded pronuciferine (338).

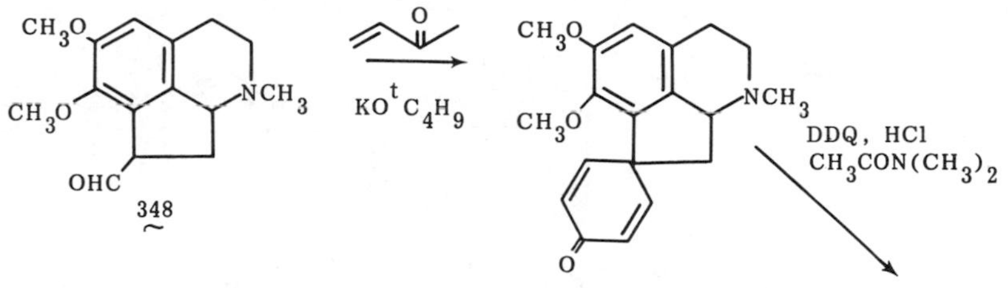

348

349 (±)-amuronine   338 (±)-pronuciferine

A phenolic alkaloid such as anonaine (350) poses a biogenetic problem because it cannot be produced from a tetrahydrobenzylisoquinoline precursor by direct phenolic coupling. Hence Barton and Cohen postulated (Scheme 46) that another skeleton was involved as an intermediate in the formation of the aporphine alkaloids. They also envisaged that an alkaloid such as roemerine (351) which has no phenolic substituent on ring D could also be produced by reduction of the intermediate dienone 352 to a dienol

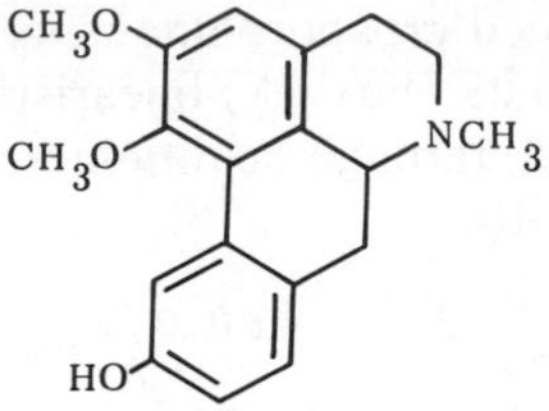

350   anonaine 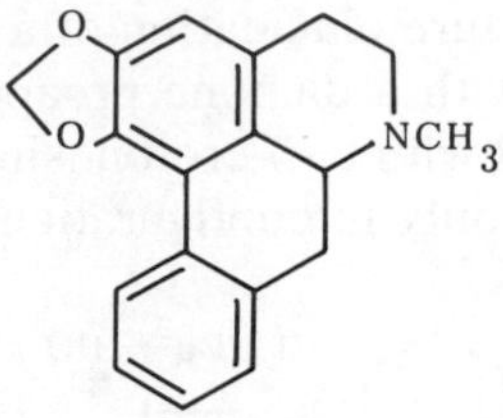   351   roemerine

353. At the time of the original proposal (1957) this intermediate skeleton, now known as a proaporphine, was not known.

The first synthesis of an alkaloid by this biogenetic approach was in 1964 by Battersby and co-workers, who obtained orientalinone (**354**) in very low yield by the potassium ferricyanide oxidation of the orientaline (**355**). Kametani carried out a similar oxidation of N-methylcoclaurine (**288**) to give glaziovine (**339**), which could be converted to pronuciferine (**338**) by methylation with diazomethane.

For the most part synthesis of proaporphines by this biogenetic approach is not very successful, and consequently there have been efforts to produce them by pseudobiogenetic routes in which somewhat less reactive intermediates are involved.

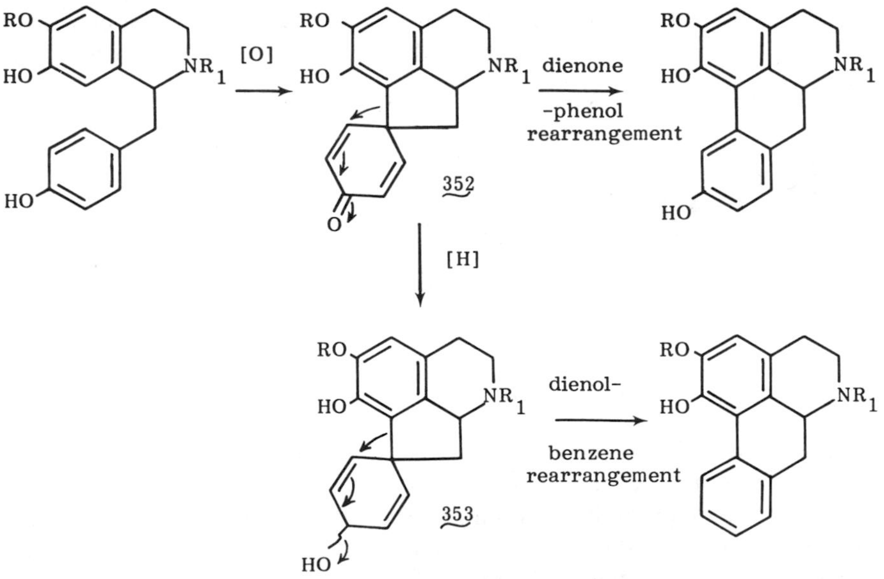

Scheme 46

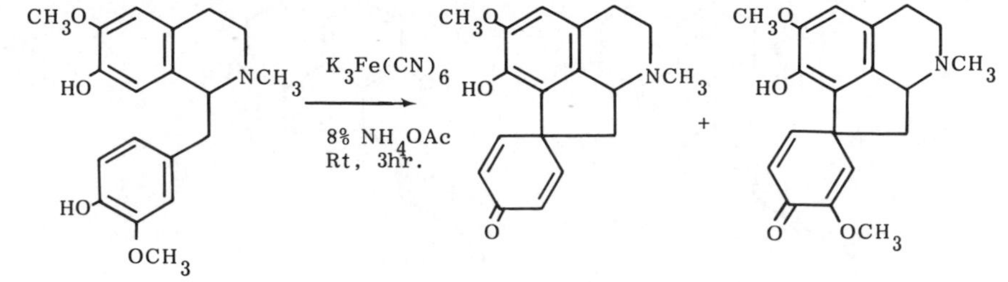

355   354   orientalinone

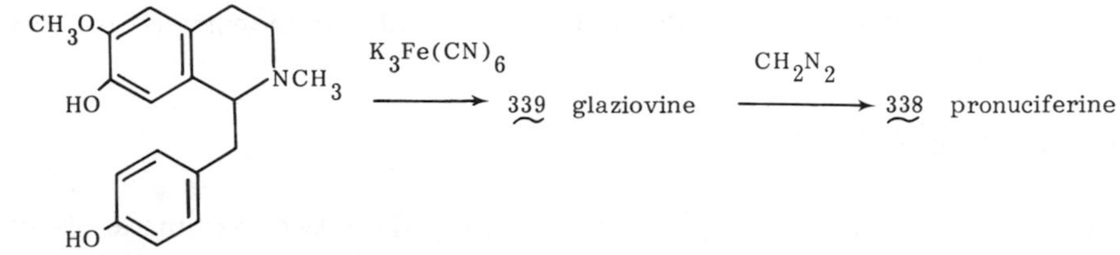

288   N-methylcoclaurine

Casagrande and Canonica have synthesized glaziovine (339) in 45% yield by diazotization of the 8-aminotetrahydrobenzylisoquinoline (356) and irradiation of the diazo intermediate. A somewhat lower yield (26%) of 339 was obtained when the 8-bromo derivative 357 was irradiated and the product debenzylated.

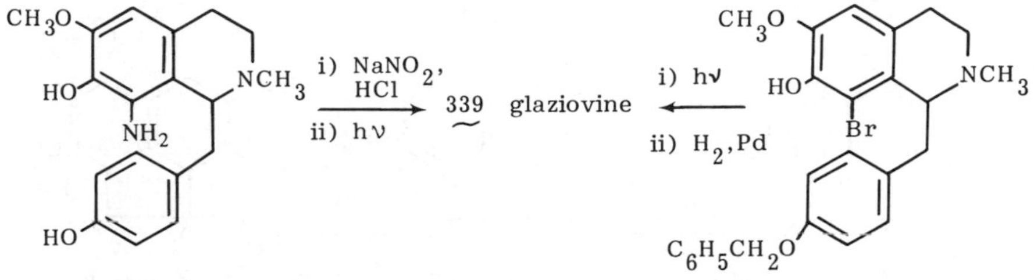

356   357

8.17.3 Absolute Configuration

The simple proaporphines such as pronuciferine are converted by acid to aporphines of known absolute configuration, but the partially reduced compounds pose a more difficult problem.

Comparison with compounds in the steroid series established that the configuration shown in 358 would have a positive Cotton effect near 334 nm, whereas 359 should show a negative effect. Since linearisine showed a pos-

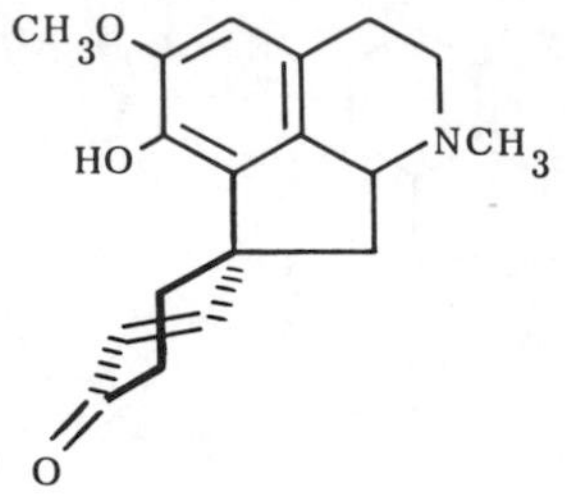

358 correct

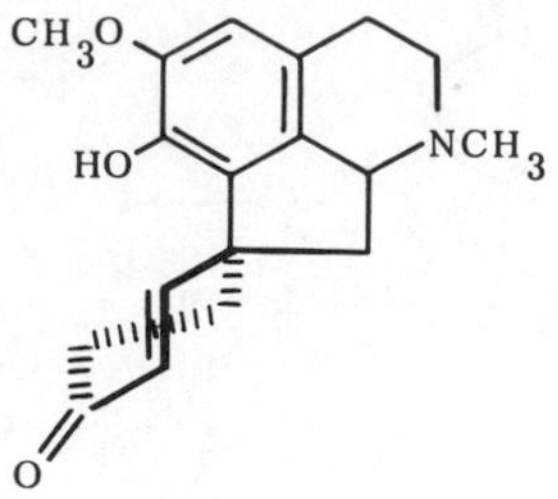

359 incorrect

itive Cotton effect, the absolute configuration **340** is demonstrated for this compound.

8.17.4 Spectral Properties

The UV spectra of the proaporphines typically show two maxima at about 230 and 285 nm (4.4 and 3.5) if a dienone system is present, and at 228, 282, and 288 nm (4.30, 3.19, and 3.22) if a partially reduced system is present.

In their mass spectra most proaporphine alkaloids show the molecular ion as a base peak with prominent ions at $M^+ - 1$ and $M^+ - 1$–28 from one pathway and at $M^+ - CH_2 = NR$ by retro Diels–Alder reaction in ring B from a second pathway (Scheme 47).

The proton NMR spectra of the dienone proaporphine alkaloids are readily assigned as indicated for crotonosine (**345**). The quite substantial cross-ring coupling constants are of interest.

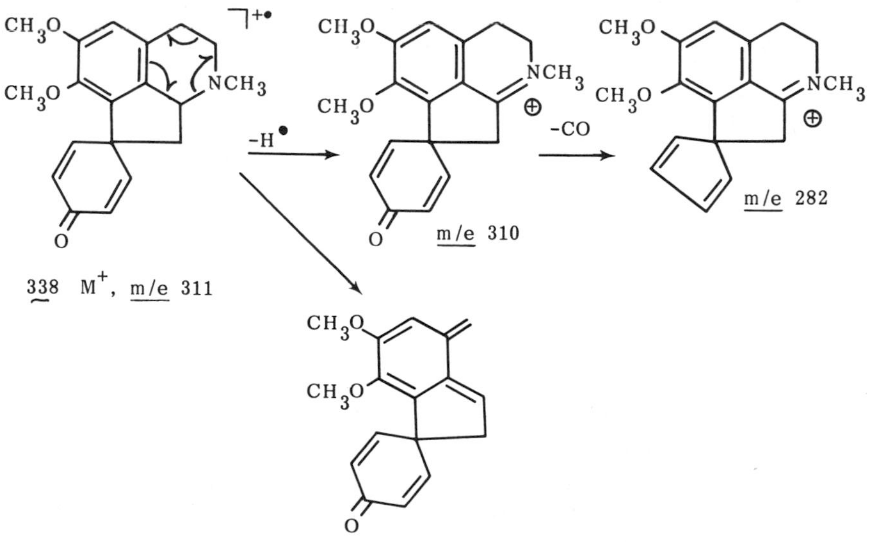

338 M^+, m/e 311

m/e 310

m/e 282

Scheme 47

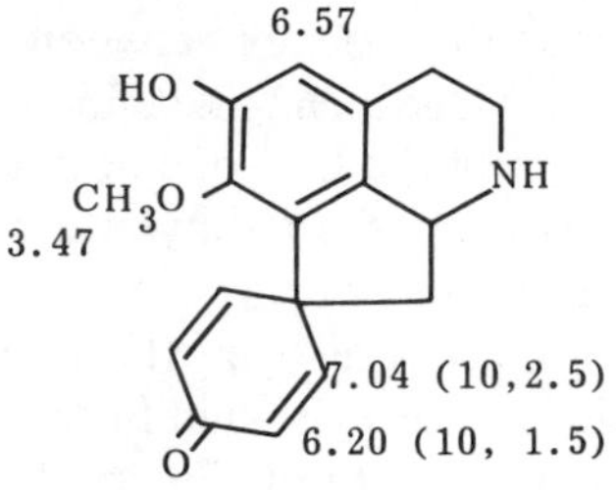

6.57

HO

CH$_3$O
3.47

NH

7.04 (10, 2.5)
6.20 (10, 1.5)

345 crotonosine

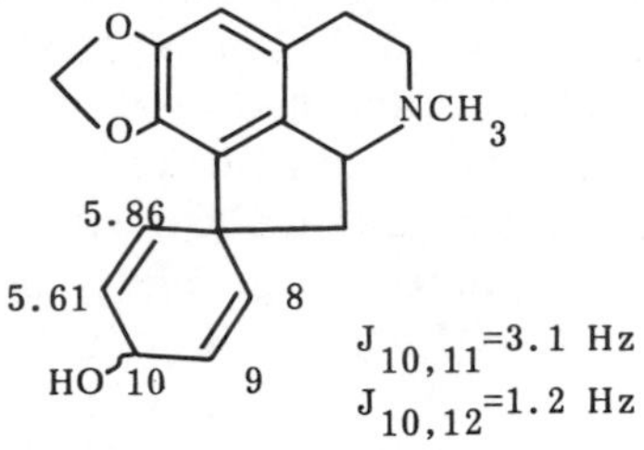

O
O

NCH$_3$

5.86

5.61

HO 10   9   8

$J_{10,11}$=3.1 Hz
$J_{10,12}$=1.2 Hz

360

The 1- and 2-methoxyl groups may be distinguished by their quite characteristic chemical shifts (1-methoxy ~ 3.8 ppm). A distinction between the 1-methoxy-2-hydroxy- and 2-methoxy-1-hydroxy arrangements can be made by acetylation, for a 3-proton is deshielded from 6.5–6.7 to 6.7–6.9 ppm by esterification of a C-2-hydroxy group.

From the coupling constants observed for the C-10 proton in roemeramine (360) it was deduced that the hydroxyl group was axial.

The ^{13}C NMR data for some representative proaporphine alkaloids in various oxidation states are shown in Table 6 and are due to Ricca and

Table 6 ^{13}C NMR Data of Some Representative Proaporphine Alkaloids

Carbon	Compound			
	361	**362**	**363**	**364**
1	141.5	140.9	140.8	141.2
2	147.6	148.0	147.9	147.6
3	110.7	110.0	110.2	109.2
3a	124.7	130.3	129.2	131.1
4	26.8	26.9	26.8	26.6
5	54.6	54.6	54.6	54.5
6a	65.2	65.3	64.6	64.7
7	46.7	43.0	48.8	50.2
7a	50.5	47.0	47.8	47.5
7b	134.8	134.0	134.5	134.3
7c	122.0	121.5	121.5	120.9
8	154.7	157.5	33.1	31.8
9	127.7[a]	126.8	35.2	30.0
10	185.5	198.9	198.5	65.5
11	126.7[a]	35.2	126.8	130.7
12	150.8	30.7	155.1	132.7
NCH$_3$	43.3	43.4	43.4	43.9
2-OCH$_3$	56.4	56.3	56.4	56.2

[a] Assignments may be reserved.

Casagrande. Of interest in these data are the differences between the chemical shifts of C-8 and C-12 in the diastereomeric pair **362** and **363**. The upfield shift of C-12 was attributed to the interaction of the hydrogen on the olefinic C-12, or the equatorial hydrogen if C-12 is part of a methylene group with the angular hydrogen on C-6a. The chemical shift of the C-7 atom is determined by the molecular geometry in ring C, and also by the diastereoisomerism at C-7. In the 11,12-dihydro series there is significant γ-gauche interaction between the axial C-11 hydrogen and the C-7 α-hydrogen; this accounts for the relatively upfield shift of C-7 in **363** and **364**.

361 (−)-glaziovine   362   363 R==O
                                         364 R=H, α-OH

8.17.5 Biosynthesis

There have been relatively few studies of the biosynthesis of proaporphine alkaloids although their importance in the biosynthesis of certain aporphine alkaloids is unquestioned.

In 1965, Barton and co-workers showed that (+)-coclaurine (**365**) labeled as indicated was incorporated into (+)-crotonosine (**345**) in *Croton linearis*. The isomer (+)-isococlaurine (**366**) was not incorporated, which tends to support the theory that coupling occurs only *ortho* or *para* to a phenolic hydroxyl group. This is surely not the major pathway to crotonosine (**345**) however, for of necessity it would involve a dimethylation–remethylation sequence.

Another highly unusual interconversion is the 0.03% incorporation of (+)-linearisine (**340**) into (+)-crotonosine (**345**), which would involve a rare N-demethylation as well as C-6a isomerization and ring D oxidation. Note that only ^{3}H-labeled precursor was used and the activity is more likely due to tritium exchange than to true incorporation.

Bhakuni and co-workers found that [2-^{14}C]tyrosine was well incorporated into crotosparine (**367**) and crotosparinine (**368**) in *Croton sparsiflorus* and both (±)-coclaurine (**365**) and (±)-norcoclaurine (**369**) were also efficient precursors. Isococlaurine (**366**) on the other hand was not incorporated. The more efficient (1.9 *vs.* 1.6%) incorporation of **365** than **369** suggests that the scheme **369** → **365** → **367** → **368** is operating. The whole area of proaporphine biosynthesis is in need of considerable further experimental work.

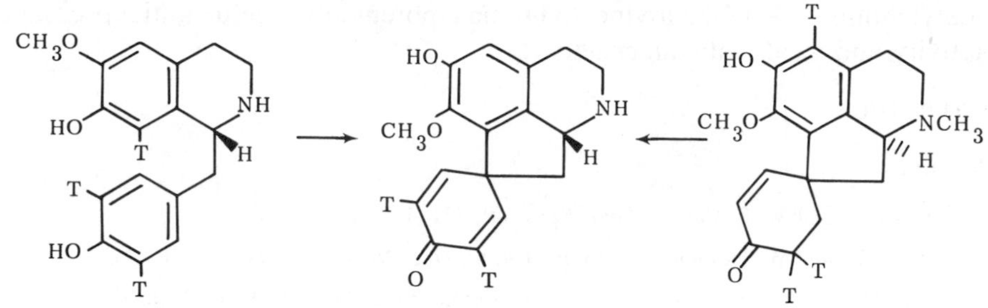

365 (+)-coclaurine   345 (+)-crotonosine   340 (+)-linearisine

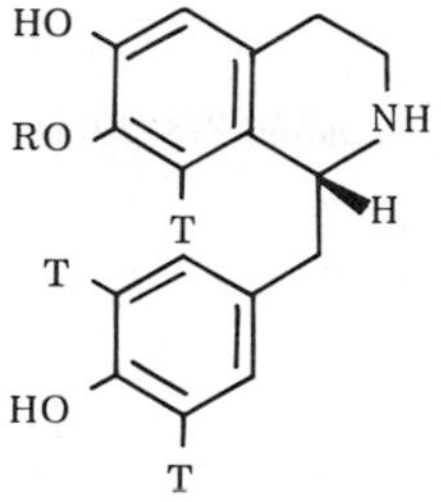

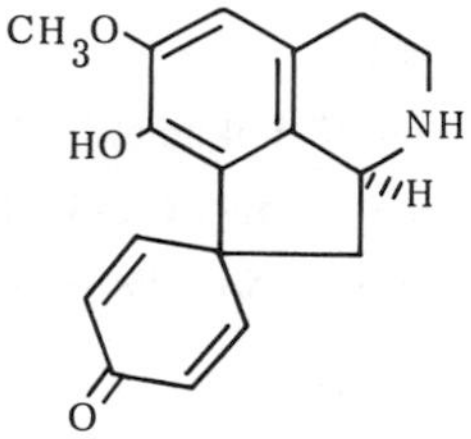

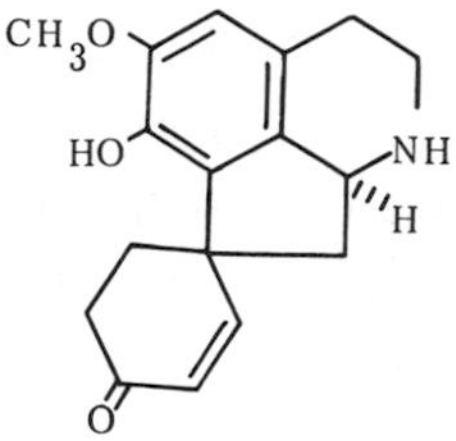

366 isococlaurine R=CH$_3$ 367 crotosparine   368 crotosparinine

369 norcoclaurine R=H

Some examples of biosynthesis of aporphines which appear to involve a proaporphine are discussed in the next section.

8.17.6 Pharmacology

(−)-Mecambrine (370) increases the motility of isolated rat or rabbit duodenum, and antagonizes the effects induced by histamine on the isolated guinea pig ileum. It is quite toxic (LD$_{50}$ 4.1 mg/kg in mice) and death is due to clonicotonic convulsions.

(+)-Pronuciferine (338) has an LD$_{50}$ of 120 mg/kg for mice and a mild local anesthetic effect. In the guinea pig ileum it has a synergistic effect with

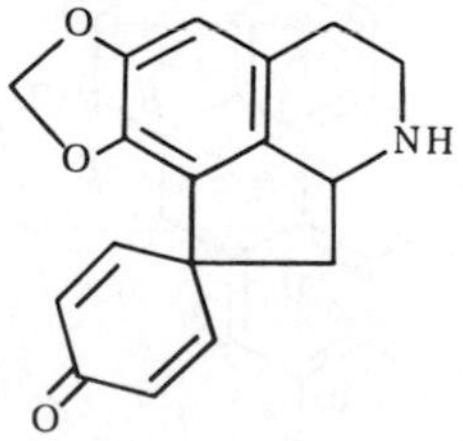

370 (−)-mecambrine

acetylcholine. (+)-Glaziovine (**339**) has potentially useful antidepressant activity and weak anticancer activity.

LITERATURE

Reviews

Stuart, K. L., and M. P. Cava, *Chem. Revs.* **68**, 321 (1968).

Bernauer, K., and W. Hofheinz, *Fortschr. Chem. Org. Naturs.* **26**, 245 (1968).

Shamma, M., *The Isoquinoline Alkaloids*, Academic Press, New York, 1972, p. 177.

Shamma, M. and J. L. Moniot, *Isoquinoline Alkaloid Research, 1972–1977*, Plenum, New York, 1978, p. 117.

Synthesis

Bernauer, K., *Experientia* **20**, 380 (1964).

Kametani, T., H. Sugi, S. Shibuya, and K. Fukumoto, *Chem. Ind. (London)* 818 (1971).

Kametani, T., and H. Yagi, *J. Chem. Soc., Sec. C* 2182 (1967).

Casagrande, C., and L. Canonica, *J. Chem. Soc. Perkin Trans I* 1647 (1975).

Spectral Properties

Baldwin, M., A. G. London, A. MacColl, L. J. Haynes, and K. L. Stuart, *J. Chem. Soc., Sec. C* 154 (1967).

Dolejs, L., *Collect. Czech. Chem. Commun.* **39**, 571 (1974).

Ricca, G. S., and C. Casagrande, *Org. Mag. Reson.* **9**, 8 (1977).

Biosynthesis

Haynes, L. J., K. L. Stuart, D. H. R. Barton, D. S. Bhakuni, and G. W. Kirby, *Chem. Commun.* 141 (1965).

Bhakuni, D. S., S. Satish, H. Uprety, and R. S. Kapil, *Phytochemistry* **13**, 2767 (1974).

8.18 APORPHINE ALKALOIDS

The aporphines are the largest group (over 100) of isoquinoline alkaloids and are represented by the general structure **371**. These alkaloids are distributed in at least 18 plant families, of which the most important are the Papaver-

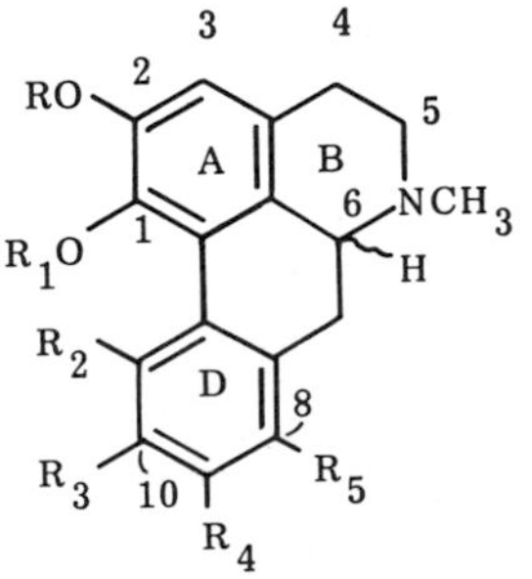

371

aceae, Annonaceae, Lauraceae, and Monimiaceae. The nitrogen atom is usually methylated; and although some noraporphines are known, they are not very stable and are often characterized as their *N*-acetyl derivatives. Aporphines are known with the C-6a stereochemistry either α or β.

The most diverse structural feature of the aporphines is the oxygenation pattern. Positions 1 and 2 are always oxygenated, either by hydroxy, methoxy or methylene dioxy groups. It is common to find further oxygen substituents at C-9, C-10, and C-11, and occasionally at C-8. It is rare to find oxygenation at C-7, except in the oxoaporphines, and even rarer to find any oxygenation in ring B. Some representatives aporphine alkaloids are shown in Figure 7.

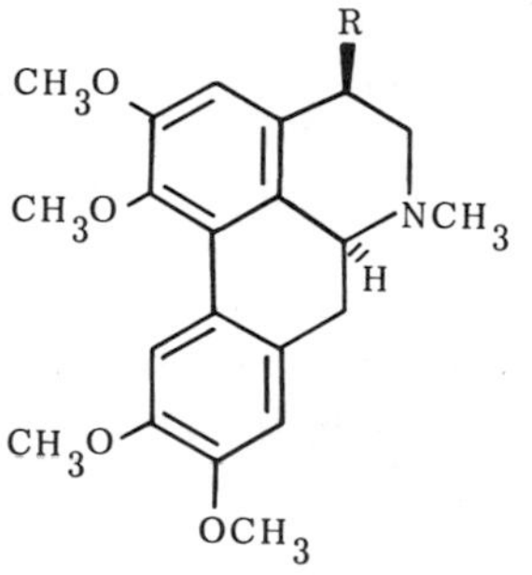

372 (+)-glaucine R=H
373 (+)-cataline R=OH

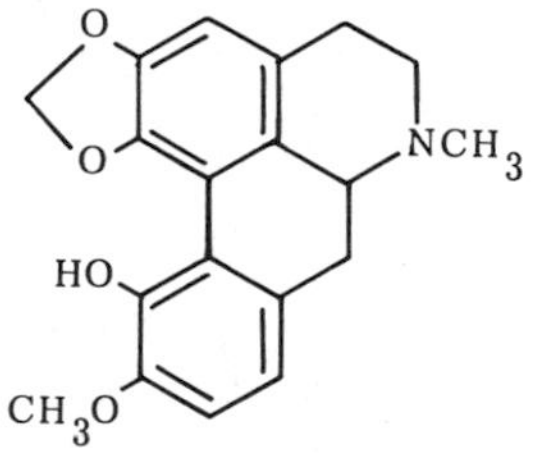

374 (+)-bulbocapnine

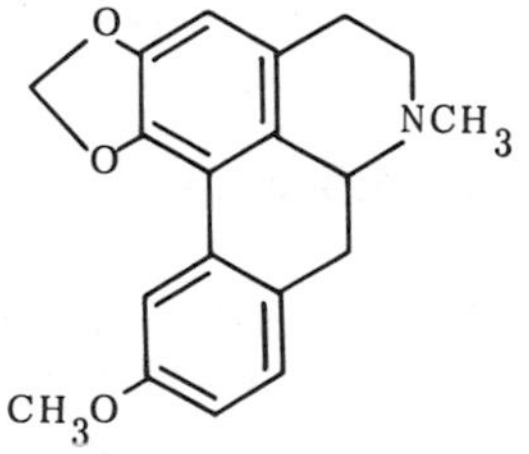

375 (−)-laureline

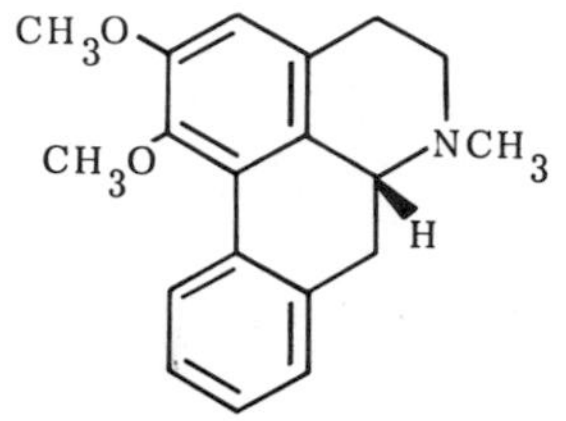

376 (−)-nuciferine

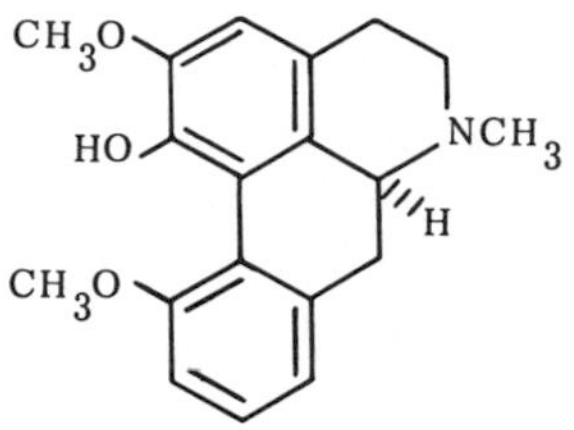

377 (+)-isothebaine

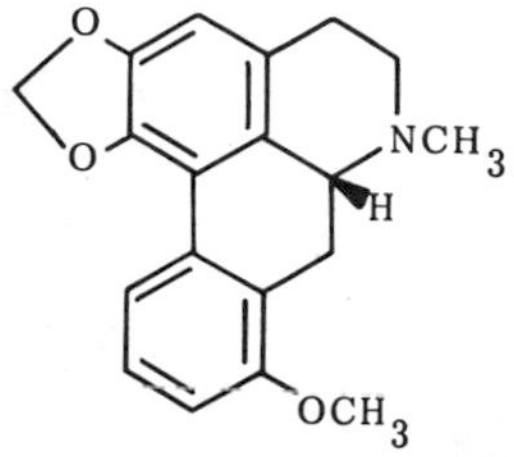

378 (−)-stephanine

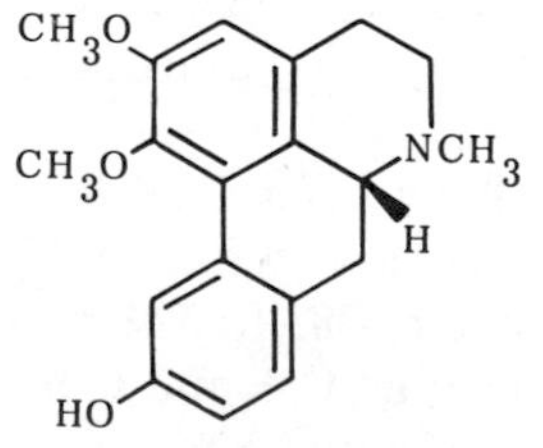

379 (+)-apoglaziovine

Figure 7 Representative aporphine alkaloids.

The first aporphine alkaloid was obtained not by isolation but as the result of chemical reaction. Thus in 1869 it was found that hot concentrated hydrochloric acid caused rearrangement of morphine (5) to apomorphine (380), which is not a natural product.

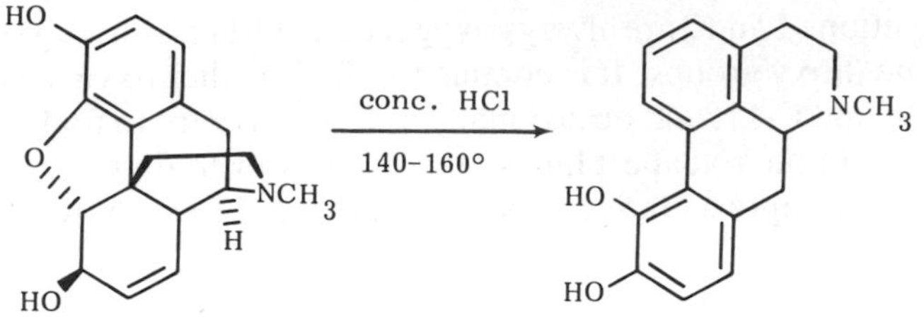

5   380   apomorphine

8.18.1   Chemistry

There are several standard reactions of aporphines worthy of mention. Oxidation of a nonphenolic aporphine alkaloid such as nuciferine (376) with potassium permanganate under mild conditions affords a phenanthrene 381,

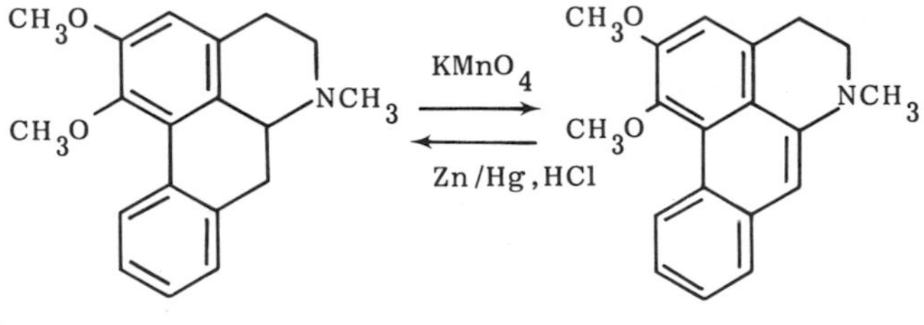

376   381

but iodine or 10% Pd-C in refluxing acetonitrile are preferable reagents. The reverse reaction can be carried out with zinc amalgam.

The phenolic groups at various positions or an aporphine nucleus display different reactivities toward methylating agents such as diazomethane and the sequence is apparently C-9 > C-10 > C-1. The methylation of isoboldine (382) is a typical example.

There are numerous non-acidic reagents which have been used for the O-demethylation of aryl ethers. The benzyl selenoate ion in hot DMF is quite highly regioselective, as shown in the conversion of 383 to 374 in 70% yield. Groups (methoxyl or methylenedioxy) at positions C-1 and C-9 are particularly susceptible to sodium–liquid ammonia reduction and a typical example is shown for nantenine (384). The importance of this reaction has been discussed previously in connection with the bisbenzylisoquinolines.

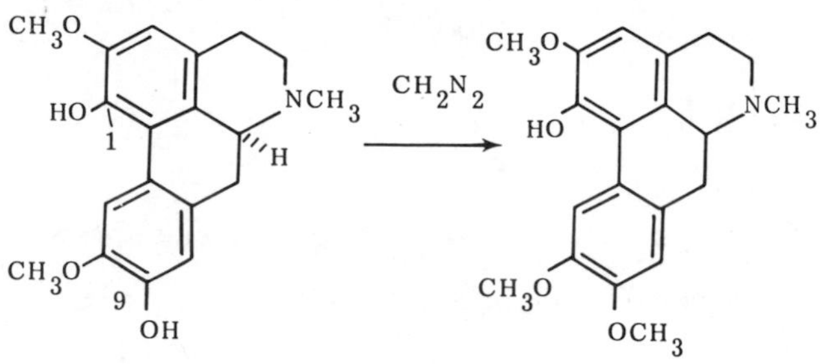

382 (+)-isoboldine

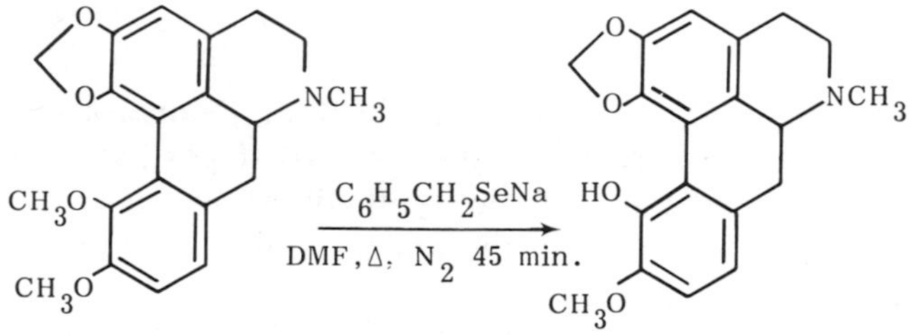

383

374 bulbocapnine

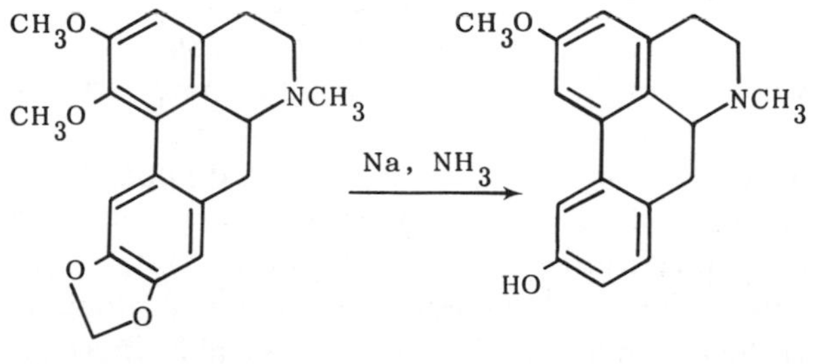

384 nantenine

8.18.2 Synthesis

The synthesis of aporphine alkaloids has been a well-investigated area in recent years and fundamentally there are two overall approaches: (a) non-biogenetically modeled and (b) biogenetically modeled.

Nonbiogenetic Synthesis

The first synthesis of an aporphine was that developed by Pschorr and Gadamer for (±)-glaucine (372). The reaction sequence begins with papaverine (183), which is nitrated at the 6'-position and through a sequence of steps transformed into the tetrahydroisoquinoline 385. The key step is then

183 papaverine

HNO$_3$

385

372 (±)-glaucine

diazotization, followed by Pschorr cyclization. Numerous syntheses of aporphine alkaloids have relied on the last two stages for formation of the tetracyclic system and most of the synthetic effort has concentrated on improving the synthetic route to the aminotetrahydroisoquinoline in order to vary the substitution pattern.

Examples of these varied approaches include (*a*) the condensation of a phenethylamine with an *o*-nitrophenylacetyl chloride followed by Bischler–Napieralski cyclization and reduction, (*b*) condensation of an *o*-nitrotoluene with a quaternary isoquinoline salt and reduction, and (*c*) the condensation of a Reissert compound such as **386** with an *o*-nitrobenzyl halide (reviewed by Popp). These fundamental routes are shown in Figure 8.

The Pschorr cyclization may also be conducted photochemically (Scheme 48), although in this instance, a morphinandicnone is also produced and this in itself can be of preparative value (Section 8.24.4).

There are two other fundamental photochemical processes for the formation of aporphines other than the photochemical Pschorr cyclization. One of these is the cyclization of a benzylidene urethane **388** in the presence of iodine and cupric acetate to give a dehydroaporphine **389** which could be reduced to (±)-nuciferine (**376**).

An alternative efficient procedure developed by Kupchan and co-workers is a photochemical dehydrohalogenation. Improved yields were obtained when the nitrogen lone pair was involved in an electron-withdrawing group such as *N*-acetyl (Scheme 49).

Biogenetic Synthesis

Two leaders in the history of biogenetic theory, Robinson and Barton, played significant roles in the area of aporphine biogenesis. The unsuccessful attempts of Schöpf and Robinson to synthesize aporphines from benzylisoquinolines have been discussed previously. Robinson indicated at that time that such transformations could be carried out if the nitrogen was not basic.

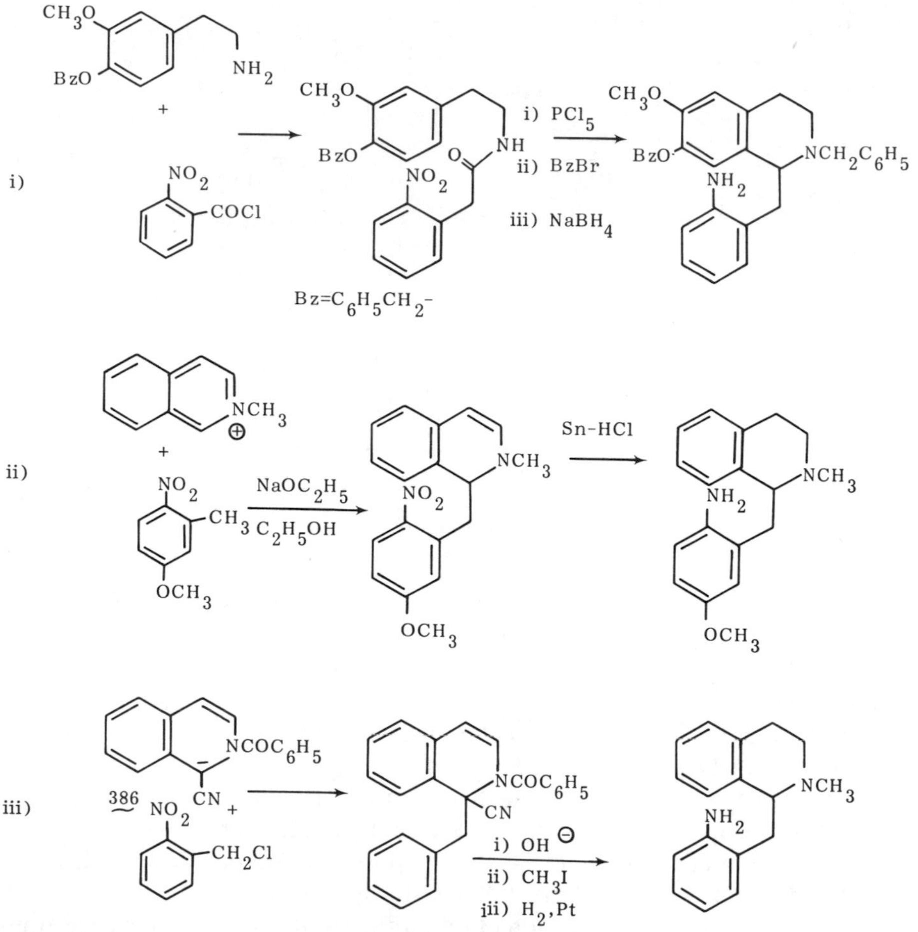

Figure 8 Fundamental routes to the aporphine skeleton involving a thermal Pschorr cyclization.

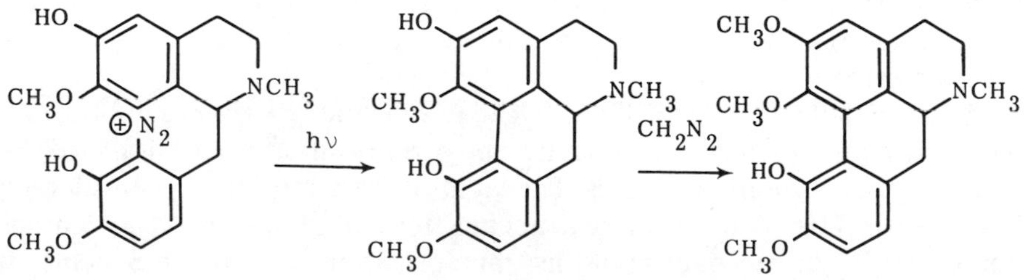

387 isocorydine

Scheme 48

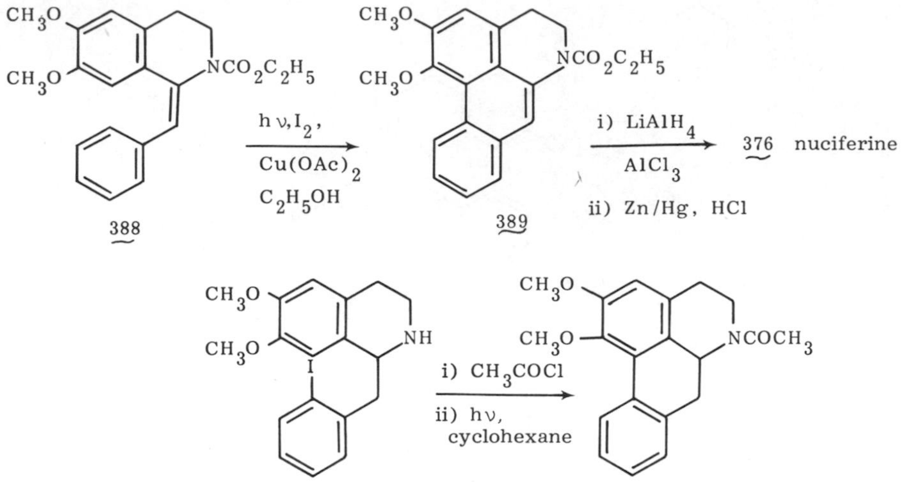

Scheme 49

In 1957, Barton and Cohen suggested that aporphines could be derived through proaporphine intermediates.

It was not until 1962 that the first successful biogenetic synthesis of an aporphine from a benzylisoquinoline was reported. Franck and Schlingloff found that oxidation of laudanosoline methiodide (**390**) with ferric cholride at 20°C for 29 hr gave **391** in 60% yield.

The most studied system is that of a nor-reticuline having either an *N*-methyl group or an *N*-carbethoxy group. Oxidation of **392** with potassium ferricyanide in dilute ammonia gave the aporphine **393** in 5–7% yield, which could be reduced to isoboldine (**382**) with lithium aluminum hydride.

The direct phenolic oxidation of reticuline (**148**) under various conditions has been studied by several groups. In no case has the yield of isoboldine (**382**) been greater than 6% and the three routes shown are equally effective. Reactions are typically carried out under very dilute conditions.

An exception to the normal dilute oxidizing conditions has been described by Franck and Tietze in which an intermediate ferric complex **394** is postulated to exist when laudanosoline (**187**) is treated with a concentrated ferric chloride solution. In dilute solution a dibenzopyrrocoline salt was produced (Section 8.15).

Oxidation of codamine (**395**) with lead tetraacetate gave a *p*-quinol acetate **396** which on treatment with acetic anhydride–sulfuric acid afforded **397** and its 4-acetoxy derivative **398**. The yield of the aporphine is considerably improved if TFA is used for rearranging the quinol acetate [e.g. thaliporphine (**399**)]. The mechanism of this transformation is shown in Scheme 50.

Kametani and co-workers have described the use of cuprous chloride and oxygen in pyridine for carrying out phenolic oxidation. Oxidation of (+)-reticuline (**182**) perchlorate with this system at room temperature for

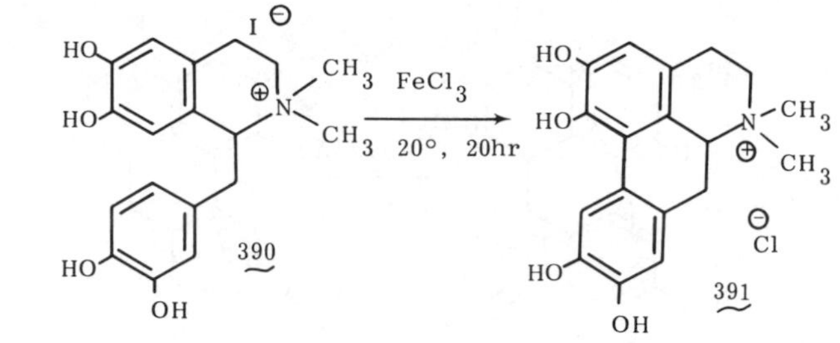

30 min gave three products, the aporphines (+)-corytuberine (**400**) (28%) and (+)-isoboldine (**382**) (8%) and the morphinandienone pallidine (**401**) (6%).

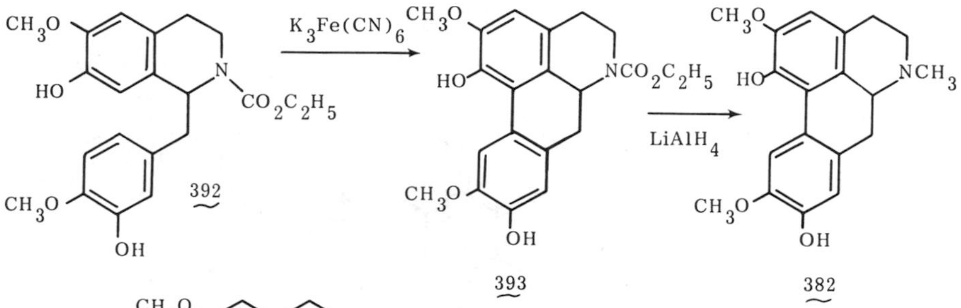

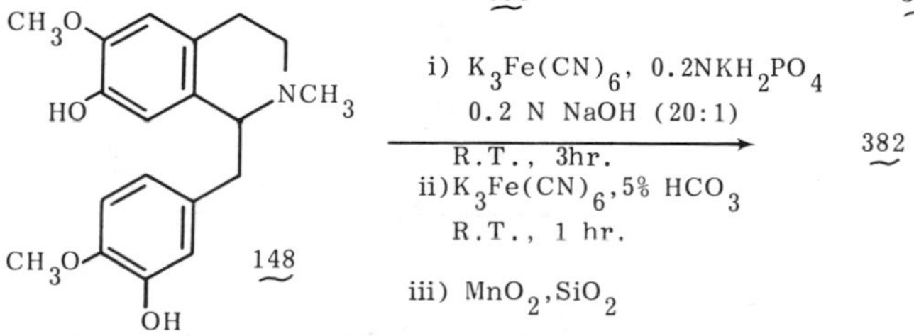

Acid treatment of (−)-orientalinone (**354**) affords two products depending on the presence or absence of water as shown in Scheme 51.

A possible biogenetic route to the aporphine system which does not involve proaporphine intermediates was disclosed by Battersby in 1971 and

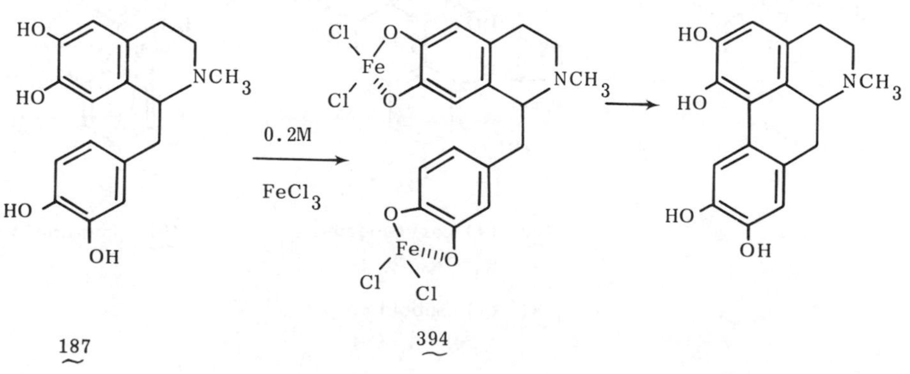

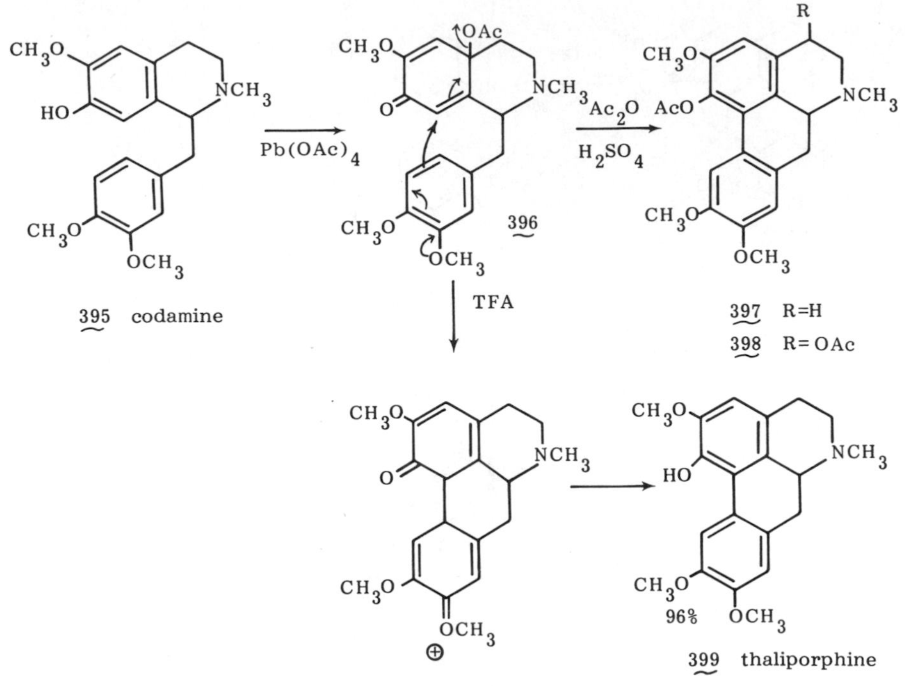

CH₃O / HO — NCH₃ (395 codamine) → Pb(OAc)₄ → 396 → TFA

396 → Ac₂O, H₂SO₄ → 397 R=H / 398 R=OAc

Scheme 50

depends on two alternative cyclizations *ortho* or *para* to the ring C phenolic group of **182**. Dienone–phenol rearrangement affords the two major substitution patterns found in the aporphines (Scheme 52).

Several remarkable biogenetic transformations of benzylisoquinolines have been discovered in recent years by Kupchan. For example laudanosine (**188**) is converted to glaucine (**372**) in 43% yield with VOF₃-TFA. In this

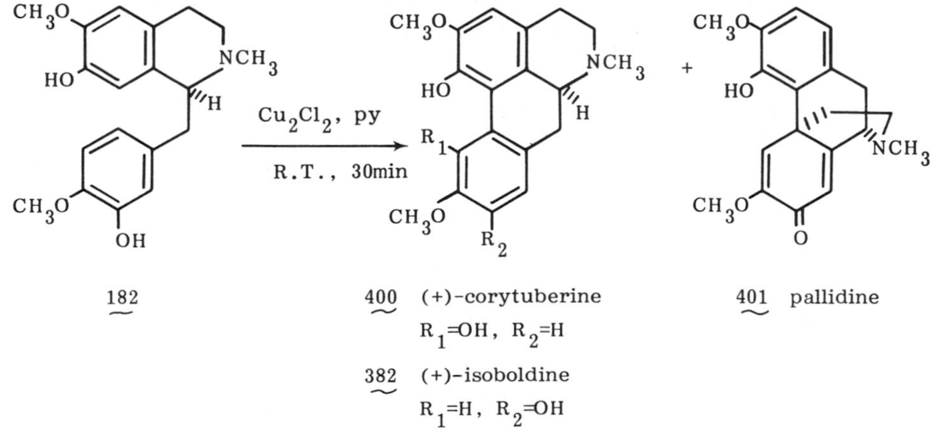

182 → Cu₂Cl₂, py, R.T., 30min →

400 (+)-corytuberine R₁=OH, R₂=H

382 (+)-isoboldine R₁=H, R₂=OH

401 pallidine

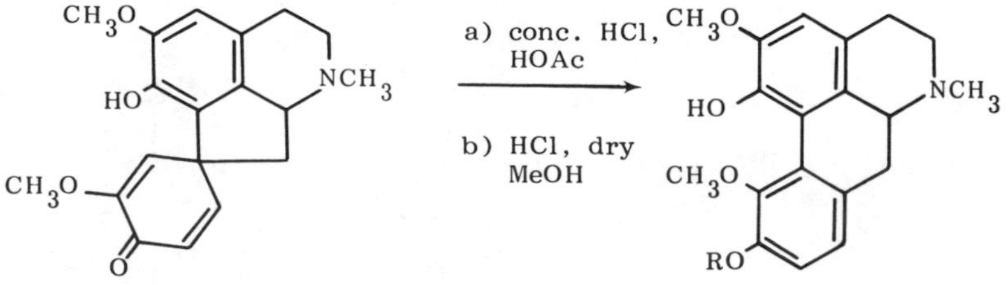

354 (-)-orientalinone

a) R=H
b) R=CH$_3$

Scheme 51

reaction the initial morphinandienone **402** rearranges to a neoproaporphine **403** by aryl migration followed by a 1,3-alkyl shift to give **372**. Two other similar reactions are discussed in the morphinandienone section.

A facile synthesis of racemic boldine (**404**) has been reported by Kupchan and co-workers (Scheme 53) and involves a neoproaporphine intermediate. Photolysis of the bromodiphenol **405** under basic conditions afforded **406** and **407**, and **406** could also be converted to **407** by further irradiation. Lithium aluminum hydride reduction then afforded boldine (**404**).

When these results (in terms of yield) are compared with the notoriously low yields obtained through phenolic oxidation, the question immediately

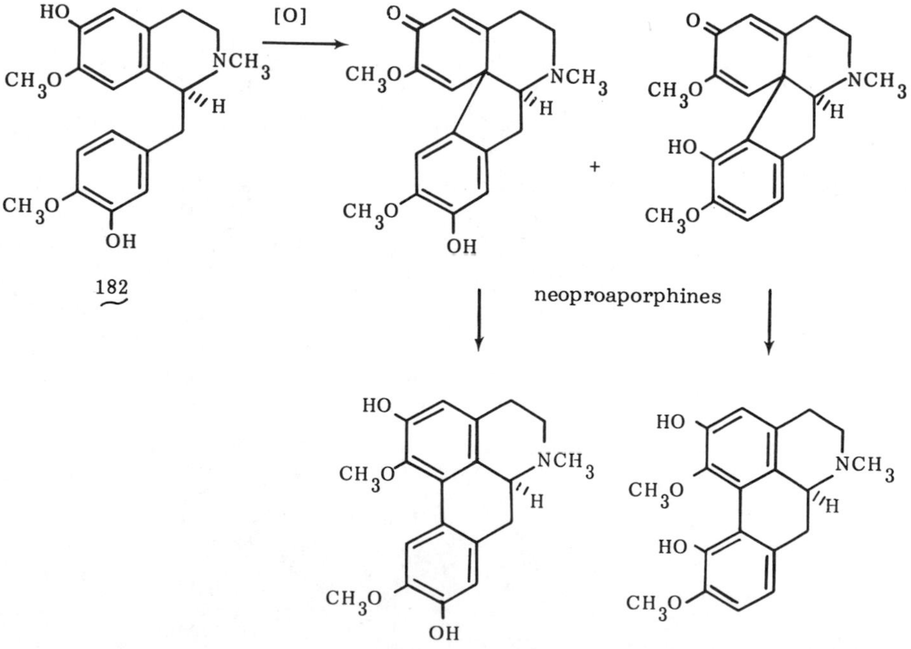

182

neoproaporphines

Scheme 52

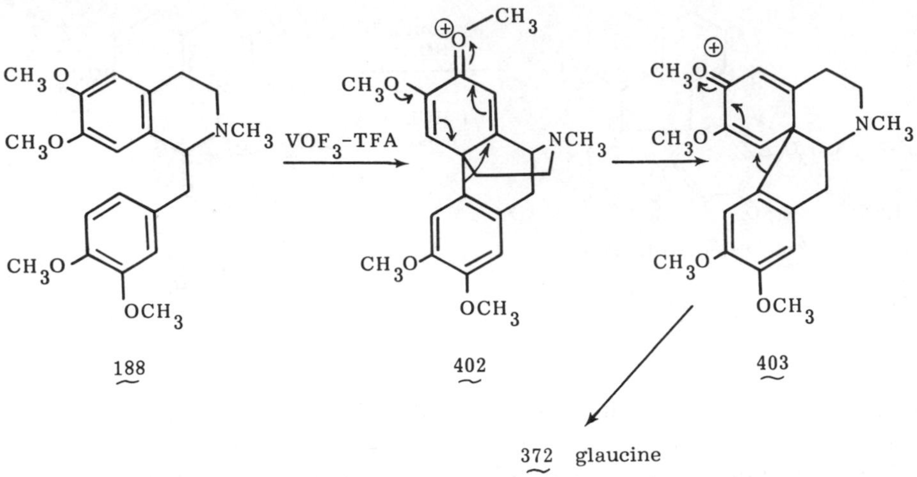

188   402   403

372   glaucine

raised is: Which of all these biogenetic approaches is the biosynthetic pathway?

8.18.3   Absolute Configuration

The stereochemistry and absolute configuration of the aporphines have been thoroughly studied. Shamma in 1960 was first to make an interesting ob-

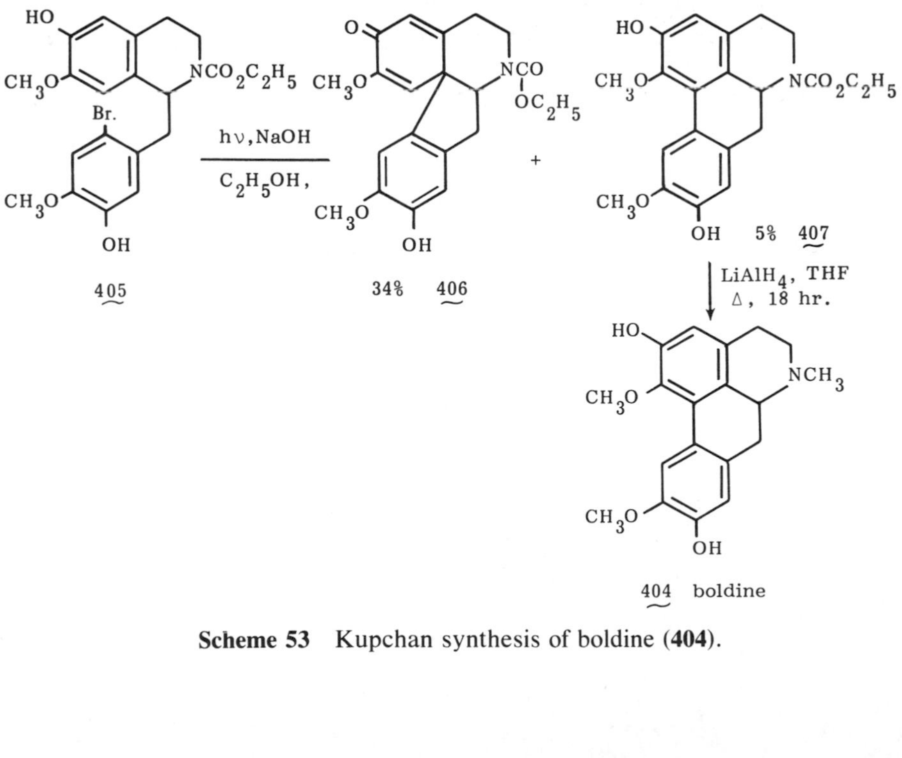

Scheme 53   Kupchan synthesis of boldine (**404**).

servation concerning the lack of planarity of the biphenyl system, and indeed two stereochemical possibilities exist as shown in **408** and **409**.

The structure determination of bulbocapnine methiodide verifies the strained nature of the biphenyl system and the severe interactions that take place. For example, the angle of twist between the two diphenyl rings is 29.9°,

<u>408</u> L–S–configuration   <u>409</u> D–R–configuration

and the distance between the oxygen at *O*-1 and *O*-11 (2.74 Å) is less than the sum of their van der Waals radii.

In spite of the complexity of their UV spectra depending on the aromatic substitution, the ORD spectra typically show a Cotton effect between 235 and 245 nm, which is diagnostic of the C-6a stereochemistry; if the Cotton effect is negative, H-6a has the β-configuration, and conversely if the Cotton effect is positive, H-6a is α.

8.18.4 Spectral Properties

More than almost any other group of isoquinoline alkaloids, the UV spectra of the aporphines are of exceptional value in disclosing structural features of the alkaloid, since they are dependent mainly on the location rather than the nature of the substituent (Table 7).

When no substituent is present at positions 10 or 11 (substitution is assumed at positions 1 and 2) the spectra show a principal peak in the region 270–280 nm (~4.3) and shoulder or smaller peak at 310–320 (~3.3).

If position 11 is free, the two maxima are observed at 282 and 303–310 nm and are of almost equal intensity (~4.2), but when position 11 is substituted the two maxima are at 268–272 nm (4.2) and 303–310 nm (~3.8) and are of unequal intensity. These differences in spectra are thought to be related to the increased strain present in the conjugated diphenyl system when both 1- and 11-substituents are present.

The mass spectra of the aporphines show no characteristic fragmentation pathways. Losses are mainly of small fragments, -Ḣ, -ĊH₃, -ȮH, and -ȮCH₃, except for a retro Diels–Alder reaction in ring B which gives a loss of -CH₂=NR (normally 29 or 43 nm if R=H or CH₃).

One major difference between the 1,2,9,10- and 1,2,10,11-substituted compounds is the intensity of the $M^+ - 1$ peak. In the former series it is

Table 7   UV Spectra of Representative Aporphine Alkaloids

	Substitution	λ_{max} (nm)	log ϵ
No 10 or 11 Substituent			
Nuciferine	1,2	230, 272, 310	4.26, 4.19, 3.32
Roemerine	1,2	235, 272, 313	4.19, 4.27, 3.59
Stephanine	1,2,8	272	4.28
Xylopine	1,2,9	230sh, 282, 307sh	4.15, 4.33, 3.74
No 11 Substituent			
Glaucine	1,2,9,10	218, 281, 302	4.58, 4.18, 4.16
Boldine	1,2,9,10	220, 284, 303, 313sh	4.6, 4.2, 4.23, 4.16
Apoglaziovine	1,2,10	218, 266, 275, 307	4.5, 4.01, 4.13, 3.96
11 Substituent			
Bulbocapnine	1,2,10,11	233, 270, 308	4.07, 3.9, 3.6
Pukateine	1,2,11	218, 272, 303	4.42, 4.10, 3.95
Isothebaine methlyether	1,2,11	273, 302	4.1, 3.92

usually the base peak, but in the latter it is less than 50% of the molecular ion.

Fundamentally there are two principal areas of interest in the proton NMR spectra (CDCl$_3$) of an aporphine or noraporphine alkaloid (Figure 9): the regions of the methoxyl groups (3.4–3.9 ppm) and the aromatic protons (6.5–8.2 ppm). The most deshielded methoxyl groups are those at C-2, C-9, and C-10 (3.8–3.9 ppm) and the most shielded is a C-1 methoxyl (3.4–3.7 ppm) (except if it is adjacent to a methylenedioxy group). An 11-methoxy group typically occurs in the region 3.6–3.8 ppm.

Additional separation of methoxyl signals is gained when benzene is used as the solvent, although this is not always easy to observe because of the problems of solubility. Meaningful shifts may be obtained by simply adding C$_6$D$_6$ to the sample. The least hindered methoxyl groups experience an upfield shift; a C-1 methoxyl therefore will experience the least shift.

The most downfield aromatic proton (H-11) is observed between 7.5 and 8.2 ppm, depending on the adjacent group at C-10. The presence of a C-10 group shields the 11-proton. The C-3 proton is almost invariably the most shielded and typically appears as a singlet in the range 6.5–6.6 ppm. Any remaining protons appear in the region 6.7–7.4 ppm. The presence of a C-11 methoxyl group causes the protons at C-8 and C-9 to have significantly different chemical shifts; however, if a C-11 hydroxy group is present the chemical shifts of H-8 and H-9 are such that overlap occurs and no coupling is observed.

As we have seen previously with the benzylisoquinolines, the placement of a phenol group can frequently be made by the addition of NaOD to the sample in d_6-DMSO, which causes quite characteristic upfield shifts. This shift is maximal (-0.85 ppm) for a proton *para* to a hydroxyl group, and as expected by resonance considerations least (0.2–0.4 ppm) for a *meta* proton. An *ortho* proton is typically shifted by 0.5 ppm.

The *N*-methyl group in a basic aporphine is typically observed around 2.5 ppm in CDCl$_3$ or d_6-DMSO. For an N-oxide dissolved in TFA, the *N*-methyl group appears considerably deshielded at about 3.5 ppm. The methylenedioxy group, when located at C-9,10, appears as a singlet (5.90 ppm), but when at C-1,2 the twisted biphenyl system induces magnetic nonequivalence between the methylene protons, which then appear as doublets ($J \simeq 1.5$ Hz) at about 5.9 and 6.1 ppm.

The ^{13}C NMR data of some representative aporphine alkaloids are shown in Table 8. Overall the data are typical of what would be expected, including the shift of two aromatic carbon atoms to below 140 ppm whenever two oxygenated aromatic carbons are adjacent. Note the downfield shift of ~3 ppm for an aromatic carbon when a phenol is converted to the corresponding methyl ether. The C-1 methoxy methyl is always the most deshielded because of the lack of planarity of the diphenyl system and the C-1 subsitutent. The chemical shifts of the C-8 and C-11a carbons should serve to distinguish between 1,2,9,10-substituted aporphines because of the shielding effect of

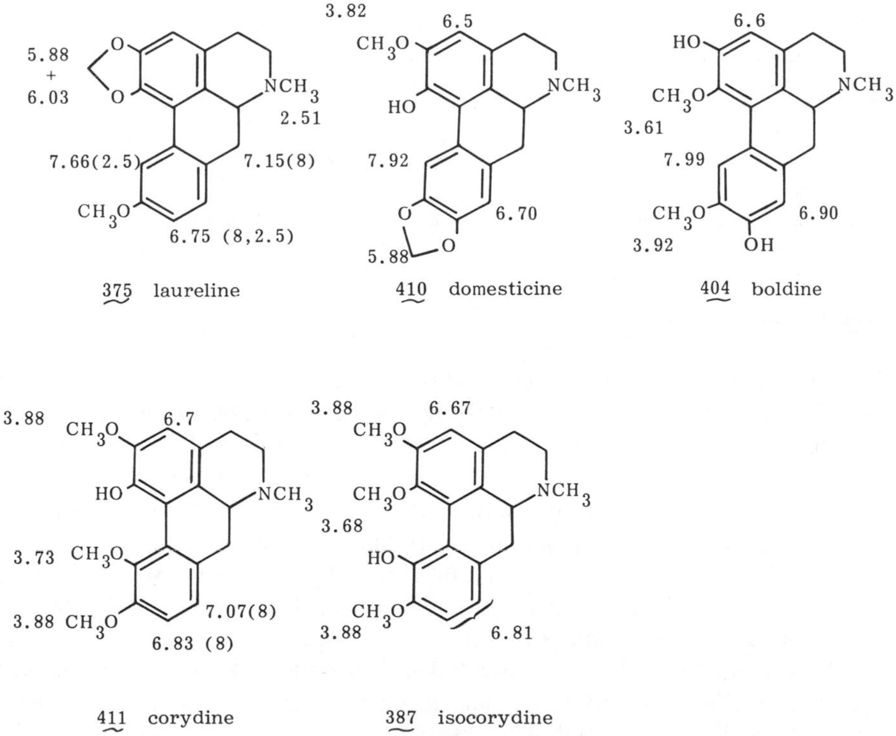

375 laureline   410 domesticine   404 boldine

411 corydine   387 isocorydine

Figure 9

Table 8 ^{13}C NMR Data for Representative Aporphine Alkaloids

	Compound			
Carbon	Nuciferine (376)	Glaucine (373)	Isocorydine (387)	Boldine (404)
1	144.6	143.9	141.7	142.0
1a	126.3	126.5	135.4	126.8
1b	128.1	128.6	129.8	125.9
2	151.4	151.5	150.8	148.1
3	110.9	110.1	110.8	113.3
3a	127.5	127.0	128.8	129.9
4	28.9	29.1	29.1	28.9
5	52.8	53.1	52.4	53.4
6a	61.9	62.3	62.6	62.6
7	34.8	34.4	35.6	34.2
7a	135.9	129.1	129.6	130.2
8	127.7	110.6	118.6	114.2
9	126.7	147.7	110.7	145.1
10	126.4	147.1	149.0	145.6
11	127.3	111.4	143.6	110.1
11a	131.6	124.2	119.8	123.6
NCH$_3$	43.5	43.4	43.6	44.0
C-1 OCH$_3$	59.7	59.8	61.7	60.2
Other OCH$_3$	59.7	55.5	55.5	56.1
	55.3	55.5	55.8	—
	—	55.7	—	—

an adjacent oxygen substituent at either C-9 or C-11 respectively. Compare for example the chemical shifts of C-8 and C-11a in 372, 387, and 404.

8.18.5 Biosynthesis

The discussion on the biogenetic synthesis of the aporphine alkaloids has raised a number of points concerning the biosynthesis of this group. In particular and depending on the orientation of phenolic and methoxy groups we might envisage any of at least five routes being in operation from a benzylisoquinoline precursor (Figure 10). *In vitro* all these routes apparently can operate; does the same situation prevail *in vivo*?

The structures of the aporphine alkaloids (+)-roemerine (351) and (+)-isothebaine (377) indicate that they are probably not derived from a tetrahydrobenzylisoquinoline by direct coupling.

Barton and co-workers investigated the formation of (+)-roemerine (351) in *Papaver dubium* L. and found that tritium-labeled (+)-N-methylcoclaurine (292) was well incorporated. The position of the hydroxy group in the precursor suggests that a proaporphine intermediate is involved (Scheme 54).

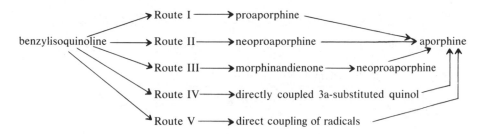

Figure 10 Biogenesis of aporphines from benzylisoquinolines.

Similarly, Battersby and co-workers showed that (+)-orientaline (**355**) was well incorporated into (+)-isothebaine (**377**) in *P. orientalis*. The formation of (+)-isothebaine (**377**) can be envisaged as occurring through (−)-orientalinone (**354**), a co-constituent, as shown in Scheme 55.

It is very difficult in some instances to say what the probable mechanism is unless variously methylated benzylisoquinoline precursors are used. For example, *N*-methyl labeled (+)-reticuline (**182**) was a precursor of bulbocapnine (**374**) in *Corydalis cava* Schweigg et Korte (Fumariaceae), but is this evidence of direct phenol coupling? Even if demethylation–remethylation is assumed not to occur, direct coupling of a 3α-substituted quinol not involving radicals is still a reasonable alternative (i.e. **412**). It is pertinent to note that the morphinandienone sinoacutine (**413**) is also a constituent of this plant. Could sinoacutine (**413**) be a key precursor *via* the intermediate neoproaporphine **414**?

Brochmann-Hanssen found that [*N*-methyl-^{14}C]reticuline (**182**) was a precursor of isoboldine (**382**), but not of magnoflorine (**415**) in *P. somniferum*. Classically, these would be regarded as the products of *ortho-para* and *ortho-ortho* coupling of reticuline (**182**). Is orientalinone (**354**) a precursor in the formation of isoboldine (**382**)? Subsequently the same group showed that [*N*-methyl-^{14}C]reticuline (**182**) was a precursor of magnoflorine (**415**) in *Aquilegia* sp.

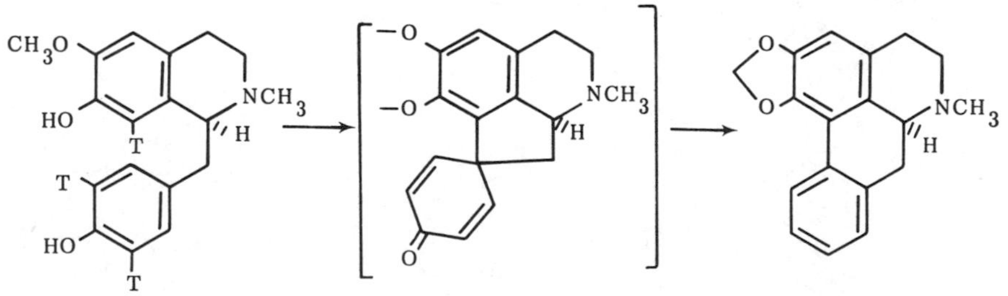

292 (+)-N-methyl-
~   coclaurine

351 (+)-roemerine
~

Scheme 54

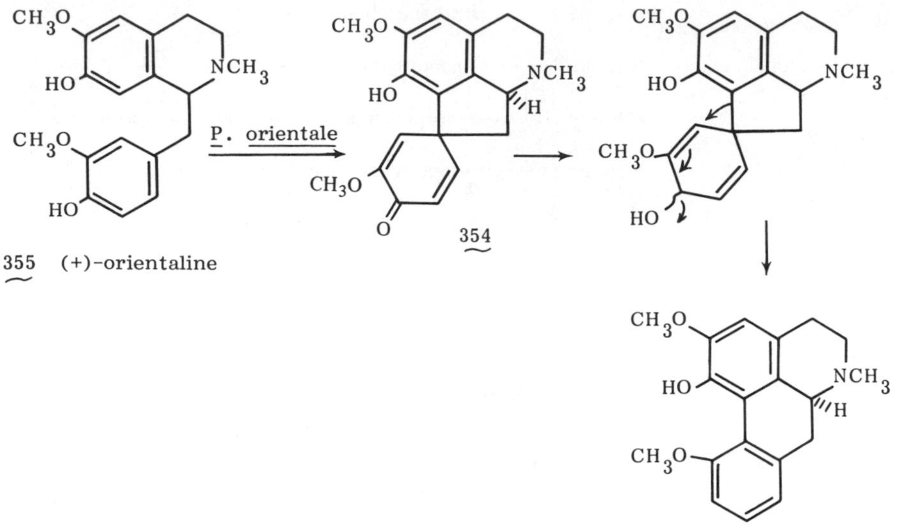

355   (+)-orientaline

354

377   (+)-isothebaine

Scheme 55

The biosynthesis of boldine (**404**) in *Litsea glutinosa* (Lour.) C.B. Rob. var. *glabraria* Hook. (Lauraceae) has also been studied in detail by Kapil and co-workers, with a quite different result.

Of several variously methylated benzylisoquinoline precursors only 4'-*O*-methylnorlaudanosoline (**416**), nor-reticuline (**239**), and reticuline (**148**)

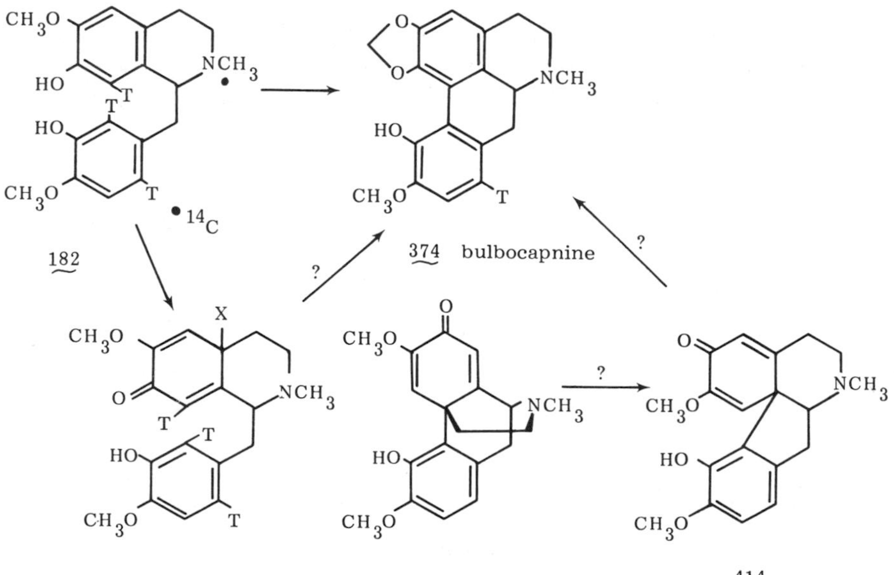

182

374   bulbocapnine

412

413   sinoacutine

414

P. somniferum

182   Reticuline   382   isoboldine

Aquilegia sp.

415   magnoflorine

were precursors. In the latter case the (+)-isomer **182** was incorporated with considerable preference into boldine. More surprising however is the high (2.0%) level of incorporation of [8-^{3}H]isoboldine (**382**) into boldine (**404**), a process that of necessity involves 2-*O*-demethylation followed by 1-*O*-methylation (Scheme 56).

Norprotosinomenine (**231**) was *not* a precursor of boldine (**404**) in this plant, which contrasts with the previously discussed work with boldine in *Dicentra eximia*.

More definitive results have been obtained by Battersby and co-workers concerning the formation of corydine (**411**) and glaucine (**322**) in *Dicentra eximia* (Ker.) Torr. (Fumariaceae). Reticuline (**148**) and orientaline (**255**) were not precursors, but 4′-*O*-methylnorlaudanosoline (**416**) and norprotosinomenine (**231**) were effective precursors. This result clearly rules out direct phenol coupling (for **411** and **148**) and a proaporphine intermediate (for **322** from **255**). According to Battersby this suggests the pathway shown in Scheme 57 involving two alternative neoproaporphine intermediates **417** and **418**. Notice that the aporphine **419** has the incorrect *O*-methylation pattern in comparison with corydine (**411**). It would be interesting indeed to know if 1-*O*-demethylation followed by 2-*O*-methylation really occurs at this point in the biosynthesis. Boldine (**404**) was a poor precursor of glaucine (**372**).

These results would suggest that boldine (**404**) is produced by two different pathways in two different plants. If this is a general trend, the biosynthesis of aporphine alkaloids may never be established. Clearly this is an area in need of considerable further study, and at this time it is not clear which, or how many, of the possible biosynthetic routes may be operating in order to produce the various aporphine alkaloids.

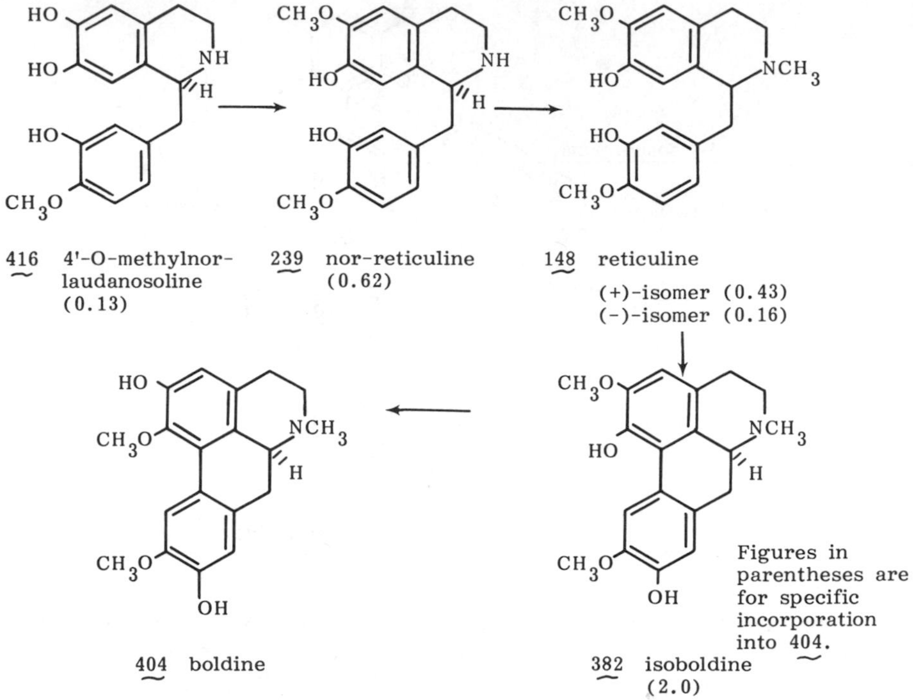

416 4'-O-methylnor-laudanosoline (0.13)

239 nor-reticuline (0.62)

148 reticuline
(+)-isomer (0.43)
(−)-isomer (0.16)

404 boldine

382 isoboldine (2.0)

Figures in parentheses are for specific incorporation into 404.

Scheme 56 Biosynthesis of boldine (**404**) in *Litsea glutinosa* var. *glabraria*.

8.18.6 Pharmacology

The aporphine alkaloids display a wide range of pharmacologic activities, although none are commercial items.

1,2-Methylenedioxyaporphine (**420**) increases arterial blood pressure, but higher doses cause strychnine-like convulsions. The methohydroxide salt has a curare-like action.

Isothebaine (**377**) increased intestinal muscle tone in rabbits and also amplified uterine contractions in the rat. Other activities observed include decreased motor activity and analgesia (mice), and an anti-inflammatory effect (rats).

Glaucine (**372**) reduced blood pressure and inhibited respiration in cats and had antitussive effects resembling codeine, but of longer duration. In rats and cats a potentially useful hypoglycemic effect was observed at 12-mg/kg doses. Dehydroglaucine has antibacterial activity.

Corydine (**411**) has central nervous system depressant and hypotensive activity and blocks transmission of nerve impulses. The corresponding 11-demethyl derivative, corytuberine, accelerates respiration and stimulates secretions.

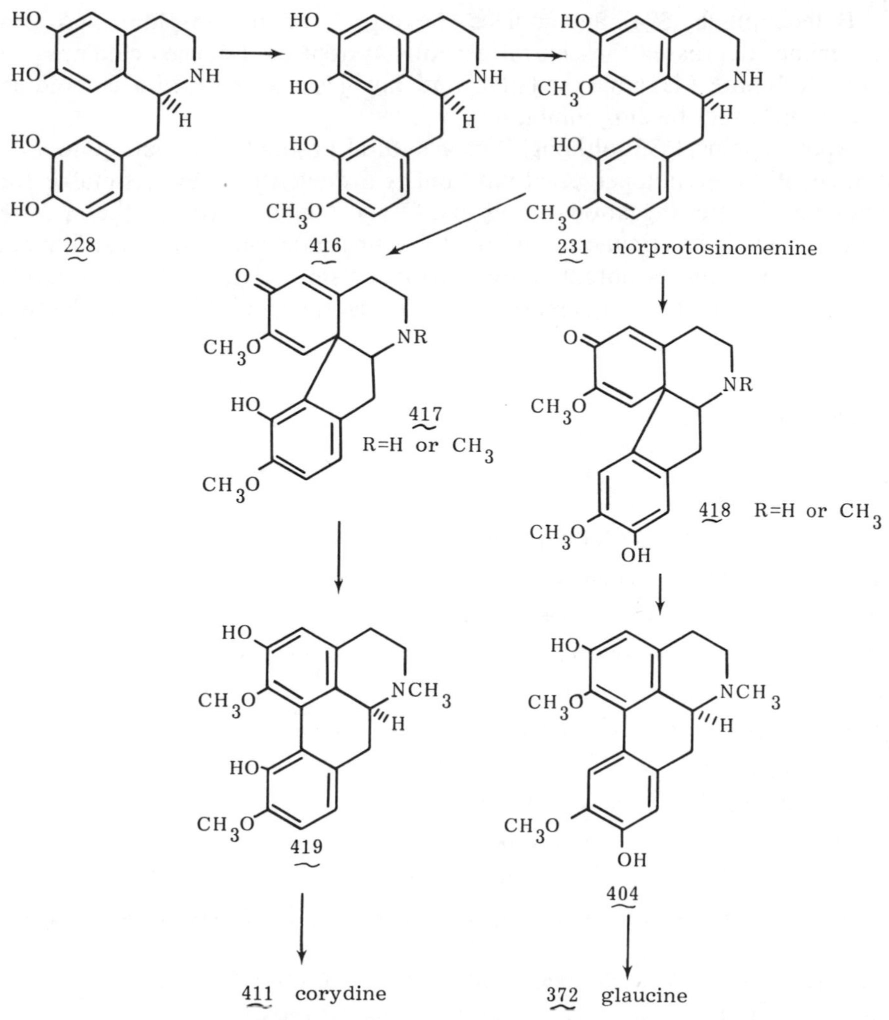

Scheme 57 Biosynthesis of corydine (**411**) and glaucine (**372**) in *Dicentra eximia*

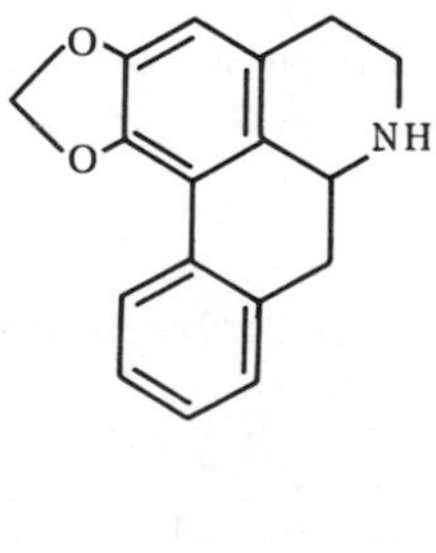

420

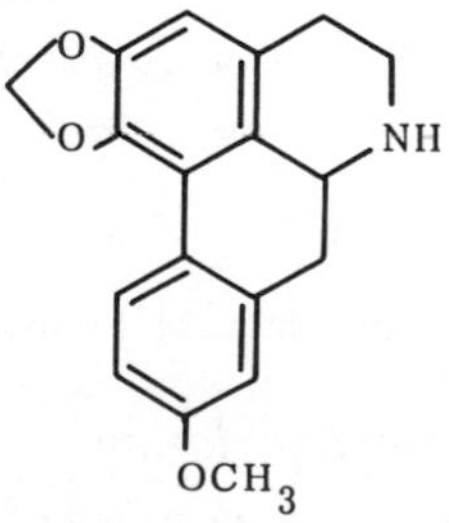

421 xylopine

Bulbocapnine (374) antagonizes the effects of apomorphine and amphetamine, depresses the central nervous system, and causes catalepsy in mice. Xylopine (421) has sedative and analgesic activity and isoboldine (382) is an insect-feeding inhibitor.

Apomorphine (380) although not a natural product has been quite well studied; it has hypotensive activity and is a powerful emetic, suitable for rapid emesis after ingestion of poisons. Of more interest from a therapeutic point of view is its stimulation of the dopaminergic system in rats and mice, and consequently its potential anti-Parkinsonism activity. Also of interest are reports that it can decrease serum prolactin levels (see also Section 9.10).

LITERATURE

Reviews

Manske, R. H. F., *Alkaloids NY* 4, 119 (1954).

Shamma, M., and W. A. Slusarchyk, *Chem. Rev.* 64, 59 (1964).

Shamma, M., *Alkaloids NY* 9, 1 (1967)

Kametani, T., *The Chemistry of the Isoquinoline Alkaloids,* Hirokawa, Tokyo, 1969.

Bentley, K. W., *Alkaloids, London* 1, 117 (1971).

Shamma, M., *The Isoquinoline Alkaloids,* Academic, New York, 1972, p. 194.

Bernhard, H. O., and V. A. Snieckus, *Alkaloids, London* 2, 114 (1972).

Shamma, M., *Alkaloids, London* 3, 139 (1973).

Snieckus, V. A., *Alkaloids, London* 4, 143 (1974).

Shamma, M., and S. S. Salgar, *Alkaloids, London* 4, 197 (1974).

Kametani, T., *The Chemistry of the Isoquinoline Alkaloids,* Vol. 2, Kikodo, Sendai, Japan, 1974.

Bernhard, H. O., and V. A. Snieckus, *Alkaloids, London* 5, 134 (1975).

Guinaudeau, H., H. Leboeuf, and A. Cavé, *Lloydia* 38, 275 (1975).

Shamma, M., *Alkaloids, London* 6, 170 (1976).

Shamma, M., *Alkaloids, London* 7, 152 (1977).

Shamma, M., *Alkaloids, London* 8, 122 (1978)..

Shamma, M. and J. L. Moniot, *Isoquinoline Alkaloid Research, 1972–1977,* Plenum, New York, 1978, p. 123.

Oxidation

Cava, M. P., A. Venkateswarlu, M. Srinivasan, and D. L. Edie, *Tetrahedron,* 28, 4299 (1972).

Methylation–O-Demethylation

Tschesche, R., P. Welzel, R. Moll, and G. Legler, *Tetrahedron* 20, 1435 (1964).

Ahmad, R., J. M. Saá, and M. P. Cava, *J. Org. Chem.* 42, 1228 (1977).

Synthesis: Nonbiogenetic

Pschorr

Tschesche, R., P. Welzel, R. Moll, and G. Legler, *Tetrahedron* **20**, 1435 (1964).

Weisbach, J. A., C. Burns, E. Macko, and B. Douglas, *J. Med. Chem.* **6**, 91 (1963).

Neumeyer, J. L., B. R. Newstadt, and J. W. Weintraub, *Tetrahedon Lett.* 3107 (1967).

Kametani, T., T. Sugahara, and K. Fukumoto, *Tetrahedron* **27**, 5367 (1971).

Popp, F. D., *Heterocycles* **1**, 165 (1973).

Photochemical

Cava, M. P., M. J. Mitchell, S. C. Havlicek, A. Lindert, and R. J. Soangler, *J. Org. Chem.* **35**, 175 (1970).

Kupchan, S. M., C.-K, Kim, and K. Miyano, *Chem. Commun.* 91 (1976).

Kupchan, S. M., and R. M. Kanojia, *Tetrahedron Lett.* 5353 (1966).

Synthesis: Biogenetic

Franck, B., and G. Schlingloff, *Ann.* **659**, 132 (1962).

Kametani, T., T. Sugahara, and K. Fukumoto, *Tetrahedron* **25**, 3667 (1969).

Chien, W. W. C., and P. Maitland, *J. Chem. Soc., Sec. C* 753 (1966).

Kametani, T., A. Kozula, and K. Fukumoto, *J. Chem. Soc., Sec. C* 1021 (1971).

Franck, B., G. Dunkelmann, and H. J. Lubs, *Angew. Chem. Int. Ed.* **6**, 1075 (1967).

Franck, B., and L. F. Tietze, *Angew. Chem. Int. Ed.* **6**, 799 (1967).

Hoshino, O., H. Hara, M. Ogawa, and B. Umezawa, *Chem. Pharm. Bull.* **23**, 2578 (1975).

Kametani, T., M. Ihara, M. Takemura, Y. Satoh, H. Terasawa, Y. Ohta, K. Fukumoto, and K. Takahashi, *J. Amer. Chem. Soc.* **99**, 3805 (1977).

Battersby, A. R., T. J. Brockson, and R. Ramage, *Chem. Commun.* 464 (1969).

Battersby, A. R., J. L. McHugh, J. Staunton, and M. Todd, *Chem. Commun.* 985 (1971).

Kupchan, S. M., V. Kameswaran, J. T. Lynn, D. K. Williams, and A. J. Liepa, *J. Amer. Chem. Soc.* **97**, 5622 (1975).

Kupchan, S. M., and C. Y. Kim, *J. Amer. Chem. Soc.* **97**, 5623 (1975).

Spectral Data

Sangster, A. W., and K. L. Stuart, *Chem. Rev.* **65**, 69 (1965).

Guinaudeau, H., M. Leboeuf, and A. Cavé, *Lloydia*, **38**, 275 (1975).

Biosynthesis

Barton, D. H. R., D. S. Bhakuni, G. M. Chapman, and G. W. Kirby, *Chem. Commun.* 259 (1966).

Battersby, A. R., R. T. Brown, J. H. Clements, and G. G. Iverach, *Chem. Commun.* 230 (1965).

Brochmann-Hanssen, E., C.-C. Fu, and L. Y. Misconi, *J. Pharm. Sci.* **11**, 1880 (1971).

Brochmann-Hanssen, E., C.-H. Chen, H.-C. Chiang, and K. McMurtney, *Chem. Commun.* 1269 (1972).

Battersby, A. R., J. L. McHugh, J. Stanuton, and M. Todd, *Chem. Commun.* 985 (1971).

Tewari, S., D. S. Bhakuni, and R. S. Kapil, *Chem. Commun.* 940 (1974).

8.19   APORPHINE DIMERS

There are three fundamental types of natural aporphine dimers known, the proaporphine–benzylisoquinoline dimers, the aporphone–benzylisoquinoline dimers and the aporphine–pavinane dimers. A fourth dimeric skeleton, a bisaporphine, has recently been synthesized.

8.19.1   Proaporphine–Benzylisoquinoline Dimers

The only proaporphine–benzylisoquinoline dimer known is pakistanamine (**422**), obtained by Shamma and co-workers from *Berberis baluchistanica* Ahrendt (Berberidaceae), where it cooccurs with the aporphine–benzyl-isoquinoline pakistanine (**423**).

On treatment with dilute hydrochloric acid, pakistanamine (**422**) under-went dienone–phenol rearrangement to give *O*-methylpakistanine (**424**). The principal proton NMR features of pakistanamine are shown in **425**.

8.19.2   Aporphine–Benzylisoquinoline Dimers

The aporphine–benzylisoquinoline dimers have been obtained from three plant families: the Berberidaceae, Ranunculaceae, and Hernandiaceae. In

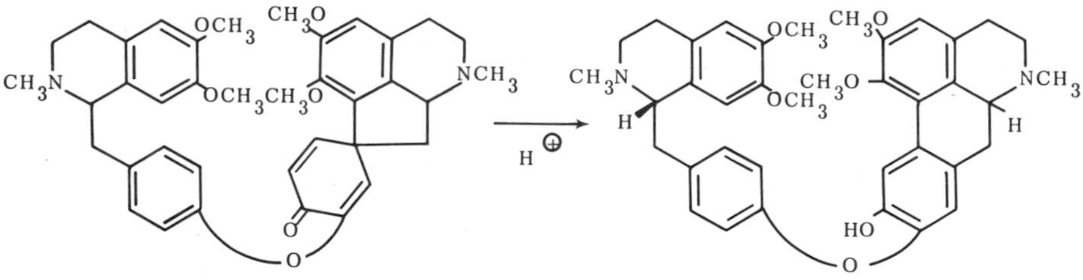

422 pakistanamine

423 pakistanine   R=H
424 R=CH$_3$

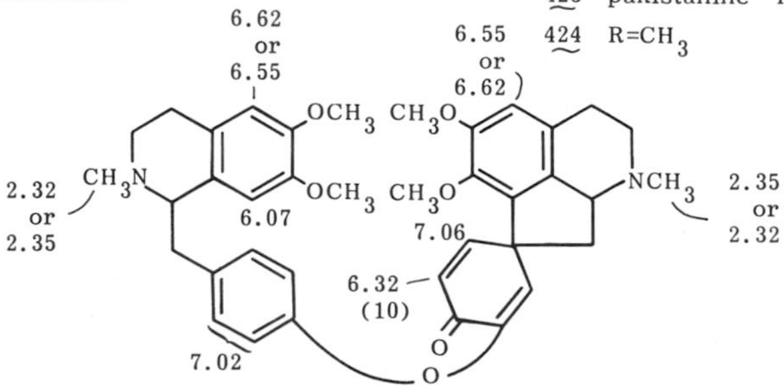

425

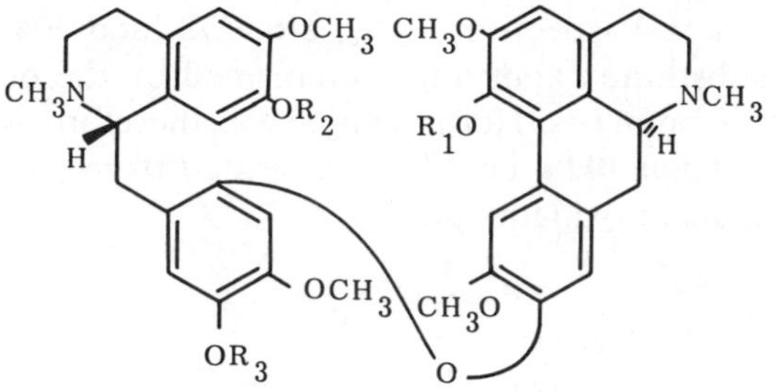

	R_1	R_2	R_3
426 thalicarpine	CH_3	CH_3	CH_3
438 pennsylvanine	CH_3	CH_3	H
439 pennsylvanamine	H	CH_3	H

Figure 11 Aporphine benzylisoquinoline dimers.

particular the genus *Thalictrum* in the Ranunculaceae has proved to be a good source of these alkaloids. The most important of this group of alkaloids is thalicarpine (426), but other representatives are shown in Figure 11.

Thalicarpine was the first obtained by Kupchan and co-workers from the roots of *Thalictrum dasycarpum* in 1963, and subsequently by Bulgarian workers in 1966.

The structure was determined in 1965 by a combination of degradation and semisynthesis. The point of attachment of the two units was initially proved by partial synthesis involving an Ullmann condensation of (+)-6'-bromolaudanosine (280) and (+)-*N*-methyllaurotetanine (427).

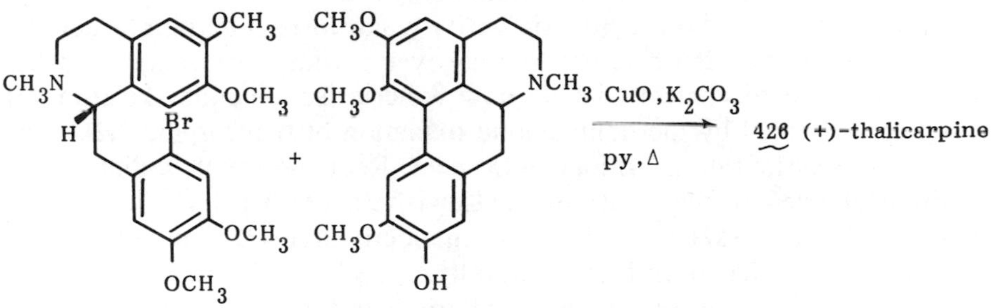

Before discussing the synthesis of thalicarpine, mention must be made of the alkaloid hernandaline (428) obtained by Cava and co-workers from *Hernandia ovigera* L. (Hernandiaceae). Crucial in the structure determination was the observation of a conjugated carbonyl at 1672 cm^{-1}, which from the proton NMR spectrum was found to be due to an aldehyde (singlet at 10.41 ppm). The UV spectrum (λ_{max} 216, 278, and 304 nm) was that of a 1,2,9,10-tetrasubstituted aporphine, confirmed by the observation of the C-11 proton at 8.20 ppm. (+)-Hernandaline was therefore assigned the structure **428**. Hernandaline (**428**) could be generated from thalicarpine by mild sodium *meta*-vanadate oxidation.

428 (+)-hernandaline

(+)-Hernandaline (**428**) is important because it is the key intermediate in the Kupchan–Liepa synthesis of the thalicarpine (**426**). Overall the strategy was to build the diaryl ether **429** appropriately substituted, from a nitrobenzylisoquinoline, **430**, then carry out a Pschorr cyclization and formylate to give hernandaline (**428**). The final isoquinoline nucleus could then be generated at the aldehyde terminus by a Reissert compound. These synthetic steps are summarized in Scheme 58.

Thalicarpine has both cytotoxic activity and *in vivo* antitumor activity against the Walker 256 carcinosarcoma over a wide dose range, and has been subjected to clinical trial. A new dimeric isoquinoline skeleton **431** has been obtained by the nonphenolic oxidation of thalicarpine (**426**) with vanadium oxyflouride in triflouracetic acid. Reduction with sodium borohydride followed by treatment with the Lewis acid, boron triflouride etherate, gave another new skeleton, a bisaporphine derivative **432**. Neither of these two new skeleta has been found naturally.

There have been no biosynthetic experiments conducted in this area, but Shamma has speculated on the biogenesis of these compounds and it is worth discussing some aspects of this problem. One question that is immediately raised concerns the formation of the aporphine nucleus; does it for example occur before or after the diphenyl ether linkage is formed?

The compounds shown in Figure 11 would seem to be produced by oxidative coupling of a 1,2,9,10-tetraoxygenated aporphine (OH at C-9) with a molecule of a benzylisoquinoline. Pakistanine (**423**) on the other hand appears to be derived from a bisbenzylisoquinoline such as **433**, which can

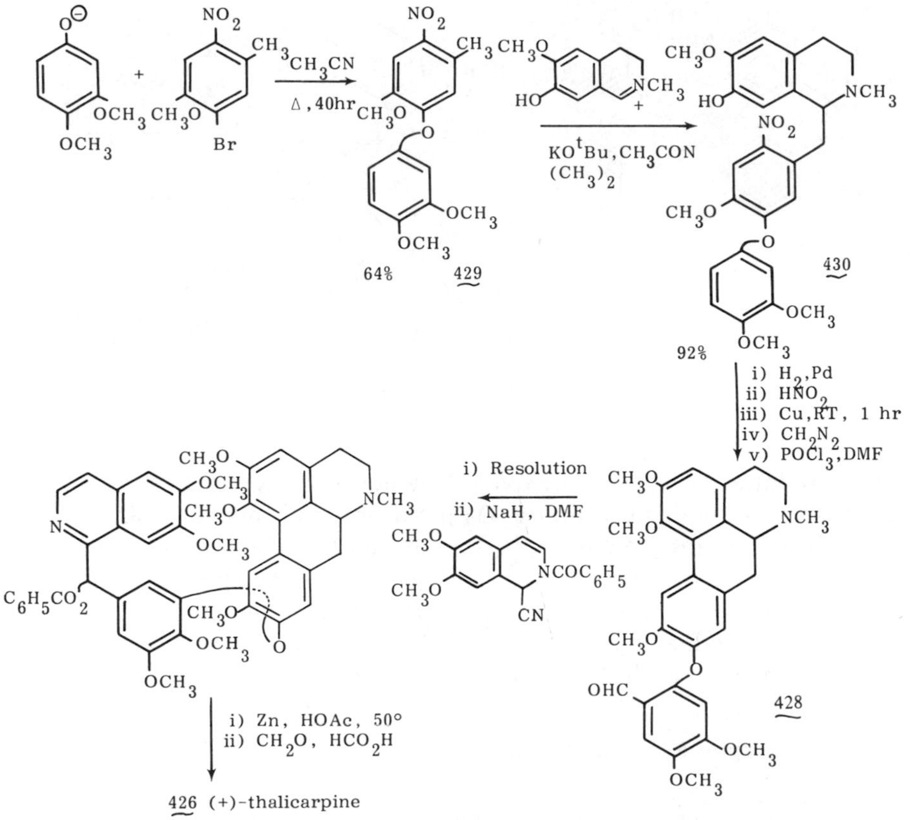

Scheme 58 Kupchan–Liepa synthesis of thalicarpine (**426**).

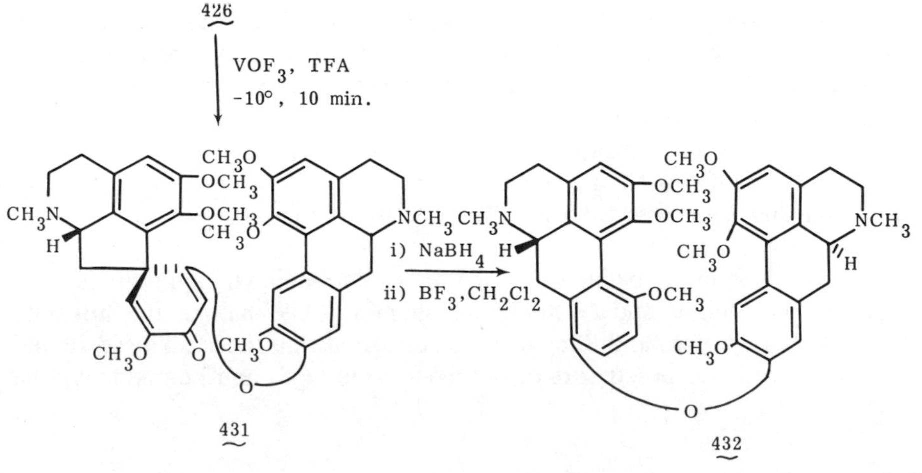

413

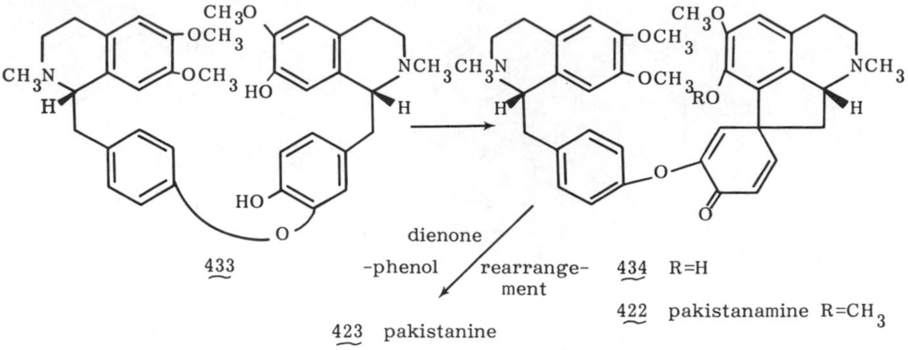

Scheme 59   Biogenesis of pakistanine (**423**).

undergo intramolecular oxidative phenolic coupling to give a proaporphine–benzylisoquinoline dimer **434** [compare pakistanamine (**422**)]. Dienone–phenol rearrangement would then afford **423** (Scheme 59). Shamma has argued that pakistanine (**423**) is the *only* aporphine–benzylisoquinoline produced in this way, because the compounds in the thalicarpine series would require bisbenzylisoquinolines derived from two tetraoxygenated reticuline-like precursors, and these are not presently known.

8.19.3   Aporphine–Pavine Dimers

Two aporphine–pavine dimeric alkaloids are known and were isolated from the giant meadow rue *Thalictrum polyganum* Muhl. (Ranunculaceae) by Shamma and Moniot in 1974. They are pennsylpavine (**435**) and pennsylpavoline (**436**).

The UV spectrum of pennsylpavine (λ_{max} 230, 280sh, 288sh, and 320sh nm) is that of a 1,2,9,10-tetrasubstituted aprophine (280, 308, and 320 nm) added to a pavine–isopavine (288 nm) system. The NMR spectrum of pennsylpavine is summarized in **437**, and the clue to the presence of a pavine system was the observation of the two bridge-head protons as doublets (J = 6 Hz) at 4.06 and 4.50 ppm.

The absolute configurations of these alkaloids were determined by the aromatic chirality method. The CD curve of 435 shows extrema at 242 (+) and 209 (−) nm attributed to the aporphine and pavine moieties respectively.

These alkaloids cooccur with two aporphine–benzylisoquinolines (+)-pennsylvanine (**438**) and (+)-pennsylvanamine (**439**) having the absolute configuration shown, and Shamma has suggested that these alkaloids could act as biosynthetic precursors of pennsylpavine (**435**) and pennsylpavoline (**436**).

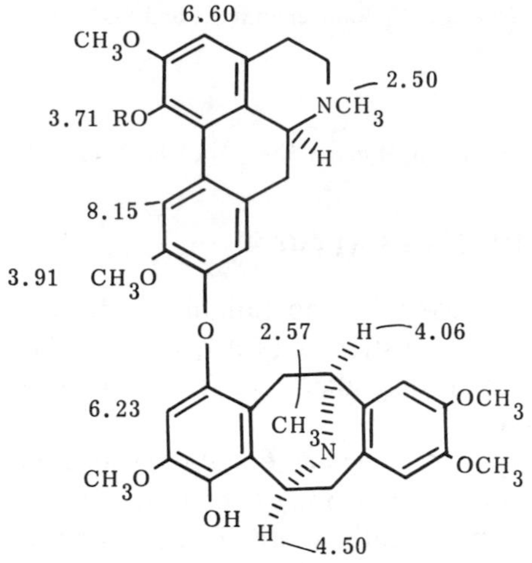

435 pennsylpavine R=CH$_3$
436 pennsylpavoline R=H

437 NMR data are for R=CH$_3$
OCH$_3$ at 3.76, 3.76.3.78, 3.78
ArH at 6.48, 6.48 and 6.52.

LITERATURE

Proaporphine–Benzylisoquinoline

Shamma, M., J. L. Monoit, S. Y. Yao, G. A. Miana, and M. Ikram, *J. Amer. Chem. Soc.* **95**, 1381 (1972).

Aporphine–Benzylisoquinoline

Shamma, M., *The Isoquinoline Alkaloids*, Academic, New York, 1972, p. 232.

Shamma, M., J. R. Moniot, S. Y. Yao, G. A. Miana, and M. Ikram, *J. Amer. Chem. Soc.* **95**, 1381 (1972).

Shamma, M. and J. L. Moniot, *Isoquinoline Alkaloid Research, 1972–1977*, Plenum, New York, 1978, p. 159.

Thalicarpine

Kupchan, S. M., K. K. Chakravarti, and N. Yokoyama, *J. Pharm. Sci.* **52**, 985 (1963).

Dutschewske, H. B., and N. M. Mollov, *Chem. Ind. (London)* 770 (1966).

Tomita, M., H. Furukawa, S.-T. Lu, and S. M. Kupchan, *Tetrahedron Lett.* 4309 (1965).

Cava, M. P., K. Besshe, B. Douglas, S. Markey, and J. A. Weisbach, *Tetrahedron Lett.* 4279 (1966).

Kupchan, S. M., A. J. Liepa, V. Kameswaran, and K. Sempuku, *J. Amer. Chem. Soc.* **95**, 2995 (1973).

Kupchan, S. M., O. P. Dhingra, V. Ramachandran, and C.-Y. Kim, *J. Org. Chem.* **43**, 105 (1978).

Aporphine–Pavine

Shamma, M., and J. L. Moniot, *J. Amer. Chem. Soc.* **96**, 3338 (1974).

8.20 OXOAPORPHINE ALKALOIDS

The oxoaporphines represent the most highly oxidized state of the aporphine skeleton. They are widely distributed (at least nine plant families) and commonly cooccur with aporphine alkaloids. The numbering system is the same as that of the aporphines.

The first oxoaporphine to be isolated (1960) was liriodenine (**440**), a bright yellow constituent of the heartwood of the tulip tree *Liriodendron tulipifera* L. (Magnoliaceae). Subsequently about 20 oxoaporphines have been isolated and, like the aporphines, beyond the perennial 1,2-dioxygenation, they exhibit a variety of oxygen substitution patterns. In this series however, there is a tendency toward 3-substitution and not 11-substitution. Some typical examples are shown in Figure 12.

Liriodenine is the most widely distributed oxoaporphine and has been obtained from at least 22 genera in eight plant families.

Shamma has summarized the chemistry that led to the assignment of the structure and mention will only be made here of the chemistry and synthesis of liriodenine (**440**).

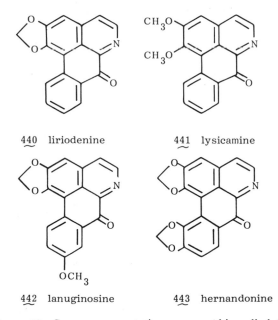

440 liriodenine   441 lysicamine

442 lanuginosine   443 hernandonine

Figure 12   Some representative oxoaporphine alkaloids.

The structure was initially confirmed by Pschorr cyclization of the amino isoquinoline **444**, but subsequently it was found that any of several aporphines could also be oxidized to **440**. The facile, efficient (30%) conversion of anonaine (**350**) into liriodenine raises the question of whether it is a natural product or an artifact.

Reduction of liriodenine (**440**) with zinc analgam affords anonaine (**350**), and the methiodide affords roemerine (**351**).

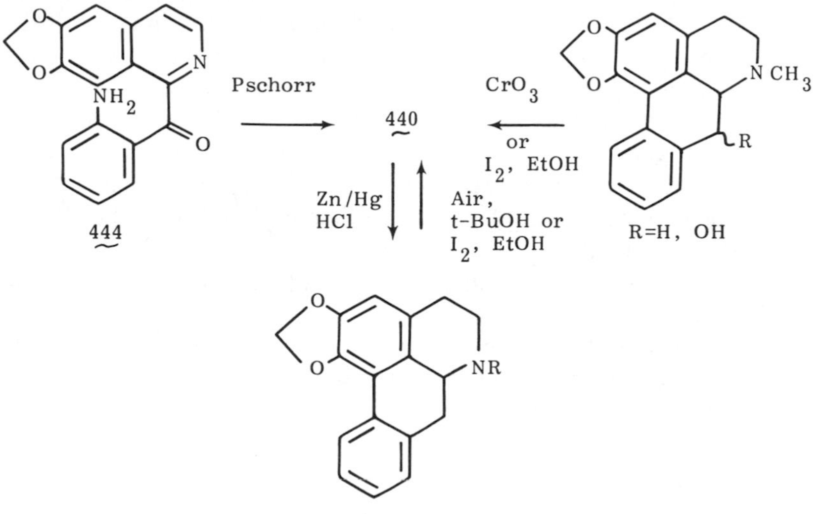

350 anonaine R=H
351 roemerine R=CH$_3$

The UV spectral data for the oxoaporphines are quite characteristic for the skeletal type and there is evidence that they may also be diagnostic for a particular oxygenation pattern. Some representative data are shown in Table 9. A 1,2-methylenedioxy derivative gives rise to a bathochromic shift

Table 9   UV Spectral Data of Some Representative Oxoaporphine Alkaloids

Compound	EtOH	
	λ_{max} (nm)	(log ϵ)
Liriodenine (**440**)	247, 268, 309, and 413 nm	(4.22, 4.13, 3.62, and 3.82)
with H$^{(+)}$	257, 277, 329, 392, and 455 nm	(4.33, 4.26, 3.67,.3.69, and 3.58)
Lysicamine (**441**)	235, 270, 307, and 400 nm	(4.47, 4.41, 3.76, and 3.94)
with H$^{(+)}$	249, 276, 306, and 453 nm	(4.33, 4.44, 3.82, and 3.58)
Cassameridine (**445**)	251, 274, 323, 353, 388, and 440 nm	(4.46, 4.40, 4.8, 3.91, 3.85, and 3.73)
with H$^{(+)}$	261, 290, 385, and 500 nm	

in the 235–250 nm band on comparison with the corresponding 1,2-dimethoxy derivative [e.g. liriodenine (440) *vs.* lysicamine (441)]. There is a substantial bathochromic shift of the longest-wavelength band on the addition of acid.

The IR spectra of oxoaporphines are typified by a band in the 1635–1660 cm^{-1} region for the 7-oxo group.

The oxoaporphines are notoriously insoluble in chloroform; indeed liriodenine crystallizes well from chloroform, consequently TFA is a common solvent to use for obtaining NMR spectral data. Representative NMR data are shown in Figure 13 for the alkaloids liriodenine (440), oxostephanine (446), lanuginosine (442), oxolaureline (447), cassameridine (445).

As with the aporphines, the C-3 proton appears at a higher field that the aromatic hydrogens and the C-11 proton is usually the most deshielded. An exception to the latter "rule" occurs in the case of oxolaureline (447), where the 10-methoxy shields the 11-proton sufficiently such that the C-8 proton is the most deshielded.

There have been no biosynthetic studies of the oxoaporphines with labeled precursors.

Liriodenine shows quite reasonable cytotoxity against the human carcinoma of the nasopharynx test system and also exhibits quite a wide range of antimicrobial activity *in vitro*.

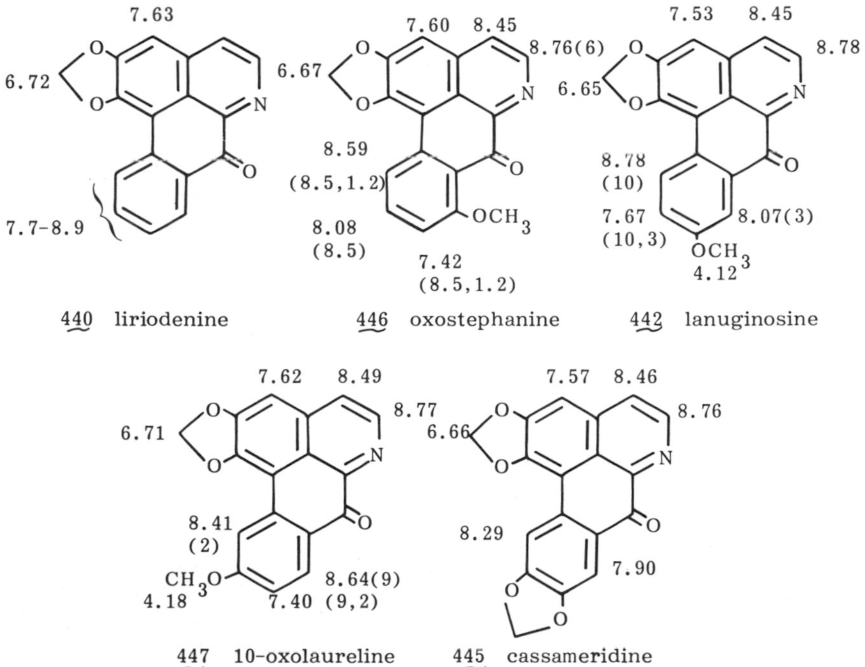

Figure 13 Proton NMR data for representative oxoaporphine alkaloids (in CF$_3$CO$_2$H).

LITERATURE

Reviews

Shamma, M. *The Isoquinoline Alkaloids*, Academic, New York, 1972, p. 245.

Shamma, M., and R. L. Castenson, *Alkaloids NY* **14**, 226 (1973).

Guinaudeau, H., M. Leboeuf, and A. Cavé, *Lloydia* **38**, 275 (1975).

8.21 DIOXOAPORPHINE ALKALOIDS, ARISTOLACTAMS, AND ARISTOLOCHIC ACIDS

Pontevedrine was isolated in 1971 from *Glaucium flavum* var. *vestitum* (Papaveraceae) and assigned the structure **448**. It gives a positive ferric chloride test, but the UV spectrum is not affected by acid or base.

In 1976, the structure of pontevedrine was revised to **449**, analogous to a compound from *S. cepharantha*. The structure was confirmed by partial synthesis from *O*-methylatheroline (**450**) by treatment with either iodine or DDQ and subsequently by total synthesis. Hydrolysis of pontevedrine (**449**) with sodium hydroxide in methanol gave **451** as the result of a benzilic acid rearrangement followed by decarboxylation.

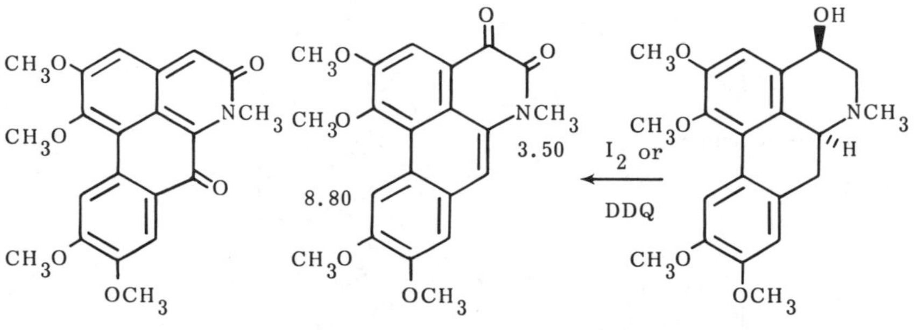

| 448 pontevedrine (original) | 449 pontevedrine (revised) | 450 O-methylatheroline |

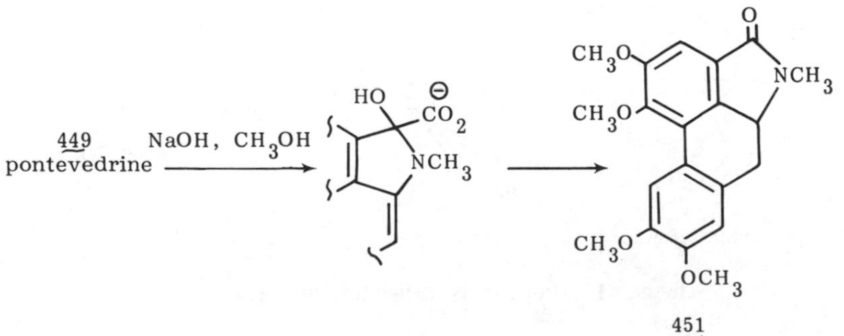

451

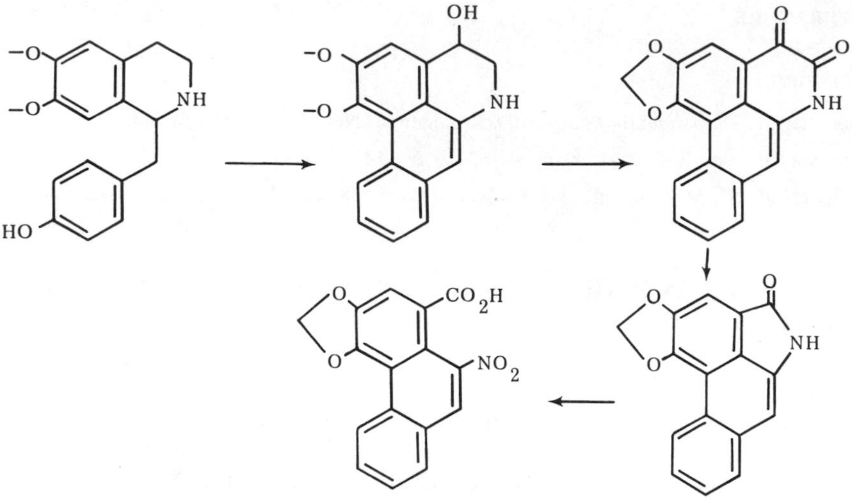
OH
O
O
NH
NH
NH
HO
O
O
NH
O
CO2H
NO2

452 aristolochic acid-II

Scheme 60 Biogenesis of aristolochic acid-II (**452**).

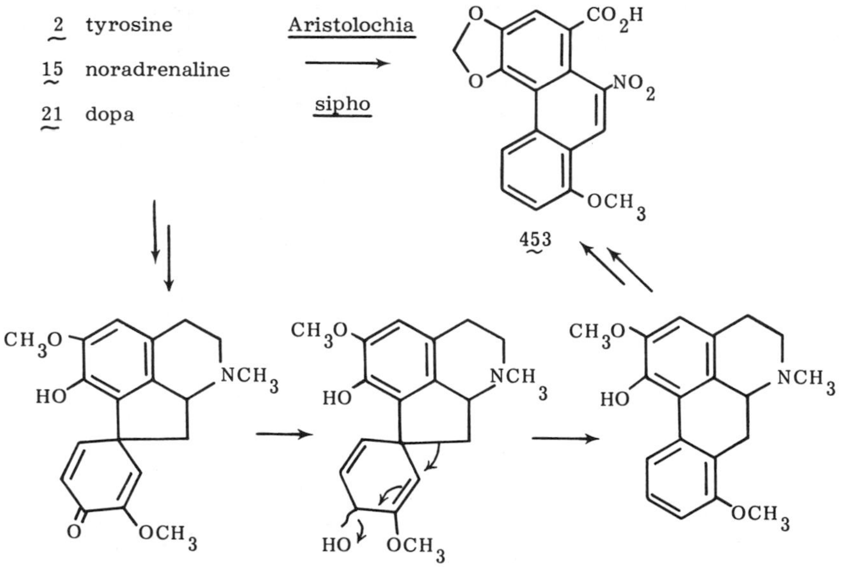
2 tyrosine
15 noradrenaline
21 dopa
Aristolochia
sipho
CO2H
NO2
OCH3
453
CH3O
HO
NCH3
OCH3
354 orientalinone

Scheme 61 Biogenesis of aristolochic acid-I (**453**).

For many years it had been thought that the several aristolochic acids (e.g. **452**) were derived from an aporphine precursor. But the recent isolations are the first biogenetic evidence for a sequence that can now be rationalized as shown in Scheme 60. No biosynthetic experiments with advanced labeled precursors have been carried out to support this biogenetic proposal.

Spenser and co-workers studied the biosynthesis of aristolochic acid-I (**453**) in *Aristolochia sipho* with simple precursors and showed that tyrosine (**2**), dopa (**21**) and noradrenaline (**15**) were precursors. The nitrogen of the nitro group was shown to be derived from the amino acid nitrogen of tyrosine. Mothes and co-workers found that norlaudanosoline (**228**) was a precursor of **453**, which would appear to contradict the incorporation of noradrenaline.

The unusual ring C substitution of aristolochic acid-I (**453**) was rationalized in terms of the dienol–benzene rearrangement shown in Scheme 61.

LITERATURE

Shamma, M., *Alkaloids, London* **6**, 183 (1976).

Shamma, M., *Alkaloids, London* **8**, 122 (1978).

Ribas, I., J. Sueiras, and L. Castedo, *Tetrahedron Lett.* 3093 (1971).

Akasu, M., H. Itoka, and M. Fujita, *Tetrahedron Lett.* 3609 (1974).

Castedo, L., R. Suau, and A. Mourino, *Tetrahedron Lett.* 501 (1976).

Castedo, L., R. Estevez, J. M. Saa, and R. Suau, *Tetrahedron Lett.* 2179 (1978).

Priestap, H. A., E. A. Ruveda, O. A. Mascaretti, and V. Deulofeu, *Anal. Assoc. Quim. Arg.* **59**, 245 (1971).

Comer, F., H. R. Tiwari, and I. D. Spenser, *Can. J. Chem.* **47**, 481 (1969).

Pailer, M., *Fortschr. Chem. Org. Naturs.* **18**, 66 (1960).

8.22 AZAFLOURANTHENE ALKALOIDS

In 1972 Cava and co-workers obtained the novel azaflouranthene alkaloids imeluteine (**454**) and rufescine (**455**) from the stems of *Abuta imene* and *A. rufescens* in the Menispermaceae. The co-occurrence of these alkaloids with homomoschatoline (**456**) suggests that they may be produced by decarbonylation of a related oxoaporphine.

A third azaflouranthene, norrufescine (**457**), has been obtained from the same sources. The proton resonances of **457** are shown.

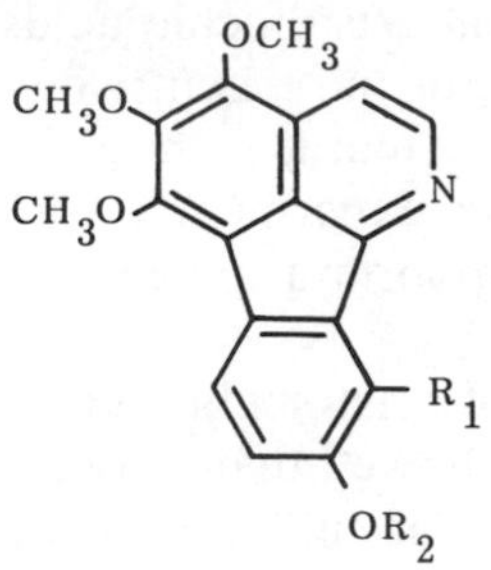

454   imeluteine $R_1 = OCH_3$
                  $R_2 = CH_3$
455   rufescine $R_1 = H$
                  $R_2 = CH_3$
457   norrufescine $R_1 = R_2 = H$

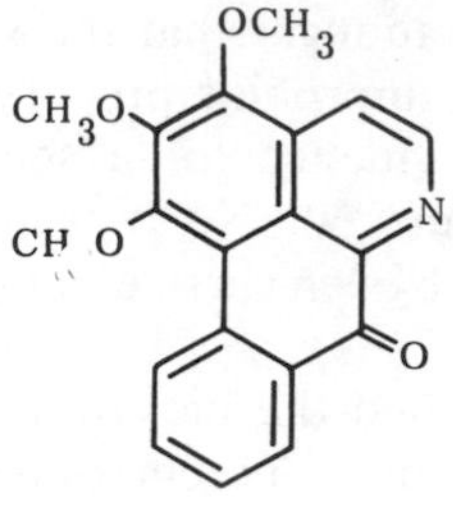

456   homomoschatoline

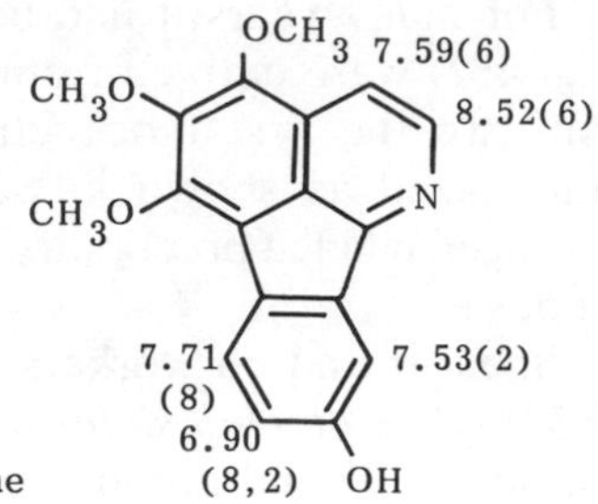

OCH_3 at
4.04, 4.08
and 4.10 ppm.

LITERATURE

Cava, M. P., K. T. Buck, and A. I. Rocha, *J. Amer. Chem. Soc.* **94**, 5931 (1972).

Cava, M. P., K. T. Buck, I. Noguchi, M. Srinivasan, M. G. Rao, and A. I. DaRocha, *Tetrahedron* **31**, 1667 (1975).

8.23   TASPINE

Taspine (**458**), on initial inspection, can hardly be classed as an isoquinoline alkaloid. It has been isolated from *Leontice alberti* Regel. and *Caulophyllum robustum* Maxim. both in the family Berberidaceae, and its unusual structure was confirmed by a biogenetic-type synthesis (Scheme 62).

8.24   MORPHINANDIENONE ALKALOIDS

There are nine main members of the morphinandienone group which can be divided into two subgroups based on their enantiomorphic nature. Morphine, codeine, neopine, thebaine, and oripavine are one group and have been obtained from *Papaver* species. The second group is comprised of sinomenine, hasubanonine, metaphenine, and protometaphenine isolated from Japanese plants of the genera *Sinomenium* and *Stephania*. Except for sinomenine the other three alkaloids in this group are discussed in the following section.

     The morphinandienones and related alkaloids are not limited in distribution to the genera mentioned above. They have been obtained from *Cassytha, Croton, Nandina, Nemuaron, Ocotea,* and *Rhigiocarya* species, and the homologous homomorphinandienones have been obtained from *Andr-*

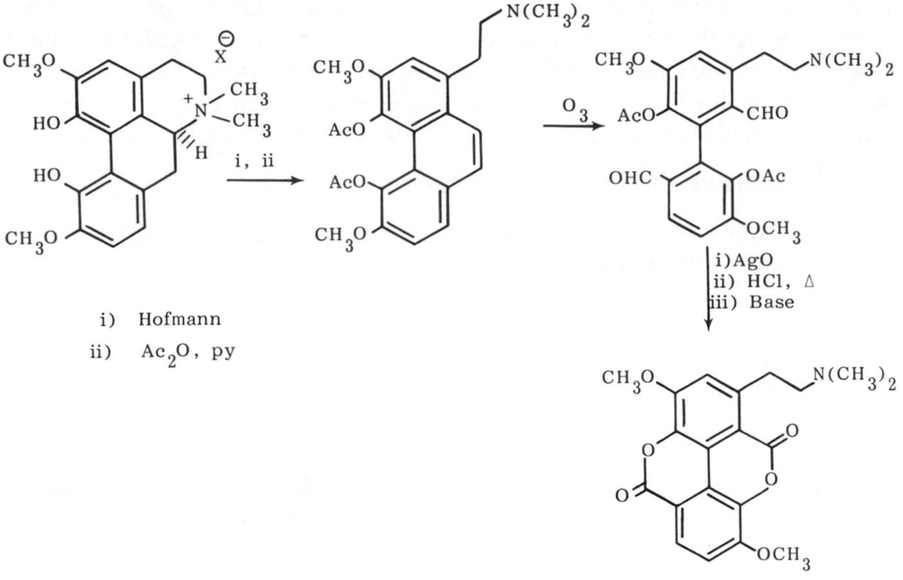

i) Hofmann

ii) Ac$_2$O, py

i) AgO
ii) HCl, Δ
iii) Base

458 taspine

Scheme 62

ocymbium, Colchicum, and *Kreysigia* species. The latter alkaloids are discussed in Section 8.35.

8.24.1 Morphine and Related Alkaloids

Historically, the most important morphinandeniones are those of the ancient drug opium. Opium is the air-dried milky exudate from incised, unripe capsules of *Papaver somniferum* or *P. alba* Mill. (Papaveraceae). It was well known in ancient Greece, and Hippocrates mentions the use of poppy juice as a cathartic, hypnotic, narcotic, and styptic. Pliny the Elder indicates the use of the seed as a hypnotic, and the latex for headaches, arthritis, and curing wounds. Dioscorides, in A.D. 77, distinguished between the latex of the capsules and a whole-plant extract. In time, the cultivation of opium spread from Asia Minor to Persia and on to India and China. The smoking of opium was first noted to be extensive in China and the Far East in the latter part of the eighteenth century. Traffic and subsequent addiction proved to be a useful source of revenue.

In Turkey *P. somniferum* var. *glabrum* has been cultivated for commericial purposes and *P. alba* in Macedonia. As the capsule turns from green to yellow, incisions are made and the dried latex removed the following day. A rich alluvial soil and fertilizers improve the yield.

It is the alkaloids of opium which are of prime importance. Over 40 alkaloids are known and several are of major significance. These are morphine, codeine, thebaine, noscapine, and papaverine. The morphine content of opium is in the range 4–21%, of noscapine 4–8%, and of the previously mentioned alkaloids 0.5–2.5%. These alkaloids occur at least partially bound to meconic acid, the presence of which can also be used to detect opium.

Isolation, Structure Elucidation, and Chemistry

The most important alkaloid of opium is morphine, which in admixture with narcotine (now noscapine) was first isolated by Derosne in 1803. One year later Seguin, in a report to the Academy of Sciences, described its isolation in pure form. A similar isolation was reported by Serturner in 1805, but it was 118 years before the structure was eventually deduced by Gulland and Robinson after considerable experimental effort.

Increasingly, "poppy straw," the dried capsule walls plus stem, has been used as a source of opiates. In 1951, 16% of the total morphine production was from straw; by 1971 this had grown to 37%. In this time morphine production rose from about 70 tons to over 170 tons per year.

Morphine and its salts are classified as narcotic analgesics, and these and other aspects of the pharmacology of these compounds will be discussed subsequently.

In 1833, Robiquet first isolated codeine from opium, where it occurs closely associated with morphine, in the concentration range 0.7–2.5%. Insufficient codeine is available by isolation, and therefore most of the codeine is produced by methylation of morphine. Several reagents have been used for this process, but the most commonly used is trimethylphenylammonium methoxide, sulfate or chloride, using either xylene or xylene–methanol as solvent.

The reverse reaction, the conversion of codeine (459) to morphine (5), has been carried out with pyridine hydrochloride (in 15–34% yield), lithium diphenylphosphide (in 61% yield), or boron tribromide at room temperature in 90% yield.

Thebaine (460) also occurs in opium and is currently a compound of great interest, because it is an alternative source for codeine, whose supply is being threatened. In particular, the plant *Papaver bracteatum* Lindl., which produces thebaine in substantial quantities with few other alkaloids, has been studied as a new crop for supplying the almost 200 tons of codeine required in a year. Structurally, it is a methyl enol ether of codeinone (461).

The structure elucidation of these alkaloids stands as a monument to the brillance of Robinson, but positive proof of the morphine skeleton did not come until synthetic endeavors were successfully completed by Gates in 1952. The synthesis is over 20 steps in length and only a brief outline will be given. Diels–Alder addition of butadiene to the unsaturated diketone 462 gave 463, which could be hydrogenated to give the keto amide 464. Standard reductions gave the amino olefin 465. The next steps were concerned with partial demethylation, and formation of the C-6 ketone 466.

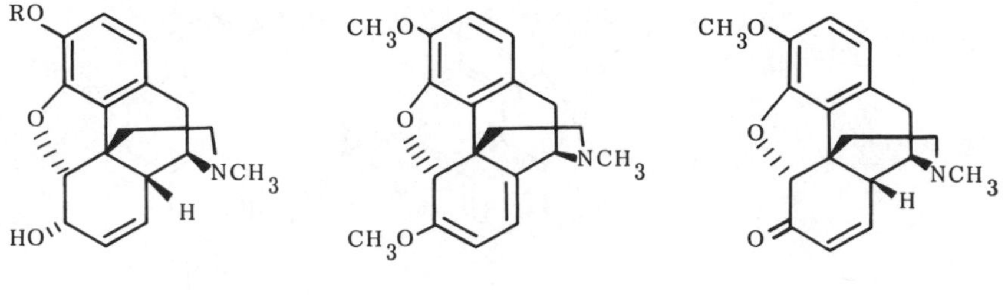

5 morphine R=H

459 codeine R=CH$_3$

460 thebaine

461 codeinone

Tribromination, elimination with C-14 epimerization, and internal displacement of the C-5 bromine by the C-4 phenolic group eventually gave 1-bromocodeinone (**467**). This compound could be debrominated and reduced with lithium aluminum hydride and then de-*O*-methylated with pyridine hydrochloride to morphine (**5**) (Scheme 63).

The stereochemistry of morphine at C-14, C-9, C-13, C-5, and C-6 could

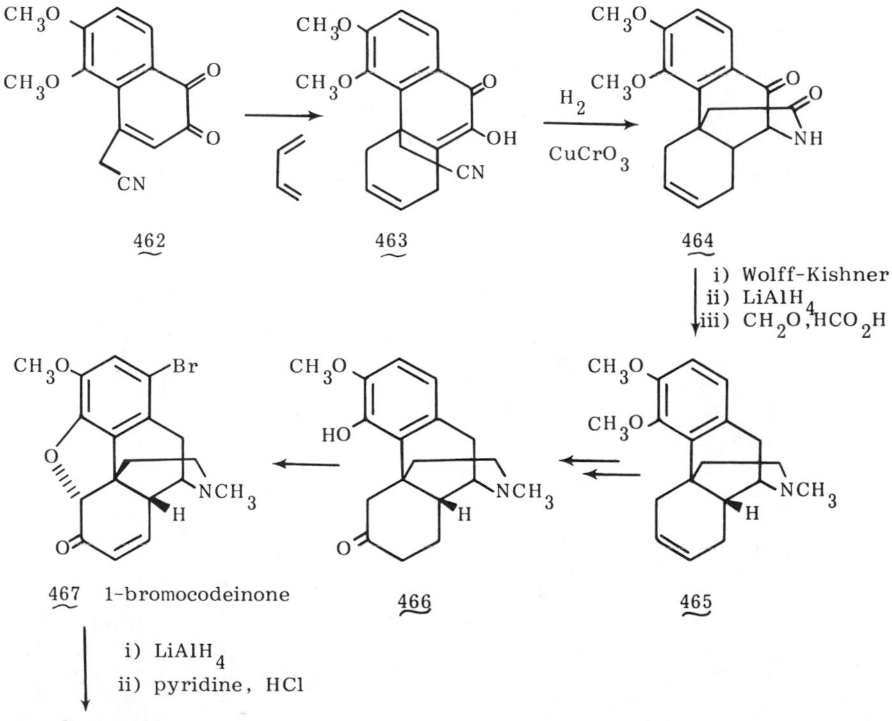

462

463

464

i) Wolff-Kishner
ii) LiAlH$_4$
iii) CH$_2$O, HCO$_2$H

467 1-bromocodeinone

466

465

i) LiAlH$_4$
ii) pyridine, HCl

5 morphine

Scheme 63

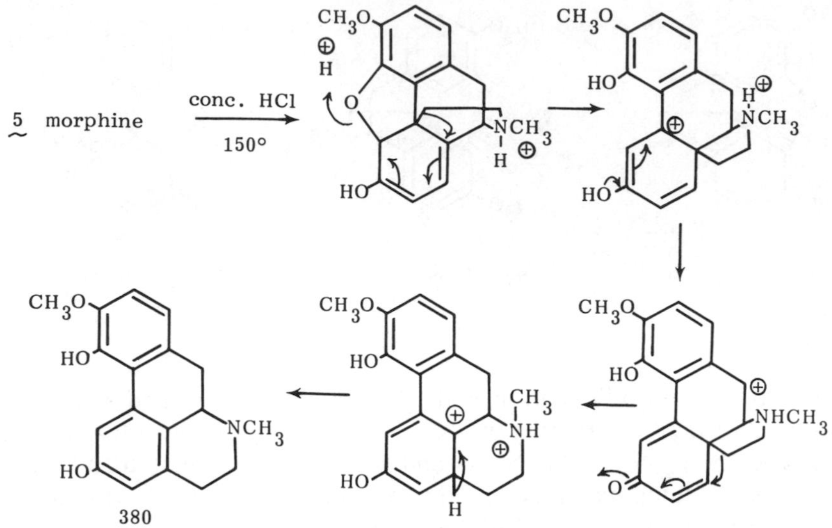

380

Scheme 64 Formation of apomorphine (**380**) from morphine (**5**).

only be deduced by X-ray crystallographic analysis, in spite of considerable chemical effort.

The action of concentrated hydrochloric acid on morphine (**5**) at 150°C produces apomorphine (**380**) by the mechanism shown in Scheme 64.

When thebaine (**460**) is heated with dilute hydrochloric acid a quite different reaction rapidly occurs to give thebenine (**468**). This rearrangement

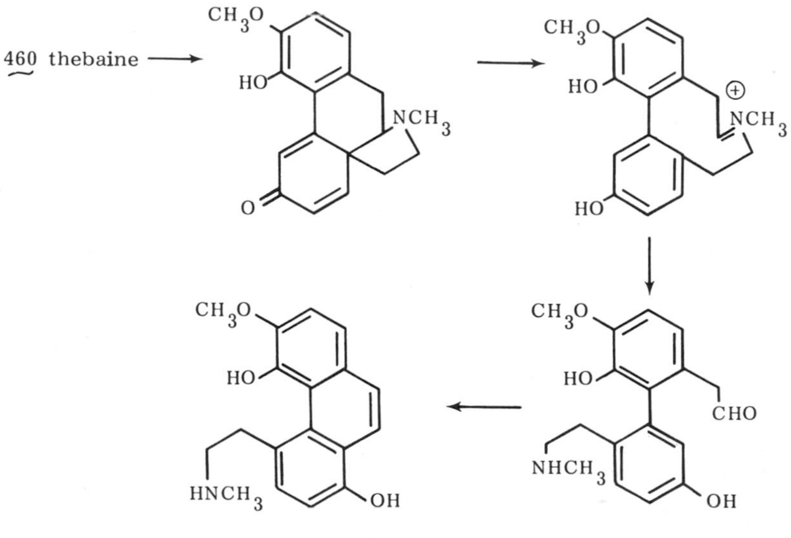

468 thebenine

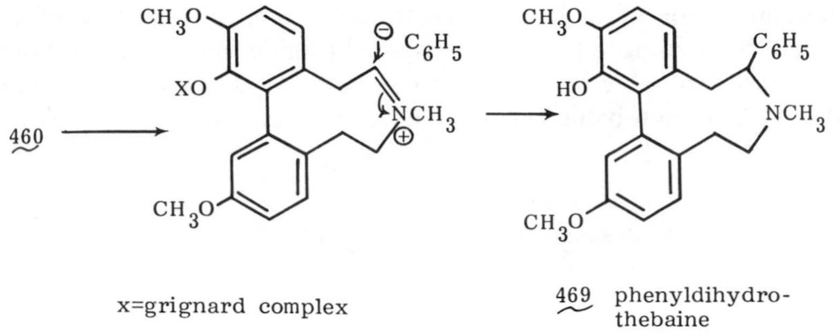

x=grignard complex

469 phenyldihydro-
      thebaine

Scheme 65

is thought to occur by fragmentation of the C-9–C-14 bond to give an iminuim species, which can be hydrolyzed and recyclized (Scheme 65).

Treatment of thebaine (**460**) with Grignard reagents gives products arising from nucleophilic attack on an iminium species to give phenyldihydro-thebaine (**469**).

Thebaine also reacts very readily with dienophiles in a Diels–Alder reaction, and literally hundreds of derivatives have been prepared from these adducts. As expected this reaction is regiospecific with dienophiles such as methyl acrylate and methyl vinyl ketone, and is quite highly stereospecific for the formation of the 7α-isomer (e.g. **470**).

Further reaction of **470** with Grignard reagents gives the alcohol **471**, also stereospecifically. O-Demethylation under alkali conditions afforded a series of phenols having analgesic activity up to 10,000 times that of morphine (e.g. **472**, R = isoamyl). It had been hoped that these compounds

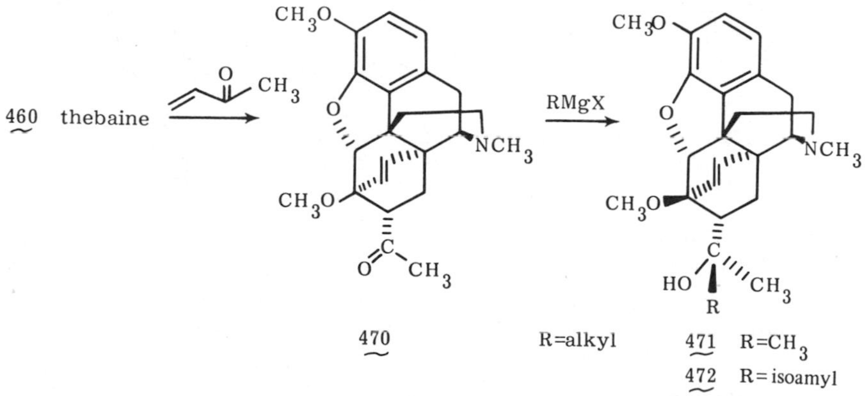

470

R=alkyl

471 R=CH$_3$
472 R= isoamyl

would separate the analgesic and respiratory depressant actions, but no such observations were made, and indeed the compounds were highly addictive.

Treatment of thebaine with aqueous sodium bisulfite at pH 4 in the presence of oxygen afforded 6-O-methylsalutaridine (**473**) in good yield, a reversal of the biosynthetic process.

The conversion of thebaine (**460**) to codeinone (**461**) can be efficiently carried out by treating **460** with a hydrogen bromide–iodine mixture in methylene chloride–butyl ether at 0°C for 7 min. Hydrolysis of the reaction mixture with sodium bicarbonate gave pure codeinone in 90% yield.

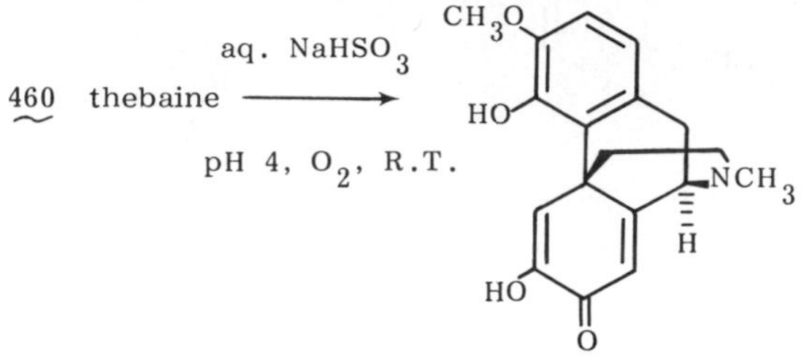

Beyerman and co-workers have recently described a synthesis of (−)-dihydrothebainone (**474**) whose conversion to thebaine (**460**) had already been described. The Birch reduction product **475** underwent exclusive acid-catalyzed cyclization to give **476**. The undesired 2-hydroxy group was selectively removed by way of the 1-phenyltetrazol-5-yl ether to afford **477** which could be reduced to **474**.

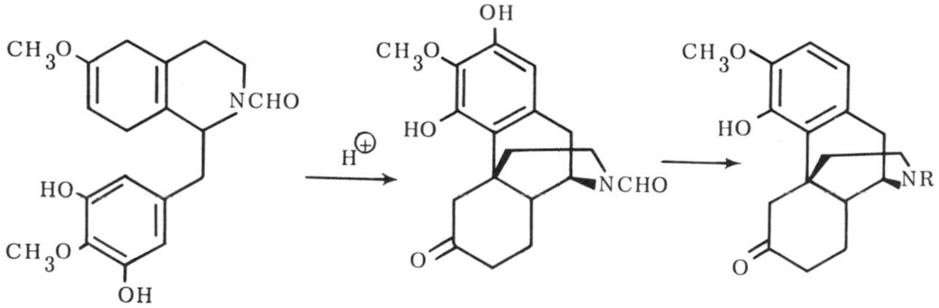

Robinson called the morphine alkaloids "the star performers in the field of molecular acrobats." Although this might now be disputed by the indole alkaloid chemists, the chemical rearrangements of the morphine alkaloids are dramatic indeed.

8.24.2 Sinomenine

(−)-Sinomenine (**478**) is the principal alkaloid of *Sinomenium acutum* (Menispermaceae) and was first studied extensively in the 1930s by Goto and collaborators in Japan. Catalytic reduction afforded a dihydro derivative, which on sodium amalgam reduction afforded (+)-dihydrothebainone (**479**), enantiomorphic with the compound from thebaine (460). It is therefore in the opposite antipodal series to the opium alkaloids and may be regarded as 7-methoxy-(+)-thebainone. Although not important as an alkaloid in its own right sinomenine affords the unique opportunity of making (+)-codeine (480) and (+)-morphine (481) available. And indeed such reactions were successfully achieved in 1954 by Goto and Yamamoto.

Brossi and co-workers have reported an improved procedure for these important compounds from (−)-sinomenine (478). Thus 478 was converted to the urethane of (+)-norcodeinone (482) which could be reduced to (+)-codeine (480) by lithium aluminum hydride. O-Demethylation with boron tribromide afforded (+)-morphine (481) which could be acetylated to (+)-heroin (483). Further mention is made of (+)-morphine in the pharmacology section.

CH$_3$O—

HO—

‴NCH$_3$

O=

OCH$_3$

478   (−)-sinomenine

CH$_3$O—

O—

O=

‴NCO$_2$C$_2$H$_5$

H

482

R$_1$O—

O—

R$_2$O‴ H

‴NCH$_3$

H

	R$_1$	R$_2$
480 (+)-codeine	CH$_3$	H
481 (+)-morphine	H	H
483 (+)-heroin	COCH$_3$	COCH$_3$
479 R=CH$_3$,optical antipode		

8.24.3   Chemistry of the Morphinandienone Alkaloids

Several reactions of the morphinandienone system are of general interest. Sodium–liquid ammonia reduction of O-methyl salutaridine (484) followed by reaction with diazomethane for example afforded (R)-(−)-laudanosine (188).

In 1965, Barton and Battersby and their respective co-workers described the biomimetic conversion of salutaridine (485) to thebaine (460). Reduction with sodium borohydride gave a mixture of diols 486, which could be converted to 460 by treatment with 0.2 M phosphate buffer at room temperature for about 3 days. The reaction proceeded more rapidly in 1 N hydrochloric acid.

A contrasting rearrangement is that of norsinoacutine (487) to the dienone 488 by treatment of the N-acetyl alcohols 489 at room temperature with 1 N hydrochloric acid.

There is also a second pathway which is open to morphinandienones such as O-methylflavinantine (490), namely their ability to rearrange to aporphine derivatives. When 490 was heated with concentrated hydrochloric

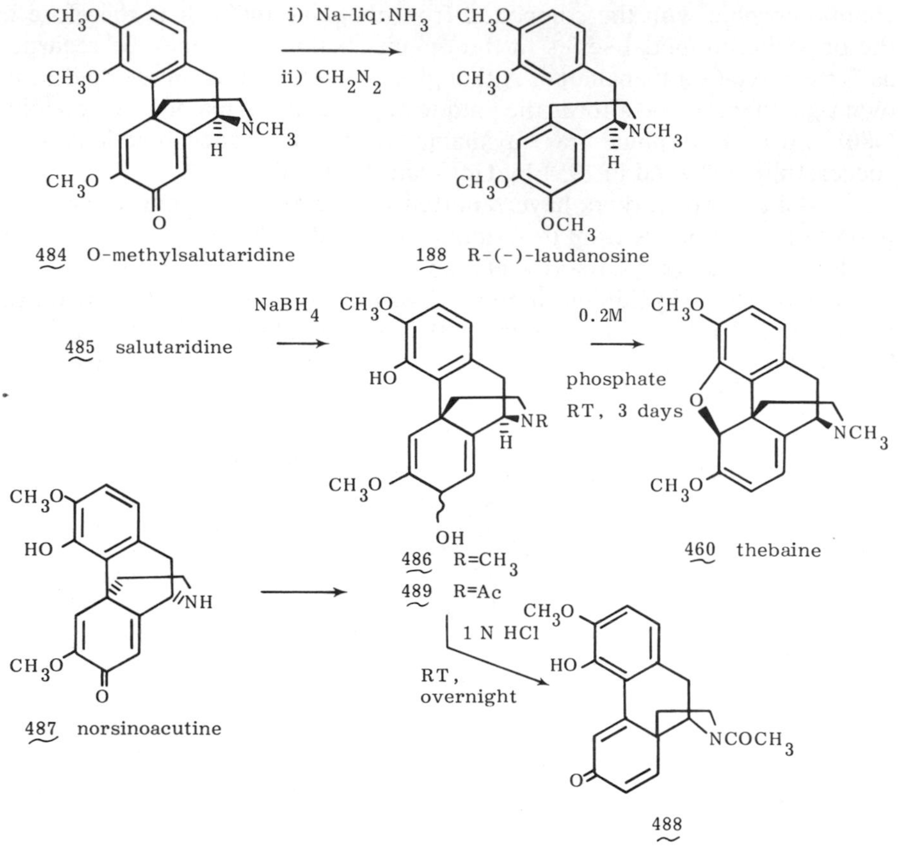

i) Na-liq.NH$_3$

ii) CH$_2$N$_2$

484　O-methylsalutaridine

188　R-(−)-laudanosine

485　salutaridine

NaBH$_4$

0.2M

phosphate

RT, 3 days

486　R=CH$_3$
489　R=Ac

460　thebaine

487　norsinoacutine

1 N HCl

RT,
overnight

488

acid, an aporphine was obtained in almost 90% yield which could be converted to glaucine (**372**).

In summary, the morphinandienones undergo two principal types of acid-catalyzed rearrangement, one that leads to the dibenzazonines, and the other to the aporphines. The difference between these two possibilities has been

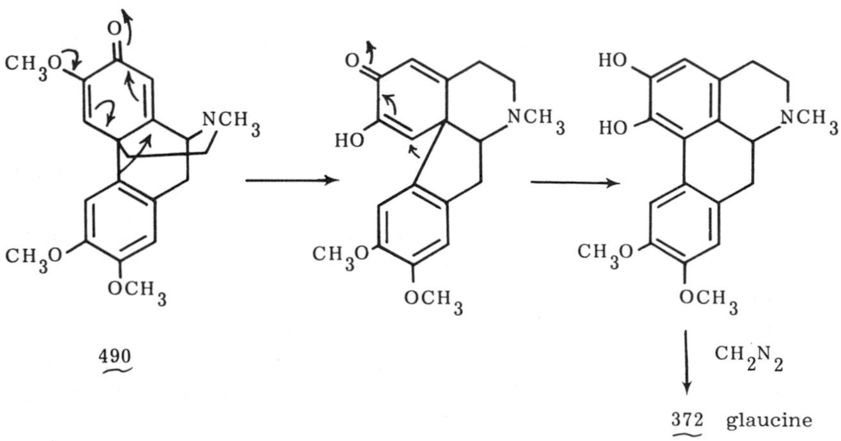

490

CH$_2$N$_2$

372　glaucine

interpreted in terms of the different degrees of involvement of the electron pair on nitrogen. In the event that this occurs to give intermediates close to **491**, stereoelectronic factors favor aryl migration to give the proerythrinadienone intermediate. When no lone-pair involvement is possible, as with the amides or reactions in BF$_3$-etherate, alkyl migration occurs to give the neospirine intermediates.

proerythrindienone
skeleton

491

These are *in vitro* interconversions; their importance in the *in vivo* biosynthesis of aporphines or benzazonines remains to be demonstrated.

8.24.4 Synthesis of Morphinandienone Alkaloids

Phenolic Oxidation

The key to any biogenetic synthesis of the morphine alkaloids from the biosynthetic precursor reticuline (**182**) (see later) is driving the oxidative coupling reaction in the direction of the *ortho-para* coupled product, salutaridine (**485**), rather than the *para-para* coupled product isosalutaridine (**492**).

Oxidation of reticuline with manganese dioxide in chloroform gave isosalutaridine (**492**) in 4% yield and salutaridine (**485**) in 0.011% yield. Low yields of **485** (% in parentheses) were also obtained with potassium ferricyanide (0.015%), potassium nitrosodisulfonate (0.0054%), and ferric chloride (0.0007%) and were determined with the aid of tritium-labeled reticuline (**182**). The major compound in these reactions is the *para-ortho* coupled isoboldine (**382**), which may be formed in yields of up to 53%.

It was not until 1975 that significant progress was made in this area, when Schwartz and Marni reported that the *N*-trifluoroacetyl derivative of nor-recticuline (**493**) could be cyclized with thallium tristrifluoroacetate to *N*-trifluoroacetylnorsalutaridine (**494**) in 11% yield. Using *N*-carbomethoxynorreticuline (**495**), the yield of the corresponding salutaridine derivative **496** was raised to 23%. Lithium aluminum hydride reduction of **496** followed by rearrangement in 1 *N* hydrochloric acid gave racemic thebaine (**460**).

Nonphenolic Oxidation

In recent years there has been considerable interest in the nonphenolic oxidation of benzylisoquinolines. For example Miller has shown that at

$$R=CO_2C_2H_5$$
i) $LiAlH_4$
ii) $1N$ HCl   R.T., 1 hr.

$(\pm)$-thebaine
460

various reagents

182	$R=CH_3$	485	$R=CH_3$
493	$R=COCF_3$	494	$R=COCF_3$
495	$R=CO_2C_2H_5$	496	$R=CO_2C_2H_5$

492   isosalutaridine

potentials near 1.1 V, laudanosine (**188**) in acetonitrile is oxidized at a platinum electrode to **497** in 50–60% yield. Further work indicated that the basic nitrogen is intimately involved in the oxidation process by anchimeric assistance.

Once again similar results are obtained with chemical oxidizing agents, and the morphinandienone may not even be the final product under these conditions. Kupchan and co-workers found that VOF_3-TFA oxidation of N-formyllaudanosine (**498**) gave the neospirinedienone **499** in good yield. That this interconversion proceeds by way of a *para-para* coupled morphinandienone intermediate **500** was demonstrated by separate experiments replacing the 6- and 7-methoxy groups with a benzyloxy group. When the 6-benzyloxy derivative **501** was oxidized, **499** was produced in 77% yield, indicating specific loss of the alkyl group from the 6-position.

The Pschorr reaction is a well-established route for the synthesis of aporphine alkaloids and has been discussed in Section 8.18.2. In 1967 however, Hey and co-workers demonstrated that morphinandienones could also be produced by similar methods and subsequently the structures of several alkaloids have been proven by synthesis using this route. For example diazotization of 6'-aminolaudanosine (**502**) at 0.5°C followed by warming to 70°C for 1 hr afforded O-methylflavinantine (**490**) in 1.4% yield.

8.24.5 Spectral Properties

The morphinandienones typically show two main UV absorption bands, one between 235 and 240 nm and the second between 275 and 290 nm, depending on the substitution. Compounds having 2,3-substitution tend to absorb closer

188

497

498 R=CH_3

501 R=$CH_2C_6H_5$

500

499

to 290 nm [e.g. flavinantine (**503**) λ_{max} 239, 286 (4.17, 3.85)], and those with 3,4-substitution closer to 275 nm [e.g. salutaridine (**485**) λ_{max} 236, 279 nm (4.23, 3.76)].

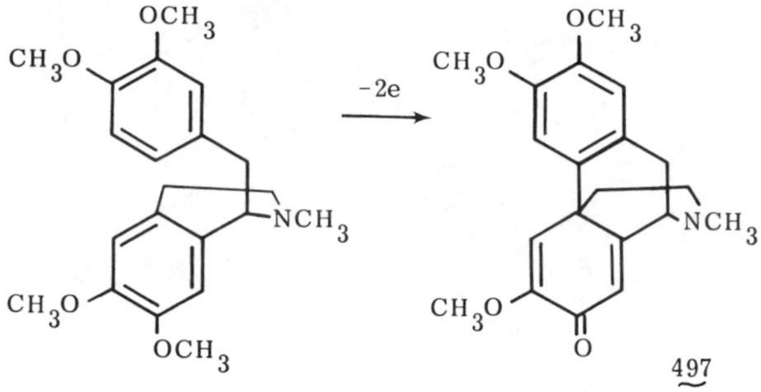

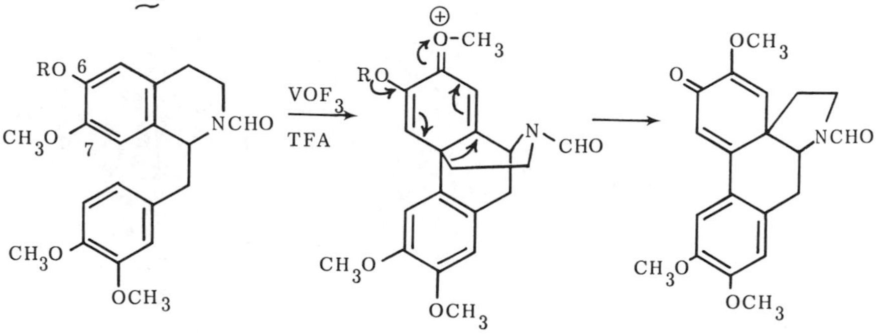

Alkaloids in the codeine–morphine series show λ_{max} at about 285 nm, and thebaine gives a similar spectrum, although the 1,3-diene system does enhance the lower-wavelength end of the spectrum. In the case of sinomenine (**478**), where there is no 4,5-bond, λ_{max} are observed at 232 and 265 nm.

The IR spectrum of morphinandienones such as salutaridine (**485**)

502

490 R=CH_3

503 R=H

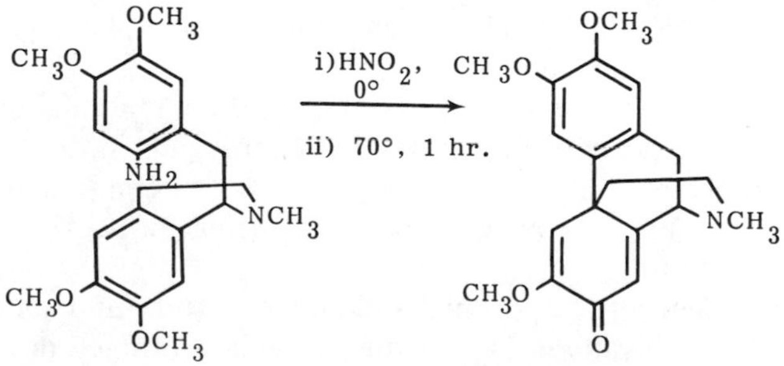

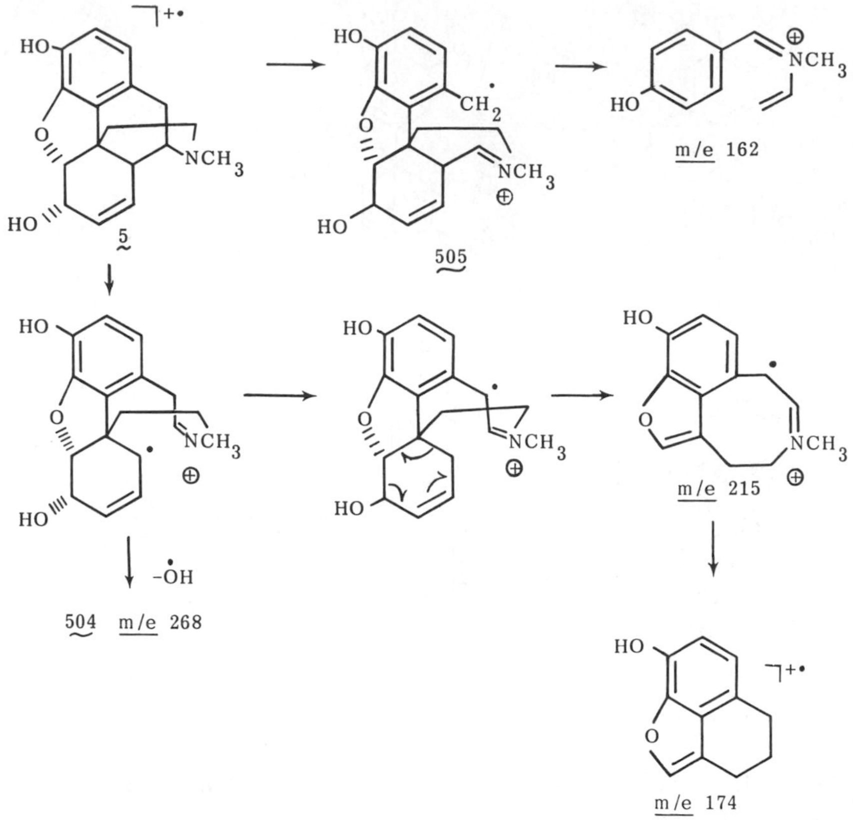

Scheme 66 Mass spectral fragmentation of morphine (**5**).

display bands at 1665, 1635, and 1615 cm^{-1} characteristic of the α-methoxy cyclohexadienone system.

The fragmentation patterns of several morphine alkaloids have been quite well studied. For morphine (**5**) the principal pathway involves the allylic and benzylic ions **504** and **505**, which further fragment as shown (Scheme 66).

When an 8,14-double bond is present as in thebaine (**460**), 13,15-cleavage is favored over 9,14-cleavage.

The distinction between B/C *cis* and B/C *trans* compounds can effectively be made by mass spectrometry. *N*-Methyl compounds in the B/C *cis* series shown an intense peak at *m/e* 59, which is absent in compounds in the B/C *trans* series. The ion *m/e* 59 is envisaged as being formed by the pathway shown in Scheme 67.

The importance of the morphinandienone system and its derivatives, particularly those alkaloids having the 4,5,-oxide bridge such as codeine (**459**) and morphine (**5**), has led to numerous studies of the NMR spectra of these compounds and some representative examples are shown in Figure 14. The geometry of ring C for codeine is favorable for both homoallylic

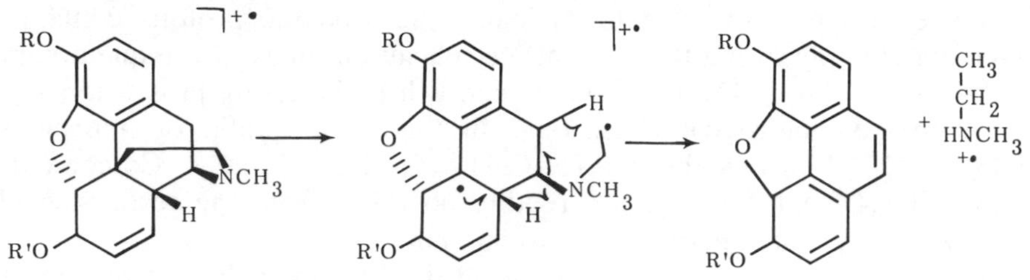

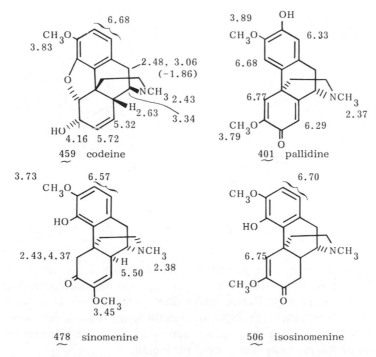

Scheme 67

($J_{6,14}$ = 2.6 Hz) and allylic ($J_{6,8}$ = 2.9 Hz) couplings, and thus this proton gives a highly complex signal.

A somewhat different spectrum is observed in the morphinandienone series for an alkaloid such as pallidine (**401**), for now the most deshielded proton is the C-5 proton which appears in the region of 6.8 ppm. Similar conclusions about the most deshielded proton allow a distinction to be made between sinomenine (**478**) and isosinomenine (**506**).

Terui and co-workers have reported the ^{13}C NMR assignments of codeine and some related alkaloids and these are shown. Some interesting features of these spectra include the substantial upfield shift of C-10 in all these compounds regarded as being due to a strong steric γ-effect of the N-methyl group.

Figure 14 Proton NMR data for some representative morphinandienone alkaloids.

Some problems found with assigning the aromatic protons could be overcome by measuring the spin lattice relaxation times of the quaternary carbons C-3, C-4, C-11, and C-12, which will be inversely proportional to the corresponding signal intensities. In this way a ratio of intensities ~4 : 2 : 2 : 1 was established for C-11 : C-12 : C-3 : C-4. Compounds having the 4,5-oxide bridge are readily observed from the resonance of C-5, which typically appears close to 90 ppm.

Carroll and co-workers have reported the ^{13}C NMR data of numerous morphine and thebaine derivatives, including morphine (5) itself.

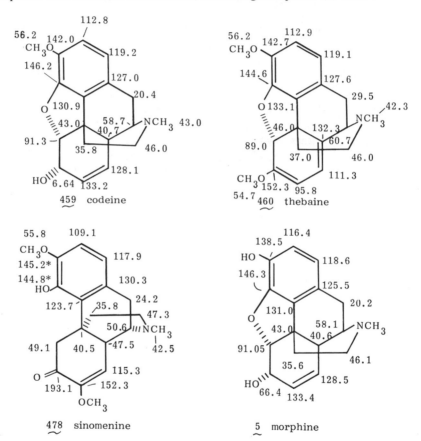

8.24.6 Biosynthesis

Robinson, in 1931, first suggested that the alkaloids of the morphine group could be derived from two units of tyrosine, *via* benzylisoquinoline intermediates. Subsequently, Barton and Cohen proposed a theory for the late stages in the biosynthetic scheme. In general terms these ideas have proven to be correct, although many unforeseen subtleties have been encountered as details of the pathway have been unraveled.

One of the classic fallacies of biosynthetic work is that a general precursor such as carbon dioxide will give a uniformally labeled product such as morphine (5). The implication here is that pool sizes of the many intermediates that form the skeleton are equal. In most instances where two or more units come together to form a single molecule, these pool sizes are not the same and consequently quite different extents of labeling of the various units by $^{14}CO_2$ will occur. In the case of morphine (5) nonuniform labeling was indeed observed by Rapoport and co-workers.

Similarly we should not assume that a molecule such as morphine which is derived from two tyrosine units will incorporate labeled tyrosine into each "half" of the molecule to the same extent. Such an expectation is reasonable only if (a) two identical units are coupling in some way or (b) a symmetrical intermediate is involved at an initial or subsequent stage.

When [2-^{14}C]tyrosine was used as a precursor, specific equal distribution of label was found at C-9 and C-16 of morphine (5) in *P. somniferum* as expected. Neubauer however found that thebaine (460) obtained from *P. bracteatum* had 64% of the activity at C-9 and 30% at C-16. This would suggest that the two C_6-C_2 units which are developed from tyrosine differ. Confirmation of this came when [1-^{14}C]dopamine labeled only C-16 of morphine (5) and codeine (459).

The classic hypothesis for morphine alkaloid formation would be to place (+)-norlaudanosoline (507) as the first "dimeric" intermediate. Indeed 507 is specifically incorporated into thebaine (460), codeine (459), and morphine (5) in *P. somniferum*. More recent work has shown that at least one other intermediate occurs before 507 in the scheme.

For some years there had been considerable speculation as to the nature of the unit combining with dopamine, and two molecules, 3,4-dihydroxyphenylacetaldehyde (168), and 3,4-dihydroxyphenylpyruvic acid (241), have featured. It will be recalled that pyruvic acid is precursor of the 1-methyltetrahydrosioquinoline alkaloids and it is therefore perhaps not surprising to discover, as Battersby and co-workers did in 1975, that the amino acid 240 was a specific precursor of morphine (5).

Decarboxylation of 240 should lead to 508 and this too was incorporated into 5. Note that this intermediate is not optically active. Stereospecific reduction then leads to (+)-norlaudanosoline (507).

If (+)-norlaudanosoline (507) is the key benzylisoquinoline, an immediate question arises as to sequencing of the O-methylation and oxidative coupling steps. When several partially O-methylated benzylisoquinolines were used as precursors only one, racemic reticuline (148) was effective. More important, (±)-reticuline (148) labeled with ^{14}C at all the methyl groups was incorporated intact into thebaine (460). This is good evidence that all methylation steps occur before phenolic coupling and that the location of the phenolic groups in the precursor is a controlling factor in the pathway.

Further work with reticuline by Brochmann-Hanssen, Battersby, and Rapoport established that there is an additional subtlety here; for the (−)-

isomer **509** is preferentially incorporated into thebaine (**460**), codeine (**459**), and morphine (**5**), while the (+)-isomer **182** accumulates.* As we shall see however in other plants it is the (+)-isomer of reticuline which is the precursor of further alkaloid groups (e.g. canadine, stylopine, beberine, narcotine, protopine).

The steps between (−)-reticuline (**509**) and thebaine (**460**) were first discussed by Barton and Cohen. Subsequently the phenol oxidation product of reticuline, the hypothetical intermediate, was synthesized by Barton *et al.* from thebaine and shown to be specifically incorporated into morphine (**5**). At about this time the same structure was isolated from *Croton salutaris* and given the name salutaridine (**485**). When [2-^{14}C]tyrosine (**2**) was used as precursor in opium, **485** could be obtained radioactively, thereby justifying its place in the biosynthetic scheme.

The remaining steps to thebaine involve closure of the 4,5-oxide bridge and loss of the C-7 oxygen. Such a transformation takes place under mild acidic conditions with the salutaridinols (**510/511**), the sodium borohydride reduction product of salutaridine (**485**). When each of the salutaridinols was labeled at C-7 with tritium and separately used as precursors of thebaine (**460**), only salutaridinol I (**510**) was incorporated. Tritium was retained indicating that oxidation to **485** had not occurred.

Rapoport has obtained a cell-free system from *P. somniferum* seedlings which can convert (−)-reticuline (**509**) to thebaine (**460**) in 10% yield, but there has been little success in developing this system to study the fundamental steps in morphine (**5**) biosynthesis.

Rapoport and co-workers have been more interested in the latter stages of the sequence, particularly the steps between thebaine (**460**) and morphine (**5**) in *P. somniferum*. Early work established that 3-*O*-demethylation of codeine (**459**) to afford morphine (**5**) is the last step. In 1972 additional data were obtained on this interconversion. Through $^{14}CO_2$ exposures studies and feeding of labeled precursors it was shown that both neopinone (**512**) and codeinone (**461**) were precursors of **459**. In addition these compounds were shown to be minor constituents of the plant. Codeine methyl ether (**513**), which was also converted to **459**, was not detected as a natural constituent of the plant (down to 0.02% of the thebaine content).

The currently accepted scheme for the biosynthesis of the morphine alkaloids in *P. somniferum* is shown in Scheme 67.

As discussed previously *P. bracteatum* is of interest as a potential commercial source of thebaine (**460**) and Hodges *et al.* have studied the pathway for thebaine formation in this species. As expected reticuline (**509**) was a precursor, but interestingly none of the demethylations to produce codeinone

---

* This is a *very* complex situation for *both* isomers are incorporated. When H-1 is labeled in the (+)-isomer no tritium is incorporated into codeine and morphine but labeling of the (−)-isomer results in substantial retention of the tritium. The conclusion is that the (+)-isomer is incorporated by way of 1,2-dehydroreticuline and that some tritum of the (−)-isomer is reversibly lost in this way.

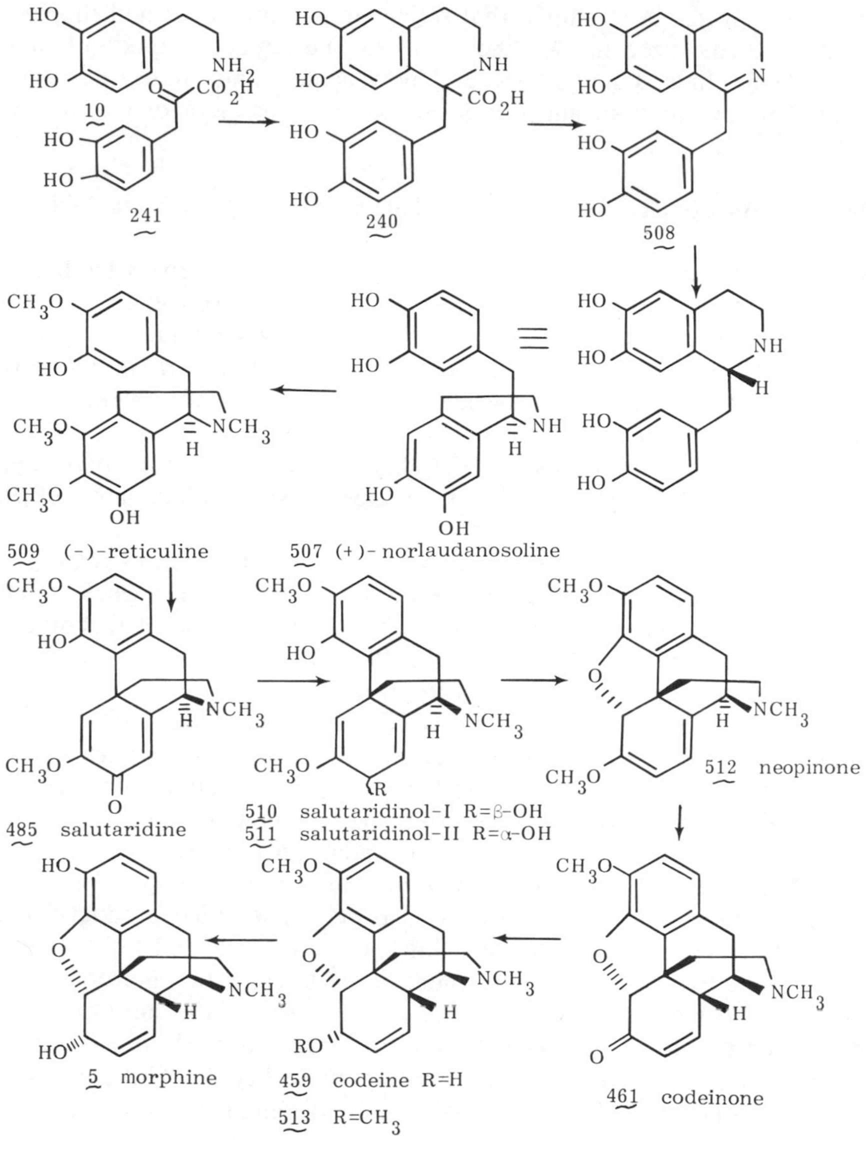

Scheme 67 Biosynthesis of morphine (**5**) in *Papaver somniferum*.

(**461**), or morphine (**5**), could be observed, hence the absence of these compounds in the plant. More recently Brochmann-Hanssen and Wunderly have confirmed these results.

A biosynthesis of some interest is the derivation of sinomenine (**478**) in *S. acutum*. This alkaloid is in the antipodal series to the *Papaver* alkaloids and consequently one would expect a derivation from (+)-reticuline (**182**) in this case. It has been shown that reticuline is a precursor and the next intermediate envisaged in the scheme is **413**, the optical antipode of salutaridine (**485**). Indeed this compound (now named sinoacutine) has been isolated from *S. acutum*, and was found to be a precursor of sinomenine (**478**).

8.24.7 Production, Translocation, and Fate of the Morphinan Alkaloids

The importance of the morphine alkaloids has led to numerous studies of the production of the alkaloids during the growing season in the various parts of the plant, and of the effects of various exogenous factors such as fertilizer, planting density, climatic conditions, and so on. Attempts to produce alkaloids from the callus tissue of *P. somniferum* and *P. bracteatum* have not been successful.

Evidence to date indicates that the alkaloids are produced, stored, and translocated in the vacuolar of the vesicles present in the latex of *P. somniferum*.

In young plants of *P. bracteatum* thebaine content increases rapidly in the roots and the concentration in the shoots increases some weeks later. During its second year of growth the highest concentration of thebaine is found in the capsule some 4–6 weeks after flowering. If budding is prevented the thebaine content remains high in the roots.

In an early chapter it was mentioned that alkaloids are not the end products of metabolism. Although details of this fascinating area are sparse and the subsequent products not well investigated, alkaloid turnover has been examined in isolated instances. One of these is morphine.

Morphine does not continuously accumulate in the latex of *P. somniferum*. The isolated latex of the opium poppy sustains alkaloid synthesis and further metabolizes morphine *in vivo* to as yet unidentified products. The alkaloids are synthesized in the vesicles and are translocated to the developing capsule by the upward movement of the latex. When morphine was fed to the stem below the developing capsules 8% of the morphine was metabolized in a few days. Translocation from the stem to the developing capsule is well established and occurs at rates up to 7 cm/hr. These transfers are most rapid during the first days of capsule development when the pericarp volume is increasing rapidly.

8.24.8 Pharmacology

Opiates

The pharmacologic actions of morphine are complex. Some specific central nervous system functions are depressed while others are stimulated. There is also some stimulation of sympathomimetic and parasympathomimetic systems.

Typically morphine produces analgesia, drowsiness, changes in mood, and mental clouding. A significant feature of morphine-induced analgesia is that it occurs without loss of consciousness. There are two factors involved in the relief of pain resulting from these analgetics; they elevate the pain threshold, and they alter the reaction of the individual to the painful experience. Drowsiness commonly occurs, the extremities feel heavy, and the body feels warm. Some patients also experience euphoria. The desire to recall this feeling of euphoria is purportedly the major cause of dependence.

When opiates are given to a patient who has not previously experienced the effects of the drug and who is not in pain, there is a sense of detachment and a feeling of anxiety or uneasiness—dysphoria. In this dysphoric state, nervousness, fear, nausea, and vomiting may occur.

In humans, death from morphine poisoning is nearly always due to respiratory arrest. Therapeutic doses of morphine in humans depress all phases of respiratory activity: respiratory rate, minute volume, and tidal exchange. The diminished respiratory volume is caused by a decrease in respiratory rate. Breathing may also be irregular and periodic.

As tolerance develops to the analgetic and euphoric effects, the respiratory center also becomes tolerant. This is the reason the addict may exhibit resistance to otherwise lethal doses of morphine. Therapeutic doses of morphine and related narcotic analgetics have no major effect on blood pressure or heart rate and rhythm in the normal recumbent patient.

Morphine has a synergistic effect with other respiratory depressants such as other opiates, barbiturates, general anesthetics, and alcohol. The depressant effects of some of the opioids may themselves be exaggerated by the phenothiazines, monoamine oxidase inhibitors, and tricyclic antidepressants.

The opiates decrease the HCl secretion in the stomach and also cause a decrease in motility. There is an increase in tone of the first part of the duodenum, which delays passage of the gastric contents for as much as 12 hr. This is probably the main reasons for the constipating effects of the morphine–like drugs. The second major effect is that the propulsive peristaltic contractions in the large intestine are decreased.

In recognizing morphine intoxication, the decrease in pupillary size is of considerable practical importance. The consensus is that this action is centrally mediated.

Morphine and its analogues are powerful depressants of the cough center and reduce the awareness of coughing. Other actions of therapeutic doses include dilation of cutaneous blood vessels. This action is thought to be due to histamine release, and this may be responsible for the itching and sweating that commonly ensues following administration of morphine.

The analgesic activity of *trans*-morphine (514) is considerably less than that of morphine. The interesting isomer of morphine, 515 which has *trans* fusion of both the B/C ring junction and of the oxygen-containing ring and

514   trans-morphine   515 trans, trans-morphine

ring C has not yet been prepared, but has been proposed by Bentley as a potent analgetic.

Codeine is a very important analgetic and antitussive drug. It is less sedative and analgetic than morphine and tolerance develops more slowly than morphine. It has less effect on the gastroinestinal tract and on the genitourinary tract than morphine. In therapeutic doses it depresses the respiratory system only slightly, but toxic doses may produce respiratory stimulation with excitement and convulsions.

Heroin, diacetyl morphine (**516**), is a highly euphoric and analgetic drug. It crosses the blood–brain barrier more rapidly than morphine and is preferred by the addict because of the initial orgiastic sensation. The euphoria produced by heroin is greater than that of morphine. Heroin is approximately 3–5 times more potent than morphine as an analgetic, and also is more potent in suppressing the cough center and causing respiratory depression.

The pharmacologic actions of hydrocodone (**517**) are midway between

516   diacetylmorphine   517   hydrocodone
      (heroin)   (dihydrocodeinone)

codeine and morphine. It shows a lower incidence of side effects and may be more effective than codeine as an antitussive. Unfortunately, it may also be more addictive than codeine.

6,7-Dihydromorphine is 6–10 times more potent as an analgetic than morphine. Its respiratory depressant action is correspondingly greater, but is less than nauseating and less constipating. The drug may also give rise to tolerance and addiction.

There has been an extensive study made of the effects of modifying groups on pharmacologic activity as shown in Table 10.

Table 10   Effects of Modifying Groups on the Analgesic Activity
of Morphine (**5**)

Modification	Increase	Decrease
3-O-Methylation		80–90%
6-O-Methylation	2-fold	
Oxidation of 6-OH	Up to 6-fold	
Reduction of 6, 7-double bond	3-fold	
Quaternization of nitrogen		Complete loss
Nitrogen ring-fission		Complete loss
Opening of 4, 5-oxide bridge		Complete loss
De-N-methyl-N-β-phenylethyl	10–15-fold	
14-Hydroxylation	Substantial	

The most potent analgesics obtained in this way are methyldihydromorphinone (**518**), 14β-acetoxydihydrocodeinone (**519**), and N-β-phenylethylnormorphine (**520**).

Two new alkaloids which have recently been obtained from *Papaver bracteatum* are 14β-hydroxycodeinone (**521**) and 14β-hydroxycodeine (**522**). These compounds had previously been synthesized for biological evaluation.

Other synthetic 14-hydroxy derivatives in which the 4,5-oxide bridge is cleaved have antitussive actvity. One of these, oxymethebanol (**523**), exhibited potent activity. Indeed, there is apparently no correlation of the factors for antitussive and analgetic activity. Both dextromethorphan and narcotine are more active than codeine in suppressing the cough reflex.

When 14-bromocodeinone (**524**) is treated with sodium borohydride, neo-

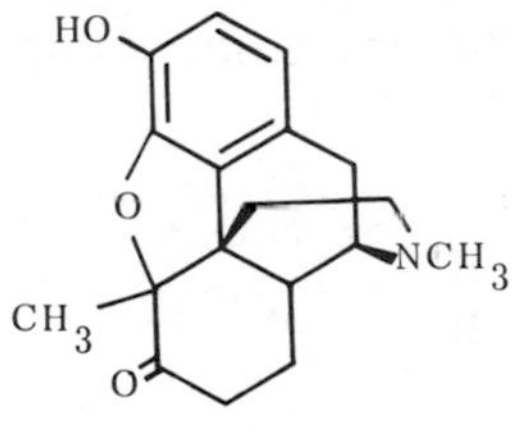

| 518 | 519 | 520 |

| 521 R=OH | 522 | 523 |
| 524 R=Br | | |

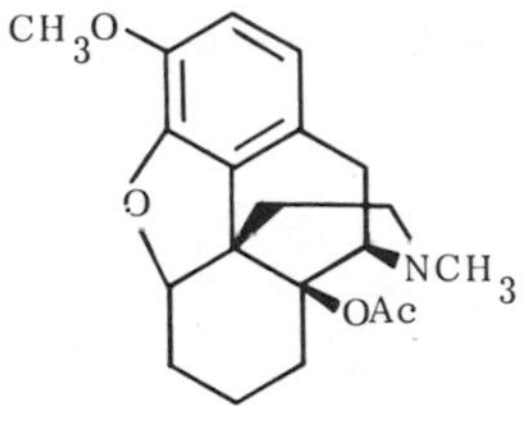

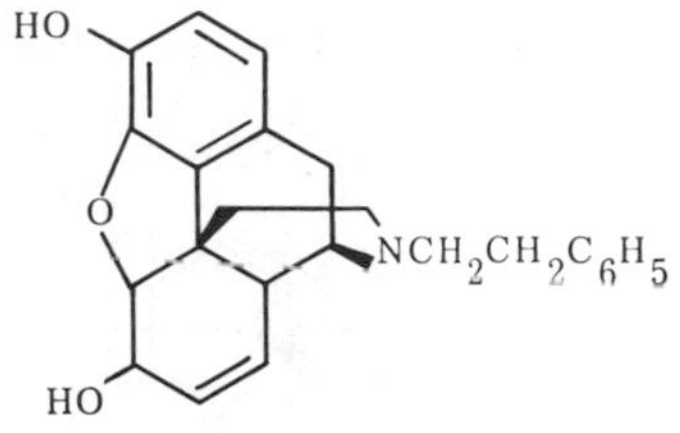

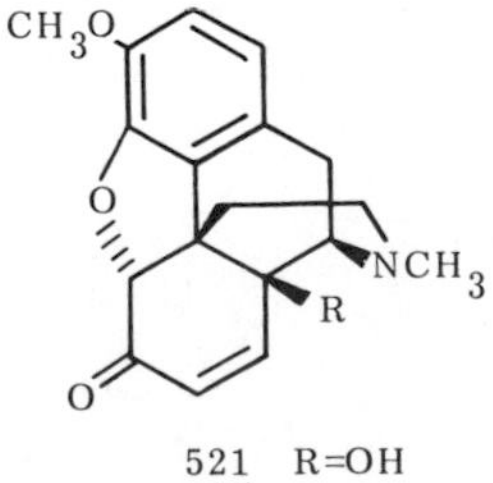

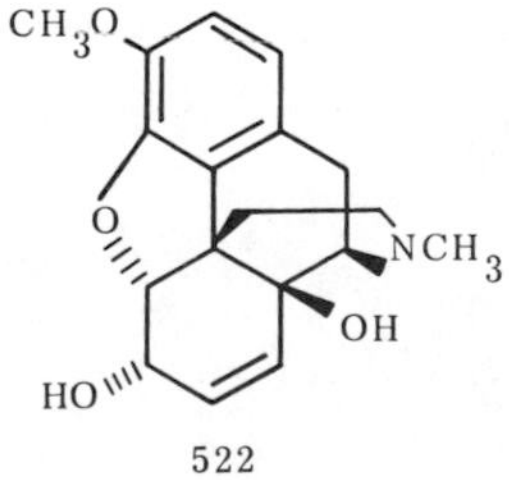

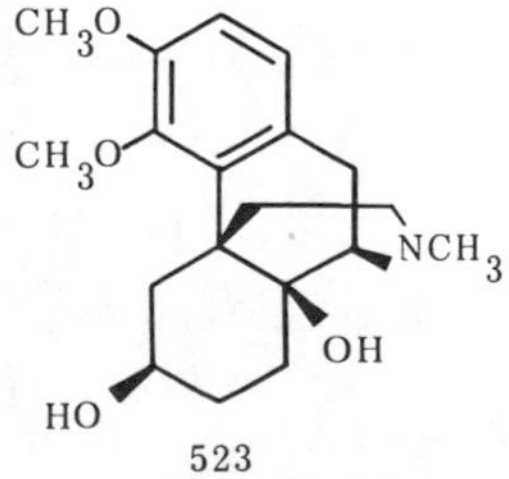

pine (525) is one of the products, together with a new isomer of codeine named indolocodeine, and assigned the structure 526. This product is thought to arise by reduction of the intermediate quaternary salt 527 derived from the carbonium ion of 524.

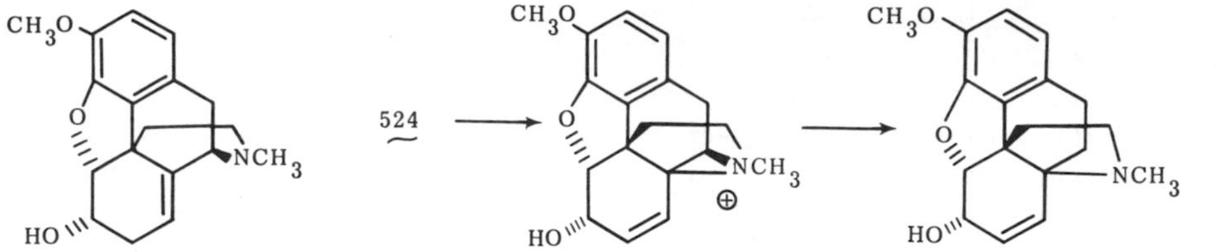

525 neopine   527   526 indolinocodeine

Opiate Antagonists

A unique feature of the structure–activity relationships of morphine (5) is that quite simple modifications produce compounds which specifically *antagonize* many of the effects of morphine. Substituting the *N*-methyl group by an *N*-allyl, *N*-propyl, or *N*-cyclopropylmethyl is sufficient to transform an agonist to an antagonist. Naloxone (528) is a pure opiate antagonist. It is characteristic that the antagonist is usually active at a markedly reduced dose compared with the agonist.

Sodium borohydride reduction of the narcotic antagonist naloxone (528) gave the 6α-hydroxy derivative identical with the principal *in vivo* metabolite.

An improved synthesis of naloxone (528) involves using vinyl chloroformate as a specific N-demethylating agent. Treatment of 3-acetyl-14-acetyloxymorphine (529) with vinyl chloroformate in dichloroethane under reflux gave 530 in quantitative yield. Hydrolysis of the protecting group

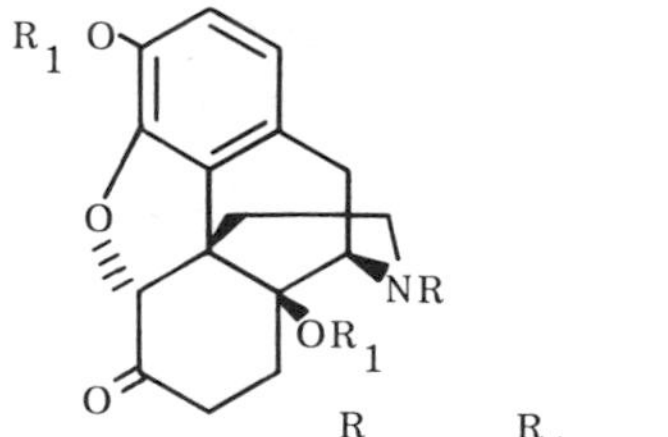

		R	R_1
528	naloxone	allyl	H
529		CH_3	CH_3CO
530		H	H
531	noroxymorphine		
		VOC	CH_3CO

$$VOC = \text{structure}$$

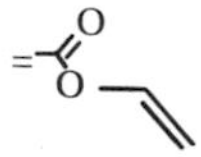

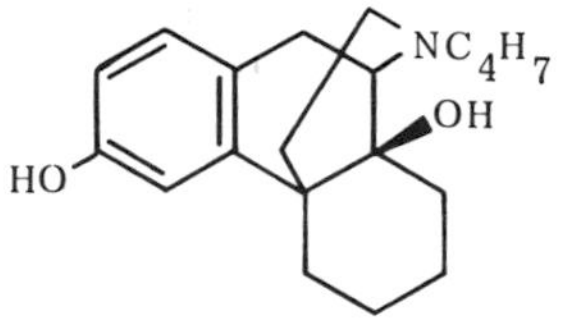

gave noroxymorphine (531), which could be N-allylated with allyl bromide in ethanol to give naloxone (528) in 70% overall yield.

The morphine derivative 532 exhibited narcotic antagonist activity comparable to naloxone, was as effective orally as parenterally, and had a long duration of action.

The pure antagonists such as naloxone do have value in the treatment of overdosed addicts, and long-acting antagonists such as naltrexone (533) may be useful in the treatment of heroin addicts.

The major human metabolite of naltrexone (533) is 6β-naltrexol (534), and a second metabolite, 2-hydroxy-3-O-methyl-6β-natrexol (535), has also been identified.

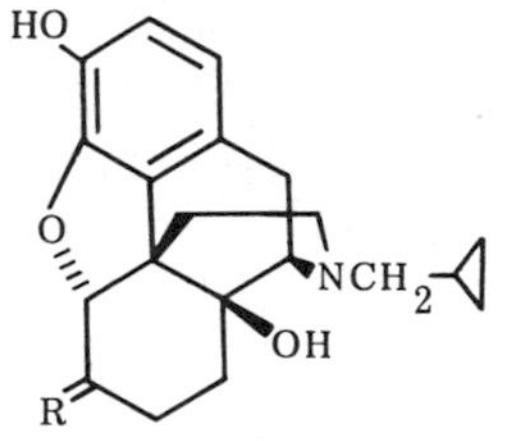

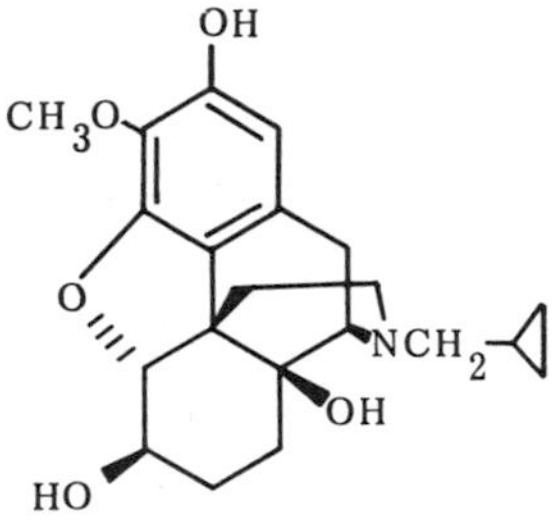

533 naltrexone R= =O

534 6β-naltrexol R=H, β-OH

535

Most antagonists also possess a certain amount of "agonist" activity. Pentazocine (536) has almost equal proportions of agonist and antagonist activity. It is a potent analgesic, but is considerably less addictive than the conventional opiates. There have been enormous, as yet unsuccessful, synthetic effects devoted to the synthesis of an analgesic free from the potential of abuse.

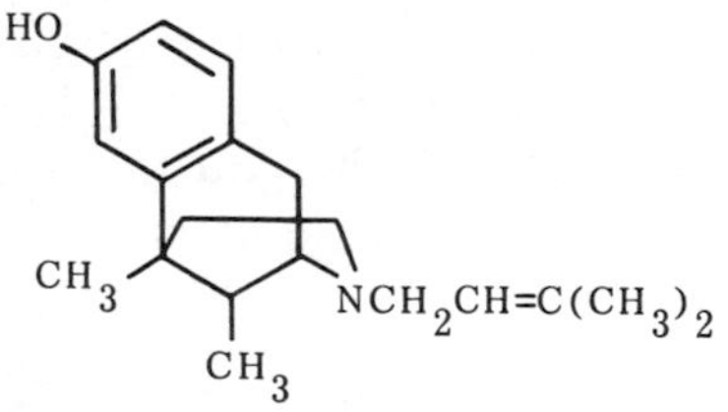

536 pentazocine

N-Allylnormorphine (537), also known as nalorphine, antagonizes all the effects of morphine, and is valuable antidote to narcotic overdose. It is a potent analgesic but has some disturbing mental effects in high doses. It is not an addictive drug.

Olofson and co-workers have also demonstrated that the conditions for hydrolysis of N-VOC and O-VOC protecting groups are quite different and have used this to good effect in the synthesis of nalorphine (**537**) from morphine (**5**). Reaction of **5** with excess vinyl chloroformate and two equivalents of the base, 1,8-bis(dimethylamino)naphthalene, gave the N-demethylated di-O-protected compound **538** in over 90% yield. The N-protecting group was selectively removed with anhydrous HBr in ether–ethanol at 25°C and the product N-allylated. Hydrolysis with aqueous acid gave nalorphine (**537**) in 83% yield.

The existence of opiate antagonists suggests that there are specific receptor sites, and the synthesis of extremely potent opiates suggests that the receptors themselves exist. Etorphine (**539**) for example is 5000 to 10,000

		R_1	R_2	R_3
5	morphine	H	H	CH_3
537	nalorphine	H	H	$CH_2CH=CH_2$
538		VOC	VOC	VOC

times more potent as an analgesic than morphine and almost as potent as LSD as a psychedelic. As a clinical agent of course it is of little value, but as a pharmacologic tool it is a very interesting compound.

The concept of a receptor site is clearly associated with the size and shape of the molecule. Such an analysis associates the stereochemical specificity of these molecules with the potential shape and characteristics of the receptor and thereby attempts to explain the activity of these compounds at the molecular level.

One of the results of the development of a receptor theory for morphine is the question of enantiospecificity of a receptor. For the case of morphine this has been investigated with (+)-morphine (**481**) by Rice and co-workers.

Two types of receptors were identified, those having high stereoselectivity which mediate morphine analgesia [i.e. (+)-morphine has little analgetic activity], and a second type of receptor possessing low stereoselectivity which mediated the syndrome of hyperexcitability and explosive motor behavior, often associated with precipitated abstinence in morphine-dependent rats. The former reaction is blocked by (−)-naloxone whereas the latter is not. Experiments with (+)-naloxone have shown that this compound has only one thousandth the activity of the (−)-isomer.

The work with (+)-morphine (481) was made possible by the development of an improved synthetic procedure from (−)-sinomenine (478). Catalytic reduction of 478 gave a mixture of C-7 stereoisomers 540 which was treated with polyphosphoric acid to afford the ketone 541. The 7,8-double bond was introduced following the method of Rapoport to yield (+)-codeinone (542). Sodium borohydride reduction followed by O-demethylation with boron tribromide gave (+)-morphine (481) in 28% overall yield.

478   (−)-sinomenine   540   541 (+)- dihydro-codeinone

i) ketalization
ii) p-TsOH
iii) MeOBr
iv) KO^tBu, R.T.
v) 5% HCl

481   (+)-morphine   542   (+)-codeinone

i) NaBH$_4$
ii) BBr$_3$

Endorphins

We have discussed in the earlier part of this section the importance and pharmacology of the opiate drugs. It is only very recently that progress has been made in the study of how these potent compounds act on the brain. The results of these studies are quite remarkable for they have disclosed the occurrence of natural compounds present in the brain having opiate activity! Collectively these opiate-like compounds are known as endorphins (a contraction of "endogenous morphine"), and all of the compounds isolated so far are polypeptide derivatives.

Methionine enkephalin, which occurs predominantly in the brain, is a simple pentapeptide, H-Tyr-Gly-Gly-Phe-Met-OH, and is part of the amino acid sequence of the pituitary peptide, β-lipotropin. Another fragment of β-lipotropin, consisting of amino acids 61 to 91, is β-endorphin. This polypeptide is the major peptide in the pituitary having opiate-like action, but is present in the brain at a level less that that of the enkephalins. Direct administration into the brain or i.v. injection produces analgesia, but tolerance and physical dependence also result.

Is β-lipotropin the storehouse for the biologically active smaller polypeptides? Subsequent work has shown that an even longer peptide, the 31K (molecular weight 31,000) form of the pituitary adrenal–stimulating hormone ACTH, may be a precursor of β-endorphin. Work with stressed rats has shown that release of ACTH into the plasma and of β-endorphin occur simultaneously from this peptide, which itself contains β-lipotropin as a subunit.

What then is the normal role of pituitary β-endorphin? It is not a source of the brain opioid peptides since excision of the pituitary does not reduce the opioid content of the brain.

The enkephalins and opiate receptors occur in similar places. Thus the periaqueductal area of the brain, a region of importance for pain perception; the nuclei of the vagus nerve, which affect respiratory function; and areas of the brain which have norepinephrine-containing neurons all have high densities of opiate receptors and enkephalins.

The only instance of an occurrence of enkephalins outside the central nervous system is in the small intestines. Remember that opiate derivatives cause constipation and are potent antidiarrhea drugs. Is peristalsis in the gut regulated by enkephalin-containing neurons?

The enkephalins are also suspected of playing a very important role in regulating emotions, for preliminary experiments have suggested that there is considerable localization of enkephalins in areas of the brain known to influence observed emotions.

There seem to be little doubt that a study of the mechanisms of the action of opiate agonists and antagonists has led to a preliminary understanding of how and where these drugs may be acting. What is more important however is that such studies have begun to unearth some of the fundamental chemical processes of the brain. These data may eventually lead to a better understanding of the chemistry and physics of psychological disease states.

LITERATURE

Reviews

Stork, G., *Alkaloids NY* **6**, 219 (1960).

Bentley, K. W., *The Isoquinoline Alkaloids,* Pergamon, Oxford, 1965, p. 79.

Swan, G. A., *An Introduction to the Alkaloids,* Wiley, New York, 1967, p. 127.

Kirby, G. W., *Science* **155**, 170 (1967).

Bentley, K. W., in S. W. Pelletier (Ed.), *Chemistry of the Alkaloids,* Van Nostrand Reinhold, New York, 1970, p. 117.

Bentley, K. W., *Alkaloids NY* **13**, 1 (1971).

Stuart, K. L., *Chem. Revs.* **71**, 47 (1971).

Kametani, T., and K. Fukumoto, *J. Het. Chem.* **8**, 341 (1971).

Gates, M., *Adv. Biochem. Psychophamacol.* **8**, 51 (1973).

Fairbairn, J. W., *Planta Med.* **29**, 26 (1976).

Synthesis

Kametani, T., and K. Fukumoto, *J. Het. Chem.* **8**, 341 (1971).

Miller, L. L., F. R. Stermitz, J. Y. Becker, and V. Ramachandran, *J. Amer. Chem. Soc.* **97**, 2922 (1975).

Kupchan, S. M., V. Kameswaran, J. T. Lynn, D. K. Williams, and A. J. Liepa, *J. Amer. Chem. Soc.* **97**, 5622 (1975).

Miller, L. L., F. R. Stermitz, and J. R. Falck, H. *J. Amer. Chem. Soc.* **95**, 2651 (1953).

Schwartz, M. A., and I. S. Marni, *J. Amer. Chem. Soc.* **97**, 1239 (1975).

Beyerman, H. C., T. S. Lie, L. Maat, H. H. Bosman, E. Baurman, E. J. M. Bijsterveld, and H. J. M. Sinnige, *Recl. Trav. Pays-Bas* **95**, 24 (1976).

Aporphine Formation

Kupchan, S. M., and C. Kim, *J. Amer. Chem. Soc.* **97**, 5623 (1975).

Spectral Properties

Bognár, R., G. Gaál, P. Kerekes, A. Lévai, S. Makleit, F. Snatzke, and G. Snatzke, *Colln. Czech. Chem. Commun.* **40**, 670 (1973).

Terui, Y., K. Tori, S. Maeda, and Y. K. Sawa, *Tetrahedron Lett.* 2853 (1975).

Wehrli, F. W., *Chem. Commun.* 379 (1973).

Carroll, F. I., C. G. Moreland, G. A. Brine, and J. A. Kepler, *J. Org. Chem.* **41**, 966 (1976).

Biosynthesis

Barton, D. H. R., and T. Cohen, *Festschrift, A. Stoll*, Birkauser, Basle, 1957, p. 117.

Rapoport, H., N. Levy, and F. R. Stermitz, *J. Amer. Chem. Soc.* **83**, 4298 (1961).

Battersby, A. R., R. Binks, and B. J. T. Harper, *J. Chem. Soc.* 3534 (1962) and references therein.

Leete, E., *J. Amer. Chem. Soc.* **81**, 3948 (1959).

Neubauer, D., *Arch. Pharm.* **298**, 737 (1965).

Battersby, A. R., R. Binks, R. J. Francis, D. J. McCaldin, and H. Ramuz, *J. Chem. Soc.* 3600 (1964).

Battersby, A. R., R. C. F. Jones, and R. Kazlauskas, *Tetrahedron Lett.* 1873 (1975).

Battersby, A. R., D. M. Foulkes, and R. Binks, *J. Chem. Soc.* 3323 (1965).

Brochmann-Hanssen, E., and B. Nielsen, *Tetrahedron Lett.* 1271 (1965).

Battersby, A. R., G. W. Evans. R. O. Martin, M. E. Warren, Jr., and H. Rapoport, *Tetrahedron Lett.* 1275 (1965).

Barton, D. H. R., G. W. Kirby, W. Steglich, G. M. Thomas, A. R. Battersby, T. A. Dobson, and H. Ramuz, *J. Chem. Soc.* 2423 (1965).

Stermitz, F. R., and H. Rapoport, *J. Amer. Chem. Soc.* **89**, 1540 (1967).

Blaschke, G., H. Parker, and H. Rapoport, *J. Amer. Chem. Soc.* **89**, 1540 (1967).

Parker, H. I., G. Blaschke, and H. Rapoport, *J. Amer. Chem. Soc.***94**, 1276 (1972).

Hodges, C. C., J. S. Horn, and H. Rapoport, *Phytochemistry* **16**, 1939 (1977).

Brochmann-Hanssen, E., and S. W. Wunderly, *J. Pharm. Sci.* **67**, 103 (1978).

Pharmacology

Bowman, W. C., M. J. Rand, and G. B. West, *Textbook of Pharmacology*, Blackwell Scientific, Oxford, 1968, p. 631.

Lewis, J. J., *Lewis' Pharmacology*, E. & S. Livingstone, Edinburgh, 1970, p. 590.

Meyers, F. N., E. Juartz, and A. Goldfein, *Review of Medical Pharmacology*, Lounge Medical Publications, Los Altos, Calif., 1970, p. 235.

Narcotic Drugs: Biochemical Pharmacology, D. H. Clouet (Ed.), Plenum, New York, 1971.

Krantz and Carr's, Pharmacologic Principles of Medical Practice, D. M. Aviado (Ed.), Williams & Wilkins, Baltimore, Md., 1972, p. 95.

Goodman, L. S., and A. Gilman, *The Pharmacological Basis of Therapeutics*, Macmillan, New York, 1975, p. 245.

Holmes, H. L., in H. L. Holmes (Ed.), *Structure-Activity Relationships of Some Conjugated Heteroenoid Compounds, Catechol Monoethers and Morphine Alkaloids,* Vol. 1, Defense Research Establishment Suffield, Ralston, Alberta, 1975, p. 453; Vol. 2 p. 1403.

Jaquet, Y., W. A. Klee, K. C. Rice, I. Ijima, and J. Minamikawa, *Science,* **198** 842 (1977).

(+)-Morphine synthesis

Iijima, I., J. Minamikawa, K. C. Rice, A. E. Jacobson, and A. Brossi, *J. Org. Chem.* **43**, 1462 (1978).

8.25   *ERYTHRINA* ALKALOIDS

The genus *Erythrina* is comprised of 107 species and is a member of the subfamily Papilionaceae in the family Leguminosae. As we shall see the skeleton of the *Erythrina* alkaloids is a biogentically interesting one.

The distribution of these alkaloids is unusual because except for some isolations from the genus *Cocculus* in the Menispermaceae, no other isolations have been reported outside the genus *Erythrina*.

8.25.1   Nomenclature and Chemistry

The nomenclature of these alkaloids is quite significant and should be clarified at this point. The name erythrinane has been given to the tetracyclic spiroamine **543**, numbered as shown.

There are three prefixes used in this series, erythroi-, erythra- and eryso-, depending on the properties of the alkaloid. The prefix eryso- indicates the presence of a phenolic group, erythroi- that ring D is a lactone, and erythra- that the parent skeleton is present. Examples of these alkaloid types are erythraline (**544**), erysodine (**545**), and β-erythroidine (**546**), and the most widely distributed are erysodine (**545**), erysovine (**547**), and erysopine (**548**).

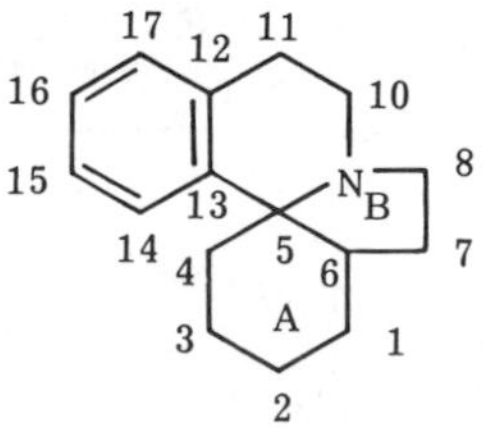

543 Erythrinane
skeleton

In many instances the phenolic groups are bound in an esterified form with sulfoacetic acid in the plant and the alkaloids are not easily extractable into organic solvent unless they are hydrolyzed by acid. Oxygenation is common in these alkaloids at carbons 3, 15, and 16, and rare elsewhere.

The structures of the major alkaloids were deduced mainly on the basis of degradation and chemical interrelationships. Methylation of erysodine (545), erysovine (547), and erysopine (548) gave an identical product, erysotrine (549), which until recently was not known as a natural product.

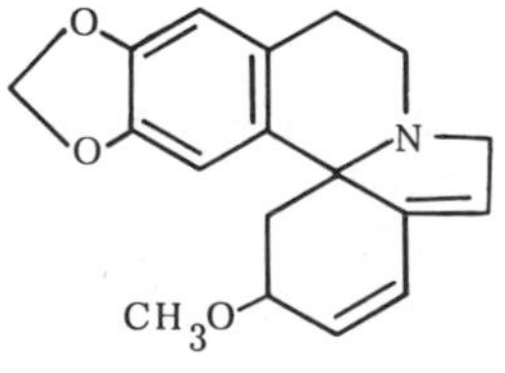

544 erythraline

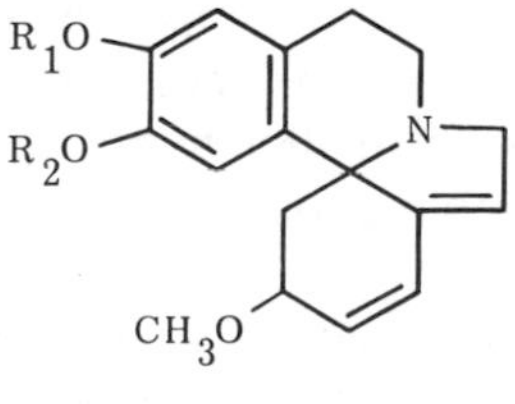

		R_1	R_2
545	erysodine	H	CH_3
547	erysovine	CH_3	H
548	erysopine	H	H
549	erysotrine	CH_3	CH_3

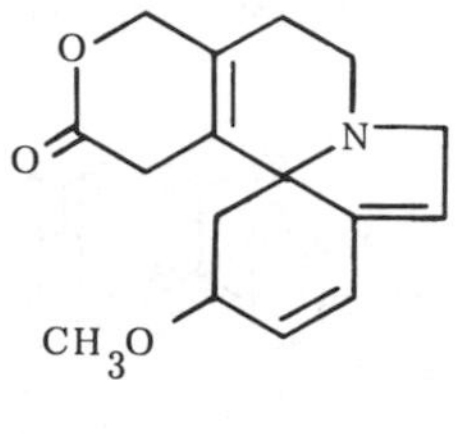

546 β-erythroidine

Oxidation of erythraline gives hydrastic acid (550) and alkali fusion affords indole (551) (from an "isoquinoline" alkaloid!). The complex data that resulted from the many degradative experiments were not rationalized until 1951, when Prelog and co-workers proposed the erythrinane skeleton. Two

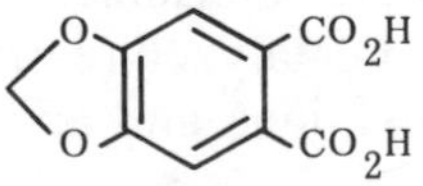

550

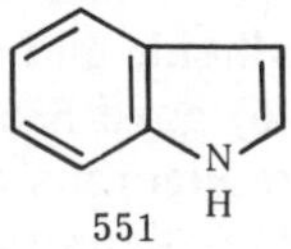

551

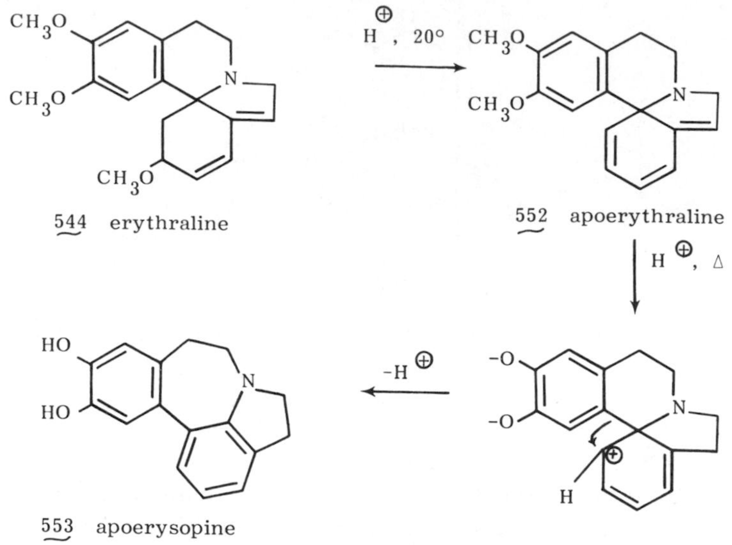

Scheme 68

reactions are of critical importance here; one is the room-temperature acid-catalyzed elimination of methanol to give apoerythraline (**552**), now showing λ_{max} 310 mm. At higher temperatures acid rearrangement leads to the biphenyl derivative apoerysopine (**553**), and rationalization of this is shown in Scheme 68.

Catalytic reduction of erythraline (**544**) gives successively a dihydro derivative **554** in which the double bond is located between C-1 and C-6, and then a stereochemically pure tetrahydro derivative **555**. The structure of

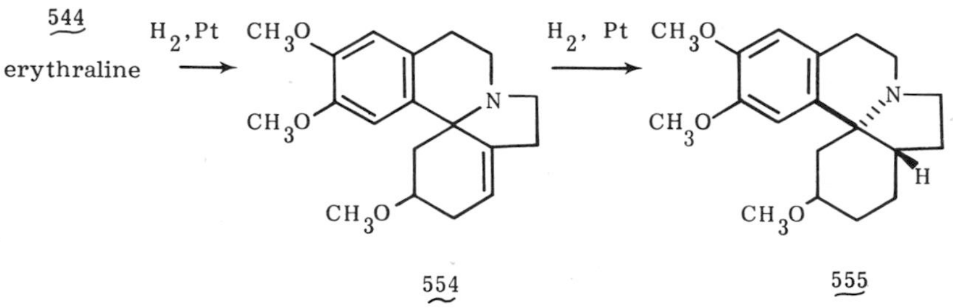

erythraline was eventually confirmed in 1958 by X-ray crystallography of the hydrobromide salt.

The structure of α- and β-erythroidines were determined by the extensive studies of Boekelheide and co-workers. Acid-catalyzed rearrangement of β-erythroidine (**546**) gave **556** under mild conditions and apo-β-erythroidine (**557**) under more vigorous conditions.

The absolute configuration at C-3 of β-erythroidine (**546**) was determined

546   556   557   apo-β-erythroidine

by Wenzinger and Boekelheide in 1963 by degradation to (3S)-3-methoxy-adipic acid, indicating C-3 in the erythroidines to be *R*. With the X-ray diffraction studies of dihydro-β-erythroidine hydrobromide by Hanson, the asymmetric center at C-5 was determined to have the *S*-configuration. Thus β-erythroidine is represented by the stereoformula **558**.

3R, 5S–

558 β-erythroidine

8.25.2 Synthesis

Over the years the *Erythrina* skeleton has been synthesized on numerous occasions, but there have been very few total syntheses of natural alkaloids reported.

The erythrinane skeleton was first synthesized by Belleau in 1953, and in the years that followed several related routes were successfully employed. Some of these approaches are shown in Scheme 69; invariably the cyclization proceeds to give the *cis*-isomer.

The first total synthesis of an *Erythrina* alkaloid was that of erysotrine (**549**), by Mondon and Nestler in 1964, and was an extension of a route first developed by Hansen. The new steps involved the introduction of a C-3 methoxyl group, and a key step is an allylic rearrangement of the C-1 hydroxyl of **559** (Scheme 70).

Mondon and Erhardt reported a biogenetic route to dihydroerysodine (**560**) involving phenol oxidative coupling of the amine **561** to give erysodienone (**562**). Sodium borohydride reduction followed by a novel acid rearrangement with hydrogen chloride in ether gave an enone **563**. Reduction to the epimeric enols and catalytic reduction of the corresponding allylic chloride gave **560** (Scheme 71).

Ito and co-workers in 1975 reported a quite different approach to the *Erythrina* skeleton, beginning with Birch reduction of the amide **564**. Acid

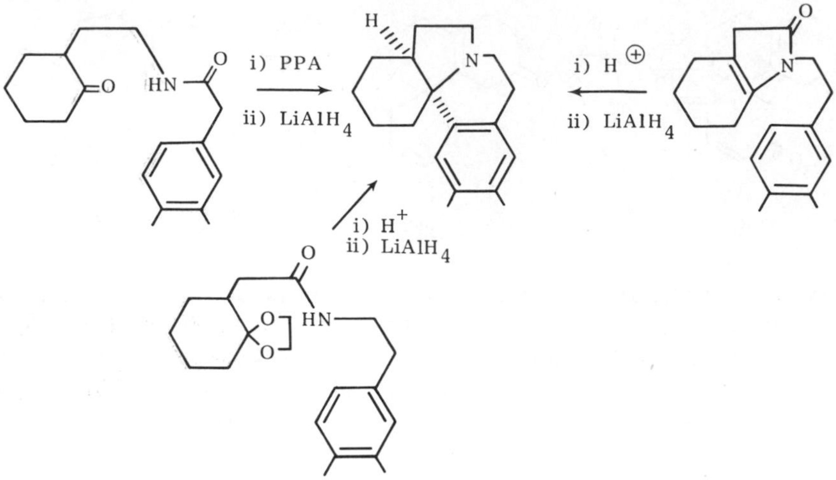

Scheme 69

hydrolysis of the product **565** gave the unsaturated ketolactam **566**, and the final cyclization was effected with formic acid to afford **567**, having the *cis* A/B stereochemistry.

Interest in recent years has centered on the biogenetic synthesis of the *Erythrina* alkaloids. In particular the stages prior to the dibenzazonine (**568**)

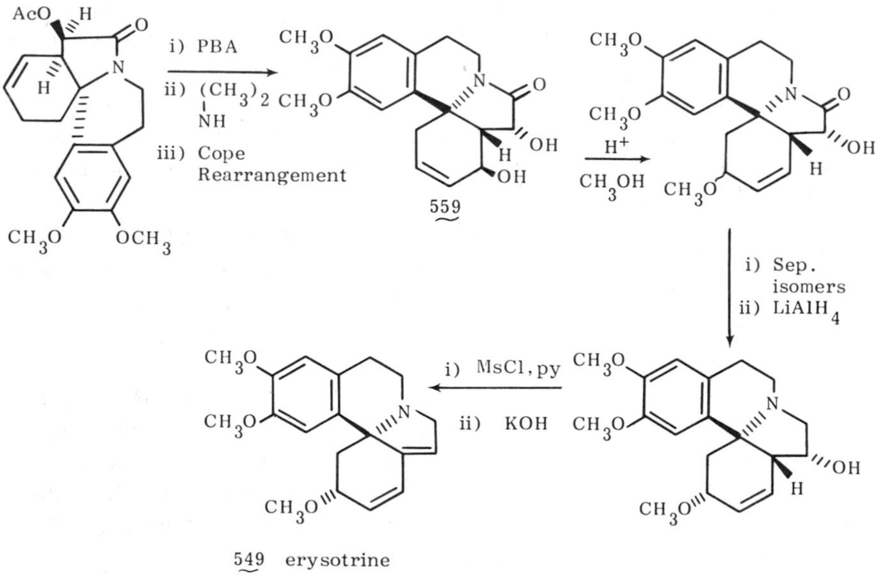

Scheme 70 Mondon–Nestler synthesis of erysotrine (**549**).

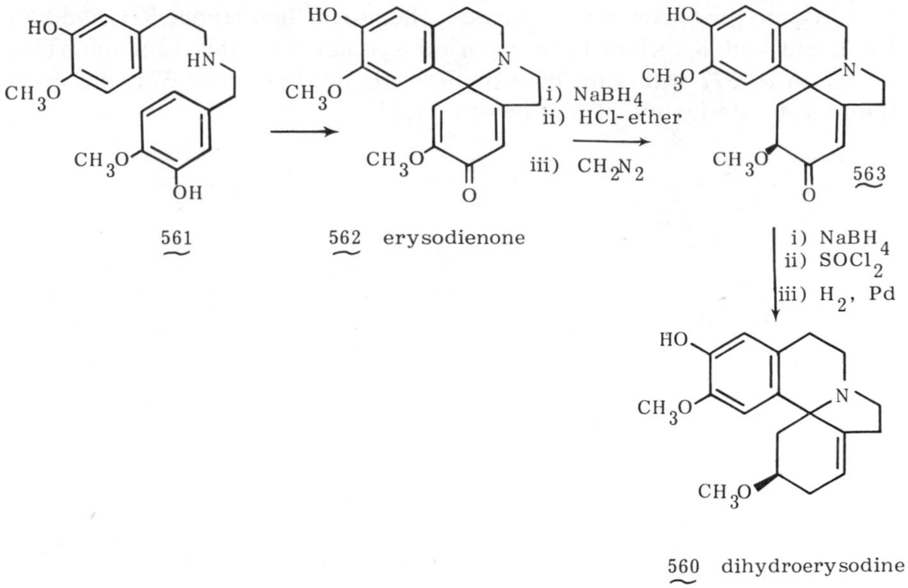

Scheme 71   Mondon–Erhardt synthesis of dihydroerysodine (**560**).

from a benzylisoquinoline, and the possible elaboration of **568** to a compound having the erythrinane skeleton.

Thus treatment of the dibenzazonine (**569**) with potassium ferricyanide in the presence of sodium bicarbonate gave the dienone **562**, and the bis-phenethylamine **570** gave the same product (the latter is probably *not* a biosynthetic process).

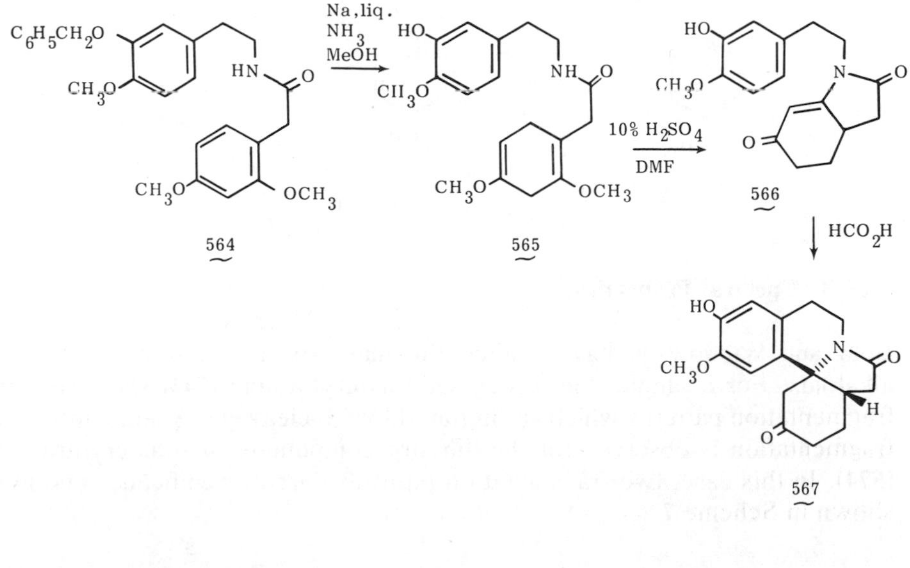

Closer to the biosynthetic scheme is the proerythradienone **571**, and both Kametani and Kupchan have reported syntheses of this skeleton. Thus oxidation of **572** with vanadium oxyflouride at -10°C gave **573**, the *N*-tri-fluoroacetyl derivative of **571**, in 40% yield.

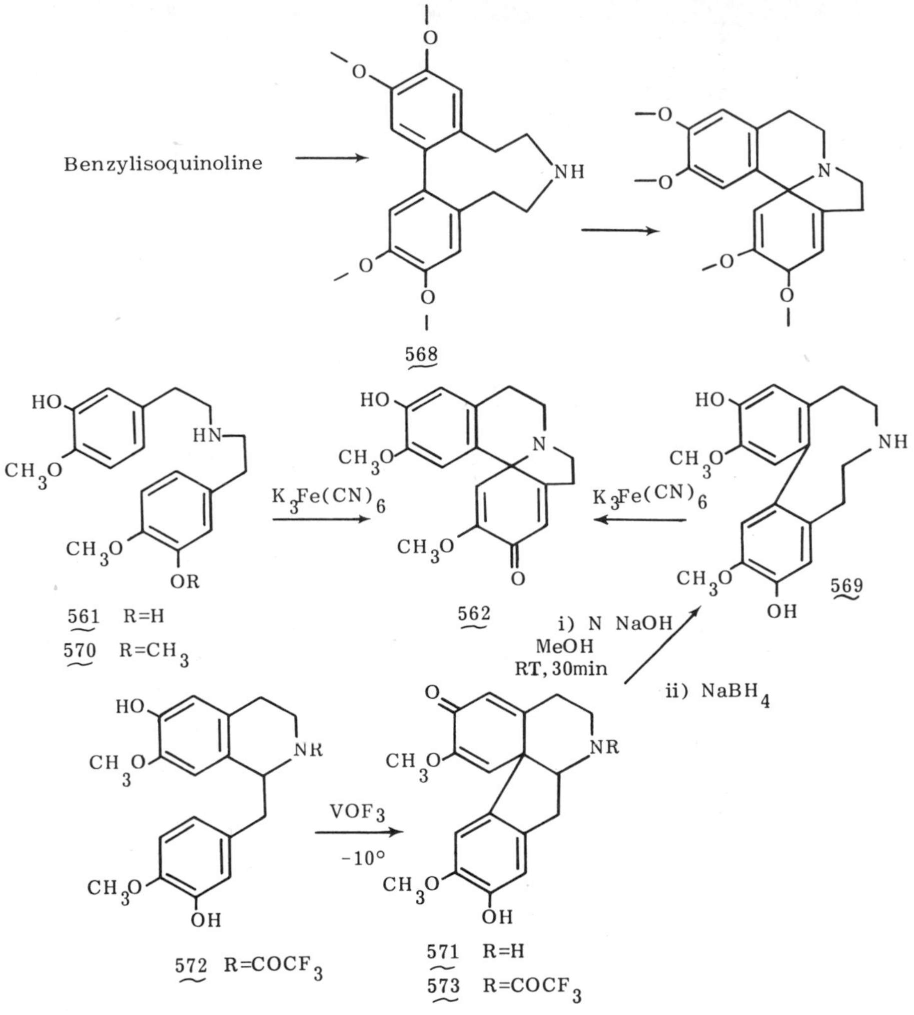

8.25.3 Spectral Properties

Boar and Widdowson have studied the mass spectra of some *Erythrina* alkaloids. For example, the dienes such as erythraline (**544**) show similar fragmentation patterns which are initiated by 5,6-cleavage. A quite different fragmentation is observed for the dihydro compounds such as erythratine (**574**). In this case, two fragmentation pathways are of significance and are shown in Scheme 72.

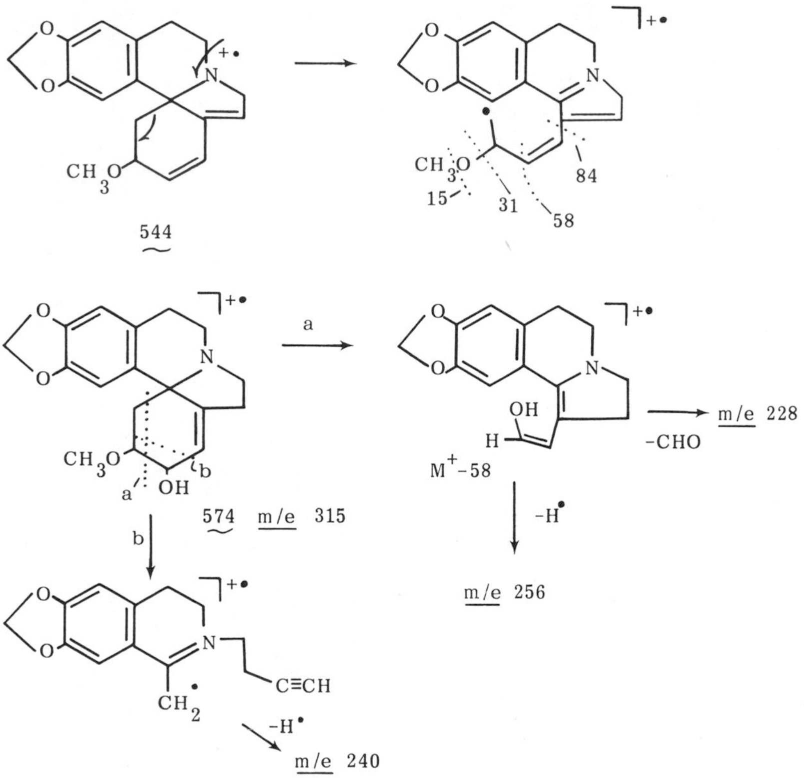

Scheme 72 Mass spectral fragmentation of the *Erythrina* alkaloids.

The NMR data of erythristemine are shown in **575** and were obtained by decoupling and INDOR techniques.

The geminal coupling constant of the 10-methylene protons (~14 Hz) indicates in solution that the nitrogen lone pair bisects the 10-methylene group. The long-range coupling between the equatorial proton at C-4 and the olefinic protons at C-1 and C-7 is of interest.

Typical data for a member of the dihydro series of compounds are shown for erythramine in **576**, and for a dienone in erysotinine (**577**).

In the erythrinane series it has been shown that the *cis*- and *trans*-erythrinanes may be distinguished by the chemical shift of the 7 and 11 protons in the two series. In the *trans* series the 14-H lies close to the axial protons at C-3 and C-1 and is consequently deshielded. Comparative data are shown in **578** and **579**, for the *cis* and *trans* systems respectively.

8.25.4 Biosynthesis

The biosynthesis of the *Erythrina* alkaloids has been investigated mainly by Barton and co-workers. [2-^{14}C]Tyrosine labeled C-8 and C-10 of β-erythro-

OCH$_3$ at 3.02, 3.21 and 3.43 (6H)

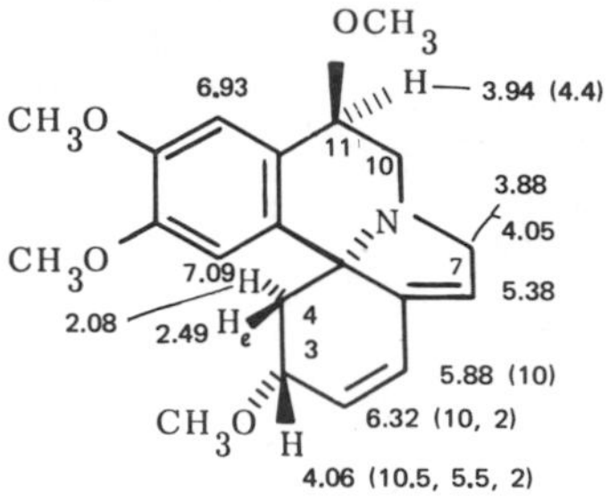

575 in C$_6$D$_6$

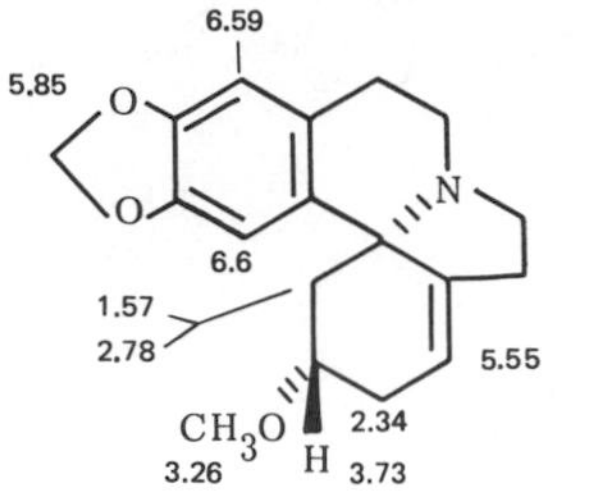

576 erythramine

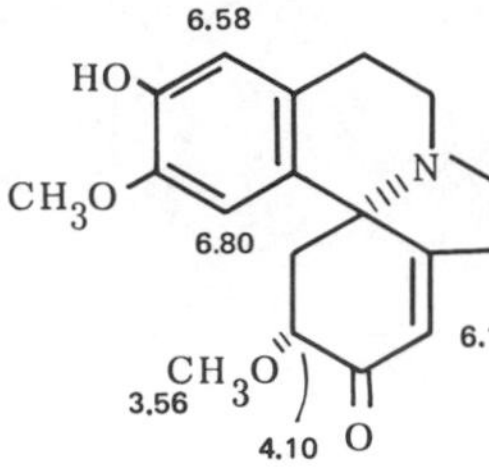

577 erysotinine

H-4a	J=10.5,-10.5
H-4e	J=-10.5,5.5
H-10a	J=4, -14
H-10e	J=4,-14

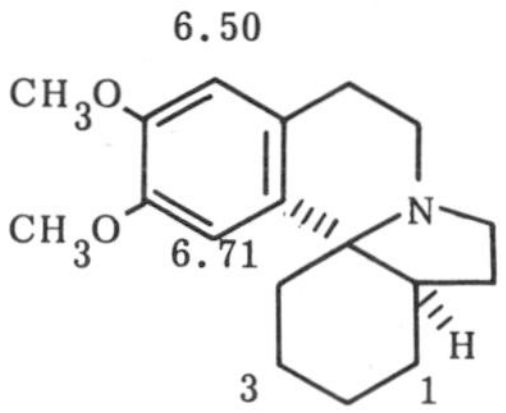

cis-
578

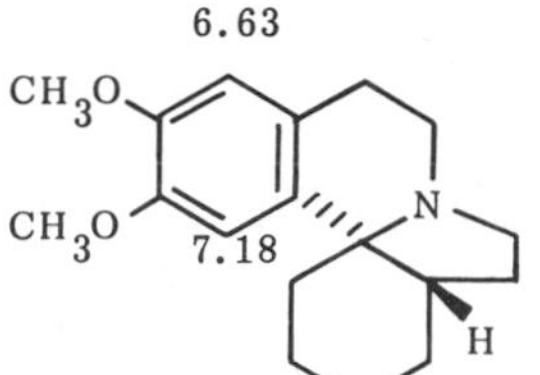

trans-
579

idine (546) in *E. berteroana*, and norprotosinomenine (231) and erysodienone (562) were precursors of erythraline (544) in *E. crista-galli*. Both the methylation pattern and the stereochemistry of 231 are critical, for the doubly labeled (+)-isomer (indicated) was a better precursor of 544 than the (−)-isomer. In addition (+)-N-nororientaline (230) and (+)-N-nor-reticuline (239) were not significant precursors of 544.

There are several pieces of evidence which indicate that a symmetrical intermediate is involved in the biosynthetic pathway; and that this intermediate has the structure 580. For example [4'-O^{14}CH$_3$]N-norprotosinomenine (231) afforded 544, labeled equally at the methoxy and methylenedioxy groups. Both singly and doubly labeled 580 were specifically incorporated into erythraline (544). These data suggested that the overall scheme for the formation of the *Erythrina* skeleton was that shown in Scheme 73.

The stages between erysodienone (562) and erythraline (544) have been investigated with several potential precursors, including labeled erysodine (545) and erysopine (548). The results indicated that the alkylation pattern in the latter stages was not important (i.e. methoxyphenol *vs.* methylene-

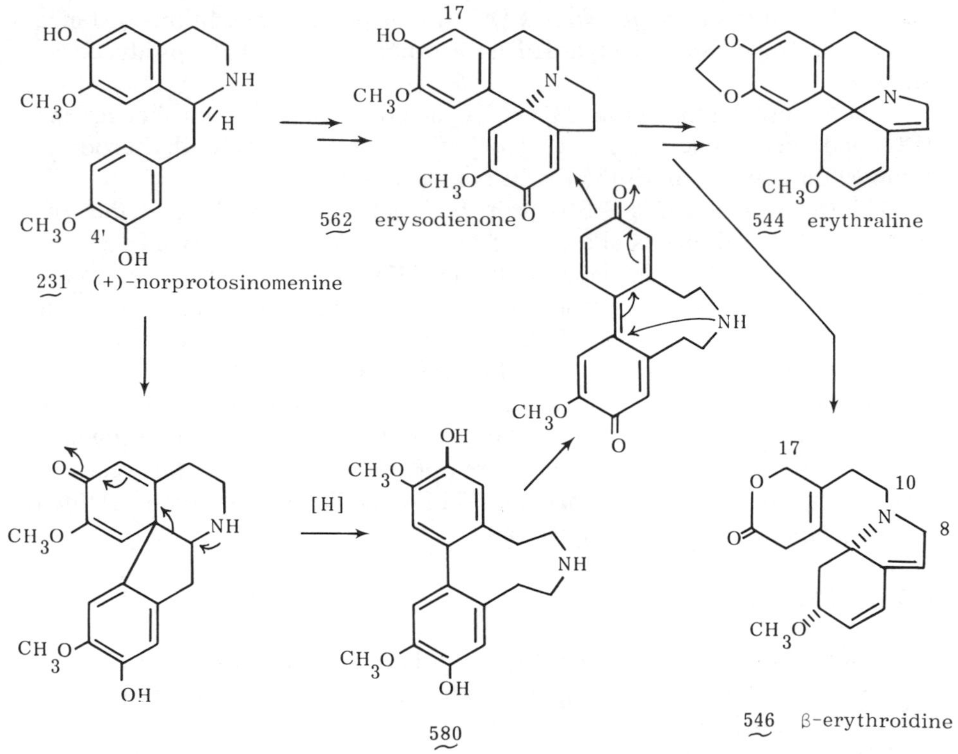

Scheme 73 Biosynthesis of the *Erythrina* alkaloids.

dioxy), and furthermore they did not delineate a unique pathway. However, it was conclusively shown that only the (−)-isomer of **562** was a precursor of erythraline (**544**) and β-erythroidine (**546**).

Similar results with **230, 239**, the amine **581**, and (+)-(**231**) were obtained by Bhakuni and co-workers when the biosynthesis of isococculidine (**582**)

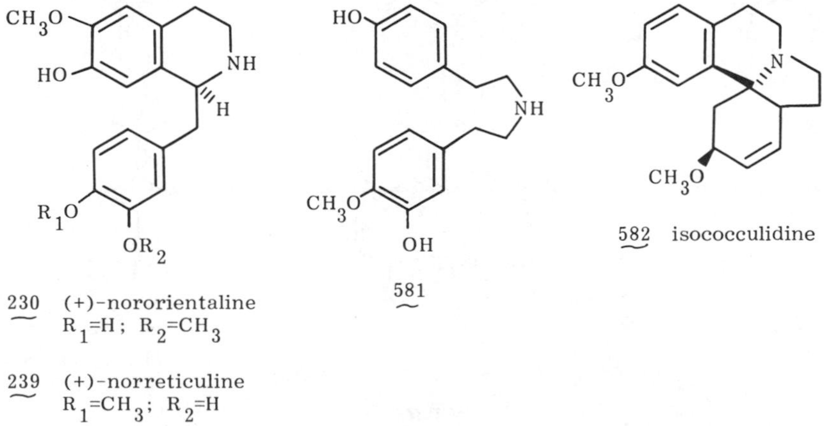

230   (+)-nororientaline
      R$_1$=H; R$_2$=CH$_3$

239   (+)-norreticuline
      R$_1$=CH$_3$; R$_2$=H

581

582   isococculidine

was studied in *Cocculus laurifolius* DC. (Menispermaceae). In this instance the benzazonine (**583**) is regarded as a crucial intermediate produced as shown.

(+)-Norprotosinomenine (**231**) is also an effective precursor of cocculine (**584**) and cocculidine (**585**) in *C. laurifolius*, and the 4′-O-methyl group is retained on conversion to the erythrinan system.

The same benazonine **583** is probably also an intermediate in the formation of **584** and **585** as shown in Scheme 74. At a more advanced stage in the biosynthetic scheme, isococculidine (**582**) was found to be a very efficient precursor (12.9% incorporation) of cocculdine (**585**), and the latter was a precursor of cocculine (**584**). This indicates that O-demethylation of the 3-O-ether may be the final step in the formation of **584**.

Derivation of the lactonic *Erythrina* alkaloids from (−)-erysodienone (**562**) has also been established, although neither the point in the sequence at which aromatic ring cleavage occurs nor the mechanism of this reaction have yet been established. Label at C-17 of erysodienone (**562**) was retained at this position in β-erythroidine (**546**).

8.25.5 Pharmacology

Cocculine (**584**) and cocculidine (**585**) have hypotensive but no central nervous system activity, and erysotrine (**549**) and isococculidine (**582**) are neuromuscular blocking agents.

The total alkaloids from the trunk bark of *E. variegata* showed smooth

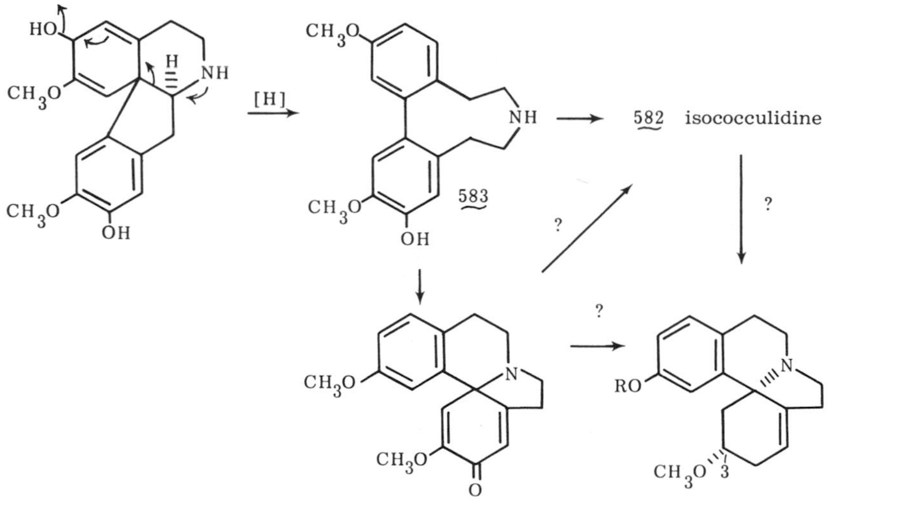

584 cocculine, R=H
585 cocculidine, R=CH_3

Scheme 74

muscle relaxant, central nervous system depressant, and anticonvulsant effects. No analgesic, anti-inflammatory, or laxative effects were observed.

LITERATURE

Reviews

Marion, L., *Alkaloids NY* **2**, 499 (1952).

Prelog, V., *Angew. Chem* **69**, 33 (1957).

Boekelheide, V., *Alkaloids NY* **7**, 201 (1960).

Hill, R. K., *Alkaloids NY* **9**, 483 (1967).

Mondon, A., in S. W. Pelletier (Ed.), *Chemistry of the Alkaloids,* Van Nostrand Reinhold, New York, 1970, p. 173.

Snieckus, V. A., *Alkaloids, London* **1**, 145 (1971).

Snieckus, V. A., *Alkaloids, London* **2**, 199 (1972).

Snieckus, V. A., *Alkaloids, London* **3**, 180 (1973).

Snieckus, V. A., *Alkaloids, London* **4**, 273 (1974).

Snieckus, V. A., *Alkaloids, London* **5**, 176 (1975).

Snieckus, V. A., *Alkaloids, London* **7**, 176 (1977).

de Silva, S. O., and V. A. Snieckus, *Alkaloids, London* **8,** 144 (1978).

Synthesis

Mondon, A., and H. J. Nestler, *Angew. Chem.* **76**, 651 (1964).

Mondon, A., and M. Erhardt, *Tetrahedron Lett.* 2557 (1966).

Stevens, R. V., and M. P. Wentland, *Chem. Commun.* 1104 (1968).

Biogenetic Synthesis

Kametani, T., T. Kobari, and K. Fukumoto, *Chem. Commun.* 288 (1972).

Barton, D. H. R., R. B. Boar, and D. A. Widdowson, *J. Chem. Soc. Sec. C* 1208 (1970).

Kametani, T., K. Takahashi, T. Honda, M. Ihara, and K. Fukumoto, *Chem. Pharm. Bull* (Japan) **20**, 1793 (1972).

Kupchan, S. M., C.-Y, Kim, and J. T. Lynn, *Chem. Commun.* 86 (1976).

Mass Spectra

Boar, R. B., and D. A. Widdowson, *J. Chem. Soc., Sec. B* 1591 (1970).

Millington, D. S., D. H. Steinman, and K. L. Rinehart, Jr., *J. Amer. Chem. Soc.* **96**, 1909 (1974).

Biosynthesis

Leete, E., and A. Ahmad, *J. Amer. Chem. Soc.* **88**, 4722 (1966).

Barton, D. H. R., R. James, G. W. Kirby, and D. A. Widdowson, *Chem. Commun.* 266 (1967).

Barton, D. H. R., R. B. Boar, and D. A. Widdowson, *J. Chem. Soc., Sec. C* 1213 (1970).

Barton, D. H. R., C. J. Potter, and D. A. Widdowson, *J. Chem. Soc. Perkin Trans. I* 346 (1974).

Barton, D. H. R., R. D. Brachs, C. J. Potter, and D. A. Widdowson, *J. Chem. Soc. Perkin Trans. I* 2278 (1974).

Bhakuni, D. S., A. N. Singh, and R. S. Kapil, *Chem. Commun.* 211 (1977).

Bhakuni, D. S., and A. N. Singh, *J. Chem. Soc. Perkin Trans. I* 618 (1978).

8.26   PROTOSTEPHANINE AND ERYBIDINE

In natural products chemistry there are very few instances where two apparently closely related compounds have quite different biosyntheses.

This is an interesting phenomenon and reemphasizes the importance of biogenesis/biosynthesis in the structure elucidation of natural products and in recognizing phytochemical differences between plant families. An example in the isoquinoline alkaloid field is apparent in the formation of the two structurally close alkaloids protostephanine (**586**) and erybidine (**587**), which differ fundamentally only in the position of an oxygen atom in one aromatic ring.

8.26.1   Chemistry and Biogenesis

Protostephanine (**586**) was obtained from *Stephania japonica* Miers. (Menispermaceae) and its structure determined as the result of extensive degradative work in the 1950s by Japanese workers. Erybidine (**587**) on the other hand was obtained relatively recently from *Erythrina bidwilli* Lindl. (Leguminosae) and is closely related to **580**, the key intermediate in the biosynthesis of the *Erythrina* alkaloids (see Section 8.25.4).

586   protostephanine   587   erybidine   580

The position of the second methoxy group in ring C of **586** is the clue that its biosynthesis is not related to that of the *Erythrina* alkaloids. Two routes have been proposed for the formation of the nucleus of **586**. One of these is a phenolic oxidative coupling of **588** and dopamine (**10**) as shown (Scheme 75) to afford **589**. The problem here is the need to postulate re-

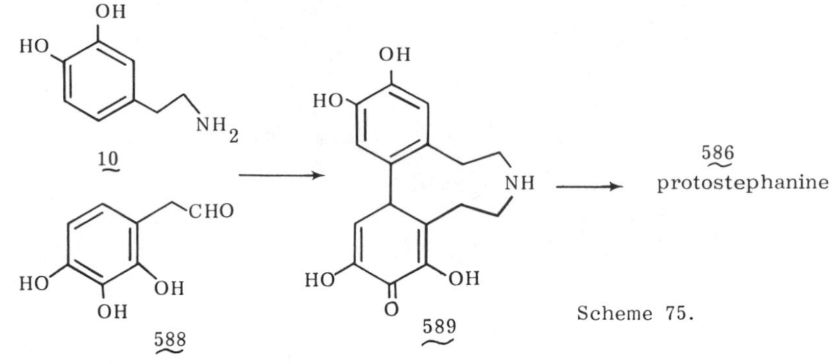

Scheme 75.

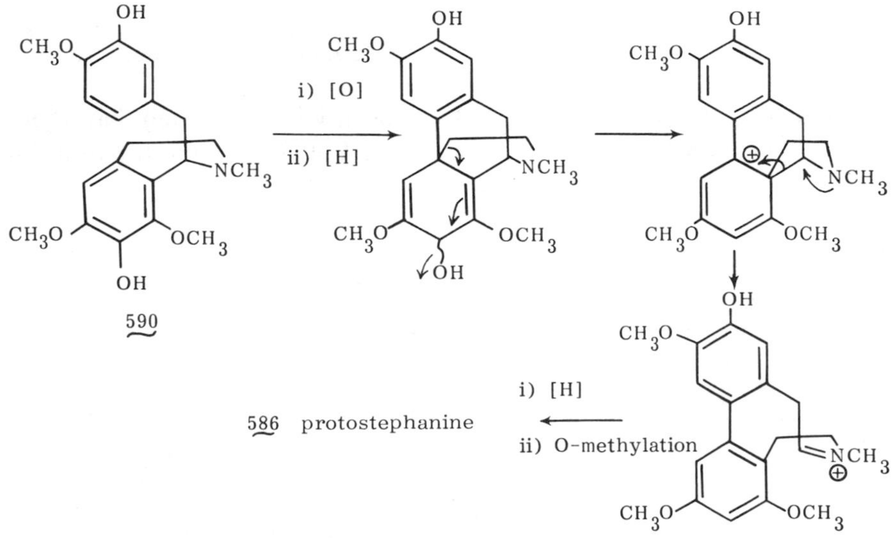

Scheme 75

ductive removal of a quinone oxygen and the unusual oxygenation pattern of the phenylacetaldehyde derivative **588**.

In 1964, Barton proposed a quite different biogenesis (Scheme 76) beginning with phenolic oxidation of an isoquinoline such as **590** and involving a dienol–benzene rearrangement as a critical step. In this scheme the intermediate is probably better written as **591**.

8.26.2 Synthesis

Acid-catalyzed rearrangement of **592** produced in steps from the isoquinoline **590** gave **593**, which could be rearranged, after a redox sequence, by heating with magnesium iodide, to a mixture of products **594** having the proto-

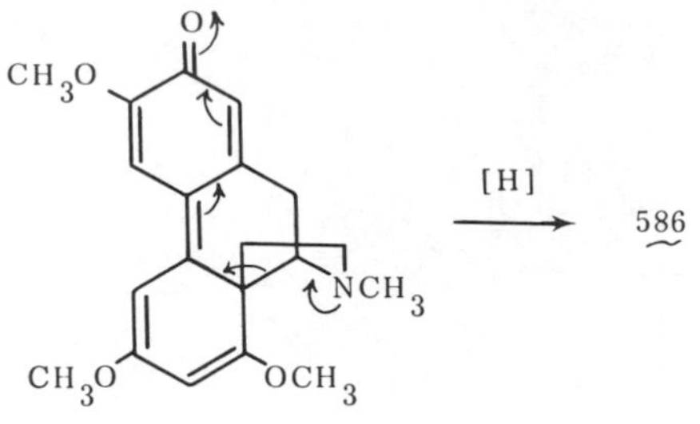

591

Scheme 76 Biogenesis of protostephanine (**586**) due to Barton.

stephanine skeleton. Treatment with diazomethane then afforded protostephanine (**586**).

No natural products containing the neospirodienone system have been
isolated, but Kupchan and Kim have demonstrated a facile route from the

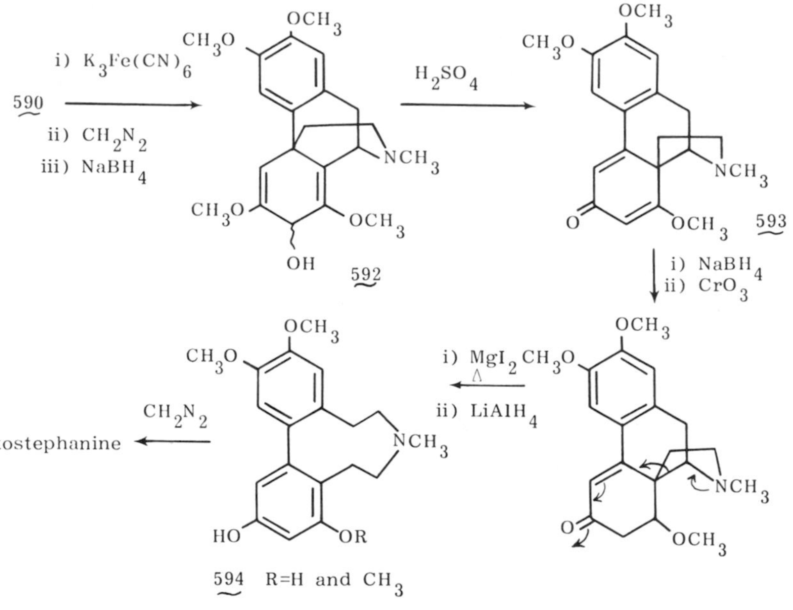

morphinandienone system to the dibenzazonine skeleton of erybidine (**587**)
which probably proceeds through a neospirinedienone intermediate.

O-Methylflavinatine (**490**), obtained in 94% yield by electrooxidative coupling of laudanosine (**188**) in HBF$_4$, when treated with boron trifluoride
etherate at room temperature followed by catalytic hydrogenation afforded
erybidine (**587**) in 85% yield. This sequence is interpreted as proceeding
from **490** by way of the neospirinedienone **594** as shown in Scheme 77.

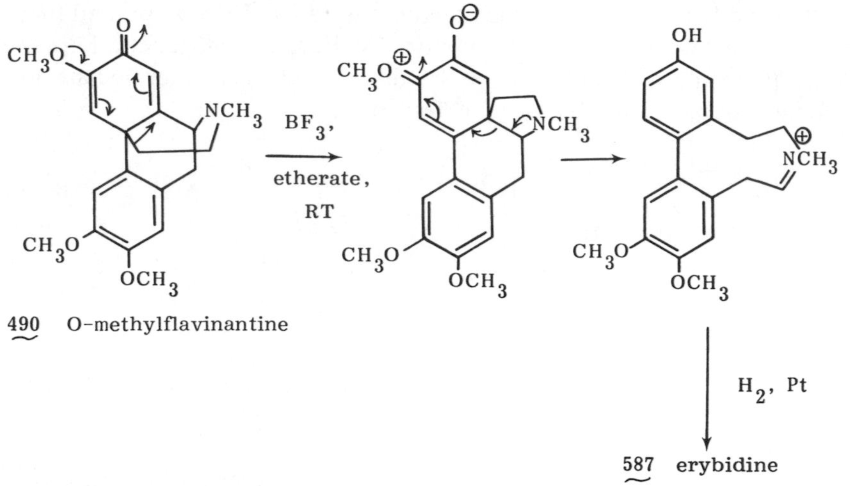

Scheme 77

8.26.3 Biosynthesis

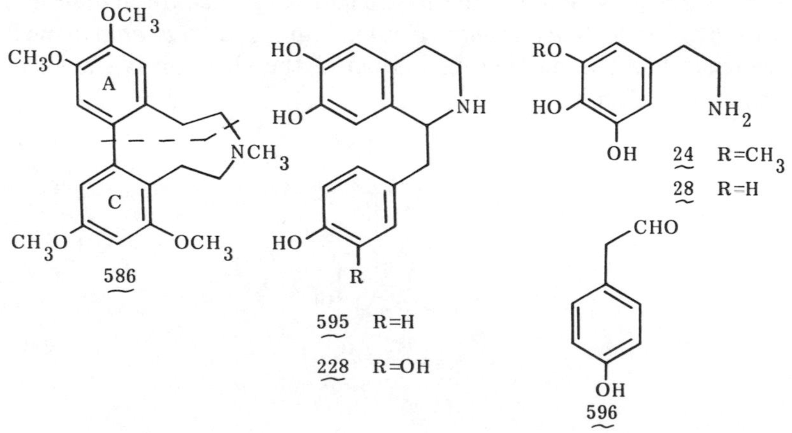

The simplicity of the structure of protostephanine (**586**) is not matched by its biosynthesis, as recent results show. Tyrosine was found to be a precursor of both C_6-C_2 units of **586**, but [2-^{14}C]dopa (**21**) and dopamine (**10**) only labeled the unit which generates ring C. Since none of the tri- or tetraoxygenated benzylisoquinoline precursors used (e.g. **228**, **595**) were effective, it was reasoned that a trioxygenated phenethylamine was condensing to form an initial benzylisoquinoline. This was supported when **28** and **24** were incorporated into **586**. The nature of the second unit which couples with the phenethylamine **24/28** is not known, but is thought to be *p*-hydroxyphenylacetaldehyde (**596**); this is discussed in Section 8.27. The derivative **597** was not a precursor, whereas **598** and **599** were (and their *N*-demethyl derivatives). Further support for Barton's biogenetic hypothesis comes from the incorporation of the dienone **600**, an intermediate in the formation of **592**, into **586** to the extent of 2.9%.

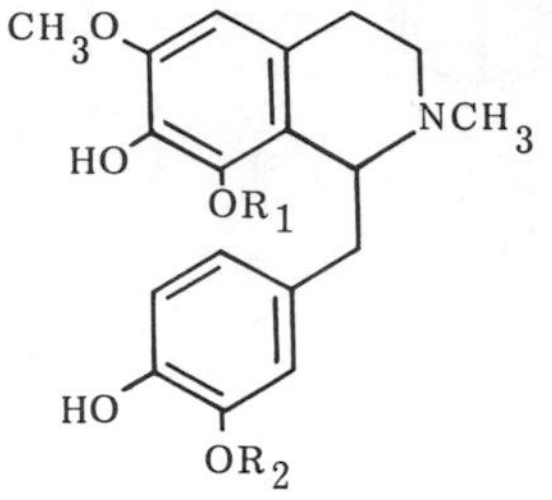

	R$_1$	R$_2$
597	CH$_3$	CH$_3$
598	H	CH$_3$
599	H	H

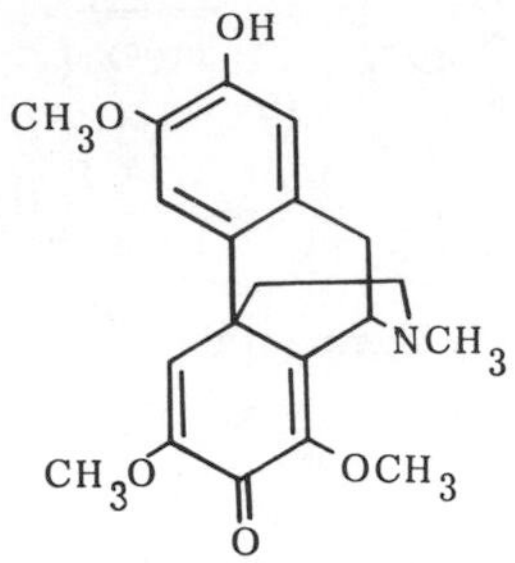

600

LITERATURE

Shamma, M., *The Isoquinoline Alkaloids*, Academic, New York, 1972, p. 418.

Barton, D. H. R., *Pure Appl. Chem.* **9**, 34 (1964).

Battersby, A. R., A. K. Bhatnagar, P. Hackett, C. W., Thornber, and J. Staunton, *Chem. Commun.* 1214 (1968).

Battersby, A. R., R. C. F. Jones, P. Kazlauskas, C. Poupat, C. W. Thornber, S. Ruchirawat, and J. Staunton, *Chem. Commun.* 773 (1974).

8.27 HASUBANAN ALKALOIDS

As far as is presently known, the hasubanan alkaloids are limited in distribution to the genera *Menispermum, Sinomenium,* and *Stephania* in the family Menispermaceae. The numbering system of the skeleton is shown in **601**, as suggested by Tomita.

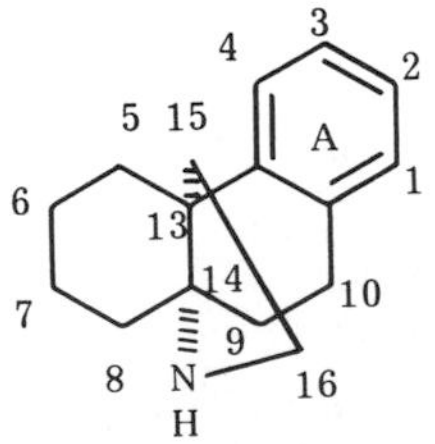

601

8.27.1 Chemistry

Hasubanonine obtained from *Stephania japonica* was originally assigned the structure **602**, because it was resistant to catalytic reduction. Subsequently it was shown to contain an α,β-unsaturated ketone and eventually the structure was modified to **149**.

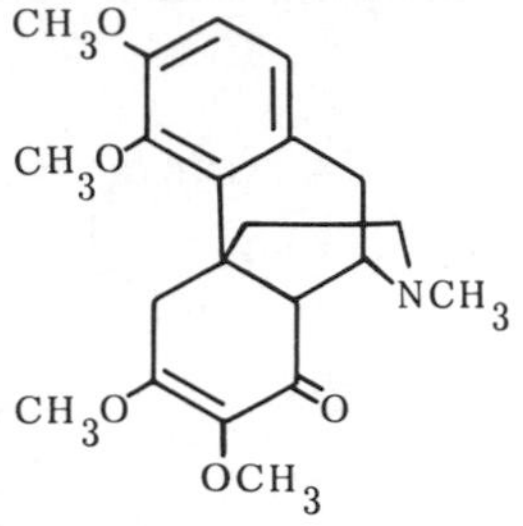

602

Some reactions of hasubanonine are worthy of mention. Sodium borohydride reduction affords an epimeric mixture of alcohols **603**, which can be reoxidized to hasubanonine (**149**) with manganese dioxide. Heating these alcohols with hydrogen bromide gave the α,β-unsaturated ketone **604** by the mechanism shown. This compound could be reduced with zinc and hydrochloric acid to the base **605**, which is important because it is antipodal by ORD to a product **606** obtained from dihydroindolinocodeinone (**607**). In

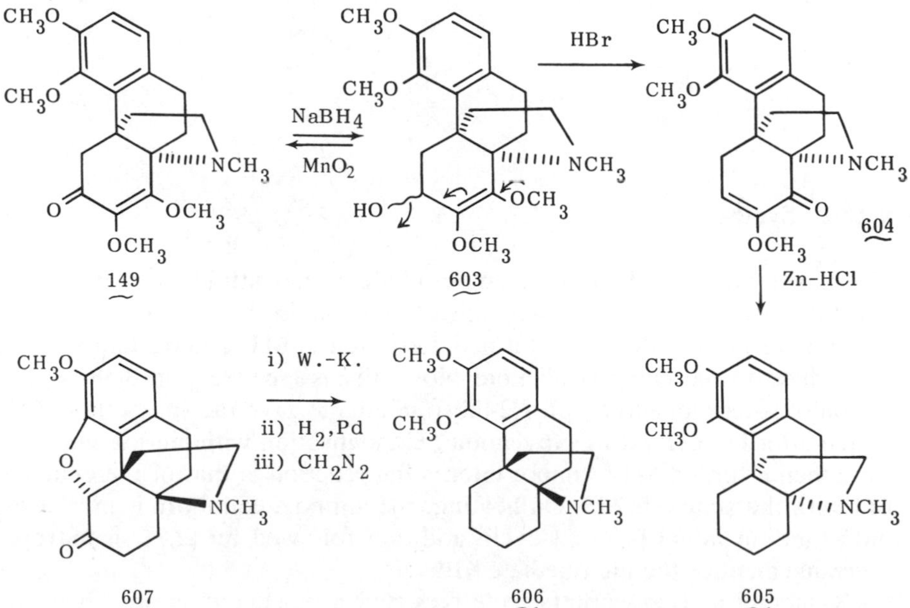

149

603

604

607

606

605

this way the absolute stereochemistry of the ethano bridge was established and relationship to the morphine alkaloids more clearly defined.

Several alkaloids in the hasubanan series have a C-8–C-10 acetal or ketal linkage and typical examples are metaphanine (**608**) from *S. japonica* and hernandifoline (**609**) from *S. hernandifolia* Walp.

On treatment with base, metaphanine (**608**) undergoes benzilic acid rearrangement to give **610**, indicating that it contains a masked α-diketone system.

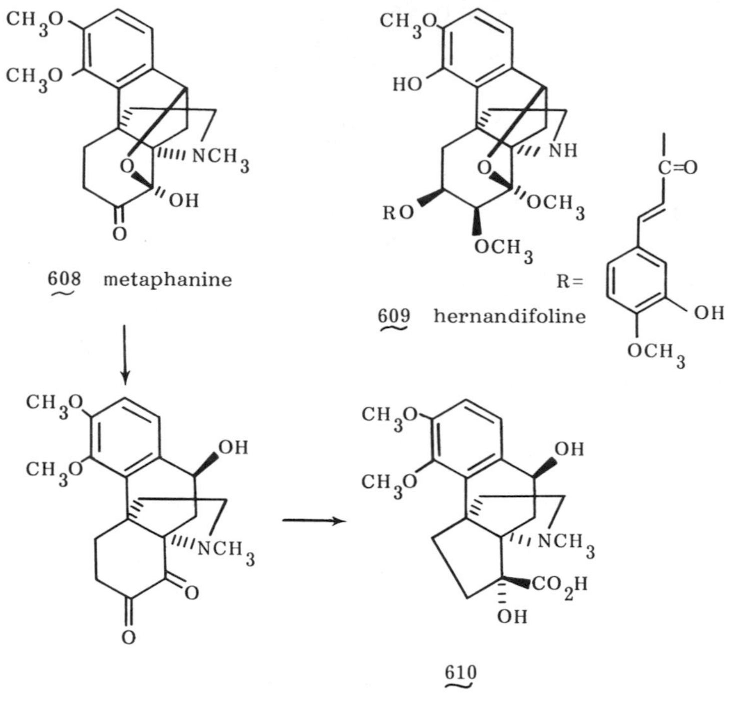

608   metaphanine

609   hernandifoline

610

8.27.2   Synthesis

The novel skeleton of the hasubanan alkaloids has attracted considerable synthetic attention and a full account is available in the review of Inubushi and Ibuka. One of the key intermediates used is **611**, analogous to an intermediate in mesembrine alkaloid biosynthesis (see Section 8.38.2). Reaction of β-tetralone **612** with 1,2-dibromoethane gave the spiroketone **613**, which afforded **611** with methylamine. Condensation with methyl vinyl ketone then afforded **614**. A more interesting reaction is that of the enamine **611** with the sulfoxide **615**. On heating, the amino alcohol **616** is produced, indicating an initial [4 + 2] cyclo addition followed by [2,3] sigmatropic rearrangement of the intermediate **617**.

Kametani and co-workers have reported a biogenetic approach to the hasubanan skeleton from reticuline (148). Reaction with trifluoroacetic anhydride followed by catalytic hydrogenation gave the amide 618, which was oxidized with vanadium oxychloride to 619. Hydrolysis afforded the enone 620.

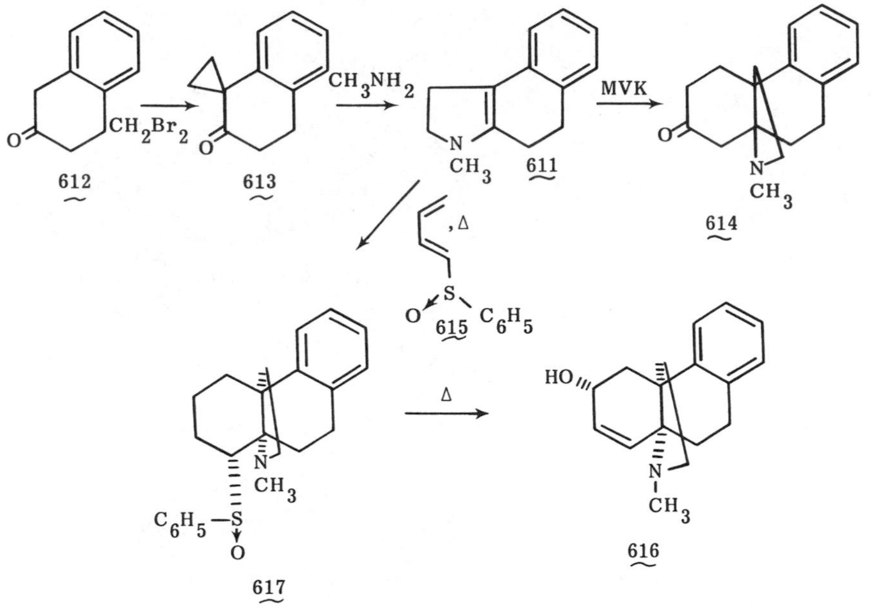

The three alkaloid types represented by cepharamine (621), hasubanonine (149), and metaphanine (608) have been synthesized by the groups of Kametani, Inubushi, and Inubushi respectively.

When 2'-bromoreticuline (622) was used in place of 148, 623 was produced and could be cyclized photochemically in the presence of sodium hydroxide and sodium iodide to 624. Hydrolysis with potassium bicarbonate followed by isomerization with methanolic hydrogen chloride gave cepharamine (621) (Scheme 78).

8.27.3 Spectral Properties

Spectroscopic definition of the hasubanan alkaloids skeleton is most effectively made by mass spectrometry, because of the highly characteristic fragments displayed by alkaloids in this series.

The spectra are dependent on the substitution in ring C. For example in the mass spectrum of the synthetic product 3,4-dimethoxy-N-methylhasubanan (623a) the initial loss is carbons 5–8, 56 mu, to give a radical ion at m/e 245 (624a).

An alkaloid such as metaphanine (608) which has a hemiketal linkage between C-8 and C-10 also shows a prominent ion corresponding to m/e 245. (m/e 229 if ring A has a methylenedioxy group). In this case the ion at m/e

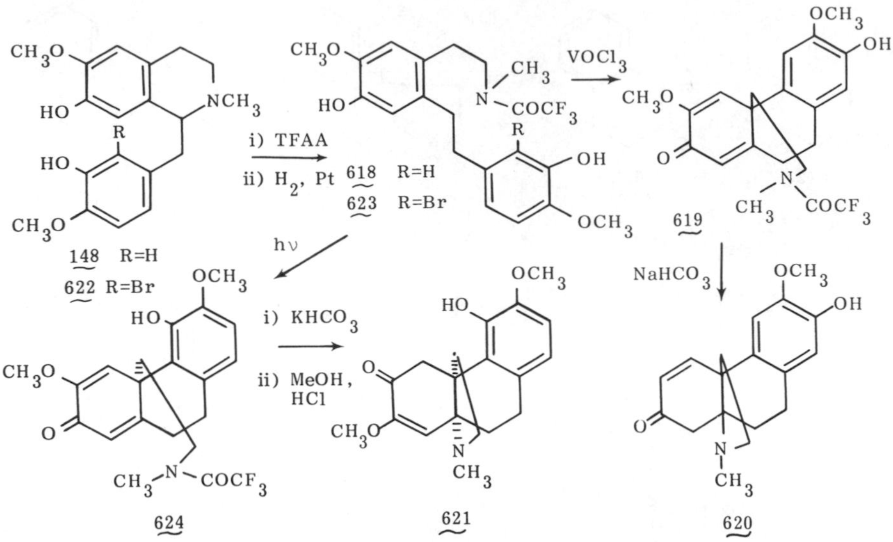

Scheme 78 Kametani systhesis of cepharamine (**621**).

245, formulated as **625**, is formed in a quite different manner as shown in Scheme 79. The source of the hydrogen which migrates to either C-13 or C-10 is not known.

8.27.4 Biosynthesis

Stephania japonica produces both the hasubanan alkaloids and protostephanine (**586**), and has therefore been an excellent plant in which to study the biosynthesis of these skeleta.

Work with labeled tyrosine and dopamine indicated that hasubanonine (**149**) and protostephanine (**586**) are derived from two different C_6-C_2 units. One of these units is a phenethylamine formed from both tyrosine and dopamine which generates ring C and the ethanamine bridge. The second unit is derived only from tyrosine. As with the biosynthesis of protostephanine (**586**) several more advanced intermediates such as trioxygenated phenethylamines and tri- and tetraoxygenated benzylisoquinolines were used. Of these only **24** and **28** were incorporated and this was specifically into ring C and the ethanamine bridge.

With the establishment of **28** as a precursor, 16 benzylisoquinolines were evaluated as precursors. Only those having the substitution of **599** were incorporated into both hasubanonine (**149**) and protostephanine (**586**). Since **599** and **626** were incorporated with either an NH or an NCH_3 group present, it is apparent that N-methylation can occur at any stage. In addition it is established that further oxygenation of ring A takes place *after* formation of the benzylisoquinoline nucleus. This explains the lack of incorporation of dopamine into this unit, and should be contrasted with the situation in

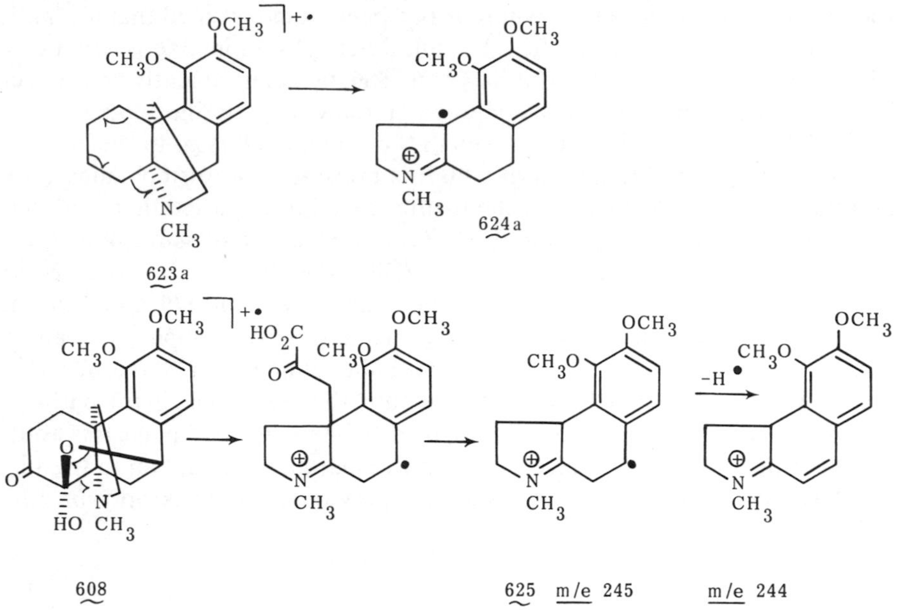

623a → **624a**

Scheme 79

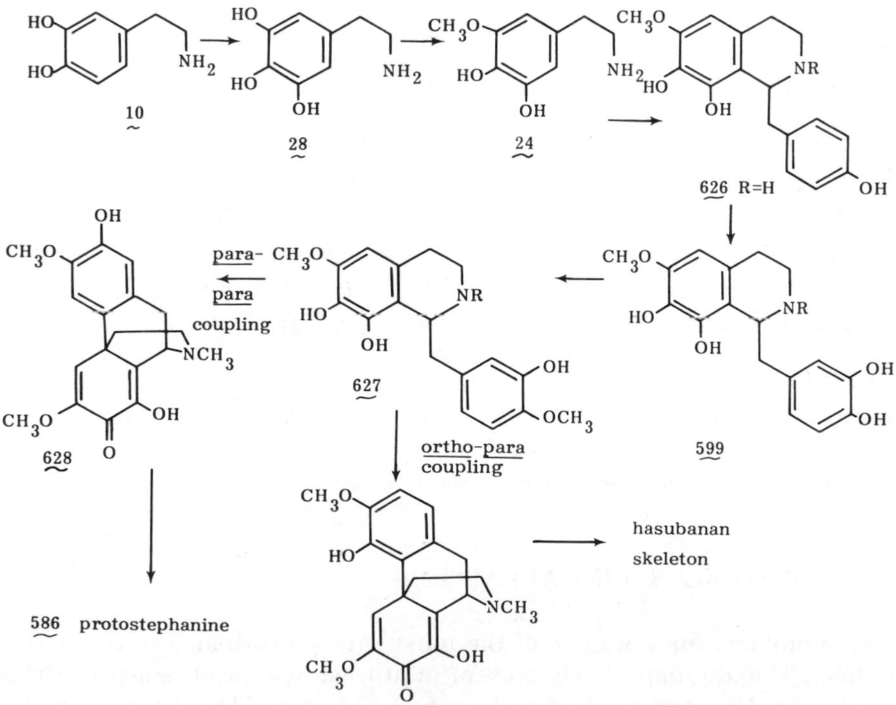

625 m/e 245   m/e 244

608

Scheme 79

10

28

24

626 R=H

para-
para
coupling

627

ortho-para
coupling

599

628   O

586   protostephanine

629

hasubanan
skeleton

Scheme 80 Biosynthesis of the protostephanine (**586**) and hasubanan skeleta.

471

the morphine alkaloid series, where it has been demonstrated that aromatic substitution does *not* occur after the initial benzylisoquinoline is produced. These data were applicable to both **149** and **586**, and consequently divergence in the biosynthetic pathway takes place at a late stage in the sequence.

Before discussing the early stages in the scheme let us go to the end and consider the oxygenation pattern which is present in protostephanine (**586**) and the hasubanan alkaloids. In the former case it is apparent that oxidative coupling has taken place *para* to the C-3 methoxy group of **586**. On the other hand in the formation of hasubanonine (**149**), coupling has occurred *ortho* to the C-4 methoxy group of **149**. The biosynthesis of **149** and **586** can therefore be envisaged as occurring by two alternative couplings of an intermediate such as **627** to afford the morphinandienones **628** and **629**. The biosynthetic pathway to this intermediate and the divergence in the pathway has been suggested to occur as shown in Scheme 80. Although this route would indicate the importance of the morphinandienone system in the biosynthesis of these skeleta, as discussed previously there is only *in vitro* evidence for this to date.

LITERATURE

Reviews

Thornber, C. W., *Phytochemistry* **9**, 157 (1970).

Bentley, K. W., *Alkaloids NY* **13**, 131 (1971)

Inubushi, Y., and T. Ibuka, *Alkaloids NY* **16**, 393 (1977).

Synthesis

Keely, S. L., A. J. Martinez, and F. C. Tahk, *Tetrahedron Lett.* 2763 (1969).

Evans, D. A., C. A. Bryan, and C. L. Sims, *J. Amer. Chem. Soc.* **94**, 2891 (1972).

Kamentani, T., T. Kobari, K. Shishido, and K. Fukumoto, *Chem. Ind. (London)* 538 (1972).

Ibuka, T., K. Tanaka, and Y. Inubushi, *Chem. Pharm. Bull.* **22**, 782 (1974).

Ibuka, T., K. Tanaka, and Y. Inubushi, *Chem. Pharm. Bull.* **22**, 907 (1974).

Biosynthesis

Battersby, A. R., R. C. F. Jones, R. Kazlauskas, C. Poupet, C. W. Thornber, S. S. Ruchirawat, and J. Staunton, *Chem. Commun.* 773 (1974).

Battersby, A. R., A. Minta, A. P. Ottridge, and J. Staunton, *Tetrahedron Lett.* 1321 (1977).

8.28 PROTOBERBERINE ALKALOIDS

The protoberberines are one of the most widely distributed of the isoquinoline alkaloid groups, being present in at least nine plant families, particularly the Annonaceae, Berberidaceae, Lauraceae, Menispermaceae, Papaverceae, and Rutaceae. Over 40 protoberberines are known and some

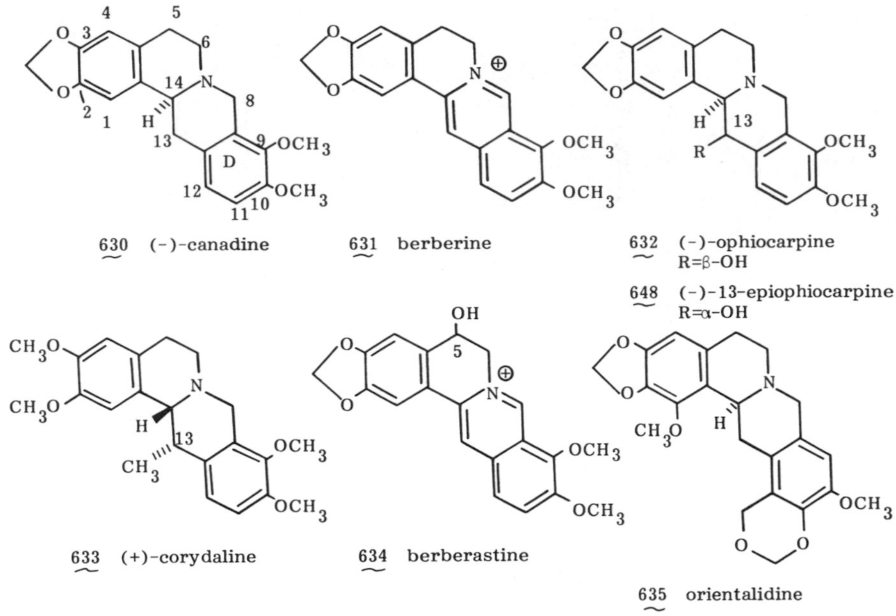

Figure 15 Some representative protoberberine alkaloids.

typical examples are shown in Figure 15. Oxygen substituents are commonly found at C-2 and C-3 and either at C-9 and C-10 or at C-10 and C-11. An alcoholic group is sometimes present at C-5 or C-13 and a methyl group may also be present at C-13. The alkaloid orientalidine (**635**) is a member of the retroprotoberberine group.

Much of the vast literature on these alkaloids has been due to the tremendous efforts of Kametani and co-workers, particularly in the area of synthesis.

8.28.1 Chemistry

The chemistry of the protoberberines, particularly that of berberine itself, is a rich and complex one (Figure 16). Lithium aluminum hydride reduces the quaternary salts to the corresponding dihydro species **636**, whereas sodium borohydride affords the corresponding tetrahydro product **637**. These reactions are reversible under mild oxidizing conditions.

In the presence of a nucleophile, attack occurs at the 8-position when the nucleophile is hydroxide and the corresponding oxoprotoberberine **638** is produced. When the nucleophile is acetone the so-called "berberine acetone"* is produced (see also the benzophenanthridine section for additional examples of this type).

___
* Berberine acetone is not a homogenous product but a mixture of three compounds.

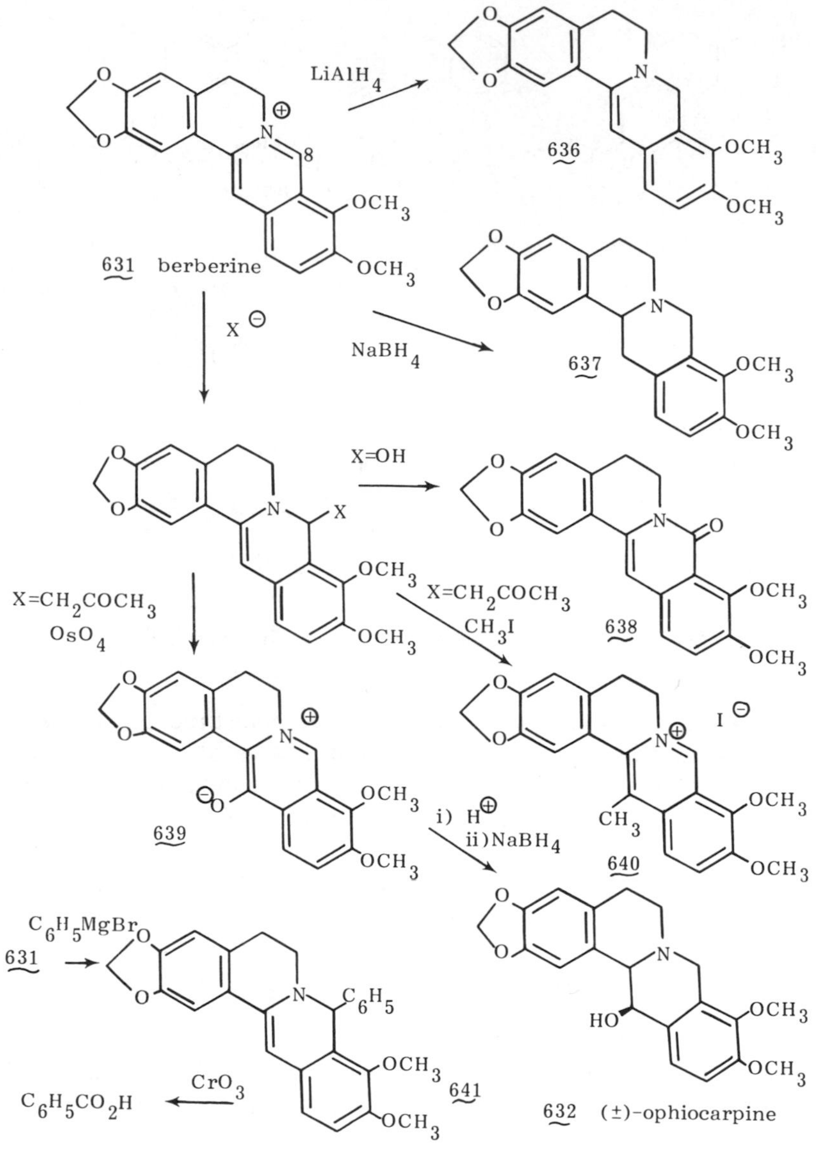

Figure 16 Chemistry of the protoberberine nucleus.

Osmium tetroxide oxidation of berberine acetone gives berberine phenol betaine (**639**), which in its protonated form can be reduced to racemic ophiocarpine (**632**) with sodium borohydride. When treated with methyl iodide the 13-methyl berberinium iodide (**640**) was produced from berberine acetone.

The 8-position is the position of attack when berberine is treated with phenyl magnesium bromide. The reaction is useful because Kuhn–Roth oxidation of **641** now leads to benzoic acid (see the biosynthesis section).

When a ketospiroisoquinoline such as **642** is irradiated in tetrahydrofuran, the berberinium derivative **643** is produced in 80% yield.

Conversely, when the protoberberine **644** is treated with alkali, the ochotensane-type alkaloid **645** is produced. But this reaction proceeds through a quinone methide intermediate so that it cannot be used for the formation of optically active spirobenzylisoquinolines.

Anionic rearrangement of the optically active methosalt of canadine (**646**) gave the corresponding optically active ochotensane derivative (**647**). The

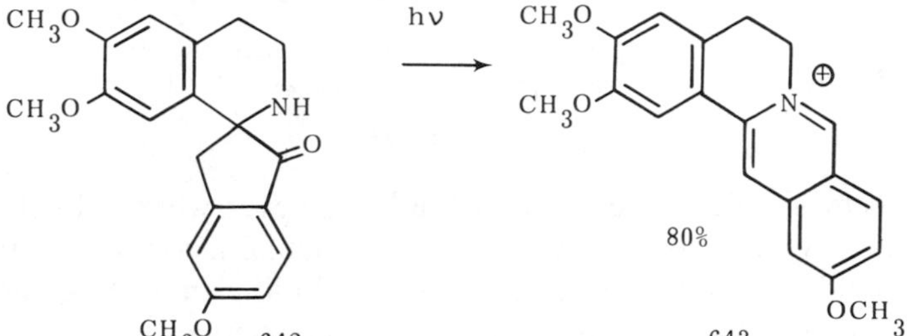

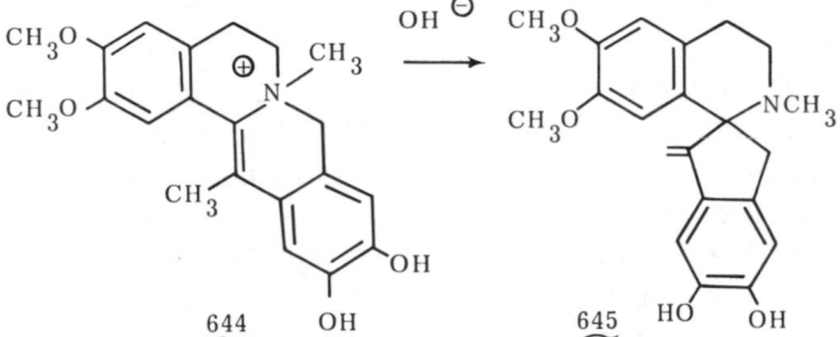

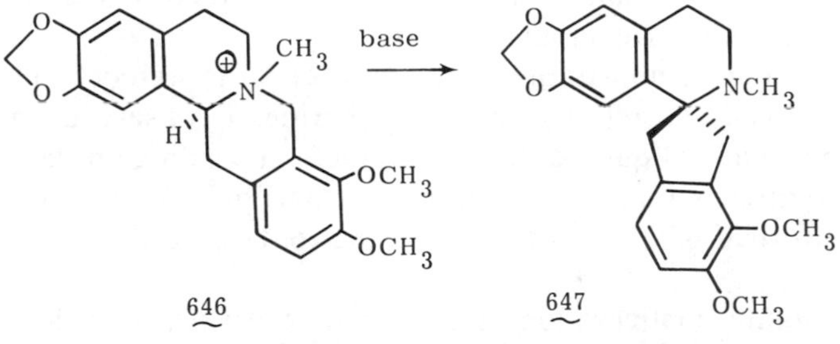

base may be butyl lithium, the ylid of DMSO, lithium aluminum hydride, or sodium bis(2-methoxyethoxy)aluminum hydride.

8.28.2 Stereochemistry

The stereochemistry of alkaloids such as ophiocarpine (**632**) and 13-epi-ophiocarpine (**648**) can be deduced from their IR spectra. Both compounds display Bohlmann bands at about 2800 cm^{-1}, indicating a *trans*-fused B/C ring juncture, but only ophiocarpine (632) has a hydrogen-bonded hydroxyl group. This indicates internal proximity of the hydroxyl proton and the nitrogen lone pair; a situation possible only when the 13-hydroxy group is *beta* and axial as in 649.

648   13 – epi ophiocarpine

649   ophiocarpine

These very simple concepts do not hold however when substituents are at positions 1, 8, or 13 (other than OH). In particular a compound may show Bohlman bands in a KBr pellet, but not in solution, indicating subtle conformational inversion, *trans* to *cis*, at the nitrogen between crystalline states and liquid. Such a phenomenon is particularly true of 1-substituted protoberberines.

The most reliable way to determine the B/C-ring junction stereochemistry is through kinetic studies of the rates of methylation of the tertiary nitrogen with methyl iodide. In general terms the *cis*-quinolizidine compounds react much faster than the corresponding *trans* isomers. Changing the substitution also changes the basicity of the nitrogen so that 10,11-substituted isomers show faster rates of *N*-methylation than 9,10-substituted isomers. When substitution is by a phenolic group, the rates of methylation are between those expected for *cis*- and *trans*-quinolizidines and consequently such data must be used with considerable caution.

An interesting demonstration of the differences in stereochemistry of two 13-methyl-substituted tetrahydroprotoberberines is observed on Hofmann degradation to afford quite different products depending on the availability of the β-hydrogen *trans* to the nitrogen lone pair. Corydaline (633) for example afforded 650, but mesocorydaline (651) gave the vinyl derivative 652.

The absolute configuration of the protoberberines was deduced when (−)-*N*-norlaudanosoline (228) of known absolute configuration afforded (−)-norcoralydine (653). The levorotatory bases have the same absolute configuration as 653 and also exhibit a negative ORD curve in the region of 240 nm.

8.28.3 Synthesis

The classic route for the formation of the protoberberines involves the condensation of a benzylisoquinoline with formaldehyde under acidic condi-

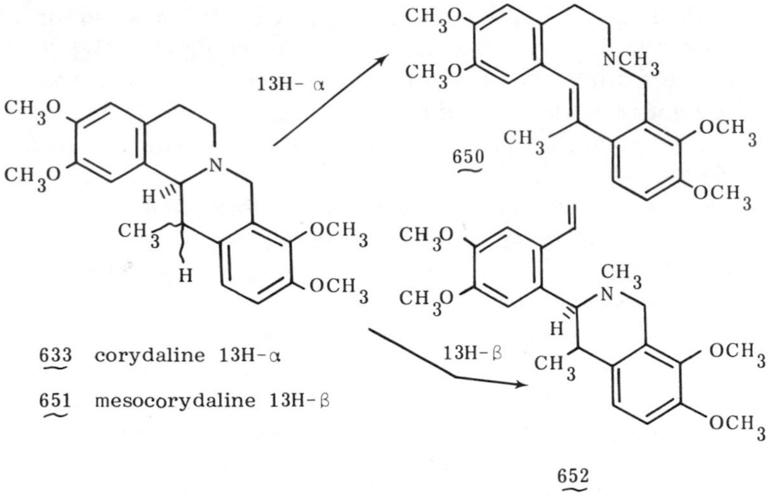

633 corydaline 13H-α

651 mesocorydaline 13H-β

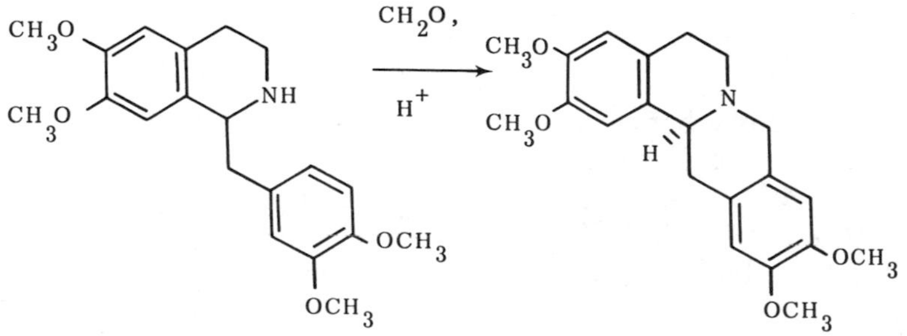

652

228 (-)-N-norlaudanosoline

653 (-)-norcoralydine

tions. This route is still the method of choice for formation of this skeleton. The major drawback is that of isomer formation, but under certain circumstances this phenomenon is pH dependent. Thus nor-reticuline (229) on reaction with formaldehyde at pH 6.3 afforded at 2:1 mixture of scoulerine (654) and coreximine (655), whereas at pH 7 only coreximine (655) is produced.

A second route for the exclusive formation of the 9,10-substituted product is to block the *para* position with a bromine substituent. In the synthesis of scoulerine (654), the bromobenzylisoquinoline 656 was condensed with

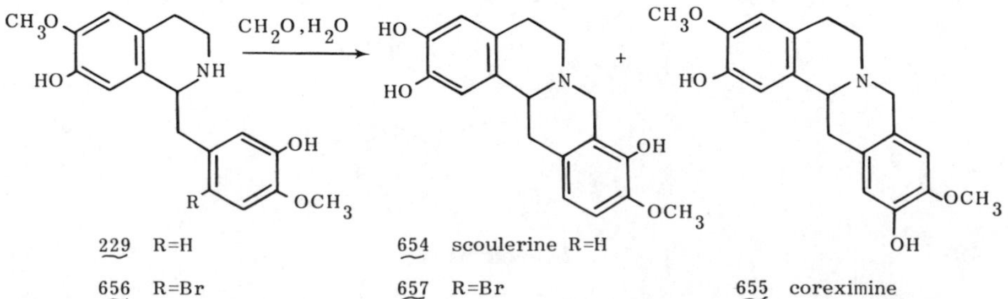

229 R=H

656 R=Br

654 scoulerine R=H

657 R=Br

655 coreximine

formaldehyde under acidic conditions to give 12-bromoscoulerine (**657**). Lithium aluminum hydride reduction then gave scoulerine (**654**).

Alternative methods for inserting the berberine bridge carbon into a benzylisoquinoline such as 3,4-dihydropapaverine (**205**) include the Vilsmeier–Haack reaction or Bischler–Napieralski cyclization of an enamide such as **658** (Scheme 81).

Kametani and co-workers have studied the formation of the protoberberine nucleus by the thermal cycloaddition of a Schiff base such as **659** to a 1-bromobenzocyclobutene. In fact the reaction took place at 100°C to give a 31% yield of the berberinium bromide **660**.

In 1925 it was demonstrated that reduction of a phthalideisoquinoline leads to an 8-oxoprotoberberine. Subsequently this reaction was used by Govindachari in the synthesis of ophiocarphine (**632**). Thus lithium aluminum hydride reduction of the phthalideisoquinoline (**661**) gave a mixture of ophiocarpine (**632**) and 13-epiophiocarpine (**648**) on workup. An alternative approach to **632** and **648** will be discussed subsequently.

8.28.4 Spectral Properties

Tetrahydroprotoberberines display an absorption maximum in the 282–289 nm range with occasionally a shoulder around 230–240 nm. Unlike the aporphine series, the 2,3,9,10-substituted compounds cannot be distinguished from their 2,3,10,11-analogues by their UV spectra.

The dihydroberberines show λ_{max} 280 and 365 nm (4.1 and 4.25), but the

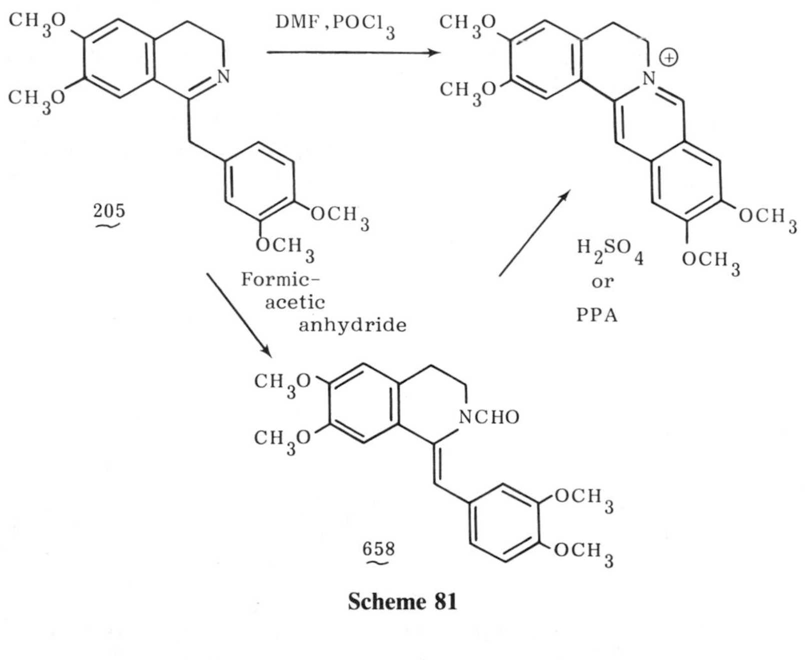

Scheme 81

659

660

661

632 13α-H
648 13β-H

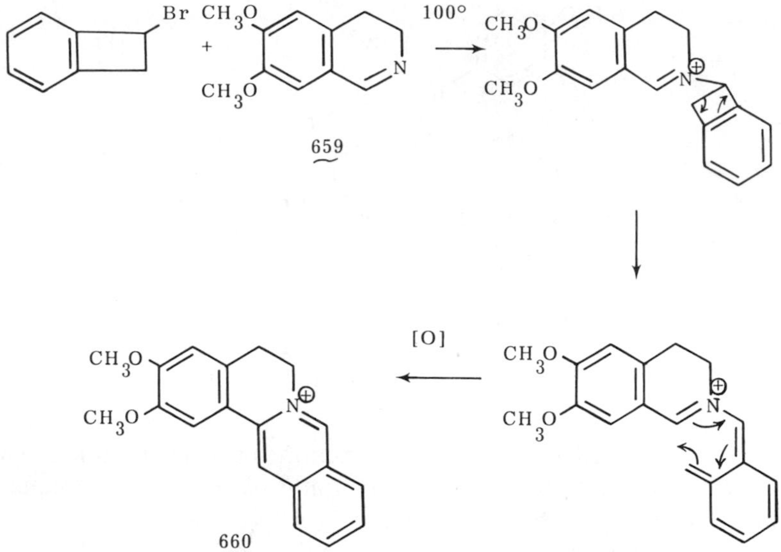

most characteristic spectra are those of the berberine salts, and these spectra can be used to distinguish between 9,10- and 10,11-substituted nuclei. For example palmatine iodide (**662**) shows λ_{max} 263, 355, and 423 nm (4.4, 4.5, and 4.0), whereas pseudopalmatine iodide (**663**) shows λ_{max} 265, 287, 310sh, 345, and 375 nm (4.3, 4.7, 4.5, 4.3, and 4.0). The two major differences are the λ_{max} of the longest-wavelength band, and the peak at 310 nm which is absent in the 9,10-substituted series.

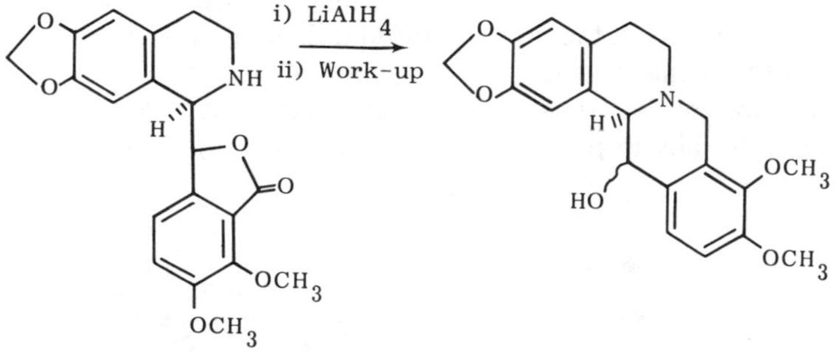

The oxygenation of ring D is best determined from two features in the proton NMR spectrum. One is the multiplicity of the aromatic protons; naively this would appear to be straightforward, but due to overlap of signals and the closeness of chemical shift, resolution may not always be an easy task. The second feature is the multiplicity of the C-8 methylene group which in C-9-unsubstituted compounds appears at a broad singlet. When the 9-

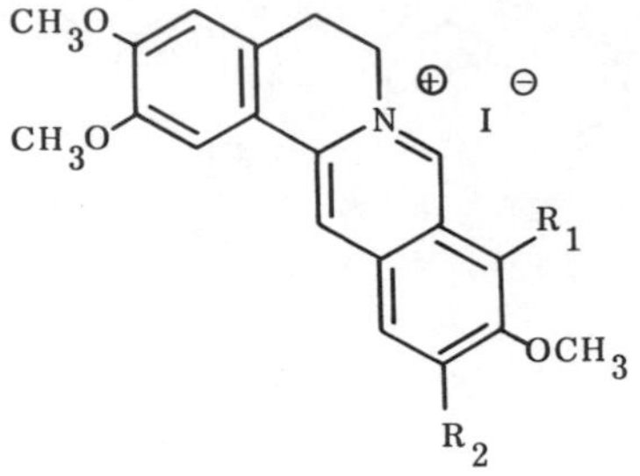

662 palmatine R_1=OCH_3; R_2=H

663 pseudopalmatine R_1=H; R_2=OCH_3

position is substituted by an oxygen function the axial proton is shifted upfield to become a doublet ($J \simeq 16$ Hz) at 3.65 ppm and the equatorial proton is deshielded to about 4.35 ppm.

Distinction between *cis*- and *trans*-quinolizidines can also be made by the CMR spectra of these compounds. Thus a 3-methoxy-1,2-methylenedioxy derivative having a *trans*-quinolizidine system shows the C-6 signal at 51.4 ppm, whereas in the *cis* compound (a 1-methoxy-2,3-methylenedioxy derivative) this signal was observed at about 48.0 ppm.

Comparison of the data for *O*-methylcapaurine (**664**) and 2,3,10,11-tetramethoxytetrahydroprotoberberine (**665**) shows the former to exist pre-

*†, indicates assignments may be reversed

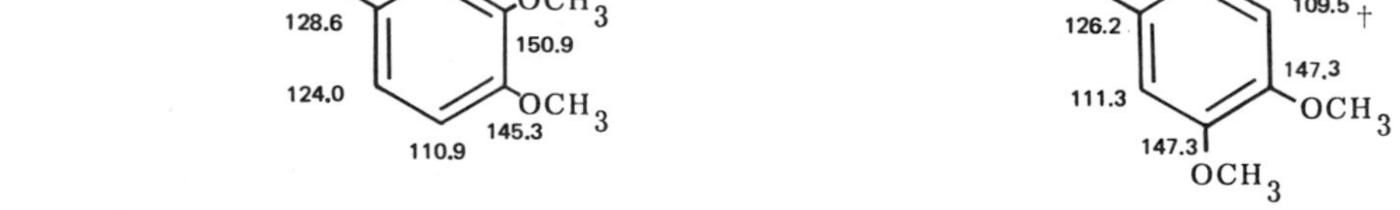

664 O-methylcapaurine

665

dominently in the *cis* form whereas the latter exists in the *trans* form. The C-13α resonance also correlates well with the stereochemistry of the B/C ring junction. A methylene at about 55 ppm is typical for the *cis*-stereochemistry, and a methylene at 59–60 ppm is characteristic of the *trans*-stereochemistry. The chemical shift of the C-8 methylene is indicative of the presence (~53–54 ppm), or absence (~57.5–58.5 ppm), of a C-9 substituent.

8.28.5 Biosynthesis

As a result of numerous studies, the biosynthesis of the berberine alkaloids has been well worked out and the currently accepted scheme is shown in Scheme 82.

The fundamental units are those for the formation of the benzylisoquinoline alkaloids with the addition of a single carbon atom which becomes C-8 of the skeleton. Thus two molecules of tyrosine (2) are involved, one proceeding to dopamine (10) *via* dopa (21), and the second to 3,4-dihydroxyphenyl pyruvic acid (241). Thus [2-^{14}C]tyrosine labels C-6 and C-13 of berberine (631) in *Hydrastis canadensis* L. (Ranunculaceae), but [1-^{14}C]dopamine (10) labels only C-6. Mannich condensation followed by decarboxylation affords norlaudanosoline (228), which on selective methylation affords reticuline (148), the now familiar precursor of so many isoquinoline alkaloids. It has been firmly established that reticuline is a precursor

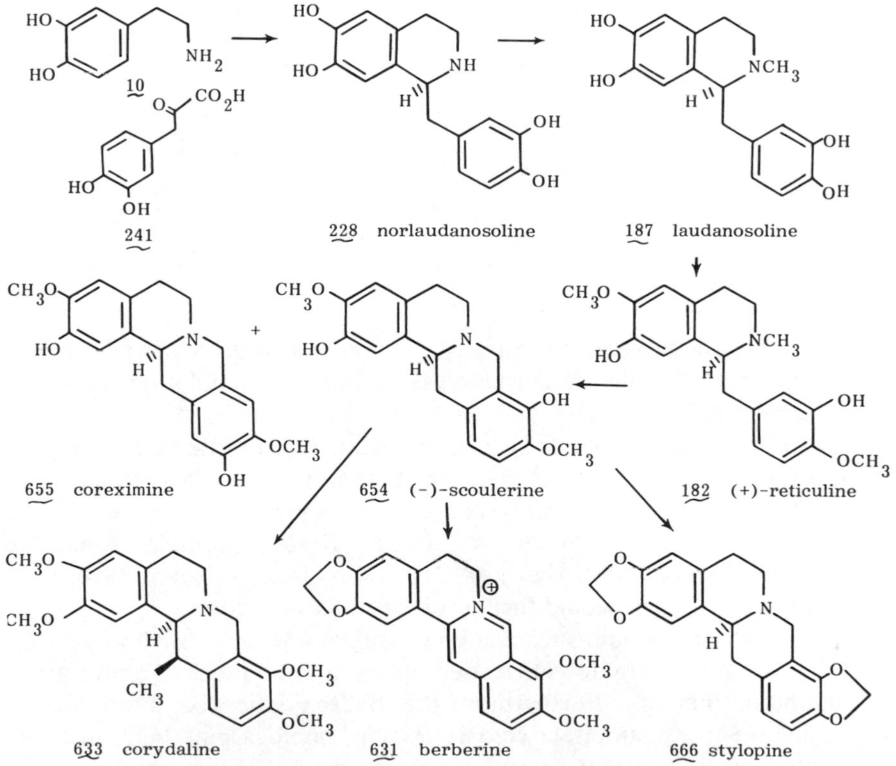

Scheme 82 Biosynthesis of the protoberberine alkaloids.

of each of the alkaloids, and that the (S)-(+)-isomer **182** is the true biosynthetic precursor, (+)-Reticuline (**182**) is therefore a precursor of berberine (**631**) in *H. canadensis*, of stylopine (**666**) in *Chelidonium majus* L. (Papaveraceae), and of coreximine (**655**) in *P. somniferum*.

The so-called "berberine bridge" carbon at C-8 of the berberine alkaloids has been shown by Spenser's group to be derived from the S-methyl group of methionine. A compound such as berberine (**631**) therefore becomes labeled at four sites after feeding [methyl-^{14}C]methionine, namely C-8, the two methoxy groups, and the methylenedioxy group. It should not be expected that this label will be distributed equally between the various sites.

The oxidation of the N-methyl group and subsequent ring closure is thought to proceed *via* the iminium species **667** to give either the 9,10-(pathway a) or 10,11-disubstituted (pathway b) series of compounds.

The methylenedioxy groups of stylopine (**666**) are derived from the corresponding methoxy phenol derivatives, so that (−)-scoulerine (**654**) is a precursor of **666** in *C. majus*.

Berberastine (**634**) has a 5-hydroxy substituent and one might suppose that this is introduced by hydroxylation of berberine (**631**). This is apparently not the case, for Monkovic and Spenser have shown that both dopamine and noradrenaline (**15**) are precursors of **634** in *Berberis japonica* Lindl. No feeding experiments with berberine (**631**) or 4-hydroxyreticuline (**668**) have been carried out which would further clarify this situation.

In contrast, an alkaloid such as ophiocarpine (**632**) having a 13-hydroxyl group has been shown by Jeffs to be derived by hydroxylation of a preformed protoberberine nucleus. Furthermore this hydroxylation is stereospecific. Scoulerine (**654**) was an effective precursor of ophiocarpine (**632**) in *Corydalis ophiocarpa*, as was the tetrahydroberberine **637**. This compound was stereospecifically labeled with tritium at the 13α and 13β positions and each

15 noradrenaline

668 4-hydroxyreticuline

631 berberine, R=H
634 berberastine, R=OH

compound evaluated as a precursor. The results demonstrated that the 13α-proton is specifically retained, indicating that hydroxylation occurs with retention of configuration.

The 13-methyl berberine alkaloids such as corydaline (**633**) are also derived specifically from reticuline (**182**), although the origin of the additional carbon atom is not known.

The protoberberines are not only interesting in themselves from a biosynthetic point of view, but are important because they act as precursors to other skeleta. In particular, as we shall see in subsequent sections, these alkaloids (e.g. **654**) are the precursors of the protopine (**669**), rhoeadine (**670**), benzophenanthridine (e.g. **671**), and phthalideisoquinolines (e.g. **672**). In addition they are thought to be the precursors of the spirobenzyl isoquinoline alkaloids, [e.g. ochotensamine (**673**)] (Scheme 83).

8.28.6 Pharmacology

There are no prescription products containing protoberberine alkaloids, but several important pharmacologic reactions have been noted.

One of the most important actions is the antimicrobial activity of berberine and derivatives, which covers the range of organisms from fungi and protozoa to bacteria, and these results have been summarized by Hahn and Ciak. Thus berberine sulfate shows good inhibitory activity at low doses against *Corynebacterium diphtheriae*, *Staphylococcus aureus*, *Xanthomonas citri*, and *Candida tropicalis*.

There has been intense interest in the anticancer activity of berberine derivatives. Berberine itself has cytotoxic activity, but the derivative coralyne chloride (**674**) displays activity against both the P-388 and L-1210 lymphocytic leukemia systems in mice. More important, this compound and the related derivative **675** have been found by Cheng and co-workers to exhibit activity over an exceptionally wide dose range. Like the antibacterial and antiprotozal activity, this is probably related to the ability of the berberine nucleus to intercalate mitochondrial DNA.

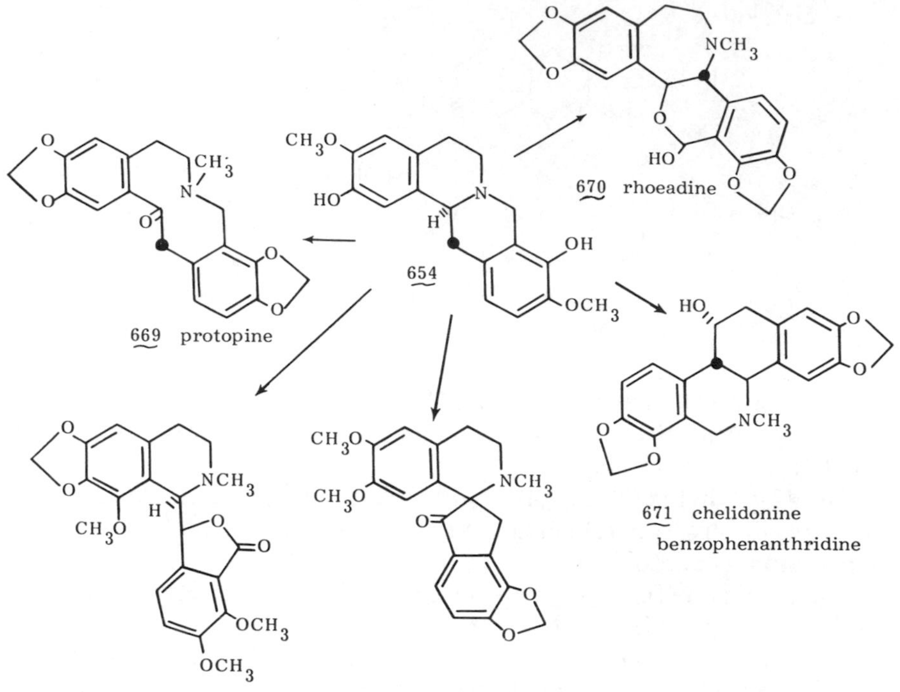

Scheme 83 Further elaboration of the protoberberine nucleus.

Berberine-containing plants also have an extensive folkloric use as antifertility remedies. Subsequent work with pure compounds has shown that activity is due to the uterine-contracting activity of this particular skeleton; canadine (**630**) for example is active *in vitro* at doses as low as 5.0×10^{-6} g/ml and berberine chloride (**631**) at 2×10^{-5} g/ml.

Canadine also has hypotensive activity and xylopinine (**676**) is a powerful adrenergic α-blocker.

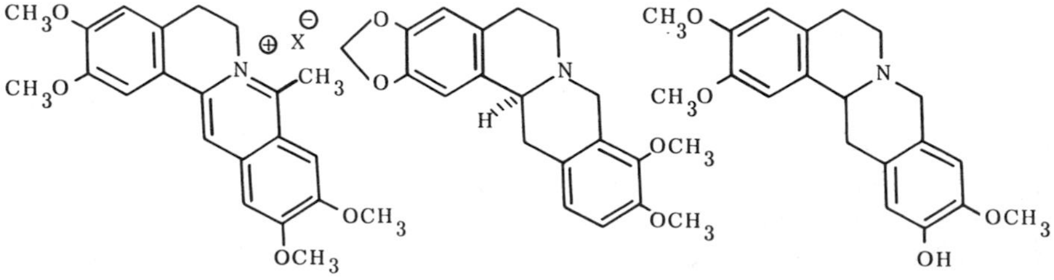

674 X = Cl 630 canadine 676 xylopinine

675 X = $C_2H_3SO_5$

LITERATURE

Reviews

Jeffs, P. W., *Alkaloids* NY **9**, 41 (1967).

Santavy, F., *Alkaloids* NY **12**, 383 (1970).

Shamma, M., *The Isoquinoline Akaloids*, Academic, New York, 1972, p. 268.

Kondo, Y., *Heterocycles* **4**, 197 (1976).

Kametani, T., M. Ihara, and T. Honda, *Heterocycles* **4**, 483 (1976).

Pai, B. R., K. Nagarajan, H. Suguna, and H. Natarajan, *Heterocycles* **6**, 1377 (1977).

Shamma, M. and J. L. Moniot, *Isoquinoline Alkaloid Research, 1972–1977,* Plenum, New York, 1978, p. 209.

Synthesis

Kametani, T., T. Kato, and K. Fukumoto, *Tetrahedron* **30**, 1043 (1974).

Kametani, T., K. Ogasawara, and T. Takahashi, *Tetrahedron* **29**, 73 (1973).

Kametani, T., T. Sugai, Y. Shoji, T. Honda, F. Satoh, and K. Fukumoto, *J. Chem. Soc. Perkin Trans I* 1151 (1977).

Govindachari, T. R., and S. Rajaduri, *J. Chem. Soc.* 557 (1957).

Biosynthesis

Gear, J. R., and I. D. Spenser, *Can. J. Chem.* **41**, 783 (1963).

Barton, D. H. R., *Proc. Chem. Soc. London* 293 (1963).

Battersby, A. R., R. J. Francis, E. A. Ruveda, and J. Staunton, *Chem. Commun.* 89 (1965).

Brochmann-Hanssen, E., C.-C. Fu, and G. Zanati, *J. Pharm. Sci.* **60**, 873 (1971).

Gupta, R. N., and I. D. Spenser, *Can. J. Chem.* **43**, 133 (1965).

Monkovic, I., and I. D. Spenser, *J. Amer. Chem. Soc.* **87**, 1137 (1965).

Jeffs, P. W., and J. D. Scharver, *J. Amer. Chem. Soc.* **98**, 4301 (1976).

Pharmacology

Kondo, Y., *Heterocycles* **4**, 197 (1976).

Hahn, F. E., and J. Ciak, *Antibiotics,* Vol. 3, D. Gottlieb, A. D. Shaw, and J. W. Corcoran (Eds.), Springer, New York, 1975, p. 577.

Amin, A. H., T. V. Sabbaih, and K. M. Abbasi, *Can. J. Microbiol.* **15**, 1067 (1969).

Zee-Cheng, K. Y., K. D. Paull, and C.-C. Cheng, *J. Med. Chem.* **17**, 347 (1974).

8.29 PROTOPINE ALKALOIDS

The protopines are distributed in the plant families Berberidaceae, Papaveraceae, Ranunculaceae, and Rutaceae and comprise about 20 members.

A typical alkaloid is protopine (**669**), one of the most widely distributed of all benzyl isoquinoline alkaloids, and the ketone group at C-14 is the characterizing feature of this group of compounds. Occasionally the 13-position is also oxygenated [e.g. 13-oxoprotopine (**677**)] or is substituted by a methyl group [e.g. corycavidine (**678**)]. Cryptopine (**679**) was first obtained from opium in 1867 and the structure deduced in 1916 by W. H. Perkin, Jr.

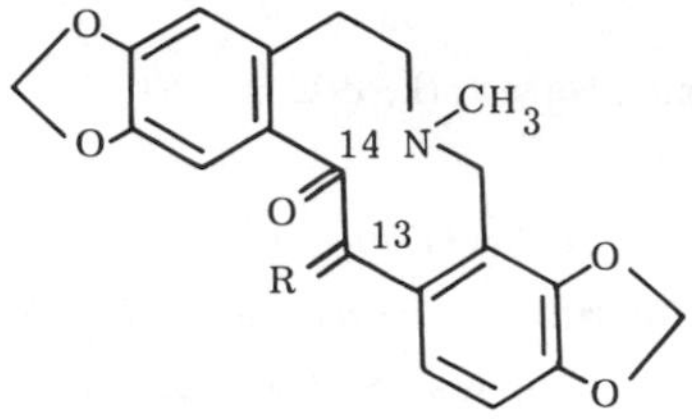

669   protopine, R=H$_2$

677   13-oxoprotopine, R=O

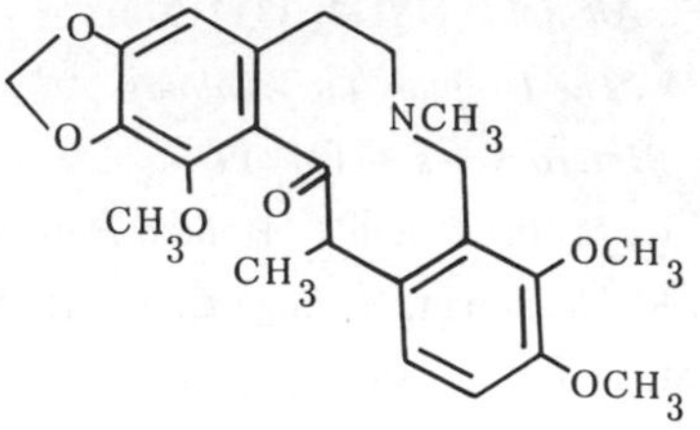

678   corycavidine

8.29.1   Chemistry and Synthesis

The relationship of the protopine to the protoberberine metho salts (**681**) with phosphorus oxychloride is shown in Scheme 84. An alternative route described by Dominquez *et al.* is the direct irradiation of a chloroform solution of cryptopine (**679**) to give the berberine species (**631**). Such interconversions are extremely useful in structure-elucidation studies.

A second interconversion of protopine (**669**) involves a transformation to sanguinarine (**682**), a benzophenanthridine alkaloid. Reaction of protopine

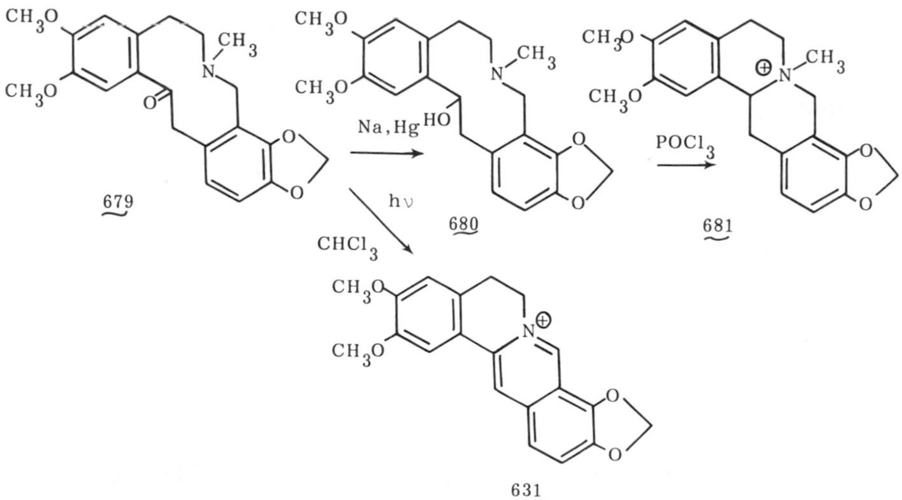

Scheme 84   Conversion of the protopine skeleton to the protoberberine skeleton.

with phosphorus oxychloride gave isoprotopine chloride (**683**), which on base treatment was converted to anhydroprotopine (**684**). These reactions were discovered by Perkin, but it was not until some 50 years later that Onda *et al.* found that irradiation of **684** gave an unstable base which could be hydrogenated to dihydrosanguinarine. Oxidation with dichlorodicyano-quinone then gave sanguinarine (**682**) (Scheme 85).

8.29.2   Spectral Properties

The protopines exhibit quite characteristic spectral properties. In the UV spectrum a maximum is observed at about 285–293 nm, with a shoulder at 232–240 nm [e.g. protopine shows λ_{max} 290 nm (4.00)]. The proton NMR values for protopine are shown in **685**. The 10-membered ring is inverting rapidly at room temperature giving rise to broad methylene signals, but cooling to $-45°C$ slows this process and well-defined peaks result. In acid the *N*-methyl singlet shifts downfield (to ~2.9 ppm) due to O-protonation and N-C-14 ring closure. The ^{13}C NMR assignments for protopine are shown in **686**.

In the mass spectrum, a retro Diels–Alder reaction gives rise to a fragment ion in which the positive charge is on the quinone methide fragment **687**, not the isoquinolone fragment **688**.

The protopines have thus far been of little interest pharmacologically. Protopine has weak uterine stimulant activity, slows the heart, and increases coronary flow, but few other effects have been observed.

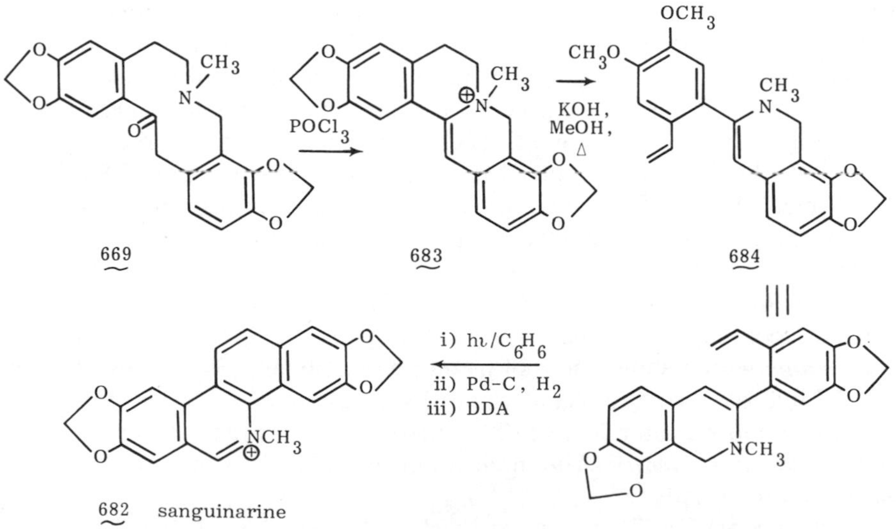

Scheme 85

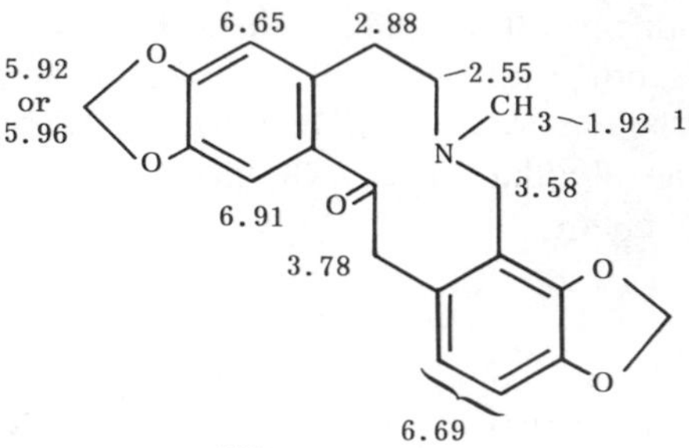

685

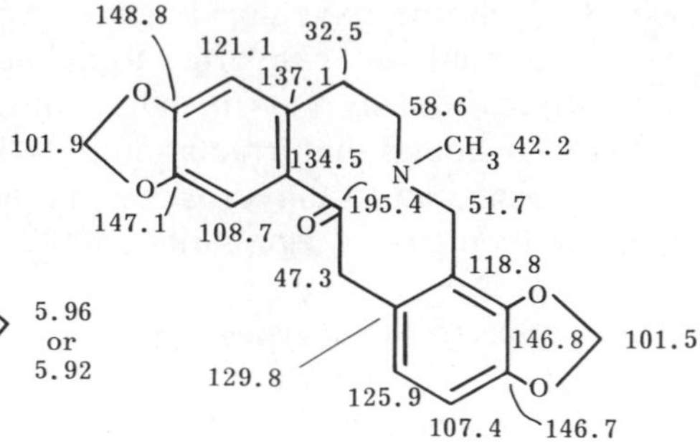

686

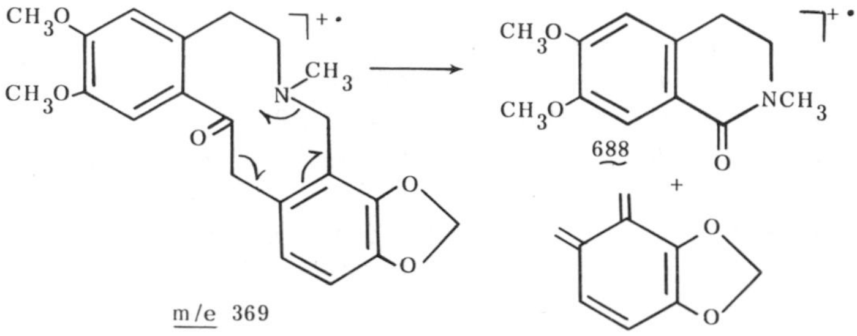

m/e 369

688

687
m/e 148

8.29.3   Biosynthesis

The biosynthesis of the protopine alkaloids has been studied concurrently with that of the tetrahydroprotoberberines and the benzophenanthridine alkaloids. The groups of Barton and Battersby have shown for example that (+)-reticuline (**182**) is a good precursor of protopine (**669**) in *Dicentra spectabilis* Lem. (Fumariaceae) and *Chelidonium majus* L. (Papaveraceae). The label from the *N*-methyl group of reticuline specifically enters C-8 of **669**. The *N*-methyl salts of reticuline are poor precursors of protopine (**669**).

A more definitive examination of the intermediacy of the tetrahydroprotoberberines in the formation of the protopines was reported in 1975 by Battersby and co-workers. (−)-(S)-Scoulerine (**654**) doubly labeled, but not the corresponding (+)-isomer, was efficiently (0.92%) incorporated into **669** in *C. majus* with complete loss of tritium originally at C-14. Methylation of the C-7 hydroxy group of scoulerine (**654**) gives isocorypalmine (**689**), and this compound was an effective (2.0% incorporation) precursor of allocryptopine (**690**) in *C. majus*. The implication (not tested) being that canadine (**630**) is an intermediate.

Further work by Takao and co-workers has confirmed that cheilanthifoline (**691**) is an intermediate between scoulerine (**654**) and stylopine (**666**), but in addition has revealed further subtleties in the latter stages of the

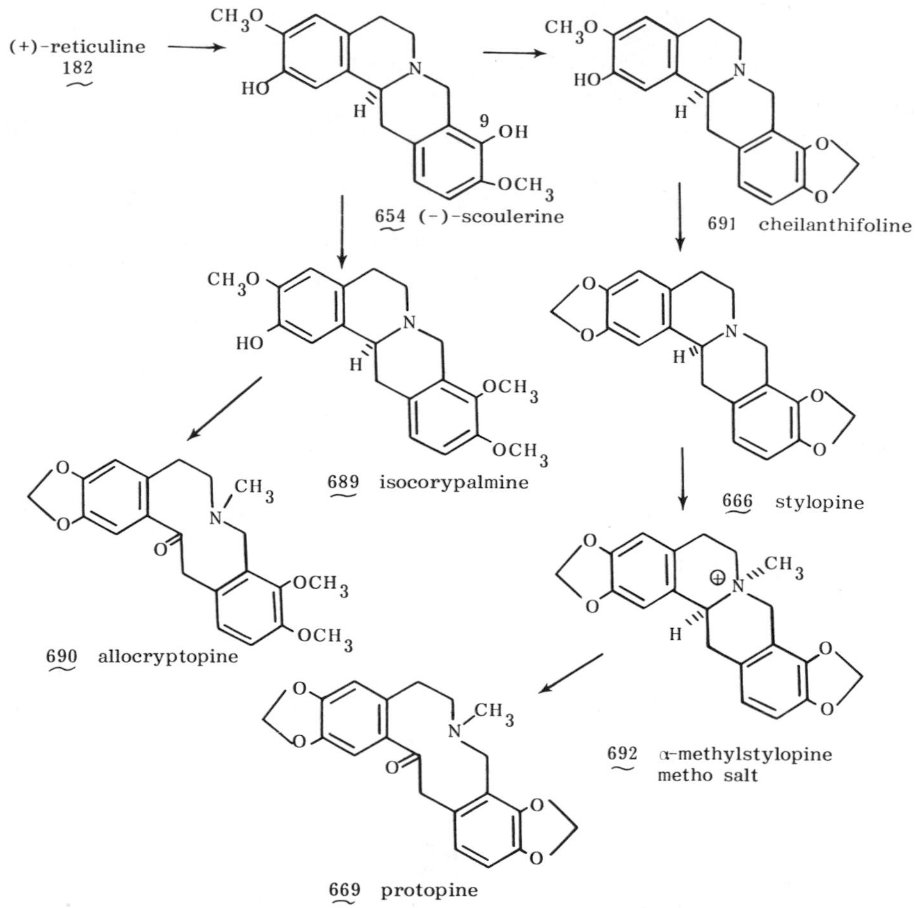

Scheme 86 Biosynthesis of protopine (**669**) and allocryptopine (**690**).

pathway. Thus protopine was formed stereospecifically from the α-metho salt of stylopine (**692**), but not the β-metho salt. This experiment was carried out using $^{13}CH_3$-labeled salts and observing the ^{13}C enrichment in the N-methyl group of **669**. The biosynthetic pathway for the formation of protopine (**669**) is shown in Scheme 86.

LITERATURE

Reviews

Shamma, M., *The Isoquinoline Alkaloids*, Academic, New York, 1972, p. 344.

Shamma, M. and J. L. Moniot, *Isoquinoline Alkaloid Research, 1972–1977*, Plenum, New York, 1978, p. 299.

Santavý, F., *Alkaloids NY*, **12**, 390 (1970).

Chemistry and Synthesis

Perkin, W. H., Jr., *J. Chem. Soc.* **109,** 815 (1916); **115,** 713 (1919).

Comin, J., and V. Delofeu, *Tetrahedron* **6,** 63 (1959).

Dominguez, X. A., J. G. Delgado, W. P. Reeves, and P. D. Garner, *Tetrahedron Lett.* 2493 (1967).

Perkin, W. H., Jr., *J. Chem. Soc.* **113,** 492 (1918).

Onda, M., K. Yonezawa, and K. Abe, *Chem. Pharm. Bull.* **17,** 2565 (1969).

Haworth, R. D., and W. H. Perkin, Jr., *J. Chem. Soc.* 1769 (1926).

Biosynthesis

Barton, D. H. R., R. H. Hesse, and G. W. Kirby, *Proc. Chem. Soc. London,* 267 (1963).

Battersby, A. R., R. J. Francis, E. A. Ruveda, and J. Staunton, *Chem. Commun.* 89 (1965).

Battersby, A. R., J. Staunton, H. R. Wiltshire, R. J. Francis, and R. Southgate, *J. Chem. Soc. Perkin Trans I.* 1147 (1975).

Takao, N., K. Iwasa, M. Kamigauchi, and M. Sugiura, *Chem. Pharm. Bull.* **24,** 2859 (1976).

8.30   RHOEADINE ALKALOIDS

With the exception of one isolation from *Bocconia frutescens*, the rhoeadines have been found only in the genus *Papaver*. There are about 30 alkaloids in the group and a typical member is rhoeadine (**670**), first obtained by Hesse

670   (+)-rhoeadine

from the red poppy *P. rhoeas* L. in 1885. Like all the other alkaloids in this series, it is dextrorotatory. The alkaloids may be detected by a quite characteristic red → brown → green color sequence with concentrated sulfuric acid.

8.30.1   Chemistry and Spectral Properties

Rhoeadine derivatives display two characteristic UV maxima, between 230 and 240 nm and between 284 and 294 nm, whose position depends on the aromatic substitution. Thus the dimethoxy compounds have a lower λ_{max} than the dimethylenedioxy derivatives such as **670**.

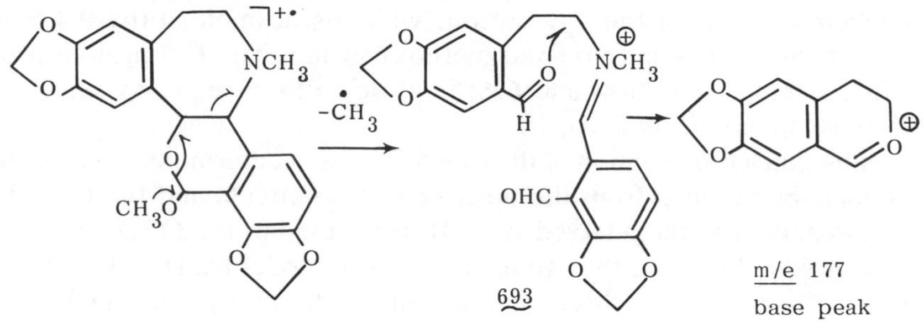

670   m/e 383   693   m/e 368   m/e 177 base peak

Scheme 87

The mass spectra of the rhoeadines having acetal group display two important fragment ions whose intensity does not change as the stereochemistry at the asymmetric centers change. This implies that the major fragmentation pathway proceeds through an achiral intermediate such as **693**, m/e 368 in Scheme 87.

The rhoeadines have three asymmetric centers, C-1, C-2, and C-14, and stereoisomers at C-1 and C-14 are known. Some of these isomers are interconvertible on acid treatment and their proton NMR spectra serve to distinguish between them. The compounds glaudine (**694**), epiglaudine (**695**), and oreodine (**696**) will be used to illustrate these reactions as discussed by

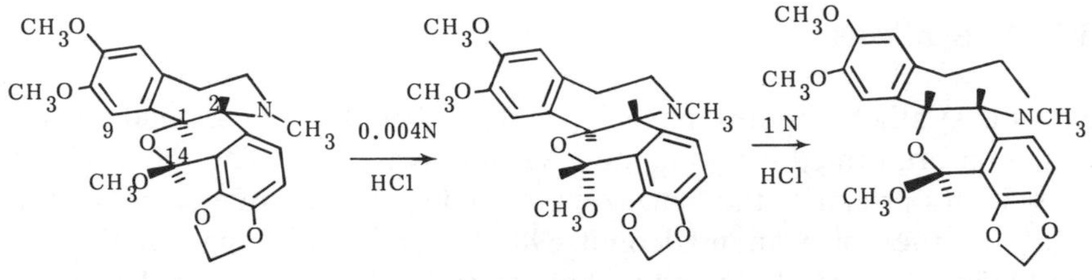

694   glaudine   695   epiglaudine   696   oreodine

Proton	δ ppm (J)		
H - 1	5.19, 9 Hz	5.57, 9 Hz	5.05, 2 Hz
H - 2	4.08, 9 Hz	4.02, 9 Hz	3.66, 2 Hz
H - 9	7.37	7.32	6.78
H - 14	5.77	5.75	5.75

Shamma and co-workers. Very dilute acid in methanol epimerizes C-14 and stronger acid epimerizes C-1 as shown. A *trans*-fused B/D ring junction results in a coupling constant of 9 Hz for the C-1 and C-2 protons. In the corresponding *cis*-fused isomer this *J* value is only 2 Hz. When the stereochemistry at C-14 is inverted as in **695** the most dramatic change is observed in the chemical shift of the C-1 proton, which is deshielded (by 0.4 ppm) because of the 1,3-diaxially related methoxy group. The C-9 proton in the *trans*-fused compounds (**694** and **695**) is close to the acetal oxygen and is therefore relatively deshielded.

The absolute configuration of the rhoeadines was determined by Shamma and Nakanishi's groups from the positive Cotton effects and the Davydov splitting observed in the allowed A → B transitions in the ORD spectra of these alkaloids. Applying the aromatic chirality model leads to the stereoformula **697** and therefore this absolute configuration for isorhoeadine. This

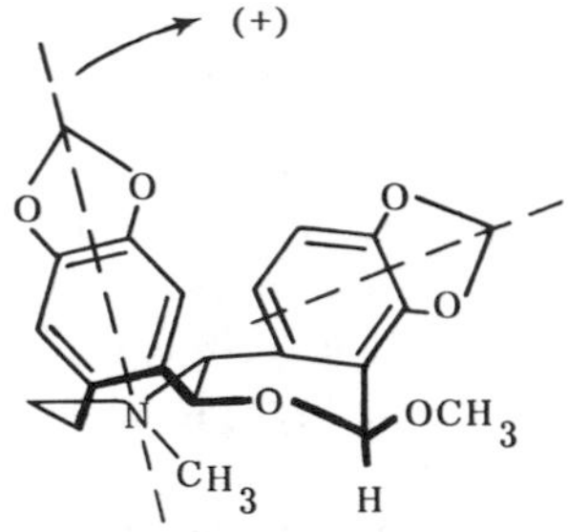

697   (+)-isorhoeadine

absolute configuration was confirmed by X-ray crystallographic analysis of rhoeagenine methiodide.

8.30.2   Synthesis

Several synthetic routes have been described for the rhoeadine nucleus based on preformed isoquinoline alkaloid precursors. The route of Irie and co-workers produces the benzazepine nucleus by rearrangement of the mesylate of the spiro aminoalcohol **698**. Oxidative cleavage of the styrene double bond in **699**, followed by borohydride reduction, afforded rhoeagenindiol (**700**) (Scheme 88).

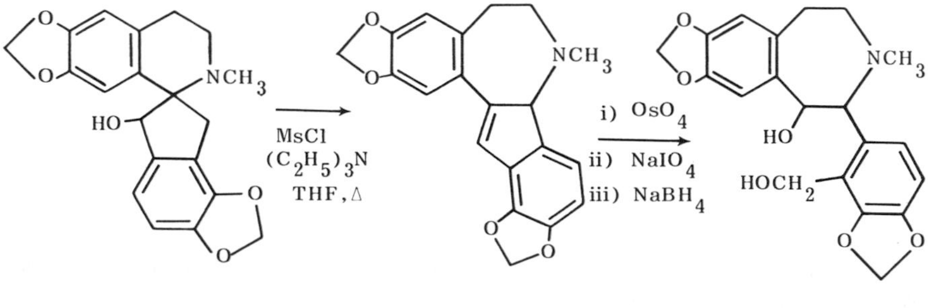

698   699   700 Rhoeagenindiol

Scheme 88   Irie and co-workers' approach to the rhoeadine nucleus.

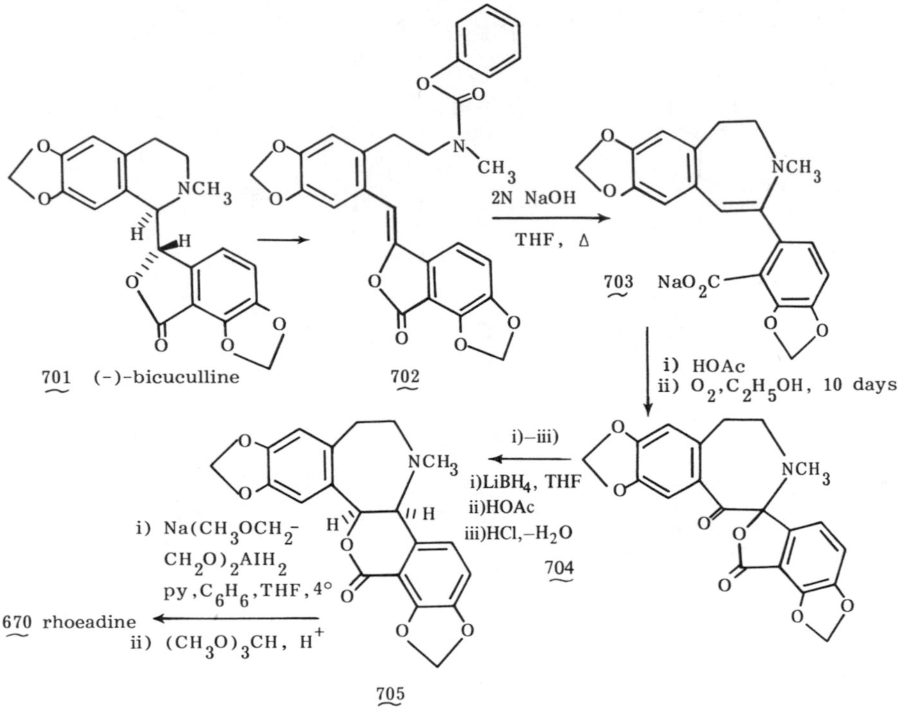

Scheme 89 Roche synthesis of rhoeadine (**670**).

A group at Hoffmann-La Roche chose a quite different approach (Scheme 89) in which the phthalideisoquinoline bicuculline (**701**) was converted into **702**. Heating with base gave the benzazepine acid **703**. The sodium salt of this acid was oxidatively cyclized to the ketospiro lactone **704**. Reduction with lithium borohydride, neutralization with acid, and dehydration of the intermediate hydroxy acid gave the lactone **705**, which was converted to rhoeadine (**670**) by standard reactions.

Prabhakar and co-workers have reported a novel photochemical route to *cis*-alpinigenine (**706**). Hofmann elimination of the methiodide of **707** afforded the *trans*-azocine **708**, which was converted to a 1,2-diol **709** and cleaved with periodate to the dialdehyde **710**. Photolysis of **710** afforded *cis*-alpinigenine (**706**) directly in 28% yield (Scheme 90).

8.30.3 Biosynthesis

The natural derivation of the rhoeadines has only recently been known in any detail. Early studies indicated that [3-^{14}C]tyrosine was utilized in the formation of each half of the molecule, but the specificity was not determined. The phthalideisoquinolines were proposed as precursors, possibly

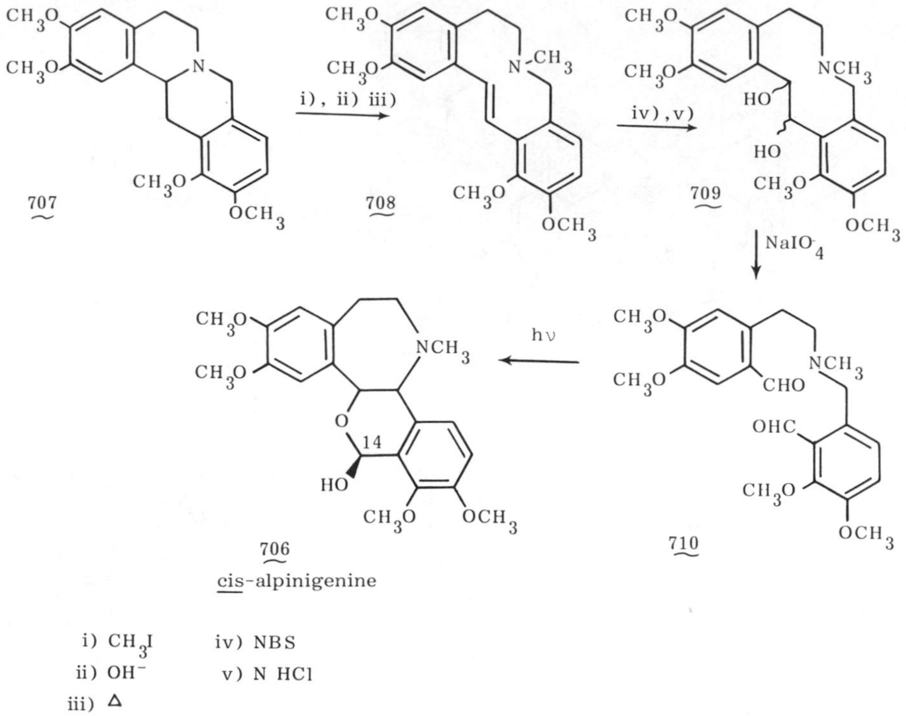

Scheme 90   Prabhakar synthesis of *cis*-alpinigenine (**706**).

in the form of their corresponding hemiacetal derivatives (e.g. **711**) (Scheme 91).

    Two groups have recently investigated the biosynthesis of the rhoead-ines. Tani and Tagahara have shown that stylopine α-metho salt (**692**) and protopine (**669**) are intermediates in the formation of rhoeadine (**670**) in *P. rhoeas*, and Ronsch has demonstrated the incorporation of the corresponding

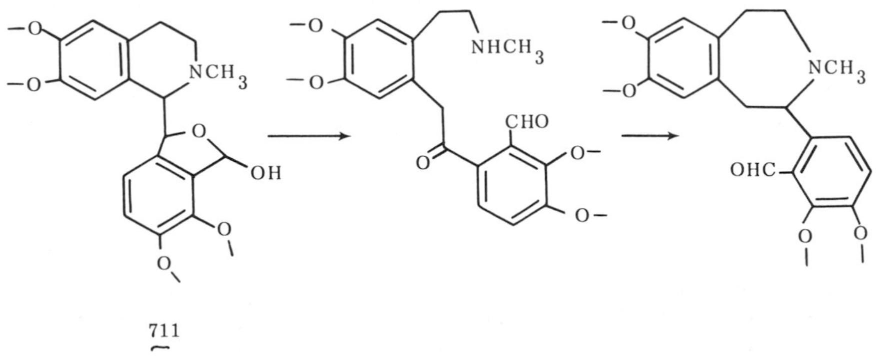

Scheme 91

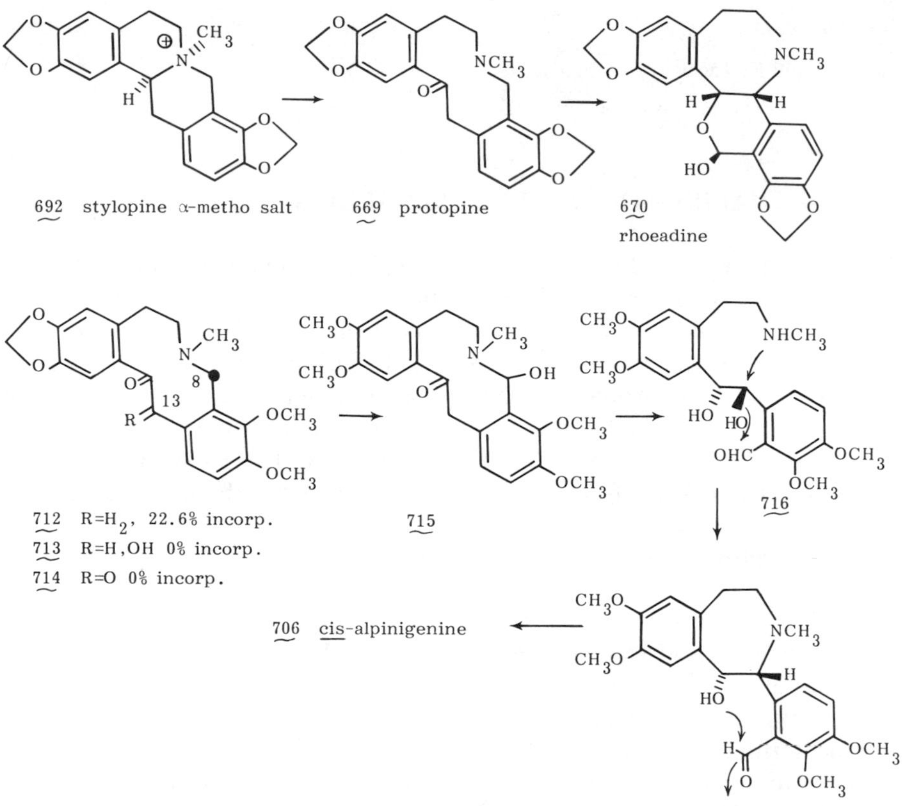

Scheme 92   Biogenesis of the rhoeadine alkaloids.

compound 712 in the tetramethoxy series into alpinigenine (706) in *P. brac-teatum*. However, neither the 13-hydroxy derivative 713 nor the 13-oxo derivative 714 were incorporated. Carbon 8 of 712 is specifically incorporated into C-14 of *cis*-alpinigenine (706), clearly demonstrating that cleavage between the N-methyl group and C-8 occurs during biosynthesis, probably by way of a carbinolamine intermediate 715. Stereospecific formation of the *trans*-dihydroxy aldehyde 716 followed by stepwise closure leads to 706 (Scheme 92).

LITERATURE

Santavý, F., *Alkaloids NY* 12, 398 (1970).

Shamma, M., J. A. Weiss, S. Pfeifer, and H. Dohnert, *Chem. Commun.* 212 (1968).

Shamma, M., J. L. Moniot, W. K. Chan, and K. Nakanishi, *Tetrahedron Lett.* 4207 (1971).

Irie, H., S. Tani, and H. Yamane, *Chem. Commun.* 1713 (1970).

Klotzer, W., S. Teitel, and A. Brossi, *Helv. Chim. Acta* 55, 2228 (1972).

Prabhakar, S., A. M. Lobo, and I. M. C. Olivera, *Chem. Commun.* 419 (1977).

Biosynthesis

Tani, C., and K. Tagahara, *Yakugaku Zasshi* **97,** 93 (1977).

Ronsch, H., *Phytochemistry* **16,** 691 (1977).

8.31   PHTHALIDEISOQUINOLINE ALKALOIDS

There are about 15 well-characterized phthalideisoquinoline alkaloids known based on the skeleton **717**, isolated mainly from the Papaveraceae and Berberidaceae. All the examples are oxygenated at C-6, C-7, C-4', and C-5', and some are also oxygenated at C-8. Carbon-1 of the tetrahydroiso-quinoline unit and carbon 9 of the phthalide moiety are the two points of asymmetry in these alkaloids, the best known alkaloids of which are (−)-α-narcotine (**672**) from *Papaver somniferum* L. (Papaveraceae) and (−)-β-hydrastine (**718**) from *Hydrastis canadensis* L. (Ranunculaceae).

8.31.1   Chemistry and Absolute Configuration

Narcotine is one of the major alkaloids of opium, *Papaver somniferum* (Papaveraceae), and was originally isolated by Derosne in 1803. The gross structure of narcotine was established by hydrolysis with dilute sulfuric acid to cotarnine (**719**) and opianic acid (**720**), and by zinc–acid reduction to hydrocotarnine (**721**) and meconine (**722**).

The stereochemistry of C-1 and C-9 of α-narcotine was established by degradation. Through a series of standard reactions α-narcotine was reduced to the benzylisoquinoline **723**, which showed a positive Cotton effect around 295 nm, indicating that the C-1 proton must be α.

The stereochemistry at C-9 is represented by the ORD of two absorptions between 320 and 335 nm and 235 and 255 nm. They are positive for the 9R-and negative for the 9S-configurations. A second Cotton effect between 280 and 300 nm reflects the configuration at C-1, which is positive for the 1R-and negative for the 1S-configuration. When C-1 and C-9 are of opposite signs, as is frequently the case, this second effect may not be observable. Most frequently though the C-1 stereochemistry controls the overall specific rotation of the compound and is 1R for levoratatory compounds. It is no-teworthy that the 1R, 9R-isomer of narcotine, so-called β-narcotine, is not a natural product.

8.31.2   Synthesis

A quite efficient synthesis of these alkaloids is a procedure developed by Hope and Robinson in 1914, in which cotarnine (**719**) is condensed with iodomeconine (**724**) and the product reduced with sodium amalgam to give racemic α-narcotine (**672**).

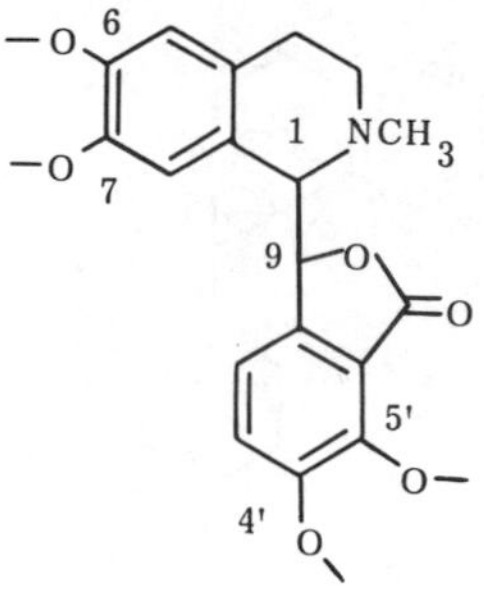

717

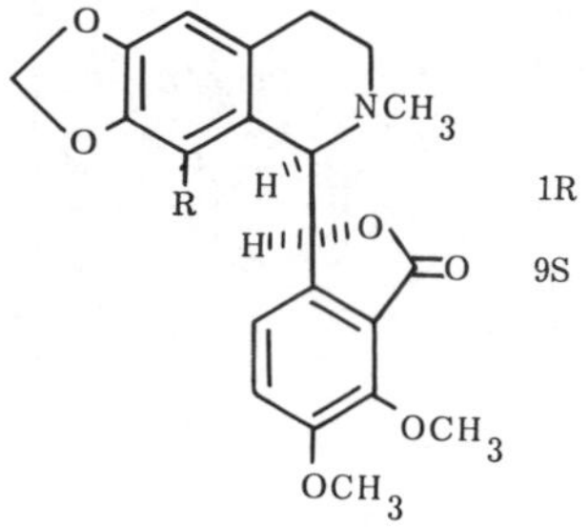

672 (−)-α-narcotine, R=OCH₃
718 (−)-β-hydrastine, R=H

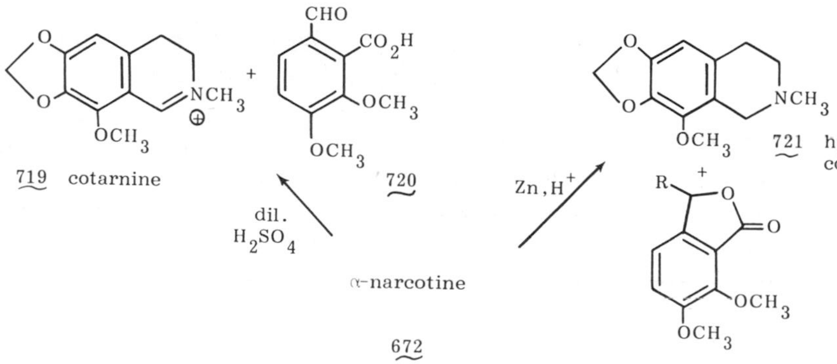

719 cotarnine

720

α-narcotine

672

721 hydro-cotarnine

722 meconine R=H
724 R=I

672
α-narcotine

i) LiAlH₄
ii) Ac₂O,py
iii) H₂, Pd–C,Δ

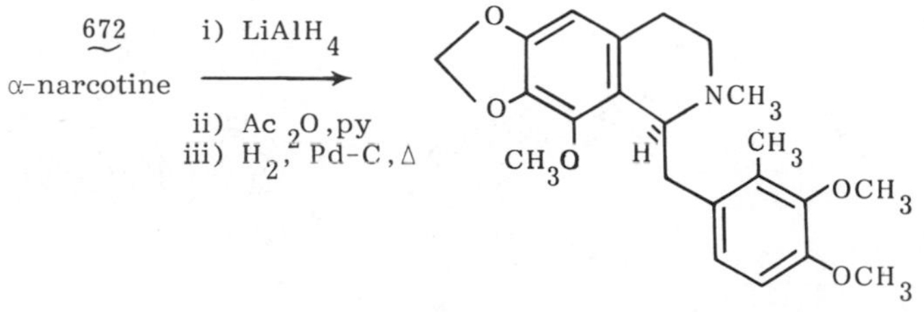

723 +ve Cotton effect

A completely different total synthetic approach has been reported by Kametani and workers involving carbene addition into the iminium bond of **725**. The reaction is considered to proceed through the aziridinium species **726**.

In 1910, Perkin and Robinson suggested that hydrastine could be formed in the plant from berberine. It was 66 years before Moniot and Shamma realized this conversion in the laboratory. Ferricyanide oxidation of ber-

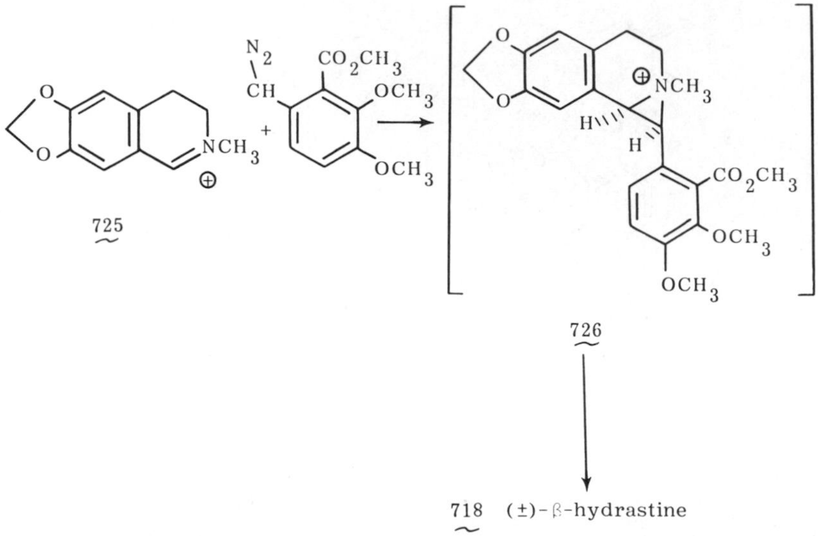

725

726

718   (±)-β-hydrastine

berine (631) gives a dimer of unknown structure which can be fragmented with methanolic hydrogen chloride to berberine chloride and δ-methoxy-berberine phenol betaine (727). The latter compound could be hydrated and N-alkylated to give 728. Sodium borohydride reduction then afforded a 1:2 mixture of racemic α-hydrastine (729) and β-hydrastine (718) respectively (Scheme 93).

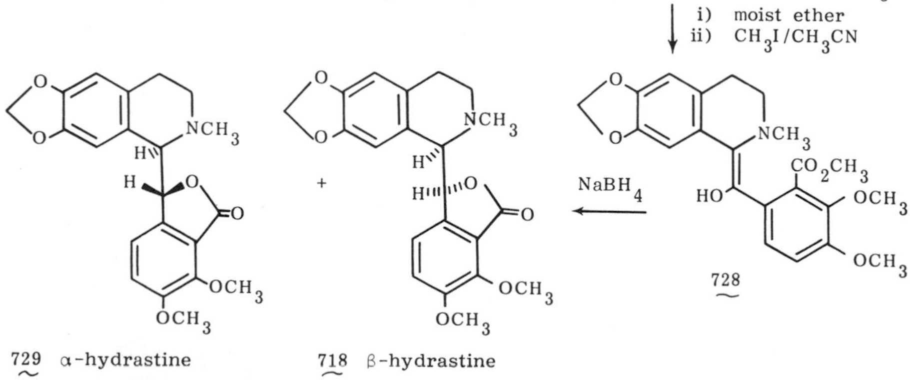

631

727

728

729 α-hydrastine

718 β-hydrastine

Scheme 93   Moniot–Shamma biogenetic transformation of berberine (631) to β-hydrastine (718).

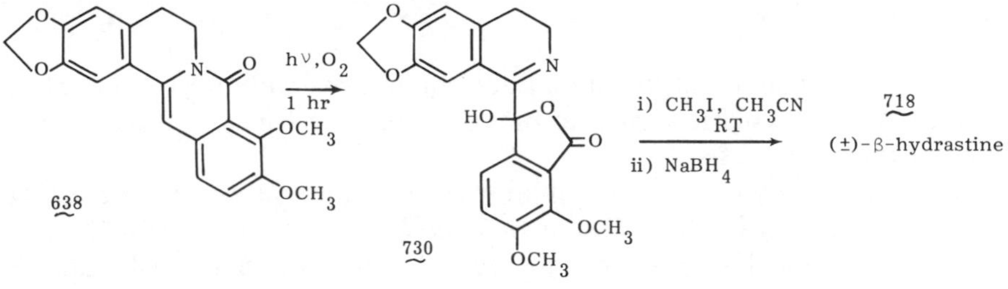

Scheme 94 Shamma synthesis of β-hydrastine (**718**).

A second partially synthetic route from the berberine skeleton reported by Shamma and co-workers involves short-term photoxidation of oxyberberine (**638**) to give the γ-lactol (**730**). N-Methylation with methyl iodide in acetonitrile followed by stereospecific sodium borohydride reduction gave racemic β-hydrastine (**718**) in 95% yield from **730** (Scheme 94).

8.31.3 Spectral Properties

Narcotine (**672**) shows a UV spectrum with maxima at 209, 291, and 309 nm (log ϵ 4.06, 3.60, and 3.69) and bicuculline (**701**) shows maxima at 225, 296, and 324 nm (log ϵ 4.57, 3.81, and 3.77).

Typical values for the NMR spectra of cordrastine I (**731**) and cordrastine II (**732**) are shown in Figure 17. Note how the coupling constant between H-1 and H-9 is independent of their relative stereochemistries.

Benzylic cleavage is a characteristic feature of the mass spectral fragmentation of phthalideisoquinolines to give an ion derived from rings A and B.

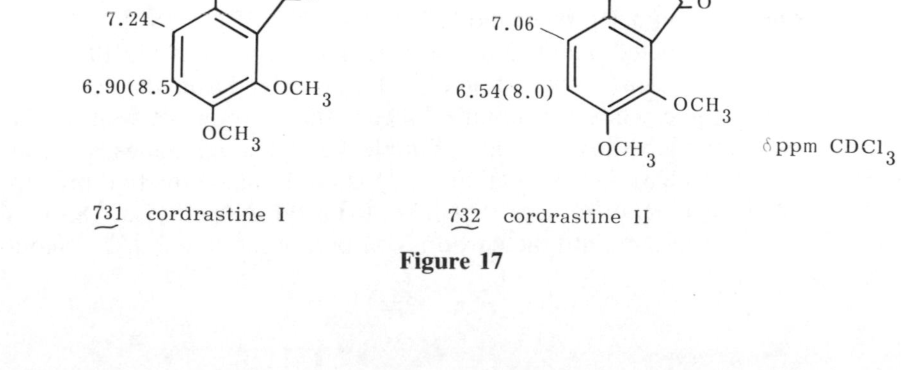

731 cordrastine I

732 cordrastine II

Figure 17

8.31.4　Biosynthesis

Robinson postulated that the phthalideisoquinolines were formed in nature by the oxidative modification of tetrahydroprotoberberines, and recent data support this hypothesis.

Complementary results were obtained when labeled tyrosines were used as precursors of hydrastine (**718**) in *H. canadensis* and narcotine (**672**) in *P. somniferum*. Thus [2-^{14}]tyrosine labeled C-1 and C-3 of **672** equally and [3-

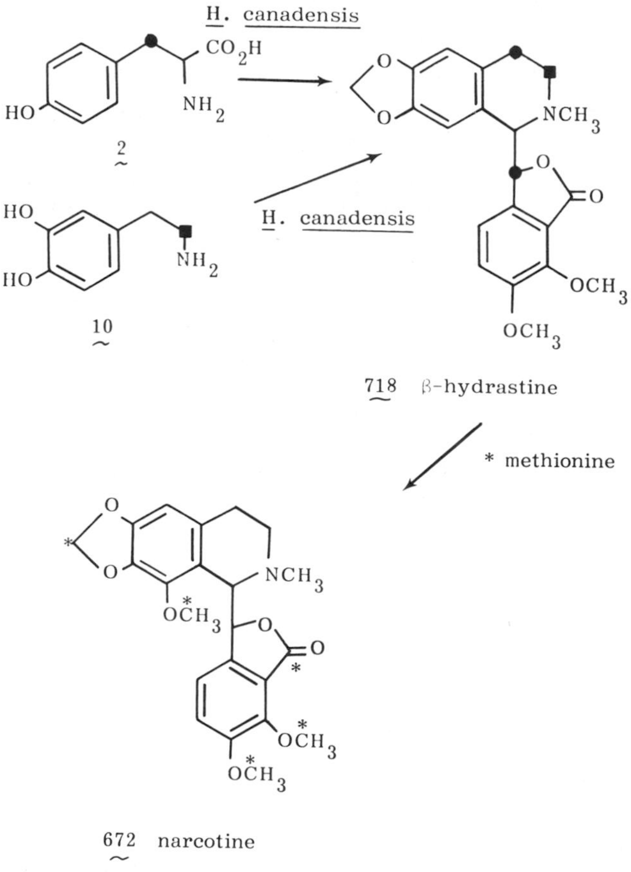

718　β-hydrastine

* methionine

672　narcotine

^{14}C]tyrosine was similarly incorporated into each half of **718**. [1-^{14}C]Dopamine however labeled only C-3 of the isoquinoline nucleus of **718**.

Evidence that the carbonyl carbon of **672** was derived from the berberine bridge carbon came when methionine labeled this carbon, as well as the methyl and methylenedioxy groups. Simple benzylisoquinolines such as norlaudanosoline (**228**) and (+)-reticuline (**182**) were also effective precursors, and it is significant to note that label from the *N*-methyl of **182** was specifically incorporated into the carbonyl carbon of narcotine (**672**). Scoulerine (654) was incorporated to the extent of 4.0% and isocorypalmine (689) 4.6% into 672. In contrast, nandinine (733) was not well incorporated, indicating the probable sequence of introduction of the two, *O*-methyl groups and formation of the methylenedioxy group. Canadine (630) was also well incorporated into narcotine (672) suggesting that it might be the next intermediate. The only additional information which is available is that the *pro*-R hydrogen at C-13 of scoulerine (654) is retained in the elaboration to 672,

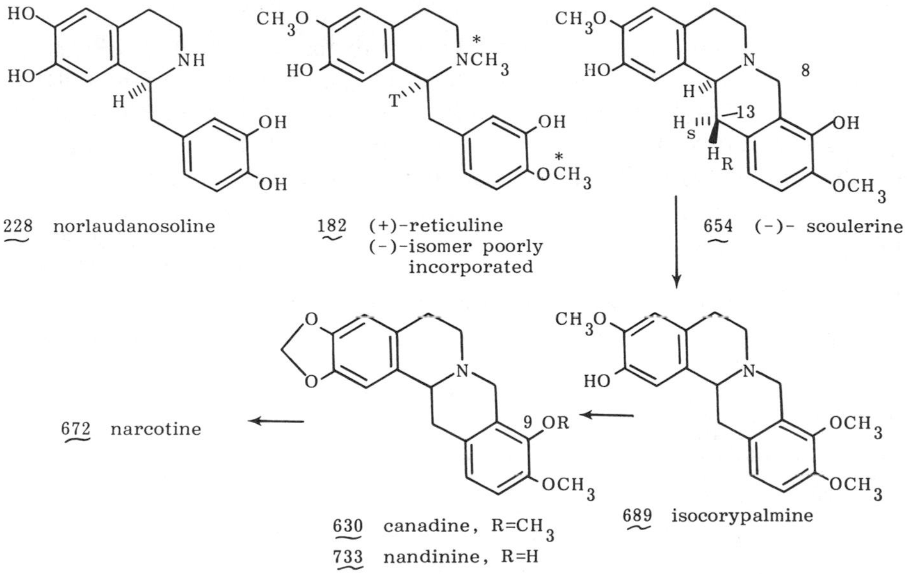

228   norlaudanosoline

182   (+)-reticuline
      (-)-isomer poorly
      incorporated

654   (-)- scoulerine

672   narcotine

630   canadine, R=CH₃
733   nandinine, R=H

689   isocorypalmine

Scheme 95   Biosynthesis of narcotine (672) in *P. somniferum*.

indicating that hydroxylation occurs with retention of configuration (Scheme 95). It would be interesting to know if in the formation of an alkaloid such as (+)-bicuculline (734), (−)-reticuline is a preferred precursor and that the *pro*-S hydrogen at C-13 of (+)-scoulerine is retained.

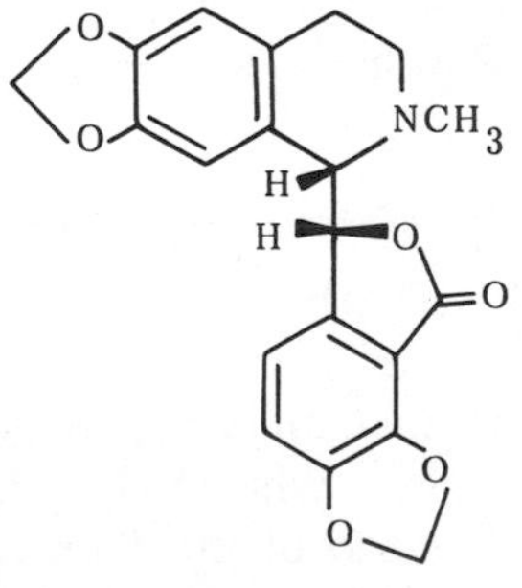

734   (+)-bicuculline

8.31.5 Pharmacology

Narcotine (672) is a mild antitussive and has a relaxant effect on smooth muscles. β-Narcotine N-oxide is a more effective antitussive than dihydrocodeine. Narcotine potentiates the mitotic effects of colchicine, although it has no action alone.

Hydrastine (718), like berberine, is used as an astringent in the inflammation of mucous membranes. It has also been used in treatment of problematic uterine hemorrhages.

Bicuculline (701) increases arterial pressure and the amplitude of cardiac contractions in the cat. The most interesting action of bicuculline (701) is its ability to antagonize the transmitter-inhibiting effect of γ-amino butyric acid in the cortex and other parts of the spinal cord. Bicuculline is also a potent competitive inhibitor of rat brain acetyl cholinesterase.

LITERATURE

Shamma, M., *The Isoquinoline Alkaloids*, Academic, New York, 1972, p. 359.

Shamma, M. and J. L. Moniot, *Isoquinoline Alkaloid Research, 1972–1977*, Plenum, New York, 1978, p. 307.

Synthesis

Hope, E., and R. Robinson, *J. Chem. Soc.* **105**, 2085 (1914).

Kametani, T., T. Honda, H. Inoue, and K. Fukumoto, *Heterocycles* **3**, 1091 (1975).

Moniot, J. L., and M. Shamma, *J. Amer. Chem. Soc.* **98**, 6714 (1976).

Shamma, M., D. Hindenlang, T.-T. Wu, and J. L. Moniot, *Tetrahedron Lett.* 6419 (1977).

Biosynthesis

Spenser, I. D., and J. R. Gear, *J. Amer. Chem. Soc.* **84**, 1059 (1962).

Battersby, A. R., and M. Hirst, *Tetrahedron Lett.* 669 (1965).

Battersby, A. R., R. J. Francis, M. Hirst, R. Southgate, and J. Staunton, *Chem. Commun.* 602 (1967).

Battersby, A. R., M. Hirst, D. J. McCaldin, R. Southgate, and J. Staunton, *J. Chem. Soc. Sec.* 2163 (1968).

Battersby, A. R., J. Staunton, H. R. Wiltshire, R. J. Francis, and R. Southgate, *J. Chem. Soc. Perkin Trans. I* 1147 (1975).

8.32 OCHOTENSANE ALKALOIDS

The ochotensane (spirobenzylisoquinoline) alkaloids are a small group of compounds, about 25 in number, which have thus far only been found in the genera *Corydalis* and *Fumaria* of the Papaveraceae. The alkaloids are characterized by the spiro linkage of tetrahydroisoquinoline unit with an indane. Two fundamental structure types have been reported, and are rep-

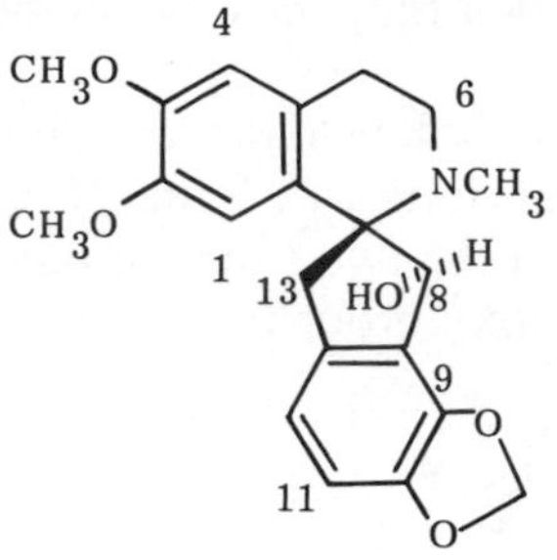

735   (-)-fumaricine 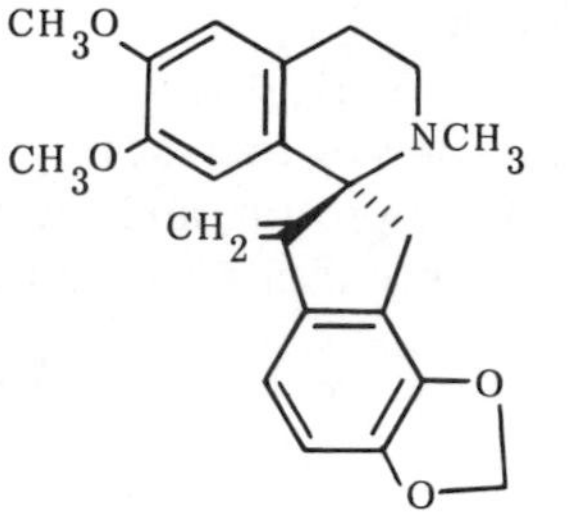   673   (+)-ochotensimine

resented by (−)-fumaricine (**735**) on one hand and (+)-ochotensimine (**673**) on the other.

8.32.1   Chemistry and Spectral Properties

Manske in 1938 had obtained three alkaloids from *Fumaria officinalis* L. and in 1940 had reported the isolation of two alkaloids from *Corydalis ochotensis* Turcz. The structures of these alkaloids were determined some 25 or more years later by the application of spectroscopic techniques.

In the proton NMR spectrum of fumariline (**736**), the substitutents and nature of the aromatic substituents as well as an N-methyl group were readily defined, together with an AB quartet for an isolated benzylic methylene. At this point two structures, **736** and **737**, could be proposed for fumariline. A

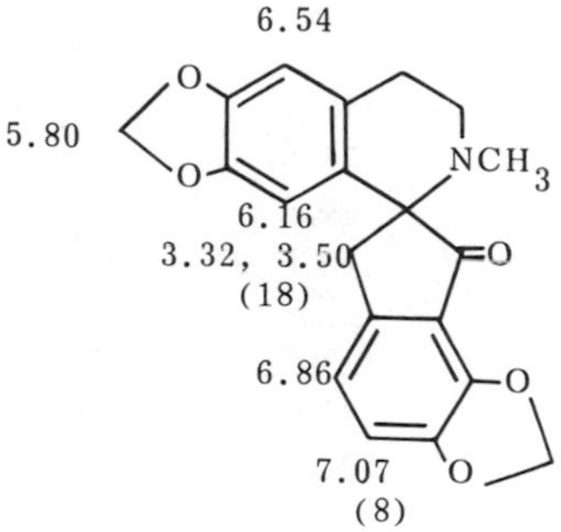

736 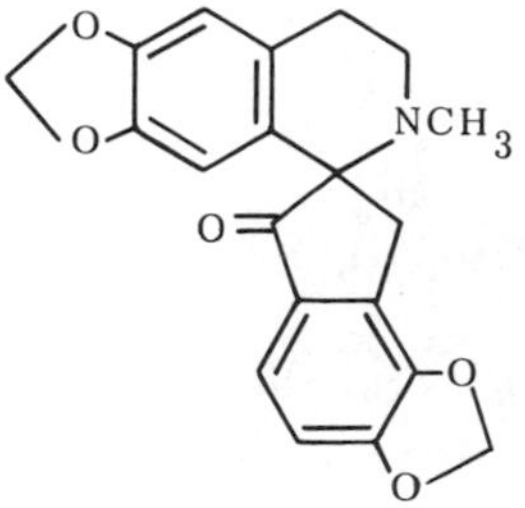   737

distinction between these was also made from the NMR spectrum. One observation was that the C-13 protons were further split by adjacent aromatic protons, and this proximity was confirmed when irradiation of the C-13 methylene caused a 19% NOE for the C-12 aromatic proton.

The NOE also proved useful in defining the stereochemistry at C-8 of fumaricine (**735**), for irradiation of the N-methyl group caused a 14%

enhancement in the C-8 singlet, showing that they must be on the same side of the molecule.

The proton NMR spectral data for ochotensimine are shown in **738**. Irradiation of one of the exomethylene protons was found to cause a 24% NOE in the C-13 proton, and irradiation of the two singlet aromatic protons indicated their proximity to the two methoxy groups.

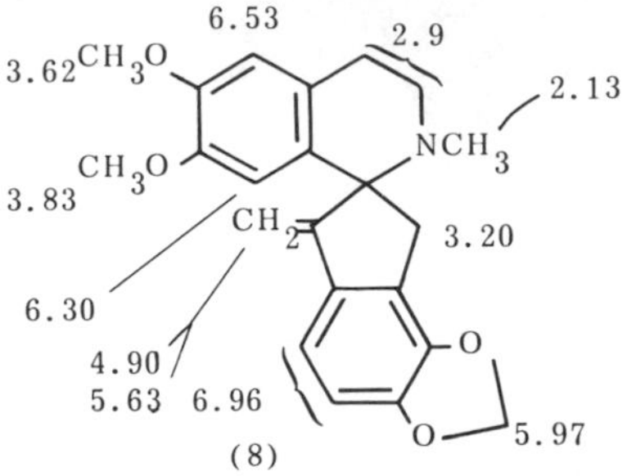

738

The ^{13}C NMR data for some representative alkaloids are shown in Table 11. The chemical shift of the C-8 resonance was found to be diagnostic of the stereochemistry of a hydroxy group at this position.

Table 11 ^{13}C NMR Data of Some Representative Ochotensane Alkaloids

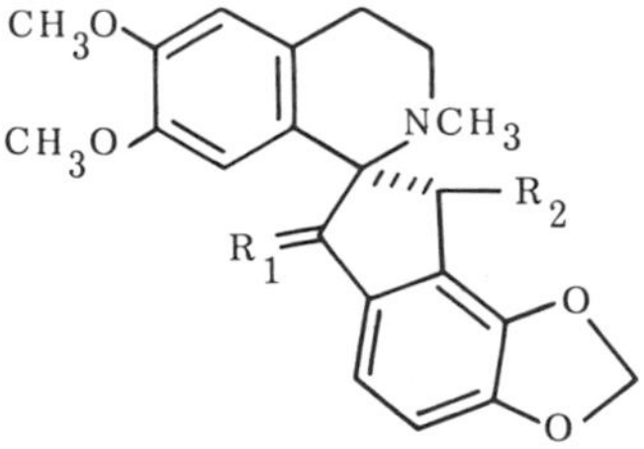

	673	**739**	**740**
1	110.5	110.7	110.7
2	147.5	147.2	148.5
3	147.7	148.9	148.6
4	110.5	112.5	111.4
4a	126.1	124.0	128.7
5	29.1	28.5	29.3
6	48.1	48.9	50.3
8	37.0	70.1	75.1
8a	123.8	132.9	134.6
9	143.0	145.0	144.4
10	148.2	154.5	154.6
11	108.0	110.4	109.5
12	113.6	119.5	119.6
12a	136.1	132.5	131.3
13	155.5	201.7	202.7
14	71.9	76.9	72.0
14a	137.2	129.7	128.7
NCH$_3$	39.0	39.7	41.9
—OCH$_2$O—	101.3	103.1	103.2
OCH$_3$	55.8, 56.1	56.0, 56.1	56.1, 56.5
13=CH$_2$	106.7		

	R$_1$	R$_2$
673 ochotensimine	CH$_2$	H
739 Raddeanone	O	β-OH
740 yenhusomidine	O	α-OH

Source: Data from D. W. Hughes *et al., Can. J. Chem.* **55**, 3304 (1977).

8.32.2 Synthesis and Biogenesis

There has been quite considerable synthetic interest in the ochotensane alkaloids and several total syntheses have been reported. McLean and co-workers investigated the condensation of indan-1,2-diones with phenethyl-amines and found that as expected the correct spiro linkage resulted. However, when condensation of 3,4-dimethoxyphenethylamine (27) and the indanone 741 was attempted, no reaction was observed, but the more nucleophilic 3-hydroxy-4-methoxyphenethylamine (23) did condense under carefully controlled conditions of acidity to give 742. Successive O- and N-methylation followed by Wittig reaction gave racemic ochotensimine (673).

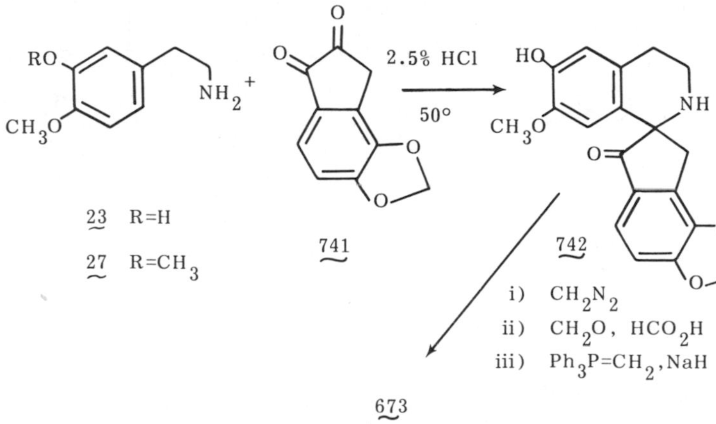

The major problem with this approach is the complex route needed to produce the indandione derivative.

Kametani and co-workers have also found that a remarkably efficient rearrangement of the benzocyclobutane derivative 743 occurs when the free base is allowed to stand in chloroform; the spiroindanone 744 is produced in 64% yield.

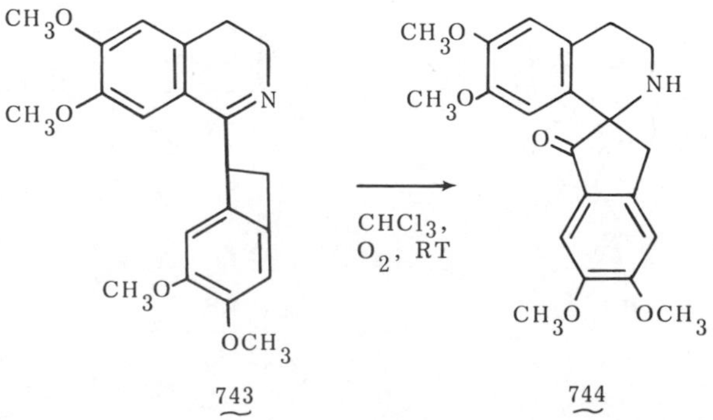

Turning to biogenetic-type synthesis it was suspected that at least the C-13 methylene derivatives of the ochotensane skeleton arise by rearrangement of the corresponding metho salts of a tetrahydro berberine as shown

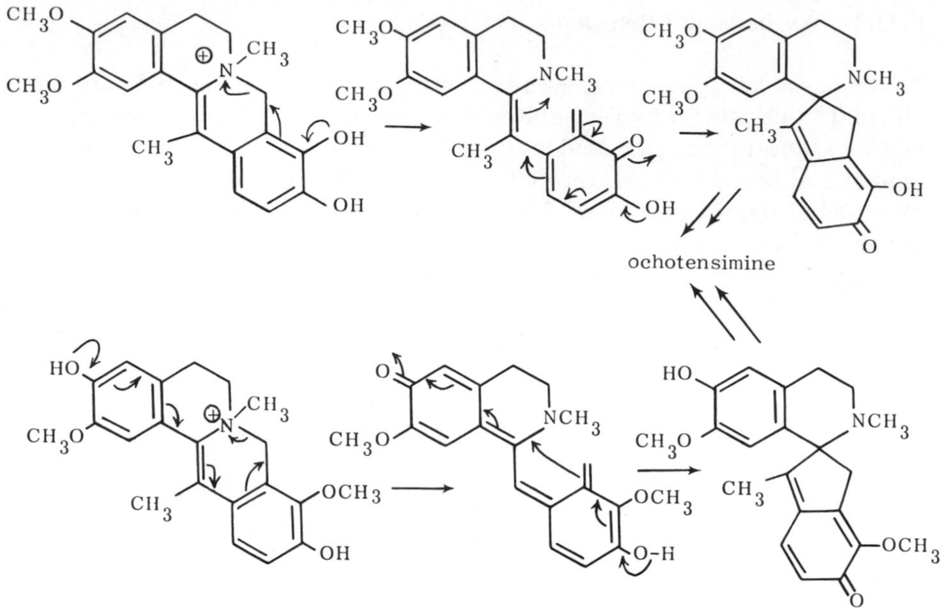

ochotensimine

Scheme 96　Biogenesis of ochotensane skeleton.

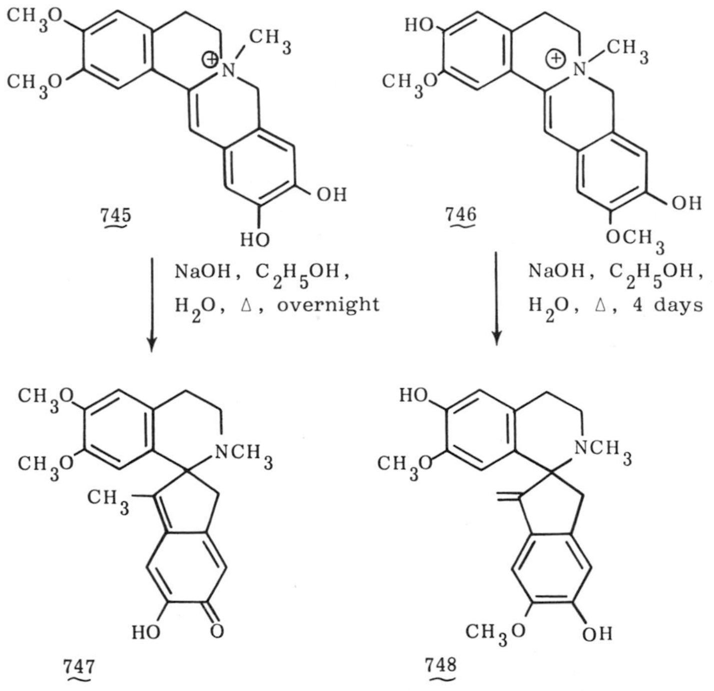

745

746

NaOH, C_2H_5OH, H_2O, Δ, overnight

NaOH, C_2H_5OH, H_2O, Δ, 4 days

747

748

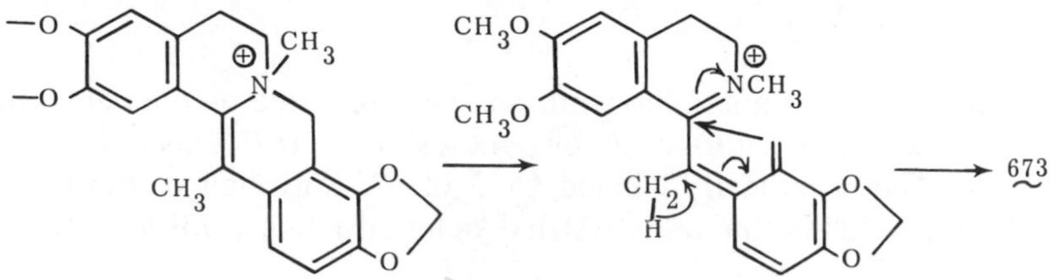

Scheme 97

in Scheme 96. An alternative hypothesis postulates the involvement of a different diphenolic protoberberine species.

In separate model experiments Shamma and co-workers demonstrated that either route could be operating. Thus treatment of the diphenol metho salts **745** and **746** afforded the corresponding ochotensane derivatives **747** and **748**. Surprisingly, the former compound preferred to exist as the quinone methide tautomer, rather than the corresponding diphenol. A phenol is not necessary for this reaction, as shown in Scheme 97.

Perhaps the most elegant *in vitro* entry into the ochotensane skeleton has been described by Manske and co-workers and involves the photolytic rearrangement of the 13-oxotetrahydroberberine metho salt **749** under basic conditions to afford **750**.

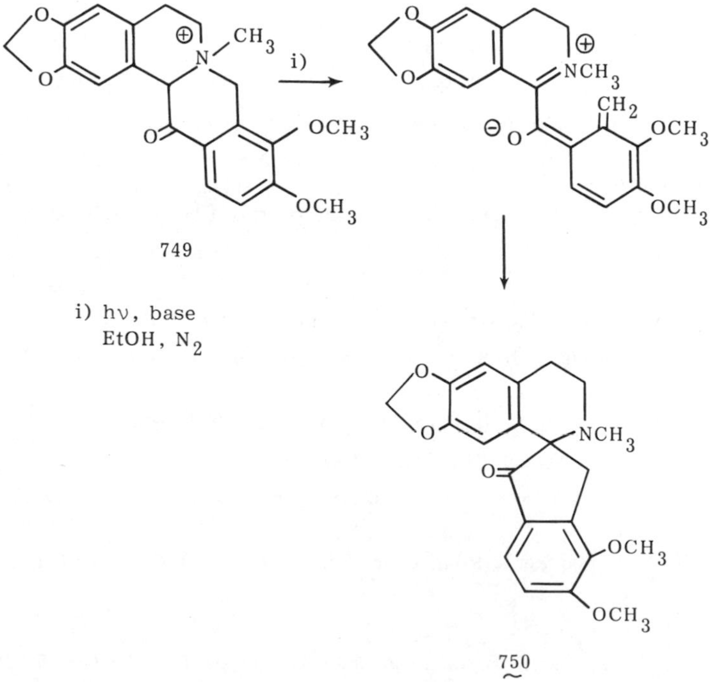

749

i) hν, base
   EtOH, N₂

750

8.32.3 Biosynthesis

There has been very limited biosynthetic work reported on the ochotensane alkaloids, and only the formation of ochotensimine (**673**) has been studied. [3-^{14}C]Tyrosine specifically labeled C-13 of **673** with half of the total ^{14}C label. [methyl-^{14}C]Methionine afforded ochotensimine (**673**) labeled at the

* from [methyl-^{14}C]-methionine

• from [3-^{14}C]-tyrosine

673 ochotensimine

C-13 methylene carbon and at C-8 (apart from the methylenedioxy and *N*- and *O*-methyl groups). This was cited as being good evidence for the intermediacy of the protoberberine skeleton in the biosynthesis.

LITERATURE

Shamma, M., *Alkaloids* NY **13**, 165 (1971).

Shamma, M., *The Isoquinoline Alkaloids,* Academic, New York, 1972, p. 380.

Shamma, M. and J. L. Moniot, *Isoquinoline Alkaloid Research, 1972–1977,* Plenum, New York, 1978, p. 325.

McLean, S., and J. Whelan, in K. F. Wiesner (Ed.), *MTP International Review of Science, Organic Chemistry,* Ser. 1, Vol. 9, *Alkaloids,* Butterworths, London, 1973, p. 161.

Synthesis

McLean, S., M.-S. Lin, and J. Whelan, *Can. J. Chem.* **48,** 948 (1970).

Kametani, T., H. Takeda, Y. Hirai, F. Satoh, and K. Fukumoto, *J. Chem. Soc. Perkin. Trans. I* 2141 (1974).

Shamma, M., and C. D. Jones, *J. Amer. Chem. Soc.* **91,** 4009 (1969).

Shamma, M., and J. F. Nugent, *Tetrahedron Lett.* 2625 (1970).

Holland, H. L., D. B. MacLean, R. G. A. Rodrigo, and R. H. F. Manske, *Tetrahedron Lett.* 4323 (1975).

Nalliah, B., R. H. F. Manske, R. Rodrigo, and D. B. MacLean, *Tetrahedron Lett.* 2795 (1973).

Biosynthesis

Holland, H. L., M. Castillo, D. B. MacLean, and I. D. Spenser, *Can. J. Chem.* **52,** 2818 (1974).

8.33 BENZO[c]PHENANTHRIDINE ALKALOIDS

The benzo[c]phenanthridine alkaloids, although known for many years, have recently been the subject of increasing interest because two members of this series show quite marked antileukemic properties.

Their distribution is limited to the Fumariaceae, Papaveraceae, and Rutaceae thus far, and they often cooccur with several other types (e.g. furoquinolines and β-indoloquinazolines). Many of these alkaloids are not basic, but are quaternary salts and can be isolated as their chlorides.

8.33.1 Chemistry and Spectral Properties

There are basically three groups of this structure type known, and two oxygenation patterns. One group is typified by alkaloids such as chelerythrine (751) and nitidine (752). These alkaloids are normally fully aromatic,

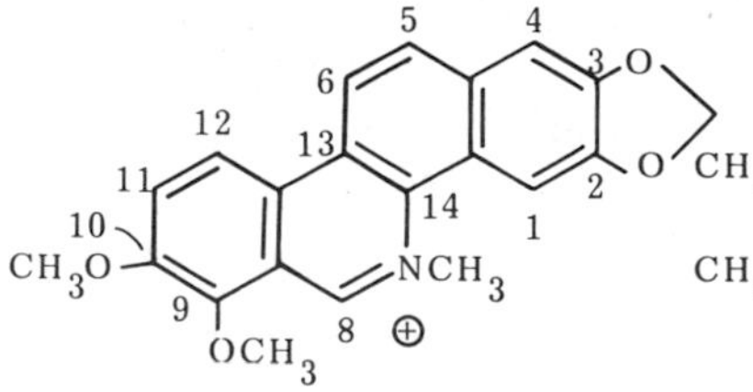

751 chelerythrine

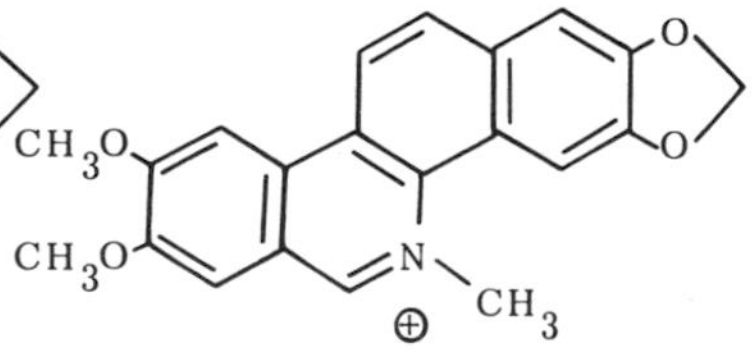

752 nitidine

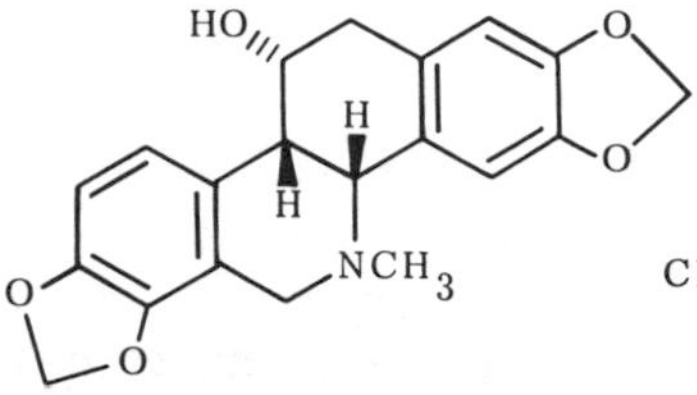

671 chelidonine

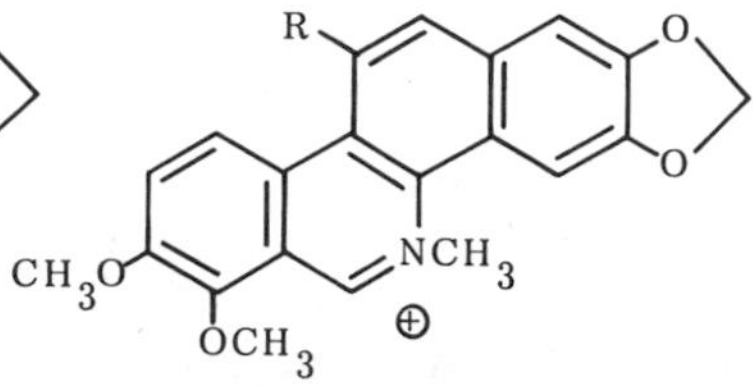

753 chelirubine, R=OCH$_3$
754 sanguinarine, R=H

although 7,8-dihydro species have been isolated. The second group are basic, have an alcohol function at C-6, and are not highly unsaturated.

The quaternary ammonium center of compounds such as sanguinarine (754) undergoes facile nucleophilic attack at C-8 even with solvents such as methanol or ethanol to give 8-alkoxydihydrosanguinarine (755).

Perhaps the most surprising alkaloids of this type are the dimeric species such as chelidimerine (756) isolated by Farnsworth and co-workers from *Chelidonium majus* and *Sanguinaria canadensis*. The structure was confirmed by synthesis when treatment of sanguinarine with acetone dicarboxylic acid gave chelidimerine (756).

754 sanguinarine $\xrightarrow[\text{C}_2\text{H}_5\text{OH}]{\text{CH}_3\text{OH}\ \text{or}}$

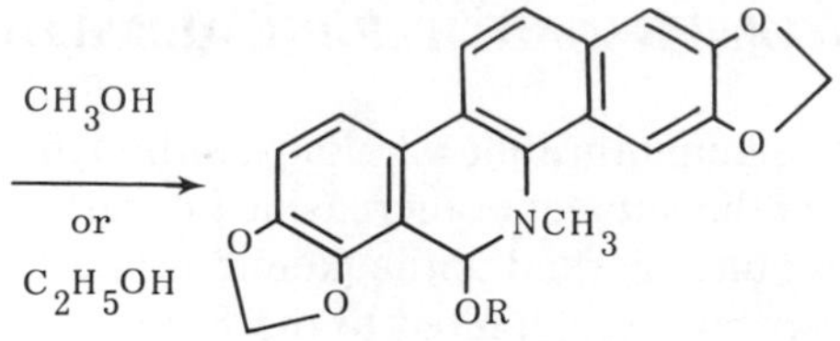

755   R=CH$_3$ or C$_2$H$_5$

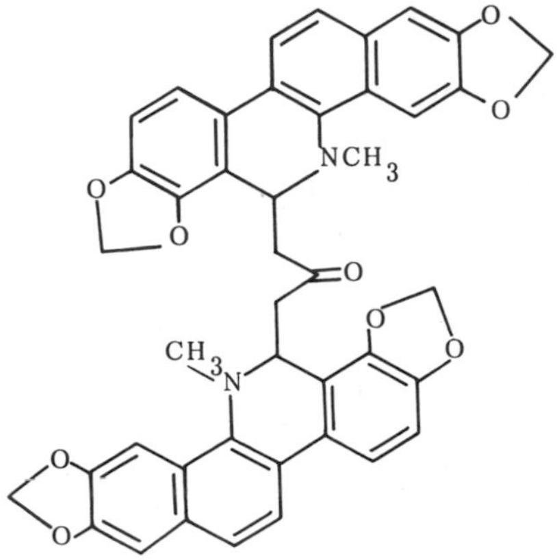

756   chelidimerine

Fagaronine was obtained as the chloride salt from the roots of *Fagara zanthoxyloides* (Rutaceae), and the UV spectrum (λ_{max} 346 nm) confirmed that fagaronine was a benzophenanthridine alkaloid containing a phenolic group. The NMR spectrum showed a quaternary *N*-methyl group (δ 5.11 ppm) and three aromatic methoxy groups (4.24, 4.11, and 4.04 ppm). The four *para*-related aromatic protons appeared at 7.66, 7.94, 8.13, and 8.36 ppm, and the two *ortho*-related protons as doublets ($J = 9$ Hz) at 8.86 and 8.16 ppm together with the iminium proton singlet at 9.97 ppm. Of the two structures that could account for this data, the correct structure was deduced by addition of base to the NMR spectrum of *N*-demethylfagaronine. Two aromatic protons were shielded under these conditions, one at 7.37 ppm by 0.63 ppm and the other at 8.68 ppm by 0.27 ppm. These data are consistent only with the structure **757** for fagaronine.

The NMR spectra of benzophenanthridines are often difficult to obtain because of solubility problems. The 8-alkoxydihydrospecies often are more

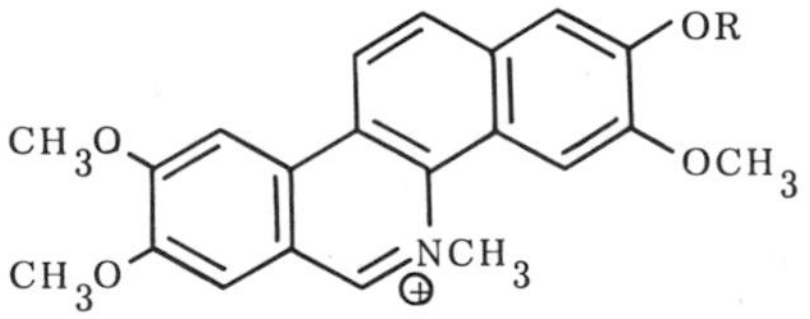

757   fagaronine, R=H
774   R=CH$_3$

7.49 7.14

7.80

7.65
7.05

3.93 CH$_3$O

CH$_3$O

3.96 OCH$_3$

3.46

7.72

NCH$_3$

H 5.55

758

amenable to study, and data for 8-methoxydihydrochelerythrine in CDCl$_3$ are shown in **758**.

8.33.2 Synthesis

There have been at least seven distinct synthetic approaches to the benzophenanthridine nucleus. Most of these routes are very complex and lengthy or are limited to a particular substitution pattern. There has been no shortage of interesting chemistry in this area; however a review of all the synthetic methods is not possible.

The first synthesis of sanguinarine (**754**) (Scheme 98) relies on the reaction between a 1,2-dihydroisoquinoline and an aldehyde to give a 4-substituted isoquinoline. Thus condensation of 2,3-methylenedioxybenzaldehyde (**759**) with the dimethyl acetal of amino acetaldehyde followed by catalytic reduction gave **760**, which was condensed with glyoxylic acid to afford **761**. A crucial intermediate in this reaction is thought to be **762**, which can undergo dehydration and rearrangement to **761**. The two remaining rings are built by condensation with 6-nitropiperonal and a quite standard Pschorr cyclization to afford **754** as the iodide.

A similar key condensation reaction was used by Dyke and Sainsbury in the synthesis of an analogue of nitidine (Scheme 99). Further reaction of the intermediate **763** was mediated not by Pschorr cyclization but by a photochemical route. In the sanguinarine series the photochemical cyclization was less efficient.

A different route was used by Stermitz and co-workers in the first synthesis of fagaronine (**757**). The route has as a crucial step a Kessar phenanthridine synthesis in which the Schiff base **764** is cyclized with sodamide in liquid ammonia.

Because fagaronine (**757**) is nonsymmetric, a suitable protecting group for the phenolic group was required. In this case an isopropyl ether group was used which could be cleaved with dimethyl sulfate, xylene, and nitrobenzene at 180°C. Reduction of the nitronaphthalene followed by condensation with *o*-bromoveratraldehyde gave **764**, and cyclization, N-alkylation, and deprotection afforded fagaronine (**757**).

The interest in the antitumor activities of compounds in the class has led

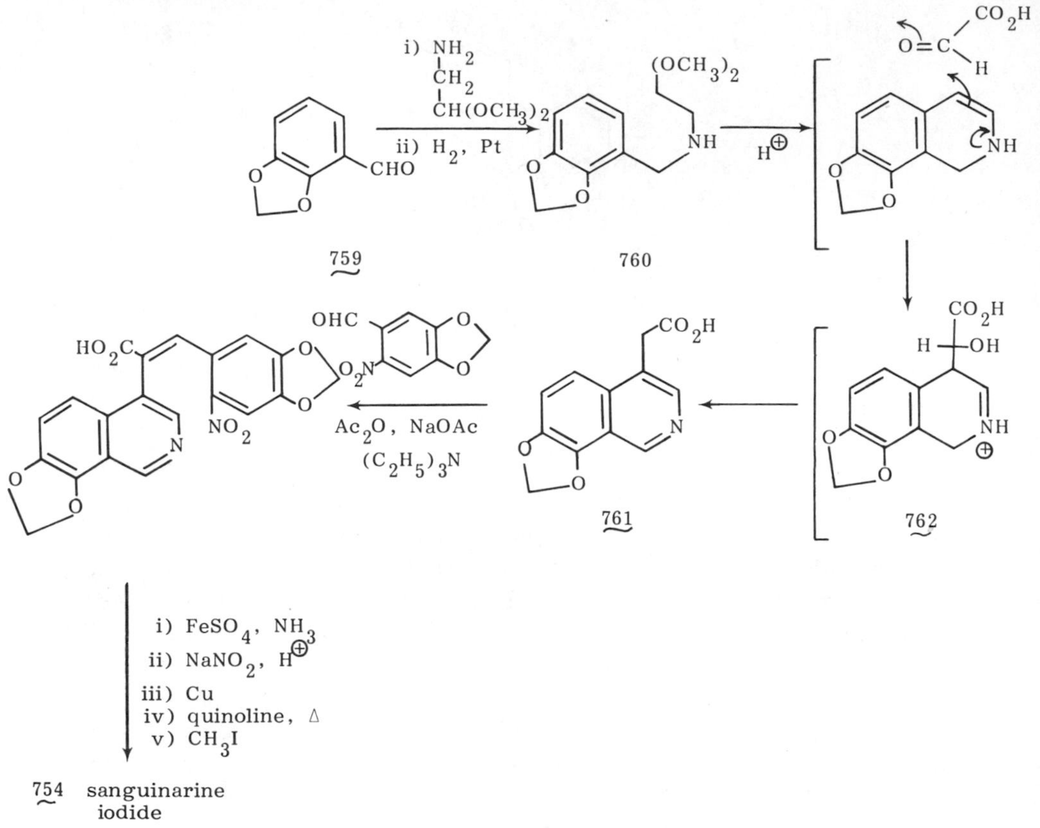

Scheme 98 Sainsbury–Duke synthesis of sanguinarine (**754**).

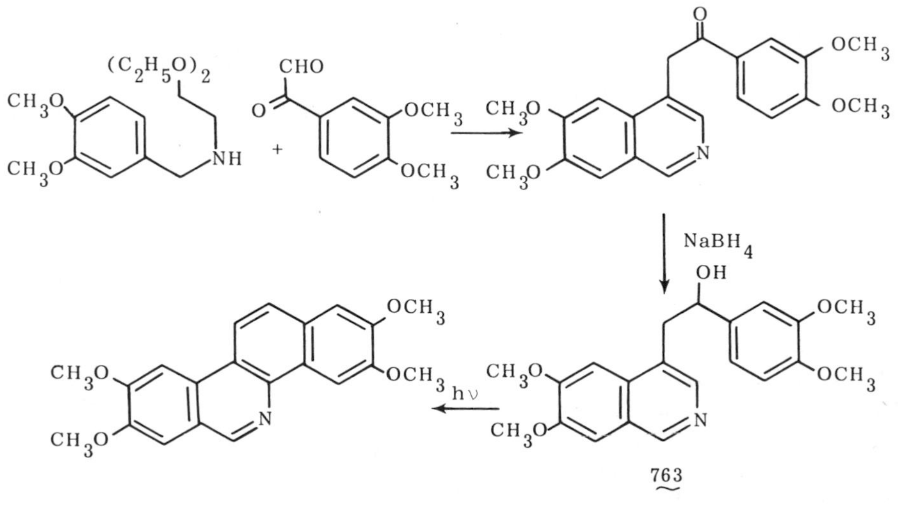

Scheme 99

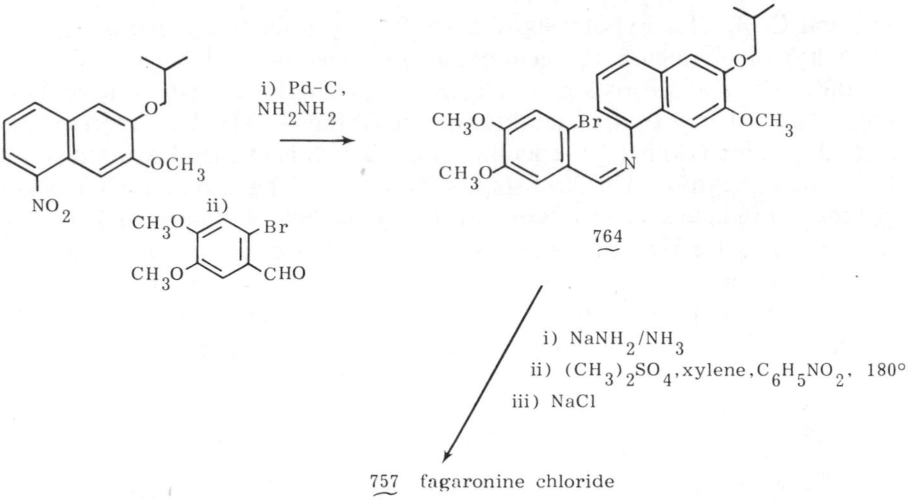

to improvements by Cheng and co-workers in a previously reported synthesis of benzo[c]phenanthridines. The key steps are synthesis of the tetralone **765**, followed by Leuckhart reaction and cyclization with phosphorus oxychloride. The synthesis is completed by aromatization with 30% palladium charcoal to give de-N-methyl nitidine (**766**) (Scheme 100).

8.33.3   Biosynthesis

The benzo[c]phenanthridines have been quite well studied from a biosynthetic aspect, and some of the more complex precursor relationships have been deduced.

It was suggested by Robinson that these alkaloids are derived by fission of the 6,7-bond of protoberberines with subsequent bond formation between

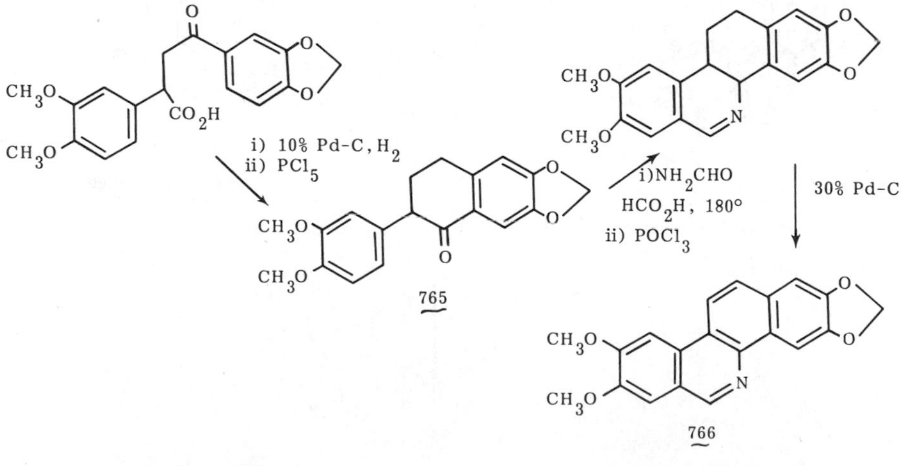

Scheme 100   Cheng synthesis of de-N-methylnitidine (**766**).

C-6 and C-13. This hypothesis was verified by a feeding experiment with [2-^{14}C]tyrosine which gave chelidonine (767) specifically labeled at C-6.

Battersby and co-workers in a series of papers have made a more thorough study of benzo[c]phenanthridine biosynthesis. The key intermediate after dopamine (10) is (+)-reticuline (182), which is cyclized oxidatively to (−)-scoulerine (654). The next step is formation of the two methylenedioxy groups to produce stylopine (666), probably *via* 768. No intermediates have been isolated for the subsequent stages, which are thought to involve 6-hydroxylation of stylopine metho salt (692) and 13,14-dehydrogenation to give 769. Subsequent rearrangement affords chelidonine (767) by cleavage to the enamine aldehyde 770 and reduction of the iminium species 771, and

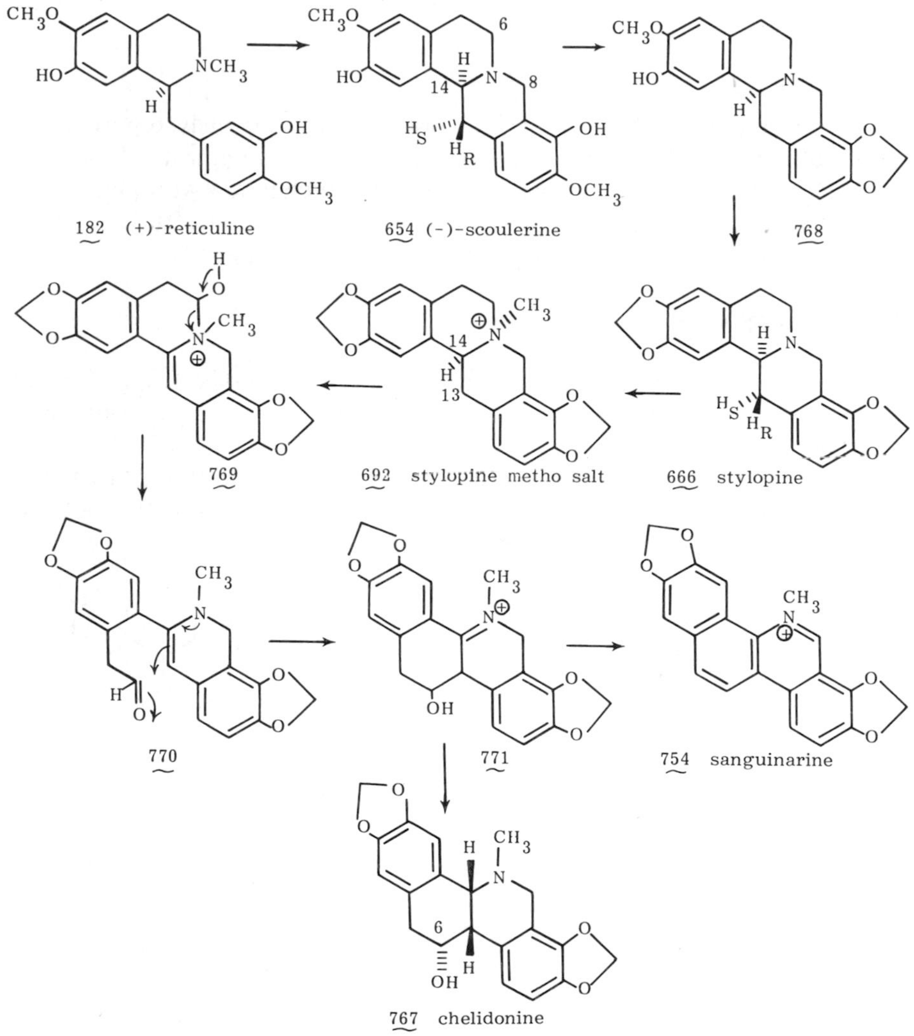

Scheme 101   Biosynthesis of benzo[c]phenanthridines.

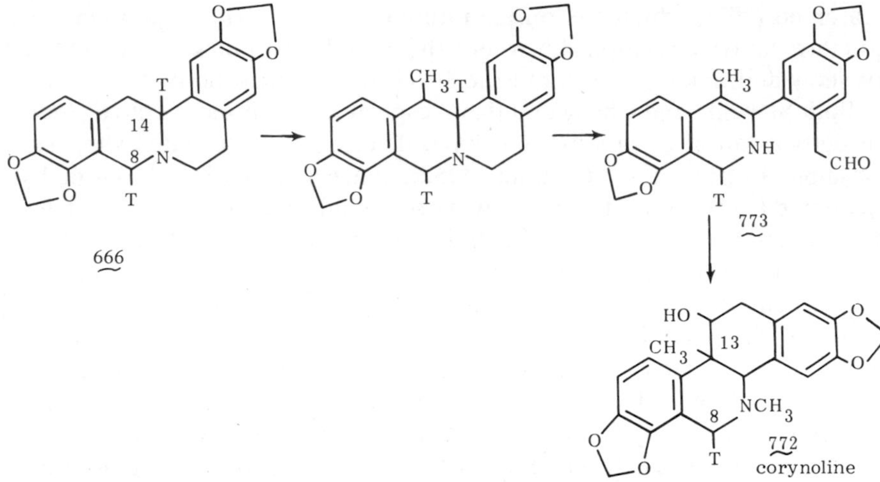

Scheme 102 Biosynthesis of the 13-methyltetrahydrobenzophenanthridines.

sanguinarine (**754**) by oxidative dehydration. In the oxidation of stylopine metho salt (**692**) a 13α-hydroxy intermediate may be involved, for the *pro*-R hydrogen at C-13 is specifically retained into chelidonine (**767**). This is probably another example of stereospecific biosynthetic hydroxylation with retention of configuration. The currently accepted theory of benzo[*c*] phenanthridine biosynthesis is shown in Scheme 101.

Takao and co-workers have shown that in *C. majus* both sanguinarine (**754**) and chelidonine (**767**) are specifically derived from the α-metho salt of the stylopine (**692**), paralleling the data obtained for protopine (**669**).

It was shown by Yagi and co-workers that [8,14-³H]stylopine (**666**) was a precursor of corynoline (**772**) in *Corydalis incisa*. The 14-³H is specifically lost and the one at C-8 is retained in **772**, suggesting that the 1,2-dihydroiso-quinoline **773** is an intermediate (Scheme 102).

8.33.4 Pharmacology

Sanguinarine chloride (**754**) and chelerythrine chloride (**751**) show antifungal, antiprotozoal, and antibacterial activity, but the principal activity of current interest is the anticancer activity of compounds in this series.

The clinical use of extracts of *Chelidonium majus* (Papaveraceae) were first described in the Russian literature in 1896, and subsequent reports include the use of chelidonine sulfate (**767**) for gastric cancer. In Europe *C. majus* has a long history of use in the treatment of warts, papillomas, and condylomas. *Sanguinaria canadensis* has a similar history and many of the isolated compounds correspond.

Sanguinarine (**754**) and chelerythrine (**751**) display cytotoxic activity and are regarded as less important than nitidine (**752**), which has both cytotoxic and antitumor activity (P-388 and lymphocytic leukemia systems), and fagaronine (**757**), which has only antitumor activity. This is an interesting structure–activity relationship where the 9,10-disubstituted compounds do not show antitumor activity, but the 10,11-disubstituted derivatives do.

Both Stermitz and Cheng and their respective collaborators have studied the derivatization of nitidine and have determined some interesting relationships. Beginning with nitidine (**752**), fagaronine (**757**) and *O*-methyl-fagaronine (**774**), Stermitz and co-workers found that almost any change in the functional groups decreased activity; even replacement of one methoxy by an ethoxy group dramatically decreased activity. The isomer of fagaronine in which the D-ring substituents are exchanged was somewhat active, but change of the phenolic group to ring A removed activity. Cheng and co-workers found that the methyl sulfate or fluorsulfonate salts were more active than the chloride of the parent quaternary compound such as nitidine.

Biochemically, sanguinarine is a quite potent inhibitor of sodium–potassium ion dependent ATPase from guinea pig brain, and fagaronine inhibits RNA-directed DNA polymerase activity from several viruses.

LITERATURE

Reviews

Manske, R. H. F., *Alkaloids NY* **4,** 253 (1954).

Manske, R. H. F., *Alkaloids NY* **7,** 430 (1959).

Manske, R. H. F., *Alkaloids NY* **10,** 485 (1968).

Santavý, F., *Alkaloids NY* **12,** 417 (1970).

Shamma, M., *The Isoquinoline Alkaloids,* Academic, New York, 1972, p. 315.

Shamma, M. and J. L. Moniot, *Isoquinoline Alkaloid Research, 1972–1977,* Plenum, New York, 1978, p. 271.

Synthesis

Dyke, S. F., B. J. Moon, and M. Sainsbury, *Tetrahedron Lett.* 3933 (1968).

Dyke, S. F., and M. Sainsbury, *Tetrahedron* **23,** 3161 (1967).

Gillespie, J. P., L. G. Amoros, and F. R. Stermitz, *J. Org. Chem.* **39,** 3239 (1974).

Zee-Cheng, K.-Y., and C. C. Cheng, *J. Het. Chem.* **10,** 85 (1973).

Cushman, M., and L. Cheng, *J. Org. Chem.* **43,** 286 (1978).

Biosynthesis

Lette, E., and J. B. Murrill, *Tetrahedron Lett.* 147 (1964).

Battersby, A. R., R. J. Francis, M. Hirst, E. A. Ruveda, and J. Staunton, *J. Chem. Soc. Perkin Trans. I* 1140 (1975).

Battersby, A. R., J. Staunton, H. R. Wiltshire, R. J. Francis, and R. Southgate, *J. Chem. Soc. Perkin Trans. I* 1147 (1975).

Battersby, A. R., J. Staunton, H. R. Wiltshire, B. J. Bircher, and C. Fuganti, *J. Chem. Soc. Perkin Trans. I* 1162 (1975).

Yagi, A., G. Nonaka, S. Nakayama, and I. Nishioka, *Phytochemistry* **16,** 1197 (1977).

Pharmacology

Cordell, G. A., and N. R. Farnsworth, *Heterocycles* **4,** 393 (1976).

Cordell, G. A., and N. R. Farnsworth, *Lloydia* **40,** 1 (1977).

Stermitz, F. R., J. P. Gillespie, L. G. Amoros, R. Romero, T. A. Stermitz, K. A. Larson, S. Earl, and J. E. Ogg, *J. Med. Chem.* **18,** 708 (1975).

Zee-Cheng, R. K. Y., and C. C. Cheng, *J. Med. Chem.* **18,** 66 (1975).

Sethi, V. S., and M. L. Sethi, *Biochem. Biophys. Res. Commun.* **63,** 1070 (1975).

8.34 PHENETHYLISOQUINOLINE ALKALOIDS

The phenethylisoquinoline alkaloids are homologues of the benzylisoquinolines, and although they are not as important as the latter group of compounds, they are nevertheless a group of increasing interest. They may be classified into six major groups: the simple 1-phenethylisoquinoline (**775**), homomorphinandenione (**776**), bisphenethylisoquinoline (**777**), homoproaporphine (**778**), homoaporphine (**779**), and homoerythrina (**780**) types, as shown in Figure 18.

The homomorphinandenione skeleton is an important intermediate in the biosynthesis of colchicine and is discussed in a separate section.

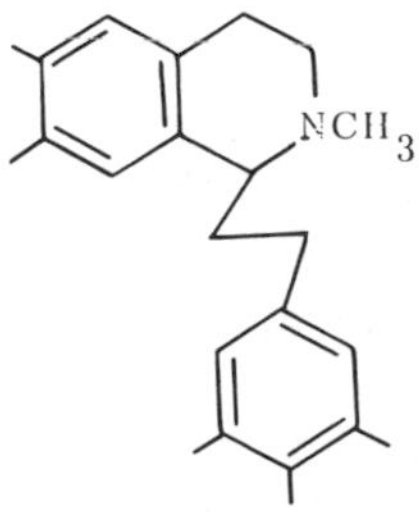

775

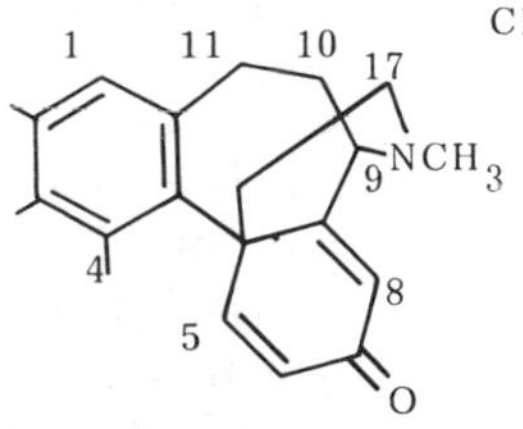

776

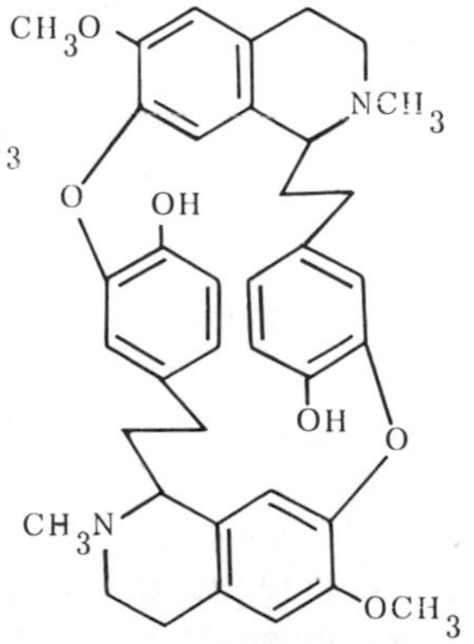

777 melanthioidine

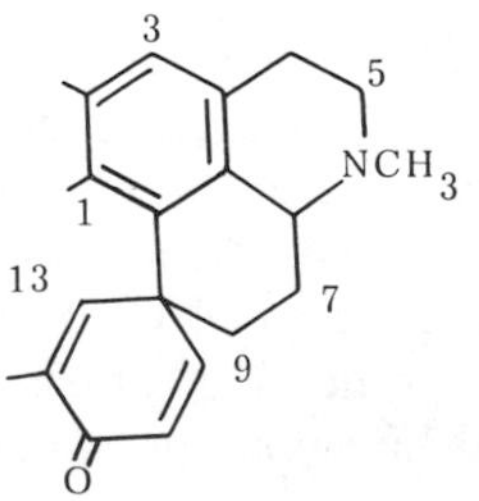

778

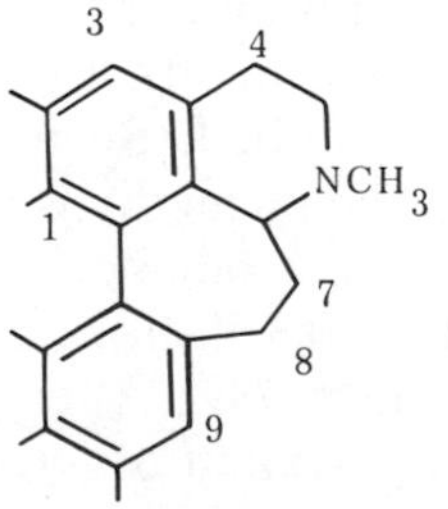

779

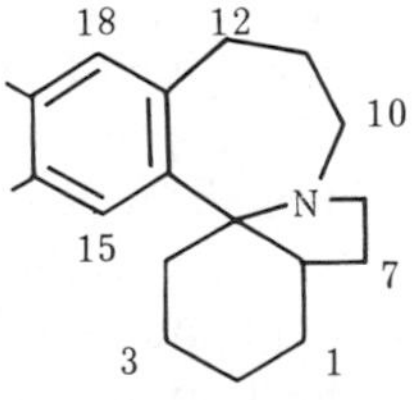

780

Figure 18 Phenethylisoquinoline alkaloid types.

From a phytochemical point of view, the phenethylisoquinolines have been obtained from six genera of higher plants: *Androcymbium, Colchicum, Kreysigia, Bulbocodium, Schelhammera,* and *Phelline,* all in the family Liliaceae.

The only example of a bisphenethylisoquinoline is melanthioidine (**777**). Several homoproaporphines have been obtained from *Kreysigia multiflora,* and typical of these is kreysiginone (**781**). This same plant is also the source of several homoaporphines, including kreysigine (**782**).

781 kreysiginone

782 kreysigine

8.34.1 Chemistry and Spectral Properties

Many of the reactions developed for the chemical conversion of the benzylisoquinolines have been applied to the phenethylisoquinolines. For example, Battersby and co-workers found that treatment of the diphenolic phenethylisoquinoline **783** with potassium ferricyanide gave the homoproaporphine multifloramine (**784**), which underwent acid-catalyzed dienone–phenol rearrangement to the homoaporphine **785**.

Although *ortho-ortho* and *ortho-para* phenolic coupling are observed in the series, it is rare that *para-para* coupling (i.e. androcymbine skeleton formation) is found. This should be contrasted with the benzyl isoquinoline series. Consequently the compound **786**, which might be expected to *para-para* couple since both *ortho* positions on ring A are blocked, still underwent *ortho* coupling to eventually afford a homoaporphine **787** in which the methoxyl group originally at C-8 had been lost.

The modified Pschorr reaction and its use in morphinandienone alkaloid synthesis has already been described (Section 8.24.4). This reaction has been applied by Kametani to the phenethylisoquinoline series and has permitted entry into the homomorphinandienone skeleton. Diazotization of the 2'-aminophenethylisoquinoline (**788**) followed by heating to 70°C for 1 hr gave the homomorphinandienone **789**. Attempted synthesis of *O*-methylandrocymbine (**790**) by the cyclization of **791** failed however, presumably due to steric crowding.

More success has been achieved with the photo-Pschorr reaction in which the intermediate diazonium salt is irradiated at low (5–10°C) temperature.

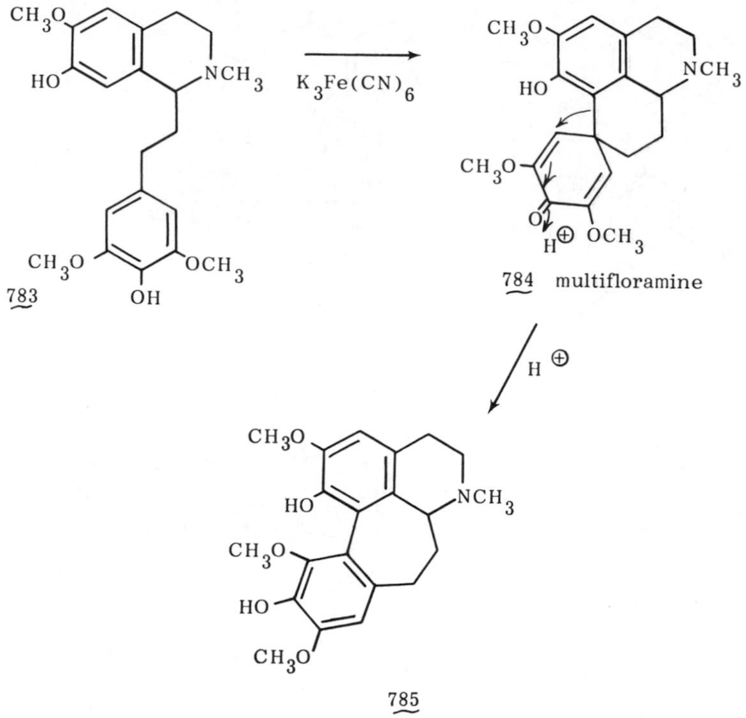

783 → $K_3Fe(CN)_6$ → 784 multifloramine

$H^{\oplus}$

785

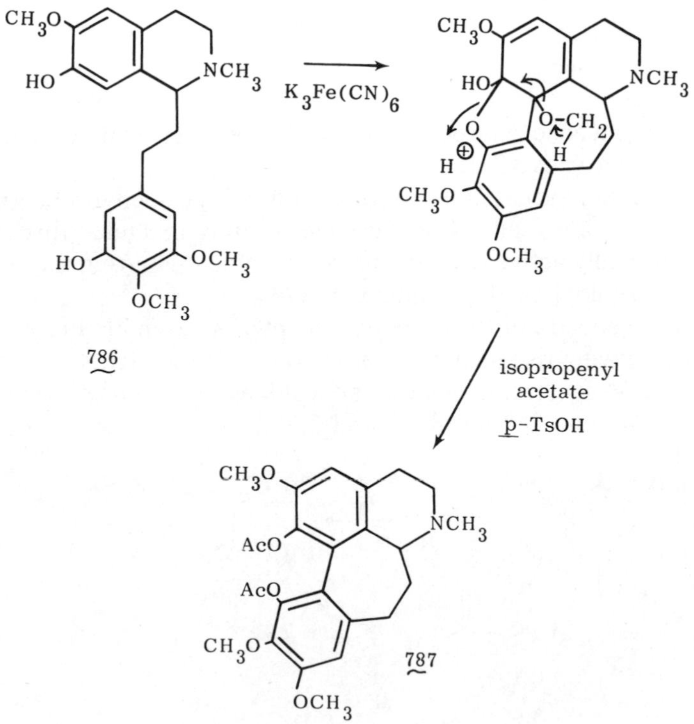

786 → $K_3Fe(CN)_6$ →

isopropenyl acetate

p-TsOH

787

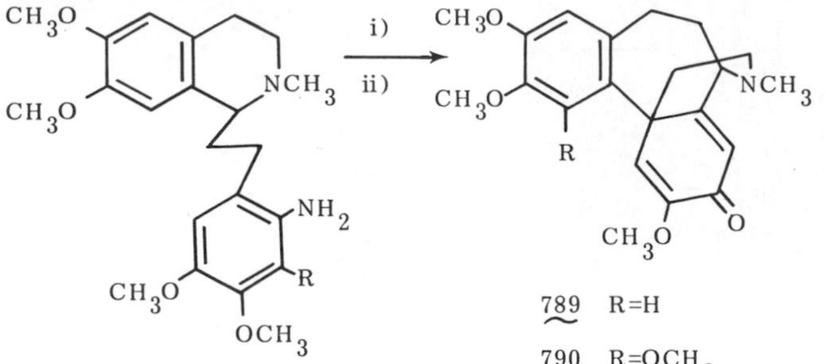

i) HNO$_2$

ii) 70°, 1 hr.

789 R=H

790 R=OCH$_3$

788 R=H

791 R=OCH$_3$

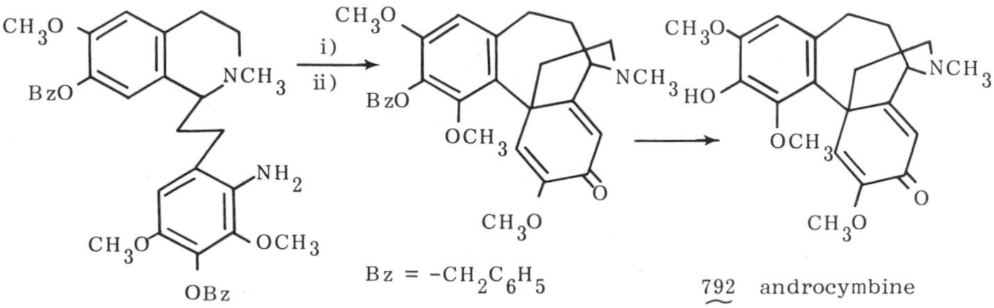

Bz = -CH$_2$C$_6$H$_5$

792 androcymbine

793

i) HNO$_2$

ii) hν

In this way racemic androcymbine (**792**) has been synthesized from the di-O-benzyl derivative **793**.

Kupchan had successfully used photolytic cyclodehydrobromination in the aporphine series, and Kametani has recently extended this to the synthesis of optically active O-methylandrocymbine (**790**) from the corresponding (−)-2′-bromophenethylisoquinoline (**794**).

The mass spectra of the homoproaporphines resemble those of the proaporphines, giving rise mainly to losses of small fragments.

In the UV spectrum, homoproaporphines such as kreysiginone (**781**) show λ$_{max}$ 243 and 287 nm (4.15 and 3.78), whereas the homoaporphines

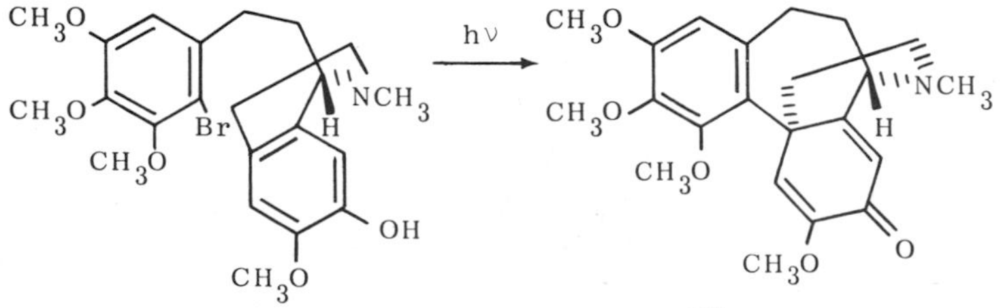

794

790

(−)-O-methylandrocymbine

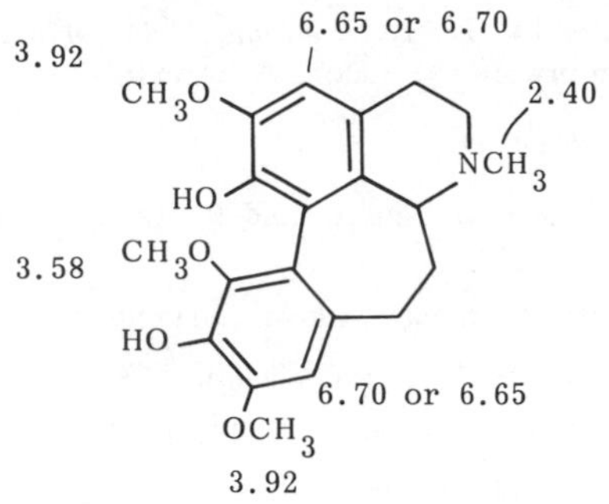

3.92

6.65 or 6.70

CH_3O

2.40

NCH_3

HO

3.58 CH_3O

HO

6.70 or 6.65

OCH_3

3.92

784 multifloramine

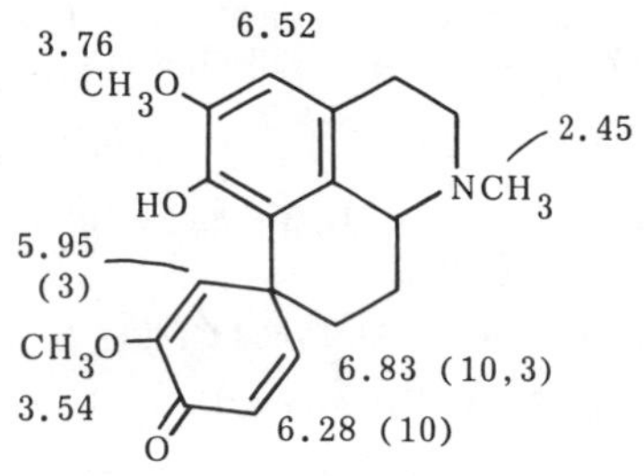

3.76   6.52

CH_3O

2.45

NCH_3

HO

5.95
(3)

CH_3O

3.54

6.83 (10,3)

6.28 (10)

O

781 kreysiginone

absorb at somewhat longer wavelengths [e.g. kreysigine (**782**)], which shows λ_{max} 257 and 293 nm (4.10 and 3.67).

Representative proton NMR data for the alkaloids multifloramine (**784**) and kreysiginone (**781**) are shown. The sizable coupling of 3 Hz between the 9- and 13-protons in **781** is noteworthy.

8.34.2 Biosynthesis

Except for androcymbine (**792**) and the homoerythrina alkaloids (discussed elsewhere), there has been minimal biosynthetic effort in this area. The homoaporphines such as **795** have been found to be derived by direct phenolic coupling. For example, the diphenolic isoquinoline **786** was incorporated into **795**, but the isomer **783** was not.

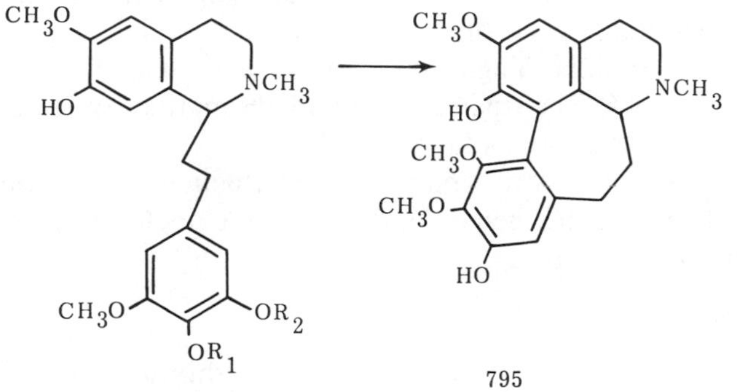

CH_3O

NCH_3

HO

CH_3O

OR_2

OR_1

CH_3O

NCH_3

HO

CH_3O

CH_3O

HO

795

783 R_1=H; R_2=CH_3
786 R_1=CH_3; R_2=H

LITERATURE

Reviews

Shamma, M., *The Isoquinoline Alkaloids*, Academic, New York, 1972, pp. 458, 469, 483.

Kametani, T., and M. Koizumi, *Alkaloids NY* **14**, 265 (1973).

Kametani, T., and K. Fukumoto, in K. F. Wiesner (Ed.), *MTP International Review of Science, Organic Chemistry*, Ser. 1, Vol. 9, *Alkaloids*, Butterworths, London, 1973, p. 181.

Synthesis

Battersby, A. R., R. B. Bradbury, R. B. Herbert, H. G. Munro, and R. Ramage, *Chem. Commun.* 450 (1967).

Battersby, A. R., E. McDonald, M. H. G. Munro, and R. Ramage, *Chem. Commun.* 934 (1967).

Kametani, T., K. Fukumoto, F. Satoh, and H. Yagi, *J. Chem. Soc., Sec. C* 3084 (1968).

Kametani, T., M. Koizumi, and K. Fukumoto, *J. Org. Chem.* **36**, 3729 (1971).

Kametani, T., Y. Satoh, and K. Fukumoto, *Tetrahedron* **29**, 2027 (1973).

Biosynthesis

Battersby, A. R., P. Bohler, M. H. G. Munro, and R. Ramage, *Chem. Commun.* 1066 (1969).

8.35   COLCHICINE

Colchicum is another of the ancient drugs of mankind, and the poisonous nature of the autumn crocus was probably known to the Greek physician Dioscorides. The name of the genus is taken from the ancient kingdom of Colchis, where the drug was used in poisonous potions. Some of the other names under which the drug is known include hermodactyl, ephemeron, and surinjan.

In Arabic writings of the sixth and seventh centuries the drug is recommended for use in the treatment of gout, but the small difference between the therapeutic and toxic doses reduced its use in classical and medieval times.

Colchicum corm is derived from the plant *Colchicum autumnale* L. (Liliaceae) native to many parts of Europe, and commercial supplies come from Poland, Czechoslovakia, Yugoslavia, and Holland. Colchicine content is in the range 0.25–0.6%. The ripe seeds are an alternative source of colchicine, where the content may be as high as 1.2%.

As well as 19 species of *Colchicum*, colchicine-type alkaloids are also present in at least 10 other genera of the Liliaceae, including *Androcymbium, Gloriosa,* and *Merendera* sp.

8.35.1   Chemistry

Colchicine (**796**), because of the amide functionality, is neutral, and darkens on exposure to light. It forms yellow crystals containing $CHCl_3$ or C_6H_6 of crystallization, and from water it gives a trihydrate. The solubility characteristics of colchicine are most unusual; it is soluble in water, alcohol, and chloroform, but only slightly soluble in petroleum ether.

In the 1940s there was intense interest in the chemistry of the B and C rings of colchicine, particularly after Dewar suggested that colchicine contained a tropolone ring. Much of the chemistry of the tropolone system was studied with colchicine. Some of the early reviews of colchicine give the details of the early chemical work and only some of transformations will be discussed here. Dilute acid on colchicine (**796**) affords the demethyl ether colchiceine **797**, which can be methylated to a mixture of colchicine **796** and isocolchicine **798**.

Treatment of colchiceine with alkaline hydrogen peroxide and methylation gave a benzenoid derivative **799**, which could be further degraded by treatment with phosphorus pentoxide, and then osmium tetroxide, to give two substituted phenanthrenes, **800** and **801**. The structures of these were confirmed by synthesis.

COCH$_3$
NH
dil. acid
CH$_2$N$_2$

796 colchicine

797 colchiceine

NHCOCH$_3$
CH$_2$N$_2$

798 isocolchicine

i) alkaline/H$_2$O$_2$
ii) CH$_2$N$_2$

NHCOCH$_3$

i) CH$_3$O
ii)

799

R$_1$
R$_2$

800 R$_1$=H; R$_2$=CHO
801 R$_1$=CHO; R$_2$=H

i) P$_2$O$_5$
ii) OsO$_4$

These data could not however distinguish between the structures **796** and **798** for colchicine and it was not until an X-ray analysis was completed in 1952 that **796** was found to be correct.

Some of the most intensely studied chemistry of colchicine has been the products of UV irradiation. There are three photoisomers of colchicine (**796**), which are known as α-, β- and γ-lumicolchicines, and have the structures **802**, **803**, and **804**. At least two of these may occur naturally, presumably as a result of the exposure of **796** to sunlight. These rearrangements were the first examples of photochemical electrocyclic reactions, and stimulated a substantial interest in the photochemistry of troponoids.

8.35.2 Synthesis

There are at least six syntheses of colchicine (**796**) or close relatives, but none are of commercial significance. That the groups of Woodward, Eschenmoser, Van Tamelen, and Scott have been involved in the synthesis of colchicine at various times indicates the importance and synthetic chal-

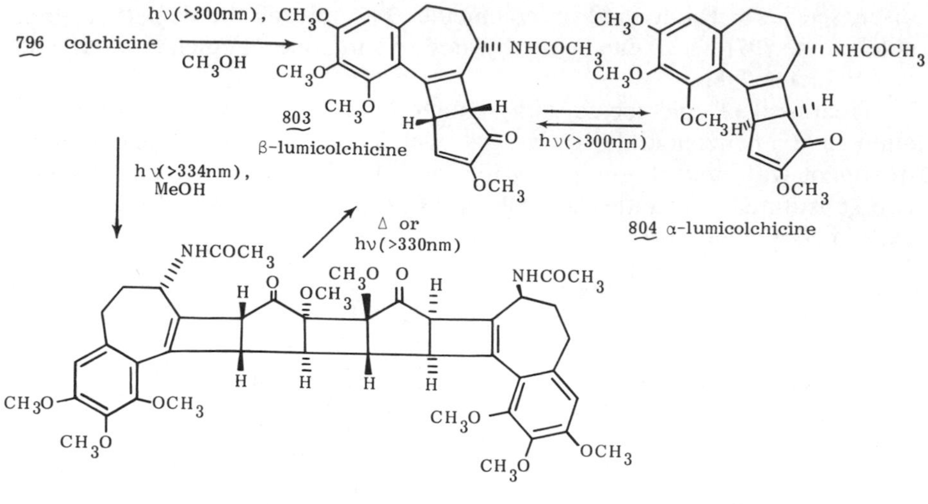
796 colchicine
hν(>300nm), CH3OH
803 β-lumicolchicine
hν(>300nm)
804 α-lumicolchicine
hν(>334nm), MeOH
Δ or hν(>330nm)

802 α - lumicolchicine

lenge presented by this molecule. Clearly these syntheses cannot be presented in detail here. Van Tamelen's and Eschemonoser's syntheses each have **805** or a close relative as an early intermediate, and **806** as a late intermediate, but quite different routes have been used to introduce the tropolone ring. The further elaboration of **806** to colchicine (**796**) involves methylation, bromination with *N*-bromosuccinimide, and displacement of

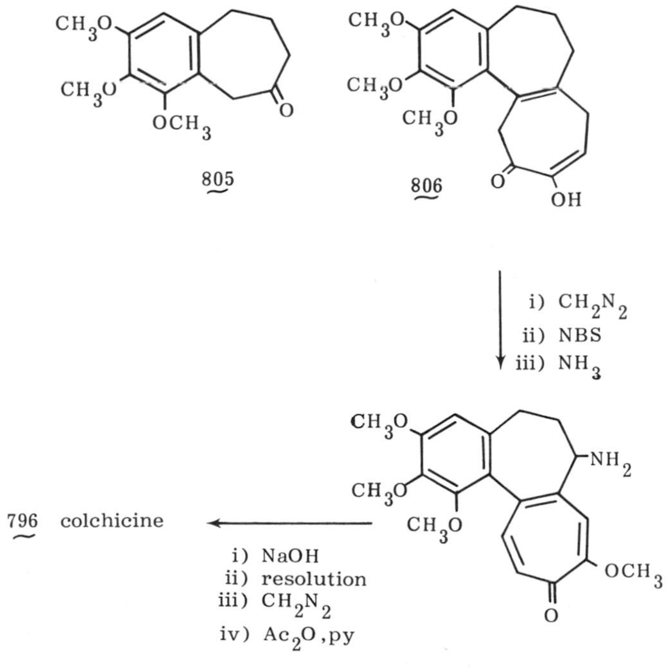
805
806
i) CH2N2
ii) NBS
iii) NH3
796 colchicine
i) NaOH
ii) resolution
iii) CH2N2
iv) Ac2O, py

the bromine with ammonia. The product was then resolved, methylated, and acetylated to give colchicine (796).

A new efficient approach to a key intermediate in the synthesis of colchicine has been reported by Tobinaga and co-workers. Anodic oxidation of 807 gave the spirodienone 808 in 80% yield. The additional carbon atom of the topolone ring was introduced by a Simmons–Smith reaction followed by oxidation of the alcohol to afford 809. Treatment of this compound with acetic anhydride–sulfuric acid at room temperature then afforded desacetamidoisocolchicine (810) in 90% yield (Scheme 103).

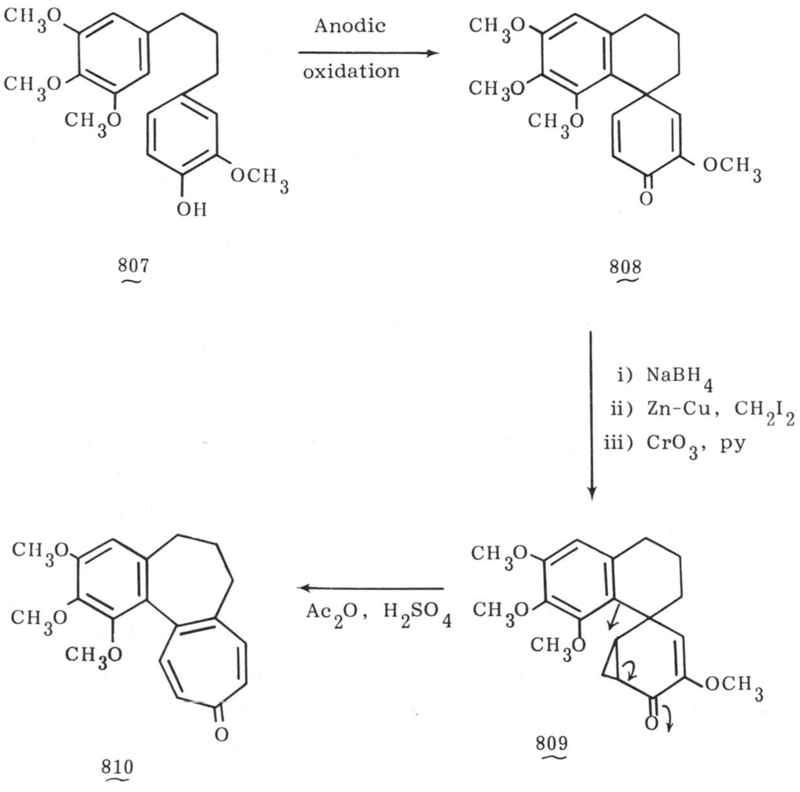

Scheme 103

The most elaborate synthesis of colchicine (796) is due to Woodward and co-workers and at this time represented significant new chemistry in the isothiazole series. Some of the key intermediates are shown in Scheme 104.

8.35.3 Biosynthesis

The novel structure of colchicine (796) has stimulated an extensive investigation of its biosynthesis by several groups. The principal investigatory group has been that of Battersby, who has summarized the work in this area.

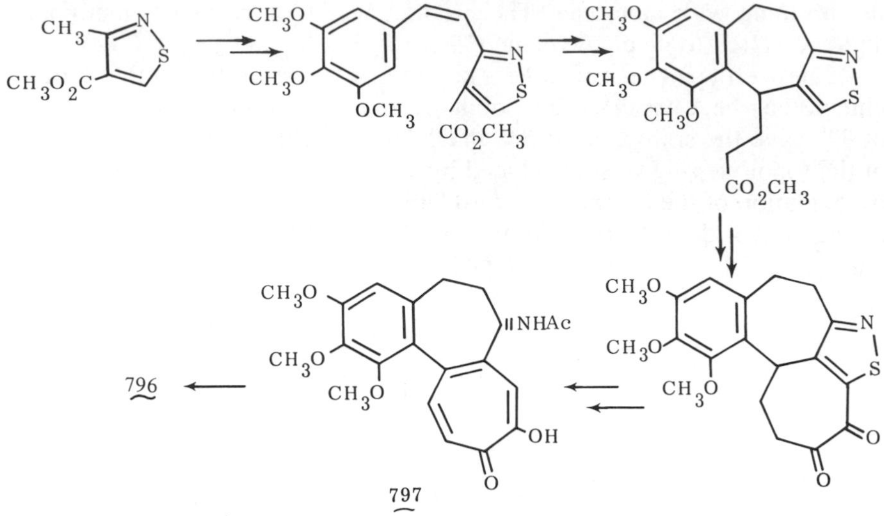

Scheme 104   Intermediates in the Woodward synthesis of colchicine (**796**).

Ring A and carbons 5, 6, and 7 of colchicine are derived from phenylalanine. The incorporation of [3-^{14}C]- and [ring-^{14}C]tyrosines was specific at C-12 and C-9 respectively, and the nitrogen of **796** was found to be derived from the amino acid nitrogen of tyrosine. However [1-^{14}C]- and [2-^{14}C] tyrosines were not specifically incorporated, indicating that carbons 1 and 2 tyrosine were lost in the biosynthesis, and that the tropolone ring system was formed by expansion of the aromatic ring with the benzylic carbon atom of tyrosine.

The key progressive steps in determining the biosynthesis of colchicine (**796**) were the isolation of androcymbine (**792**) from *Androcymbium melanthioides,* and the demonstration that it was well and specifically incorporated into **796**. But what of the derivation of this compound? Autumnaline (**811**), obtained from a *Colchicum* sp. and labeled specifically at C-9, was next evaluated and found to be an efficient precursor, thereby establishing the sequence autumnaline (**811**), *O*-methylandrocymbine (**790**), and finally colchicine (**796**). Variously substituted phenethylisoquinoline precursors were also evaluated but were poorly incorporated, suggesting that the particular substitution pattern of autumnaline (**811**) was correct for the true precursor of colchicine. The steps between phenylalanine (**1**) and autumnaline (**811**) are still under study. Variously substituted cinnamic acids were evaluated as precursors, but except for cinnamic acid, none were effective precursors and it has been suggested that reduction of the double bond takes place before hydroxylation of the aromatic ring. Present indications are that the route shown in Scheme 105 is probably correct to the point of *O*-methylandrocymbine (**790**). Beyond this compound no intermediates are known, and the mechanism for ring expansion to the tropolone is conjecture.

In one of the first uses of ^{13}C-enriched precursors in plant biosynthesis Battersby and co-workers showed that [1-^{13}C]-autumnaline (**811**) specifically labeled C-7 of colchicine (**796**).

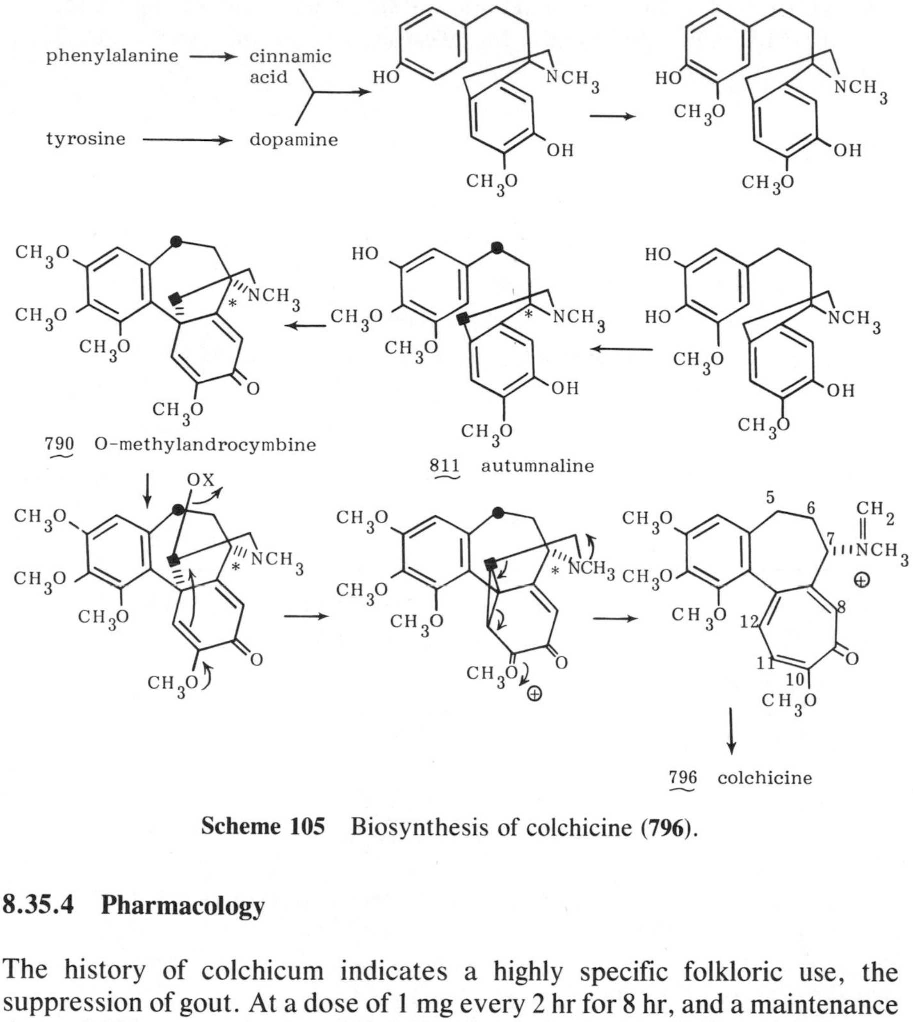

phenylalanine ⟶ cinnamic acid

tyrosine ⟶ dopamine

790   O-methylandrocymbine

811   autumnaline

796   colchicine

Scheme 105   Biosynthesis of colchicine (**796**).

8.35.4   Pharmacology

The history of colchicum indicates a highly specific folkloric use, the suppression of gout. At a dose of 1 mg every 2 hr for 8 hr, and a maintenance dose of 500 μg twice a day, colchicine is still regarded as the most effective treatment.

In 1889, it was reported that colchicine had the ability to arrest mitosis at metaphase. Until the 1930s and 1940s this startling observation provoked little interest. But at that time two potential areas for the use of colchicine were explored. One was the possibility of using colchicine as an anticancer agent, and the second was evaluation of the genetic effects produced in plants.

Although colchicine effectively regressed tumors in dogs and mice, two problems were evident; the inhibition of cell division is not specific for malignant cells, and the required dose is borderline toxic.

Colchicine has the ability to artificially induce polyploidy (multiple chromosome groups). Typical effects are larger flowers, pollen grains, and stomata. In practically every alkaloid-containing plant where colchicine has been used to induce polyploidy, the alkaloid content has been remarkably increased, sometimes by up to 150%.

LITERATURE

Reviews

Cook, J. W., and J. D. London, *Alkaloids NY* **2**, 261 (1952).

Wildman, W. C., *Alkaloids NY* **6**, 247 (1960).

Fell, K. R., and D. R. Ramsden, *Lloydia* **30**, 123 (1967).

Wildman, W. C., and S. W. Pelletier (Ed.), *Chemistry of the Alkaloids,* Van Nostrand Reinhold, New York, 1970, p. 199.

Synthesis

Van Tamelen, E. E., T. A. Spencer, Jr., D. S. Allen, Jr., and R. L. Orvis, *Tetrahedron* **14**, 8 (1961).

Schreiber, J., W. Leimgruber, M. Pesaro, P. Schudel, T. Threfall, and A. Eschenmoser, *Helv. Chim. Acta* **44**, 540 (1961).

Woodward, R. B., *The Harvey Lectures, Series 59,* Academic, New York, 1965, p. 31.

Sunagawa, G., T. Nakamura, and J. Nakazawa, *Chem. Pharm. Bull.* **10**, 291 (1962).

Scott, A. I., F. McCapra, R. L. Buchanan, A. C. Day, and D. W. Young, *Tetrahedron* **21**, 3605 (1965).

Kotani, E., F. Miyazaki, and S. Tobinaga, *Chem. Commun.* 300 (1974).

Biosynthesis

Battersby, A. R., T. A. Dobson, D. M. Foulkes, and R. B. Herbert, *J. Chem. Soc. Perkin Trans. I* 1730 (1972).

Battersby, A. R., R. B. Herbert, L. Pijewska, R. Santavy, and P. Sedmera, *J. Chem. Soc. Perkin Trans. I* 1736 (1972).

Battersby, A. R., R. B. Herbert, E. McDonald, R. Ramage, and J. H. Clements, *J. Chem. Soc. Perkin Trans. I* 1741 (1972).

Battersby, A. R., P. W. Sheldrake, and J. A. Milner, *Tetrahedron Lett.* 3315 (1974).

8.36 *CEPHALOTAXUS* AND HOMOERYTHRINA ALKALOIDS

The Japanese plum-yew, *Cephalotaxus harringtonia* (Forbes) K. Koch var. *harringtonia* in the family Cephalotaxaceae, is the source of a very interesting group of alkaloids: interesting for two reasons, the novel structure and the potent antileukemic activity of some members of this series.

Alkaloids were first detected in this genus in 1950, but no alkaloids were isolated until Paudler obtained cephalotaxine in 1963. The alkaloids belong to the homoerythrina series and structurally similar alkaloids have been obtained from *Schelhammera* sp. and *Phelline comosa* Labill. (Liliaceae).

The most interesting alkaloids are the complex esters of cephalotaxine (**812**), typical examples of which are harringtonine (**813**) and homoharringtonine (**814**). The esters are not crystalline and until recently X-ray analysis

812 cephalotaxine, R=H

813 harringtonine, R=-C-CCH$_2$CH$_2$C(CH$_3$)$_2$

814 homoharringtonine, R=-C-C(CH$_2$)$_3$C(CH$_3$)$_2$

had been limited to the methiodide and *p*-bromobenzoate derivatives of cephalotaxine. When an X-ray analysis of cephalotaxine itself was made it was found that the crystal contained two independent cephalotaxine molecules in different conformations.

8.36.1 Synthesis of Cephalotaxine

The major problem in bringing these alkaloids to clinical trial has been paucity of material. Consequently considerable effort has been put into their synthesis.

Two groups have reported on the synthesis of cephalotaxine (**812**), and Weinreb and Semmelhack have collaborated to summarize the results of their respective groups. The unique spiro rings and benzazepine nucleus pose some interesting problems which each group solved in different ways. The Weinreb–Auerbach synthesis proceeds in eight steps from 1-prolinol (**815**) and 3,4-methylenedioxyphenylacetyl chloride (**816**), and involves the enamine **817** as a crucial intermediate. The initial N-acylated product **818** was cyclized with a Lewis acid catalyst and reduced to give the enamine **817**.

The remaining ring and formation of the spiro center were introduced by way of an unusual dicarbonyl compound. Acylation of **817** with the mixed anhydride from ethyl chloroformate and pyruvic acid gave **819**, which could be cyclized with magnesium methoxide to demethylcephalotaxinone (**820**). This compound exists solely as **820** and not as the tautomer **821**. Under equilibrating conditions, treatment of **820** with dimethoxypropane in the presence of *p*-toluenesulfonic acid afforded a keto enol ether (cephalotaxinone) which could be reduced stereospecifically with sodium borohydride to racemic cephalotaxine (**812**) (Scheme 106).

The synthesis of Semmelhack adopts a quite different approach and the key steps are shown in Scheme 107. The crucial intermediates are the aza-spiro[4,4]nonene derivative **822** and the nitrobenzene sulfonate ester **823**. Alkylation of **822** with the sulfonate ester in the presence of diisopropyl-ethylamine gave **824** in good yield.

Several methods were successful for the completion of the synthesis of cephalotaxine (**812**), but the most spectacular is a base-mediated photolysis in liquid ammonia to give cephalotaxinone (**825**) in 49% yield, by way of the anion radical **826**.

8.36.2 Homoerythrina Alkaloids

The term homoerythrina refers to a small group of compounds having the *Erythrina* skeleton but with an additional carbon atom in ring C. These

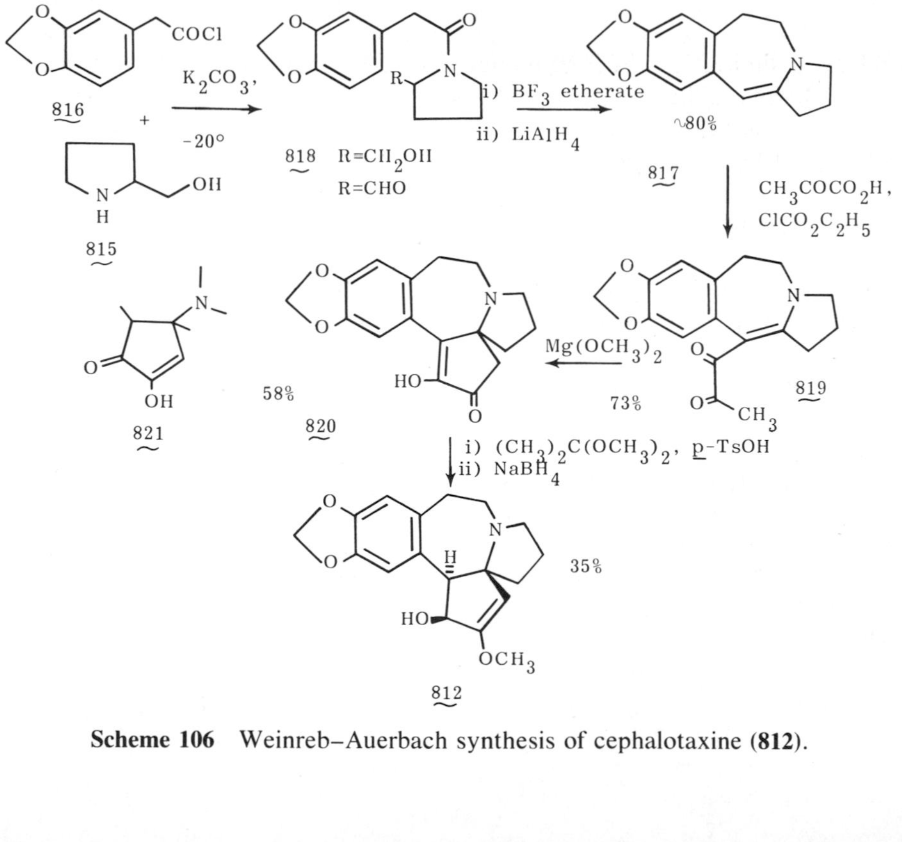

Scheme 106 Weinreb–Auerbach synthesis of cephalotaxine (**812**).

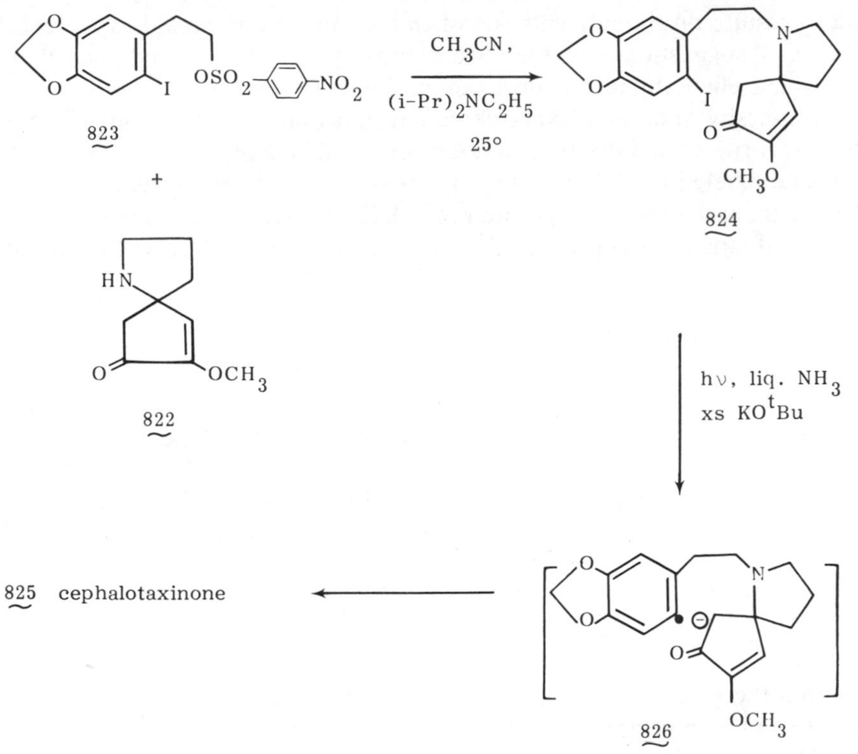

Scheme 107 Semmelhack synthesis of cephalotaxinone (**825**).

alkaloids occur in the genus *Schelhammera* and in *Cephalotaxus harringtonia*. A typical alkaloid is schelhammeridine (**827**).

Potier and co-workers described several alkaloids of this structure type from *Phelline comosa* in the Liliaceae, including **828**. Such isolations pose some interesting taxonomic problems.

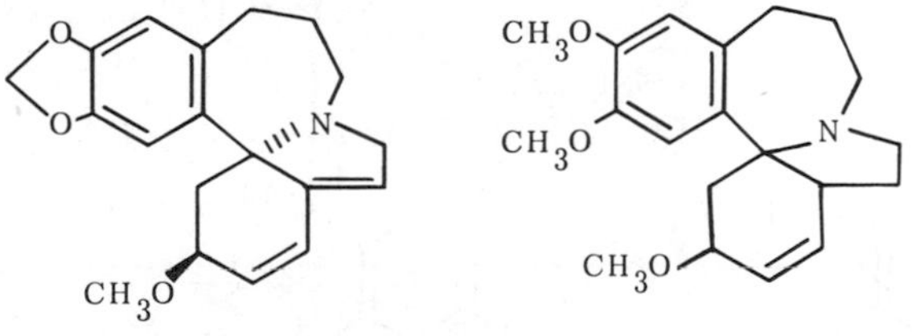

827 schelhammeridine 828

8.36.3 Biosynthesis

The *Cephalotaxus* alkaloids co-occur with alkaloids of the homoerythrina type such as schelhammeridine (**827**), and two groups have reported preliminary results on the biosynthesis of this group of compounds as a result of Powell's suggestion that they were biosynthetically related, possibly involving a phenethyltetrahydroisoquinoline derivative **829**.

Battersby and co-workers demonstrated that only one unit of tyrosine was incorporated into the homoerythrina schelhammeridine (**827**) which became labeled at C-2. Although [1-^{14}C]- and 2-^{14}C]phenylalanines and [2-^{14}C]cinnamate were incorporated into **827**, the site of labeling was not determined, and it was proposed that **830** is a key intermediate in the formation

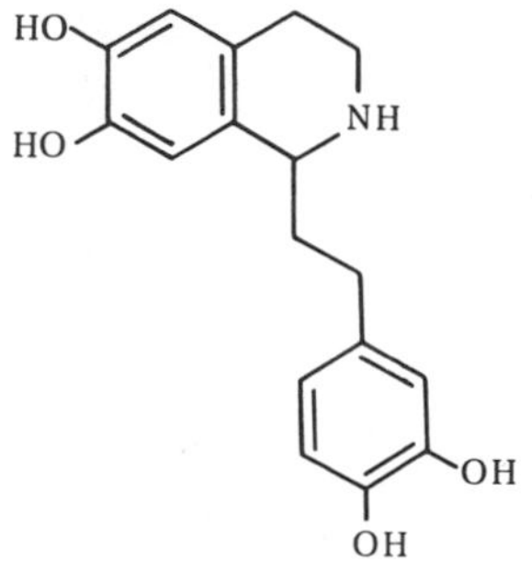

829 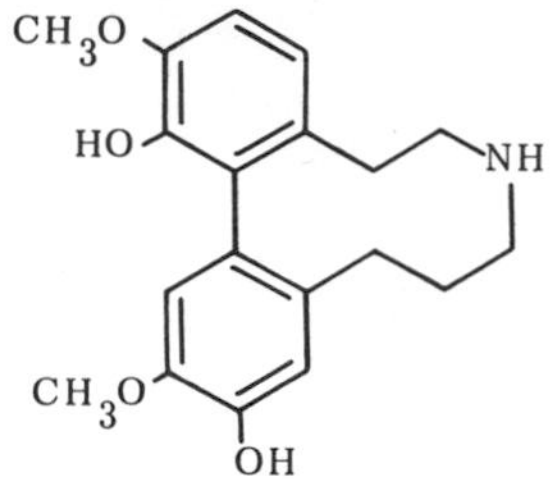   830

of both the *Cephalotaxus* and homoerythrina alkaloids (compare *Erythrina* alkaloid biosynthesis).

The main thrust in studying the biosynthesis of the *Cephalotaxus* alkaloids has come from Parry's laboratory. In contrast to Battersby's data, Parry and Schwab found that both [2-^{14}C]- and [3-^{14}C]tyrosines were good precursors, and that in the latter case 100% of the activity was at C-4 and C-11. On Battersby's scheme [3-^{14}C]tyrosine would lead to labeling at C-11 and C-6.

Further results by Parry, using [ring-^{14}C]L-tyrosine (**2**) as a precursor and degrading cephalotaxine (**812**) to 4,5-methylenedioxy phthalic acid (**831**),

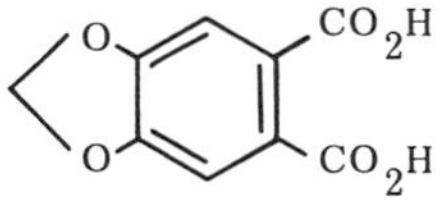

831 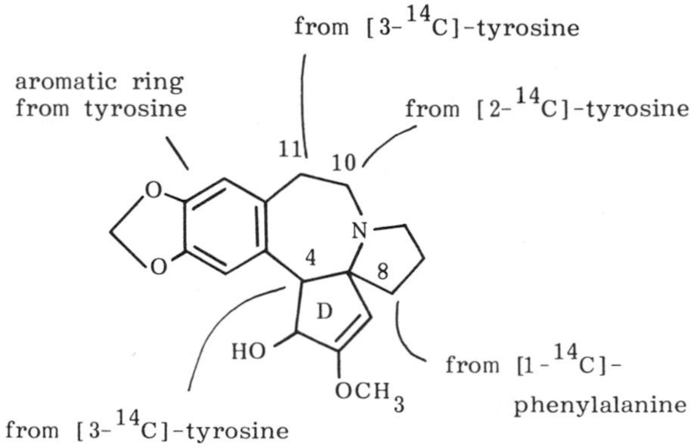   812

indicated that 90% of the total activity was in this fragment. The C_3 moiety of two tyrosine units is therefore incorporated, but only one aromatic ring. It was also found that when [2-^{14}C]tyrosine was used as a precursor, the level of incorporation into cephalotaxine was lowered, but the specificity of incorporation at C-10 increased.

Previously, phenylalanine had been found to be a very poor precursor, but repetition of this experiment indicated that C-8 of **812** was specifically derived from C-1 of phenylalanine. This result is analogous to those obtained for other phenethyl isoquinoline alkaloids. In contrast to experiments with colchicine however, cinnamic acid was not a precursor of cephalotaxine (**812**). No further data are available on this fascinating biosynthetic problem in which it appears that ring D of **812** is produced from phenylalanine by the loss of one carbon atom from the aromatic ring. There have been no experiments with [ring-^{14}C]phenylalanine that would substantiate this however.

Some of the key results in this area are summarized in **812**.

LITERATURE

Reviews

Weinreb, S. M., and M. F. Semmelhack, *Accts. Chem. Res.* **8**, 158 (1975).

Findlay, J. A., in K. F. Wiesner (Ed.), *MTP International Review of Science, Organic Chemistry* Ser. 2, Vol. 9, *Alkaloids*, Butterworths, London, 1976, p. 23.

Isolation

Paudler, W. W., G. I. Kerley, and J. McKay, *J. Org. Chem.* **28**, 2194 (1963).

Powell, R. G., D. Weisleder, C. R. Smith, and J. A. Wolff, *Tetrahedron Lett.* 4081 (1969).

Powell, R. G., *Phytochemistry* **11**, 1467 (1972).

Delfel, N. E., and J. A. Rothfus, *Phytochemistry* **16**, 1595 (1977).

Synthesis

Weinreb, S. M., and J. Auerbach, *J. Amer. Chem. Soc.* **97**, 2503 (1975).

Semmelhack, M. F., B. P. Chong, R. D. Stauffer, T. D. Rogerson, A. Chong, and L. D. Jones, *J. Amer. Chem. Soc.* **97**, 2507 (1975).

Biosynthesis

Battersby, A. R., E. McDonald, J. A. Milner, S. R. Johns, J. A. Lamberton, and A. A. Sioumis, *Tetrahedron Lett.* 3419 (1975).

Parry, R. J., and J. M. Schwab, *J. Amer. Chem. Soc.* **97**, 2555 (1975).

Schwab, J. M., M. N. T. Chang, and R. J. Parry, *J. Amer. Chem. Soc.* **99**, 2368 (1977).

8.37   AMARYLLIDACEAE ALKALOIDS

The limited taxonomic distribution of alkaloids is classically represented by the alkaloids of the Amaryllidaceae. No other alkaloids occur in this plant family and these alkaloids have not yet been found in any other plant family. Over 100 alkaloids have been found in this series, almost all in the past 20 years.

For the most part they are derived by the elaboration of a nucleus of 15 carbon atoms, an aromatic C_6-C_1 unit, and a reduced aromatic C_6-C_2 unit. Almost incredibly there are eight well-established ring systems (Figure 19). Some typical examples are lycorine (**832**), galanthamine (**833**), crinine (**834**), tazettine (**835**), narciclasine (**836**), and montanine (**837**). A biogenetic numbering is used for these skeleta.

8.37.1 Isolation and Chemistry

Lycorine (**832**) is the most widely distributed alkaloid in this family and is easy to isolate because of its insolubility in chloroform. The structure was deduced by chemical degradation and X-ray crystallography of dihydrolycorine hydrobromide.

Galanthamine (**833**) is a crystalline alkaloid obtained commercially from *Galanthus woronwii* Losinsk., and has also been isolated from *Lycoria squamingera* Maxim., *Leucojum vermum* L., *G. nivalis* L., and several *Ungernia* species. The structure was determined by Barton and Kirby in 1962 and subsequently confirmed by X-ray crystallography. Lycoramine (**838**), first studied in the 1930s, is dihydrogalanthamine, a relationship that was easily established by catalytic reduction.

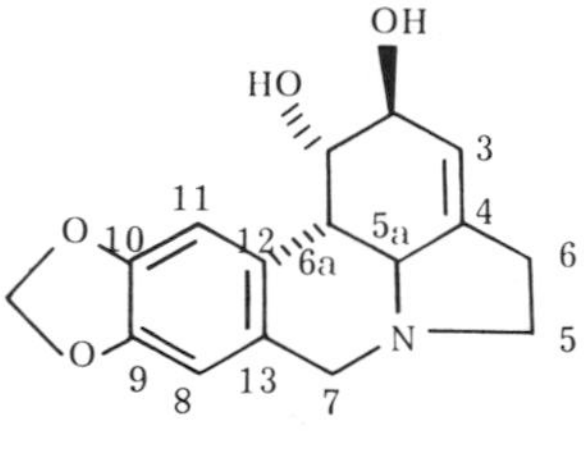

832 lycorine

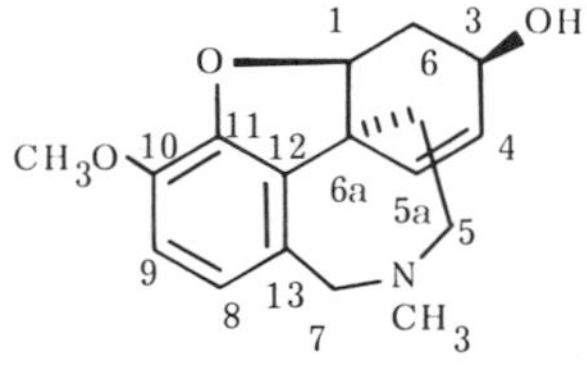

833 galanthamine

838 lycoramine 4,5a-H2

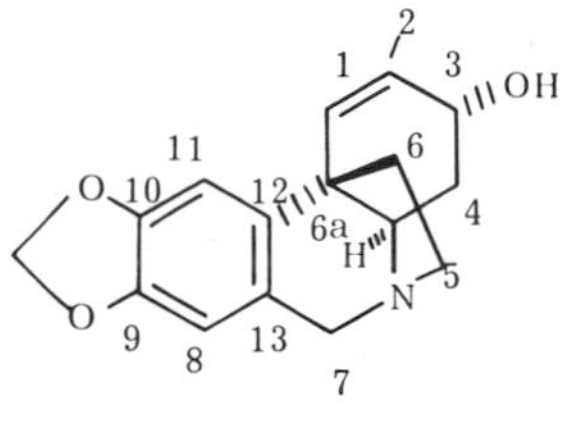

834 crinine

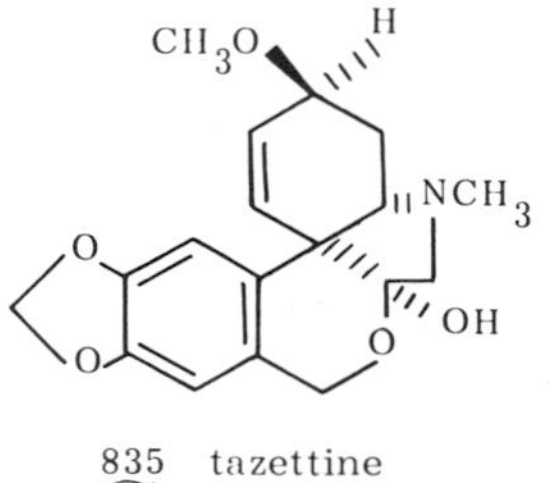

835 tazettine

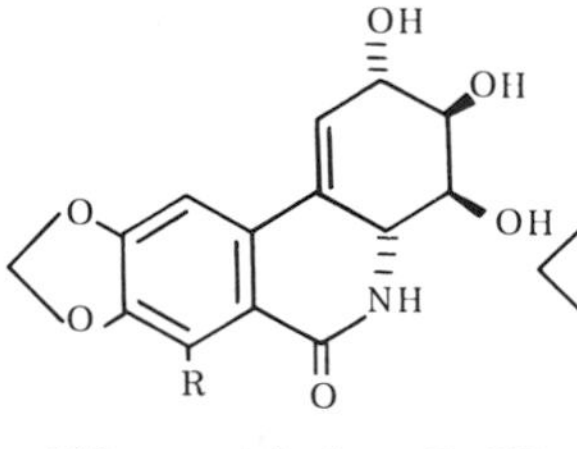

836 narciclasine, R=OH

851 margetine, R=H

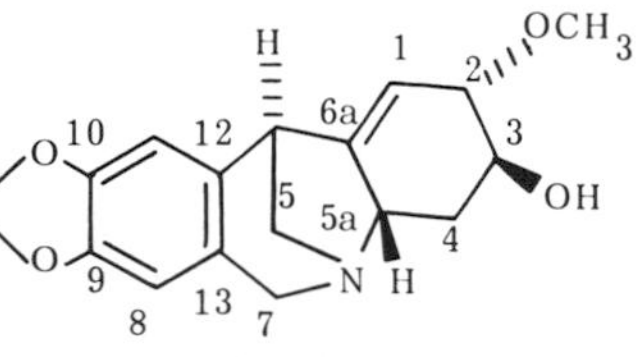

837 montanine

Figure 19 Some representative Amaryllidaceae alkaloids.

Crinine (**834**) occurs frequently in *Crinum* species, and its structure was established by Wildman in 1960. The allylic alcohol was proven by manganese dioxide oxidation to oxocrinine (**839**). N-Methylation of this compound followed by treatment with alkali gave an optically inactive dienone **840**. This demonstrates the ease with which some of the skeletal interrelationships occur in the Amaryllidaceae alkaloids.

Several of the crinane-type alkaloids have oxygenation at C-6 and C-7 and typical examples are haemanthamine (**841**) and haemanthidine (**842**), which have been obtained from over 20 genera of the Amaryllidaceae. It is of interest that these alkaloids are in the opposite antipodal series to crinine. (**834**). Treatment of haemanthamine (**841**) with acid afforded **843**. This established both the stereochemistry of the C-6 hydroxy group and the *cis*-fusion of the pyrrolidine ring to the reduced aromatic system. In solution, haemanthidine exists as an equilibrating mixture of the C-7 epimers.

Haemanthidine (**842**) on oxidation with manganese dioxide (benzylic hydroxyl group) affords the corresponding 7-oxo compound **844**. The

841 haemanthamine, R=H$_2$
842 haemanthidine, R=OH, H
844 R = =O

843

application of Bredt's rule to this system indicates that although the carbonyl is adjacent to nitrogen, the compound is still basic and the carbonyl group functions as a ketone not a lactam. Thus lithium aluminum hydride reduction gives the parent compound **842**, not haemanthamine (**841**).

The versatility of haemanthamine (**841**) is demonstrated further by its reaction with mesyl chloride in pyridine and mild hydrolysis to give iso-haemanthidine (**845**). In simple terms this reaction may be considered as proceeding stereospecifically by way of the mechanism shown in Scheme 108.

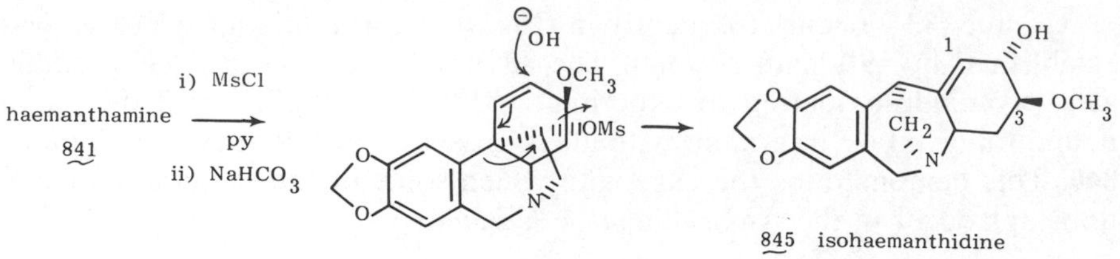

Scheme 108

Quite a different nucleus, the [2]benzopyrano[3,4g]indole system, is found in tazettine (**835**), an artifact produced from pretazzetine (**846**) during workup with base, alumina, or prolonged standing. Thus **846** is the major alkaloid of *Sprekelia formosissima* L., but is converted to tazettine on standing. Oxidation with manganese dioxide gave a lactone **847** which could also be produced from **842** by lactonization and N-methylation. Clearly in this reaction sequence the 6-oxo system in behaving more like an amide.

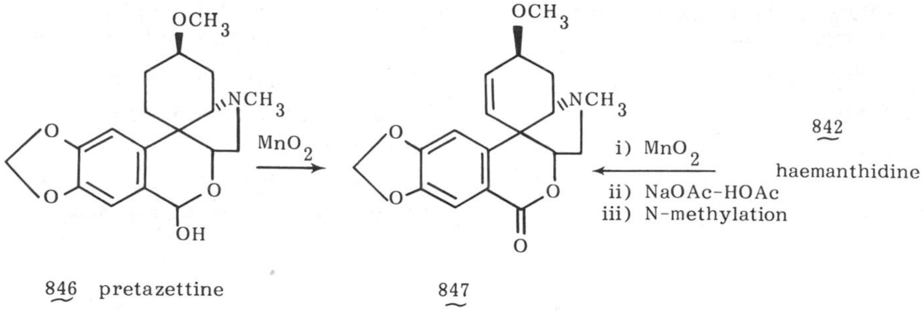

846   pretazettine   847

Yet another structural variant is montanine (**837**) from *Haemanthus montanus* native to S. Africa, an alkaloid in the methanomorphanthridine series. Oppenhauer oxidation gave the aminophenol **848** by β-elimination of the amino group in the intermediate ketone **849**, and tautomerization. A related alkaloid, cherylline (**850**), was recently obtained from a number of *Crinum* species.

Groups in Italy and Japan have investigated several members of the Amaryllidaceae, showing anticancer activity. The alkaloids responsible for this activity are narciclasine (**836**), margetine (**851**), and narciprimine (**852**). The latter compound is an artifact. Treatment of narciclasine with dilute acid gave narciprimine (**852**), the structure of which was proven by photolytic synthesis (Scheme 109). The structure and stereochemistry of narciclasine (**836**) were recently confirmed by X-ray crystallography.

Clivia miniata produces a number of alkaloids based on the [2]benzopyrano[3,4g]indole skeleton, including clivonine (**853**).

Two alkaloids which are apparently biosynthetic end points or degradation products are belladine (**854**) and ismine (**855**).

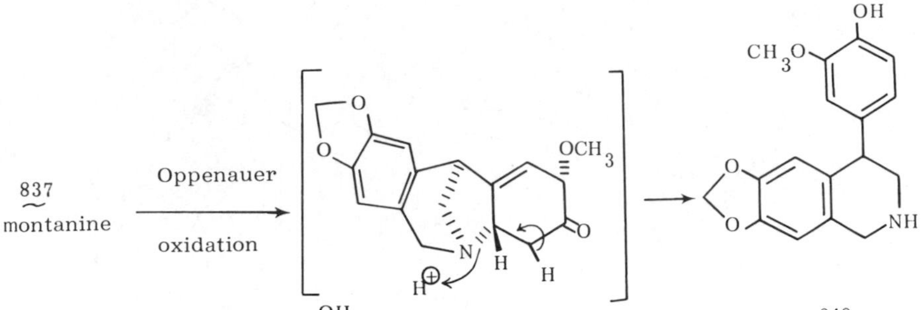

837
montanine

Oppenauer
oxidation

849

848

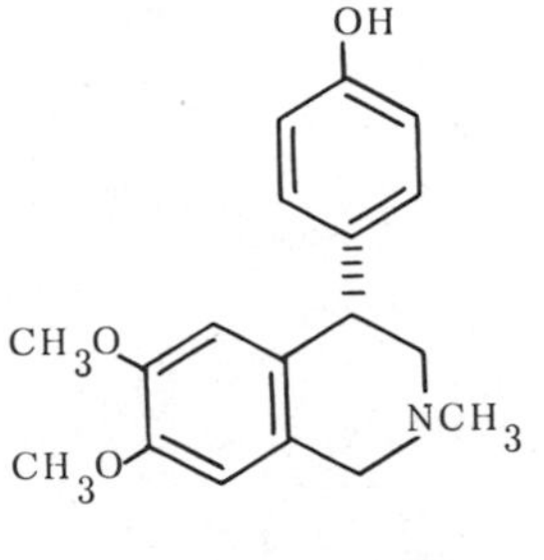

850   cherylline

8.37.2   Synthesis

Stevens has provided an excellent review of the synthesis of Amaryllidaceae alkaloids and thus only limited examples will be given here.

The first synthesis in this series was that of (±)-crinane (**856**) by Wildman in 1956 and the basic strategy (Scheme 110) involves Michael addition to

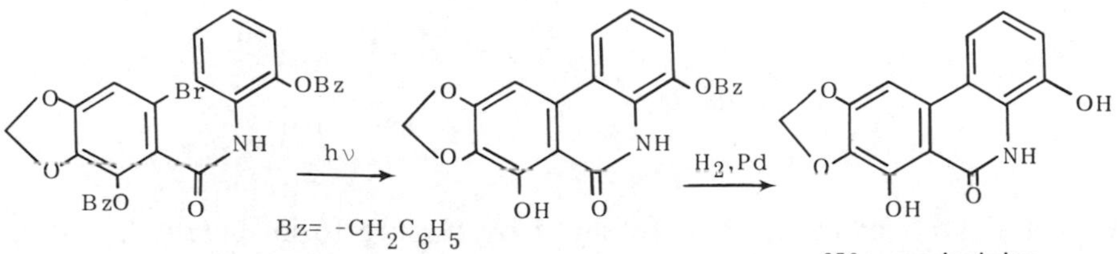

Bz= -CH₂C₆H₅

852   narciprimine

Scheme 109   Mondon–Krohn synthesis of narciprimine (**852**).

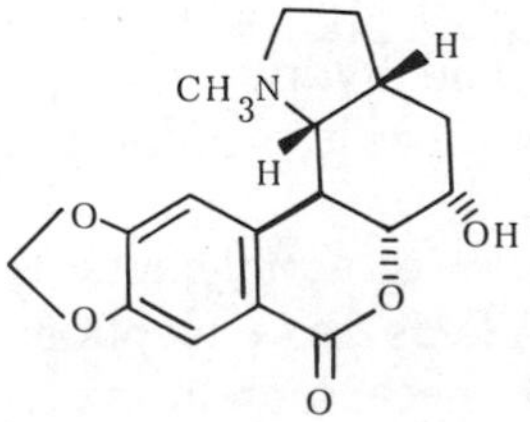

853   clivonine

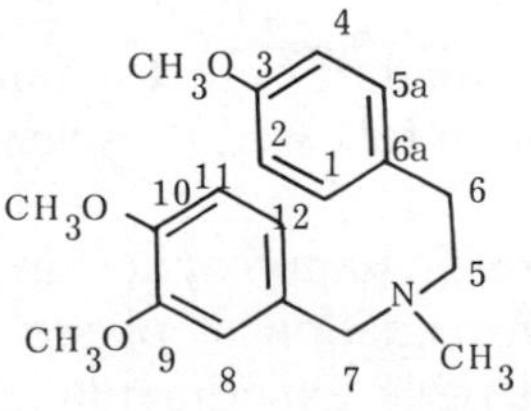

854   belladine

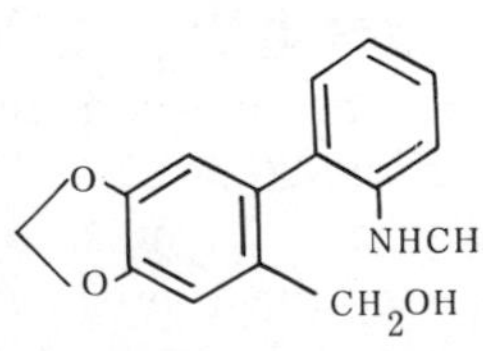

855   ismine

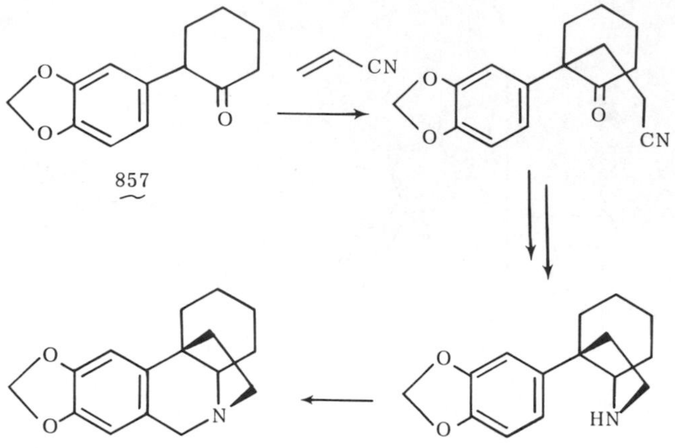

857

856   (±)-crinane

Scheme 110

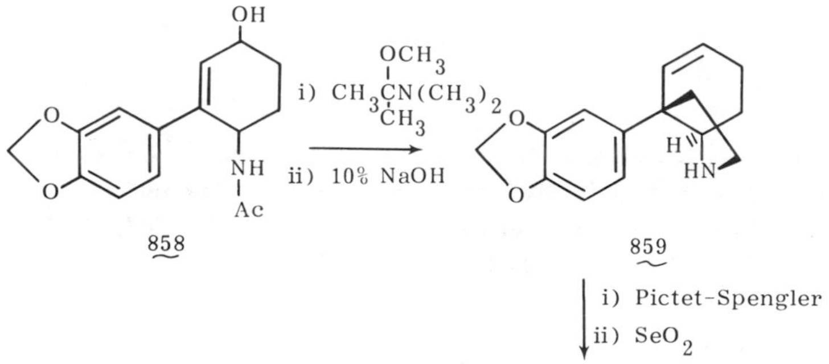

858

859

i) Pictet–Spengler

ii) SeO$_2$

834   (±)-crinine

the 2-arylcyclohexanone (**857**) followed by closure to an octahydroindole derivative and Pictet–Spengler cyclization.

Since this time several syntheses in the crinane series have appeared and many new routes developed. The Muxfeldt synthesis involves as a key intermediate **858**, which can be elaborated to **859** by the Meerwein–Eschenmoser reaction, base treatment and hydride reduction. Pictet–Spengler condensation followed by selenium dioxide oxidation gave (±)-crinine (**834**).

Perhaps the most interesting approaches are those of a biogenetic type involving the one-electron oxidation of phenols. Yields are exceptionally variable and often considerable experimentation is necessary in order to maximize yields. Thus although **860** gave narwedine (**861**) in only 1.4% yield, oxidation of **862** with basic potassium ferricyanide proceeded in 81% yield.

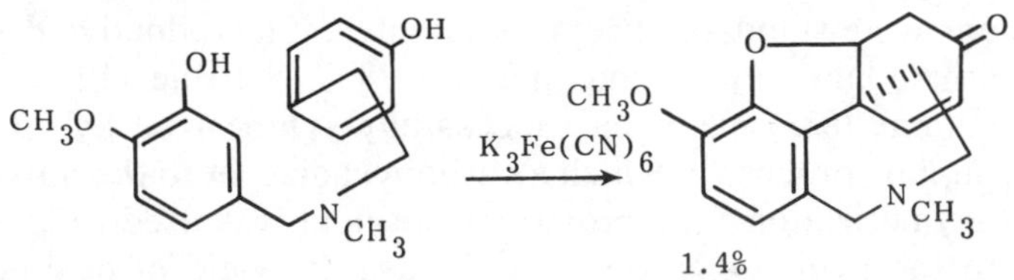

860 → 861 narwedine 1.4%

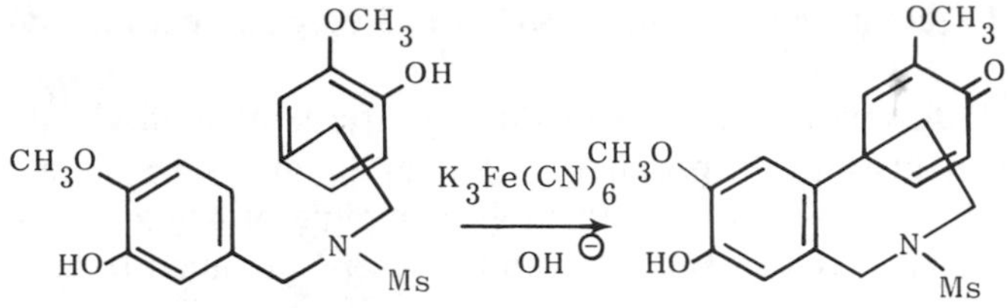

862 → 81%

Unfortunately, the resistance of the *N*-mesyl group to hydrolysis under mild conditions prevented further elaboration of this scheme. Overall the most successful system appears to be to use trifluoroacetyl group for protection of the nitrogen and vanadium oxychloride for coupling, and an example is in the synthesis of maritidine (**863**) as shown in Scheme 111.

A chiral synthesis of (+)-maritidine (**863**) has been achieved by Yamada and co-workers beginning with Schiff base **864** from veratraldehyde and L-tyrosine methyl ester. In this instance phenolic coupling of the trifluoroacet-amide derivative **865** was accomplished with thallium tris-trifluoroacetate in 67% yield. Treatment with methanolic ammonia and removal of the N-protecting group gave the enone **866** as a *single* diastereoisomer. The sub-

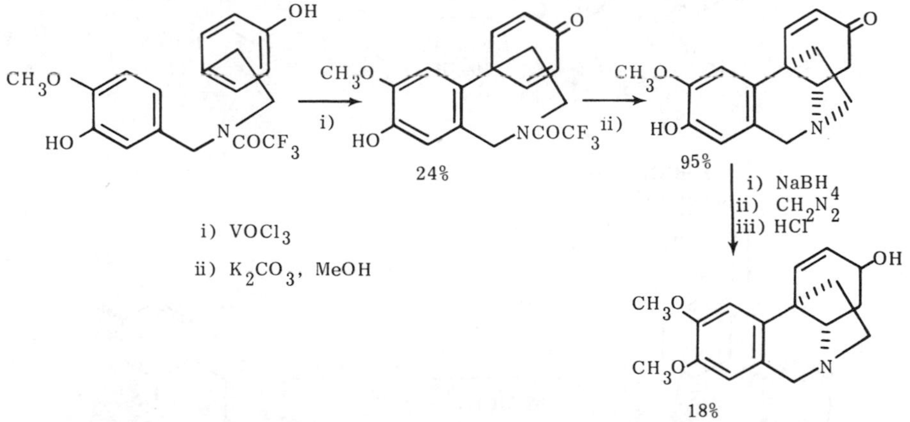

i) VOCl$_3$

ii) K$_2$CO$_3$, MeOH

i) NaBH$_4$
ii) CH$_2$N$_2$
iii) HCl

863 maritidine 18%

Scheme 111

sequent steps were standard except for an interesting reductive decyanation reaction with sodium–liquid ammonia at −78°C (Scheme 112).

Galanthamine (833) has been successfully synthesized by an oxidative phenolic coupling process in which a carbonyl adjacent to the nitrogen again prevents polymerization. The bromo substitutent was used to inhibit coupling *para* to the hydroxy group. In this case the yield of desired product is a quite respectable 40%. Lithium aluminum hydride reduction gave a mixture of racemic galanthamine (833) and 3-epigalanthamine (864) (Scheme 113).

Yamada has extended his work using L-tyrosine methyl ester to induce asymmetry, this time to produce (+)-galanthamine (865), the optical antipode of natural galanthamine. Phenolic coupling of 866 afforded 867 which could be transformed to 868. Subsequent steps involved reductive decyanation, hydrolysis of the acetyl group, and reductive elimination of the phenolic group (Scheme 114).

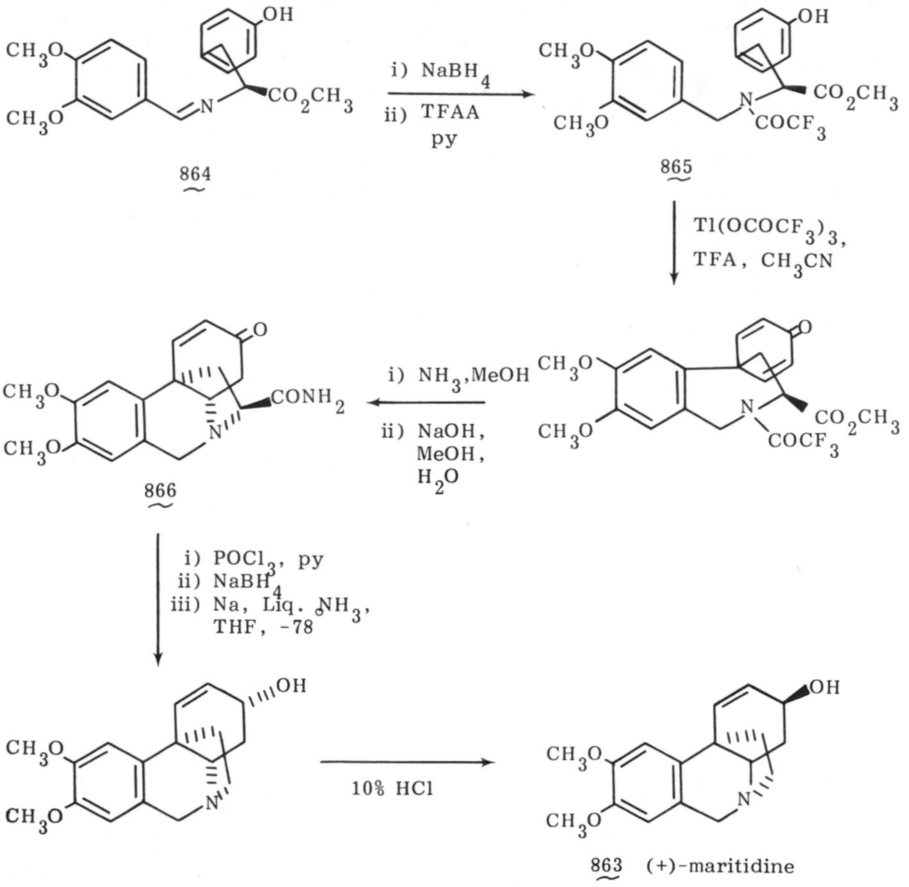

Scheme 112 Yamada synthesis of (+)-maritidine (863).

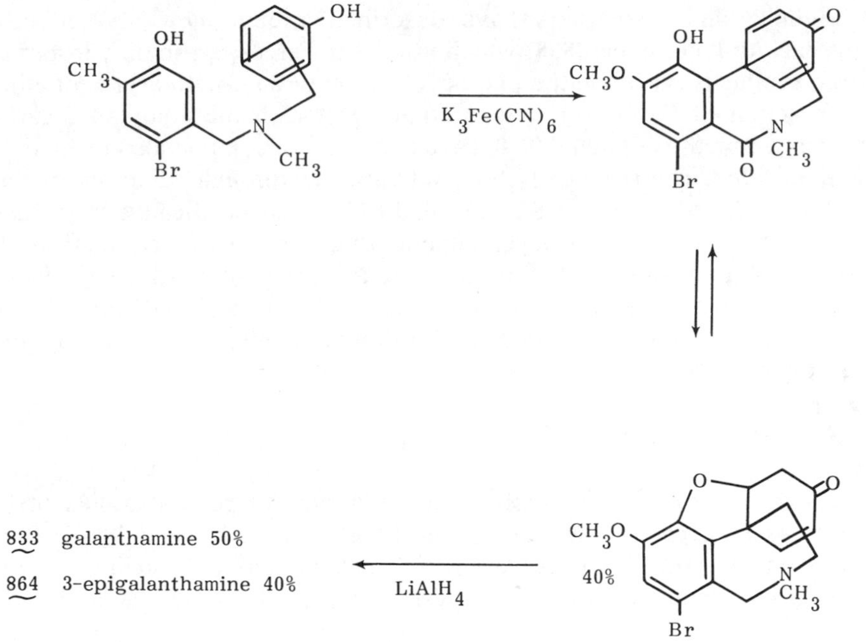

Scheme 113   Kametani synthesis of (±)-galanthamine (**833**).

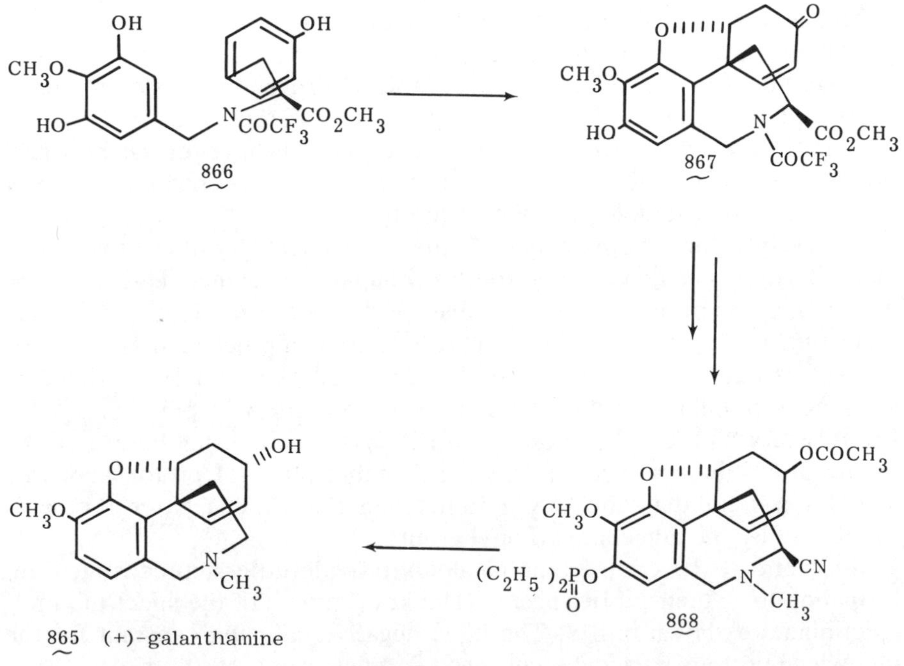

- **833** galanthamine 50%
- **864** 3-epigalanthamine 40%

Scheme 114   Yamada synthesis of (+)-galanthamine (**865**).

865 (+)-galanthamine

Schultz and co-workers have described a quite new and effective approach to lycoramine (838) which utilizes a stereospecific dihydrobenzofuran synthesis involving the photocyclization of an α-keto aryl enol ether.

Reaction of the epoxy keto urethane 869 with the anion of 5-carbomethoxy-2-methoxy phenol (870) gave a mixture of compounds in which 871 predominated. This reaction is thought to proceed through a diketone enolate such as 872. Irradiation of 871 afforded 873 stereospecifically. With these stereo centers fixed, the azacycloheptane ring was constructed to afford the ketone 874. In lycoramine the oxygen functionality is at the adjacent position to that in 874 and consequently a 1,2-carbonyl transposition was required. The resulting ketone was reduced with lithium aluminum hydride to afford lycoramine (838) (Scheme 115).

8.37.3 Spectral Data

The UV spectra of some representative Amaryllidaceae alkaloids of the major skeletal types show quite similar maxima (λ_{max} 240 and 288–295 nm) based on the tetrasubstituted aromatic nucleus. The only change is with those alkaloids having a δ-lactone system, which show maxima at 268 and 308 nm.

The most important peaks in the mass spectrum of lycorine (832) are at m/e 226 and 227 and are envisaged as being produced by initial hydrogen ion transfer followed by retro Diels–Alder process (Scheme 116).

Montanine (837), M^+301, shows the substantial loss of a methoxy radical to give an ion at m/e 270, followed by decyclization and aromatization to give an ion at m/e 223 (Scheme 117).

Tazettine (835), M^+331, undergoes initial hydrogen in transfer and retro Diels–Alder reaction to give the base peak m/e 247 (Scheme 118).

The NMR spectra of the Amaryllidaceae alkaloids have for the most part proved crucial in determining the structure and stereochemistry of new alkaloids. Some examples will illustrate this.

In the spectrum of clivonine (853) the stereochemistry of the five asymmetric centers was determined from the coupling constants. Thus the C-5a proton (4.06 ppm) appears as a doublet of doublets with $J_{5a,6a} = 12.5$ Hz indicating a *trans* ring junction and a *cis* relationship between H-5 and H-5a. The proton at C-1 (2.87 ppm) also appeared as a doublet of doublets showing a coupling constant with the C-6a proton (3.23 ppm) of 9.5 Hz (*trans*) and with the C-2 proton of 5.8 Hz (*cis*). One unusual feature of the spectrum of clivonine acetate was the fact that the C-11 proton appeared at a lower field than the C-8 proton, even though this latter proton is deshielded by the adjacent carbonyl group.

In contrast, the crinine-type alkaloids provide little stereochemical information from their NMR spectra. The key features of the spectrum of 3-epicrinine are shown in 875. The large negative coupling constant for the methylene protons adjacent to nitrogen is noteworthy.

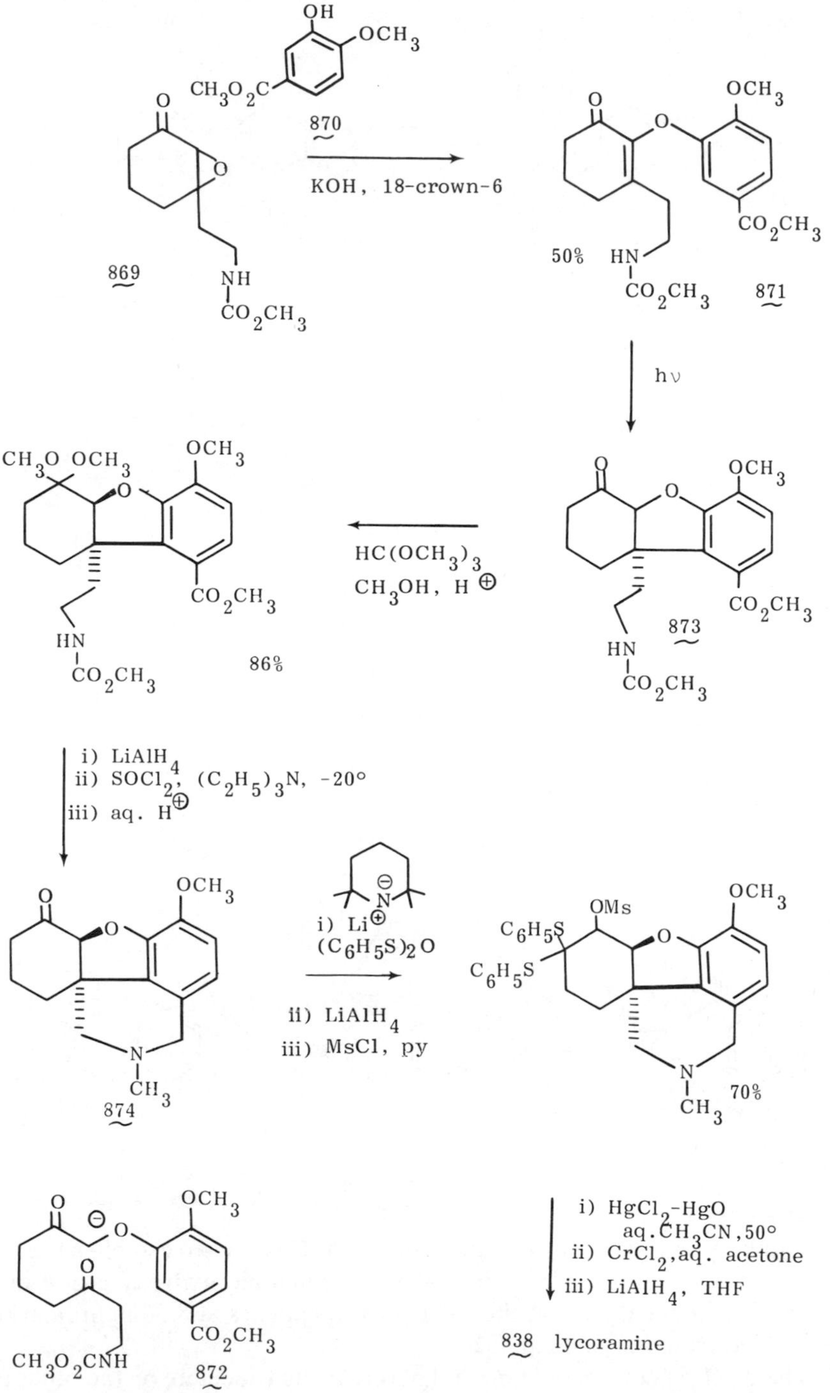

Scheme 115 Schultz and co-workers' synthesis of lycoramine (**838**).

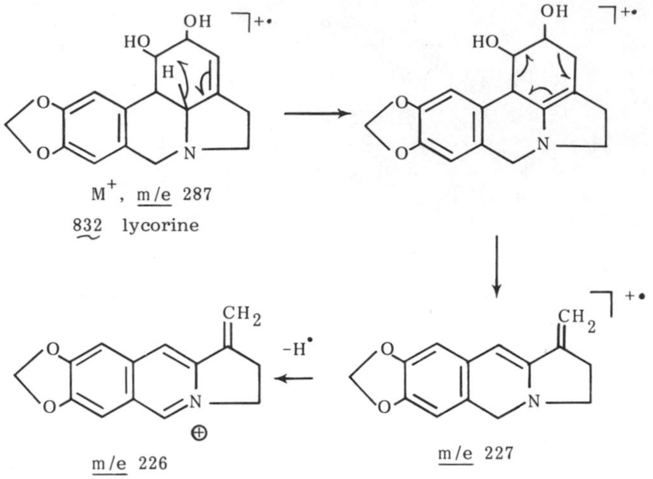

Scheme 116

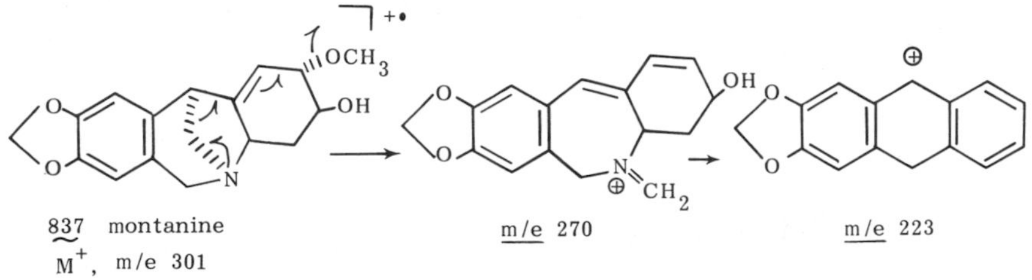

Scheme 117

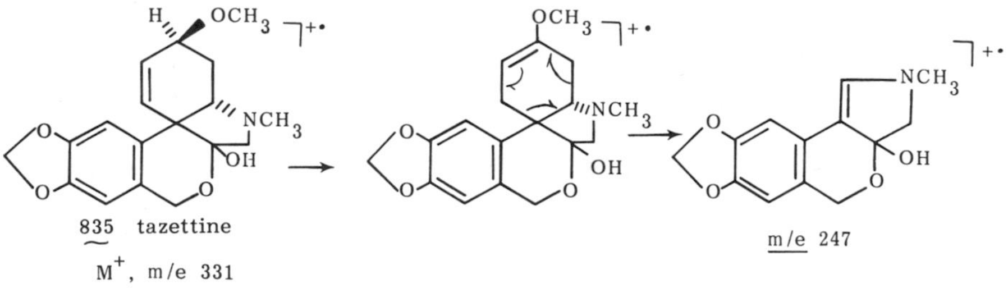

Scheme 118

Similarly, the spectrum of narciprimine (**852**) serves to establish the structure. In the structural isomer in which the phenolic hydroxy group in ring C is at C-1 rather than C-4, the C-11 proton appears at 8.50 ppm, markedly deshielded from its value in **852**.

The NMR spectrum of *O*-methyl narciclasine triacetate on the other hand was used to make the stereochemical assignments shown in **876**. The small coupling constant between H-2 and H-3 was initially interpreted in terms

875 3-epicrinine   852 narciprimine   876 O-methylnarciclasine
                                                         triacetate

of a *cis* stereochemistry. But the coupling constant between vicinal diace-
tates is known to be reduced when one acetate lies in the plane of one of
the coupling protons.

The ^{13}C NMR data for some representative Amaryllidaceae alkaloids are
shown in Table 12.

8.37.4 Biosynthesis

The biosynthesis of the Amaryllidaceae alkaloids has been studied mainly
by Barton, Battersby, Kirby, and Fuganti and their respective co-workers.
Barton postulated that the alkaloids of the Amaryllidaceae were biogenetically derived from norbelladine (**877**) by way of various phenolic coupling
processes (Scheme 119). The overall correctness of the proposal was established by Battersby in the early 1960s using both singly and doubly labeled
norbelladine (**877**). In particular [5,7-^{14}C$_2$]norbelladine (**877**) was incorporated as an intact unit into galanthamine (**833**) and haemanthamine (**841**),
and a more advanced precursor **878** was also specifically incorporated into
galanthamine (**833**) without scrambling of an *N*-methyl label. In an analogous
experiment, doubly labeled **879** was incorporated into haemanthamine (**841**)
at C-5 and the methylenedioxy group. The methylation sequence of norbelladine is crucial in determining the precursor relationships; thus **879** was
not incorporated into galanthamine (**833**), although *N*-methylnorbelladine
(**880**) was.

With this preliminary information in hand, two areas of interest remain,
the mode of formation of norbelladine (**877**) and the stereospecificity of the
elaboration of the various individual alkaloids.

The origin of the C$_6$-C$_2$-N-C$_1$-C$_6$ system has been clearly established.
The C$_6$-C$_2$ unit is specifically derived from tyrosine (**2**) and the C$_6$-C$_1$ specifically from phenylalanine (**1**). The latter amino acid is not *para*-hydroxyl-
ated to tyrosine in Amaryllidaceous plants. Thus [3-^{14}C]tyrosine (**2**) specifically labels C-6 (and not C-7) of lycorine (**832**), haemanthamine (**841**),
and tazettine (**835**). Conversely [3-^{14}C]phenylalanine (**1**) and [3-^{14}C]*p*-cou-

Table 12 ^{13}C NMR of Some Representative Amaryllidaceae Alkaloids

	Compound					
Carbon	Crimine (834)	Haemantha-mine (841)	Galanthamine (833)	Montanine (837)	Tazettine (835)	Clivonine (853)
1	129.5[a]	128.0[a]	88.1	113.1	130.9	70.2
2	127.2[a]	127.2[a]	30.0	79.9	128.8	34.0
3	63.5	80.0[b]	62.2	68.5	72.8	29.5[a]
4	33.5	29.5	126.0	33.1	27.0	68.5
5	54.8	61.5	54.3	57.3	62.1	53.9
5[a]	63.0	62.7	126.8	60.9	70.3	82.8
6	44.7	73.0[b]	34.0	45.9	102.1	32.0[a]
6[a]	44.7	50.0	48.2	154.0	50.5	34.0
7	58.6	63.3	60.5	58.7	65.4	164.7
8	140.0	106.9	121.6	107.2[b]	104.1	109.5
9	132.5	146.5[c]	110.5	146.7[d]	146.7	146.8
10	147.0	147.0[c]	145.5	145.9[d]	146.7	152.5
11	96.0	103.3	144.0	106.7[b]	109.5	107.8
12	138.0	135.0	132.7	124.6	125.8	141.2
13	116.0	126.9	129.5	132.5	128.2	119.0
—OCH₂O—	99.3	101.0	—	101.1	101.0	102.7
OCH₃	59.3	56.0	56.0	55.4	56.0	—
NCH₃	—	—	42.2	—	42.3	46.0

[a-d] Assignments may be reversed.
Source: Data from W. O. Crain, Jr. *et al.*, *J. Amer. Chem. Soc.* **93**, 990 (1971).

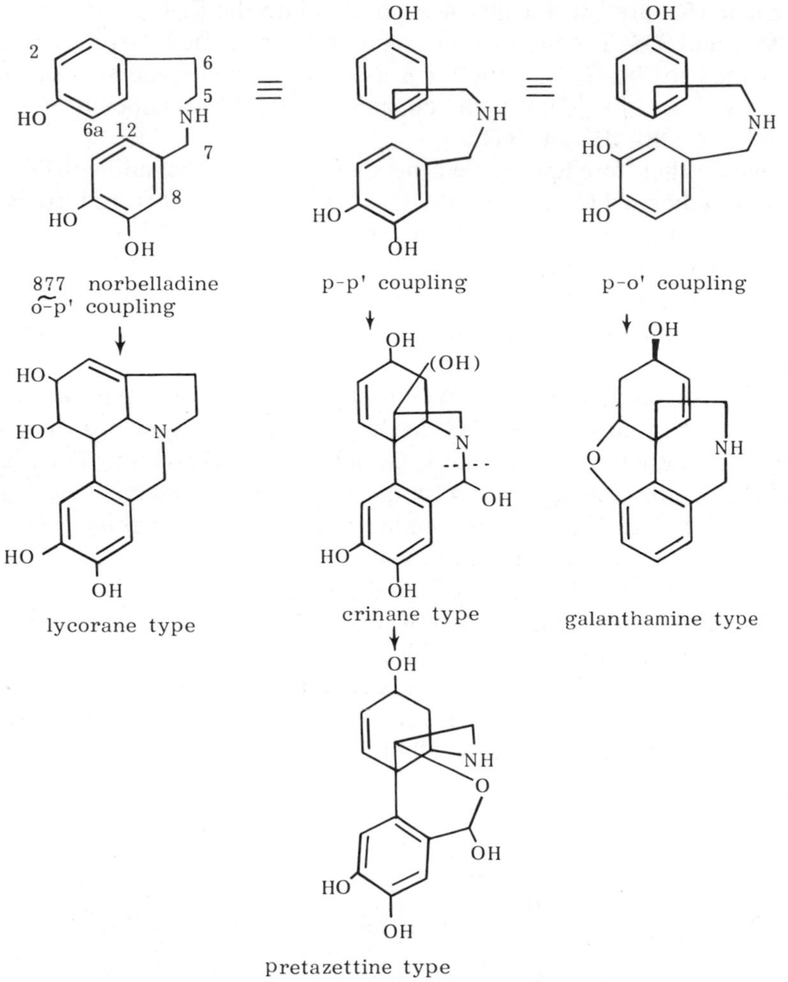

877 norbelladine
ŏ–p' coupling

p–p' coupling

p–o' coupling

lycorane type

crinane type

galanthamine type

pretazettine type

Scheme 119 Biogenetic relationships of the major Amaryllidaceae skeleta.

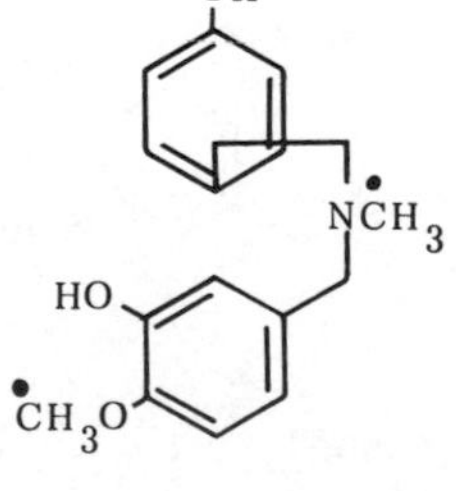

878

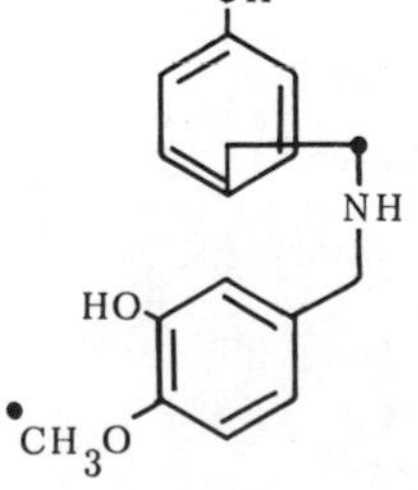

879

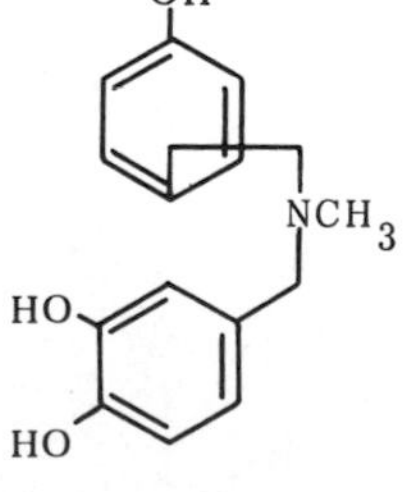

880

maric acid (58) are specifically incorporated into the C_6-C_1 unit (i.e. C-7) of 832, 841, and 835. The next step is apparently aromatic hydroxylation prior to cleavage to the C_6-C_1 unit, for p-hydroxybenzaldehyde was not incorporated whereas [7-^{14}C]protocatechualdehyde (180) was specifically incorporated into both 832 and 841.

It would therefore appear that the sequence to norbelladine (877) is that shown in Scheme 120 in which the C_6-C_2 unit is derived from tyrosine and the C_6-C_1 unit from phenylalanine via caffeic acid (59).

The further rearrangement of initially formed alkaloid skeleta has also been studied and, to give one example, it has been demonstrated that the sequence haemanthamine (841) → haemanthidine (842) → tazettine (835) operates in one direction only.

In the conversion of O-methylnorbelladine (889) having tritium at C-2, 6a, 8, and 11 to norpluviine (881) in the lycorane series, two tritium atoms were specifically lost at C-12 and C-6a, where phenyl coupling occurs. Two plausible mechanisms to account for this result are reasonable. One of these, shown in Scheme 121, involves a nine-membered ring biphenyl 882 as an intermediate. This product would be comparable to the 10-membered ring biphenyl involved in *Erythrina* alkaloid formation (Scheme 73). The tritium at C-2 in norpluviine (881) is stereospecifically oriented as shown and is of interest because on conversion to lycorine (832) in *Zepheranthes candida* and *Narcissus* daffodils this tritium is retained. Thus C-2 hydroxylation is stereospecific and occurs with inversion of configuration. This appears to be the only example of a biological hydroxylation taking place with an *overall*

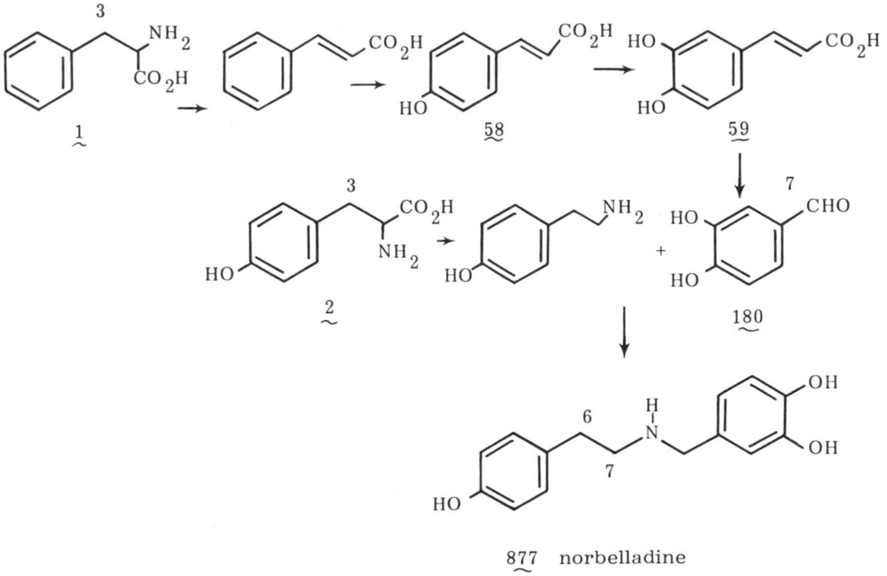

Scheme 120   Formation of norbelladine (877) in Amaryllidaceae plants.

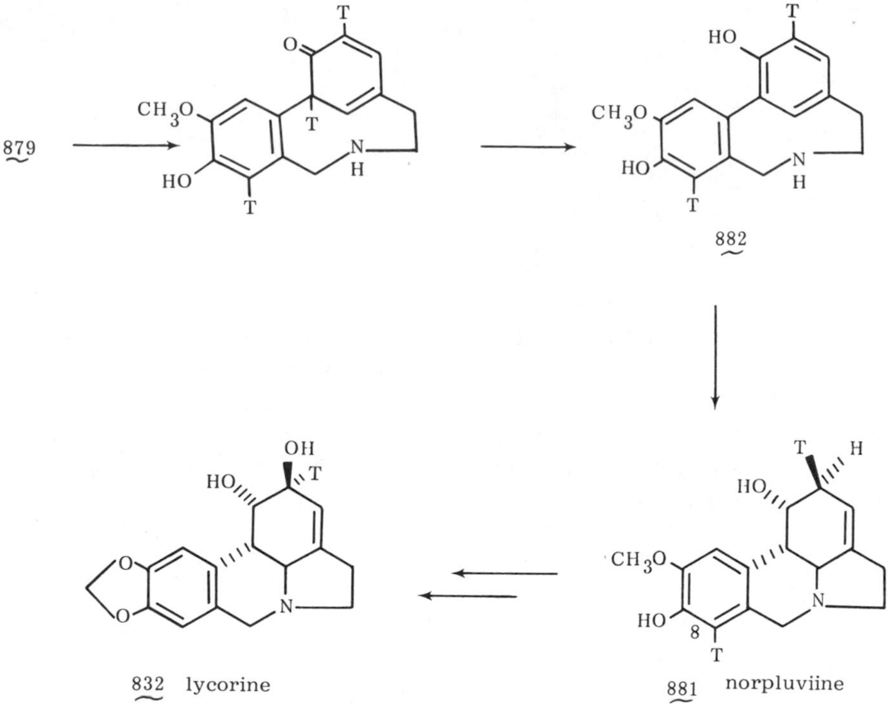

Scheme 121

inversion of configuration. It may well indicate that the process is not a direct hydroxylation but rather that it involves several steps.

In the biosynthesis of haemanthamine (**841**) a hydroxyl group is introduced at C-6 and has the α-stereochemistry. Two groups, those of Kirby and Battersby, have examined the stereospecificity of this process. The conclusion was that hydroxylation occurs with retention of configuration by removal of the *pro*-R hydrogen. Thus Kirby found that from [2S, 3S- ³H] tyrosine (**2**), haemanthamine (**841**) retained 87% of the tritium at C-6. The stage at which this hydroxylation occurs has also been investigated. Since [3-³H]crinine (**834**) but not 3-epicrinine (**875**) was incorporated, this reaction occurs *after* formation of the nucleus and is dependent on the C-3 stereochemistry. The point at which the *o*-methoxyphenol is converted to a methylenedioxy is not known.

Haemanthamine (**841**) is a precursor of haemanthidine (**842**) and further work has established that the benzylic hydroxylation process is also sterospecific, although it is not known which hydrogen is lost or whether hydroxylation proceeds with inversion or retention of configuration.

The group of alkaloids represented by clivonine (**853**) or lycorenine (**883**) have been suggested to be derived from norpluviine (**881**) along the lines shown in Scheme 122. In substantiation of this, doubly labeled norpluviine

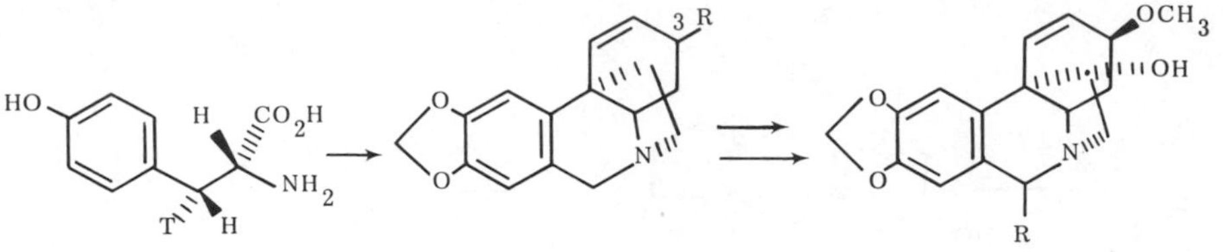

2S-[3S-³H]-tyrosine

| 834 | crinine, R=β-OH |
| 875 | 3-epicrinine, R=α-OH |

| 841 | haemanthamine, R=H |
| 842 | haemanthidine, R=OH |

(881) was incorporated into lycorenine (883) without loss of label, but pluviine (884) was not incorporated, indicating that the phenolic group in 881 is necessary for further reaction at the benzylic position. It has also been shown that the hydrogen introduced in the formation of norbelladine (877), the *pro*-7R hydrogen, is the same one lost in the hydroxylation of C-7 of norpluviine (881).

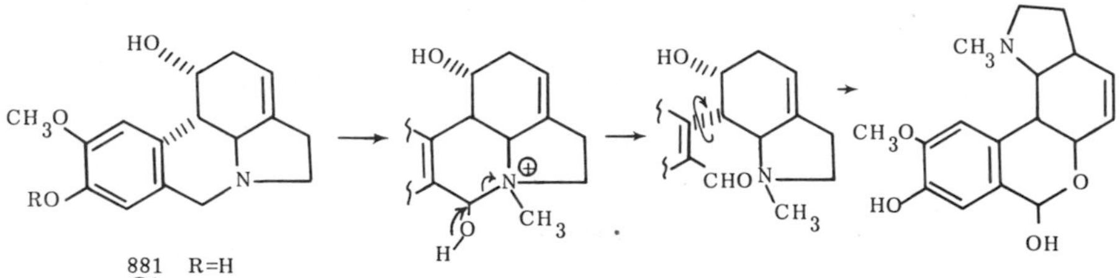

| 881 | R=H |
| 884 | R=CH₃ |

883 lycorenine

Scheme 122 Biogenesis of lycorenine (883).

The limited work on the biosynthesis of montanine (837) from norbelladine (877) has given conflicting results, Fuganti's group indicating that [6R-³H]O-methylnorbelladine (879) retained tritium on incorporation into 837, with Wildman finding that there is a loss of tritium during this reaction. The conversion of haemanthamine (841) *in vitro* into a montanine-like compound proceeds by loss of the 6α-hydroxy group. Wildman found, in confirmation with others, that haemanthamine (841) in the same plant had also lost ³H specifically at C-6.

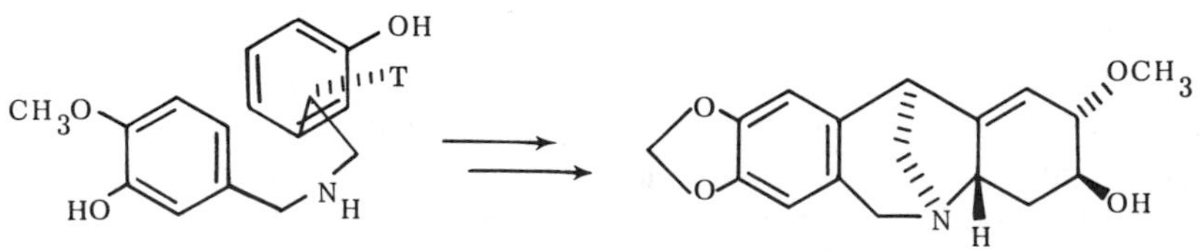

879 [6R-³H]-O-
methyl norbelladine

837 montanine

The norbelladines **885** and **886** were incorporated into narciclasine (**836**). From **885**, 75% of the original tritium was retained, and this was interpreted as occurring by random protonation at C-4 during the formation of a crinine intermediate, followed by stereospecific hydroxylation. Note that C-2 reduction and hydroxylation is completely stereospecific. The intermediacy of a crinine-type intermediate was established when crinine (**834**) [but not 3-epicrinine (**875**)] was incorporated into narciclasine (**836**). Interestingly, only the (+)-isomer of crinine is incorporated. The mechanism of loss of the C-2 unit of the crinane percursor is not known.

885   886   836   narciclasine

Ismine (**855**) occurs in several Amaryllidaceae plants (e.g. *Spreckelia formosissima*) with the more familiar alkaloids such as haemanthamine (**841**).

Although it was ignored for many years it is now clear that this alkaloid is derived from the crinane skeleton by elimination of the ethano bridge at a late stage. Thus noroxomaritidine (**887**), triply labeled with ^{3}H at C-7 and C-11 and ^{14}C at C-5, was incorporated into ismine (**855**) in *S. formosissima* with complete loss of ^{14}C label and 50% of the tritium label from C-7 of **887**. Further work demonstrated that the *pro*-S proton at C-7 was preferentially removed on oxidative cleavage of the C-7-N bond. These results parallel the 7-hydroxylation of haemanthamine (**841**) to haemanthidine (**842**).

887   noroxomaritidine   855   ismine

8.37.5   Pharmacology

Galanthamine (**833**) has been the subject of recent interest because of its powerful cholinergic activity and a report that it exhibits analgesic activity comparable to morphine. The alkaloid has been used in Russia in the treatment of myasthenia gravis, myopathy, and diseases of the nervous system. Several derivatives have been evaluated for their anticholinesterase activity and also as antibacterial agents and central nervous system depressants.

Pretazettine (**846**) has been used in combination with DNA-binding and alkylating agents in the treatment of the Rauscher leukemia virus.

Several other Amaryllidaceae alkaloids also display anticancer activity, including narciclasine (**836**) and oxidation product **888** of lycorine (**832**).

Lycorine (**832**) and lycoricidine (**889**) possess plant growth inhibiting

888

889   lycoricidine

properties, and narwedine (**861**) from *Narcissus cyclamineus* enhances the pharmacological effects of caffeine in mice.

Jimenez and co-workers have examined the inhibition of protein synthesis in eukaryotic cells by various Amaryllidaceae alkaloids. The most potent of the common alkaloids were dihydrolycorine, haemanthamine, lycorine, narciclasine, and pretazettine.

LITERATURE

Reviews

Cook, J. W., and J. D. London, *Alkaloids NY* **2**, 331 (1952).

Wildman, W. C., *Alkaloids NY* **6**, 289 (1960).

Wildman, W. C., *Alkaloids NY* **11**, 307 (1968).

Wildman, W. C., in S. W. Pelletier (Ed.), *Chemistry of the Alkaloids*, New York, 1970, p. 151.

Snieckus, V. A., *Alkaloids, London* **2**, 185 (1972).

Snieckus, V. A., *Alkaloids, London* **3**, 169 (1973).

Jeffs, P. W., in K. F. Wiesner (Ed.), *MTP International Review of Science, Organic Chemistry*, Ser. 1, Vol. 9, *Alkaloids*, Butterworths, London, 1973, p. 273.

Fuganti, C., *Alkaloids, NY* **15**, 83 (1975).

Snieckus, V. A., *Alkaloids, London* **4**, 266 (1974).

Snieckus, V. A., *Alkaloids, London* **5**, 170 (1975).

Reed, J. N. and V. A. Snieckus, *Alkaloids, London* **8**, 137 (1978).

Synthesis

Stevens, R. V., in J. Ap. Simon (Ed.), *The Total Synthesis of Natural Products,* Vol. 3, Wiley-Interscience, New York, 1977, p. 439.

Wildman, W. C., *J. Amer. Chem. Soc.* **78,** 4180 (1956).

Muxfeldt, H., R. S. Schneider, and J. B. Mooberry, *J. Amer. Chem. Soc.* **88,** 3670 (1966).

Schwartz, M. A., and R. Holton, *J. Amer. Chem. Soc.* **92,** 1090 (1970).

Yamada, S., K. Tomioka, and K. Koga, *Tetrahedron Letts.* 57 (1976).

Kametani, T., K. Yamaki, H. Tagi, and K. Fukumoto, *J. Chem. Soc., Sec. C* 2602 (1969).

Tomioka, K., K. Shimizu, S. Yamada, and K. Koga, *Heterocycles* **6,** 1752 (1977).

Tomioka, K., K. Koga, and S. Yamada, *Chem. Pharm. Bull.* **25,** 2681 (1977).

Schultz, A. G., Y. K. Lee, and M. H. Berger, *J. Amer. Chem. Soc.* **99,** 8065 (1977).

Biosynthesis

Battersby, A. R., R. Binks, S. W. Brener, H. M. Fales, W. C. Wildman, and R. J. Highet, *J. Chem. Soc.* 1595 (1964).

Barton, D. H. R., G. W. Kirby, J. B. Taylor, and G. M. Thomas, *J. Chem. Soc.* 4545 (1963).

Wildman, W. C., H. M. Fales, and A. R. Battersby, *J. Amer. Chem. Soc.* **84,** 681 (1962).

Zulalian, J., and R. J. Sukadolnik, *Proc. Chem. Soc.* 422 (1964).

Fales, H. M., and W. C. Wildman, *J. Amer. Chem. Soc.* **86,** 294 (1964).

Kirby, G. W., and H. P. Tiwan, *J. Chem. Soc., Sec. C* 676 (1966).

Heimer, N. E., and W. C. Wildman, *J. Amer. Chem. Soc.* **89,** 5265 (1967).

Bruce, I. T., and G. W. Kirby, *Chem. Commun.* 207 (1968).

Fuganti, C., and M. Mazza, *Chem. Commun.* 1196 (1971).

Fuganti, C., and M. Mazza, *Chem. Commun.* 1388 (1971).

Kirby, G. W., and J. Michael, *J. Chem. Soc. Perkin Trans. I* 115 (1973).

Fuganti, C., J. Staunton, and A. R. Battersby, *Chem. Commun.* 1154 (1971).

Fuganti, C., and M. Mazza, *J. Chem. Soc. Perkin Trans. I* 954 (1973).

Battersby, A. R., J. E. Kelsey, J. Staunton, and K. E. Suckling, *J. Chem. Soc. Perkin Trans. I* 1609 (1973).

Harken, R. D., C. P. Christensen, and W. C. Sildman, *J. Org. Chem.* **41,** 2450 (1976).

Wildman, W. C., and B. Olesen, *Chem. Commun.* 551 (1976).

Pharmacology

Bolssier, J., G. Combes, and J. Pagny, *Ann. Pharm. Fr.* **18,** 888 (1960).

Kametani, T., C. Seino, K. Kamaki, S. Shibuya, K. Fukumoto, K. Kigasawa, F. Satoh, M. Huragi, and T. Hayasaka, *J. Chem. Soc., Sec. C* 1043 (1971).

Morlock, E. B., and L. Goldman, U.S. Pat. 3,673,177 (1962); *Chem. Abstr.* **77,** 88753w (1972).

Furusawa, E., N. Suzuki, S. Funisawa, and J. Y. B. Lee, *Proc. Soc. Exp. Biol. Med.* **149,** 771 (1975).

Mondon, A., and K. Krohn, *Chem. Ber.* **108,** 445 (1975).

Jimenez, A., A. Santos, G. Alonso, and D. Vasquez, *Biochem. Biophys. Acta* **425,** 342 (1976).

8.38 MESEMBRINE AND RELATED ALKALOIDS

Mesembrine (**890**) is one of several closely related alkaloids isolated from the genus *Sceletium* in the family Aizoaceae. Stimulation of general interest in this group of compounds was preceded by reports of the use of "Channa" by natives in Africa as a narcotic drug. There are three fundamental structures, mesembrine, joubertiamine (**891**) and a new *Sceletium* isolate.

  Alkaloids in the mesembrine series are characterized by the *cis* stereochemistry of the perhydroindole system **892**, and since this stereochemistry is also found in alkaloids of the crinane (**856**) skeleton, these groups are quite closely related from a synthetic point of view.

  The ^{13}C NMR shift assignments for mesembrenone are shown in **893**.

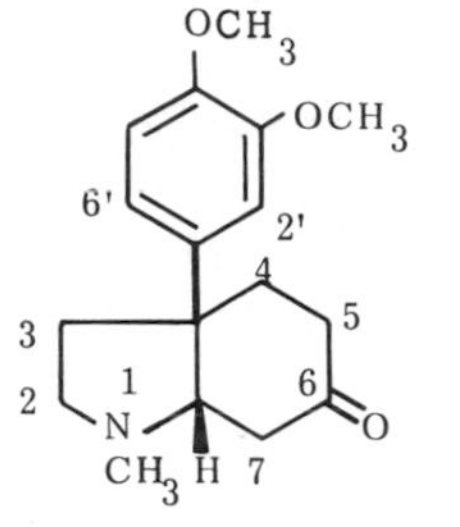

890 (−)-mesembrine 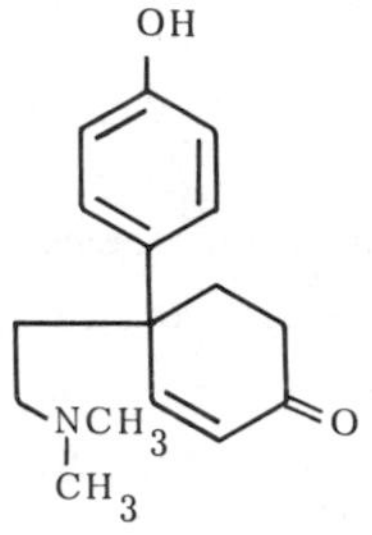 891 joubertiamine 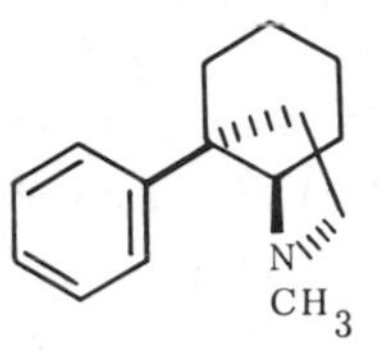 892 mesembrane 856 crinane
                     nucleus   nucleus

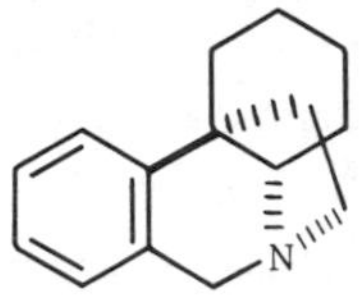

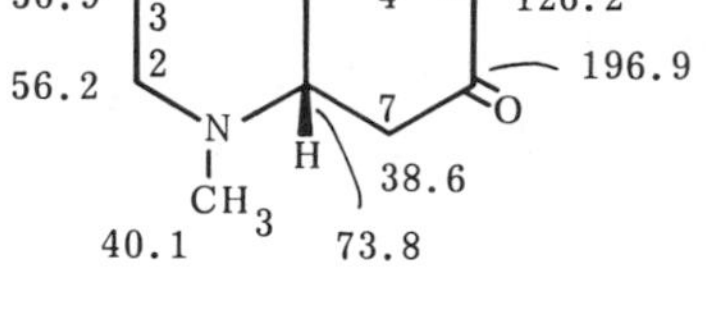

893 mesembrenone

8.38.1 Synthesis

Several novel syntheses of mesembrine and related compounds have been reported since Popelak and co-workers' first efforts in 1960 on the synthesis of the mesembrane nucleus. One of the new approaches developed almost simultaneously in three separate laboratories in 1968 is the now familiar cyclopropyl imine to pyrrolidine enamine rearrangement. Thus heating the imine **894** with either aqueous HBr or ammonium chloride gave **895**, which on reaction with methyl vinyl ketone or β-chlorovinyl methyl ketone in the presence of acid gave racemic mesembrine (**890**). Joubertiamine (**891**) may be obtained from a mesembrine-like intermediate **896** by conversion to the methiodide salt **897** followed by elimination of the β-amino group, with concomitant O-demethylation.

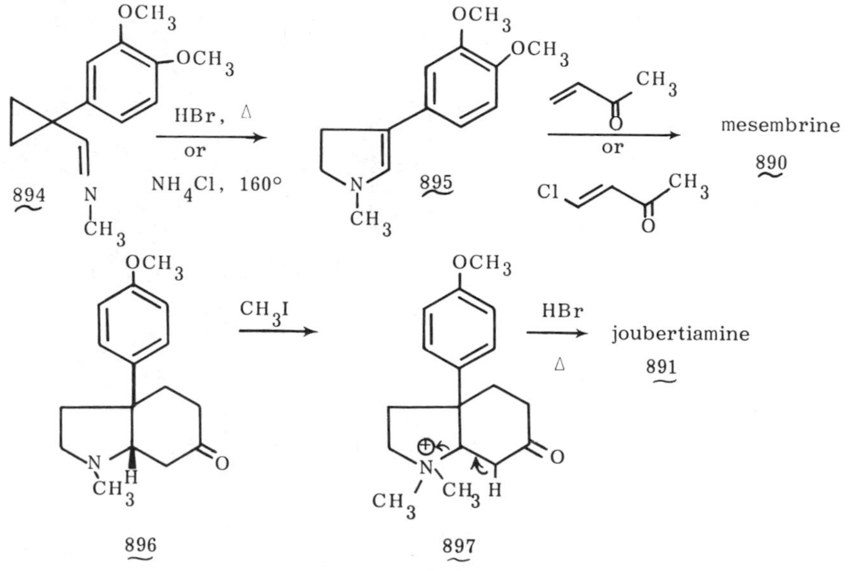

Yamada and Otani have reported an interesting asymmetric synthesis of unnatural (+)-mesembrine (**898**) from the formamido derivative **899**. Condensation with L-proline pyrrolidide (**900**) followed by reaction with methyl vinyl ketone and acid hydrolysis gave the optically active cyclohexanone **901** in 38% yield. Acid hydrolysis then afforded (+)-mescmbrine (**898**) (Scheme 123).

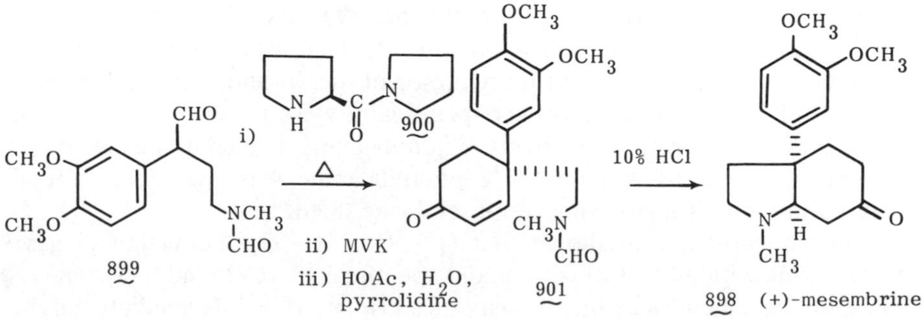

Scheme 123

Before turning attention to the biosynthesis of the mesembrane alkaloids mention must be made of a new skeletal alkaloid type, represented by Sceletium alkaloid A₄ (**902**) obtained from *S. namaquense* and *S. strictum* by Jeffs and co-workers. The structure of this alkaloid was suggested from the spectral data and subsequently confirmed by X-ray crystallographic analysis. Some of the proton NMR spectral data are shown in **903**. Forbes and co-workers have reported a synthesis of **902** which involves the key steps of

902 Sceletium
     alkaloid A₄

903

a thermal rearrangement of a cyclopropylimine followed by annulation with a methyl vinyl ketone analogue.

8.38.2 Biosynthesis

The similarity between the structures of the mesembrine-type alkaloids and the crinane alkaloids initially stimulated speculation that these alkaloids were biosynthetically related. Perhaps surprisingly this has turned out not to be the case.

This area has been studied for many years by Jeffs and co-workers, but the pathway is so complex that only recently have even the broad outlines of the pathway become evident.

In 1967, Jeffs reported that mesembrine (**890**), like the Amaryllidaceae alkaloids, was derived from phenylalanine (**1**) (C_6 unit) and tyrosine (**2**) (C_6-C_2 unit). Thus methionine was a precursor of the *N*- and *O*-methyl groups, and 2- and 3-^{14}C-labeled tyrosines specifically were incorporated into C-2 and C-3 of mesembrine respectively. Phenylalanine labeled in the side chain was not a precursor, but [ar-U-^{14}C]phenylalanine was specifically incorporated into the aromatic ring of mesembrine (**890**).

Subsequently it was shown that [1'-^{14}C, 2',6'-^{3}H₂]phenylalanine gave mesembrine without loss of tritium and specifically at C-2' and C-6'. On this basis any scheme proceeding by way of a crinane-type intermediate can be excluded since one tritium would be lost under these circumstances. This experiment provides other valuable information too, because it also defines the point of coupling of C-1 of the phenylalanine unit.

As a result of these experiments the Scheme 124 was proposed to account for the formation of the mesembrine alkaloids. The O-methylnorbelladine derivative **879** which had previously been shown to be a precursor of mesembrinol (**904**) in *Sceletium strictum* is a key intermediate in this scheme.

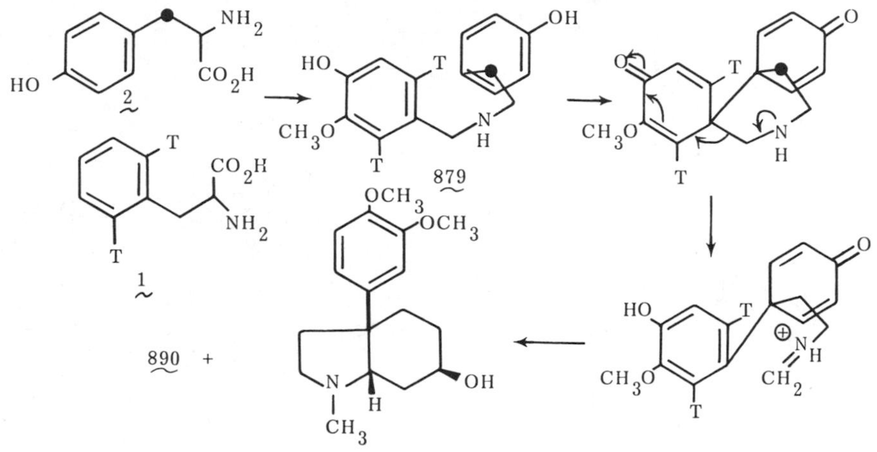

Scheme 124   Initial proposal for the biogenesis of mesembrine (**890**) and mesembrinol (**904**).

Additional experiments confirmed the route tyrosine → tyramine → N-methyltyramine but also showed that labeled O-methylnorbelladine (**879**) was not incorporated intact. This result eliminated Scheme 124 as a reasonable biogenetic possibility and it was only comparatively recently that much progress was made on this vexing problem.

In 1974, Jeffs and co-workers proposed a new biogenetic scheme (Scheme 125) for the *Sceletium* alkaloids. Crucial in this scheme is the concept that since alkaloids such as mesembrenone (**893**), Sceletium alkaloid A$_4$ (**902**), and N-formyltortuosamine (**905**) co-occur in *S. namaquense,* they are probably biogenetically related.

To account for the previously observed data, **906** was proposed as an intermediate and this compound was suggested to have two fates. One (pathway a) proceeds through cleavage of the phenyl ring to β-ring carbon bond followed by loss of the original three-carbon side chain from phenylalanine and cyclization to sceletenone (**907**). This compound is regarded as the key intermediate in the formation of the mesembrenone (**893**) and joubertiamine (**891**) alkaloid types.

The alkaloids such as Sceletium alkaloid A$_4$ (**902**) are proposed to be derived by pathway b, in which the three-carbon side chain from the *p*-

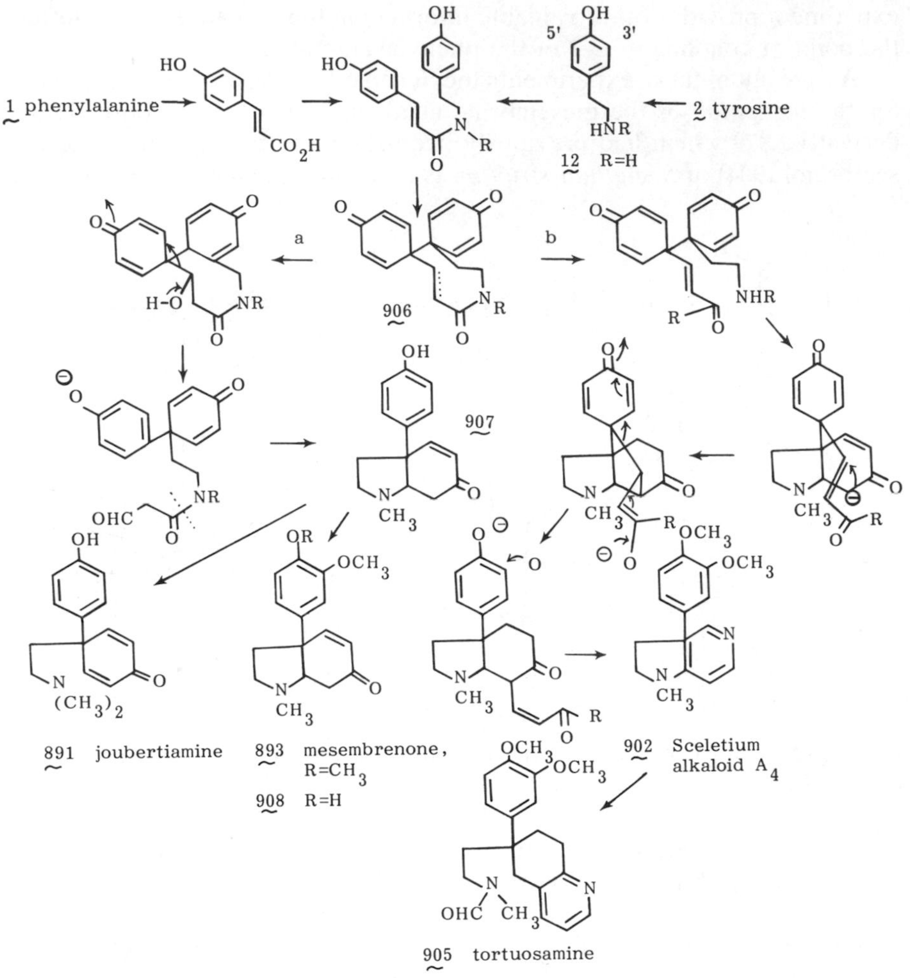

Scheme 125   Jeffs proposal for the biogenesis of *Sceletium* alkaloids.

coumaric acid precursor migrates to carbon 3 of the tyramine-derived unit. As yet there is no experimental evidence to support this proposal.

Examination of the beginning and very late stages in the scheme has established the route cinnamic acid, 4'-hydroxycinnamic acid, and 2,3-dihydro-4'-hydroxycinnamic acid. 3',4'-Dioxygenated cinnamic acids were not incorporated, suggesting that the C-3' oxygen of mesembrine (**890**) was introduced late in the scheme as indicated.

In agreement with this, sceletenone (**907**) was a good precursor (2% incorporation) of mesembrenol (**904**) and 4'-O-demethylmesembrenone (**908**) (58%) and mesembrenone (**893**) (63%) were exceptional precursors of mesembrenol (**904**).

The situation is more complicated than this however, for in 1976 Jeffs reported a very surprising result. When [1-^{14}C, 3',5'-^{3}H$_2$]tyramine (12) (or N-methyltyramine) was used as a precursor, 50% of the original ^{3}H activity was lost, the remaining activity being equally distributed between C-5 and C-7 of 904. Although this establishes that β-protonation of the enolate at C-

904

7 after conjugate addition of the nitrogen is a stereospecific process, Scheme 125 would predict intact incorporation of ^{3}H activity. There must therefore remain some subtlety in the biosynthetic scheme to be discovered.

LITERATURE

Isolation and Synthesis

Popelak, A., and G. Lettenbauer, *Alkaloids NY* 9, 467 (1967).

Arndt, R. R., and P. E. J. Kinger, *Tetrahedron Lett.* 3237 (1970).

Stevens, R. V., and M. P. Wentland, *J. Amer. Chem. Soc.* 90, 5580 (1968).

Keely, S. L., Jr., and F. C. Tank, *J. Amer. Chem. Soc.* 90, 5584 (1968).

Curphey, T. J., and H. L. Kim, *Tetrahedron Lett.* 1441 (1968).

Yamada, S., and G. Otani, *Tetrahedron Lett.* 1133 (1971).

Jeffs, P. W., T. Capps, D. B. Johnson, J. M. Karle, N. H. Martin, and B. Rauckman, *J. Org. Chem.* 39, 2703 (1974).

Forbes, C. P., G. L. Wenteler, and A. Wiechers, *Tetrahedron* 34, 487 (1978).

Biosynthesis

Jeffs, P. W., W. C. Archie, R. L. Hawks, and D. S. Farrier, *J. Amer. Chem. Soc.* 93, 3752 (1971).

Jeffs, P. W., H. F. Campbell, D. S. Farrier, and G. Molina, *Chem. Commun.* 228 (1971).

Jeffs, P. W., H. F. Campbell, D. S. Farrier, G. Ganguli, N. H. Martin, and G. Molina, *Phytochemistry* 13, 933 (1974).

Jeffs, P. W., J. M. Karle, and N. H. Martin, *Phytochemistry* 17, (1978).

Jeffs, P. W., D. B. Johnson, N. H. Martin, and B. S. Rauckman, *Chem. Commun.* 82 (1976).

8.39 IPECAC ALKALOIDS

Another of the very old medicinals is ipecacuanha, the "little wayside plant that causes vomiting." Derived from the root of *Cephaelis ipecacuanha* (Biot.) A. Rich. (Rubiaceae) native to Brazil, it was brought to Europe in the late seventeenth century, where its fame spread, particularly as a result of its successful use in the treatment of dysentry at the time of Louis XIV.

Two types of ipecac are of commercial significance, the Rio or Brazilian, and the Cartegene from Nicaragua or Panama.

Total imports into the United States are in the order of 25,000–30,000 lb annually, mainly from Brazil. The Rio variety of ipecac usually contains about 2% total alkaloids, 60–70% of which consists of emetine and 25% of cephaeline.

Emetine was one of the alkaloids isolated by Pelletier in the early nineteenth century (1817), but was not obtained in pure form until 1894. The correct molecular formula, was established by Pyman some 20 years later, and it was he who was responsible for establishing the interrelationship of emetine and some of the cooccurring alkaloids such as cephaeline, psychotrine, and *O*-methylpsychotrine.

The extensive degradations necessary to establish the structure of emetine are well described by Openshaw and it would be pointless to rediscuss them at this point.

The elaborate degradative work could not distinguish among three possible structures for emetine as represented by **909**. Robert Robinson noted

909

that the structure in which bond 2 is correct could be inferred on biogenetic grounds, namely the notion of prephenic acid first advanced by Woodward to explain the formation of strychnine. As we shall see subsequently the biosynthesis of emetine (**910**) is quite different from that suggested by Robinson.

$\underset{\sim}{910}$ emetine

8.39.1 Synthesis and Chemistry

The pharmacologic significance of emetine (910) coupled with the interesting structure and stereochemical problems, has resulted in at least a dozen syntheses. Indeed of the more complex alkaloids, emetine ranks as one of the most synthesized alkaloids.

The first synthesis developed was that of Eustigneeva and Preobrazhenskii in 1950. The route is nonstereospecific and has been subjected to several modifications. A vastly improved, stereospecific synthesis was reported by Battersby and Turner and involves the keto lactam 911 as a key intermediate, which can easily be converted to the α,β-unsaturated lactam 912. This compound was condensed with malonic ester to give the thermodynamically more stable *trans* isomer 913. Bischler–Napieralski cyclization and catalytic reduction afforded the ester 914, having three of the four asymmetric centers of emetine (910) (Scheme 126)

Only two syntheses of emetine or its derivatives are of commercial significance. One of these is the first Roche synthesis of 2,3-dehydroemetine (915), and the second is the Openshaw–Whittaker synthesis of emetine (910) itself.

The Roche synthesis (Scheme 127) involves the preparation of the tricyclic ketone 916 as a key intermediate. This compound could be converted to the δ-unsaturated acid 917 and condensed with homoveratrylamine 918 to afford the amide 919. By appropriate steps 919 can be converted into both emetine (910) and 2,3-dehydroemetine (915), but is principally of use for the latter compound.

The most startling synthesis of emetine is the Openshaw–Whittaker approach, in which the tricyclic ketone is produced stereospecifically in a very interesting condensation reaction. Condensation of the keto amine 920 with the iminium chloride at room temperature gave an essentially quantitative conversion to ketone 916. This racemic product could be resolved with (−)-camphor-10-sulfonic acid to give levorotatory 916 in almost quantitative

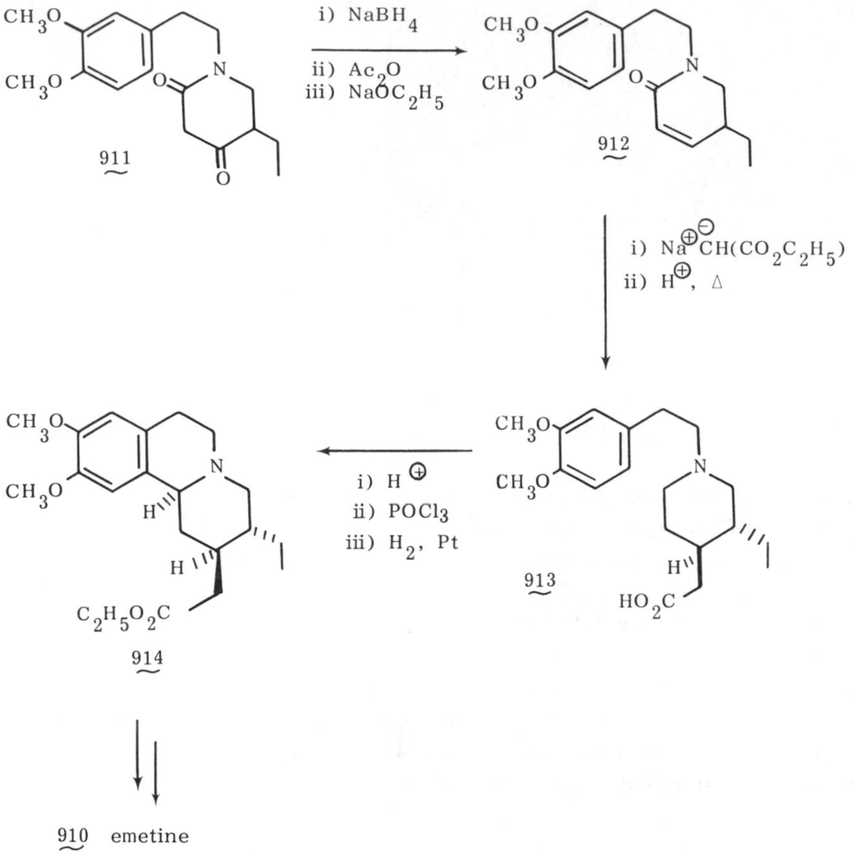

Scheme 126   Battersby–Turner synthesis of emetine (**910**).

yield! This amazing transformation proceeds through an immonium species which effectively permits epimerization at C-1 of (+)-**916**. Condensation of optically active **916** by a Wadsworth–Emmons procedure with the phosphorane **921** followed by catalytic reduction afforded the amine ester **922** (Scheme 128). The route to emetine from this point is a well-established one, although the Burroughs–Wellcome group found that when the ester **922** is condensed with homoveratrylamine (**918**), the reaction is markedly accelerated in the presence of 2-hydroxypyridine.

The commercial preparation of emetine therefore presents no unusual problems in procedures. However, exposure during handling of the drug should be limited, for in susceptible individuals severe conjunctivitis, epidermal inflammation, and asthma attacks may result.

There are several other important alkaloids of ipecac root, including cephaeline (**923**), psychotrine (**924**), and emetamine (**925**); however a more interesting alkaloid from ipecac root is ipecoside. This compound was first

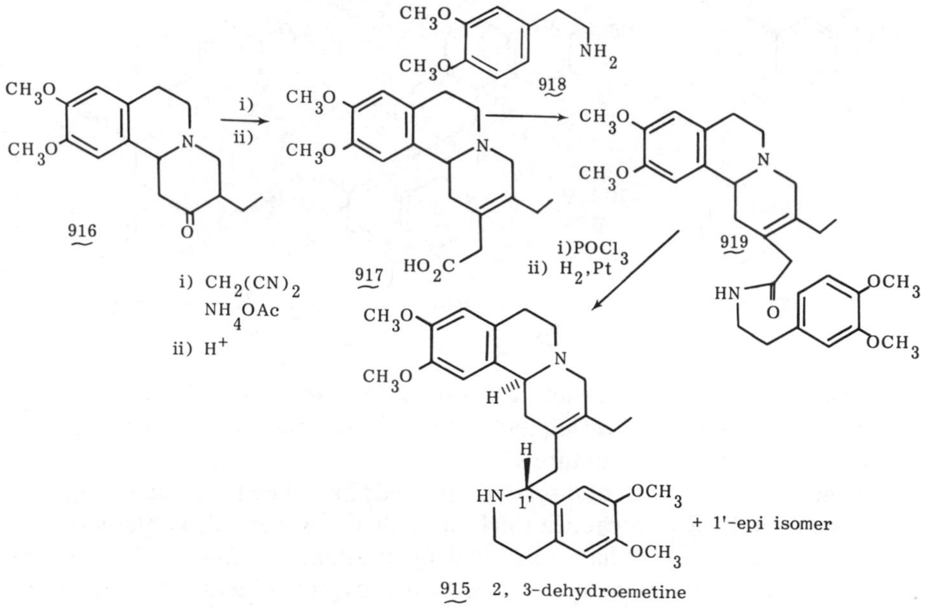

Scheme 127 Roche synthesis of 2,3-dehydroemetine (**915**).

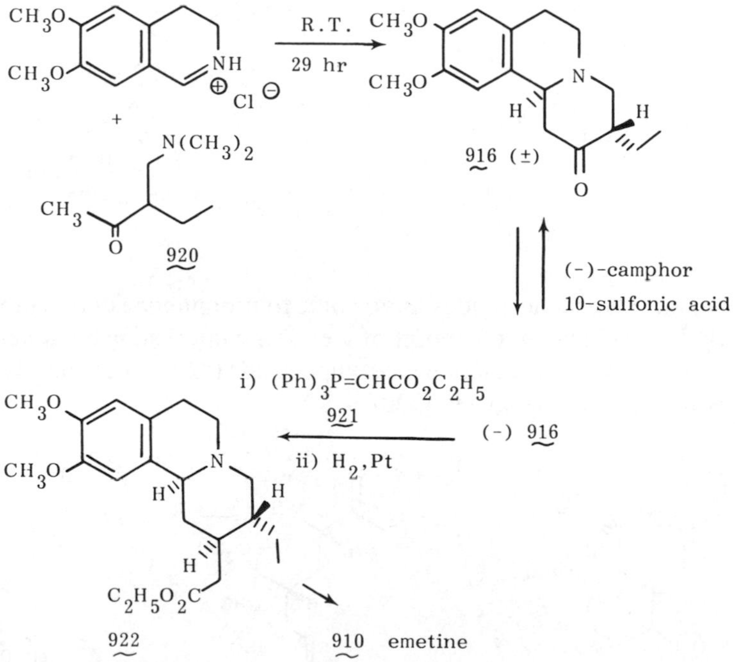

Scheme 128 Openshaw-Whittaker synthesis of (−)-emetine (**910**).

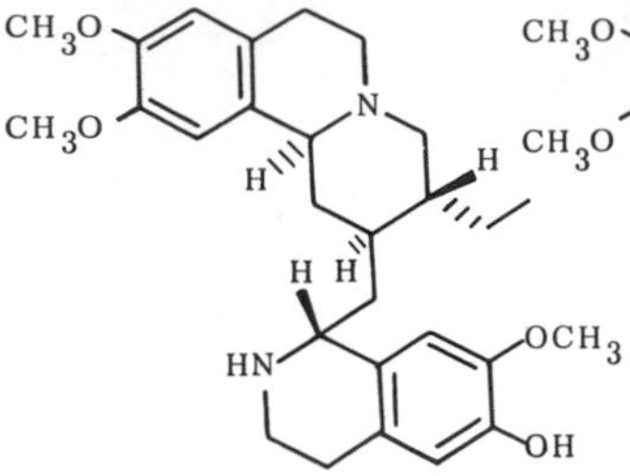

923   cephaeline

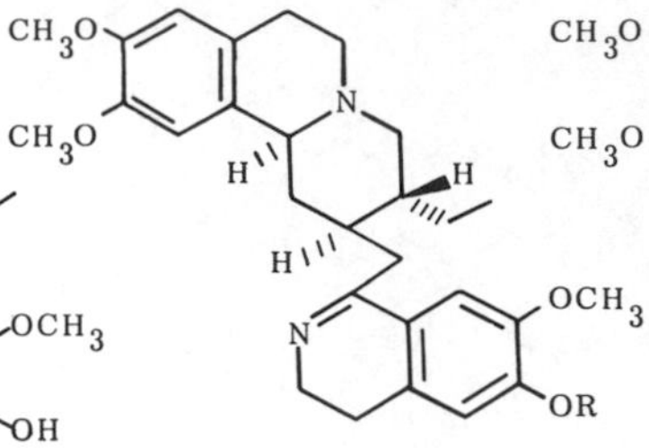

924   psychotridine, R=H
931   R=CH₃

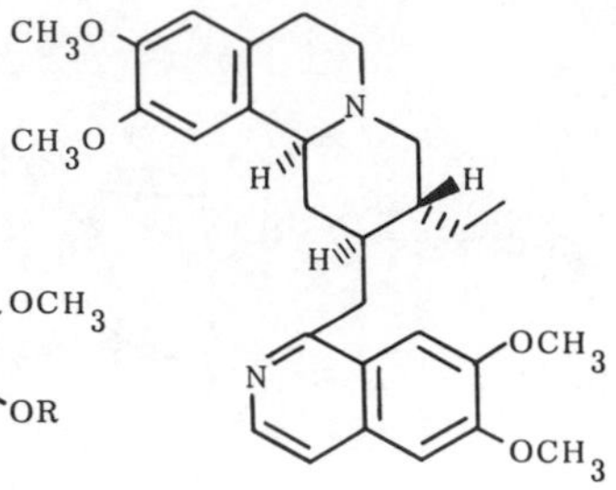

925   emetamine

obtained in 1952 and the structure deduced to be **926** on the basis of several degradative reactions. Subsequently the C-1 stereochemistry was revised and ipecoside has the structure **927**.

Emetine, cephaeline, and related alkaloids have been obtained from several plants in the Rubiaceae. In 1964 these alkaloids were also obtained from the root bark of the Indian medicinal plant *Alangium lamarckii* Thwaites (Alangiaceae). This plant has in addition provided several new alkaloids, including a novel lactam alangiside (**928**) obtained from the fruit.

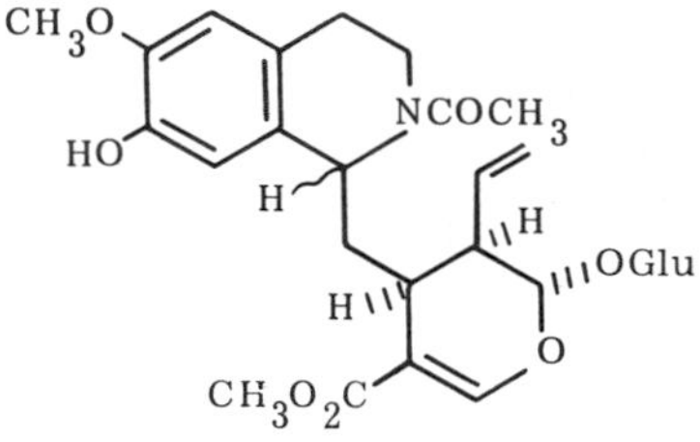

926   H-1 α

927   H-1 β

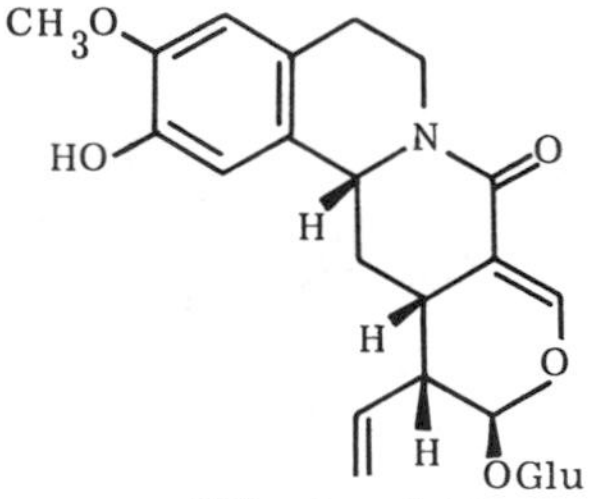

928   alangiside

Nitrogenous glycosidic amides have come to prominence in recent years, particularly from a biosynthetic point of view. One interesting biogenetically modeled transformation is that of deacetylipecoside (**929**) under slightly basic conditions to demethylalangiside (**930**).

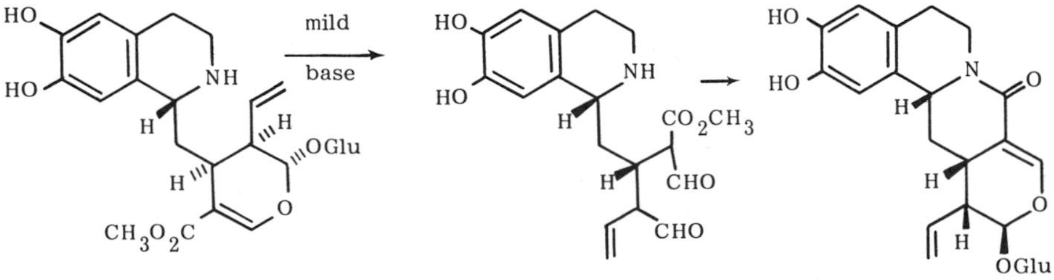

929   deacetylipecoside

930   demethylalangiside

8.39.2   Spectral Properties

The UV spectra of the ipecac alkaloids are relatively simple, giving rise to maxima at about 230 (log ϵ 4.23) and 280 nm (3.87), and as expected stereo-isomerization has little effect on the spectra. O-Methylpsychotrine (**931**) shows λ_{max} 244, 290, 306, and 360 nm and is consistent with the addition of 3,4-dihydro- and tetrahydroisoquinoline chromophores.

In the mass spectrum the pattern of cleavage is dependent on the un-saturation of ring D. Cephaeline (**923**) gives rise to two intense fragments

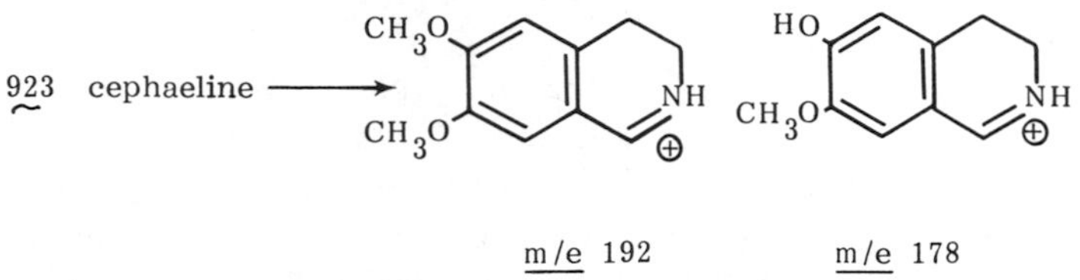

923   cephaeline

m/e 192   m/e 178

at m/e 178 and 192. The former ion is derived by benzylic cleavage of the lower tetrahydroisoquinoline ring, and the latter by a retro Diels–Alder fragmentation of the C ring with an internal hydrogen transfer.

Very little work has been done on the complex NMR spectra of the ipecac alkaloids beyond the straightforward assignments, but the ^{13}C NMR data for emetine are shown in **932**.

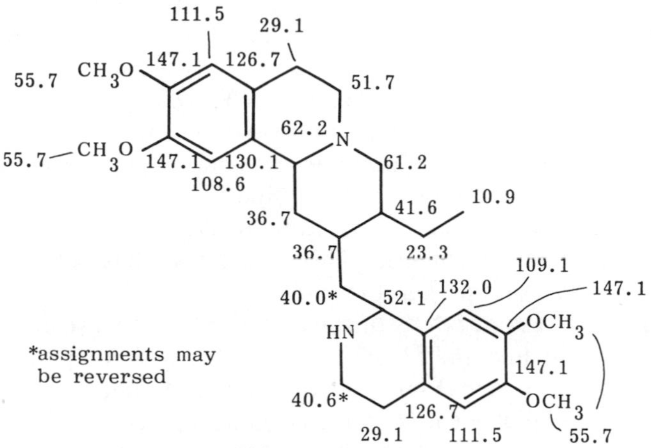

*assignments may
be reversed

932

8.39.3   Biosynthesis

The biosynthesis of emetine is quite different from that of any of the other isoquinoline alkaloids. Fundamentally it is derived from dopamine and the monoterpene secologanin, and since this area is very closely related to monoterpenoid indole alkaloid biosynthesis, discussion is deferred to Chapter 9.

8.39.4 Pharmacology

Emetine exhibits a number of profound pharmacologic effects which have been summarized by Brossi and by Grollman and Jarkovsky. Principal among these are the antiamebic and anticancer activities.

Various groups of South American Indians have used ipecac root for the treatment of amebic dysentery since long before their contact with the European civilizations.

Amebic dysentery is caused by the microrganism *Entamoeba histolytica*. That emetine was responsible for the activity of ipecac roots in *E. histolytica* infections was first shown in 1912, and its use in the clinical treatment of amebic dysentery is well established and highly effective. Emetine shows no antibacterial activity, but does show clinical antiviral activity. The main toxic effects of emetine therapy are cardiotoxicity, muscle weakness, and gastrointestinal problems, particularly diarrhea, nausea, and vomiting. These toxic side effects are often observed at therapeutic doses.

The antitumor properties were first described in 1918 by Lewishon and in 1919 by Van Hoosen. Subsequently, dehydrometine proved effective in the treatment of chronic granulocytic leukemia and other malignancies. Testing in mice confirmed the anticancer activity, but further clinical testing with advanced carcinomas was not successful. Dehydroemetine did give rise to partial remissions, and Grollman has reviewed the renewed interest in the antiviral and antitumor activities of emetine and related compounds.

Emetine has a profound irreversible effect on DNA synthesis and essentially produces total inhibition at 1 μM concentration; it also rapidly inhibits the cell-free incorporation of amino acids into protein, and appears to be quite general inhibitor of protein synthesis in animal cells, plants, cell-free yeast extracts, and *E. histolytica*. Emetine does not, however, penetrate the cell wall of yeast or bacteria. This is a classic case of the difference between eukaryotes and prokaryotes.

There have been some interesting investigations of the structure–activity relationships of emetine and its derivatives. Isoemetine (epimeric at C-1′), 1′,2′-dehydroemetine, and *N*-methylemetine are all inactive, but 2,3-dehydroemetine retains biological activity. Although other alkyl groups can be substituted at C-3, the *trans* stereochemistry is critical for activity.

LITERATURE

Reviews: General

Janot, M. M., *Alkaloids NY* **3**, 363 (1953).

Manske, R. H. F., *Alkaloids NY* **7**, 419 (1960).

Openshaw, H. T., in S. W. Pelletier (Ed.), *The Chemistry of the Alkaloids,* Van Nostrand Reinhold, 1970, p. 85.

Brossi, A., G. V. Parry, and S. Teitel, *Alkaloids NY* **13,** 189 (1971).

Shamma, M., *The Isoquinoline Alkaloids, Chemistry and Pharmacology,* Academic, New York, 1972, p. 427.

Robinson, R., *Nature (London)* **162,** 524 (1948).

Battersby, A. R., and H. T. Openshaw, *J. Chem. Soc.* 3207 (1949).

Brossi, A., *Pure Appl. Chem.* **19,** 171 (1969).

Reviews: Biosynthesis

Battersby, A. R., *Pure Appl. Chem.* **14,** 124 (1967).

Cordell, G. A., *Lloydia* **37,** 219 (1974).

Reviews: Pharmacology

Grollman, A. P., and Z. Jarkovsky, in J. W. Corcoran, and F. E. Hahn (Eds.), *Antibiotics,* Vol. 3, *Mechanism of Action of Antimicrobial and Antitumor Agents,* Springer-Verlag, New York, 1975, p. 420.

Manno, B. R., and J. E. Manno, *Clin. Toxicol.* **10,** 221 (1977).

Grollman, A. P., *Ohio State Med. J.* **66,** 257 (1970).

Synthesis

Eustigneeva, R. P., and N. A. Preobrazhenskii, *Tetrahedron* **4,** 223 (1958).

Battersby, A. R., and J. C. Turner, *Chem. Ind. (London)* 1324 (1958).

Brossi, A., M. Baumann, and O. Schnider, *Helv. Chim. Acta* **42,** 1515 (1959).

Openshaw, H. T., and N. Whittaker, *J. Chem. Soc.* 1449, 1461 (1963).

8.40 PHENANTHROINDOLIZIDINE ALKALOIDS

Tylophora asthmatica, Wight et Arn. (Asclepiadaceae) is a perennial climber indigenous to eastern and southern India and Burma. It is noted for its asthmatic, antidysenteric, and emetic properties, and indeed is known as Indian ipecac.

Hooper, in 1891, was the first to isolate a crystalline alkaloid from this plant and subsequently examination of *Tylophora brevipes* gave an alkaloid similar to that obtained by Hooper. In 1935, Ratnagiriswaran and Venkatachalam became the first to begin chemical studies on the alkaloids of *T. asthmatica.* They isolated two alkaloids, tylophorine (**933**) and tylophorinine (**934**), determined their molecular formulas, and noted that there were considerable dermatitic problems in working with these compounds.

Since this time several other genera in the Asclepiadaceae, including *Antitoxicum, Vincetoxicum, Cyananclus,* and *Pergularia,* have been found to contain alkaloids of this type and presently about seven alkaloids in this groups are known.

Some of the newer representative members of this group are tylophor-inidine (**935**), tylocrebrine (**936**), isotylocrebrine (**937**), and pegularinine (**938**).

The compounds *O*-methyltylophorinidine (**939**) and pegularinine (**938**) show opposite (but not equal) optical rotations. Since **939** has the C-13a-H and the C-14-OH *trans* diaxial, pegularinine should have either structure **938** or its mirror image. The observation that deoxypegularinine (**940**), also a

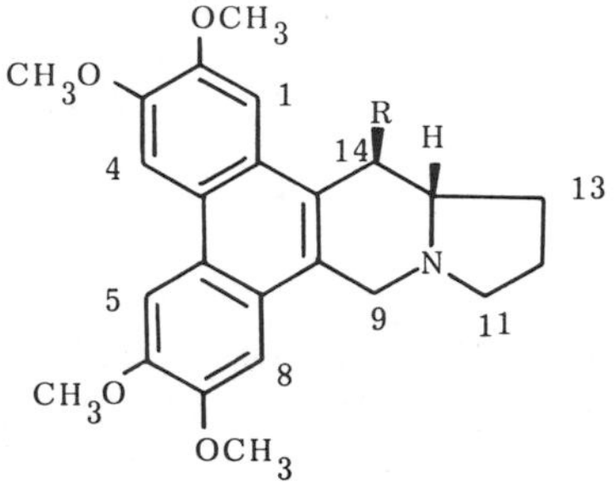

933 tylophorine, R=H
934 tylophorinine, R=OH

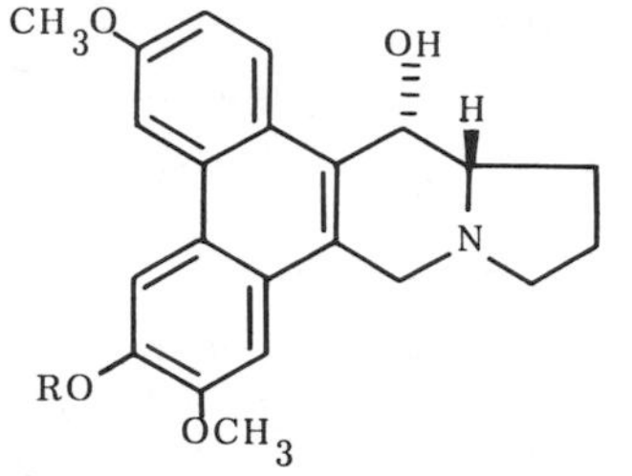

935 tylophorinidine, R=H
939 R=CH₃

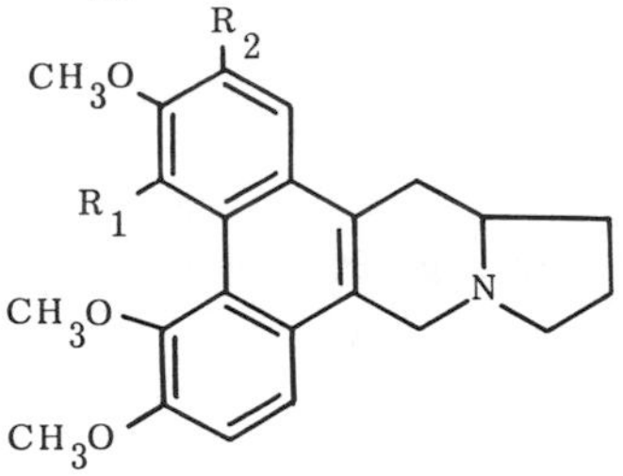

936 tylocrebrine,
     R₁=H, R₂=OCH₃

937 isotylocrebrine,
     R₁=OCH₃, R₂=H

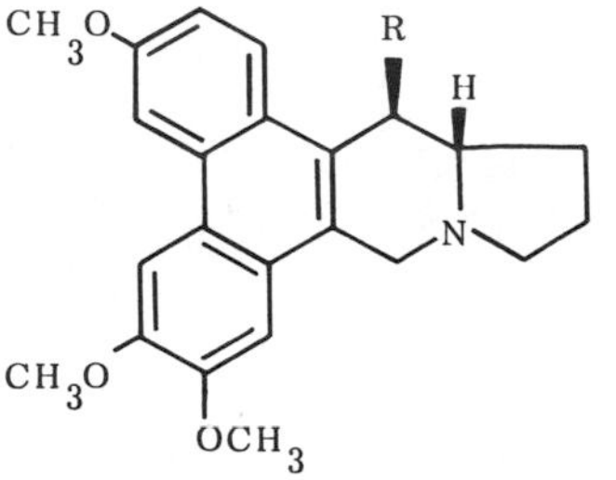

938 pegularinine, R=OH

940 R=H

natural product, has a negative Cotton effect in the region of 270 nm indicates a C-13a absolute stereochemistry the same as in tylophorine (**933**).

Typically these alkaloids show λ_{max} 258, 286, 211, 341, and 360 nm (4.3, 3.9, 3.5, 3.1, and 2.6).

8.40.1 Synthesis

Liepa and Summons have reported an interesting synthesis of tylophorine (**933**) which relies on a nonoxidative aromatic coupling as the key step. Condensation of veratraldehyde with potassium laevulinate followed by esterification and Michael addition of 3,4-dimethoxybenzyl cyanide gave **941** in about 38% overall yield. Catalytic hydrogenation permitted double internal ring formation to afford the lactam **942**. Nonphenolic oxidative coupling of this compound with vanadium oxyfluoride at 0°C gave the phenanthrene **943**, which could be reduced to tylophorine (**933**) (Scheme 129).

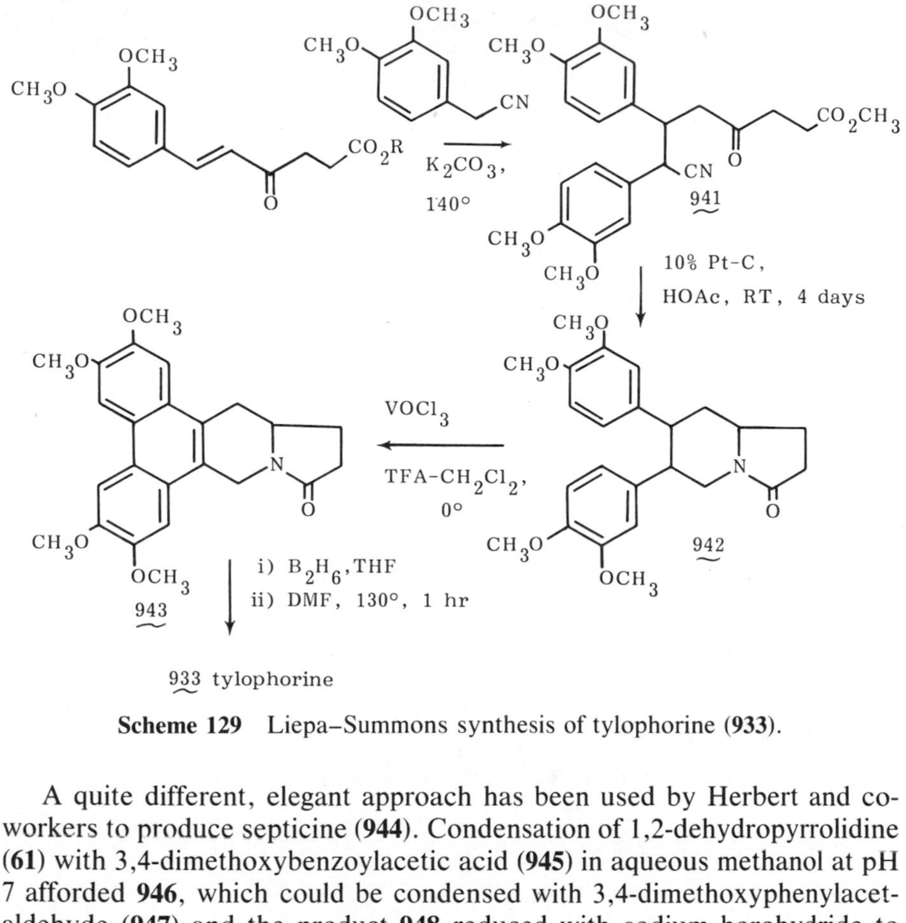

Scheme 129 Liepa–Summons synthesis of tylophorine (**933**).

A quite different, elegant approach has been used by Herbert and co-workers to produce septicine (**944**). Condensation of 1,2-dehydropyrrolidine (**61**) with 3,4-dimethoxybenzoylacetic acid (**945**) in aqueous methanol at pH 7 afforded **946**, which could be condensed with 3,4-dimethoxyphenylacet-aldehyde (**947**) and the product **948** reduced with sodium borohydride to give septicine (**944**) in 24% yield. The mechanism of the crucial step, namely formation of the diene **949**, is shown in Scheme 130.

8.40.2 Biosynthesis

At first sight the phenanthroindolizidine skeleton appears to pose a difficult biogenetic problem, but closer examination, and the knowledge that the alkaloid septicine (**944**) co-occurs in *T. asthmatica,* sheds some light on the probable pathway. For example, by breaking at the double bond in **944** to give **950** we see that formally at least, these structures can hypothetically

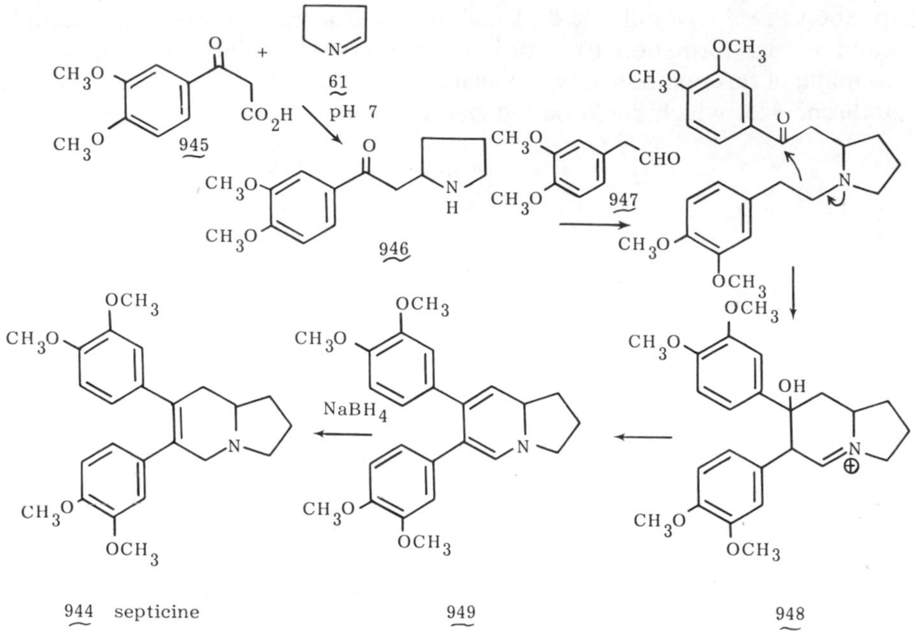

Scheme 130 Herbert and co-workers' synthesis of septicine (**944**).

be derived from two C_6-C_2 units and ornithine. These ideas have been confirmed by biosynthetic studies.

The derivation of ring E was established by feeding experiments with ornithine, and the remainder of the molecule was found to be derived from phenylalanine and tyrosine. Both [2-^{14}C]tyrosine (**2**) and [2-^{14}C]phenylalanine (**1**) were used as precursors of tylophorine (**933**) in *T. asthmatica,* and it was conclusively shown that ring A and carbons 14 and 14a of **933** are

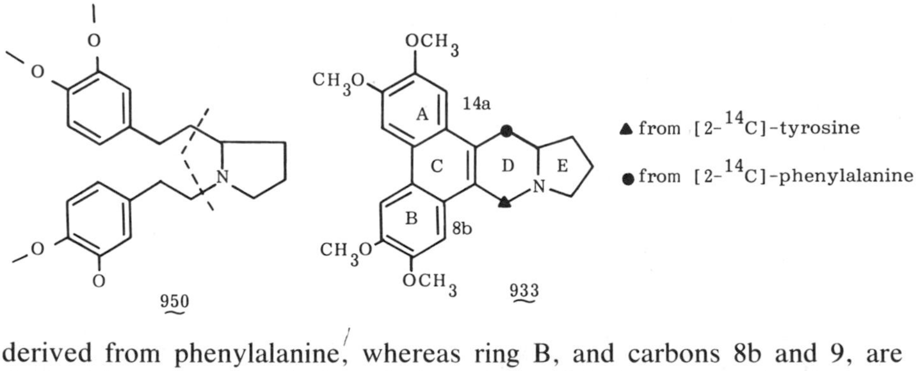

derived from phenylalanine, whereas ring B, and carbons 8b and 9, are derived from tyrosine. Hydroxylation of phenylalanine to tyrosine does not apparently occur in this plant. In confirmation of this, [2-^{14}C]cinnamic acid (**42**) was specifically incorporated into C-14 (but not C-9) of tylophorine (**933**).

With these results in hand from the Indian workers, Herbert and co-workers turned their attention to more advanced precursors, and established that the 2-phenacyl pyrrolidine system **951** is an important precursor for tylophorinine (**934**).

As an extension of this work, the 6,7-diphenylhexahydroindolizines **952**,

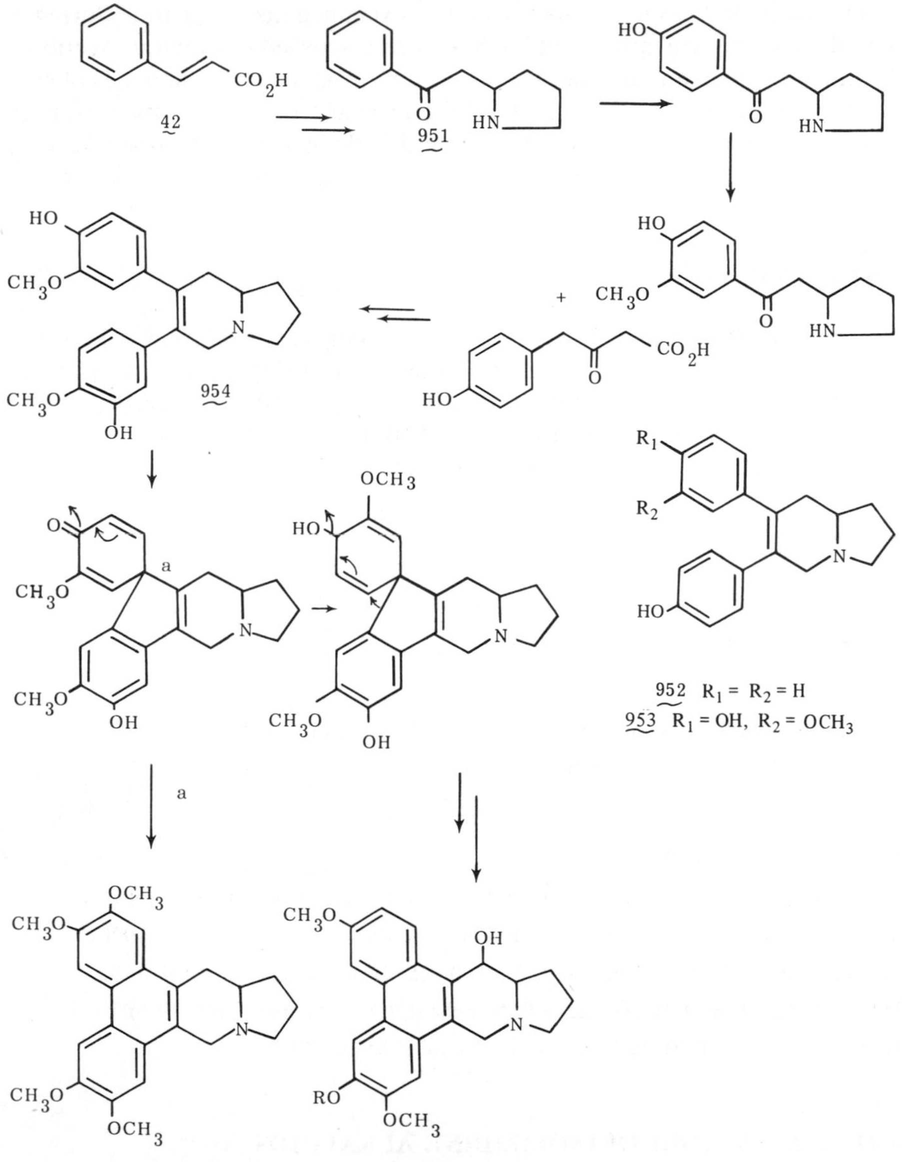

933 tylophorine

934 tylophorinine, $R=CH_3$
935 tylophorinidine, R=H

Scheme 131 Biosynthesis of the *Tylophora* alkaloids.

953 and 954 were examined as precursors, each being doubly labeled. Only 953 and 954 were incorporated into tylophorine (933), tylophorinine (934), and tylophorinidine (935) in *T. asthmatica,* suggesting that the biosynthetic pathway does indeed involve a dioxygenated 2-phenacyl pyrrolidine precursor. The incorporation of 954 into all three alkaloids also demonstrates that during the formation of tylophorinine (934) an oxygen atom from ring A in 954 is lost. Such a loss can be explained as occurring during a dienol–benzene rearrangement, as has been discussed in previous sections. Tylophorine (933) on the other hand would be produced by a dienone–phenol rearrangement of ring A in which all the aromatic oxygen atoms present in 954 are retained. On this basis Herbert and Jackson have postulated Scheme 131 for the biosynthesis of the phenanthroindolizidine alkaloids of *T. asthmatica.*

8.40.3 Pharmacology

Tylophorine (933) is toxic to *Paramecium caudatum* at a dilution of 1 : 50,000 and is lethal to frogs at 0.4 mg/kg. It has a paralyzing action on the heart muscles, but is probably better noted for its cytotoxicity. Tylocrebrine (936) shows antitumor activity against the L-1210 lymphocytic leukemia system.

All of these alkaloids have vesicant activity.

LITERATURE

Review

Govindachari, T. R., *Alkaloids NY* **9,** 517 (1967).

Synthesis

Liepa, A. J., and R. E. Summons, *J. Chem. Soc. Chem. Commun.* 826 (1977).

Herbert, R. B., F. B. Jackson, and I. T. Nicholson, *Chem. Commun.* 450 (1976).

Biosynthesis

Leete, E., *Biosynthesis* **5,** 148 (1977).

Mulchandani, N. B., S. S. Iyer, and L. P. Badheka, *Phytochemistry* **8,** 1931 (1969).

Mulchandani, N. B., S. S. Iyer, and L. P. Badheka, *Phytochemistry* **10,** 1047 (1971).

Mulchandani, N. B., S. S. Iyer, and L. P. Badheka, *Phytochemistry* **15,** 1697 (1976).

Herbert, R. B., F. B. Jackson, and I. T. Nicholson, *Chem. Commun.* 865 (1976).

Herbert, R. B., and F. B. Jackson, *Chem. Commun.* 955 (1977).

8.41 PHENANTHROQUINOLIZIDINE ALKALOIDS

Cryptopleurine (955), a phenanthroquinolizidine alkaloid, was first obtained from *Cryptocarya pleurosperma* Wh. and Fr. (Lauraceae), and subsequently as the main cytotoxic principle of *Boehemeria cylindrica* (L.) Sw. (Urtica-ceae), where it cooccurs with the acetamide derivative **956** and the seco-phenanthroquinolizidine **957**. Cryptopleurine shows no antitumor activity even though it is one of the most cytotoxic alkaloids known.

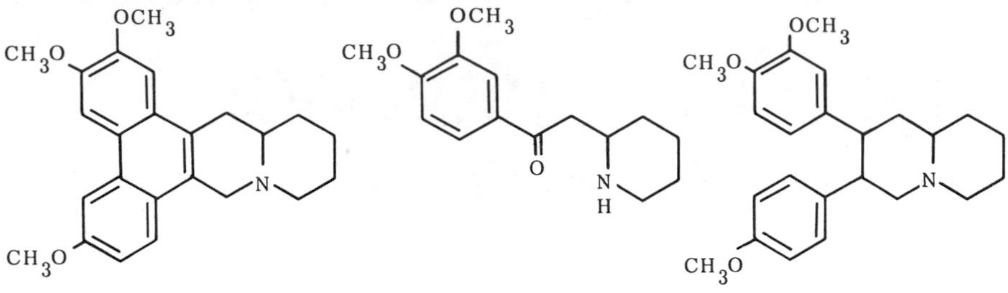

955 cryptopleurine   956   957

Several syntheses of cryptopleurine have been described, and the most effective involves anodic oxidation of the quinolizidinone **958** to give oxo-cryptopleurine (**959**), which can be easily reduced to **955**.

No studies of the biosynthesis of cryptopleurine have been made, but the cooccurrence with **956** and **957** indicates that it probably takes place

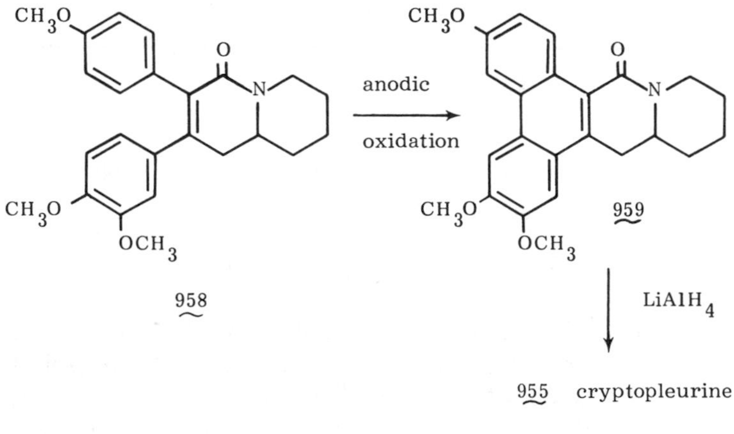

958   anodic oxidation →   959

LiAlH$_4$

955 cryptopleurine

along the same lines as that of the *Tylophora* alkaloids, but utilizing lysine rather than ornithine.

LITERATURE

Farnsworth, N. R., N. K. Hart, S. R. Johns, J. A. Lamberton, and W. Messmer, *Aust. J. Chem.* **22**, 1805 (1969).

Kotani, E., M. Kitazawa, and S. Tobinaga, *Tetrahedron* **30**, 3027 (1974).

chapter 9

ALKALOIDS DERIVED FROM TRYPTOPHAN

There has been a tremendous interest in those alkaloids derived from the amino acid tryptophan (**1**) during the past 20 years. So much interest in fact that this is probably now the largest group of alkaloids. Indole alkaloids occur in both fungi and plants, where the families Apocynaceae, Rubiaceae, and Loganiaceae are particularly good sources. This group of compounds has proved to be a source of abundant challenges to many types of chemist; to the natural product chemist interested in structure elucidation and chemical correlation, to the synthetic organic chemist because of the diverse structural and stereochemical problems, and to the organic chemist interested in the biosynthesis of these alkaloids because of the intriguing rearrangements that occur.

There are far too many compounds in this group for a complete discussion to be made of all of these; consequently only the major structure types can be discussed. With all of these alkaloid types it is not surprising to find that they show extremely diverse range of pharmacologic activities, many of clinical significance.

Tryptophan is the biosynthetic precursor of all of these alkaloids, but except for the most simple alkaloids it is rare that it is the sole carbon source. Often as many carbons are provided by another carbon source, such as a monoterpene unit. It is the structure diversity resulting from this process which has been responsible for the tremendous interest in indole alkaloid biosynthesis during the past 15 years or so.

9.1 THE SIMPLE BASES

There are several important alkaloids derived by simple modification of tryptamine (**2**). For the most part these compounds are not particularly limited in their distribution. Indole (**3**) for example occurs in the flowers of a variety of plants, including many *Jasminum* and *Citrus* species, *Narcissus jonquilla* L., and *Chimonanthus fragrans* Lindl.

The origin of indole in plants is uncertain, since it is apparently in an enzyme-bound form in the formation of tryptophan, and can also be a degradation product of tryptophan.

The N_b-methyl derivative of L-tryptophan, also known as abrine (4), has so far only been isolated from jequirity seeds, *Abrus precatorius* L. (Leguminosae).

Curiously though the N_b-methyl betaine of tryptophan, hypaphorine (5), occurs widely in the seeds of *Erythrina* species, and in some instances it is the predominating alkaloid. In the seeds of *E. acanthocarpa* E. May for example, it occurs to the extent of 5.8%.

The corresponding methyl ester of N_b-dimethyl tryptophan (6) is the major alkaloid of *Pultenaea altissima* F. Muell. ex Benth. (Leguminosae).

1 tryptophan, $R=CO_2H$ 3 indole
2 tryptamine, $R=H$

4 abrine   5 hypaphorine   6

Gramine (7) was first obtained from chlorophyll-deficient barley mutants and is also a constituent of normal sprouting barley (*Hordeum vulgare* L.). It is of interest because it is an important synthetic intermediate in the formation of indole acetic acid (8), tryptamine (2), and tryptophan (1) (Scheme 1).

Gramine contains one less carbon in the side chain than tryptamine, and has therefore proved to be an interesting compound for biosynthetic study. Is it for example derived from indole by addition of a C_1-N unit or from tryptophan by loss of one of the side-chain carbon atoms? If the latter is correct, which carbon atom is retained and what is fate of the nitrogen of tryptamine?

The pioneering work in this area was carried out by Marion. Tryptophan was shown to be a precursor, and when [3-^{14}C]- and [ring-2-^{14}C]tryptophans were used in admixture specific intact incorporation was observed. In the biosynthetic process therefore the bond between the 2- and 3-positions of the side chain is broken, and the 3-carbon retained.

Several intermediates were postulated in which the 3-methylene group of the side chain was oxidized; examples include indolyl-3-pyruvic acid,

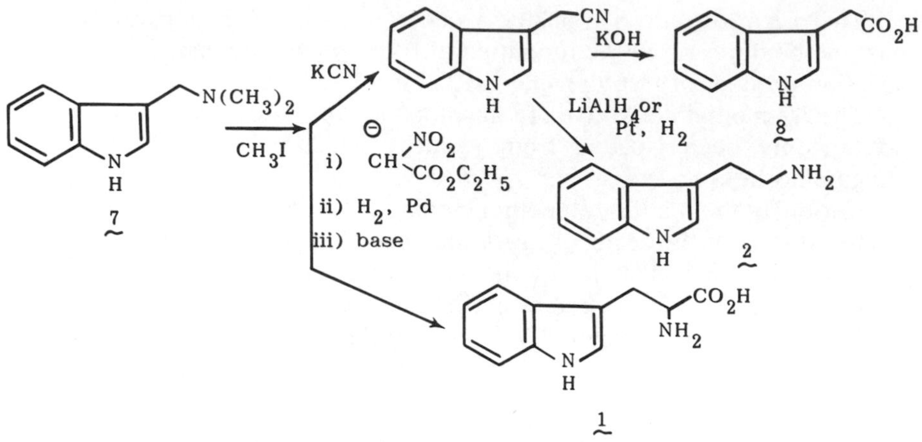

Scheme 1　Chemical modifications of gramine (**7**).

indolyl-3-acrylic acid, and indolyl-3-carboxaldehyde. These possibilities were eliminated when Leete and O'Donovan showed that [3-^{14}C,3-^{3}H$_2$]tryptophan was incorporated without loss of ^{3}H label into gramine.

Wenkert has suggested that loss of the side chain involves as an initial step condensation with pyridoxal phosphate (**9**) to give a Schiff base **10**. This intermediate can now undergo fission between C-2 and C-3 of tryptophan to yield a 3-methylene indole iminium species **11** with loss of C-2. It is then envisaged (Scheme 2) that amination occurs to sequentially form 3-

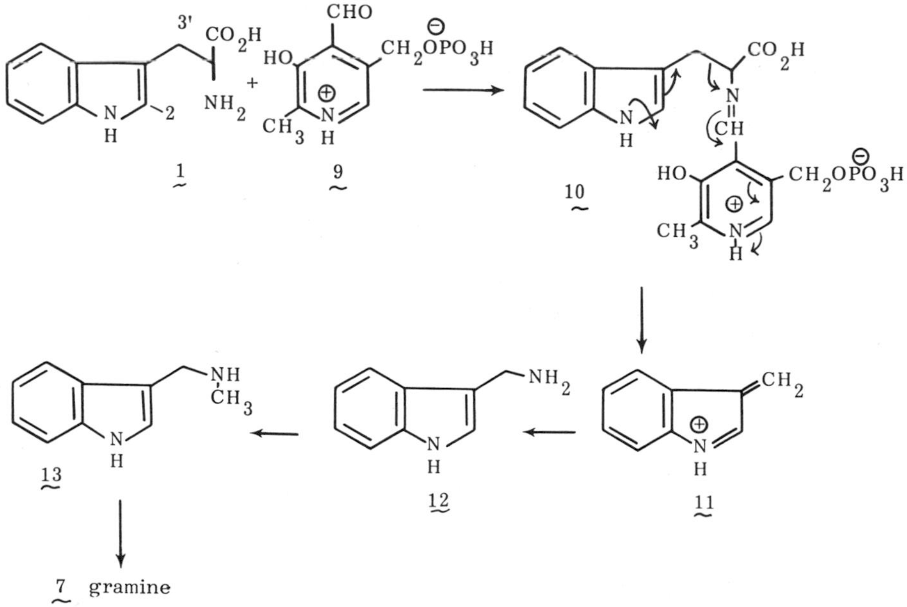

Scheme 2　Biosynthesis of gramine (**7**) in barley seedlings.

aminomethylindole (12), 3-methylaminomethylindole (13), and gramine (7). The origin of the nitrogen atom is not known.

Polyalthenol (14) is an indolosesquiterpene from *Polyalthia oliveri* Engl., where it co-occurs with aporphine alkaloids. The ^{13}C NMR data are shown and the base peak in the mass spectrum was observed at *m/e* 130 (11).

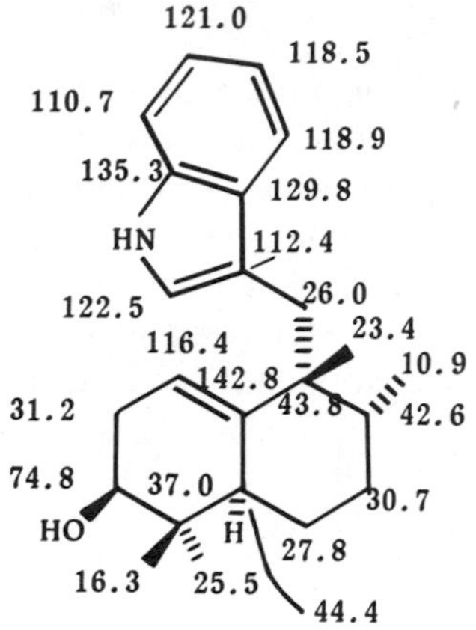

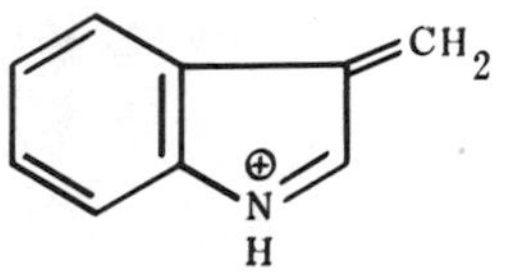

9.2   SIMPLE TRYPTAMINE DERIVATIVES

Tryptamine (2) is a widespread constituent of both plants and fungi, and is also present in several edible fruits, including plums, eggplants, and tomatoes. Clearly though tryptamine is more familiar as a synthetic and biosynthetic intermediate.

N-Methyltryptamine (15), also known as dipterine, has not been found widely, but does occur in *Girgensohnia diptera* Bge. and *Arthrophytum leptocladum* Popov in the Chenopodiaceae.

N,N-Dimethyltryptamine (16) on the other hand is more widely occur-

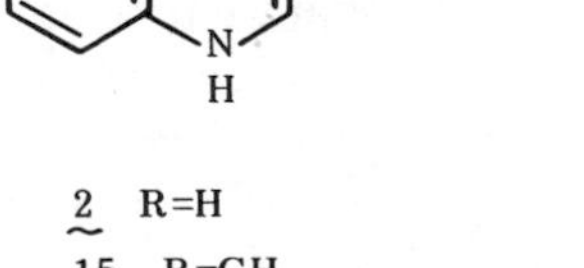

ring. It was first obtained as one of the hallucinogenic constituents of the seeds and seed pods of *Piptadenia peregrina* Benth. (Leguminosae), a plant used as a narcotic snuff by certain American Indian tribes. The leaves of *Prestonia amazonica* (Benth.) Macbride (Apocynaceae) are used by some Colombian and Peruvian Indians for their hallucinogenic properties and

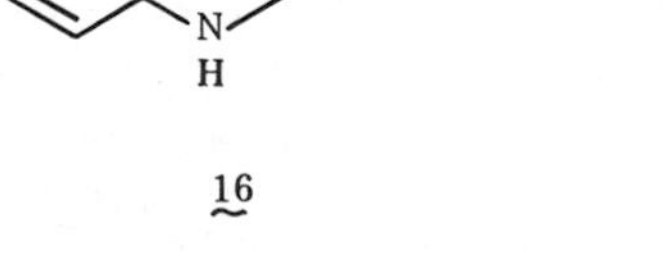

N,*N*-dimethyltryptamine was obtained in this case also. Similar reports concern the roots of the Brazilian plant *Mimosa hostilis* Benth.

In humans *N*,*N*-dimethyltryptamine causes rapidly developing, but brief, hallucinations when injected intramuscularly.

9.2.1 Psilocin and Psilocybin

Mexican Indian tribes have, since pre-Columbian times, used hallucinogenic and narcotic drugs in their religious rituals and this use continues today. Botanical investigation of some of these drugs has established that they are mushrooms, principally of the genus *Psilocybe* and occasionally of the genus *Stropharia*. Subsequently psilocybin (**17**) was isolated and shown to be the hallucinogenic principle of *Psilocybe mexicana* Heim and several other *Psilocybe* species. Psilocin (**18**) is usually less abundant in these mushrooms

$\underline{17}$ psilocybin, R=OPO_3H_2

$\underline{18}$ psilocin, R=H

than is psilocybin (**17**), but the reverse is true in *Psilocybe baeocystis* Singer and Smith.

Psilocybin (**17**) affords psilocin (**18**) and one equivalent of phosphoric acid on hydrolysis, and the UV spectrum is typical of a 4-hydroxyindole (λ_{max} 268, 285, and 294). The structures were confirmed by a synthesis beginning with 4-benzyloxyindole (**19**). Using a standard sequence, **19** was condensed with oxalyl chloride, the product aminated with dimethylamine, and then reduced with lithium aluminum hydride to afford 4-benzyloxy-*N*,*N*-dimethyltryptamine (**20**). Catalytic removal of the protecting group gave psilocin (**18**) which could be converted to psilocybin (**17**).

Ingestion of the fungi results in intoxication and hallucination, and qualitatively the effects are described as similar to those of mescaline and lysergic acid diethylamide (see Section 9.10.11).

i) $(COCl)_2$

ii) $HN(CH_3)_2$

iii) $LiAlH_4$

$\underline{19}$

$\underline{20}$

$\xrightarrow{Pd, H_2}$ $\underline{18}$ psilocin

i) $(C_6H_5CH_2O)_2POCl$

ii) Pd, H_2

$\underline{17}$ psilocybin

It has been demonstrated that tryptophan is a precursor of psilocybin (**17**) in *Psilocybe sempervira*, but details of the hydroxylation process have not been studied.

9.2.2 5-Hydroxytryptamine and Derivatives

5-Hydroxytryptamine (**21**), serotonin, is an important mammalian neurohormone, but as well as in brain tissue it has also been detected in blood, and in stomach, intestine, and lung tissues. The mechanism of action of some psychotomimetic drugs appears to be due to an interference with brain 5-hydroxytryptamine. High concentrations of 5-hydroxyindoleacetic acid (**22**), a metabolite of 5-hydroxytryptamine, is diagnostic of certain types of intestinal tumors.

The topical pharmacology of 5-hydroxytryptamine (**21**) is somewhat confusing. It was first isolated from *Mucuna pruriens* DC (Leguminosae) as a skin irritant, acting by promoting histamine release, and has been suggested to be the irritant of *Urtica dioica* L., the stinging nettle. However, it also occurs in numerous fruits, including bananas, tomatoes, plum avocado, eggplant, pineapple, and passion fruit, which certainly are not noted for their irritancy.

Several syntheses of 5-hydroxytryptamine (**21**) are available involving all the well-established syntheses of tryptamine derivatives. A typical example begins with 5-benzyloxyindole (**23**) and utilizes oxalyl chloride and ammonia to afford 5-benzyloxyindoleglyoxylamide (**24**) followed by lithium aluminum hydride and catalytic reduction. An alternative route involves the Grignard reaction of 5-benzyloxyindole (**23**) with ethyleneimine to afford 5-benzyloxytryptamine (**25**) directly.

5-Methoxy-*N,N*-dimethyltryptamine (**26**) is present in the bark of *Piptadenia peregrina* Benth and is the principal hallucinogenic constituent of the *Virola* snuffs of certain Amazonian Indian tribes.

5-Hydroxy-*N,N*-dimethyltryptamine (**27**), also known as bufotenine, has been obtained from several interesting sources. The seeds of *P. peregrina* have been used for centuries as a ceremonial hallucinogenic snuff. In the Caribbean, where it is known as cohoba, inhalation is through a bifuricated tube. *P. colubrina* Benth. seeds are used in Brazil for a similar purpose. The respective contents of **27** are 0.94% in *P. peregrina* seeds and 2.1% in *P. colubrina* seeds, although whether **27** is a true hallucinogenic principle in these snuffs remains to be established.

Bufotenine (**27**) is also a constituent of several fungi, particularly *Amanita* species, but is most famous as a constituent of the skin secretion of the toad *Bufo vulgaris* Laur. First obtained in 1893, it was characterized in 1934, and has been synthesized by the standard routes described previously for other simple tryptamines.

The tryptamide **28** occurs in a *Thermoactinomyces* species (strain TM-64). Borreline (**29**) is one of the more interesting simple plant indole alkaloids

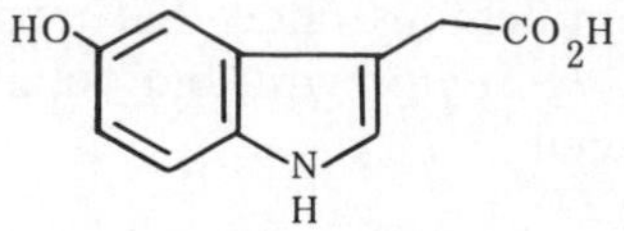

$$HO\text{—}\overset{\displaystyle}{\underset{N,H}{\text{indole}}}\text{—}CO_2H$$

22

23 $\xrightarrow{\text{i) }(COCl)_2\text{ ii) }NH_3}$ 24

i) C_2H_5MgBr

ii) $\overset{CH_2}{\underset{CH_2}{|}}\diagdown NH$

iii) Pd, H_2

i) LiAlH$_4$

ii) Pd, H$_2$

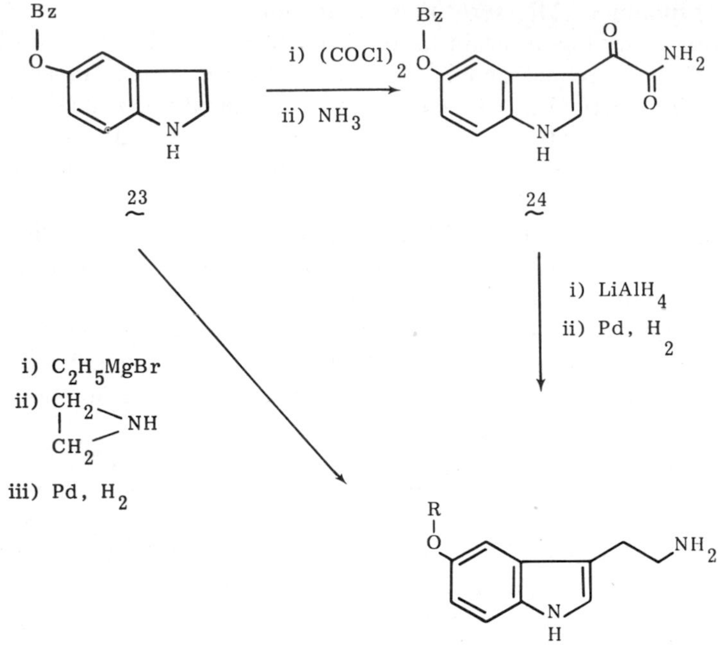

$Bz = CH_2C_6H_5$

21 5-hydroxytryptamine, R=H
25 R=Bz

to be obtained recently. Isolated in 0.3% yield from a *Borreria* sp. (Rubiaceae), the structure was determined by a combination of spectral and chemical evidence. The UV spectrum (λ_{max} 216, 245, 262, and 304 nm) suggested a 3-acylindole, and this was confirmed by an ion in the mass spectrum at *m/e* 144 (**30**). The ^{13}C NMR spectral data are shown in **31**, and the structure was confirmed by single-crystal X-ray analysis. A possible biosynthesis has not been discussed.

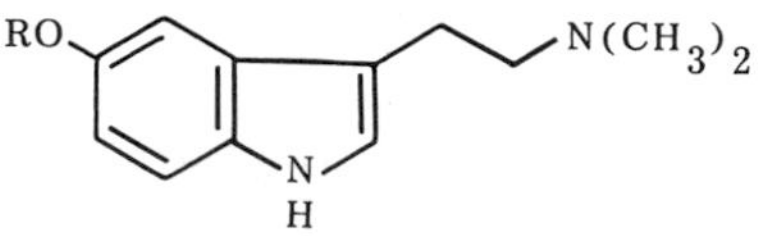

$$RO\text{—}\overset{\displaystyle}{\underset{N,H}{\text{indole}}}\text{—}N(CH_3)_2$$

26 R=CH$_3$
27 bufotenine, R=H

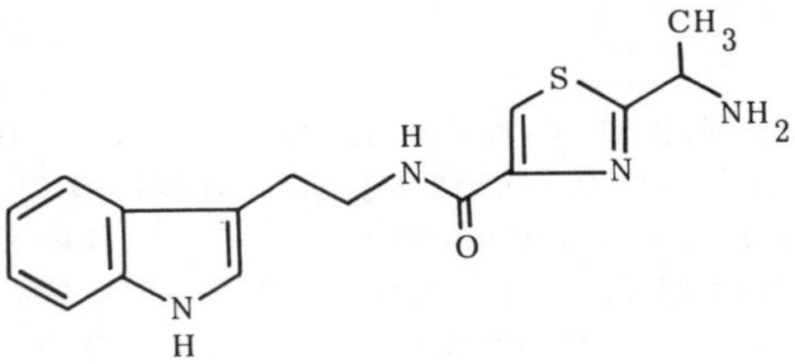

28

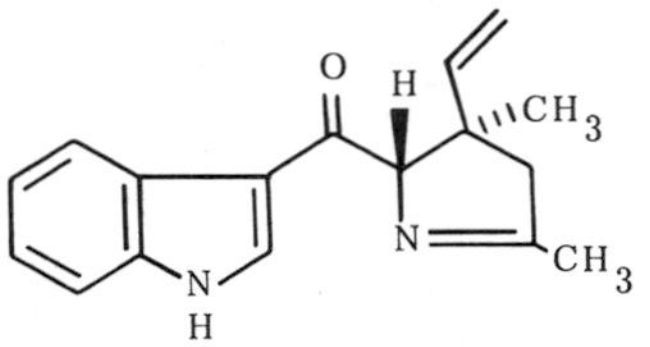

29 borreline

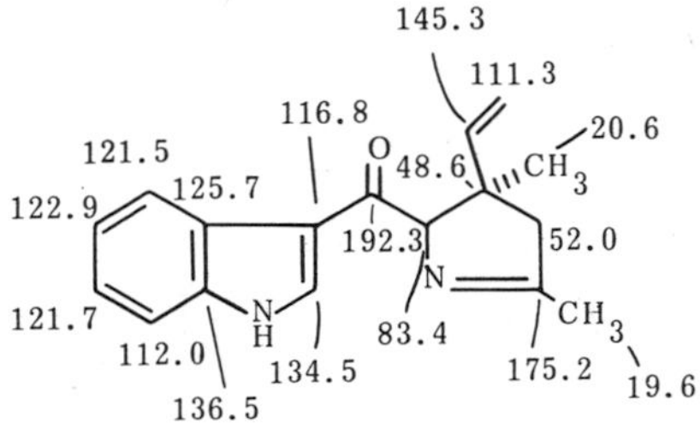

31

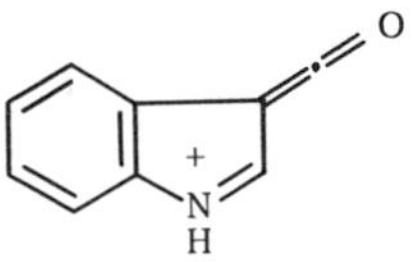

30 m/e 144

LITERATURE

Review

Saxton, J. E., *Alkaloids, NY* **8**, 1 (1965).

Borreline

Jössang, A., H. Jaquemin, J. L. Pousset, A. Cavé, M. Damak, and C. Riche, *Tetrahedron Lett.* 1219 (1977).

9.3 PYRROLNITRIN

Pyrrolnitrin (**32**) was obtained from the fungus *Pseudomonas aureofaciens* by Arima and co-workers and has been the subject of numerous studies. The Fujisawaka Pharmaceutical Co. has had an extensive synthetic program in this area with the aim of potentiating activity, and there has been substantial interest in the synthesis of the parent compound.

9.3.1 Synthesis

One interesting synthesis of pyrrolnitrin begins with condensation of the toluene derivative **33** with diethyloxalate and reaction of the product successively with chloroacetone and sulfuric acid–methylene chloride at $-15°C$ to afford **34**. Reaction of **34** with ammonia then gave the 3-phenyl pyrrole **35**. Alkyl groups on a pyrrole nucleus are particularly susceptible to halogenation which can lead under optimum conditions to a trichloro derivative. Hydrolysis of such a group leads to a carboxylic acid, and pyrrole carboxylic acids are well known for their ease of decarboxylation. Chlorination of **35** with sulfuryl chloride in acetic acid gave an intermediate tetrachloro derivative which was hydrolyzed to afford the diacid **36** after acidification. Thermal decarboxylation of the diacid yielded pyrrolnitrin (**32**).

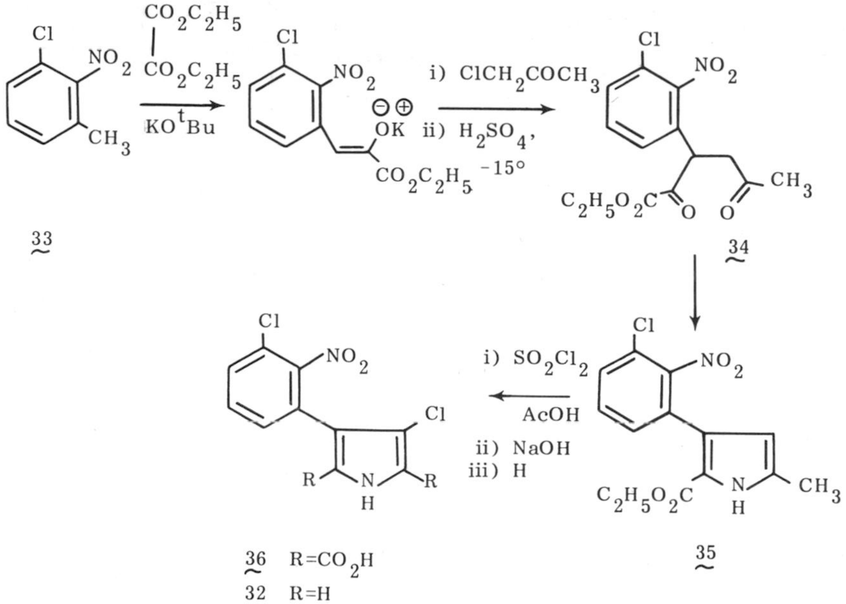

9.3.2 Biosynthesis

The novel structure of pyrrolnitrin (**32**) has prompted examination of the biosynthesis of this compound by several interesting approaches.

Gorman and co-workers established in 1967 that tryptophan was a key carbon source, and these same workers proposed a novel cleavage between N-1 and C-2 in order to account for the observed results. In this scheme N-1 becomes the nitrogen in the nitro group and the amino acid nitrogen ends up in the pyrrole ring. The experiments using ^{15}N-labeled tryptophans which thoroughly established this point were discussed in Section 2.3.3.

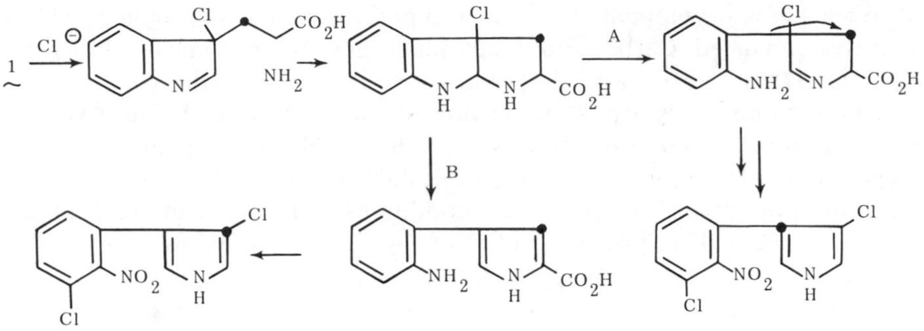

Scheme 3   Biogenesis of pyrrolnitrin (**32**).

At this point two hypothetical pathways could be proposed in which in one instance (pathway A) there is a 1,2-aryl shift, and in the second route (pathway B) the cleavage is direct and no shift of an aryl unit occurs (Scheme 3). The validity of pathway A was also established with a stable isotope, in this case ^{13}C. In order to understand the experiment, consider the alternative fates of carbon 3 in the side chain of tryptophan. By pathway A this carbon is attached to an aryl group, but from pathway B this carbon is attached to chlorine. In essence then the experiment is to examine the origin of C-3 and C-4 in pyrrolnitrin.

The assignments of C-3 (111.7 ppm) and C-4 (115.3 ppm) were made unambiguously, after reduction of the nitro group gave an amino compound in which the signal at 115.3 ppm was now shifted to 118.6 ppm.

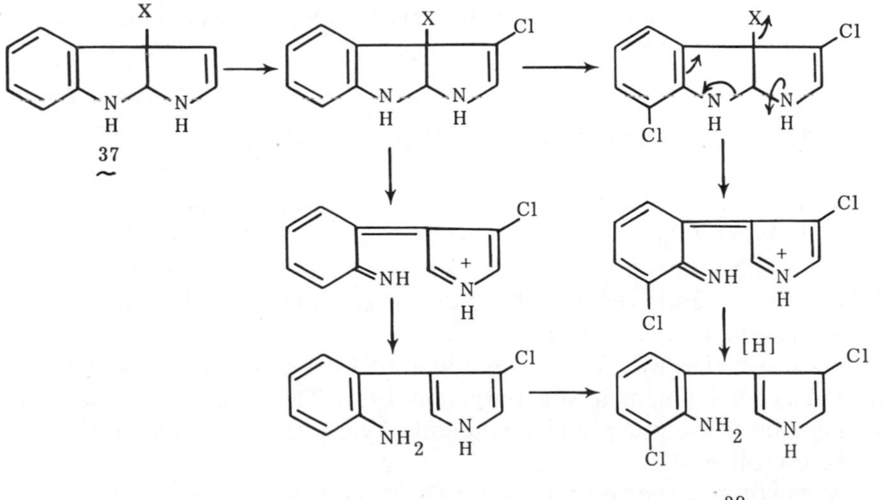

When [3-^{13}C]tryptophan was used as a precursor, only the signal at 111.7 ppm was enhanced. Carbon 3 of tryptophan therefore becomes C-3 of pyrrolnitrin (32) and no 1,2-aryl shift is involved in the biosynthesis.

Another interesting aspect of pyrrolnitrin biosynthesis is the observation made by Gorman and co-workers that both D- and L-tryptophans were incorporated. L-Tryptophan was found to inhibit pyrrolnitrin production and was also a poor precursor under these conditions. D-Tryptophan on the other hand was an excellent precursor of pyrrolnitrin both on its own and in the presence of an excess of unlabeled L-tryptophan.

However, when [2-^{3}H]tryptophans were used as precursors, in the L-series tritium was partially (~70%) retained, but in the D-series it was essentially lost. No explanation for this result is available as yet.

Two possibilities for the point of halogenation are likely. One is the involvement of the enamine 37, which can be chlorinated once or twice, or alternatively aromatic chlorination of the 3-chloro-4-(2-aminophenyl)pyrrole (38) to give the aminopyrrolnitrin (39) which co-occurs with 32.

LITERATURE

Imanaka, H., M. Kousaka, G. Tamura, and K. Arima, *J. Antibiot. (Tokyo) Ser. A* **18**, 207 (1965).

Gosteli, J., *Helv. Chim. Acta* **55**, 451 (1972).

Biosynthesis

Hammill, R., R. Elander, J. A. Mabe, and M. Gorman, *Antimicrob. Agts. Chemother.* 388 (1967).

Floss, H. G., P. E. Manni, R. L. Hammill, and J. A. Mabe, *Biochem. Biophys. Res. Commun.* **45**, 781 (1971).

Floss, H. G., *Lloydia* **35**, 399 (1972).

Martin, L. L., C.-J. Chang, H. G. Floss, J. A. Mabe, E. W. Hagaman, and E, Wenkert, *J. Amer. Chem. Soc.* **94**, 8942 (1972).

9.4 PHYSOSTIGMINE AND RELATED COMPOUNDS

Physostigma venenosum Balf. (Leguminosae) is a perennial woody climber native to West Africa. In 1845, Daniell described the use of the bean by the natives as an ordeal poison. Although the alkaloid content of the beans is quite low (~0.2%), several alkaloids have been isolated, and of these, physostigmine is the most important.

Physostigmine was first isolated in 1865 by Vee and its gross structure 40 deduced by Stedman and Barger in 1925. The alkaloid is notable for having a urethane group which is readily hydrolyzed with aqueous base to afford eseroline (41).

As might be expected for such a strained system, fragmentation of the ring system is quite facile. Thus eseroline methiodide when subjected to

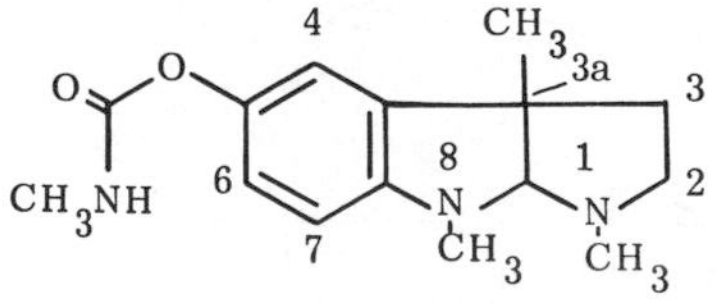

40 physostigmine 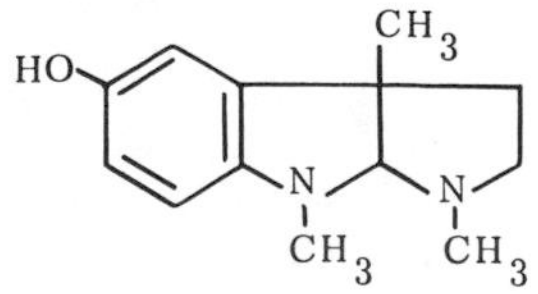   41 eseroline

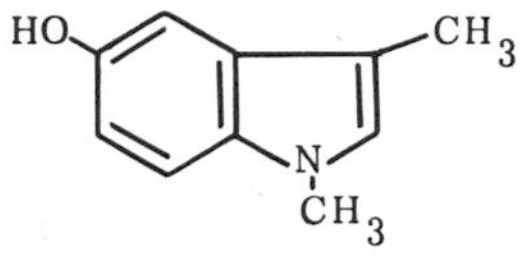

42 physostigmol 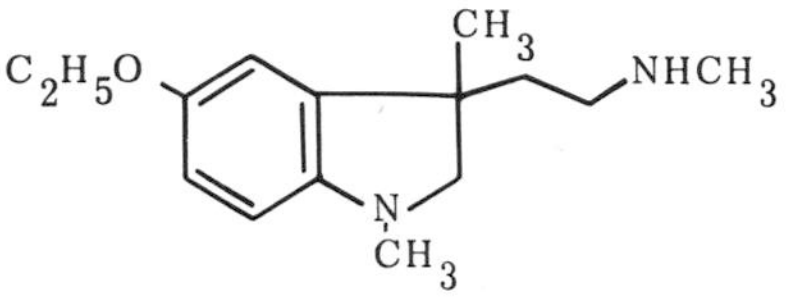   43

Hofmann elimination affords physostigmol (42) and the ethyl ether of eseroline is reduced catalytically to a ring-opened dihydro species 43.

9.4.1 Synthesis

Several syntheses of physostigmine have appeared, the first by Julian and Pikl in 1935 and a second by Harley-Mason and Jackson. The latter synthesis is of interest because it proceeds from 44 to eseroline (41) in one step using potassium ferricyanide as an oxidizing agent.

Classically and based on biogenetic principles one would like to develop a synthesis involving the internal cyclization of a 5-hydroxytryptamine derivative. For the most part however, such reactions have been observed as aberrations of other schemes and are not synthetically useful.

When N-acetyl tryptamine (45) is condensed with 3,3-dimethylallylbromide, 46 is the major product, and reaction of N-acetyltryptophan ethyl ester 47 with t-butyloxychloride-triethylamine affords the 2,3-dihydro[2,3-b]indole 48.

The most recent synthesis (Scheme 4) involves an interesting rearrangement of a cycloprop[b]indole. Irradiation of the Reissert compound 49 gave the cycloprop[b]indole 50, which on treatment with base in aqueous ethanol afforded 51 in 72% yield. This rearrangement is thought to proceed as shown by way of the indolenine 52. Further elaboration of the C-ring to afford eseroline methyl ether (53) was accomplished by standard methods.

9.4.2 Absolute Configuration

It had been assumed on steric grounds that the B/C ring junction in physostigmine was *cis* and this was verified when a nuclear Overhauser effect

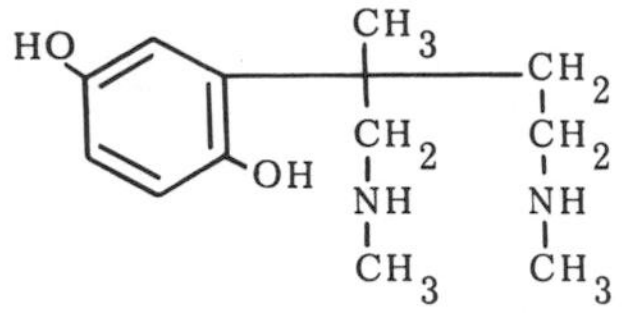

44

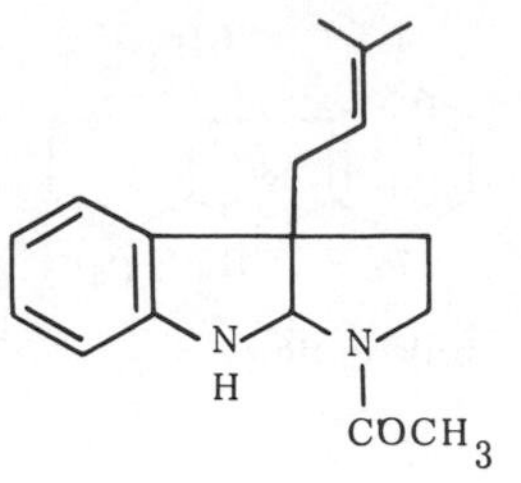

46

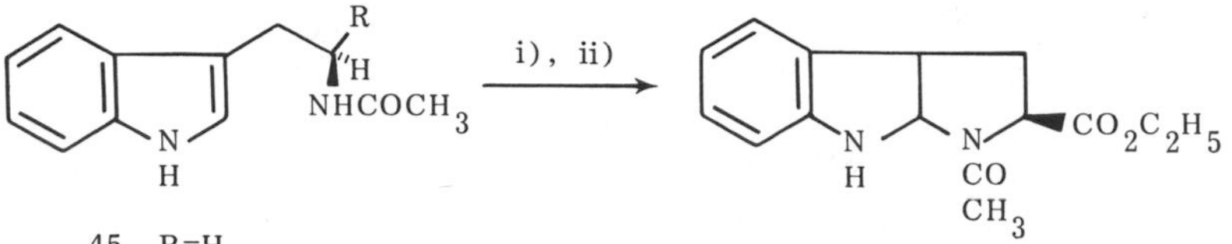

45   R=H

47   R=CO$_2$C$_2$H$_5$

48

i) (CH$_3$)$_3$COCl

ii) Ru–Al$_2$O$_3$, H$_2$

(NOE) was observed between the 3-methyl group and the 8-proton. The absolute configuration of physostigmine was determined independently by two groups in 1969 to be that shown in **54**. In both cases physostigmine was degraded to the hydroxyindoline **55**, which was then degraded further by different methods, in one case to the amino acid **56** and in the other case to the oxindole **57**. In each instance comparison was made with synthetic compounds of known absolute configuration.

49

50

51

52

53

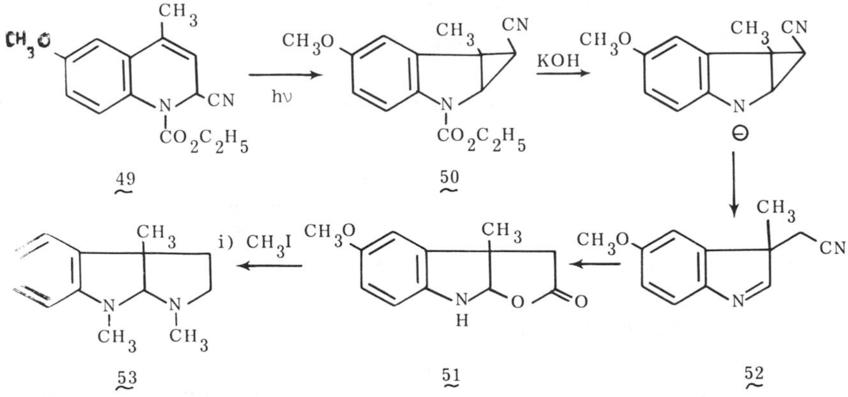

Scheme 4   Ikeda synthesis of eseroline methyl ether.

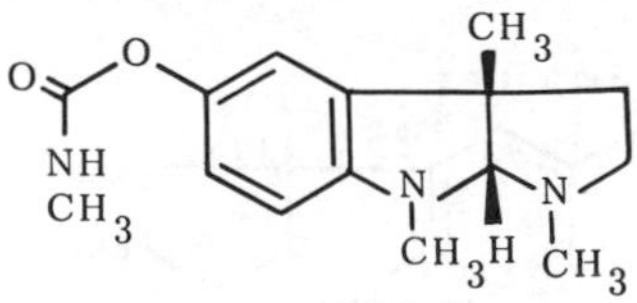

54

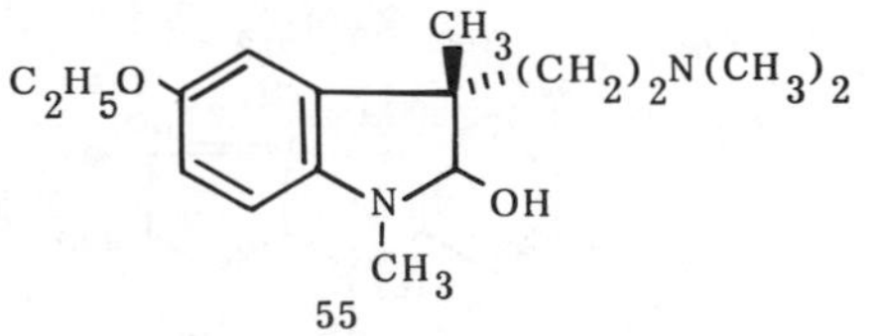

55

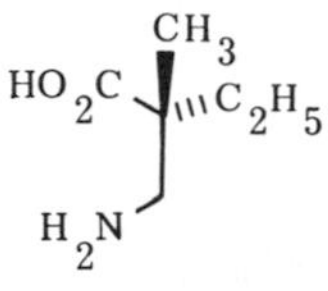

56

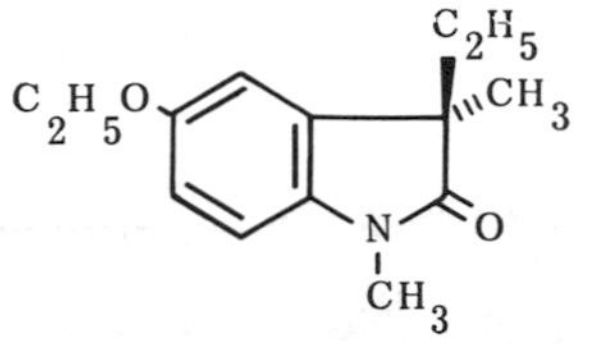

57

9.4.3 Detection and Estimation

The reaction of physostigmine with ammonia has been one of the official qualitative tests for physostigmine for many years. The product has a quite characteristic blue color and has been formulated as the phenoxazone **58**.

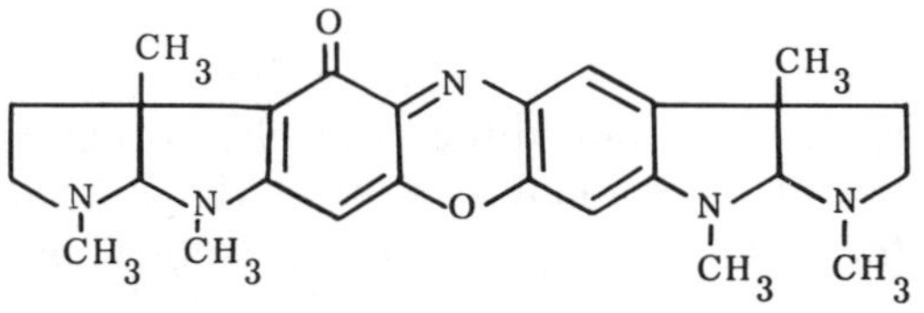

58

Until recently most of the methods for the quantitative assay of physostigmine were nonspecific. Quite recently a method has been described which involves reaction with sodium nitrite in acid solution. The resulting yellow absorbance is quantitated at 417 nm and is not affected by decomposition products.

9.4.4 Spectral Properties

The ^{13}C NMR data of physostigmine and some related compounds have recently been determined by Robinson and co-workers and are shown in **59** together with the corresponding data for physovenine in **60**. As expected the most pronounced shift differences are at C-2 and C-8α followed by γ-

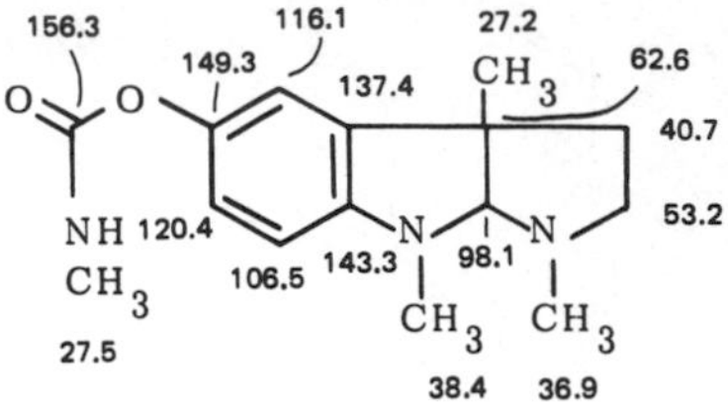

59

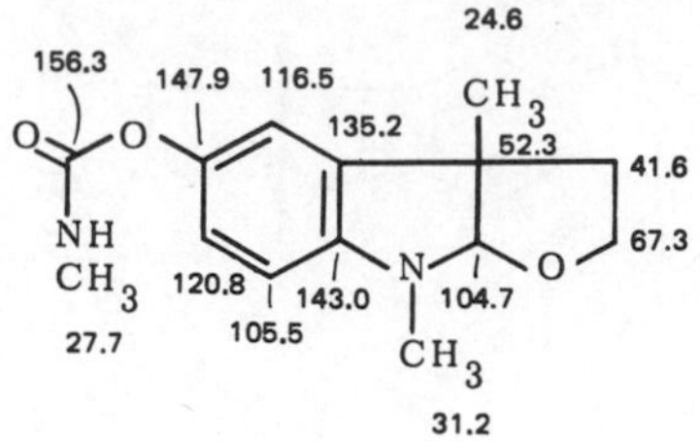

60

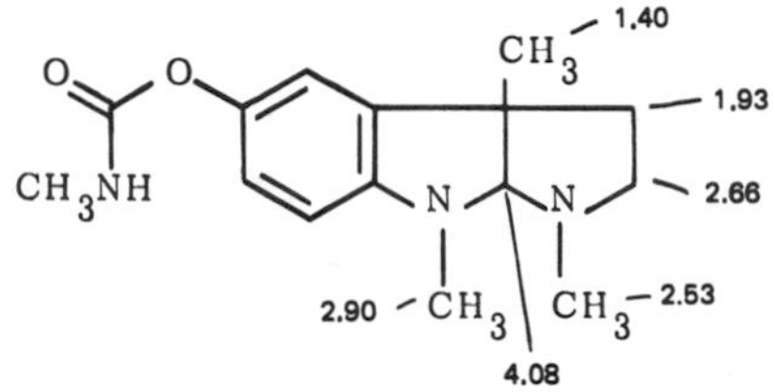

δppm TMS

61

effect differences at N-8 and C-3α methyl groups. The proton NMR data are shown in **61**.

9.4.5 Biosynthesis

No details are known of the biosynthesis of physostigmine (**54**). Biogenetically though it would be predicted to be derived from tryptophan and methionine, with the latter probably also being responsible for the C-3 methyl group.

9.4.6 Pharmacology

Physostigmine is a reversible cholinesterase inhibitor, and in the form of a 0.25% ointment the salicylate salt is used in the treatment of glaucoma. Typically for a compound having a cholinesterase inhibitor activity, physostigmine also stimulates such secretions as saliva and perspiration, and acts to increase tone and peristalsis of the gastrointestinal tract. The urethane group is essential for activity since eseroline (**41**) is inactive. The N-8 methyl group and N-1 and its methyl group are not necessary however, as shown by activity in the erythrocyte acetylcholinesterase system *in vitro*.

It has been suggested that these results support the notion that the indolinium cations such as **62** are the active species. Indeed the indoline derivative **63** was 100 times more active than physostigmine as a cholinesterase inhibitor.

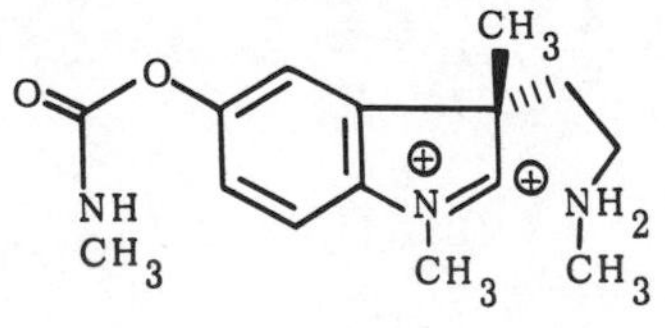

62 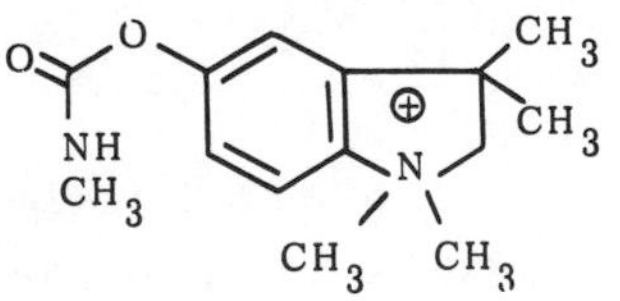   63

9.4.7 Geneserine

For many years geneserine was regarded as the N_1-oxide of physostigmine (**64**). However several factors appeared to mitigate against this, among these was the nonhygroscipic character, the low water solubility, and the absence of $M^+ - 17$ or $M^+ - 18$ ions in the mass spectrum. When the NMR spectrum of geneserine was examined, the N_1-methyl group had the same chemical shift as in physostigmine, indeed the only major differences were a downfield shift of about 0.63 ppm for the C-8α proton and an upfield shift of 0.22 ppm in the C-3 methyl resonance. On this basis and by comparison with N-methyltetrahydro-1,2-oxazine (**65**), geneserine has been assigned the structure **66**.

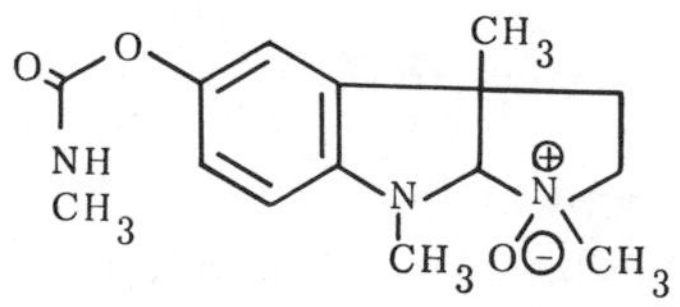

64   geneserine
      (old structure)

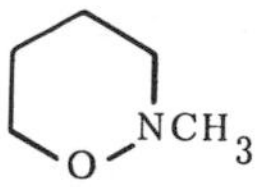

65

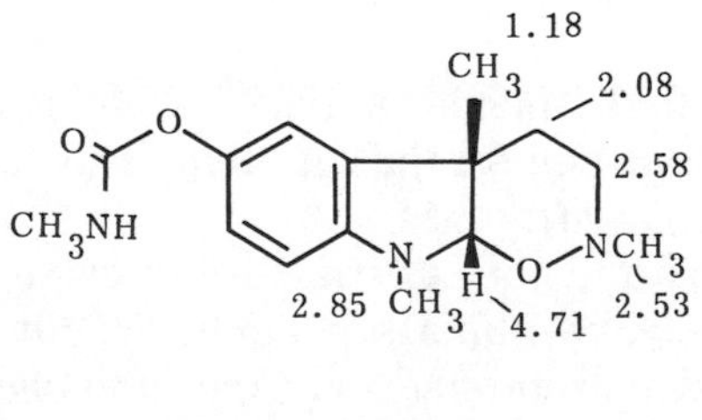

66   geneserine
      (new structure)

LITERATURE

Reviews

Coxworth, E., *Alkaloids, NY* **8,** 27 (1965).

Robinson, B., *Alkaloids, NY* **10,** 383 (1968).

Robinson, B., *Alkaloids, NY* **13,** 213 (1971).

Synthesis

Julian, P. L., and J. Pikl, *J. Amer. Chem. Soc.* **57,** 755 (1935).

Harley-Mason, J., and A. H. Jackson, *J. Chem. Soc.* 3651 (1954).

Ikeda, M., S. Matsugashita, and Y. Tamura, *J. Chem. Soc. Perkin Trans I* 1770 (1977).

Spectral Properties

Longmore, R. B., and B. Robinson, *Chem. Ind. (London)* 622 (1969).

Hill, R. K., and G. R. Newkome, *Tetrahedron*, **25,** 1249 (1969).

Crooks, P. A., B. Robinson, and O. Meth-Cohn, *Phytochemistry* **15,** 1092 (1976).

Pharmacology

Long, J. P., and C. J. Evans, in A. Burger (Ed.), *Drugs Affecting the Peripheral Nervous System*, Vol. 1, Marcel Dekker, New York, 1967.

9.5   THE OLIGOMERS OF TRYPTAMINE

There are several alkaloids known which are derived by the polymerization of tryptamine units. These compounds are of quite diverse botanical origin and have some interesting structures.

9.5.1   Calycanthine and Chimonanthine

The simplest compounds have been isolated from plants in the family Ca- lycanthaceae, a small family comprised only of the genera *Calycanthus* and *Chimonanthus*. The most prominent alkaloid, calycanthine (**67**), was ob- tained from several species of *Calycanthus*, and the related plant *Chimon- anthus fragrans* afforded the dimeric tryptamine derivative chimonanthine (**68**).

Calycanthine (**67**) was first isolated in 1905, but in spite of strenuous efforts and several proposed structures the data could still not be interpreted in terms of a unique molecular array.

It was not until 1960 that the correct structure of calycanthine was de- duced, almost simultaneously, by Robertson and by Woodward, the former using X-ray analysis of the dihydrobromide, and the latter chemical methods.

Calycanthine (**67**) has not been synthesized, but an interesting synthesis of chimonanthine (**68**) has been reported by Hendrickson. The sodium salt

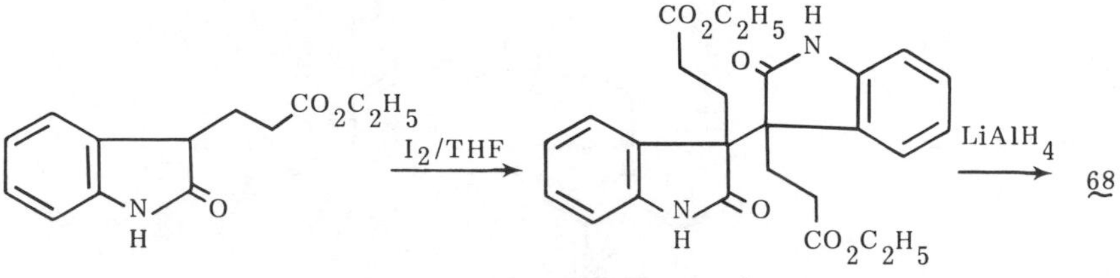

69 70

Scheme 5 Hendrickson synthesis of chimonanthine.

of the oxindole **69** on oxidation with iodine in tetrahydrofuran dimerized to a mixture of diastereoisomers **70**. Reduction of one of these with lithium aluminum hydride in THF gave racemic chimonanthine (**68**) (Scheme 5).

Schütte has suggested that in fact chimonanthine (**68**) is an intermediate in the biosynthesis of calycanthine (**67**) as shown in Scheme 6. It is probably more reasonable however to go directly from the diimine **71** to calycanthine (**67**), rather than involving chimonanthine (**68**).

9.5.2 Hodgkinsine

The next highest oligomer is a trimer, hodgkinsine (**72**), isolated from *Hodgkinsonia frutescens* F. Muell., a plant in the family Rubiaceae quite unrelated to the Calycanthaceae.

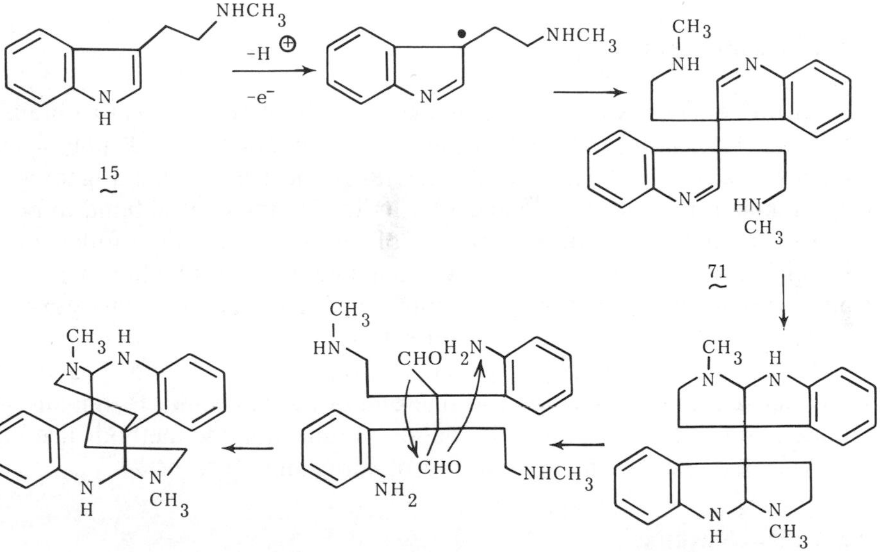

67 calycanthine 68 chimonanthine

Scheme 6 Biogenesis of chimonanthine (**68**) and calycanthine (**67**).

72 hodgkinsine

The compound was originally thought to be an isomer of calycanthine, but subsequent work established the molecular formula $C_{33}H_{38}N_6$. The structure was solved both by chemical means (Manchester group), and by X-ray analysis of the trimethiodide derivative (Australian group). The CD spectrum of hodgkinsine is interesting because the components T_1 and T_2 are of opposite absolute configuration and consequently the only contributor to the CD spectrum is the unit T_3.

The mass spectral fragmentation of hodgkinsine ($M^+ - 518$) was similar to that of chimonanthine, where the C-3'–C-3" bond is broken to yield ions at m/e 172 and 173 from the lower unit and ions at m/e 345 and 344 from the upper two components.

9.5.3   Quadrigemines

The quadrigemines A and B are minor alkaloids of *H. frutescens* obtained by Parry and Smith. They are tetramers of tryptamine and display a molecular ion at m/e 690 ($C_{44}H_{50}N_8$). The mass spectrum of quadrigemine A gave fragment ions at m/e 345 and 344, indicating the central bond to be C-3-to-C-3' linked. Hofmann elimination of the tetramethiodide followed by sodium borohydride reduction gave an indole-indoline identical with that from hodgkinsine except for optical rotation. Quadrigemine A therefore has the structure **73** except for stereochemistry.

Quadrigemine B yielded fragment ions at m/e 172, 173, 516, and 517, indicating a chimonanthine unit at one end of the molecule. Hofmann degradation and sodium borohydride reduction following the methods used for hodgkinsine indicated the structure of quadrigemine B to be **74**.

9.5.4   Psychotridine

Psychotria beccarioides Wernh. in the family Rubiaceae is a shrub native to New Guinea and has afforded the alkaloid psychotridine, a pentamer of

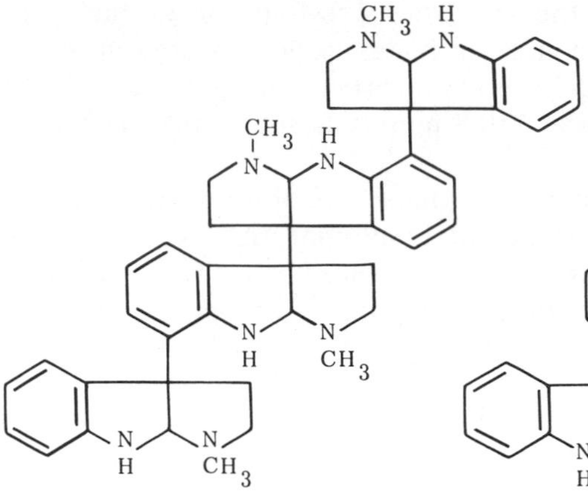

73 quadrigemine A 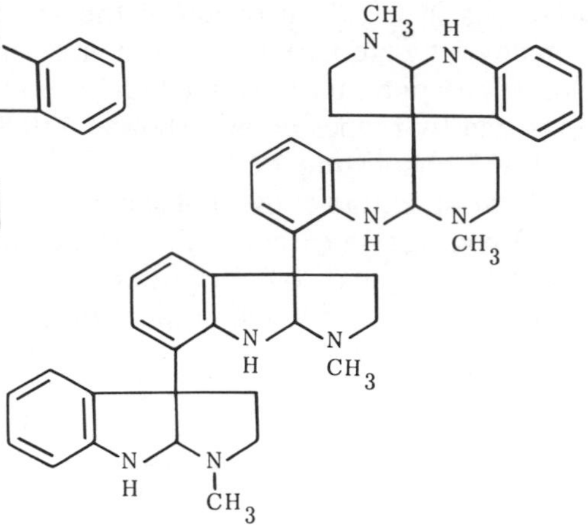   74 quadrigemine B

tryptamine having the molecular formula $C_{55}H_{62}N_{10}$. So far this is the most complex alkaloid in this series.

The mass spectrum of psychotridine gave ions at m/e 172, and m/e 344 (base peak), 345, and m/e 516 (72%) and 517 with a molecular ion at m/e 862. The two important ions at m/e 516 and 344 indicate a C-3a to C-3a' linkage between the second and third units in a sequence of five N_b-methyl-tryptamine units. If only the standard hodgkinsine linkages occur, the other units are fixed in their orientation and psychotridine has the structure **75**.

The ^{13}C NMR spectrum of psychotridine was compared with hodgkinsine and although not all the carbon resonances could be assigned, the number

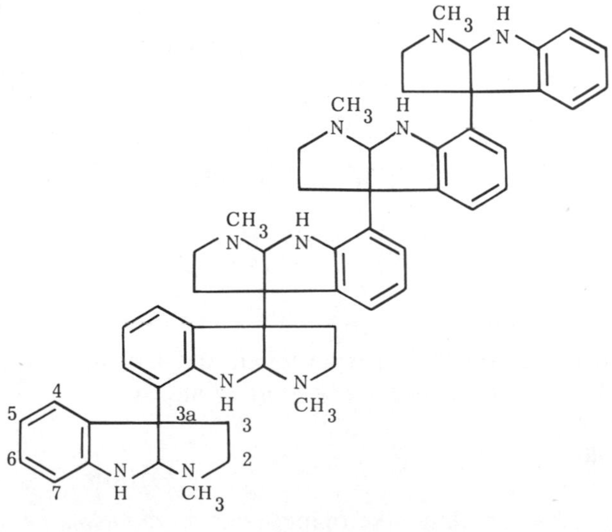

75 psychotridine

of signals in a certain region of the spectrum was found to be highly informative. Quaternary C-3a resonances in hodgkinsine appeared in the region 60–64 ppm, unsubstituted C-7 aromatic carbons in the range 108–109 ppm, and two types of 7a carbon at 150.8 ppm (C-7 substituted) and 151.5 ppm (C-7 unsubstituted).

In the spectrum of psychotridine five signals were observed in the region 60–63 ppm for 3a carbons, a resonance at 148.9 ppm (three C-7a carbons) and at 150.7 ppm (two C-7a carbons). Although these data are not definitive for the structure of psychotridine, they do support the proposed structure.

LITERATURE
Review

Manske, R. H. F., *Alkaloids, NY* **8,** 581 (1965).

Hodgkinsine

Fridrichsons, J. P., M. F. Mackay and A. McL. Matheson, *Tetrahedron* **30,** 85 (1974).

Quadrigemines

Parry, K. P., and G. F. Smith, *J. Chem. Soc., Perkin Trans I*, 1678 (1978).

Psychotridine

Hart, N. K., S. R. Johns, J. A. Lamberton, and R. E. Simmons, *Aust. J. Chem.* **27,** 639 (1974).

9.6   DIKETOPIPERAZINES DERIVED FROM TRYPTOPHAN

Some diketopiperazines were discussed in Chapter 8, but there are also several distinct classes of diketopiperazines derived from tryptophan. These compounds include the echinulins, the brevianamides, the sporidesmins, and certain compounds that exhibit tremorigenic activity.

One of the simplest of these compounds is **76** from *Aspergillus ruber*,

76

which is probably derived from tryptophan and leucine. The other compounds in this series are considerably more complex.

9.6.1   Echinulin

Echinulin (**77**) was the first diketopiperazine to be isolated carrying an isoprenylated tryptophan residue, and was initially obtained by Quilico from

77   echinulin

Aspergillus echinulatus in 1943. The structure was determined in 1964 by chemical degradation, but Birch made a very significant contribution to this effort using biogenetic reasoning after making a study of the biosynthetic precursors.

Biosynthesis

Birch evaluated the structure in terms of three structural elements: L-tryptophan, L-alanine, and mevalonic acid. Using the last two precursors it was shown that echinulin contained three isoprene units and that [1-^{14}C]alanine was specifically incorporated into the diketopiperazine ring.

Subsequently it was shown that *cyclo*-L-alanyl-L-tryptophyl (78) was also efficiently incorporated into echinulin (77). Since the Cotton effects for echinulin (77) and 78 in the region 200–300 nm were the same, the stereochemical integrity of the precursor units into the product was established. The implication was also made that the isoprenylation reactions occur as late stages in the biosynthesis. Based on the observed tendencies of the indole nucleus to electrophilic substitution, the location of substituents in echinulin are not surprising. However, the orientation of the C-2 substituent and the sequence of isoprene substitutions are of some interest.

Allen has prepared a cell extract of *A. amstelodami* which catalyzes a dimethylallylpyrophosphate-mediated isoprenylation of 78. The product was identified as the 2-substituted derivative 79, and was later isolated as preechinulin from *A. chevalieri*. Doubly-labeled preechinulin 79 was incorporated intact into echinulin (77) (Scheme 7).

The mechanism of introduction of the C-2 substituent is not clear. Model reactions with 3-methylindole appear to eliminate direct C-2 substitution or rearrangement from the 3-position and favor N-alkylation as the initial step. For example the *N*-alkyl derivative 80 is converted by acid into a mixture of 81 and 82.

Aspergillus amstelodami produces a number of metabolites which are only substituted at the 2-position of the indole nucleus, and which are elaborated in the side chain and the dioxopiperazine ring; some examples are neoechinulin A (83), neoechinulin B (84), and cryptoechinulin C (85). The same organism also produces the corresponding compounds such as neoechinulins C (86) and D (87) and neoechinulin (88) which are substituted at

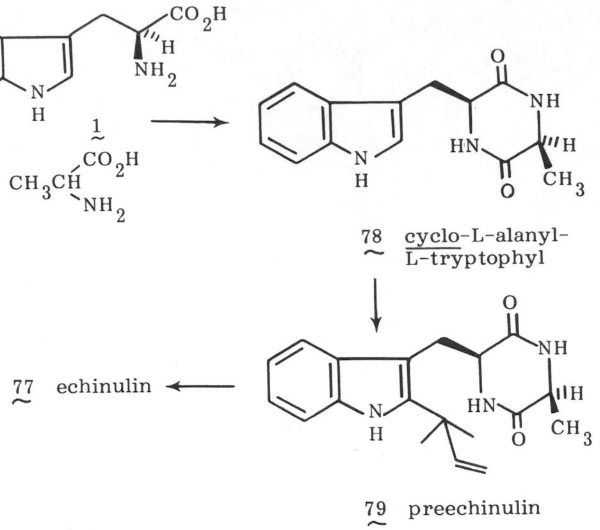

Scheme 7   Biosynthesis of echinulin (**77**).

C-6. *Aspergillus ruber* produces isomeric compounds (e.g. **89**) called iso-echinulins, where substitution is at C-5.

Experiments with [3R-^{3}H]- and [3S-^{3}H]L-tryptophans, doubly labeled with ^{14}C, showed that the 3S-proton was specifically removed on incorporation into neoechinulin (**88**), indicating that elimination occurs by *cis*-elimination of the 2- and 3-hydrogens in the tryptophan side chain.

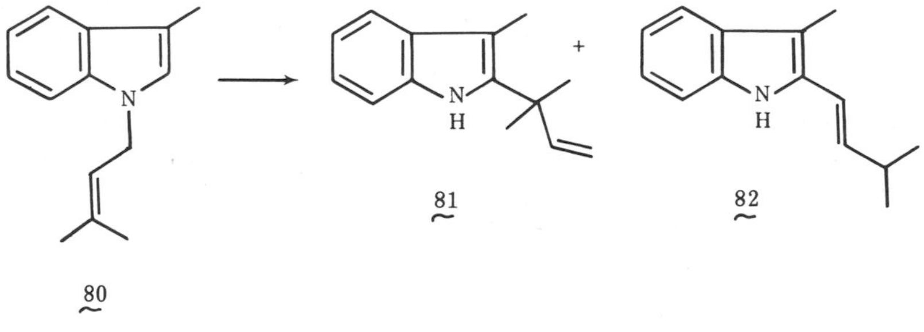

A quite interesting new approach to the "isoprene" side chain of echinulin (**77**) has been the work of the Italian group using leucine as a precursor. [4,5-^{3}H$_2$, U-^{14}C]L-Leucine was incorporated into echinulin (**77**) with loss of about half of the tritium and this was followed by studies with 4R[5-^{13}C]leucine (**90**) and 4S[5-^{13}C]leucine (**91**). The labeling of echinulin from these products is shown in **92**.

These results are in agreement with a derivation of mevalonic acid (**93**) from leucine in which the 3-methyl group of MVA arises from the *pro*-4S

79 preechinulin

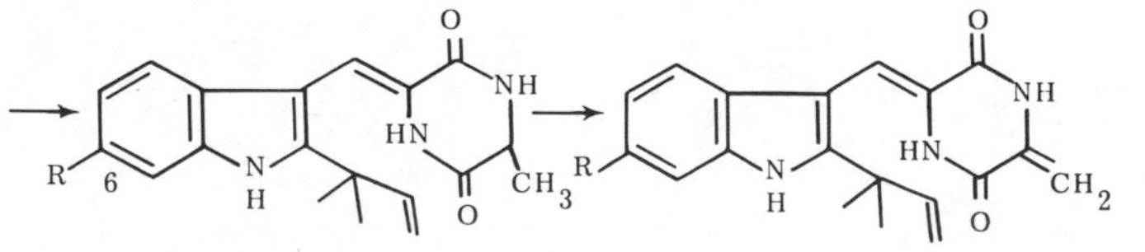

83 neoechinulin A, R=H
87 neoechinulin D, R=-CH$_2$CH=C(CH$_3$)$_2$

84 neoechinulin B, R=H
86 neoechinulin C, R= -CH$_2$CH=C(CH$_3$)$_2$

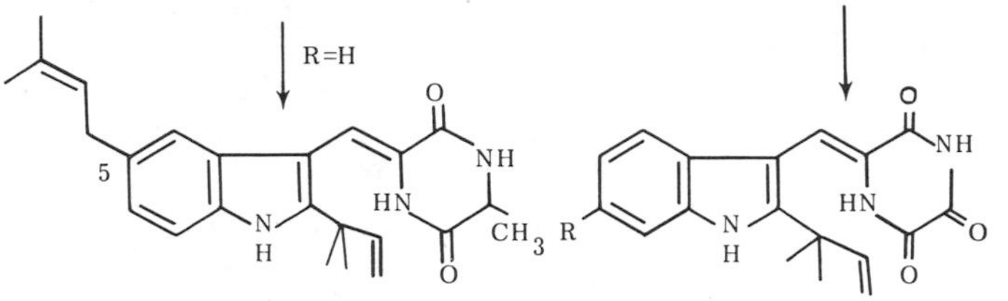

89

85 cryptoechinulin C, R=H
88 neoechinulin, R=-CH$_2$CH=C(CH$_3$)$_2$

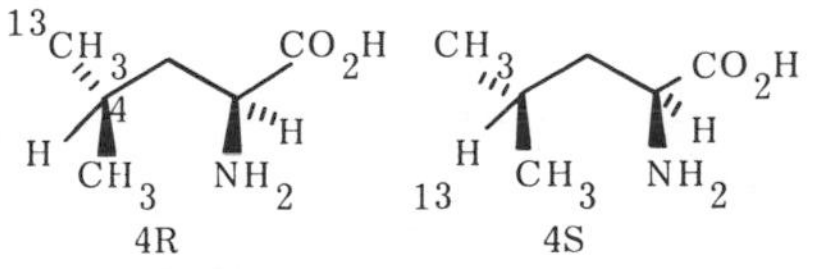

90

91

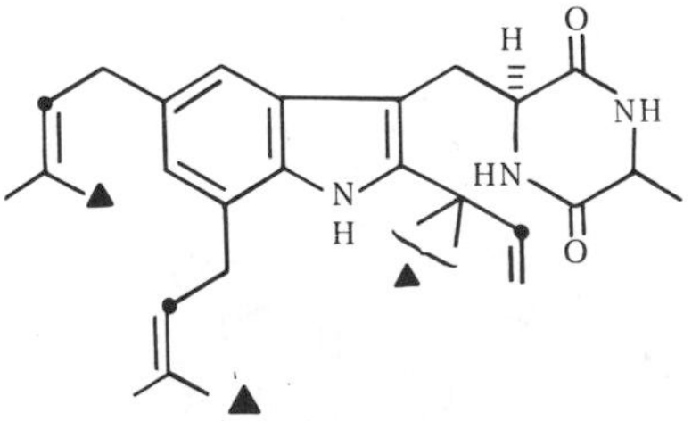

92

• from 4R-[5-^{13}C]-leucine

▲ from 4S-[5-^{13}C]-leucine

methyl group of leucine and the C-4 methylene group of MVA from the *pro*-4R methyl group of leucine as shown in Scheme 8.

An additional structure type derived from the neoechinulin B framework is represented by the alkaloids **94** and **95**, also obtained from *A. amstelodami*. The compounds are envisaged as being derivatives of the neoechinulins and auroglaucine (**96**) by a Diels–Alder cycloaddition.

Synthesis

The first total synthesis of echinulin (**77**) was reported by Kishi in 1971. Heating the alkylated aniline **97** with zinc chloride afforded the aniline **98**, which on Fischer cyclization with **99** gave the indole **100**. Reduction of the ester to the alcohol, oxidation to the aldehyde, and a Wittig reaction yielded **101**, which could be converted to the corresponding gramine derivative **102**

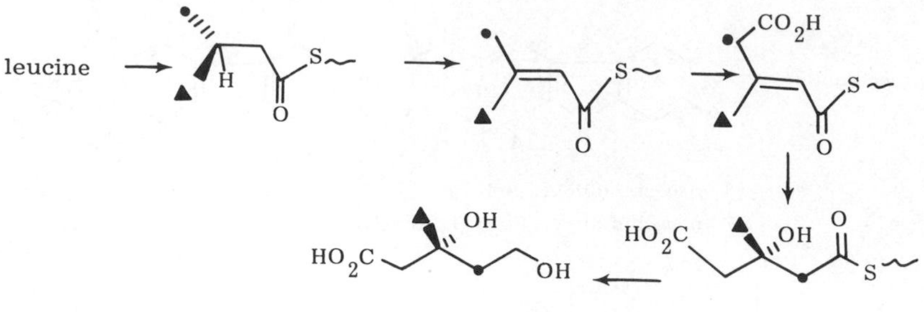

$$\underset{93}{}\quad MVA$$

● from $4R-[5-^{13}C]$-leucine

▲ from $4S-[5-^{13}C]$-leucine

Scheme 8   Biosynthesis of mevalonic acid (**93**) from leucine.

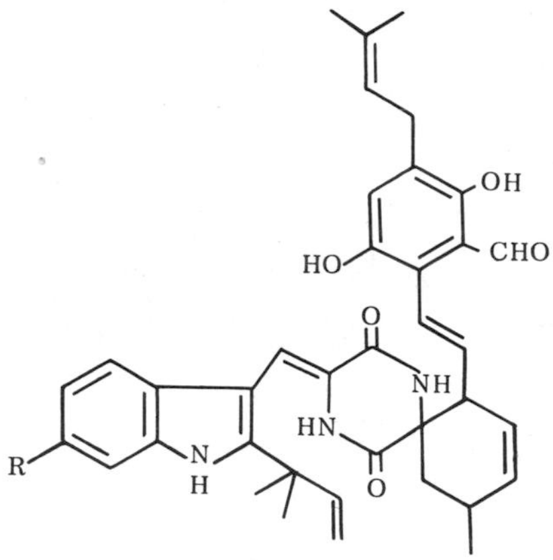

94 R=H

95 R=

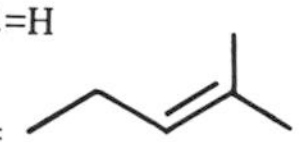

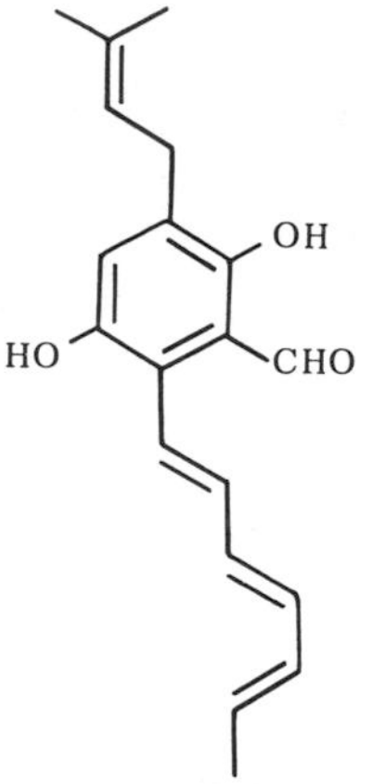

96   auroglaucine

and condensed with **103** to yield **104**. Hydrolysis and decarboxylation afforded echinulin as a mixture of *cis* and *trans* isomers (Scheme 9).

LITERATURE

Quilico, A., and L. Panizzi, *Chem. Ber.* **76**, 348 (1943).

Birch, A. J., G. E. Blance, S. David, and H. Smith, *J. Chem. Soc.* 3128 (1961).

Birch, A. J., and K. J. Farrar, *J. Chem. Soc.* 4277 (1963).

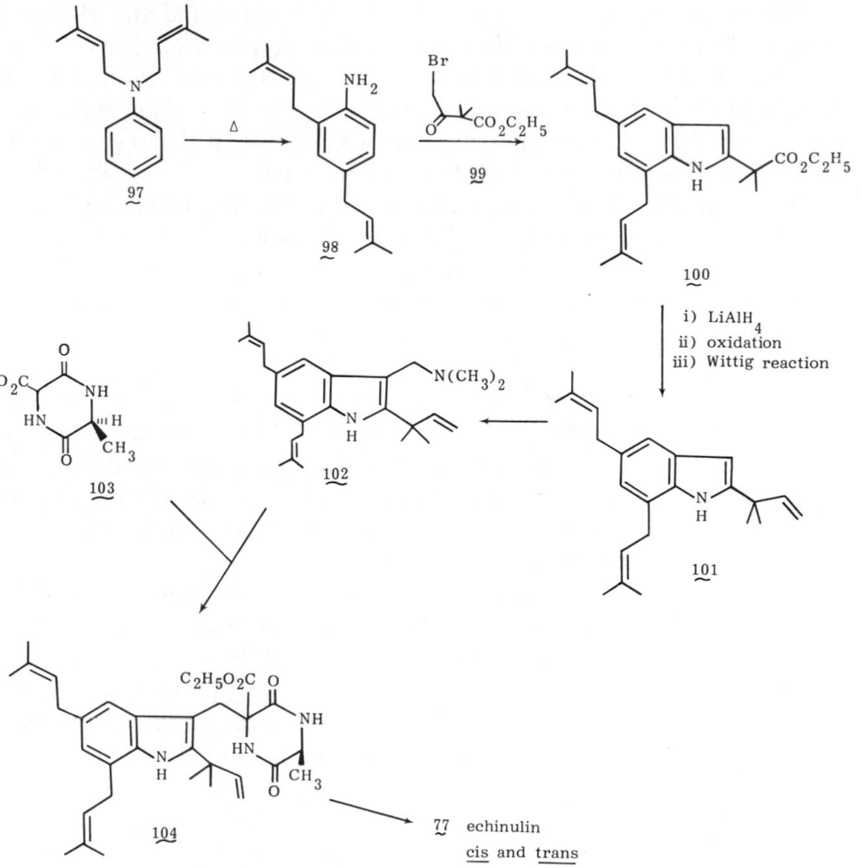

Scheme 9 Kishi synthesis of echinulin (**77**).

MacDonald, J. C., and G. P. Slater, *Can. J. Microbiol.* **12**, 455 (1966).

Slater, G. P., J. C. MacDonald, and R. Nakashima, *Biochemistry* **9**, 2886 (1970).

Allen, C. M., *Biochemistry* **11**, 2154 (1922).

Allen, C. M., *J. Amer. Chem. Soc.* **95**, 2386 (1973).

Cardillo, R., C. Fuganti, D. Ghiringhelli, and P. Graselli, *Chem. Commun.* 778 (1975).

Takamatsu, N., S. Inoue, and Y. Kishi, *Tetrahedron Lett.* 4665 (1971).

Cardillo, R., C. Fuganti, D. Ghiringhelli, P. Graselli, and G. Gatti, *Chem. Commun.* 474 (1977).

9.6.2 Brevianamides and Austamides

Brevianamides

In 1969, Birch and co-workers reported the isolation of a new group of neutral compounds from *Penicillium brevi-compactum* and named them brevianamides.

Brevianamide A has a 3-indoxyl UV spectrum and the IR spectrum suggested the presence of a diketopiperazine, containing a single exchangeable NH. Proton NMR data established the presence of two methyl groups attached to a quaternary, saturated carbon, and since there were no 3- or 6-protons of the diketopiperazine nucleus present in the region 4.5 to 3.7 ppm, these positions must be further substituted.

The available evidence was rationalized after it was shown that all of the carbon atoms could be accounted for by three precursors: tryptophan, proline (106), and an isoprene unit. These data together with biogenetic reasoning led to the probably molecular structure 105 for brevianamide A. Some of the principal mass spectral fragments of brevianamide A (105) are shown in Scheme 10.

The second isolate, brevianamide E (107), was a dihydroindole also containing a diketopiperazine unit. The proton NMR spectrum indicated the presence of an isoprene unit having a gem-dimethyl group and a vinyl moiety. Unlike brevianamide A, brevianamide E gave 1 mole of proline (106) on acid hydrolysis. Reduction with zinc–acetic acid gave the indole 108, showing an important fragment ion at m/e 198.

Biosynthetically, [3'-^{14}C]tryptophan (1), [2-^{14}C]mevalonic acid (93), and [U-^{14}C]L-proline (106) were incorporated into brevianamides A and E, and in addition cyclo-L-[3'-^{14}C]tryptophyl-L-prolyl (109) was incorporated into brevianamide A (105) and was found to be a co-metabolite. The compound 109 accumulates in A. ustus and could be the next biosynthetic intermediate, although this has not been established. Brevianamide E (107) would be

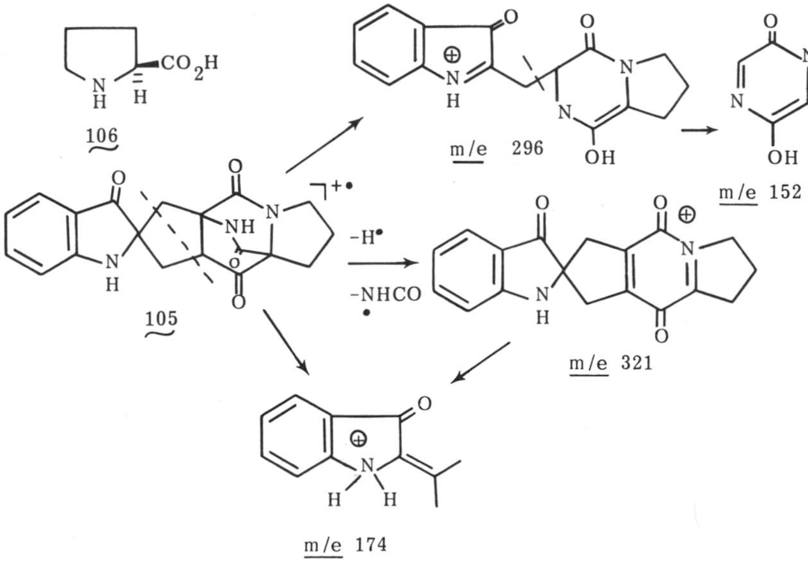

Scheme 10   Mass spectral fragmentation of brevianamide A (105).

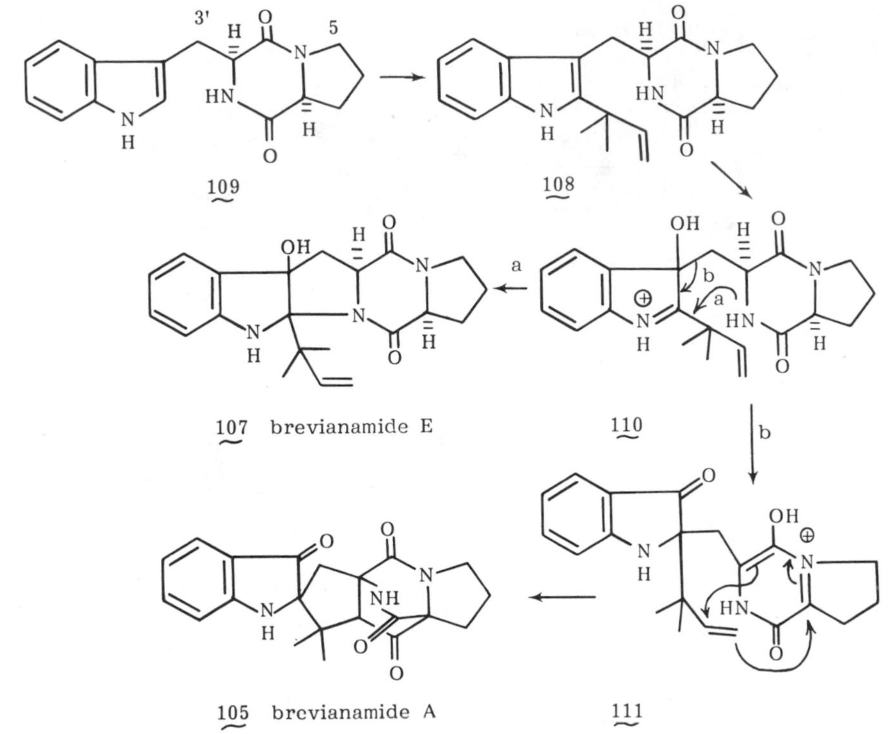

Scheme 11 Biogenesis of the brevianamides A (**105**) and E (**107**).

produced from the 3-hydroxyindolenine **110** by intramolecular attack of the amide nitrogen as shown. Brevianamide A on the other hand could be produced from the same hydroxyindolenine by migration of the bulky 3-substituent to give an oxindole **111**. Oxidation of the dioxopiperazine followed by an internal Diels–Alder reaction affords **105** (Scheme 11).

Deoxybrevianamide E (**112**) has been synthesized by the route shown in Scheme 12. Aerial oxidation converts this compound to brevianamide E (**107**).

Austamides

Very closely related to the brevianamides are the austamides obtained from *Aspergillus ustus*, and some representative structures are **113** and austamide (**114**), the toxic principle. These are novel compounds because they contain a new nitrogen-to-carbon bond from the isoprene unit to the tryptophan nitrogen.

The isolation of **113** is important because it is regarded as a biosynthetic intermediate in the formation of austamide (**114**). The previously encountered intermediate **79** could also be involved in the formation of **114**, but another more reasonable explanation is the involvement of the compound

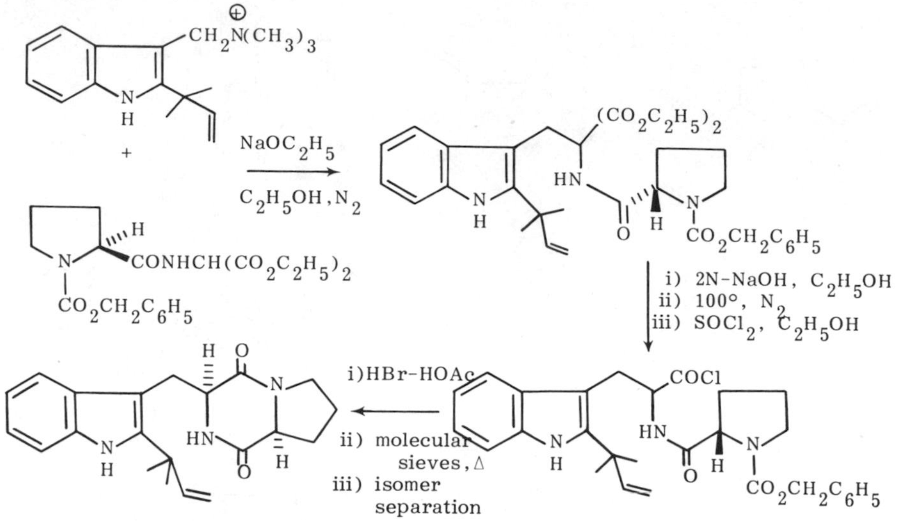

Scheme 12 Ritchie–Saxton synthesis of deoxybrevianamide E (**112**).

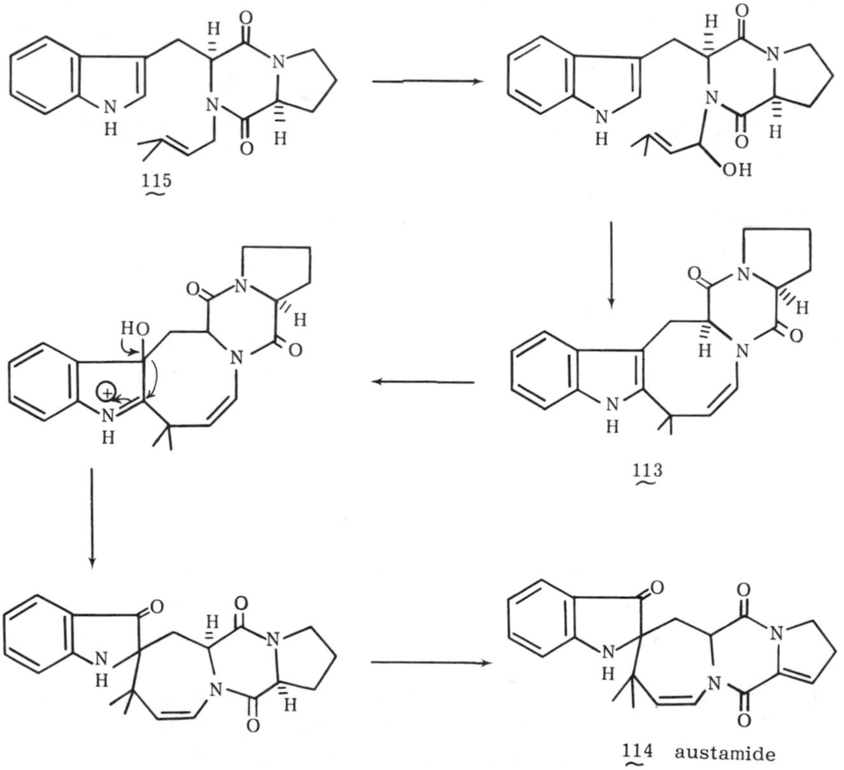

112 deoxybrevianamide E

Scheme 13 Biogenesis of austamide (**114**).

114 austamide

115, where the isoprene unit is originally attached to the tryptophan nitrogen (Scheme 13). Hydroxylation followed by attack from C-2 of the indole nucleus would lead to **113**. Oxindole formation and dehydrogenation in the proline ring affords austamide (**114**).

The proton NMR spectral data of austamide are shown in **116**.

116

LITERATURE

Birch, A. J., and J. J. Wright, *Chem. Commun.* 644 (1969); *Tetrahedron* **26**, 2329 (1970).

Ritchie, R., and J. E. Saxton, *Chem. Commun.* 611 (1975).

Steyn, P. S., *Tetrahedron Lett.* 3331 (1971); *Tetrahedron* **29**, 107 (1973).

9.6.3   Sporidesmins and Related Compounds

The fungus *Pithomyces chartarum* occurs in certain farm areas in New Zealand and causes facial eczema in sheep grazing on infected pastures. The main active principle, sporidesmin A (**117**), was first obtained and characterized in 1963. Subsequently several other closely related compounds were obtained from the same fungus and these compounds have been divided into two groups: those in which the sulfur function is modified [e.g. sporidesmin A (**117**), sporidesmin E (**118**), and sporidesmin G (**119**)] and those in which the indolopyrrolo pyrazine nucleus has been modified [e.g. sporidesmin B (**120**) and sporidesmin H (**121**)].

The structure of sporidesmin A (**117**) was deduced by chemical and spectroscopic techniques and confirmed by X-ray analysis. The absolute stereochemistry indicated by circular dichroism studies is the same as that in the gliotoxin series (Chapter 8).

Chemistry and Synthesis

Much of the chemistry of this system parallels that in the gliotoxin series, but the occurrence of compounds with additional sulfur bridge atoms also

117 sporidesmin A, n=2   120 sporidesmin B, R_1=Cl, R_2=OH
118 sporidesmin E, n=3   121 sporidesmin H, R_1=H, R_2=Cl
119 sporidesmin G, n=4

introduces some new chemical factors. For example, treatment of sporidesmin E (**118**) with one equivalent of triphenylphosphine affords sporidesmin A (**117**) by specific removal of the central sulfur atoms. Alternatively, treatment of sporidesmin A (**117**) or sporidesmin E (**118**) with dihydrogen disulfide or hydrogen polysulfide affords sporidesmin G (**119**).

Kishi and co-workers have reported a total synthesis of sporidesmin A which is similar to that used for the preparation of dehydrogliotoxin and is shown in Scheme 14. The aryl thio acetal group was used to establish the two sulfur linkages. Closure of the amide nitrogen to the indole 2-position with introduction of the acetyl group at C-3 was accomplished with iodosobenzene diacetate.

Biosynthesis

Towers and Wright investigated the formation of the sporidesmins from unlabeled amino acids and concluded that tryptophan, alanine, and methionine were involved. As yet there is no information concerning sporidesmin biosynthesis from labeled alanine or methionine, but variously labeled tryptophans have been used to study the stereochemistry of the tryptophan side-chain hydroxylation.

Feeding experiments with [1'-^{14}C], [2'-^{14}C], and [3'-^{14}C]tryptophans established that all the side-chain carbons are retained in sporidesmin A (**117**), and experiments using [3'S-^{3}H]- and [3'R-^{3}H]tryptophans, in admixture with [3-^{14}C]tryptophan, demonstrated that the *pro*-3R proton is specifically removed on hydroxylation. As observed in Chapter 8 for hydroxylation of the phenylalanine/tyrosine side chain, hydroxylation proceeds with retention of configuration.

Beyond these very preliminary results nothing is known of the involvement of other intermediates or the states at which the various substituent groups are introduced.

Several biogenetic routes to sporidesmin A have been proposed and one of these, shown in Scheme 15, involves *cyclo*-L-alanyl-L-tryptophyl (**78**), the intermediate involved in echinulin (**77**) biosynthesis.

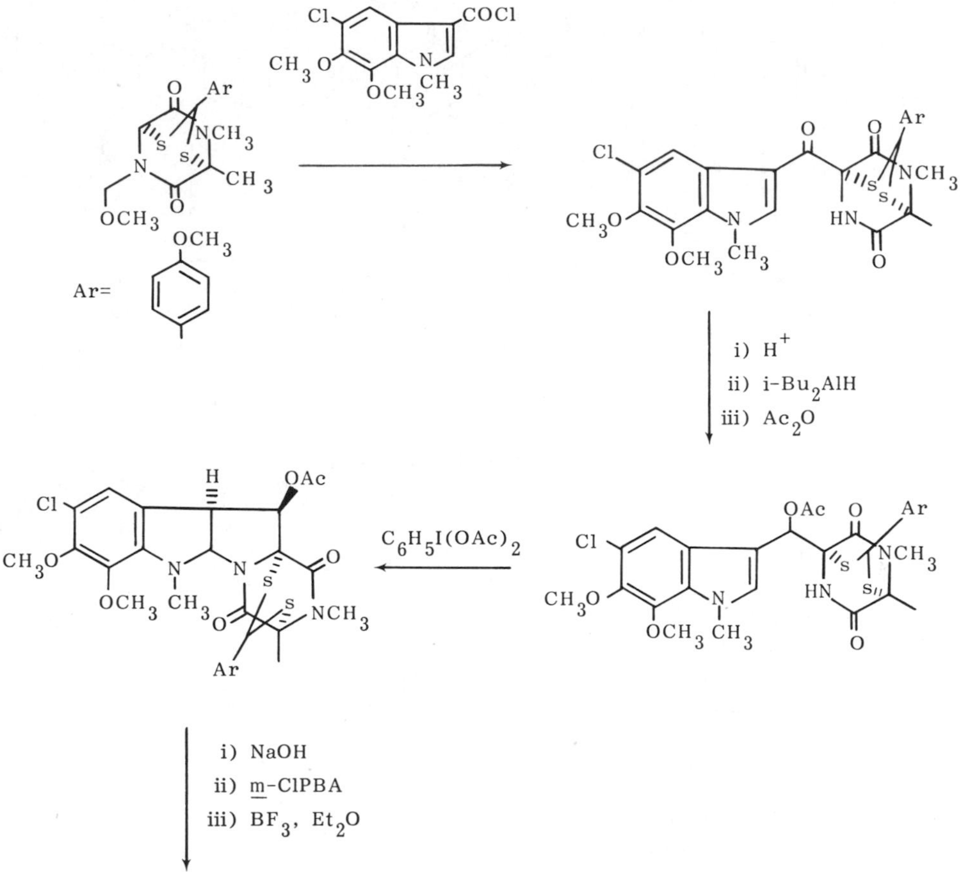

117 sporidesmin A

Scheme 14 Kishi synthesis of sporidesmin A (**117**).

Pharmacology

Sporidesmin E (**118**) is cytotoxic at 0.04 ng/ml, but is considerably (~100 times) less active than gliotoxin as an antibacterial agent *in vitro*. The LD_{50} is about 250 mg/kg in mice and death is due to pulmonary edema. In rats the LD_{50} is about 8 mg/kg and for sheep about 0.75 mg/kg. Hepatic lesions are observed in mice, rats, guinea pigs, rabbits, and sheep, and are lethal.

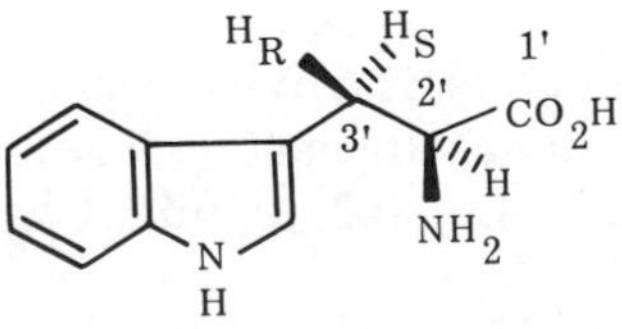

1

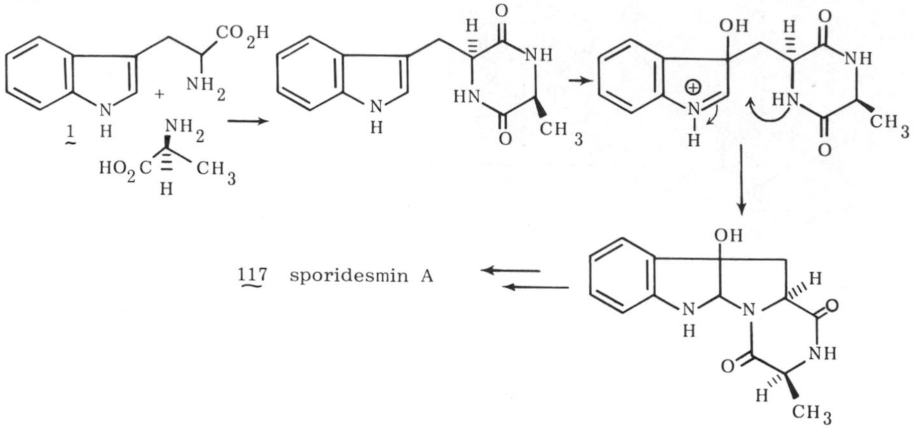

Scheme 15   Biogenesis of sporidesmin A (**117**).

LITERATURE

Taylor, A., S. Kadis, A. Ciegler, and S. J. Ajl (Eds.), in *Microbial Toxins*, Vol. 7, Academic, New York, 1971, p. 337.

Sammes, P. G., *Fortschr. Chem. Org. Naturs.* **32**, 51 (1975).

Kishi, Y., S. Nakatsuka, T. Fukuyama, and M. Havel, *J. Amer. Chem. Soc.* **95**, 6493 (1973).

Vining, L. C., and J. L. C. Wright, *Biosynthesis* **5**, 240 (1977).

9.6.4   Miscellaneous Diketopiperazines

Closely related to the sporidesmins are the fungal dimeric indole alkaloids such as verticillin A (**122**) from *Verticillium* sp. and chaetocin (**123**) from *Chaetomium minutum.*

Chaetomium sp. were found to be closely associated with pastures on which sheep grew slowly or not at all in both Nova Scotia and New Zealand, and the fungi were consequently cultured and investigated. Degradative and spectral data suggested the structure **123** for chaetocin, and this was confirmed by X-ray analysis. The circular dichroism data of both this compound and verticillin A are antipodal to the sporidesmins. The ^{13}C NMR data for chaetocin are shown in **124**.

Chaetocin has a MIC of 0.001 μg/ml against penicillin-resistant *Staphlococcus aureus*.

Verticillin A has antimicrobial activity against gram-positive bacteria, but is not active against gram-negative bacteria or fungi. It shows no antiviral activity, but is cytotoxic (ED$_{50}$ 0.2 μg/ml). Chaetocin has no antiviral activity.

Oxaline (**125**) was the main alkaloid obtained by cultivating *Penicillium oxalicum* on maize meal. The structure was solved by X-ray crystallography.

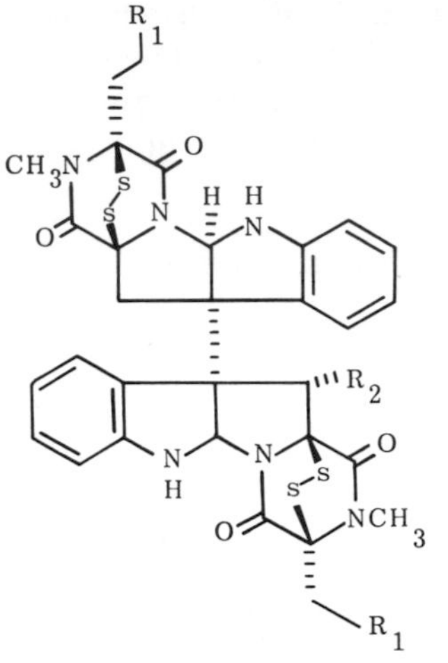

122 verticillin A, R_1=H, R_2=OH
123 chaetocin, R_1=OH, R_2=H

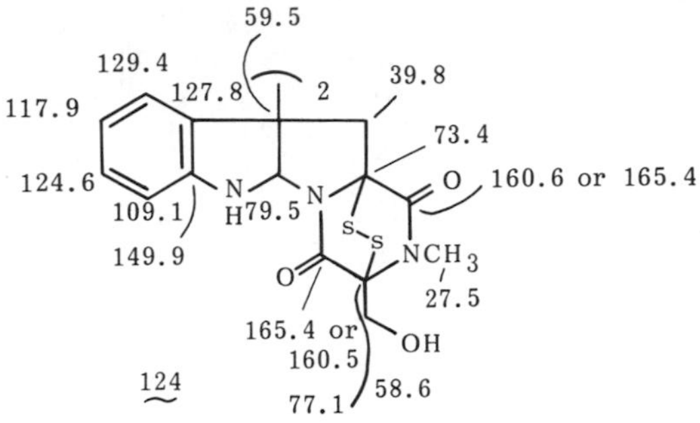

It is a very interesting compound from a structural point of view for several reasons: it has an isoprene unit inverted at the 3-position, it contains both dehydro-tryptophan and histidine units, but most amazingly it contains a carbon (C-2) attached to *three* nitrogen atoms. An investigation of the origin of this additional nitrogen would be of interest.

Penicillium roquefortii Thom. is a source of ergoline alkaloids, but recently Polonsky and co-workers obtained the compound roquefortine (**126**)

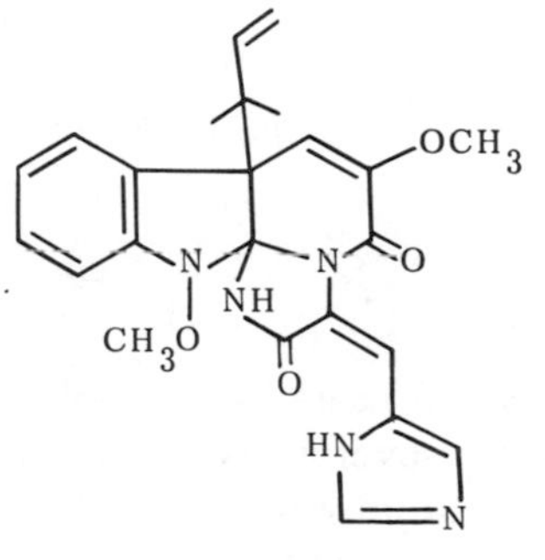

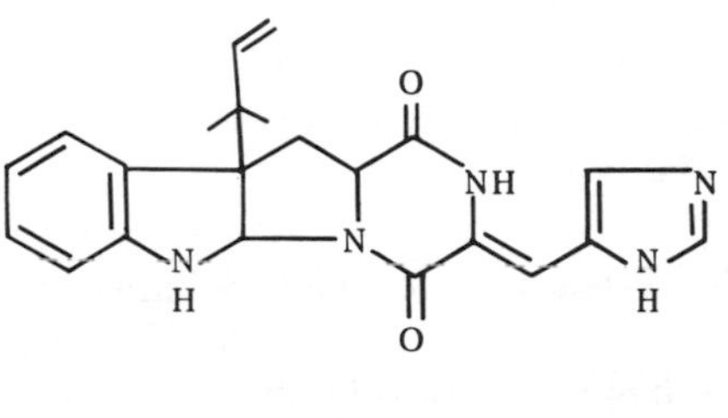

126 roquefortine

125 oxaline

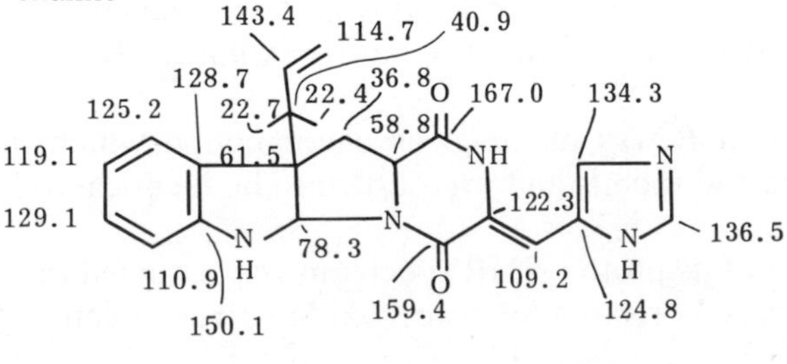

from this fungus. The structure clearly bears considerable resemblance to that of oxaline (**125**). The ^{13}C NMR data of roquefortine are shown.

Surugatoxin (**127**) is a mydriatic isolated as the toxin occurring in the gut of the gastropod *Babylonia japonica*. The occurrence is quite unique

127   surugatoxin

since the toxin is isolated only from specimens living in Suruga Bay off the southeast coast of Honshu. The structure was deduced by X-ray analysis.

Nothing is known of the biosynthesis of this interesting metabolite.

LITERATURE

Nagel, D. W., K. G. R. Pachler, P. S. Steyn, P. L. Wessels, G. Gafuer, and G. J. Kruger, *Chem. Commun.* 1021 (1974).

Scott, P. M., M. A. Merrien, and J. Polonsky, *Experientia* **32,** 140 (1976).

Kosuge, T., H. Zenda, A. Ochiai, N. Masaki, M. Noguchi, S. Kimura, and Nanita, *Tetrahedron Lett.* 2545 (1972).

9.6.5   Tremorigenic Indole Alkaloids

There has been considerable interest in the past 5 years in metabolites that produce tremors in mice and rabbits. As a result, several new and interesting compounds have been described; they include fumitremorgin B (**128**) from *A. fumigatus* Fres. and *A. caespitosus*, verruculogen (**129**) from *Penicillium verruculosum* Peyronel and *A. caespitosus*, tryptoquivaline (**130**) and tryptoquivalone (**131**) from *A. clavatus*, and paxilline (**132**) from *Penicillium paxilli* Bainier.

Fumitremorgin B was obtained as a compound which induces strong tremors in mice and rabbits and was shown to be identical to lanosulin from *P. lanosum*.

On the basis of its proton NMR spectrum was assigned the structure **128**, and it is very closely related to the peroxide verruculogen (**129**).

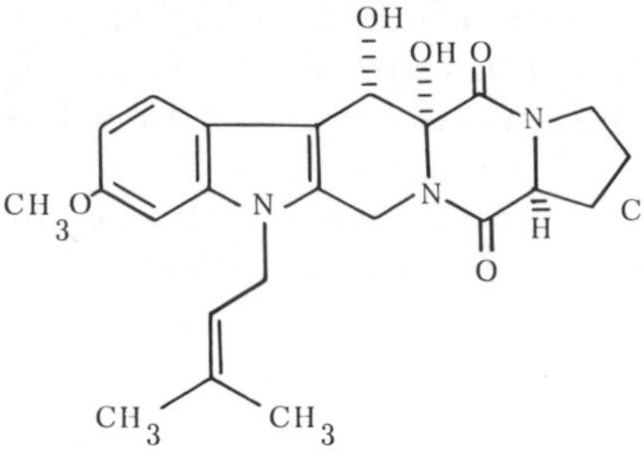

128 fumitremorgin B

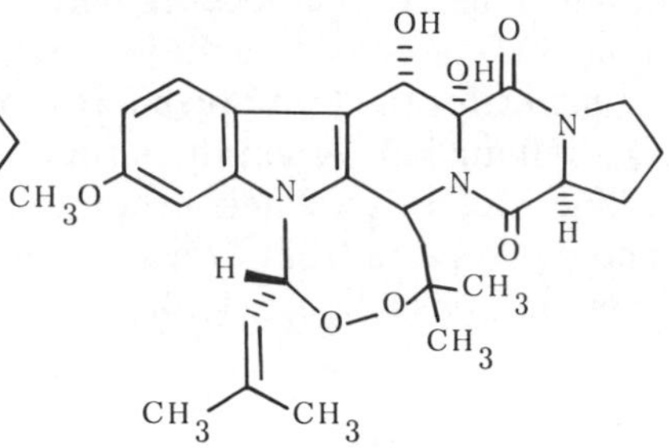

129 verruculogen

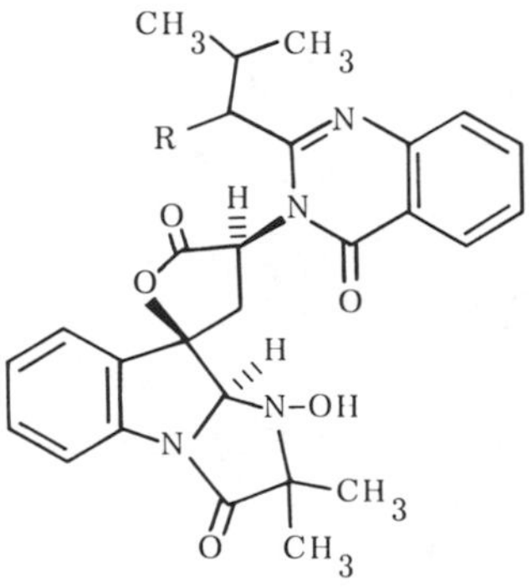

130 tryptoquivaline, R=-OCOCH$_3$, β-H

131 tryptoquivalone, R= =O

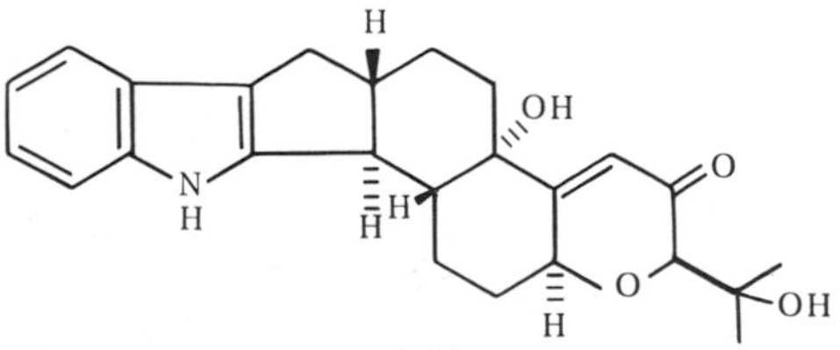

132 paxilline

The UV spectrum of verruculogen showing λ_{max} 226, 277, and 295 nm is typical of the 6-methoxyindole system, and the principal proton NMR assignments are shown in **133**. The novel peroxide structure was determined by single-crystal X-ray analysis. The ED$_{50}$ for the tremor response is 0.39 mg/kg i.p. in mice, and the LD$_{50}$ is 2.4 mg/kg.

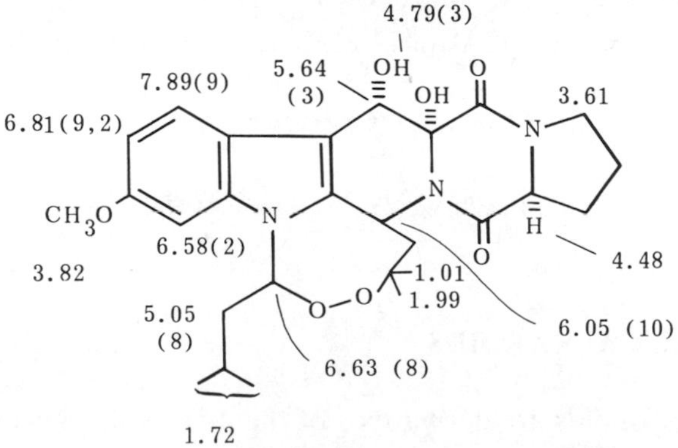

133 verruculogen

Tryptoquivaline (**131**) co-occurs with cytochalasin E (Section 6.8).

Paxilline (**132**) is considerably less active than verruculogen, having an ED$_{50}$ of 25 mg/kg for the tremorgenic response, and is also considerably less toxic (LD$_{50}$ 150 mg/kg). Again the structure was determined by X-ray analysis. Paxilline is closely related structurally to two compounds obtained by Arigoni and co-workers from *Claviceps paspali*. These compounds are paspaline (**133a**) and paspalicine (**133b**).

133a paspaline   133b paspalicine

There have been no studies on the biosynthesis of these compounds. The biogenesis of the amide alkaloids may be envisaged as occurring from tryptophan and proline and either two units of mevalonate in the case of verruculogen (**129**), or tryptophan, anthranilic acid, valine, and methyl alanine in the case of tryptoquivaline.

The biogenesis of paxilline is less clear; the relationship to paspaline indicates that a methyl group is lost and suggests an isoprenoid biogenesis as shown in **133a**. The major problem biogenetically appears to be the formation of the remaining five carbon atoms which do not have the isoprenoid skeleton.

LITERATURE

Yamasaki, N., K. Sasago, and K, Miyaki, *Chem. Commun.* 408 (1974).

Yamasaki, M., H. Fumjimoto, T. Akiyama, U. Sankawa, and Y. Iitaka, *Tetrahedron Lett.* 27 (1975).

Fayos, J., D. Lokensgard, J. Clardy, R. J. Cole, and J. W. Kirksey, *J. Amer. Chem. Soc.* **96**, 6785 (1974).

Clardy, J., J. P. Springer, G. Büchi, K. Matsuo, and R. Wightman, *J. Amer. Chem. Soc.* **97**, 663 (1975).

9.7 HARMALA ALKALOIDS

The Harmala alkaloids are derivatives of the β-carboline system (**134**), numbered as shown. This system, particularly in its 1,2,3,4-tetrahydro form (**135**), occurs frequently in the indole alkaloid series and there are literally

134   β-carboline

135   tetrahydro-β-
      carboline

hundreds of monoterpenoid-derived indole alkaloids containing this unit as part of their skeleton.

In β-carboline itself the pyridine nitrogen is weakly basic, but in the tetrahydro derivative this piperideine nitrogen is now quite strongly basic. The indole NH is of course acidic in both cases.

Most of the harmala alkaloids are substituted at C-1 by a methyl group, and the parent compound having this structure is known as harman (**136**). These alkaloids have been found in several plant families, including the Leguminosae, Malpighiaceae, and Rubiaceae, but most famously are associated with the seeds of *Peganum harmala* L. (Rutaceae), where they cooccur with quinazoline alkaloids such as vasicine (Chapter 7).

The alkaloid content of the seeds of *P. harmala* is in the range 2–3%, and principal alkaloids of interest are harman (**136**), harmine (**137**), and harmaline (**138**). Harmine is probably also the active principle of the narcotic

136   harman, R=H
137   harmine, R=OCH$_3$

138   harmaline

drug yage (*Banisteria caapi* Spruce), and causes tremors when injected intracerebrally.

9.7.1   Synthesis and Chemistry

Several approaches have been developed for the formation of this ring system. One, utilizing the Fischer indole synthesis, involves cyclization of the product from the condensation of ·2-methyl-3-hydrazinopyridine (**139**) and cyclohexanone (**140**), followed by a palladium–charcoal dehydrogenation to afford harman (**136**). An alternative route uses a Pictet–Spengler type of condensation, involving the reaction of acetaldehyde and 6-methoxytryptamine (**141**) at pH 6.7 to give tetrahydroharmine which can be dehydrogenated to harmine (**137**). A recent modification of this approach by Spenser and co-workers using glyoxal in place of acetaldehyde permits the synthesis of harmaline (**138**) by acid-catalyzed dehydration. It is pertinent to note that harmaline exists in the imine form **138** rather than as the corresponding enamine **142**.

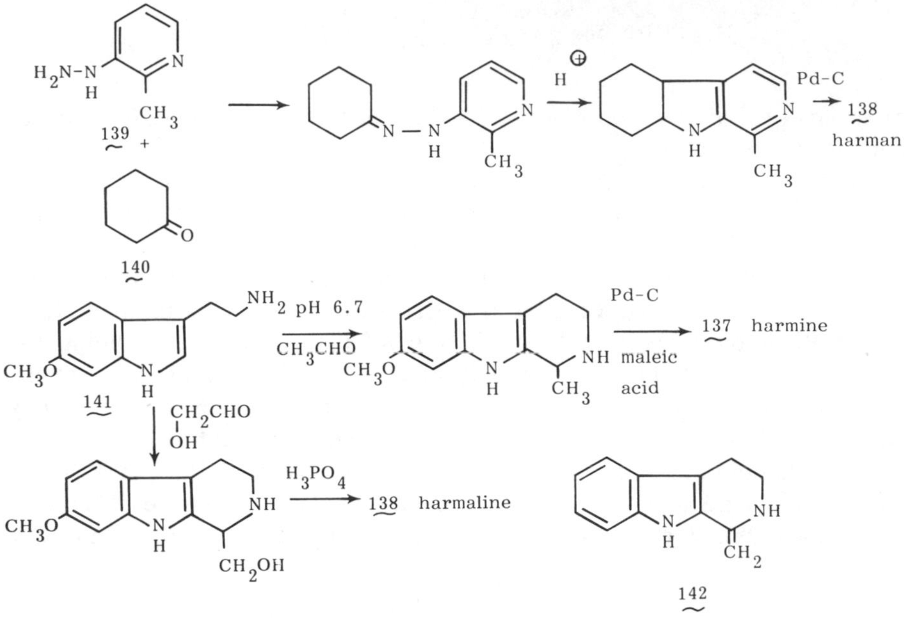

Much of the original chemistry of the harmine/harmaline system was first studied by Perkin and Robinson, and some of this is worth mentioning because it shows the classic differences between benzenoid and pyridine nuclei. Oxidation of harmine (**137**) with chromic acid for example affords harminic acid (**143**), which as a pyrrole carboxylic acid, readily undergoes decarboxylation to afford apoharmine (**144**). Oxidation with nitric acid on the other hand affords 4-methoxy-3-nitrobenzoic acid (**145**). In this case nitration has deactivated the benzene ring and allowed oxidation of the pyridine nucleus.

The enamine–imine tautomerization of harmaline was mentioned previously, and since harmaline forms both *N*-acetyl **146** and mono-*N*-methyl **147** derivatives, this might be taken in favor of the enamine structure. However the UV spectrum, which shows λ_{max} 260, 344, and 376 nm, indicates that this is not so. An equilibrium process in which the enamine form is a minor but chemically active contributor is therefore responsible for the formation of the N_b-substituted derivatives.

9.7.2 Brevicolline

Brevicolline (**148**) from *Carex brevicaria* (Cyperaceae) is an unusual harmala type of alkaloid having an *N*-methyl pyrrolidine ring attached at the 4-po-

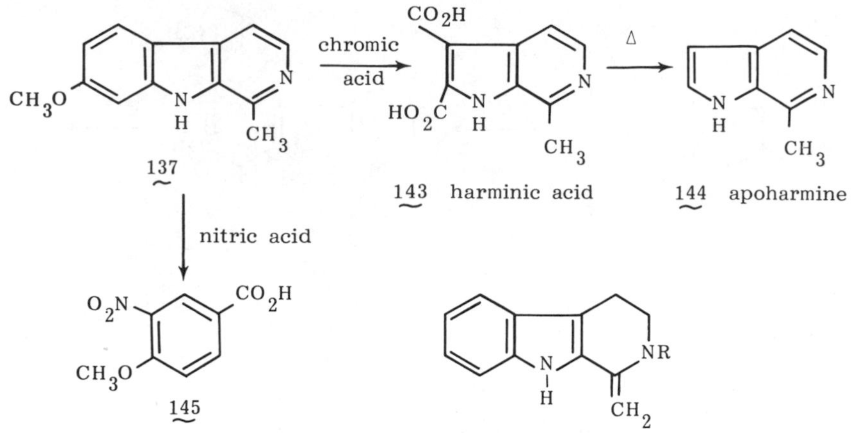

146 R=COCH$_3$
147 R=CH$_3$

sition of 1-methyl harman. A simple nonbiogenetic synthesis of **148** begins with the condensation of the indole derivative **149** with *N*-methyl pyrrole to give **150**, which could be reduced to the 2′-substituted tryptamine **151**. The *N*-acetyl derivative of **151** was cyclized and oxidized photochemically to brevicolline (**148**) (Scheme 16).

Biogenetically brevicolline (**148**) is probably derived as shown in **152**.

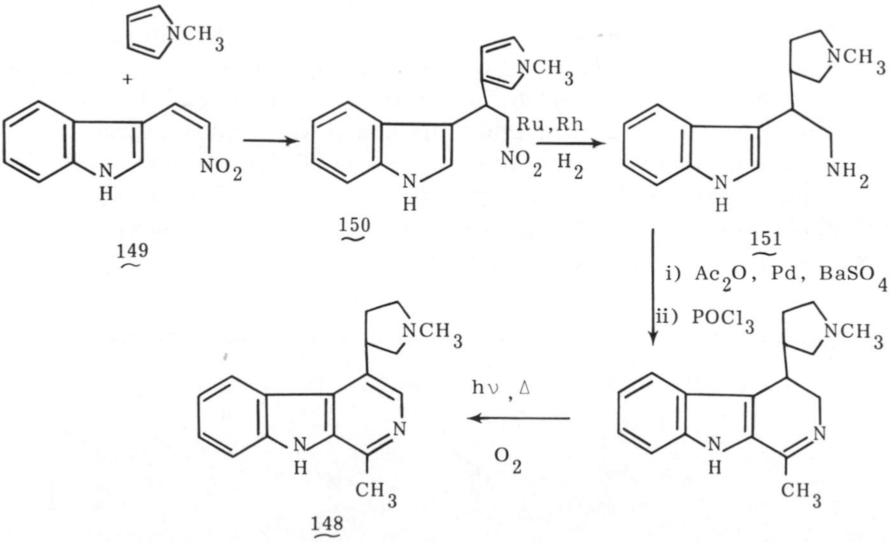

Scheme 16 Synthesis of brevicolline (**148**).

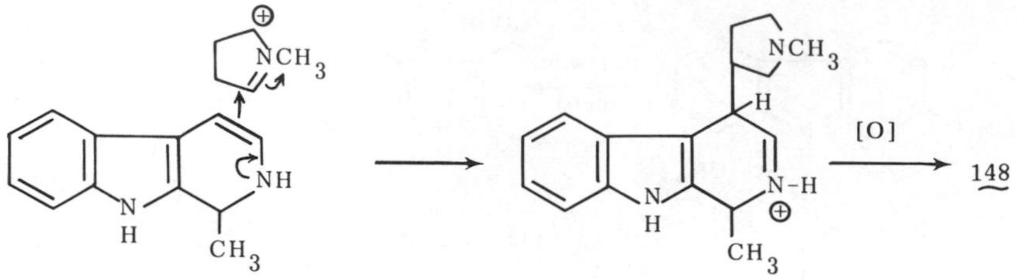

152

LITERATURE

Marion, L., *Alkaloids NY* **2,** 369, 499 (1952).

Manske, R. H. F., *Alkaloids NY* **8,** 47 (1965).

Telezhentskaya, M. V., and S. Yu Yunusov, *Chem. Nat. Compd.* **13,** 613 (1978).

9.8 CARBAZOLE AKLALOIDS

Whereas most groups of indole alkaloids have been known for many years, the carbazole alkaloids are a relatively recent addition and this area has been well reviewed by Kapil. Although these alkaloids are included in this tryptophan-derived alkaloid chapter, there is at present no evidence to support this classification.

Their distribution is at present limited to the family Rutaceae, and the Indian curry leaf plant *Murraya koenigii* Spreng. is a prolific producer of these compounds. Other sources have included *Glycosmis pentaphylla* (Retz.) DC. and *Clausena heptaphylla* Wt. and Arn.

Three basic types of carbazole alkaloid have been recognized; all are based on the parent carbazole nucleus, but they differ in the number of carbon atoms attached to nucleus. The groups are better analyzed however in terms of the number of isoprene units linked to an indole nucleus. Thus the first group, represented by murrayanine (**153**), has an indole unit plus one isoprene unit, the second group [e.g. heptaphylline (**154**)] has an indole unit plus two isoprene units, and the third group [e.g. mahanimbine (**155**)] has an indole unit plus three isoprene units.

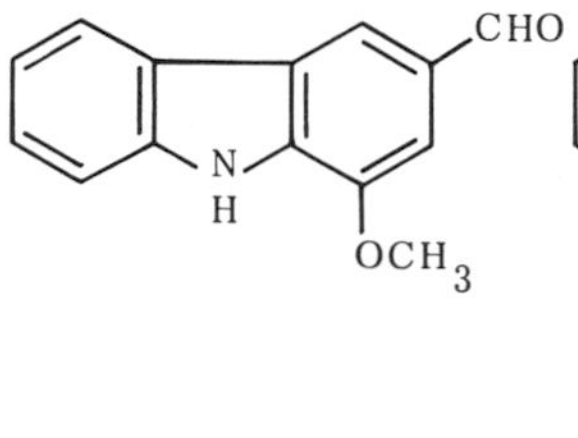

153 murrayanine 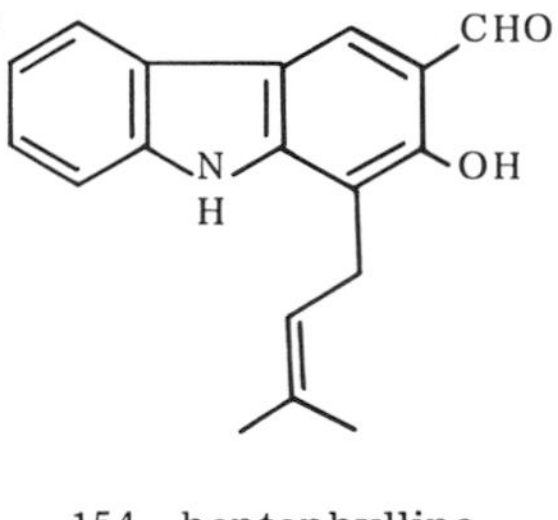   154 heptaphylline

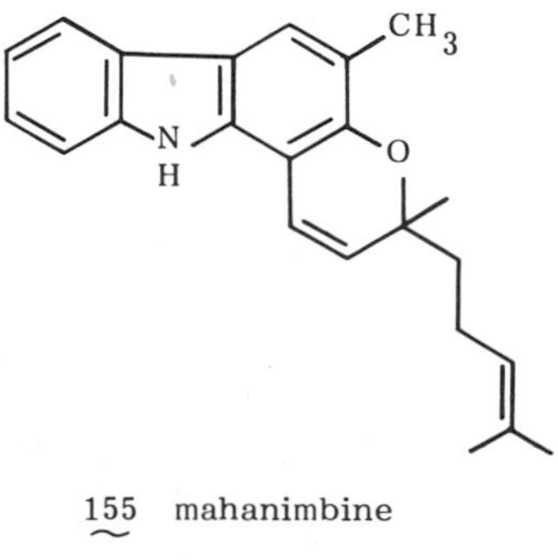

155 mahanimbine

9.8.1 Chemistry and Synthesis

Murrayanine (153) was obtained from the stem bark of *M. koenigii* and was found to yield *N*-methyl and 2,4-dinitrophenyl hydrazone derivatives. The UV spectrum (λ_{max} 238, 247, 274, 289, and 335 nm) was similar to that of 3-formylcarbazole, but after potassium borohydride reduction was similar to that of 1-methoxycarbazole. The possible structures were distinguished from the NMR spectrum, which showed two deshielded aromatic protons at 8.09 ppm and four at 7.39 ppm, in favor of structure 153. This structure has been confirmed by synthesis.

Heptaphylline (154) was obtained from the roots of *C. heptaphylla* and again the UV spectrum was similar to that of carbazole-3-aldehyde and this was confirmed by the IR spectrum. The NMR spectrum showed the presence of two olefinic methyl groups at 1.66 and 1.83 ppm, an olefinic proton (*t*, *J* = 6 Hz) at 5.35 ppm, and a methylene doublet at 3.60 ppm, suggesting a 3,3'-dimethylallyl unit attached to the carbazole nucleus. A phenolic group was placed between the formyl group and the dimethylallyl unit since reaction with polyphosphoric acid gave the isomeric chroman 156. The structure was confirmed by synthesis from the phenolic carbazole aldehyde 157.

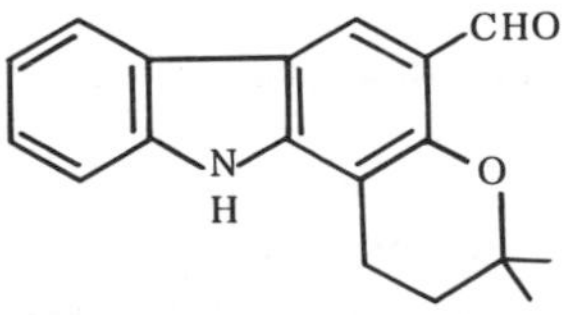

156

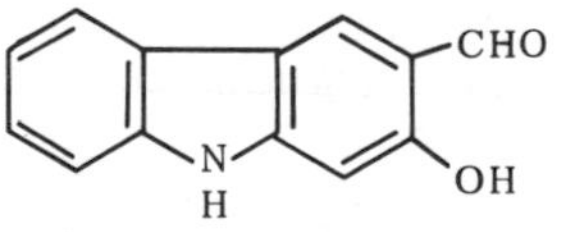

157

The most complex alkaloids are those derived from indole and three isoprene units. Mahanimbine (155) from the stem bark of *M. koenigii* was the first alkaloid of this type to be isolated. Its UV spectrum was similar to that of girinimbine (158), showing λ_{max} 239, 288, 328, 343, and 357 nm.

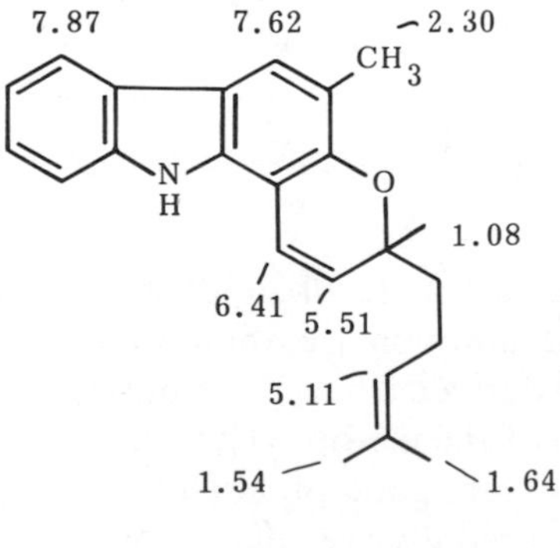

155   mahanimbine

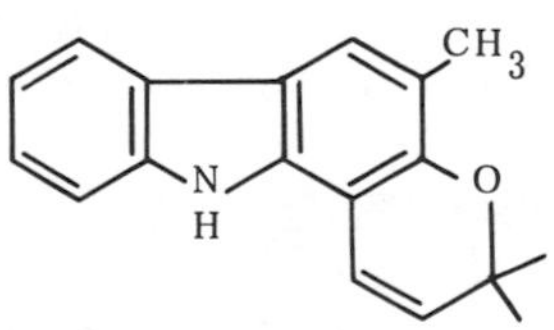

158   girinimbine

As in the synthesis of heptaphylline (154), the carbazole 159 is a key intermediate, and Kureel and co-workers have provided an efficient synthesis of 155. Ullmann condensation of 2-nitrobromobenzene with 4-bromo-2-methylanisole (160) gave the biphenyl 161, which could be cyclized to the carbazoles 162 and 163 with triethyl phosphite. Demethylation of the major component afforded 159. Condensation with citral can be carried out in the presence of anhydrous ferric chloride or pyridine with 2% benzoic acid to afford mahanimbine (155) (Scheme 17).

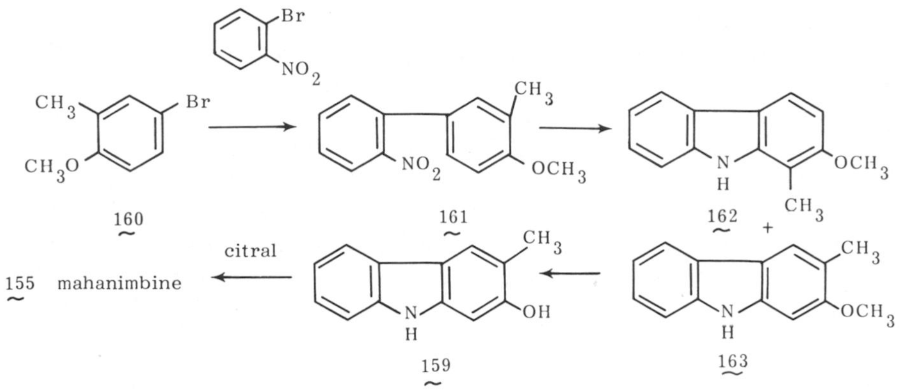

Scheme 17   Synthesis of mahanimbine (155).

When mahanimbine (155) was dissolved in benzene and shaken with silica gel for 48 hr, bicyclomahanimbine was produced and assigned the structure 164.

The reaction is regarded as proceeding through cyclomahanimbine (165), which may also be produced from 155 by reaction with acid in chloroform solution (Scheme 18). Both 164 and 165 have been obtained from *M. koenigii*, but probably are best regarded as artifacts rather than true natural products. Some proton NMR spectral values are indicated for these compounds.

The structure of bicyclomahanimbine was subsequently revised. The revision was stimulated by chemical work in the cannabinoid series. Acid treatment of cannabichromene (166) affords cannabicyclol (167), and by analogy bicyclomahanimbine is regarded as having the structure 168.

9.8.2   Biosynthesis

Information on the biosynthesis of the carbazole alkaloids is extremely limited. Kapil and Popli showed that the additional carbon atom on the carbazole ring of koenigicine (169) was not derived from [^{14}C-methyl]methionine, but that as expected the carbon atoms of the methoxyl groups were labeled.

[2-^{14}C]- and [2-^{3}H]Mevalonic acids were good precursors of koenigicine (169) and mahanimbine (155), but no degradative results are available. These

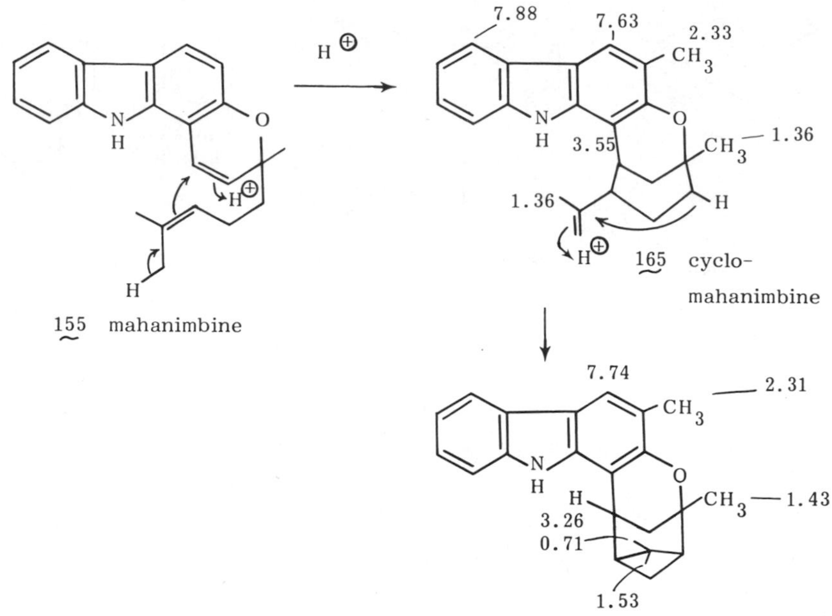

155 mahanimbine

165 cyclo-
mahanimbine

164 bicyclomahanimbine

Scheme 18

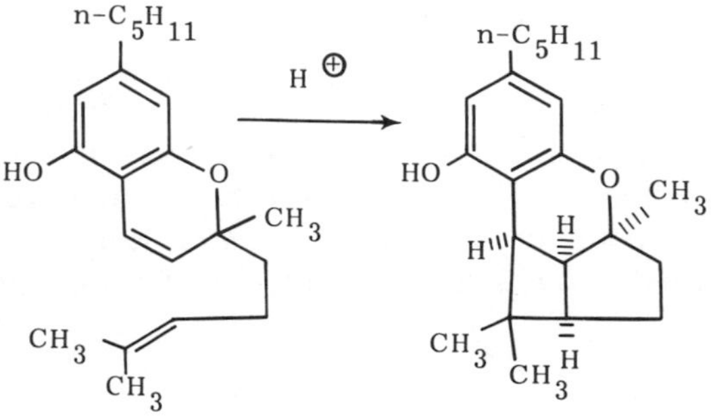

166 cannabichromene

167 cannabicyclol

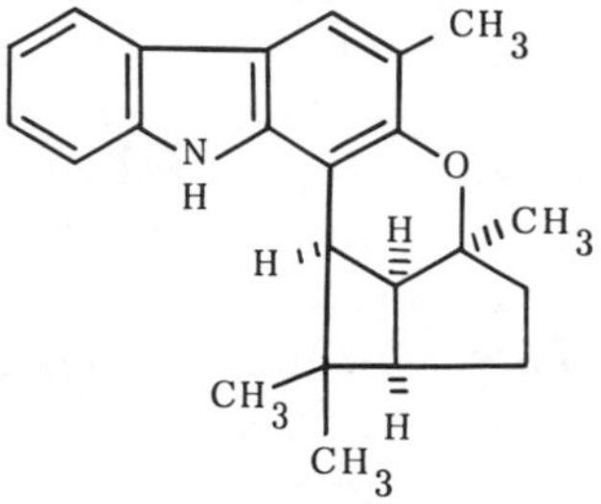

168 bicyclomahanimbine
(revised)

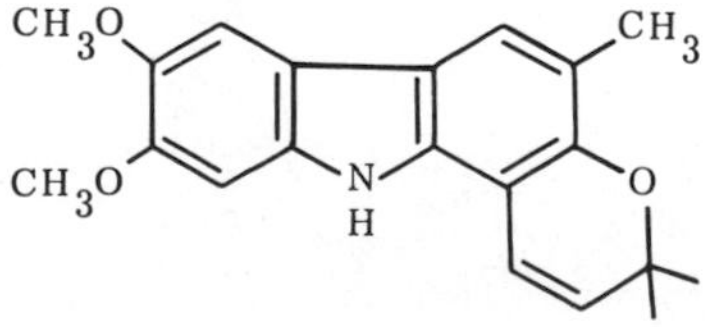

169 koenigicine

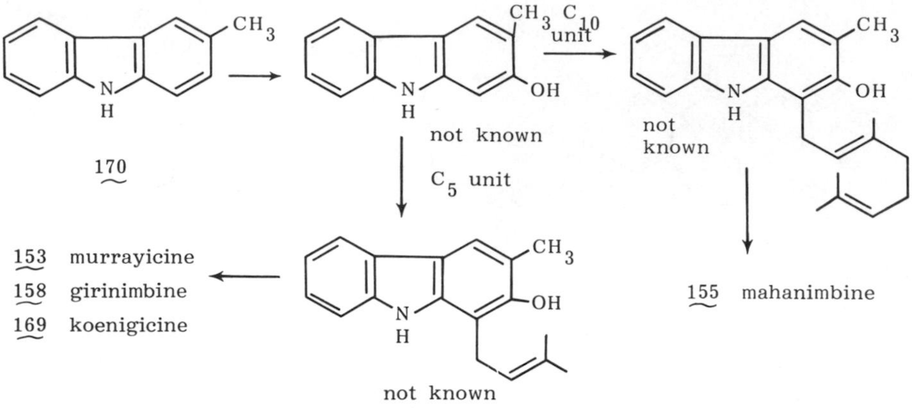

Scheme 19 Biogenesis of carbazole alkaloids.

results, together with the evidence from isolations, prompted Kapil to propose that 3-methylcarbazole (**170**) is the key biosynthetic intermediate in the
formation of the carbazole alkaloids and that this compound may then be
the object of hydroxylation and prenylation reactions (Scheme 19). None
of the simple prenylated biogenetic precursors are known, and there is no
information as to the mechanism of formation of 3-methyl carbazole (**170**).
This compound has been obtained recently from *Clausena heptaphylla* Wight
et Arn.

It is perhaps worth mentioning that no other tryptophan-derived indole
alkaloids co-occur with these carbazole derivatives, so the possibility exists
that tryptophan biosynthesis is blocked at the intermediate stage of indole
(see Chapter 2), thus permitting prenylation to occur at the 2-position. This
compound 2-(3-methyl-2-butenyl)indole (**171**) has not yet been found naturally, but the 6-isomer **172** is a constituent of *Richardia sinuata* (Rubiaceae)
and 6-(3-methylbuta-1,3-dienyl)indole (**173**) is a constituent of *Monodora
tenuifolia* Benth. (Annonaceae).

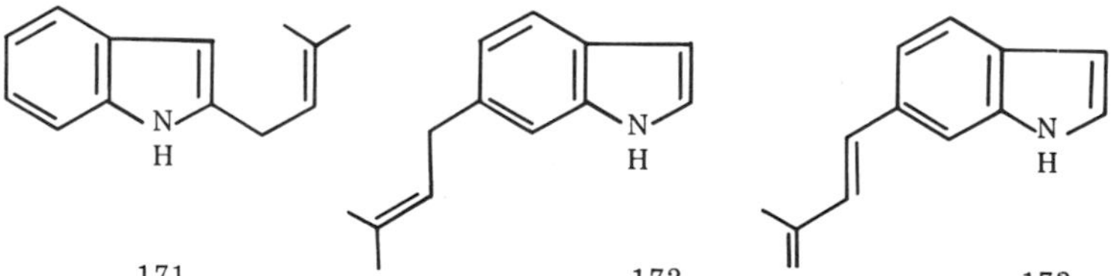

LITERATURE

Kapil, R. S., *Alkaloids, NY* **13**, 273 (1971).

Chakraborty, D. P., and B. K. Chowdhury, *J. Org. Chem.* **33**, 1265 (1968).

Kureel, S. P., R. S. Kapil, and S. P. Popli, *Chem. Commun.* 1120 (1969).

9.9   CANTHIN-6-ONES

The canthin-6-ones are a small, yet interesting group of alkaloids having the skeleton **174**. The parent compound, canthin-6-one, was first isolated from *Pentaceras australis* Hook F. (Rutaceae) in 1952. Subsequently it has been obtained from several *Zanthoxylum* species, also in the Rutaceae, and from a number of plants in the Simaroubaceae, a botanically close family to the Rutaceae.

9.9.1   Chemistry and Synthesis

Hydrolysis with aqueous ethanolic alkali gave the *cis*-acrylic acid derivative **175**, which was easily converted to canthin-6-one (**174**) by alkali. Prolonged heating with alkali gave the *trans*-acrylic acid **176**, which could not be recyclized. Vigorous zinc–acetic acid reduction gave the deoxy compound **177**.

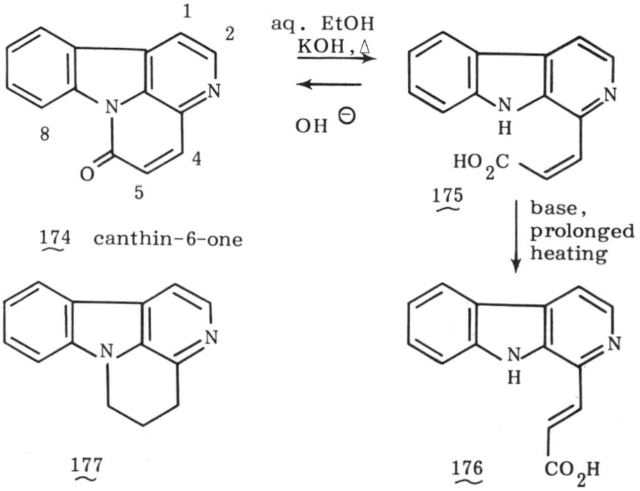

Closely related to canthin-6-one is 5-methoxy canthin-6-one (**178**), which was demethylated with hydrogen bromide–acetic acid to the phenol **179**. Reaction of the phenol with *o*-phenylenediamine gave the hydroxyquinazoline **180**, thereby defining the 1,2-arrangement of the oxygen substituents. The phenol **179** was synthesized by condensation of diethyloxalate with the dilithium derivative of 1-methyl β-carboline (**136**).

Some other canthin-6-ones which have been obtained recently include the 1-methoxy, 4-methoxy, and 8-hydroxy derivatives.

A more recent synthesis by Mitscher and co-workers involves a Doebner reaction on β-carboline-1-carboxaldehyde (**181**), or less directly by malonic ester condensation on **181**, to afford **182** followed by copper-catalyzed thermal decarboxylation to canthin-6-one (**174**).

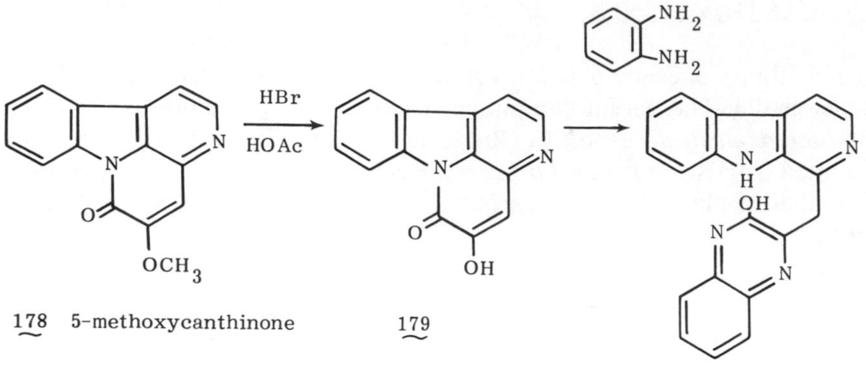

178 5-methoxycanthinone   179

180

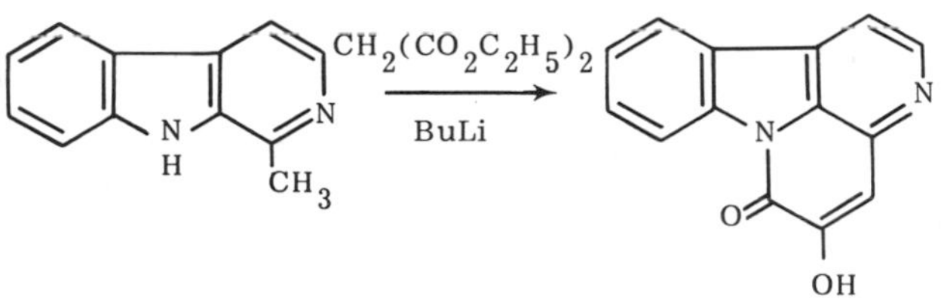

136   179

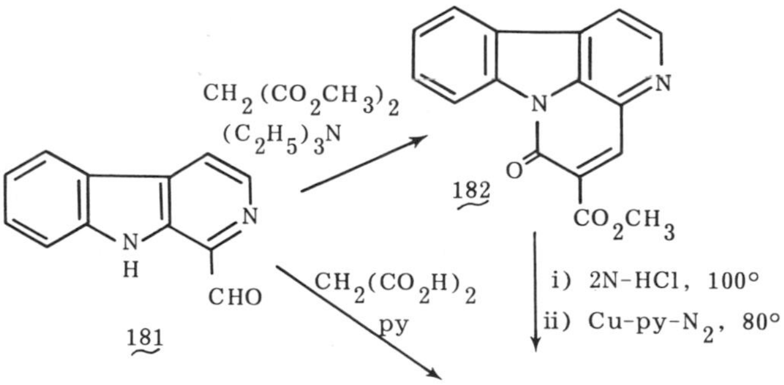

182

181

canthin-6-one

174

9.9.2 Spectral Properties

The most characteristic spectral feature of the canthin-6-ones is their UV spectra. Canthin-6-one shows λ_{max} 248 (log ϵ 4.51), 257 (4.48), 268 (4.36), 300 (4.25), 346 (4.16), 364 (4.38), and 381 nm (4.36). But 8-hydroxycanthin-6-one, which has a hydrogen-bonded hydroxyl, shows λ_{max} 253 (4.11), 270

Table 1 Proton NMR Spectral Data of Canthin-6-one Derivatives

	H-1	H-2	H-4	H-5	H-8	H-9	H-10	H-11
				δ (ppm)	(Coupling Constant, Hz)			
Canthinone (**174**)	7.76	8.73	6.89	7.92	8.47	7.40	7.55	7.88
	(5.1)	(5.1)	(9.8)	(9.8)	(7.1, 1.9, 1.2)	(7.1, 7.1, 1.2)	(7.1, 7.1, 1.9)	(7.1, 1.2)
1-Methoxycanthinone	—	8.46	6.82	7.93	8.64	7.59	7.42	8.15
			(10)	(10)	(8.0)			(8.0)
5-Methoxycanthinone (**178**)	8.14	8.78	7.4	—	8.55	7.58	7.80	8.40
	(5.2)	(5.2)			(7.2, 1.9)			(7.2, 1.9)
8-Hydroxycanthinone	8.08	8.78	6.97	8.12	—	7.00	7.56	7.97
	(4.9)	(4.8)	(10.6)	(10.6)		(7.8)	(7.8)	(7.8)

(4.07), 294sh (3.52), 330 (3.88), 340 (3.86), and 399 nm (3.86) and is only changed to a more typical canthin-6-one type of spectrum on the addition of base.

In the IR spectrum canthin-6-ones show strong bands at about 1670 and 1635 cm^{-1} in the carbonyl region.

The mass spectrum of canthin-6-one shows M$^+$ at m/e 220, with important successive losses of 28, 28, and 25 mass units.

Some representative proton NMR spectral data of canthin-6-one derivatives are shown in Table 1. The substitution is normally readily established by the absence of a particular coupling in the spectrum in comparison with canthin-6-one (174) itself. The 2-proton is invariably the most deshielded, and if a 1-proton is present the coupling constant is about 5 Hz. The 4- and 5-protons, if both are present, typically appear around 6.9 and 7.9 ppm with a coupling constant of about 10 Hz.

9.9.3   Pharmacology

Canthin-6-one (174) shows antimicrobial activity against *Staphylococcus aureus* and *Mycobacterium smegmatis*, but the 5-methoxy analogue is devoid of activity.

9.10   ERGOT ALKALOIDS

The ergot alkaloids are one of the most important groups of indole alkaloids from a pharmaceutical point of view.

Ergot is the dried sclerotium of the filamentous fungus *Claviceps purpurea* (Fries) Tulasne (Hypocreaceae), which grows parasitically on rye and other gramineaceous crop plants.

9.10.1   Historical Perspective

The history of ergot in Europe is quite exceptional and has been eloquently described by Barger. The first authenticated reports of the effects of ergot appear in the writings of Chou Kaning in 1100 B.C., when its use in obstetrics was described. The Egyptians, the Greeks, and the Romans probably all knew of ergot; but through the eighteenth century botanists considered ergot to be a "super" rye with an enlarged kernel. It was not until 1764 that von Munchhausen first recognized ergot as a fungus.

Ergot is probably the oldest known mycotoxin and was commonly ingested in the form of bread made from ergot-infested flour. In A.D. 944, in the Aquitane region of France, 20,000 people, about half the population, died of ergot poisoning. All through the Middle Ages ergot plagues ravaged continental Europe, but in contrast to other mycotoxicoses, the typical forms of ergotism have been eradicated. Ergot has thus changed from a dreaded toxin to a highly important source of pharmaceuticals.

Two types of toxicity were prevalent in Europe and were collectively known as ergotism. One of the toxicities is known as "St. Anthony's Fire," a gangrenous infection of the extremities which resulted in a bloodless, often dramatic, loss of blackened limbs. This is the potent vasoconstriction action of the ergot alkaloids in operation. A second form of toxicity was a state of delirium and hallucinations, and in more severe cases convulsions.

In 1582, Adam Lonicer described in his *Kreuterbuch* a preparation of ergot which produced strong uterine contractions. Its use as an oxytoxic became widespread in France, Germany, and the United States until the number of stillborn babies rose to a point where the Medical Society of New York instituted an enquiry. As a result it was recommended in 1824 that the drug only be used in the control of postpartum hemorrhage.

The chemical investigation of ergot was begun by Vauquelin in 1816. The first alkaloid preparation was obtained by the French pharmacist Tanret in 1875, but it was 78 years before Stoll and Hofmann identified one component of this mixture as ergocristine. The first homogeneous alkaloid isolated was ergotamine, obtained in 1918 by Stoll. It took Stoll and co-workers 35 years to identify this compound.

As the importance of ergot increased, the production of wild ergot in the field in Spain and Portugal became insufficient to meet world requirements. Currently the annual world production of lysergic acid is in excess of 12,000 kg at a total cost of about $50 million.

Fundamentally, there are four procedures to obtain ergot alkaloids: (*a*) isolation from the crude drug grown parasitically in the field, (*b*) extraction from saprophytic cultures, (*c*) partial synthesis, and (*d*) total synthesis.

Ergot may be produced parasitically on rye and Gröger has reviewed this and the other methods of commercial ergot production. In-field cultivation by innoculation is still carried out in several countries in central Europe and numerous attempts to improve the yield of alkaloids by strain selection have been made. In spite of these efforts this is not a workable process for large-scale production of the non-peptide alkaloids.

The most successful commercial technique is that of saprophytic culture. This work was begun in 1881, but it was nearly 80 years before the process became a commercial success. In 1960, Chain and co-workers were the first to report the formation of lysergic acid derivatives, particularly the α-hydroxyethylamide, from cultures of *Claviceps* sp. on an industrial scale. This is the technique that is commonly used, and *Claviceps paspali* Stevens and Hall has proved to be the most productive fungus. Yields of up to 2 g/liter of lysergic acid α-hydroxyethylamide have been obtained, and the study of mutants of *C. paspali* could lead to even more productive strains. Some mutants of *C. purpurea* have been produced by cobalt irradiation and alkaloid yields, mainly ergonovine and ergotoxine, of up to 400 mg/liter claimed. Even higher yields of ergotamine have been obtained from submerged cultures of *C. purpurea* strains parasitic to Triticale. However, the peptide alkaloids are still largely obtained from ergot cultivated in the field.

There have been two medicinally important alkaloids of ergot, ergotamine and ergonovine, but continuing and developing research in this area has led to at least one other pharmaceutical product, and this will be discussed subsequently.

9.10.2 Alkaloid Structure Types

There are three main groups of ergot alkaloids, whose structures indicate that they are biogenetically related:

1. The clavine type (e.g. elymoclavine, agroclavine, and chanoclavine-I).
2. The water-soluble lysergic acid derivatives (e.g. ergonovine).
3. The water-insoluble lysergic acid derivatives (e.g. ergotamine, ergocornine, and ergocryptine).

All these compounds have a central tetracyclic ring system **183**, known as the ergoline nucleus and numbered as shown. In some of the simplest clavine alkaloids only two or three of the rings may be present. Invariably the N-6 amine is methylated and there is a carbon atom attached at C-8. The latter may be in any oxidation state ranging from a methyl group to a carboxylic acid, but the majority of alkaloids are derivatives of an acid, where usually they are present in an amide linkage. Unsaturation in ring D is also quite common, typically being either at 8,9 or at 9,10. A molecule containing some of these structural features and which is the fundamental building block of many ergot aklaloids is *d*-lysergic acid (**184**).

d-Lysergic acid (**184**) and its derivatives undergo epimerization at C-8 under a variety of conditions to afford the corresponding *d*-isolysergic acid derivative **185** (Scheme 20). Epimerization is more difficult in the 9,10-di-

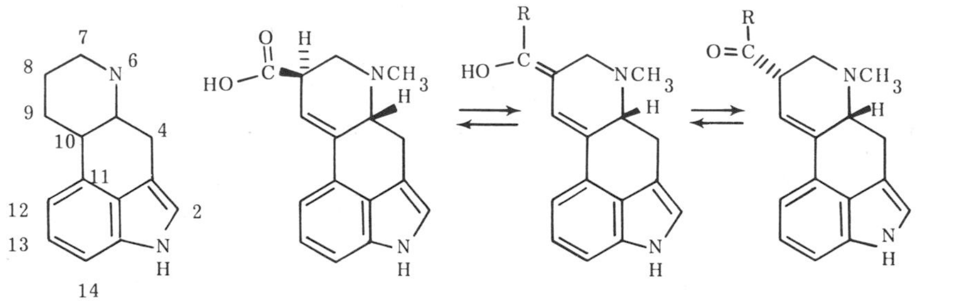

183 ergoline nucleus

184 *d*-lysergic acid

185 *d*-isolysergic acid derivative

Scheme 20

hydro series. As a result of epimerization, the *d*-lysergic acid derivatives (weakly dextrorotatory or levorotatory) are frequently found with the corresponding *d*-isolysergic compound (strongly dextrorotatory). These epimers are easily separated chromatographically, the isolysergic acid derivatives being less polar. Compounds in the natural series which are C-8 diastereoisomers have names ending in -inine.

The classic difference between the lysergic acid and isolysergic acid series occurs in their pharmacologic activity. Compounds in the lysergic acid series display potent activity, whereas the isolysergic acid derivatives are weakly active or inactive.

Another type of isomer of lysergic acid is paspalic acid (**186**), the major

186   paspalic acid

alkaloid of strains of *C. paspali* from Portugal. Stability is a problem here because isomerization to lysergic acid occurs with ease under either neutral or alkaline conditions.

9.10.3   Clavine Alkaloids

Over 32 clavine (ergoline) and secoergoline alkaloids have been obtained since Abe isolated agroclavine (**187**) from ergot growing on *Agropyrum semicostatum* Nees (Gramineae). In these alkaloids the C-17 carbon atom is in a lower oxidation state than lysergic acid and the double bond may be at either the 8,9- or 9,10-positions or it may not be present at all. In some instances the D-ring is cleaved; this is the secoergoline series of compounds. Some typical representatives of these compounds are shown in Figure 1, and other examples will be discussed in the presentation of the biosynthesis of ergot alkaloids.

For many years the clavine alkaloids appeared to be limited in their distribution to *Claviceps* species, but in 1961 three clavine alkaloids were obtained from *Aspergillus fumigatus*, and subsequently agroclavine (**187**), elymoclavine (**188**), and chanoclavine-I (**192**) were obtained from the same source. More recently, clavine alkaloids have been obtained from several *Penicillium* and *Aspergillus* species.

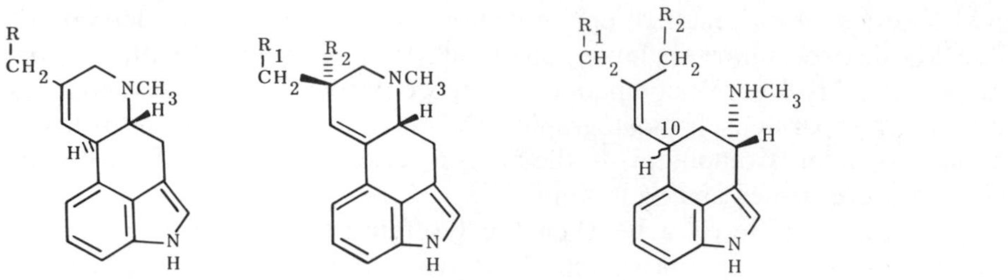

187 agroclavine, R=H
188 elymoclavine, R=OH

		R_1	R_2
189	lysergol	OH	H
190	setoclavine	H	OH
191	penniclavine	OH	OH

		R_1	R_2
192	chanoclavine-I 10α-H	OH	H
193	chanoclavine-II 10β-H	OH	H
194	isochanoclavine-I 10α-H	H	OH

Figure 1 Some representative clavine alkaloids.

The mycelium of *Penicillium roqueforti*, as well as roquefortine (**126**), also contains isofumigaclavine A (**195**), but from *Penicillium concavorugulosum* a different, quite new structure type was obtained. These compounds are the rugulovasines A and B, having the structures **196** and **197**, as determined by X-ray analysis. Interconversion of the diastereoisomers occurs on warming in polar solvents and has been suggested to take place by way of the intermediate **198**, formed through a reverse Mannich reaction.

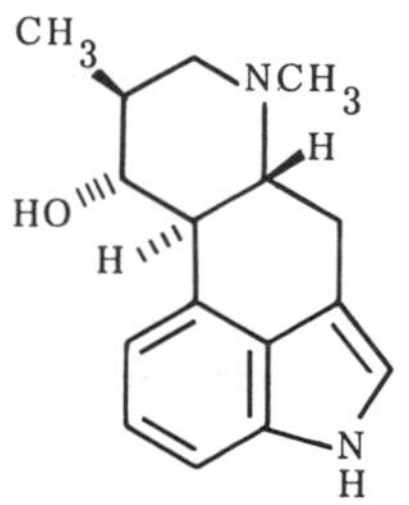

195 isofumigaclavine A

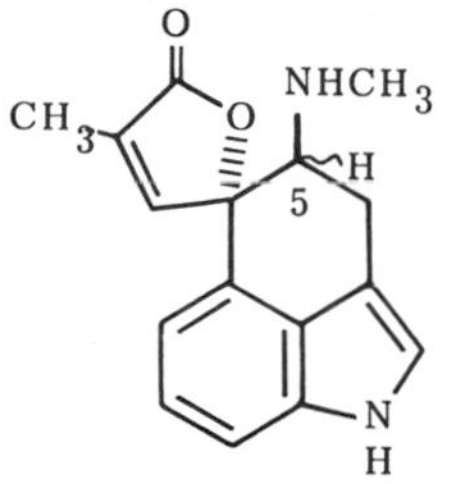

196 rugulovasine A H-5β

197 rugulovasine B H-5α

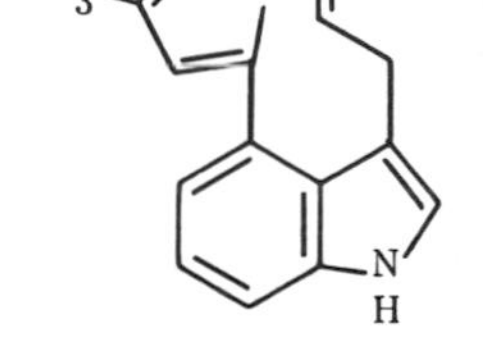

198

Perhaps the most startling discovery however was that some higher plants also contain ergoline alkaloids.

9.10.4 Simple Lysergic Acid Amides

Although lysergic acid diethylamide (LSD) (**199**) is not a natural product, several simple amides are known from natural sources, and one of these is of considerable commercial importance.

The simplest amide derivative is lysergic acid amide (**200**), obtained initially as a hydrolysis product of ergot alkaloids and subsequently as one of the main alkaloids of *Paspalum distichum* L. It is now known as ergine (**200**) and the C-8 epimer as erginine (**201**), in accordance with the established nomenclature of these alkaloids.

The carbinolamine **202** has been obtained from *Claviceps paspali*. In weakly acidic solution, fragmentation occurs to yield ergine and acetaldehyde.

Ergonovine was isolated almost simultaneously in 1935 in four different laboratories; as ergometrine by Dudley and Mair, as ergotocin by Kharasch and Legault, as ergobasine by Stoll and Burckhardt, and as ergostetrine by Thompson. In Europe (except Switzerland) the alkaloid is known as ergometrine; in Switzerland it is known as ergobasine, and in the United States ergonovine is the *United States Pharmacopeia* (USP)-adopted name.

Ergonovine is a water-soluble ergot alkaloid and this property can be used in the isolation of the alkaloid. The filtrate, after removing the water-insoluble material, is extracted with ethylene dichloride to afford crude ergonovine, which is recrystallized as the maleate salt from methanol–ether and then methanol. This procedure is not applicable when water-soluble pigments are present.

The structure of ergonovine (**203**) was deduced in 1935 by Jacobs and Craig, who found that alkaline hydrolysis yielded lysergic acid (**184**) and L-(+)-2-aminopropanol. The yield from ergot however was not sufficient to meet demand and several partial syntheses have been described.

The first partial synthesis of ergonovine was developed by Stoll and Hofmann and involves the condensation of *d*-isolysergic acid hydrazide with L-2-amino-1-propanol to afford ergonovinine (**204**); ergonovine is produced by treatment with alkali. The problem with this synthesis is that in the formation of the hydrazide racemization occurs and the crude product must be resolved with di(*p*-toluyl)tartaric acid.

199 lysergic acid diethylamide, $R_1=R_2=C_2H_5$

200 ergine, $R_1=R_2=H$

202 $R_1=CHCH_3$, $R_2=H$
       $\overset{|}{OH}$

203 ergonovine, $R_1=-\overset{\overset{CH_3}{|}}{C}HCH_2OH$, $R_2=H$

201 erginine, R=H

204 ergonovinine, $R-\overset{\overset{CH_3}{|}}{C}HCH_2OH$

The Pioch process involves the reaction of the mixed anhydride of lysergic and trifluoroacetic acids at room temperature with L-2-amino-1-propanol. Variable quantities of the corresponding amino ester are also produced.

The Garbrecht process utilizes the mixed anhydride of *d*-lysergic and sulfuric acids with L-2-amino-1-propanol. The reaction in this instance does not produce any of the amino ester. The mixed anhydride is produced from the lithium salt of *d*-lysergic acid with the sulfur trioxide complex of dimethylformanide at room temperature. These synthetic routes are summarized in Scheme 21.

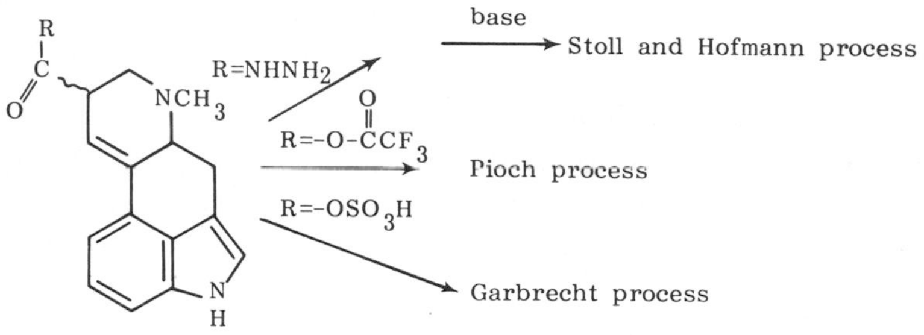

Scheme 21   Partial synthesis of ergonovine (**203**).

9.10.5   Peptide Ergot Alkaloids

The most commercially significant ergot alkaloids belong to the "peptide" group. Hydrolysis of these alkaloids gives lysergic acid, proline, a second amino acid, an α-keto acid, and one equivalent of ammonia. The second amino acid is either L-phenylalanine, L-leucine, or L-valine; and pyruvic, dimethylpyruvic, or α-keto butyric acids are the α-keto acid. A typical example is ergotamine (**205**), which on hydrolysis yields lysergic acid (**184**), ammonia, proline (**106**), pyruvic acid, and L-phenylalanine. No keto acid or ester moiety was detected in the peptide, and consequently an α-amino acid group was thought to be present which was hydrolyzed to an α-keto acid and ammonia.

After several reactions the structure of the peptide moiety was proposed to be **211**, which is the cyclized form of the more easily recognized internally acylated polypeptide **212**.

Ergotamine was first isolated by Stoll in 1918, but the complexities of the nucleus and of the peptide moiety precluded complete determination of the structure until 1951. This structure was confirmed some 10 years later by synthesis (see later).

The importance of ergotamine tartrate as a therapeutic agent has led to a number of efforts designed to improve the yield from the crude drug. In one of these, the drug is treated with tartaric acid until no more alkaloids

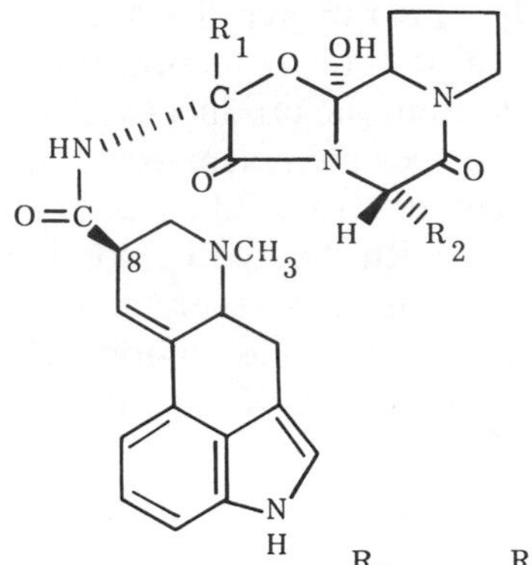

		R_1	R_2
205	ergotamine	CH_3	$CH_2C_6H_5$
206	ergocristine	$CH(CH_3)_2$	$CH_2C_6H_5$
207	ergokryptine	$CH(CH_3)_2$	$CH_2CH(CH_3)_2$
208	ergocornine	$CH(CH_3)_2$	$CH(CH_3)_2$
209	ergosine	CH_3	$CH_2CH(CH_3)_2$
210	dihydroergosine	CH_3	$CH_2CH(CH_3)_2$, 9, 10α–H_2
218	ergosinine	CH_3	$CH_2CH(CH_3)_2$, 8–epi

are removed. After basification the mixture is extracted with trichloroethylene, and ergotamine tartrate precipitated by the addition of methanolic tartaric acid.

It has been estimated that 95% of the commercial production of the peptide ergot alkaloids is obtained by the extraction of ergot cultivated in the field.

The synthesis of the peptide portions of any of the peptide ergot alkaloids poses a problem because of the lability of the α-amino acid grouping and the formation of the cyclol structure. It was discovered however that if certain structure conditions are met, cyclol formation occurs spontaneously and gives a quite stable product. The key to a successful synthesis was determined to be introduction of the α-amino group in the last step.

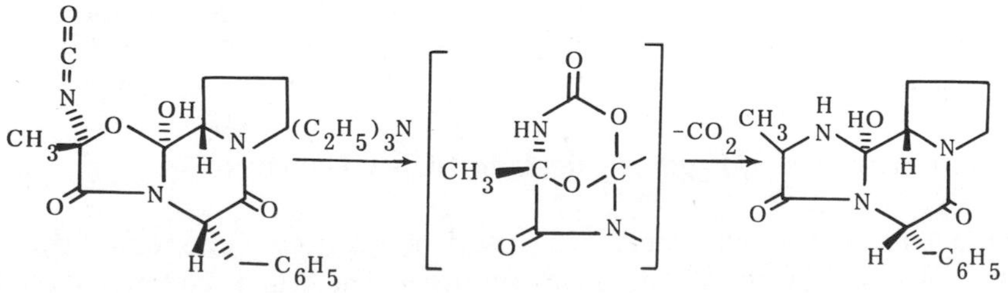

This approach was first used successfully by Hofmann and co-workers in the synthesis of ergotamine. The half-ester acid chloride **213**, when treated with *cyclo*-L-phenylalanyl-L-prolyl (214) in pyridine, gave the acylated diketopiperazine 215 (Scheme 22). Catalytic reduction (Pd–C/H$_2$) removed the benzyl protecting group, whereupon the product cyclized spontaneously to form the cyclol ester as a mixture of stereoisomers differing in configuration at C-2. The ester was converted to a mixture of amines 216 by degradation. Treatment of one of these with the hydrochloride of lysergic acid chloride in the presence of tributylamine gave ergotamine (205). An improved procedure uses the optically active acid chloride 213, and also permits determination of the C-12′ stereochemistry in the synthetic intermediate 216 and hence in the natural product.

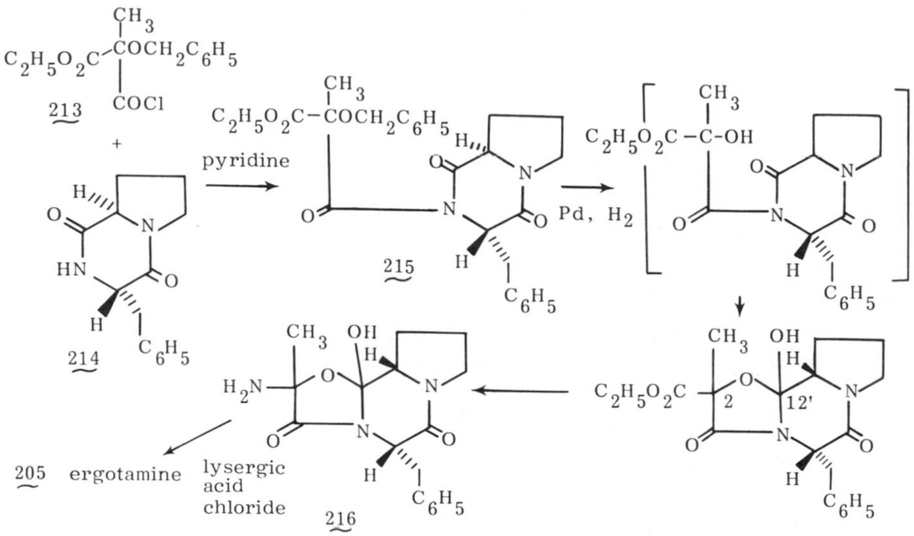

Scheme 22 Stoll and co-workers' partial synthesis of ergotamine (205).

Ergotoxine was originally obtained from ergot by Kraft and by Barger and Carr and was subsequently demonstrated by Stoll and Hofmann to be a mixture of three peptide ergot alkaloids: ergocristine (206), ergocornine (208), and ergocryptine (207). The alkaloids themselves have no clinical applications, but a mixture of 9,10-dihydro derivatives possesses strong sympathicolytic activity. The mesylate of dihydroergotoxine has been used as a peripheral vasodilator and hypotensive, and in combination with papaverine and sparteine is an effective medication for vascular disorders in the aged.

9.10.6 Occurrence of Ergot Alkaloids in Higher Plants

Ololiuqui is an hallucinogenic drug used in religious ceremonies by the Indians of several Central American countries. The drug is comprised of the seeds of two plants, *Ipomoea violacea* L. and *Rivea corymbosa* (L.) Hall f., both in the family Convolvulaceae. Hofmann and Tscherter obtained several clavine alkaloids from these seeds, including elymoclavine (**188**), chanoclavine-I (**192**), and lysergic acid amide (**200**), the principal hallucinogenic constituent of ololiuqui. As many as seven alkaloids have been identified from 500 mg of *I. violacea* var. "Pearly gates" seeds, and the major of these is chanoclavine-I acid (**217**).

217   chanoclavine - I
      acid

Subsequent work with other *Ipomoea* species has afforded several amide derivatives, including ergine (**200**), erginine (**201**), and ergonovine (**203**). The peptide ergot alkaloids ergosine (**209**) and its 8-epimer ergosinine (**218**) were obtained from *Ipomoea argyrophylla* Vatke, and this of particular interest.

To date, ergoline alkaloids have only been isolated from plants in the Convolvulaceae, and in particular the genera *Ipomoea*, *Argyreia*, *Rivea*, and *Stictocardia* are fruitful sources. It should be emphasized that these are true alkaloids of the plant, *not* artifacts from an associated parasitic fungus. The yields do not however compare with those from fungal sources, and it seems unlikely that plants will be an economic source of these alkaloids.

9.10.7 Detection and Separation of Ergot Alkaloids

Ergot alkaloids give characteristic color reactions with sulfuric acid, but the best detecting reagent is Van Urk's reagent, acidified *p*-dimethylamino-benzaldehyde, which gives a characteristic blue color. The reaction has been standardized (measurement at 590 nm) for the quantitative assay of these alkaloids. The alkaloids separate well with mixtures of chloroform:ethanol (90:10 and 95:5) on Silica gel G.

9.10.8 Synthesis of the Ergoline Nucleus

Considering the amount of effort that has been exerted toward the synthesis of other alkaloids (e.g. camptothecine, Section 9.13.2), it is amazing that so few synthetic procedures for the formation of the ergoline nucleus are known. Perhaps however, this is due in part to the ready availability of lysergic acid by fermentation.

If we consider the structure of lysergic acid, we see that it contains only two asymmetric centers, one double bond, one tertiary nitrogen, and an indole nucleus. It really does not seem to be *that* challenging; however the system is sufficiently difficult that only two syntheses of lysergic acid (185) itself have been reported.

The first synthetic approach to the ergoline nucleus was a synthesis of racemic dihydrolysergic acid (219) by Uhle and Jacobs and completed in 1945. The main steps in this synthesis are shown in Scheme 23, and although the parent ring system was obtained without difficulty, the subsequent reductive manipulations proved very difficult.

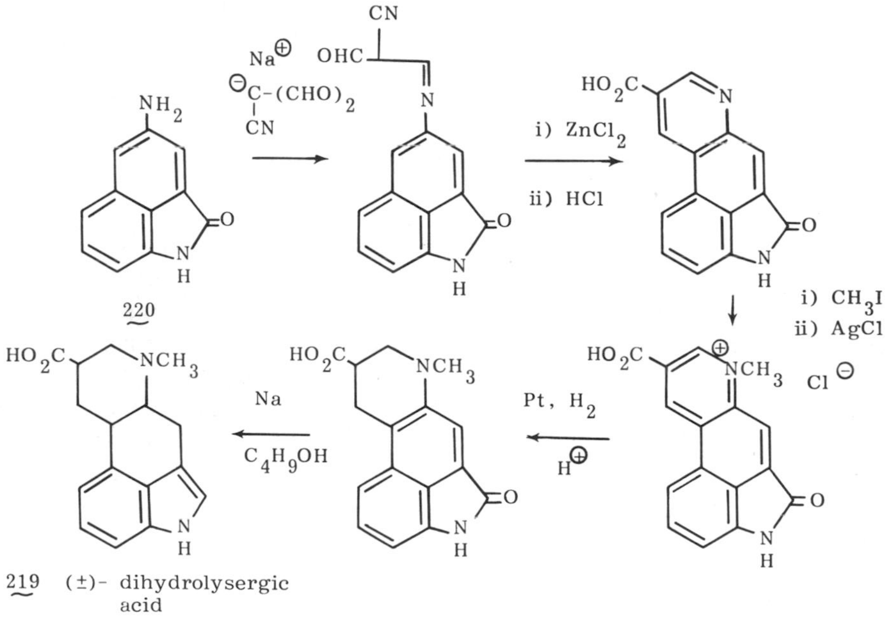

Scheme 23 Uhle–Jacobs synthesis of (±)-dihydrolysergic acid (219).

An improved synthesis of (+)-dihydrolysergic acid and later of the *d*-(−)-isomer was achieved by Stoll and Rutschmann some years later and also begins with the amino amide 220 as shown in Scheme 24. The dihydronorlysergic acid mixture 221 was N-methylated and the dihydrolysergic acids resolved with L-norephedrine to afford *d*-(−)-dihydrolysergic acid (222).

Several other synthetic approaches which made use of the naphthostyril or benz[c,d]indoline system could not be applied to the synthesis of lysergic acid or its derivatives, because of the failure to introduce the double bond into the 9,10 position. Fundamentally the problem involves the facile isomerization of the ergoline system to a benzindoline 223 under acidic conditions.

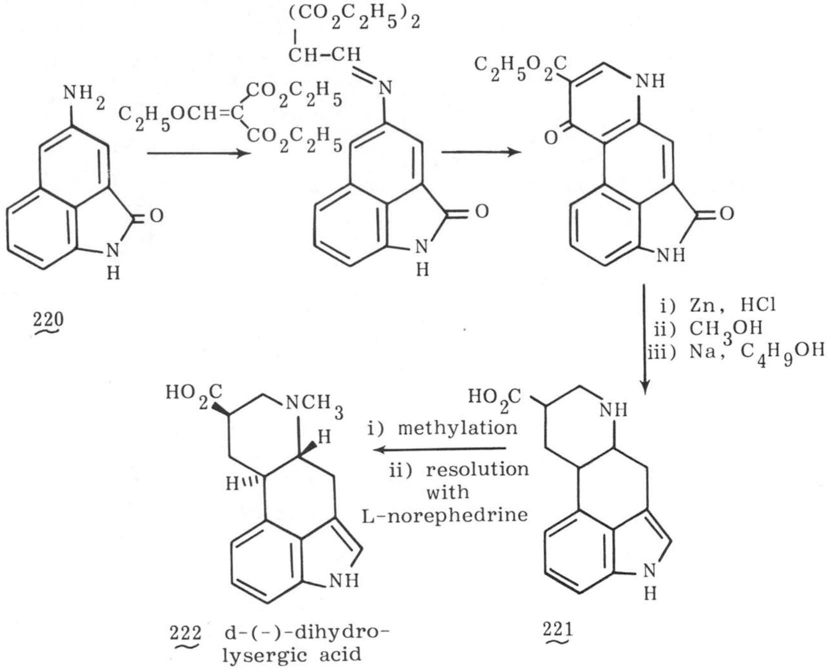

Scheme 24 Synthesis of *d*-(−)-dihydrolysergic acid (**222**).

For the synthesis of lysergic acid it was therefore necessary to develop a quite new approach, and this was initially due to Kornfeld and co-workers at Eli Lilly and Company. The indoline–indole dehydrogenation step is the last in the sequence. Friedel–Crafts cyclization of the acid chloride **224** followed by α-bromination afforded **225**. Alkylation of methylaminoacetone ketal **226** with **225** gave **227**, which was hydrolyzed, condensed, and N-acetylated (to prevent oxidation) to yield the α,β-unsaturated ketone **228**. Sodium borohydride reduction followed by conversion to the chloride and displacement with cyanide gave the nitrile **229**. Methanolysis of the nitrile, hydrolysis of the *N*-acetyl group, and dehydrogenation with deactivated Raney nickel afforded racemic lysergic acid (**184**) (Scheme 25).

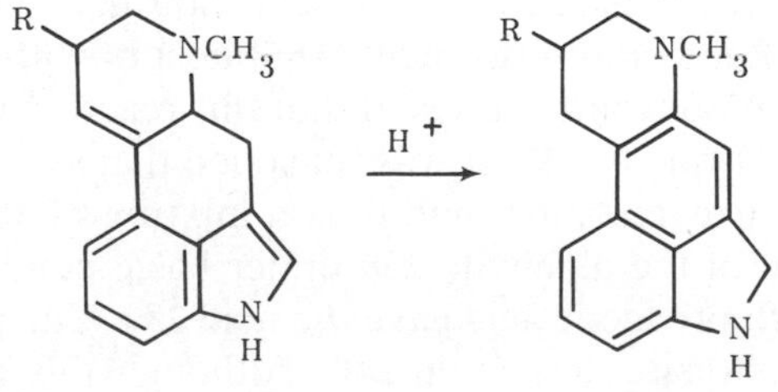

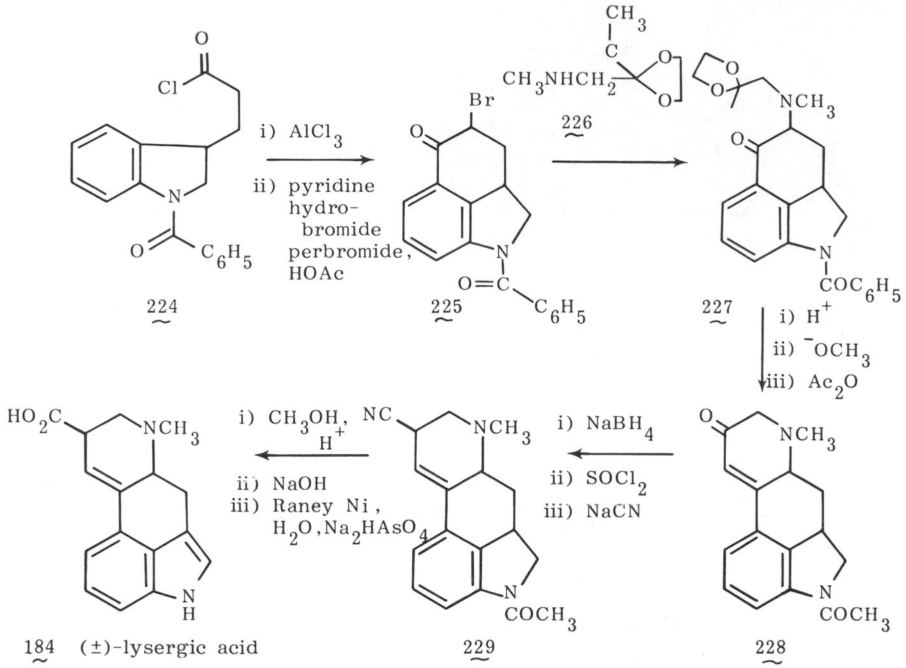

Scheme 25   Kornfeld and co-workers' synthesis of (±)-lysergic acid (**184**).

A quite different synthetic approach to lysergic acid (**184**) has been reported by Julia and co-workers and involves as a key step the intramolecular addition of an enolate anion to a benzyne. Condensation of methyl 6-methyl nicotinate (**230**) with 5-bromoisatin (**231**) gave **232**, which was reduced in turn with zinc–acetic acid and diborane to afford the indoline **233**. N_a-Acetylation, N_b-quaternization, and sodium borohydride reduction gave the α,β-unsaturated ester **234** as a mixture of *cis* and *trans* isomers. The *cis* isomer was treated with sodium amide–liquid ammonia to yield the *N*-acyl indoline **235** by way of the benzyne enolate **236**. This compound was synthesized from lysergic acid in order to confirm the identity, but was not itself carried through the additional steps to lysergic acid (**184**) (Scheme 26).

Ramage and co-workers have reported the only other total synthesis of lysergic acid (**184**). It is based on an observation that the epimerization of lysergic acid (**184**) and isolysergic acid (**185**) occurs with racemization (i.e. inversion at C-5). Woodward suggested that this reaction proceeded through on opening of the D-ring to **237**. It was reasoned that synthesis of **237** would lead to spontaneous cyclization, and thus a mixture of **184** and **185**.

Wittig reaction of the aldehyde **238** under basic conditions followed by treatment with trifluoroacetic acid gave the acid **239**. Curtius degradation led to the *p*-toluenesulfonate derivative **240**. Although this compound did not cyclize spontaneously, methylation under Eschweiler–Clark conditions gave a mixture of **241** and **242**. Methanolysis of the mixture **241** afforded **243** as a mixture of C-8 epimers in which the C-8 β-substituted product predominated (3:1). This compound had previously been converted to (+)-lysergic acid (**184**) (Scheme 27).

In the extensive synthetic work by Floss, Cassady, and co-workers on prolactin inhibitors, a new synthesis of the ergoline nucleus (9,10-dihydro series) was developed from the tricyclic ketone **244**. In five steps this compound was converted to the isomeric ketone **245**, which was transformed by ethyl α-(bromomethyl)acrylate and methylamine to the 5,10-dehydro species **246**. Sodium cyanoborohydride reduction gave **247** having the C,D-*trans*-stereochemistry. Hydrolysis, esterification, and manganese dioxide oxidation then afforded methyl (±)-dihydrolysergate (**248**) (Scheme 28).

Several clavine alkaloids have been synthesized from lysergic acid (**184**), and these include penniclavine (**191**) and elymoclavine (**188**). A key intermediate in many of these syntheses is methyl 10α-methoxy-Δ^{8,9}-lysergide (**249**). This compound is readily prepared from methyl lysergate (**250**), by oxidation with mercuric acetate in methanol followed by alkaline sodium borohydride reduction. The benzylic methoxyl group can be removed reductively. For example in the synthesis of elymoclavine (**188**), **249** was reduced with lithium aluminum hydride to afford 10α-methoxyelymoclavine (**251**), which could be oxidized with manganese dioxide and reduced with lithium aluminum hydride–aluminum chloride to **188** (Scheme 29).

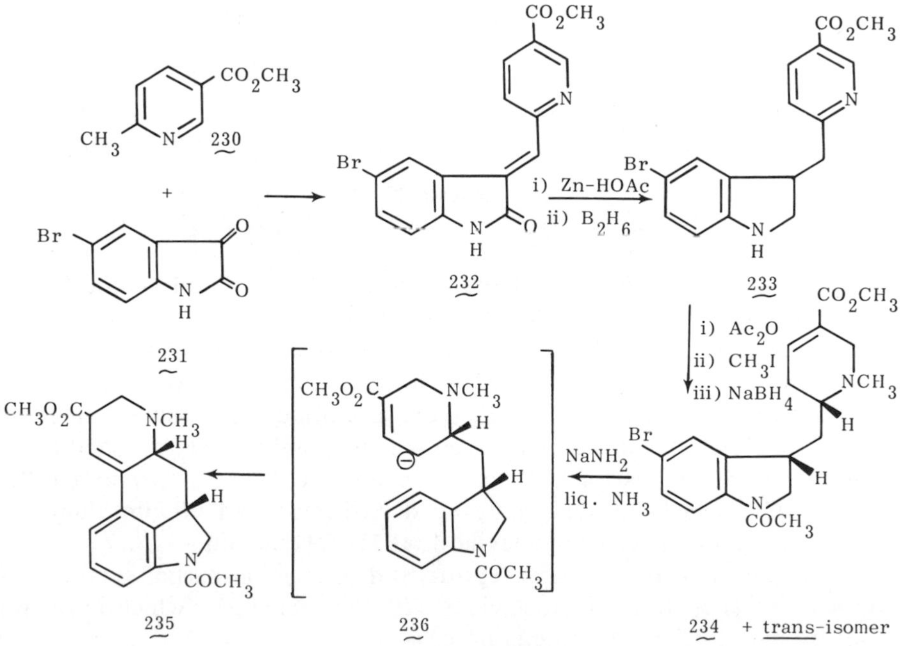

Scheme 26

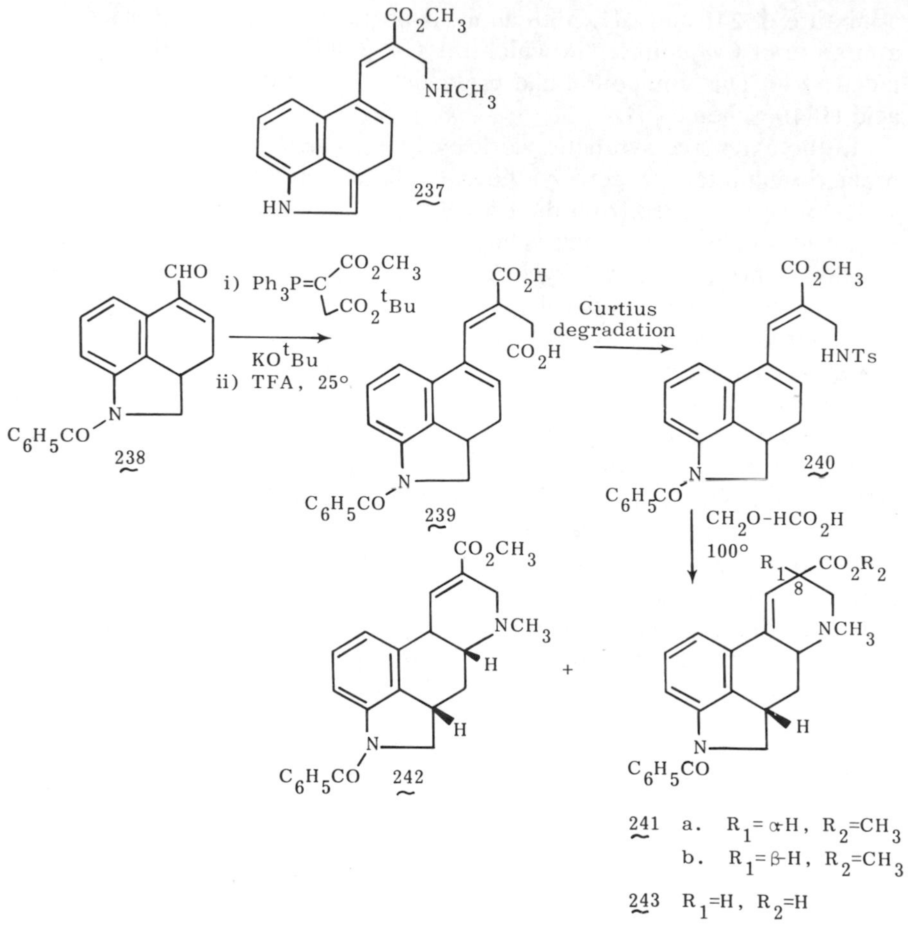

Scheme 27

Pleininger and co-workers have reported the only total synthesis of a chanoclavine, namely chanoclavine-I (192). The key intermediate is the nitro ketone (252), prepared in several steps from β-naphthol. Treatment of the phenylhydrazone derivative of 252 with aluminum in aqueous ethanol and reaction with ethyl chloroformate gave a mixture of biurethane derivatives 253. Ozonolysis afforded a dialdehyde which spontaneously cyclized to the compound 254. Wadsworth–Emmons reaction with the phosphorane 255 yielded 256, which on acid-catalyzed dehydration and lithium aluminum hydride reduction gave chanoclavine-I (192) (Scheme 30).

Floss and co-workers have reported an interesting partial synthesis of agroclavine (187) from chanoclavine-I (192) (Scheme 31), which in overall concept mimics the biosynthetic process.

9.10.9 Spectral Properties

The $\Delta^{8,9}$- and $\Delta^{9,10}$-ergolenes are easily distinguished by their UV spectra. Thus agroclavine (187) shows typical indole maxima [λ_{max} 284 (log ε 2.88) and 293 nm (3.81)], whereas ergotamine (205) shows a 4-vinyl indole UV spectrum [λ_{max} 318 nm (log ε 3.86)].

The proton NMR spectral data of chanoclavine-I (192) and isochano-clavine-I (194) show an interesting difference between the vinyl protons and the allylic methyl groups in each compound.

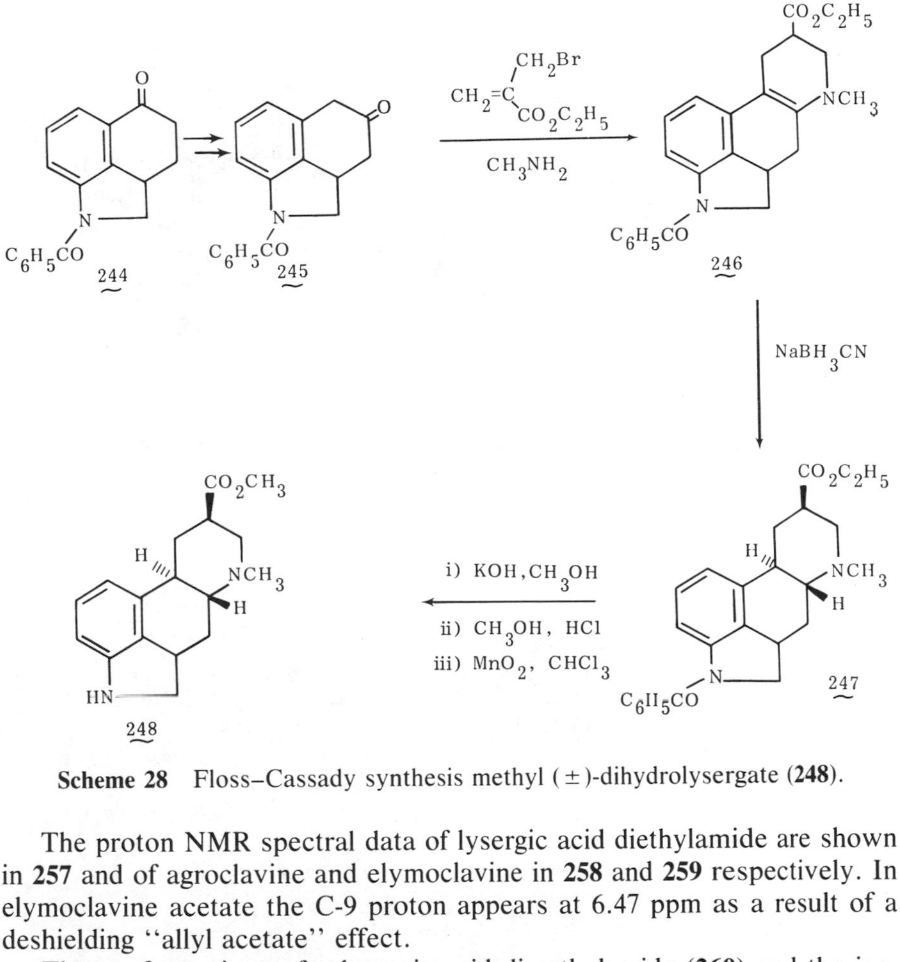

Scheme 28 Floss–Cassady synthesis methyl ($\pm$)-dihydrolysergate (248).

The proton NMR spectral data of lysergic acid diethylamide are shown in 257 and of agroclavine and elymoclavine in 258 and 259 respectively. In elymoclavine acetate the C-9 proton appears at 6.47 ppm as a result of a deshielding "allyl acetate" effect.

The conformations of D-lysergic acid dimethylamide (260) and the iso-lysergic acid isomer (261) have been determined by analysis of a 220-MHz proton NMR spectrum. Ring D was found to adopt a half-chair conformation in which both the N-methyl and C-8 amide functions are pseudoequatorial in each isomer. The dramatic difference between the pharmacologic activ-

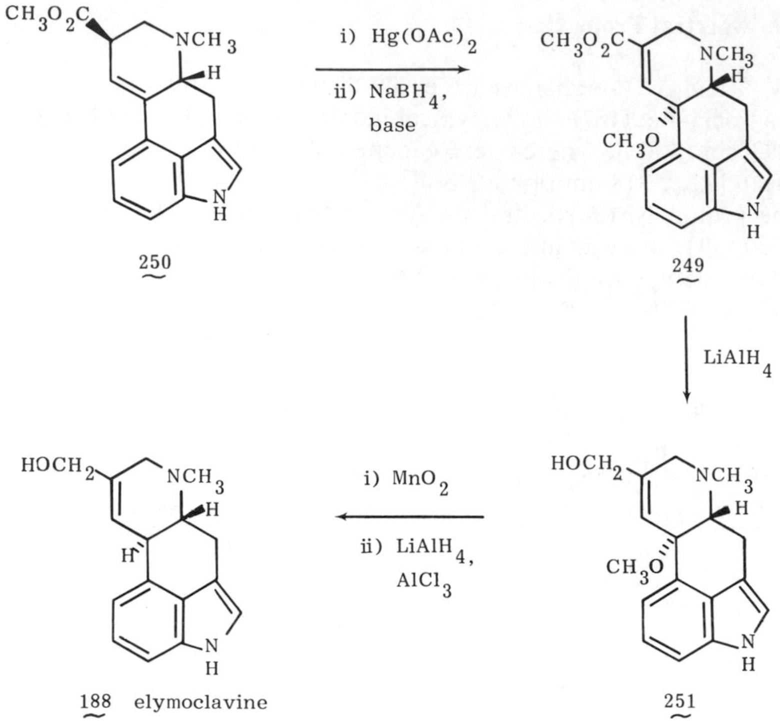

Scheme 29 Bach–Kornfeld synthesis of elymoclavine (**188**).

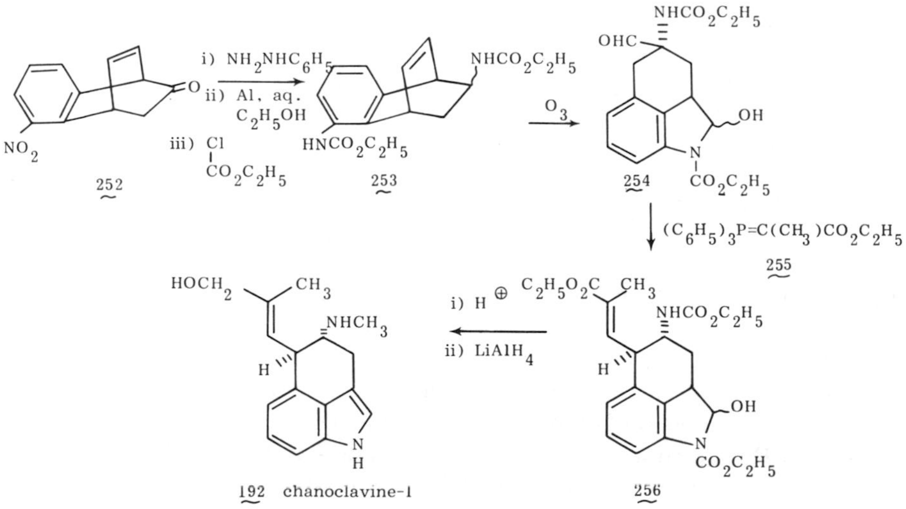

Scheme 30 Pleininger and co-workers' synthesis of chanoclavine-I (**192**).

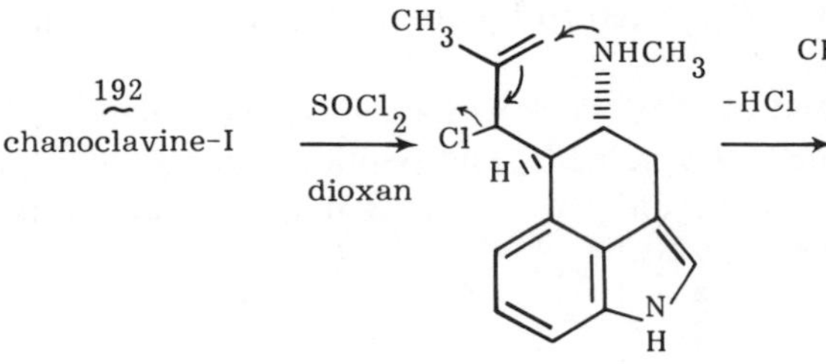

192
chanoclavine-I

$\xrightarrow[\text{dioxan}]{\text{SOCl}_2}$

187 agroclavine

Scheme 31

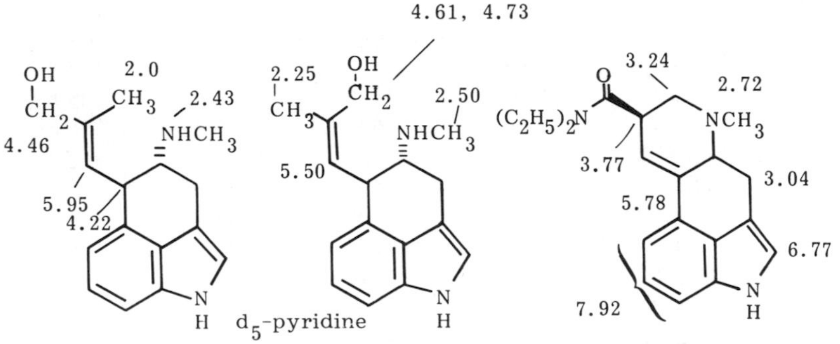

4.61, 4.73

chanoclavine-I

isochanoclavine-I
d$_5$-pyridine

257

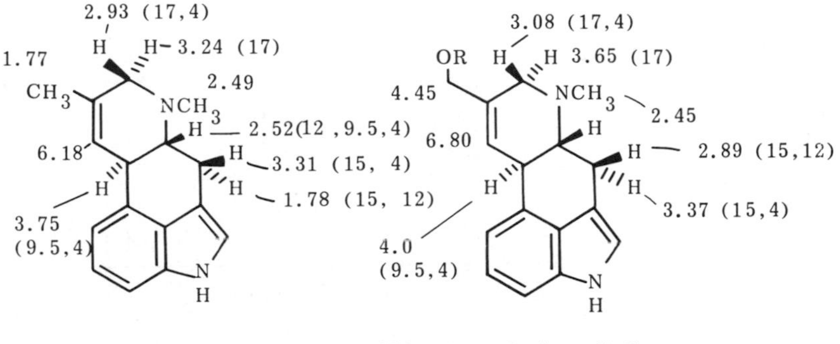

258 agroclavine

259 elymoclavine, R=H
262 elymoclavine acetate, R=Ac

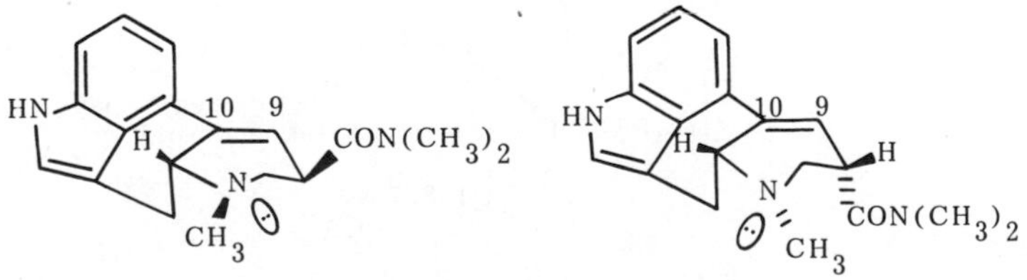

260 lysergic acid dimethylamide

261 isolysergic acid dimethylamide

ities of these isomers was suggested to be due to the α- or β-orientation of the nitrogen lone pair.

Wenkert and Floss and their co-workers have described an extensive ^{13}C NMR analysis of clavine, lysergic acid, and peptide ergot alkaloid derivatives. The data for some representative ergot alkaloids are shown in Table 2.

Table 2 ^{13}C NMR Data of Representative Ergot Alkaloids

	Agroclavine (**187**)	Elymoclavine Acetate (**262**)	Lysergic Acid Methyl Ester (**250**)	Ergonovine (**203**)	Ergotamine (**205**)
C-2	118.3	117.9	118.2	119.1	119.4
C-3	111.2	111.3	110.2	108.9	108.8
C-4	26.4	26.4	26.9	26.8	26.6
C-5	63.6	63.4	62.6	62.6	62.4
C-7	60.2	56.2	54.6	55.5	55.1
C-8	131.9	130.9	41.8	42.8	42.5
C-9	119.4	124.8	117.6	120.1	118.3
C-10	40.8	40.5	136.0	135.0	136.0
C-11	131.9	131.3	127.6	127.4	127.1
C-12	112.0	112.2	112.0	111.0	111.0
C-13	122.0	122.6	122.9	122.4	122.2
C-14	108.4	108.7	109.4	109.0	110.2
C-15	134.0	133.4	133.7	133.7	133.8
C-16	126.6	126.1	125.9	125.8	125.9
C-17	19.9	66.2	172.4	171.2	174.3
N—CH$_3$	40.2	40.5	43.4	43.4	43.4
CH$_3$	—	20.6	51.9	17.4	—
C—O	—	170.7	—	—	—
NCH	—	—	—	46.4	—
OCH$_2$	—	—	—	64.4	—

Peptide Portion of Ergotamine

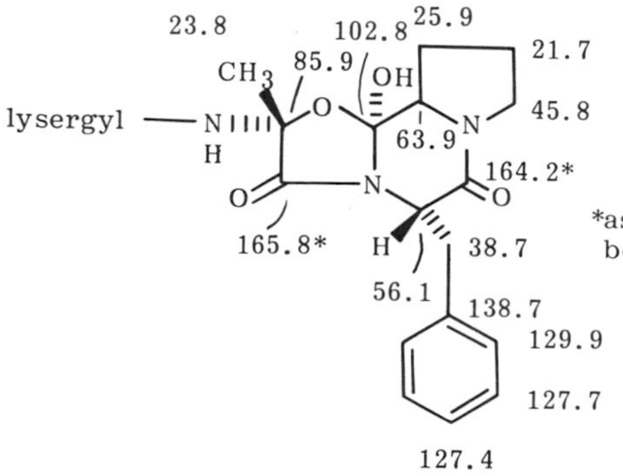

*assignments may be reversed

9.10.10 Biosynthesis

Numerous proposals for the natural formation of ergot alkaloids, in particular the ergoline nucleus, were made following the structure determination of lysergic acid. But Mothes and co-workers are credited with the suggestion that tryptophan condensed directly with an isoprene unit at C-4 as shown in 263. A similar proposal was also made independently by Birch in 1958.

263   264

The first experimental evidence for the involvement of tryptophan was obtained by Mothes using [3-^{14}C]tryptophan, who showed that the lysergic acid moiety of ergonovine (203) and the peptide ergot alkaloids were labeled by the precursor. This work was carried out with ergot-infected rye plants and was confirmed subsequently by feeding experiments with saprophytic cultures; incorporations into elymoclavine (188) were in the range 10–39%. Since that time, only saprophytic cultures have been used to study the biosynthesis of ergot alkaloids.

[1-^{14}C]Tryptophan was not incorporated into the alkaloids, in agreement with the hypothesis, but the intact incorporation of the remainder of the side chain of tryptophan has been established by feeding experiments with the nitrogen atom labeled, and with [2-^{14}C, 2-^{3}H] and [3-^{14}C, 3-^{3}H]tryptophans.

Gröger, Birch, and Ramstad showed at about the same time that mevalonic acid (93) was a good (9–23% specific incorporation) precursor of the ergoline system. In particular [2-^{14}C]mevalonate (93) was specifically incorporated into C-17 of elymoclavine (188) and agroclavine (187). This apparently "direct" incorporation conceals a number of subtleties, as we shall see. [Methyl-^{14}C, ^{3}H]Methionine was incorporated into the clavine alkaloids specifically at the N-methyl group, and consequently all the carbon and nitrogen atoms of the clavine nucleus are accounted for.

With these data on hand, attention was turned to the details of the biosynthetic process, and it was only then that the complexities of ergot alkaloid biosynthesis were unveiled. The recent review by Floss presents these data in a complete, yet succinct manner.

Both D- and L-tryptophans are incorporated into the ergoline nucleus, the former with loss of the amino nitrogen and presumably by way of indole pyruvic acid. L-Tryptophan is incorporated without loss of this nitrogen and this is interesting because the configuration at C-5 in the alkaloids corresponds to the present in a D-amino acid. At some point therefore inversion of configuration at C-2 from tryptophan occurs with retention of the hydrogen. With the fundamental precursors established as 3R-mevalonic acid, L-tryptophan, and methionine, attention can now be turned to the mechanistic pathway of ergoline nucleus formation.

Feeding experiments with labeled N-methyl tryptamine (15) and tryptophan (1) established that the first step in the pathway is isoprenylation of tryptophan. The mechanism of this substitution at C-4 of tryptophan (not the most likely position, *vide infra* echinulins) is unknown in spite of numerous *in vitro* model studies. However, 4-(γ,γ-dimethylallyl)tryptophan (264) is a good precursor of the clavines in *C. paspali* and was later isolated from ergot cultured under anerobic conditions or in the presence of ethionine. More recently, an enzyme catalyzing the condensation of L-tryptophan and γ,γ-dimethylallylpyrophosphate has been obtained. Studies with this enzyme using model compounds established that the substitution probably occurs directly by a "close encounter" theory in which C-4 is the only site on tryptophan made available by the enzyme.

192   chanoclavine   →   187   agroclavine   →   188   elymoclavine   →   187   lysergic acid

Scheme 32   Biogenetic sequence to lysergic acid due to Rochelmeyer.

Let us turn now our attention to the more intimate details of the pathway between γ,γ-dimethylallyltryptophan (263) and lysergic acid (184). Two early hypotheses for this biosynthesis were those of Rochelmeyer (Scheme 32) and Abe (Scheme 33). The major differences concern the involvement of the chanoclavine series of compounds. One scheme suggests that the chanoclavines are at the beginning of the scheme, the other that they are at the end of the pathway.

lysergaldehyde   →   188   elymoclavine   →   187   agroclavine   →   192   chanoclavine
      ↓
184   lysergic acid

Scheme 33   Biogenetic sequence to lysergic acid due to Abe.

Extensive work, principally by Agurell and Ramstad, established some of the biosynthetic relationships of the alkaloids. Initially, the sequence agroclavine → elymoclavine → penniclavine was established, and elymoclavine was separately shown to be a precursor of lysergic acid. Somewhat surprisingly chanoclavine was not a precursor and its involvement was not regarded as significant at this time. When [2-^{14}C, 2-^{3}H]mevalonate was used as a precursor of agroclavine (187), the ^{3}H/^{14}C ratio was the same as that in the starting material, thereby eliminating lysergaldehyde as a precursor of agroclavine (**187**).

Although many mechanisms for the conversion of **264** to agroclavine were postulated at this time, they involved for the most part allylic carbonium ions and subsequently have been shown to be overly simplistic.

Perhaps the most important discovery was the definition of the role of the chanoclavines. This was initiated by the isolation of chanoclavine-I (**192**), chanoclavine-II (**193**), and isochanoclavine-I (**194**), and the question was immediately raised as to the exact nature of the "chanoclavine," which was a precursor of agroclavine. One point worth reiterating here is that of the importance of isolating biogenetically significant compounds.

The real test came when these newly isolated chanoclavines were used as precursors. Only chanoclavine-I (**192**) was a good (up to 40%) precursor of elymoclavine (**188**), and this suggests that chanoclavine-I (**192**) cannot be interconverted to chanoclavine-II (**193**) and isochanoclavine-I (**194**).

192 chanoclavine-I   187 agroclavine   188 elymoclavine

193 chanoclavine-II   194 isochanoclavine-I

A most interesting result was observed when [2-^{14}C]mevalonic acid was used as a precursor of secoergolines, for the label was consistently found in the methyl group, and in the tetracyclic series (e.g. **187** and **188**) the C-17 carbon atom was labeled. In the conversion of chanoclavine-I to agroclavine (**187**) therefore, a *cis,trans* isomerization of the 8,9-double bond must occur in the process of closing ring D.

These results obtained by Arigoni were confirmed by Floss and co-workers, who showed with [17-^{14}C]- and [7-^{14}C]chanoclavine-I that the hydroxymethyl group of chanoclavine-I (**192**) appeared at C-7 in agroclavine (**187**).

Conversely the *C*-methyl group of **192** gave rise to C-17 in agroclavine (**187**) and elymoclavine (**188**).

With these results in hand, a scheme could be proposed in which chano-clavine-II (**193**) and isochanoclavine-I (**194**) are by-products of the main pathway which proceeds through chanoclavine-I → agroclavine → elymo-clavine, and involves *cis,trans* isomerization in the formation of **187** and **192**.

Having some of the elements of the final stages apparently deduced, attention in several laboratories was turned to the next major challenge, the formation of ring C of chanoclavine-I (**192**) from 4-(γ-γ-dimethylallyl) tryptophan (**264**).

From previous work in other systems it is well established that C-2 of mevalonate corresponds to C-4 of isopentenylpyrophosphate and that in the subsequent conversion to dimethylallylpyrophosphate by isopentenylphos-phate isomerase this label appears in the E-methyl group of DMAPP. In chanoclavine-I (**192**) however this label appears in the Z-methyl group. Does IPP isomerase have a different stereochemistry in ergot alkaloid bio-synthesis than in all other known systems, or is a second *cis-trans* isomer-ization involved between **93** and **188**?

It will be recalled (Chapter 2) that in both the IPP isomerase and prenyl transferase reactions the *trans* double bond is formed by loss of the *pro*-4S hydrogen of mevalonate. In the ergot series it was shown that the *pro*-4R hydrogen of mevalonate was retained to the extent of 70% into elymoclavine (**188**). The loss of the label from this position will be discussed subsequently. The key experiment to prove the second *cis,trans* isomerization was carried out by Arigoni and co-workers, who synthesized 4-(γ,γ-dimethy-lallyl)tryptophan[Z-^{14}CH$_3$] (**264**) and showed that C-7 of elymoclavine (**188**) contained 98.5% of the incorporated activity. Thus either no or an even number of *cis,trans* isomerizations occur during biosynthesis.

By experiments with various desoxychanoclavine-I derivatives which showed no incorporation into clavine alkaloids, it was established that hydroxylation of the E-methyl group occurred before closure of the C-ring. At this point 4-(4′-hydroxy-3′-methyl-2-butenyl)tryptophan (**265**) would ap-

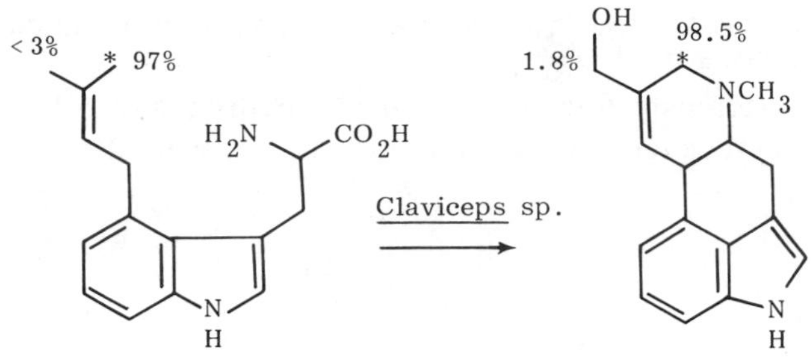

264 (Z-^{14}CH$_3$)-dimethyl-allyltryptophan

188 elymoclavine

pear to be a key intermediate, and initial biosynthetic results suggested that this was indeed the case.

However, when E-265 was incorporated into elymoclavine (188) and the product degraded, label was found at C-17 and not C-7. Even more distracting was the fact that the agroclavine (187) obtained was inactive. But this was only a prelude to an experiment involving both [E-^{14}CH$_3$]- and [Z-^{14}CH$_3$]265, which gave rise to elymoclavine (188) labeled in identical fashion, with 97% of the activity at C-7.

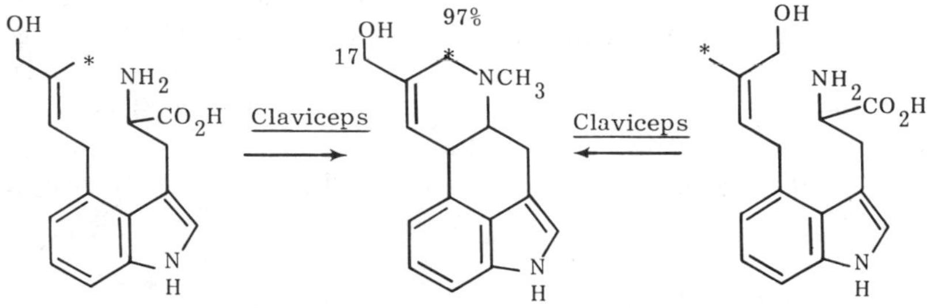

265
E-isomer

188

265
Z-isomer

Since chanoclavine-I (192) is a well-established intermediate, it is clear that neither E- nor Z-isomers of 265 are precursors of agroclavine and elymoclavine and that the organism is metabolizing the E-isomer to elymoclavine as though the hydroxyl group were not present. The incorporation of the Z-isomer has been suggested to occur by way of isomerization to the E-isomer prior to incorporation.

Further experiments, although not having a direct bearing on the intermediate steps in the pathway, do suggest another possibility. These results involve incorporation of the 5-hydrogens of mevalonate.

Both Arigoni's and Floss' group showed that the principal clavine alkaloids of ergot, agroclavine (187), elymoclavine (188), and chanoclavine-I (192) were produced with stereoselective retention of the pro-5R hydrogen and loss of the pro-5S hydrogen of mevalonate. In addition, chanoclavine-

II (193), having the cis-5,10 stereochemistry, also retained the pro-5R hydrogen of MVA.

With this stereospecific loss of a pro-5S hydrogen of MVA independent of the C-10 stereochemistry it is tempting to postulate 266 (no stereochem-

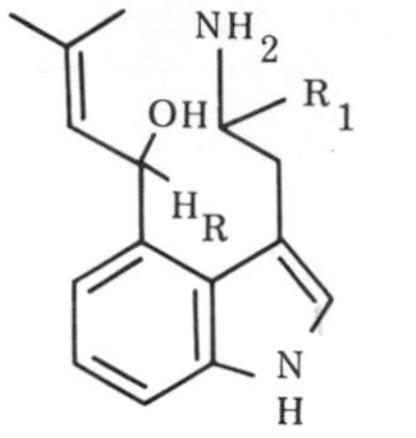

266 R₁=H or CO₂H

istry defined) as a key intermediate in the formation of chanoclavine-I (192). We will return to the possible stereochemistry of this compound subsequently.

When chanoclavine-I specifically labeled with tritium at C-17 was used as a precursor of elymoclavine (188), half of the tritium activity was lost, suggesting that chanoclavine-I aldehyde (267) is an intermediate. This was confirmed with [17-³H, 4-¹⁴C]267, which retained all the activity in the iso-lated elymoclavine (188), and with [3'-³H₂]mevalonate, which showed the presence of two deuteriums in chanoclavine-I and one in elymoclavine (188). The proton NMR spectrum of the latter compound established that the deuterium was stereospecifically at the pro-7S position. Isochanoclavine-I aldehyde (268) has not as yet either been used as a precursor or detected in Claviceps sp.

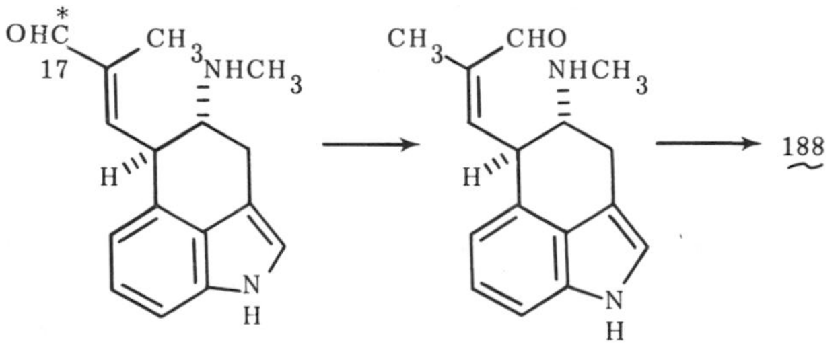

267 chanoclavine-I   268 isochanoclavine-I
     aldehyde   aldehyde

Let us return to the question of the 30% loss of the pro-4R proton of mevalonate on incorporation into C-9 of elymoclavine (188). Further experiments by Floss and co-workers established that the retention of this hydrogen in elymoclavine can vary between 40 and 80%, and is dependent on the rate of alkaloid production. There is another subtlety involved here though, for several experiments by both Floss and Arigoni have established that there is an intermolecular transfer of the C-9 hydrogen under these conditions. For example with [2-^{13}C]mevalonate and [4-^{2}H$_2$]mevalonate used as simultaneous precursors, chanoclavine-I (**192**) contained only M$^+$ and M$^+$ + 1 species in the mass spectrum. However, elymoclavine (**188**) gave M$^+$, M$^+$ + 1, *and* M$^+$ + 2 species, the latter containing *both* ^{13}C and ^{2}H in the same molecule.

This quite exceptional result has been interpreted in terms of the recycling of the C-9 hydrogen from molecule to another molecule as shown in Scheme 34. The EnzX-^{2}H complex must exchange ^{2}H for ^{1}H slowly under these conditions, thereby allowing EnzX-^{2}H to be involved in the initial reaction with a second chanoclavine-I molecule. There are other mechanisms which have also been postulated for this process.

The point in the pathway at which *N*-methylation takes place is at present unknown. But there is evidence that this occurs before or during the formation of ring C. For example, although *N*-norchanoclavine-I (**269**) is a natural product, it is produced *from* chanoclavine-I (**192**), not *vice versa*. One mechanism suggested for this step is shown subsequently.

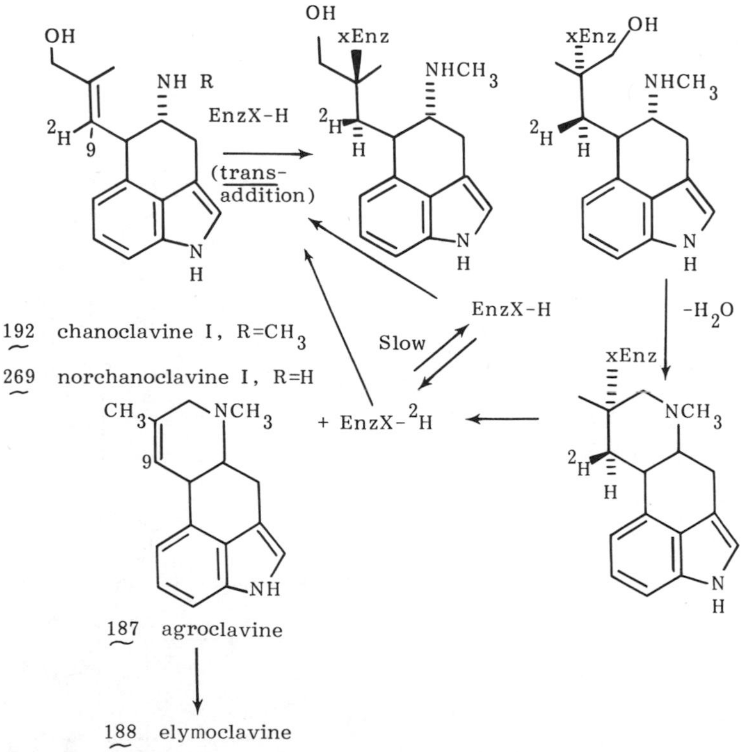

192 chanoclavine I, R=CH$_3$

269 norchanoclavine I, R=H

187 agroclavine

188 elymoclavine

Scheme 34 Postulated mechanism for the recycling of the C-9 hydrogen of chanoclavine-I (**192**).

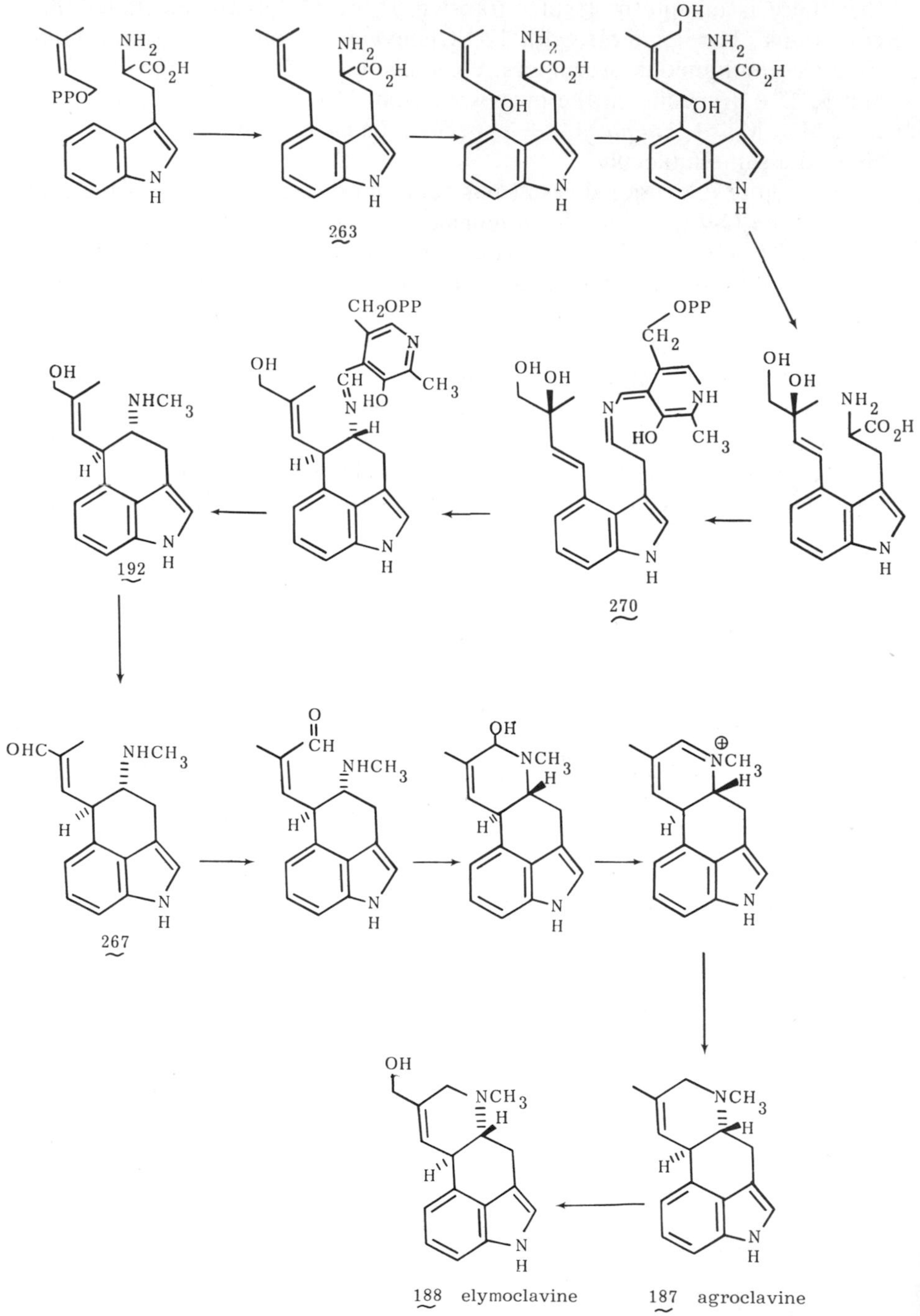

Scheme 35 Biogenesis of the clavine alkaloids.

As Floss has pointed out, in spite of the tremendous efforts that have been put into studying ergoline alkaloid biosynthesis, a unifying scheme is still not possible at this time. Scheme 35 explains much of the available evidence and also indicates how formation of a Schiff base 270 with pyridoxal phosphate could be important in the closure of the C-ring.

Such work has taken us only to the point of a $\Delta^{8,9}$-compound. How is lysergic acid (184), which has a $\Delta^{9,10}$-bond, produced? Elymoclavine (188) is a well-established precursor of lysergic acid (184) and it can safely be concluded that the double-bond isomerization takes place after formation of elymoclavine (188). It has been established also that oxidation of the alcohol to the aldehyde is the next step, and since paspalic acid (186) was well converted to lysergic acid derivatives, it was suggested that this is the point at which isomerization occurs. Floss has indicated that since this isomerization (i.e. of 186 to 184) occurs in vitro (albeit slowly), this is not unequivocal proof that the incorporation of 186 is due to an enzymatic process.

Attempts to prepare either lysergaldehyde (271) have so far proved unsuccessful, and consequently the true intermediate nature of these compounds has not been evaluated. An alternative mechanism (Scheme 36) postulates the 1,2-addition of coenzyme A across the enol form of $\Delta^{8,9}$-lysergaldehyde (272) to give an intermediate hemiacetal 273, which can be

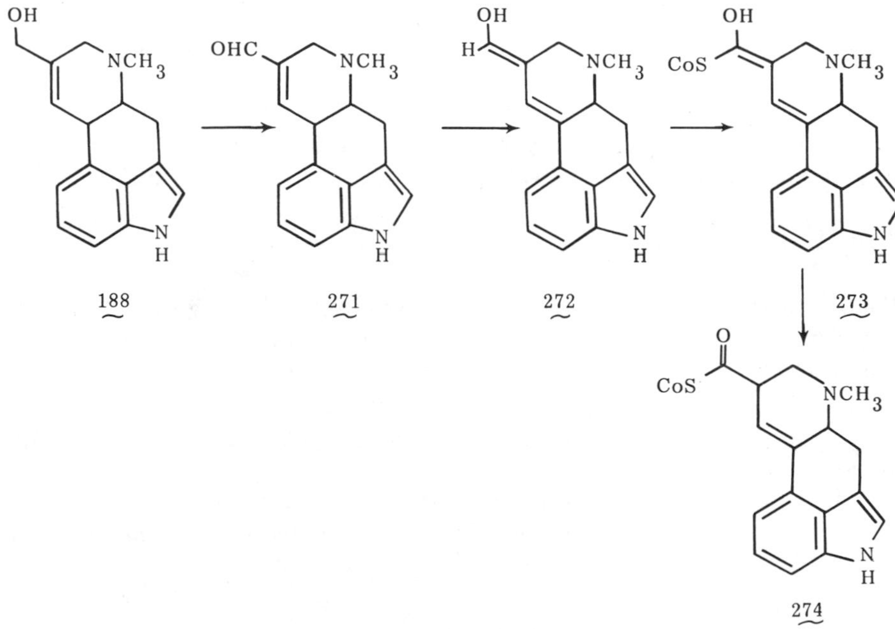

Scheme 36

oxidized to lysergic acid coenzyme A (274). This scheme has the advantage that activation of the acid group will permit more ready amide formation, which is the characteristic next step.

The formation of the lysergic amide derivatives has also been well studied, but the route is not clear as yet. L-Alanine is an effective precursor of the side chain of ergonovine (203) and may be incorporated by way of L-alaninol, although this is not at present defined.

Although an attractive biogenetic scheme involving lysergylalanine (275)

203   R=CH$_2$OH
275   R=CO$_2$H

as a key intermediate in the formation of both ergonovine (203) and ergotamine (205) has been proposed by Agurell, 275 has not yet been isolated or detected in a *Claviceps* sp.

There are similar problems involved when attention is turned to the formation of the peptide moiety of ergotamine (205) or the ergotoxine group of compounds, and comparatively little work has been done in this area.

9.10.11   Pharmacology

Pharmacologically there are three main actions of the ergot alkaloids: (*a*) peripheral, (*b*) neurohormonal, and (*c*) adrenergic blockade. The two most important peripheral effects are smooth muscle contraction, typified by vasoconstriction, and uterotonic effects. The well-known use of ergot alkaloids in obstetrics is dependent on the contractile effects on the uterine smooth muscle.

The neurohormonal effects are typified by serotonin and adrenaline antagonism, and the use of ergot alkaloids as sympathicolytic agents relies on their adrenolytic action.

Clavine Alkaloids

There are no clavine alkaloids which are presently prescription products, but a number of derivatives have interesting pharmacologic activities, and these are discussed in the section dealing with prolactin release and the ergot alkaloids. Of the parent alkaloids, agroclavine (**187**) is a potent uterine stimulant and several other clavine alkaloids exhibit ecbolic properties.

Ergonovine

Ergonovine (**203**) is a powerful uterine contractant but with low vasoconstrictor action. It is the drug of choice in the treatment of postpartum hemorrhage and acts very rapidly. The typical oral dose of the maleate salt is 0.2 mg. An alternative product of somewhat longer duration of action is methyl ergonovine (**276**), which is also used the maleate salt.

A third product in this series is methysergide, having the structure **277**.

	R_1	R_2
203 ergonovine	$-NH-\underset{\underset{CH_3}{\vert}}{CH}CH_2OH$	H
276 methylergonovine	$-NH-\underset{\underset{CH_2CH_3}{\vert}}{CH}CH_2OH$	H
277 methysergide	$-NH-\underset{\underset{CH_2CH_3}{\vert}}{CH}CH_2OH$	CH_3

Unlike the other products, this compound is used as a cranial vasodilator in the prophylactic treatment of migraine headaches. It is contraindicated in pregnancy because of the potent uterine stimulant actions it possesses.

A recent area of sustained interest has been in the many new partially synthetic derivatives which have been prepared. Some of these compounds are of substantial interest. The N',N'-diethylurea (**278**) is a potent serotonin antagonist which can be used in the treatment of migraine and hypertension. The acetyl amino ergoline **279** has specific oxytoxic activity comparable to that of ergonovine, and the piperazine derivative **280** is a potent stimulator which may be useful in the treatment of psychoses and schizophrenia.

Ergotamine Tartrate

Ergotamine tartrate is used as a specific analgesic in the treatment of migraine. The typical dose is 2 mg orally and relief is often quite dramatic. A

278 279 280

preferred treatment for longer duration is dihydroergotamine methane sulfonate, in which the 9,10-double bond is hydrogenated. Caffeine is an important adjunct in migraine therapy, where it acts to constrict cerebral blood vessels and reduce blood flow. In addition, compounds in the dihydro series have a reduced stimulating action on both smooth muscle and the uterus.

The ergotoxine alkaloids, particularly ergocornine, inhibit implantation of the ovum in female rats and mice. Although the mechanism of action is not known, such actions may open up new areas of drug development.

Lysergic Acid Diethylamide (LSD)

Lysergic acid diethylamide (LSD) is, without doubt, the most notorious and controversial ergot alkaloid derivative, and it is impossible here to even begin to discuss the legend that has grown around this simple derivative of lysergic acid.

Lysergsaurediethylamid was first synthesized in 1938 by Albert Hofmann in the laboratories of Sandoz AG in Basel. Its potent pharmacologic effects became known on Friday, April 16, 1943, when Hofmann recorded in his laboratory notebook that he "was forced to stop my work in the laboratory in the middle of the afternoon and to go home." As he lay at home in "a dazed condition" he described how "there surged upon him an uninterrupted stream of fantastic images of extraordinary plasticity and vividness and accompanied by an intense, kaledioscope-like play of colors." Hofmann was convinced that he had accidentally ingested some of the compound with which he had been working and consequently the following Monday, April 19, he prepared a solution containing 0.25 mg of LSD and deliberately ingested it. After about 40 minutes he wrote of "difficulty in concentration, visual disturbances, marked desire to laugh. . . ." As he was riding home he found that the symptoms were much stronger than previously. He reported: "I had great difficulty in speaking coherently, my field of vision swayed before me. . . . I had the impression of being unable to move from the spot. . . ." After reaching home a doctor arrived, but the symptoms continued for 6 hours, at which point Hofmann described how "all objects appeared in unpleasant, constantly changing colors, the predominant shades being sickly green and blue. . . . A remarkable feature was the manner in which all acoustic perceptions were transformed into optical effects."

Hofmann had consumed what is now regarded as five times the normal effective dose of LSD.

The first report on LSD in the scientific literature came from Zurich in 1947, and in 1953 LSD became an investigational new drug. In the years 1953–1966, considerable quantities of LSD was distributed for biochemical and animal behavior research, but in 1966 the responsibility for distribution was turned over by Sandoz to the federal government.

Absorption of LSD from the gastrointestinal tract is rapid, hence its effectiveness by the oral route. Tolerance develops rapidly and daily doses are ineffective in 3 to 4 days. LSD is a sympathomimetic agent and typical early symptoms are dilated pupils, an increase in salivation, and an elevated temperature and blood pressure.

The hallucinogenic properties of LSD are probably due to an inhibition of a basic brain stem mechanism which integrates sensory input. It apparently occupies serotonin receptor sites and potentiates noradrenaline systems.

Ergot Alkaloids and Prolactin Release

The principal clinical uses of the ergot alkaloids have been described in the various individual alkaloid discussions, and these have principally been, of the muscular, vascular, and central nervous system type. Another major area in which the ergot alkaloids have an important action is on the various processes that are controlled by the hypothalamic–hypophyseal system.

For example, ergot alkaloids inhibit lactation in animals and in humans, the latter effect having first been observed in 1676. In addition, ergot alkaloids inhibit nidation in rats (i.e. implantation of the fertilized egg). These effects have been traced to the ability of certain ergot alkaloids to inhibit the release of the hormone prolactin. But why is this so important?

Prolactin is a mammalian hormone responsible for mammary growth and milk production and is secreted by the adenohypophysis. However, the evidence available suggests that prolactin is necessary for the induction and growth of chemically-induced mammary tumors in mice. Ergocornine (208) significantly reduced the size and incidence of these lesions and, like ergocryptine (207), it is also a potent inhibitor of nidation, lactation, and prolactin release. Although the corresponding 8,10-dihydro compounds are of comparable activity, the C-8 epimers are devoid of activity.

Substitution at the 2-position by a halogen reduced toxicity and central nervous system side effects, and in particular 2-bromo-α-ergocryptine (281) has been a compound of considerable recent interest and is now commer-

cially available. This compound inhibits lactation and decreases serum prolactin levels in women but more recently has been studied in the treatment
of both male and female sexual disorders. It has been shown to enhance
sexual libido in women, restore menstrual cycles in dysmmenorrhaeic
women, increase sexual libido in men, restore spermatogenesis, and raise
testosterone levels. The remission of pituitary gland tumors has also been
observed.

The peptide moiety of the ergot alkaloids is not necessary for prolactin
inhibition to be observed. Both ergonovine (203) and methylergonovine (276)
show activity, as does agroclavine (184), but the most interesting work has
come in the area of the simple ergolines. This area has been reviewed by
Cassady and Floss.

The most important of these simple ergoline derivatives are the compounds VUFB 6605 (282) and VUFB 6683 (283) developed by Semonsky
and co-workers in Prague, and lergotrile (284) developed by Kornfeld and
Clemens at Eli Lilly and Company.

281   2-bromo-α-ergokryptine

282   VUFB 6605, R=H

284   Lergotrile, R=Cl

283   VUFB 6683

The Purdue group however established some interesting structure–activity
relationships for prolactin inhibition in this series: (a) the tetracyclic ergoline
nucleus, a basic nitrogen at C-6 with an alkyl substituent and the stereochemistry at C-5, C-8, and C-10 are critical for significant prolactin inhibition; and (b) substitution at C-2, C-7, C-9, or N-1 or at N-6 with an electron-
releasing group reduced or abolished activity.

LITERATURE

Reviews

Barger, G., *Ergot and Ergotism,* Gurney and Jackson, London, 1931.

Hofmann, A., *Die Mutterhornalkalide,* Enke Verlag, Stuttgart, 1964.

Stoll, A., and A. Hofmann, *Alkaloids NY* **8**, 726 (1965).

Stoll, A., and A. Hofmann, in S. W. Pelletier (Ed.), *Chemistry of the Alkaloids,* Van Nostrand Reinhold, New York, 1970, p. 267.

Bove, F. J., *The Story of Ergot,* Karger Verlag, Basel, 1970.

Gröger, D., *Fortschr. Chem. Forsch.* **6**, 159 (1966).

Gröger, D., S. Kadis, A. Ciegler, and S. J. Ajl (Eds.), in *Microbial Toxins,* Vol. 7, Academic, New York, 1972, p. 321.

Floss, H. G., *Tetrahedron* **32**, 873 (1976).

Synthesis

Uhle, F. C., and W. A. Jacobs, *J. Org. Chem.* **10**, 76 (1945).

Stoll, A., and J. Rutschmann, *Helv. Chim. Acta* **33**, 67 (1950).

Kornfeld, E. C., E. J. Fornefeld, G. B. Kline, M. N. Mann, D. E. Morrison, R. E. Jones, and R. B. Woodward, *J. Amer. Chem. Soc.* **78**, 3987 (1956).

Julia, M., F. Le Goffic, J. Igolen, and M. Baillarge, *Tetrahedron Lett.* 1569 (1969).

Armstrong, V. M., S. Coulton, and R. Ramage, *Tetrahedron Lett.* 4311 (1976).

Crider, A. M., J. M. Robinson, H. G. Floss, and J. M. Cassady, *J. Med. Chem.* **20**, 1473 (1977).

Bach, N. J., and E. C. Kornfeld, *Tetrahedron Lett.* 3225 (1974).

Pleininger, H., W. Lehnert, D. Mangold, D. Schmalz, A. Volkl, and J. Westphal, *Tetrahedron Lett.* 1827 (1975).

Spectral Data

Bach, N. J., H. E. Boaz, E. C. Kornfeld, C.-J. Chang, H. G. Floss, E. W. Hagaman, and E. Wenkert, *J. Org. Chem.* **39**, 1272 (1974).

Biosynthesis

Ramstad, E., *Lloydia* **31**, 327 (1968).

Thomas, R., and R. A. Bassett, *Progr. Phytochem.* **3**, 46 (1972).

9.11 MONOTERPENOID-DERIVED INDOLE ALKALOIDS

This is the major group of indole alkaloids and numbers about 1000 members of widely differing structure and pharmacologic activity. Their common thread, as we shall see in the section on biosynthesis, is the derivation from a single precursor derived by the joining of an amino acid, tryptophan, and a monoterpenoid, secologanin. The variety of structures in this group appears to be unbounded and space permits a discussion of only some of these. More complete reviews of structures are to be found in the works of Hesse, and Gabetta and Mustich, and of chemistry in the Manske *Alkaloids* and Chemical Society Specialist Periodical Reports Series.

The alkaloids of principal therapeutic importance are shown in Table 3, together with their sources.

Table 3 Some Monoterpenoid Indole Alkaloids of Pharmacologic Significance

Compound	Source	Pharmacologic Activity
Ajmalicine	*Catharanthus roseus*	Hypotensive
Ajmaline	*Rauvolfia* sp.	Antiarrythmic
Camptothecine	*Camptotheca acuminata*	Anticancer
Deserpidine	*Rauvolfia canescens*	Hypotensive
Ibogaine	*Tabernanthe iboga*	Psychotomimetic
Leurocristine	*Catharanthus roseus*	Anticancer
Quinidine	*Remijia* sp.	Cardiac depressant
Quinine	*Cinchona* sp.	Antimalarial
Rescinnamine	*Rauvolfia* sp.	Hypotensive
Reserpine	*Rauvolfia vomitoria*	Hypotensive
Strychnine	*Strychnos nux vomica*	CNS depressant
Vincaleukoblastine	*Catharanthus roseus*	Anticancer
Vincamine	*Vinca major*	Hypotensive

9.11.1 Occurrence of Indole Alkaloids in Higher Plants

Simple indole alkaloids are of quite wide distribution (at least 35 plant families). In contrast, the more complex indole alkaloids are essentially limited in their distribution to the Apocynaceae (principally), the Rubiaceae, and the Loganiaceae, with only isolated examples of their occurrence in the families Alangiaceae, Annonaceae, and Nyssaceae. This is in marked contrast to the benzylisoquinoline alkaloids, whose complex members are considerably more widespread.

Not all members of the Apocynaceae contain indole alkaloids however; some, such as *Holarrhena* and *Funtumia,* are noted for their steroidal alkaloids, and others, such as *Strophanthus,* contain cardiac glycosides. There is some evidence that these compound classes are mutually exculsive in the Apocynaceae.

9.11.2 Biogenetic Classification of Indole Alkaloids

Unlike the organization of any other alkaloid group, the biogenesis of the monoterpenoid indole alkaloids will be mentioned before the individual groups of alkaloids are discussed. A fuller discussion of aspects of the biogenesis and biosynthesis of these fascinating and complex alkaloids follows in a subsequent section.

The fundamental building blocks for the monoterpene indole alkaloids are tryptamine (tryptophan) and the iridoid, secologanin (**285**). The units combine to form (*in vivo*) strictosidine (**286**), the nitrogenous glycoside which is the key intermediate in the biosynthetic elaboration which subsequently occurs.

285   secologanin

286   strictosidine

In 1971, Kompis, Hesse, and Schmid discussed an interesting new approach to the classification of indole alkaloids based on the established biosynthesis. In this way the alkaloids were divided into five classes and within each class several subclasses were developed.

The main skeleta are subdivided as shown in Table 4, which is an ab-

Table 4   Biogenetic Classification of Indole Alkaloids

Class 1 Alkaloids containing a nonrearranged secologanin skeleton

1.1	Strictosidine group	1.11	Picraline group
1.2	Corynantheine group	1.12	Corymine group
1.3	Vallesiachotamine group	1.13	Ajamline group
1.4	Adifoline group	1.14	Perakine group
1.5	Talbotine group	1.15	Oxindole group
1.6	Stemmadenine group	1.16	Pseudoxindole group
1.7	Mavacurine group	1.17	Condylocarpine group
1.8	Cinchonamine group	1.18	Akuammicine group
1.9	Sarpagine group	1.19	Strychnine group
1.10	Peraksine group		

Class 2 Alkaloids containing an opened secologanin skeleton

2.1	Secodine group

Class 3 Alkaloids containing a rearranged secologanin skeleton

3.1	Quebrachamine group	3.6	Vindolinine group
3.2	Aspidospermine group	3.7	Pleiocarpine group
3.3	Schizophylline group	3.8	Kopsine group
3.4	Vincamine group	3.9	Oxindole group
3.5	Schizozygine group		

Class 4 Alkaloids containing a rearranged secologanin skeleton

4.1	Fruticosine group

Class 5 Alkaloids containing a rearranged secologanin skeleton

5.1	Catharanthine group
5.2	Pseudooxindole alkaloids

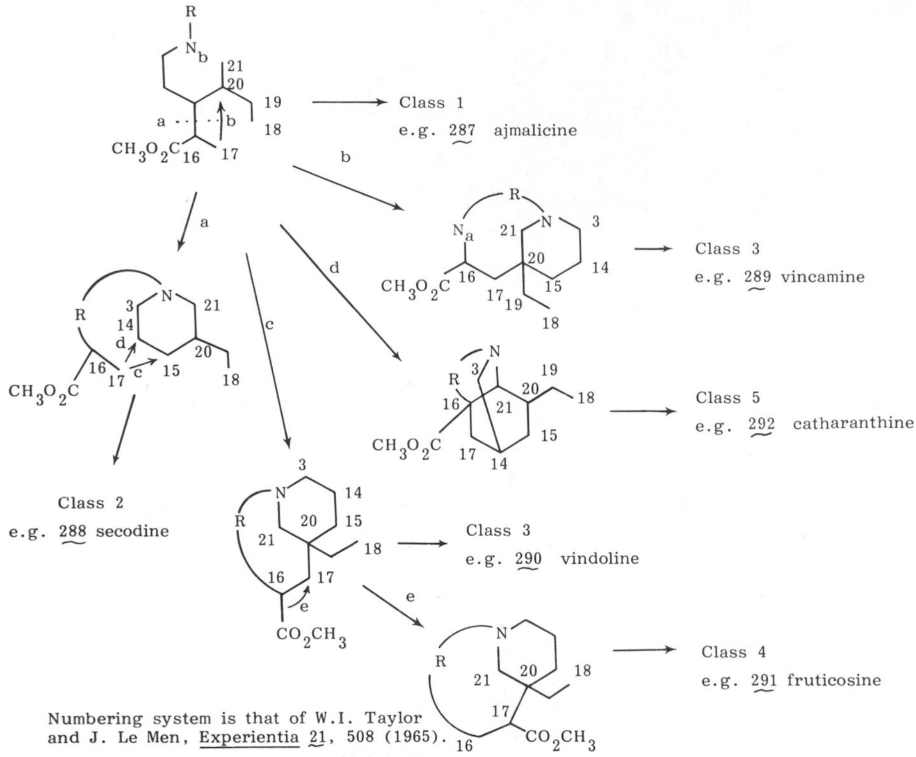

Numbering system is that of W.I. Taylor and J. Le Men, *Experientia* 21, 508 (1965).

breviated version of that developed by the Swiss group. The rationale for this elaborate classification is shown in Scheme 37. Note that the numbering system of the terpenoid indole alkaloids is internally consistent, thus carbon 3 in one skeleton is the same carbon, numbered 3, in a quite different skeleton. It is immaterial that a myriad of complex rearrangements may have taken place in converting one skeleton to another, or that their formation may be from a third, different skeleton. This system was devised by Le Men and Taylor and is used by almost all workers in the field. The usefulness of this approach will become apparent as more skeleta are discussed.

9.12 NITROGENOUS GLYCOSIDES AND RELATED COMPOUNDS

Without doubt the most significant indole alkaloids to have been isolated in recent years are the nitrogenous glycosides, in particular strictosidine (**286**), the key intermediate in the biosynthesis of all indole alkaloids.

9.12.1 Isolation and Chemistry

Strictosidine was not the first nitrogenous glycoside to be obtained; cordifoline (**293**) and deoxycordifoline (**294**) from *Adina cordifolia* Benth & Hook.

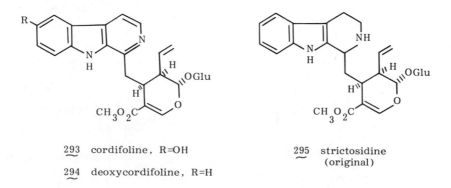

Scheme 37 Biogenetic classification of indole alkaloids. [After I. Kompis *et al.*, *Lloydia* **34**, 269 (1971).]

must claim that distinction. However, the isolation of strictosidine (**295**) by Smith from *Rhazya stricta* Decne. and *R. orientalis* A. DC. as an unstable amorphous gum began what continues to be a most exciting period of indole alkaloid chemistry, and further discussion is to be found in the section on the biosynthesis of indole alkaloids. There is only limited discussion here on the controversies surrounding this area, which have recently been clarified.

Shortly after Smith obtained strictosidine (**295**), Battersby obtained two nitrogenous glycosides, vincoside (**296**) and isovincoside (**297**), from *Catharanthus roseus* G. Don, and also succeeded in producing these alkaloids *in vitro* by the condensation of tryptamine (**2**) and secologanin (**285**) at pH 6.2. Clearly one of these compounds (**296** or **297**) should be identical with strictosidine, and this was found to be isovincoside. Three groups deduced the C-3 stereochemistry of vincoside (**298**) and strictosidine (**286**) and there is now no question as to the veracity of these assignments, particularly since the absolute configuration has recently been confirmed by X-ray analysis.

These compounds are quite unstable, but in characteristically different ways. Vincoside (**298**) decomposes on evaporation of a methanol solution to yield vincosamide (**299**), whereas strictosidine (**286**) prefers to decompose to vallesiachotamine (**300**). The latter alkaloid was first obtained by Walser and Djerassi, but its significance was not realized for several years. Strictosidine (**286**) will also undergo decomposition to strictosamide (**301**), but vincoside (**298**) does not yield the 3-epivallesiachotamines (**302**). Since the early isolation work several further interesting nitrogenous glycosides have been obtained, and this area has been reviewed by Brown and Kapil. Some of these compounds include 5α-carboxystrictosidine (**303**) from *R. orientalis*,

original assignment

correct assignment

vincoside   **296**   3α – H   **298**   3β – H

isovincoside **297**   3β – H   **286**   3α – H
                                              strictosidine

299   vincosamide, 3β – H   **300**   vallesiachotamines, 3α – H

301   strictosamide, 3α – H   **302**   3β – H

5α-carboxystrictosamide (**304**) and 5α-carboxyvincosamide (**305**) from *Adina rubescens,* and rubenine (**306**) from the same plant.

Not all of the novel alkaloids in this series are glycosides, as recent isolation work has shown. *Anthocephalus cadamba* Miq. for example has

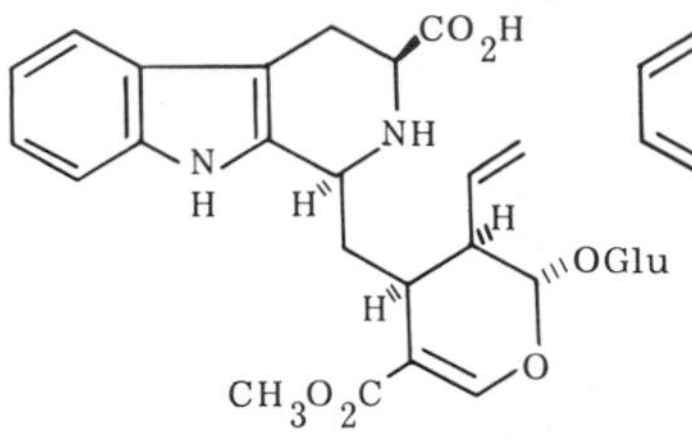

303	304 3α - H
	305 3β - H
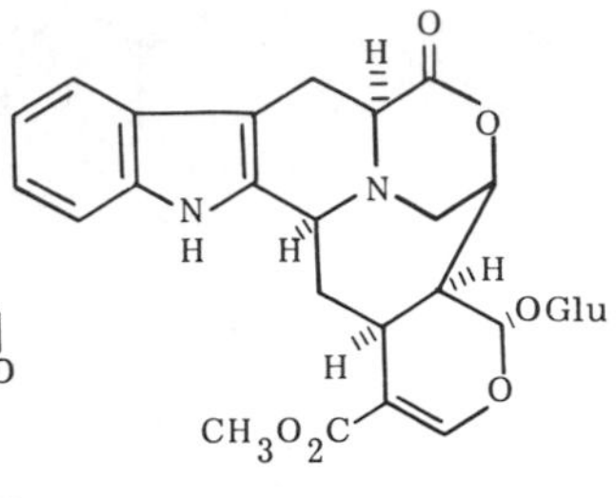	306 rubenine

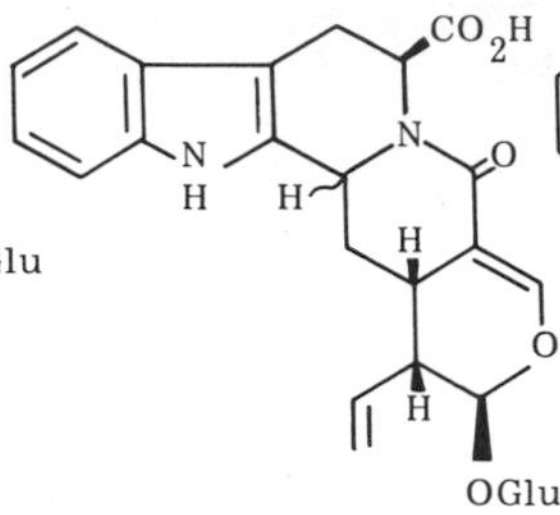

afforded cadamine (**307**), and naufoline (**308**) has been obtained from *Nau-clea latifolia* Sm.; *Pauridiantha lyalli* (Baker) Bremek. has yielded several alkaloids, of which lyadine (**309**) is just one example; and *Alstonia constricta* F. Muell. has yielded alstonilidine (**310**). All of these alkaloids are relatively simple derivatives of the strictosidine system.

Comparison of the ^{13}C NMR data for the acetylated vincoside and stric-tosidine shows somewhat surprisingly that C-3 epimerization has little effect on C-6 and C-15. Thus C-6 is shifted less than expected because of the

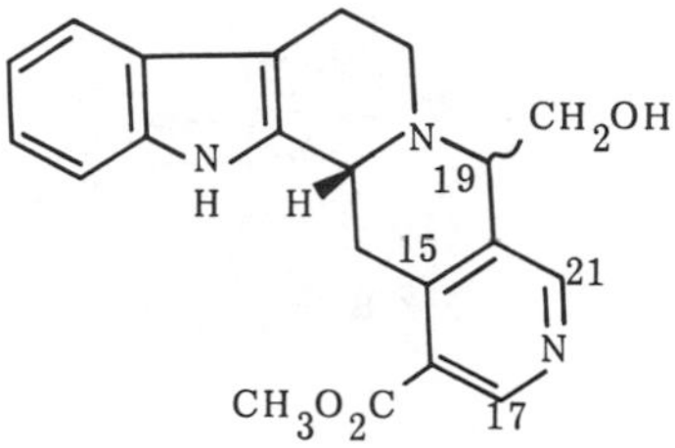

307 cadamine

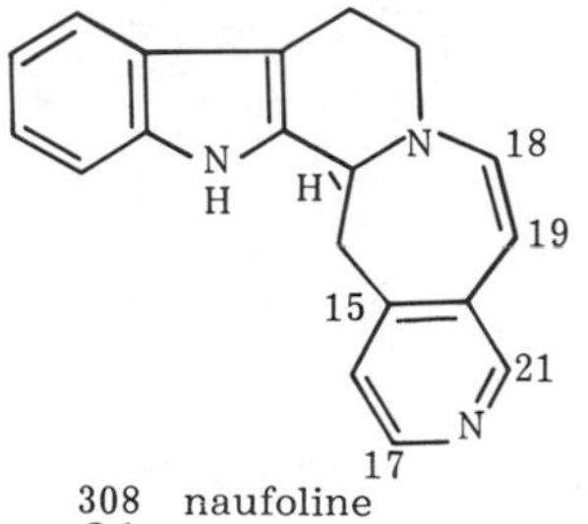

308 naufoline

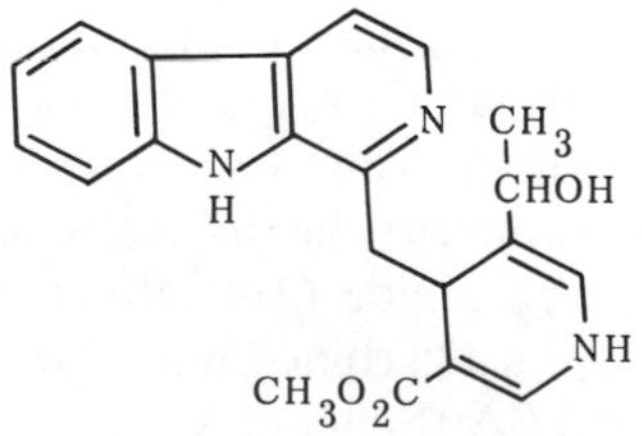

309 lyadine

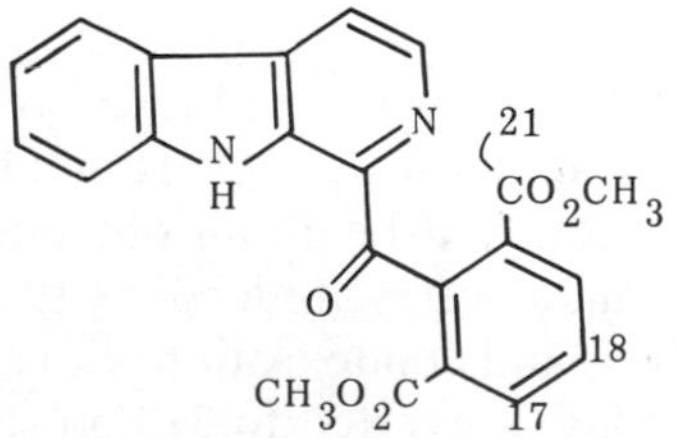

310 alstonilidine

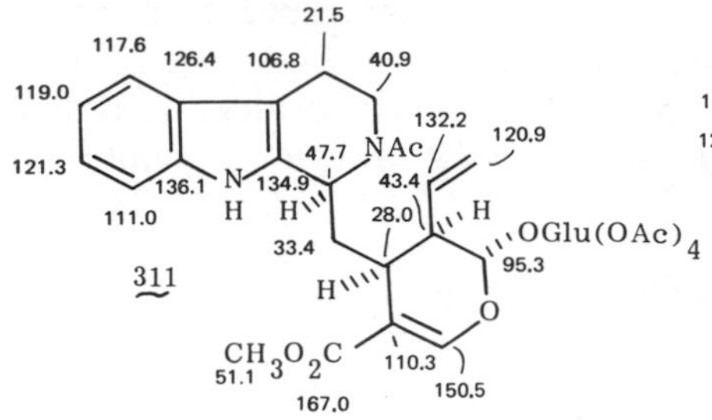

sp^2 nature of C-21, which reduces nonbonded interactions on H-6 with a C/D *cis* conformation. These and other comparisons with model compounds indicated that the C-3α H isomer strictosamide tetraacetate (**301**) prefers the *cis*-quinolizidone structure, whereas vincosamide tetraacetate (**302**) prefers the *trans*-quinolizidone conformation.

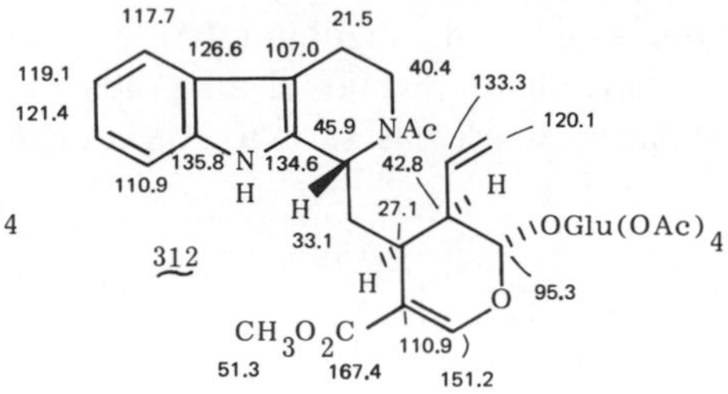

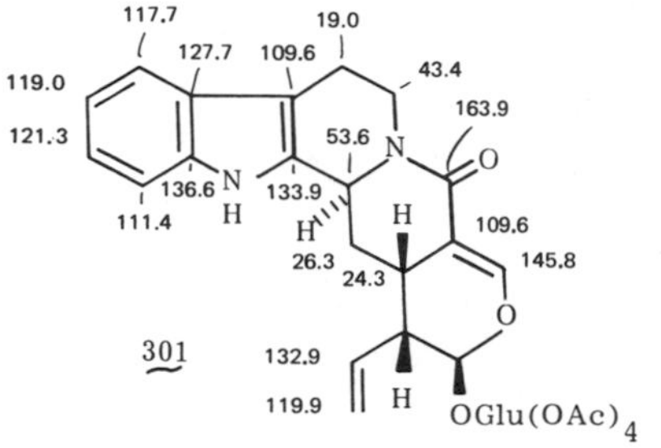

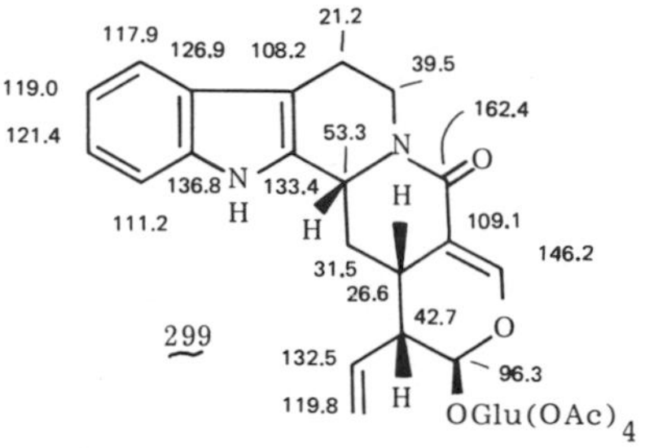

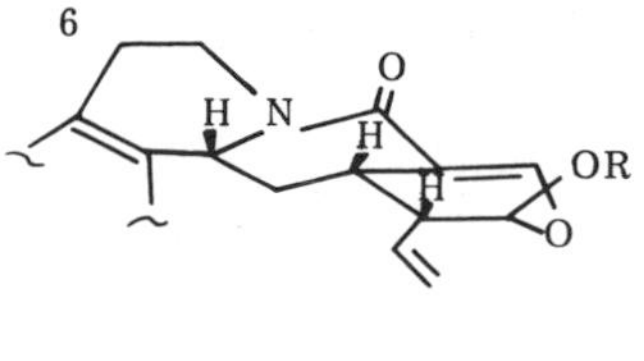

C-3βH   <u>trans</u>   C/D 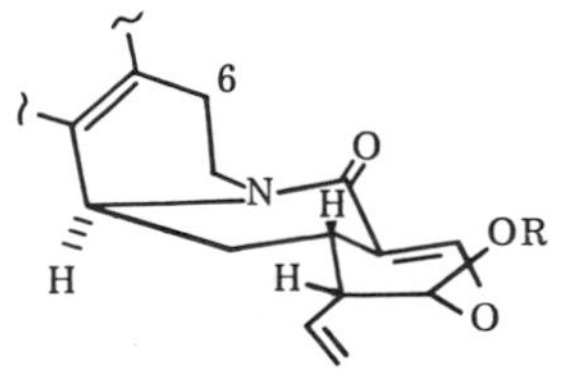   C-3αH   <u>cis</u>   C/D

9.12.2 Biomimetic Chemistry

The availability of the nitrogenous glycosides has led to some interesting biomimetic chemistry of these systems, mainly by Brown and co-workers.

Dihydroangustine (**313**) is an alkaloid obtained from *Strychnos angustiflora* Benth. and bears an obvious overall similarity to the strictosamide/vincosamide series. Indeed 18,19-dihydrovincosamide (**314**) after enzymic hydrolysis and condensation with ammonia gave a carbinolamine **315**, which was readily converted to **313** on standing in TFA (Scheme 38).

The enigma of why the C-3β isomer of the nitrogenous glycosides [i.e. vincoside (**298**)] was a precursor of the C-3α Corynanthe alkaloids was

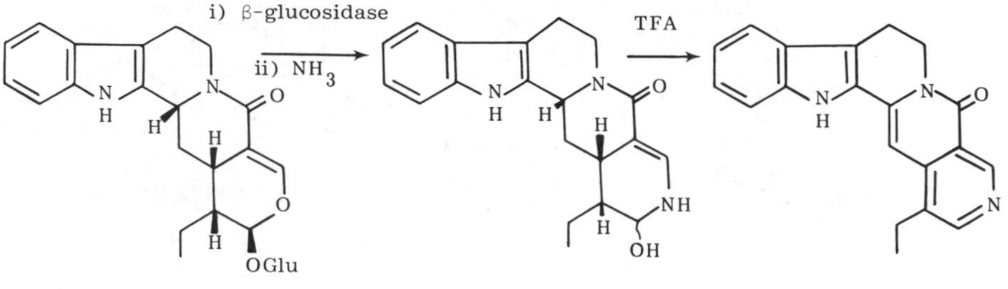

314   315   313   dihydroangustine

Scheme 38

rationalized by Brown in terms of a novel intermediate dihydromancunine (**316**). This intermediate was simply produced by exposing dihydrovincoside (**317**) to β-glucosidase at pH 5. Dihydromancunine (**316**) could be easily transformed to a dihydroisositsirikine (**318**) (Scheme 39). Another synthetic scheme beginning with *N*-benzyldihydrovincoside (**319**) gave dihydrocorynantheine (**320**).

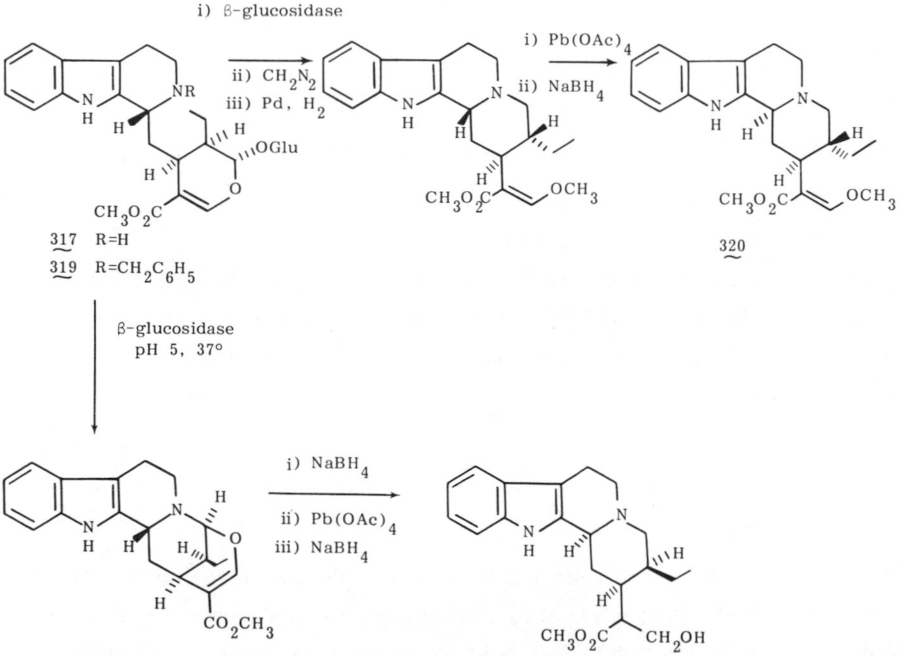

317   R=H
319   R=CH$_2$C$_6$H$_5$

320

316   318   a dihydroisositsirikine

Scheme 39

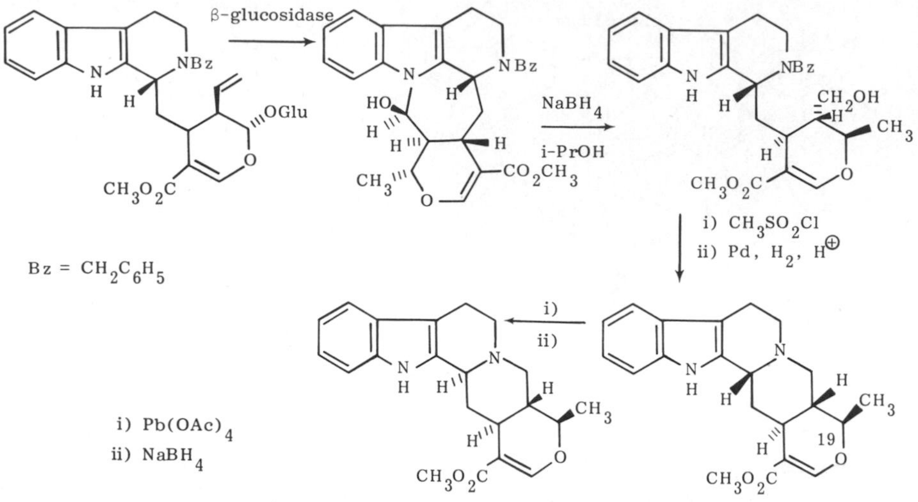

321 19-epiajmalicine 3-iso-19-epiajmalicine

Scheme 40 Biogenetic synthesis of 19-epiajmalicine (**321**).

One biomimetic synthesis worthy of mention is that of 19-epiajmalicine (**321**) (Scheme 40).

LITERATURE

Review

Brown, R. T., and R. S. Kapil, *Alkaloids NY* **17**, 546 (1979).

Structure Elucidation

Smith, G. N., *Chem. Commun.* 912 (1968).

Battersby, A. R., A. R. Burnett, and P. G. Parsons, *J. Chem. Soc., Sec. C* 1193 (1969).

DeSilva, K. T. D., G. N. Smith, and K. E. H. Warren, *Chem. Commun.* 905 (1971).

Blackstock, W. P., R. T. Brown, and G. K. Lee, *Chem. Commun.* 910 (1971).

Kennard, O., P. J. Roberts, N. W. Isaacs, F. H. Allen, W. D. S. Motherwell, K. H. Gibson, and A. R. Battersby, *Chem. Commun.* 899 (1971).

Mattes, K. C., C. R. Hutchinson, J. P. Springer, and J. Clardy, *J. Amer. Chem. Soc.* **97**, 6270 (1975).

Biomimetic Chemistry

Brown, R. T., A. A. Charalambides, and H. T. Cheung, *Tetrahedron Lett.* 4837 (1973).

Brown, R. T., C. L. Chapple, and A. A. Charalambides, *Chem. Commun.* 756 (1974).

Brown, R. T., C. L. Chapple, R. Platt, and S. K. Sleigh, *Chem. Commun.* 1829 (1976).

Brown, R. T., and C. L. Chapple, *Chem. Commun.* 740 (1974).

9.13 CAMPTOTHECINE

The rare Chinese ornamental tree *Camptotheca acuminata* Decsne, also known as "tree of joy" and "tree of love," is the only member of the genus *Camptotheca* in the family Nyssaceae.

No folklore has been reported for this plant, but in 1958 an extract of the leaves was found to exhibit very high antitumor activity in the L-1210 system, and subsequently extracts of the fruits, twig, stem bark, and root bark also exhibited activity. These samples were obtained mainly from southern California, where the plant was introduced in 1927.

9.13.1 Isolation and Spectral Properties

Fractionation for the principle responsible for this activity led Wall and co-workers to camptothecine (**322**), a pale yellow crystalline compound, whose structure was deduced by X-ray analysis of the iodoacetate derivative. As expected the aromatic system is planar and the lactone ring occupies a boat conformation.

The UV spectrum showed λ_{max} 220, 254, 290, and 370 nm and IR spectrum indicated the presence of hydroxyl (3440 cm^{-1}), δ-lactone (1760–1745 cm^{-1}), lactam (1660 cm^{-1}), and aromatic (1610 and 1588 cm^{-1}) groups. Some of proton NMR data for camptothecine are shown in **323** and the ^{13}C NMR data in **324**.

Camptothecine does not form stable salts with acids and the lactone ring is cleaved reversibly on treatment with base.

Camptotheca acuminata has also afforded 10-hydroxycamptothecine (**325**), 10-methoxycamptothecine (**326**), and 9-methoxycamptothecine (**327**) has been obtained from *Mappia foetida* Miers. (Icacinaceae). Camptothecine (**322**) is obtained only in very low (0.005%) yield from *C. acuminata* and a better source (up to 1%) was found to be *M. foetida*. More recently the alkaloid was obtained together with 10-methoxycamptothecine (**326**) as an antiviral constituent of *Ophiorrhiza mungos* L. (Rubiaceae) and also from *Ervatamia heyneana* (Wall.) T. Cooke (Apocynaceae).

9.13.2 Synthesis

The importance of camptothecine (**322**) as an anticancer agent and its scarcity from natural sources led to extensive investigations concerning the synthesis of camptothecine itself as well as of various derivatives. Indeed these efforts have resulted in no less than 11 distinct total syntheses, to make camptothecine the most synthesized of the complex alkaloids. Clearly space does not permit a discussion of these efforts in any great detail and the reviews of Schultz, and Shamma and St. Georgiev, provide excellent discussions of the early efforts.

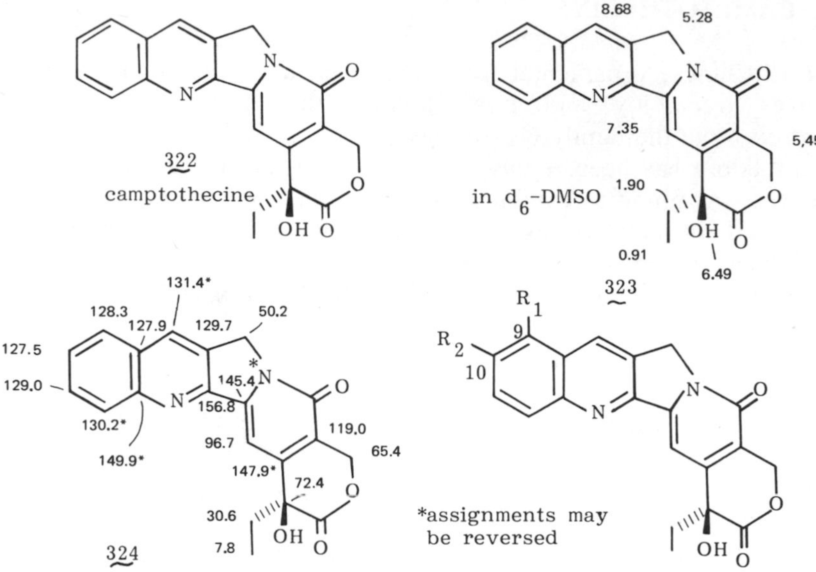

322
camptothecine

in d$_6$-DMSO

323

324

*assignments may
be reversed

data from Hutchinson et. al.
J.Amer.Chem.Soc. 97, 5609 (1974)

325   R$_1$=H; R$_2$=OH
326   R$_1$=H; R$_2$=OCH$_3$
327   R$_1$=OCH$_3$; R$_2$=H

The first synthesis of camptothecine was due to Stork and Schultz, and some of the steps are shown in Scheme 41. A key step is the annelation of ring E with the anion of the carbonate ester **328** on the unsaturated lactam **329**. This provides a compound possessing all the carbon atoms of **322**, without the C-20 oxygen, requiring a series of reductions and oxidations to reach camptothecine (**322**).

The Winterfeldt synthesis of camptothecine (**322**), the biogenetic aspects of which are discussed subsequently, inserts the ethyl group and the 20-hydroxy groups as late stages in the biosynthesis. Thus this synthesis begins with compounds in the indole series that can be oxidatively rearranged to the quinolone with potassium *t*-butoxide in the presence of oxygen. DIBAL selectively reduces the unsaturated ethyl ester of **330** to give an alcohol diester **331**, which spontaneously lactonized and decarboxylated on treatment with trifluoroacetic acid to afford **332** (Scheme 42).

The most effective synthesis of camptothecine (**322**) is one developed by Bradley and Büchi (Scheme 43) and involves as a key step the condensations of the pyrrolo[3,4-*b*]quinoline **333** with the acids **334** to afford **335**. Cyclization to the pyridone **336** was achieved by successive treatment with boron trifluoride etherate and heating in toluene in the presence of a trace of trifluoroacetic acid. This compound had previously been converted by Dan-

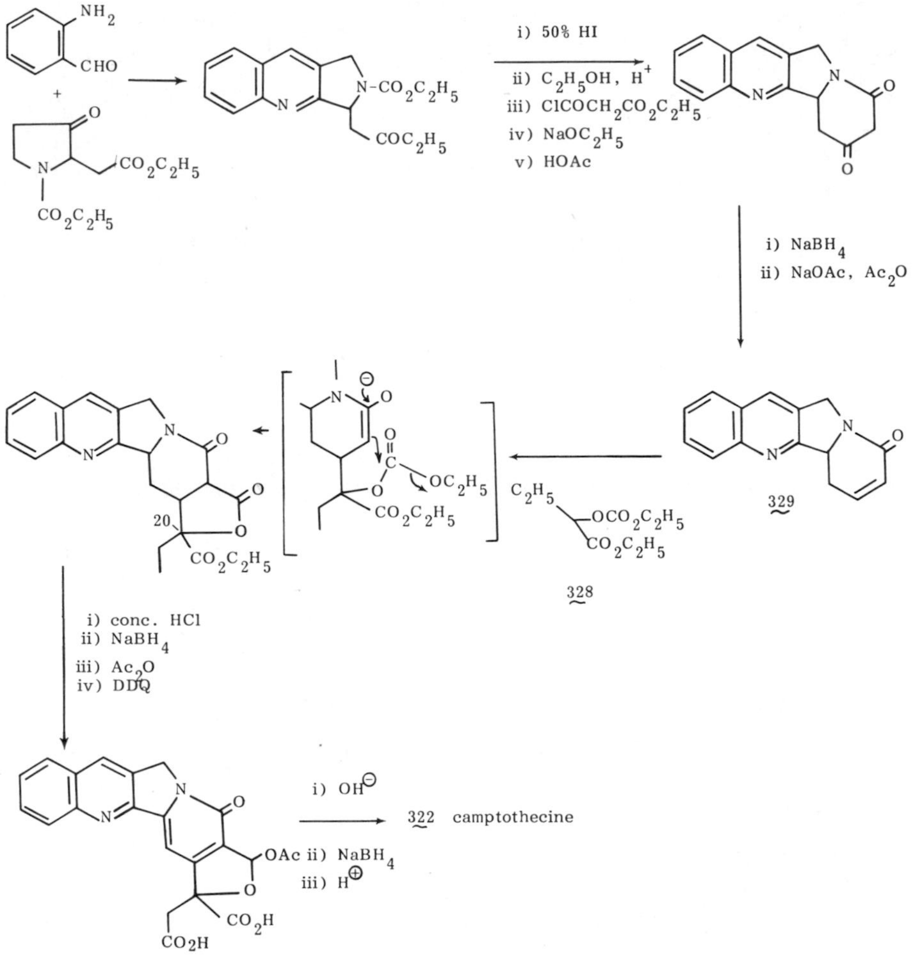

Scheme 41   Stork–Schultz synthesis of camptothecine (**322**).

ishefsky and co-workers to 20-desethyl-20-desoxycamptothecine (**332**) by condensation with paraformaldehyde.

9.13.3   Biogenesis

Biogenetically, camptothecine (**322**) was regarded by Wenkert as being formed from an alkaloid such as isositsirikine (**337**), and in particular a mechanism was suggested for the rearrangement to the pyrrolo[3,4-*b*]quinoline system. This rearrangement, which bears some similarity to the formation of the quinoline ring of quinine, has been investigated by several groups from a synthetic point of view.

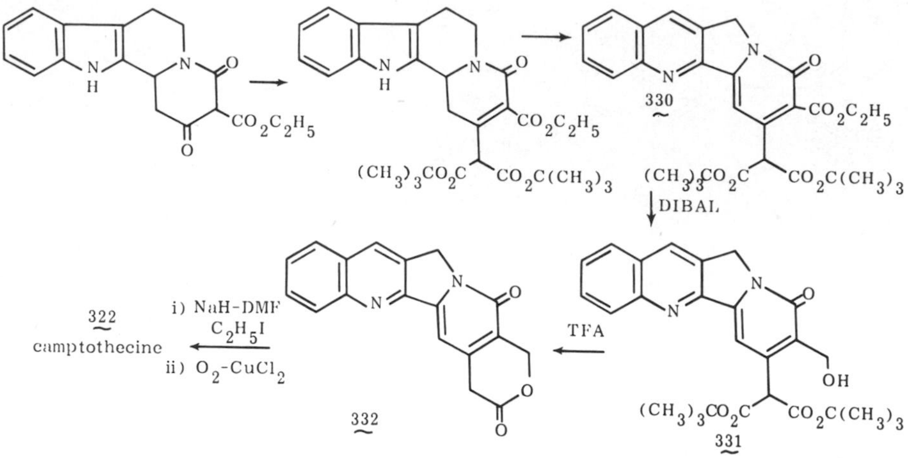

Scheme 42 Winterfeldt synthesis of camptothecine (**322**).

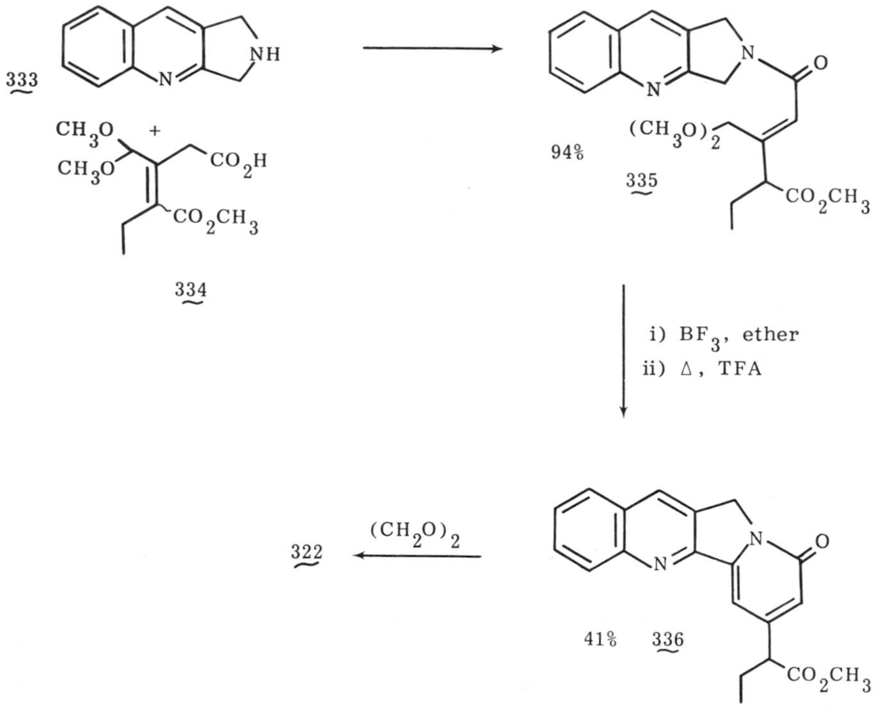

Scheme 43 Bradley–Büchi synthesis of camptothecine (**322**).

Winterfeldt and co-workers for example in their total synthesis found that autoxidation of **338** in DMF in the presence of potassium *t*-butoxide gave the quinolone **339** in high yield. This compound could be chlorinated and oxidized to the pyridone **340**.

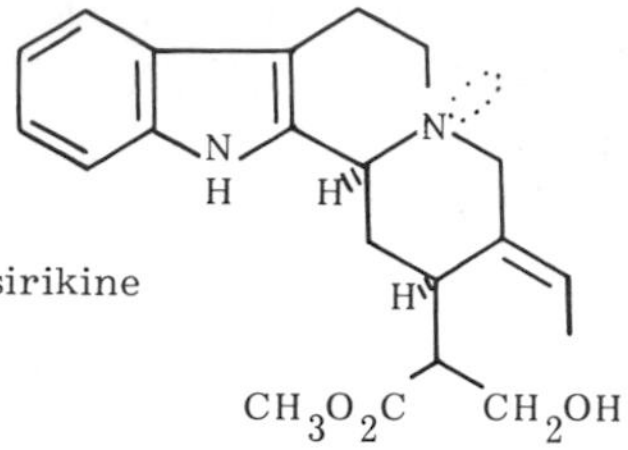

337   isositsirikine

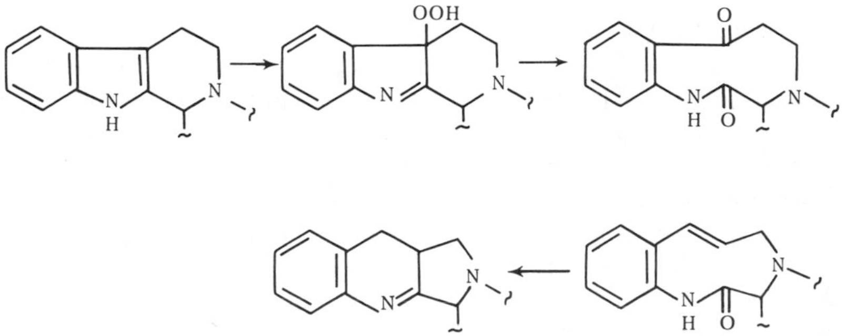

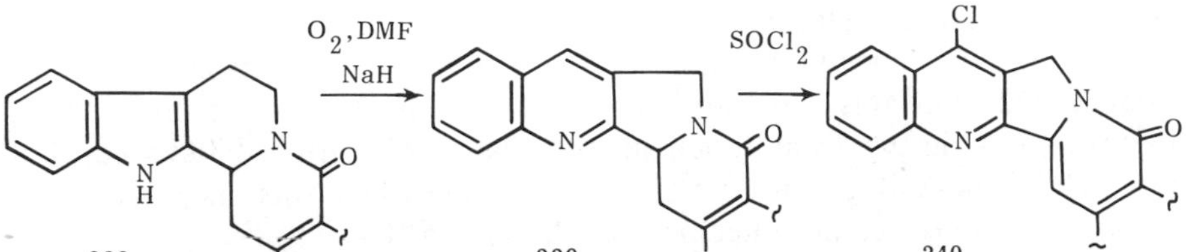

338   339   340

The biogenesis of camptothecine became clearer after Smith and co-workers had deduced the correct structure of strictosamide (**301**) and vincosamide (**299**). It occurred to both Hutchinson and the author that such a structure bore an exceptionally close skeletal resemblance to the camptothecine nucleus except for the B/C ring system. Hutchinson subsequently went on to investigate this idea both *in vitro* and *in vivo*.

For example, periodate oxidation of 18,19-dihydrostrictosamide tetraacetate (**341**) followed by cyclization of the intermediate keto amide with triethylamine gave the quinolone **342**. The corresponding 3β-isomer **343** reacted more slowly and incompletely.

The pyridone ring could be introduced into either **341** or **342** by reaction with DDQ, the former giving **344**.

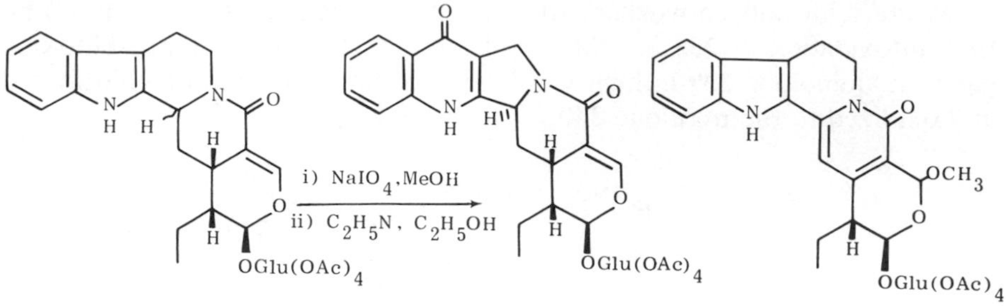

i) NaIO$_4$.MeOH

ii) C$_2$H$_5$N, C$_2$H$_5$OH

341   3α–H

343   3β–H

342

344

9.13.4   Biosynthesis

Some *in vitro* biogenetic work concerning the rearrangements of compounds in the vincoside lactam–strictosamide series was described in the preceding section. But what is the *in vivo* evidence for such a scheme?

Biosynthetic data on even the most simple precursors were not forthcoming until 1974, when Hutchinson and co-workers reported that tryptophan (**1**), mevalonic acid (**93**), and secologanin (**285**) were incorporated at low levels into camptothecine (**322**). An improved incorporation was obtained using a mixture of vincoside (**298**) and strictosidine (**286**). But the best precursors were the lactams corresponding to the 18,19-dihydro series. Thus strictosamide (**301**) was incorporated to the extent of 2.0% and 18,19-dihydrostrictosamide (**345**) was incorporated to the extent of 4.7%. The corresponding 3R-isomer vincosamide (**299**) was a poor precursor of camptothecine (**322**), and this was the first published indication that a nitrogenous glycoside having the 3S stereochemistry was crucial in the biosynthesis of any indole alkaloid group. [5-^{13}C]Strictosamide (**301**) was specifically incorporated to give camptothecine (**322**) labeled at C-5.

In confirmation that further reaction takes place through the lactam in-

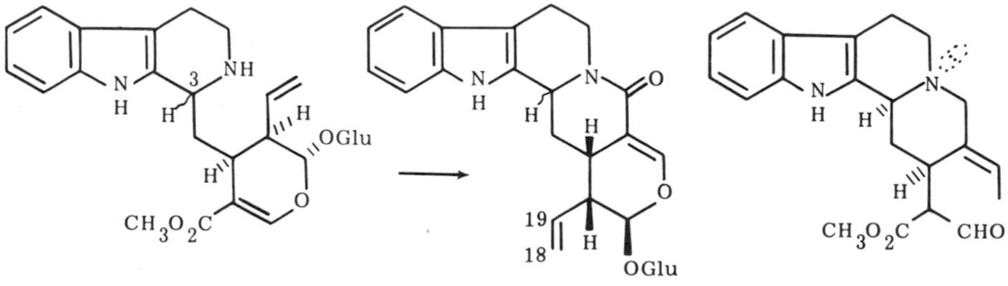

286   3α – H

298   3β – H

299   3β – H

301   3α – H

345   3α – H, 18, 19-dihydro-

346   geissoschizine

termediates rather than the more usual rearrangement to the Corynanthe alkaloids, geissoschizine (**346**) was found to be a poor precursor of camptothecine (**322**).

Rapoport and Sheriha have been more interested in the early amino acid and terpenoid precursors and have established tryptophan, tryptamine, mevalonic acid, and geraniol as moderate precursors.

At this point no precise scheme for the formation of camptothecine (**322**) can be written, but the good incorporation of **345** suggests that 18,19-hydrogenation is an early biosynthetic step. The remainder of the pathway may follow that shown in Scheme 44.

9.13.5 Pharmacology

Camptothecine is a potent anticancer agent in rodents. It shows exceptional activity in the L-1210 lymphocytic leukemia test system in mice, a test which is highly discriminating between the many natural compounds that show activity in P-388 test system.

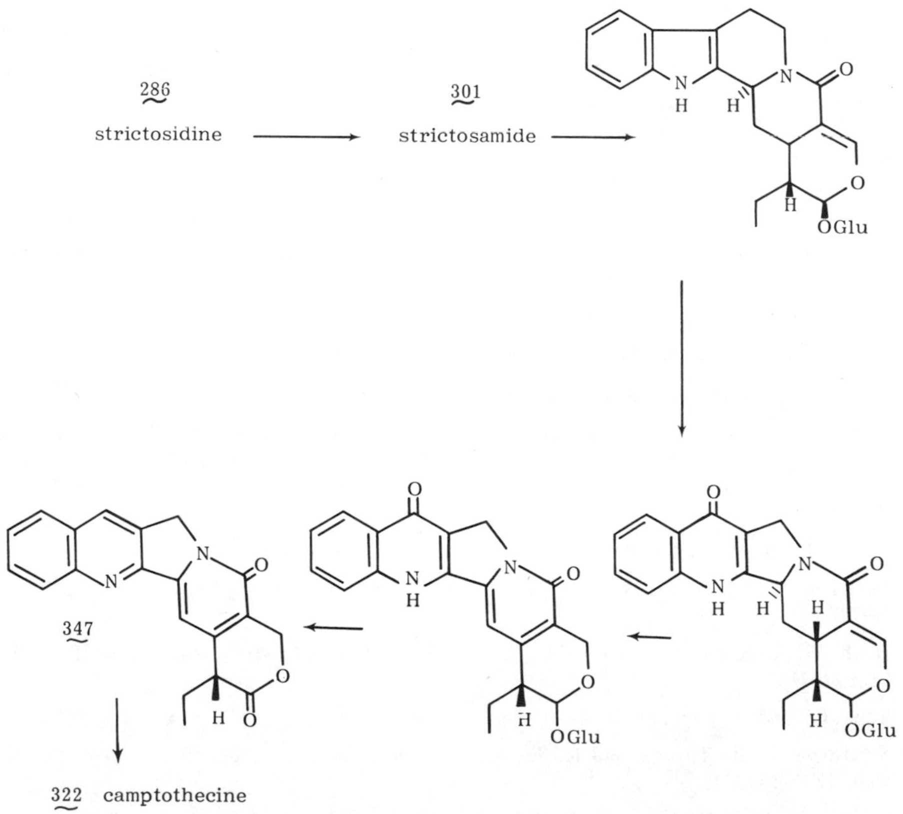

Scheme 44 Biogenesis of camptothecine (**322**).

In a preliminary clinical evaluation of camptothecine some positive responses were noted, particularly in patients with advanced gastrointestinal carcinoma. Additional clinical trials demonstrated that hematopoietic depression, diarrhea, alopecia, and cystitis were among the undesired side effects, and little clinical benefit was observed. Surprisingly, a more extensive evaluation has not been carried out, especially since the compound is in clinical use in the People's Republic of China.

Camptothecine is a potent cytotoxic agent and acts by inhibiting nucleic acid synthesis in HeLa and L-1210 cells. Thus DNA and RNA (particularly ribosomal RNA) synthesis are inhibited 50% by 5 μM camptothecine, although protein synthesis is unaffected up to 100 μM. The inhibition of nucleic acid synthesis is reversible.

Total cellular DNA in HeLa cells remains unchanged after treatment with camptothecine, but single strand breaks are induced and the average molecular weight is lower. The drug has no effect on nucleic acid synthesis in rat liver, brain mitochondria, or *E. coli*.

Camptothecine inhibits replication of vaccinia virus and cleaves intracellular viral DNA. It is not effective on poliovirus, where RNA is the template for nucleic acid synthesis.

20-Deoxycamptothecine (**347**) and 10-methoxycamptothecine (**326**) are almost as effective as **322** in inhibiting RNA synthesis in HeLa cells, and as expected both are potent anticancer agents *in vivo*.

LITERATURE

Reviews

Perdue, R. E., Jr., R. L. Smith, M. E. Wall, J. L. Hartwell, and B. J. Abbott, Tech. Bull. No. 1415, Agricultural Research Service, U.S. Department of Agriculture, 1970.

Schultz, A, G., *Chem. Revs*. **73**, 385 (1973).

Shamma, M., and V. St. Georgiev, *J. Pharm. Sci.* **63**, 163 (1974).

Danieli, B., and G. Palmisano, *Fitoterapia* **45**, 87 (101).

Horwitz, S. B., in J. W. Corcoran and F. E. Hahn, (Eds.), *Antibiotics,* Vol. 2, *Mechanism of Action of Antimicrobial and Antitumor Agents,* Springer-Verlag, New York, 1975, p. 48.

Synthesis

Stork, G., and A. G. Schultz, *J. Amer. Chem. Soc.* **93**, 4074 (1971).

Volkman, R., S. Danishefsky, J. Eggler, and D. M. Solomon, *J. Amer. Chem. Soc.* **94**, 3631 (1972).

Boch, M., T. Korth, J. M. Nelke, D. Pike, H. Radunz, and E. Winterfeldt, *Chem. Ber.* **105**, 2126 (1972).

Tang, C., and H. Rapoport, *J. Amer. Soc.* **94**, 8615 (1972).

Sugasawa, T., T. Toyoda, and K. Sasakura, *Tetrahedron Lett.* 5109 (1972); *Chem. Pharm. Bull.* **22**, 771 (1974).

Meyers, A. I., R. L. Nolen, E. W. Collington, T. A. Narwid, and R. C. Strickland, *J. Org. Chem.* **38**, 1974 (1973).

Kende, A. S., T. J. Bentley, R. W. Draper, J. K. Jenkins, M. Joyebix, and J. Kubo, *Tetrahedron Lett.* 1307 (1973).

Shamma, M., D. A. Smithers, and V. St. Georgiev, *Tetrahedron*, **29**, 1949 (1973).

Wani, M. C., H. F. Campbell, G. A Brine, J. A. Kepler, M. E. Wall, and S. G. Levin, *J. Amer. Chem. Soc.* **94**, 3531 (1972).

Richman, J. E., *Diss. Abstr. Int. B.* **36**, 243 (1975).

Bradley, J. C., and G. H. Büchi, *J. Org. Chem.* **41**, 699 (1976).

Shanghai Fifth Pharmaceutical Plant, *K'o Hsueh T'ung Pao* **21**, 40 (1976); *Chem. Abstr.* **84**, 122100n (1976).

Biosynthesis

Hutchinson, C. R., A. H. Heckendorf, P. E. Daddona, E. Hagaman, and E. Wenkert, *J. Amer. Chem. Soc.* **96**, 5609 (1974).

Sheriha, G., and H. Rapoport, *Phytochemistry* **15**, 505 (1976).

9.14   CORYNANTHE ALKALOIDS

There are few alkaloids in this group but two of these, geissoschizine (**346**) and corynantheine aldehyde (**348**), are of considerable biosynthetic significance. A mixture of corynantheine and dihydrocorynantheine was first isolated by Janot and Goutarel from the bark of *Pseudocinchona africana* A. Chev. in 1938. The structure was deduced in the mid-1950s.

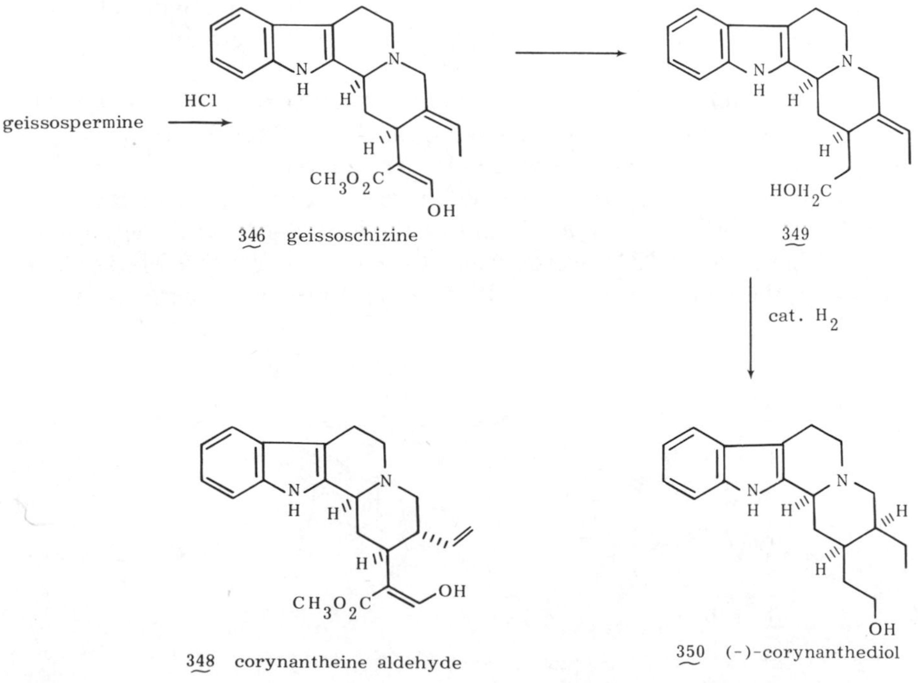

346   geissoschizine

349

348   corynantheine aldehyde

350   (-)-corynanthediol

Geissoschizine (**346**) was not isolated as a natural product until relatively recently. First isolated as part of geissospermine, a dimeric alkaloid obtained in 1887 from the Brazilian *Geissospermum laeve* (Vellozo) Baillon, the structure was determined by decarbomethoxylation and standard reactions to geissoschizol (**349**). Catalytic reduction gave (−)-corynanthediol (**350**), a compound of known absolute configuration.

Geissoschizine (**346**) was first obtained from *Catharanthus roseus* and has subsequently been isolated from several plants. The ^{13}C NMR data are shown in **351**.

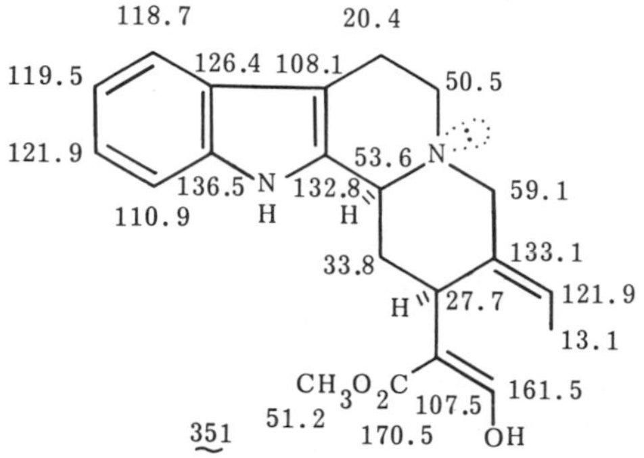

9.14.1   Chemistry

The biosynthetic importance of this skeleton has led to a number of *in vitro* studies of biogenetic-type interconversions and only a very few of these will be discussed.

Goutarel and co-workers have recently investigated the reactivity of the aldehyde ester group of corynantheine aldehyde (**348**) toward the 18,19-vinylic double bond when activated by mercuration.

In the presence of mercuric acetate in warm acetic acid the mercuronium ion derived from **348** gave a mixture of three products on demercuration. Two of these, **287** and **321**, were derived by nonstereoselective Markovnikov attack by the enolic oxygen at C-19. The third product is derived by anti-

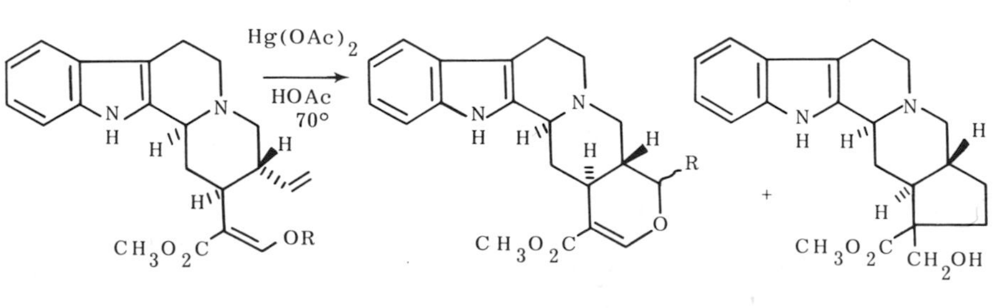

348   R=H

353   R=CH$_3$

287   ajmalicine, R=α-CH$_3$

321   19-epiajmalicine, R=β-CH$_3$

352

Markovnikov attack of the C-16 methine at C-18, leading to the unnatural condensation product **352**.

Corynantheine (**353**) in the presence of mercury(II) trifluoroacetate in aqueous THF gave a mixture of cyclic ketals which could be treated with polyphosphoric acid to yield ajmalicine (**287**) and 19-epiajmalicine (**321**).

Goutarel has also investigated the Prins reaction of corynantheine (**353**). Thus treatment of corynantheine with 0.5 N hydrochloric acid at 100°C gave a mixture of three products, the epimeric dihydroxyyohimbanes **354** and **355** and 19β-chloro-β-yohimbine (**356**), formed as shown in Scheme 45.

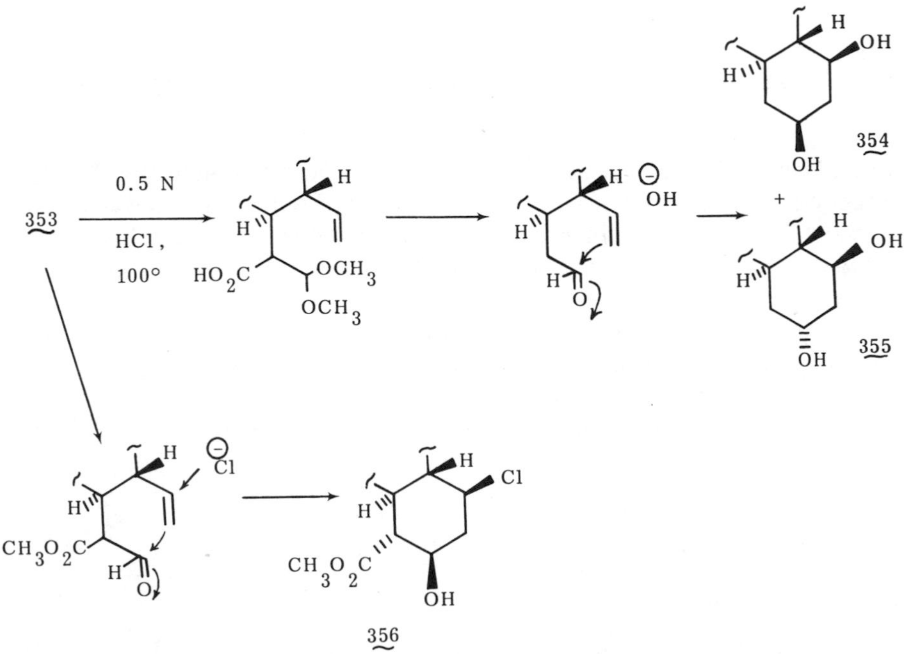

Scheme 45

9.14.2 Synthesis

The first total synthesis of a corynantheine type alkaloid was reported by Van Tamelen in 1958. Using a scheme identical in many steps to the ajmalicine synthesis (see Section 9.15), the keto ester **357** was produced. The vinyl side chain was generated by the thermal decomposition of the tosylhydrazone **358** in the presence of base to produce the desired vinyl compound **359** and an unwanted *trans*-ethylidene derivative. Formylation of **359** in the usual way followed by treatment with diazomethane gave *dl*-corynantheine (**353**) (Scheme 46).

A particularly effective synthesis of optically active corynantheidine has been developed by Szantay. Condensation of **360** with methyl cyanoacetate

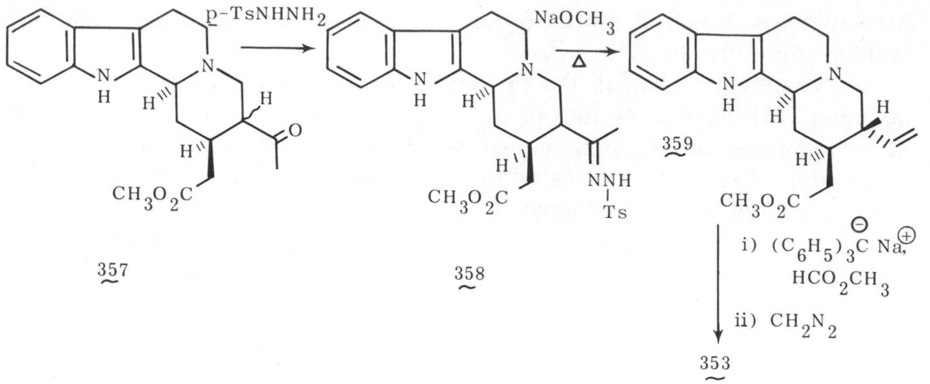

Scheme 46 Van Tamelen synthesis of *dl*-corynantheine (**353**).

gave **361** in 66% yield, which could be converted by sodium borohydride reduction and methanolysis to afford the diester **362**. Controlled low-temperature lithium aluminum hydride reduction afforded the aldehyde ester **363**, which could be methylated to give **364** (Scheme 47). The racemate was resolved with dibenzoyl-*d*-tartaric acid.

An interesting approach to the introduction of the C-15 carbon side chain used by Ziegler involves the Eschenmoser rearrangement of the allylic alcohol **365** with the dimethylacetal of dimethylacetamide (**366**) to give a mixture of the amides **367**. Hydrolysis and esterfication gave **368**, which could

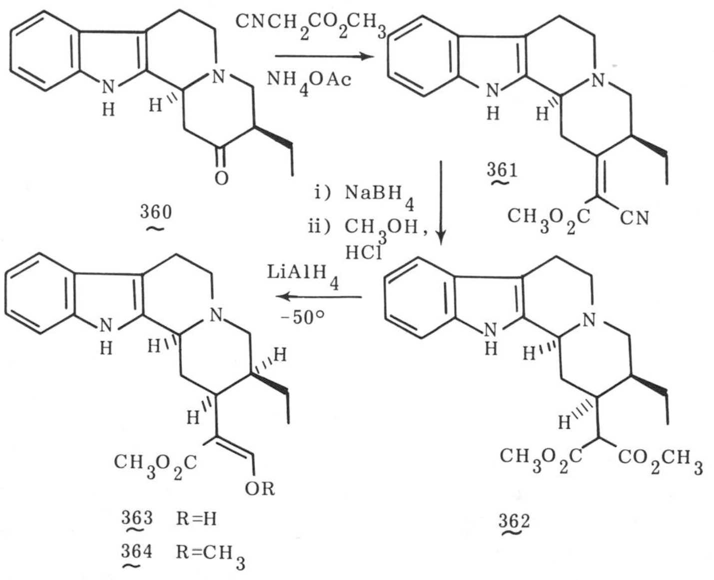

Scheme 47 Szantay synthesis of corynantheidine (**364**).

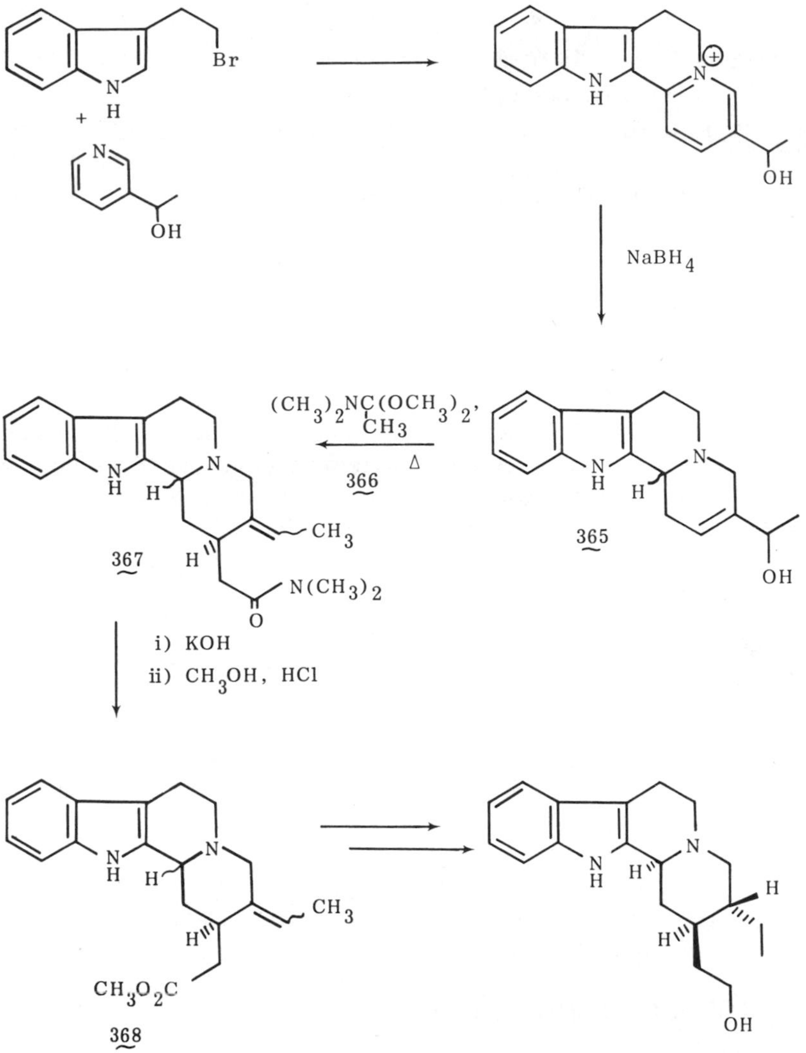

Scheme 48 Ziegler synthesis of dihydrocorynantheol (**369**).

be elaborated in an unexceptional way to dihydrocorynantheol (**369**) and its C-3 epimer (Scheme 48).

LITERATURE

Chemistry

Boivin, J., M. Pais, and R. Goutarel, *Tetrahedron* **33**, 305 (1977).

Pais, M., L. A. Djakoure, F.-X. Jarreau, and R. Goutarel, *Tetrahedron* **33**, 1449 (1977).

Synthesis

van Tamelen, E. E., and I. G. Wright, *J. Amer. Chem. Soc.* **91,** 7349 (1969).

Szantay, Cs., and M. Barczai-Beke, *Tetrahedron Lett.* 1405 (1968).

Wenkert, E., K. G. Dave, R. G. Lewis, and P. W. Sprague, *J. Amer. Chem. Soc.* **89,** 6741 (1967).

Ziegler, F. E., and J. G. Sweeney, *Tetrahedron Lett.* 1097 (1969).

9.15   AJMALICINE AND RELATED COMPOUNDS

There is a group of alkaloids formerly known as heteroyohimbanes, derived by the cyclization of a geissoschizine/corynantheine aldehyde.

The most important member of this group is ajmalicine (**287**), which like the yohimbine derivatives has been found to occur in several stereoisomeric forms, the most widespread of which is 19-epiajmalicine (**321**). Other examples are akuammigine (**370**) and mayumbine (**371**).

287 ajmalicine, R= β-H	370 akuammigine, R=β-H
321 19-epiajamalicine, R=α-H	371 mayumbine, R=α-H

9.15.1   Isolation and Stereochemistry

Ajmalicine (**287**) was first isolated from Yohimbe bark, and later from *R. serpentina* and several other *Rauvolfia* species. It is one of the principal alkaloids of *Catharanthus roseus*.

The configuration at C-3 of these alkaloids is a persistant stereochemical problem, and of this more will be said later. The coupling constants between H-19 and H-20 are influenced by their relative stereochemistry, but at least one center must be of known configuration if the other is to be deduced.

The stereochemistry at C-19 in the ajmalicine and corresponding oxindole series can be deduced from the chemical shift of H-19. In the 19β,20α-series, this signal appears at 4.2–4.5 ppm, whereas in the 19α,20β-series it is consistently in the range 3.7–3.8 ppm.

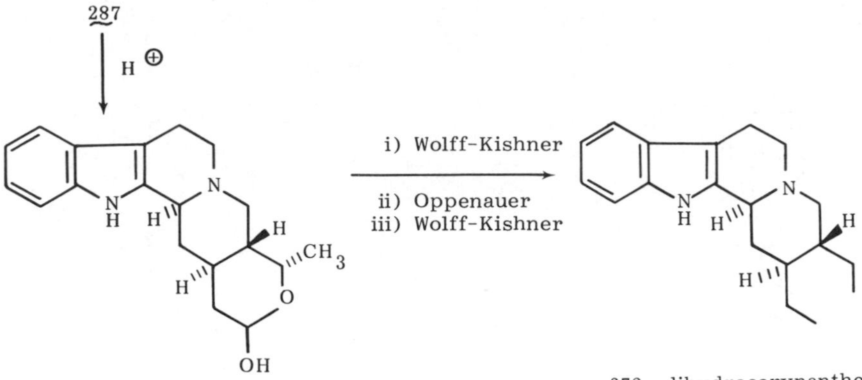

287

H ⊕

i) Wolff-Kishner

ii) Oppenauer
iii) Wolff-Kishner

372 dihydrocorynantheane

Scheme 49

The absolute configuration was deduced by conversion to dihydro-corynantheane (**372**) as shown (Scheme 49).

9.15.2 Synthesis

Van Tamelen reported a synthesis of ajmalicine (**287**) which confirmed the structure assignment. Condensation of the acetyl triester **373** with tryptamine (**2**) in the presence of formaldehyde gave **374**. Cyclization was effected by phosphorus oxychloride and the iminium species **375** subjected to catalytic reduction. Removal of the carboxyl group from the β-ketonic ester and sodium borohydride reduction at low temperature afforded the lactone **376**. The final carbon atom was introduced at C-16 with tritylsodium–methyl formate and the product subjected to an acyl lactone rearrangement under acidic conditions. The resulting acetal **377** on elimination of methanol yielded ajmalicine (**287**) (Scheme 50).

The work of Brown on the biomimetic synthesis of Corynanthe alkaloids was mentioned earlier and one route to 19-epiajmalicine ((**321**) (Scheme 40) was discussed. A slightly different approach involves the conversion of secologanin tetraacetate (**378**) to elenolic acid (**379**), which can be transformed as shown (Scheme 51) to a mixture of ajmalicine (**287**), 19-epi-ajmalicine (**321**), and tetrahydroalstonine (**380**). A similar synthetic route from elenolic acid by van Tamelen had afforded only ajmalicine (**287**). Epimerization at C-20 and C-19 can occur by tautomerization to the ene imine **381**.

9.15.3 Pharmacology

Ajmalicine has been used for the treatment of circulatory disorders, and 10-methoxyajmalicine (**382**) exhibits hypotensive and vasodilator actions.

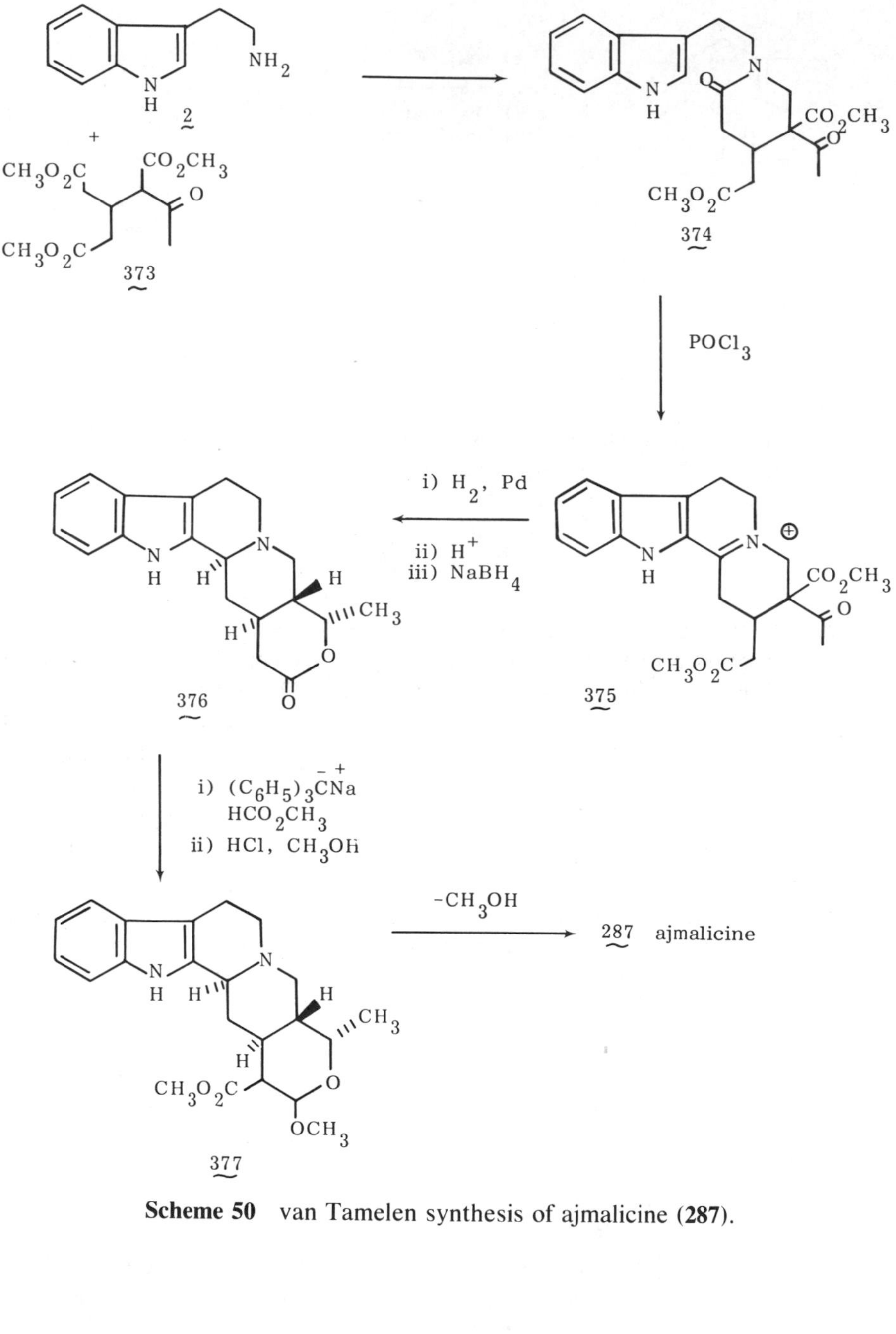

Scheme 50 van Tamelen synthesis of ajmalicine (**287**).

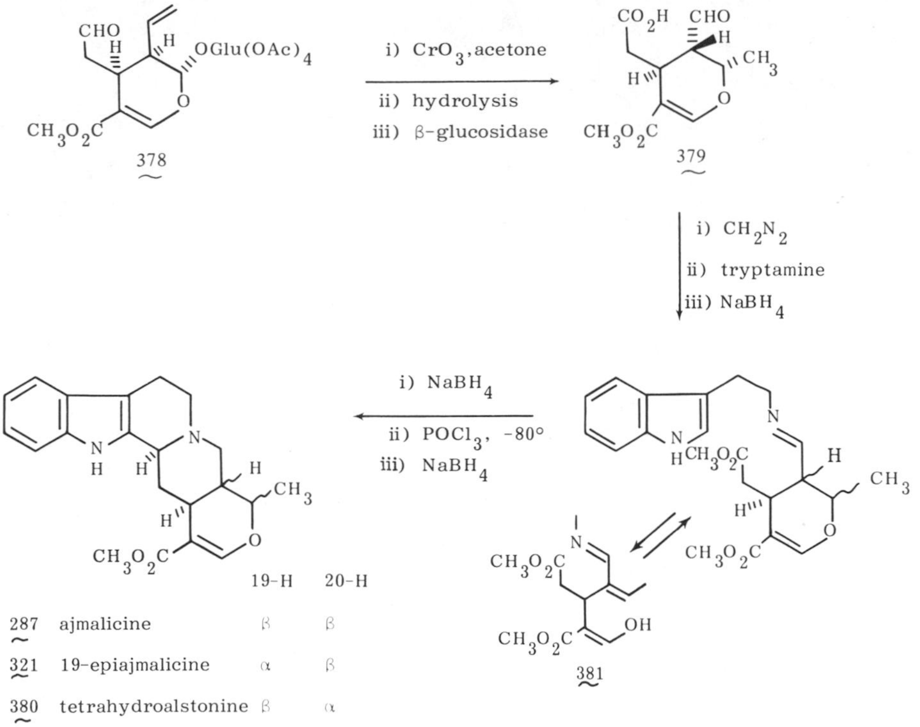

CHO

H   H

OGlu(OAc)$_4$

CH$_3$O$_2$C

378

i) CrO$_3$,acetone

ii) hydrolysis

iii) β-glucosidase

CO$_2$H CHO

H

H   CH$_3$

O

CH$_3$O$_2$C

379

i) CH$_2$N$_2$

ii) tryptamine

iii) NaBH$_4$

i) NaBH$_4$

ii) POCl$_3$, −80°

iii) NaBH$_4$

N

H CH$_3$O$_2$C

H

CH$_3$

O

CH$_3$O$_2$C

N

CH$_3$O$_2$C

CH$_3$O$_2$C   OH

381

N

H   H

H

CH$_3$

O

CH$_3$O$_2$C

19-H 20-H

		19-H	20-H
287	ajmalicine	β	β
321	19-epiajmalicine	α	β
380	tetrahydroalstonine	β	α

Scheme 51

LITERATURE

van Tamelen, E. E., and C. Placeway, *J. Amer. Chem. Soc.* **83**, 2594 (1961).

Brown, R. T., C. L. Chapple, D. M. Duckworth, and R. Platt, *J. Chem. Soc. Perkin Trans. I* 160 (1976).

Mackeller, F. A., R. C. Kelly, E. E. van Tamelen, and C. Dorshel, *J. Amer. Chem. Soc.* **95**, 7155 (1973).

9.16 OXINDOLE ALKALOIDS

There is an important group (over 40 alkaloids) of monoterpene alkaloids containing the oxindole (**383**) nucleus and these alkaloids are typically found to cooccur with their corresponding corynantheoid or ajmalicinoid analogues.

Two typical examples are rauvoxine (**384**) and rauvoxinine (**385**), which were found in the leaves of *Rauvolfia vomitoria* and could be produced in low yield from reserpiline (**386**). This compound was also formed on the acid-catalyzed equilibration of either alkaloid, thereby indicating the parent compounds to be C-7 epimers in the 3β-H series.

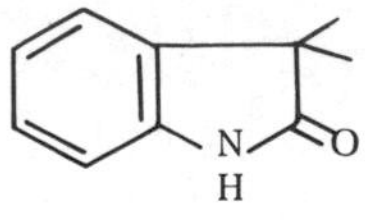

383

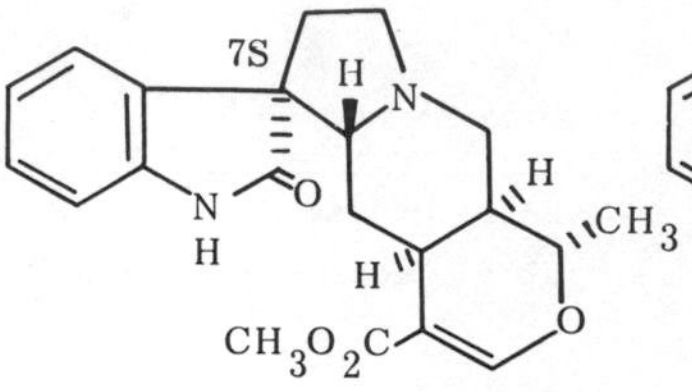

384 rauvoxine

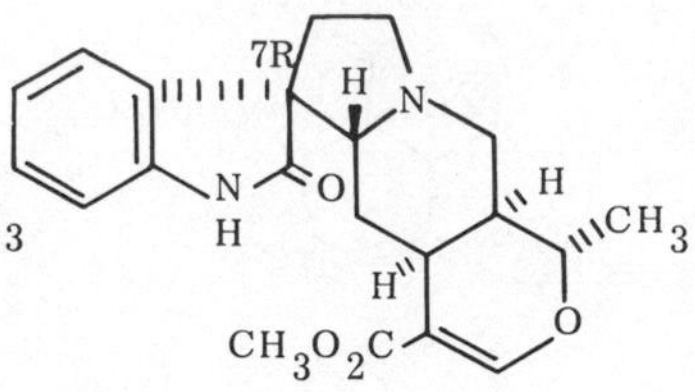

385 rauvoxinine

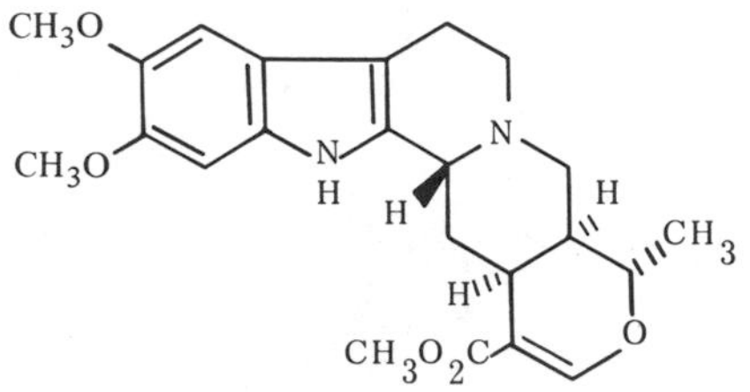

386 reserpiline

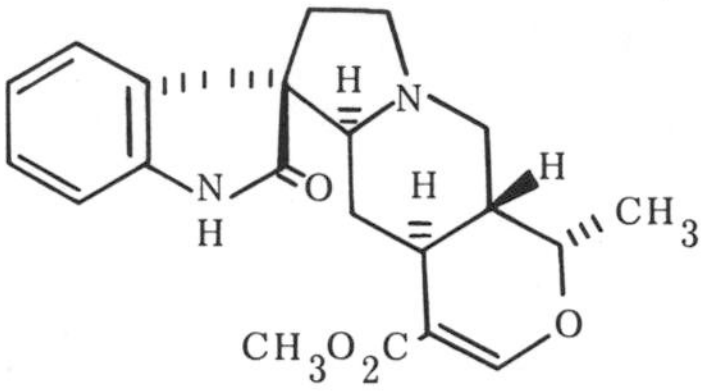

395 mitraphylline

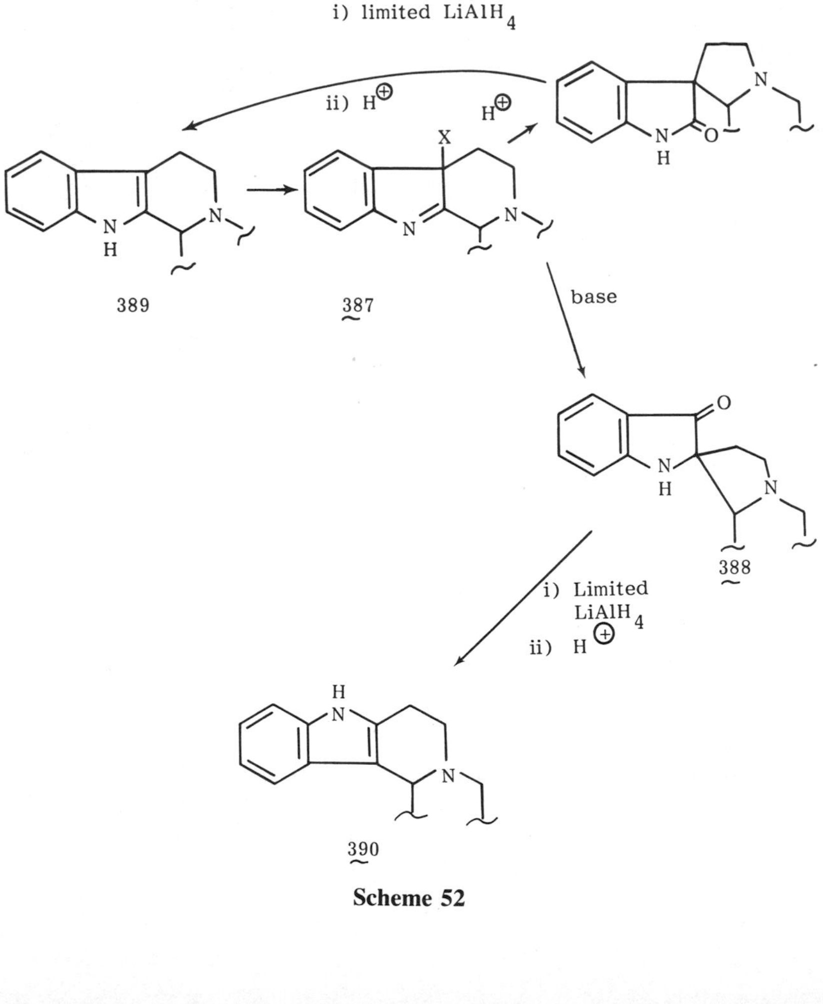

i) limited LiAlH$_4$

ii) H$^{\oplus}$

H$^{\oplus}$

389

387

base

388

i) Limited LiAlH$_4$

ii) H$^{\oplus}$

390

Scheme 52

Yohimbinoids or ajmalicinoids may be rearranged to oxindoles by oxidation with *t*-butyl hypochlorite or lead tetraacetate to afford the corresponding indolenine **387** followed by acid rearrangement. Base rearrangement yields a pseudoindoxyl **388** (Scheme 52).

Reduction of oxindoles with a limited amount of lithium aluminum hydride followed by treatment with acid regenerates the indole **389**. Pseudoindoxyls are converted to isoindoles **390** under analogous conditions.

The mass spectra of these oxindole alkaloids are quite characteristic since they lack the crucial C-2-C-3 bond of the tetrahydro-β-carboline alkaloids such as ajmalicine (**287**).

The latter alkaloid for example typically shows a prominent $M^+ - 1$ ion by loss of the C-3 hydrogen; and characteristic ions in the low mass range of *m/e* 184 (**391**), *m/e* 170 (**392**), *m/e* 169 (**393**), and *m/e* 156 (**394**), the base peak.

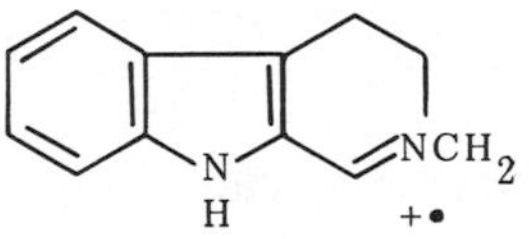

391   *m/e* 184 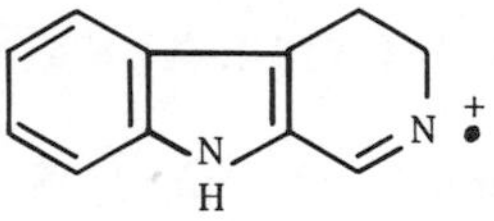   392   *m/e* 170 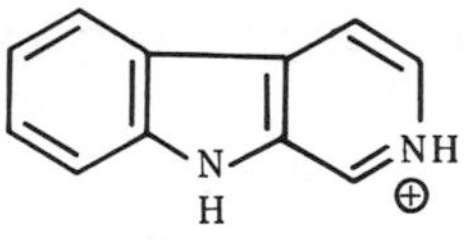   393   *m/e* 169

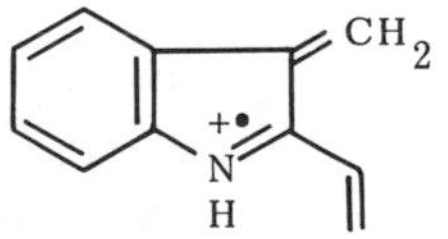

394   *m/e* 156

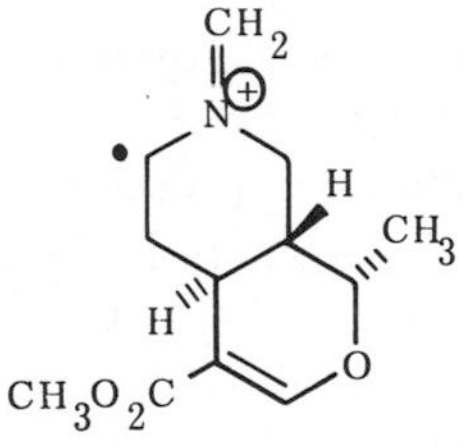

396   *m/e* 223

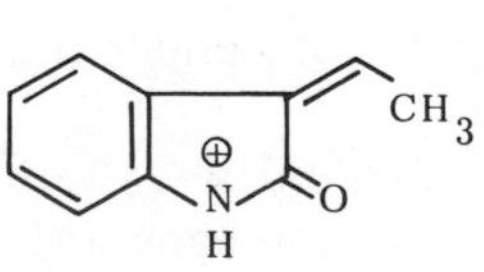

397 *m/e* 159 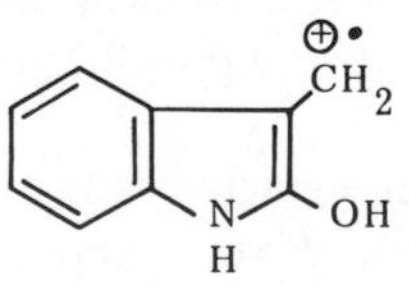   398   *m/e* 146 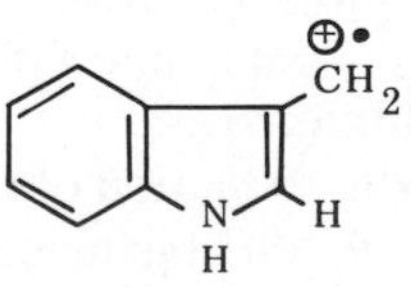   399 *m/e* 130

Mitraphylline (395) on the other hand does not show loss of a single hydrogen; rather the most intense ion is at m/e 223, which is attributed to 396. Ions characteristic of the indole nucleus are found at m/e 159 (397), 146 (398), and 130 (399). Absent are ions such as 391 and 392, typical of the β-carboline nucleus.

Mitraphylline (395) has a weak depressant effect and hypotensive activity.

LITERATURE

Joule, J. A., *Alkaloids, London* **1**, 150 (1971).

Saxton, J. E., *Alkaloids NY* **8**, 59 (1965).

Shellard, E. J., *Planta Med. Phytother.* **7**, 179 (1973).

9.17   YOHIMBINE AND RELATED ALKALOIDS

The bark of the tree *Corynanthe yohimbe* K. Schum. (Rubiaceae), indigenous to the Cameroons and the French Congo, has been used in the treatment of arteriosclerosis, and is also said to be an aphrodisiac. The principal alkaloid is yohimbine (400), which co-occurs with several stereoisomers and was first isolated by Speigel in 1900.

Yohimbine and its derivatives have been obtained from several plant genera in the Apocynaceae (*Rauvolfia, Amsonia, Vallesia, Aspidosperma,* and *Catharanthus*), Loganiaceae (*Gelsemium, Strychnos*), and Euphorbiaceae (*Alchornea*).

9.17.1   Stereochemistry

Yohimbine has five asymmetric carbon atoms (C-3, C-15, C-16, C-17, and C-20), and since these centers are common to several groups of compounds, let us consider them in a little detail.

Vigorous Oppenauer oxidation of yohimbine (400) and β-yohimbine (401) afforded yohimbone (402), indicating that these alkaloids have the same configuration at positions 3, 15, and 20. Two other series were also established when α-yohimbine (403) and alloyohimbine (404) gave alloyohimbone (405) and pseudoyohimbine (406) gave pseudoyohimbone (407). The three series are therefore known as normal, allo, and pseudo.

Yohimbine (400) and pseudoyohimbine (406) differ only at C-3 since Wolff–Kishner reduction of both yohimbone (402) and pseudoyohimbone (407) gave yohimbane (408) (C-3 epimerization occurs under the strongly basic conditions).

Catalytic reduction of sempervirine (409), an alkaloid of *Gelsemium sempervirens* Ait. (Loganiaceae), gave racemic alloyohimbane (410), which could be resolved to give the natural (−)-isomer. This compound should

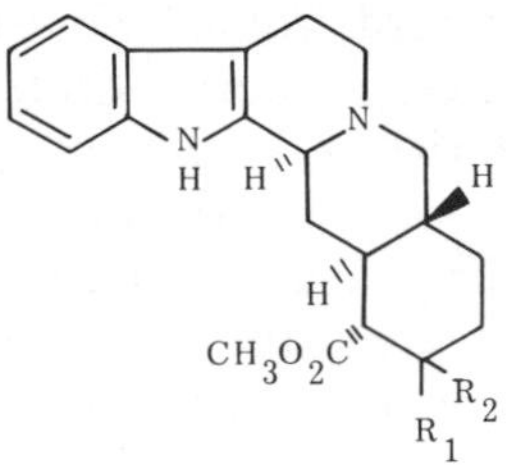

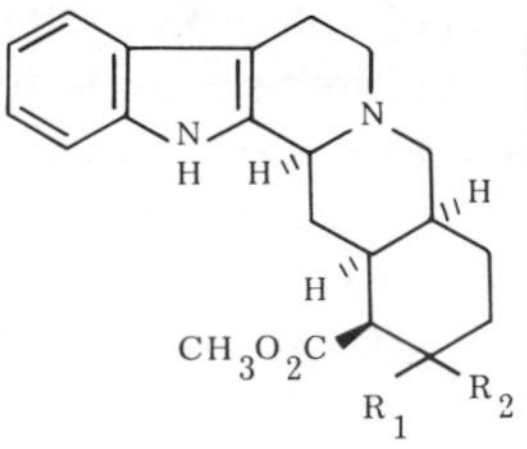

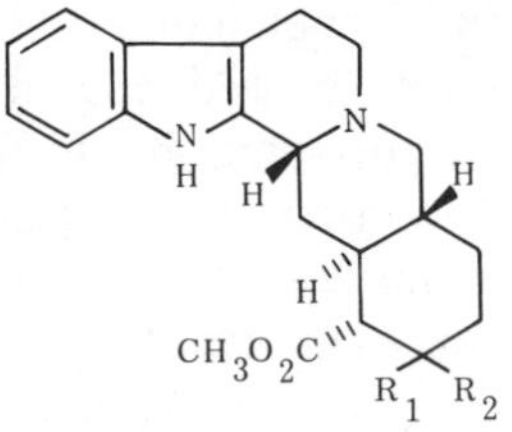

400 yohimbine, $R_1 = \alpha\text{-OH}$, $R_2 = H$
401 β-yohimbine, $R_1 = \beta\text{-OH}$, $R_2 = H$

403 α-yohimbine, $R_1 = \alpha\text{-OH}$, $R_2 = H$
404 alloyohimbine, $R_1 = \beta\text{-OH}$, $R_2 = H$

406 pseudoyohimbine, $R_1 = \alpha\text{-OH}$, $R_2 = H$

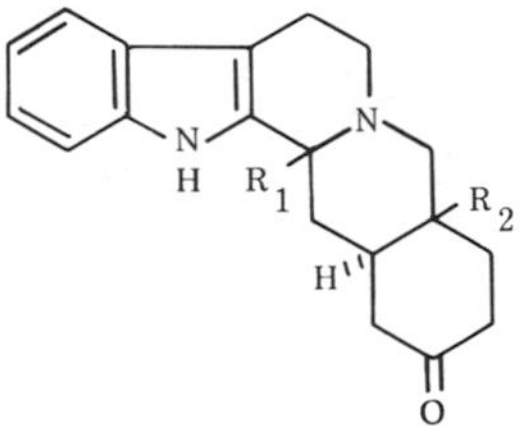

402 yohimbone, $R_1 = \alpha\text{-H}$, $R_2 = \beta\text{-H}$
405 alloyohimbone, $R_1 = \alpha\text{-H}$, $R_2 = \beta\text{-H}$
407 pseudoyohimbone, $R_1 = \beta\text{-H}$, $R_2 = \beta\text{-H}$

therefore have *cis* hydrogen atoms at C-3, C-15, and C-20. These data establish the relative stereostructures for yohimbane, alloyohimbane, and pseudoyohimbane.

A fourth stereoisomeric possibility is the 3-epiallo series, which was isolated in the form of 3-epi-α-yohimbine (**412**) *after* it has been synthesized.

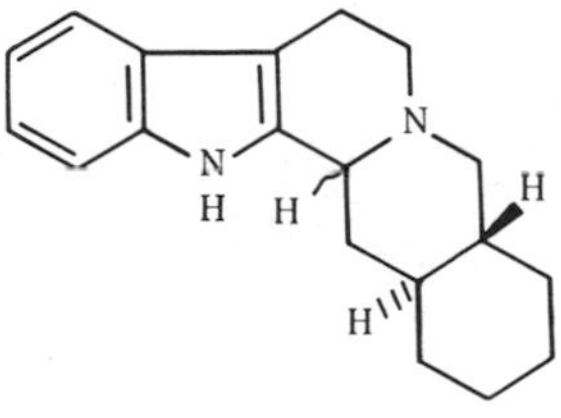

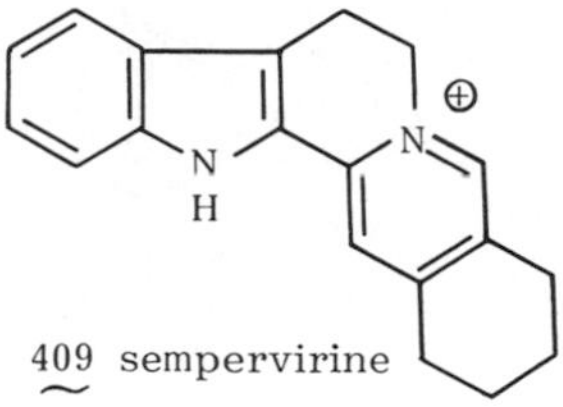

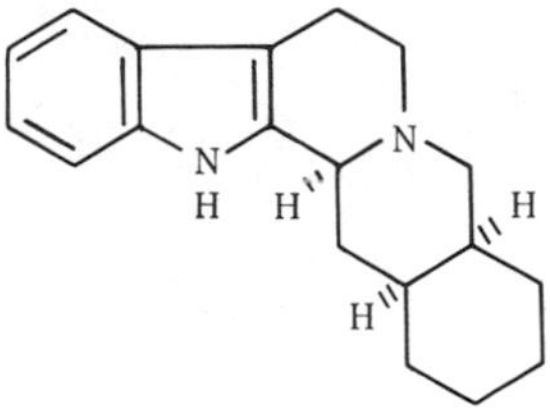

408 yohimbane, 3α-H
411 pseudoyohimbane, 3β-H

409 sempervirine

410 alloyohimbane

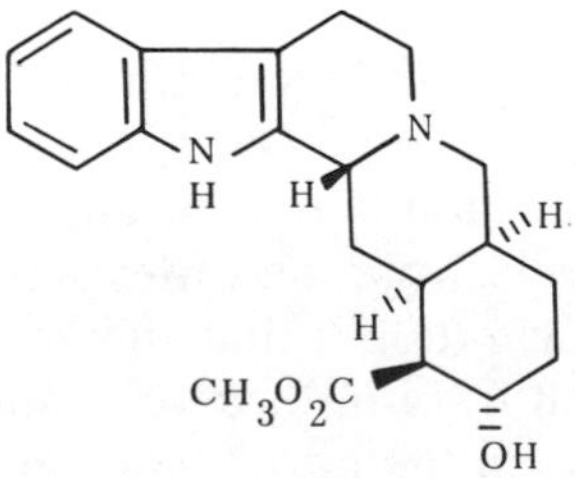

412 3-epi-α-yohimbine

Mild Oppenaeur oxidation of yohimbine (**400**) affords yohimbinone (**413**), which can be reduced to β-yohimbine (**401**) by sodium borohydride. The latter compound should therefore have an equatorial C-17 hydroxy group, with the corresponding group in yohimbine being axial. Corynanthine can be epimerized to yohimbine on treatment with base and therefore should have an axial carbomethoxy group at C-16. This suggests that yohimbine has the complete stereochemistry shown in **400** and corynanthine that shown in **414**.

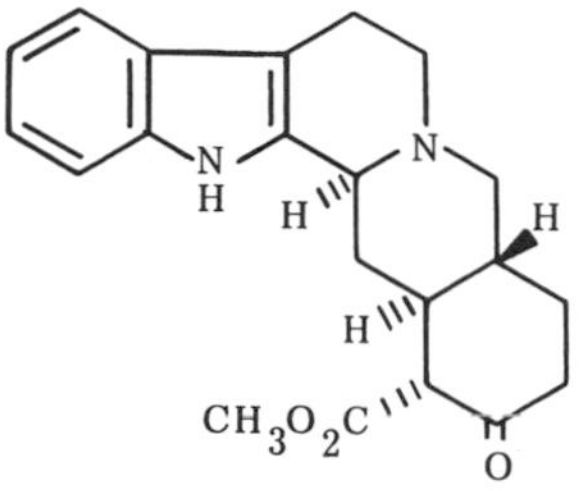

413   yohimbinone 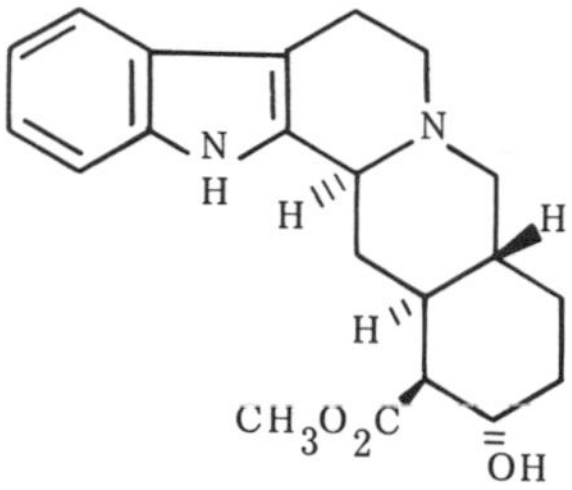   414   corynanthine

Pseudoyohimbine and β-yohimbine have been suggested to have the structures **406** and **401** respectively. Alloyohimbine and α-yohimbine have the hydrogens at C-3, C-15, and C-20 *cis*, but α-yohimbine on treatment with *p*-toluenesulfonyl chloride gave an internal quaternary salt **415**, formed by displacement of a C-17 toluenesulfonyloxy group. Alloyohimbine does

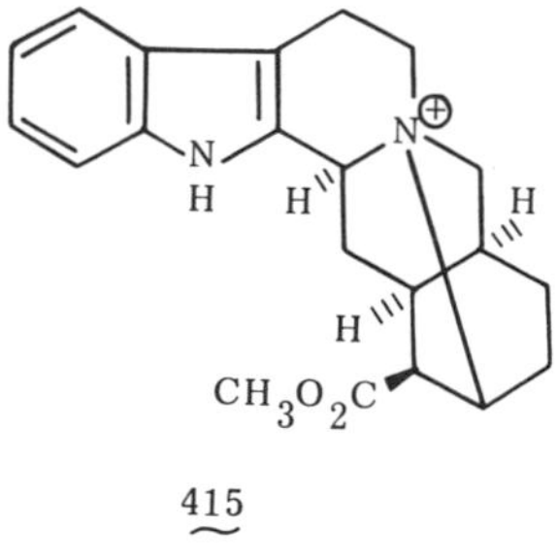

415

not undergo this reaction, and consequently these compounds are epimers at C-17.

The only remaining question is that of absolute configuration, and this was determined by molecular rotation differences in the yohimbine series. The correct absolute configuration is that shown in **400** for yohimbine. Note that the stereochemistry at C-15 in these compounds does not change and is the same as that at C-15 in the nitrogenous glycoside and corynantheine aldehyde series.

9.17.2 Synthesis

The Czech group adopted a strategy that involved the formation of the A, B, and C rings followed by a Wadsworth–Emmons reaction to afford **416**. On treatment with base, internal condensation occurred to afford an intermediate cyclohexenone, which could be catalytically reduced to yohimbone (**402**) (Scheme 53).

The Stork synthesis forms the stereocenters of the D- and E-rings in a decahydroisoquinoline **417**, followed by alkylation with tryptophyl bromide to afford **418** (Scheme 54). In the conversion to yohimbine the kinetically controlled product is formed initially, and this is oxidized to a 3,4-iminium species, which can then be reduced to the more stable 3α-H stereochemistry.

The Kametani group has reported a number of quite different synthetic strategies to the yohimbane nucleus and also a synthesis of yohimbine (**400**) itself. Some examples of the synthetic approaches are shown in Figure 2.

Condensation of the pyrrolidine enamine **419** with methyl 3-oxo-4-pentenoate (**420**) gave 15,16-dehydroyohimbinone (**421**) in 17% yield, which could be reduced catalytically to yohimbinone (**422**). Sodium borohydride

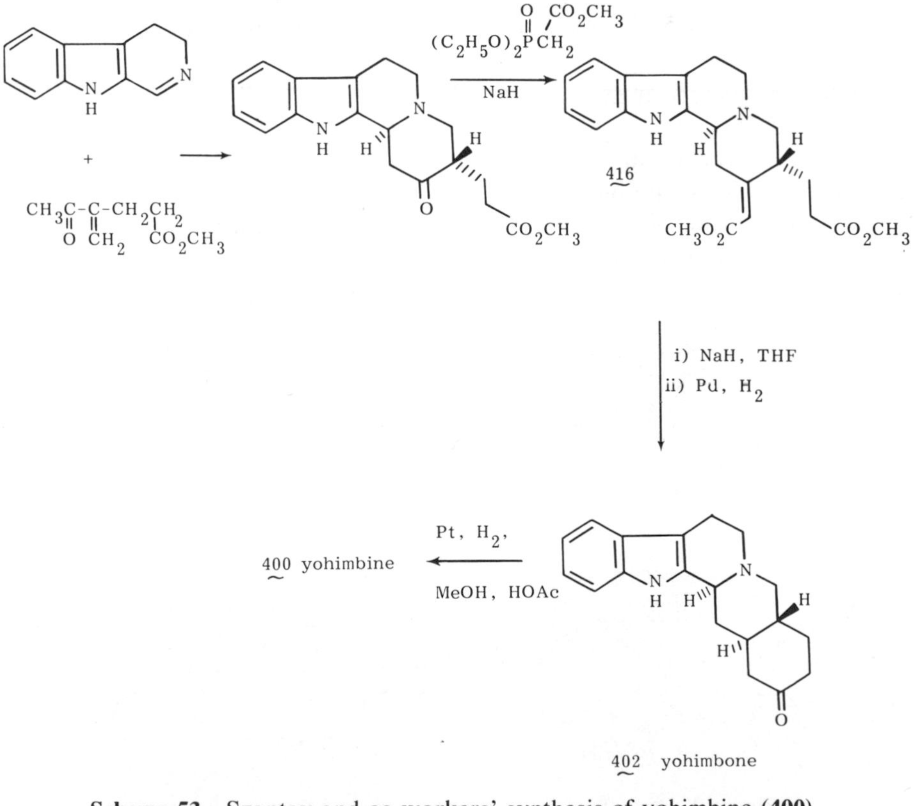

Scheme 53   Szantay and co-workers' synthesis of yohimbine (**400**).

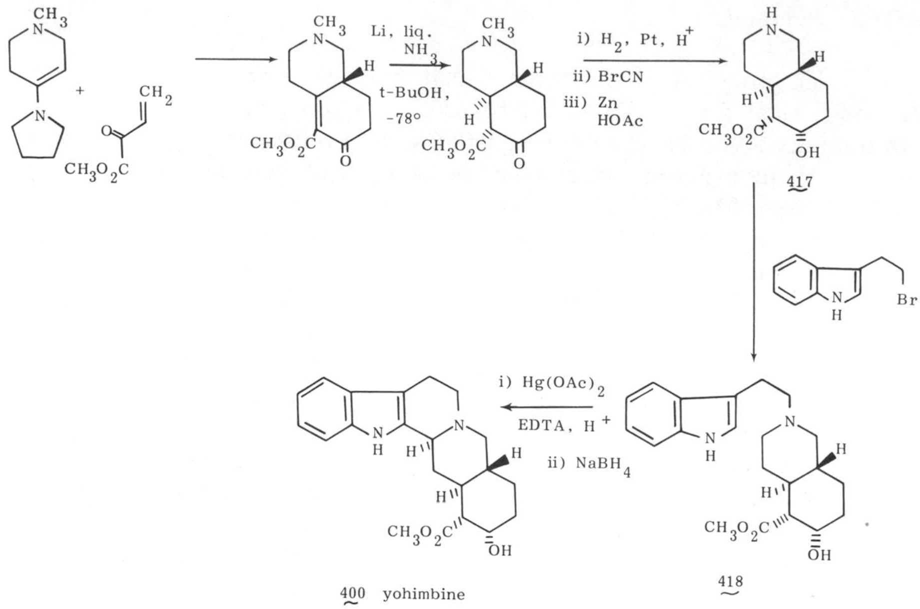

Scheme 54   Stork–Darling synthesis of yohimbine (**400**).

reduction of yielded a mixture of yohimbine (**400**) and β-yohimbine (**401**) (Scheme 55).

There has been quite reasonable interest in the microbiological 18-hydroxylation of yohimbine derivatives, since this position is required oxygenated in the reserpine nucleus (Section 9.18). A patent has been obtained for this process using a *Streptomyces* species, but the method is not of commercial significance.

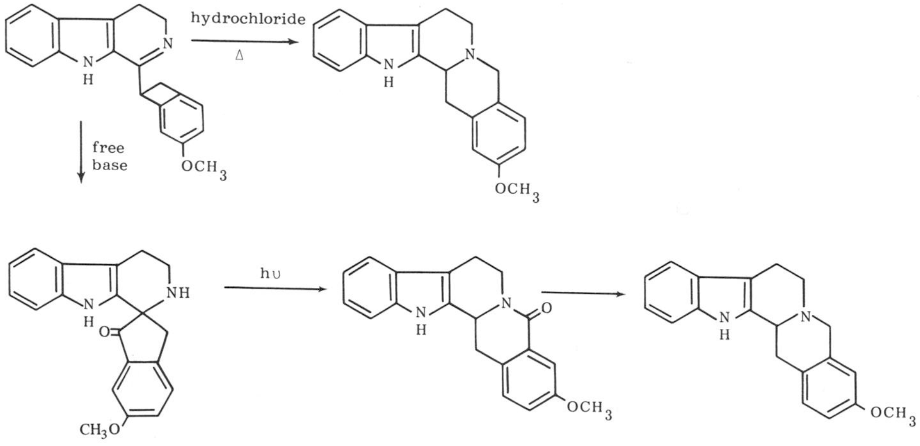

Figure 2   Kametani approaches to the yohimbane nucleus.

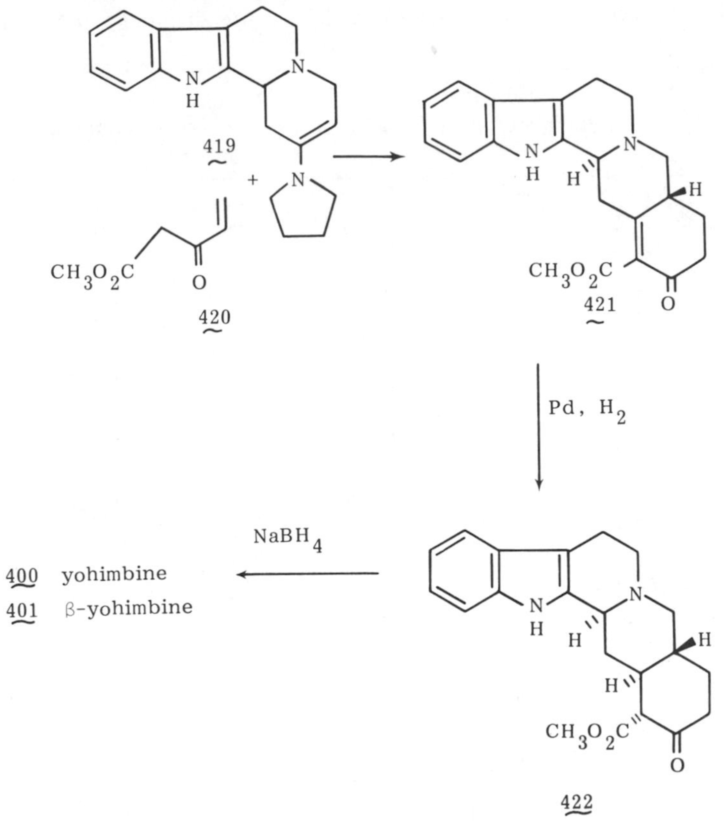

Scheme 55 Kametani total synthesis of yohimbine (**400**).

9.17.3 Spectral Properties

The ^{13}C NMR data for some representative heteroyohimbane and yohimbane alkaloids are shown in Table 5. Dramatic differences on changing the configuration at C-3 (e.g. yohimbine to pseudoyohimbine) were observed for C-3, C-6, C-21, and to a lesser extent C-15. In the *cis*-quinolizideine form of pseudoyohimbine, C-2 exerts a γ-effect on C-15 and C-21 and γ-reciprocal effects on C-6 and C-21. The resonance of C-21 is therefore particularly shielded in pseudoyohimbine.

The allo stereochemistry of α-yohimbine (**403**) confirms the *trans*-quinolizideine configuration by its similarity to the data of yohimbine. On the other hand reserpine, in the epi allo series, is shown by the shift of the C-6 and C-15 resonances to be a *cis*-quinolizideine.

The chemical shift difference of C-21 between pseudoyohimbine (**406**) and α-yohimbine (**403**) is due to a modified β-effect from C-19 and the shift of C-14 by additional γ-effects from C-17 and C-19.

Table 5 ^{13}C NMR Data of Some Representative Ajmalicinoid and Yohimbinoid Alkaloids

	Ajmalicine 287	Tetrahydro-alstonine 380	Yohimbine 400	Pseudo-Yohimbine 406	α-Yohimbine 403	Reserpine 423
C-2	134.0	134.4	134.3	134.0	134.3	130.2
C-3	59.8	52.6	59.8	53.7	60.1	53.6
C-5	52.7	53.3	52.1	50.7	53.2	51.1
C-6	21.3	21.7	21.5	16.4	21.7	16.7
C-7	106.1	107.6	107.5	105.9	108.1	107.7
C-8	126.6	126.9	127.0	127.2	127.1	121.9
C-9	117.3	117.8	117.7	117.2	117.9	118.2
C-10	118.4	119.0	118.8	118.1	119.1	108.7
C-11	120.5	120.9	120.8	129.1	121.1	155.8
C-12	110.6	110.6	110.6	111.1	110.6	95.0
C-13	135.9	135.8	135.8	135.5	135.7	136.1
C-14	32.1	34.2	33.8	32.2	27.6	24.1
C-15	30.1	31.2	36.4	32.4	37.9	32.2
C-16	106.5	109.3	52.6	52.4	54.6	51.6
C-17	154.5	155.5	66.9	66.6	66.0	77.8
C-18	14.5	18.4	31.4	30.9	33.2	77.7
C-19	73.3	72.3	23.1	23.0	24.5	29.6
C-20	40.2	38.3	40.2	39.5	36.4	33.8
C-21	56.2	56.0	61.0	51.5	60.4	48.8
CO	167.3	167.8	175.1	172.9	172.4	172.5
OCH$_3$	50.6	51.0	51.7	51.2.	51.8	51.6

Source: Data from E. Wenkert *et al.*, *J. Amer. Chem. Soc.* **98**, 3645 (1976).

Some general chemical shift ranges for the various stereochemical arrays of these systems were deduced by Wenkert and co-workers and are summarized in Table 6.

Table 6   Typical Shift Ranges (δ, ppm) for Ajmalicinoid and Yohimbinoid Alkaloids

	Normal or Allo	Pseudo or Epiallo
C-3	60 ± 1	5.35 ± 0.5
C-6	21.5 ± 0.5	16.5 ± 0.5

9.17.4   Pharmacology

Yohimbine and its derivatives exhibit a number of pharmacologic activities, although the purported aphrodisiac properties have yet to be substantiated. A gold–isoquinoline–yohimbine complex has been suggested for the treatment of rheumatic disease. Yohimbine derivatives also exhibit hypotensive and cardiostimulant activities.

LITERATURE

Reviews

Saxton, J. E., *Alkaloids NY* **7**, 1 (1960).

Manske, R. H. F., *Alkaloids NY* **8**, 694 (1965).

Monteiro, H. J., *Alkaloids NY* **11**, 145 (1968).

Stereochemistry

Bartlett, R., N. J. Dastoor, J. Hrbek, W. Klyne, H. Schmid, and G. Snatzke, *Helv. Chim. Acta* **54**, 1238 (1971).

Synthesis

van Tamelen, E. E., M. Shamma, A. W. Burgstahler, J. Wolinsky, R. Tamm, and P. E. Aldrich, *J. Amer. Chem. Soc.* **91**, 7315 (1969).

Szantaÿ, C., K. Honty, and L. Toke, *Tetrahedron Lett.* 4871 (1971).

Stork, G., and S. E. Darling, *J. Amer. Chem. Soc.* **86**, 1961 (1964).

Kametani, T., Y. Hivai, M. Kajwara, T. Takahashi, and K. Fukumoto, *Chem. Pharm. Bull.* **23**, 2634 (1975).

9.18 *RAUVOLFIA* ALKALOIDS

The genus *Rauvolfia* is comprised of about 150 species distributed throughout the tropics and subtropics of the world, and is typically found in the rain forests and tropical savannas. *R. serpentina* (L.) Benth. ex Kurz, the most important member of this genus, ascends to elevations of 4500 feet in Assam and *R. tetraphylla* L. has a similar habitat in the tropical Americas.

9.18.1 History

Rauvolfia is one of the ancient medicines, and authenticated reports of the use of *R. serpentina* data back to 1000 B.C. The drug was used in Ayurvedic medicine as a cure for dysentery, as a remedy for snake bite, and as a febrifuge. The hypotensive activity of the root of *R. serpentina* was first reported in 1933, but it was not until 1952 that a group at CIBA in Basel isolated the principal active constituent, reserpine. As reserpine (**423**) became an important clinical agent, the indigenous supplies became depleted and cultivation did not develop rapidly enough. At this point India began what became a series of embargoes designed to limit supplies of the drug. This uncertainty of continuing supplies lead to the development of *R. vomitoria* Afz., an African species, and *R. tetraphylla* L., a Central American species, as alternative commercial sources.

The search for other alkaloids having similar effects which could replace reserpine continued in the 1950s and afforded two compounds, rescinnamine (**424**) and deserpidine (**425**), which were introduced into clinical use. This search for improved drugs having a modified reserpine structure still continues, and some of the most recent efforts will be discussed subsequently.

		R_1	R_2
423	reserpine	OCH_3	3,4,5-trimethoxybenzoyl
424	rescinnamine	OCH_3	3,4,5-trimethoxycinnamoyl
425	deserpidine	H	3,4,5-trimethoxybenzoyl

9.18.2 Structure and Stereochemistry

The alkaloid content of *R. serpentina* root is typically in the range 0.8–1.3%, and in addition to reserpine (0.1%) several other alkaloids cooccur, including rescinnamine, serpentine, yohimbine, ajmalicine, and ajmaline.

The structure of reserpine was deduced on the basis of some elegant degradative studies and further work also deduced the assignment of the six asymmetric centers. The structure was proved by Woodward's legendary synthesis.

Alkaline hydrolysis of reserpine affords trimethoxy benzoic acid and reserpic acid (**426**). The latter compound contains both aliphatic and aromatic carboxyl groups, a carboxylic acid, and a hydroxy group. The stereorelationship of these last two groups was deduced when a γ-lactone was produced on treatment with acetic anhydride–pyridine.

Detosylation of methyl reserpate tosylate with collidine gave an anhydroreserpate derivative **427**, which was hydrolyzed and decarboxylated on treatment with acid to the ketone **428**. This proved the contiguous nature of the ring-E substituents (Figure 3).

Crucial in determining the stereochemistry in the reserpine series was a correlation with the yohimbine series, and this was achieved when the tosylate of methyl deserpidate gave α-yohimbine (**403**). Only one center was inverted during these experiments, that at C-3.

The stereochemistry of the D-E ring juncture was deduced by the formation of an internal quaternary salt **430**, on treatment of methyl 3-iso-reserpate tosylate (**429**) as shown.

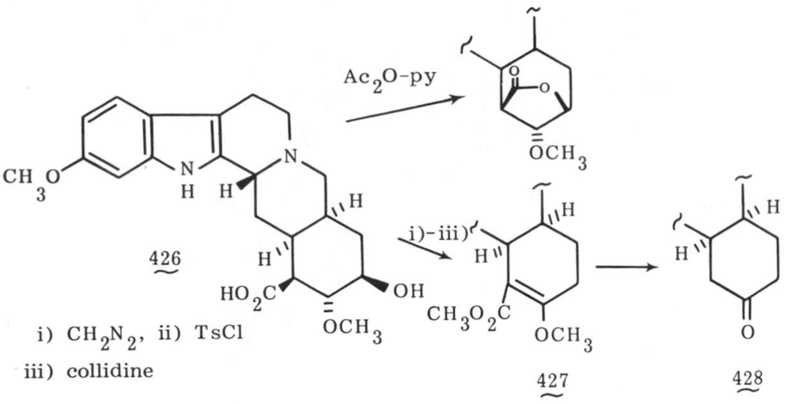

Figure 3 Some degradative reactions of reserpine (**423**).

The C-17 stereochemistry of reserpine was also deduced by internal salt formation: namely that treatment of 3-epi-α-yohimbine (**412**) with tosyl chloride in pyridine the salt **431** results, formed by displacement of a C-17 tosylate with inversion of configuration. The groups at C-16 and C-17 must therefore be *trans* in **412** and reserpine. Reserpine therefore has the complete stereostructure represented by **423**.

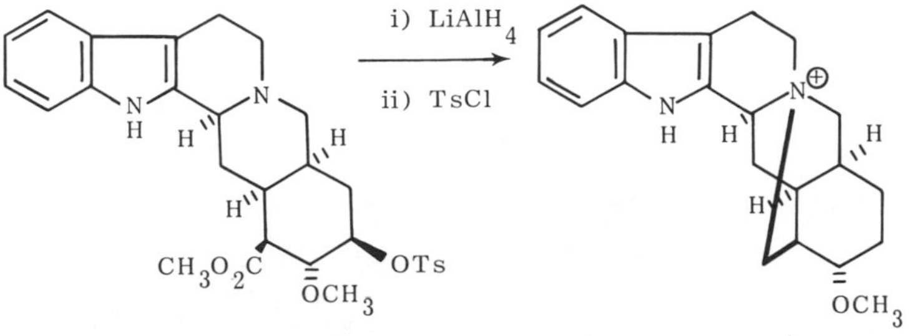

429 430

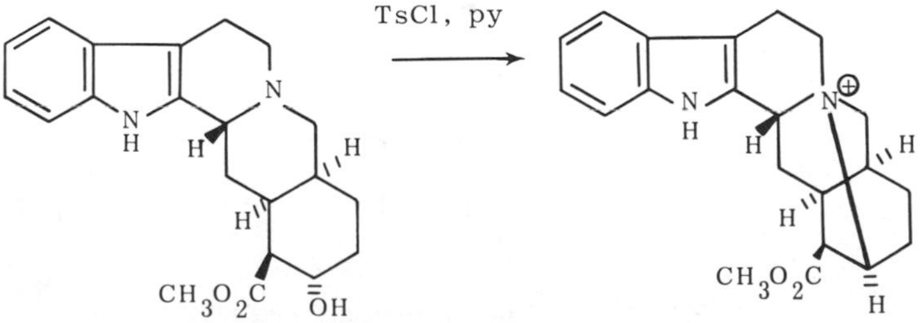

412 431

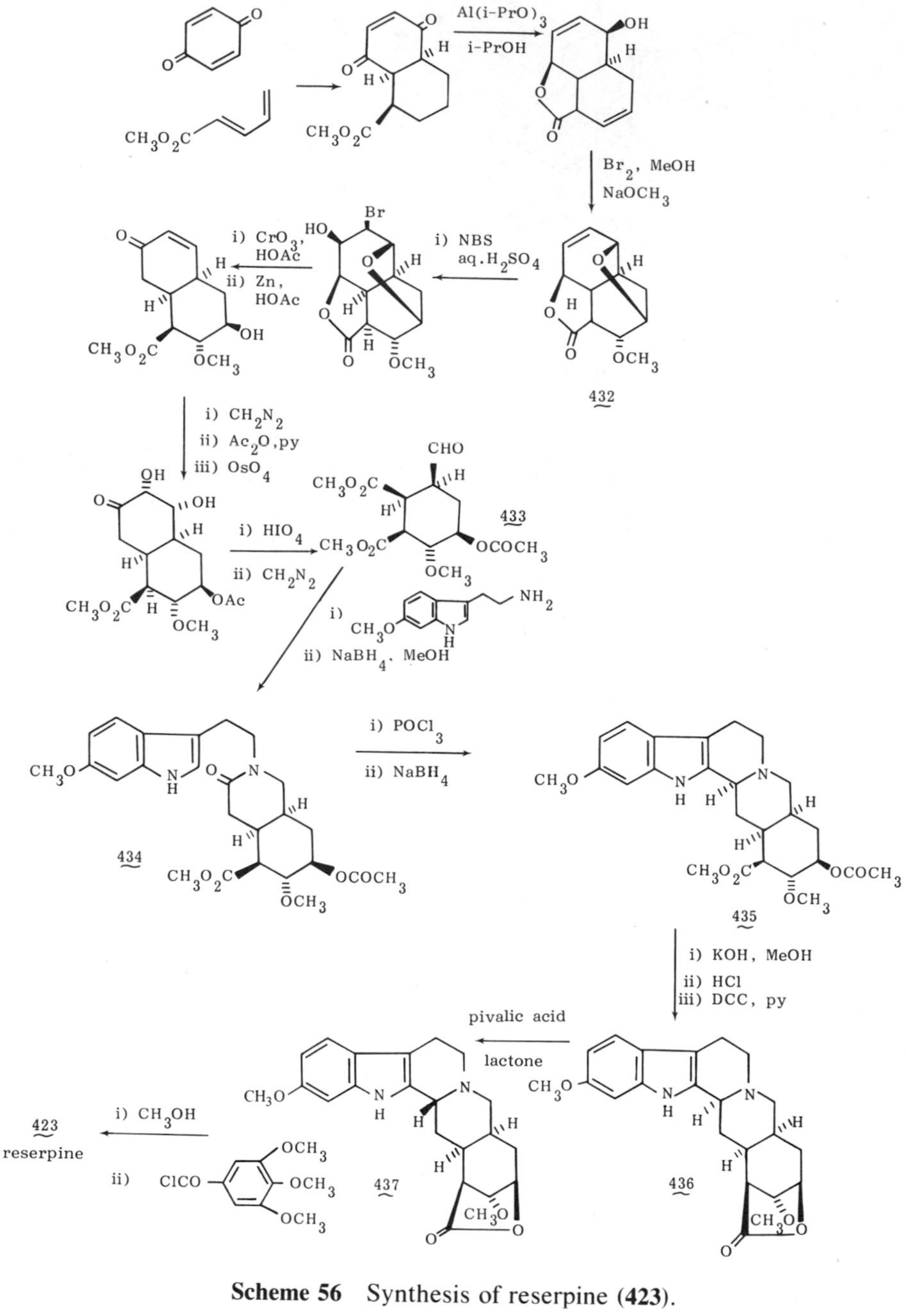

Scheme 56 Synthesis of reserpine (**423**).

9.18.3   Synthesis

The Woodward synthesis of reserpine is notable for the elegance with which the stereocenters not only are introduced, but are maintained in the subsequent reactions. The details of each step will not be discussed, but a few pertinent points should be noted.

The first key intermediate is the tetracyclic compound **432**, which incorporates five of the six stereo centers of reserpine. The most efficient, although not the original, route to this compound is shown in Scheme 56. The next key compound is the aldehyde diester **433**, which can then be condensed with 6-methoxytryptamine (**141**) to give the amide ester **434** after sodium borohydride reduction. This compound can be cyclized with phosphorus oxychloride and the intermediate imine reduced with sodium borohydride to give the more stable 3α-H derivative **435**. In order to convert to the less stable 3β-form, the *cis*-diequatorial substituents in ring E were intramolecularly joined in the form of a lactone **436**. In this conformation C-3 is now thermodynamically unstable and can be converted to the more stable 3β stereochemistry by heating with pivalic acid in xylene to give reserpic acid lactone **437**. Standard reactions converted this to reserpine.

Synthetic reserpine is now available at close to the price of the natural material as a result of some quite extensive modifications of the original Woodward synthesis. For example, optical resolution is performed at an early stage (what is the pharmacology of enantio reserpine?), and the trimethoxybenzoyl group is introduced at an earlier point (e.g. in **438**).

The stereochemistry at C-3 is also introduced in a different way; namely the $\Delta^{3,4}$-quaternary perchlorate is reduced with zinc and perchloric acid in THF–acetone to give reserpine exclusively.

9.18.4 Pharmacology

The central nervous system effects of reserpine are due to catecholamine and 5-hydroxytryptamine (serotonin) depletion in the brain. It acts to produce sedation and tranquilization, and together with the other alkaloids is useful in the treatment of hypertension.

R. serpentina and its constituents are used in combination with several groups of compounds, particularly thiazides and the *Veratrum* alkaloids in the various stages of more severe hypertension.

The onset of action of reserpine is quite slow, and is independent of the route of administration. Drug action is prolonged and effects may still be observed after 4–5 days.

A number of side effects may be observed, including drowsiness, bradycardia, excessive salivation, nausea, diarrhea, and increased gastric secretion. The most serious side effect is depression, which is also a contraindication. Weight gain often occurs on prolonged administration and a number of hormonal effects have been noted.

Reserpine is a stimulator of prolactin release (ergot alkaloid section), and recent evaluations of women over 50 have indicated a substantial increase in the incidence of breast cancer after prolonged usage of reserpine-containing drugs.

Reserpine is a potent uterine stimulant and is therefore also contraindicated during pregnancy.

Rescinnamine exhibits similar pharmacologic properties to reserpine but with lower side effects. Dihydrorescinnamine, in which the double bond of the cinnamic ester moiety has been reduced, is claimed to be less toxic than either reserpine or rescinnamine.

Over the years numerous different ester groups have been tried at C-18 in order to potentiate hypotensive activity and/or lower the sedative effects. Efforts to increase the hypotensive activity by changing the esters at C-16 and C-18 have largely been unsuccessful; reserpine and deserpidine are the most active compounds. Sedative activity could be affected however and syrosingopine (**439**), which is hypotensive in humans at ~3 mg orally (reserpine 0.3 mg), has only minimal activity. On the other hand, **440** has no hypotensive activity but is a potent tranquilizer. The 18-ethers of methyl reserpate (e.g. **441**) also possess marked sedative activity and the onset of action is within minutes. Antihypertensive activity is restored if the chain is lengthened or a second ether introduced, e.g. **442**, which has no sedative activity.

Methyl-18-ketoreserpate (**443**) is not antihypertensive, but does have about one-fifth the sedative activity of reserpine. A more interesting compound is the pyrrolidino derivative **444**, which has curare activity. The mode

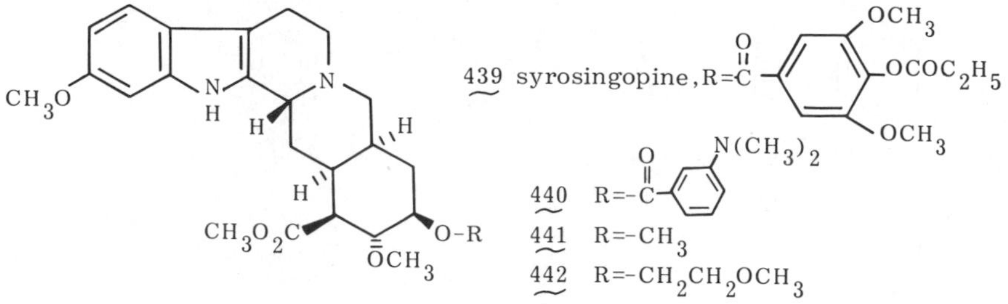

439 syrosingopine, R=$\overset{\overset{\text{O}}{\|}}{\text{C}}$

440 R=-$\overset{\overset{\text{O}}{\|}}{\text{C}}$

441 R=-CH$_3$

442 R=-CH$_2$CH$_2$OCH$_3$

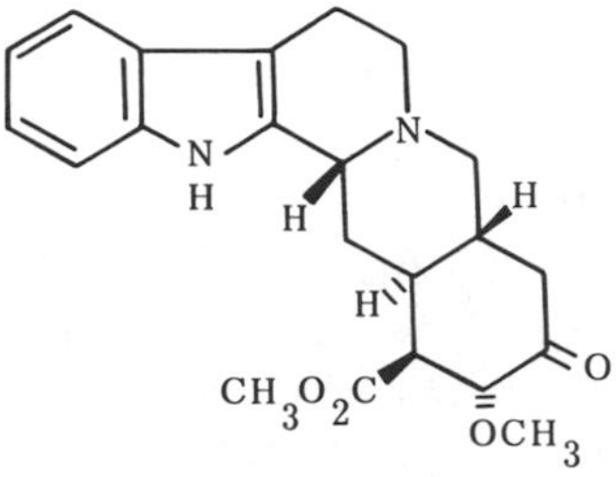

443

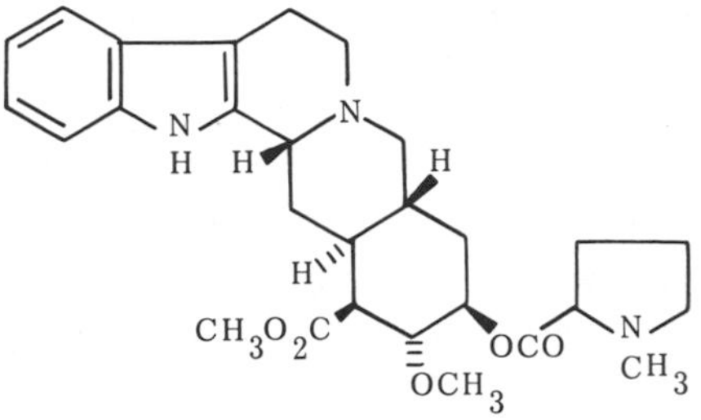

444

of activity is quite interesting because it has a slow onset of activity and this activity persists for several hours.

LITERATURE

Reviews

Chatterjee, A., S. C. Pakrashi, and G. Werner, *Fortschr. Org. Chem. Naturs.* **13**, 346 (1956).

Woodson, R. B., H. W. Youngken, E. Schlittler, and J. A. Schnieder, *Rauvolfia: Botany, Pharmacognosy, Chemistry and Pharmacology*, Little Brown, Boston, 1957.

Lucas, R. A., *Progr. Med. Chem.* 146 (1963).

Schlittler, E., *Alkaloids NY* **8**, 287 (1965).

Synthesis

Woodward, R. B., F. E. Bader, H. Bickel, A. J. Frey, and R. W. Kierstead, *Tetrahedron* **2**, 1 (1958).

9.19 AJMALINE–SARPAGINE ALKALOIDS

The sarpagine and ajmaline structures are represented by the skeleta **445** and **446.** Both skeleta contain a corynanthe-type nucleus in which C-5 and C-16 are joined and in the sarpagine type C-17 is additionally linked to C-7.

These groups are quite widespread and representatives have been obtained from the genera *Aspidosperma, Catharanthus, Picralima, Rhazya,* and *Strychnos,* among others.

445   sarpagine skeleton   446   ajmaline skeleton

9.19.1   Chemistry and Synthesis

Ajmaline is an unusual alkaloid because it contains a carbinolamine moiety and this imparts some interesting chemistry into the molecule.

The structure of ajmaline (447) was not put on a firm basis until 1956, when the structure was deduced by Woodward and Schenker. It was another variant of the yohimbinoid skeleton, and as such represented a further challenge to the then prevalent biogenetic hypothesis of indole alkaloids from prephenic acid.

Anhydrous ajmaline is not easy to obtain because of the ease of methanol solvate formation. Above its melting point ajmaline is converted to isoajmaline (448), a process that can more easily be achieved by heating in ethanolic alkali. The carbinolamine moiety of ajmaline behaves, because of Bredt's rule, more like a tertiary amine, except of course that many of the reactions are better explained in terms of the species chanoajmaline (449): for example the formation of isoajmaline (448), the formation of an oxime, and a reaction with Tollen's reagent. Reduction with sodium borohydride gave dihydro-chanoajmaline (450), which could be converted to 21-deoxyajmaline (451). Lead tetraacetate oxidation of 451 gave the aldehyde 452, which yielded an alcohol 453 by standard reactions. The tosylate of this alcohol underwent fragmentation on heating in collidine to yield a compound 454 in the corynanthe series, which could readily be correlated with compounds of known stereochemistry.

The remaining stereochemical issue was therefore the C-17 hydroxy group, and this could be deduced by reduction of 21-deoxyajmalone (455), which gave only 17-epi-21-deoxyajmaline (456) (attack from least hindered side) and from the proton NMR spectrum, in which in the normal series $J_{16,17}$ is ~2 Hz, but in the 17-epi series is ~9 Hz.

There is also a group of alkaloids which contain a 16-carbomethoxy group; one example is quebrachidine (457), a constituent of Aspidosperma quebrachoblanco Schlecht. The stereochemistry of the C-17 hydroxy group was deduced when the lithium aluminum hydride product of quebrachidine gave an acetonide derivative 458. The compound is useful because it permits assignment of the C-16 stereochemistry to some of the sarpagine alkaloids.

Sarpagine has the structure 459, and the stereochemistry was deduced by correlation with the alcohols in the 21-deoxyajmaline series. There are

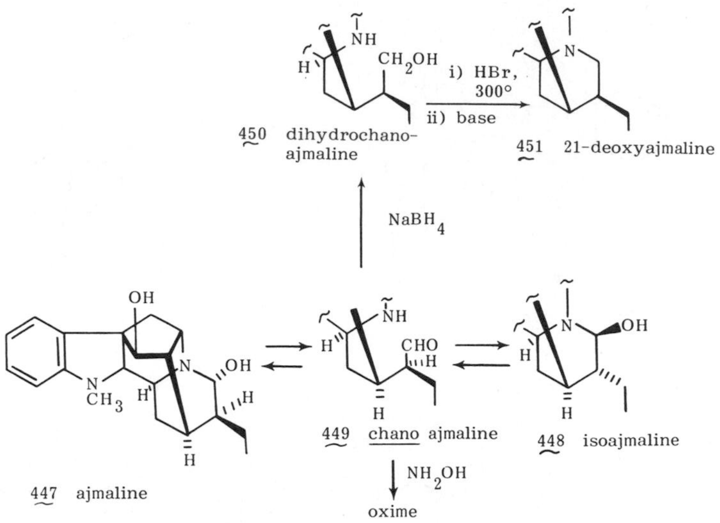
NH
CH2OH
H
i) HBr, 300°
ii) base
N
450 dihydrochano-ajmaline
451 21-deoxyajmaline
NaBH4
OH
N
OH
H
N
CH3
H
447 ajmaline
NH
CHO
H
449 chano ajmaline
NH2OH
oxime
N
OH
448 isoajmaline

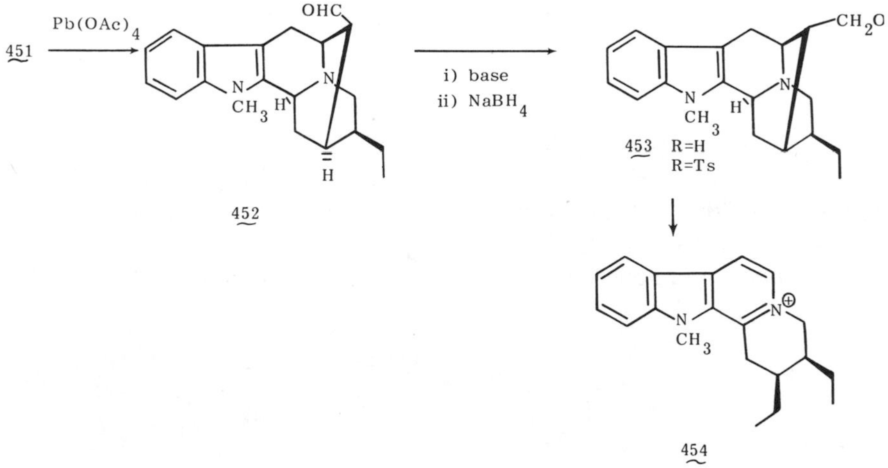
451
Pb(OAc)4
OHC
N
N
CH3
H
452
i) base
ii) NaBH4
CH2O
N
N
CH3
H
453 R=H
R=Ts
N⊕
N
CH3
454

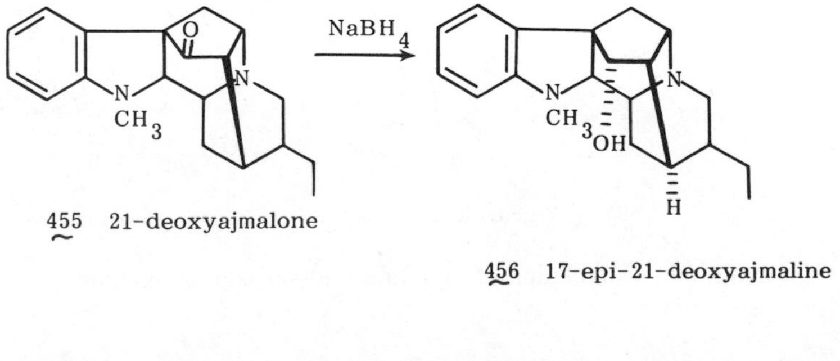
O
N
N
CH3
455 21-deoxyajmalone
NaBH4
N
N
CH3
OH
H
456 17-epi-21-deoxyajmaline

457   quebrachidine   458

459   sarpagine

more alkaloids in this series which contain a 16-carbomethoxy group, and examples are polyneuridine (**460**) and akuammidine (**461**). The relative stereochemistries of these alkaloids was deduced by relationship with quebrachidine (**457**) as shown in Scheme 57.

The other major subgroup in the sarpagine series is one in which the C-

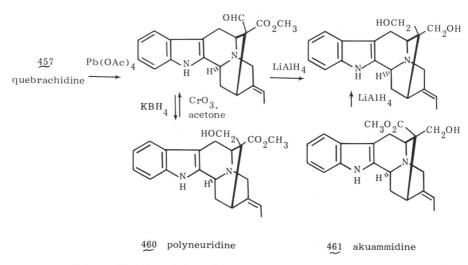

460   polyneuridine   461   akuammidine

Scheme 57   Correlation of ajmaline and sarpagine alkaloids.

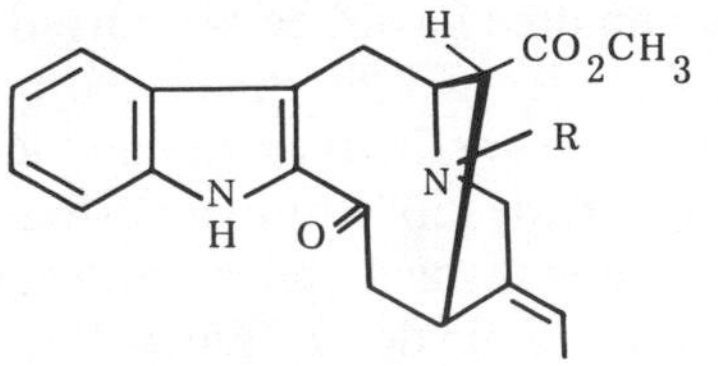

462 vobasine, R=CH₃
463 perivine, R=H

3–N-4 bond is cleaved. Vobasine (**462**) was the first of these alkaloids to be studied and perivine (**463**) is an important alkaloid of *C. roseus*. The stereochemistry at C-16 is determined from the NMR spectrum. Typically the carbomethoxy group appears at δ 3.88 ppm in the 16-epi series, but is markedly shielded to ~2.50 ppm in the normal series.

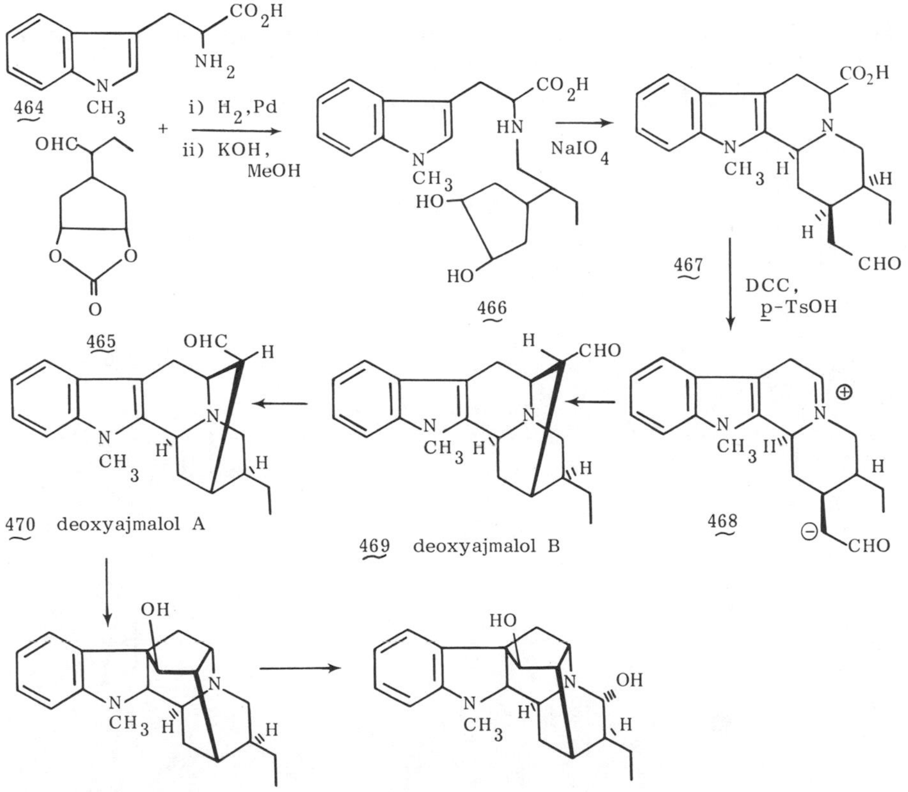

Scheme 58 van Tamelen–Oliver synthesis of ajmaline (**447**).

Three syntheses of ajmaline have been reported, and only the biogenetic type synthesis of van Tamelen will be discussed in detail.

Reductive alkylation of N-methyltryptophan (464) with the aldehyde 465 followed by hydrolysis gave the diol amino acid 466. Cleavage with periodate liberated two aldehydes, one of which underwent spontaneous Pictet–Spengler cyclization to give the tetracyclic derivative 467. Oxidative decarboxylation with DCC and p-toluenesulfonic acid afforded an intermediate iminium 468 species, which could be cyclized directly to deoxyajmalal B (469). Reductive cyclization allowed partial isomerization of 469 to the C-16 epimer deoxy-ajmalal A (470) and formation of 21-deoxyajmaline (451). The C-21 hydroxy group was introduced by an oxidative process involving phenylchlorofor-mate to afford ajmaline (447) (Scheme 58).

An elegant total synthesis of 16-decarbomethoxyvobasine (471) by Lan-glois and Potier involves as a key step the Claisen rearrangement of an *ortho* ester to introduce an acetic ester unit at C-15 (Scheme 59).

9.19.2 *Picralima* Alkaloids

The seeds of *Picralima nitida* Th. and H. Dur. have been used by natives in West Africa as an antipyretic. Chemical studies begun in 1926 have af-forded a number of interesting alkaloids related to akuammiline (472) but

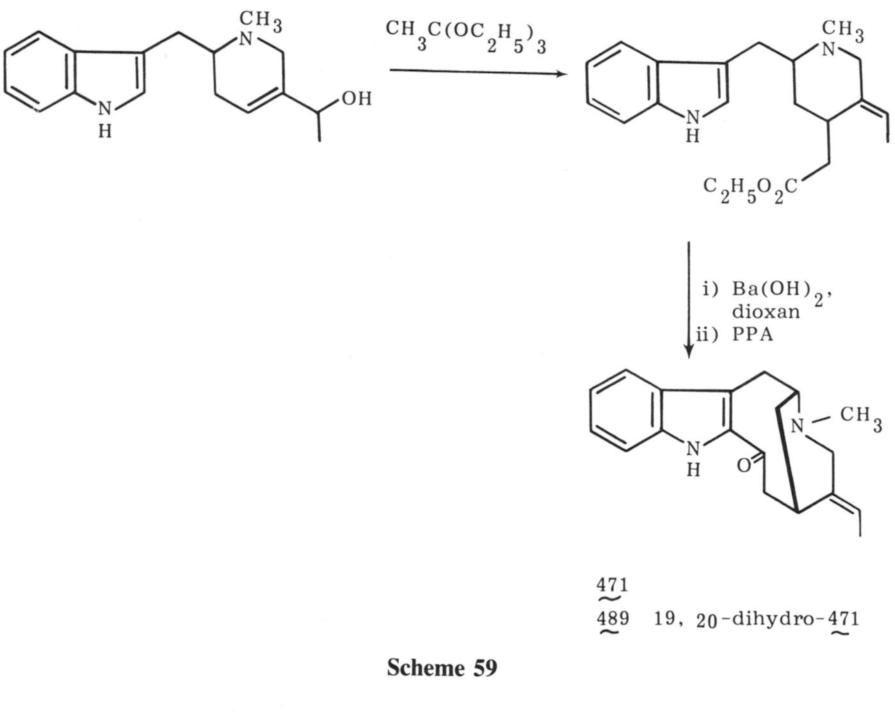

471

489   19, 20-dihydro-471

Scheme 59

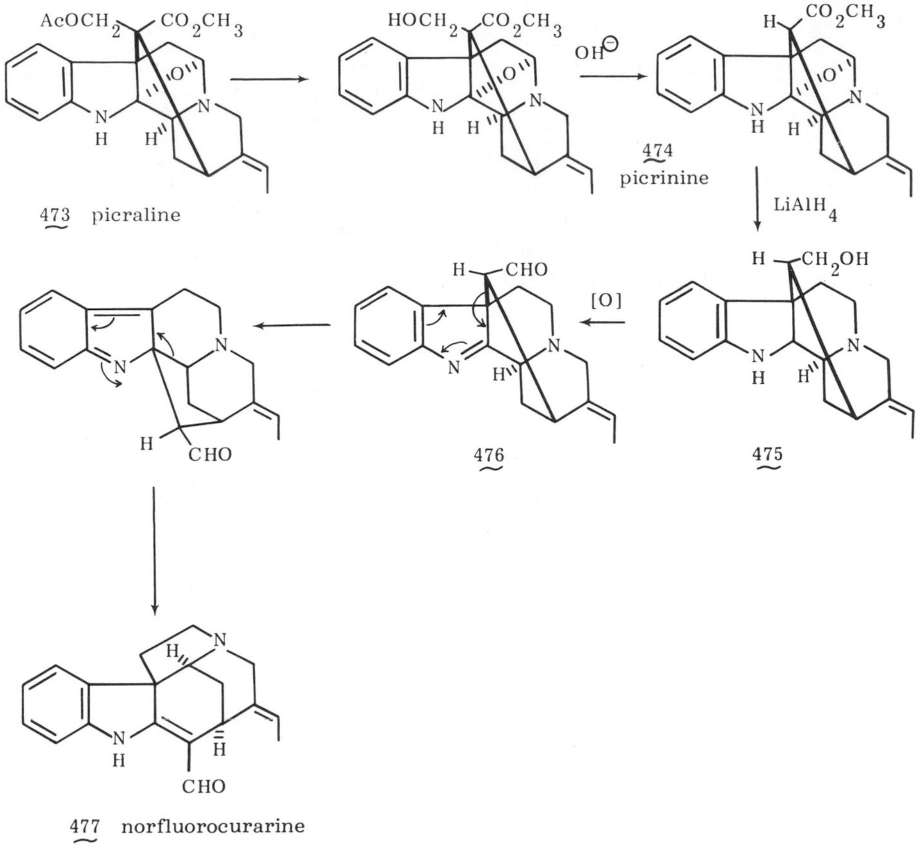

Scheme 60

containing a C-2–C-6 ether bridge. The most important of these alkaloids is picraline (**473**).

Acid hydrolysis of picraline (**473**) followed by retro aldolization gave picrinine (**474**), which was reduced by lithium aluminum hydride to the alcohol **475**. Oppenauer oxidation gave an indolenine aldehyde **476** and a rearranged product in the *Strychnos* series, norfluorocurarine (**477**). By an analogous series of reactions, deacetylakuammiline (**478**) gave akuammicine (**479**). The mechanism of the interesting rearrangement, which could be a biogenetic model, is shown in Scheme 60.

9.19.3 Spectral Properties

Compounds in the sarpagine series show a strong $M^+ - 1$ peak in the mass spectrum which could not, for steric reasons, have originated from either

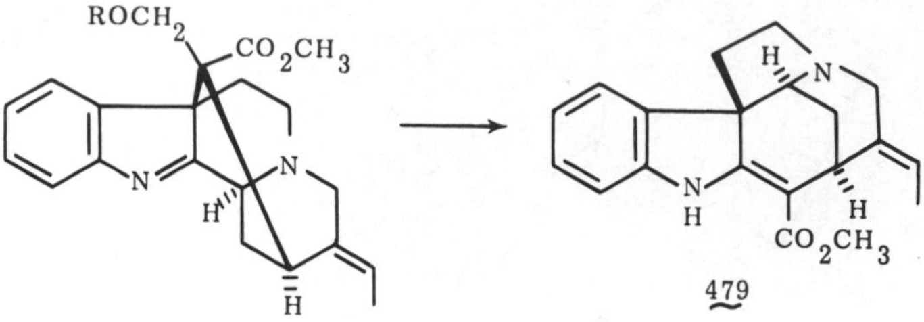

479

472 akuammiline, R=Ac

478 deacetylakuammiline, R=H

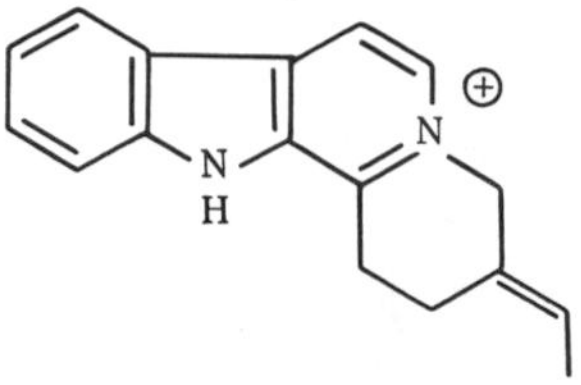

480 m/e 248

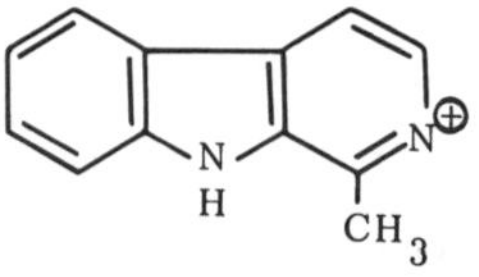

481 m/e 181

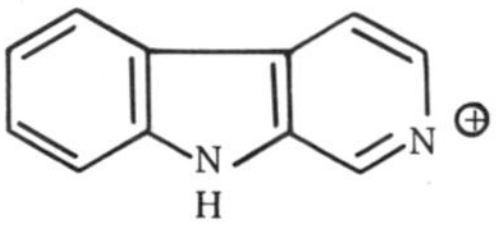

482 m/e 168

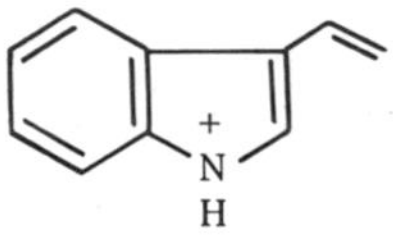

483 m/e 142

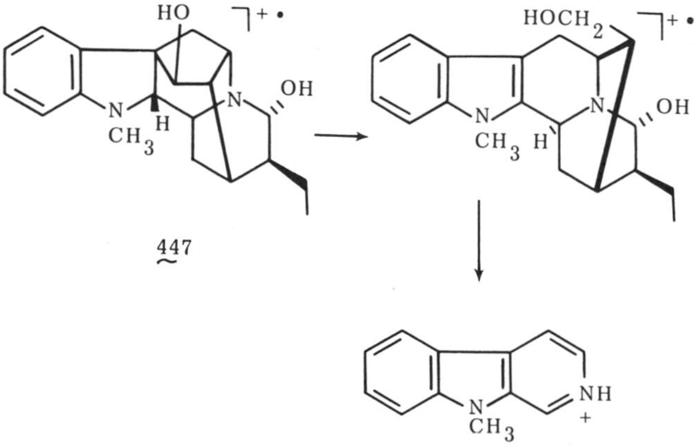

447

484 m/e 183

C-5, C-6, or C-21, and is regarded as coming from C-3. In addition, there are several simple fragmentary losses of CH_3, H_2O, CH_3OH, and where appropriate, CO_2CH_3. The fragmentation pathway to a β-carboline is quite different and gives rise in the sarpagine series to ions at m/e 248 (480), m/e 181 (481), m/e 168 (482), and m/e 142 (483).

The stereochemistry at C-2 in the ajmaline series can be determined by mass spectrometry. When the C-2 hydrogen is cis to the C-17 bridge, migration to C-17 occurs, giving intense ions at m/e 182 and 183 (484), the latter analyzing for $C_{12}H_{11}N_2$.

The mass spectrum of vobasine (462) shows a pronounced base peak at m/e 180 (485), assumed to occur by homolytic fission of the 5,6-bond, followed by β-cleavage as shown in 486.

The proton NMR spectral data for dehydrovoachalotine are shown in 487. The ^{13}C NMR data for vobasine are shown in 488.

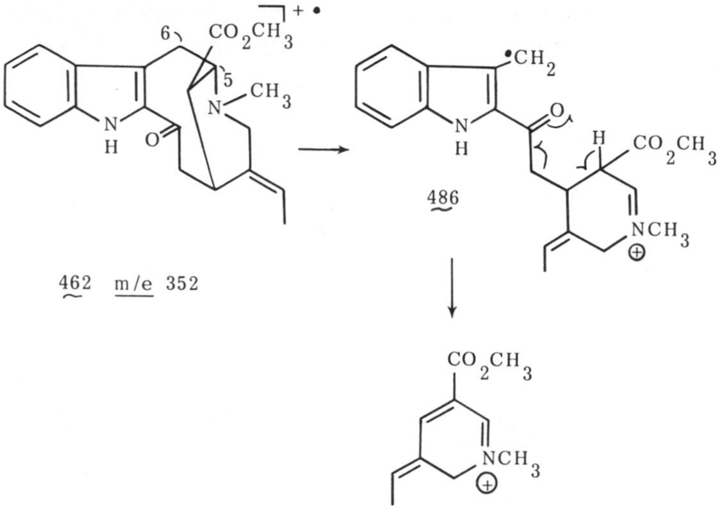

462   m/e 352

486

485   m/e 180

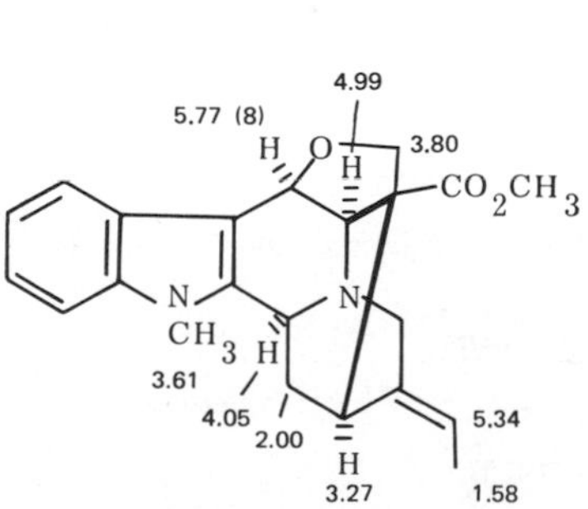

487   dehydrovoachatoline

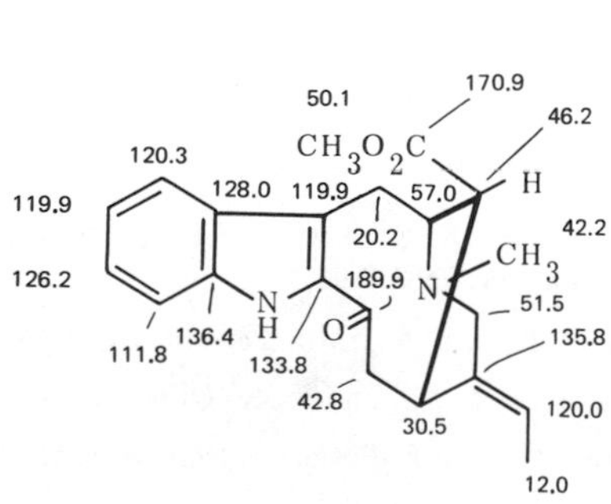

488   vobasine

9.19.4 Pharmacology

Ajmaline has recently attracted attention because of its coronary dilating and antiarrythmic effects. A number of derivatives of ajmaline have been compared in their effectiveness and toxicities, and found to be more potent than the parent compound.

"Neoajmaline" on the other hand is a central nervous system stimulant, a vasodilator, and uterine stimulant. It also lowered blood pressure and in high doses caused death through respiratory arrest.

Isoajmaline (**448**) after initial central nervous system stimulation causes depression and death due to lowered blood pressure, but in contrast to neoajmaline,* it is a uterine relaxant.

Decarbomethoxydihydrovobasine (**489**) from *Hazuntia silicifola* shows vasodilating and hypotensive activities, and related compounds such as the *N*-alkyl dregamines (e.g. **490**) exhibit antiviral activity.

Several of the *Picralima* alkaloids display interesting pharmacologic activity. Thus akuammine (**491**) augments the hypertensive effects of ad-

490

491   akuammine

renaline and has local anesthetic action almost equal to that of cocaine. Akuammidine (**461**) is said to be three times more active than cocaine.

LITERATURE

Reviews

Taylor, W. I., *Alkaloids NY* **8**, 789 (1965).

Taylor, W. I., *Alkaloids NY* **11**, 41 (1968).

Synthesis

Masamune, S., S. K. Ang, C. Egli, N. Nakatsuka, S. K. Sarkar, and Y. Yasunari, *J. Amer. Chem. Soc.* **89**, 2506 (1967).

Mashimo, K., and Y. Sato, *Tetrahedron Lett.* 905 (1969).

van Tamelen, E. E., and L. K. Oliver, *J. Amer. Chem. Soc.* **92**, 2136 (1970).

Langlois, Y., and P. Potier, *Tetrahedron* **31**, 419 (1975).

---

* The structural relationship of "neoajmaline" and ajmaline has never been fully determined.

Pharmacology

Chatterjee, M. L., and M. S. De, *Nature (London)* **200,** 1067 (1963).

Femmer, K., G. Gabsch, and K. Braun, *Pharmazie* **31,** 36 (1976).

Grimm, I., W. Sziegoleit, H. Riedel, and W. Förster, *Pharmazie* **31,** 44 (1976).

9.20 *CINCHONA* ALKALOIDS

The alkaloids of *Cinchona* have proved over the years to be of considerable scientific and therapeutic importance. The most recent account of these alkaloids is the review of Solomon, but Turner and Woodward's review is exceptionally interesting.

9.20.1 History

The history of these alkaloids in Western medicine dates back well over 300 years, although the indigenous natives do not appear to have used *Cinchona* bark. Its use in the treatment of malaria was first recorded by a Jesuit priest in 1633. The genus name was given by Linnaeus in 1742, who derived the name from that of a consort of the Spanish Viceroy of Peru in the 1630s. As the medicinal properties became widely known, extensive harvesting of Cinchona in the eighteenth and early nineteenth centuries led to substantial supply problems from South America, where *Cinchona* is indigenous to the tropical Andes. Determined efforts by the Dutch in the East Indies, particularly Java, permitted them to become essentially the main supplier of quinine. There is little doubt that quinine was one of the key factors in the colonization of Africa and Asia by the British and Dutch.

Until World War II, quinine was the only specific antimalarial remedy. Then, as Java became a battleground and supplies of quinine were cut off, it became critical to find a substitute. It was this that stimulated the tremendous synthetic program to develop a new antimalarial and led to the demise of quinine for general use. More recently another alkaloid of *Cinchona,* quinidine, has found use in the treatment of certain heart conditions.

Cinchona is the dried stem or root bark of various *Cinchona* species, mainly *C. officinalis* L., *C. succirubrum* Pav., *C. calisaya* Wedd., and *C. ledgeriana* Moens. The number of cultivated hybrids however often makes a definitive identification difficult.

The history of the alkaloids of *Cinchona* bark is quite an unusual one. The first crystalline product was obtained by Gomes, a Lisbon physician, who named it cinchonine. Pelletier and Caventou first obtained pure cinchonine in 1820, and subsequently they also obtained quinine. By 1884 twenty-five alkaloids had been described; however in the next 94 years only six additional alkaloids have been isolated, the last in 1941. Of these, 13 are of unknown structure.

The alkaloids only occur in two genera, *Cinchona* and *Remijia,* and typically are white crystalline solids capable of forming two types of salts, a monosalt which is sparingly soluble in water, and a bis salt which is highly water soluble.

There are eight major alkaloids of *Cinchona* (**492–499**), which are actually four pairs of stereoisomers. These alkaloids are comprised of two parts: a quinoline nucleus and a quinuclidine moiety. The numbering system shown in **500** is due to Rabe. A prefered numbering system is that shown in **501**, which is biogenetically based.

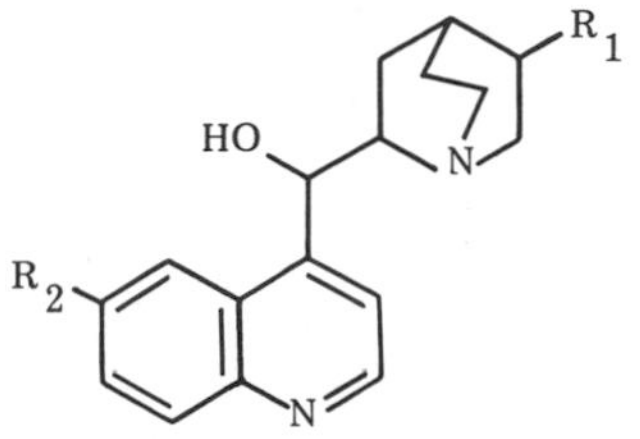

		R_1	R_2		
492	cinchonine	$-CH=CH_2$	H	cinchonidine	496
493	quinidine	$-CH=CH_2$	OCH_3	quinine	497
494	dihydrocinchonine	$-CH_2CH_3$	H	dihydrocinchonidine	498
495	dihydroquinidine	$-CH_2CH_3$	OCH_3	dihydroquinine	499

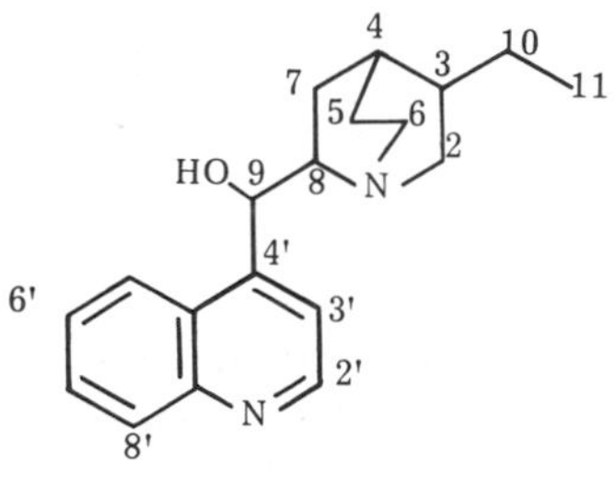

500 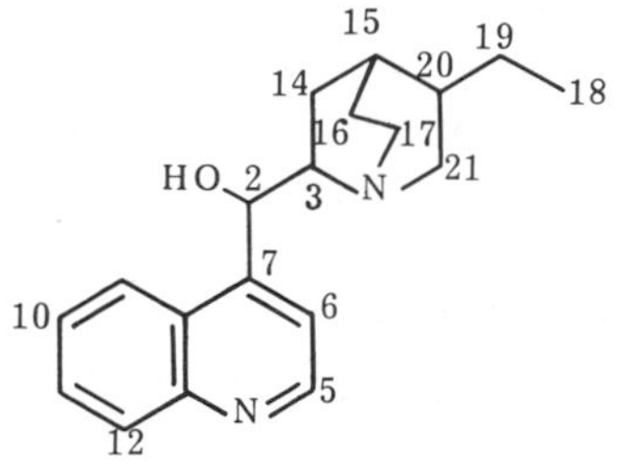   501

The eight major alkaloids can also be divided into two groups based on their optical rotation; thus cinchonine and quinidine and their dihydroderivatives are dextrorotatory, whereas cinchonidine and quinine and their dihydroderivatives are levorotatory. It is unfortunate that the nomenclature is not consistent and that cinchonine was not named cinchonidine, in line with quinidine.

Quinine was first obtained in pure form by Pelletier and Caventou in 1820, although Fourcroy may have obtained crude quinine as early as 1792.

Besides Pelletier and Caventou, the history of this compound is checkered with the names of great chemists. Elemental analyses were carried out by Liebig, by Laurent, and finally by Skraup. The rearrangement to the cinchonatoxines was discovered by Pasteur. Konigs and Skraup developed much of the early chemical work. Rabe deduced the plane structure. Prelog and Rabe were responsible for developing the stereochemistry. There is the legend of W. H. Perkin, who tried in 1856 to synthesize quinine from allyltoluidine and potassium dichromate, and who in the process founded a whole dyestuffs industry. Rabe first synthesized the dihydroalkaloids, followed by Woodward and Doering, who reported the first synthesis of quinine itself. Woodward, Robert Robinson, and Battersby speculated on the biosynthesis of the skeleton.

Quinidine was first obtained in 1833 and subsequently in 1847. The compound was named by Pasteur in 1853. It is a minor constituent of many cinchona barks and can be obtained from the quinine sulfate mother liquors. A preferred commercial source is *Cuprea* bark, *Remijia pedunculata* Fluckiger (Rubiaceae), where it is frequently accompanied by as much as 25–30% of dihydroquinidine. Quinidine may also be produced by the isomerization of quinine.

9.20.2   Structure Elucidation and Chemistry

The structure elucidation of the cinchona alkaloids began with the potash distillations carried out by Gerhardt in 1842 and concluded with Rabe's demonstration that a secondary and not a tertiary alcohol group was present. Details of the reactions that led to the structure of quinine are discussed in the reviews of Turner and Woodward and of Solomon.

Mention should also be made of the action of acid on these compounds. In 1853, Pasteur observed that on mild acid treatment, cinchonine or cinchonidine gave a keto amine **502**, a member of the so-called "toxine" compounds. These compounds are intermediates in some of the syntheses of the quinuclidine nucleus.

In compounds such as quinine (**497**) which contain both a vinyl group and a methoxy group, the vinyl group is more susceptible to acid attack than the latter group and is capable of undergoing a variety of rearrangements. These include double-bond isomerization to 19,20-, Grob fragmentation, and

497

502   cinchotoxine

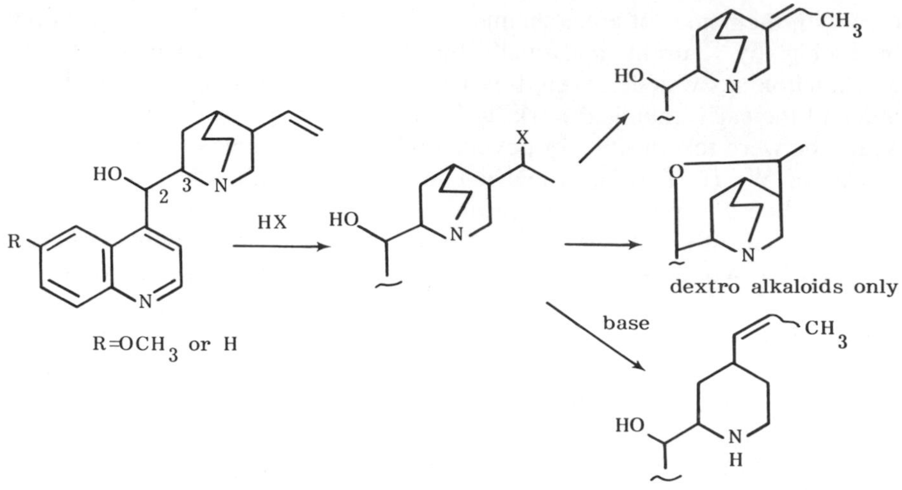

Scheme 61

cyclic ether formation, and are shown in Scheme 61. The formation of a cyclic ether is specific for compounds in the *dextro* series in which the C-3 and C-20 substituents are *cis*.

The interconversion of quinine (**497**) to quinidine (**493**) is best carried out by a modified Oppenauer reaction (KO*t*-Bu and benzophenone), followed by treatment under Meerwein–Ponndorf conditions (isopropanol). Note that because of epimerization at C-3, quinine and quinidine on oxidation give the same equilibrium mixture of ketones.

The remaining matter to be discussed is that of stereochemistry. Cinchonidine and quinidine are stereoisomers of cinchonine and quinine, respectively. Cinchonine and quinidine are dextrorotatory and cinchonidine and quinine are levorotatory. There are four asymmetric centers present in these compounds, at C-2, C-3, C-15, and C-20, and some evidence for their relationship can be deduced from some of the reactions previously discussed.

For example, both cinchonine (**492**) and cinchonidine (**496**) give rise to the same cinchotoxine. These compounds must therefore have the same configurations at C-15 and C-20.

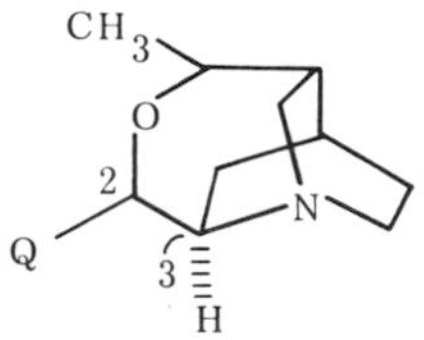

cyclic ether from
cinchonine

Reduction of the C-2 hydroxy group to give a methylene group in cinchonine does not give the same product as cinchonidine. These alkaloids therefore differ in stereochemistry. Remembering that cinchonine can be converted to a cyclic ether, whereas cinchonidine cannot, serves to define the stereochemistry at C-3 (but not C-2) in these compounds.

All the alkaloids are identical in configuration except at C-3 and C-2, and therefore four isomers are possible in each series. The specific rotations of these compounds are:

Cinchonine	$+224°$	Quinidine	$+254°$
2-Epicinchonine	$+120°$	2-Epiquinidine	$+102°$
2-Epicinchonidine	$+63°$	2-Epiquinine	$+43°$
Cinchonidine	$-111°$	Quinine	$-158°$

Clearly the contributions from C-3 and C-2 are positive in the most strongly dextrorotatory compounds. The absolute configuration can be deduced by correlation with $(-)$-ephedrine (503) and $(+)$-pseudoephedrine (504), which indicate that quinine and cinchonidine, like ephedrine, have the *erythro* configuration. The stereochemistries of the four major compounds are therefore as shown in Figure 4.

The hydroxy group at C-2 in quinine can be selectively removed by irradiation in acid solution; sodium amyl alcohol also carries out this reaction, but also reduces the B-ring.

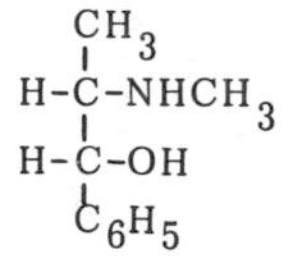

$$\begin{array}{c} CH_3 \\ | \\ H-C-NHCH_3 \\ | \\ H-C-OH \\ | \\ C_6H_5 \end{array}$$

503 (-)-ephedrine

$$\begin{array}{c} CH_3 \\ | \\ H-C-NHCH_3 \\ | \\ HO-C-H \\ | \\ C_6H_5 \end{array}$$

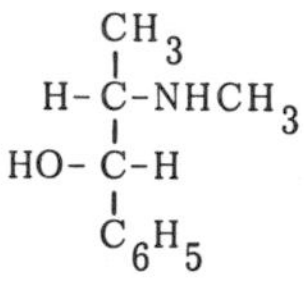

504 (+)-pseudoephedrine

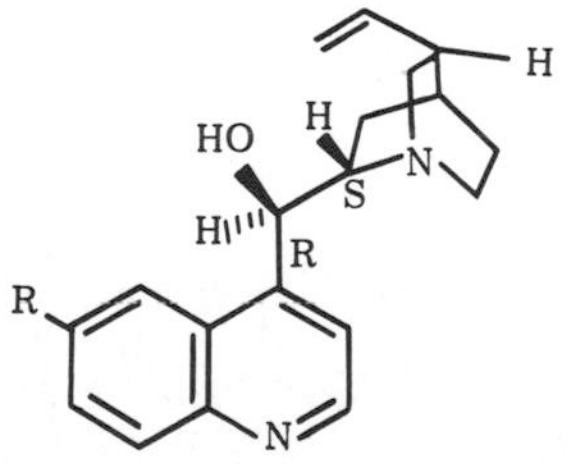

496 cinchonidine, R=H
497 quinine, R=OCH$_3$

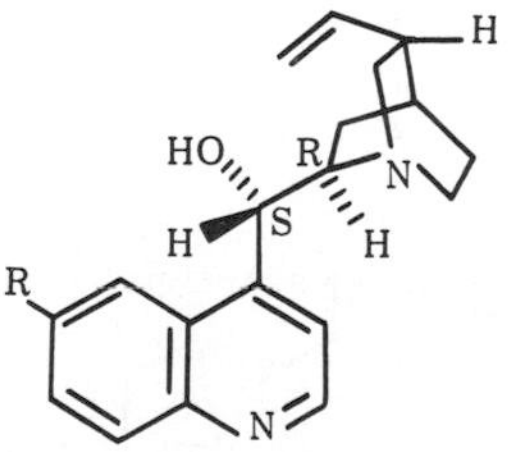

492 cinchonine, R=H
493 quinidine, R=OCH$_3$

Figure 4

There have been some interesting retrobiomimetic transformations of the *Cinchona* alkaloids, and one of these is shown in Scheme 62. In the demethoxy series the hexahydroderivative **505** on treatment with base gave an aniline **506**, which with lithium *t*-butoxide–benzophenone (modified Oppenauer) followed by treatment with acidified ethanol gave the indole **507**. Lithium aluminum hydride reduction then afforded dihydrocinchonamine (**508**).

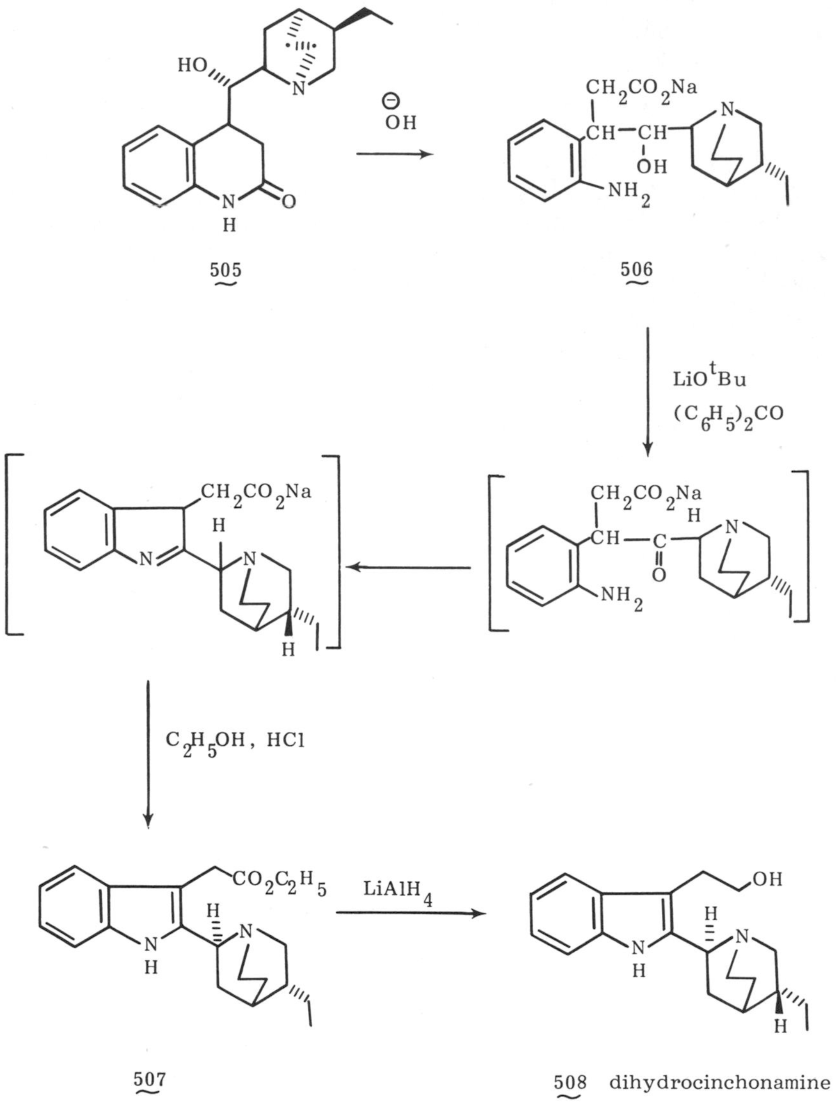

Scheme 62

9.20.3 Cinchonamine and Related Alkaloids

A number of the alkaloids of *Cinchona* and *Remijia* do not contain a quinoline nucleus but rather an indole or indoline nucleus; some examples include cinchonamine (**509**) and quinamine (**510**).

The most striking structural feature of these alkaloids is the quinuclidine nucleus, which is substituted in the same way as that in cinchonidine and related alkaloids.

Quinamine (**510**) was first isolated from *Cinchona* species in 1872, and cinchonamine (**509**) was obtained from *Remijia purdieana* in 1881. These alkaloids were not regarded as important by most investigators during the period when the structural and chemical work on the quinoline alkaloids was being carried out, and it was not until 1950 that Witkop deduced the structure of quinamine. In the same year Prelog and co-workers determined the structure of cinchonamine (**509**). Lithium aluminum hydride reduction of quinamine yields cinchonamine, and chromic acid oxidation of quinamine and cinchonamine affords 5-vinylquinuclidine-2-carboxylic acid (**511**).

509 cinchonamine   510 quinamine   511

These are the only examples where the quinuclidine nucleus of a cinchona alkaloid survives oxidative cleavage.

The stereochemistry of cinchonamine (**509**) was deduced by partial synthesis from cinchonidine (**496**) as described previously. The stereochemistry of quinamine (**510**) does not appear to be known in detail.

9.20.4 Synthesis

The first attempt to synthesize quinine was made in 1856 in the "rough laboratory" at the home of W. H. Perkin, Sr., who treated allyl toluidine with potassium dichromate. Quinine was not the product, but subsequent experiments with aniline led to the founding in Manchester of the great dyestuffs industry.

Perhaps this experiment may seem a little naïve to the uninitiated, but it illustrates an important point which the pharmaceutical industry consistently overlooks: namely that nature produces a wider range of structures and with more imagination than any group of synthetic organic chemists could ever devise. Furthermore, nature has a knack of doing things selectively, and as we shall see in the case of quinine and quinidine, stereochemistry is frequently the key to specific drug reaction.

More rational attempts at the synthesis of quinine and its derivatives were not to brought to fruition until some 75 years later by Rabe. He had worked on the degradation of quinine derivatives for many years and began to investigate the use of intermediates obtained in these reactions for the partial synthesis of compounds in the dihydro series.

Quinine itself was first synthesized by Woodward and Doering in 1944, and is a classic achievement in synthetic organic chemistry. The quinoline portion was derived from the ethyl ester of 6-methoxyquinoline 4-carboxylic acid (**512**) and the terpene portion from a stereospecifically constructed 3,4-disubstituted piperidine **513**, also known as homomeroquinene. Some of the key steps in this synthesis are shown in Scheme 63, and include the cleavage of the *cis*-decahydroisoquinolone **514** by basic ethyl nitrite and the Claisen condensation to yield the cinchonatoxine skeleton in the form of quinotoxine (**515**).

The subsequent steps had been worked out previously by Rabe in 1911. Treatment with hypobromous acid gave the *N*-bromo derivative **516**, which can be cyclized to quinidinone (**517**). Reaction with sodium isopropoxide in toluene affords a mixture of quinine (**497**) and quinidine (**493**).

The Roche group have described a modern approach to the synthesis of quinine which differs from the Woodward synthesis in several important aspects, although the overall strategy is quite similar.

A similar homomeroquinene derivative **518** is again the key intermediate, although the synthetic route is somewhat different (Scheme 64). Alkylation with the lithium salt of 4-methyl-6-methoxyquinoline (**519**) gave **520**, which was reduced with disobutylaluminum hydride to the alcohol **521**. On heating, dehydration occurs and internal cyclization takes place to afford a deoxyquinidine/deoxyquinine mixture (**522**). The oxygen at C-2 was introduced with molecular oxygen in the presence of base.

Taylor and Martin have reported a total synthesis of quinine (**497**), which also passes through the vinylquinoline (**523**). Condensation of the ylid **524** with *N*-acetyl-3(R)-vinyl-4(S)-piperidine acetaldehyde (**525**) afforded an intermediate olefin **526**, which was hydrolyzed *in situ* to **523**. Spontaneous intramolecular Michael addition then occurred to yield (38%) a mixture of deoxyquinine and deoxyquinidine (**522**), which was converted by base-catalyzed hydroxylation to a mixture of quinine (**497**) and quinidine (**493**) (Scheme 65).

An alternative synthesis of the quinine nucleus involves condensation of 6-methoxyquinoline 4-carboxaldehyde (**527**) with the quinuclidone derivative **528** (Scheme 66).

Yamada and co-workers have reported a synthetic approach to the cinchonamine skeleton which is interesting because it is possibly biomimetic in nature. Sodium borohydride reduction of the lactam alcohol **529** effectively cleaved the N-C-5 bond to give **530** in 90% yield.

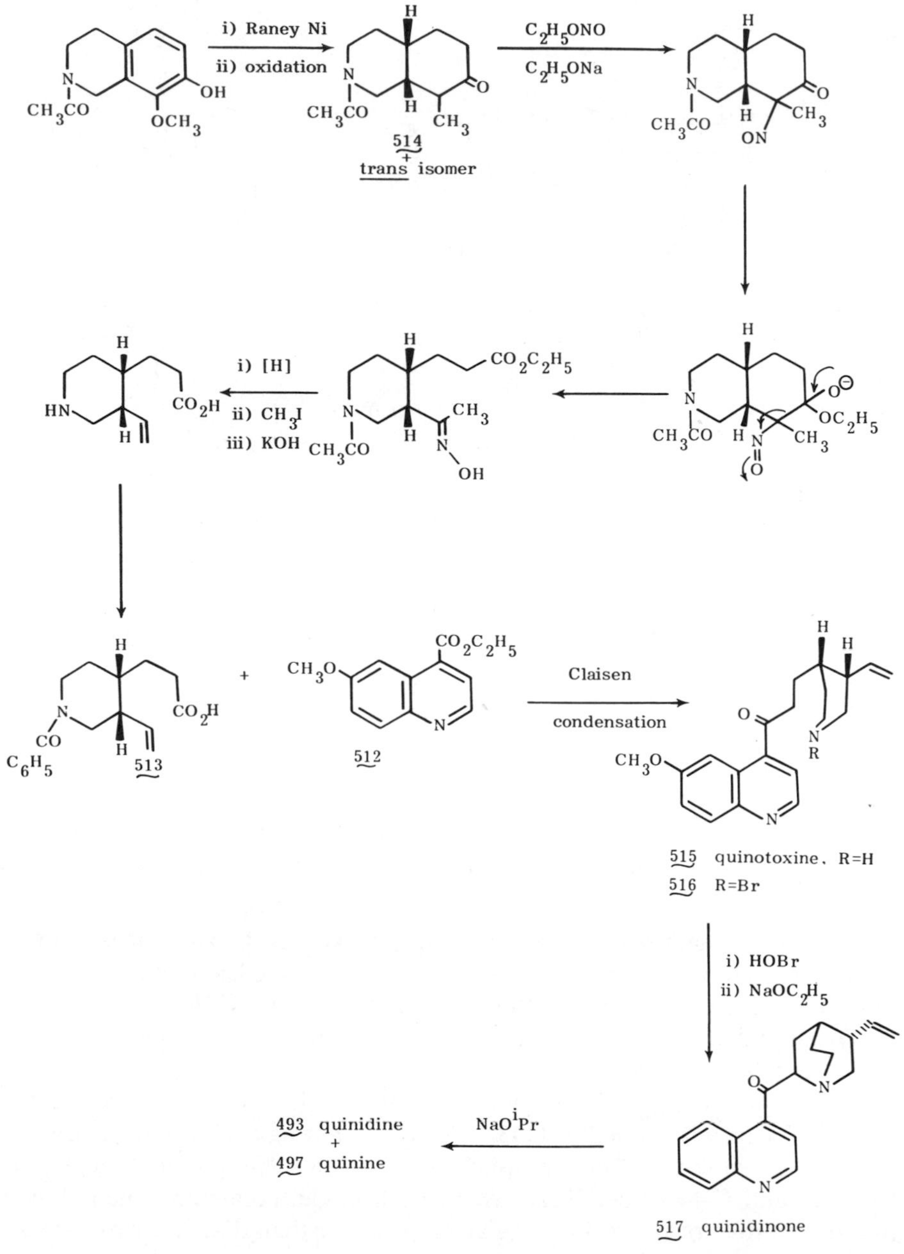

Scheme 63 Woodward–Doering synthesis of quinine (**497**).

715

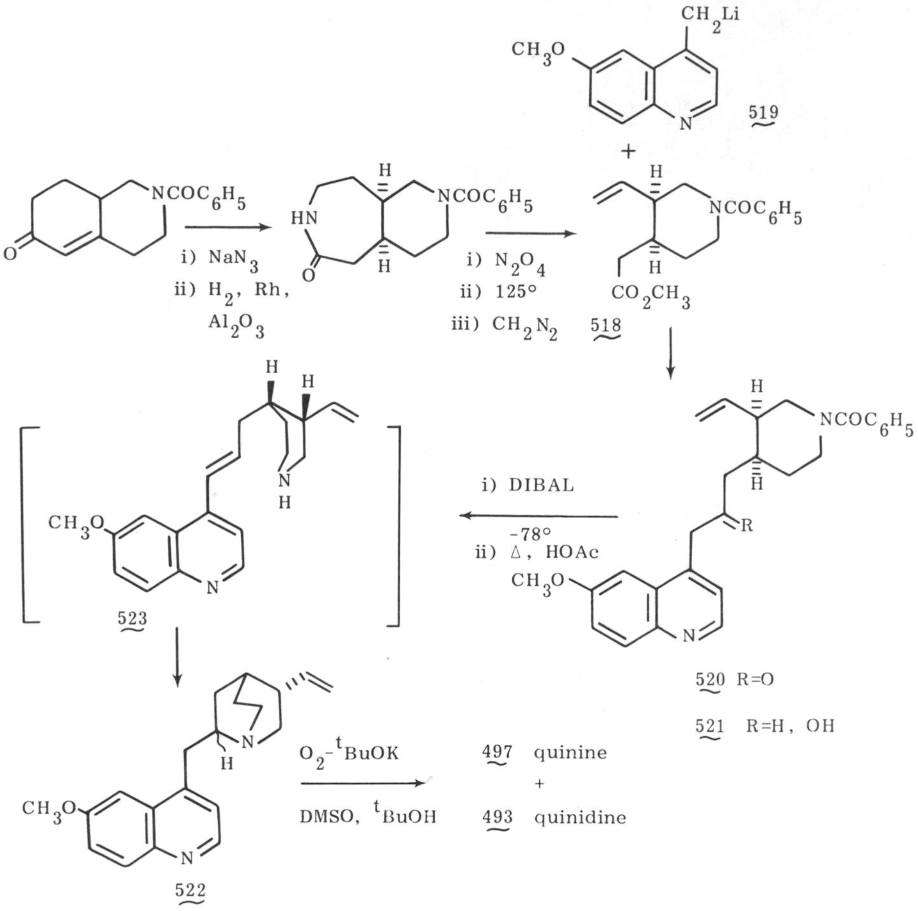

Scheme 64 Roche synthesis of quinine (**497**) and quinidine (**493**).

One of the main metabolites of quinidine (**493**) is the 20-hydroxy derivative **531**, and Carroll and co-workers have described a synthesis of this metabolite and determined its stereochemistry by ^{13}C NMR.

Osmium tetroxide–sodium periodate oxidation of **532**, obtained from quinidine (**493**) as shown, was treated with vinyl magnesium bromide at 25°C to afford a mixture of C-20 epimers **531** and **533**. In their ^{13}C NMR spectra on comparison with quinidine (**493**), the isomer **531** having the C-20 hydroxyl *syn* to C-16 shows a 5.68-ppm upfield shift and a 0.83 ppm downfield shift for C-16 and C-14 respectively. Work with model compounds had shown that the γ-effect of the 20-hydroxy group is larger than that of a vinyl group.

9.20.5   Spectral Properties

The UV spectra of cinchonine (**492**) and quinine (**497**) serve to readily distinguish the two compound groups. Cinchonine (**492**) shows a typical quin-

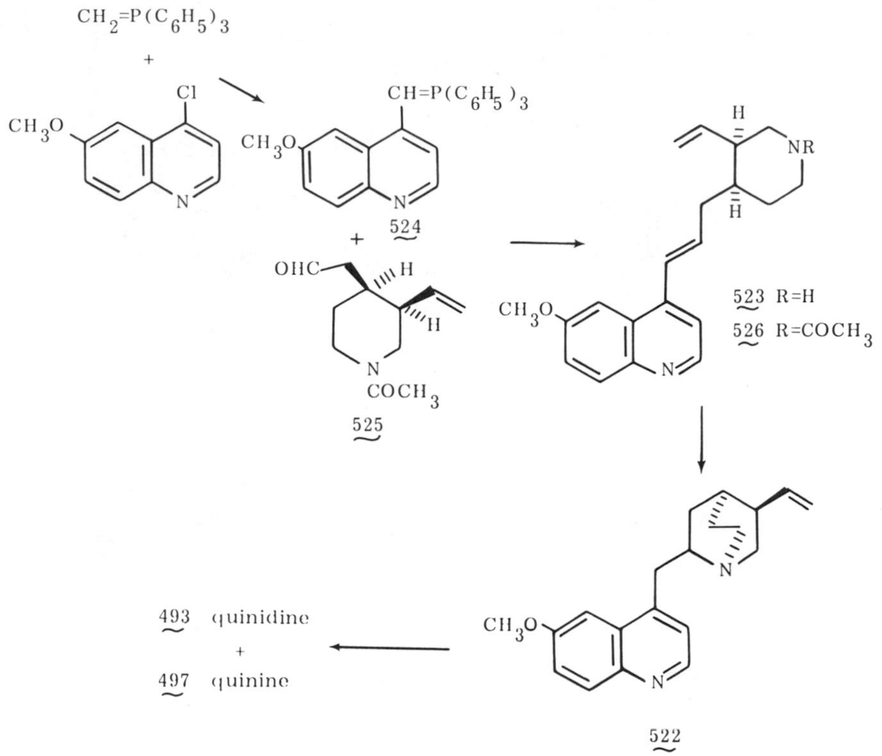

Scheme 65 Taylor–Martin synthesis of quinine (**497**) and quinidine (**493**).

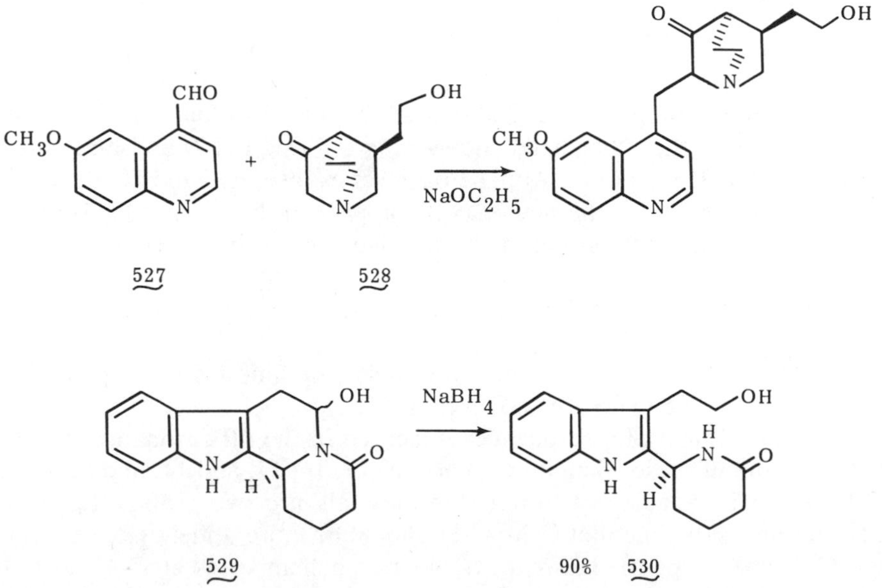

Scheme 66 Coffen–McEntee approach to the quinine nucleus.

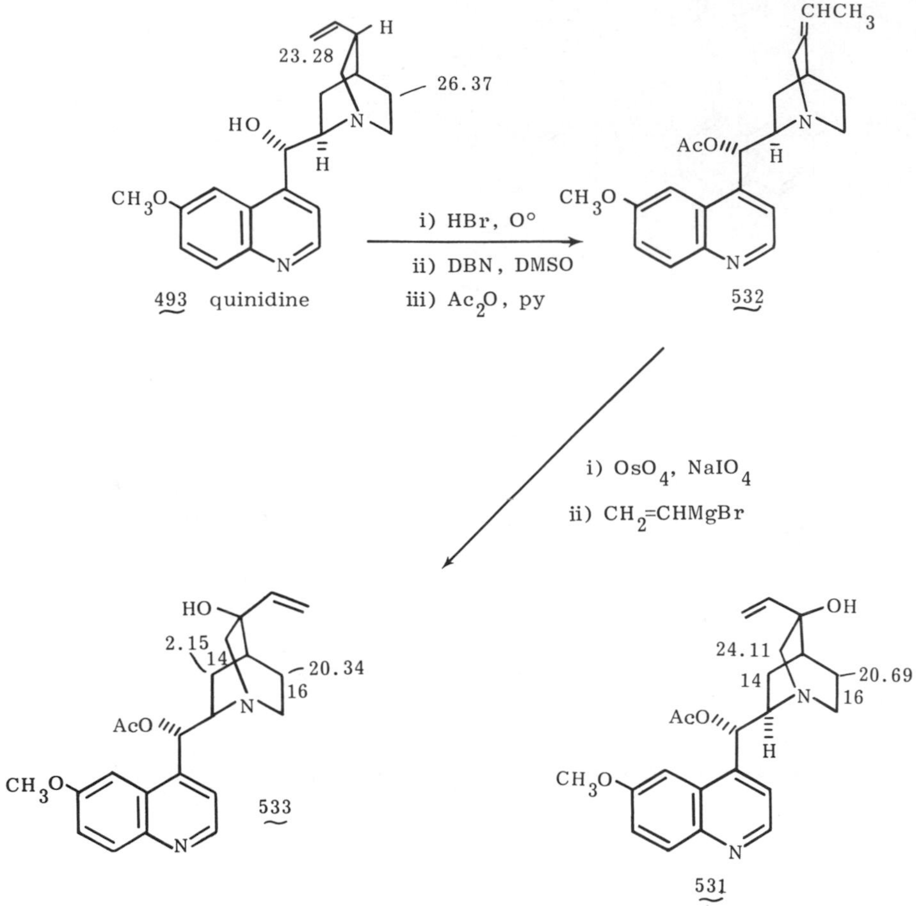

oline spectrum with maxima at 280 (log ε 3.7), 300 (3.6), and 315 nm (3.4). Quinine (497) on the other hand shows λ_{max} 275 (log ε 3.7) and bifurcated maxima at 320–355 nm (3.8). At pH 1 each series lose the band at ~280 nm; in the cinchonine series the new maxima appear in the range 305–315 nm, but appear at 315 and 350 nm in the quinine series, the latter peak being more intense.

Some of the proton NMR assignments of natural dihydroquinine in chloroform are shown in 534.

The ^{13}C NMR data for cinchonidine (496), quinine (497), 2-epiquinine (535), and quinidine (493) are shown in Table 7.

There are four triplet resonances observed in the off-resonance decoupled spectrum of these alkaloids for carbons 21, 16, 17, and 14, and because C-21 and C-17 are adjacent to nitrogen they fall into two groups. Because of the tertiary substituent at C-20, C-21 should be more upfield (55–56 ppm) than C-17 (40–43 ppm). However if the stereochemistries at C-3 and C-2 are inverted (e.g. quinidine), C-17 does not experience a marked γ-effect, and thus has a normal chemical shift.

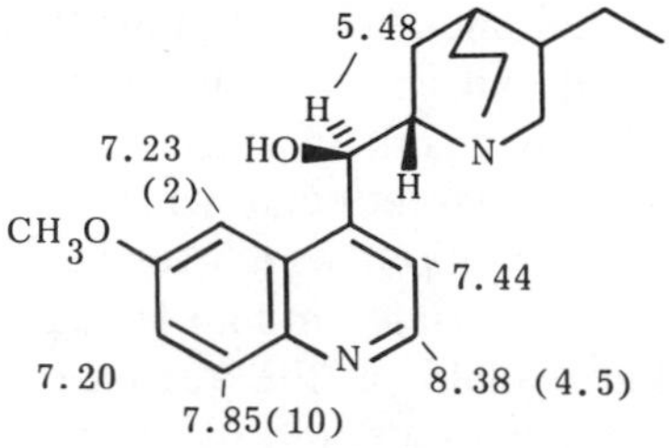

5.48

H

7.23 HO

(2)

CH₃O

N

H

7.44

7.20

N

8.38 (4.5)

7.85(10)

534

For carbons 16 and 14 a distinction may be made theoretically on the grounds that C-14 is α to two tertiary groups (C-15 and C-3), whereas C-16 is α to only one of these. Carbon-16 is therefore expected to be downfield of the C-14 resonance.

9.20.6 Pharmacology

Quinine and the *Cinchona* alkaloids display a wide variety of pharmacological properties. Quinine is toxic to many bacteria and other unicellular

Table 7 ^{13}C NMR Data for Representative Cinchona Alkaloids (δ, ppm, CDCl₃)

Carbon	Cinchonidine (496)	Quinine (497)	2-Epiquinine (535)	Quinidine (493)
2	71.46	71.51	71.17	71.51
3	60.24	59.85	. 61.32	59.55
5	149.75	147.01	147.30	147.06
6	122.85	121.09	121.04	121.09
7	149.75	148.33	144.26^a	148.18
8	125.19	126.43	127.95	126.33
9	118.10	101.40	102.47	101.25
10	126.38	157.44	157.25	157.34
11	128.78	118.30	119.87	118.25
12	129.46	130.89	131.33	130.94
13	147.84	143.67	144.64^a	143.63
14	21.19	21.44	24.92	20.76
15	27.76	27.74	27.76	28.44
16	27.46	27.46	27.08	26.24
17	43.04	43.00	40.55	49.42^a
18	114.13	114.08	114.33	114.23
19	141.61	141.66	141.18	140.54
20	39.77	39.76	39.62	39.96
21	56.81	56.86	55.25	49.85^a

Source: Data from C. G. Moreland *et al., J. Org. Chem.* **39,** 2413 (1974).

^{a} Tentative assignments.

organisms. It is also a local anesthetic of considerable duration. Orally it may cause gastric pain and nausea, and subcutaneous or intramuscular injections are painful and may result in sterile abscesses. Other important activities are the cardiovascular effects, which will be discussed subsequently, uterine muscle stimulation, which results in an oxytocic effect, a curare-like effect on skeletal muscle, and an analgesic effect comparable to salicylate. The alkaloid is available as the free base and as several salts; the most commonly used are the sulfate and the dihydrochloride.

Quinine is rarely employed alone as an antimalarial, being largely replaced by synthetics. It finds current use in combination with 8-aminoquinoline for the treatment of resistant strains of *Plasmodium falciparum* and for the relief of nocturnal leg cramps.

Malaria is again becoming "a major threat to the health of mankind" and $1\frac{1}{2}$ million people died from this disease in 1976. Strains of malaria resistant to the synthetic antimalarials are blamed for this alarming situation.

Toxicity from quinine is usually due to overdose or hypersensitivity. Cinchonism is the term normally applied to these effects, which include tinnitus, vomiting, diarrhea, fever, and respiratory depression.

Quinidine is more effective than quinine in its action on cardiac muscle. The major clinical uses of quinidine (as the sulfate salt) are the prevention or abolition of certain cardiac arrythmias. It is employed much less frequently than preparations of digitalis. The usual dose is 100–200 mg one to four times a day. The gluconate and polygalactouronate derivatives are available for oral use, but the alginate salt has fewer gastric side effects.

A number of galenical preparations of cinchona are used as bitter tonics and stomachics. Decoctions and acid infusions are occasionally used in gargles. Totaquine, a mixture containing not less than 70% and not more than 80% of total anhydrous crystallizable cinchona alkaloids, is used as an antiprotozoan where the purified product is not available.

LITERATURE

Reviews

Turner, R. B., and R. B. Woodward, *Alkaloids NY* **3**, 1 (1953).

Henry, T. A., *The Plant Alkaloids*, J. & A. Churchill, London, England, 1949, p. 418.

Solomon, W., in S. W. Pelletier (Ed.), *The Chemistry of the Alkaloids*, Van Nostrand Reinhold, New York, 1970, p. 301.

Synthesis

Woodward, R. B., and W. E. Doering, *J. Amer. Chem. Soc.* **66**, 849 (1944).

Uskoković, M., J. Gutzwiller, and T. Henderson, *J. Amer. Chem. Soc.* **92**, 203 (1970).

Gutzwiller, and M. Uskoković, *J. Amer. Chem. Soc.* **100**, 576 (1978) and following references.

Taylor, E. C., and S. F. Martin, *J. Amer. Chem. Soc.* **94**, 6218 (1972).

Coffen, E. L., and T. E. McEntee, *Chem. Commun.* 539 (1971).

Yamada, S., K. Murato, and T. Shioiri, *Tetrahedron Lett.* 1605 (1976).

Carroll, F. I., A. Philip, and M. C. Coleman, *Tetrahedron Lett.* 1757 (1976).

Spectral Data

Williams, T., R. G. Pitcher, P. Bommer, J. Gutzwiller, and M. Uskoković, *J. Amer. Chem. Soc.* **91,** 1871 (1969).

Moreland, C. G., A. Philip, and F. I. Carroll, *J. Org. Chem.* **39,** 2413 (1974).

9.21 *STRYCHNOS* ALKALOIDS

The genus *Strychnos* in the family Loganiaceae comprises a number of small trees and climbing shrubs native to Africa, Asia, and South America. The Asian species are sources of strychnine and brucine, whereas the South American species are better known as the source of certain types of curare, where the active constituents are dimeric *Strychnos* alkaloids.

The most important Asian *Strychnos* species are *S. nux vomica* L. and *S. ignatii* Berg. The seeds of these species contain 2–3% of total alkaloids, of which approximately one-half of the first drug and two-thirds of the second is strychnine. In each case the next most important constituent is brucine.

Few alkaloids have put as much fear into the hearts of man as strychnine. It is widely known as a poison, although in reality it is only moderately toxic, and its complex chemistry proved a fruitless challenge for many brilliant chemists for over a century.

9.21.1 Structure Elucidation and Chemistry

Strychnine was first isolated (1817) from *S. ignatii* Berg. by Pelletier and Caventou, and these workers were also responsible for the first isolation (1819) of brucine. Structural investigations were begun by Hanssen and Tafel and continued by Leuchs, Perkin, Robinson, and Wieland. Finally the structure of strychnine was established by Robinson and co-workers in 1946 and subsequently confirmed by X-ray crystallographic analysis, and a brilliant total synthesis by Woodward.

A discussion of the complex chemical work that led to the structure of strychnine is beyond the scope of this book. However, a number of interesting reactions are worthy of mention because they serve to point out the ease with which critical information can now be gathered spectroscopically.

Strychnine (**536**) contains one olefinic double bond, which is easily reduced catalytically. Since strychnine and dihydrostrychnine (**537**) have similar pK_a values (7.37 and 7.45 respectively), this double bond is not conjugated to nitrogen. In sharp contrast, treatment of strychnine with Raney nickel affords neostrychnine, an *isomer* of strychnine having a pK_a 3.8 and deduced to be the enamine **538**. Unlike most enamines however, protonation occurs on nitrogen and not the β-carbon, since the resulting immonium

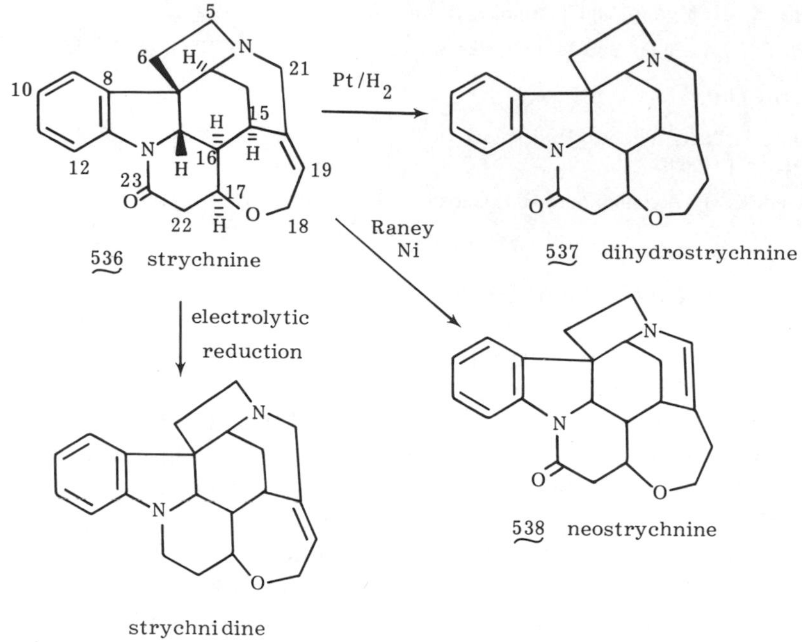

536 strychnine

537 dihydrostrychnine

538 neostrychnine

strychnidine

species would violate Bredt's rule. Mild hydrolysis of strychnine effects cleavage of the amide bond to form strychnic acid, which may be recyclized to strychnine with acid.

Some of the degradation products of strychnine are quite amazing. Alkaline pyrolysis affords indole (**3**) and 20% nitric acid gives 5,7-dinitroindole-2,3-dicarboxylic acid (**539**); hot alcoholic potassium hydroxide affords tryptamine (**2**)! Distillation from zinc dust yields carbazole (**540**) (Scheme 67).

In the presence of air, oxidation of strychnine (**536**) with copper salts affords pseudostrychnine (pK_a 5.6), containing one additional oxygen atom.

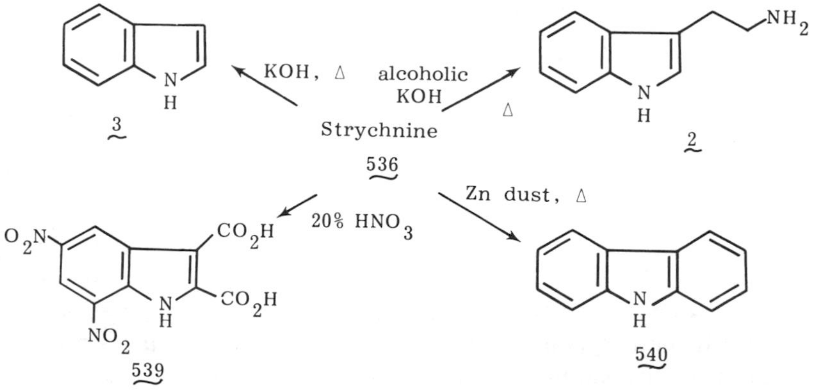

Scheme 67 Some degradations of strychnine (**536**).

Pseudostrychnine is an equilibrium mixture of the carbinolamine **541** and the keto amine **542**. The latter on Baeyer–Villiger-type oxidation affords the diamide strychnone (**543**), presumably by way of the lactone derivative **544**. This reaction was very important in deducing the structure of strychnine because it indicated that only one carbon atom was present between the indole nucleus and the tryptamine nitrogen.

9.21.2 Synthesis

The synthesis of strychnine was and is a synthetic achievement of almost unparalleled brilliance. Here is the master truly at work and certain aspects of the synthesis were clearly conceived along the then current biogenetic postulates. Mention will only be made here of the important or unusual steps in the sequence.

The tryptamine **545** on condensation with ethyl glyoxylate afforded a Schiff's base **546** that could be cyclized in novel fashion by treatment with toluenesulfonyl chloride in pyridine to **547**. After reduction of the imine and N-acetylation the next step is a biogenetically modeled ozonolysis of the catechol dimethyl ether system followed by cyclization to afford **548**. Dieckmann cyclization of the two ester groups could not be effected on the N-tosyl species and it was consequently necessary to form the N-acetyl derivative, which smoothly underwent cyclization. However, this compound existed in the enol form **549,** and it was therefore necessary to remove the oxygen by Raney nickel reduction of the corresponding benzyl thioenol ether and reduction of the double bond. The product was converted to the more thermodynamically stable form **550** and the acid resolved with quinine. The (−)-isomer was shown to be identical to a degradation product of strychnine.

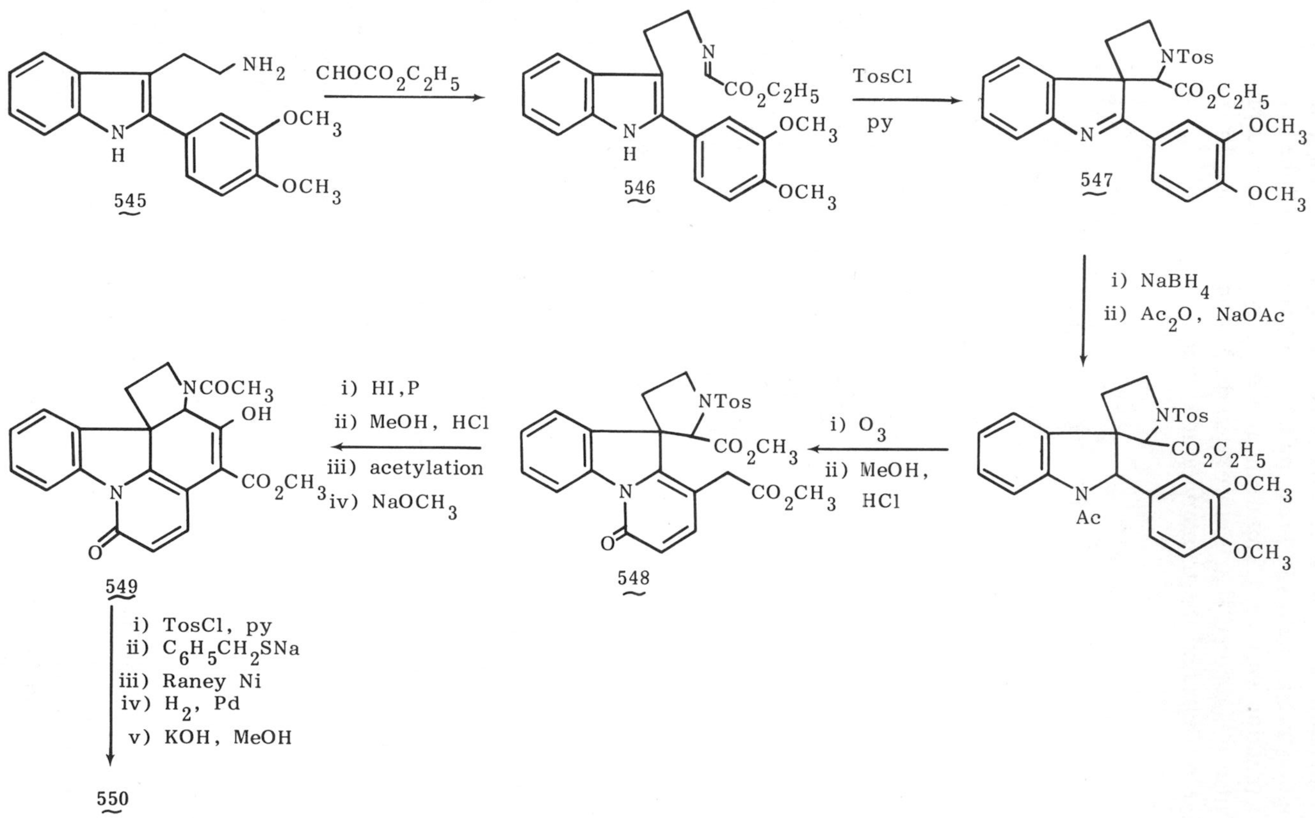

NH2
CHOCO2C2H5
OCH3
OCH3
N
H
545
N
CO2C2H5
OCH3
OCH3
N
H
546
TosCl
py
NTos
CO2C2H5
OCH3
OCH3
N
547
i) NaBH4
ii) Ac2O, NaOAc
NTos
CO2C2H5
OCH3
OCH3
N
Ac
i) O3
ii) MeOH, HCl
NTos
CO2CH3
CO2CH3
N
O
548
i) HI,P
ii) MeOH, HCl
iii) acetylation
iv) NaOCH3
NCOCH3
OH
CO2CH3
N
O
549
i) TosCl, py
ii) C6H5CH2SNa
iii) Raney Ni
iv) H2, Pd
v) KOH, MeOH
550

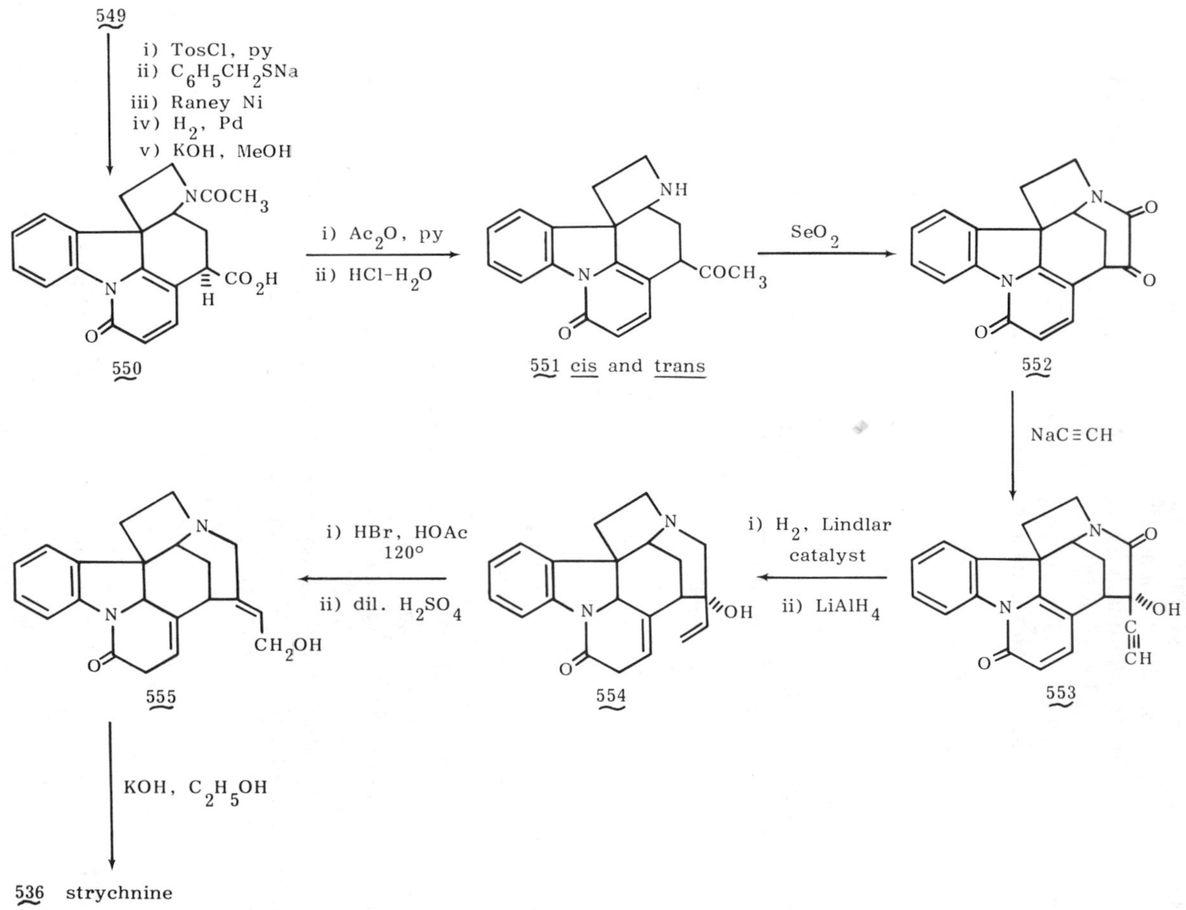

Scheme 68 Woodward synthesis of strychnine (536).

725

Treatment with acetic anhydride–pyridine followed by acid hydrolysis of the intermediate enol acetate gave the *cis* and *trans* isomers of the methyl ketone **551**. Selenium dioxide oxidation afforded the α-keto amide **552**. The final two carbons of strychnine were inserted by reaction with sodium acetylide to yield **553**. Controlled reduction of the acetylene to the alkene, followed by lithium aluminum hydride, unexpectedly gave a product **554** in which not only had the amide been reduced, but the pyridone had also undergone partial reduction. After 1,3-hydroxyl shift to afford the isomeric allylic alcohol **555**, alcoholic base yielded strychnine (**536**) (Scheme 68).

9.21.3 Alkaloids of Calebash Curare

The curare of certain South American Indian tribes was mentioned in Section 8.13 in connection with bisbenzylisoquinoline alkaloids.

Typically, the species of *Strychnos* used are *Strychnos toxifera* F. Schomb. and *S. castelneana* Baill., and upward of 40 dimeric alkaloids have been detected by partition chromatography on cellulose.

The dimeric Calebash-curare alkaloids are derived from the monomeric units Wieland–Gumlich aldehyde (caracurine VII) (**556**) and 2,16-dihydro-norfluorocurarine (**557**) or their N_b-metho salts. Wieland–Gumlich aldehyde (**556**) had previously been obtained by degradation of strychnine in 1932.

Caracurine VII methiodide (**558**) on heating with sodium acetate–acetic acid undergoes self-condensation by opening of the hemicacetal ring and attack of the indole nitrogen by the newly liberated aldehyde of a second molecule. Deprotonation then affords the most important Calebash-curare alkaloid C-toxiferine (**559**). A similar reaction with Wieland–Gumlich aldehyde itself affords the ditertiary base caracurine V (**560**), and this compound on heating in acetic acid under nitrogen may be transformed into bisnortoxiferine (**561**). The reaction may be reversed by brief warming with methanolic hydrochloric acid.

The other minor variations in these compounds are shown by the oxidation level of C-18 and C-18' and typical examples are C-alkaloid H (**562**) and C-dihydrotoxiferine (**563**). Each of these alkaloids together with C-toxiferine (**559**) represent the three major structure types of the quaternary Calebash-curare alkaloids.

C-Toxiferine (**559**), C-dihydrotoxiferine (**563**), and C-alkaloid H (**562**) undergo acid-catalyzed hydrolysis under nitrogen to give the corresponding monomeric species, and these were critical in determining the structures of the parent alkaloids.

9.21.4 Spectral Properties

There have been several reports of the ^{13}C NMR data for the simple *Strychnos* alkaloids, but the most comprehensive data are those of Verpoorte *et al*. The data for strychnine (**536**), brucine (**564**), pseudostrychnine (**541**),

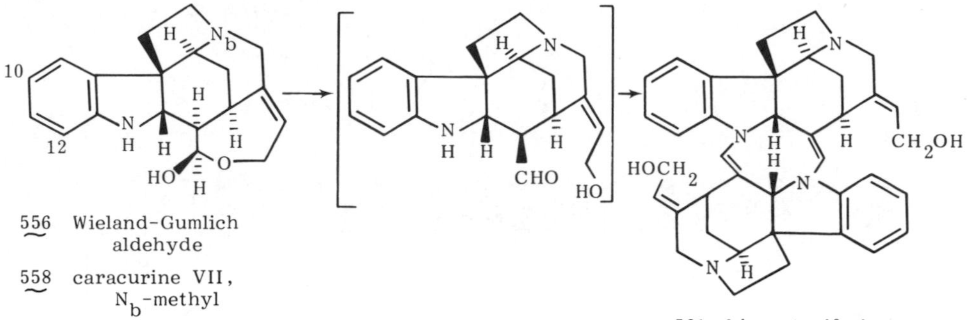

556 Wieland–Gumlich aldehyde

558 caracurine VII, N_b-methyl

561 bisnortoxiferine

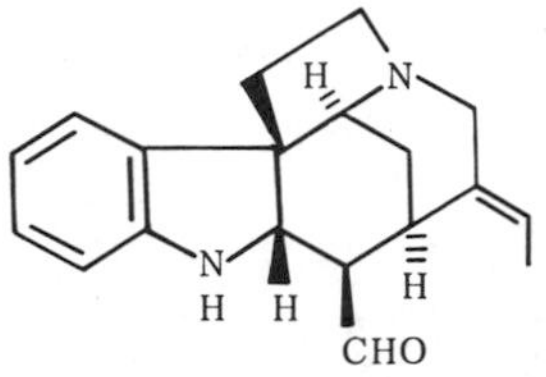

557 2, 16-dihydronorfluoro-curarine

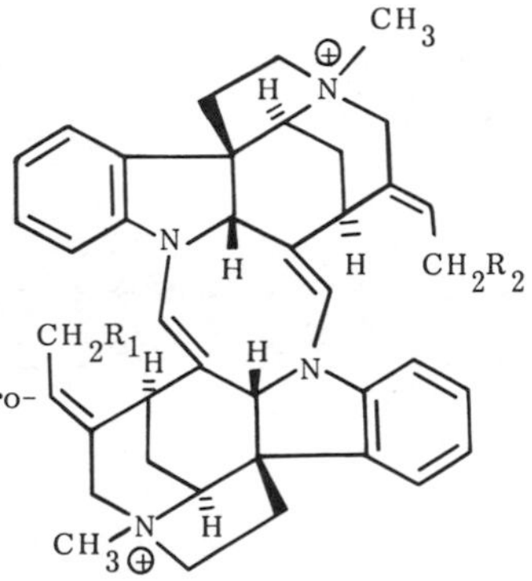

560 caracurine V

559 C-toxiferine, $R_1 = R_2 = OH$
562 C-alkaloid H, $R_1 = OH$, $R_2 = H$
563 C-dihydrotoxiferine $R_1 = R_2 = H$

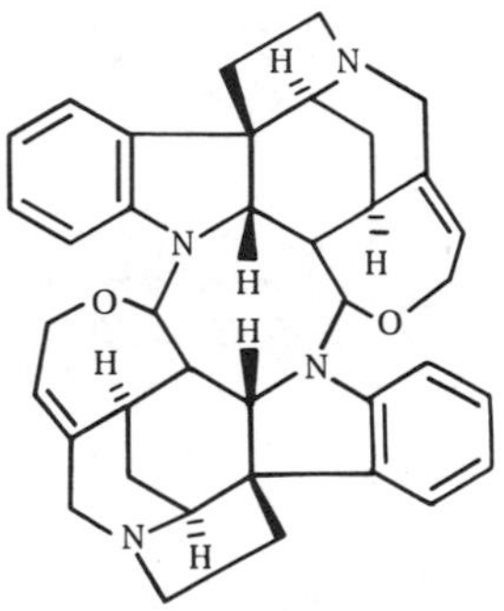

icajine (**565**), bisnordihydrotoxiferine (**566**), and alcuronium (**567**) are shown in Table 8.

The most important difference on the introduction of the 16-hydroxy group (strychnine *vs.* pseudostrychnine) is a downfield shift of 31.5 ppm for the C-16 proton.

9.21.5 Pharmacology

The poisonous properties of *Nux vomica* have been known since the sixteenth century, when it was introduced in Germany as a rat poison.

Pharmacologically, strychnine excites all portions of the central nervous system. It is a powerful convulsant and death results from asphyxia. Strychnine has no therapeutic uses in Western medicine, but Chinese workers have described the use of strychnine nitrate in the treatment of chronic aplastic anemia. Many rodenticides formerly contained strychnine as the principal active component, but these are becoming increasingly rare. Accidental poisoning of children however still occurs, and Diazepam is commonly used to treat cases of strychnine poisoning.

The Calebash-curare alkaloids are noted for their paralytic activity and

Table 8 ^{13}C NMR Data of the *Strychnos* Alkaloids

	Strychnine (536)	Brucine (564)	Pseudostrychnine (541)	Icajine (565)	Bisnordihydro toxiferine (566)	Alcuronium (567)
C-2	60.1	60.3	60.1	58.7	72.3	70.6
C-3	60.1	59.9	91.6	193.5	68.0	75.8
C-5	42.8^a	42.3	39.7	45.5	42.5	38.1
C-6	50.2	50.1	50.0^a	47.3	52.8^a	61.6
C-7	51.9	51.9	51.0	54.8	54.3	53.2
C-8	132.6	132.6	—	133.4	137.2	134.0
C-9	122.3	105.7	127.1	126.1	119.1	121.6
C-10	124.3	146.2	124.2	124.1	122.6	124.4
C-11	128.6	149.3	128.4	128.0	128.2	130.8
C-12	116.2	101.1	115.8	115.5	107.0	109.5
C-13	142.2	136.0	—	140.3	146.1	146.0
C-14	26.8	26.8	35.2	41.5	24.5	21.5
C-15	31.5	31.5	33.6	35.5	29.8	30.2
C-16	48.2	48.3	48.2^a	46.4	117.8	114.9
C-17	77.5	77.8	77.5	77.9	129.9	133.0
C-18	64.6	64.6	64.9	65.4	12.9	57.2^a
C-19	127.7	127.2	126.9	130.2	115.8	130.5
C-20	140.3	140.6	138.9	141.5	141.1	134.0
C-21	52.7	52.7	52.4	62.4	54.8^a	62.6
C-22	42.3^a	42.3	42.5	—	—	43.0
C-23	169.4	168.9	—	—	—	167.2
N—CH$_3$	—	—	—	39.4	—	—
O—CH$_3$	—	56.2, 56.4	—	—	—	—
N—CH$_2$—CH=CH$_2$	—	—	—	—	—	57.6^a, 130.5
						125.5

Source: Data from R. Verpoovte *et al.*, *Org. Mag. Reson.* **9**, 567 (1977).
^{a} Tentative assignments

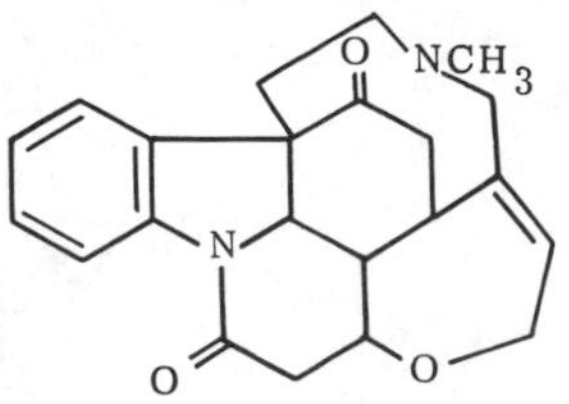

565 icajine

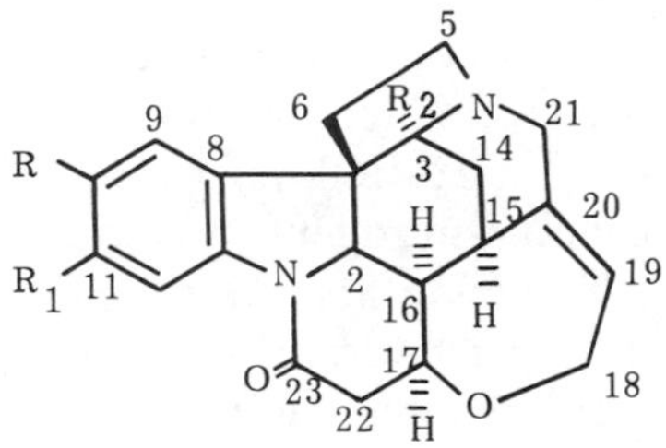

536 strychnine, R_1=H, R_2=H

564 brucine, R_1=OCH$_3$, R_2=H

541 pseudostrychnine, R_1=H, R_2=OH

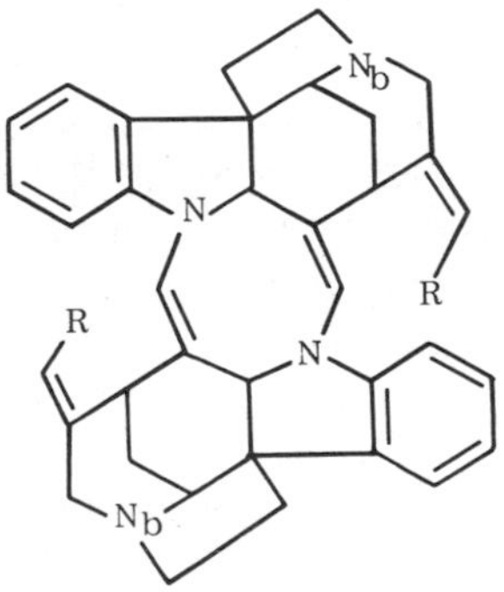

566 bisnordihydrotoxiferine, R=CH$_3$

567 Alcuronium, R=CH$_2$OH, N$_b$ -CH$_2$-CH=CH$_2$

are regarded as the most paralytic natural or synthetic neuromuscular block-ing agents. The typical assay involves injection of the alkaloid solution into the tail vein of mice and measuring the time that elapses before head drop, abolition of righting reflexes, and death or recovery.

The isolated alkaloids show a wider range of paralytic activity, and at least seven alkaloids are more active than *d*-tubocurarine. C-Toxiferine (**559**) for example is about 10 times more active, and C-alkaloid G (**568**) and C-alkaloid E (**569**) are 100 times more active than *d*-tubocurarine, being active at 1 μg/kg.

Further hydroxylation at C-18 increases the polarity and also the potency of the alkaloids, and those having a C-2-C-2′ ether bridge are particularly active.

In their paralytic activity in the rabbit head-drop test, C-toxiferine (**559**) is the most active, showing activity at 0.004 mg/kg (dose for *d*-tubocurarine is 0.15 mg/kg). Some of these data are shown in Table 9.

The most potent curarizing alkaloids have little effect on blood pressure

568 C-alkaloid G, R=H, R_1-OH
569 C-alkaloid E, R=R_1=OH
570 C-curarine, R=R_1=H

571 C-calebassine

in the cat, but one alkaloid, C-alkaloid B (structure unknown), lowers blood pressure significantly before any paralysis occurs.

C-Curarine (570), C-toxiferine (559), and the synthetic compound *N,N'*-diallyl bisnortoxiferine have been used clinically to induce paralysis. Toxiferine at a dose of 2 mg i.v. produces paralysis for 50–160 min in a 70 kg human and is probably the most useful specific curarizing agent. Advantages include no depression of blood pressure, no histamine liberation, and an absence of bronchoconstriction. The diallyl bis nor derivative (alcuronium) is of interest because of its short (15–20 min) reproducible paralytic action. It is a compound in quite wide clinical use in Europe.

Table 9 Pharmacologic Activity of the Calebash-Curare Alkaloids

Alkaloid	18-CH₂R	18'-CH₂R'	Mouse Head Drop	LD_{100} (μg/kg)	Duration of Paralysis (min)
C-toxiferine group					
C-toxiferine (559)	OH	OH	9	23	12
C-curarine group					
C-curarine (570)	H	H	30	50	4
C-alkaloid G (568)	OH	H	5	12	7
C-alkaloid E (569)	OH	OH	4	8	18
C-calebassine					
C-calebassine (571)	H	H	240	320	3

9.21.6 Akuammicine

Akuammicine (479) is a β-anilinoacrylate alkaloid obtained from several apocynaceous plants, including *Picralima nitida* and *Catharanthus roseus* L., although it may well be an artifact of isolation.

One of the remarkable reactions of akuammicine (479) is its facile degradation in methanol at 80°C to the biogenetically interesting betaine 572. At higher temperatures further decomposition to the carbazole 573 and 2-ethyl pyridine (574) occurs (Scheme 69).

Acid hydrolysis of akuammicine (479) affords the indolenine 575, which exists in equilibrium with the iminium salt 576 (Scheme 70). Hydride

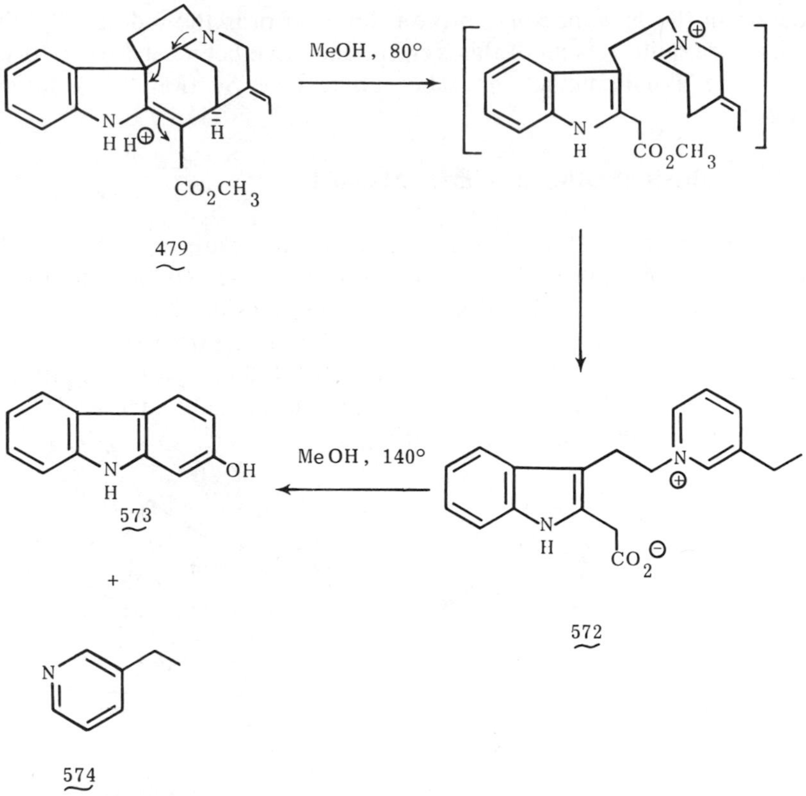

Scheme 69 Degradation of akuammicine (**479** in methanol.

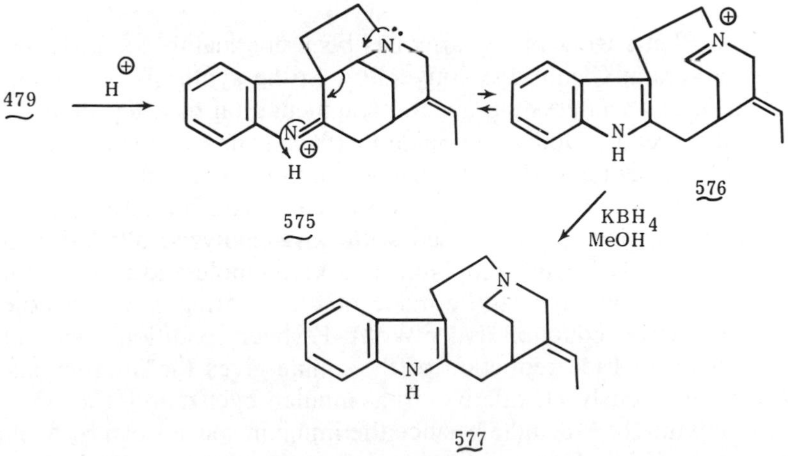

Scheme 70

731

reduction in the presence of a proton donor affords the indole **577**. These reactions are quite typical of this system and have been used to good value in both the akuammicine and tabersonine (see Section 9.23) series of alkaloids.

9.21.7 Synthesis of Other *Strychnos* Alkaloids

The two fundamental skeleta of *Strychnos* alkaloids are represented, as we have discussed previously, by the alkaloids tubifoline (**578**) and condyfoline (**579**). The skeletal differences (location of the ethyl side chain) might suggest that quite different synthetic plans would be necessary to produce these skeleta. In fact this is not so, for the synthetic (and biosynthetic) difference is only whether oxidative cyclization of an intermediate such as **580** takes place at the 3-position or the 21-position.

578　tubifoline　　　　　　579　condyfoline

580

The synthetic work in this area has been due mainly to Harley-Mason and co-workers at Cambridge University. In the synthesis of **578** and **579** use is made of an interesting general fragmentation of a tetrahydro β-car boline discovered by Dolby and Sakai in 1964. Thus treatment of the hexa-hydroindoloindolizine (**581**) with α,α'-dichlorobutyric anhydride gave an amide ester **582**, which could be hydrolyzed and oxidized to the keto amide **583**. The fourth ring is now closed with base-catalyzed alkylation of the ketone to afford **584**. This compound is a keto amide and is a very useful intermediate because use can be made of the differing carbonyl function-alities. Successive reduction under Wolff–Kishner conditions and and lith-ium aluminum hydride reduction of the amide gives the intermediate **580** mentioned previously. Oxidative transannular cyclization (Pt, O$_2$) then affords a mixture of **578** and **579** since the iminium species can be formed at either C-3 or C-21. This reaction was first observed by Schumann and Schmid.

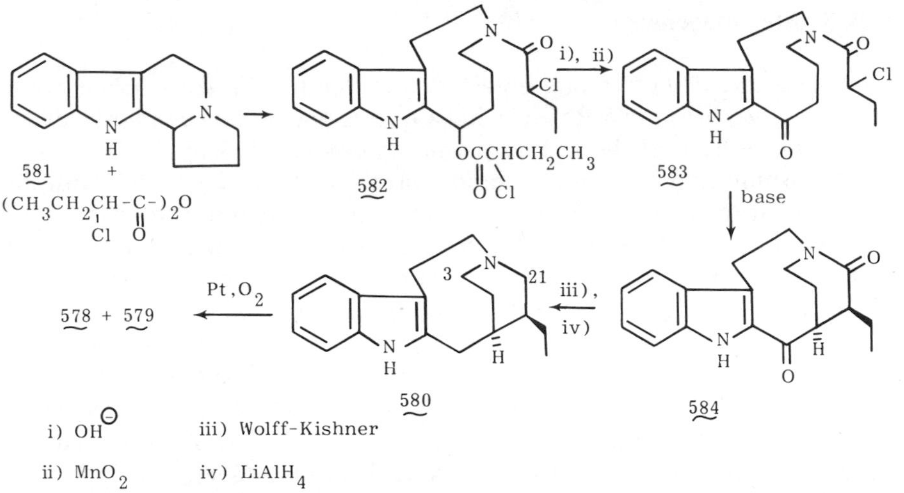

i) OH$^{\ominus}$ iii) Wolff-Kishner

ii) MnO$_2$ iv) LiAlH$_4$

In the 19, 20-dehydro series, Crawley and Harley-Mason have reported a synthesis of norfluorocurarine (**477**) (Scheme 71). Again the sequence begins with **581** and a methoxybromobutyric anhydride is used to permit subsequent introduction of the double bond. The aldehyde was introduced using a Wadsworth-Emmons procedure and the lactam amide reduced with aluminum hydride to afford **585**. Considerable problems were encountered with the selective reduction of the amide carbonyl without reducing the 19,20-double bond. Oxidation and acid hydrolysis of **585** afforded norfluorocurarine (**477**).

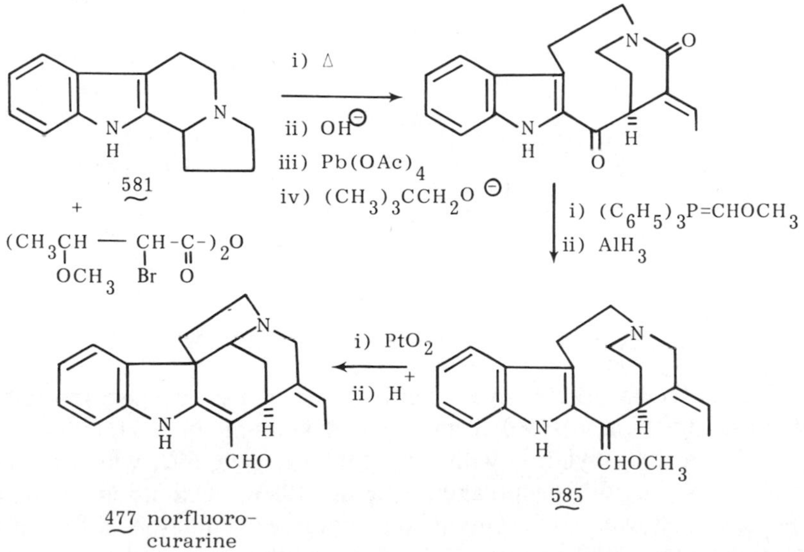

Scheme 71 Crawley–Harley–Mason synthesis of norflurocurarine (**477**).

9.21.8 Stemmadenine

Stemmadenine (586) was first obtained from the rare Brazilian plant *Stemmadenia donnell-smithii* (Rose) Woodson. It is one of the key intermediates in the biosynthesis of the *Aspidosperma* and iboga alkaloids.

Potassium permanganate oxidation gave condylocarpine (587) with liberation of formaldehyde. The condylocarpine was identical to the natural material. In the NMR spectrum of condylocarpine two singlets at about 4.0 ppm were assigned to the C-15 and C-21 protons. Note that condylocarpine (587) has a strong positive rotation, whereas akuammicine (479) has a strongly negative rotation. This is highly diagnostic of the two skeleta, because the stereochemistry at C-15 is predetermined biosynthetically.

Vallesia dichotoma produces a wide range of alkaloids, many of which are of biogenetic significance; stemmadenine (586) is one of these, and precondylocarpine (588) is another. Precondylocarpine (588), which may be regarded as a cyclized oxidation product of stemmadenine (586), had an indolenine UV spectrum (λ_{max} 280 nm) and a two-proton singlet for a primary alcohol group. Treatment with base gave condylocarpine (587) by a modified retro-aldol reaction, and each alkaloid gave 16-methylene aspidospermatidine (589) on lithium aluminum hydride reduction.

586 stemmadenine

587 condylocarpine

588 precondylocarpine

589

Kutney and co-workers have recently reported a stereospecific synthesis of 16-epistemmadenine (590) from dihydroakuammicine (591). N-Formylation followed by alkylation with formaldehyde gave 592, which on methanolysis afforded dihydropreakuammicine (593). Oxidation with lead tetraacetate followed by sodium borohydride reduction afforded 16-epistemmadenine (590) (Scheme 72). This synthesis served to define the stereochemistry at C-16 in stemmadenine (594) itself.

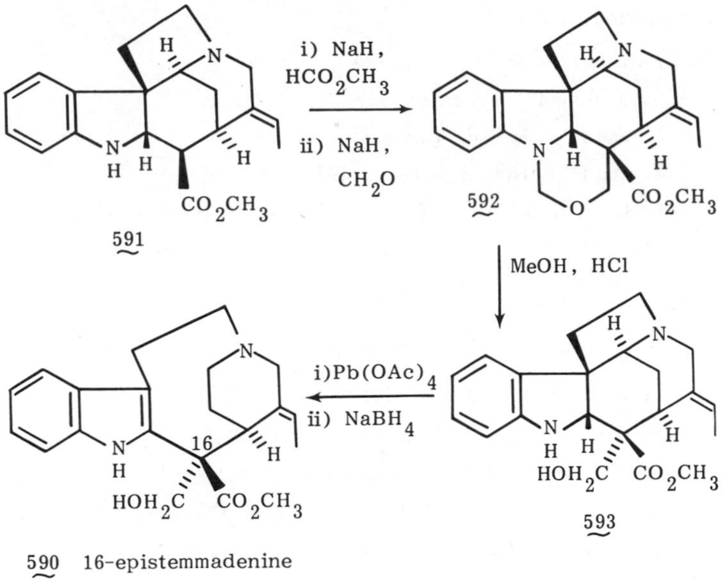

Scheme 72 Kutney and co-workers' synthesis of 16-epistemmadenine (**590**).

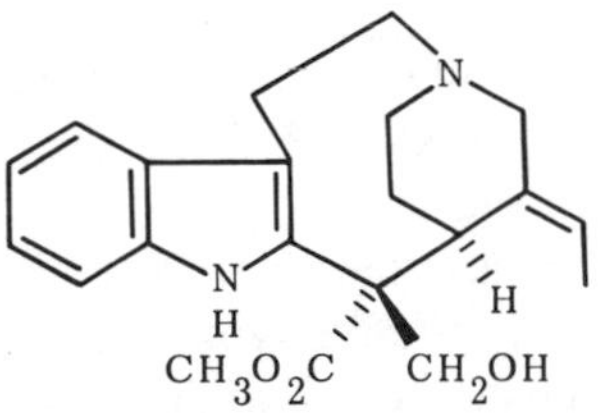

594   stemmadenine

LITERATURE

Reviews: Strychnine

Holmes, H. L., *Alkaloids NY* **1,** 376 (1950).

Holmes, H. L., *Alkaloids NY* **2,** 513 (1952).

Hendrickson, J. B., *Alkaloids NY* **6,** 179 (1960).

Smith, G. F., *Alkaloids NY* **8,** 591 (1965).

Taylor, W. I., *Indole Alkaloids, an Introduction to the Enamine Chemistry of Natural Products,* Pergamon, Oxford, 1966, p. 73.

Swan, G. A., *An Introduction to the Alkaloids,* Wiley, New York, 1967, p. 256.

Battersby, A. R., and H. F. Hodson, *Alkaloids NY* **11,** 189 (1968).

Reviews: Calebash Curare

Bernauer, K., *Fortschr. Chem. Org. Nature* **17**, 183 (1959).

Kaner, P., H. Schmidt, and P. Waser, *Farmaco* **15**, 126 (1960).

Battersby, A. R., and H. F. Hodson, *Quart. Rev.* **14**, 77 (1960).

Battersby, A. R., and H. F. Hodson, *Alkaloids NY* **8**, 515 (1965).

Gorman, A., M. Hesse, H. Schmid, P. G. Waser, and W. H. Hopf, *Alkaloids, London* **1**, 200 (1971).

Spectral Data

Verpoorte, R., P. J. Hylands, and N. G. Bissett, *Org. Mag. Reson.* **9**, 567 (1977).

Synthesis of Other Strychnos Alkaloids

Dolby, L. J., and S. Sakai, *J. Amer. Chem. Soc.* **86**, 1890 (1964).

Schumann, D., and H. Schmid, *Helv. Chim. Acta* **46**, 1966 (1963).

Dadson, B. A., J. Harley-Mason, and G. H. Foster, *Chem. Commun.* 1233 (1968).

Dadson, B. A., and J. Harley-Mason, *Chem. Commun.* 665 (1969).

Crawley, G. C., and J. Harley-Mason, *Chem. Commun.* 685 (1971).

9.22   SECODINE ALKALOIDS

In the section on the biosynthesis of monoterpenoid indole alkaloids the significance of the dihydropyridine **595** having an acrylic ester unit at C-2 of the indole nucleus is discussed.

Fundamentally, we can say that in all probability a very close relative of this compound is involved in the biosynthetic steps between stemmadenine (**586**) and the two major groups of alkaloids in the *Aspidosperma* and iboga series. Although probably not a biosynthetic process, these skeleta are formally derived by the Diels–Alder cyclization of the acrylic ester unit with a dihydropyridine in either of two ways (Scheme 73).

This type of cyclization was proposed in the early 1960s to account for the formation of these alkaloid skeleta, even though no alkaloids having this skeleton were known at that time.

Two critical isolations in 1968 changed this scene, for at this time the monomeric alkaloid **596** was isolated from *Tabernaemontana cumminsii* and the secamines, a group of dimeric alkaloids, were obtained from *Rhazya stricta* Decsne. and *R. orientalis* A. DC.

The secamines, of which tetrahydrosecamine (**597**) is an example, were characterized by an extremely intense base peak at either *m/e* 124 or 126 having the structures **598** and **599** respectively.

This observation led to the analysis of other alkaloid fractions of compounds having such intense base peaks and resulted in the isolation of a group of monomeric alkaloids (e.g. **600** and **601**), an additional group of dimeric alkaloids, the presecamines having the UV spectrum of an indole *and* a β-anilinoacrylic ester **602**.

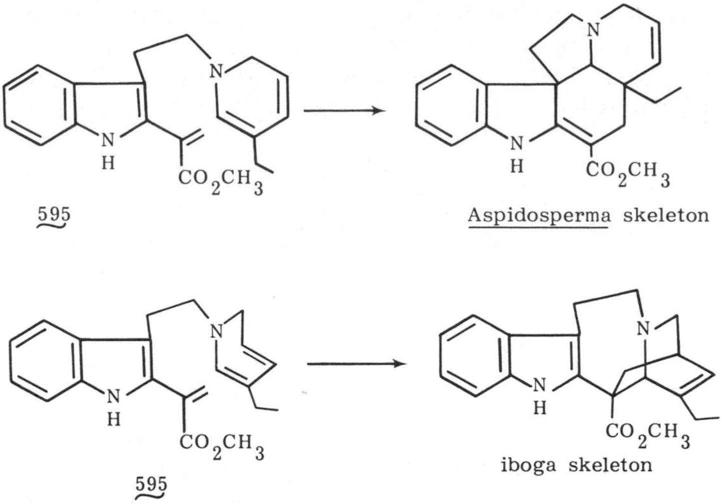

595

Aspidosperma skeleton

595

iboga skeleton

Scheme 73 Alternative cyclizations of the dihydropyridine (**595**).

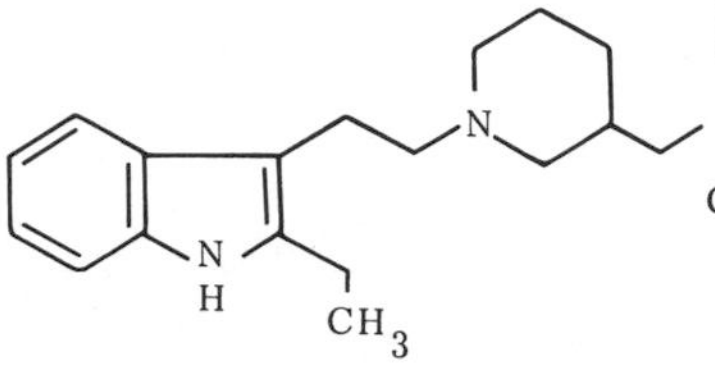

CH$_3$O$_2$C

16

16'

596

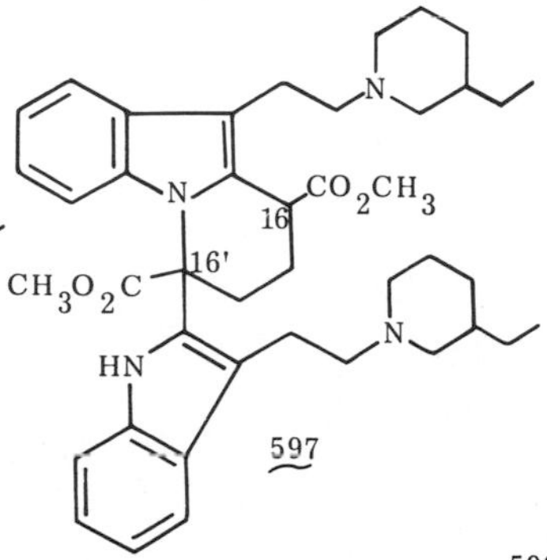

597

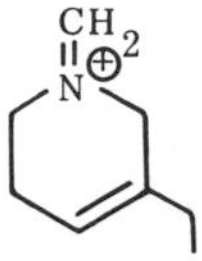

598

m/e 124

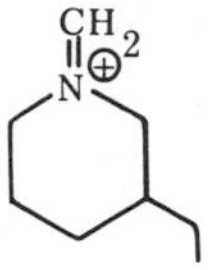

599 m/e 126

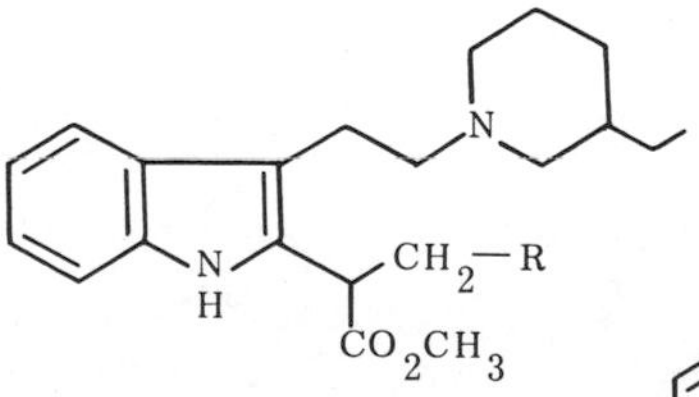

600 R=H

601 R=OH

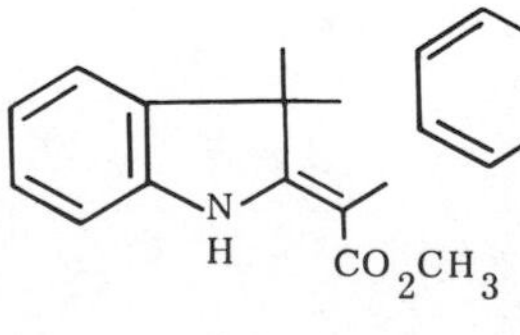

16'

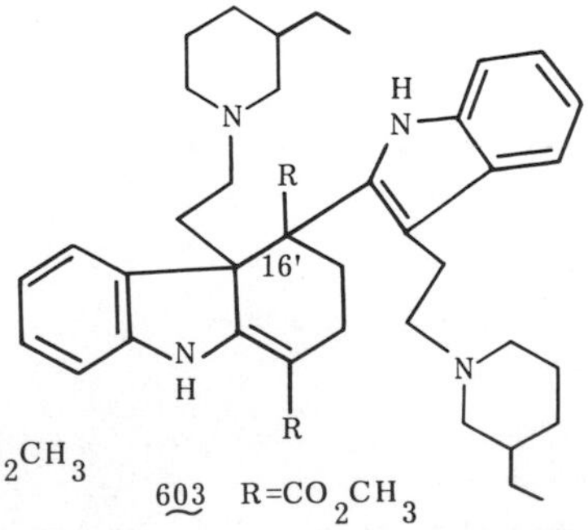

603 R=CO$_2$CH$_3$

602 β-anilinoacrylate unit

Tetrahydropresecamine (**603**), in spite of being dimeric, gave a mass spectrum at quite low probe temperatures which essentially consisted of two ions, at m/e 340 and m/e 126 (**599**). Biogenetically the m/e 340 ion can best be rationalized as having the structure **604**, formed by a retro Diels-Alder

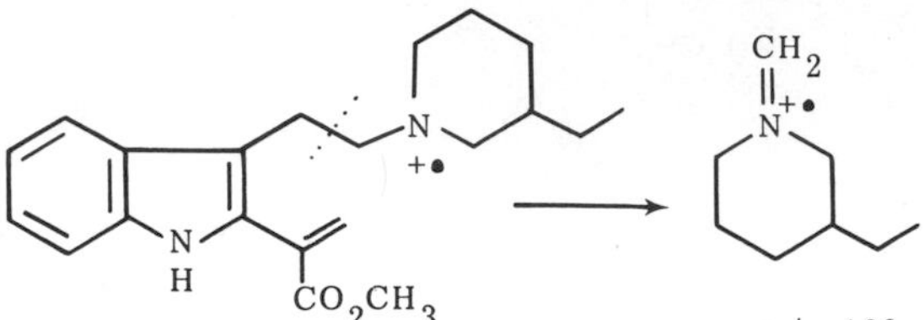

m/e 126

<u>604</u> m/e 340

reaction. The two units of the dimer are apparently identical from the simplicity of the mass spectrum, and since one of the indole units remains unreacted in the starting material (UV spectrum), it is the acrylic ester portion of this unit that is the ethylenic product in the retro Diels–Alder reaction.

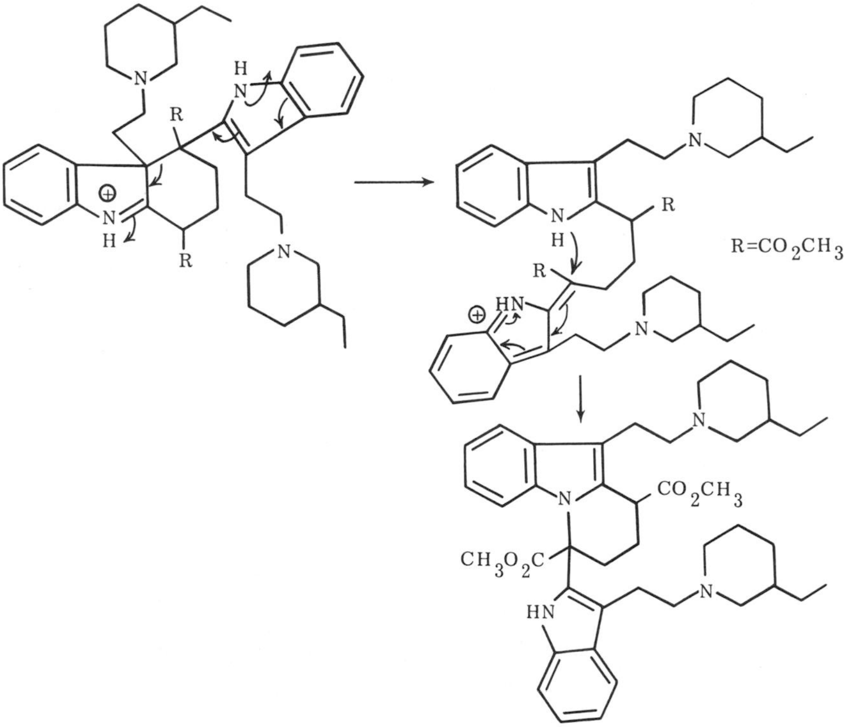

$R=CO_2CH_3$

Scheme 74   Mechanism of the rearrangement of the presecamines to the secamines.

The name presecamine was given to this compound class after a quite remarkable rearrangement was observed. In dilute hydrochloric acid at room temperature tetrahydro*pre*secamine (**603**) was quantitatively rearranged to tetrahydrosecamine (**597**). Moreover the reaction was kinetically controlled since only one diastereoisomer (at C-16) was produced. The mechanism of this reaction is shown in Scheme 74.

The retro Diels–Alder reaction to give a monomer could also be carried out by sublimation. The trapped product was identified as **605** from its spectral data and from a number of simple chemical reactions Figure 5. The compound **606** formed from presecamine itself was given the name secodine.

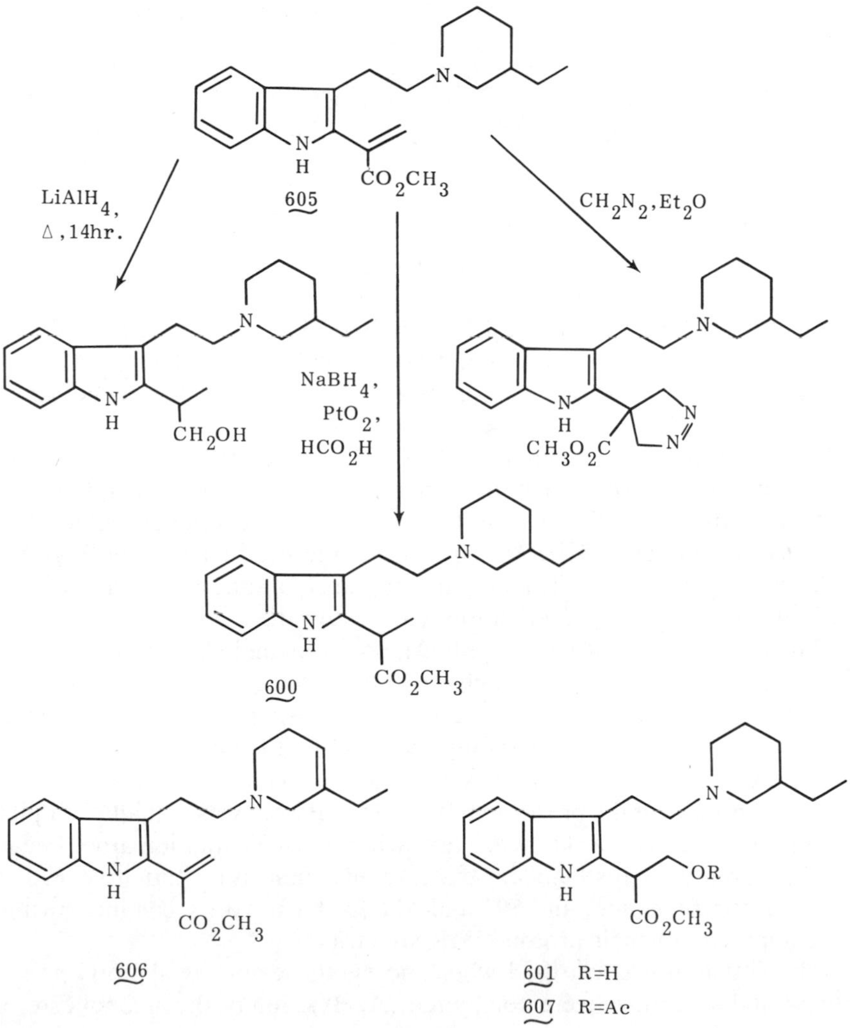

Figure 5

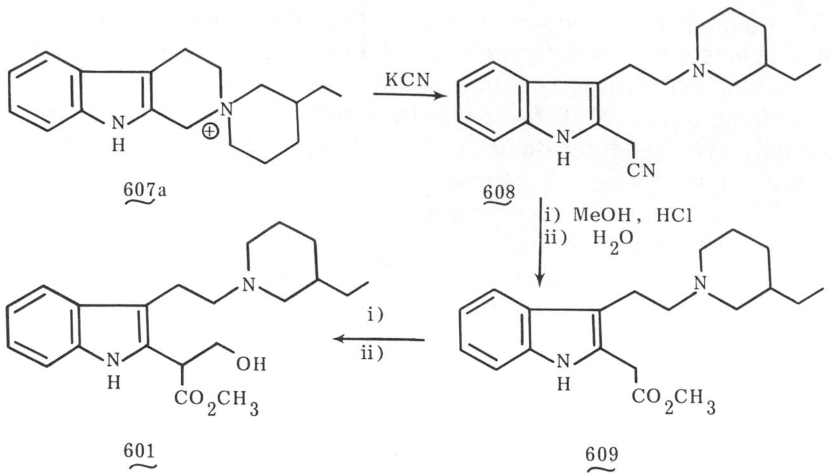

Scheme 75

In the condensed phase at 0°C a Diels–Alder reaction occurred to give tetrahydropresecamine (**603**) as a mixture of C-16′ diastereoisomers, and consequently the relationship was established between the secodines, the presecamines, and the secamines.

At about this time the alcohol ester **601** was obtained from *R. orientalis* and in an attempt to establish the primary alcohol function, acetylation with acetic anhydride–pyridine was attempted. Rather than the expected acetate ester **607** having an indolic UV spectrum, the product isolated was tetrahydropresecamine (**603**)! Thus in one step acetylation, elimination of acetic acid, and Diels–Alder dimerization had occurred.

The alcohol ester **601** was synthesized by cyanide cleavage of the quaternary ammonium salt **607a** to give **608** (see also quebrachamine synthesis in Section 9.23). Methanolysis gave the corresponding methyl ester **609**, which was formylated and carefully reduced to **601** (Scheme 75). This completed a total synthesis of the three alkaloid skeleta.

The synthesis of the presecamines and secamines was actually completed before the structure was known. This rather unusual situation arose because of the inability to distinguish structure **603**, and an alternative **610**, for tetrahydropresecamine, and **597** and **611** for tetrahydrosecamine, owing to the complexity of their proton NMR spectra.

The situation was clarified when the syntheses of the skatolyl presecamines* and secamines were completed. Acetylation of the alcohol ester **612**

---

* Skatole is the trivial name for 3-methyl indole.

gave the acetate ester **613** with no elimination to the acrylic ester **614** occurring unless a tertiary base such as triethylamine was present. The acrylic ester **614** underwent Diels–Alder condensation to give a mixture of the diastereomeric skatolyl presecamines **615**. Acid rearrangement afforded the skatolyl secamines **616** (Scheme 76). The aliphatic region of the proton NMR spectra of these compounds was extremely complex and not in agreement

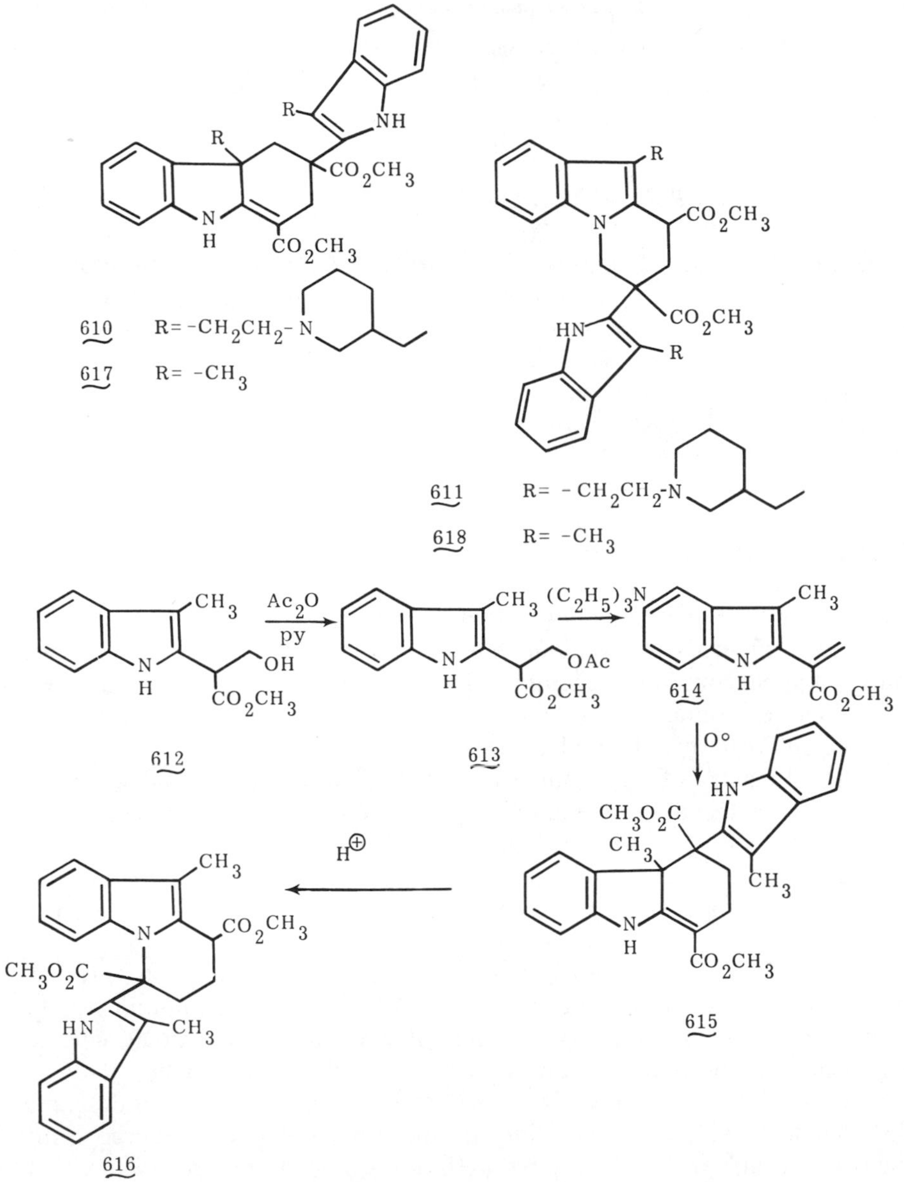

Scheme 76

with the concept of two AB systems expected for the structures **617** and **618**.

LITERATURE

Evans, D. A., G. F. Smith, G. N. Smith, and K. S. J. Stapleford, *Chem. Commun.* 859 (1968).

Cordell, G. A., G. F. Smith, and G. N. Smith, *Chem. Commun.* 189 (1970).

Brown, R. T., G. F. Smith, K. S. J. Stapleford, and D. A. Taylor, *Chem. Commun.* 190 (1970).

Cordell, G. A., G. F. Smith, and G. N. Smith, *Chem. Commun.* 191 (1970).

Cordell, G. A., G. F. Smith, and G. N. Smith, *J. Ind. Chem. Soc.* **55,** 1083 (1978).

9.23   *ASPIDOSPERMA* ALKALOIDS

This group of alkaloids is one of the fundamental skeleta of the monoterpenoid indole alkaloids, and it is not surprising to find that simple members of the series such as (−)-tabersonine (**619**) and (+)-1,2-dehydroaspidospermidine (**620**) are quite widely distributed.

The basic skeleton is that shown in **621**, but it may be modified by the formation of additional carbon–carbon bonds, as shown in the vindolinine (**622**), aspidofractinine (**623**), aspidoalbine (**624**), and kopsine (**625**) skeletal types.

9.23.1   Pentacyclic Alkaloids

The "parent" *Aspidosperma* alkaloid is tabersonine (**619**), which was first isolated from seeds of *Amsonia tabernaemontana* Walt. The alkaloid was only obtained from *Catharanthus roseus* when the seedlings were examined. It has not been obtained from the mature plant, where vindoline (**626**) is the major *Aspidosperma* alkaloid.

The chemistry of tabersonine is quite typical of the many alkaloids in this series that contain a β-anilinoacrylate (**602**) chromophore and which characteristically give λ_{max} 300 (log ϵ 4.03) and 328 nm (4.19) (Figure 6). These compounds have the same chromophore as akuammicine (**479**) and give rise to the same IR absorptions at 1610 and 1665 cm^{-1} for the unsaturated system with a sharp NH band at 3400 cm^{-1}.

Catalytic reduction of tabersonine (**619**) yields the 14,15-dihydro product, which is also known as vincadiffomine (**627**). Lithium aluminum hydride reduction affords a mixture of the alcohol **628** and the alkane **629**, whereas zinc–sulfuric acid reduction gives the 2,16-dihydro product **630**.

Acid hydrolysis of either **619** or **627** gave the corresponding indolenines **631** and **620**, which on potassium borohydride reduction undergo Smith cleavage to afford the corresponding indoles **632** and **633** respectively. In this process hydride is introduced at C-21 as shown in Scheme 76.

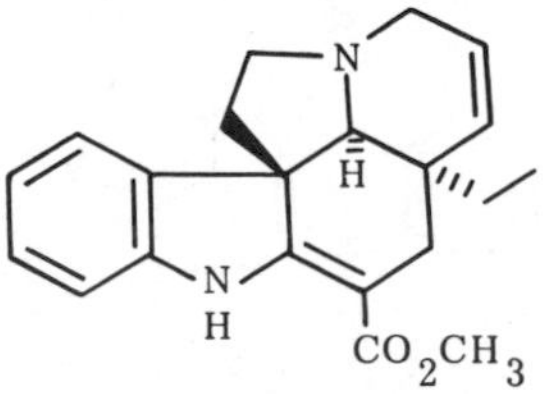

619 (−)-tabersonine

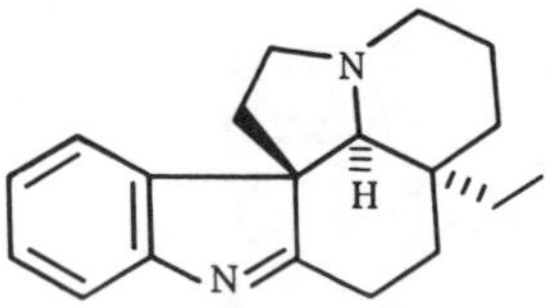

620 1,2-dehydro-
     aspidospermidine

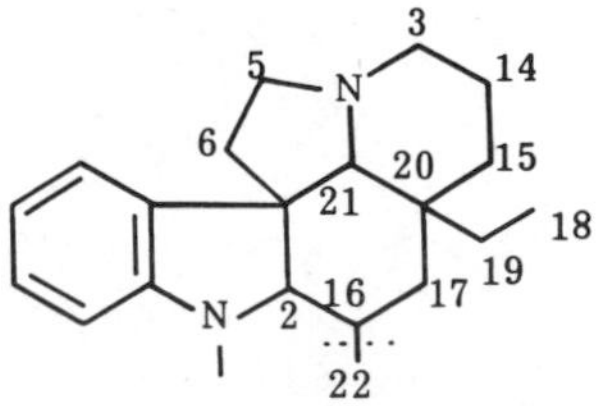

621 aspidospermine
     type

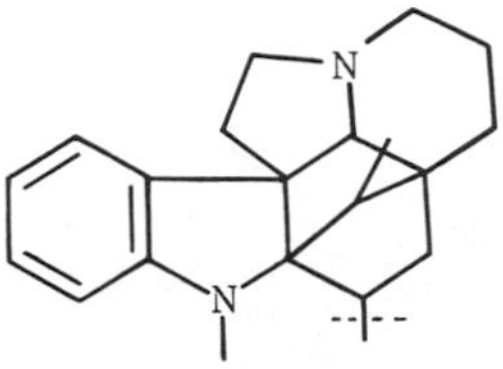

622 vindolinine
     type

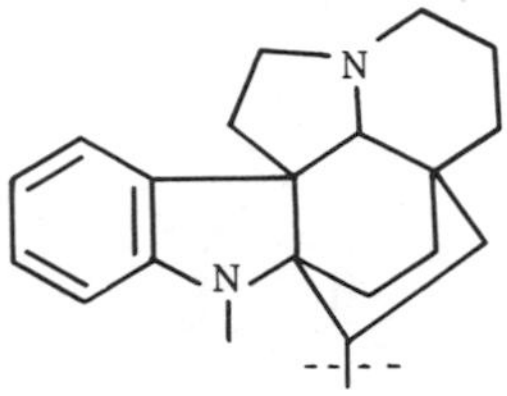

623 aspidofractinine
     type

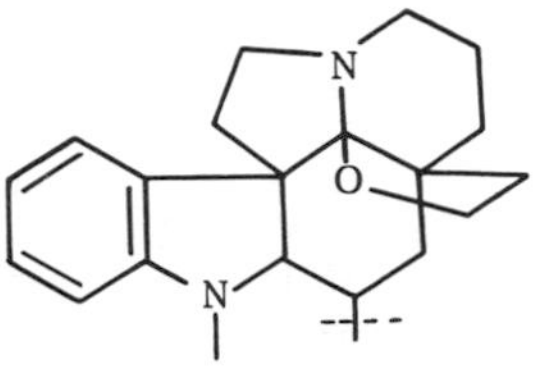

624 aspidoalbine type

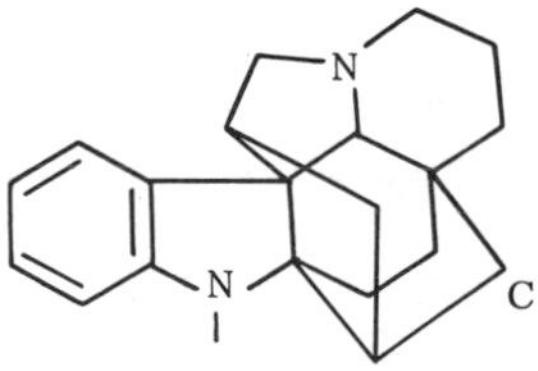

625 kopsine type

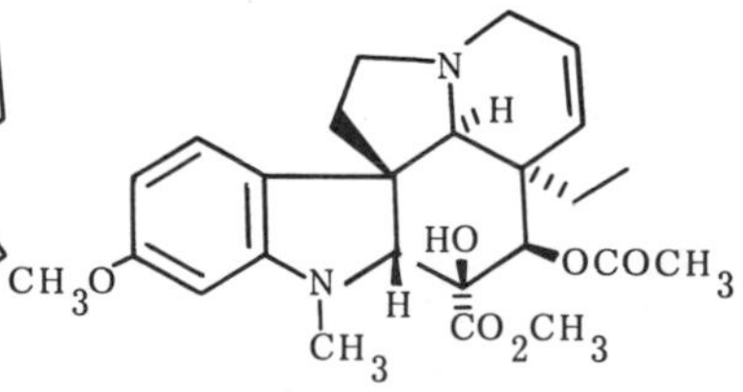

626 vindoline

The indoles **632** and **633** are compounds in the quebrachamine series which contain only one asymmetric center, and since both enantiomers are known from natural sources such a conversion allows the absolute configuration at C-20 to be determined.

Two types of mass spectral fragmentation are observed for the simple *Aspidosperma* alkaloids depending on the oxidation state of C-2 and C-16. If the compound is a β-anilinoacrylic ester, retro Diels-Alder reaction occurs in the C-ring to give a species that fragments predominantly through the 5,6-bond to afford two stable ions such as **634** and **635** from vincadifformine (**627**) (Scheme 77). The quaternary ethyl side chain is also lost easily to afford a radical ion of mass 29 plus any substituents. By careful observation of these ions, considerable information can be gleaned concerning the nature and location of substituents.

The second major fragmentation pathway (Scheme 78) occurs in the 2,16-dihydro series, where an alternative retro Diels-Alder reaction occurs to expel C-16 and C-17 with any substituents; this is followed by homolytic cleavage of the C-5–C-6 bond as shown.

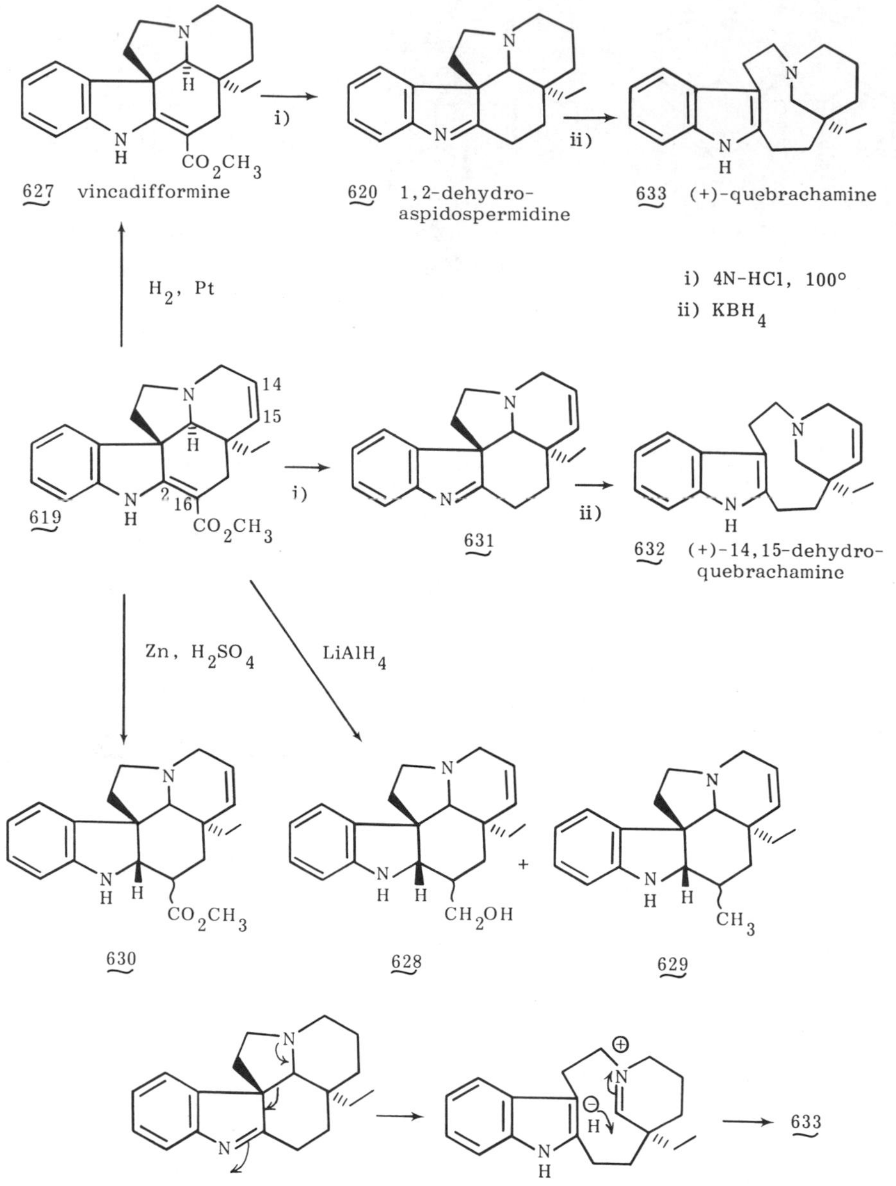

Figure 6 Chemical transformations of tabersonine (**619**).

Some of the *Aspidosperma* alkaloids obtained recently include (+)-3-oxominovincine (**636**) from *Tabernaemontana riedellii*, vincoline (**637**) from *C. roseus* and *Melodinus balansae* Baill., and cathovaline (**638**) from *C. lanceus* (Boj.) Pichon and *C. ovalis* Mgf.

Potier and co-workers have described a synthesis of deacetylcathovaline

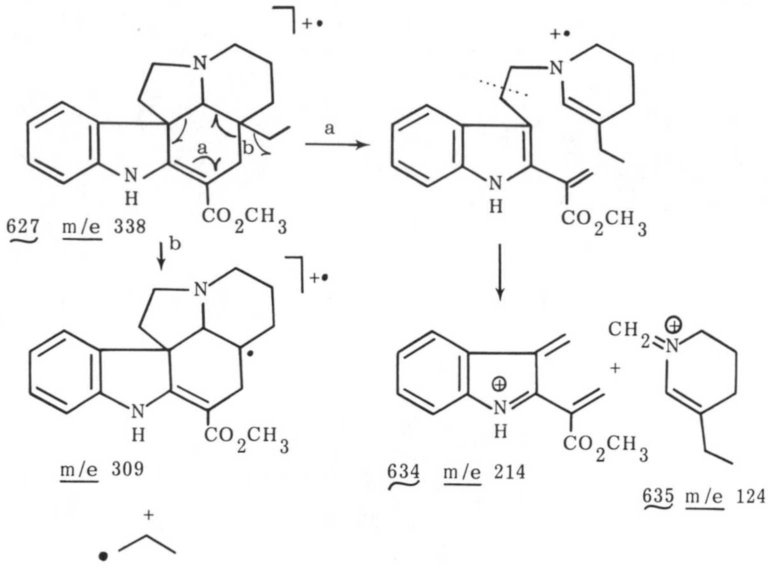

Scheme 77

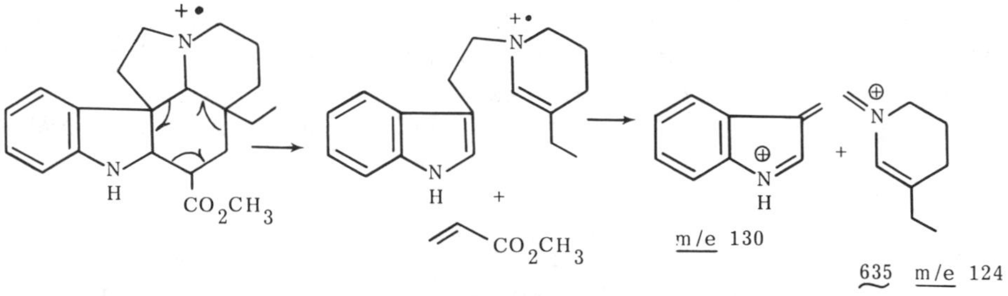

Scheme 78

(639) from the N-oxide of deacetylvindorosine ((640) by a modified Polonovskii reaction (Scheme 79).

The most novel *Aspidosperma* alkaloid to be obtained in the past few years is the neutral alkaloid rhazinilam (641) from *Rhazya stricta* and *Aspidosperma quebrachoblanco*. The structure was confirmed by an imaginative partial synthesis from (+)-1,2-dehydroaspidospermidine (620) by treatment with *m*-chloroperbenzoic acid followed by ferrous sulfate to give (−)-rhazinilam (641) in 30% yield.

9.23.2 Aspidofractinine Group

The aspidofractinine group of alkaloids is derived from the aspidospermine nucleus by closure of C-18 with C-2; they are therefore hexacyclic, and

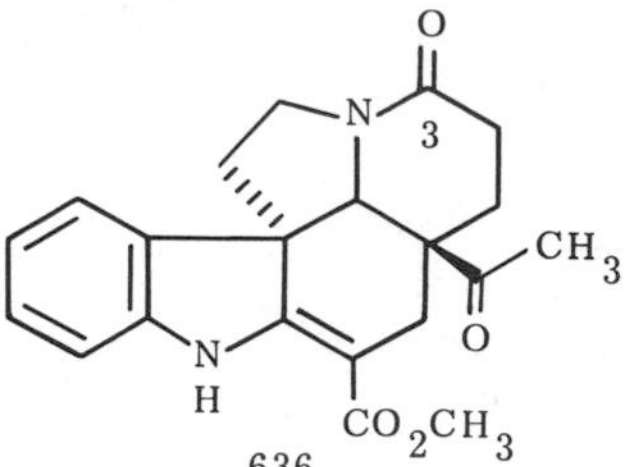

636

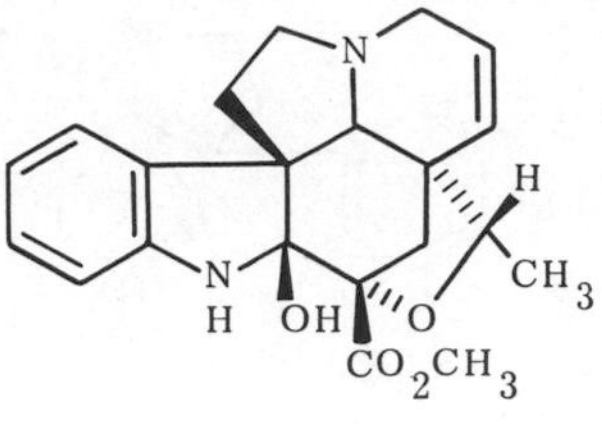

637   vincoline

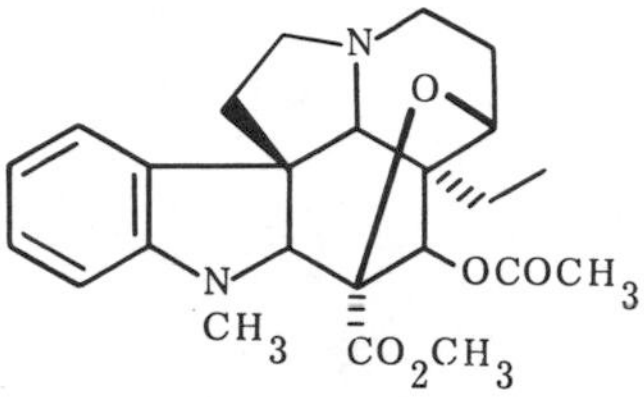

638   cathovaline

aspidofractinine (**642**) from *A. refractum* Mart. is a typical representative. A further complexity is found in alkaloids of the kopsane type, in which C-16 is joined to C-6. These groups of alkaloids are at present limited to the genera *Aspidosperma, Hunteria, Kopsia,* and *Pleiocarpa.*

The hexacyclic nature of the alkaloids was determined by a combination

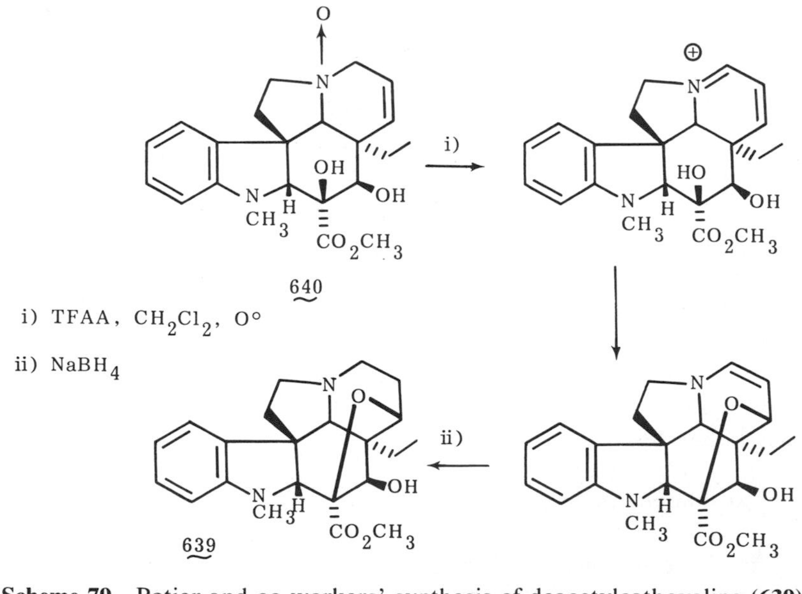

640

i) TFAA, CH_2Cl_2, 0°

ii) $NaBH_4$

639

Scheme 79   Potier and co-workers' synthesis of deacetylcathovaline (**639**).

620 (+)-1, 2-dehydro-
     aspidospermidine

641 rhazinilam

of high-resolution mass spectrometry and the absence of olefinic double bonds. The major mass spectral fragmentation pathways of these alkaloids are shown in Scheme 80. Pathway B gives an intermediate identical with that obtained from vincadifformine (**627**), but is dependent on the stereochemistry at C-16 for the H transfer to produce the ethyl side chain.

A correlation of (−)-minovincine (**643**) with (−)-aspidofractinine (**642**) was achieved when the former alkaloid was heated to 105°C in the presence of 3 N hydrochloric acid to give (−)-19-oxoaspidofractinine (**644**) in high yield. Clemmensen reduction of afforded (−)-aspidofractinine (**642**) (Scheme 81).

9.23.3 Vindolinine

Vindolinine is one of the major alkaloids of the genus *Catharanthus* and on the basis of spectral data was assigned the structure **645**. Two pieces of evidence were critical in assigning the structure. One of these was a three-proton doublet at about 0.85 ppm for the methyl group, and a "doublet" at 3.85 ppm thought to be indicative of a C-2 proton.

The structure for vindolinine and other alkaloids in the series stood for many years until the ^{13}C NMR spectrum of vindolinine was evaluated. These data indicated a quaternary carbon at 81.4 ppm and only three (C-16, C-19, and C-21) aliphatic methine carbons. Hence the C-19 carbon, rather than being linked to C-6, is linked to C-2. The ^{13}C assignments for vindolinine are shown in **646**.

9.23.4 Absolute Stereochemistry

The structure of (−)-aspidospermine-N_b-methiodide (**647**) was determined by X-ray crystallography, but its absolute stereochemistry remained un-

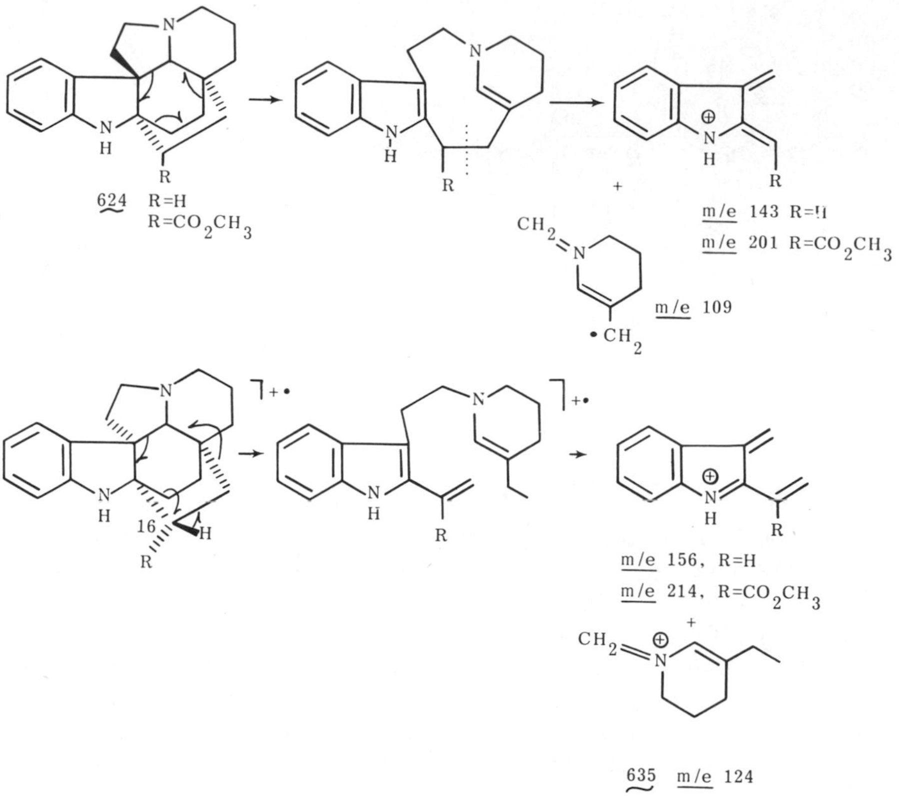

$CH_2 \overset{\cdots}{=} N$ m/e 109

· CH₂

<u>624</u> R=H
R=CO₂CH₃

m/e 143 R=H
m/e 201 R=CO₂CH₃

m/e 156, R=H
m/e 214, R=CO₂CH₃

$CH_2 = \overset{\oplus}{N}$

<u>635</u> m/e 124

Scheme 80 Mass spectral fragmentation of aspidofractinine alkaloids.

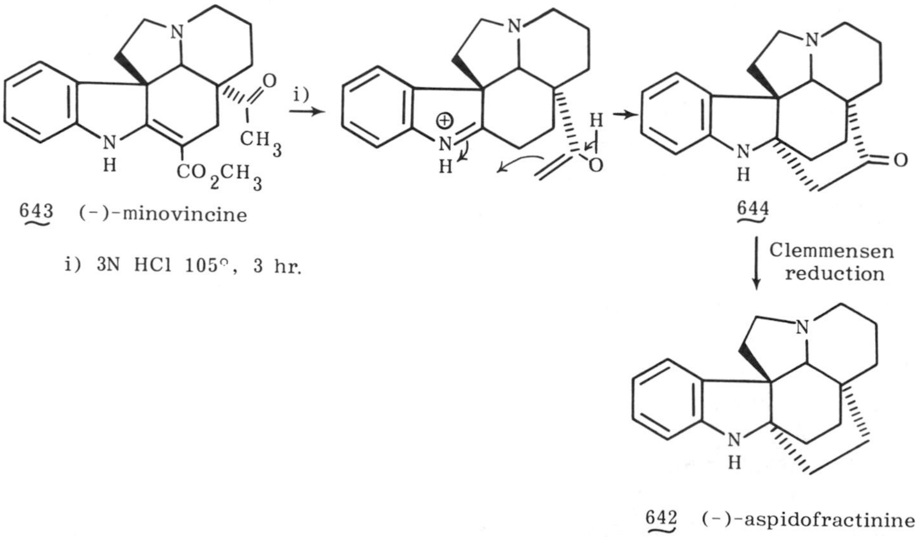

<u>643</u> (−)-minovincine

i) 3N HCl 105°, 3 hr.

<u>644</u>

Clemmensen
reduction

<u>642</u> (−)-aspidofractinine

Scheme 81 Correlation of minovincine (**643**) and aspidofractinine (**642**).

748

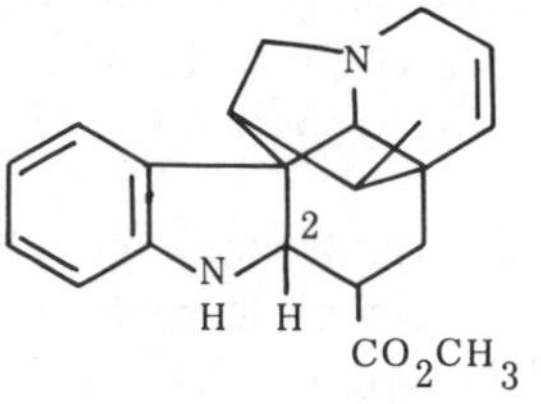

645   vindolinine
(old structure)

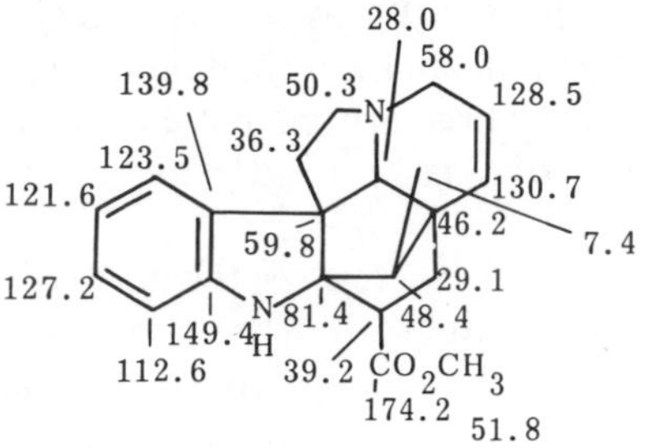

646

known until that of the related compounds cleavamine (**648**) and *N*-acetyl pseudoaspidospermidine (**649**) had been deduced. The optical rotatory dispersion curves with maxima at 264 and 237 nm are critical in defining the absolute stereochemistry.

The absolute configuration in which C-6 is β in an *N*-acetyl dihydroindole is typified by a positive Cotton effect in the region 262–278 nm and a trough in the region 224–247 nm depending on the UV chromophore. This rule also applies to the *Strychnos* alkaloids such as *N*-acetyltubifolidine (**650**). A weak Cotton effect is also observed in the region 285–290 nm, but the sign is not related to the absolute stereochemistry. The sign of the α_D is *not* diagnostic of the absolute configuration in this series, for both (−)-aspidospermine (**651**) and (+)-aspidospermidine (**652**) have the same absolute stereochemistry.

9.23.5   Synthesis

There has been an enormous effort in recent years in the area of *Aspidosperma* alkaloid synthesis, and numerous successes have been achieved.

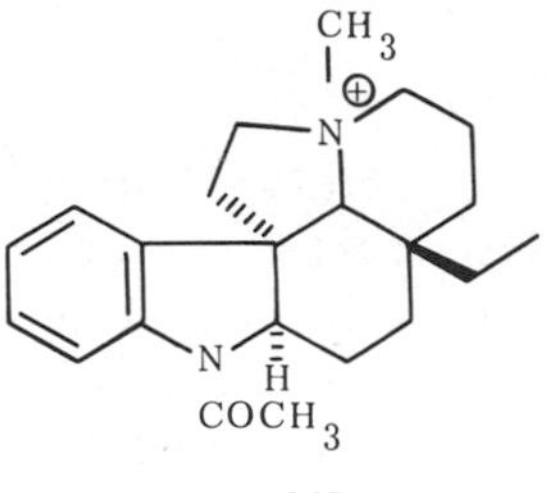

647

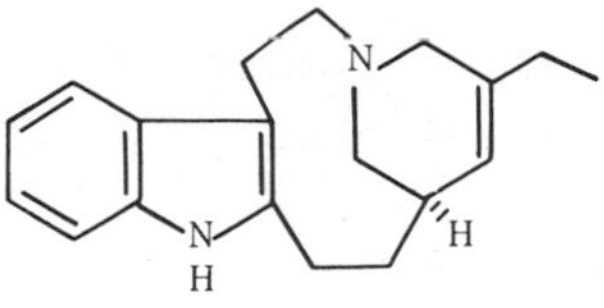

648

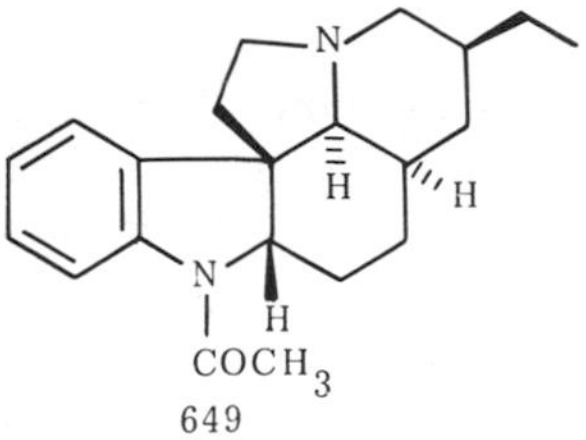

649

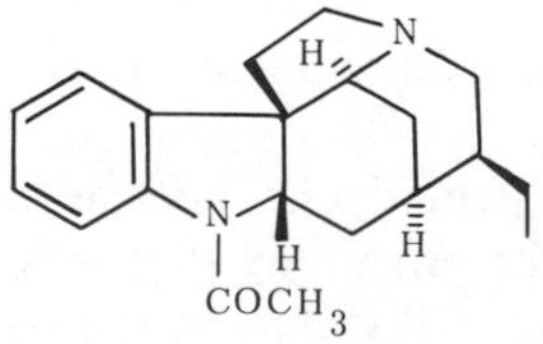

650   N-acetyltubifolidine 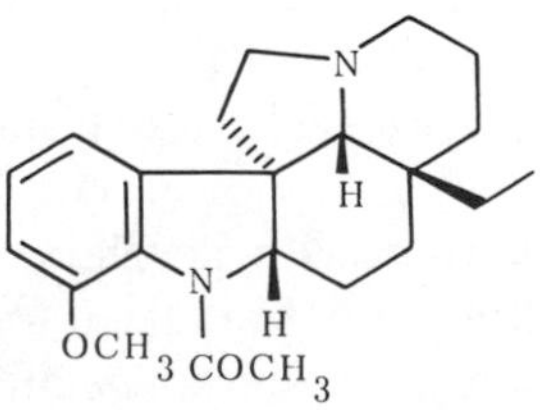   651   (-)-aspidospermine 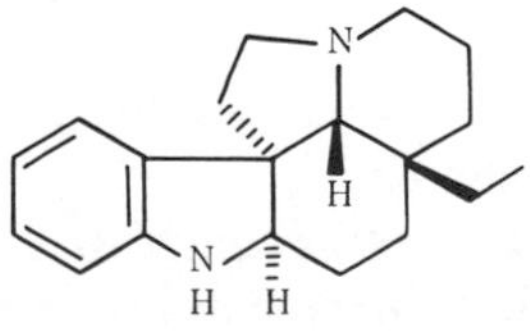   652   (+)-aspidospermidine

Tabersonine (**619**), as we have discussed previously, is one of the fundamental *Aspidosperma* alkaloids, and is consequently an important synthetic target. Three groups have described synthetic procedures to **619** and the routes of Ziegler and Takano are shown in Schemes 82 and 83 respectively.

One of the key steps is the formation of the quaternary mesylate **653**, which can be attacked at the α-position by a nucleophile such as cyanide to give the quebrachamine derivative **654**. This type of cleavage reaction, developed by Harley-Mason, is quite standard and has been used on several occasions in indole alkaloid synthesis.

The route developed by Takano offers an alternative synthesis of the quaternary mesylate **653**, and the approach of Le Men is to elaborate a 3-oxovincadifformine by introducing the 14,15-double bond through phenyl selenation.

Ziegler has also reported a synthesis of minovine (**654**) which involves the use of the Diels–Alder reaction in a key step analogous to the biogenetic formation of the β-anilinoacrylate nucleus from a C_3-acrylic ester. Thus heating **655** and **656** in methanol gave a regiospecific mixture of isomers at the ester group having the structure **657**. Debenzylation and a double alkylation step then affords minovine (**654**) (Scheme 84). The instability of the dihydropyridine corresponding to **656** precludes the use of this synthesis for the formation of the 14,15-dehydro compounds.

The parent compound aspidospermidine (**652**) was the subject of intensive synthetic work by the groups of Stork and Ban, who carefully studied the formation of the isomers of **658**. Treatment of the *O*-methoxyhydrazone of **658** with formic acid gives the product of Fischer indole cyclization, aspidospermidine (**652**).

The key compound in any synthetic endeavors is vindoline (**626**), because it constitutes one half of the antitumor alkaloid vincaleukoblastine, and both Kutney and Büchi have reported syntheses of this important *Aspidosperma* alkaloid.

Büchi had earlier reported a synthesis of vindorosine (**659**), but the reaction sequence failed at an early stage in the 6-methoxytryptamine series which could have afforded vindoline (**626**). The problem was resolved when the 6-tosyloxy derivative of the *trans* acetamide (**660**) afforded the desired dihydroindole **661** in 89% yield. After sequential removal of the protecting groups, *O*-methylation and removal of the *N*-acetyl group with triethyloxonium fluoroborate and sodium bicarbonate gave the diamine **662**. This compound is a β,β'-diamino ketone and therefore potentially has two positions susceptible to epimerization. The most stable configuration is the one shown.

When this amine is condensed with acrolein, a pentacyclic ketone is produced which is now beautifully functionalized for the remaining synthetic steps. The α,β-unsaturated ketone is stereospecifically alkylated at the α-position to afford **663**. The ketone is then α-carbomethoxylated with di-

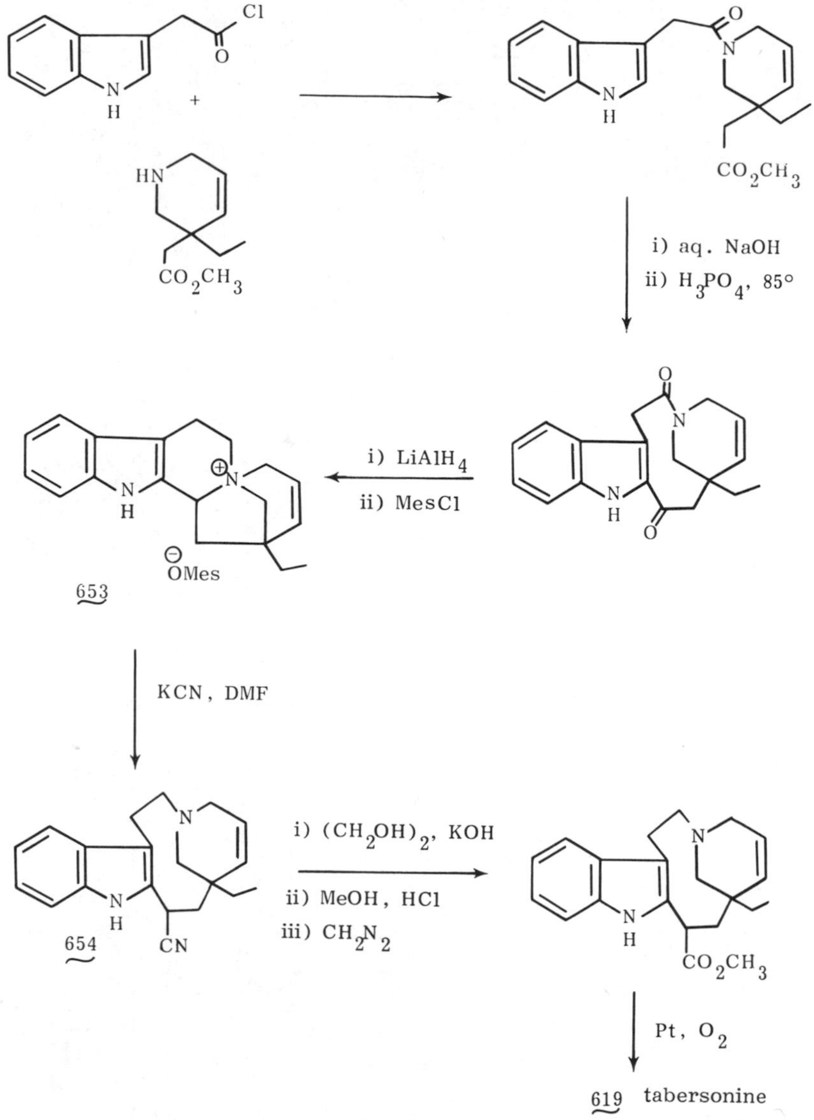

Scheme 82   Ziegler–Bennett synthesis of tabersonine (**619**).

methyl carbonate and sodium hydride and the C-16 hydroxy group introduced stereospecifically with oxygen–hydrogen peroxide under basic conditions to yield **664**. This reaction is regarded as proceeding by peroxide
anion attack C-17 and subsequent nucleophilic attack by a carbanion at C-
16 to give as intermediate hydroxyepoxide intermediate that is isomerized
to the α-hydroxy ketone **664**. The C-17 ketone was also reduced stereospecifically using sodium bis(2-methoxyethoxy)aluminum hydride at −20°C

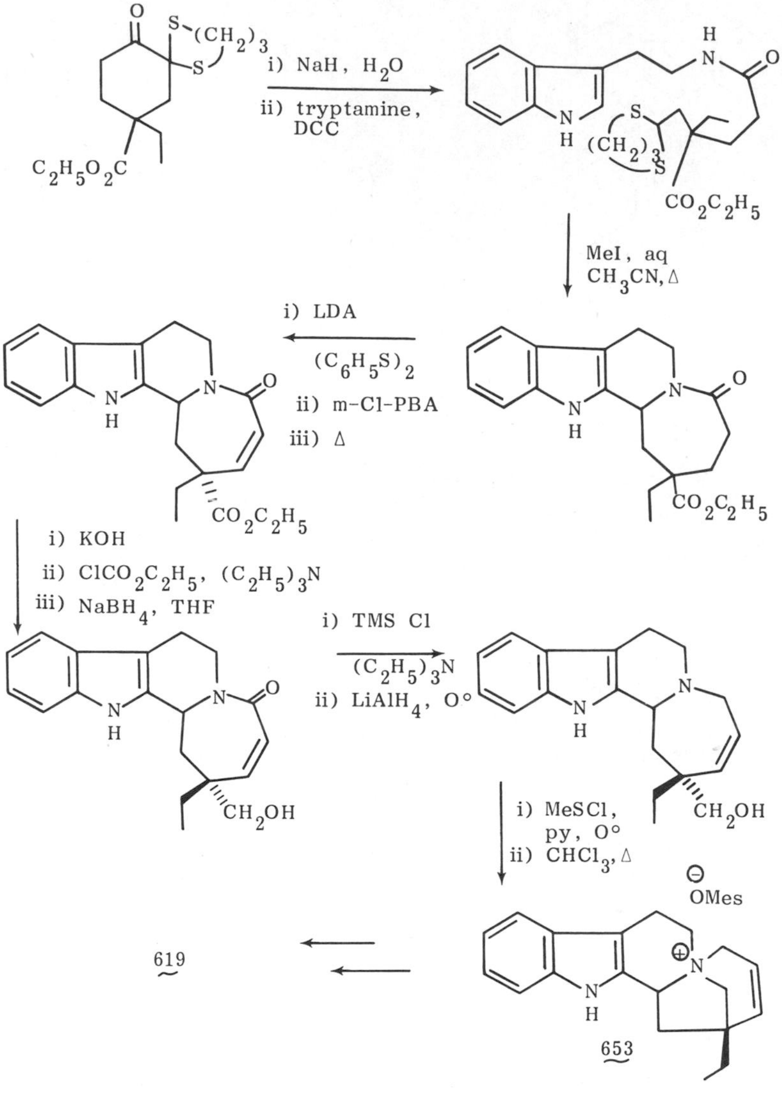

Scheme 83 Takano synthesis of tabersonine (**619**).

after prior addition of aluminum chloride. The exclusive formation of the β-hydroxy isomer is due to complexation of the N-4 group with aluminum chloride on the β-face of the molecule. Acetylation of the C-17 alcohol gave vindoline (**626**) (Scheme 85).

A second synthesis of vindoline has been described by Kutney and co-workers, and it extends the previous synthetic work on 11-methoxyminovine (**665**). α-Hydroxylation of the β-ketoester **666** using the method of Büchi

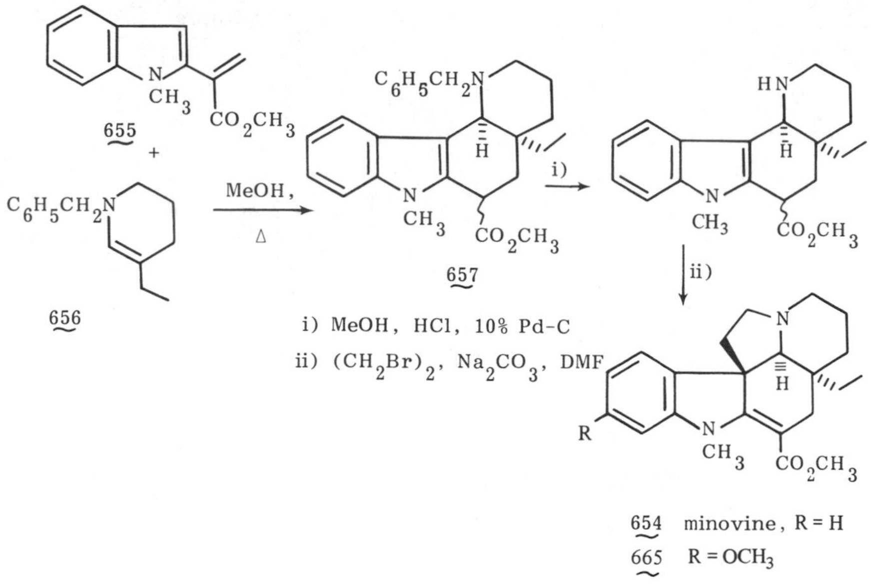

Scheme 84 Ziegler–Spitzer synthesis of minovine (**654**).

followed by controlled lithium aluminum hydride reduction and acetylation gave 14,15-dihydrovindoline (**667**).

Reaction of **667** with mercuric acetate in refluxing dioxan gave the ether lactam **668**, in one step. The intermediates in this interesting reaction are the six-membered lactam and the α,β-unsaturated lactam, which then undergoes Michael attack at the β-position to afford **668**. The ether bridge was opened by removal of a proton α- to the lactam carbonyl with the anion of triphenylmethane, and the tertiary hydroxyl group acetylated to afford **669**. The lactam carbonyl was removed by conversion to the imine ether with Meerweins reagent followed by borohydride reduction and treatment with moist silica gel to yield vindoline (**626**) (Scheme 86).

Ban and co-workers have described several successful synthetic efforts on various types of *Aspidosperma* alkaloids, and only a few of these studies will be described here.

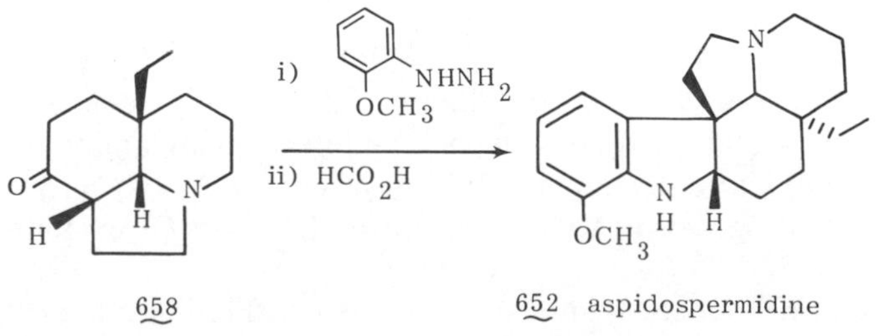

658

652 aspidospermidine

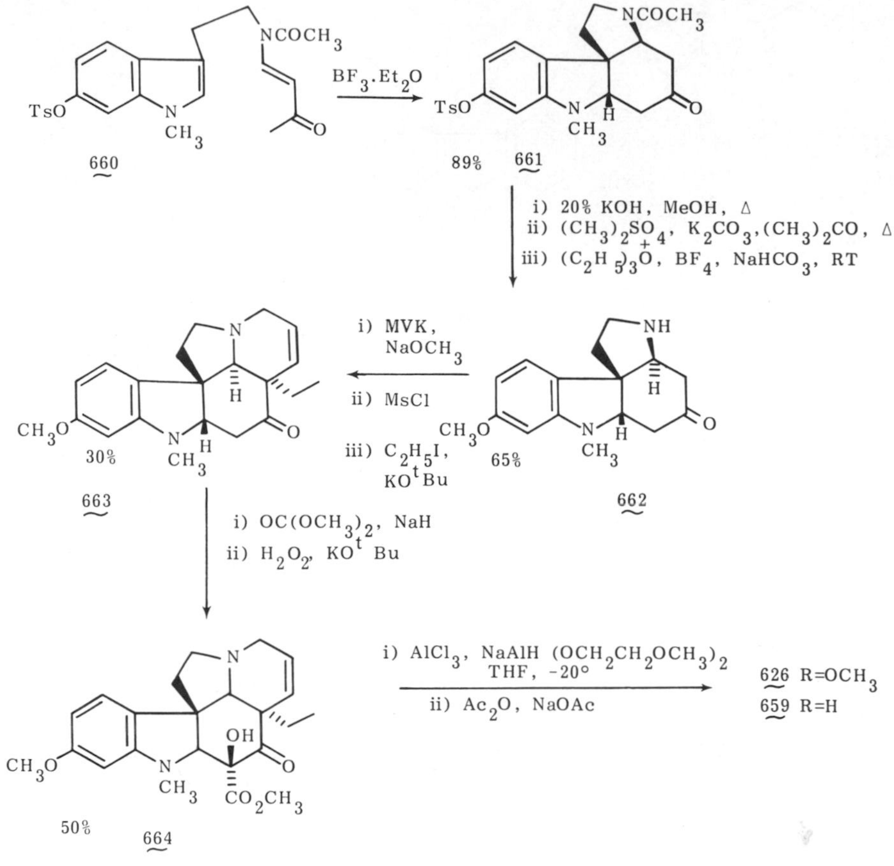

Scheme 85 Büchi synthesis of vindoline (**626**).

The rearrangement of the minovincine system to the aspidofractinine skeleton was described previously. Ban and co-workers have used this approach in a synthesis of aspidofractinine (**642**) involving the keto lactam **670** as a key intermediate.

The direct conversion of **670** to the parent compound aspidofractinine (**642**) met only with failure and consequently another method was sought for introduction of the two-carbon unit.

The key reaction is the Diels–Alder addition of nitroethylene to the diene **671**, which as expected gave a product **672** derived by closure of the most stable radicals. Catalytic reduction gave an intermediate amine which could be diazotized to afford the olefin **673**. Removal of the N-tosyl group followed by catalytic hydrogenation completed the synthesis of (−)-aspidofractinine (**642**) (Scheme 87).

The use of **670** for the synthesis of the aspidofractinine nucleus led Ban to investigate methods to introduce an ethyl side chain at C-20. Reaction with methyl vinyl sulfone using lithium diisopropyl amide as the base gave the Michael addition product **674** in 82% yield. Desulfurization with Raney nickel gave the residual ethyl side chain and again standard reactions afforded *N*-acetyl aspidospermidine (**675**) (Scheme 88).

Early syntheses of the *Aspidosperma* nucelus had expended considerable effort in carefully establishing the C-D-E ring stereochemistry. Le Men and workers, elaborating on the efforts of Harley-Mason, have found that this is perhaps not as critical as was once thought and they have described quite effective syntheses of both aspidospermidine (**652**) and vincadifformine

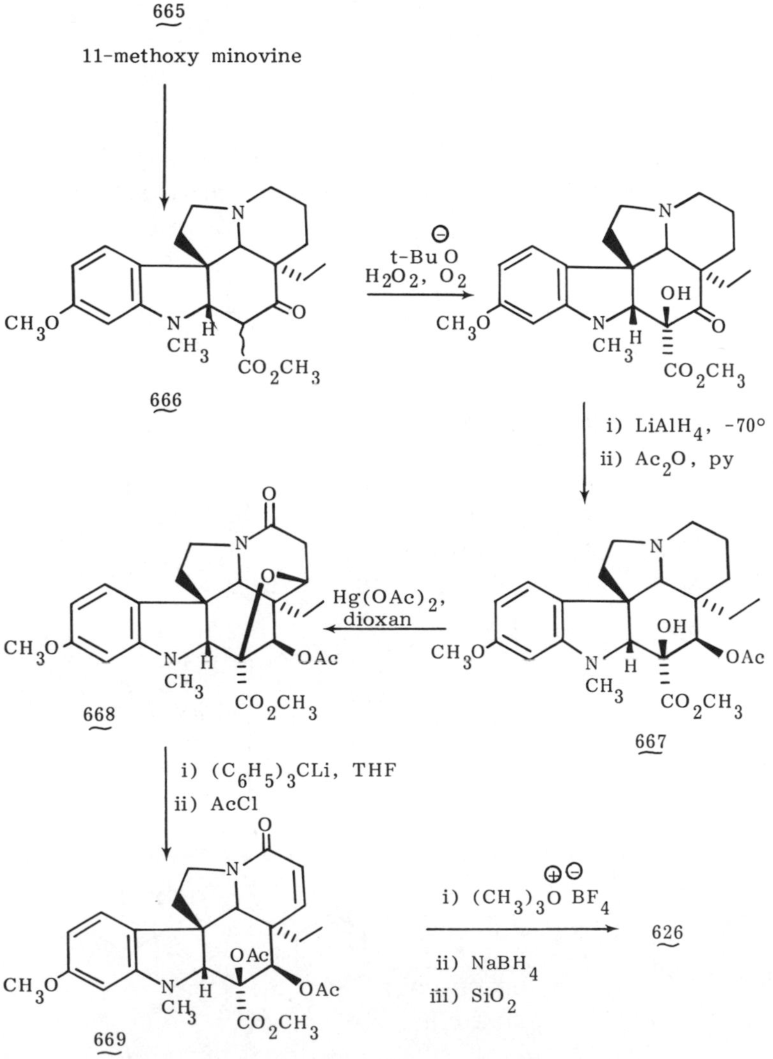

Scheme 86 Kutney synthesis of vindoline (**626**).

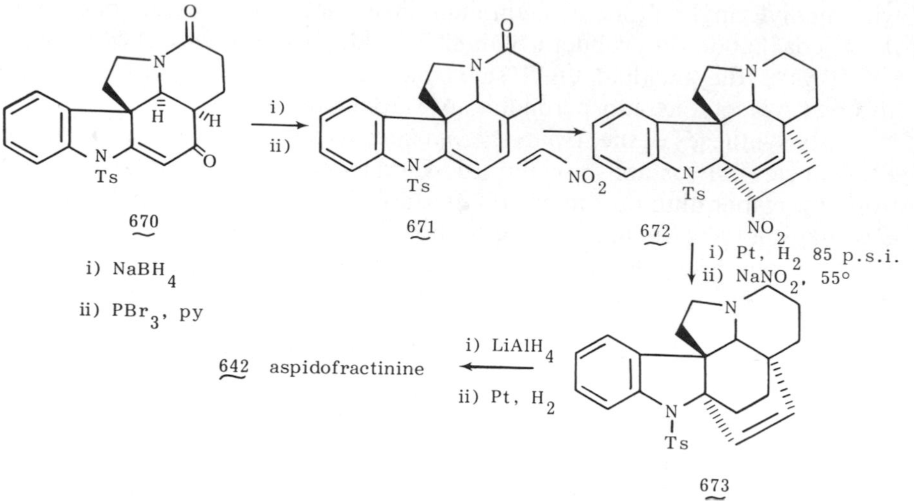

Scheme 87 Ban synthesis of aspidofractinine (**642**).

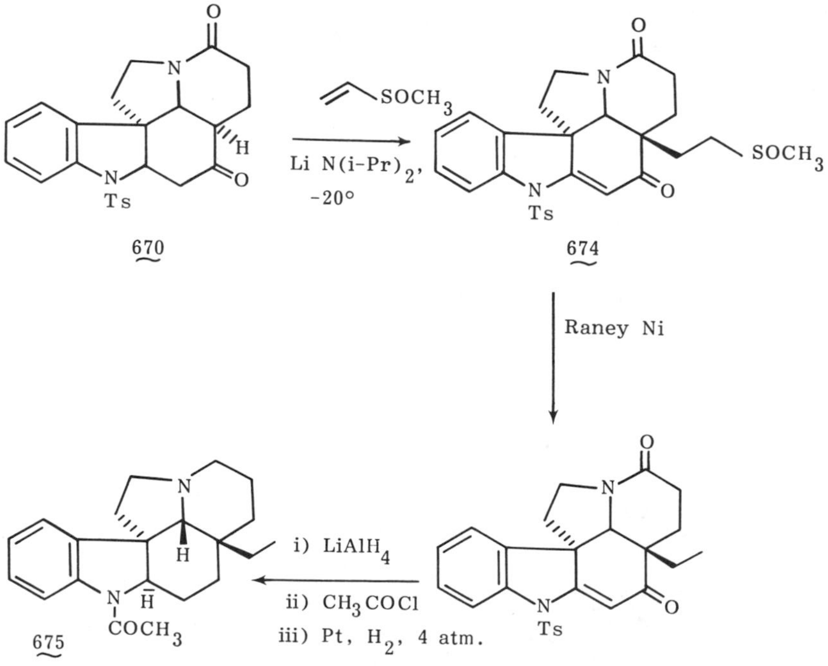

Scheme 88 Ban synthesis of N-acetylaspidospermidine (**675**).

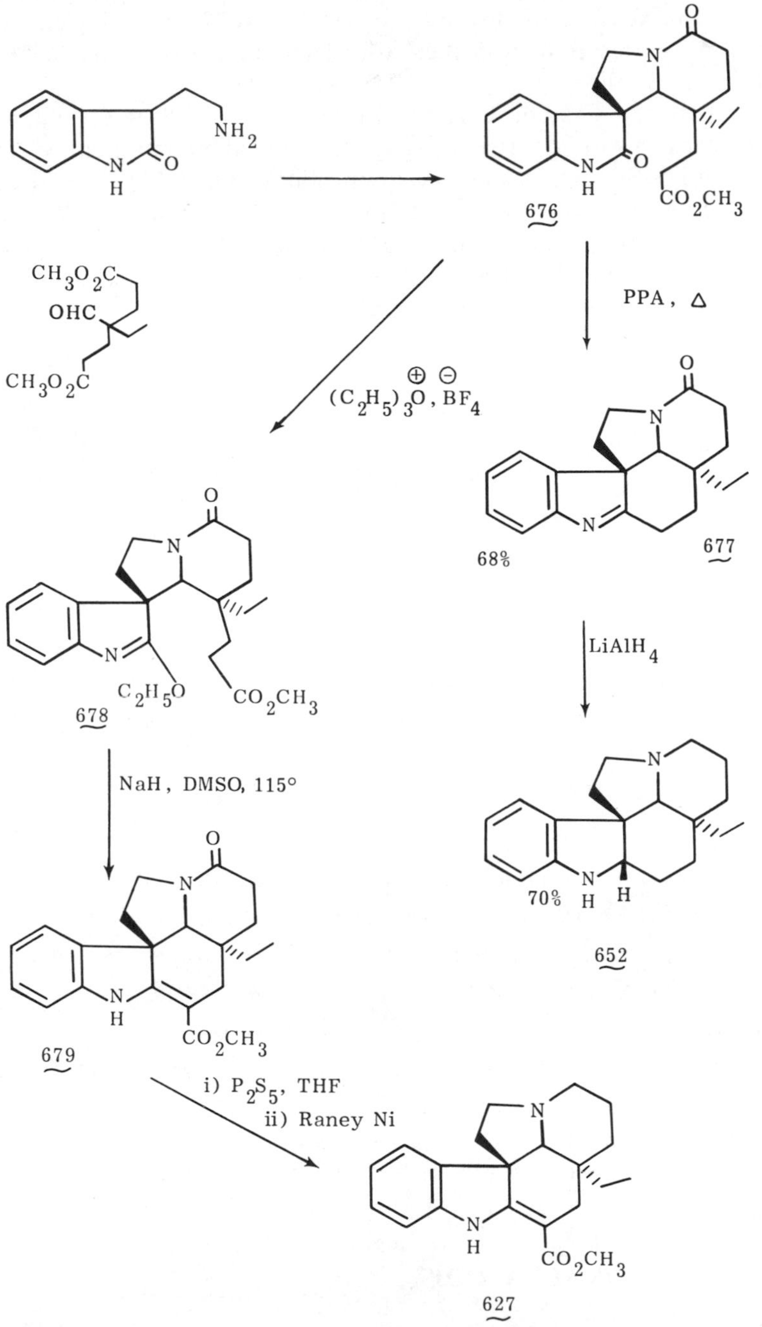

Scheme 89 Le Men synthesis of aspidospermidine (**652**) and vincadifformine (**627**).

(627). The key intermediate is **676**, which on treatment with polyphosphoric acid gave 3-oxo-1,2-dehydroaspidospermidine (**677**) in 68% yield. Lithium aluminum hydride reduction then afforded aspidospermidine (**652**) in 70% yield.

Alternatively, C-2 and C-16 of **676** could be joined by formation of the imino ether **678** with Meerwein's reagent followed by treatment with sodium hydride in DMSO, which gave 3-oxovincadifformine (**679**). The 3-oxo group was selectively reduced by conversion to the thiolactam and Raney nickel reduction to give vincadifformine (**627**) (Scheme 89). One of the complications in this route is that the use of the ethyl version of Meerwein's reagent allows some ester exchange to take place.

LITERATURE

Reviews: General

Saxton, J. E., *Alkaloids NY* **7**, 1 (1960).

Gilbert, B., *Alkaloids NY* **8**, 336 (1965).

Gilbert, B., *Alkaloids NY* **11**, 205 (1968).

Cordell, G. A., *Alkaloids NY* **17**, 200 (1979).

Reviews: Synthesis

Wenkert, E., *Acc. Chem. Res.* **1**, 78 (1968).

Kutney, J. P., in K. F. Wiesner (Ed.), *MTP International Review of Science, Organic Chemistry*, Ser. 1, Vol. 9, *Alkaloids*, Butterworths, London, 1973, p. 23.

Winterfeldt, E., *Progr. Chem. Org. Nat. Prods.* **31**, 469 (1974).

Synthesis

Ziegler, F. E., and G. B. Bennett, *J. Amer. Chem. Soc.* **93**, 5930 (1971).

Takano, S., S. Hatakeyama, and K. Ogasawara, *J. Amer. Chem. Soc.* **98**, 3022 (1976).

Levy, J., J.-Y. Laronze, J. Laronze, and J. Le Men, *Tetrahedron Lett.* 1579 (1978).

Ziegler, F. E., and E. B. Spitzner, *J. Amer. Chem. Soc.* **95**, 7146 (1973).

Ohnuma, T., T. Oishi, and Y. Ban, *Chem. Commun.* 301 (1973).

Büchi, G. E., K. E. Matsumoto, and H. Nishimura, *J. Amer. Chem. Soc.* **93**, 3299 (1971).

Ando, M., G. Büchi, and T. Ohnuma, *J. Amer. Chem. Soc.* **97**, 6880 (1975).

Kutney, J. P., U. Bunzhi-Trepp, K. K. Chan, J. P. deSouza, Y. Fujise, T. Honda, J. Katsube, F. K. Klein, A. Leutwiler, S. Morehead, M. Rohr, and B. R. Worth, *J. Amer. Chem. Soc.* **100**, 4220 (1978).

9.24 *MELODINUS* ALKALOIDS

The genus *Melodinus* was subjected to analysis for alkaloids in the mid-1960s, and although some typical *Aspidosperma* alkaloids such as 19-acetoxytabersonine (**680**) were isolated, the major alkaloids were a new type of indole alkaloid in which ring B had expanded to become six-membered with a concomitant contraction of the C-ring.

Meloscine (**681**) shows an acylaniline UV chromophore (λ_{max} 229, 263 nm) and five olefinic protons, including three for a vinyl group. The complexity of these patterns indicated that the vinyl group was attached to a quaternary center, whereas the *cis*-olefinic bond has a single methylene adjacent.

Another variation on this structure type is meloscandonine (**682**). Unlike the other *Melodinus* alkaloids, no vinyl group was observed. Instead, a doublet methyl group was found at 1.12 ppm together with a cyclopentanone carbonyl at 1750 cm^{-1}. The stereochemistry of the methyl group was established by examining the proton NMR spectra of the acetate derivatives of the dihydromeloscandonines. In one of these, **683**, the coupling constant

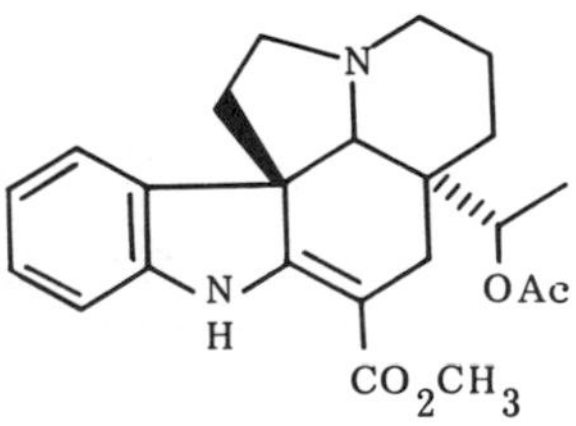

680

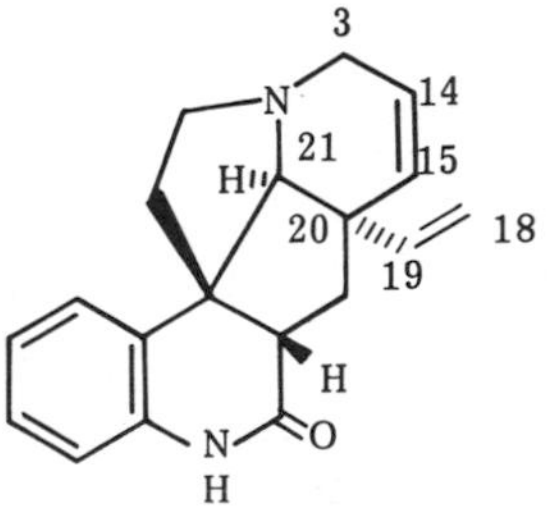

681   meloscine

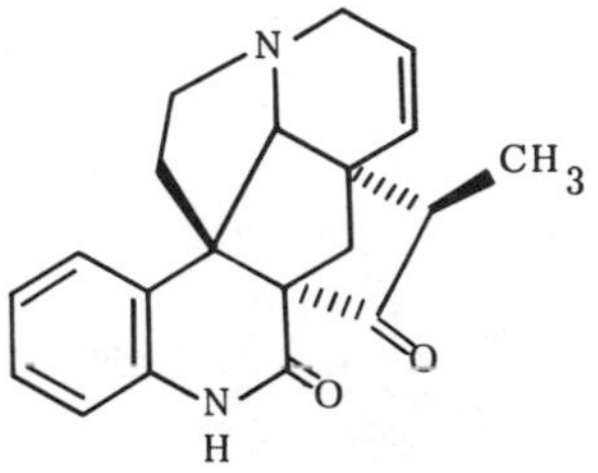

682   meloscandonine

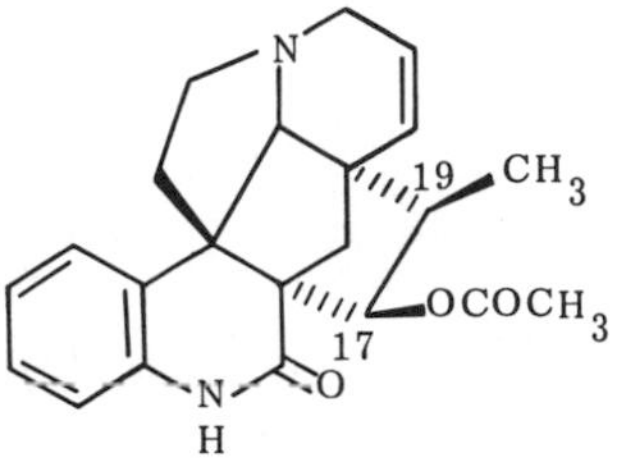

683

between the acetate methine proton at C-17 was shown to be 7.4 Hz, indicating a *cis* relationship between H-17 and H-19. The epimeric acetate displayed an acetate singlet at 1.23 ppm, indicating it to be held proximate to the shielding area of the aromatic nucleus.

The ^{13}C NMR spectral data of meloscandonine and meloscine are shown in **684** and **685**.

The biogenesis of these alkaloids has been suggested by Wenkert and co-workers to involve 18,19-dehydrotabersonine (**686**), which can then undergo stereospecific electrophilic hydroxylation at C-16, and addition of water at N-1 and C-2 to give **687**. Rearrangement then gives scandine (**688**),

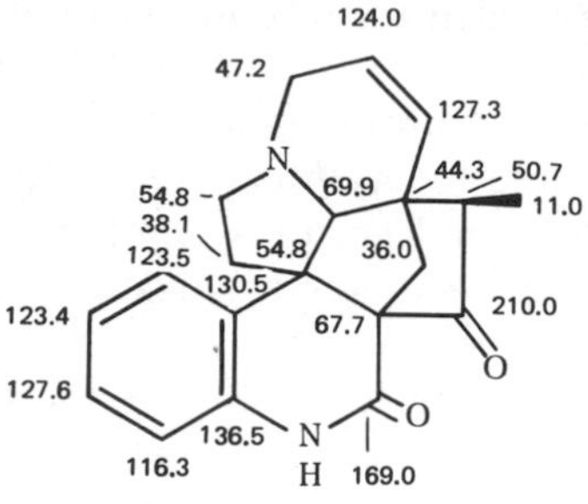

684 meloscandonine

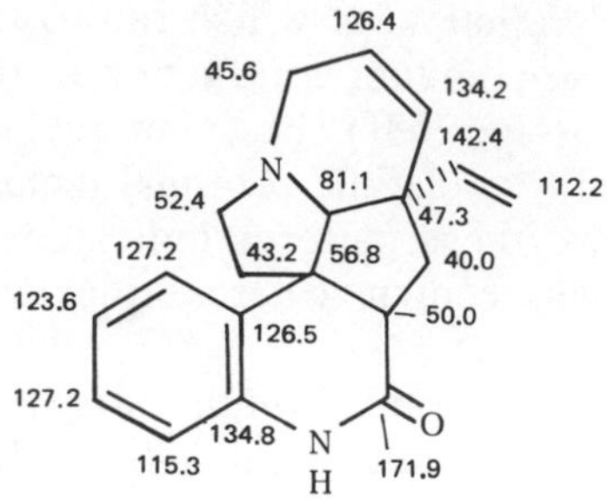

685 meloscine

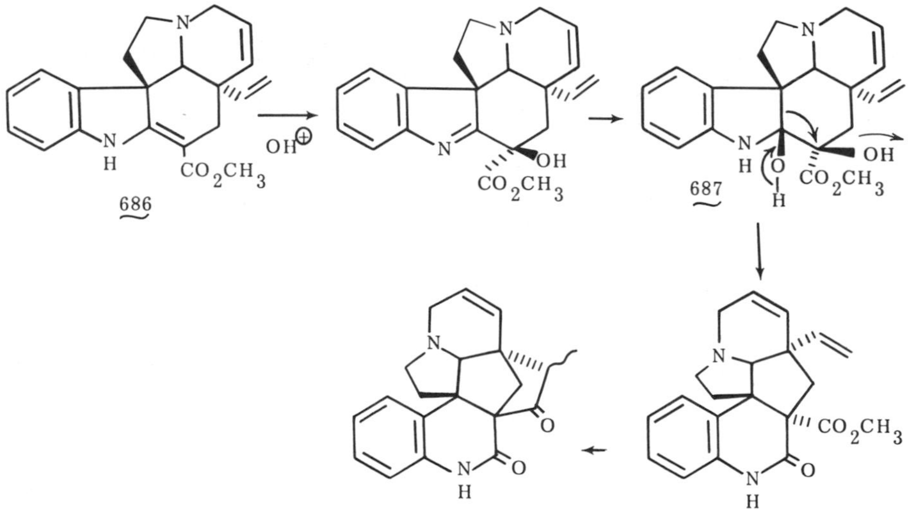

686

687

682/689

688 scandine

which can be cyclized reductively to meloscandonine (**682**) and 19-epimelo-scandonine (**689**).

LITERATURE

Gilbert, B., *Alkaloids NY* **11**, 206 (1968).

Cordell, G. A., *Alkaloids NY* **17**, 200 (1979).

Daudon, M., M. H. Mehri, M. M. Plat, E. W. Hagaman, and E. Wenkert, *J. Org. Chem.* **41**, 3275 (1976).

9.25   IBOGA ALKALOIDS

9.25.1   Isolation and Chemistry

The roots of the West African plant *Tabernanthe iboga* are used by natives to combat sleep and hunger (central nervous system stimulation) and the principal alkaloid, ibogaine, was first obtained at the turn of the century.

Ibogamine (**690**), an alkaloid whose structure was determined by Taylor in 1957, is a typical example of the skeleton. But other alkaloids in the series are further substituted at C-16 by a carbomethoxy group [e.g. coronaridine (**691**)] or have an aromatic methoxy group [e.g. voacangine (**692**)].

These alkaloids are particularly common in the genera *Voacanga*, *Tabernaemontana*, *Ervatamia*, *Pandaca*, and *Conopharyngia*. Unlike any other indole alkaloids they are particularly susceptible to oxidation at the C-7 position to yield an hydroxyindolenine (e.g. **693**), which may rearrange to a 3-oxindole [e.g. iboluteine (**694**)]. *In vitro* this reaction is base-catalyzed.

690   ibogamine

691   coronaridine, R=H
692   voacangine, R=OCH$_3$

693

694   iboluteine

The *cis*-16,21 stereochemistry of iboluteine (**694**) was deduced when the tosylated oxime **695** underwent a second-order Beckmann rearrangement in refluxing pyridine to give the aniline **696** and the amino ketone **697**. The latter was subjected to Von Braun degradation and lithium aluminum hydride reduction to afford the decahydroisoquinolone **698**, whose ORD curve was studied.

Eglandine (**699**) from *Gabunia eglandulosa* and its 10-methoxy N-oxide derivative **700** from *Ervatamia heyneana* are examples of a new iboga alkaloid type. The isoxazoline group can be cleaved reductively with lithium aluminum hydride to give the parent iboga alkaloid, but with a hydroxymethyl group at C-16 (i.e. **701**).

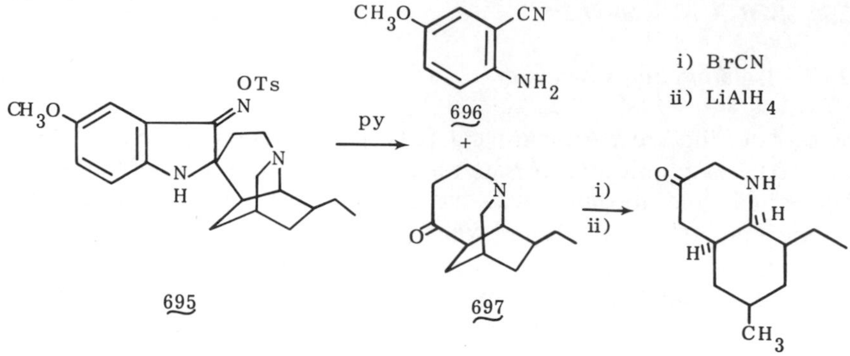

695   696   697   698

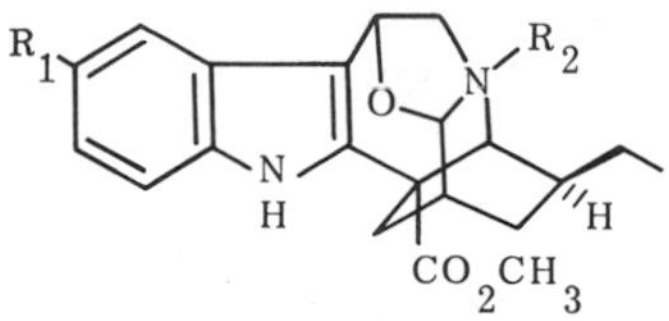

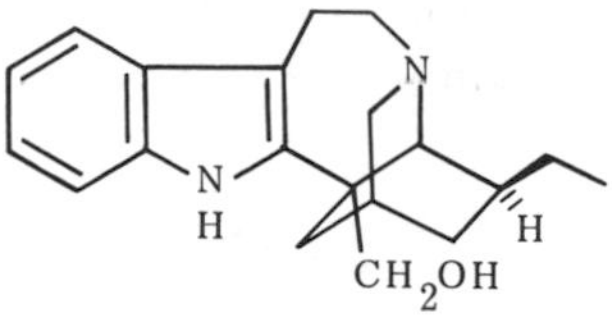

699 eglandine, R₁=H, R₂not present
699 eglandine, R_1=H, R_2not present

700 R_1=OCH$_3$, R_2=O

701

Two alkaloids in this series, catharanthine (**702**) and a 16,21-secoderivative cleavamine (**648**), are of particular significance.

Catharanthine is the iboga alkaloid which has been studied biosynthetically in *Catharanthus roseus*, where it is one of the major alkaloids. Cleavamine (**648**) is a product of the reductive cleavage of the antitumor dimeric alkaloid vincaleukoblastine (Section 9.29.1), and may also be prepared by similar reaction of catharanthine (**702**). This skeleton has yet to be obtained from natural sources.

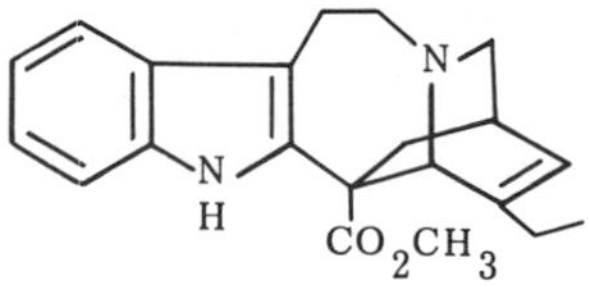

702 catharanthine

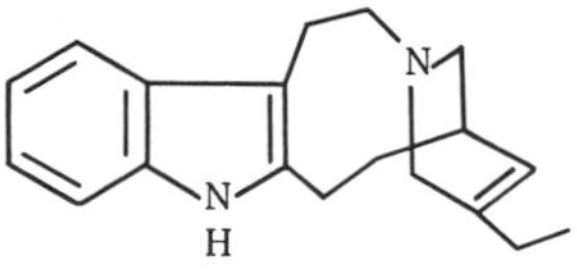

648 cleavamine

Catharanthine (**702**) has also proved to be important because of a number of facile rearrangements which occur. For example, catharanthine (**702**) is rearranged to pseudocatharanthine (**703**) on brief heating in glacial acetic acid. The mechanism is regarded as that shown in Scheme 90. Note that when the structure of pseudocatharanthine is redrawn as **704**, it resembles an *Aspdosperma*-type alkaloid in which the ethyl side chain has been shifted

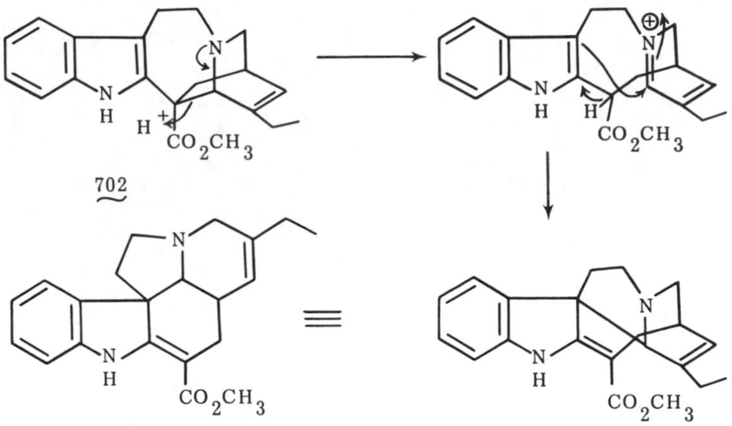

Scheme 90

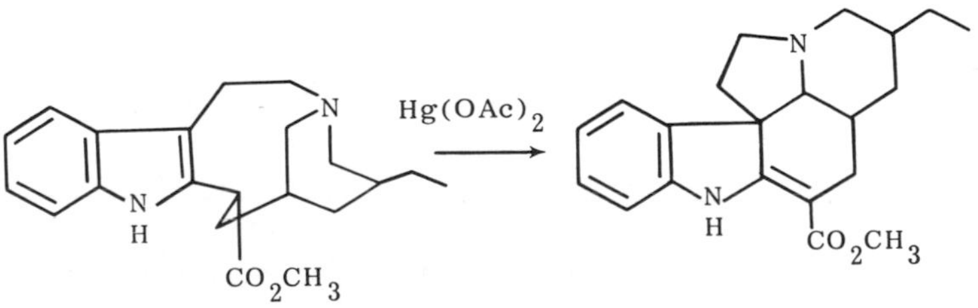

to C-14.* A similar reaction occurs with 16β-carbomethoxydihydrocleavamine (**705**) to afford dihydropseudocatharanthine (**706**) on treatment with mercuric acetate.

On heating catharanthine–N-oxide (**707**) at 40–60°C, [2,3]sigmatropic rearrangement occurs to give quantitatively the isoxazolidine **708**. In the presence of a nucleophile (e.g. acetate) a different rearrangement occurs, to give the novel pentacyclic product **709** (Scheme 91).

When dihydrocatharanthine (**710**) is refluxed with glacial acetic acid it is converted mainly to the C-20 epimer coronaridine (**691**), presumably by way of the intermediate **711**, which can equilibrate C-20 through the corresponding enamine (Scheme 92).

9.25.2 Absolute Stereochemistry

The absolute stereochemistry of catharanthine (**712**) was determined by X-ray crystallographic studies on cleavamine methiodide (**713**) and is shown.

* Indeed the preferred name is pseudotabersonine.

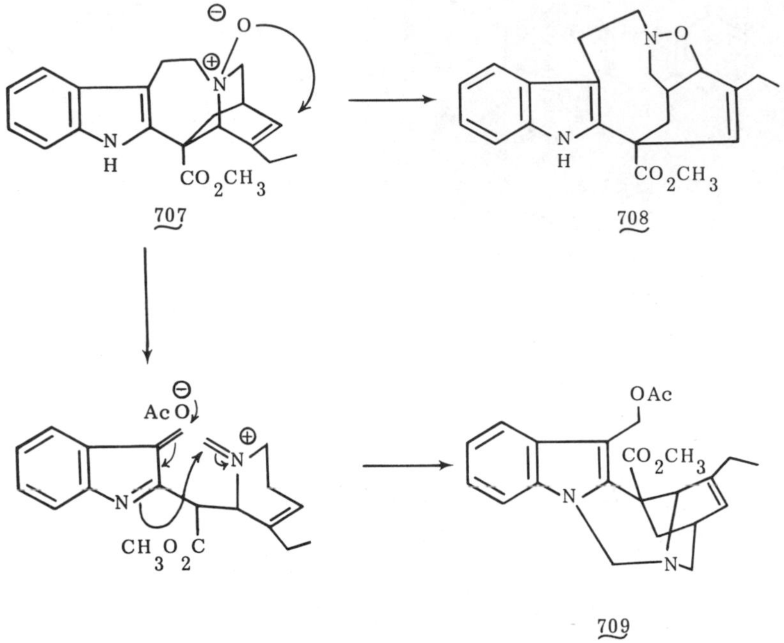

Scheme 91 Rearrangements of catharanthine–N-oxide (**707**).

However, the absolute stereochemistry of the more widespread iboga alkaloids such as coronaridine and ibogamine was not known, even though they had been chemically interrelated with catharanthine.

The CD curves of compounds having a C-16 carbomethoxyl group and lacking an aromatic methoxyl substituent show three groups of Cotton

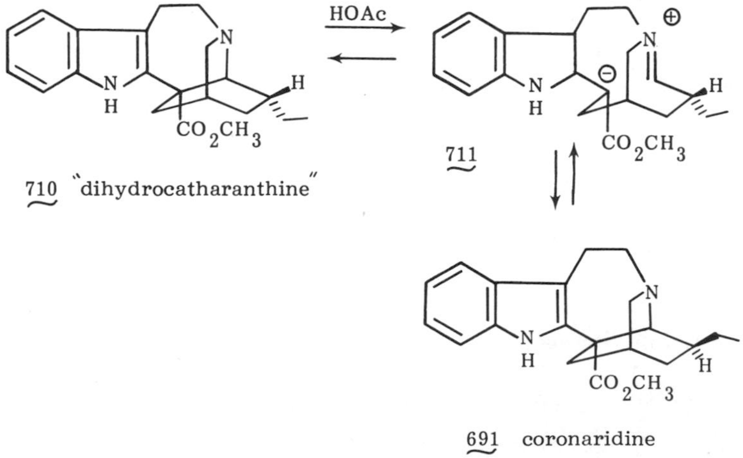

Scheme 92 C-20 epimerization of dihydrocatharanthine (**710**).

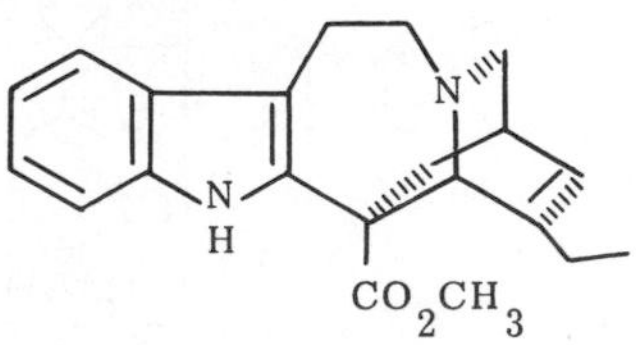

712 catharanthine 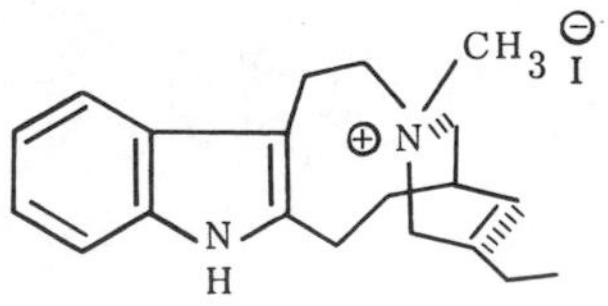   713 cleavamine methiodide

effects. Coronaridine (**691**) for example shows a triplet of partially overlapping negative bands between 270 and 295 nm, a triplet of positive bands in the range 240–250 nm, and a short-wavelength negative band at 215 nm. Catharanthine showed bands at the same wavelengths, but the sign of each band was reversed.

Thus (+)-catharanthine (**712**) and (−)-coronaridine (**691**) have the opposite absolute configuration. Indeed, (+)-catharanthine (**712**) is to date the only natural representative of this class. These conclusions concerning absolute configuration were substantiated by an X-ray crystallographic analysis of (+)-coronaridine hydrobromide.

9.25.3 Synthesis

The tremendous interest in the dimeric indole alkaloids of the vincaleukoblastine type stimulated synthetic investigation of the iboga skeleton as well as of the *Aspidosperma* system discussed previously.

The first total synthesis in the area was that of ibogamine (**690**) by Büchi and co-workers, and the same group subsequently also reported the first synthesis of velbanamine (**714**) and catharanthine (**712**). Quite different successful approaches to the skeleton have been described by Kutney and Nagata and their respective co-workers. These synthetic efforts have been well reviewed by Kutney.

The strategy developed by Kutney involves as a key step the reductive

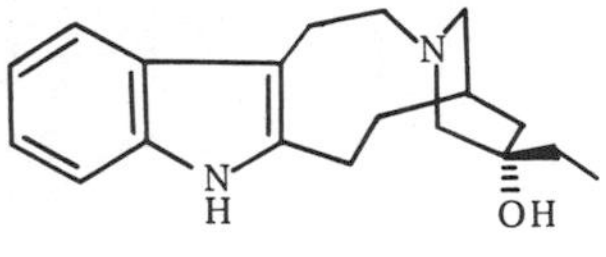

714 velbanamine

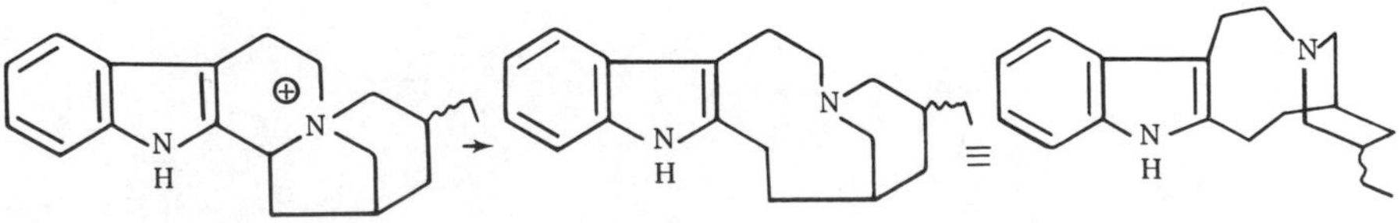

714a

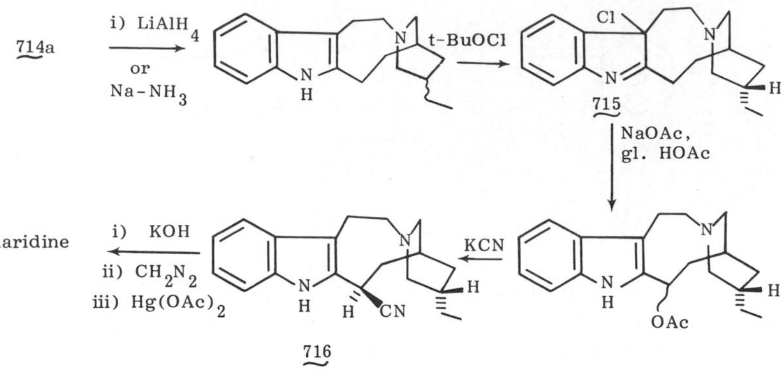

Scheme 93 Kutney and co-workers' synthesis of (±)-coronaridine (**691**).

cleavage of an appropriately substituted quaternary species to directly afford the cleavamine system as shown. In practice this reaction could be carried out with either lithium aluminum hydride or sodium–liquid ammonia.

The iboga synthesis was completed by treating the chloroindolenine of 20β-dihydrocleavamine (**715**) with sodium acetate in glacial acetic acid followed by further direct reaction with potassium cyanide to afford 16-cyano-20β-dihydrocleavamine (**716**), which could be hydrolyzed and oxidized to give racemic coronaridine (**691**) (Scheme 93).

A quite different synthesis of ibogamine (**690**) developed by Trost involves as a key step a silver–palladium-catalyzed olefin arylation, and proceeds in four steps from the dienol acetate **717** in 17% overall yield (Scheme 94). Beginning with an optically active dienol ester, a synthesis of optically active **690** was achieved.

The synthetic routes developed by Büchi and Kutney have more recently

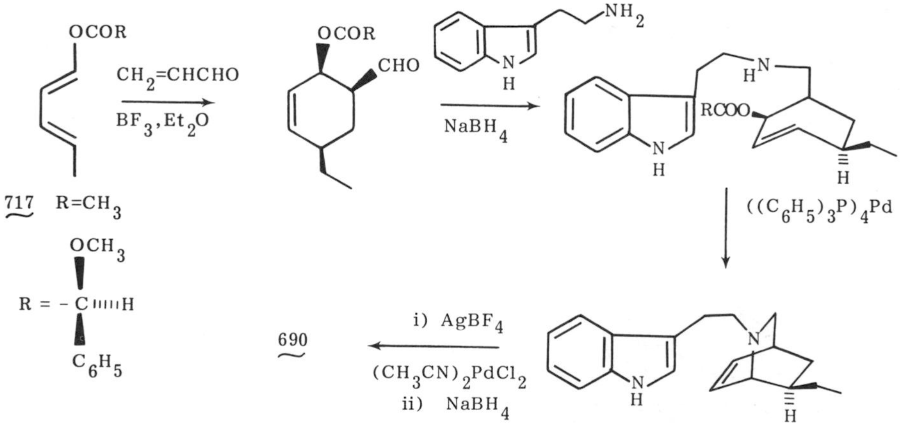

Scheme 94 Rosenmund synthesis of 20-epiibogamine (**717**).

been extended to syntheses of velbanamine (714) and catharanthine (712). In the Büchi synthesis the quinuclidine 718 is a key intermediate, which can be transformed in two steps to the desethyl iboga derivative 719. Deprotection followed by aldol cleavage then gave 720. Standard transformations afforded velbanamine (714).

The intermediate 719 could also be converted to 721, which is now set up for the familiar introduction of the additional substituent at C-16 by way of a chloroindolenine intermediate (Scheme 95).

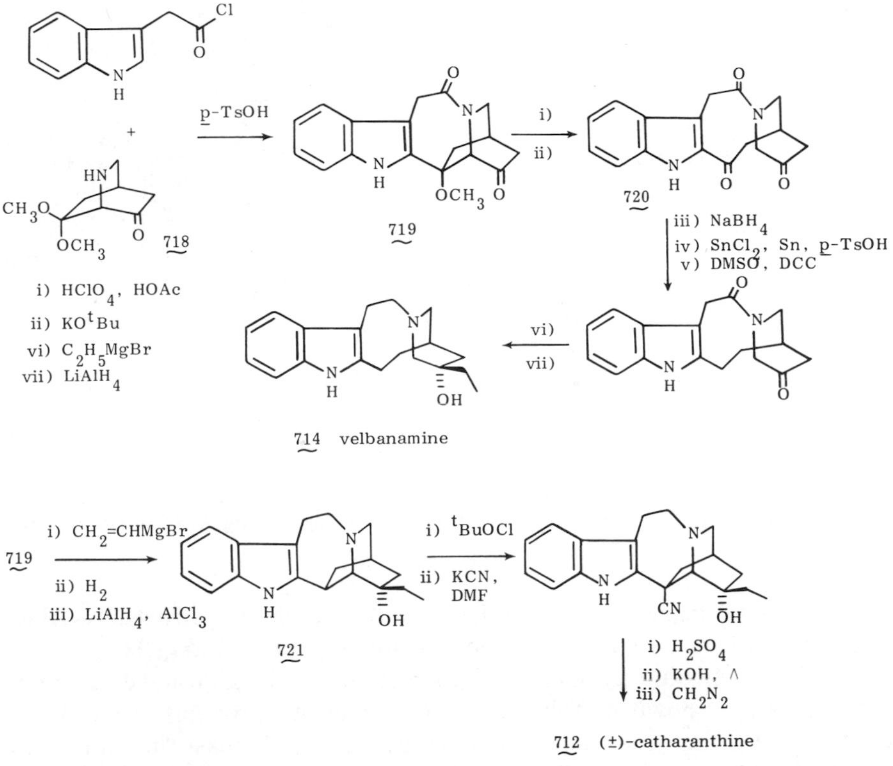

Scheme 95 Büchi synthesis of velbanamine (719) and catharanthine (712).

9.25.4 Spectral Properties

The mass spectrum of ibogamine (690) shows a group of peaks at m/e 122, 124, 136, and 149 from the piperidine unit, and a second group at m/e 156, 175, 251, 265, and 280 containing the indole nucleus. These pathways are shown in Scheme 96.

One interesting point concerning the mass spectra of tabersonine (619), catharanthine (712), and pseudocatharanthine (704) is that they are essen-

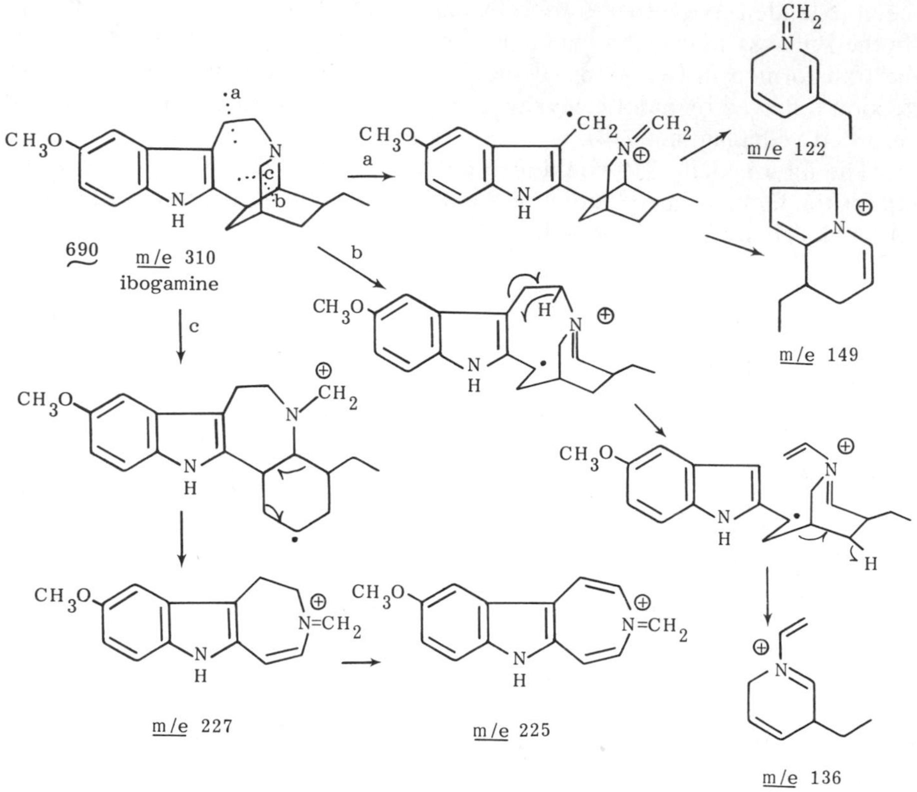

Scheme 96 Typical mass spectral fragmentation of the iboga alkaloid skeleton.

tially identical, because as an initial step they undergo retro Diels–Alder reaction to afford virtually the *same* intermediate radical ion **722**.

The ^{13}C NMR shift assignments for some quebrachamine and cleavamine derivatives are shown in Table 10. The former are probably derived biosynthetically from the *Aspidosperma* nucleus, whereas the latter are obtained from the iboga skeleton.

As in the Corynanthe and yohimbe series of alkaloids, the chemical shift of C-6 is useful in defining the stereochemistry at C-16 in a cleavamine-type

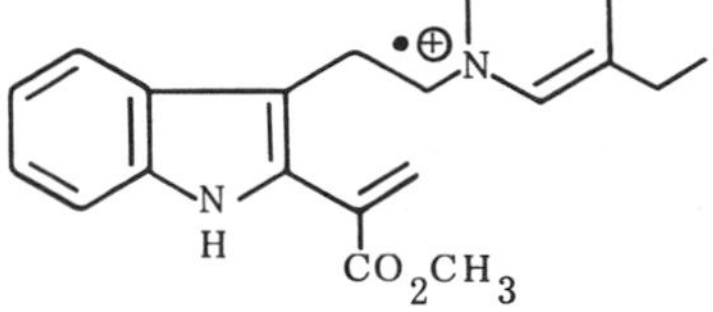

Table 10 ¹³C NMR Data for Some Representative Quebrachamine and Cleavamine Derivatives

	Quebrachamine (633)	Vincadine (723)	Cleavamine (648)	16β-Carbomethoxy-15,20β-Dihydro-cleavamine (705)	enantio-16β-Carbomethoxy velbanamine (724)
C-2	139.7	135.2ᵃ	139.2	135.0	c
C-3	54.9	55.0	53.5ᵃ	50.6	50.9
C-5	53.2	52.7	53.8ᵃ	52.5	52.0
C-6	21.7	21.8	26.1	22.1	22.4
C-7	108.3	109.4	109.5	109.5	109.2
C-8	128.6	127.7	128.5	127.8	127.7
C-9	117.1	117.4	117.6	117.4	117.5
C-10	118.4	118.5	118.5	118.6	119.0
C-11	119.9	120.6	120.3	120.7	121.4
C-12	109.9	110.5	109.8	110.5	111.1
C-13	134.5	134.9ᵃ	135.2	135.0	134.0
C-14	22.6	22.3	35.3	32.8ᵃ	30.5
C-15	33.4	33.9	122.3	36.7ᵇ	39.5
C-16	22.4	37.8	22.4	39.0	39.3
C-17	34.7	38.6	34.1	36.3ᵇ	36.1
C-18	7.8	7.4	12.6	11.3	6.9
C-19	32.0	30.6	27.6	27.3	32.6
C-20	36.9	37.9	140.4	33.1ᵃ	71.2
C-21	56.6	56.7	55.1	61.3	66.1
C—O	—	176.2	—	176.1	175.2
OCH₃	—	52.0	—	52.1	52.2

Source: Data from E. Wenkert *et al.*, *Helv. Chim. Acta* **59**, 2711 (1976).
ᵃ,ᵇ Assignments may be reversed.
ᶜ Signal not detected.

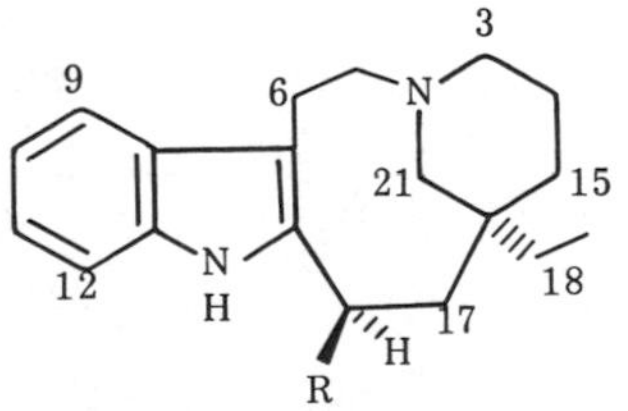

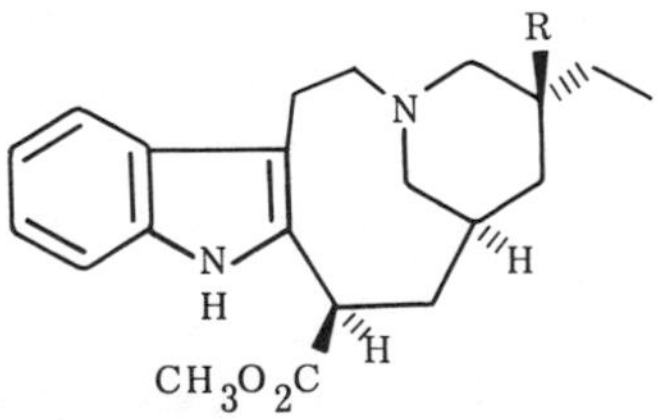

633 (+)-quebrachamine, R=H

723 vincadine, R=CO₂CH₃

705 16β-carbomethoxy-15,20β-dihydrocleavamine, R=H

724 16β-carbomethoxyvelbanamine, R=OH

compound. Thus C-6 in compounds where the C-16 substituent is β resonates at 22 ± 0.7 ppm, but at 26.2 ± 0.3 ppm when this substituent is α.

The chemical shift of C-14 serves to readily distinguish between the quebrachamine and cleavamine derivatives since in the former group of compounds this is a low-field methylene (22.5 ppm), but in the latter is a methine carbon at somewhat higher field (30–34 ppm).

The γ-effect observed for C-18 in the quebrachamine series (~8 ppm) is also diagnostic provided that a C-20 substituent in the cleavamine series can be eliminated.

The data were used to describe the preferred conformers of these structures. Of the four *a priori* possibilities, two were eliminated because of severe strain or thermal instability, leaving two choices, **725** and **726**. All

of the compounds indicated in Table 10 revealed a four-line multiplet for the C-21 aminomethylene because of the chemical shift difference between the amino methylene protons. This effect is observed when a hydrogen is disposed *trans*-diaxial to the nitrogen electron pair; such a situation occurs only in conformer **725**.

9.25.5 Pharmacology

Several iboga alkaloids are central nervous system stimulants, as evidenced by their antagonism to reserpine catalepsy. In addition, several of these compounds cause hypotension and bradycardia, and ibogaline was the most active of these. Coronaridine has diuretic and cytotoxic activity and catharanthine has hypoglycemic activity.

The acyl derivatives of 10-methoxyibogamine (ibogaine) exhibit analgesic and anti-inflammatory activity. 10-Methoxyibogamine itself is regarded as the principal hallucinogenic constituent of *Tabernanthe iboga* root, and has been placed in the same category as LSD by the U.S. Food and Drug Administration.

LITERATURE

Reviews: General

Taylor, W. I., *Alkaloids NY* **8**, 203 (1965).

Taylor, W. I., *Alkaloids NY* **11**, 79 (1968).

Reviews: Synthesis

Kutney, J. P., in K. F. Wiesner (Ed.), *MTP International Review of Science, Organic Chemistry*, Ser. 1, Vol. 9, *Alkaloids*, University Park Press, Baltimore, Md., 1973, p. 27.

Kutney, J. P., in J. P. ApSimon (Ed.), *The Total Synthesis of Natural Products*, Wiley-Interscience, New York, 1977, p. 273.

Absolute Stereochemistry

Bláha, K., Z. Koblicová, and J. Trojanek, *Tetrahedron Lett.* 2763 (1972).

Kutney, J. P., K. Fuki, A. M. Treasurywala, J. Fayos, J. Clardy, A. I. Scott, and C. C. Wei, *J. Amer. Chem. Soc.* **95**, 5407 (1973).

Synthesis

Büchi, G., D. L. Coffen, K. Kocsis, P. E. Sonnet, and F. E. Ziegler, *J. Amer. Chem. Soc.* **87**, 2073 (1965); *J. Amer. Chem. Soc.* **88**, 3099 (1966).

Büchi, G., P. Kulsa, and R. L. Rosati, *J. Amer. Chem. Soc.* **90**, 2448 (1968).

Büchi, G., P. Kulsa, K. Ogasawara, and R. L. Rosati, *J. Amer. Chem. Soc.* **92**, 999 (1970).

Kutney, J. P., R. T. Brown, E. Piers, and J. R. Hadfield, *J. Amer. Chem. Soc.* **92**, 1709 (1970).

Kutney, J. P., W. J. Cretney, P. LeQuesne, B. McKague, and E. Piers, *J. Amer. Chem. Soc.* **92**, 1712 (1970).

Trost, B. M., S. A. Godleski, and J. P. Genet, *J. Amer. Chem. Soc.* **100**, 3930 (1978).

9.26 BIOGENETIC INTERCONVERSION OF MONOTERPENOID INDOLE ALKALOIDS

The many varied skeleta of the monoterpenoid indole alkaloids have stimulated an almost unending stream of studies aimed at the interconversion of these skeleta along biogenetic lines. Once again this is a vast area and only select examples will be given here; other examples are mentioned throughout this discussion of monoterpene indole alkaloids.

It is generally assumed that all indole alkaloids having a center of asymmetry at C-15 have the same absolute stereochemistry. This stereochemistry being derived from position 5 in secologanin (**285**). The minor aberrations in this idea which have been observed from time to time have largely been ignored.

The first example of such a novel occurrence was the isolation of (±)-akuammicine from *Picralima nitida* by Edwards and Smith. This compound should be a mixture of the components **479** and **727** [i.e. (−)-akuammicine], which is usually encountered, and (+)-akuammicine, in which the configuration at C-15 is now antipodal. Le Men and co-workers some years later isolated (+)-20-epilochneridine (**728**), and clearly this too must have C-15 inverted.

Scott and Yeh have investigated this phenomenon *in vitro* and have found that heating (−)-19,20-dihydroakuammicine (**729**) in degassed methanol at 90°C (see Akuammicine, Section 9.21.6, for the rationale for this particular

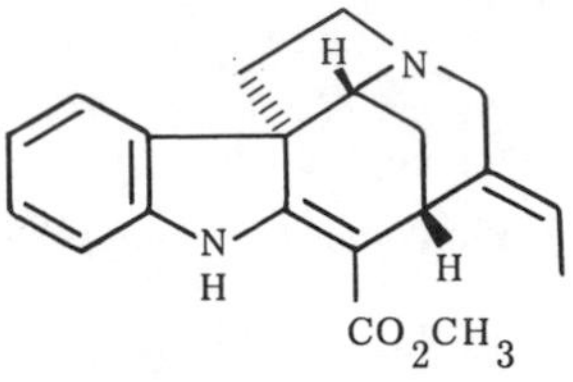

727   (+)-akuammicine 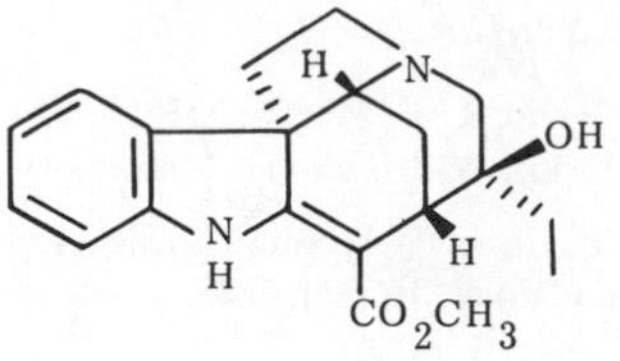   728   (+)-20-epi-lochneridine

temperature) gave an approximately 4:1 mixture of starting material and the isomer **730**, which now is in the (+)-series of compounds. This is an equilibrium process, since **730** gave a similar mixture under these conditions (Scheme 97). The mechanism proposed for this reaction involves C-16 protonation, C-3–C-7 cleavage, and loss of a C-15 proton to give the enamine

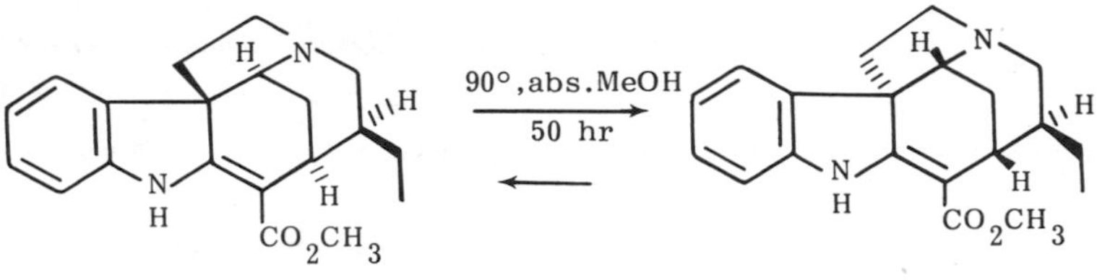

90°, abs.MeOH
50 hr

4 parts   730

729   (-)-19,20-dihydroakuammicine

731. Fragmentation of the C-15–C-16 bonds permits formations of a *seco*-intermediate **732**, which now permits racemization at C-15 to occur. Whether this process occurs *in vivo* remains to be determined with appropriately (C-15) labeled precursors.

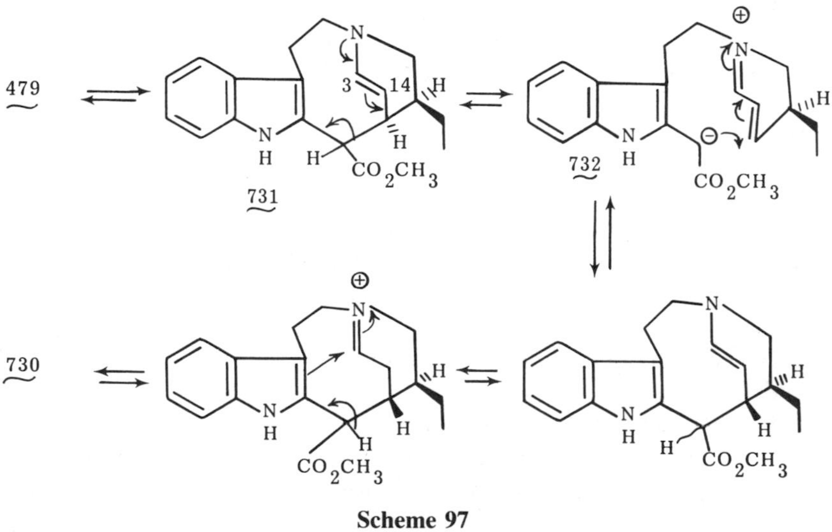

479

731

732

730

Scheme 97

The most controversial area of indole alkaloid biosynthesis without doubt has been the *in vitro* conversions of alkaloids such as stemmadenine (**594**), tabersonine (**619**), and geissoschizine (**345**) to alkaloids of biogenetic significance as reported by Scott and Qureshi. These interconversions were postulated to proceed through dehydrosecodine (**595**).

Tabersonine (**619**) after a 16-hr reflux in glacial acetic acid was said to be converted to 12% catharanthine (**712**), 28% pseudocatharanthine (**704**), and a small quantity of **619**. Previous work by the Eli Lilly group had shown that catharanthine (**712**) after only a 2½-hr reflux in glacial acetic acid gave racemic pseudocatharanthine (**704**) in 45% yield.

However, subsequent work by groups at Manchester and Paris failed to reproduce these results; instead three bases, allocatharanthine (**733**), dihydroallocatharanthine (**734**), and acetoallocatharanthine (**735**), were isolated from the tabersonine (**619**) reaction and acetylstemmadenine (**736**) (55% yield) when stemmadenine (**594**) was refluxed for 50 hr in glacial acetic acid.

Subsequently Scott agreed that allocatharanthine (**733**) was indeed a product. A number of other transformations were also carried out at this time on a silica gel surface at 150°C, although the mechanisms of these transformations have not been established. When *O*-acetyl-19,20-dihydro-preakuammicine (**737**) was heated in methanol at 80°C for 15 min, a 9:1

619   tabersonine   712   catharanthine   704   pseudocatharanthine

733   allocatharantine   734   dihydroallocatharanthine, R=H
                                 735   acetoallocatharanthine, R=OAc

594   stemmadenine, R=H
736   O-acetylstemmadenine, R=Ac

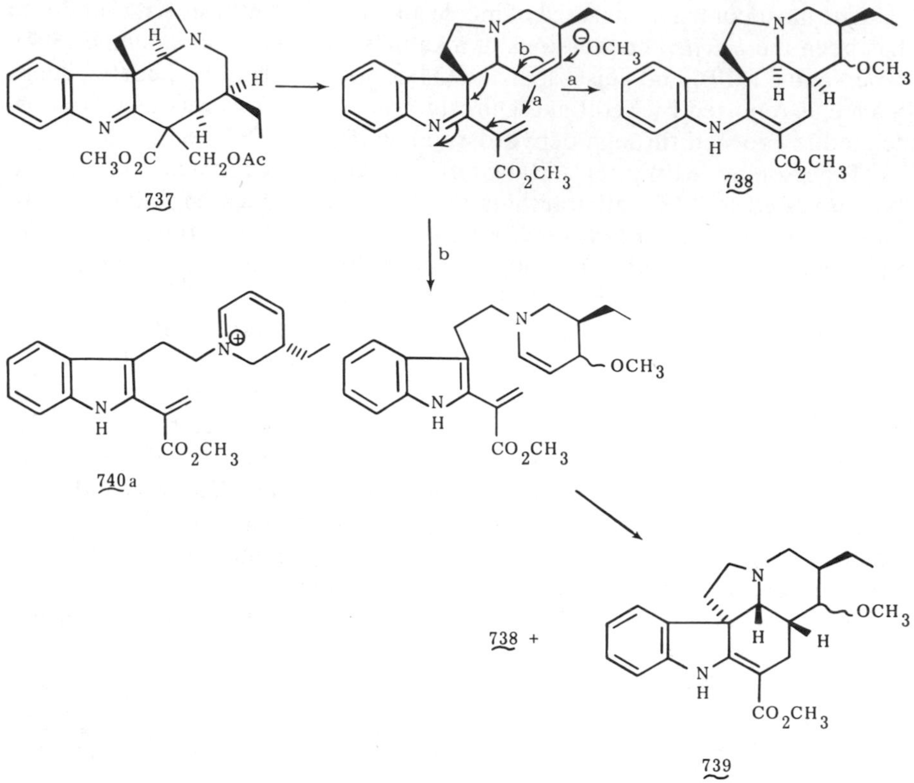

Scheme 98

mixture of (+)- and (−)-15-methoxy-15,20-dihydropseudocatharanthine (**738** and **739**) was produced in 3.5% yield. The protonated dehydrosecodine **740a** is regarded as an intermediate, but other explanations are also possible (Scheme 98).

Some other transformations of interest are the oxidation of stemmadenine (**594**) to preakuammicine (**740**), precondylocarpine (**588**), akuammicine (**479**), and condylocarpine (**587**) in the presence of oxygen and a catalyst, and the formation of the pyridinium species **741**, on heating allocatharanthine (**733**) in methanol at 140°C. The structure was proved by sodium borohydride reduction to tetrahydrosecodine (**600**).

Previous workers had found that only condylocarpine was produced on the oxidation of stemmadenine (**594**). One explanation for this may be the optical purity of C-16 in the stemmadenine used, for it is postulated that one C-16 diastereoisomer may give rise to the precondylocarpine system and the other to the preakuammicine system.

One final example concerns a novel indole alkaloid skeleton which was obtained synthetically prior to its isolation from a natural source.

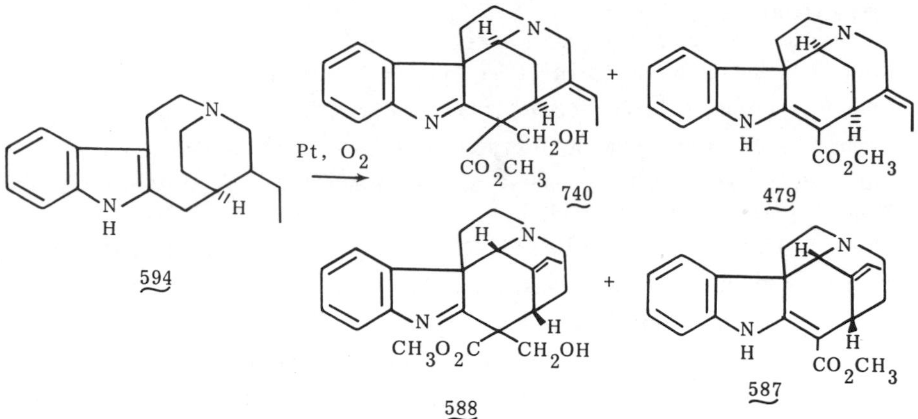

594 740 479 588 587

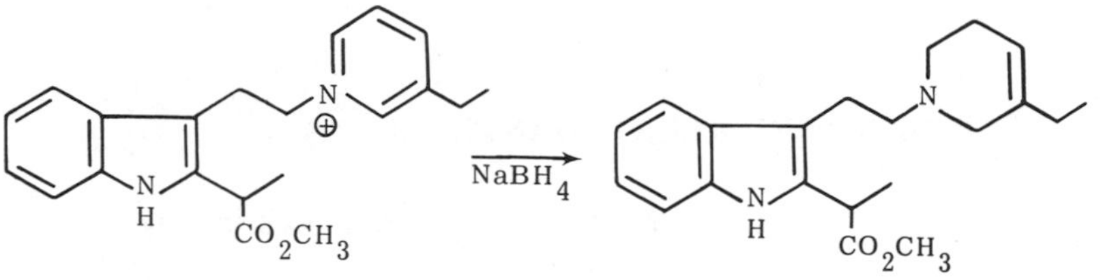

741 600

Andranginine (**742**) is a racemic alkaloid isolated by Potier and co-workers from *Craspidospermum verticillatum* Boj. var. *petiolare*. By coincidence the alkaloid had been synthesized previously by Scott and Wei, who found that heating precondylocarpine acetate (**743**) at 100°C gave **742** in 28% yield by the mechanism shown in Scheme 99.

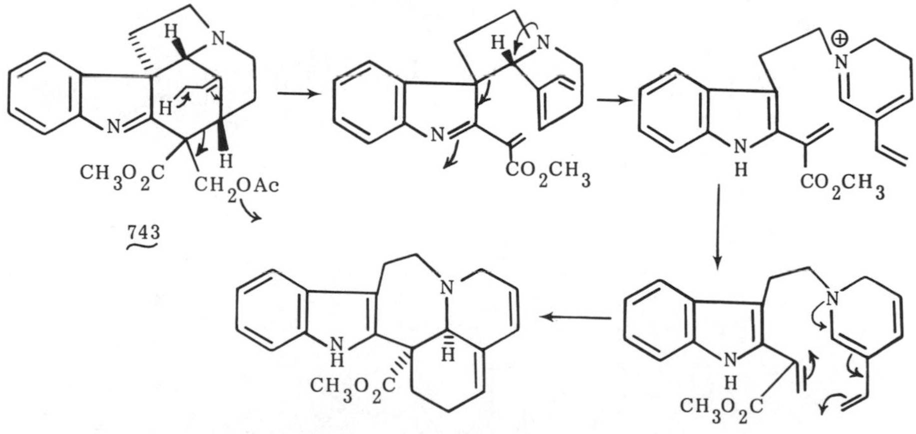

743

742 andranginine

Scheme 99 Biogenetic synthesis of andranginine (**742**).

LITERATURE

Scott, A. I., and C. L. Yeh, *J. Amer. Chem. Soc.* **96**, 2273 (1974).

Qureshi, A. A., and A. I. Scott, *Chem. Commun.* 945 (1968).

Gorman, M., N. Neuss, and N. J. Cone, *J. Amer. Chem. Soc.* **87**, 93 (1965).

Brown, R. T., J. S. Hill, G. F. Smith, K. S. J. Stapleford, J. Poisson, M. Muqunet, and N. Kunesch, *Chem. Commun.* 1475 (1969).

Brown, R. T., J. S. Hill, G. F. Smith, and K. S. J. Stapleford, *Tetrahedron* **27**, 5217 (1971).

Muquet, M., N. Kunesch, and J. Poisson, *Tetrahedron* **28**, 1363 (1972).

Scott, A. I., and C. C. Wei, *J. Amer. Chem. Soc.* **94**, 8263 (1972).

Scott, A. I., and A. A. Qureshi, *J. Amer. Chem. Soc.* **91**, 5874 (1969).

Kan-Fan, C., G. Massiot, A. Ahond, B. C. Das, H.-P. Husson, P. Potier, A. I. Scott, and C. C. Wei, *Chem. Commun.* 164 (1974).

9.27   PANDOLINE AND RELATED COMPOUNDS

Until quite recently there were no natural compounds having the pseudo-catharanthine or more correctly pseudo-*Aspidosperma* skeleton.

The first isolation was of pandoline (**744**) from *Pandaca calcarea* and the structure was determined from the spectral data. Since an ethyl group was observed but no carbinolamine was present, the hydroxy group should be located at C-14 or C-20. The ^{13}C NMR spectral data indicated a quaternary oxycarbon at 71.0 ppm which shifted to 82.0 ppm on acetylation. Three methylene carbons were also shifted by this reaction, indicating the structure for pandoline.

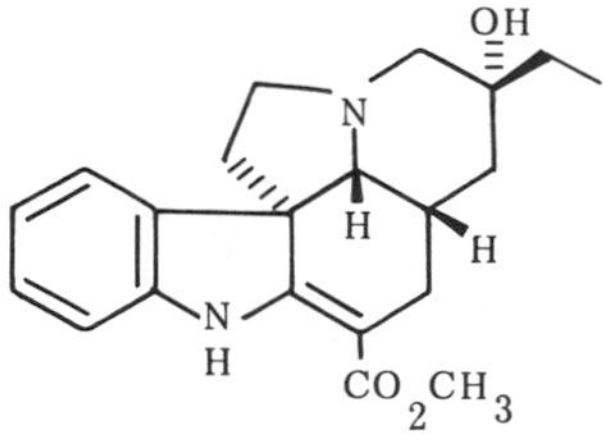

744   pandoline 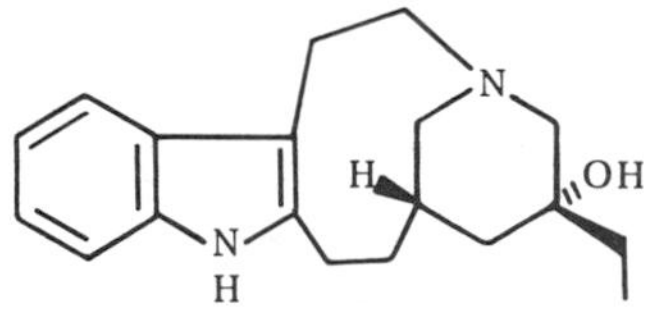   745   (−)-velbanamine

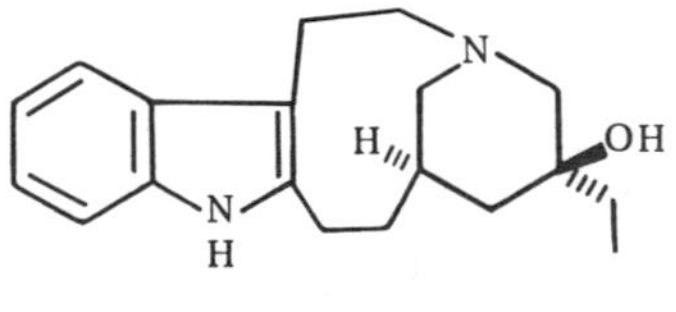

746   (+)-velbanamine

The 250-MHz PMR spectrum established the *cis* nature of the C-D-ring junction with a broad singlet at 2.86 ppm. Pandoline had a positive α_D and consequently should have the absolute stereochemistry shown.

Pandoline was subsequently obtained from *Melodinus polyadenus* (Baillon) Boiteau and *Ervatamia obtusiuschula* Mgf., but the problem of the C-20 stereochemistry remained. This point was resolved by conversion to (−)-velbanamine (745), which could be correlated with (+)-velbanamine (748).

The isolation of this alkaloid has been followed by several further reports of related alkaloids, including (+)-pseudotabersonine (704) from *Pandaca caducifolia* Mgf., capuronidine (747) from *Capuronetta elegans* Mgf., and iboxyphylline (748) and ibophyllidine (749) from *Tabernanthe* species.

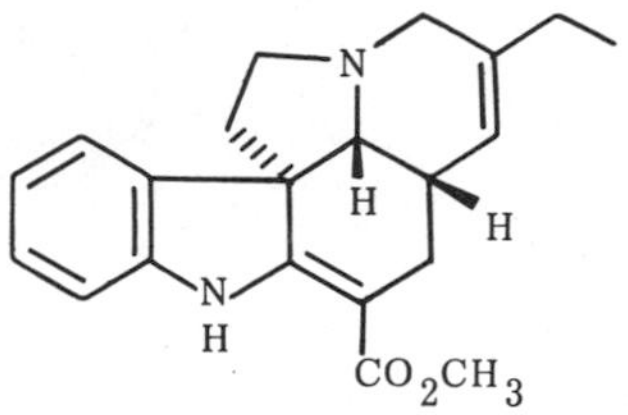

704 pseudotabersonine

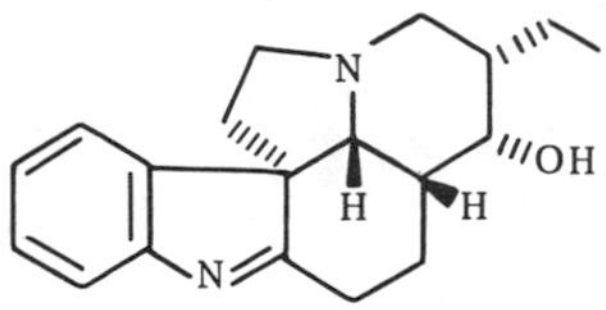

747 capuronidine

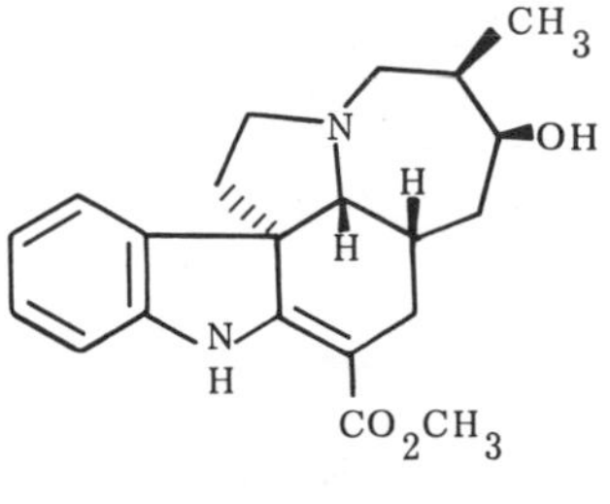

748 iboxyphilline

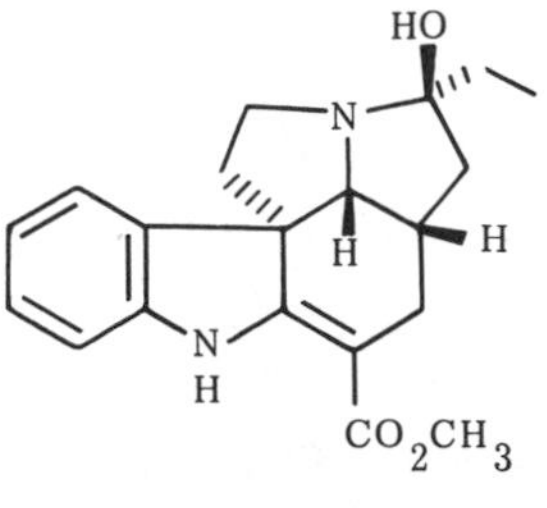

749 ibophyllidine

LITERATURE

Le Men, J., G. Lukacs, L. Le Men-Oliver, J. Levy, and M. J. Hoizey, *Tetrahedron Lett.* 483 (1974).

Bruneton, J., A. Cavé, E. W. Hagaman, N. Kunesch, and E. Wenkert, *Tetrahedron Lett.* 3567 (1976).

9.28 BISINDOLE ALKALOIDS

Some bisindole alkaloids derived from tryptamine have already been encountered in Section 9.5.1. In this section brief mention will be made of some of the bismonoterpenoid indole alkaloids which have been obtained in recent years. Examples of these alkaloids have already been discussed in the sections on the secodine alkaloids, the *Catharanthus* alkaloids, and the *Strychnos* alkaloids. The major bisindole alkaloid groups and their sources are shown in Table 11, and some examples of the alkaloids are shown in Figure 7.

There has been considerable debate as to the natural occurrence of the bisindole alkaloids. Some of these species are possibly formed during iso-

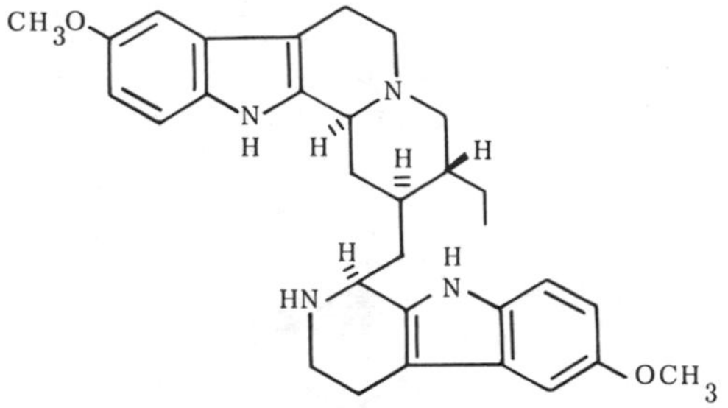

750 isocinchophyllamine

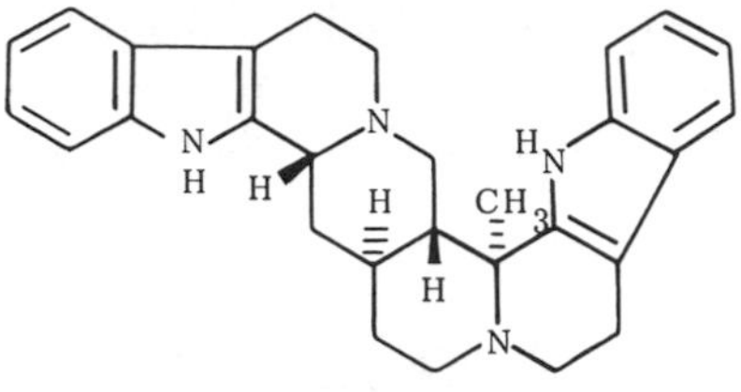

751 roxburghine D

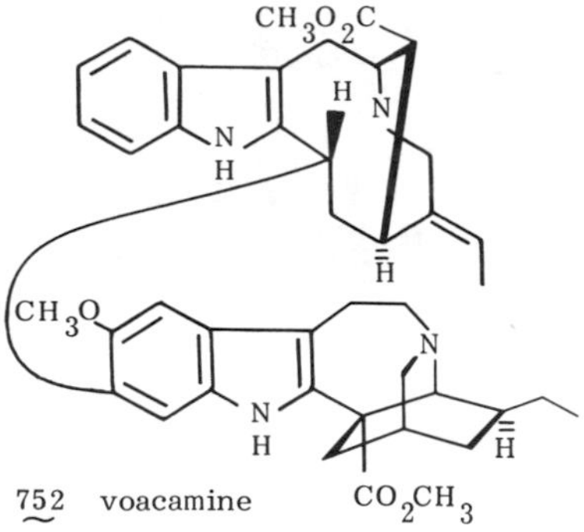

752 voacamine

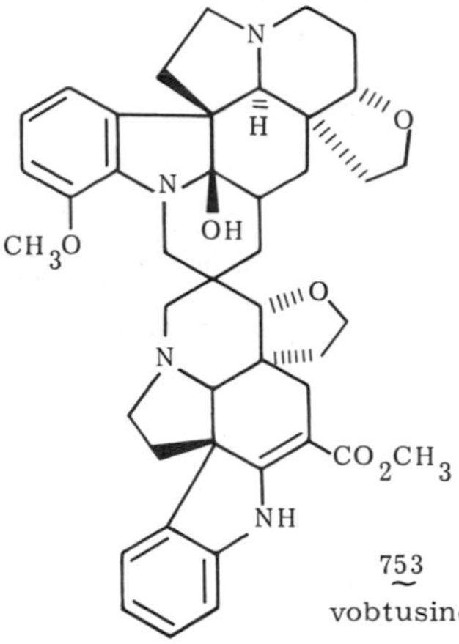

753 vobtusine

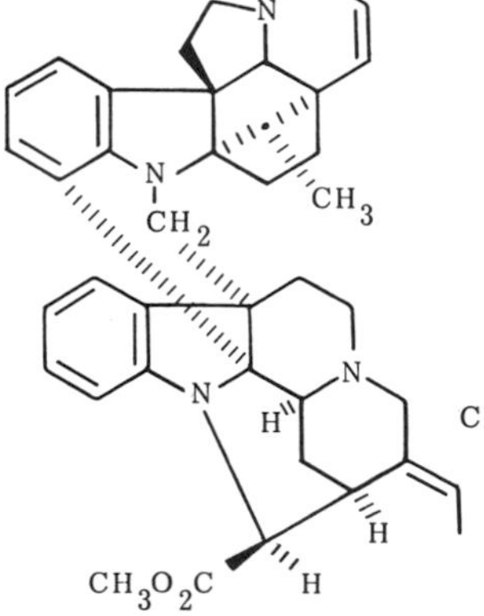

754 pycnanthine

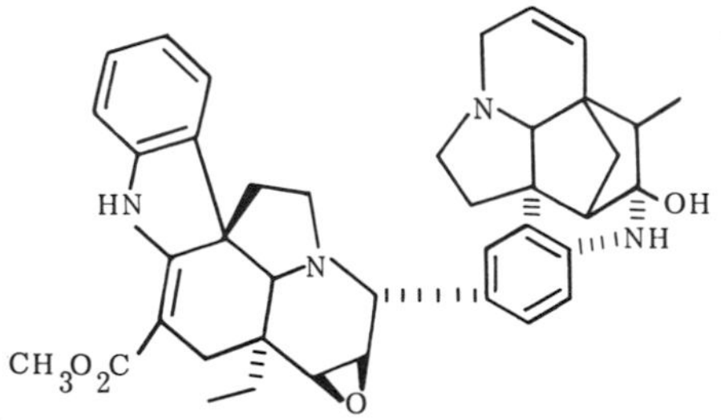

755 scandomeline

Figure 7 Some representative bisindole alkaloids.

Table 11 Summary of Bisindole Alkaloids and Their Isolations

Compound Type	Source
Calycanthine, chimonanthine	*Calycanthus* sp., *Chimonanthus* sp.
Calebash curare	*Strychnos* sp.
Tubulosine	*Alangium* sp., *Pogonopus* sp., *Cassinopsis* sp.
Cinchophyllamine	*Cinchona ledgeriana*
Aspidosperma–velbanamine	*Catharanthus* sp.
Roxburghines	*Uncaria gambier*
Haplophytine	*Haplophytum cimicidium*
Geissospermine	*Geissospermum* sp.
Secamines	*Rhazya* sp., *Amsonia* sp.
Vobasine–iboga	*Callichilia, Conopharyngia, Gabunia, Stemmadenia, Tabernaemontana,* and *Voacanga* sp.
Pleiocarpamine–vindolinine type	*Pleiocarpa* sp.
Villalstonine type	*Alstonia* sp.
Kopsane–eburna	*Pleiocarpa* sp.
Strictamine–eburna	*Hunteria umbellata*
Vobtusine type	*Callichilia* sp., *Voacanga* sp.
Melodinus–aspidosperma	*Melodinus* sp.

lation, but without doubt the vast majority of these are true natural products. The presecamines were observed in freshly extracted plant material and an enzyme preparation from *C. roseus* has been obtained which catalyzes the formation of the *Catharanthus* dimers from the monomeric alkaloids.

9.28.1 Spectral Properties

The UV spectra of indole alkaloids are, as we have seen, very important in determining the structural class to which the alkaloid belongs.

On the basis of their UV spectra, the bisindole alkaloids can be divided into two major groups, those in which the two chromophores are isolated and those in which some orbital overlap occurs, leading to modification of the chromophore. Each of these two groups can be further subdivided into groups where the two "monomer" chromophores are the same or are different.

The chromophore of a bisindole alkaloid can frequently be identified by the superposition of the two parent monomer chromophores (e.g. the presecamines), which are the sum of an indole and a β-anilinoacrylate.

Minor differences are frequently observed and are usually due to steric effects of the introduction of substituents into the aromatic nucleus.

The color reactions with ceric ammonium sulfate can frequently rationalized in terms of the two chromophores, although sometimes this is not possible for reasons which are not clear. The unique nature of the color reaction, pink, gray, dark brown, and so on, however, is sometimes useful in determining bisindole alkaloids in complex mixtures.

The proton NMR spectra of the bisindole alkaloids are typically very complex, and except in unusual circumstances only the signals of functional groups are usually identified. The aliphatic region of the spectrum is most often mountainous with no discernible coupling constants or readily identified signals.

The mass spectra of the bisindole alkaloids are usually complex (one classic exception is the presecamines), but considerable information can be obtained if the fragmentation pattern of the monomeric units is known. Frequently fragments are lost from one monomer unit, and these losses may be very helpful in eliminating certain positions for the linking of the two monomers.

The higher temperature needed to volatalize the bisindole alkaloids may also lead to some complications in their mass spectra. If a dimer contains a carbomethoxyl group and a basic nitrogen in a favorable steric arrangement, inter- or intra-transmethylation may occur. In this process intramolecular ions at $M^+ + 14$ and $M^+ + 15$ are observed, the former being most prominent. In some instances these ions are more significant than the true parent ion. The voacamine alkaloids and the bisindole alkaloids of *Catharanthus* sp. are notorious in this respect.

9.28.2   Isolation and Chemistry

The *Cinchona* alkaloids were discussed previously because of the therapeutic significance of quinine and quinidine. The leaves of *Cinchona ledgeriana* are the source of some biogenetically interesting bisindole alkaloids, although the quinoline alkaloids are not present in any dimeric form.

The isolated alkaloids are cinchophyllamine and isocinchophyllamine. The latter was originally assigned the structure **756**, but this was recently corrected by X-ray crystallographic analysis, which demonstrated the alternative structure **750**. The relationship of cinchophyllamine to isocinchophyllamine is at present unclear.

756   isocinchophyllamine
       (old structure)

The roxburghines A–E are an interesting group of alkaloids obtained by Merlini and co-workers from *Uncaria gambier* Roxb. They all have the same molecular formula and are in fact stereoisomers. Roxburghine D is a typical example and the structure was elegantly established by spectroscopic and degradative means and total synthesis. Spectroscopy established the stereochemistry at C-3 and C-20, but the problem of C-19 remained and was resolved by synthesis.

Addition of methyl *t*-butyl malonate to the unsaturated ketone **757** gave a single product, **758**. Treatment with TFA, tryptomine, dicyclohexylcarbodiimide, and finally acid gave the lactam **759** stereospecifically. Diisobutylaluminum hydride reduction of in glyme at $-70°C$ afforded roxburghine D (**751**), thereby establishing the structure (Scheme 100).

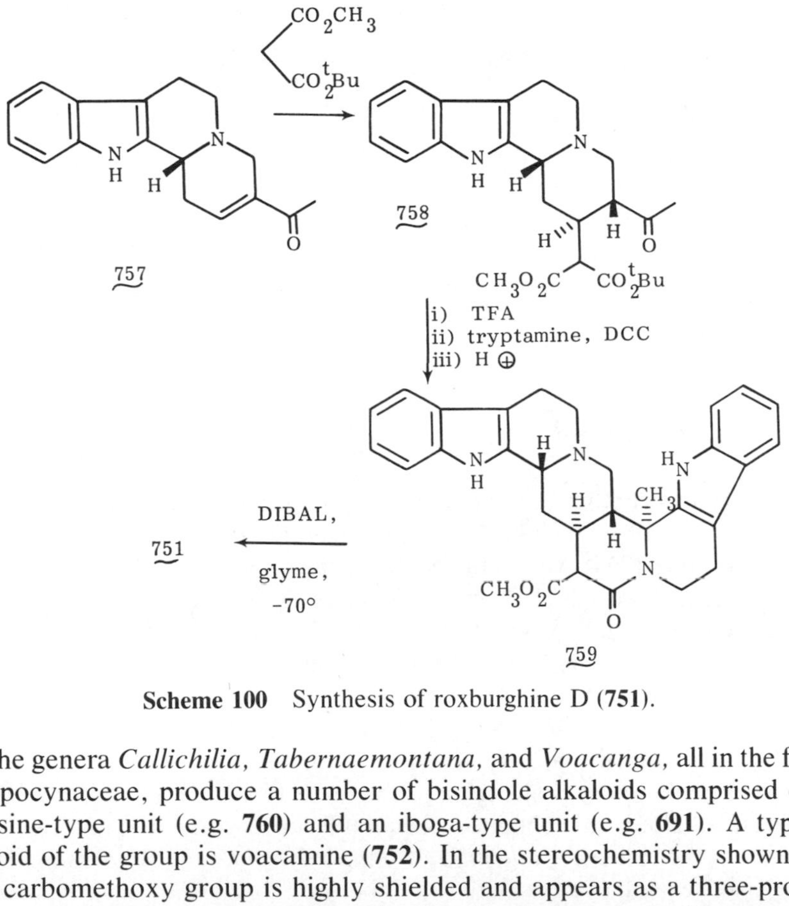

Scheme 100 Synthesis of roxburghine D (**751**).

The genera *Callichilia*, *Tabernaemontana*, and *Voacanga*, all in the family Apocynaceae, produce a number of bisindole alkaloids comprised of a vobasine-type unit (e.g. **760**) and an iboga-type unit (e.g. **691**). A typical alkaloid of the group is voacamine (**752**). In the stereochemistry shown the C-16 carbomethoxy group is highly shielded and appears as a three-proton singlet at 2.63 ppm. In the epimer this signal is observed at 3.53 ppm.

Some members of this series, such as tabernamine (**761**) from *Tabernaemontana johnstonii* Pichon, are weakly cytotoxic in the P-388 cell culture system. Cleavage of tabernamine under acidic conditions afforded ibogamine

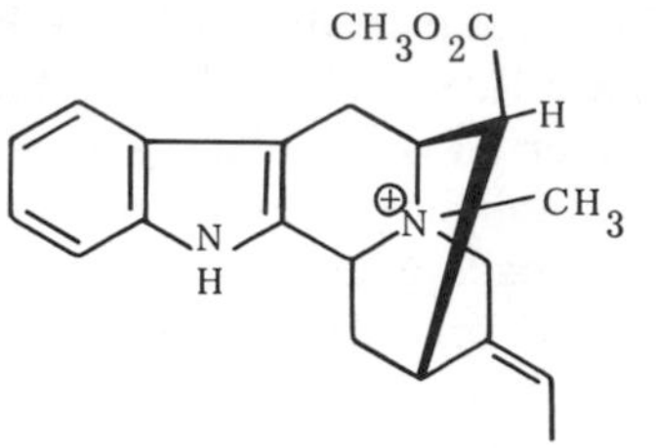

760   vobasinol 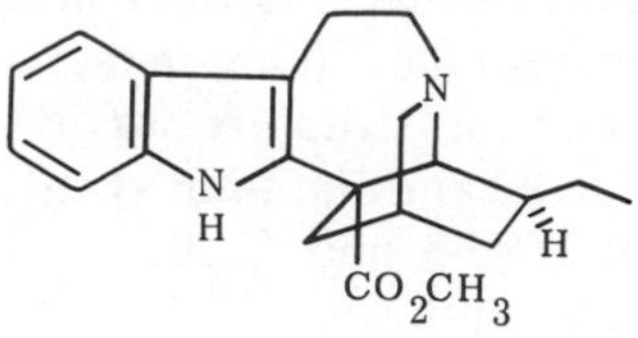   691   coronaridine

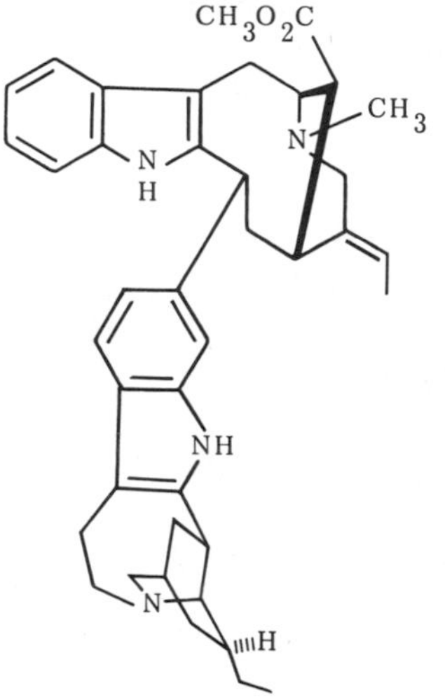

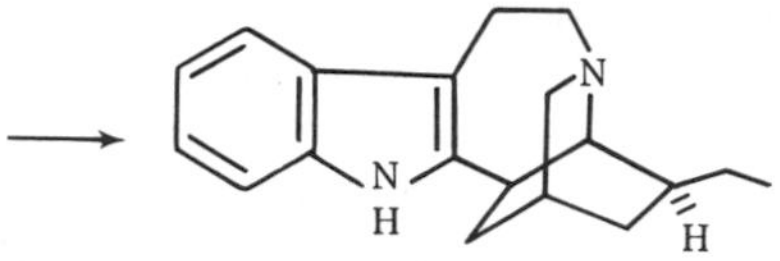

690   ibogamine

\+

760

761   tabernamine

(**690**), and the structure of the dimer was confirmed by the reverse reaction, a condensation of vobasinol (**760**) and ibogamine (**690**).

Pycnanthine (**754**) was obtained from the root bark of *Pleiocarpa pyc-nantha* (K. Schum.) Stapf. var *pycnantha* Pichon and initially caused a problem because the UV spectrum (λ_{max} 267, 309, and 325 nm) was not the summation of any two chromophores. Subsequently, this extended chromophore, which is quite characteristic of the group, was shown to be due to the spatial interaction of two indoline chromophores.

Vobtusine and the apparently closely related alkaloid callichiline have proved to be the most challenging problems in bisindole alkaloid chemistry.

These alkaloids are obtained, frequently in reasonable quantities, from species in the genera *Callichilia, Conopharyngia,* and *Voacanga* in the Apocynaceae and over a dozen members of this group are known. The structures are particularly problematic because the two monomeric units do not separate from the whole compound either chemically or mass spectrometrically. This makes intellectual dissection of the molecule and establishment of the bonds between the two units extremely difficult.

Since the UV spectrum is that of vincadifformine plus an *N*-alkyl-7-methoxy indoline, the coupling between the two components must not involve the aromatic nuclei, but does involve the *N*-alkylindoline.

The monomeric alkaloid beninine (**762**) co-occurs with vobtusine and provided some useful structural clues; in particular the ion *m/e* 138 (**763**) was observed from both compounds.

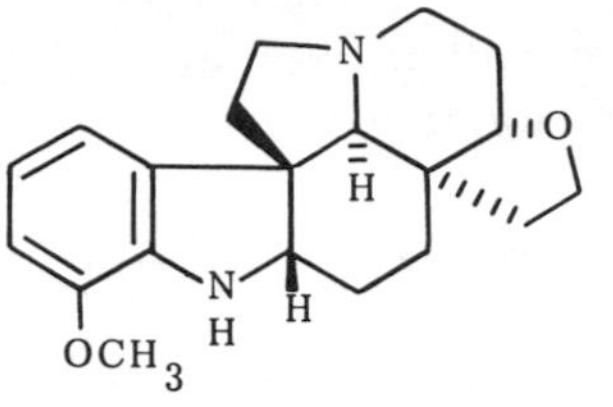

762　beninine

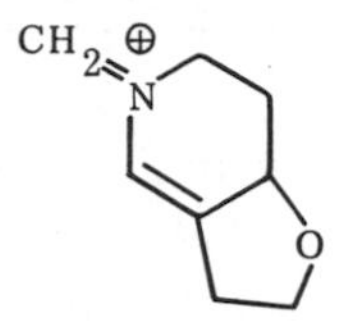

763　*m/e* 138

Recently the dibromo derivative of vobtusine was examined by X-ray analysis and the structure was deduced to be **753**. Note the failure of the carbinolamine moiety of **753** to undergo hydride reduction, which led to the

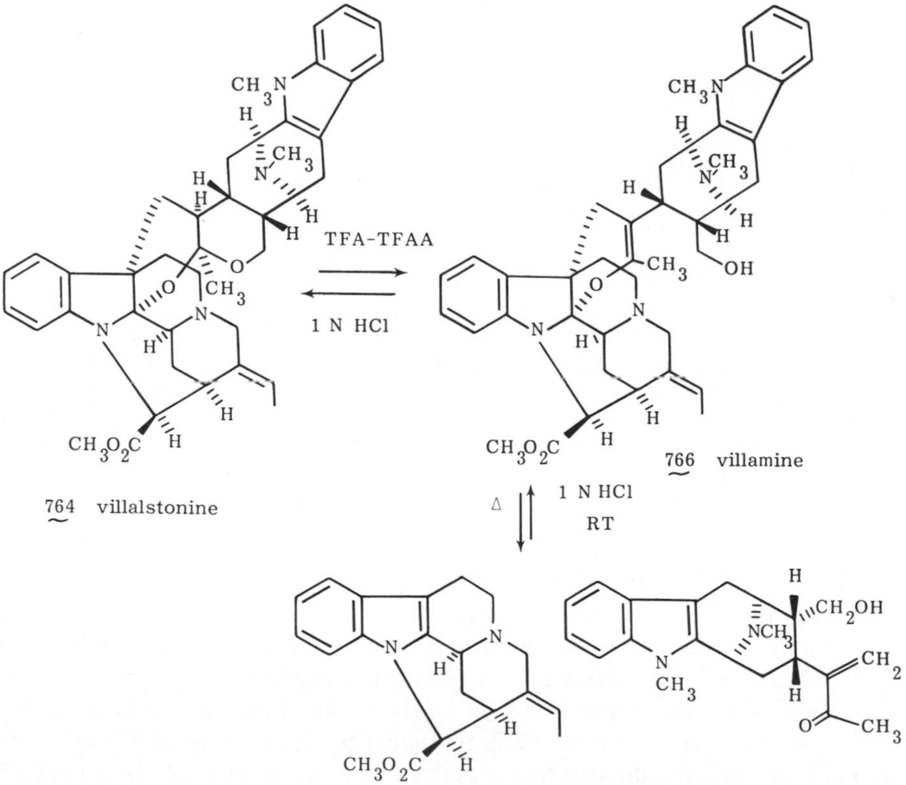

764　villalstonine

TFA-TFAA

1 N HCl

766　villamine

Δ ⇅ 1 N HCl / RT

765　pleiocarpamine

767　macroline

original proposal of a tertiary alcohol at C-16'. Wenkert and co-workers have discussed the ^{13}C NMR spectra of vobtusine and related compounds.

Another genus of the Apocynaceae which produces bisindole alkaloids is *Alstonia*, and a typical member is villalstonine (**764**). These alkaloids can be cleaved with acid. For example on standing in 70% perchloric acid **764** gives (+)-pleiocarpamine (**765**), but the other "half is destroyed."

The key compound in the structure determination of villalstonine was villamine (**766**), formed by treatment with trifluoroacetic acid–trifluoroacetic acetic anhydride. The reverse reaction can be carried out with 1 *N* hydrochloric acid.

The mass spectrum of villamine was virtually the sum of the two component monomeric halves, pleiocarpamine (**765**) and macroline (**767**), and the retro Diels–Alder reaction could also be carried out by thermolysis. The reverse reaction was found to occur at room temperature after 18 hr in 0.2 *N* hydrochloric acid to give villalstonine in 40% yield.

LITERATURE

Reviews

Gorman, A. A., M. Hesse, H. Schmid, P. G. Waser, and W. H. Hopff, *Alkaloids, London* **1**, 200 (1970).

Reviews: Alstonia Alkaloids

Saxton, J. E., *Alkaloids NY* **12**, 207 (1970).

Saxton, J. E., *Alkaloids NY* **14**, 167 (1973).

Vobtusine

Gorman, A. A., V. Agwada, M. Hesse, U. Renner, and H. Schmid, *Helv. Chim. Acta* **49**, 2072 (1966).

Lefebvre-Soubeyran, O., *Acta Cryst., Sect. B.* **29**, 2855 (1973).

Rolland, Y., N. Kunesch, J. Poisson, E. W. Hagaman, F. M. Schell, and E. Wenkert, *J. Org. Chem.* **41**, 3270 (1976).

9.29 *CATHARANTHUS* ALKALOIDS

The Madagascan plant *Catharanthus roseus* is probably the most significant alkaloid-containing plant to be studied in the last 25 years.

The genus *Catharanthus* is comprised of eight species, six of which are native to Madagascar. Now *C. roseus* is pantropical and is widely cultivated and grown as an herbaceous border. There has been considerable confusion in the literature concerning the name of the plant and the name *Vinca rosea* Linn. is often seen. This name is incorrect and should not be used. The genus *Vinca* is quite distinct and produces a different array of alkaloids (see Section 9.30).

Catharanthus roseus is notable because of the number of alkaloids that have been isolated (over 90), and because two of these, vincaleukoblastine (VLB), and leurocristine (VCR), are among the two most important clinically useful anticancer agents known.

9.29.1 Bisindole Alkaloids of *Catharanthus roseus*

Two groups, unknown to each other, became interested in the folklore reports of the hypoglycemic activity of the plant. Although these reports could not be substantiated, the Canadian group found that certain of the alkaloid fractions produced a peripheral granulocytopenia and bone marrow depression in rats. The group at Eli Lilly found that alkaloid extracts of this plant were effective against the P-1534 leukemia in mice.

The first important compounds to be isolated by Svoboda were vincaleukoblastine (**768**) and leurosine (**769**), but other fractions showed even higher activity, and subsequently leurocristine (**770**) and leurosidine (**771**)

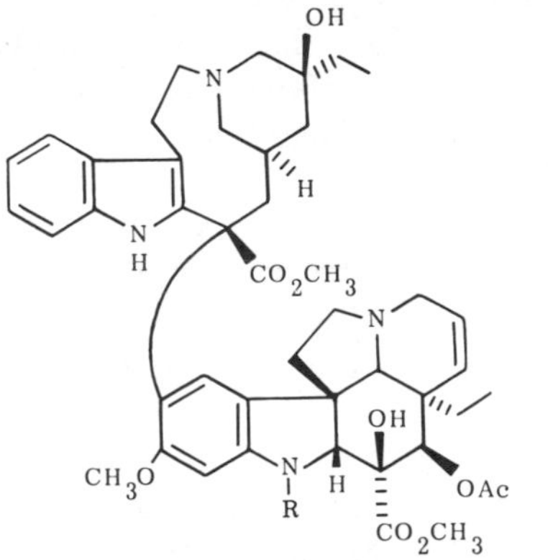

768   vincaleukoblastine, R=CH$_3$
770   leurocristine, R=CHO

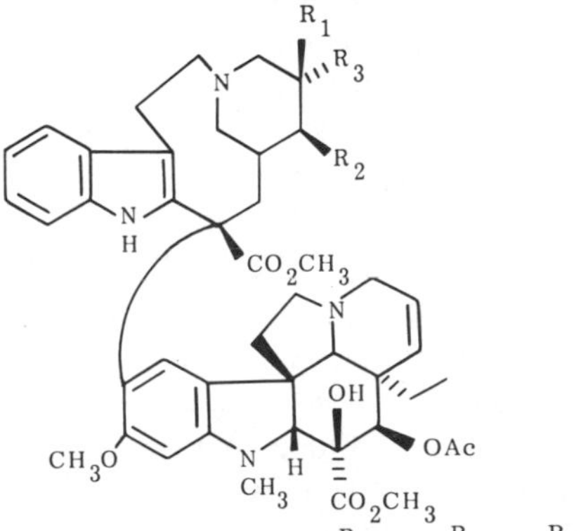

	R$_1$	R$_2$	R$_3$
769 leurosine		$-O-$	C$_2$H$_5$
771 leurosidine	C$_2$H$_5$	H	OH
774 vincadioline	OH	α-OH	C$_2$H$_5$

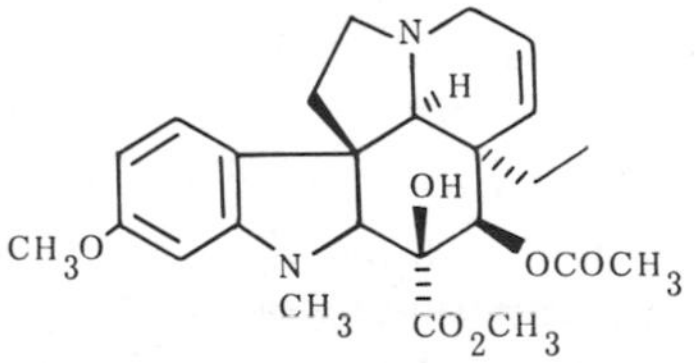

626   vindoline

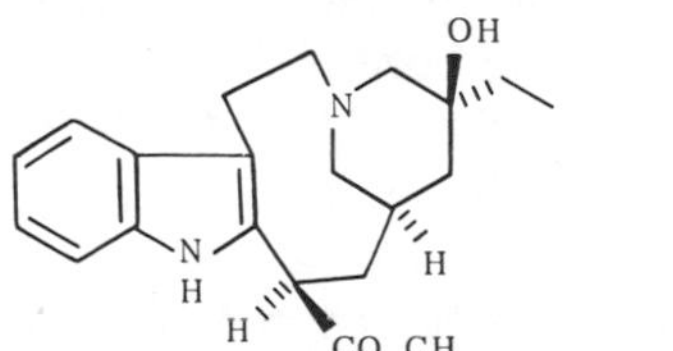

724   16β-carbomethoxyvelbanamine

were obtained. Leurocristine has become one of the most critical compounds in the chemotherapeutic armamentarium against human cancerous states. The two important alkaloids VLB and VCR are relatively minor alkaloids and yields are typically 1 g and 20 mg/1000 kg respectively. The techniques developed for the commercial production of these alkaloids stand as a milestone in natural products chemistry, and should be regarded with the same awe and respect that a complex synthesis receives.

The structure of VLB (768) was deduced by a combination of spectral data and degradative evidence which indicated that the molecule was composed of two parts, vindoline (626) and 16β-carbomethoxyvelbanamine (724). The position of the tertiary hydroxyl group was determined by mass spectrometric examination, and the location of carbon–carbon bond linking the two units by reductive cleavage in the presence of deuterium. The stereochemistry and absolute configuration of VLB were obtained by X-ray analysis of the methiodide salt.

9.29.2   Partial Synthesis

The low yields of these two alkaloids from natural sources has stimulated several approaches to improving their availability. One of these has been to develop chemical methods for the conversion of VLB to VCR, since the former is more abundant. The Lilly workers have found that *Streptomyces albogriseolus* is capable of carrying out a specific N-demethylation, thereby permitting N-formylation to be carried out. The preferred method however is to oxidize the *N*-methyl group with chromium trioxide in acetone at low temperature to give VCR directly.

There are two other important approaches to the synthesis of the dimers which converge at the final stage, the joining of the two monomeric units. One approach is totally synthetic and involves the formation of the two monomeric units. A second method is to modify the corresponding natural monomeric alkaloids prior to joining the units.

In either of these two routes the crucial step is that of joining the two monomeric units. It is only relatively recently that this has been achieved by Potier and co-workers and by Kutney and co-workers. These approaches are based on a modification of the Polonovskii reaction of vindoline with an iboga alkaloid N-oxide. For example in the synthesis of 15′,20′-dehydrovincaleukoblastine (772), catharanthine N-oxide (707) is treated with vindoline (626) in methylene chloride–trifluoroacetic anhydride at −50°C followed by sodium borohydride reduction. The yield is on the order of 50%.

A critical aspect of this synthetic work was the discovery that the configuration at C-16′ could be defined by circular dichroism. This was important because although only the natural stereochemistry is produced at low temperature, under different conditions (room temperature or +42°C) a mixture of natural and unnatural isomers was produced in which the unnatural isomer predominated. The unnatural isomers are devoid of antileukemic activity and it is therefore of prime importance to be able to maximize the yield of the natural isomers. The rationalization of the difference in yields as the temperature is varied is an interesting point. At low temperatures the reaction is regarded as proceeding through a concerted mechanism (Scheme 101), whereas the high-temperature reaction is thought to occur by a nonstereospecific reaction on the iminium species 773 (Scheme 102). Sev-

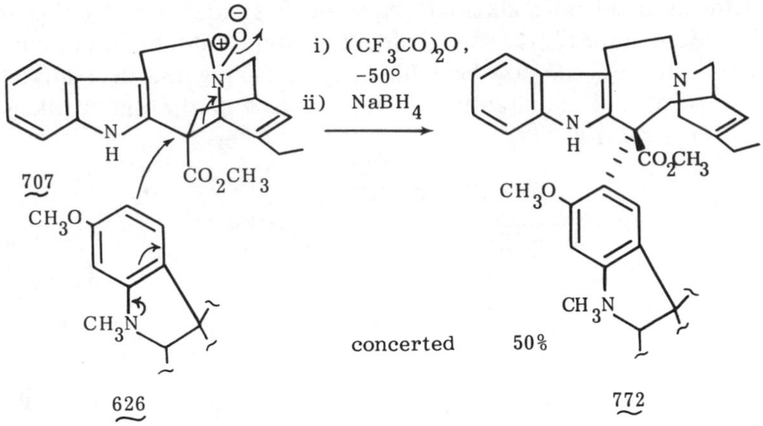

Scheme 101

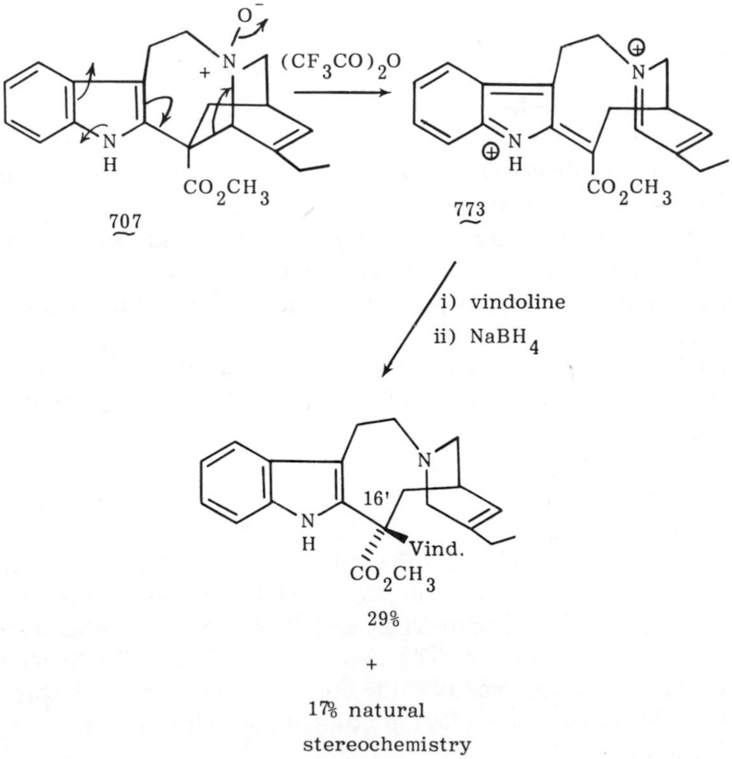

i) vindoline

ii) NaBH$_4$

29%

+

17% natural
stereochemistry

Scheme 102

eral of the antileukemic alkaloids have been synthesized by this method including leurosidine (771) and vincaleukoblastine (768), but more important, substantial progress can now be made in synthesizing new derivatives, since vindoline (626) and catharanthine (712) are two of the major alkaloids of *Catharanthus roseus*.

leurosidine   vincaleukoblastine

9.29.3 Spectral Properties

The dimeric indole alkaloids of *Catharanthus roseus* are a true challenge to the natural product chemist. Not only are the spectral data extremely complex, but the compounds are only rarely crystalline and typically are quite unstable. In addition, ions at 14 and/or 28 mu *above* the true molecular ion frequently make even accurate determination of the molecular weight a frustrating task.

Although much information can be gained from the proton NMR spectrum of these compounds, particularly with regard to establishing the nature of the vindoline moiety, it is the quite recent developments in the area of ^{13}C NMR spectroscopy which have revolutionized this area. The spectral data for vincaleukoblastine (768), leurosine (769), and vincadioline (774) are shown in Table 12. These data permitted the rapid assignment of the gross structure to vincadioline (774) by the shift of C-15' upfield by ~9 ppm. As expected C-3', C-19', and C-21' all show γ-shifts of about 5 ppm, substantiating the assignment of a hydroxyl group to C-15'. Several other instructive points are worthy of mention. The consistency of the resonances for the vindoline protion are notable, as is the considerably reduced deshielding of C-15' and C-20' in leurosine (769) in comparison with the dihydroxy compound vincadioline (774).

Table 12 ^{13}C NMR of Some Bisindole *Catharanthus* Alkaloids

	Vindoline Portion				Indole Portion		
Carbon	VLB^a (768)	Leurosine^b (769)	Vincadioline^a (774)	Carbon	VLB	Leurosine	Vinca- dioline
C-2	83.3	83.3	83.4	C-2	131.4	130.7	131.6
C-3	50.2	50.2	50.4	C-3	48.0	42.3	43.2
C-5	50.2	50.2	50.4	C-5	55.8	49.6	55.6
C-6	44.6	44.5	44.6	C-6	28.2	24.6	28.5
C-7	53.2	53.1	53.3	C-7	117.0	116.7	116.7
C-8	122.6	123.0	123.0	C-8	129.5	129.1	129.4
C-9	123.5	123.4	123.6	C-9	118.4	118.1	118.5
C-10	121.1	120.4	120.6	C-10	122.1	122.2	122.3
C-11	158.0	157.6	158.1	C-11	118.7	118.4	118.9
C-12	94.2	94.0	94.2	C-12	110.4	110.3	110.5
C-13	152.5	152.8	152.7	C-13	135.0	134.6	134.9
C-14	124.4	124.3	124.5	C-14	30.1	33.5	39.2
C-15	129.9	129.7	130.0	C-15	41.4	60.3	75.2
C-16	79.7	79.5	79.6	C-16	55.8	55.3	55.8
C-17	76.4	76.2	76.5	C-17	34.4	30.7	32.8
C-18	8.3	8.3	8.4	C-18	6.9	8.6	6.2
C-19	30.8	30.7	30.9	C-19	34.4	28.0	29.2
C-20	42.7	72.6	42.7	C-20	69.4	59.9	71.3
C-21	65.5	65.5	65.7	C-21	64.2	54.0	60.3
COOC̲H₃	170.8	170.7	170.0	C̲OOCH₃	174.0	174.1	174.8
COOC̲H₃	52.1	52.1	52.2	COOC̲H₃	52.3	52.3	52.4
ArOC̲H₃	55.8	55.7	55.8				
NC̲H₃	38.3	38.2	38.3				
OC̲OCH₃	171.6	171.4	171.4		δ (ppm) from TMS in CDCl₃		
OCOC̲H₃	21.1	21.0	21.1				

^{a} Data from D. E. Dorman and J. W. Paschal, *Org. Mag. Reson.* **8**, 413 (1976).
^{b} Data from E. Wenkert *et al.*, *Helv. Chim. Acta* **58**, 1560 (1975).

9.29.4 Pharmacology

Although several of these bisindole alkaloids display antileukemic activity, only two of these are presently in clinical use. Vincaleukoblastine (VLB) (**768**), but more particularly leurocristine (VCR) (**770**), have proved to be extremely valuable single agents or combination regimens in the treatment of many forms of human neoplastic disease.

A critical or comprehensive discussion of the clinical usefulness of VLB is beyond the scope of this book, but some brief generalizations are warranted. The principal use of VLB is in the management of Hodgkin's disease (68% overall response rate) either alone or in combination with chlorambucil. Other uses have included the treatment of resistant choriocarcinona and

lymphosarcoma. In combination with the antibiotics bleomycin and acti-
nomycin D, VLB has proved to be effective in the treatment of testicular
tumors.

As a single agent, VCR is highly active in the treatment of childhood
leukemia (ACL), and is also a critical component of several combination
regimens both for ACL and non-Hodgkin's lymphomas. It is part of the
famous MOPP (mechlorethamine, oncovin, procarbazine, and prednisone)
combination used in the treatment of Hodgkin's disease.

LITERATURE

Reviews

Taylor, W. I., and N. R. Farnsworth, (Eds.), *The* Catharanthus *Alkaloids, Botany, Chemistry
and Pharmacology*, Marcel Dekker, New York, 1973.

Synthesis

Potier, P., N. Langlois, Y. Langlois and F. Gueritte, *J. Chem. Soc., Chem. Commun.* 670
(1975).

Kutney, J. P., T. Hibino, E. Jahngen, T. Okutani, A. H. Ratcliffe, A. M. Treasurywala, and
S. Wunderly, *Helv. Chim. Acta* **59**, 2858 (1976).

Mangeney, P., R. Z. Andriamialisoa, N. Langlois, Y. Langlois and P. Potier, *J. Amer. Chem.
Soc.* **101**, 2243 (1979).

9.30   *VINCA* ALKALOIDS

The genus *Vinca* is comprised of six species native to the Mediterranean
region of Europe and to western Asia. The most important species are *Vinca
major* L. and *Vinca minor* L.

Vinca minor was first examined by Lucas in 1859, who obtained an
alkaloid fraction, but until 1953 only four *Vinca* alkaloids were known. After
the isolation of reserpine many apocynaceaous plants were examined for
alkaloids and now over 90 *Vinca* alkaloids are known. Approximately 30 of
these occur in other closely related plants, but only about 10 in the genus
Catharanthus. Characteristically, the *Vinca* species do not show anticancer
activity and no dimeric alkaloids have been isolated and characterized from
Vinca species.

Plants of the genus *Vinca* have a quite extensive folklore literature, *V.
minor* has been used as a remedy for toothache and hypertension and as a
carminative, vomitive, and astringent. *V. major* has been used in France as
an abortifacient, an astringent, and a tonic. The total alkaloids of both *V.
minor* and *V. major* produce marked hypotension in animals and this was
traced to the alkaloid vincamine (**775**). Figure 8 gives some of the representative alkaloids of *Vinca* species; they range from *Corynanthe*-type alkaloids, through to those in the *Strychnos* and *Aspidosperma* series. Notable

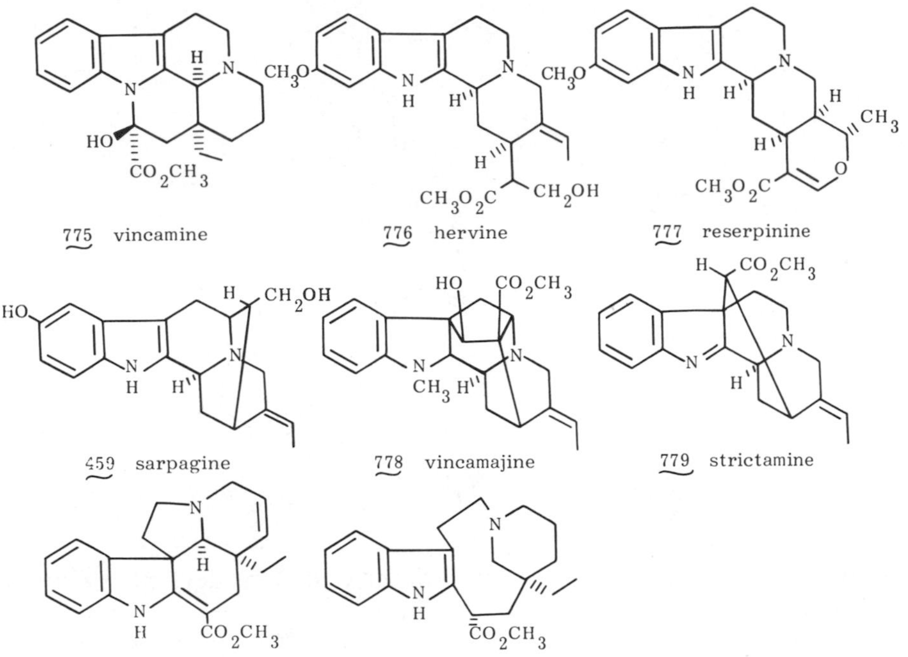

775 vincamine; 776 hervine; 777 reserpinine; 459 sarpagine; 778 vincamajine; 779 strictamine; 619 (-)-tabersonine; 723 vincadine

Figure 8 Some representative *Vinca* alkaloids.

by their absence are alkaloids of the iboga type. This would seem to be the simplest explanation as to why the genus *Vinca* contains none of the dimeric alkaloids typical of *Catharanthus*.

Vincamine (**775**) is the most important *Vinca* alkaloid, where it may be present in yields up to 2–3%. Treatment with acid gave (−)-eburnamonine (**780**), the optical antipode of the major alkaloid of *Hunteria eburnea* Pichon. Acetylation of vincamine (**775**) does not lead to an acetyl derivative but rather apovincamine (**782**).

9.30.1 Synthesis

Several synthetic approaches to the Eburna nucleus of vincamine (**775**) have been described, and depending on the individual synthetic goal quite different problems are posed.

Many of the nonbiogenetic approaches have employed tryptamine as the indole template in condensation with an appropriately functionalized di- or tricarbonyl system. A simple example is the synthesis of eburnamonine (**780**) by Barlett and Taylor (Scheme 103), the key step of which involves condensation of β-ethyl-β-formyl adipic acid (**783**) with tryptamine (**2**) to afford the lactam **782**.

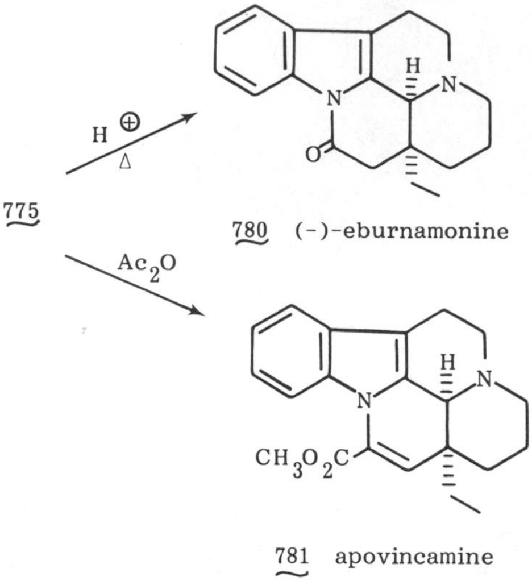

780   (−)-eburnamonine

781   apovincamine

A more elaborate synthetic scheme (Scheme 104) was devised by Gibson and Saxton in order to reach racemic vincamine (**775**).

A highly efficient synthesis of (±)-eburnamine (**784**) by Schlessinger and co-workers involves alkylation of the dianion of **785** with methyl bromoacetate to complete the carbon skeleton, which can then be elaborated by standard methods (Scheme 105).

9.30.2   Biomimetic Interconversion of the *Aspidosperma* to the Eburna Skeleton

About half of the vincamine currently used is partially synthesized from tabersonine and considerable effort has been expended in studying this interconversion, particularly by Le Men and co-workers in France.

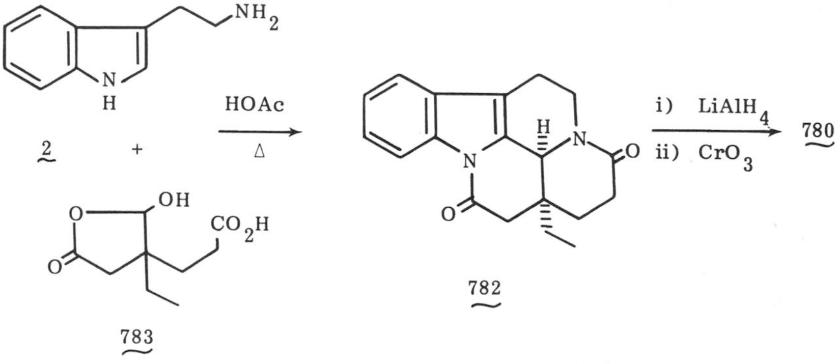

Scheme 103   Bartlett–Taylor synthesis of eburnamonine (**780**).

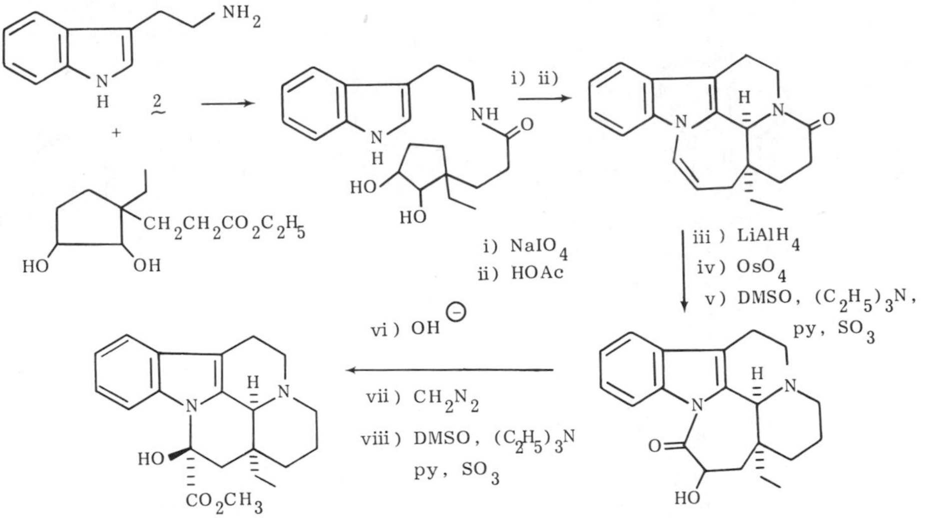

Scheme 104 Gibson–Saxton synthesis of vincamine (**775**).

The *Aspidosperma* skeleton can also be converted to the Eburna skeleton directly. For example, the N-oxide of (+)-1,2-dehydroaspidospermidine (**786**) was rearranged to (+)-eburnamine (**784**) with triphenylphosphine in aqueous acetic acid at 0°C in 42% yield. The hydroxyindolenine **787** is regarded as an intermediate.

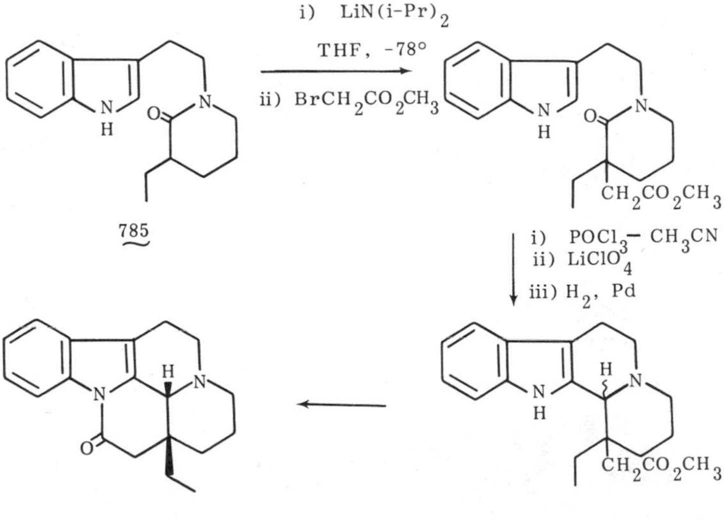

Scheme 105 Schlessinger and co-workers' synthesis of eburnamonine (**780**).

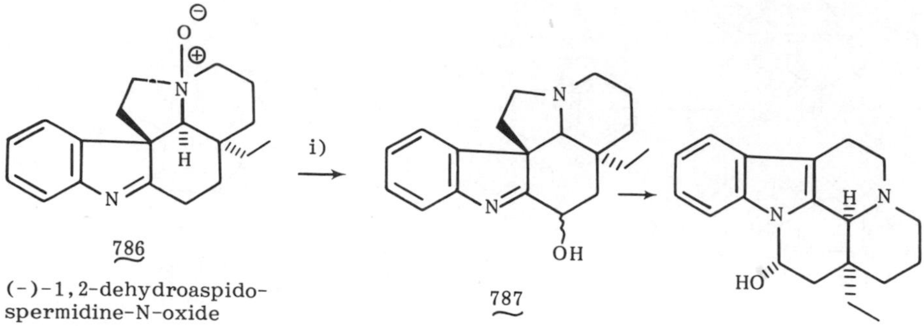

786

(−)-1,2-dehydroaspido-
spermidine–N–oxide

787

784 (+)-eburnamine

i) $(C_6H_5)_3P$, aq. HOAc,

0°, 48 hr

When (−)-vincadifformine (**627**) was treated with lead tetraacetate in benzene, the acetoxyindolenine **788** was produced in about 50% yield. Treatment with trifluoroacetic acid in chloroform at 0°C followed by reaction with sodium acetate in acetic acid gave vincamine (**775**) in 36% overall yield. An improved procedure involving the reaction of (−)-vincadifformine (**627**) with p-nitroperbenzoic acid gave **789**, which could be rearranged to (+)-vincamine (**775**) in 66% yield. This is the method of choice for the formation of vincamine from tabersonine.

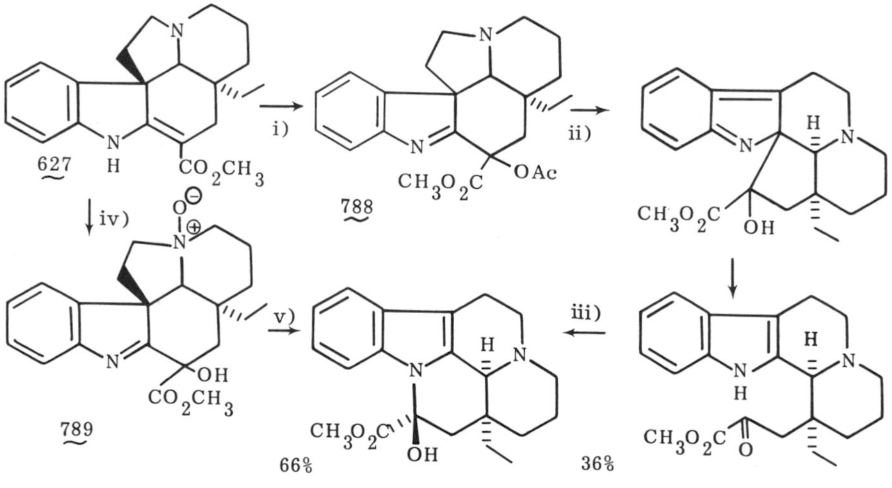

i) Pb(OAc)$_4$

ii) TFA

iii) NaOAc, HOAc

iv) p-NO$_2$PBA

v) $(C_6H_5)_3P$, HOAc

9.30.3 Spectral Properties

The mass spectra of the Eburna alkaloids are quite characteristic (Scheme 106) and the major fragmentation pathway is initiated by a retro Diels-Alder reaction in the C-ring. The resulting radical ion may undergo two pathways as shown to afford the major fragment ions at m/e 249 and 208 for eburnamenine (**789**) and at m/e 308 and 267 for apovincamine (**781**).

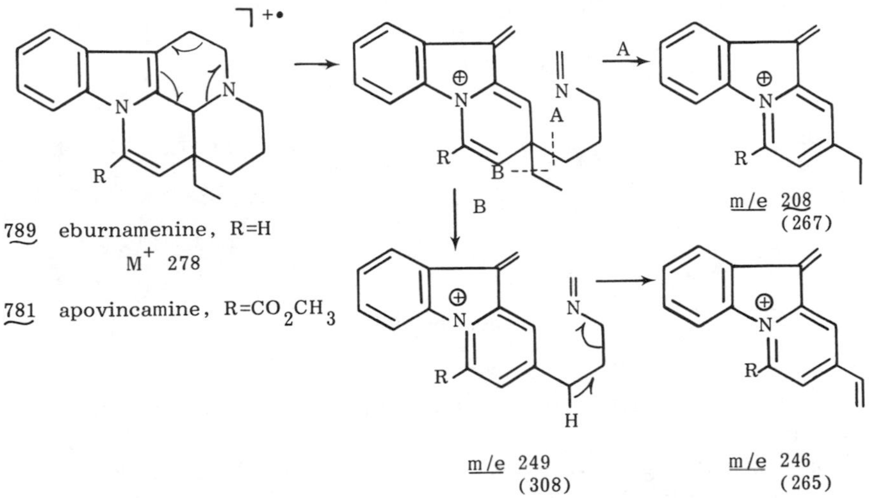

789   eburnamenine, R=H
      M^+ 278

781   apovincamine, R=CO$_2$CH$_3$

Scheme 106 Mass spectral fragmentation of eburnamenine (**789**) and apovincamine (**781**).

In the more complex alkaloids such as vincarodine (**790**), the only Eburna-type alkaloid from a *Catharanthus* species, this pathway is still prominent, but other pathways (e.g. Scheme 107) also become important.

The NMR spectra of the Eburna alkaloids bear many resemblances to those in the *Aspidosperma* series (quaternary ethyl group) and also the *Corynanthe* series (Ar–CH–N system), and by possessing both these characteristics are in themselves distinctive. The proton NMR assignments of vincarodine in d_5-pyridine are shown in **791**. The ^{13}C NMR assignments of vincarodine are shown in **792**.

9.30.4 Pharmacology

The total alkaloids of *V. minor* show a pronounced hypotensive effect which was not blocked by atropine and only partly blocked by nicotine. A 1% solution of the total alkaloids was comparable in anesthetic effect to 0.7% cocaine.

In mammals at 6–7.5 mg/kg a curare-like effect was observed, but at lower (1–3 mg/kg) doses only the hypotensive effects were observed. Clinically, the crude drug was active at 3.0 g/day orally in 50% of patients with

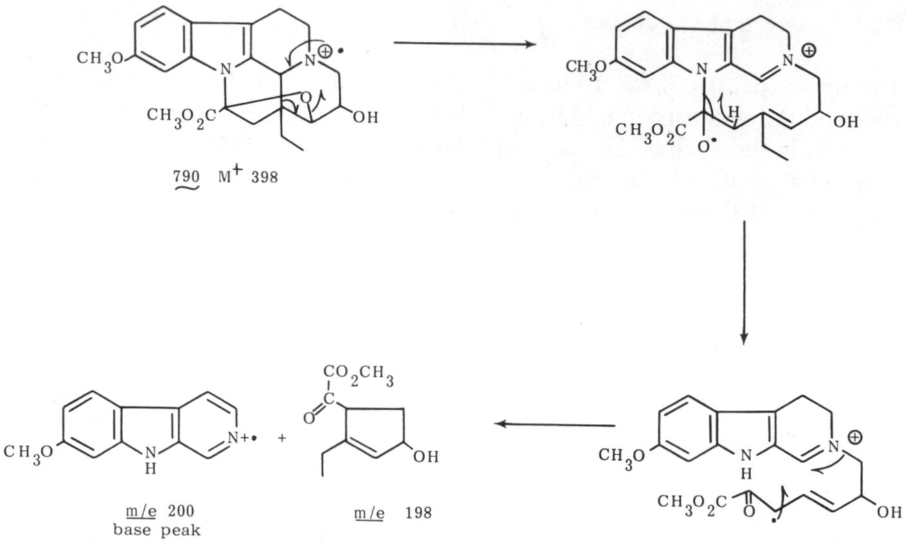

CH_3O

CH_3O_2C OH

790   M^+ 398

CH_3O

CH_3O_2C $O^•$ OH

CO_2CH_3

CH_3O $N^{+•}$ +

m/e 200
base peak

OH

m/e 198

CH_3O

CH_3O_2C O   OH

Scheme 107

threatening cerebral hemorrhage, and in 80% of patients with varying stages of hypertension.

The pharmacologic properties of vincamine were first studied by Raymond-Hamet in 1954. In mice and rats at 5 mg/kg, blood pressure is lowered by 18–48%. The i.v. LD_{50} in mice is 75 mg/kg.

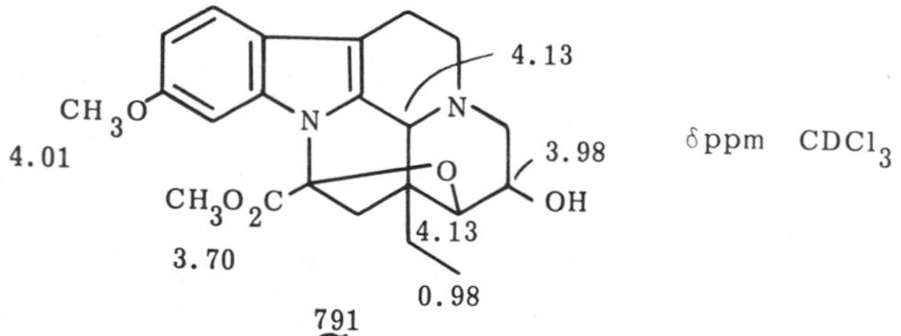

CH_3O

4.13
4.01
CH_3O_2C
3.98
δ ppm $CDCl_3$
O
OH
4.13
3.70
0.98
791

125.3   110.9
118.6   18.4   133.2
109.6   50.1   56.1
55.3
CH_3O H
137.8 N   45.4*
96.2
156.3
CH_3O_2C 66.3
90.5   OH
46.1*   43.9
25.7   9.3   82.0

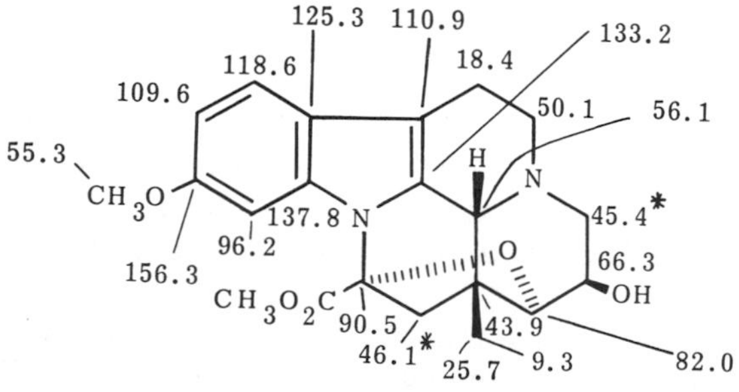

792

Clinically at 5–20 mg per patient p.o. daily, **775** was effective in 16 of 31 patients with hypertension and is claimed to be useful in cerebral angiospastic syndromes. Its real value as a hypotensive agent has not been well documented in the literature. Side effects include heart depression, respiratory depression, and sedative effects.

Numerous vincamine derivatives have been prepared in an attempt to modify the effects. The ethers of vincaminediol are potent muscle relaxants, but more interestingly apovincamine derivatives are longer-acting than vincamine itself.

The *N*-acetyl aspartate of vincamine is a potent cerebral vasodilator and vincamine 2-oxoglutarate is used in the treatment of anoxia.

LITERATURE

Reviews

Taylor, W. I., and N. R. Farnsworth (Eds.), *The* Vinca *Alkaloids, Botany, Chemistry and Pharmacology*, Marcel Dekker, New York, 1973.

Taylor, W. I., *Alkaloids NY* **8**, 249 (1965).

Taylor, W. I., *Alkaloids NY* **8**, 269 (1965).

Taylor, W. I., *Alkaloids NY* **11**, 126 (1968).

Synthesis

Bartlett, M. F., and W. I. Taylor, *J. Amer. Chem. Soc.* **82**, 5491 (1960).

Gibson, K. H., and J. E. Saxton, *Chem. Commun.* 1490 (1969); *J. Chem. Soc. Perkin Trans I* 2776 (1972).

Wenkert, E., J. S. Bindra, and B. Chauncy, *Synth. Commun.* **2**, 285 (1972).

Pfäffli, P., W. Oppolzer, R. Wenger, and H. Hauth, *Helv. Chim. Acta* **58**, 1131 (1975).

Hermann, J. L., G. R. Kieczykowski, S. E. Normandin, and R. H. Schlessinger, *Tetrahedron Lett.* 801 (1976).

Biomimetic Synthesis

Mauperin, P., J. Lévy, and J. Le Men, *Tetrahedron Lett.* 999 (1971).

Hugel, G., B. Gourdier, J. Lévy, and J. Le Men, *Tetrahedron Lett.* 1597 (1974).

Hugel, G., J. Lévy, and J. Le Men, *C.R. Acad. Sci. Paris, Ser. C.* **274**, 1350 (1972).

9.31 MONOTERPENOID INDOLE ALKALOIDS LACKING THE TRYPTAMINE BRIDGE

There are several indole alkaloids which, although they contain a monoterpene unit in a readily recognizable form, lack the typical two-carbon tryptamine bridge. Some examples include uleine (**793**), pericalline (**794**), ervatamine (**795**), and ellipticine (**796**) (Figure 9). Although they are limited

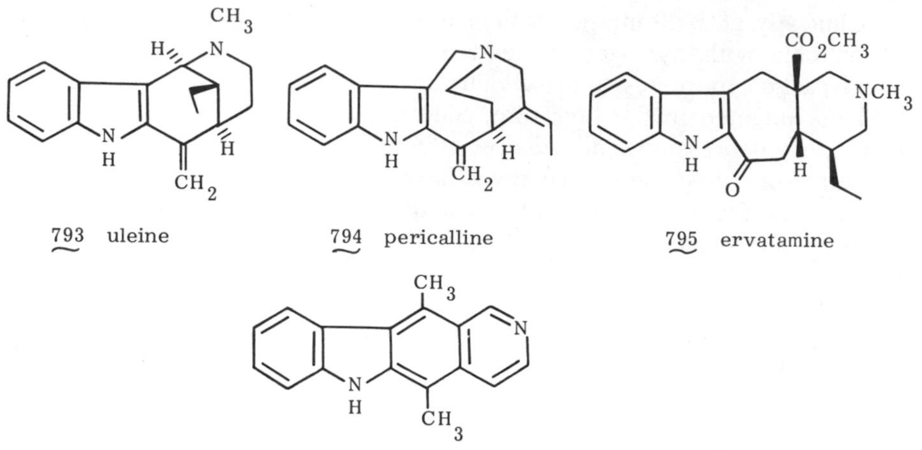

793   uleine   794   pericalline   795   ervatamine

796   ellipticine

Figure 9   Monoterpene indole alkaloids lacking the tryptamine bridge.

in distribution to certain genera in the Apocynaceae, within those genera they are often widely distributed.

9.31.1   Uleine and Pericalline

Uleine (**793**) was first obtained from *Aspidosperma ulei* Mgf. The UV spectrum (λ_{max} 309 nm) indicated an α-vinyl indole nucleus, which from the NMR spectrum was an exocyclic methylene group (singlets at 5.27 and 4.98 ppm). A one-proton doublet (J = 2Hz) at 4.07 ppm is ascribed to a proton on the single bridging carbon between the indole nucleus and N_b. The structure was determined by degradation and has been confirmed by several syntheses.

Treatment of the pyrrolidine enamine of 3-ethyl-4-propionylpyridine (**797**) with phenyl diazonium chloride gave **798**, which underwent Fischer indole cyclization with phosphoric acid to **799**. N-Methylation followed by borohydride reduction gave **800**, which could be cyclized to a mixture of dasycarpidone (**801**) and 20-epidasycarpidone (**802**). Wittig reaction gave uleine (**793**) (Scheme 108).

Büchi and co-workers have reported an elegant stereospecific synthesis of uleine (**793**). Condensation of 3-formylindole (**803**) with 1-aminohexan-3-one (**804**) under Mannich conditions gave **805**, which was formylated and treated with potassium acetylide to give **806**. The acetylenic group was rearranged with mercuric acetate to the α-acetoxy methyl ketone **807**. Elimination of acetic acid followed by catalytic reduction gave the ketone **808**, which was cyclized with boron trifluoride etherate and the N-formyl group reduced with lithium aluminum hydride to give uleine (**793**) (Scheme 109).

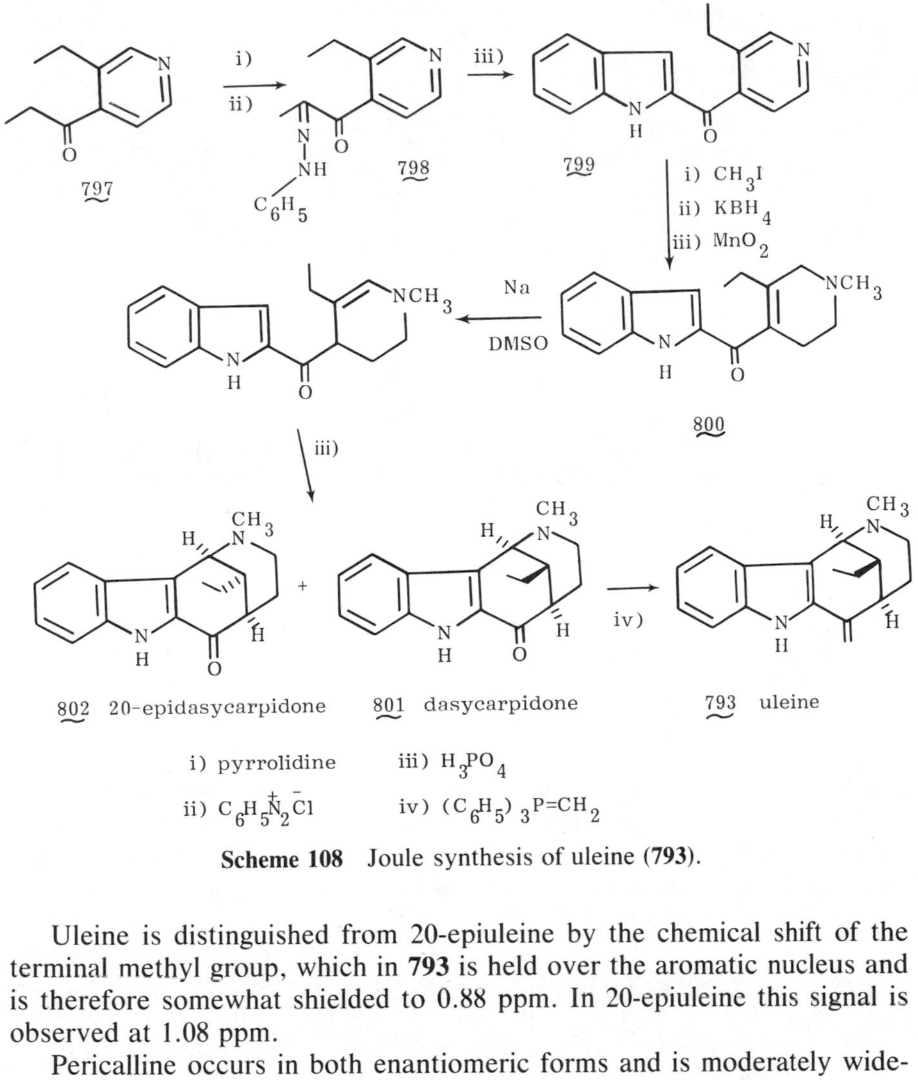

Scheme 108 Joule synthesis of uleine (**793**).

Uleine is distinguished from 20-epiuleine by the chemical shift of the terminal methyl group, which in **793** is held over the aromatic nucleus and is therefore somewhat shielded to 0.88 ppm. In 20-epiuleine this signal is observed at 1.08 ppm.

Pericalline occurs in both enantiomeric forms and is moderately widespread, since isolations include several *Aspidosperma* sp., *Vallesia dichotoma* Ruiz et Pav., *Catharanthus roseus, Schizozygia caffaeoides* (Boj.) Baill., *Conopharyngia holstii* Stapf., and *Ervatamia heyneana.*

Pericalline shows an AB pattern (4.27, 4.47 ppm), indicating a methylene group between the indole nucleus and the basic nitrogen. Also observed were ethylidene and exomethylene groups. Degradative work established the structure **794**, but no syntheses have been reported as yet.

9.31.2 Ellipticine and Olivacine

Ellipticine (**796**) and olivacine (**809**) are two highly aromatic alkaloids having the 6H-pyrido[4,3-*b*]carbazole nucleus (**810**); they differ in the position of

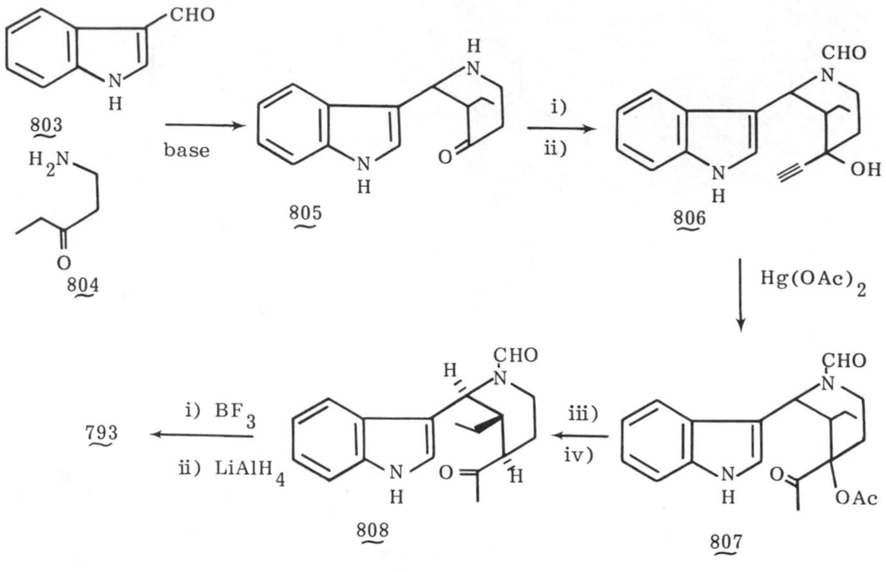

i) HC₂COCH₃   iii) Δ

ii) K⊕ ⊖C≡CH   iv) H₂, Pd

Scheme 109   Büchi synthesis of uleine (**793**).

one of the methyl groups on the aromatic nucleus. The two other natural compounds known in this series are 9-methoxyellipticine (**811**) and 9-methoxyolivacine (**812**). Interest in these compounds has been stimulated by the demonstration of potent *in vivo* antileukemic activity.

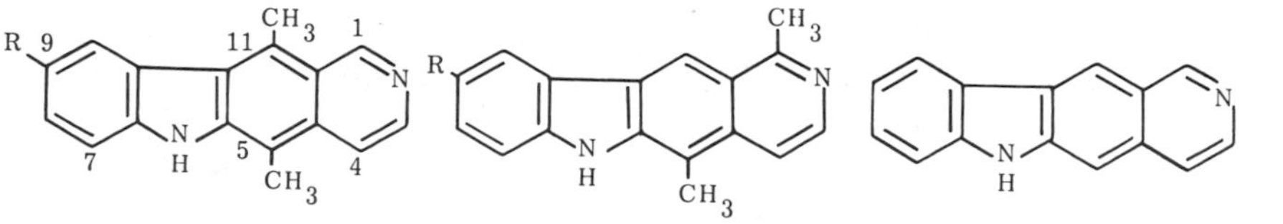

796   ellipticine, R=H   809   olivacine, R=H   810   pyrido [4,3-b] carbazole

811   R=OCH₃   812   R=OCH₃

Ellipticine was first obtained from *Ochrosia elliptica* Labill (Apocynaceae) in 1959, and subsequently from at least eight other *Ochrosia* species as well as *Aspidosperma subincanum* Mart. and *Bleekeria vitiensis* (Mgf.) A. C. Smith.

Olivacine (**809**) was isolated from *Aspidosperma olivaceum* Muell, in 1958 and subsequently several *Aspidosperma*, *Peschiera*, and a *Tabernaemontana* species.

The complex UV spectrum [λ_{max} 238 (log ϵ 4.36), 276 (4.74), 287 (4.90), 295 (4.88), 333 (3.71), 343 (3.47), 384 (3.61), and 401 nm (3.58)] and the molecular formula ($C_{17}H_{14}N_2$) of olivacine indicated a highly aromatic structure. A related base guatambuine (**813**) often co-occurs with olivacine and the latter could be converted to **813** by catalytic reduction of olivacine methiodide. Much of the structure work was therefore carried out with guatambuine (**813**). This alkaloid has a carbazole UV spectrum (λ_{max} 242, 262, and 294 nm) and unlike olivacine is a strong base in which the basic nitrogen is not part of the chromophore. The structure of **814** was determined by Hofmann degradation and reduction to followed by a second Hofmann degradation–reduction sequence to **815**.

Synthesis

The structure of ellipticine (**796**) was originally confirmed by an amazing synthesis developed by Woodward and co-workers. 3-Acetylpyridine (**816**) was condensed with indole (**3**) in the presence of zinc chloride to afford the bisindolylpyridylethane (**817**). Acetylation of the pyridine nucleus under reducing conditions gave the diacetyl pyridine derivative (**818**), which on pyrolysis at 200°C and 10^{-4} mm yielded ellipticine (**796**) in 2% yield (Scheme 110). Ellipticine (**796**) now ranks as one of the most synthesized alkaloids and no fewer than 15 approaches have been described. Only some of the more interesting and efficient of these will be described here.

Strategically one can approach ellipticine in several ways. Since it contains an isoquinoline unit (C and D rings), one can imagine synthetic routes involving standard isoquinoline-forming reactions. One can also envisage approaches where the reactivity of the indole 2- and 3-positions toward

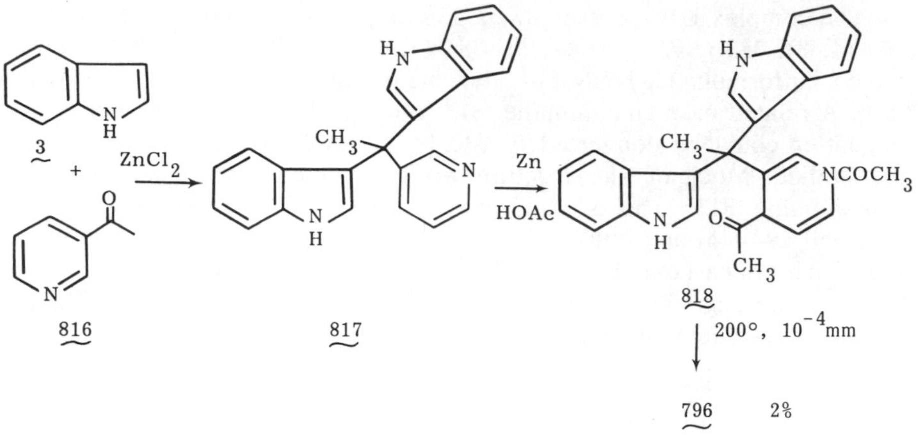

Scheme 110   Woodward synthesis of ellipticine (**796**).

electrophiles is utilized. In addition, although nothing is known of the bio-synthesis, one can imagine biogenetic approaches to both ellipticine and olivacine. All of these methods have been used successfully.

A route that has been used to synthesize ring A-modified ellipticines was developed by Cranwell and Saxton and modified by Loder. Condensation of indole with hexane-2,5-dione (**819**) gave the 1,4-dimethylcarbazole (**820**),

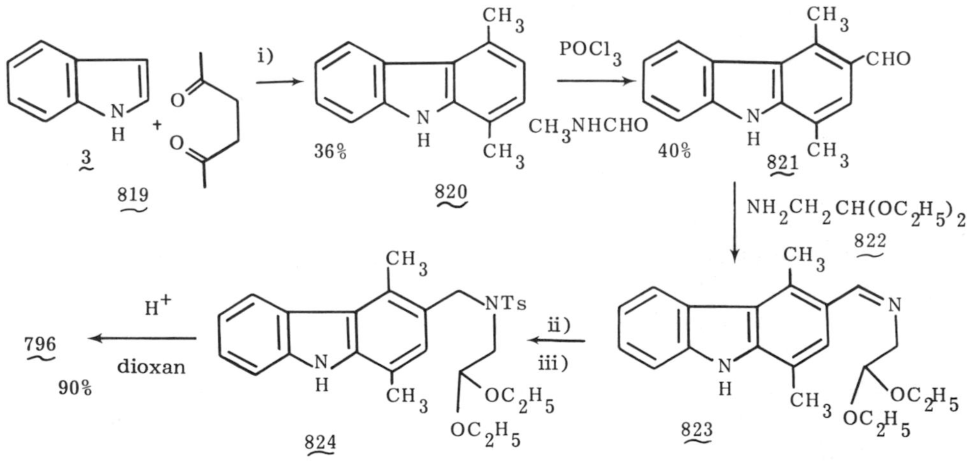

i) dry HCl

ii) H$_2$, Pt

iii) TsCl, py

Scheme 111

which was preferentially formylated at the 3-position by the Vilsmeier reagent. The product **821** could be condensed with the diethyl acetal of amino acetaldehyde (**822**) to afford a Schiff base **823**, which could be cyclized directly to ellipticine (**796**) with 90% phosphoric acid (Scheme 111).

Jackson and co-workers have reported an improved conversion of the carbazole aldehyde **821** to **796**. Condensation with the diethylacetal of amino acetaldehyde followed by reduction and tosylation afforded the tosylate **824** in 93% yield, and acid hydrolysis gave **796** in 90% yield.

A synthesis developed by Potier is of interest for some of the chemistry involved, particularly the use of the Polonovskii reaction to complete cyclization of ring C.

Condensation of allylic alcohol **825** with methylorthopropionate in butyric acid gave the ester **826**, which after hydrolysis was treated with methyl lithium to give the ketone **827**. Fischer indole cyclization with polyphosphoric acid of the phenylhydrazone of **827** gave an indole **828**.

The N-oxide of **828** in trifluoroacetic anhydride at 0°C gave, in 92% yield, the hexahydroellipticine (**829**), convertible with palladium–charcoal to **796** in 35% yield. The overall yield from **825** is of the order of 19% (Scheme 112).

An entirely new approach which leads to an intermediate requiring C-D ring closure has been described by Kozikowski and Haven. Crucial in this synthesis (Scheme 113) is the Diels–Alder addition of an acrylic acid moiety to an oxazole followed by elimination of the oxygen bridge to give a pyridine. Thus heating the oxazole **830** in acetic acid–acrylonitrile at 145°C gave the indolyl–ethylpyridine **831** in 16% yield, which can be converted to **796** as shown.

A quite different synthetic approach (Scheme 114) has been used by

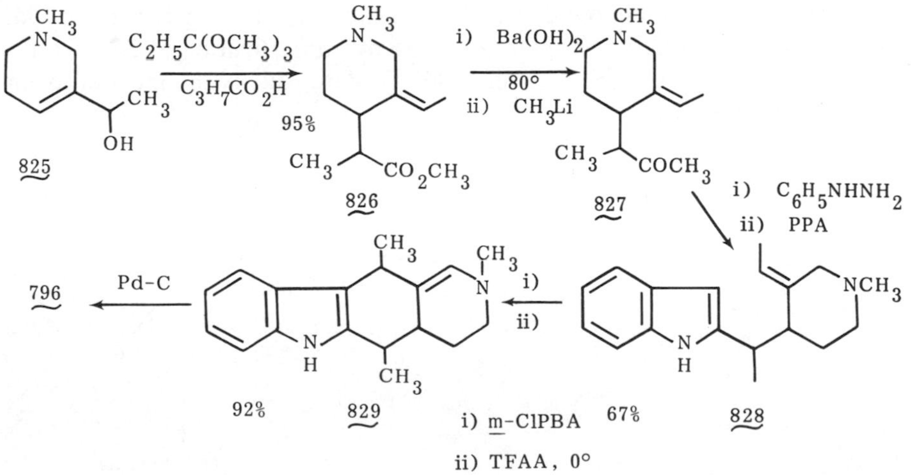

Scheme 112

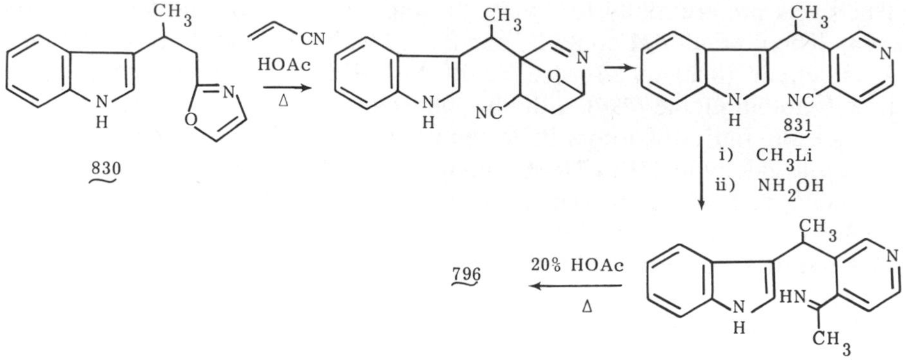

Scheme 113   Kosikowski–Hovan synthesis of ellipticine (**796**).

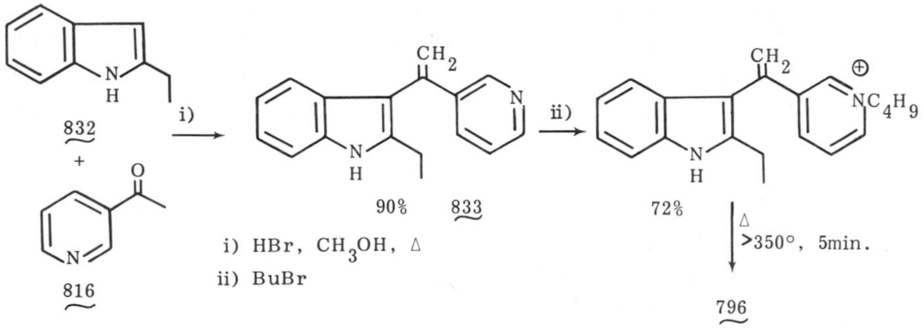

Scheme 114   Bergman–Carlsson synthesis of ellipticine (**796**).

Bergman and Carlsson. It had been shown previously by these workers that condensation of 2-ethyl indole (**832**) with 3-acetylpyridine (**816**) in acetic acid gave **833**. This compound has not been successfully transformed to **796**. N-Alkylation with butyl bromide followed by *rapid* heating at 350°C for 5 min gave **796** in about 70% overall yield. To date, this is the most efficient synthesis of **796**.

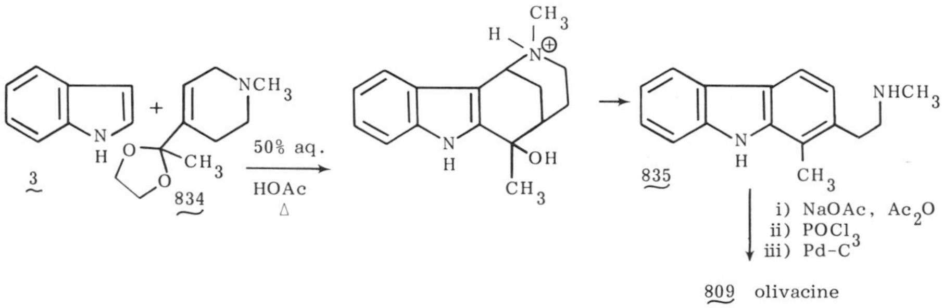

Scheme 115   Besselievre–Husson synthesis of olivacine (**809**).

The most dramatic synthesis of olivacine is that due to Besselievre and Husson (Scheme 115). Heating indole (3) and the piperideine **834** in aqueous acetic acid for 56 hr afforded **835** in 74% yield. Acetylation and Bischler-Napieralski cyclization with phosphorus oxychloride and dehydrogenation of the product with palladium–charcoal gave olivacine (**809**) in 45% yield.

Biosynthesis

There have been no extensive biosynthetic studies on the formation of ellipticine (**796**) and olivacine (**809**), but there have been some interesting biogenetic speculations. Potier and co-workers for example have suggested a route (Scheme 116) which unifies the formation of the uleine alkaloids with those in the ellipticine series by way of two alternative cyclizations derived as shown from a stemmadenine derivative. The olivacine series has a somewhat more complex biogenesis (Scheme 117) involving cleavage of the N_b-C-21-bond followed by rotation of the 15,20-bond and cyclization to an intermediate which can be further elaborated to olivacine (**809**).

There is no experimental evidence to support this scheme except the circumstantial evidence that the uleine and ellipticine/olivacine alkaloids do occur together in some *Aspidosperma* species.

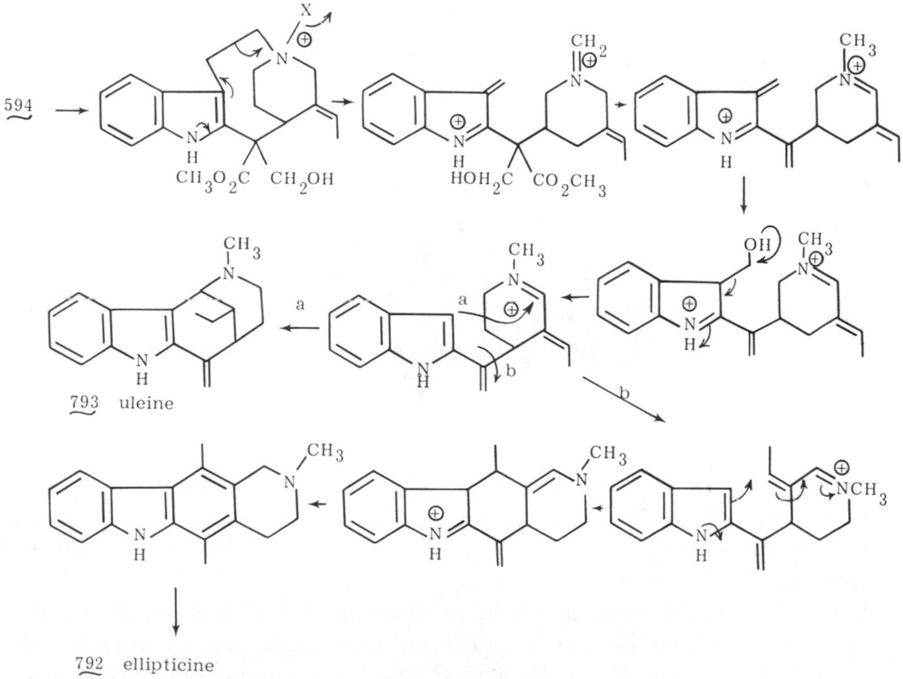

Scheme 116 Biogenesis of ellipticine (**796**) and uleine (**793**) from stemmadenine (**594**).

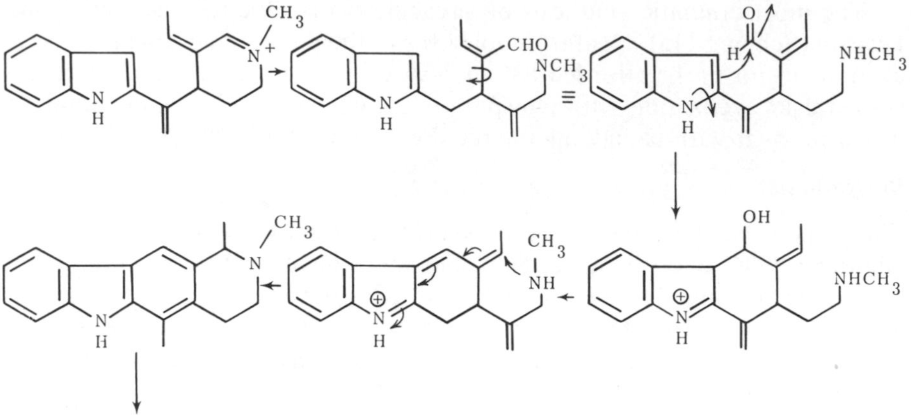

<u>809</u> olivacine

Scheme 117 Biogenesis of olivacine (**809**).

Pharmacology

Ellipticine (**796**) and olivacine (**809**) display potent activity against the L-1210 lymphocytic leukemia system in mice, and many derivatives have also been studied. 9-Methoxyellipticine (**811**) and 9-hydroxyellipticine (**836**) were the most potent derivatives and subsequently other compounds found to display activity include 2-methyl-9-hydroxyellipticine (**837**) and 2,6-dimethyl-9-hydroxyellipticine (**838**). The second compound was active at one-

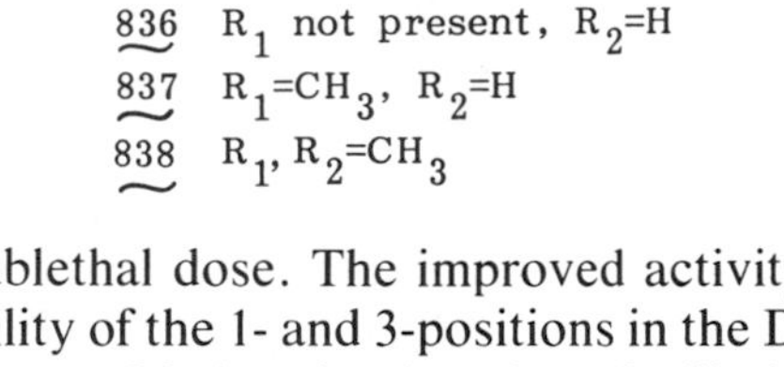

<u>836</u> R_1 not present, R_2=H
<u>837</u> R_1=CH$_3$, R_2=H
<u>838</u> R_1, R_2=CH$_3$

hundredth of the sublethal dose. The improved activity may be due to the improved susceptibility of the 1- and 3-positions in the D-ring to nucleophilic attack by thiol groups. Of the simple salts of ellipticine the lactate and hydrochloride are particularly active.

In mammals, the main metabolic product of ellipticine is 9-hydroxyellipticine (**836**), although this reaction dose not proceed by way of an NIH shift.

Ellipticine is toxic to L-1210 and Don (Chinese hamster) cells *in vitro*. In synchronous Don cells, similar inhibitions of RNA, protein, and DNA synthesis were observed.

9-Methoxyellipticine (**811**) is a powerful inhibitor of DNA synthesis comparable to cytosine arabinoside. However, RNA and protein synthesis were also strongly inhibited in resting stage cells. The alkaloid intercalates between the base pairs of DNA at low and high ionic strength, and causes an unwinding of 6° in the helix for each intercalated molecule. This intercalative binding can also be studied by UV spectroscopy. At 100:1 mole ratio of DNA to ellipticine the spectra is analogous to that of ellipticine in 0.01 N dilute hydrochloric acid.

In large mammals ellipticine at doses of 27–40 mg/kg produces irreversible cardiovascular and respiratory depression and at even lower doses (3–12 mg/kg) hemolysis was a significant problem in rhesus monkeys, although this could be prevented when ellipticine was administered in citrate buffer.

The clinical activity of 9-methoxyellipticine lactate against acute myeloblastic leukemia and Hodgkin's disease was studied by a French group in 1970. Of the 12 patients treated having acute myeloblastic leukemia, three were considered as complete remissions.

The study of ellipticine derivatives as potential clinical agents is clearly an area of considerable promise for the future.

9.31.3 Ervatamine and Related Alkaloids

One of more interesting new indole alkaloid structure types to be obtained in recent years is represented by ervatamine (**795**), a constituent of *Ervatamia orientalis* (Apocynaceae).

The structure of ervatamine posed some interesting biogenetic problems because there are *three* carbon atoms between the indole nucleus and the basic nitrogen atom.

Three routes have been described for the formation of the ervatamine skeleton. One of these (Scheme 118) involves the gramine alkylation of 4-piperidone ester to afford the amino ketone **839**. 2-Acylation, base-cata-

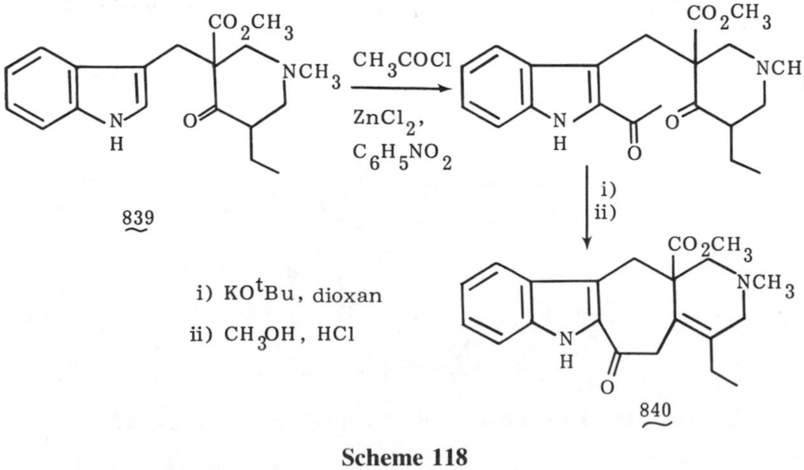

Scheme 118

lyzed condensation, and dehydration afforded 15,20-dehydroervatamine (**840**).

A biogenetic synthesis of the ervatamine skeleton has been described by Potier and co-workers and involves a modified Polonovskii reaction on the N-oxide of vobasine (**462**). Thus treatment with trifluoroacetic anhydride followed by sodium borohydride reduction of an intermediate iminium

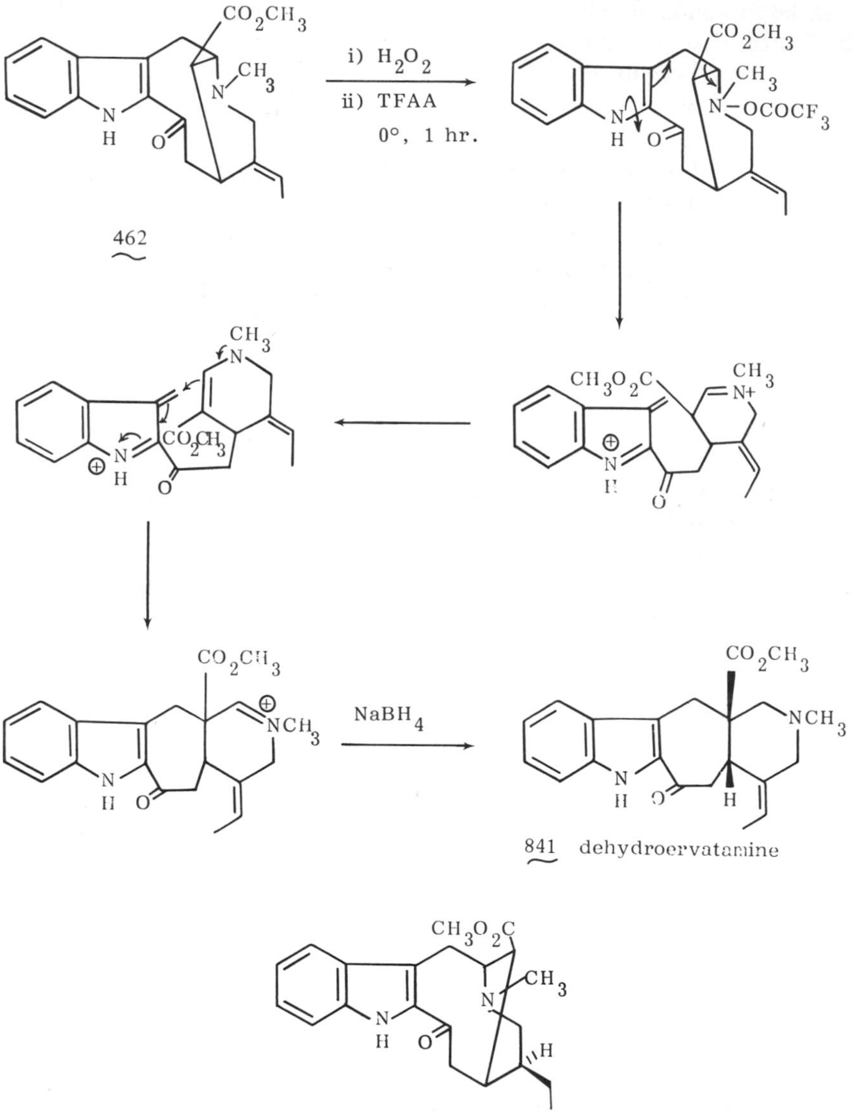

462

841 dehydroervatamine

842 tabernaemontanine

Scheme 119 Partial synthesis of dehydroervatamine (**841**).

species afforded 19, 20-dehydroervatamine (**841**) stereospecifically (Scheme 119). Tabernaemontanine (**842**) has been converted to ervatamine (**795**) by an analogous series of reactions.

6-Oxosilicine (**843**) is an interesting alkaloid in the ervatamine series because it contains an α,β-diacylindole skeleton. It was synthesized from **844** and **845** as shown in Scheme 120.

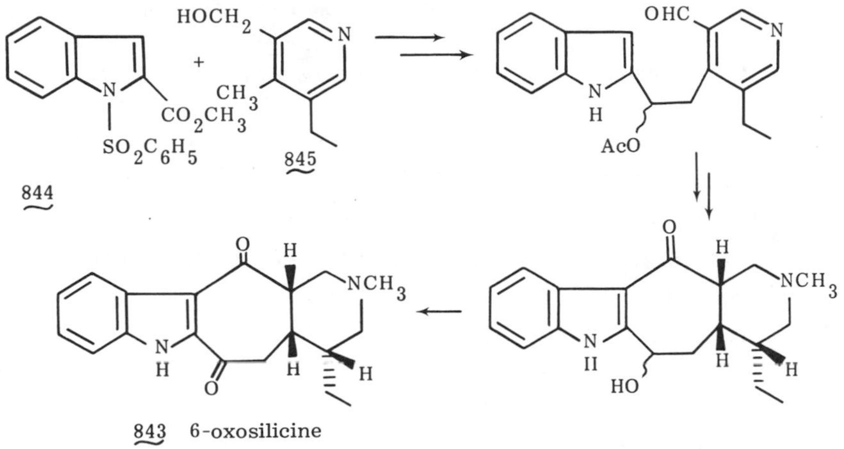

843 6-oxosilicine

Scheme 120

LITERATURE

Reviews

Gilbert, B., *Alkaloids NY* **8**, 474 (1965).

Cordell, G. A., *Alkaloids NY* **17**, 200 (1979).

Uleine–Pericalline

Wilson, N. D. V., A. Jackson, A. J. Gaskell, and J. A. Joule, *Chem. Commun.* 584 (1968).

Büchi, G. H., S. J. Gould, and F. Naf, *J. Amer. Chem. Soc.* **93**, 2492 (1971).

Ellipticine–Olivacine
Synthesis

Woodward, R. B., G. A. Iacobucci, and R. A. Hochstein, *J. Amer. Chem. Soc.* **81**, 4434 (1959).

Cranwell, P. A., and J. E. Saxton, *J. Chem. Soc.* 3842 (1962).

Dalton, L. K., S. Demerac, B. C. Elmer, J. W. Loder, J. M. Swan, and T. Teitel, *Aust. J. Chem.* **20**, 2715 (1967).

Jackson, A. H., P. R. Jenkins, and P. V. R. Shannon, *J. Chem. Soc. Perkin Trans. I* 1698 (1977).

Langlois, Y., N. Langlois, and P. Potier, *Tetrahedron Lett.* 955 (1975).

Kozikowski, A. P., and N. M. Hasan, *J. Org. Chem.* **42**, 2039 (1977).

Besselievre, R., and H.-P. Husson, *Tetrahedron Lett.* 1873 (1976).

Biogenesis

Kunesch, G., C. Poupat, N. van Bac, G. Henry, T. Sevenet, and P. Potier, *C. R. Acad. Sci. Paris, Ser. C* **285**, 89 (1977).

Pharmacology

Hartwell, J. L., and B. J. Abbott, *Advan. Pharmacol. Chemother.* **7**, 117 (1969).

Hayat, M., G. Mathe, M.-M. Janot, P. Potier, N. Dat-Xuong, A. Cavé, T. Sevenet, C. Kan-Fan, J. Poisson, J. Miet, J. Le Men, F. Le Goffic, A. Gouyette, A. Ahond, L. K. Dalton, and T. A. Conners, *Biomedicine* **21**, 101 (1975).

Herman, E. H., D. G. Chadwick, and R. M. Mahtre, *Cancer Chemother. Rept. Part 1* **58**, 637 (1974).

Mathe, G., M. Hayat, F. De Vassal, L. Schwarzenberg, M. Schneider, J. R. Schlumberger, C. Jasmin, and C. Rosenfeld, *Rev. Eur. Etud. Clin. Biol.* **15**, 451 (1970).

Ervatamia *Alkaloids*

Knox, J. R., and J. Slobbe, *Tetrahedron Lett.* 2149 (1971); *Aust. J. Chem.* **28**, 1843 (1975).

Langlois, Y., and P. Potier, *Tetrahedron* **31**, 423 (1975).

Husson, A., Y. Langlois, C. Riche, H.-P, Husson, and P. Potier, *Tetrahedron* **29**, 3095 (1973).

9.32   THE BIOSYNTHESIS OF INDOLE ALKALOIDS

There are few areas of natural product chemistry which have been as en-
thusiastically studied as has the biosynthesis of indole alkaloids. The area
has consistently been at the forefront when new techniques have been developed and perhaps because of this it has been an area of considerable
competition between rival research groups. For the most part this has produced order out of chaos and has completely refuted the original ideas of
indole alkaloid biosynthesis. It must be said in retrospect however, that
experimental haste has resulted in several major errors being perpetrated.

There have been numerous reviews of this area over the years but few
are comprehensive.

9.32.1   Early Work

With a group of alkaloids as diverse as the indoles, it was a long time before
the unifying threads of unique scheme became evident. The early hypotheses
centered on explaining the structures of strychnine (**536**) and yohimbine
(**400**), but as more diverse molecular skeleta were unraveled, so the need
to develop a more elaborate scheme became apparent. It should be remem-
bered that one of the key points in working with indole alkaloids, particularly
from a biosynthetic or biogenetic point of view, is that many plants contain
more than one of the classes of alkaloid in the Schmid scheme. *Catharanthus
roseus* G. Don, which is the most investigated indole alkaloid–containing

846   353   536

plant, contains representatives of classes 1, 3, and 5. Because of this it has been the plant of choice of the study of indole alkaloid biosynthesis.

It is not pertinent to discuss the old theories for the formation of indole alkaloids. Rather let us begin in the early 1960s when the "monoterpene hypothesis" was developed quite independently by Thomas and Wenkert.

The class 1 alkaloids in the Schmid classification, such as corynantheine (353), have the skeleton 846 for the nontryptamine portion of the alkaloids. This skeleton is exactly the same as that in some of the cyclopentane monoterpenoid glycosides such as verbenalin (847) and genipin (848). Moreover the *seco*-cyclopentane unit found in compounds such as swertiamarin (849) and gentiopicroside (850) was thought to be more than coincidental. In

847   verbenalin   848   genipin   849   swertiamarin   850   gentiopicroside

addition, it was noted that the stereochemistry in the appropriate position (C-5) in these compounds was the same as that consistently found at C-15 in the indole alkaloids.

Thomas suggested that the unit 846 was derived from two units of mevalonate by way of a cyclopentane and a secocyclopentane precursor, although at the time these compounds were not known. Wenkert discussed the structure diversification of alkaloids having the unit 846 to those of some of the other classes, principally those in the *Aspidosperma* (Class 3) and iboga (Class 5) groups.

Almost as important was the suggestion by both Wenkert and Levy that an intermediate having the secodine skeleton was a key compound in the formation of *Aspidosperma* and iboga alkaloids (Figure 10). A simple schematic representation can be drawn to illustrate the relationship of the cory-

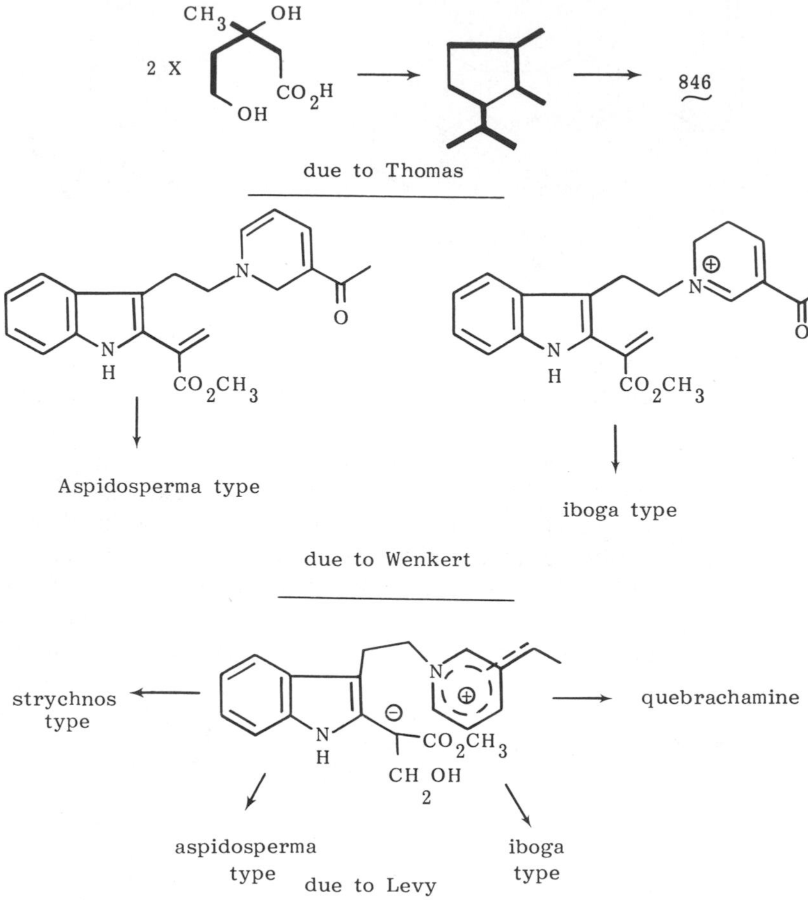

Figure 10 Biogenetic ideas for the monoterpenoid indole alkaloids as postulated in the early 1960s.

nanthe skeleton represented by unit **846** to the *Aspidosperma* (**851**), iboga (**852**), and yohimbe (**853**) skeleta.

The initial work was aimed at establishing the utility of mevalonate, and this was achieved when [2-^{14}C]mevalonate (**93**) was incorprated into serpentine (**854**), vindoline (**626**), and catharanthine (**712**) in *C. roseus*. A veritable flood of feeding experiments with labeled mevalonates followed and completely established the specific derivation of the nontryptamine unit from two units of mevalonate (Figure 11).

Attention then focused on the next intermediate, which was soon established to be geranyl pyrophosphate for each of the major alkaloid classes.

The next question was a crucial one, for potentially the monoterpene units could be rearranging prior to incorporation into the various diverse

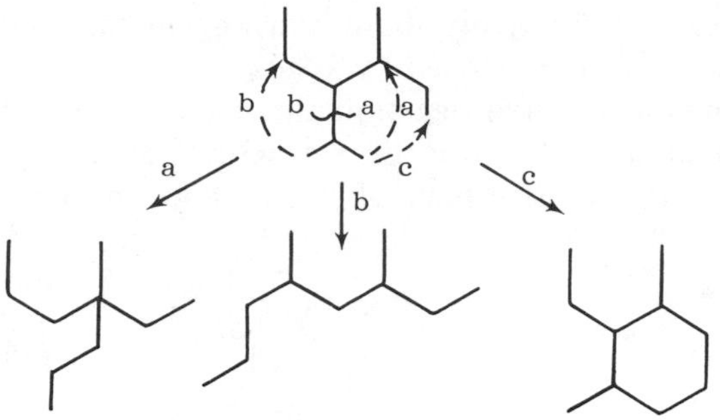

851
Aspidosperma
type

852 iboga type

853 yohimbe
type

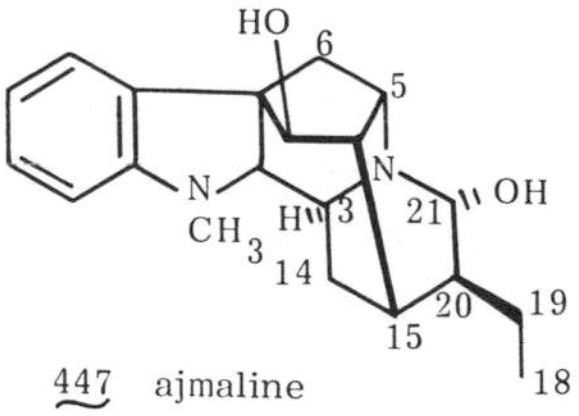

447 ajmaline

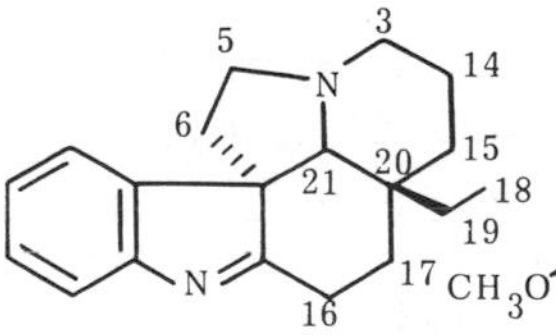

620 (+)-1,2-dehydro-
aspidospermidine

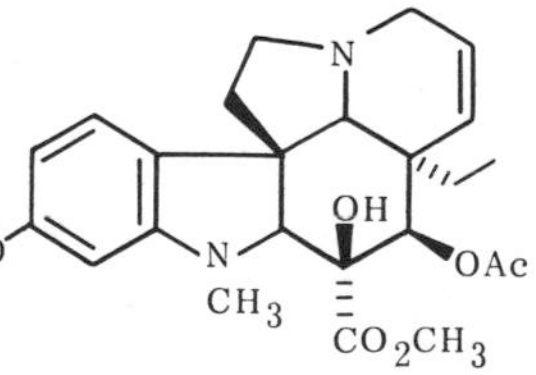

626 vindoline

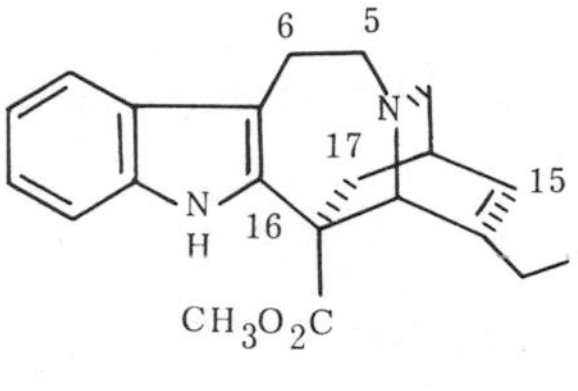

712 catharanthine

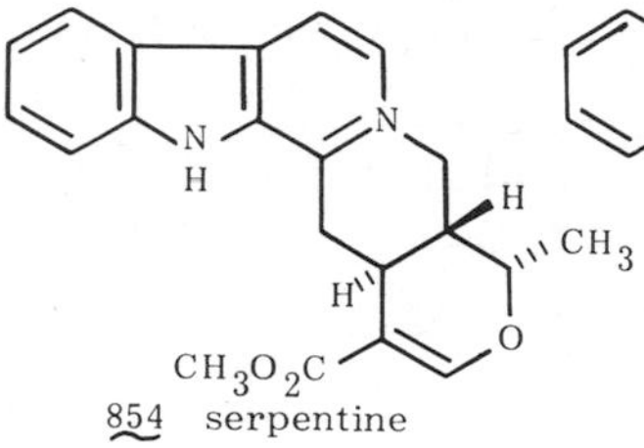

854 serpentine

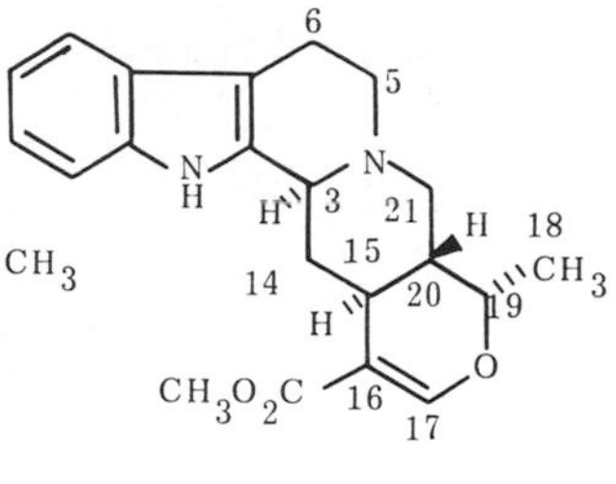

287 ajmalicine

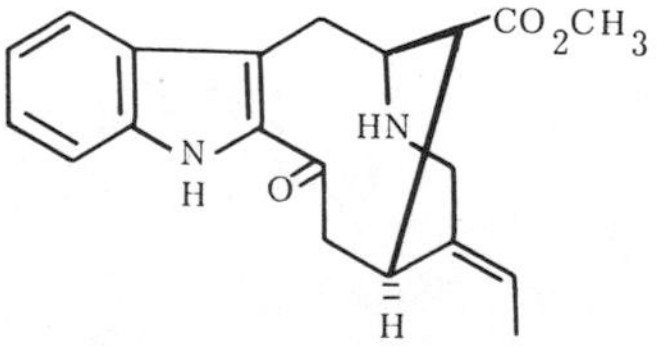

463 perivine

Figure 11 Indole alkaloids commonly examined biosynthetically.

alkaloid skeleta. Alternatively these rearrangements could be taking place at the alkaloid level as proposed by Wenkert.

At this time the tetrahydroisoquinoline monoterpene glycoside ipecoside (855) was isolated, and it was this structural evidence which led to the suggestion that the monoterpene glycoside loganin (856) gave an aldehyde

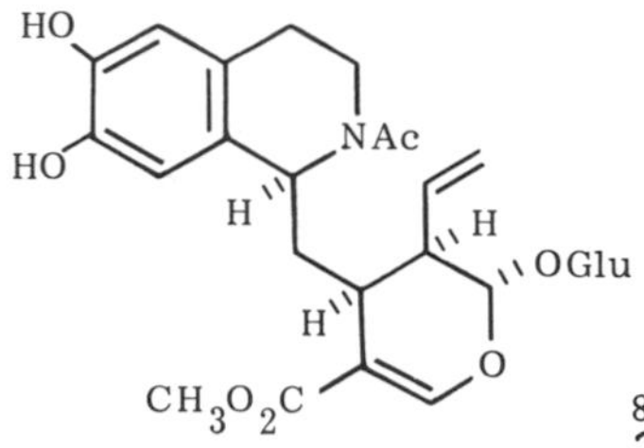

855   ipecoside

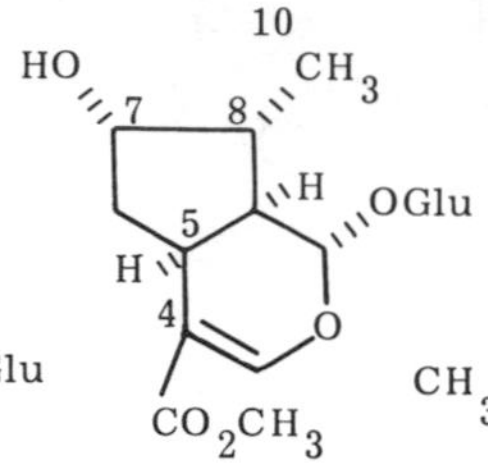

856   loganin

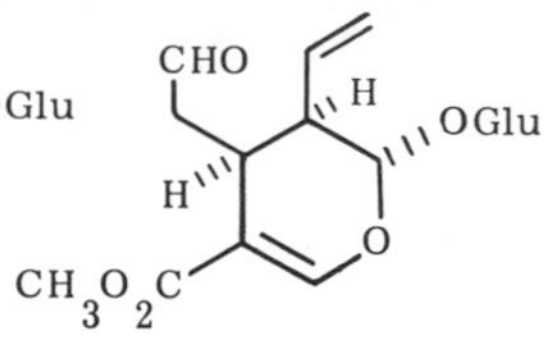

285   secologanin

that combined with tryptamine to yield the many diverse indole alkaloid skeleta.

Loganin (856) was specifically incorporated into all the major alkaloid skeleta and also into ipecoside (855) in *Cephaelis ipecauanha*. The specifically labeled mevalonates and variously labeled geraniols were also incorporated into loganin (856) in *Menyanthes trifoliata*, and this sets up the simple scheme mevalonate → indole alkaloids. In Chapter 2 the formation of loganin (856) from geraniol was discussed more completely, and we will not dwell on this point further here, but rather turn our attention to the next major intermediate, secologanin (285).

This compound was first obtained by the mild alkaline hydrolysis of the lactol ester foliamenthin (857), and was later shown by dilution analysis to

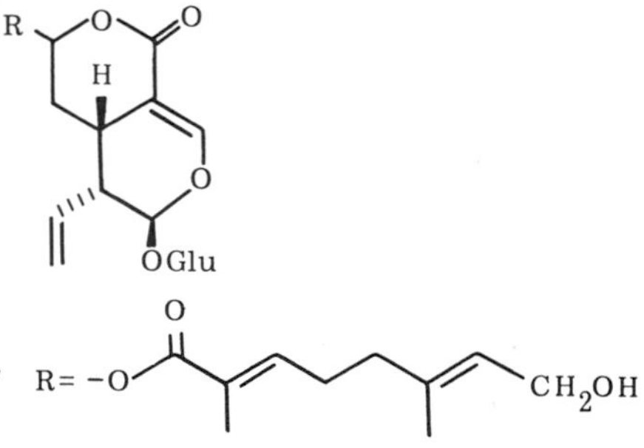

857   R= −O

854   R=H

be present in *C. roseus*. Subsequently, it has been isolated from several plant species.

The key experiments, however, were those involving the feeding of secologanin (285) to *C. roseus* and examining the alkaloids. Good incorporation into the three major classes was observed and in addition sweroside (858), the lactol corresponding to secologanin, was incorporated into vindoline (626). The possible mechanism of formation of secologanin (285) from loganin (856) will be discussed subsequently.

Turning to the other portion of the indole alkaloids it has been found that both tryptamine and tryptophan are effective precursors of the major classes, and therefore the origin of the complete carbon and nitrogen framework is established for the indole alkaloids (Figure 10).

9.32.3 The Nitrogenous Glycosides

The structure determination of cordifoline (293), an alkaloid of *Adina cordifolia*, provided the first example of the secoiridoid unit in combination with tryptophan. But the true intermediate in indole alkaloid biosynthesis was discovered by accident. Smith had been investigating why on storage certain alkaloid fractions of *Rhazya orientalis* lost basic material to a neutral fraction, and found that this was due to the formation of the alkaloid vallesiachotamine (300). It was reasoned that this compound was being pro-

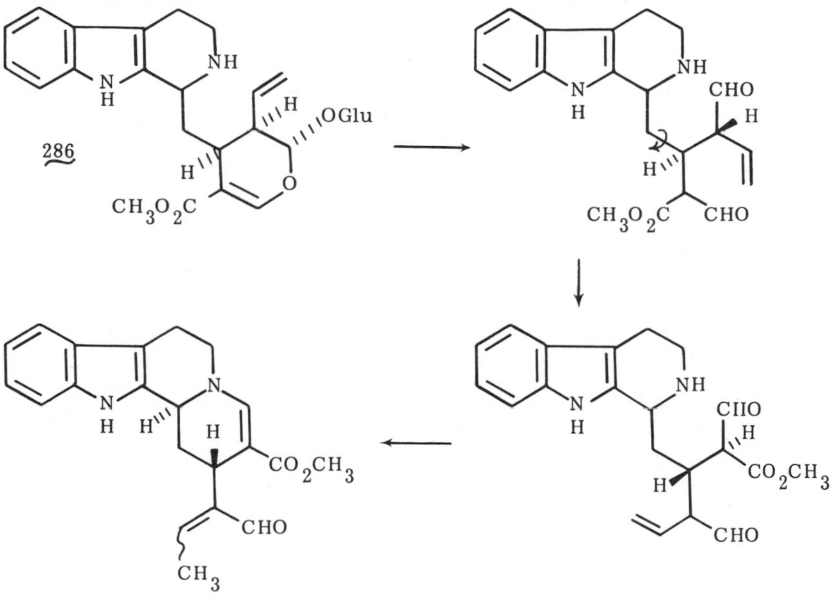

300 vallesiachotamines

duced from a nitrogenous glycoside precursor by the mechanism shown. Smith therefore undertook to isolate this precursor and succeeded in obtaining strictosidine (295).

Battersby and co-workers subsequently obtained evidence by dilution analysis for the presence of both C-3 epimers of 295 in *C. roseus*. Vincoside

(296) and isovincoside (297) were later isolated from *C. roseus* as their hydrochlorides and assigned the stereochemistries shown. More important from a biosynthetic point of view, only vincoside (296) was incorporated into vindoline (626), ajmalicine (287), and catharanthine (712). Isovincoside (297) was not a precursor. Note that vincoside (296) was assigned the same

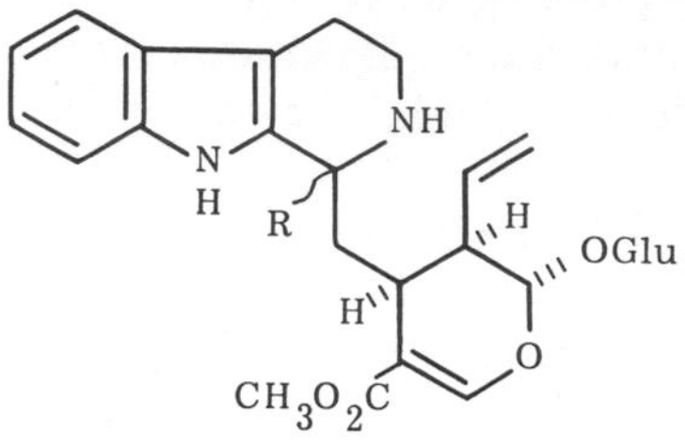

296   vincoside, R=α- H
297   isovincoside, R=β- H
      original assignments

stereochemistry as occurs at C-3 in the *Corynanthe* alkaloids geissoschizine (345) and ajmalicine (287).

At this point the Manchester group conclusively demonstrated that iso-vincoside and strictosidine were identical and that the C-3 stereochemistry of vincoside was incorrect. The key transformation was a chemical corre-lation of strictosidine (286) with dihydroantirhine acetate (859), whose

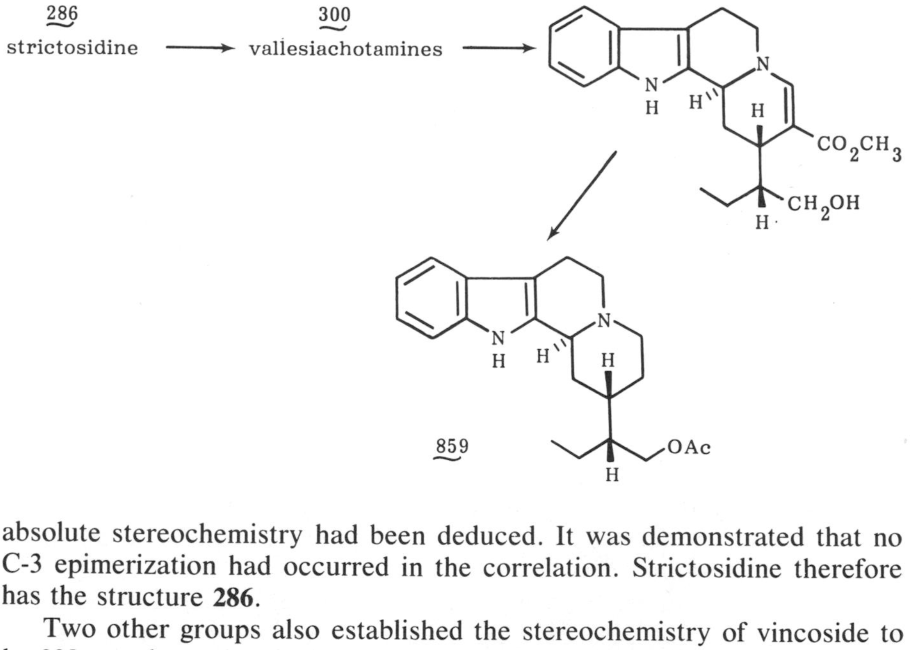

286   300
strictosidine ⟶ vallesiachotamines ⟶

absolute stereochemistry had been deduced. It was demonstrated that no C-3 epimerization had occurred in the correlation. Strictosidine therefore has the structure 286.

Two other groups also established the stereochemistry of vincoside to be 298, one by a chemical correlation of vincosamide, the second by an X-

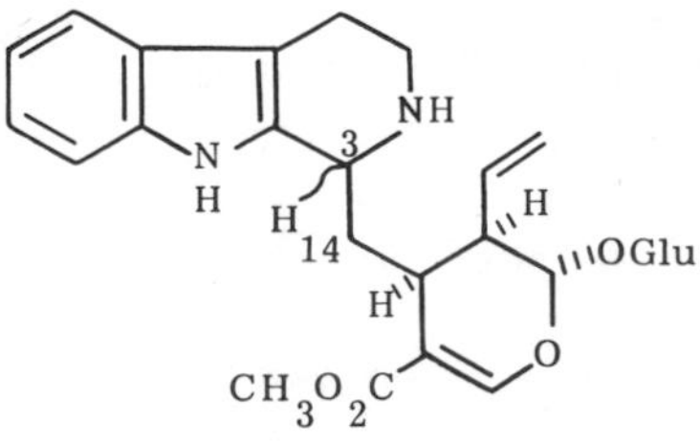

286 strictosidine, 3α-H
298 vincoside, 3 β-H

ray analysis of an ipecoside derivative. Since ipecoside had been used previously in a correlation with vincoside by molecular rotation differences, the structure of ipecoside was revised to **860** and that of vincoside to **298**.

A prior result now became of critical significance, for [7-³H]loganin, which specifically labels the C-3 proton of vincoside (**298**), was retained on

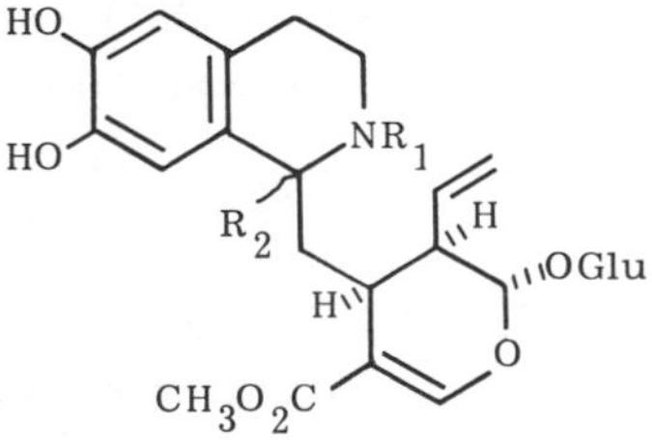

860 ipecoside, R_1=Ac, R_2= α-H
864 desacetylipecoside, R_1=R_2= β- H
865 desacetylisoipecoside, R_1=H, R_2=α-H

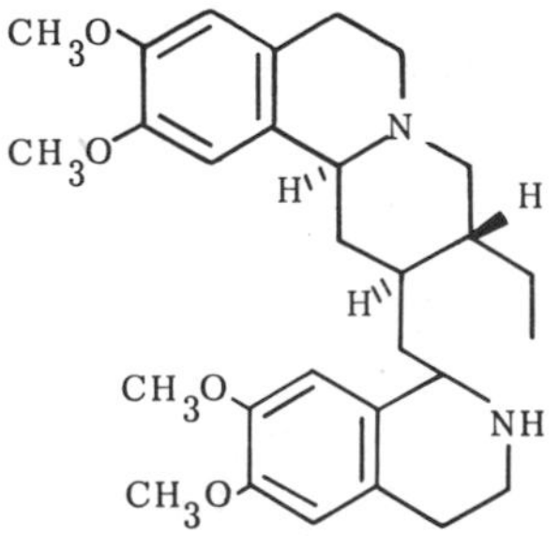

863 cephaeline

incorporation into the alkaloids. In the conversion of vincoside (**298**) to the alkaloids therefore, a C-3 epimerization occurs with specific retention of the C-3 proton.

The fate of the strictosidine (isovincoside) (**286**), which had been used as a potential precursor by Battersby and co-workers, was never determined. This situation remained until quite recently, when several groups demonstrated that much of the early work with vincoside as precursor was incorrect.

The first experimental evidence for the possible involvement of the 3α-H isomers was obtained by Hutchinson and co-workers, who were studying the biosynthesis of the camptothecine nucleus. These workers showed conclusively that strictosamide (**301**), but not vincosamide (**299**), was a specific precursor of camptothecine (**322**). It was presumed at this point that strictosidine (**286**) was the precursor of **322**. Subsequently, the same group established this hypothesis experimentally using [14-³H₂]strictosidine (**286**) and vincoside (**298**). The former compound was a considerably better precursor

301   strictosamide, 3α-H
299   vincosamide, 3β-H

322   camptothecine

of camptothecine (322) than the latter. At the time when this experiment was run it was thought that possibly the *Camptotheca* alkaloids, being isolated from a different plant family, were perhaps an exception to the involvement of vincoside (298) in indole alkaloid biosynthesis.

Zenk was interested in the biosynthesis of ajmalicine (287), particularly the enzymes and their cofactors in cell-free enzyme systems and in the cell suspension cultures of *C. roseus*.

In studying these systems strictosidine (286) was obtained as a transient intermediate, but vincoside (298) was not detected. Was this because all the vincoside (298) was being used to produce the alkaloids, leaving only strictosidine? Or is the enzyme that catalyzes the condensation of tryptamine (2) and secologanin (285) stereospecific for one isomer only? Using an enzyme preparation from *C. roseus* cells at pH 6.5 in the presence of δ-D-gluconolactone, which inhibits ajmalicine (287) biosynthesis, 40% of the activity from [2-^{14}C]tryptamine was incorporated into strictosidine (286). When vincoside (298) was added as carrier, no labeled product was isolated, thereby demonstrating the stereospecific formation of strictosidine (286) by the enzyme system. Cell-free systems of *Amsonia tabernaemontana, Rhazya orientalis, Rhazya stricta,* and *Vinca minor* gave identical results, and the name strictosidine synthetase was given to the enzyme system.

The next step was to prove that strictosidine (286) is indeed the key precursor. In *C. roseus* plants, 286 was well incorporated into ajmalicine (287), vindoline (626), and catharanthine (712), whereas vincoside (298) was not incorporated. Similarly, 286 was a precursor of indole alkaloids in *Amsonia, Cinchona, Rhazya, Stemmadenia, Uncaria,* and *Vinca* species.

More recently these results have been extended to other indole alkaloid–producing systems, not only those having the C-3α stereochemistry but also those with a C-3β H. In *Rauvolfia canescens* α-yohimbine (403) and reserpiline (386) were studied, and in *Mitragyna speciosa* mitragynine (861) and speciociliatine (862) were examined. In all instances strictosidine (286) was a precursor but vincoside (298) was not.

When [3-^{3}H]strictosidine (286) was used the tritium was retained on

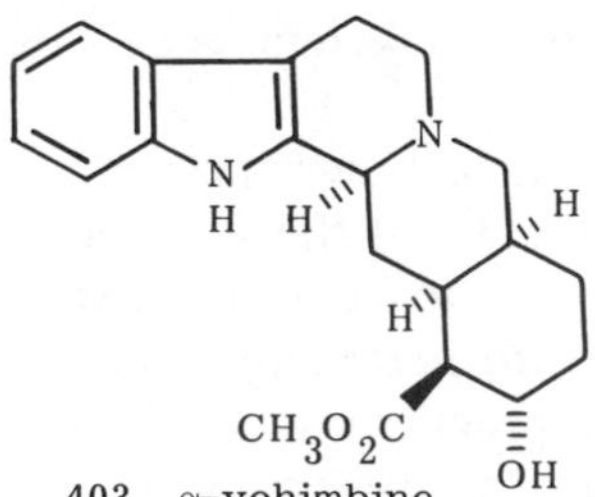

403   α-yohimbine

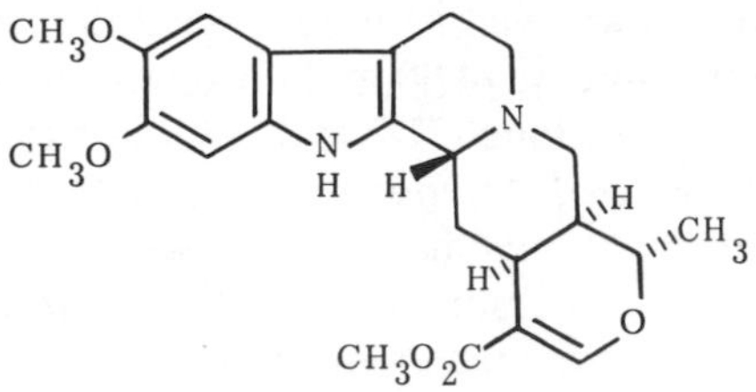

386   reserpiline

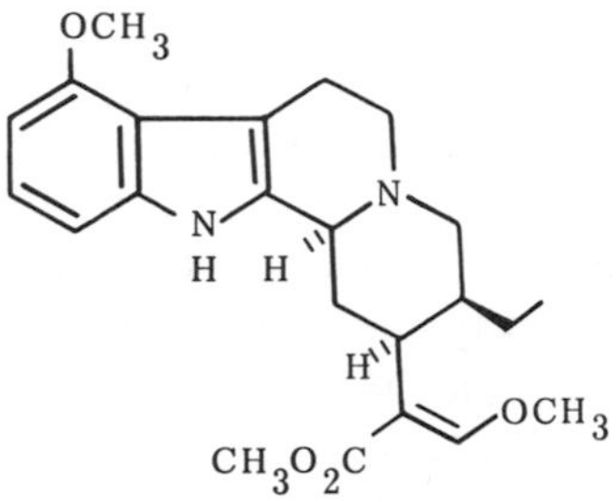

861   mitragynine

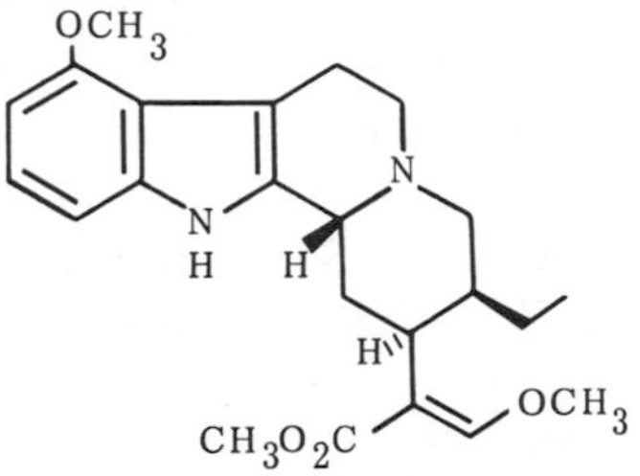

862   speciociliatine

conversion into the 3α-H alkaloids **403** and **861**, but completely lost in the 3β-H alkaloids.

It is clear that strictosidine (**286**) is the ubiquitous precursor of indole alkaloids and that as suggested previously, vincoside, which is definitely not a precursor, is probably not a natural product.

At this point a pertinent question is: What is the nature of the precursor of cephaeline (**862**) in *Cephaelis ipecacuanha*? Is it desacetylipecoside (**864**), as had previously been claimed by Battersby, or is desacetylisoipecoside (**865**) the precursor? The latter situation would parallel the indole alkaloid series.

9.32.3   Elaboration of the Nitrogenous Glycosides

The prime focus of speculative comment on the biosynthesis of indole alkaloids has been the elaboration of the nitrogenous glycosides. In other words how are various co-occurring indole alkaloid skeleta interrelated biosynthetically? Many workers suggested that overall the scheme was: corynanthe → strychnos → aspidosperma → iboga.

In essence this has been shown to be correct, but most of the details have still to be worked out, for many of the intermediates remain to be identified, and as we shall see little is known about the stereospecificity of most of the processes involved.

Geissoschizine (**345**) is one of the earliest alkaloids in the pathway and has been obtained from *C. roseus*, as well as other sources. Feeding ex-

periments showed it to be a precursor of the major skeletal types and also
of akuammicine (479) and coronaridine (691).

Catharanthine (712) and vindoline (626) are the classic representatives of
the iboga and *Aspidosperma* skeleta respectively and the next postulated
intermediate stemmadenine (594) was also shown to be a precursor of the
alkaloids. Tabersonine (619), a very simple *Aspidosperma* alkaloid, was also
labeled by stemmadenine (594) and was itself a precursor of vindoline (626).

The possible reversibility of some of these transformations was also
investigated. Tabersonine (619) was an apparent precursor of catharanthine
(712), but the reverse was not true. As we shall see subsequently these
results are unlikely to be proven correct in the final analysis of this field.

Other corynanthe-type precursors have also been used, but whether they
are true intermediates remains to be definitively established. Ajmalicine
(287) for example was a precursor of 712 and 626, presumably by reversal
to geissoschizine (345), although this interconversion has not been demonstrated *in vitro*. Corynantheine aldehyde (348), the vinyl isomer of geissoschizine (345), was a precursor of catharanthine (712) and vindoline (626)
in seedlings but not in plants.

More recently several groups have reported on the formation of some
of the early intermediates in geissoschizine (345) and ajmalicine (287)
biosynthesis.

Using *Catharanthus* cell suspensions in the absence of reduced pyridine
nucleotide an intermediate accumulated which was identified as cathenamine
(866). Also, when borohydride was added to capture any intermediate iminium species, sitsirikine (867) and 16-episitsirikine (868) were obtained, indicating species 869 in the formation of ajmalicine (287).

869   4, 21-dehydrocorynantheine   867   sitsirikine, 16β-H
         aldehyde   868   16-episitsirikine, 16α-H

The currently accepted scheme for the formation of geissoschizine (345)
and ajmalicine (287) in *C. roseus* is shown in Scheme 121. After strictosidine
the first step is regarded as hydrolysis by a β-glucosidase followed by
opening of the lactol ring and rotation of the 15,20-bond to afford 870.
Condensation, followed by elimination of water yields 4,21-dehydrocory-
nantheine aldehyde (869) which can then be deprotonated and the aldehyde
attack the dienamine at the γ-position to yield cathenamine (866). In the
presence of NADPH 20,21-reduction occurs to afford either ajmalicine (287)

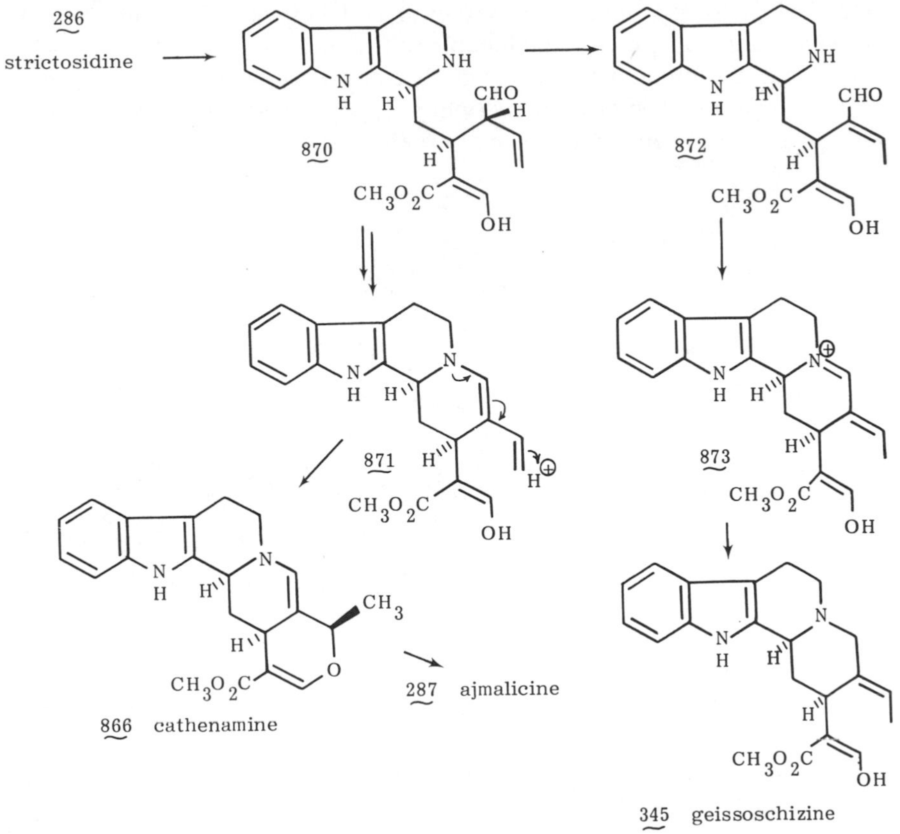

Scheme 121 Biosynthesis of ajmalicine (**287**) and geissoschizine (**345**).

or 20-epiajmalicine (tetrahydroalstonine) (**380**). The mechanism of formation of geissoschizine (**345**) is unknown; it may be produced by δ-protonation of the dienamine **871**, followed by 4,21-reduction. Alternatively, if the intermediate aldehyde **870** has a finite lifetime, isomerization to the aldehyde **872** could occur, and condensation would then lead to **873**. Whatever the stage of the isomerization, it is a stereocontrolled process since only one isomer has ever been found in all the compounds containing an ethylidene residue.

The oxindole alkaloids such as mitraphylline (**395**) have been shown to be derived from the corresponding indole precursor, in this case ajmalicine (**287**). The oxindole alkaloids may not be as biosynthetically inert as was once thought..

9.32.4 Elaboration of Dehydrogeissoschizine

Dehydrogeissoschizine (**873**) is potentially a highly reactive intermediate and is one of the main ''acrobats'' in the biosynthetic scheme. Four quite reasonable reactions are shown in Scheme 122. Reaction between N_a and C-16 yields the pleiocarpamine (**765**) skeleton and between C-7 and C-16 the formyl strictamines (**874**). Attachment by C-2 and cleavage of the C-2–C-3 bond yields an iminium species, which can be reduced to stemmadenine (**594**) or undergo an internal nucleophilic attack to afford preakuammicine (**740**), a strychnos-type alkaloid. Yet another alternative involves hydroxy-

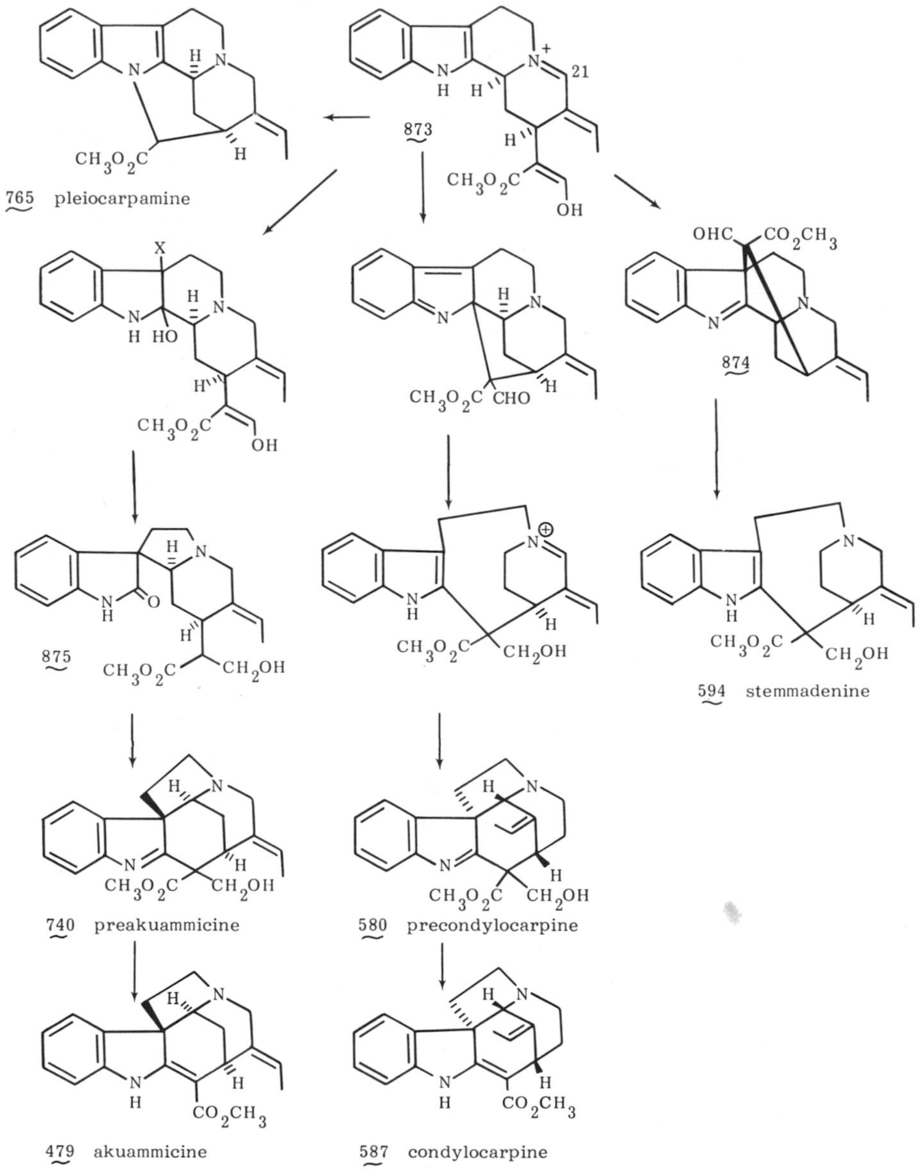

Scheme 122 Biosynthesis of the main indole alkaloid skeleta from dehydrogeissoschizine (**873**).

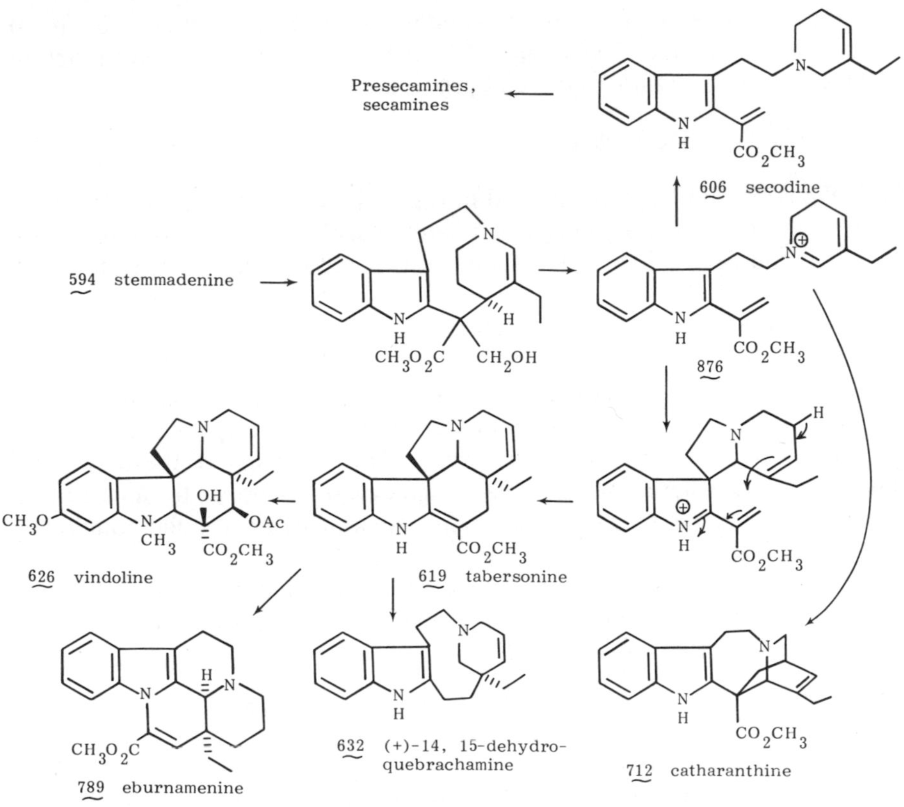

Presecamines, secamines

606 secodine

594 stemmadenine

876

626 vindoline

619 tabersonine

789 eburnamenine

632 (+)-14, 15-dehydro-quebrachamine

712 catharanthine

Scheme 122 (*continued*).

lation at C-2 followed by rearrangement to the oxindole **875**. This compound can then undergo internal condensation to give preakuammicine (**740**).

Stemmadenine is the next key intermediate in the formation of the *Aspidosperma* and iboga types. In sequential isolation studies with seedlings, Scott and co-workers found that geissoschizine (**245**) was produced initially, followed by stemmadenine (**594**), akuammicine (**479**), and tabersonine (**619**), and finally the iboga alkaloids.

Although stemmadenine (**594**) was isolated from *Stemmadenia donnell-smithii*, it was regarded as a rare alkaloid, and its biosynthetic role was not realized for many years. The steps between stemmadenine (**594**) and tabersonine (**619**) and catharanthine (**712**) are not known with any degree of certainty. Some aspects can be explained by the involvement of a *seco*-intermediate such as **876**. This compound may be reduced directly to secodine (**606**) or undergo two different reactions, either between C-7 and C-3, to eventually give the *Aspidosperma* nucleus or a Diels–Alder type of process to afford catharanthine (**712**) in the iboga series.

The *Aspidosperma* alkaloids such as tabersonine (**619**) are precursors of the Eburna skeleton (e.g. **789**) and probably of the quebrachamine skeleton (e.g. **632**). Geissoschizine (**345**), stemmadenine (**594**), and tabersonine (**619**) were incorporated into vincamine (**775**) in *Vinca minor*. The latter reaction *in vitro* is an important commercial process.

9.32.5   The Secodine Skeleton

Both Wenkert and Levy postulated in the early 1960s a C_7-C_3 intermediate in the formation of the *Aspidosperma* and iboga alkaloids. It was some years later that the first evidence for such a novel indole alkaloid skeleton was obtained.

The first evidence for a *seco*-C_7-C_3 intermediate was obtained by the isolation from *Rhazya* species of the dimeric indole alkaloids, the secamines (e.g. **597**). The first monomer to be isolated was decarbomethoxytetra-hydrosecodine (**596**) from *Tabernaemontana cumminsii* and the various secodine derivatives from *Rhazya* species. The chemistry of these systems is discussed elsewhere, but as we have said before, evidence for a particular biosynthetic scheme often comes from the isolation of a particular compound or structure type.

Radioactive secodines were incorporated into vindoline (**626**) in *C. roseus*, but other experiments have met with mixed success and much further work is needed in this area.

9.32.6   Stereoselectivity of Indole Alkaloid Interconversions

With the fundamental scheme of indole alkaloid biosynthesis established, some attention has been paid to the stereoselectivity of the various interconversions.

The data so far indicate that if we consider the alkaloids ajmalicine (**287**), catharanthine (**712**), and vindoline (**626**), then tritium from loganin labeled at positions 1, 5, 7, and 8 is specifically retained into the alkaloids, which become labeled at C-21, 15, 3, and 19 respectively.

Typically of terpene biosynthesis, the *pro*-4R proton of mevalonic acid is retained into loganin (**856**) at positions 9 and 5, but only the proton at C-5 of loganin, which becomes H-15 in the alkaloids, is retained subsequently. [1-^{3}H]Geraniol (**877**) labels position 21 of the alkaloids with an overall loss of half the activity. The loss takes place in the formation of

$\underline{877}$ geraniol

loganin (**856**), and consequently half is specifically retained from loganin (**856**) into the alkaloids. Thus the NADPH-mediated reduction of 4,21-dehydrocorynantheine aldehyde (**819**) is stereospecific, and in the double-bond isomerization of stemmadenine (**594**), it is this proton that is specifically removed.

When [6,8-³H]loganin (**856**) was used as precursor of vindoline (**626**), the activity from C-8 was retained at C-9 and from C-6 it was found that a stereoselective loss of the *pro*-6R hydrogen of loganin took place. Some support for this comes from studies with the [5R-³H]- and [5S-³H]mevalonates which label C-1 and C-6 of loganin (**856**). The *pro*-5R proton of mevalonate labels the *pro*-6S position of loganin (**856**). The data showed that the *pro*-5R hydrogen of mevalonate is lost in the formation of catharanthine (**712**) and the *pro*-5S hydrogen is lost on incorporation into vindoline (**626**). It is considered likely that these stereospecific losses occur from C-14 at the point involving the cleavage of stemmadenine (**594**) and the recyclization to the *Aspidosperma* and iboga skeleta.

One of the points which has remained ignored is that relating to the two antipodal series observed in the *Aspidosperma* and iboga alkaloids. Typical examples are (+)- and (−)-quebrachamine (**633**) and (**878**), (−)-tabersonine (**619**) and (+)-minovincinine (**879**), and (+)-vincadifformine (**880**) in the *Aspidosperma* series, and catharanthine (**712**) and ibogamine (**690**) in the

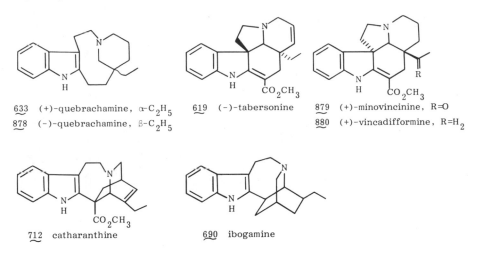

633 (+)-quebrachamine, α-C₂H₅
878 (−)-quebrachamine, β-C₂H₅

619 (−)-tabersonine

879 (+)-minovincinine, R=O
880 (+)-vincadifformine, R=H₂

712 catharanthine

690 ibogamine

iboga series. It is essential that the stereochemical conclusions reached about catharanthine (**712**) biosynthesis be examined for an antipodal iboga alkaloid, and similarly for an alkaloid in the (+)-*Aspidosperma* series.

These optical antipodes can occur in the same plant, for *Amsonia tabernaemontana* yields both (+)-vincadifformine (**880**) and (−)-tabersonine (**619**).

The C-16 stereochemistry of stemmadenine (**594**) was unknown for many years and was recently established by Kutney and co-workers from the synthesis of 16-epistemmadenine (590). However, there is no evidence that only one stemmadenine (594) is produced naturally, and it may be that specific enzyme attack on each of the diastereoisomers at C-16 of stem-madenine is important in the formation of the antipodal *Aspidosperma* and iboga alkaloids.

9.32.7 Yohimbine

The early biogenetic ideas of Barger suggested that the yohimbe class of alkaloids were the progenitors for the corynanthe and *Strychnos* alkaloids. Ironically these alkaloids would appear to be at the end of a pathway (Scheme 123) involving the corynanthe alkaloids as intermediates. An alternative postulate due to Inouye suggests that secoiridoids of the type 881 are involved in a condensation reaction with tryptamine (2) to afford the yohimbe alkaloids. The recent demonstration that strictosidine (286) is a precursor of α-yohimbine (403) in *Rauvolfia canescens* supports the suggested scheme.

9.32.8 Ajmaline–Sarpagine Alkaloids

Van Tamelen proposed that alkaloids of the ajmaline–sarpagine group are formed by the attack of C-16 on an electrophilic iminium species at C-5 formed by oxidative decarboxylation of a 5-carboxygeissoschizine type (882) as shown in Scheme 124.

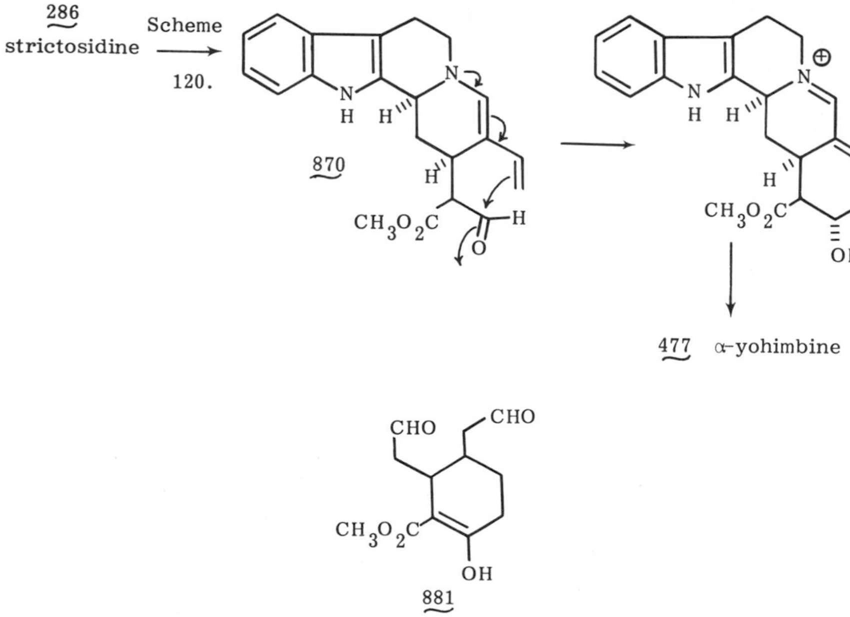

Scheme 123   Biogenesis of α-yohimbine (447) from strictosidine (286).

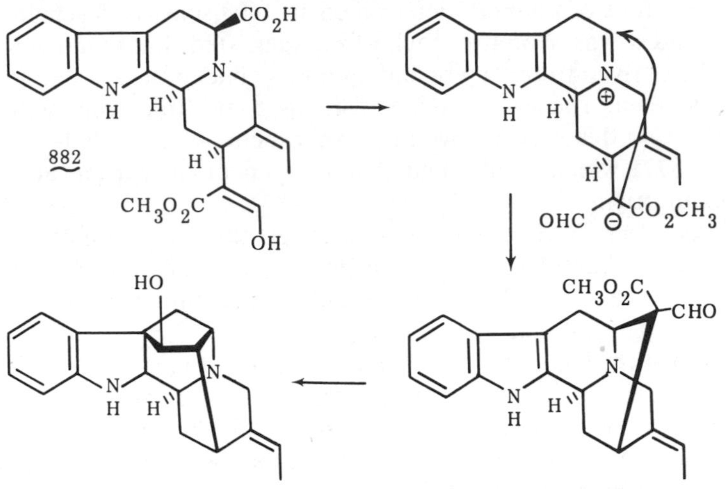

ajmaline type sarpagine type

Scheme 124 Biogenesis of ajmaline–sarpagine alkaloids.

In recent years several compounds have been isolated which might have some bearing on the biosynthesis. The first of these is 5α-carboxystricto-sidine (**883**) and the second is adirubine (**884**). The latter alkaloid is of interest because it demonstrates that at least rearrangement to the corynanthe skeleton can occur from **883** in the amino acid series. No biosynthetic studies

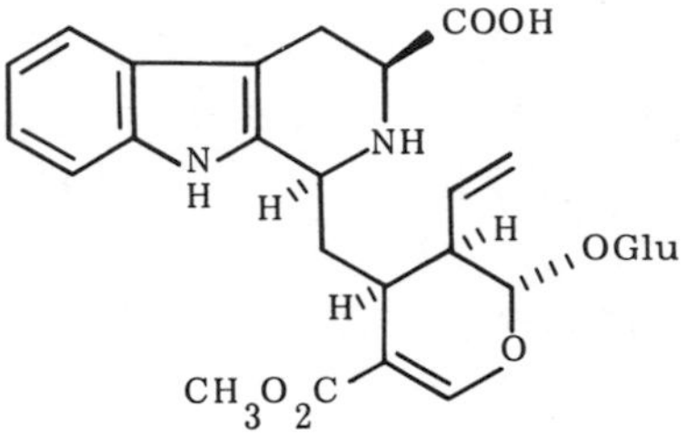

883 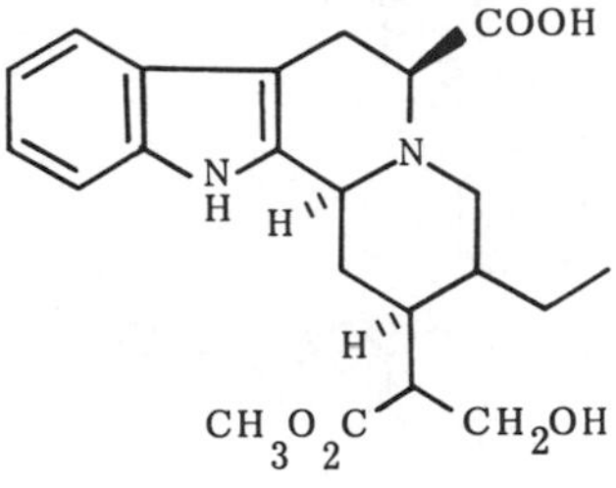 884

with these types of compounds have been conducted, but it will be recalled that the formation of a sarpagine intermediate by a cyclization similar to that in Scheme 124 has been used in a total synthesis of ajmaline (**447**).

9.32.9   Quinine

It was mentioned previously that quinine (**497**), in spite of having a quinoline nucleus, is actually a monoterpene indole alkaloid. Let us examine the evidence for this unusual transformation.

The structure of quinine, determined in 1909, clearly fascinated many chemists and it was Woodward who first suggested that quinine could potentially be derived from tryptophan, and this concept was elaborated upon by van Tamelen. The first evidence for this came from work with [2-^{14}C] tryptophan and this was followed by experiments with [2-^{14}C]- and [3-^{14}C] geraniols (877) which established that the remaining carbon atoms were terpenoid in origin.

Activity from [7-^{3}H]loganin (856) was retained into quinine (497), and labeled strictosidine (286) was also a precursor, indicating that rearrangement takes place at the alkaloid level and with retention of the C-3 proton of 286. Corynantheal (885) was also a precursor.

The fate of the indole and amino acid nitrogen atoms was examined with ^{15}N-labeled tryptophans. The indole nitrogen was found to be specifically retained as the quinoline nitrogen, whereas ^{14}C label from the indole C-2 position was specifically at C-2, the carbon atom between the quinoline and quinuclidine systems. Some of the late stages have also been investigated, but the results are as not a clear cut as one would like. The currently accepted pathway for the biosynthesis of the *Cinchona* alkaloids is shown in Scheme 125.

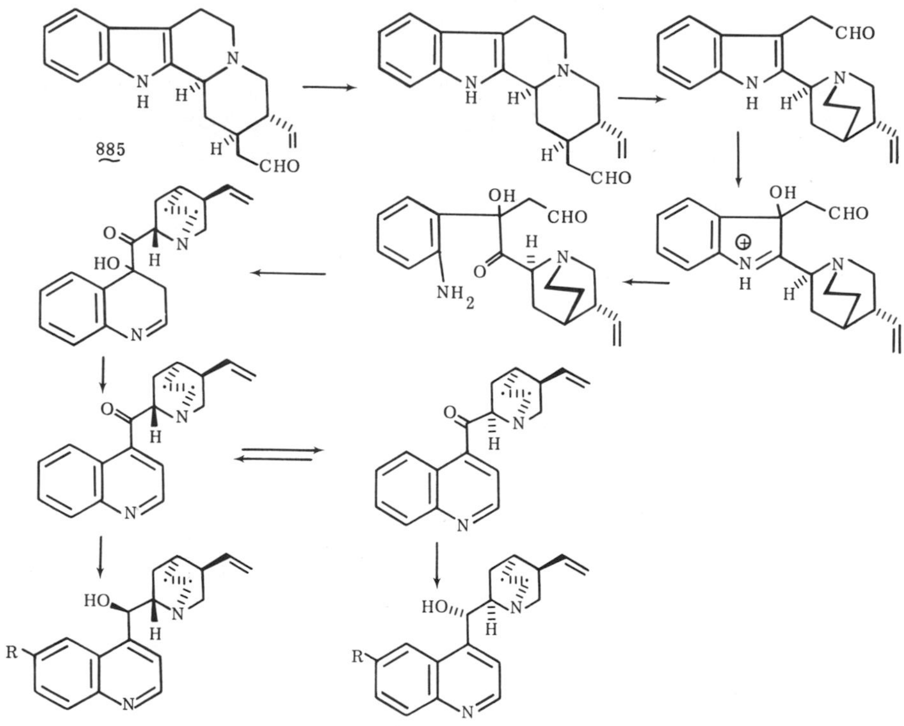

496   cinchonidine, R=H
497   quinine, R=OCH$_3$

492   cinchonine, R=H
493   quinidine, R=OCH$_3$

Scheme 125   Biosynthesis of *Cinchona* alkaloids.

9.32.10 Uleine–Pericalline

The uleine–pericalline group (Section 9.31.1) are a biosynthetically interesting series of alkaloids which are characterized by having only one carbon
atom between the indole nucleus and N_b. The remaining carbon atoms appear to have skeleta akin to those of the akuammicine (**479**) or condylocarpine (**587**) type.

Two of these alkaloids, pericalline (**494**) and uleine (**793**), have been
studied biosynthetically, but with mixed results. Pericalline (**794**) is derived
from tryptophan by specific retention of C-3 and loss of C-2. Stemmadenine
(**594**) was also a precursor, and consequently it was proposed that specific
loss of C-5 of stemmadenine was a late process in the pathway. No feeding
studies have been reported with labeled mevalonate, monoterpenes, or
iridoids.

3-Aminomethyl indole (**886**), variously labeled tryptophans, stemmadenine (**594**) and vallesamine (**887**), were not precursors of uleine (**793**) in
Aspidosperma pyricollum. However, the plant may not have been producing
uleine (**793**) at the time.

One of the problems with proposing any unifying scheme for the biosynthesis of these alkaloids is that when 16,17-dihydrosecodin-17-ol (**888**)

886

887   vallesamine

888

was used as a precursor the ester carbonyl group was specifically incorporated into the exomethylene of pericalline (**794**). The secodine substrate
is probably not a true biosynthetic intermediate, and one must consider
whether results obtained with such a compound should be used to develop
biosynthetic schemes.

Potier and co-workers proposed a scheme for the formation of pericalline
(**794**) from stemmadenine (**594**) involving a concerted Polonovskii fragmentation and loss of the ester group as carbon dioxide (Scheme 126). A similar
scheme was proposed by Kutney and co-workers to explain their results
with the secodine precursor, except that in this case C-17 of the stemmadenine derivative was lost as formaldehyde.

The biogenesis of uleine (**793**) has been discussed in Scheme 116. In this
scheme C-3 of the side chain of tryptophan is specifically loss in uleine
biosynthesis, whereas C-2 is retained.

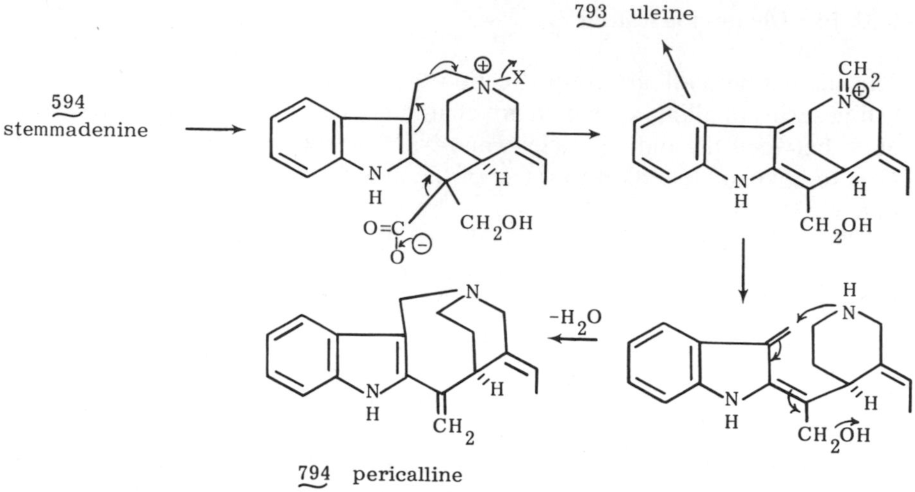

Scheme 126 Biogenesis of pericalline (**794**).

LITERATURE

Reviews

Battersby, A. R., *Quart. Revs.* **15**, 259 (1961).

Taylor, W. I., *Science* **153**, 954 (1966).

Gröger, D., K. Stolle, and K. Mothes, *Arch. Pharm. (Weinheim)* **300**, 393 (1967).

Gröger, D., in K. Mothes, and H. R. Schütte (Eds.). *Biosynthese der Alkaloide*, VEB Deutsche Verlag der Wissenschaften, Berlin.

Scott, A. I., *Acc. Chem. Res.* **3**, 151 (1970).

Sundberg, R. J., *The Chemistry of Indoles*, Academic, New York, 1970, p. 269.

Battersby, A. R., *Alkaloids, (London)* **1**, 31 (1971).

Parry, R. J., in N. R. Farnsworth, and W. I. Taylor (Eds.), *The Catharanthus Alkaloids, Botany, Chemistry and Pharmacology* Marcel Dekker, New York, 197.

Cordell, G. A., *Lloydia* **37**, 219 (1974).

Early Work

Woodward, R. B., *Nature* **162**, 155 (1948).

Robinson, R., *The Structural Relations of Natural Products*, Clarendon, Oxford. 1955.

Monoterpenoid Hypothesis

Wenkert, E., *J. Amer. Chem. Soc.* **84**, 98 (1962).

Thomas, R., *Tetrahedron Lett.* 544 (1961).

Money, R., I. G. Wright, F. McCapra, E. S. Hall, and A. I. Scott, *J. Amer. Chem. Soc.* **90**, 4144 (1968).

Battersby, A. R., R. T. Brown, J. A. Knight, J. A. Martin, and A. O. Plunkett, *Chem. Commun.* 346 (1966).

Leete, E., and S. Ueda, *Tetrahedron Lett.* 4915 (1966).

Brechbuler-Bader, S., C. J. Coscia, P. Loew, C. von Szczepanski, and D. Arigoni, *Chem. Commun.* 136 (1968).

Battersby, A. R., E. S. Hall, and R. Southgate, *J. Chem. Soc., Sec. C* 721 (1969).

Battersby, A. R., A. R. Burnett, and P. G. Parsons, *J. Chem. Soc., Sec. C* 1187 (1969).

Nitrogenous Glycosides

Smith, G. N., *Chem. Commun.* 912 (1968).

Battersby, A. R., A. R. Burnett, and P. G. Parsons, *J. Chem. Soc. Sec. C* 1193 (1969).

DeSilva, K. T. D., G. N. Smith, and K. E. H. Warren, *Chem. Commun.* 905 (1971).

Blackstock, W. P., R. T. Brown, and G. K. Lee. *Chem. Commun.* 910 (1971).

Kennard, O., P. J. Roberts, N. W. Isaacs, F. H. Allen, W. D. S. Motherwell, K. H. Gibson, and A. R. Battersby, *Chem. Commun.* 899 (1971).

Mattes, K. C., C. R. Hutchinson, J. P. Springer, and J. Clardy, *J. Amer. Chem. Soc.* **97**, 6270 (1975).

Hutchinson, C. R., A. H. Heckendorf, P. E. Daddona, E. Hagaman, and E. Wenkert, *J. Amer. Chem. Soc.* **96**, 5609 (1974).

Heckendorf, A. H., and C. R. Hutchinson, *Tetrahedron Lett.* 4153 (1977).

Stöckigt, J., and M. H. Zenk, *Chem. Commun.* 646 (1977).

Rueffer, M., N. Nagakura, and M. H. Zenk, *Tetrahedron Lett.* 1593 (1978).

Elaboration of the Nitrogenous Glycoside

Battersby, A. R., and E. S. Hall, *Chem. Commun.* 793 (1969).

Qureshi, A. A., and A. I. Scott, *Chem. Commun.* 948 (1968).

Husson, H.-P., C. Kan-Fan, T. Sevenet, and J.-P. Vidal, *Tetrahedron Lett.* 1889 (1977).

Stöckigt, J., M. Rueffer, M. H. Zenk, and G.-A. Hoyer, *Planta Med.* **33**, 188 (1978).

Elaboration of Dehydrogeissoschizine

Scott, A. I., P. B. Reichardt, M. B. Slaytor, and J. G. Sweeney, *Bioorg. Chem.* **1**, 157 (1971).

Kutney, J. P., J. F. Beck, V. R. Nelson, and R. S. Sood, *J. Amer. Chem. Soc.* **93**, 255 (1971).

Kutney, J. P., J. F. Beck, C. Ehret, G. Poulton, R. S. Sood, and M. D. Westcott, *Bioorg. Chem* **1**, 194 (1971).

Rueffer, M., C. Kan-Fan, H.-P. Husson, J. Stöckigt, and M. H. Zenk, *J. Chem. Soc., Chem. Commun.* 1016 (1979).

Stereoselectivity

Battersby, A. R., T. C. Byrne, R. S. Kapil, J. A. Martin, T. G. Payne, D. Angoni, and P. Loew, *Chem. Commun.* 951 (1968).

Quinine Biosynthesis

Kowanka, N., and E. Leete, *J. Amer. Chem. Soc.* **84**, 4919 (1962).

Leete, E., and J. N. Wemple, *J. Amer. Chem. Soc.* **91**, 2698 (1969).

Battersby, A. R., and E. S. Hall, *Chem. Commun.* 194 (1970).

Battersby, A. R., and R. J. Parry, *Chem. Commun.* 31 (1971).

Uleine Biosynthesis

Kutney, J. P., V. R. Nelson, and D. C. Wigfield, *J. Amer. Chem. Soc.* **91**, 4278 (1969).

Kutney, J. P., J. F. Beck, and G. B. Fuller, *Heterocycles* **1**, 5 (1973).

Potier, P., and M. M. Janot, *C. R. Acad. Sci. Paris, Ser. C* **276**, 1727 (1973).

Kutney, J. P., *Heterocycles* **4**, 429 (1976).

chapter 10

ALKALOIDS DERIVED FROM HISTIDINE

Histidine (**1**) and the corresponding amine histamine (**2**) are the most widely distributed compounds containing the imidazole nucleus. In spite of this broad occurrence however, few alkaloids are known whose biogenesis can be attributed to histidine (**1**). Of those that are known, essentially nothing has been deduced of their biosynthesis.

L-Histidine forms a complex with copper(II) and this can be used to resolve racemic histidine. L-Histidinato-L-threonatocopper(II) has been detected in human serum.

Some examples of alkaloids containing the imidazole nucleus, and plausibly derived from histidine, are ergothioneine (**3**) from ergot, spinacin (**4**) from the liver of the shark *Acanthia vulgaris*, glochidine (**5**) and glochicidine (**6**) from *Glochidion* species, casimiroedine (**7**) from *Casimiroa edulis* La Llave et Lejarza (Rutaceae), and pilocarpine (**8**) from various *Pilocarpus* species (Rutaceae).

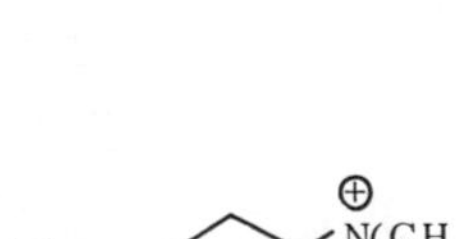

1 histidine

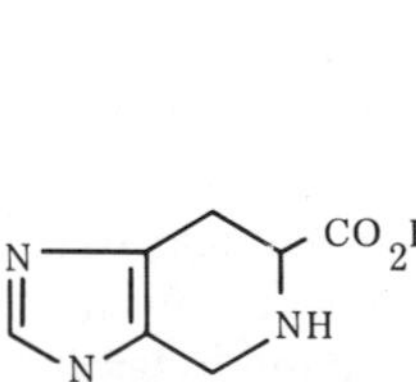

2 histamine

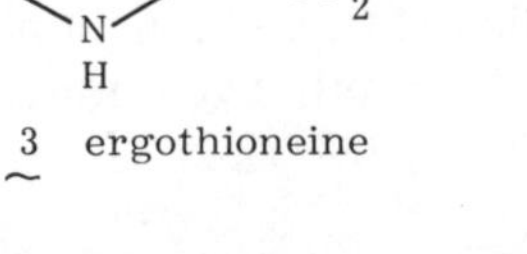

3 ergothioneine

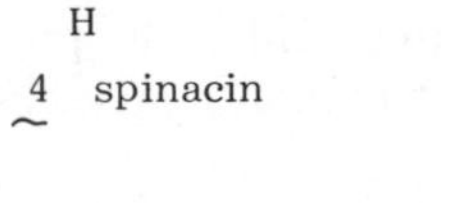

4 spinacin

833

Spinacin (**4**) would correspond to compounds in the tetrahydronorharman series derived from tryptophan (Chapter 9). In this respect it is perhaps surprising to find that there are no alkaloids in this series, like emetine (Chapter 8) or geissoschizine (Chapter 9), which are derived by condensation of the amino acid or the free amine with secologanin (**9**).

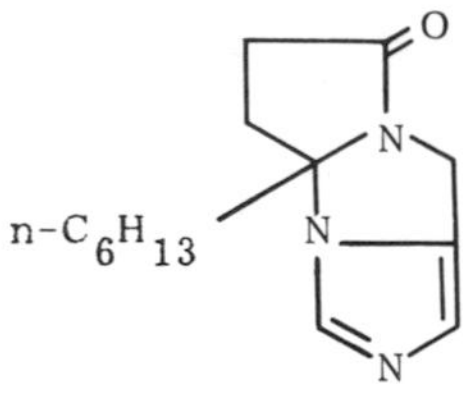

5 glochidine

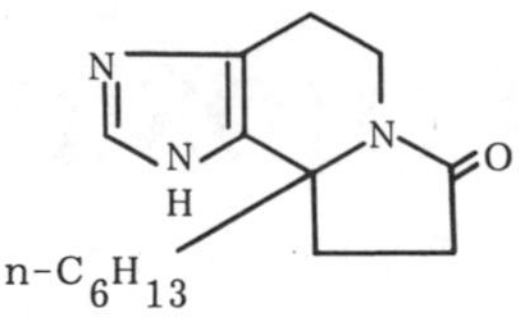

6 glochicidine

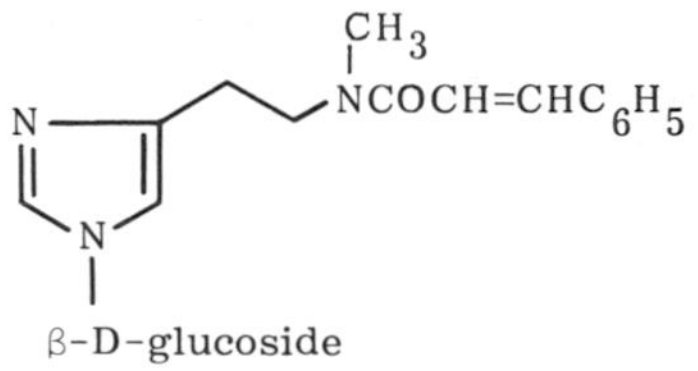

β-D-glucoside

7 casimiroedine

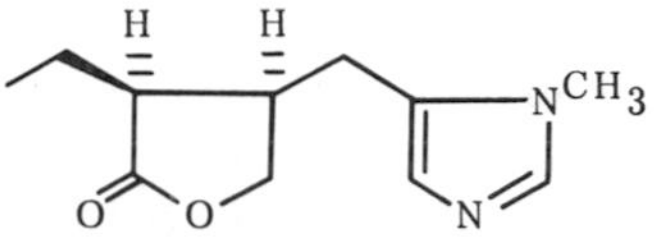

8 pilocarpine

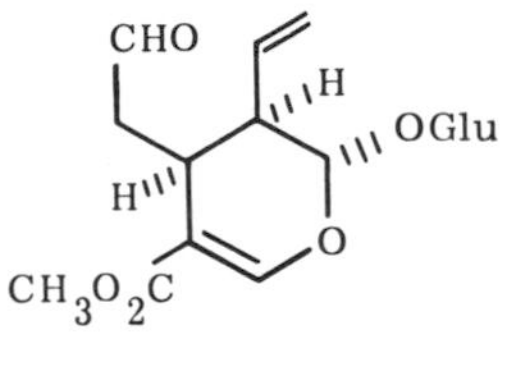

9 secologanin

10.1   CASIMIROEDINE

Although the application of degradative, spectroscopic, and X-ray techniques afforded the structure of casimiroedine, the *cis* or *trans* nature of the cinnamide moiety was not determined.

This stereochemical point in casimiroedine (**7**) was confirmed by total synthesis from the chloromercury derivative **10**. Glycosylation and treatment with methanolic methylamine gave casimidine (**11**), and reaction of **11** with *trans*-cinnamic acid in the presence of EEDQ (*N*-ethoxycarbonyl-2-ethoxy-1,2-dihydroquinoline) gave casimiroedine (**7**) in 70% yield.

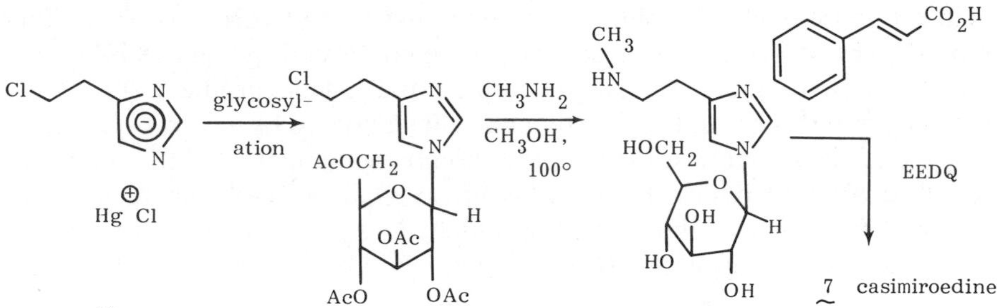

10   11   7   casimiroedine

10.2   PILOCARPINE

Jaborandi is the name applied to the leaves of various species of *Pilocarpus* (Rutaceae) native to S. America. The crude drug was used to promote salivation, but subsequently was also shown to act on the pupil of the eye and on the sweat glands.

The alkaloid responsible for these effects is pilocarpine (**8**), an unstable, hygroscopic, viscous oil. First obtained by Gerrard in 1875 from *P. jaborandi* Holmes, the structure of pilocarpine was deduced by Jowett in 1903.

The principal commercial source is *Pilocarpus microphyllus* Stapf, the so-called Maranham jaborandi. To avoid conversion to the unwanted iso-pilocarpine (**12**), the leaves are processed as rapidly as possible in the following way. The alkaloids are converted to their salts and the leaves defatted with petroleum ether. Alkalinization and extraction with benzene gives the crude alkaloid mixture, from which the pilocarpine can be obtained as the nitrate salt.

When heated with acid or alkali, pilocarpine (**8**) is readily isomerized to isopilocarpine (**12**). But on treating either **8** or **12** with permanganate, the same products—ammonia, methylamine, acetic acid, propionic acid, iso-pilopic acid (**13**), and homoisopilopic acid (**14**)—are obtained. The latter

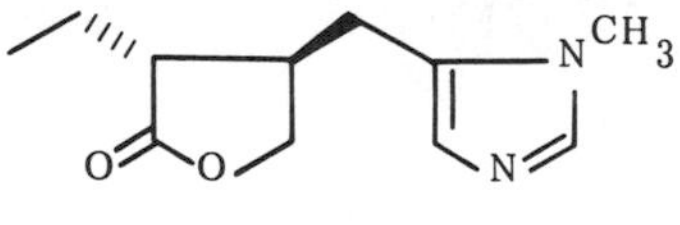

12   isopilocarpine

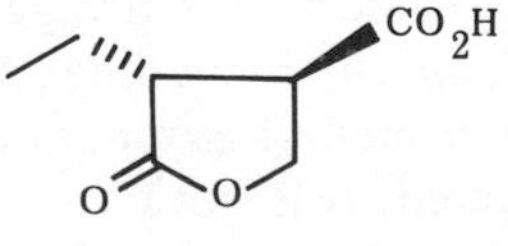

13   isopilopic acid 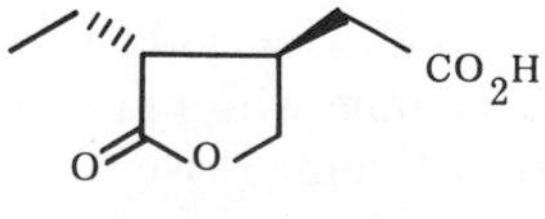   14   homoisopilopic acid

acid can more readily be obtained by ozonolysis and hydrolysis from either **8** or **12**. The structures of these acids were confirmed by synthesis.

Jowett first suggested that pilocarpine and isopilocarpine were stereoisomers, and this has subsequently been substantiated by several syntheses.

One of these syntheses began with homoisopilopic acid (**14**), which was readily converted to the α-hydroxy ketone **15**. Oxidation to the keto aldehyde followed by condensation with formaldehyde and ammonia gave pilocarpidine and isopilocarpidine (**16/17**), which could be methylated to **8** and **12**.

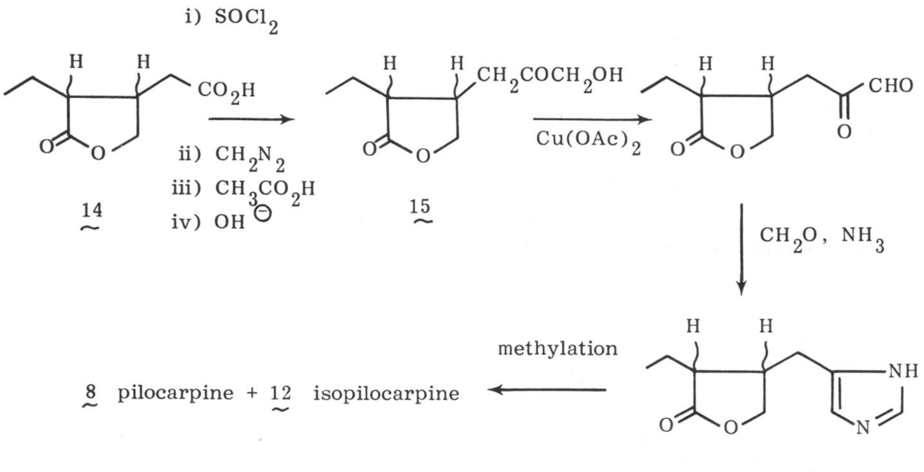

The configuration of the iso series was confirmed by successive electrolysis and lithium aluminum hydride reduction of **14** to the (−)-diol (**18**). The presence of optical activity in this diol is proof of the *trans* configuration of this, and the previous, compounds.

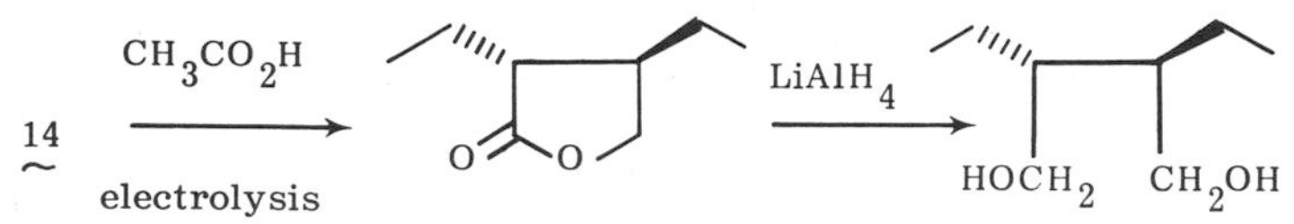

A recent synthetic effort has been a modified route to homopilopic acid (**19**). Michael addition of diethylethylmalonate to 2-oxo-5-ethoxy-2,5-dihydrofuran (**20**) followed by treatment with HBr–acetic acid gave **21**, which as the corresponding ester could be hydrogenated and hydrolyzed to **19**.

Only very preliminary data are available on the biosynthesis of pilocarpine (**8**). Methionine was a precursor of the *N*-methyl group, but threonine, acetate, and histidine (**1**) were not incorporated. It is not clear whether the lack of incorporation is real or due to the precursor not reaching the site of synthesis.

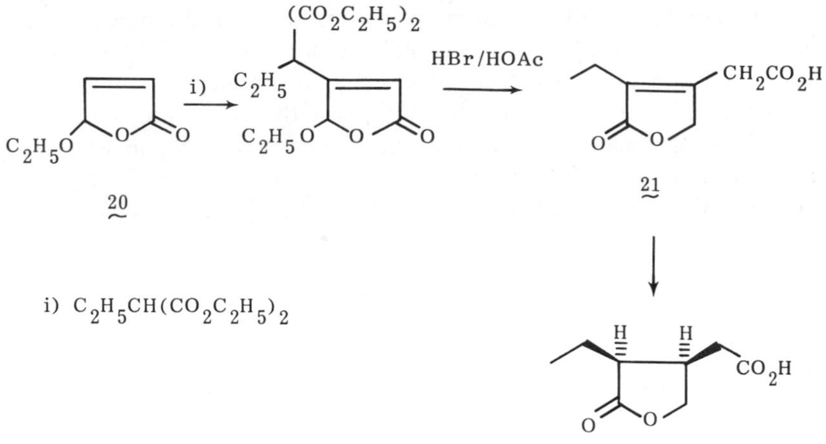

i) $C_2H_5CH(CO_2C_2H_5)_2$

Pilocarpine stimulates the parasympathetic nerve endings, thereby increasing salivatory, gastric, and lachrymal secretions. The hydrochloride and nitrate salts are used in the treatment of glaucoma. The ability of quaternary pilocarpine derivatives to act as acetylcholine antagonists has been investigated, and the salt 22 was found to be the most potent.

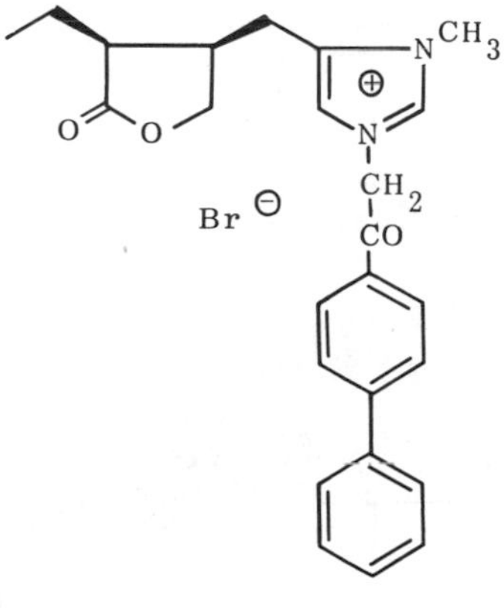

10.3 ALKALOIDS RELATED TO PILOCARPINE

From *Cynometra ananta* in the Leguminosae, three alkaloids, anantine (23), cynometrine (24), and cynodine (25), have been isolated whose structures bear an obvious similarity to pilocarpine.

Cynometrine (24), $C_{16}H_{19}N_3O_2$, showed the presence of a five-membered ring lactam in the IR (ν_{max} 1689 cm^{-1}), and from the proton NMR two N-

methyl groups (2.86 and 3.43 ppm) were observed. A doublet at 5.03 ppm typical of a benzylic alcohol indicated the presence of only one adjacent proton. The imidazole 5-proton appeared as a doublet ($J = 1$ Hz) at 5.93 ppm, but the 2-proton was in the region of the remaining five aromatic protons. The mass spectrum showed a molecular ion at 285 amu, and principal fragment ions at m/e 179 and 108, formulated as **26** and **27** respectively. Hydrolysis of cynodine (**25**) gave benzoic acid and cynometrine.

The ^{13}C NMR values for cynometrine are shown in **28**.

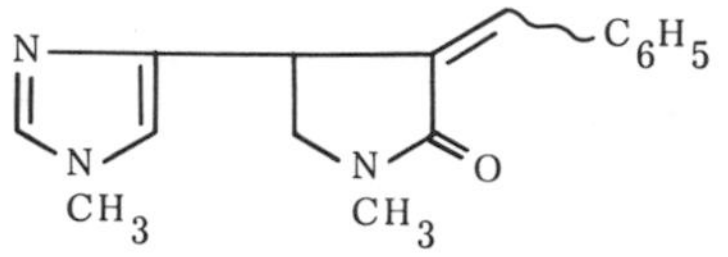

23   anantine

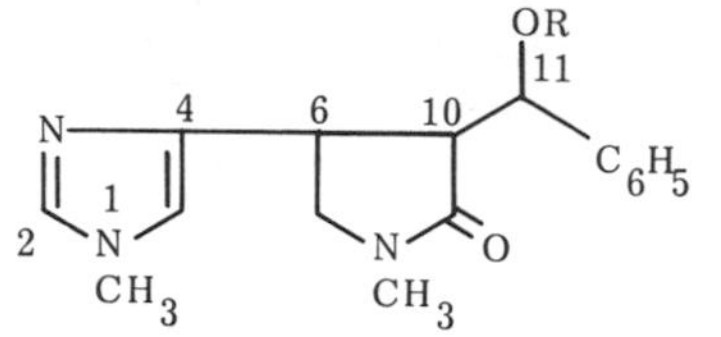

24   cynometrine, R=H

25   cynodine, R=COC$_6$H$_5$

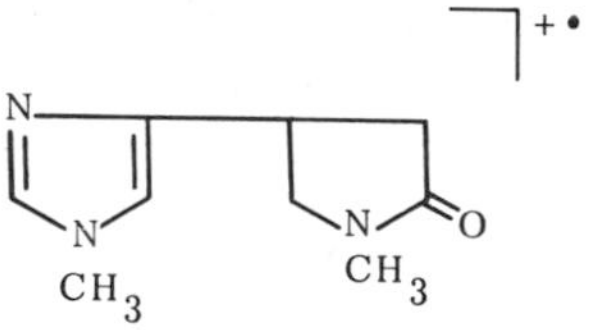

26   m/e 179

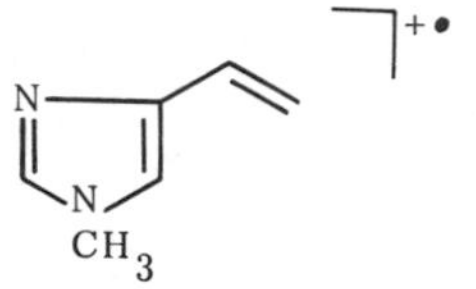

27   m/e 108

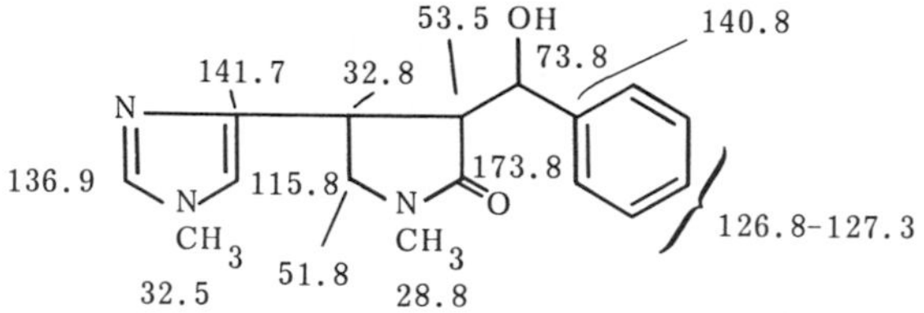

28

Several other alkaloids are also known which are very similar to pilocarpine (**8**), for example pilosine (**29**) and pilosinine (**30**). The pilosine originally isolated from *P. microphyllus* and *P. jaborandi* has been shown to be a 1:1 mixture of (+)-pilosine (**29**) and (+)-isopilosine (**31**).

These alkaloids and pilocarpine (**8**) have been synthesized by Link and Bernauer. The imidazole aldehyde **32** was condensed with diethyl succinate to give **33** which could be reduced with lithium borohydride to racemic

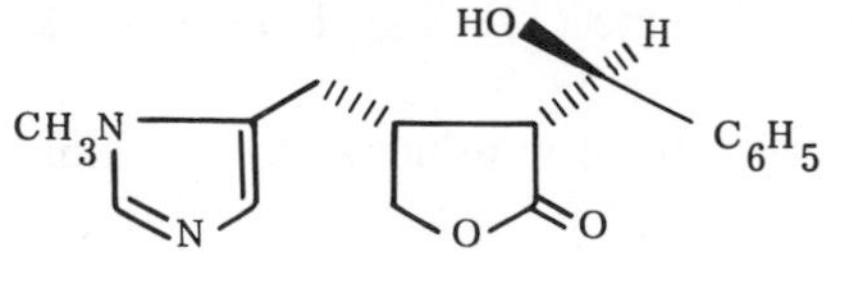

29 (+)-pilosine

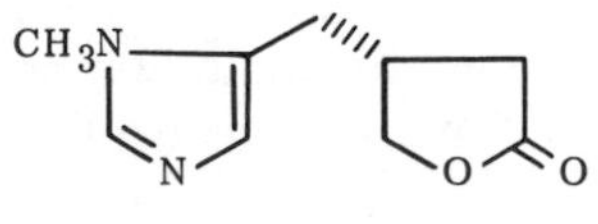

30 pilosinine

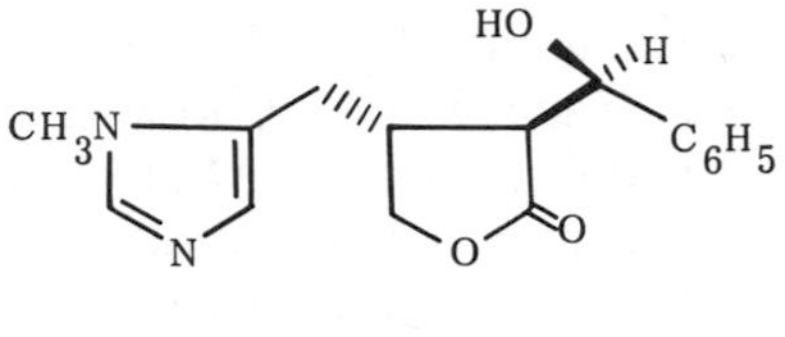

31 (+)-isopilosine

pilosinine (30) and resolved. Benzoylation and catalytic reduction of (+)-pilosinine (30) gave a mixture of (+)-isopilosine (31) and (−)-epiisopilosine (34).

Acetylation and catalytic reduction of optically active 30 followed by elimination gave 35, which on further catalytic reduction gave (+)-pilocarpine (8) and (+)-isopilocarpine (12). The configuration at C-3 in 30 is the therefore established as R, and since retroaldolization of 31 or 34 gave (+)-30, these too must have the 3R-configuration (Scheme 1). The absolute

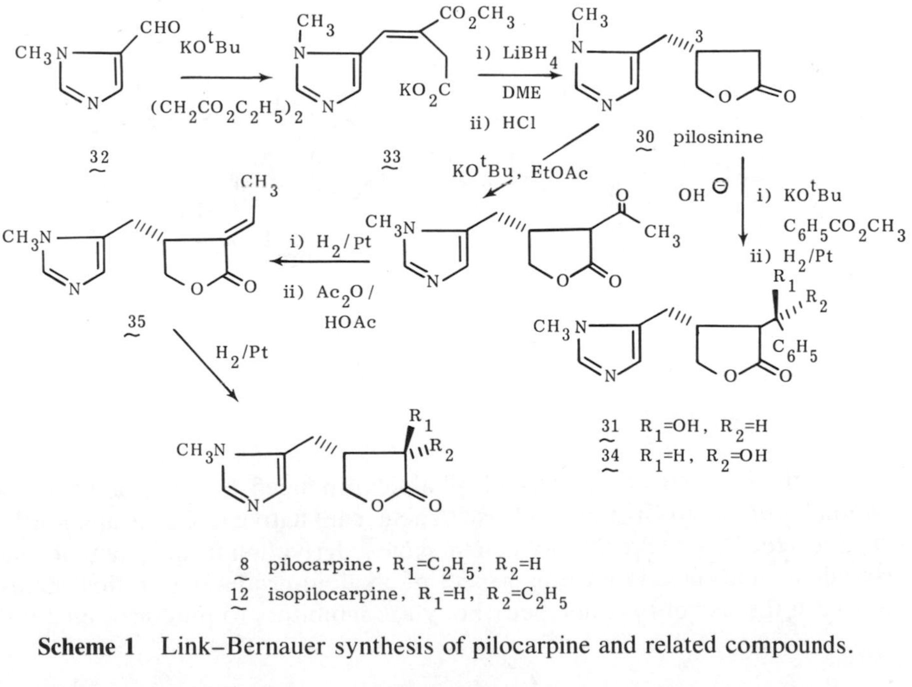

32

33

30 pilosinine

35

31 R$_1$=OH, R$_2$=H
34 R$_1$=H, R$_2$=OH

8 pilocarpine, R$_1$=C$_2$H$_5$, R$_2$=H
12 isopilocarpine, R$_1$=H, R$_2$=C$_2$H$_5$

Scheme 1 Link–Bernauer synthesis of pilocarpine and related compounds.

configuration (2S,3R,6R) of (+)-isopilosine (**31**) was determined by X-ray analysis.

Racemic anantine (**23**) has recently been synthesized beginning with 4-formyl-*N*-methyl imidazole.

10.4   MISCELLANEOUS ALKALOIDS

Further novel alkaloids apparently derived from histidine have been isolated in recent years, and some examples are cypholophine (**36**) from *Cypholophus friesiamus* (Urticaceae), dolichotheline (**37**) from *Dolichothele sphaerica* (Cactaceae), cinnamoyl histamine (**38**) from several plants in the Sterculi-aceae, and longistrobine (**39**) from *Macrorungia longistrobus* C.B. Cl. (Acanthaceae).

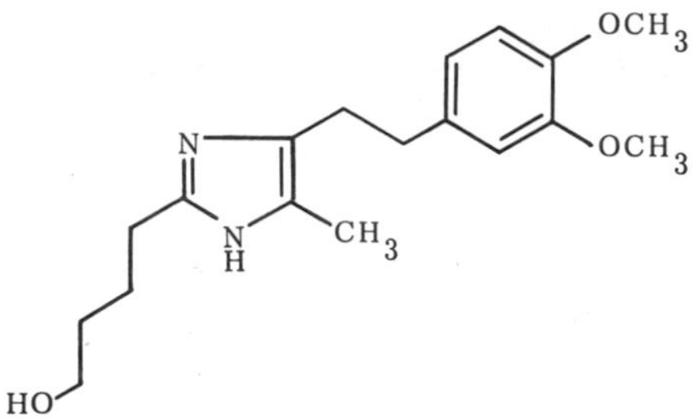

36   cypholophine 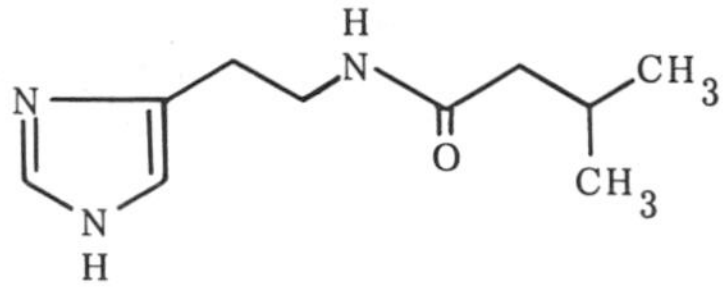   37   dolichotheline

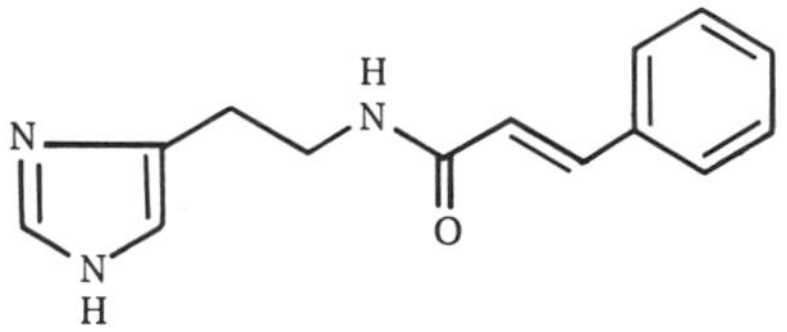

38 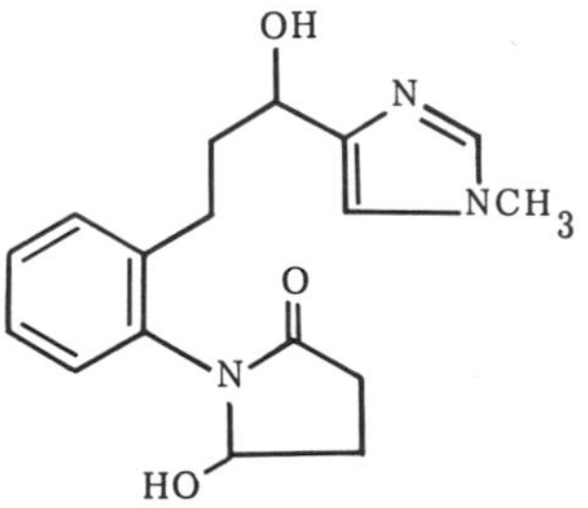   39   longistrobine

10.4.1   Dolichotheline

Dolichotheline (**37**) is an unusual alkaloid produced by the cactus *Doli-chothele sphaerica* Britton and Rose (Cactaceae) native to Texas and north-ern Mexico. Biogenetically, we can imagine a derivation from histamine and isovaleric acid. One interesting aspect of alkaloid production in this cactus has been the use of histidine decarboxylase inhibitors to produce unnatural

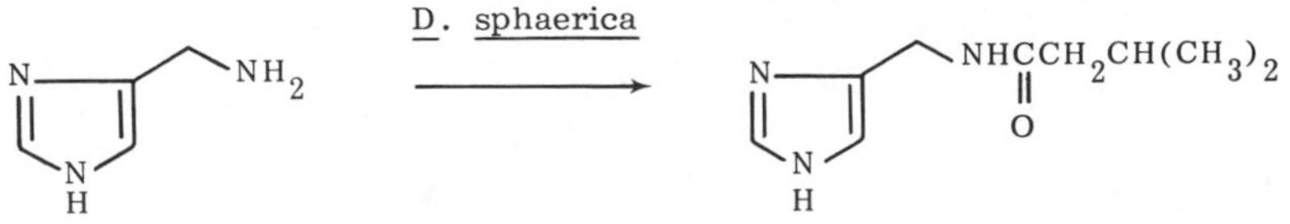

D. sphaerica

40

41

analogues. Thus α-methylhistidine in the presence of aminomethylimidazole (**40**) gave **41**, at the expense of dolichotheline (**37**).

10.4.2 Longistrobine and Isolongistrobine

Several imidazole alkaloids have been isolated from *Macrorungia longistrobus,* including isolongistrobine (**42**), dehydroisolongistrobine (**43**), macrorine (**44**), and isomacrorine (**45**).

Isolongistrobine and dehydroisolongistrobine were originally assigned the structures **46** and **47** respectively, based on oxidation and zinc dust distillation which yielded isomacrorine (**45**). Subsequently Wuonola and Woodward synthesized **42** and **43** from the imidazole **46**, in which the aniline **47** is a key intermediate (Scheme 2). Longistrobine (**39**) is the *N*-methyl imidazole isomer of **42**.

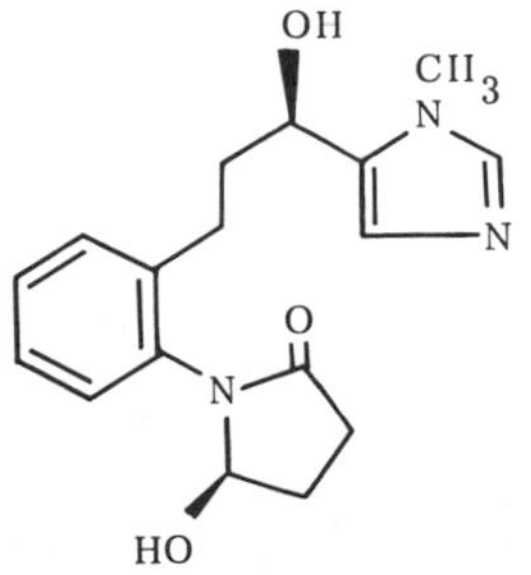

42   isolongistrobine

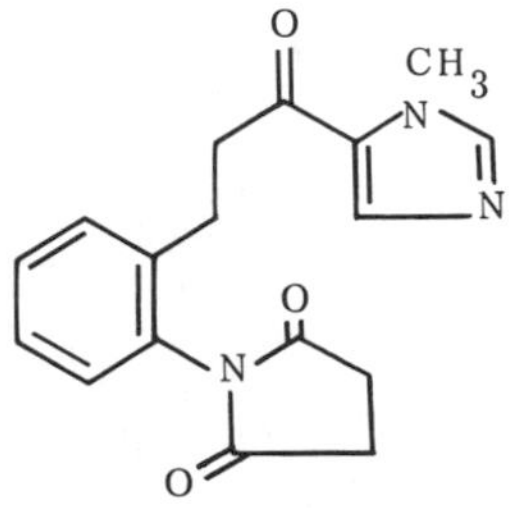

43   dehydroisolongistrobine

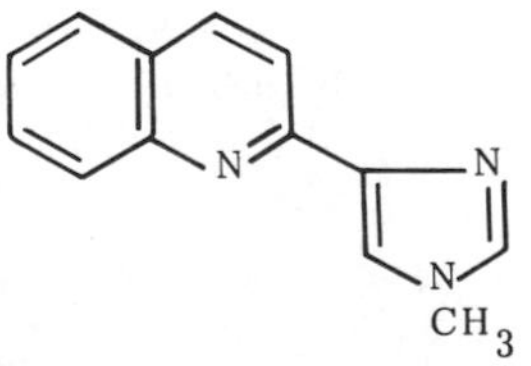

44   macrorine

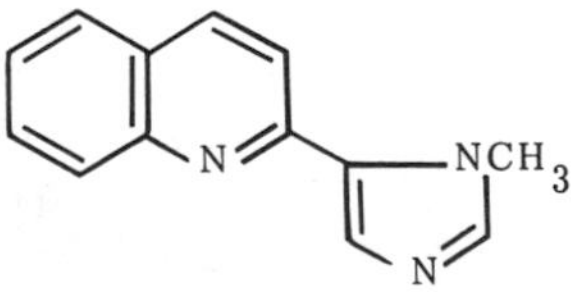

45   isomacrorine

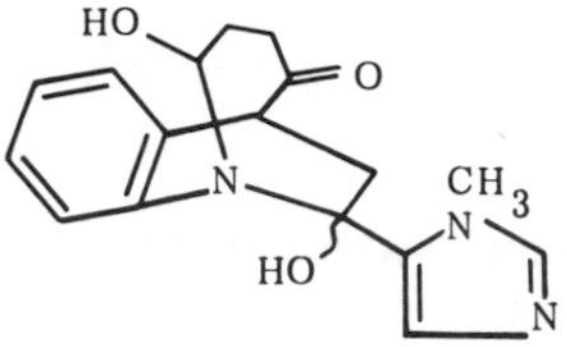

46   isolongistrobine
     (incorrect)

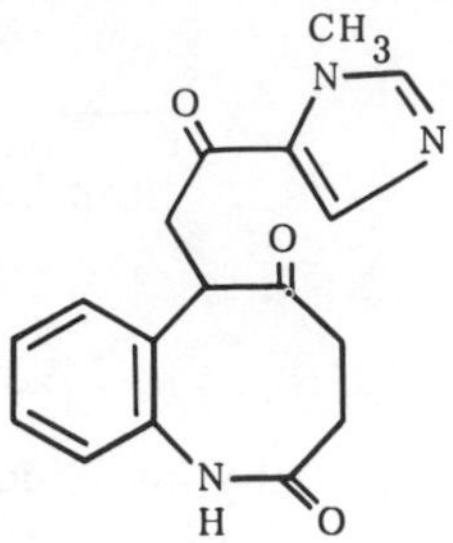

47   dehydroisolongistrobine
     (incorrect)

10.4.3   Zoanthoxanthin

There is a group of marine animals related to the sea anemones and the corals. Chemical studies on this group are rare, but one member, *Parazoanthus axinellae,* has afforded zoanthoxanthin (**48**), whose structure was established by X-ray analysis. Note the presence once again of an amino histidine unit in this product, which can formally be dissected and recombined in terms of two histidine-derived units as shown in Scheme 3.

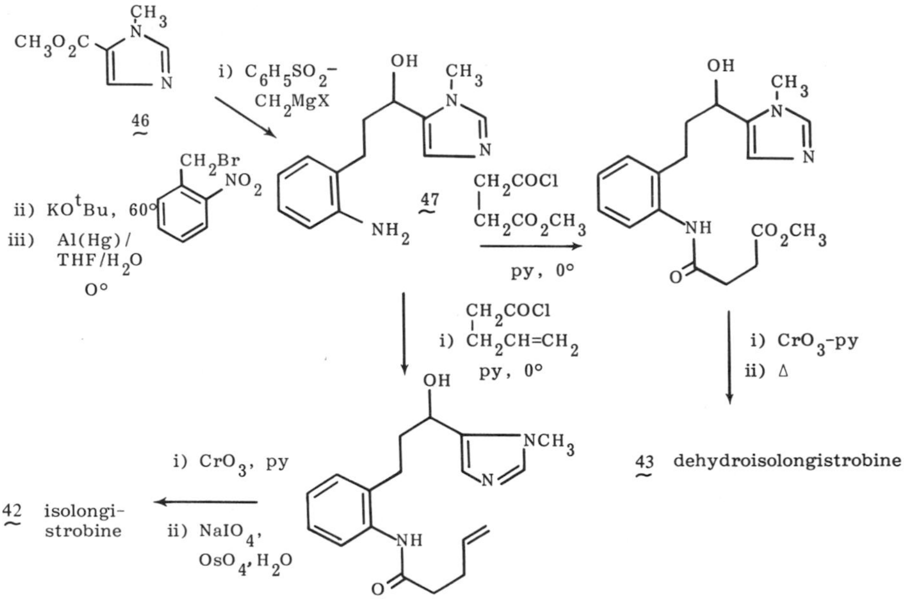

Scheme 2   Wuonola–Woodward synthesis of isolongistrobine (**39**) and dehydroisolongistrobine (**42**).

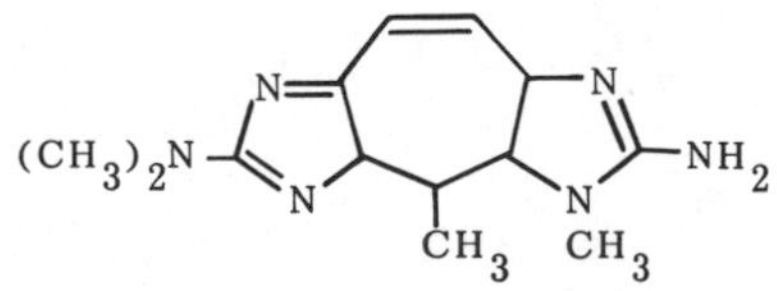

48 zoanthoxanthin

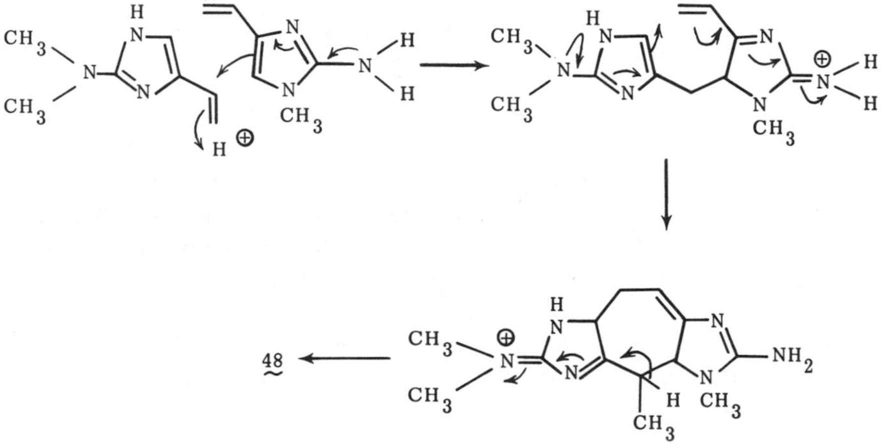

Scheme 3 Biogenesis of zoanthoxanthin (48).

10.4.4 Dibromophakellin

Several other alkaloids are known from marine sources which contain an imidazole nucleus, and among these are dibromophakellin (49) and oroidin (50) from the sponge *Phakellia flabellata*.

The structure of dibromophakellin was deduced from the spectral data and by X-ray crystallography. The UV spectrum showed λ_{max} 233 and 281 nm typical for a 2-acylpyrrole, and the IR spectrum bands at 1675 (C=N) and 1635 cm^{-1} (C=O). The proton and ^{13}C NMR data for dibromophakellin are shown in 51 and 52 respectively.

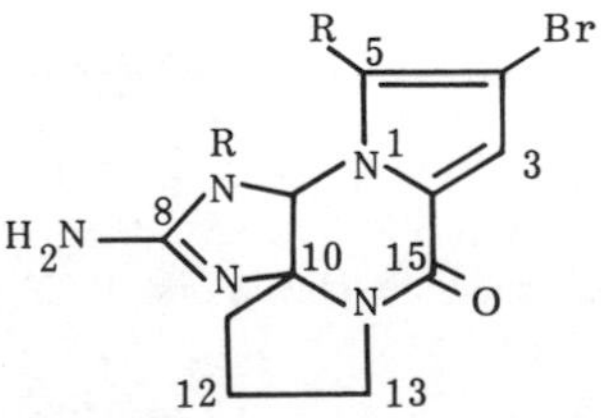

49 dibromophakellin, R=Br
54 monobromophakellin, R=H

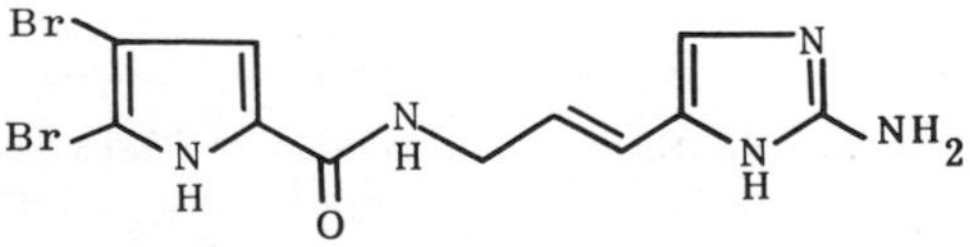

50 oroidin

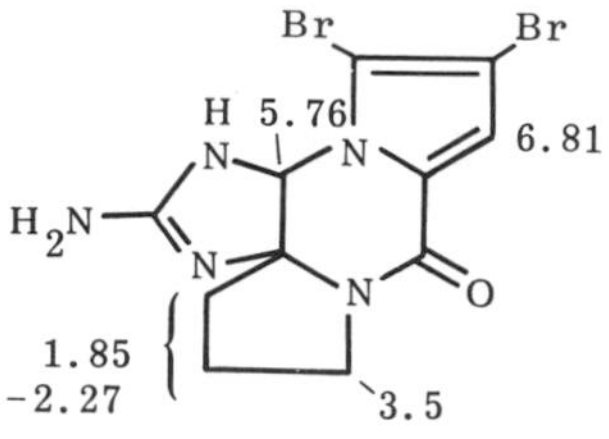

51

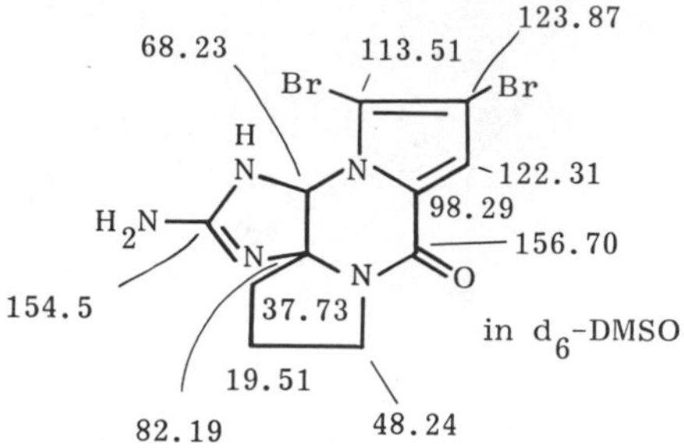

52

One confusing aspect of this work was the low pK_a (<8) value of dibromophakellin (49) compared with other guanidine derivatives (pK_a 13.4). The X-ray analysis however established that the aminoimidazoline ring is twisted and cannot become planar on protonation. Resonance within the cation is therefore considerably reduced.

Oroidin (50) and dibromophakellin (49) are probably biosynthetically related through the intermediate 53, suggesting an overall derivation of both compounds from proline and histidine.

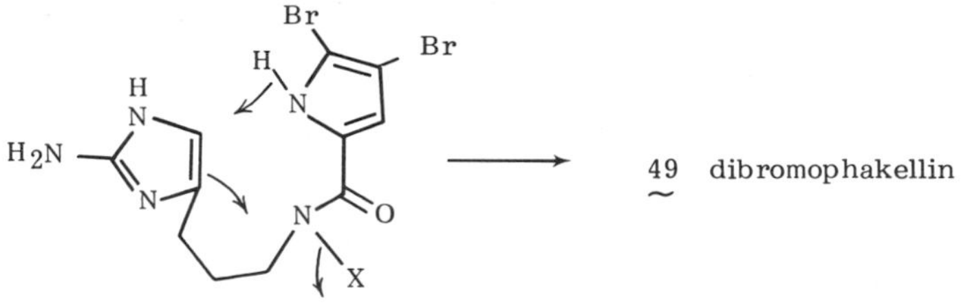

49   dibromophakellin

53

The hydrochlorides of dibromophakellin (49) and monobromophakellin (54) exhibit mild antibacterial activity against *B. subtilis* and *E. coli*.

LITERATURE

Reviews

Battersby, A. R., and H. Y. Openshaw, *Alkaloids NY* **3**, 202 (1953).

Luckner, M., *Secondary Metabolism in Plants and Animals*, Academic, New York, 1972, p. 282.

Snieckus, V. A., *Alkaloids, London* **1**, 455 (1971).

Snieckus, V. A., *Alkaloids, London* **2**, 271 (1972).

Snieckus, V. A., *Alkaloids, London* **3**, 299 (1973).

Snieckus, V. A., *Alkaloids, London* **4**, 395 (1974).

Snieckus, V. A., *Alkaloids, London* **5**, 265 (1975).

Casimiroedine

Raman, S., J. Reddy, W. N. Lipscomb, A. L. Kapoor, and C. Djerassi, *Tetrahedron Lett.* 357 (1962).

Panzica, R. P., and L. B. Townsend, *J. Amer. Chem. Soc.* **95**, 8737 (1937).

Pilocarpine and Related Alkaloids

Swan, G. A., *Introduction to Alkaloids*, Wiley-Interscience, New York, 1967, p. 188.

Preobrazhenskii, N. A., V. A. Preobrazhenskii, A. F. Wompe, and M. N. Shchukina, *Chem. Ber.* **66**, 1536 (1933).

Brochmann-Hanssen, E., M. A. Nunes, and C. K. Olah, *Planta Med.* **28**, 1 (1975).

Link, H., and K. Bernauer, *Helv. Chim. Acta* **55**, 1053 (1972).

Tchissambou, L., M. Benechie, and F. Khuong-Huu, *Tetrahedron Lett.* 1801 (1978).

Isolongistrobine

Arndt, R. R., S. H. Eggers, and H. Jordaan, *Tetrahedron* **25**, 2767 (1969).

Wuonola, M. A., and R. B. Woodward, *J. Amer. Chem. Soc.* **95**, 284 (1973).

Wuonola, M. A., and R. B. Woodward, *J. Amer. Chem. Soc.* **95**, 5098 (1973).

Wuonola, M. A., and R. B. Woodward, *Tetrahedron* **32**, 1085 (1976).

Dibromophakellin

Sharma, G., and B. Magdoff-Fairchild, *J. Org. Chem.* **42**, 4118 (1977).

chapter 11

ALKALOIDS DERIVED BY THE ISOPRENOID PATHWAY

Several examples of alkaloids containing mevalonate-derived units have been discussed previously; they include the furoquinoline alkaloids, the indole alkaloids related to echinulin, and all the monoterpenoid indole alkaloids. But there are many alkaloids derived almost exclusively from a terpene unit which remain to be described.

These alkaloids can be conveniently divided into groups based on the number of mevalonate units involved; thus the hemi- and monoterpenes are treated first, followed by the sesquiterpenes, the diterpenes, the triterpenes, and the steroids. These alkaloids are wide-ranging in their occurrence, structures, and biological activity, and there are no unifying threads, save their origin from mevalonate.

11.1 HEMITERPENOID ALKALOIDS

The classic hemiterpenoid alkaloids (i.e. those containing a single isoprene unit) are the furoquinoline alkaloids (Chapter 7) and the echinulin and ergot alkaloids (Chapter 9). But there are several other alkaloids which apparently contain simple isoprene units. For example, *Alchornea* species in the Euphorbiaceae produce alchorneine (**1**) and the guanidine **2**, which is known as pterogynine.

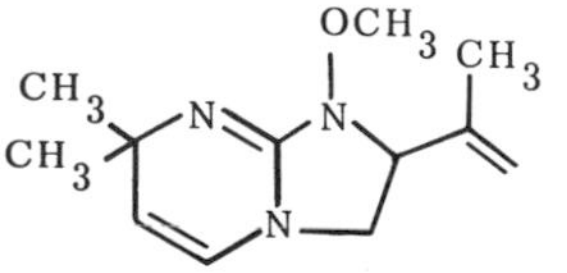

1   alchorneine

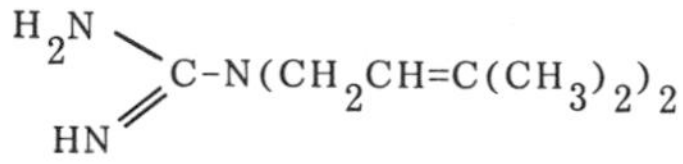

2   pterogynine

11.2 MONOTERPENOID ALKALOIDS

Before discussing some of the better established monoterpene alkaloids, brief mention should be made of chaksine (3), a guanidine alkaloid from *Cassia lispidula* Vahl. (Leguminosae), where the linear monoterpene unit is clearly seen.

Chaksine has a number of interesting pharmacologic effects, including respiratory paralysis in mice, vasoconstriction in rats, and contraction of the guinea pig ileum.

Arenaine (4) from *Plantago arenaria* Decne. (Plantaginaceae) is a guanidine derivative of the acid 5.

11.2.1 Chemistry

The remaining monoterpene alkaloids may be classified into two categories based on whether they are derived from a monoterpene before or after loganin (6) in the biosynthetic sequence (see Chapter 9). Those derived by some of the many possible condensation reactions of secologanin (7) with ammonia give rise to some very interesting structures.

Examples of the first group include β-skytanthine (8) and actinidine (9), and cantleyine (10), bakankoside (11), gentianine (12), gentioflavine (13), and probably 14 from *Antirrhinum majus* L. (Scrophulariaceae) are examples of the second general group.

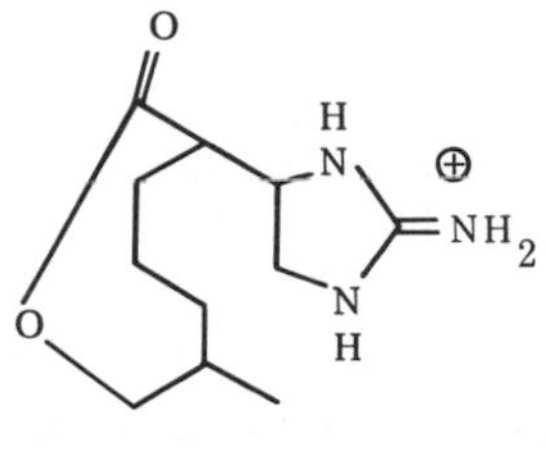

3 chaksine 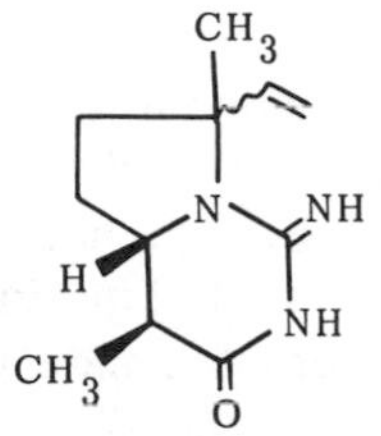   4 arenaine 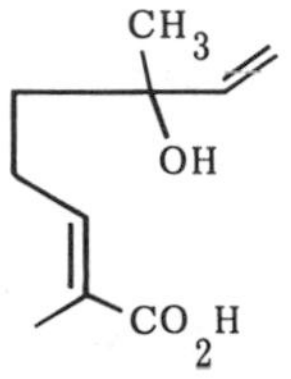   5

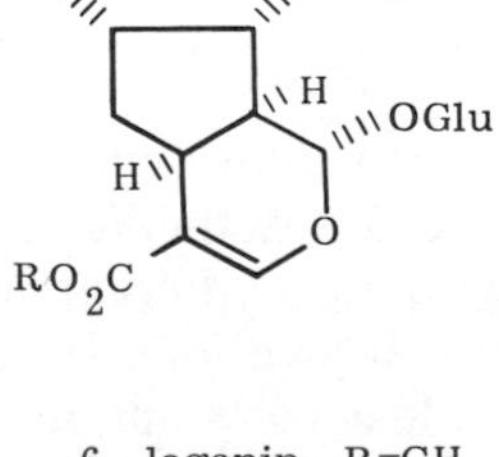

6 loganin, R=CH₃

35 R=H

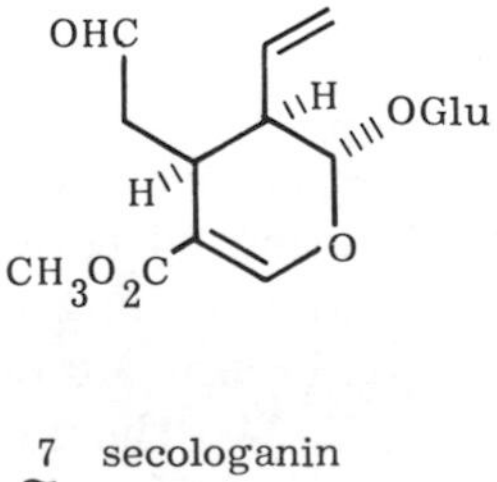

7 secologanin

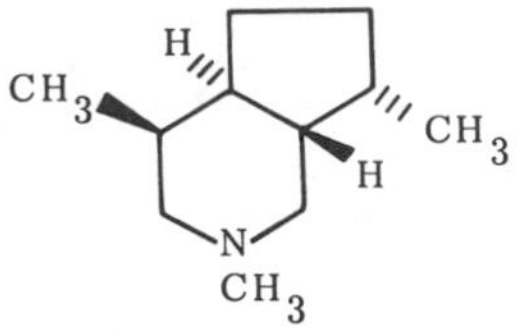

8   β-skytanthine 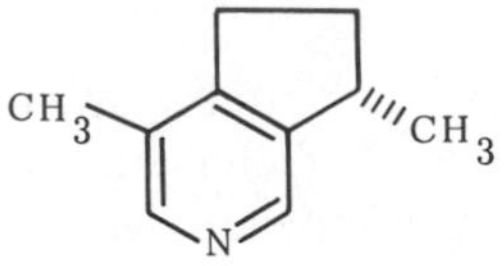   9   actinidine 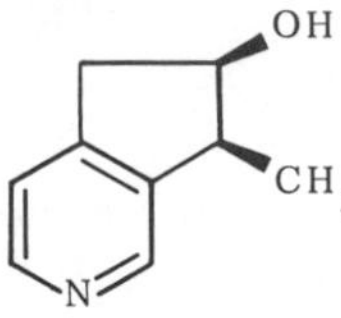   10   cantleyine

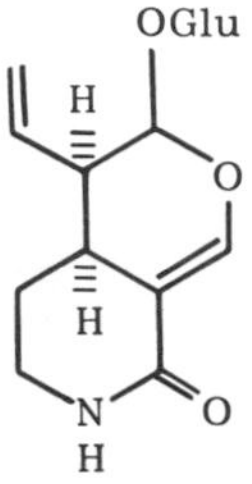

11   bakankoside 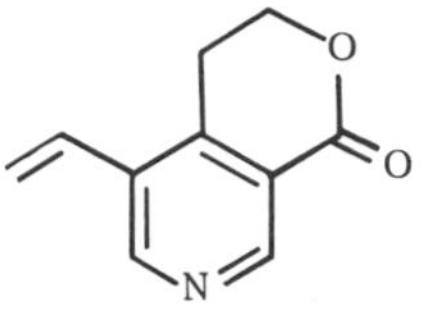   12   gentianine 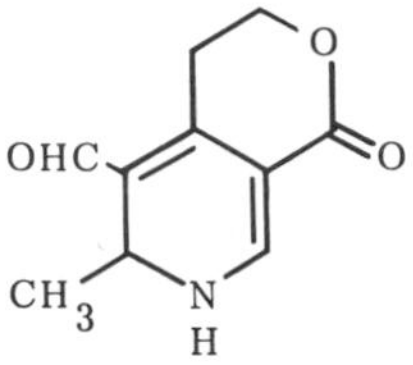   13   gentioflavine

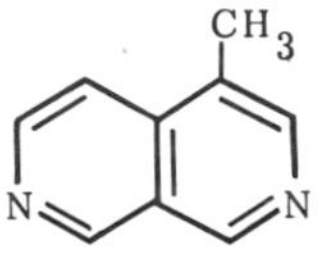

14

β-Skytanthine (**8**) was isolated in 1961 from the Chilean shrub *Skytanthus acutus* Meyern. (Apocynaceae) and at about this time a related alkaloid, actinidine (**9**), was obtained from *Actinidia polygama* Franch. (Actinidiaceae). Dehydrogenation of β-skytanthine gave actinidine (**9**), the structure of which was based on permanganate oxidation to 5-methylpyridine-3,4-dicarboxylic acid (**15**) and synthesis (Scheme 1). The intermediate **16** was also available optically active from *d*-pulegone (**17**), and consequently the configuration of natural actinidine could be determined.

A different synthesis of *d*-actinidine (**9**) from *d*-pulegone (**17**) involves the [1,5]-sigmatropic migration of hydrogen in the vinyl ketene **18** to give an α,β,γ,δ-unsaturated aldehyde **19**. Condensation with hydroxylamine affords **9** directly (Scheme 2).

In the mass spectrum, skytanthine (**8**) shows characteristic ions at *m/e* 84 (**20**), 58 (**21**), and 44 (**22**) (Scheme 3). Actinidine (**9**), because of the presence of the aromatic ring, displays quite a different mass spectrum. Two losses of methyl groups are observed to give ions at *m/e* 132 (**23**) (allylic loss) and *m/e* 117 (**24**). There is also a loss of 28 mu, including the nitrogen atom to an ion *m/e* 89 formulated as **25** (Scheme 4). The mechanism of this loss is not known.

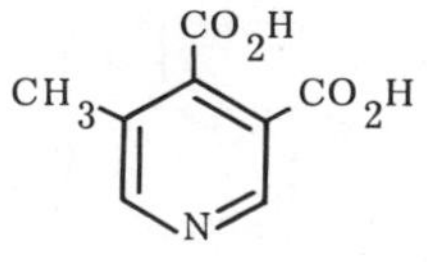

CO$_2$H

CH$_3$ ... CO$_2$H

N

15

CH$_3$

O

CH$_3$ ... C$_2$H$_5$O$_2$C

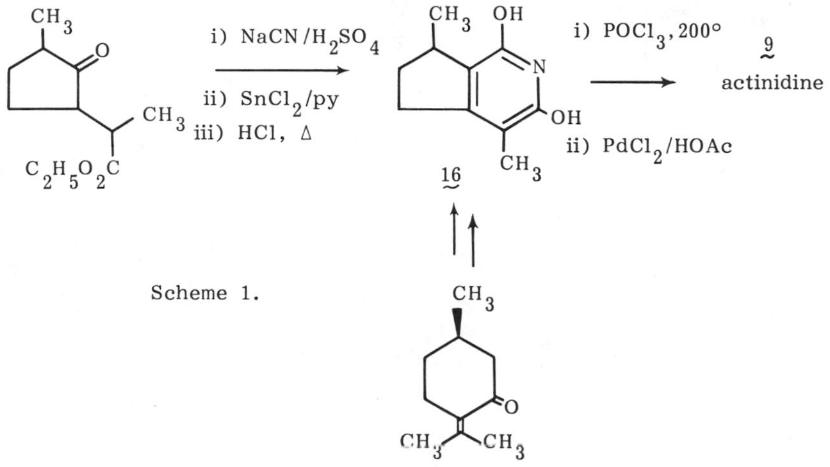

i) NaCN/H$_2$SO$_4$

ii) SnCl$_2$/py
iii) HCl, Δ

CH$_3$ OH

N

OH

CH$_3$

16

i) POCl$_3$, 200°

ii) PdCl$_2$/HOAc

9

actinidine

Scheme 1.

CH$_3$

O

CH$_3$ ... CH$_3$

17   d-pulegone

Scheme 1

17
d-pulegone

CH$_3$ CO$_2$H

CH$_3$
CH$_3$

i) COCl$_2$

ii)

N

N

110°

CH$_3$ C=O

H
CH$_2$

CH$_3$

18

CH$_3$ CHO

CH$_3$

19

NH$_2$OH

Δ

90%

9   (+)-actinidine

Scheme 2.

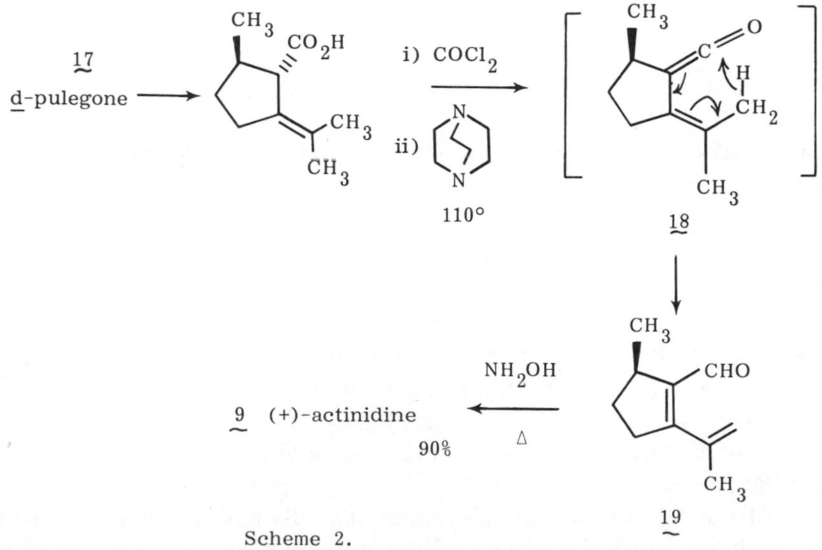

Scheme 2

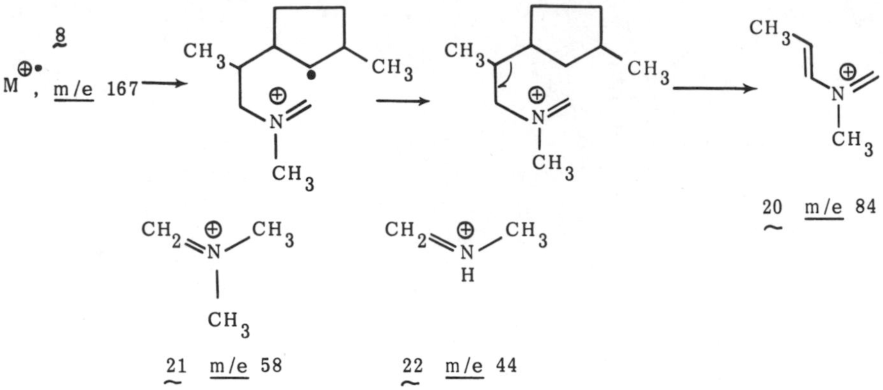

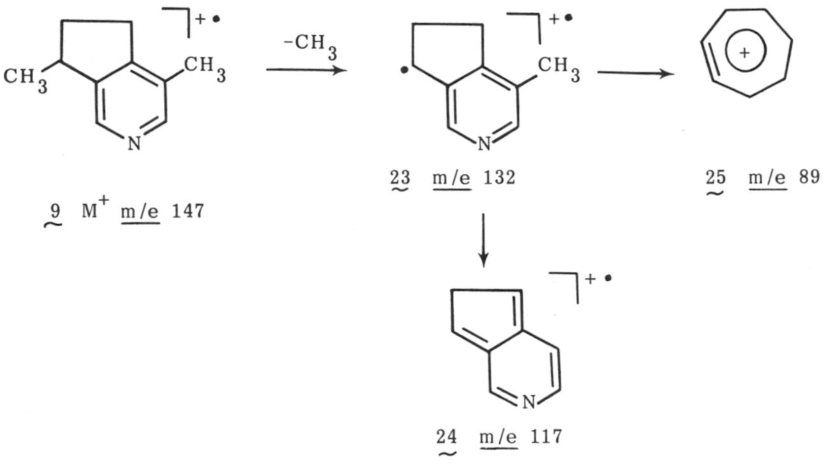

Schemes 3 and 4   Mass spectral fragmentation of skytanthine (**8**) and actinidine (**9**).

Cantleyine (**10**), first obtained from *Jasminum* species and subsequently from *Cantleya corniculata* (Icacinaceae), is the classic monoterpene alkaloid derived from an iridoid. In the absence of ammonia *Cantleya corniculata* does not afford cantleyine (**10**), and it was found that loganin (**6**) on treatment with ammonia at room temperature gave **10** in high yield.

The mass spectrum of cantleyine shows two principal fragmentation pathways of the hydroxycyclopentane ring, giving rise to four important ions (Scheme 5).

One of the best known monoterpene alkaloids is gentianine (**12**), which since first being isolated in 1944 has been obtained from a number of plants, particularly *Gentiana* species.

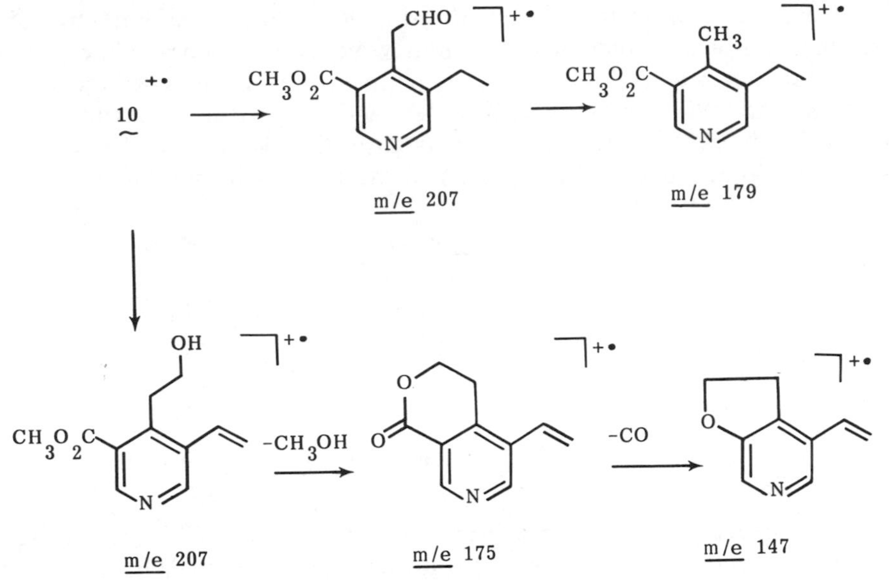

Scheme 5

The structure was deduced by degradation and synthesis. The NMR spectrum of gentianine shows the two pyridine α-protons at 9.11 and 8.80 ppm, the vinylic protons at 5.76, 5.59 and 6.80 ppm, and the two methylene groups at 4.55 and 3.09 ppm.

Once again the isolation of gentianine is common after ammonia is used during workup, and in this case two iridoids, swertiamarin (**26**) and gentiopicroside (**27**), are probably responsible for many of the isolations of **12**.

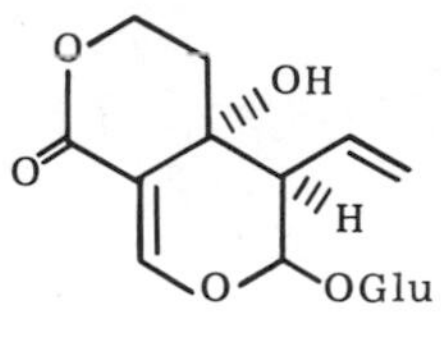

26 swertiamarin

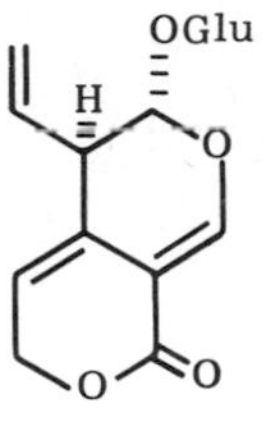

27 gentiopicroside

Another interesting *Gentiana* alkaloid is gentioflavine (**13**). The NMR spectrum displayed an aldehyde proton at 10.1 ppm, two methylene groups at 4.35 and 3.0 ppm, two-proton singlets at 8.45 ppm (NH) and a very low field "olefinic" proton at 8.8 ppm. A secondary methyl group was observed from a doublet at 1.3 ppm and an associated highly deshielded methine proton at 5.2 ppm.

Bromination of gentioflavine with bromine water gave a derivative **28**, and reaction of this compound with Raney nickel gave gentianidine (**29**), a well-characterized alkaloid. Gentioflavine therefore was assigned the structure **13**. The yellow color (λ_{max} 235, 298, and 410 nm) indicates that conjugation is from the nitrogen atom through to the aldehyde carbonyl. Presumably this accounts for the stability of the 1,2-dihydropyridine system.

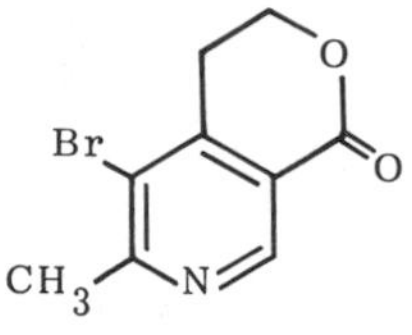

28 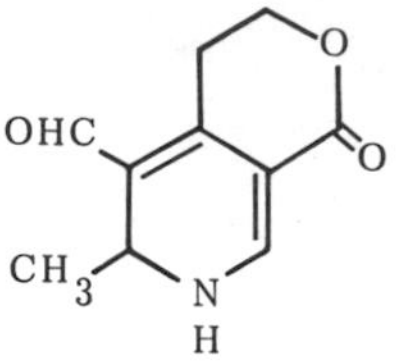   13 gentioflavine 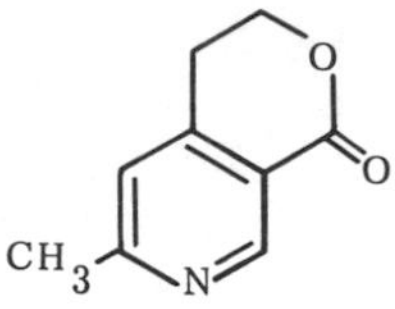   29 gentianidine

Gentiocrucine is an alkaloid from *Gentiana cruciata* Gilib. which was originally assigned the structure **30**. Further work corrected this structure to **31**, the first stable primary enamide system, synthetic or natural.

One of the problems with the original structure was that gentiocrucine gave *two* 2,4-dinitrophenylhydrazones. The PMR spectrum was quite complex, and from the CMR spectrum this was shown to be due to the presence of two isomers, since 14 carbon resonances were observed (**32** and **33**).

Gentiocrucine was synthesized by the route shown in Scheme 6.

A related compound, enicoflavine (**34**), has been obtained from *Enicostemma hyssopifolium* Willd. as a mixture of isomers, and now we can see the relationship of gentiocrucine to the monoterpene alkaloids. For at room

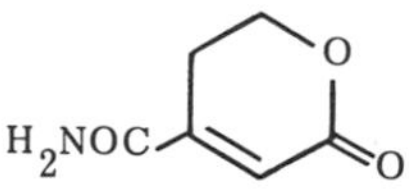

30 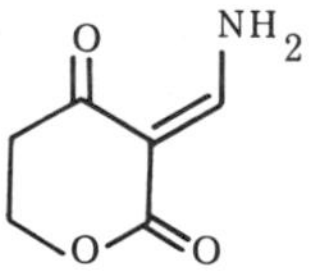   31 cis and trans

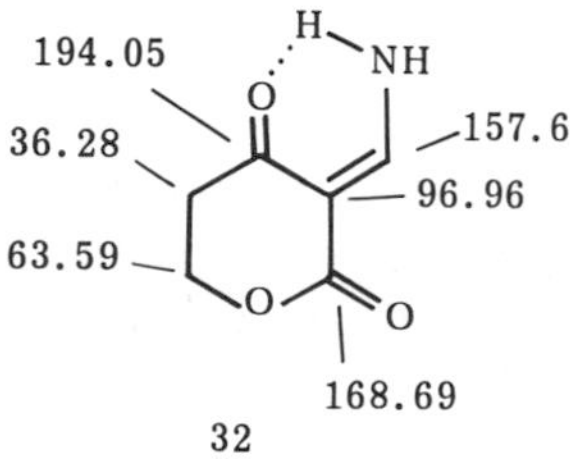

32 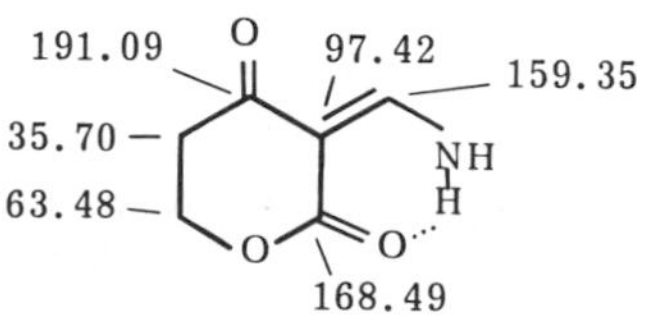   33

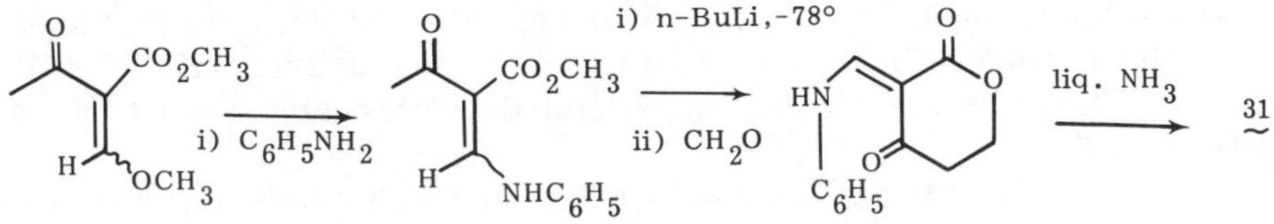

Scheme 6 Synthesis of gentiocrucine (**31**).

temperature, enicoflavine decomposed to gentiocrucine (**31**) and crotonaldehyde by a retro-aldol process. The monoterpene skeleton of **34** is shown by heavy lines.

The problem that some, or most, of the monoterpenoid alkaloids may be formed as a result of the addition of ammonia to a performed iridoid or

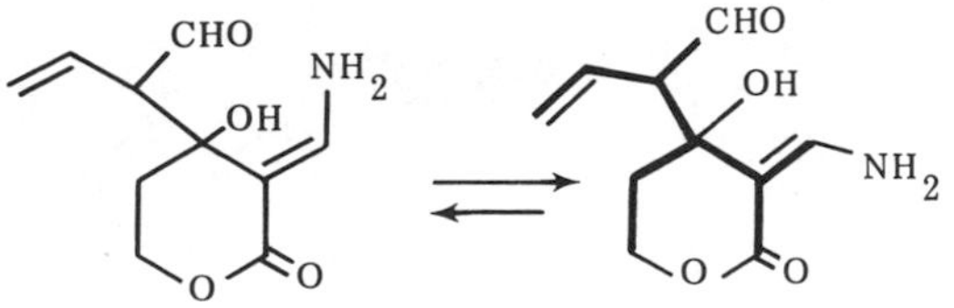

34

secoiridoid also has its benefits. For it means that by investigating the formation of the iridoids, the biosynthesis of the monoterpene alkaloids is being elucidated at the same time. A discussion of the formation of loganin (**6**) and secologanin (**7**) was included in the discussion of the indole alkaloids.

11.2.2 Biosynthesis

One of the earliest studies on the biosynthesis of this group of alkaloids was of actinidine (**9**). Lysine, aspartic acid, and quinolinic acid were not precursors, but acetate, mevalonate, and geranylpyrophosphate were incorporated. The monoterpenoid pathway is therefore confirmed.

Neither loganin (**6**) nor actinidine (**9**) were precursors of skytanthine derivatives in *Tecoma stans* Juss., indicating that the branching point for the formation of the skytanthine-type alkaloids is prior to loganin, and that oxidation of the piperidine ring to a pyridine is not a reversible process.

On the other hand, both loganin (6) and loganic acid (35) were precursors of gentiopicroside (27), the projected precursor of gentianine (12). [5,9-^{3}H, 3,7,11-^{14}C]Loganic acid (35) was incorporated with loss of half of the tritium label into 27.

Gentioflavine (13) is a sufficiently active metabolite to be a precusor of both gentianine (12) and gentianidine (29).

13   gentioflavine

29   gentianidine

12   gentianine

A more interesting aspect of the monoterpene alkaloids, particularly those derived from a secoiridioid precursor, is their biogenesis, since this aspect has not been investigated experimentally.

The structure diversification of the monoterpene alkaloids can be accounted for in terms of the many alternative cyclizations of the ester trialdehyde 36, which is the ring-opened aglycone of secologanin. These ideas are expanded in Scheme 7, where the principal alkaloids are shown.

An important distinction should be made here between the diverse indole alkaloids, where rearrangement takes place at the alkaloid level, and the monoterpene alkaloids, where rearrangement appears to occur at the iridoid level.

11.2.3   Pharmacology

Much of the original interest in gentianaceous plants arose because of the widespread indigenous use of gentian in Europe. At the turn of the century *Gentiana* spp. were current in at least 20 pharmacopaeias as tonics.

Gentianine (12) is the most widely studied of this group of alkaloids. It is not particularly toxic and in low doses exhibits a central nervous system stimulant action. It also exerts hypotensive, anti-inflammatory, and muscular relaxant actions. The reputed tonic effects of gentian are probably due to the hypotensive and muscle relaxant actions.

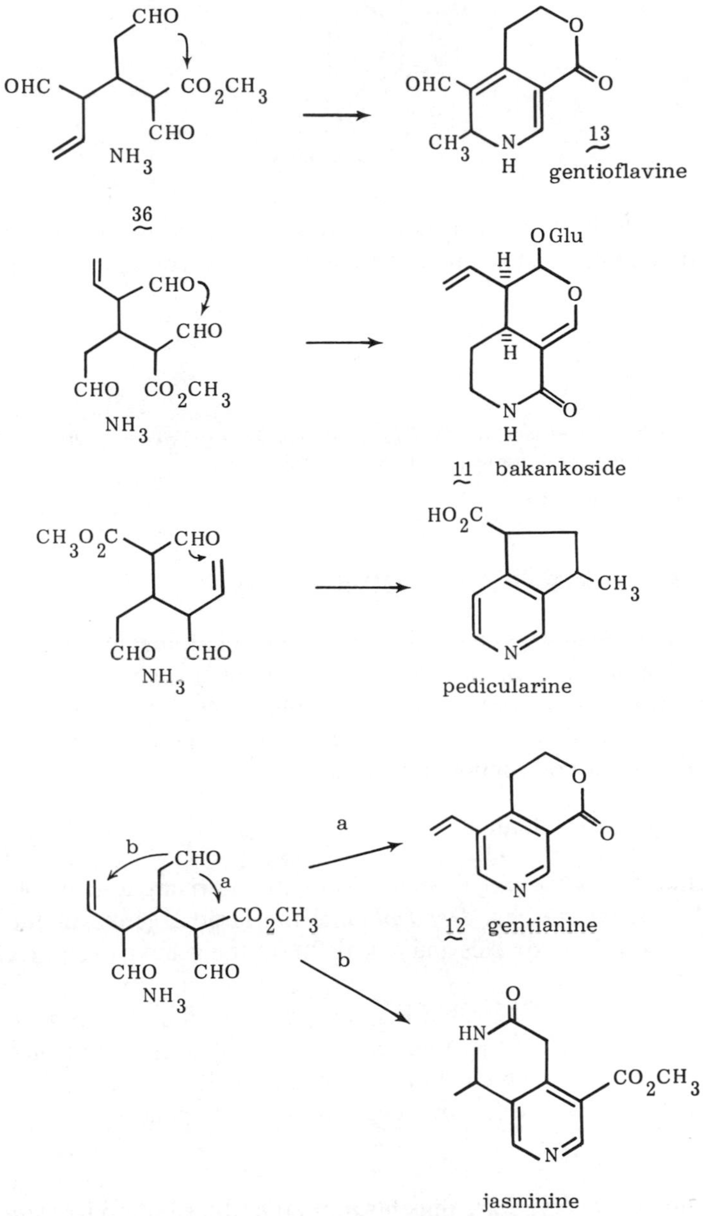

Scheme 7 Biogenesis of the monoterpene alkaloids derived from secologanin.

37   tecomine

Skytanthine (**8**) is tremorigenic, tecomine (**37**) shows hypoglycemic activity, and actinidine (**9**) is a potent attractant for several species of Felidae.

LITERATURE

Reviews

Snieckus, V., in K. F. Wiesner (Ed.), *MTP International Review of Science, Organic Chemistry*, Ser. 2, Vol. 9, *Alkaloids*, Butterworths, London, 1976, p. 191.

Cordell, G. A., *Alkaloids NY* **16**, 432 (1977).

11.3   SESQUITERPENE ALKALOIDS

There are three major classes of sesquiterpene alkaloid, the dendrobine (**38**) type, the alkaloids of *Nuphar* such as deoxynupharidine (**39**), and the complex celastraceous alkaloids such as maytoline (**40**). They do not represent a biogenetically simple group, and cannot be discussed in the same organized way as can the monoterpenoid alkaloids.

11.3.1   *Dendrobium* Alkaloids

The Orchidaceae is the largest plant family, comprising at least 900 genera and nearly 35,000 species. *Dendrobium*, the largest genus of the family, consists of about 1100 species and nearly 200 of these have been investigated for alkaloids.

The Chinese drug "Chin-Shih-Hu" contains the plant *Dendrobium nobile* Lindl. (Orchidaceae), and is used as a tonic. The principal alkaloidal constituent is dendrobine, which although isolated in 1932, was not structurally defined until 1964 after extensive degradation and spectral work.

Chemistry

The alkaloids of this type are thus far limited in distribution to *Dendrobium* species, even though the structurally related picrotoxinin (**41**)-type compounds are obtained from plants in the Menispermaceae. *Dendrobium* species also give rise to other alkaloids [e.g. crepidine (**42**) from *D. crepidatum* Griff.], which appears to be derived from shikimate, acetate, and possibly ornithine.

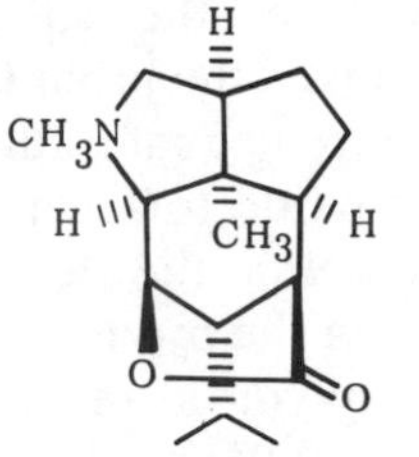

38 dendrobine

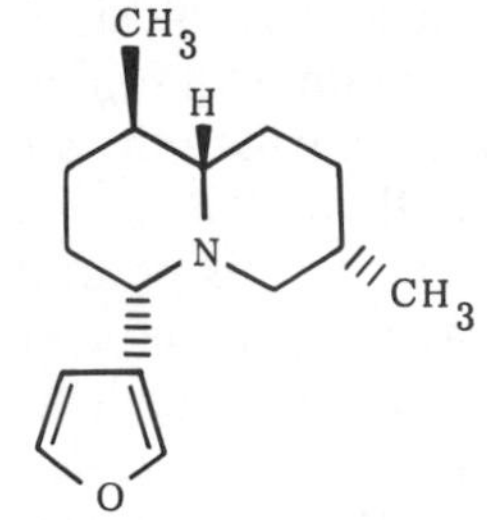

39 deoxynupharidine

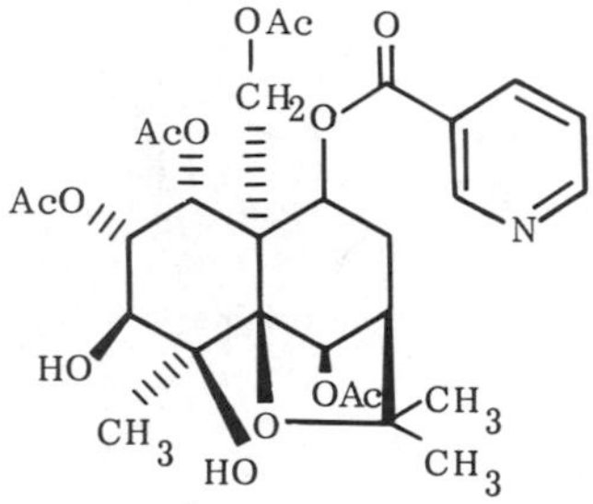

40 maytoline

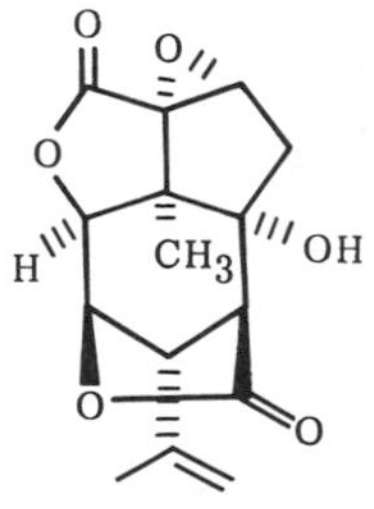

41 picrotoxinin

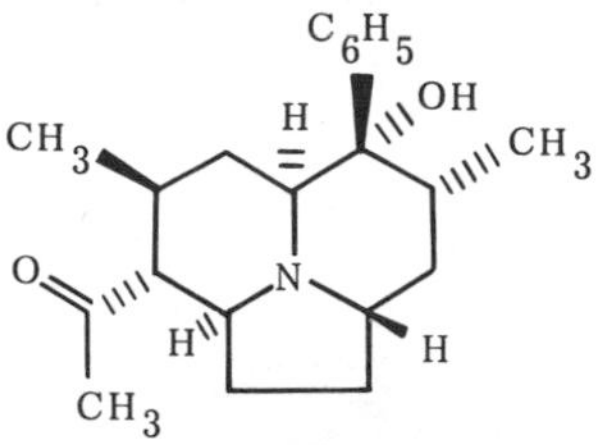

42 crepidine

Several syntheses of the dendrobine nucleus have been achieved; the one described here is due to Kende and co-workers. The all-*cis* nature of the three main rings plus the stereochemistry of lactone makes dendrobine an almost globular molecule. Such stereochemical features can be used to good effect in developing synthetic strategies.

Four of the seven asymmetric carbons of dendrobine are introduced early in the synthesis by construction of the pyrrolidine hydrindone network. A Diels-Alder reaction of the quinone **43** with butadiene provides two of these. After protection of the diosphenol, the *cis*-disubstituted olefin **44** is cleaved and the aldehydes allowed to undergo aldol condensation. One of the two products is **45** which is now set up for reductive amination to afford **46** having the required stereochemistry. The *cis* stereochemistry of the hydrindan system effectively limits the stereochemical possibilities for reductive amination. Reduction of the ketone **46** with lithium aluminum hydride followed by acid hydrolysis and dehydration gives the α,β-unsaturated ketone **47**. Reaction with lithium divinylcuprate gives the product **48** of conjugate addition, in which the vinyl group has entered equatorially. The two-carbon unit is then cleaved with ruthenium tetroxide to a carboxylic acid and methylated. At this point it is necessary to define the sterochemistries of the carbomethoxyl and isopropyl groups. One of the products on reaction of **49** with methoxide–methanol is **50**, and borohydride reduction of this compound gives the axial alcohol. Chromatography on silica resulted in spontaneous cyclization to dendrobine (**38**) (Scheme 8).

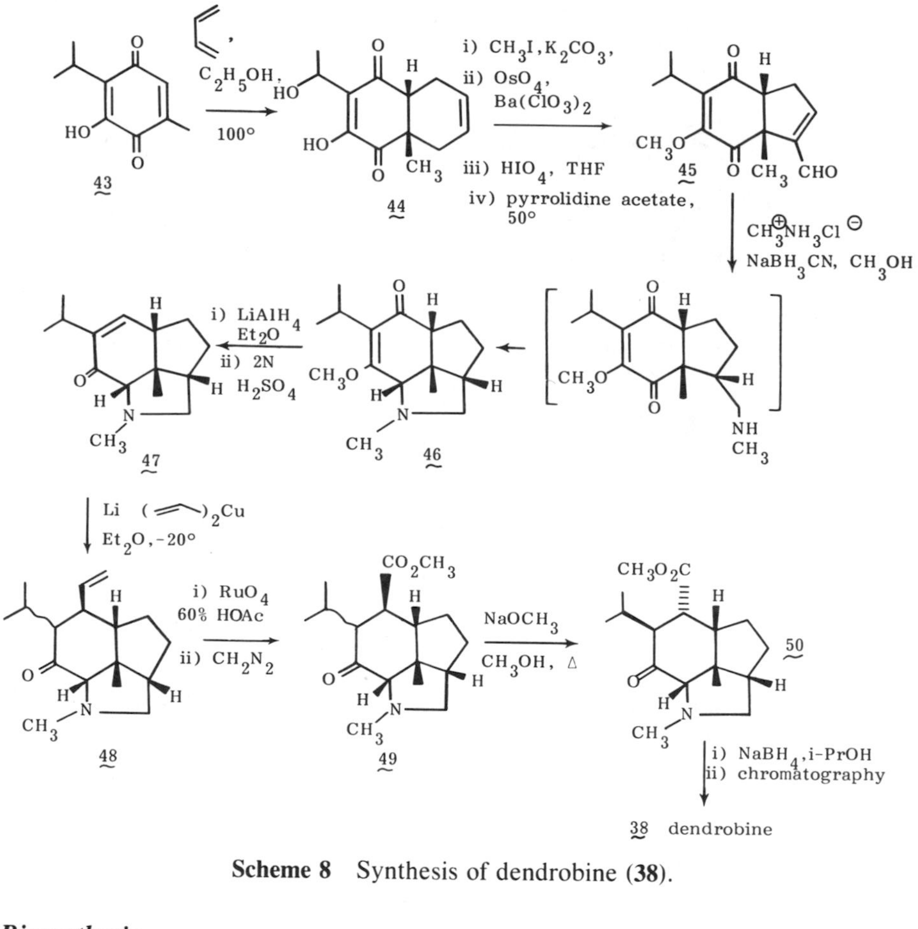

Scheme 8   Synthesis of dendrobine (**38**).

Biosynthesis

The *Dendrobium* alkaloids have been classed as sesquiterpenes, but what is the evidence for this? Most of the biosynthetic information concerning these alkaloids has come from studies not of dendrobine, but of tutin (**51**) or picrotoxinin (**41**).

The first biosynthetic experiments were reported in 1966 when it was

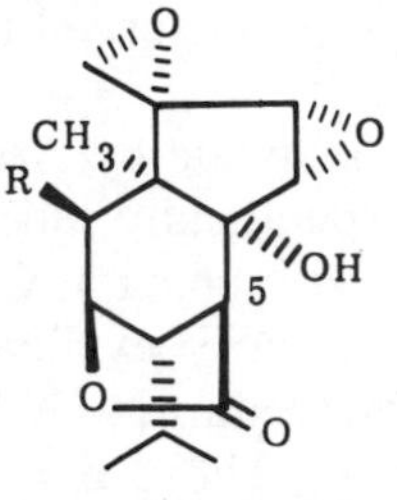

51 tutin, R=OH
53 coriamyrtin, R=H

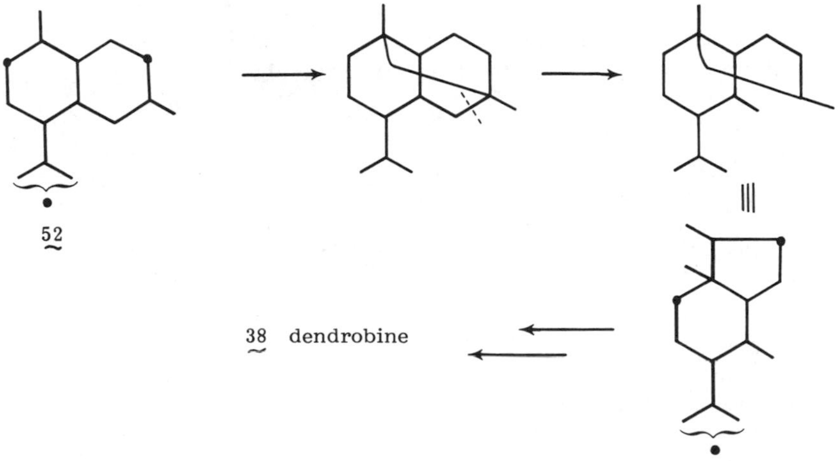

52

38 dendrobine

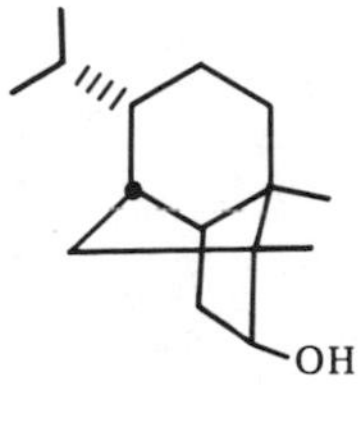

54 copaborneol

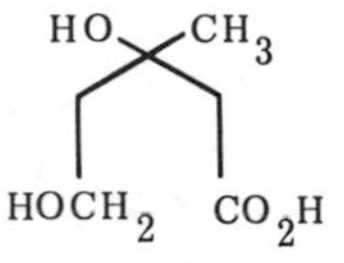

55 mevalonic acid

Scheme 9

shown that dendrobine (**38**) was specifically derived from three mevalonate units. On this basis the Japanese workers suggested Scheme 9 to explain the observed specificity from the cadalane skeleton **52**. Further studies with tutin (**51**) and coriamyrtin (**53**) showed that not all three mevalonate units were incorporated equally, probably indicating different pool sizes for IPP and DMAPP. Labeling of the terminal carbons of the isopentenyl unit was almost equal. Normally this result is only observed in the iridoid series of compounds.

The next significant observation was made by Arigoni and co-workers, who showed that copaborneol (54), specifically labeled as shown, was a precursor of tutin (51), 90% of the activity being at C-5.

Further information concerning the early cyclization steps and intermediate hydride shifts were obtained by feeding [4-^{14}C]mevalonate (55), [4R-^{3}H] mevalonate, and [5-^{3}H]mevalonate. These experiments showed that probably cis,trans-farnesol (56) was a key precursor, proceeding via a 1,3-hydride shift in the intermediate germacrene 57 to 58. Crucial in this interpretation was that from [5-^{3}H$_2$]mevalonate one-sixth of the tritium was specifically lost, and that activity was located at C-3, C-5, C-8, and C-11 in dendrobine (38) (Scheme 10). Additional work established that it is the pro-5R proton which migrates to the isopropyl side chain.

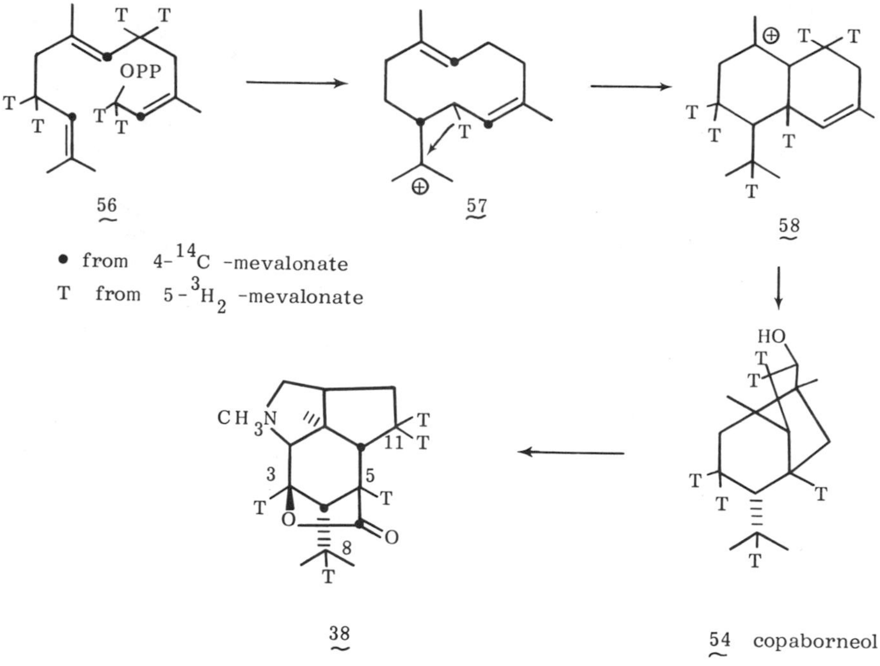

• from   4-^{14}C -mevalonate

T   from   5-^{3}H$_2$ -mevalonate

Scheme 10

Pharmacology

Dendrobine exhibits a number of quite marginal pharmacologic actions. It lowers blood pressure and has weak antipyretic and analgesic actions. In large doses cardiac activity is reduced. Intravenous injection produces convulsions.

11.3.2   The *Nuphar* Alkaloids

The rhizomes of the water lilies *Nuphar japonicum* DC. and *N. luteum* Sibth. (Nymphaceae) were first shown to contain alkaloids in 1931. The structures of the alkaloids have proved to be very interesting and were initially determined by degradative reactions. There are two classes of these alkaloids, a monomeric group of minimum general formula $C_{15}NO$, and a group of dimeric alkaloids having a minimum formula $C_{30}N_2S$.

Monomeric Alkaloids

Deoxynupharidine (**39**), an oil, is one of the most important of the monomeric alkaloids obtained from *N. japonicum*. The carbon skeleton was deduced by Hofmann degradation and hydrogenation to a saturated product containing a tetrahydrofuran ring. Palladium-catalyzed dehydrogenation of deoxynupharidine followed by oxidation with alkaline permanganate gave pyridine 2,5-dicarboxylic acid.

The relative configurations of the methyl and furyl groups have been established in several ways. In one of these, two-stage Hofmann degradation of deoxynupharidine gave a furanodiene, which on permanganate oxidation gave (−)-2-methyl adipic acid (**59**) having the R configuration. These results coupled with X-ray analysis established the absolute configuration to be that shown in **39**.

39

59   (-)-(R)-α-methyl–
      adipic acid

Although several of the monomeric alkaloids of *Nuphar* are quinolizidines, a number of bases have been isolated which contain a piperidine side chain; one example is nuphamine (**60**) from *N. japonicum*. The relative configurations of the groups on the piperidine ring were deduced when **60** gave (−)-deoxynupharidine (**39**) and (−)-7-epi-deoxynupharidine (**61**) on catalytic hydrogenation.

As expected, the initial steps in the mass spectral fragmentation of deoxynupharidine (**39**) is rupture of the four bonds β to the nitrogen. The most important of these alternative ions is **62**, which undergoes further loss of an ethyl group to **63** and retro Diels–Alder reaction to **64**, or loss of the piperidine ring in various ways as shown (Scheme 11).

Thiodeoxynupharidinols such as **65** display fungicidal activity.

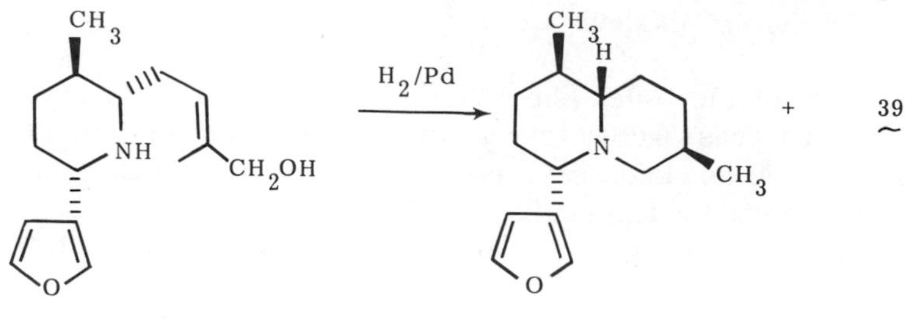

60 nuphamine

61

Dimeric Alkaloids

There are now at least 10 dimeric sulfur-containing alkaloids known, whose distribution is limited to *N. luteum* and a subspecies *N. luteum* subsp. *macrophyllum*.

Thiobinupharidine is a crystalline solid having an NMR spectrum showing two C-CH$_3$ groups (0.91 ppm, $J = 5$ Hz), six furan protons, two equatorial C-6 protons, six CH-N protons, and a pair of doublets ($J = 11.5$ Hz) at 2.32 ppm attributed to a -CH$_2$S group. The absence of additional *C*-methyl groups suggests that the sulfur is involved in a linkage between two nupharidine units through the methyl group at C-7.

The structure **66** proposed for thiobinupharidine has been confirmed by X-ray analysis of the dihydrobromide dihydrate derivative.

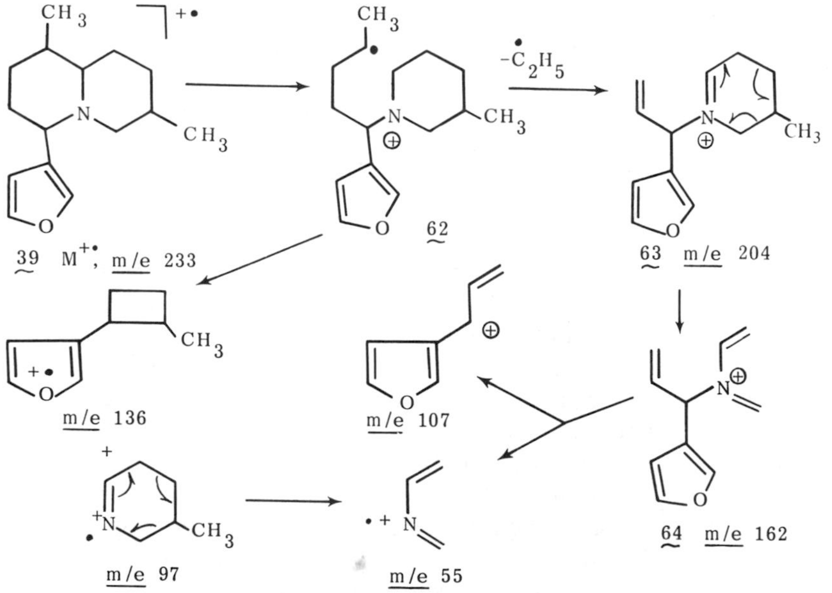

Scheme 11 Mass spectral fragmentation of deoxynupharidine (**39**).

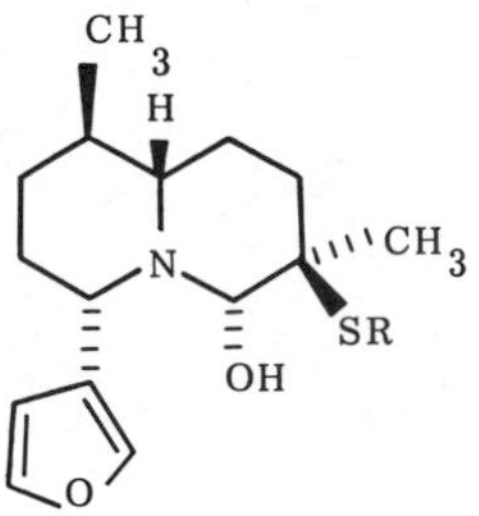

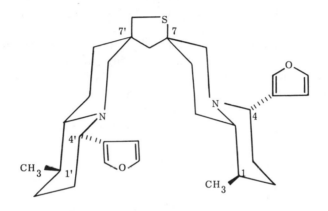

65

R=alkyl, alkenyl and
cyclohexyl

66   thiobinupharidine

A number of the dimeric *Nuphar* alkaloids display weak antibacterial activity.

Castoramine

Closely related to the *Nuphar* alkaloids is castoramine from the Canadian beaver. First isolated by Lederer, who assigned the molecular formula $C_{15}H_{23}NO_2$, the alkaloid was investigated further and shown to form a mono-acetate and be easily convertible to "deoxynupharidine." Since the NMR spectrum showed only one secondary methyl group, the hydroxyl group should be on the C-1 or C-7 methyl group. The alkaloid is generally considered to have the structure **67**, but this has never really been proven.

11.3.3   Celastraceae Alkaloids

This is a relatively new group of alkaloids in terms of structure analysis, although their existence has been known for 40 years. The first alkaloids were isolated in 1947 from the spindle tree *Euonymus europaeus* L. and in

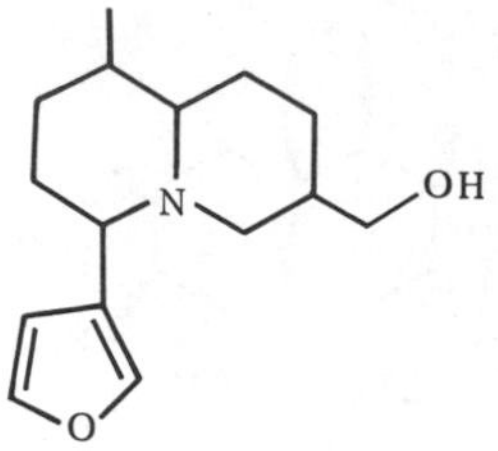

67   castoramine

1950 from the thunder god vine, *Tripterygium wilfordii* Hook. Yet another plant that has yielded sesquiterpene alkaloids is Khat, *Catha edulis* Forsk., although the stimulatory action is due to norpseudoephedrine (Chapter 9). Most of these alkaloids have been obtained from the seeds of the appropriate plant.

These unrelated isolations did not really gel until Kupchan and co-workers determined the structures of maytoline (**40**) and maytine (**68**) from *Maytenus ovatus*. Since this time over 20 alkaloids have been added to this group; some are monoesters, others are macrocyclic diesters.

Evonine (**69**) is the major alkaloid of *E. europaeus*. Hydrolysis gave five equivalents of acetic acid, evoninic acid (**70**), and an alcohol evoninol (**71**).

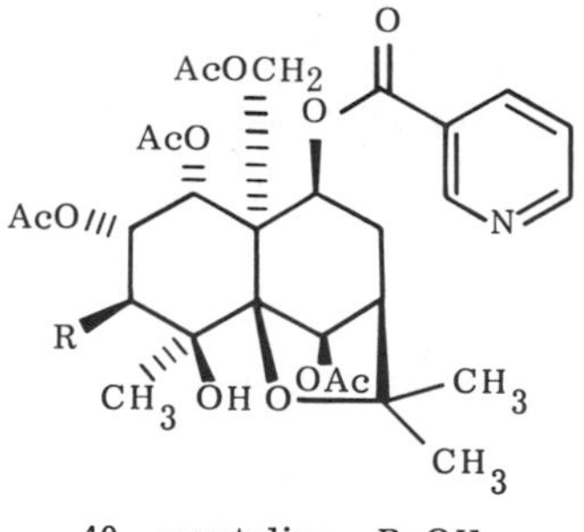

40   maytoline, R=OH
68   maytine, R=H

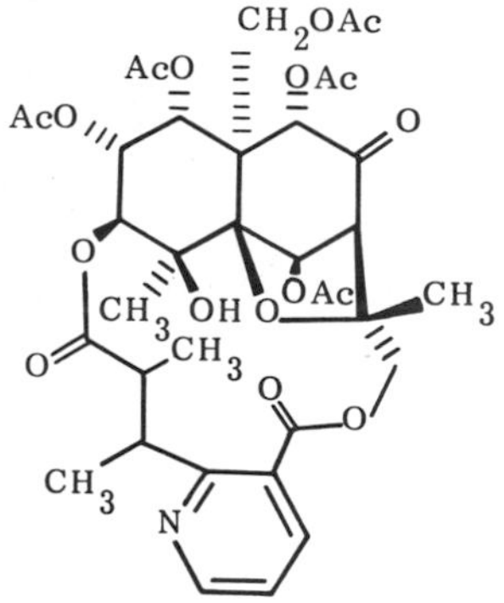

69   evonine

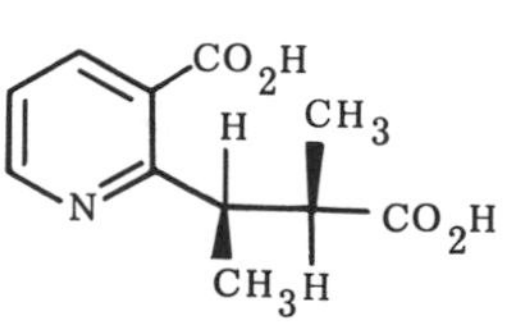

70   evoninic acid

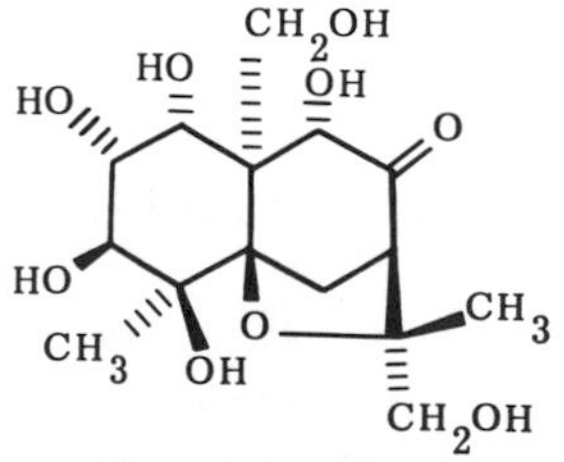

71   evoninol

There are several important points about this group of compounds; (*a*) that these alkaloids are limited in distribution to the Celastraceae, and (*b*) there are no other groups of sesquiterpenes which contain such highly oxygenated skeleta. Even the sterochemistries of some of the parent alcohols have some interesting features. The C-4 methyl group is always axial and with few exceptions the C-1, C-2, C-3, and C-6 hydroxyl groups are equatorial, axial, and equatorial respectively.

Very little biosynthetic information is available, although the pyridine ring is derived from nicotinic acid. The biosynthesis of the five-carbon side chain on the nicotinic acid residue is unknown.

11.3.4 Miscellaneous Alkaloids

Confirmation of the structure of fabianine (**72**) from *Fabiana imbricata* Ruiz and Pav. came from a facile synthesis beginning with the pyrrolidine enamine of pulegone (**73**) (Scheme 12). Some of the proton NMR data are shown in **74**.

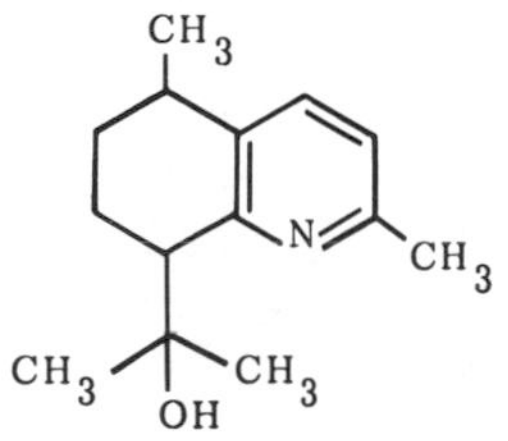

72   fabianine

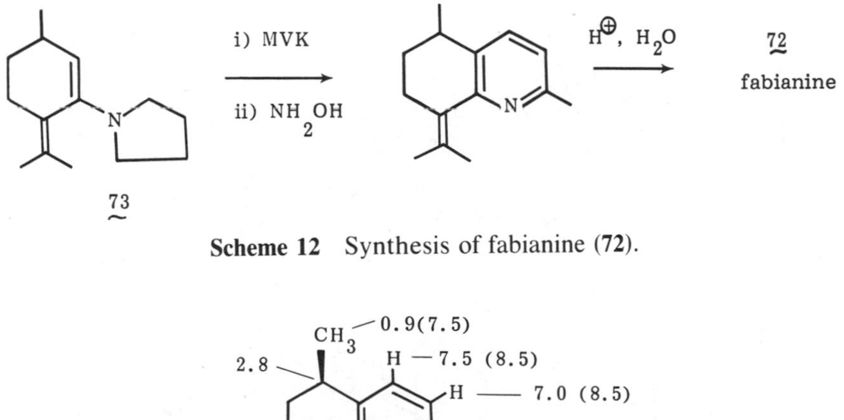

Scheme 12   Synthesis of fabianine (**72**).

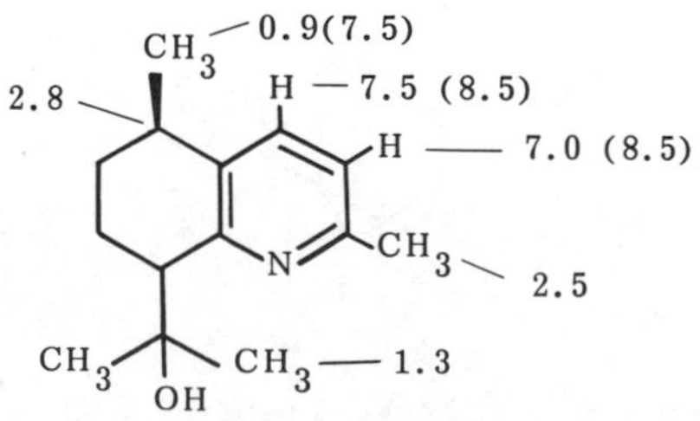

74

Guaipyridine (**75**) was obtained together with patchouli pyridine (**76**) from *Pogostemon patchouli* Pellet. (Labiateae). Isomerization of guaiol (**77**) gave **78**, which by ozonolysis afforded a diketone **79**. Condensation with hydroxylamine followed by elimination of water then gave **75**, identical to the natural product. The principal NMR assignments are shown in **80**.

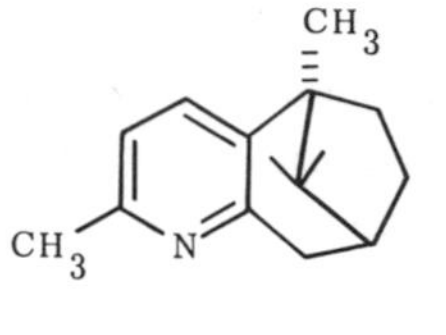

76 patchouli pyridine

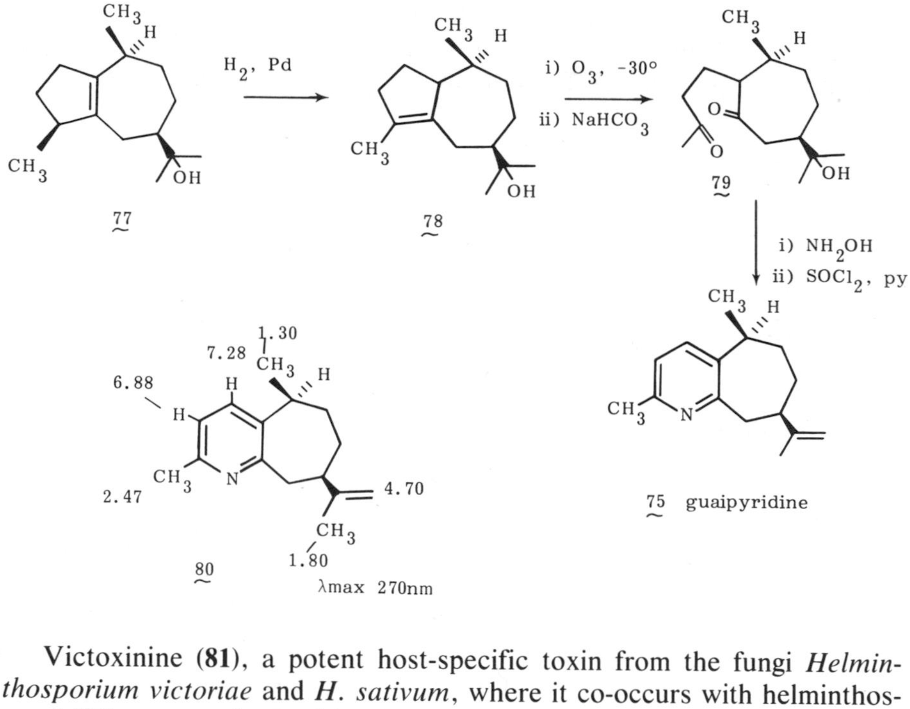

77

78

i) O$_3$, -30°

ii) NaHCO$_3$

79

i) NH$_2$OH

ii) SOCl$_2$, py

75 guaipyridine

80

λmax 270nm

Victoxinine (**81**), a potent host-specific toxin from the fungi *Helminthosporium victoriae* and *H. sativum*, where it co-occurs with helminthosporal (**82**), was synthesized from prehelminthosporal (**83**). Pulchellidine (**84**), an alkaloid having anti-inflammatory activity from *Gaillardia pulchella*

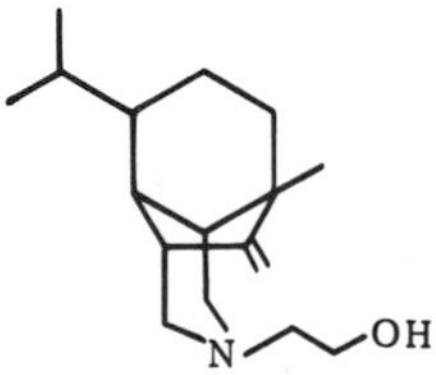

81 victoxinine

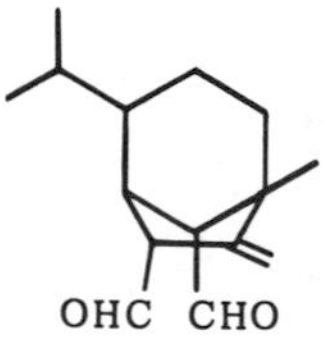

82 helminthosporal

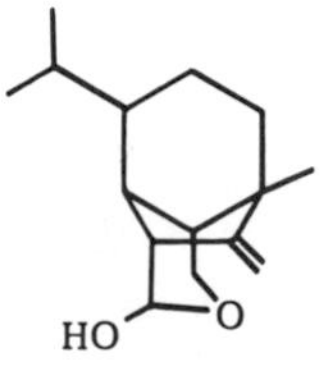

83 prehelminthosporal

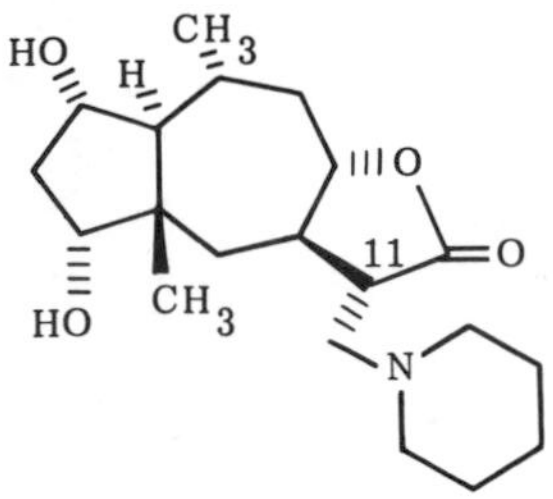

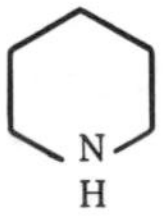

84 pulchellidine

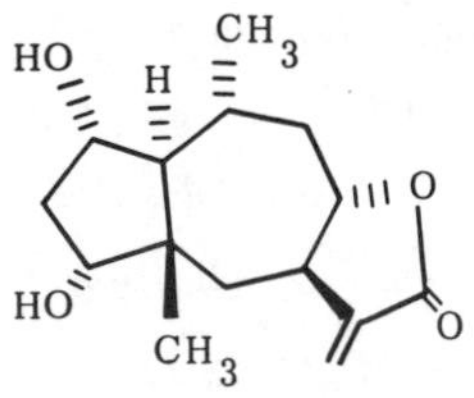

85 neopulchellidine, 11-epi

86

87 pulchelline

Fouger. (Compositae), co-occurs with neopulchellidine (**85**). However, only **84** was formed *in vitro* by the Michael addition of piperidine **86** to the sesquiterpene lactone pulchelline (**87**).

LITERATURE

Reviews: General

Edwards, O. E., in W. I. Taylor and A. R. Battersby (Eds.), *Cyclopentanoid Terpene Derivatives*, Marcel Dekker, New York, 1969, p. 357.

Snieckus, V., in K. F. Wiesner (Ed.), *MIP International Review of Science, Organic Chemistry*, Ser. 2, Vol. 9, *Alkaloids*, Butterworths, London, 1976, p. 191.

Dendrobine

Porter, L. A., *Chem. Revs.* **67**, 441 (1967).

Synthesis

Inubushi, Y., T. Kikuchi, T. Ibuka, K. Tanaka, I. Saji, and K. Tokano, *Chem. Commun.* 1252 (1972); *Chem. Pharm. Bull.* **22**, 349 (1974).

Yamada, K., M. Suzuki, Y. Hayakawa, K. Aoki, H. Nakamuro, H. Nagase, and Y. Hirata, *J. Amer. Chem. Soc.* **94**, 8278 (1972).

Kende, A. S., T. J. Bentley, R. A. Mader, and D. Ridge, *J. Amer. Chem. Soc.* **96**, 4332 (1974).

Biosynthesis

Cordell, G. A., *Chem. Revs.* **76**, 425 (1976).

Yamazaki, M., M. Matsuo, and K. Arai, *Chem. Pharm. Bull.* **14**, 1058 (1966).

Biollaz, M., and D. Arigoni, *Chem. Commun.* 633 (1969).

Turnbull, K. W., W. Acklin, D. Arigoni, A. Corbella, P. Gariboldi, and G. Jommi, *Chem. Commun.* 598 (1972).

Corbella, A., P. Gariboldi, and G. Jommi, *Chem. Commun.* 600 (1972).

Corbella, A., P. Gariboldi, and G. Jommi, *Chem. Commun.* 729 (1973).

Nuphar Alkaloids

Wrobel, J. T., *Alkaloids, NY* **9**, 441 (1967).

Wrobel, J. T. *Alkaloids, NY* **16**, 181 (1977).

Celastraceae Alkaloids

Smith, R. M. *Alkaloids, NY* **16**, 215 (1977).

11.4 DITERPENE ALKALOIDS

Although not of commercial or therapeutic significance, the diterpene alkaloids have recently proved to be an area of great interest. Since 1955 tremendous efforts have been made to elucidate the structure and synthesize the most important members of this group of compounds.

Diterpene alkaloids occur predominently in the Ranunculaceae, particularly the genera *Aconitum* and *Delphinium*. They have also been obtained from the genera *Garrya* in the Garryaceae, *Inula* in the Compositae, *Spivaea* in the Rosaceae, and *Anopterus* in the Escalloniaceae. Interest in this group of compounds arises from the use of *Aconitum* roots and leaves (so-called Monkshood) as poisons and in native remedies for the treatment of gout, hypertension, neuralgia, rheumatism, and as a local anesthetic. *Aconitum* species are some of the most poisonous plants known. It is the C_{19}-ester alkaloids which are responsible for toxicity, acting by slowing the heart rate and lowering the blood pressure.

The diterpene alkaloids are conveniently divided into three general classes: (*a*) a group of relatively simple C_{20}-alkaloids having little oxygenation; (*b*) a group of C_{19}-alkaloids substituted by many hydroxyl or methoxyl groups, where some of the hydroxyl groups are esterified; and (*c*) the *Erythrophleum* alkaloids.

11.4.1 C_{20}-Diterpene Alkaloids

The C_{20}-alkaloids often occur as esters (acetate, benzoate), but are relatively nontoxic. The two principal skeleta are the veatchine (**88**) and atisine (**89**)

88 veatchine skeleton   89 atisine skeleton

types. *Garrya* species produce the veatchine skeleton, while the atisine skeleton is found in the alkaloids of *Aconitum* and *Delphinum* species.

Veatchine (**90**) from *Garrya veatchii* Kellogg is a strong base (pK_a 11.5) containing an exomethylene, a C-CH$_3$, and a secondary hydroxy group. Selenium dehydrogenation gave 1-methyl-7-ethylphenanthrene (**91**) and 1-methyl-7-ethyl-3-azaphenanthrene (**92**). Lithium aluminum hydride reduction gave a diol **93**, indicating hydrogenolytic cleavage of an N-C-O linkage. Pyrolysis of **90** gave **94**, which could be reduced and alkylated to the diol **93**, and osmium tetroxide oxidation then gave garryine (**95**).

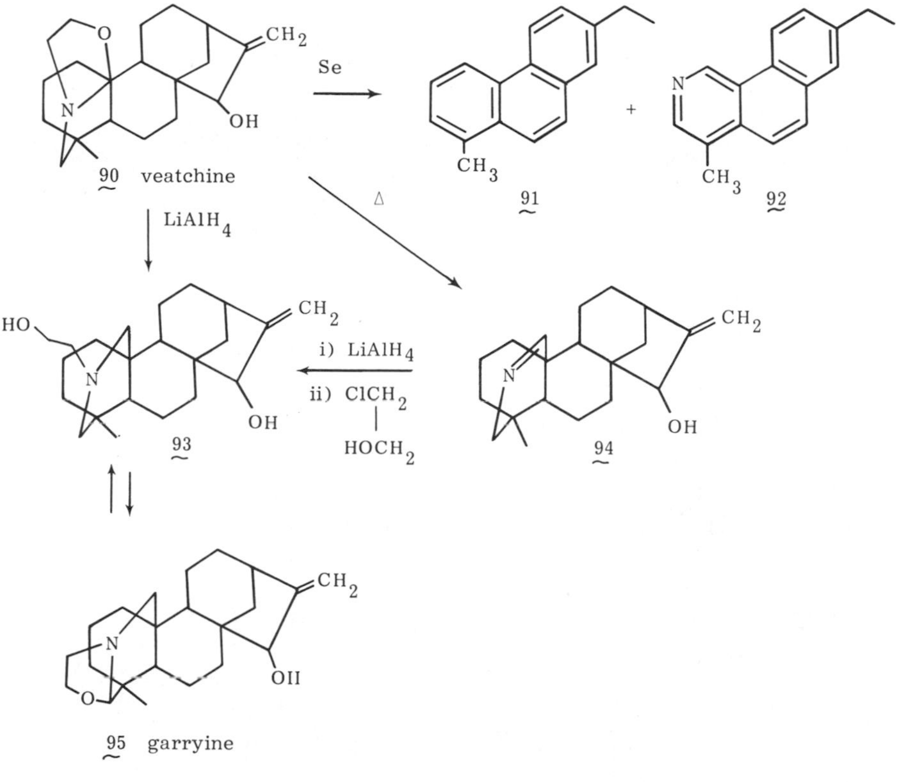

95 garryine

The close structural relationship between atisine and the *Garrya* alkaloids greatly facilitated the structure elucidation of these alkaloids. Atisine (**96**) itself is isomeric with veatchine (**90**) and garryine (**95**), and is the principal alkaloid of *Aconitum heterophyllum* Wall. Like veatchine (**90**) it contains a quaternary C-methyl group (0.70 ppm, s), and exocyclic methylene (5.0 ppm, 2H, m), and a secondary alcohol. Many of the chemical reactions, particularly the reductive ones, are quite analogous with those of veatchine.

The main differences in chemical reaction are the products of selenium dehydrogenation, which were 1-methyl-6-ethylphenanthrene (**97**) and 1-methyl-6-ethyl-3-azaphenanthrene (**98**). On this basis it was suggested that veatchine (**90**) and atisine were isomers in ring D, and that the latter had the structure **96**. In an interesting series of reactions, atisine (**96**) and veatchine (**90**) were correlated through the amide ester **99**.

The facile isomerization of veatchine (**90**) to garryine (**95**) is paralleled by a similar isomerization of atisine (**96**) to isoatisine (**100**) under mild basic conditions. Further proof of the relative configurations of oxazolidine moieties came from partial synthetic work with the imine **101** from atisine as shown in Scheme 13.

The unusual symmetry of the bicyclo[2,2,2]octane system (rings C and

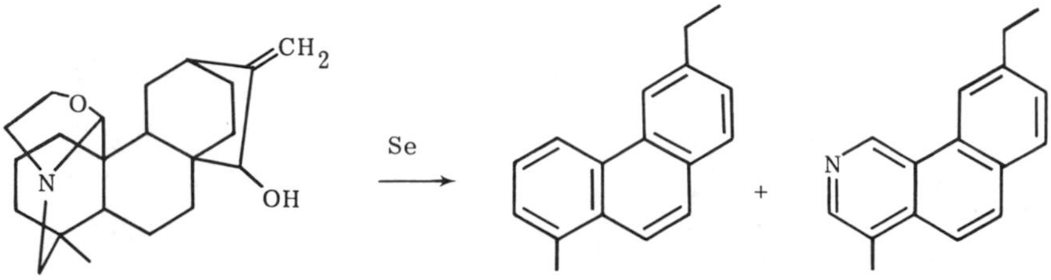
CH2
O
N
OH
Se
+
N

96 atisine 97 98

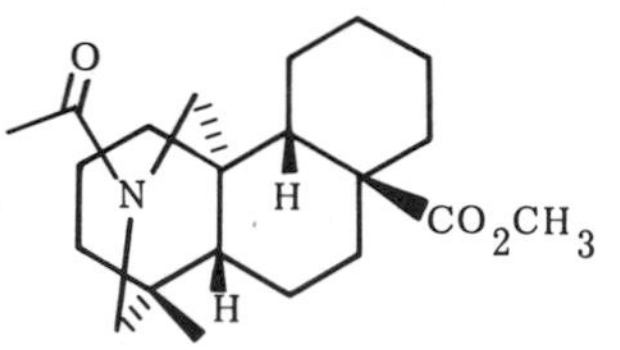
O
N
H
H
CO2CH3

99

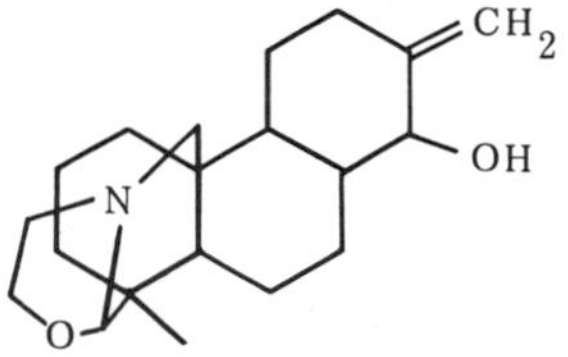
CH2
OH
N
O

100 isoatisine

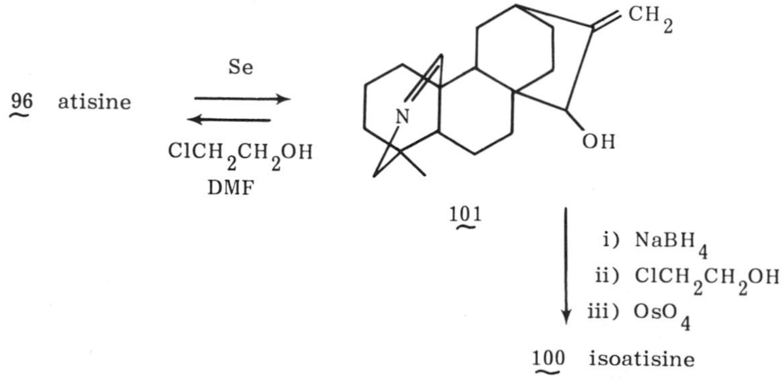
96 atisine
Se
ClCH2CH2OH
DMF
CH2
N
OH
101
i) NaBH4
ii) ClCH2CH2OH
iii) OsO4
100 isoatisine

Scheme 13 Interrelationship of atisine (**96**) and isoatisine (**101**).

D) present in atisine poses an interesting problem for the location of the allylic alcohol, which could potentially be on the bridge either *cis* or *trans* to the piperidine ring (i.e. it could be either **102** or **103**). Because the corresponding ketone gives two products on reduction, location of this group

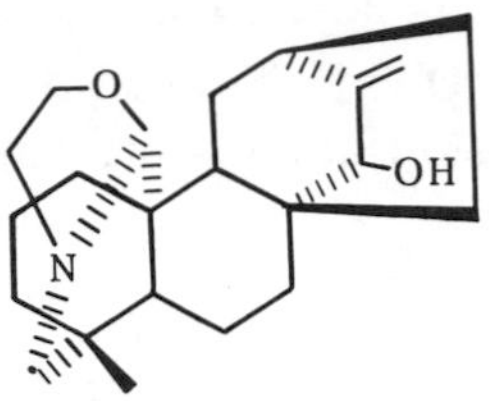

102 cis

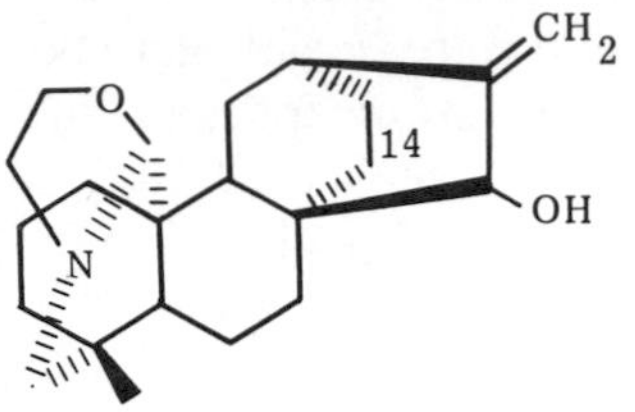

103 trans

at 14, a very crowded position, is eliminated and atisine has the structure **104**.

The stereochemistry of the hydroxyl group at C-15 in these various compounds has been demonstrated in an interesting way. Garryfoline (**105**) on treatment with mineral acid at room temperature is isomerized to cuachichicine (**106**); veatchine (**107**) is stable to these conditions. Since **105** and **107** differ only in the stereochemistry at C-15, the reaction is stereochemically controlled and involves a 15,16-hydride shift (**108**). Veatchine (**107**) therefore has a pseudoaxial hydroxyl group with regard to ring D. Since atisine also undergoes this reaction, it must be epimeric to veatchine at C-15 and have the stereochemistry **104**.

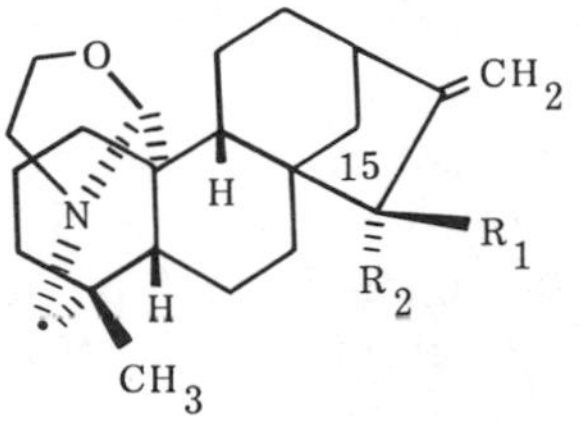

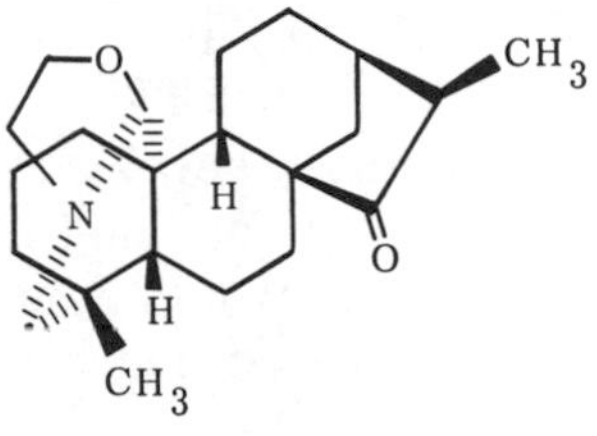

105 garryfoline, R_1=OH; R_2=H   106 cuachichincine
107 veatchine, R_1=H, R_2=OH

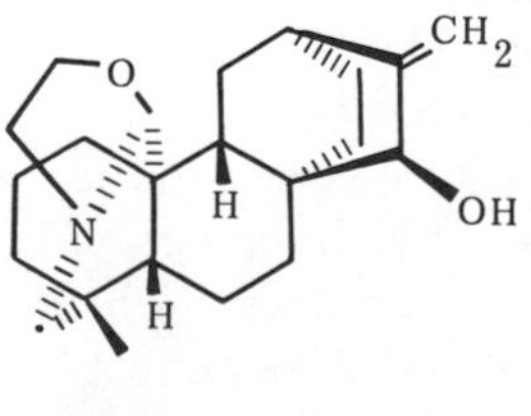

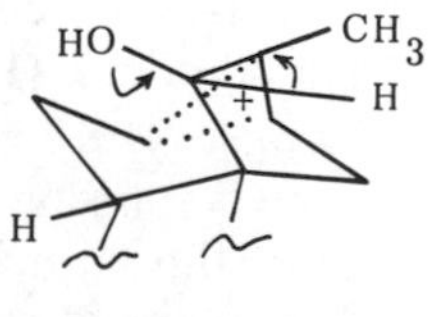

104 atisine

108

One interesting point is that in the NMR spectrum of atisine at room temperature, the C-4 methyl group is observed as two unequal signals in the ratio 2 : 1, which above 85°C occur as a singlet. Thermodynamic calculations indicated an energy barrier of 19.7 kcal/mole between these two species. Two peaks are also observed in the region 3.4–4.0 ppm for the C-20 proton, and were suggested to be signals due to the C-20 α- and β-protons of two interconverting species. On addition of D_2O both the C-20 H and C-4 methyl signals coalesced, indicating the stabilization of a highly polar, possibly even ionic species such as 109.

109

Garryine (95) has been correlated with (−)-kaurene (110) as shown in Scheme 14, and the absolute configuration of atisine was deduced by conversion to the phenol 111, which was enantiomeric with a compound 112 derived from podocarpic acid (113).

The first total synthesis of atisine was completed in 1963 by Nagata et al. and proceeds in 35 steps from 114. Some of the crucial intermediates are shown in Scheme 15. One of the intermediates in this synthesis prior to the vinyl ketone 115 was also used by Nagata and co-workers for a synthesis of veatchine (107). Key in this synthesis is the rearrangement of the brosylate 116 to the bridge ketone 117.

Another approach developed by Wiesner and co-workers develops the A, B, C, and piperidine rings and then forms the D-ring. Some of the intermediates are shown in Scheme 16. The compound 118 is the principal synthetic target, which had been elaborated previously by Masamune to veatchine (107).

These are the principal synthetic routes to the atisine–veatchine general group of compounds, and although many improvements have been made (see reviews) no dramatically new approaches have been described.

In the years since many of the early structures were determined there have been numerous new structures elucidated, but few having the veatchine skeleton. Some examples include denudatine (119) from *Delphinium denudatum* Wall., spiradine G (120) from *Spivaea japonica* L. (Rosaceae), kobusine (121) from several Japanese *Aconitum* species, and several dimeric alkaloids, including staphisagrine (122) from *Delphinium staphisagria* L.

The structure of the latter alkaloid was determined with the aid of the ^{13}C NMR data for atisine (123) and veatchine (124) as shown.

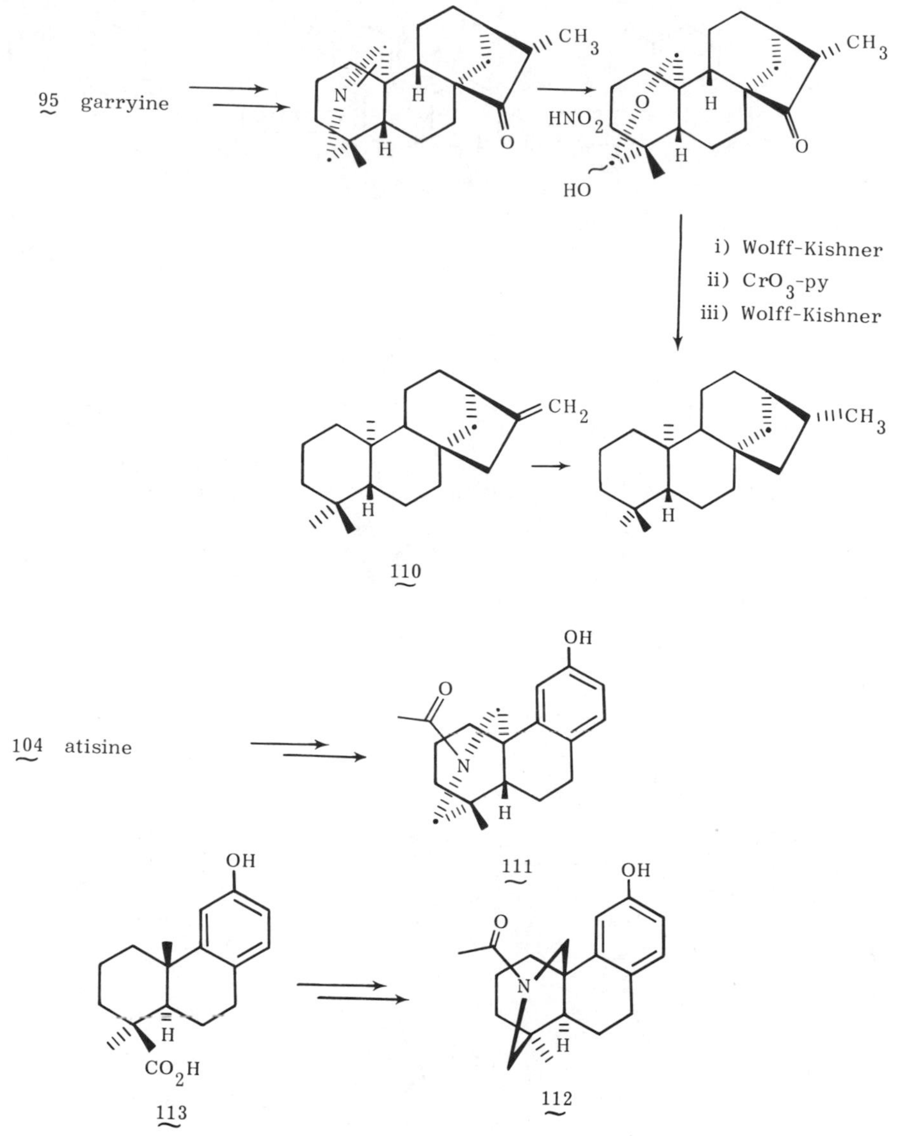
95 garryine
CH3
N
H
H
O
HNO2
O
HO
i) Wolff-Kishner
ii) CrO3-py
iii) Wolff-Kishner
CH2
110
104 atisine
OH
O
N
111
OH
H
CO2H
113
112

Scheme 14

11.4.2 C_{19}-Diterpene Alkaloids

The C_{19}-diterpene alkaloids are classically represented by aconitine, first isolated by Geiger in 1833. They are the highly toxic basic esters and 2–5 mg may be lethal in humans. Two general structure types are found, and are represented by aconitine (**125**) and lycoctonine (**126**) on one hand, and

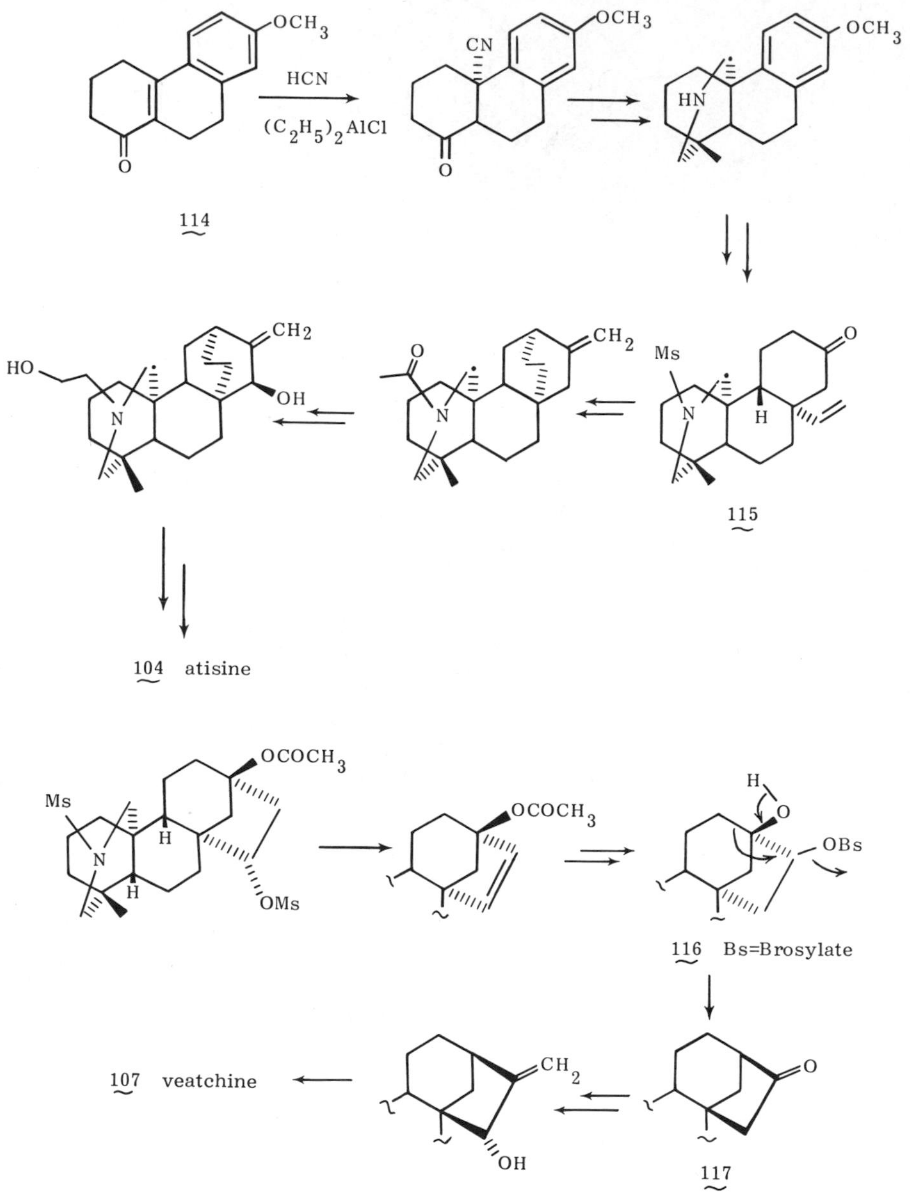

HCN
$(C_2H_5)_2AlCl$
114
115
104 atisine
116 Bs=Brosylate
117
107 veatchine

Scheme 15

a minor type [e.g. heteratisine (**127**)] having a modified lycoctonine skeleton and a lactone ring.

Major differences between the aconitine and lycotonine types are the presence of a 7-oxygenated group in the latter, and the oxygen substituent at C-6 is α in the aconitine series, but β in the lycoctonine series.

The structures and absolute configurations of the parent alkaloids have been confirmed by X-ray crystallography, and many of the recently isolated alkaloids have been chemically correlated with the parent compounds.

A new chemical method for determining the stereochemistry of the C-6 substituent involves neutral permanganate oxidation. In the 6α-methoxyl series N-dealkylation occurs, but in the 6β-series a 19-lactam results. This has been explained in terms of the orientation of attack by the permanganate ion, and the steric availability of the α-hydrogens for oxidative loss (i.e. the direction in which iminium formation can occur).

In a group of alkaloids with so many functional groups it is not surprising to find that there are several interesting chemical reactions which occur.

For example, heating delphinine (128) or any other alkaloid with a C-8

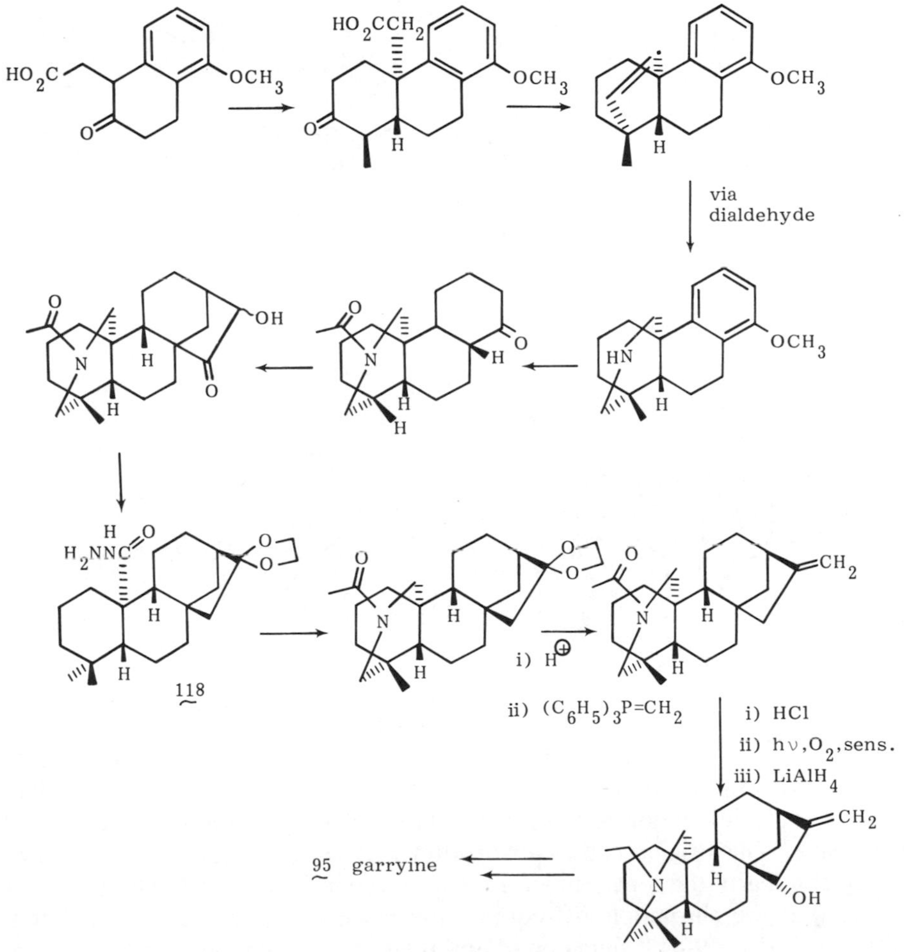

Scheme 16 Synthesis of garryine (95).

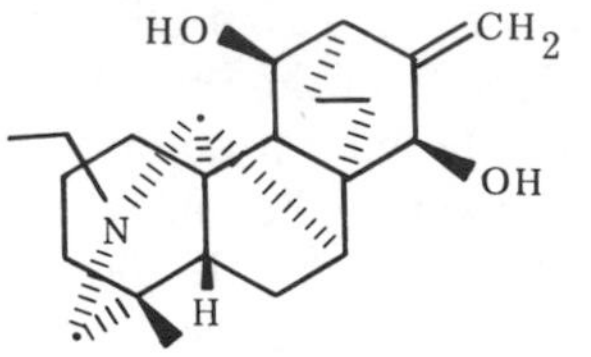

119   denudatine

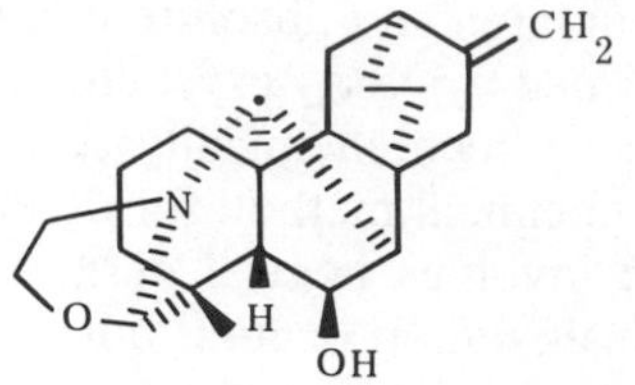

120   spiradine G

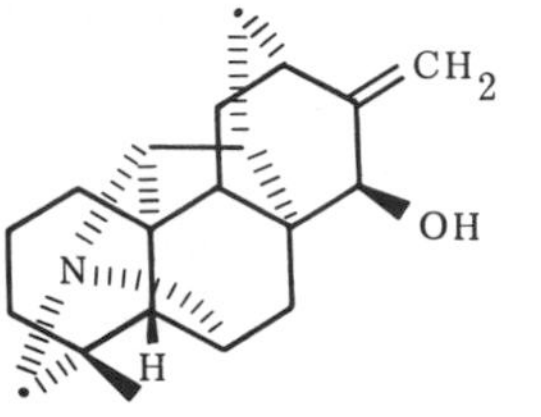

121   kobusine

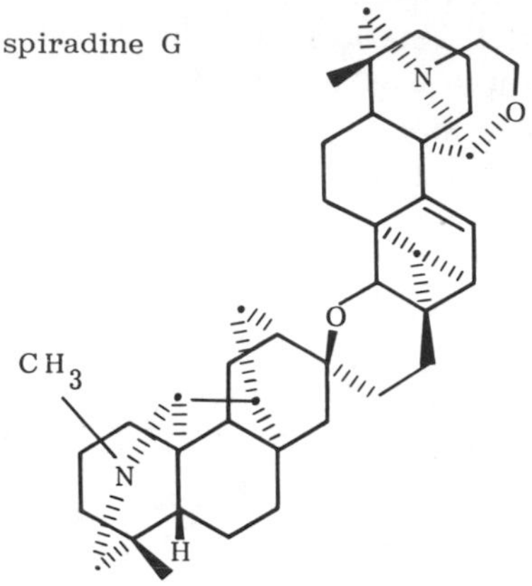

122   staphisagrine

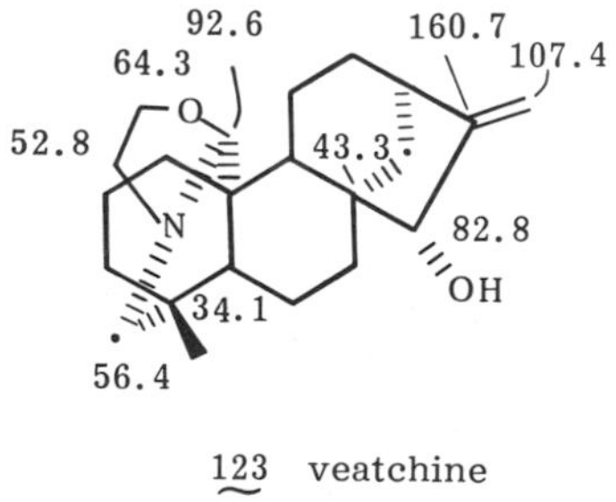

123   veatchine

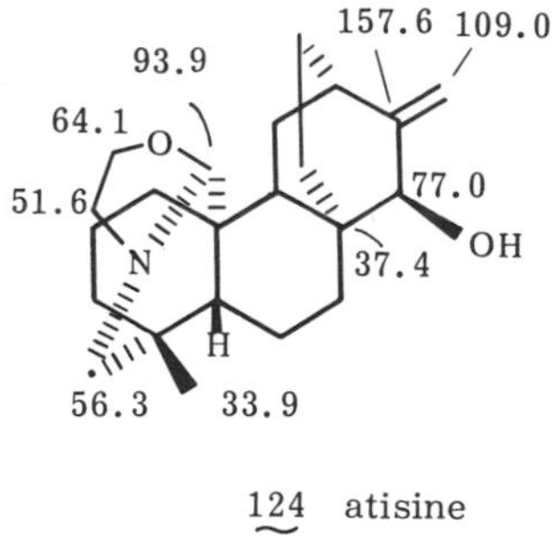

124   atisine

acetate group (and no C-15 substituent) causes loss of acetic acid to give an olefin **129**. The UV spectra of these derivatives are unusual because they show an absorption between 235–245 nm ($\epsilon \sim 6000$). This absorption disappears on acidification or if an *N*-acetyl derivative is used rather than the free base. The origin of this absorption is still somewhat of a mystery, although clearly the lone pair of a basic nitrogen is involved. There is evidence that the chromophore requires the involvement of the lone pair on nitrogen, the C-19–C-7 sigma bond and the C-8–C-15 double bond as shown in **130**.

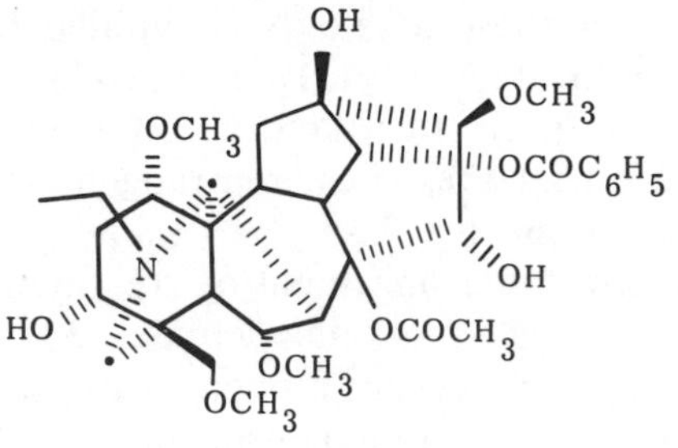

125 aconitine

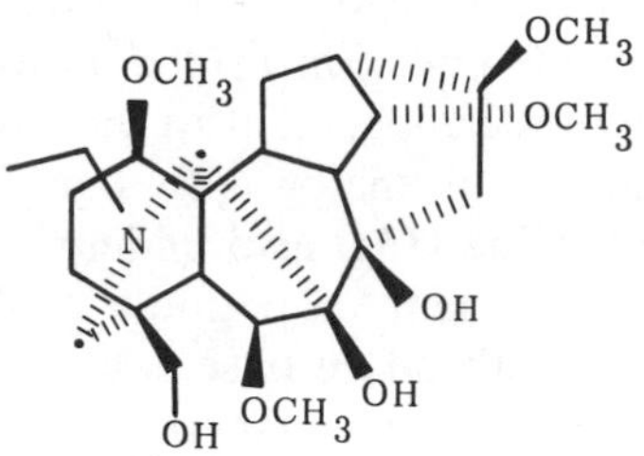

126 lycotonine

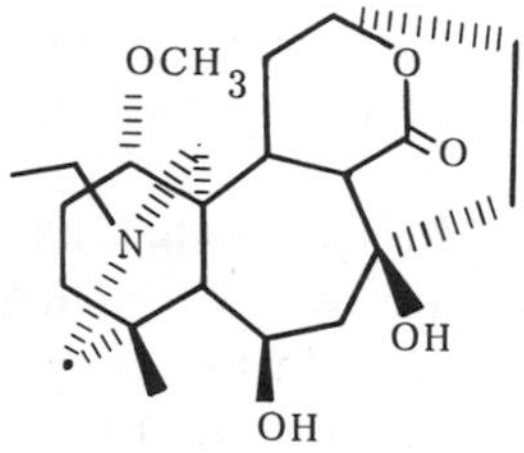

127 heteratisine

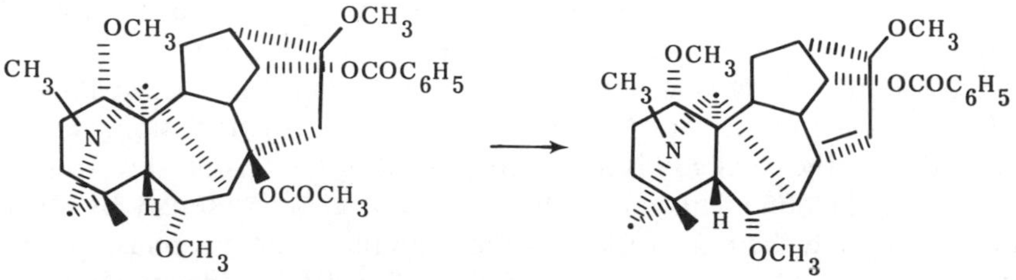

128 delphinine

129

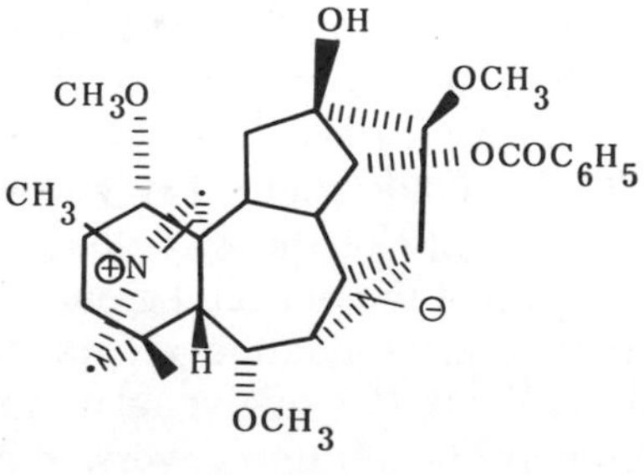

130

Another common rearrangement of these alkaloids is typified by the chromium trioxide–pyridine oxidation of oxonine (131). The product is the result of oxidation of the 14-hydroxy group to a ketone followed by an acyloin rearrangement to the isomeric triol 132. These rearrangements also occur on chromatography over basic alumina.

Oxonine (131) also undergoes a novel rearrangement on treatment with one equivalent of periodic acid to give a seco-ketoaldehyde 133. When heated with dilute base in the presence of oxygen, a phenol 134 was produced. The mechanism proposed for this interesting rearrangement is shown in Scheme 17. Aconitine (104) and delphinine (128) were correlated by this method.

In spite of enormous effort the structure of lycoctonine (126) could not be deduced by chemical methods, but was resolved after X-ray analysis of a derivative. Many of the alkaloids isolated subsequently have been related to derivatives of lycoctonine.

The best known lactone diterpene alkaloid is heteratisine (135) from *A. heterophyllum*, whose structure was determined by a combination of X-ray analysis and chemical and spectral properties.

Pyrolysis of heteratisine acetate (136) affords pyroheteratisine (137) in 90% yield. An unusual feature of this reaction is that dehydroheteroatisine (138) does not undergo this reaction. Several mechanisms have been proposed to account for the apparent involvement of the 6α-hydrogen in the reaction. The most plausible of these is shown in Scheme 18. As yet it has not been verified by deuterium-labeling studies.

The first total synthesis of an alkaloid of the lycoctonine type was reported by Wiesner and associates in 1974.

Talatisamine (139) was obtained from *Aconitum talassicum* Popov. in 1940 and the structure was established by chemical correlation. A crucial step in the synthesis of 139 is a rearrangement of the atisine skeleton to the lycoctonine skeleton. This reaction had been studied previously by Overton and by Ayer and their co-workers. One example is the gas-phase pyrolysis of the β-tosylate 140 derived from atisine (104), which yielded the keto olefin 141 in 70% yield.

The synthesis developed by Wiesner begins with the keto amide 142 having the atisine skeleton, which in a number of standard steps was converted to the tosylate 143. The chromium trioxide–pyridine, sodium borohydride sequence is necessary to invert the alcohol configuration at C-15. When heated at 180°C the amide olefin 144 was produced in 40% yield, together with an equal quantity of the 8,15-double bond isomer. Lithium aluminum hydride reduction of the *N*-acetyl group, deketalization, and stereospecific sodium borohydride reduction set up the molecule for a final oxidative coupling of C-19 with C-8 to give talatisamine (139) (Scheme 19).

The atisine–lycoctonine sketal interconversion has been proposed as a biogenetic process, with C-7 to C-20 bond formation occurring subsequent to the B-C ring rearrangement.

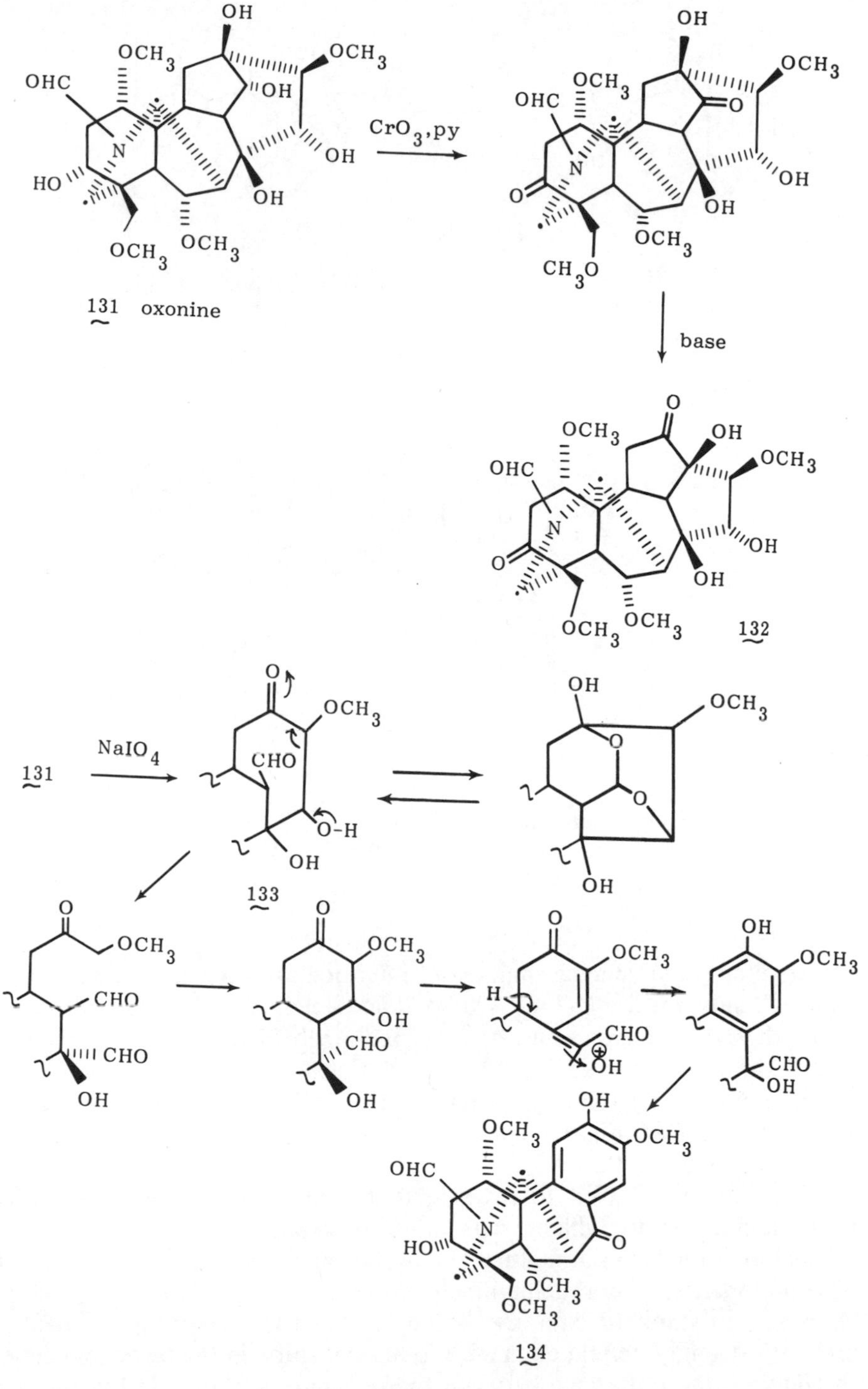
OH
OCH3
OHC
OCH3
OH
N
OH
HO
OH
OCH3
OCH3
131 oxonine
CrO3,py
OH
OCH3
OHC
OCH3
O
N
OH
O
OH
OCH3
CH3O
base
O
OCH3
OH
OHC
OCH3
N
OH
O
OH
OCH3
OCH3
132
131
NaIO4
O
OCH3
CHO
O-H
OH
133
OH
OCH3
O
O
OH
O
OCH3
CHO
CHO
OH
O
OCH3
OH
CHO
OH
O
OCH3
H
CHO
OH
OH
OCH3
CHO
OH
OH
OCH3
OCH3
OHC
N
HO
O
OCH3
OCH3
134

Scheme 17

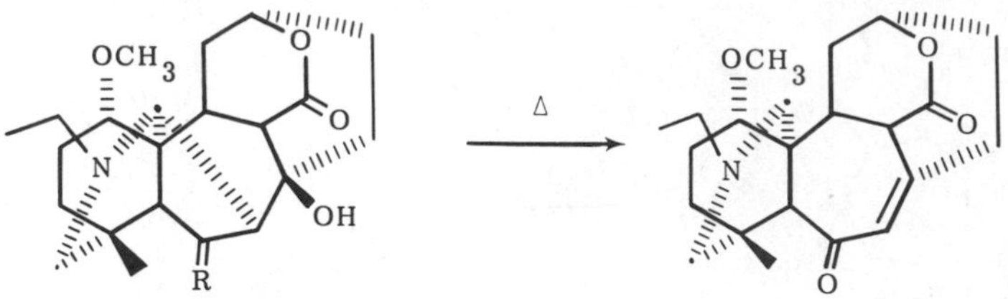

135   R=H,β-OH

136   R=H, β-OCOCH$_3$

138   R= =O

137   pyroheteratisine

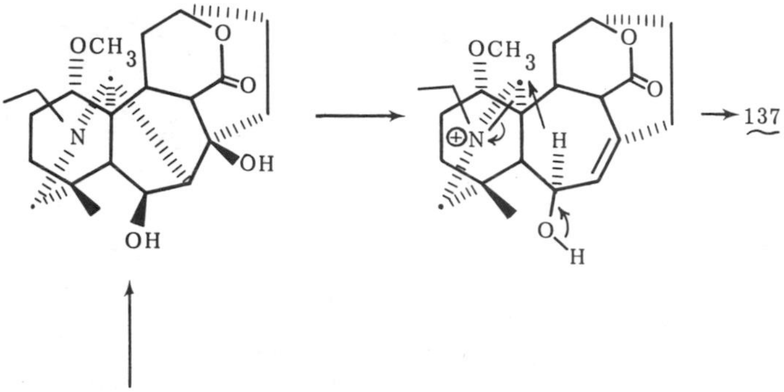

135   heteratisine

Scheme 18

An alternative scheme implies an intermediate such as napelline (**145**) which already has the C-7 to C-20 bond and belongs to the veatchine group rather than the atisine group. A suggested mechanism for this interconversion is shown in Scheme 20.

Pelletier and Djarmati have reported the ^{13}C NMR spectra of several *Aconitum* and *Delphinium* alkaloids, and some of the principal resonances are shown in Table 1.

The data were used to investigate the anomalous UV spectrum of pyrodelphinine (**129**), which it was known from previous work was removed on addition of acid. In a mixture of chloroform : acetic acid (1 : 8) a stepwise protonation of pyrodelphinine occurred.

Based on standard data, the original shift of C-8 is strongly downfield and C-15 strongly upfield compared to normal shifts in the terpenoid series. In addition, the difference between the resonances (30.3 ppm) in the nonprotonated state is intermediate between those of an unconjugated double bond (10–20 ppm) and those of a conjugated double bond (up to 40 ppm). However, in the protonated species the difference between the C-8 and C-15 resonances is reduced to 20.1 ppm.

Thus the structure of pyrodelphinine is a resonance hybrid involving the lone pair on nitrogen, the C-17–C-7 σ bond and π-electon pair of the C-8–C-

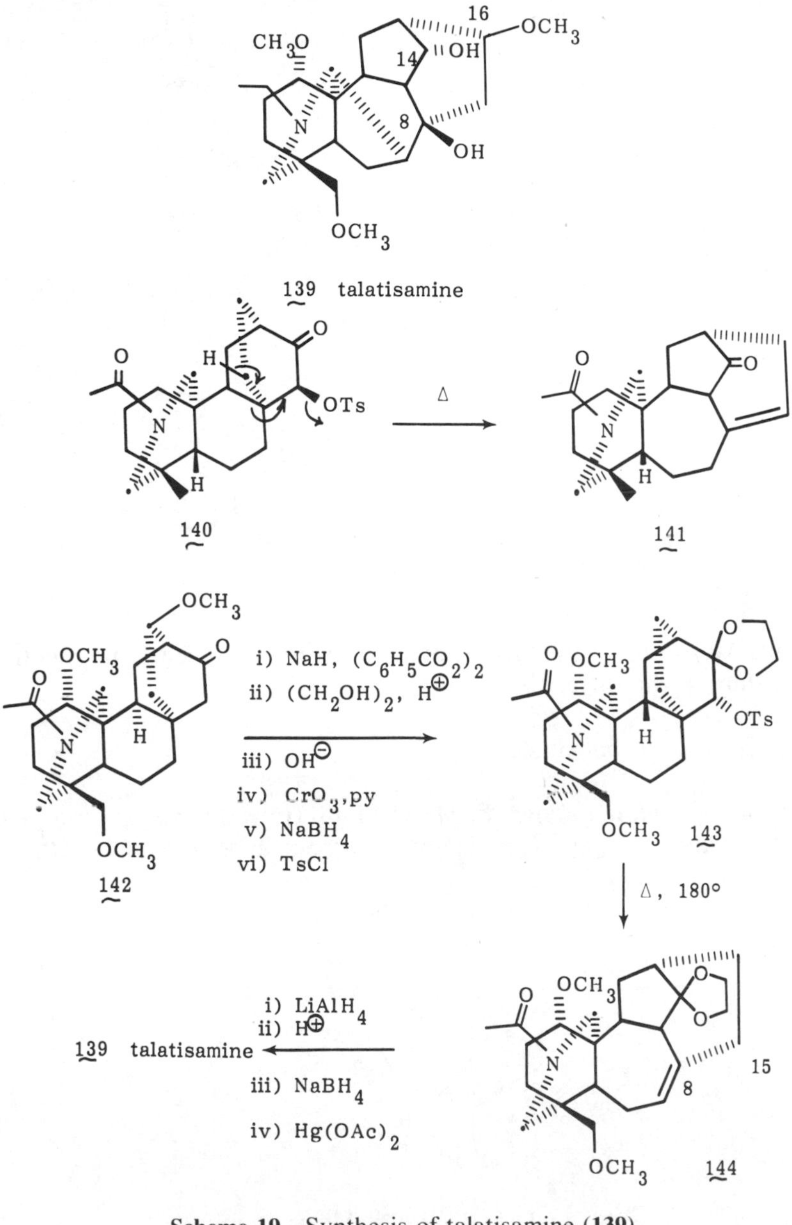

139 talatisamine

140   141

Scheme 19 Synthesis of talatisamine (**139**).

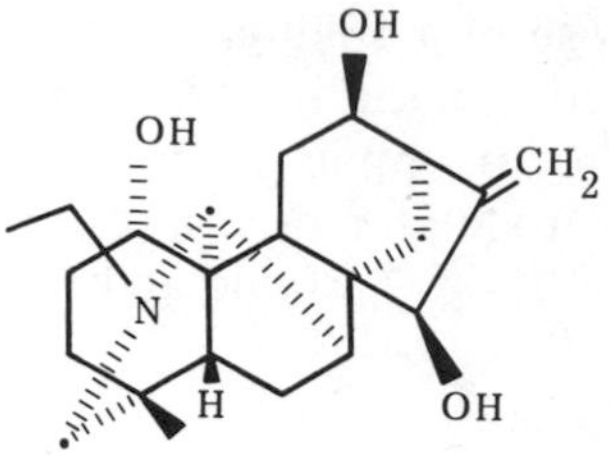

145   napelline

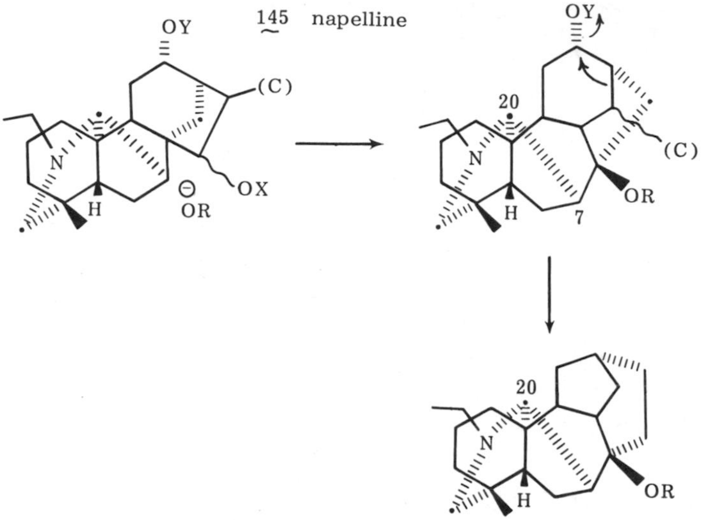

Scheme 20

15 bond. More important this resonance hybrid represents the electronic ground state of the molecule.

11.4.3   *Erythrophleum* Alkaloids

Erythrophleum chlorostachys (F. Muell.) Bail. (Leguminosae) is a large tree native to northern Australia and is noted for being highly toxic to both cattle

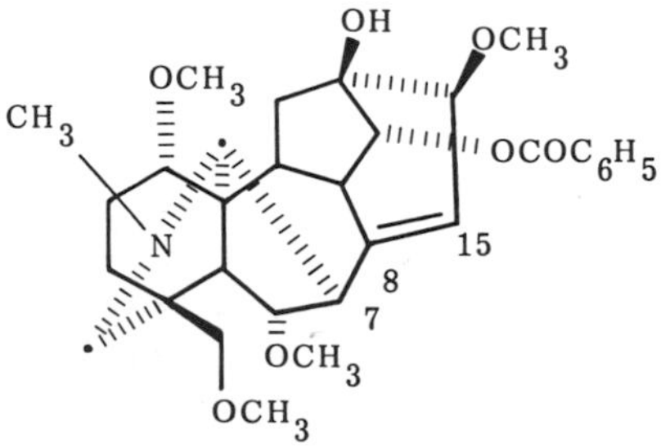

129   pyrodelphinine

Table 1 Carbon-13 NMR Shifts for Some Representative Diterpene Alkaloids

Carbon	Aconitine	Delphinine	Pyrodelphinine	Pyrodelphinine in Acetic Acid (1 : 8 Mole Ratio)
1	83.4	84.9	86.1	81.6
2	36.0	26.3	25.3	23.6
3	70.4	34.7	35.3	31.8
4	43.2	39.3	40.0	40.0
5	46.6	48.8	48.5	48.7
6	82.3	83.0[b]	83.6	82.9[c]
7	44.8[a]	48.2	50.4	47.0
8	92.0	85.4	146.6	142.4
9	44.2[a]	45.1	47.6	47.0
10	40.8	41.0	46.7	45.4
11	49.8	50.2	51.9	51.5
12	34.0	35.7	38.4	36.6
13	74.0	74.8	77.7	77.8
14	78.9	78.9	79.1	77.8
15	78.9	39.3	116.3	122.3
16	90.1	83.7[b]	83.6	82.8[c]
17	61.0	63.3	78.6	76.5
18	75.6	80.2	80.3	78.9
19	48.8	46.1	56.5	58.6
N—CH$_2$ (—CH$_3$)	46.9	42.3	42.7	43.5
| CH$_3$	13.3	—	—	—
1'-CH$_3$	55.7	56.1	56.5	55.5
6'-CH$_3$	57.9	57.6	58.1	58.6
16'-CH$_3$	60.7	58.6	57.1	57.4
18'-CH$_3$	58.9	58.9	59.2	59.2
—C=O	172.2	169.4	—	—
CH$_3$	21.3	21.4	—	—
—C=O	165.9	166.0	168.0	—
C$_6$H$_5$	129.8	130.4	130.5	—
	129.6	129.6	130.0	—
	128.6	128.4	128.1	—
	133.2	132.8	132.7	—

[a-c] Assignments may be reversed.
Source: Data are from S. W. Pelletier and Z. Djarmati, *J. Amer. Chem. Soc.* **98**, 2626 (1976).

and sheep. Two leaves are said to be fatal for a goat. *E. guineense* G. Don, an arboreal of central Africa, is a notorious ordeal poison. In both instances diterpene alkaloids are considered to be the toxic constituents.

A typical alkaloid of *Erythrophleum* species is cassaine (**146**), showing λ_{max} 215 nm (ϵ 20,000) and having a 4-carbomethoxy group. Not all of the alkaloids are α,β-unsaturated esters; some, such as norcassamidide (**147**), are amide derivatives involving *N*-methyl ethanolamine and co-occur with their corresponding ester derivatives. The esters are interconverted to the corresponding amides simply by passage over alumina.

Two points of stereochemical interest concern the C-14 methyl group

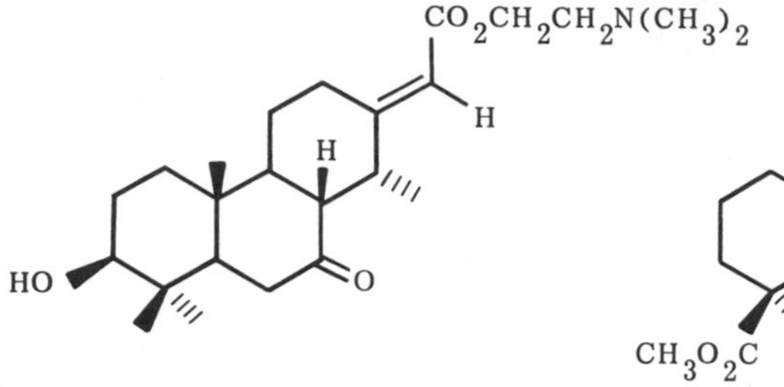

146   cassaine 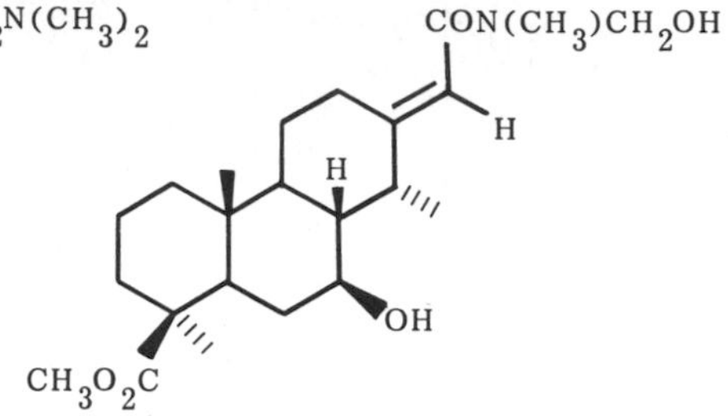   147   norcassamidide

and the unsaturated ester. In the proton NMR spectrum of **148** (a natural isomer), the C-14 proton appears at 2.77 ppm, whereas in the C-14 isomer **149** this proton is strongly deshielded to 4.30 ppm. This is to be expected for a carboxyl group *cis* to the equatorial C-14 proton.

Representatives of this alkaloid group (e.g. **150**) display potent cytotoxic activity. An interesting structure–activity relationship is that the parent alcohol **151** is 1000 times less cytotoxic than **150**.

These alkaloids also exhibit potent cardiotonic effects in the isolated rabbit papillary muscle system. One of the most active compounds in this respect is **152**, having an ester at the 3-position and oxygenation at C-7.

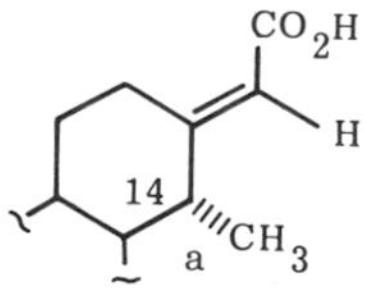

148 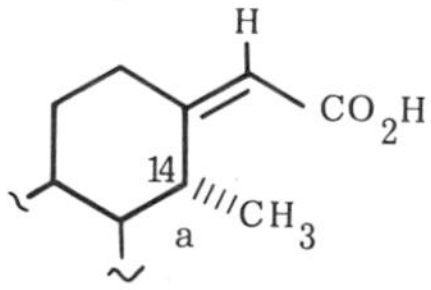   149

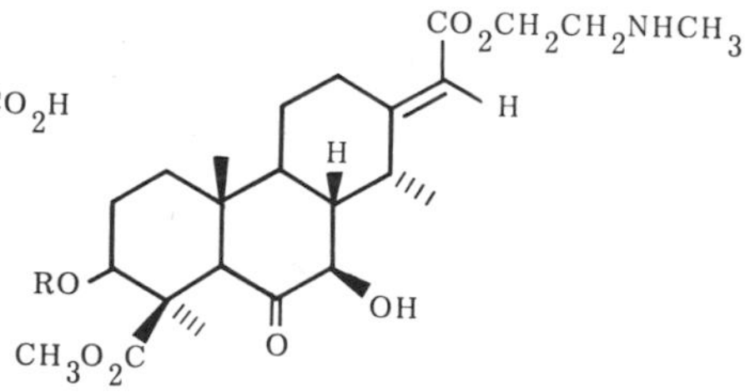

150   R=Ac
151   R=H

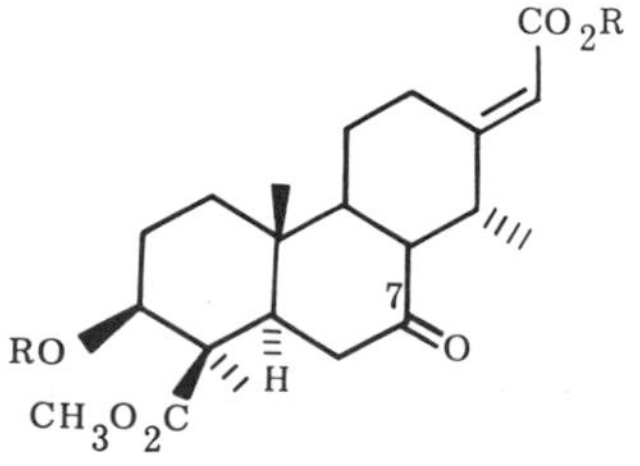

152

R= $CH_2CH_2N(CH_3)_2$

LITERATURE

Weisner, K., and Z. Valenta, *Fortschr. Chem. Org. Naturs.* **16**, 26 (1958).

Stern, E. S., *Alkaloids NY* **7**, 473 (1960).

Pelletier, S. W., *Quart Revs.* **21**, 526 (1967).

Pelletier, S. W., and L. H. Keith, *Alkaloids NY* **12**, 2 (1970).

Pelletier, S. W., and L. H. Keith, in S. W. Pelletier (Ed.), *Chemistry of the Alkaloids*, Van Nostrand Reinhold, New York, 1970, p. 503.

Keith, L. H., and S. W. Pellitier, in S. W. Pelletier (Ed.), *Chemistry of the Alkaloids*, Van Nostrand Reinhold, New York, 1970, p. 549.

Edwards, O. E., *Alkaloids, London* **1**, 343 (1971).

Pelletier, S. W., and L. H. Wright, *Alkaloids, London* **2**, 247 (1972).

Pelletier, S. W., and S. W. Page, *Alkaloids, London* **3**, 232 (1973).

Pelletier, S. W., and S. W. Page, in K. F. Wiesner (Ed.), *MTP International Review of Sciences, Organic Chemistry*, Ser. 1, Vol. 9, *Alkaloids*, Butterworths, London, 1973, p. 319.

Pelletier, S. W., and S. W. Page, *Alkaloids, London* **4**, 323 (1974).

Hauth, H., *Planta Med.* **25**, 201 (1974).

Pelletier, S. W., and S. W. Page, *Alkaloids, London* **5**, 230 (1975).

Pelletier, S. W., and S. W. Page, *Alkaloids, London* **6**, 256 (1976).

Pelletier, S. W., and S. W. Page, in K. F. Wiesner (Ed.), *MTP International Review of Science, Organic Chemistry*, Ser. 2, Vol. 9, *Alkaloids*, Butterworths, London, 1976, p. 53.

Pelletier, S. W., and S. W. Page, *Alkaloids, London* **8**, 219 (1978).

C_{20}-Diterpene Alkaloids: Synthesis

Nagata, W., T. Sugasawa, M. Narisada, T. Wakabayashi, and Y. Hayase, *J. Amer. Chem. Soc.* **85**, 2342 (1963); *J. Amer. Chem. Soc.* **89**, 1483 (1967).

Nagata, W., M. Narisuda, T. Wakabayishi, and T. Sugasawa, *J. Amer. Chem. Soc.* **86**, 929 (1964); *J. Amer. Chem. Soc.* **89**, 1499 (1967).

Masamune, S., *J. Amer. Chem. Soc.* **86**, 288, 290 (1964).

Valenta, Z., K. Wiesner, and C. M. Wang, *Tetrahedron Lett.* 2437 (1964).

Beames, D. J., J. A. Halleday, and L. N. Mander, *Aust. J. Chem.* **25**, 138 (1972).

C_{19}-Diterpene Alkaloids

Synthesis

Wiesner, K., T. Y. R. Tsai, K. Huber, S. E. Bolton, and R. Vlahov, *J. Amer. Chem. Soc.* **96**, 4990 (1974).

Rearrangements

Johnson, J. P., and K. H. Overton, *J. Chem. Soc. Perkin Trans. I*, 1490 (1972).

Ayer, W. A., and D. P. Deshpande, *Can. J. Chem.* **51**, 77 (1973).

Erythrophleum *Alkaloids*

Dalma, G., *Alkaloids NY* **4**, 265 (1954).

Morin, R. B., *Alkaloids NY* **10**, 287 (1968).

Loder, J. W., and R. H. Nearn, *Tetrahedron Lett.* 2497 (1975).

11.5 *DAPHNIPHYLLUM* ALKALOIDS

Daphniphyllum macropodum (Daphniphyllaceae) is a tall tree common to Japan, where it is known as "yuzuriha."

An alkaloid mixture was first obtained in 1909, but no structures were determined until 1966, when daphniphylline hydrobromide was subjected to X-ray crystallography. At least 19 alkaloids have been isolated and characterized from the bark, leaves, and fruit of *D. macropodum*, and more recently six new alkaloids have been obtained from *D. teijsmanni* Zoll. This is now therefore a significant group of alkaloids. At the present time these alkaloids may be divided into five subgroups.

The major group is the daphniphylline type, represented by the parent compound daphniphylline (**153**) and its desacetyl derivative, codaphniphylline (**154**).

Daphniphylline consists of two units, an amine and an acetal, joined through a straight chain of three carbon atoms containing an α-acetoxyketo group. Basic hydroylsis followed by periodate cleavage gave an acetal acid **155** and an unstable aldehyde **156**, which was reduced (NaBH₄) and acetylated to **157**.

Treatment of daphniphylline with 6N hydrochloric acid at 80°C gave deacetylisodaphniphylline (**158**) quantitatively. Acetylation gave isodaphniphylline (**159**), which contained a primary acetate group (δ 3.99 ppm, 2H, dd, *J* = 11 Hz). These data suggested that the diketone diol which gave rise to isodaphniphylline was **160**.

The *Daphniphyllum* alkaloids display two prominent ions at *m/e* 286 (**161**) and 272 (**162**), and Scheme 21 has been suggested to explain their formation.

Secodaphniphylline (**163**) is an isomer of deacetyldaphniphylline but contains an NH group. The structure was established by NMR comparison with known alkaloids and the main features are illustrated on the structure.

The structure of yuzurimine (**164**), a major alkaloid of *D. macropodum*, was determined by X-ray analysis and reflects a skeleton having only 22 carbon atoms. Hydrolysis in methanolic sodium hydroxide gave a dideacetyl derivative **165** and further reaction with thionyl chloride in pyridine gave a cyclic sulfite **166**, indicating the steric proximity of the primary and secondary hydroxyl groups. The carbinolamine nature of the third hydroxy group was established by alkaline treatment of yuzurimine methiodide to give the keto amine **167**.

The ketone carbonyl frequency in this compound was observed at 1633 cm⁻¹, indicative of considerable transannular interaction between the lone

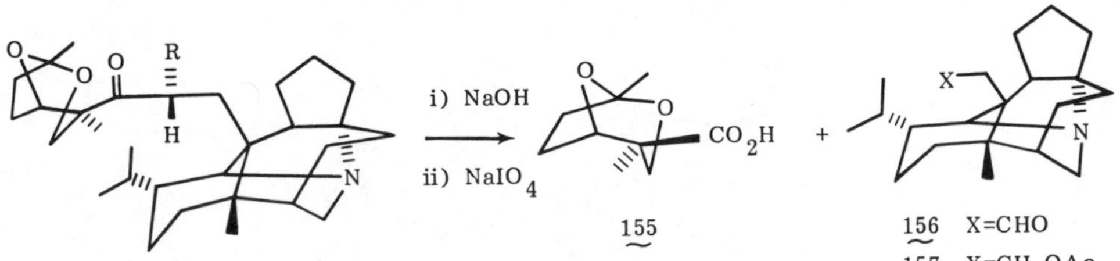

i) NaOH

ii) NaIO$_4$

155

156 X=CHO
157 X=CH$_2$OAc

153 daphniphylline, R=OCOCH$_3$
154 codaphniphylline, R=H

153 daphniphylline $\xrightarrow{\text{H}^{\oplus}}$

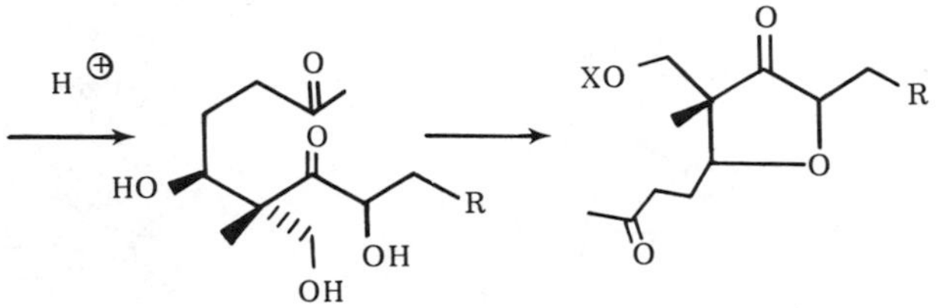

160 R=amine
     moiety

158 X=H
159 X=COCH$_3$

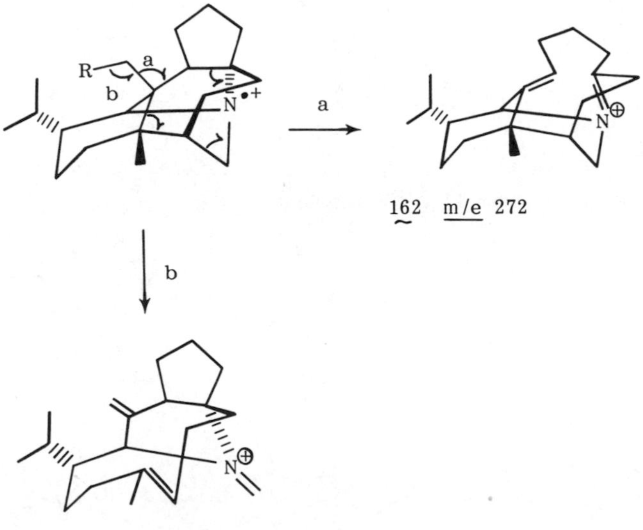

162 m/e 272

161 m/e 286

Scheme 21

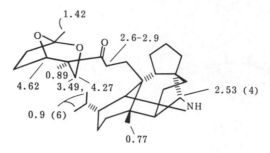

163   secodaphnilline

164   yuzurimine, R=COCH₃
165   R=H
166   R= >S=O

167

168   from [5-¹⁴C]-mevalonate

codaphnilline

Scheme 22   Biogenesis of the ketal moiety of the *Daphniphyllum* alkaloids.

pair on nitrogen and the carbonyl group. No reduction of this carbonyl group is observed on reaction with sodium borohydride in refluxing THF for 21 hr!

These are representative of the alkaloid skeleta found in *D. macropodum*, and their novelty has stimulated some interesting biosynthetic experimentation.

Daphniphylline (**153**) and co-daphniphylline (**154**) are the major alkaloids of the bark and leaves of *D. macropodum* and were each labeled by mevalonate (**55**) to the extent of 0.13%. Degradation indicated a derivation **168** from six mevalonate units; that is, these are triterpene alkaloids *not* diterpene alkaloids as had originally been suggested. More important, labeled squalene (**169**) was incorporated into both **153** and **154**. When [4RS-^{3}H$_2$]mevalonate was used, five of the six possible tritium labels were retained.

These alkaloids are comprised of an amine and a ketal moiety. Biosynthetically, few details of any intermediates are known, but there have been some interesting biogenetic schemes proposed. One of these involves a possible formation of the ketal moiety as shown in Scheme 22; another involves the formation of the amine moiety (Scheme 23). Note the proposed intermediacy of secodaphniphylline (**163**).

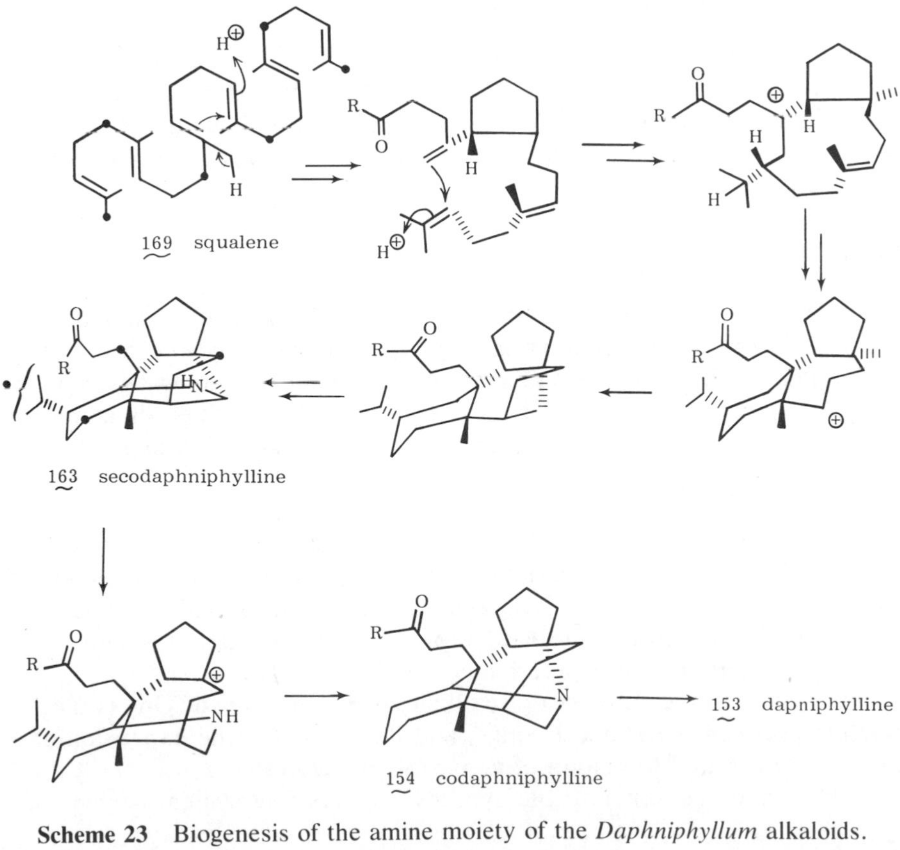

Scheme 23   Biogenesis of the amine moiety of the *Daphniphyllum* alkaloids.

There is little evidence for either of these schemes, particularly with regard to the stages of introduction of ketal formation. As further degradation results are reported, these biogenetic schemes will become clearer.

D. macropodum as a decoction has been used as a vermicide and asthma remedy for years. Yuzurimine (**164**) has muscle relaxant activity.

LITERATURE

Yamamura, S., and Y. Hirata, *Alkaloids NY* **15**, 41 (1975).

Yamamura, S., and Y. Hirata, in K. F. Wiesner (Ed.), *MTP International Review of Science, Organic Chemistry*, Ser. 2, Vol. 9, *Alkaloids*, Butterworths, London, 1976, p. 161.

Edwards, O. E., *Alkaloids, London* **1**, 343 (1971).

Pelletier, S. W., and S. W. Page, *Alkaloids, London* **3**, 232 (1973).

Pelletier, S. W., and S. W. Page, *Alkaloids, London* **4**, 323 (1974).

Pelletier, S. W., and S. W. Page, *Alkaloids, London* **6**, 256 (1976).

Pelletier, S. W., and S. W. Page, *Alkaloids, London* **8**, 243 (1978).

11.6   STEROIDAL ALKALOIDS

There are four major groups of steroidal alkaloids which have been obtained from plants: (*a*) the *Veratrum* alkaloids, (*b*) the *Solanum* alkaloids, (*c*) the *Holarrhena* and *Funtumia* alkaloids, and (*d*) the *Buxus* alkaloids. In addition, there are two groups of alkaloids obtained from animals and these will be discussed initially.

11.6.1   Steroidal Alkaloids from Animal Sources

Alkaloids from animal sources are becoming increasingly common. Unknown a few years ago there are now many examples, and the two groups of steroidal alkaloids are particularly interesting from a structural viewpoint. These compounds occur in secretions of the skin glands of certain amphibia and appear to protect the skin against fungal and bacterial infection.

Salamandra *Alkaloids*

One of the more unusual methods of alkaloid isolation is used for the *Salamandra* alkaloids, which are obtained by evacuation of the skin glands of the animal after anaesthesia. The crude product is separated by extraction with ethanol followed by chromatography. The main alkaloid is samaderin (**170**), whose structure was determined by X-ray analysis.

Two structural features of these alkaloids are of interest. One is the *cis* junction between the A and B rings, and the second is the expansion of a steroidal ring A and formation of an isoxazolidine system.

A biogenetic approach to the synthesis of this ring system begins with

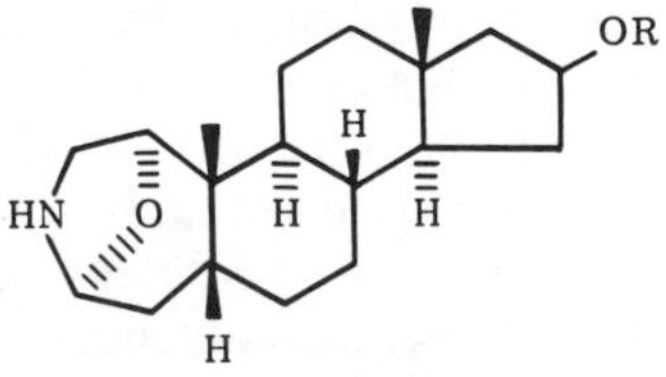

170 samaderin R=H
       R=COCH$_3$

the androstene derivative **171**. Catalytic hydrogenation introduced the *cis*-ring junction and cleaved the epoxide to afford **172**, which after a Schmidt reaction (HN$_3$, H$_2$SO$_4$) afforded two lactams, one of which was **173**. Reduction with lithium–ethylamine in the presence of 2-methylpropan-2-ol yielded a key intermediate for synthetic elaboration (Scheme 24).

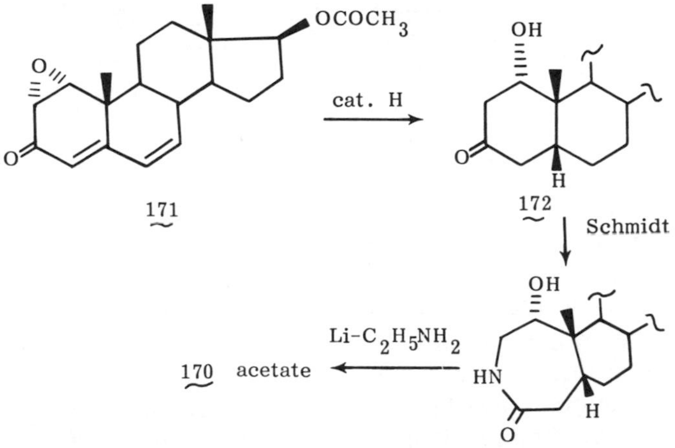

Scheme 24 Biogenetic-type synthesis of the *Salamandra* alkaloid series.

Biosynthetically, these alkaloids are formed, like other steroids, from mevalonate *via* cholesterol. The ring-A enlargement possibly proceeds after a fission between carbon atoms 2 and 3. The nitrogen atom that is inserted is derived from glutamine.

Phyllobates *Alkaloids*

The Indians native to the river forests of Colombia have used secretions of the skin from a highly colored frog. *Phyllobates aurotaenia*, known as *kokoi*, as a poison for blow darts. Investigations began in 1871 on the chemical nature of this material, but it was not until Witkop and his associates brought all the modern physical methods to bear on this problem that much progress was made.

The first structure to be determined, by X-ray analysis, was that of batrachotoxinin A (**174**). This compound could be correlated with batrachotoxin (**175**) by hydrolysis of the latter to **174** and a pyrrole carboxylic acid (**176**). Because the parent compound contains a pyrrole nucleus capable of electrophilic substitution, **175** gives a red color with Ehrlich's reagent and a blue color with dimethylaminocinnamaldehyde.

174   R=H, batrachotoxinin A

175   R= , batrachotoxin

176

The novelty of the skeleton of batrachotoxin (*batrachos* is Greek for frog) has been the subject of considerable synthetic interest and the first successful approach was reported in 1973. Clearly there are two important regions in the molecule to be constructed: (*a*) the 3,9-ether bridge and (*b*) the heterocyclic ring between C-13 and C-14.

Epoxidation of the dihydroxyketone **177**, partial acetylation, and oxidation gave a ketone **178**, which was osmylated and acetylated to give **179**. Treatment with sodium iodide in acetic acid, followed by dehydration with DDQ, gave the dienone **180**. Epoxidation of **180** gave the 6α,7α-epoxide, which could be catalytically reduced to the internal hemiketal **181** after treatment with methanol. Acetylation of the 7-alcohol and a standard series of reactions gave the 14β,18-diol **182**. Oxidation of the primary alcohol with dimethyl sulfoxide–acetic anhydride to an aldehyde was followed by condensation with methylamine and reduction to the amine **183**. N-Chloroacetylation, treatment with acid, and base-catalyzed cyclization gave **184**. Hydrolysis and reacetylation gave the 11α,20(S)-diacetate, which on dehydration with thionyl chloride in pyridine, followed by reduction and hydrolysis, afforded batrachotoxinin A (**174**) (Scheme 25).

Batrachotoxin and its close relatives are presently of interest as a tool for investigating the control of membrane permeability, because they are extremely potent, selective, and essentially irreversible activators of so-called "sodium channels" in excitable membranes. This action is specifically antagonized by tetrodotoxin, which acts to lower sodium conductance.

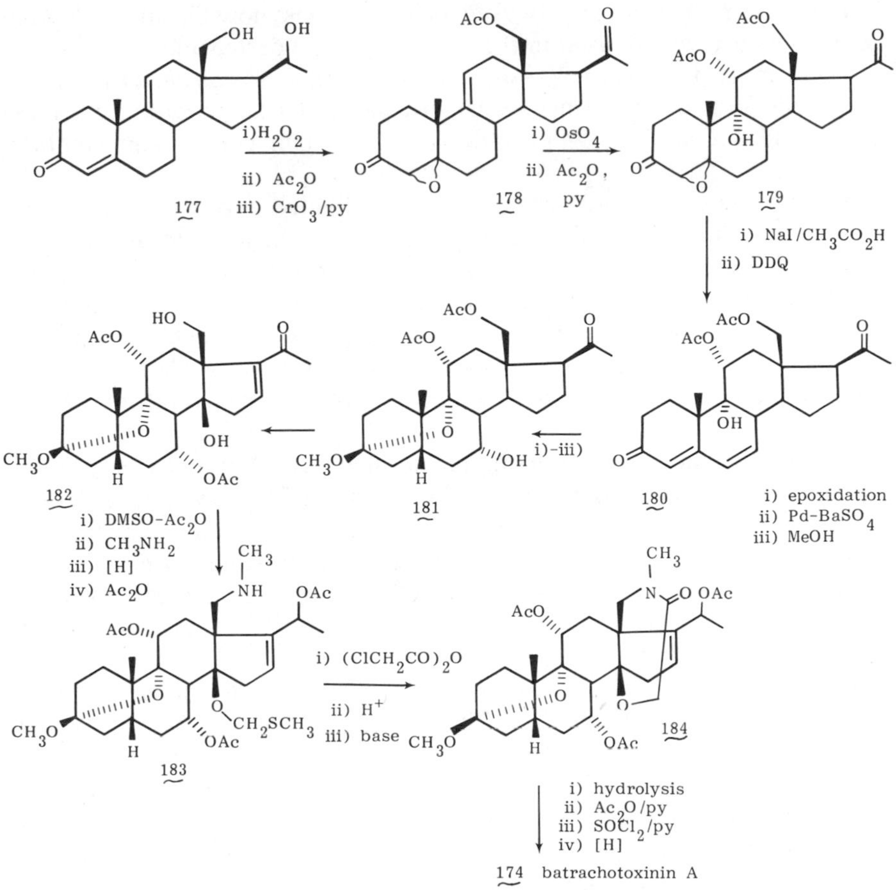

OH OH

i)H$_2$O$_2$
ii) Ac$_2$O
iii) CrO$_3$/py

177

AcO O

178

i) OsO$_4$
ii) Ac$_2$O, py

AcO
AcO
OH

179

i) NaI/CH$_3$CO$_2$H
ii) DDQ

HO
AcO

O OH OAc

CH$_3$O H

182

i) DMSO-Ac$_2$O
ii) CH$_3$NH$_2$
iii) [H]
iv) Ac$_2$O

AcO
AcO O

CH$_3$O H OH

181

i)-iii)

AcO
AcO OH

180

i) epoxidation
ii) Pd-BaSO$_4$
iii) MeOH

CH$_3$
NH OAc

AcO

CH$_3$O O CH$_2$SCH$_3$
H OAc

183

i) (ClCH$_2$CO)$_2$O
ii) H$^+$
iii) base

CH$_3$
N O OAc

AcO

CH$_3$O O OAc
H

184

i) hydrolysis
ii) Ac$_2$O/py
iii) SOCl$_2$/py
iv) [H]

174 batrachotoxinin A

Scheme 25 Partial synthesis of batrachotoxinin A (**174**).

Batrachotoxin is one of the most potent cardiotoxins known and produces ventricular fibrillation and tachycardia at the 2 nM level in the cat heart.

11.6.2 Steroidal Alkaloids from Plant Sources

Veratrum *Alkaloids*

Veratrum species have been noted for their pharmacologic activity for well over 300 years. In 1672, the great chronicler of New England lore, John Josselyn, said of the white Hellebore, *Veratrum viride* (Liliaceae): "the powder of the root is good for toothache. . . . the root sliced thin and boyled in vinegar, is very good against herpes milliaris."

The plant was used by several tribes of American Indians to kill lice, as a catarrh remedy, against rheumatism, and as a hypotensive. Until about

1950 the veratrums were mainly used as insecticides; now the alkaloids from these plants are widely used in the treatment of hypertension.

There are two species of *Veratrum* of importance: *Veratrum viride* Aiton, green hellebore, collected in the eastern parts of Canada and the United States, and *V. alba* L., white hellebore, found in central and southern Europe.

The *Veratrum* alkaloids are conveniently divided into two quite distinct groups, according to their degree of oxygenation. The jerveratrum alkaloids contain one to four oxygen atoms and usually occur as the free alkamine or as mono-D-glycosides. Typical examples of this class are jervine (185) and veratramine (186). The ceveratrum alkaloids are very highly oxygenated and contain seven to nine oxygen atoms. They occur in the plant with some of the hydroxy groups esterified; examples are the protoveratrines A and B (187) and (188).

Jervine does not form an oxime of the hindered 11-carbonyl group, but does have a long-wavelength n-π^* band at 360 nm (ϵ 70). O,N-Diacetyljervine on acetolysis in the presence of BF_3 gave a nitrogen-free dienone 189.

The chemistry of jervine has a number of interesting features, two of which are a hydrogenation product (Pd/10% aqueous acetic acid) having the dienone structure 190, and a Wolff–Kishner reduction product formulated as the conjugated diene 191.

More challenging intellectually is the product derived from jervine (185) by treatment with mineral acid and known as isojervine. A conjugated carbonyl is present (ν_{max} 1640, 1690 cm^{-1}), but the UV spectrum now only shows a weak band at 252 nm. Catalytic hydrogenation gave a dihydro derivative which showed typical λ_{max} value at 238 nm for an unsaturated ketone. After considerable chemical and spectroscopic effort the compound was formulated as 192, produced by rearrangement as shown in 193.

The synthesis of jervine was reported by Masamune *et al.* in 1967, and that of veratramine by Johnson in the same year.

Sodium borohydride reduction of the acetate of 194 and bromination gave a mixture of bromides. The higher-melting-point bromide was treated with the pyrrolidine enamine of 1-acetyl-(3S)-methyl-5-piperidone (195) to afford 196. Reduction with lithium–ethylamine and hydrogenation (H_2–Pt) gave a $\Delta^{13,17}$-derivative which was epoxidized to 197. Treatment with base and elimination of water gave diacetyl-5,6-dihydro-11-deoxojervine (198) by intramolecular displacement at C-17. After chromium trioxide oxidation to introduce the 11-keto group, standard reactions gave jervine (185) (Scheme 26).

The synthesis of jervine illustrates an important point about the biogenetic relationships of the veratramine, jervine, and cevanine bases. The first-named group of compounds may be represented in skeletal fashion by the structure 199. If union (by an ether bridge) is made (e.g. join a), the jervanine bases result. If however the union is between C-18 and the basic nitrogen,

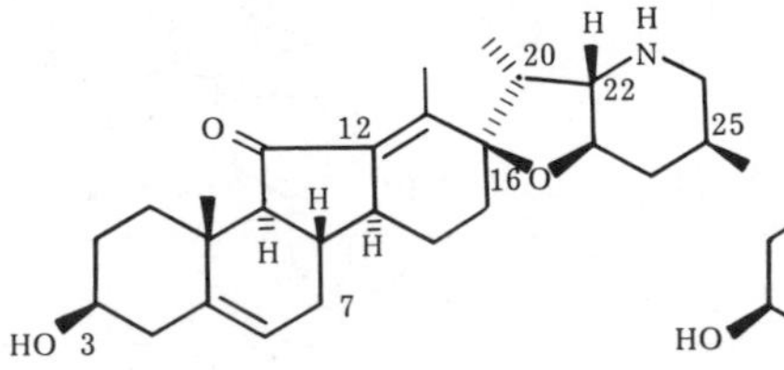

185 jervine

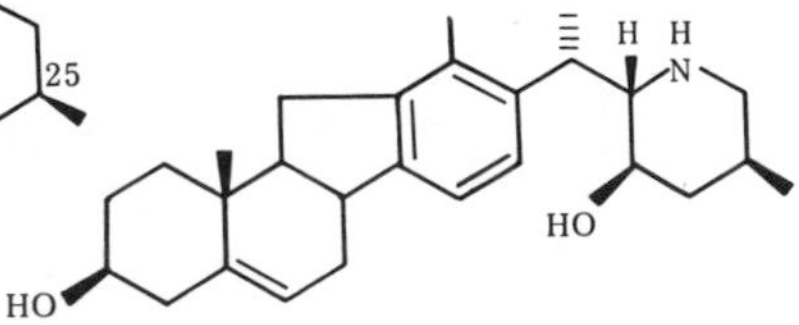

186 veratramine

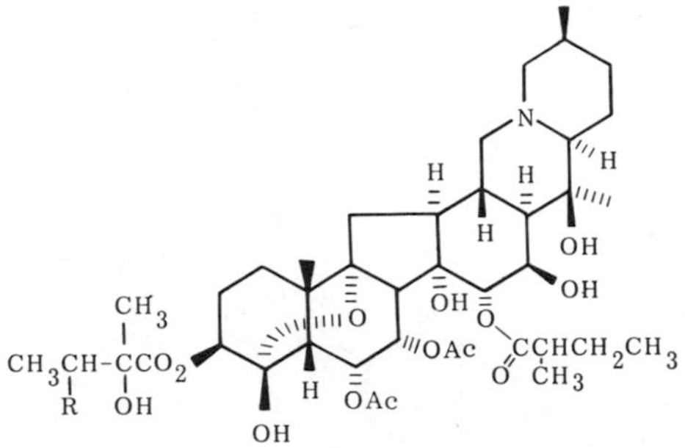

187 protoveratrine A, R=H
188 protoveratrine B, R=OH

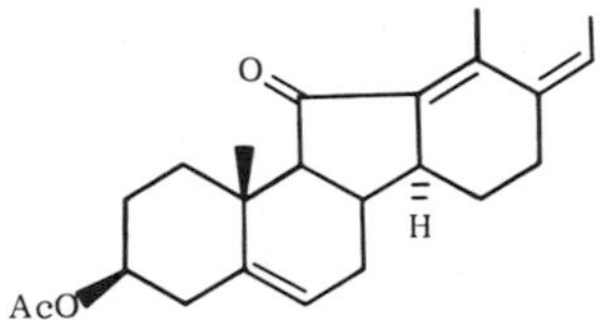

189

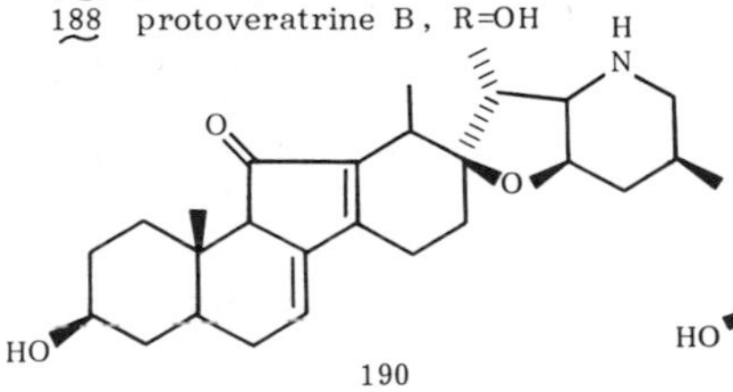

190

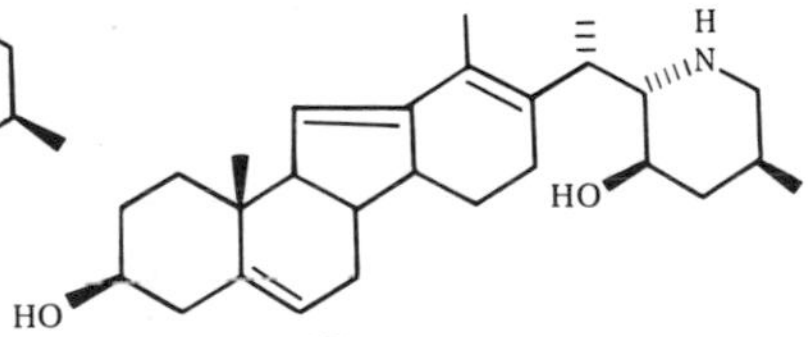

191

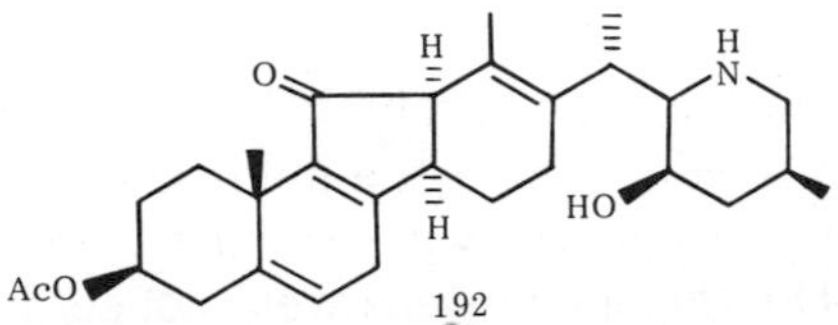

192

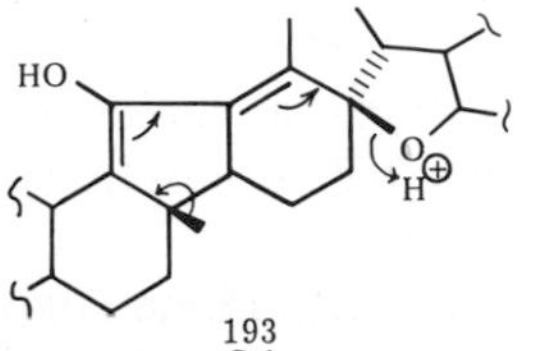

193

the cevanine bases are produced. These ideas have recently been applied to the total synthesis of verticine (**200**), a member of the Ceveratrum group of compounds.

The cevane skeleton is represented by structure **201**. The biogenetic derivation by N-to-C-18-bound linkage of a veratramine-type alkaloid is reflected in the numbering system as shown.

The more complex members of this series, having many sites of oxygenation and esterification, are found in the *Veratrum* species discussed previously. A simpler group of compounds having this skeleton occur in *Fritillaria* species in the family Liliaceae, and a well-studied example is verticine (**200**).

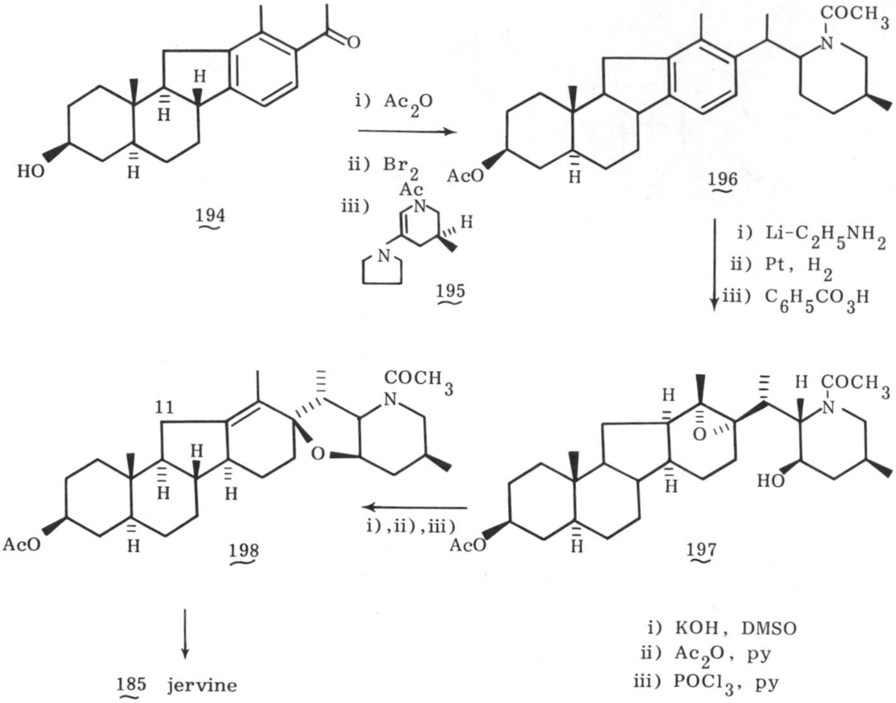

Scheme 26 Masamune synthesis of jervine (**185**).

Treatment of the C-nor-D-homosteroid derivative **202** with the lithium derivative of 5-methyl pyridine with tosyl chloride in pyridine followed by treatment with triethylamine, and then sodium borohydride, gave a mixture of two olefins **203**. Catalytic (Pt/H_2) reduction gave a mixture of C-25 epimers, the β-isomer of which when reacted with sodium naphthalene afforded deoxoverticinone (**204**). The remaining steps were concerned with the stereospecific introduction of the 6α-hydroxy group.

Chromium trioxide oxidation followed by bromination of the ketone with pyridine hydrobromide perbromide and elimination of HBr with lithium

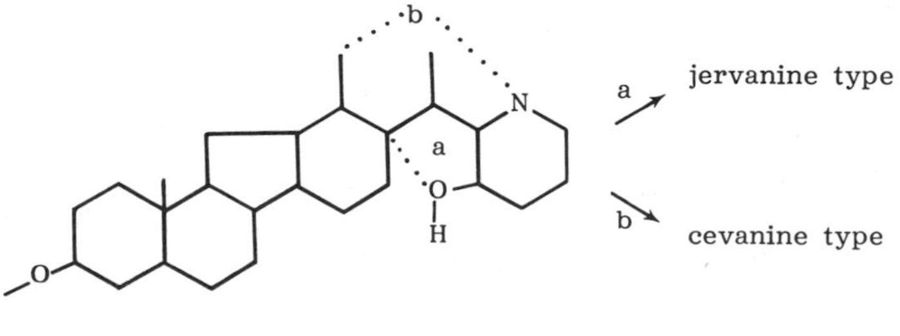

199 veratramine type

200 verticine

201 cevane skeleton

bromide gave a mixture of 1,2- and 4,5-dehydro ketones **205**. The former, unwanted product was reduced and recycled while the latter was ketalized (isomerization to the 5,6-dehydro derivative) to **206**. Oxidative hydroboration gave a mixture of 6-alcohols **207**, the α-isomer of which was hydrolyzed and reduced to verticine (**200**) (Scheme 27).

The structure of cevine (**208**), a typical representative of the polyoxygenated *Veratrum* alkaloids, was deduced only after many years of complex degradative work. Note that cevine (3β-H) is not a natural product, but rather an isomerization product of veracevine having the 3α-H stereochemistry of **209**. Treatment of veracevine (**209**) with base gave cevagerine (**210**), capable of forming an orthoester **211** in the D-ring, thereby indicating the stereochemistry of hydroxy groups at positions 12, 14, and 17. When this orthoacetate was treated with sulfuric acid, a different orthoacetate **212** resulted, which now involved the hydroxy groups at C-9, C-12, and C-14. The relative stereochemistries of the hydroxyl groups at positions 16, 17, and 20 became clear when cevine consumed only two equivalents of periodate, which are involved in cleavage of the 3,4- and 12,14-bonds.

One other reaction of these systems is of interest. Heating the 3,4,16-triacetate of cevine in methanol specifically removed with 16-acetyl group. This unusual reaction is regarded as proceeding with intramolecular assistance of the tertiary nitrogen and a 1,3-diaxial involvement of the axial 20-hydroxyl group as shown in **213**.

The most important alkaloids of *Veratrum* are the hypotensive polyesters of protoverine (**214**). Protoverine (**214**) itself contains a vicinal triol since formic acid is produced on periodate cleavage. Acetonide formation prevented periodate cleavage. In this series both the 7-esters and the 16-esters are hydrolyzed very easily, indicating 1,3-diaxial participation of hydroxyl groups at positions 14 and 20 respectively.

Oxidation of the 16-hydroxyl group of germine (**215**) to a ketone and treatment with base gave a cross-conjugated diosphenol (λ_{max} 330 nm) having

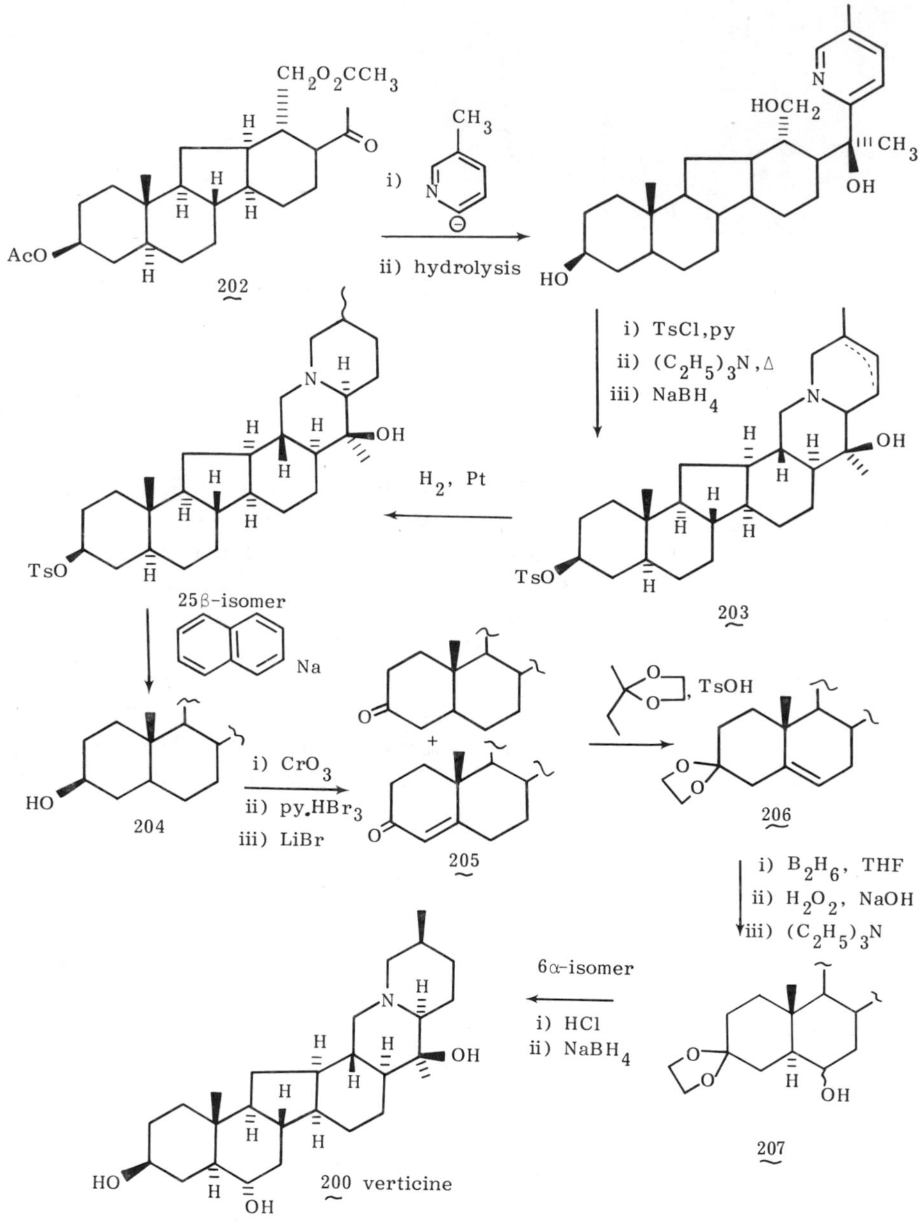

Scheme 27 Kutney synthesis of verticine (**200**).

the structure **215a**, and formed by double elimination of the tertiary hydroxyls at 14 and 20.

Protoverine (**214**) contains one more hydroxyl group than germine and the location of this group was determined when periodate cleavage of the methanolysis product of protoveratrine A (**187**) indicated the *cis* relationship

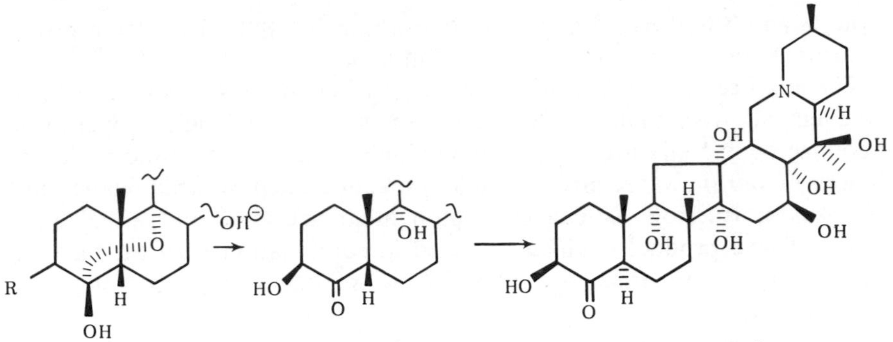

208 cevine, R=α-OH
209 veracevine, R=β-OH

210 cevagenine

i) Ac$_2$O, HClO$_4$
ii) 5% KOH, MeOH

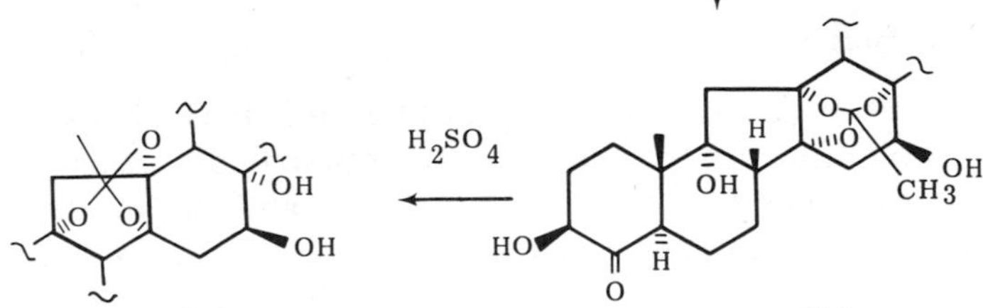

212

H$_2$SO$_4$

211

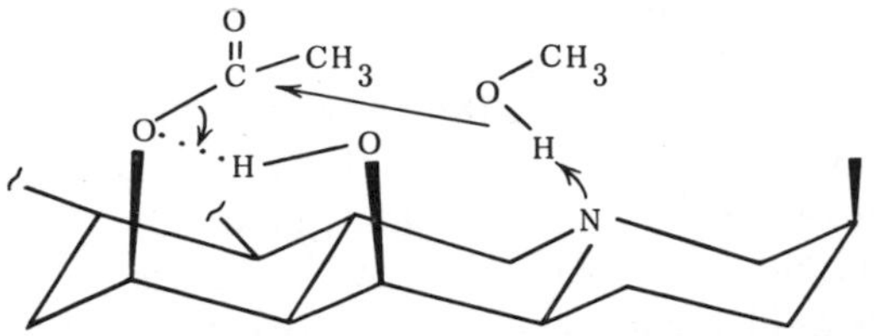

213

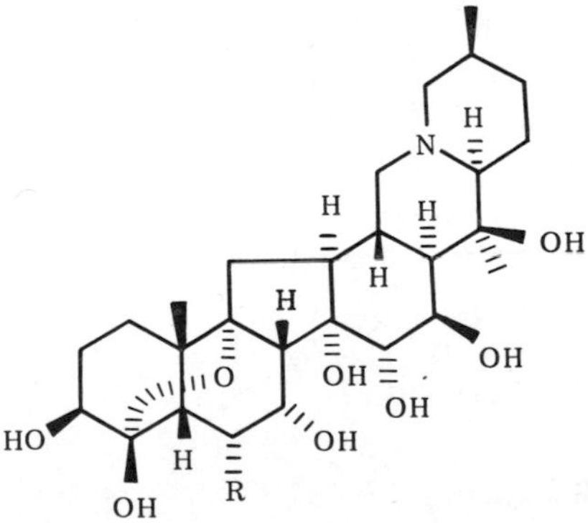

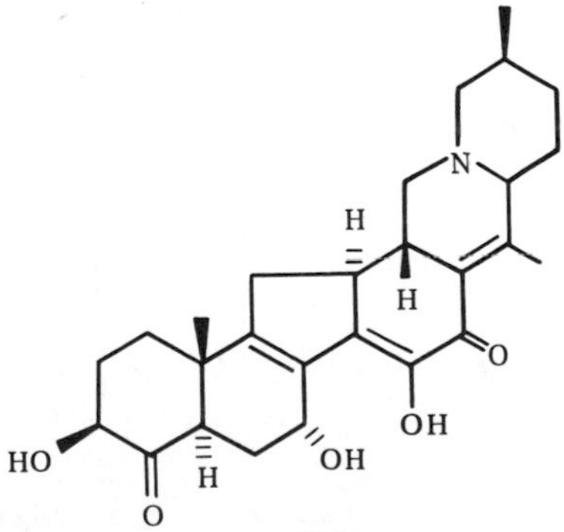

214 protoverine, R=OH
215 germine, R=H

215a

of the 6- and 7-hydroxy groups. Protoveratrine B (188) differs from proto-veratrine A only within the ester group at C-3.

The mixture of protoveratrines was first isolated by Salzberger in 1890, and eventually reexamined by Craig and Jacobs in 1943. The protoveratrine of commerce is a mixture of the two alkaloids, although several other prep-arations of *Veratrum* are also available, including the powdered whole root, the alkaloid fraction (so-called cryptenamine acetate), and an alkaloid extract. These products, which are used in combination with the alkaloid fractions of *R. serpentina*, are useful in the treatment of various stages of hypertension.

Intravenous injection of the protoveratrines produces pronounced bra-dycardia and a fall in blood pressure by stimulation of vagal afferents. The major problem with the protoveratrines is the small margin between the therapeutic and side effects. The latter are usually nausea and vomiting.

Solanum *Alkaloids*

The Solanaceae yields several varieties of steroidal alkaloids derived by subsequent bond formations of a pregnane or modified pregnane derivative having a methyl piperidine whose α-position is joined to the 20-position of a steroid moeity. There are two structure types, the spirosolanes and the solanidines.

The spirosolanes such as tomatidine (216) and solasodine (217) have the basic nitrogen involved in an oxaazaspirane unit, which constitutes the original steroid side chain.

In 1935, Fisher demonstrated that the expressed juice of tomato plants inhibited the growth of a *Fusarium* fungus. In 1948, Fontaine and co-workers isolated and crystallized the fungistatic agent, naming it "tomatine." The same workers established that tomatine, now known as α-tomatine (218), was a glycosidal alkaloid comprising an aglycone tomatidine (216) and a tetrasaccharide, β-lycotetraose. Degradation of tomatidine gave 3β-acetoxy-5α-pregn-16-en-20-one (219), and this compound has also been used as a

216   tomatidine, R=H
217   solasodine, $\Delta^{5,6}$, R=H
218   tomatine, R=β-lycotetraose

starting material for the synthesis of **216**. The diacetate **220**, available from **219** by way of the 16β,17α-bromohydrin, was condensed with the lithium salt of 5-methyl pyridine at -40°C to give a mixture of pyridyl carbinols **221**, which could be dehydrated to the vinyl pyridine derivative **222**. Hydrolysis of the ester groups and catalytic hydrogenation afforded dihydrotomatidine (**223**), the N-chloro derivative of which on treatment with base gave tomatidine (**216**) in high yield (Scheme 28).

Tomatine, like most saponins, forms an insoluble 1 : 1 complex with cholesterol. The complex is quite stable but can be dissociated with acid or dimethyl sulfoxide. Spectrophotometric methods for the determination of tomatine include the products of treatment with concentrated sulfuric acid, silicotungstic acid, or phosphomolybdic acid.

In *Solanum* species α-tomatine is accompanied by other steroidal glyco-alkaloids, but in the genus *Lycopersicon* tomatine is often the only alkaloid present. Alkaloid content in tomato plants is highly variable depending on the stage of growth but is typically 0.9–1.9% in the leaves, 0.3–0.6% in the

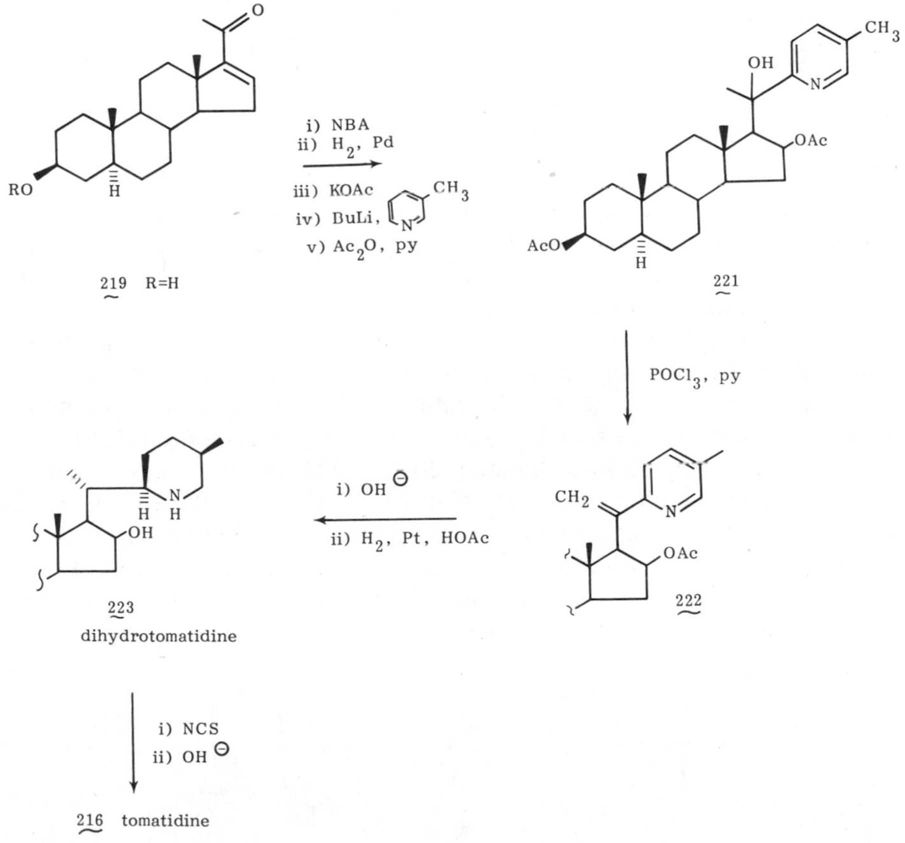

Scheme 28   Synthesis of tomatidine (**216**).

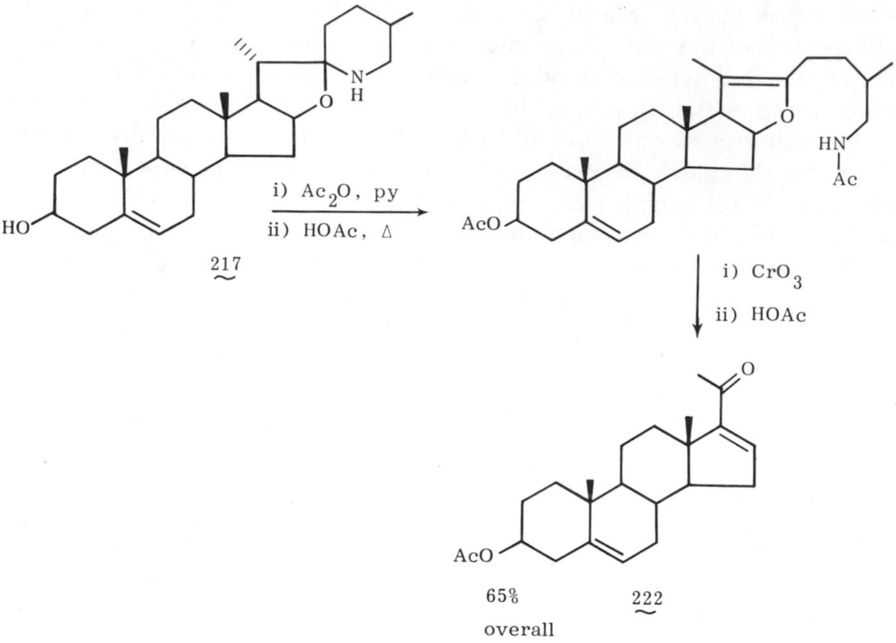
HO
N
H
O
i) Ac_2O, py
ii) HOAc, Δ
217
AcO
O
HN
Ac
i) CrO_3
ii) HOAc
O
AcO
65%
overall
222

Scheme 29

stems and roots, and 0.04% in ripened fruits. When the fruits are removed during active growth, the alkaloid levels of the remaining plant parts may be raised up to fivefold. Synthesis of tomatine occurs in the roots, and translocation then takes place throughout the plant.

The need for new high-yield sources of precursors of the biologically active steroid hormones has prompted an extensive search of the genus *Solanum* for solasodine (**217**). The rationale for this interest concerns the reaction sequence shown in Scheme 29, whereby solasodine (**217**) is converted to 3β-acetoxy-5,16-pregnadien-20-one (**222**) in 65% overall yield.

The facile interconversion of the *N*-nitroso derivatives of the spirosolanes to the steroidal sapogenins by treatment with acetic acid is of considerable use in the determination of the C-25 stereochemistry. A typical example is the conversion of *N*-nitrososolasodine (**223**) to diosgenin (**224**). In addition, the spirosolanes show stereoisomerism at C-22. A typical example is the

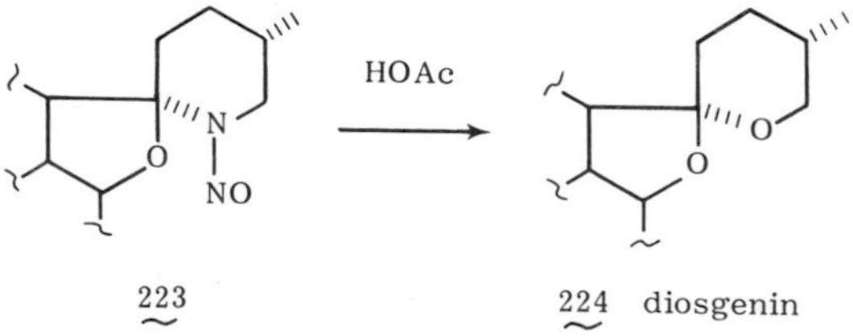
N
O
NO
HOAc
O
O
223
224 diosgenin

dissimilar nature of the ORD curves of the *N*-nitroso derivatives of tomatidine and solasodine. This has been confirmed by X-ray analysis of the
hydroidodide of tomatidine, which indicated that tomatidine had the structure **225**, and that solasodine could be represented by **226**.

The second major group of *Solanum* alkaloids are the solanidanes, which

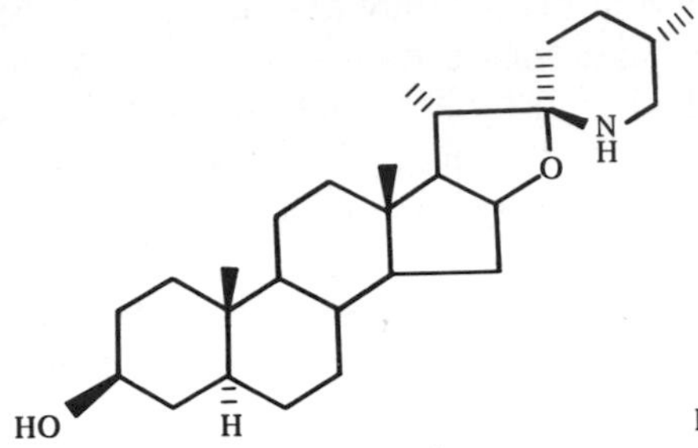

225 tomatidine

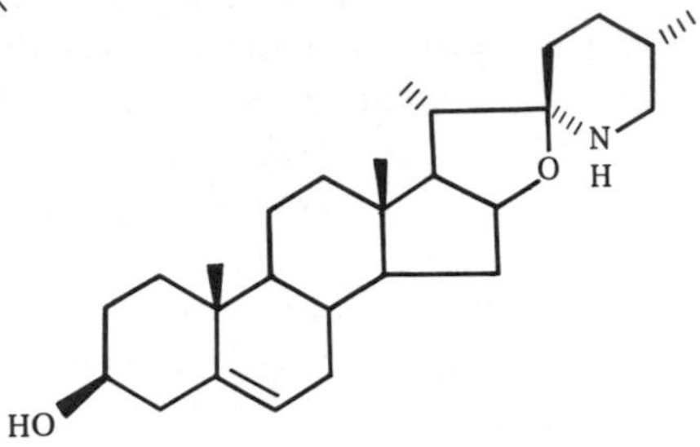

226 solasodine

also occur as glycoalkaloids. An example is solanidine (**227**), whose structure
was confirmed through the lactam **228** by reductive (Zn, acetic acid) condensation of the nitro keto ester **229**. The imine ester **230**, which is also
produced in the reaction, can be converted to the lactam with sodium
borohydride.

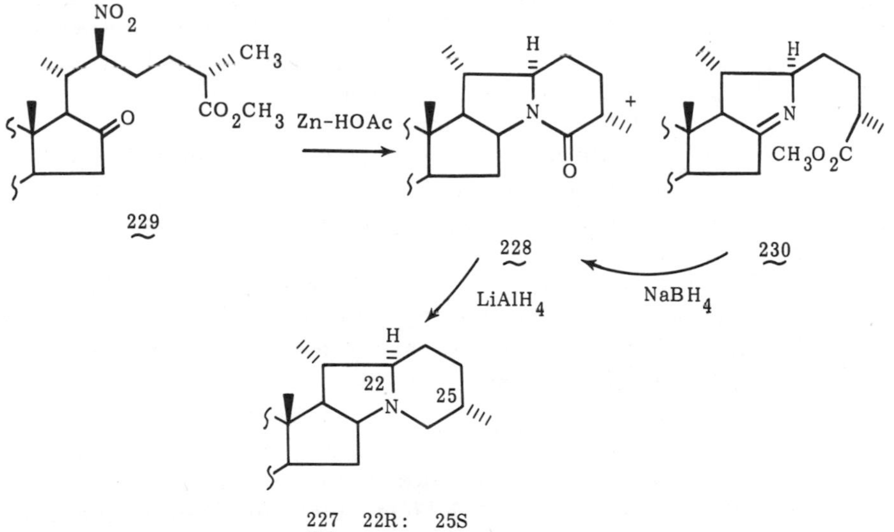

229

228   230

LiAlH$_4$ NaBH$_4$

227 22R: 25S

Tomatine partially inhibits the growth of several types of gram-positive
and gram-negative bacteria and animal pathogenic fungi. It is more active
against plant pathogenic and human dermatophytic fungi and *Candida
albicans*, but it has no effect on human pathogenic *Actinomyces*. The *in vivo*
activity of tomatine is extremely limited.

Tomatine appears to have some antifeeding effects for Colorado beetle and potato beetle larvae by leaf infiltration, but this has no use in the field. The pure alkaloid is not very toxic orally (single dose of 1 g/kg), but is toxic by i.v. injection (LD_{50} 18 mg/kg). Death is thought to be due to a massive decrease in blood pressure.

The molecular basis for the toxicity of tomatine may be due to its ability to complex with 3β-hydroxy steroids, and this is quite analogous to the action of the polyene antibiotics, which complex with membrane sterols and alter membrane permeability. An ointment of tomatine has been marketed for use in the treatment of mycotic dermatosis, but its clinical efficacy remains in doubt.

Solanine is a mitotic poison and inhibits human plasma cholinesterase. Solasodine and its glycosides have bradycardiac activity similar to that of veratramine.

Steroidal Alkaloids of the Apocynaceae

The family Apocynaceae is noted for the presence of indole alkaloids (Chapter 9) and cardenolides. But another group of compounds has, over the years, also received considerable attention. Quite simply they are derivatives of the pregnane (231) nucleus having an amino substituent at either C-3 or C-20 or both.

Three genera are of importance, *Holarrhena*, *Funtumia*, and *Malouetia*, and a fourth, *Chonemorpha*, has also received attention. A taxonomically unrelated genus *Pachysandra* in the family Buxaceae also produces alkaloids of this structure type. Some representative examples include funtumine (232), irehine (233), kurchessine (234), and conessine (235).

Many of the reactions used to characterize these alkaloids involve correlation in some way with the corresponding pregnane derivative. Amino groups in the 3-position are relatively easy to convert to the 3-keto group and the amino group can be reintroduced by oximation and lithium aluminum hydride reduction.

Conversion of a 3β-hydroxysteroid to a 3α-amino steroid when 5,6- is unsaturated can be effected by tosylation and displacement with sodium azide in hexamethylphosphotriamide (HMPT). Reduction with lithium aluminum hydride gives the 3α-amino-Δ^5-steroid, but reduction with sodium borohydride gives the 3α-amino-5α-H derivative (Scheme 30).

The corresponding 3β-compounds are obtained from the 3α,5α-cyclo-6β-hydroxy steroids by treatment with hydrazoic acid in benzene in the presence of boron trifluoride, followed by lithium aluminum hydride reduction (Scheme 31).

Several of the alkaloids have a rather unusual substitution pattern in the A ring, namely a 4-oxygen substituent, and an example is terminaline (236) from *Pachysandra terminalis* Sieb. and Zucc. The chemical shift of the 19-methyl group is characteristic of the 4-hydroxy-group stereochemistry. In

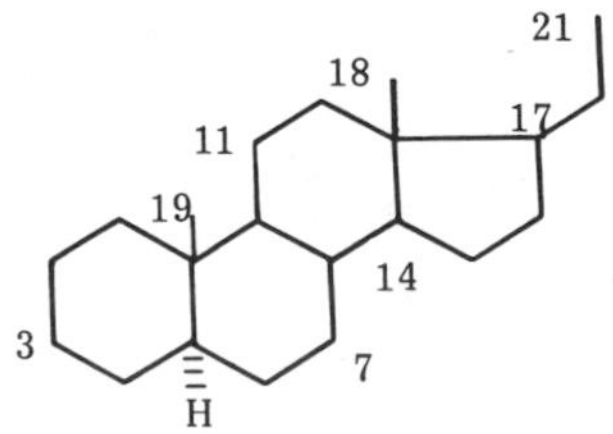

231 5α-pregnane

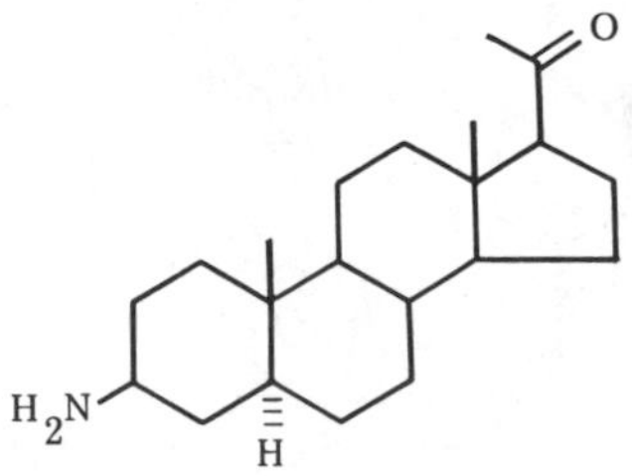

232 funtumine

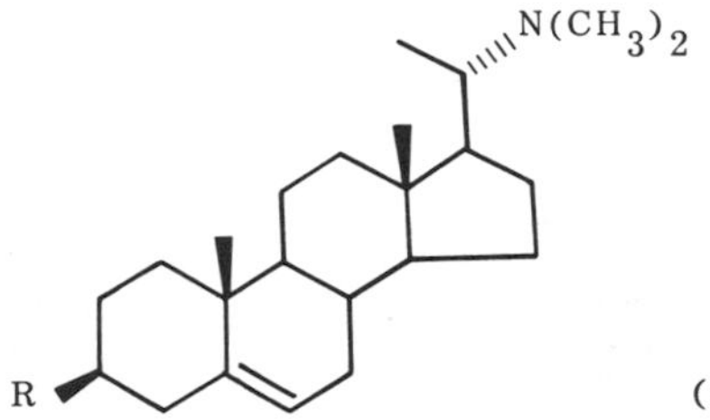

233 irehine, R=OH
234 kurchessine, R=N(CH$_3$)$_2$

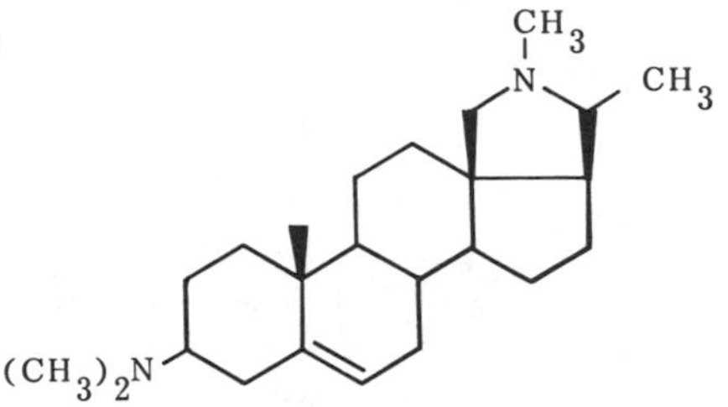

235 conessine

the 4β-series this signal appears at 1.02 ppm, but at 0.82 ppm in the 4α-series.

The structure of conessine (**235**) was elucidated by Haworth and co-workers in 1953 by an extensive series of degradative experiments, and has been confirmed by several syntheses. One of these, developed by Corey, solves the problem of functionalizing the C-18 group by photochemical means. Thus irradiation of the N-chloro compound **237** gave the isomeric C-18 chloride **238**, which could be readily cyclized with base.

A second route to the introduction of the pyrrolidine ring begins with a compound in the 18-nor series such as **239**. Reductive β-cyanation with potassium cyanide and ammonium chloride gave a mixture of C-13 stereo-isomers. Protection of the ketone, lithium aluminum hydride reduction, and acid hydrolysis gave an intermediate imine which could easily be reductively methylated (Scheme 32).

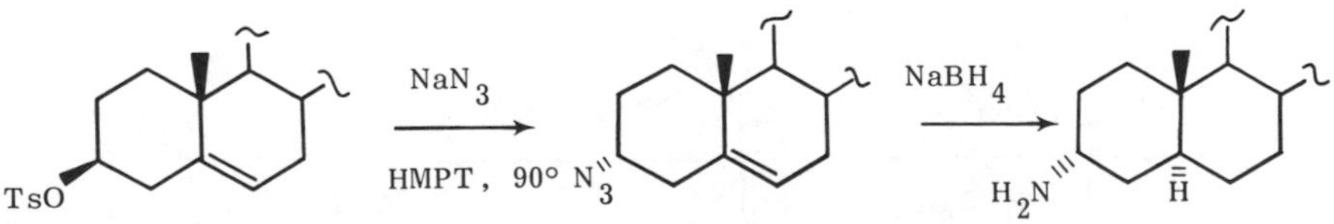

Scheme 30

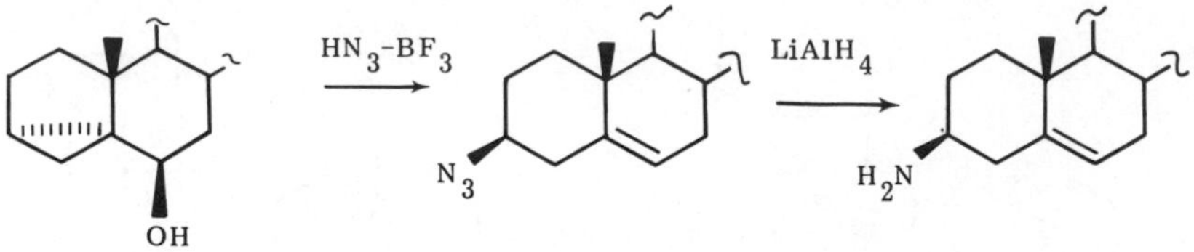

Scheme 31

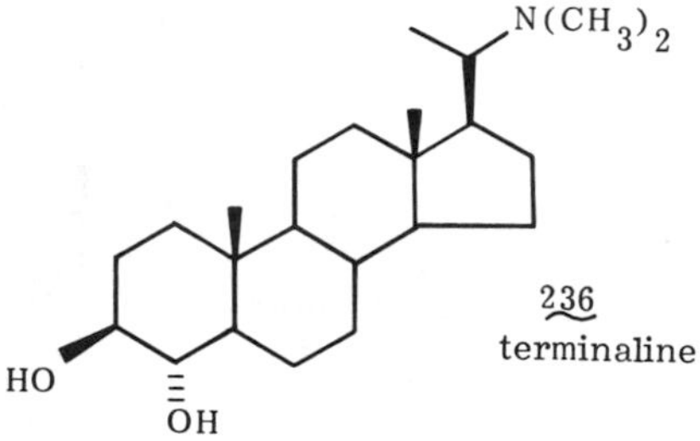

236
terminaline

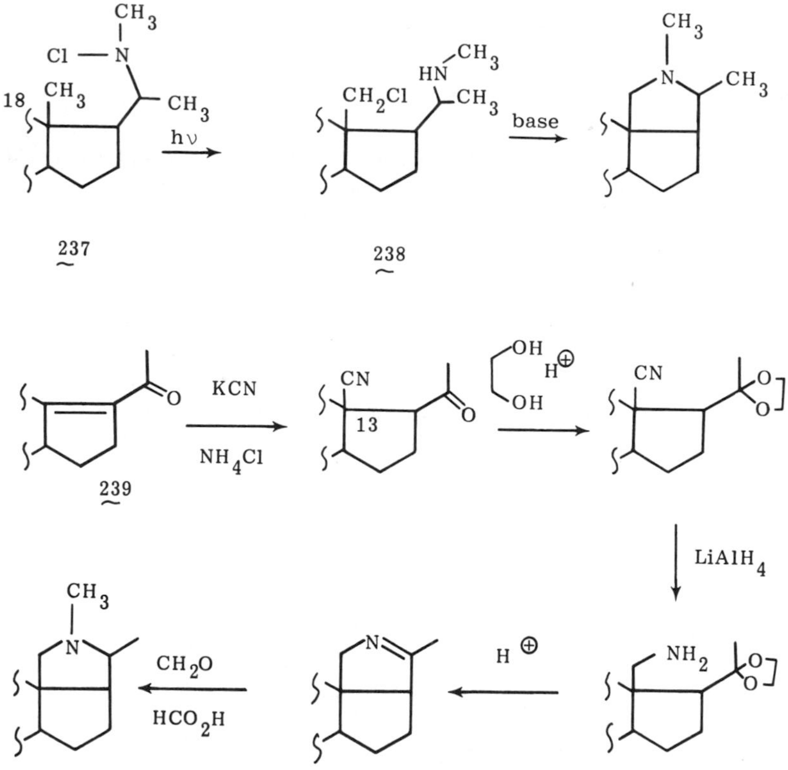

237

238

239

Scheme 32

The interest in these alkaloids over the past 20 or so years is due to the availability of normally inaccessible 18-substituted steroids, particularly those related to the adrenocortical hormone, aldosterone. For example the pyrroline **240** is readily available from the corresponding *N*-chloro compound and treatment with two equivalents of *p*-nitroperbenzoic acid gave keto oxime **241** probably *via* **242** and **243**. Standard reactions then afford the 20-epimeric lactone mixture **244**.

An alternative suggested source of the hormones is paravallarine (**245**), the main alkaloid of *Paravallaris microphylla* Pitard.

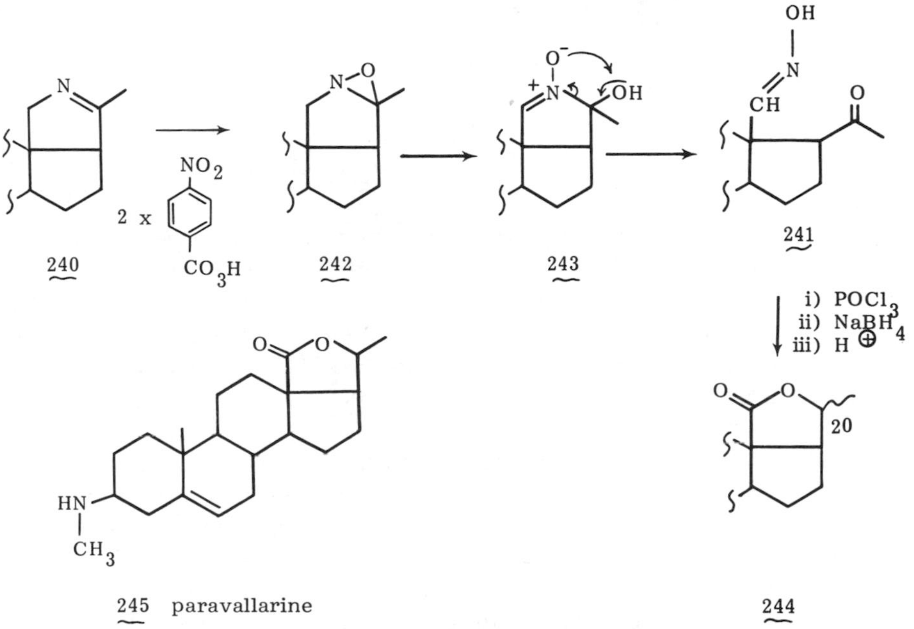

240   242   243   241

i) $POCl_3$
ii) $NaBH_4$
iii) $H^{\oplus}$

245   paravallarine   244

A reaction of some biogenetic interest in this area is the rearrangement of the mesylate of a 12β-hydroxyconanine (i.e. **246**) to give a C-nor-D-homoconanine **247** with lithium aluminum hydride or sodium borohydride under anhydrous conditions. The reaction proceeds with migration of the 18α-proton to C-13 as shown.

A compound such as **248** having a 12α-mesyloxy group on treatment with lithium aluminum hydride however gives a 12,13-dehydro-20-aminopregnane derivative **249**; this is the Grob fragmentation.

Buxus Alkaloids

Buxus sempervirens L., the common boxwood, and a member of the plant family Buxaceae, is noted for its extensive folk literature. Since ancient times extracts of boxwood have been used against various dermatitic problems. In the nineteenth century the plant gained a considerable reputation as an antimalarial, and as late as 1957 a patent was issued to Merck and Co. for use of the alkaloid mixture in the treatment of tuberculosis.

Chemical investigations began in 1830, but it was not until 1949 that pure alkaloids were obtained, and in the period 1958–1961, great strides were made in the structure elucidation of this group. The crucial work was that of Kupchan and Brown which led to the structure of cyclobuxine. These authors recognized that this was a "new class of steroidal alkaloids . . . intermediate . . . between the lanosterol and cholesterol-type steroids." The group is characterized by substitution with a methyl, methylene, hydroxy methyl, or alkoxymethyl at C-4, and at C-14 by a methyl group. Structurally, the alkaloids are a complex group and an elaborate system has been developed for naming them, where a suffix letter is used to designate the substitution by alkyl groups of nitrogen atoms at C-3 and C-20. Thus a derivative with the suffix A has two methyl groups each at the C-3 and C-20 nitrogen atoms, whereas one with the suffix D has only one methyl group at each. The two basic structural types are derivatives of 9β,19-cyclo-4,4, 14α-trimethyl-5α-pregnane, and derivatives of *abeo*-9-(10 → 19)-4,4,14α-trimethyl-5α-pregnane. Some representatives are shown in Figure 1.

There have been considerable problems involved in determining the stereochemistry of the C-4 methyl or hydroxymethyl groups. The 4-monomethyl derivatives were originally assigned as β, through a combination of molecular rotation data, NMR spectra, and other evidence. In 1966, these assignments were reversed based on comparison of the chemical shifts of the cyclopropyl protons. These appear at a lower field in the 4β-methyl and 4,4-dimethyl series than in the 4α-methyl series. The *Buxus* alkaloids having a hydroxymethyl group at C-4 were assigned stereochemistry such that this group was β. Additional evidence was also inferred from the facile hydrolysis of an amide group at C-3, and of an acetoxymethyl at C-4, by a basic amino

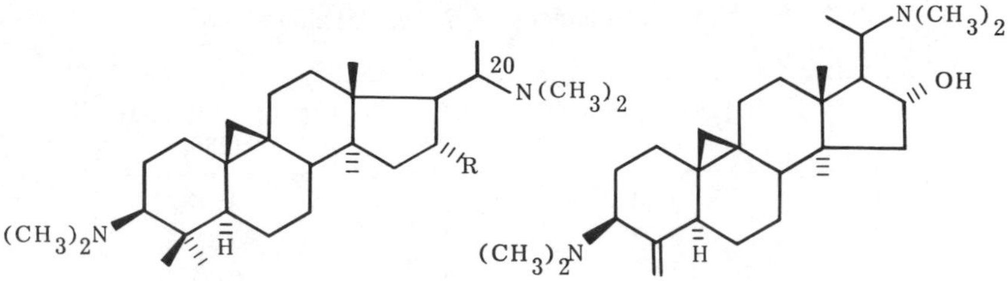

250 cycloprotobuxine, R=H

251 cyclovirobuxine, R=OH

252 cyclobuxine

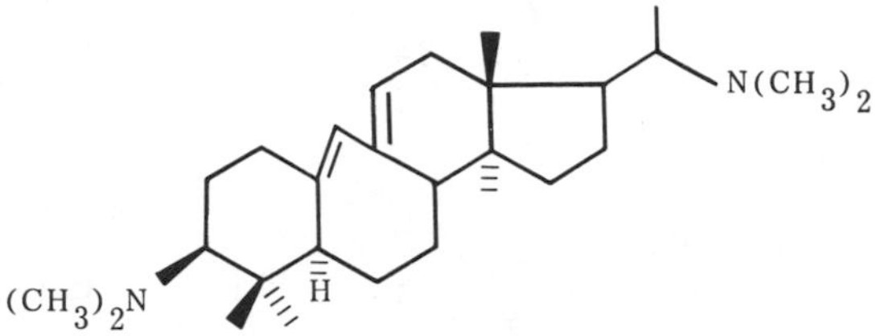

253 buxamine

Figure 1 Some representative *Buxus* alkaloids.

group at C-3. In 1975 however, ^{13}C NMR evidence was accumulated to show that this stereochemical assignment should be reversed.

The ^{13}C NMR spectral data for cyclobuxidine F (**254**) are shown in Table 2 in comparison with the corresponding data for cycloartanol (**255**) and cycloprotobuxine F (**256**). The amino methine resonances of the latter compound are assigned at 61.3 ppm (C-3) and 59.6 ppm (C-20). In **254** both these carbon atoms are in a γ-relationship to a hydroxyl group and consequently are shielded (by about 2.3 and 2.6 ppm respectively). Observe also how the γ-effect of the C-20 dimethyl amino group shields the C-21 resonance.

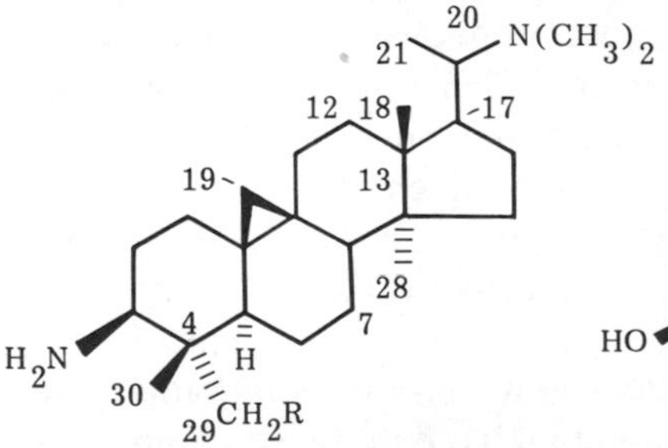

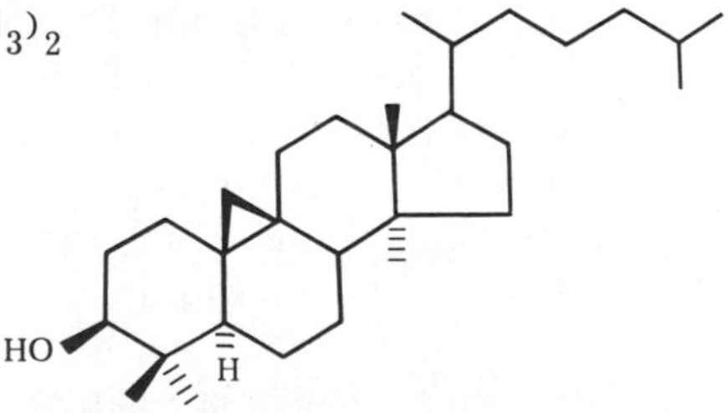

254 cyclobuxidine F, R=OH

256 cycloprotobuxine F, R=H

255 cycloartanol

Table 2 ^{13}C NMR Resonances of *Buxus* Alkaloids and Cycloartanol

	Cyclobuxidine F (254)	Cycloartanol (255)	Cycloprotobuxine F (256)
C-1	31.4	31.9	31.0
C-2	32.7	30.3	32.5
C-3	59.0	78.5	61.3
C-4	42.0	40.3	39.7
C-5	44.8	47.0	47.8
C-6	20.9	21.0	21.3
C-7	25.9	28.0	26.9
C-8	47.9	47.8	47.8
C-9	19.0	20.0	19.7
C-10	25.9	26.0	26.0
C-11	25.9	26.0	26.0
C-12	34.6	35.5	35.1
C-13	44.8	45.1	44.1
C-14	47.2	48.7	48.9
C-15	44.8	32.8	32.5
C-16	79.0	26.5	26.3
C-17	62.5	52.2	50.6
C-18	19.0^a	17.9^b	18.2^c
C-19	30.4	29.8	29.5
C-20	57.0	36.0	59.6
C-21	9.6	18.3^b	9.3
C-28	20.9^a	19.3^b	19.2^c
C-29	73.7	25.4	25.8
C-30	9.6	14.0	14.0
N$_b$—CH$_3$	2 × 40.6	—	2 × 39.7

$^{a-c}$ Assignments may be reversed.

The most important aspect however was that in **254** there were substantial changes in the methyl resonances C-29 and C-30 on introduction of a hydroxyl group at C-29 (or C-30). Distinction between these two was made on the basis of assignments in cycloartanol (**255**), where the signal at 14.0 ppm is assigned to the C-30 axial methyl group. Consequently hydroxylation has occurred on C-29, resulting in a γ-effect (shielding of 4.4 ppm) on the C-30 signal. In cyclobuxidine-F and the compounds related to it, the hydroxy methyl group at C-4 is therefore in an equatorial, α-configuration.

Aminoglycosteroids This is a relatively new class of compound in which an amino sugar is linked to a steroidal genin. To date these compounds are restricted in distribution to the *Holarrhena* species of Asia, and are frequently 14β-hydroxylated pregnane derivatives such as holacurtine (**257**) or aminoglycoside derivatives of cardenolides such as mitiphylline (**258**), a derivative of digitoxigenin.

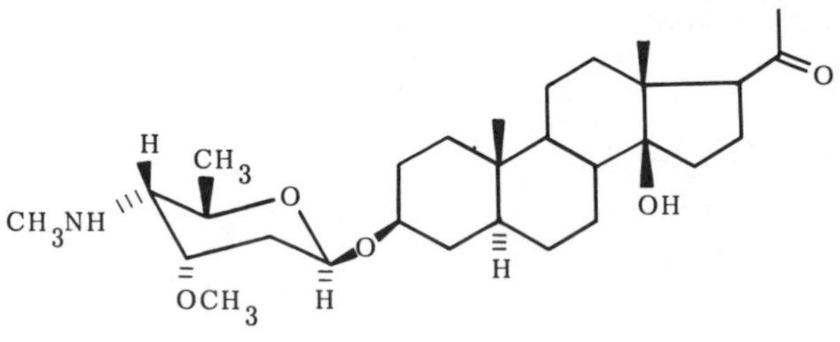

257   holacurrhine

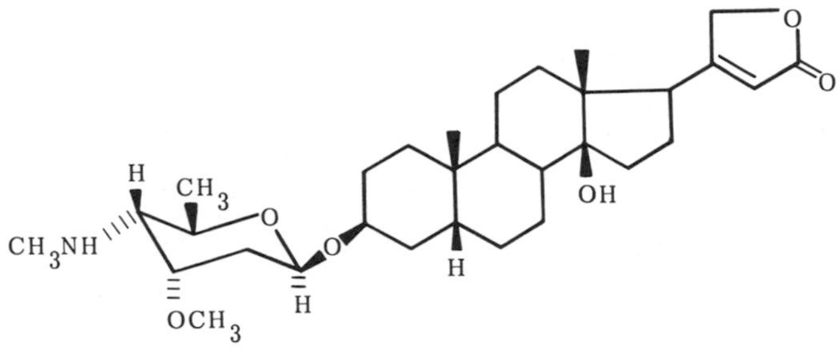

258   mitiphylline

Spectral Data of Steroidal Alkaloids

Mass Spectra Budzikiewicz has evaluated the mass spectra of several types of steroidal alkaloid, and deduced many of the principal fragmentation processes.

When an aliphatic basic nitrogen is present, fragmentation occurs between the carbon atoms α and β to the nitrogen. A typical example in the steroid alkaloid area is of a 20-amino pregnane which characteristically gives fragmentation of the 17,20-bond as shown in **259**. The ion *m/e* 72 (**260**) accounts for over 80% of the total ion current.

The spectrum of a 3-dimethylaminopregnane shows characteristic ions at *m/e* 84 (**261**) and 110 (**262**), which are rationalized by alternative 3,4- or 2,3-fragmentation as shown. Careful observation of the changes in the mass of these ions can be extremely useful in determining or demonstrating the position of substitution in rings A and B or a compound of this type.

The conessine type of alkaloid having the 18,20 imino linkage shows a prominent loss of 15 mu (the 21 carbon atom), and an intense ion at *m/e* 71 (**263**), produced by successive cleavage of the 17,20- and 13,18-bonds.

Solanidine (**227**) shows prominent fragment ions at *m/e* 204 and 150, the latter being the base ion. These ions have been formulated as **264** and **265** respectively, produced by the mechanisms shown in Scheme 33.

Alkaloids of the protoverine type which have hydroxyl groups display very few fragment ions; indeed in many only the ion *m/e* 112 (**266**) has any reasonable stability. This can be envisaged to arise from the F ring as shown.

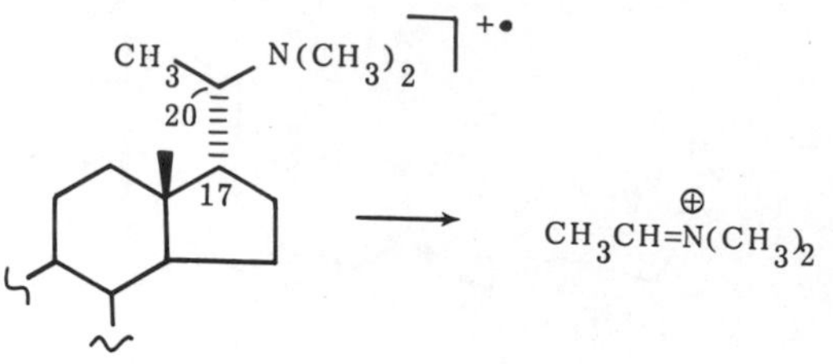

259 260 m/e 72

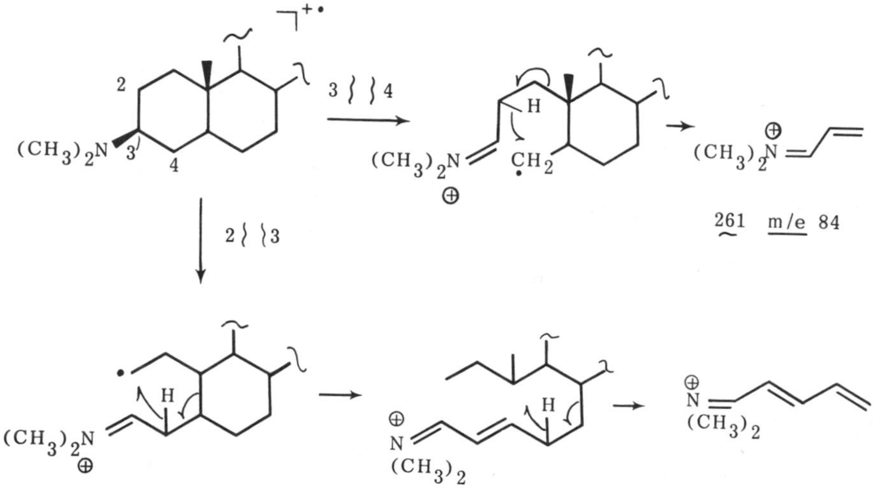

261 m/e 84

262 m/e 110

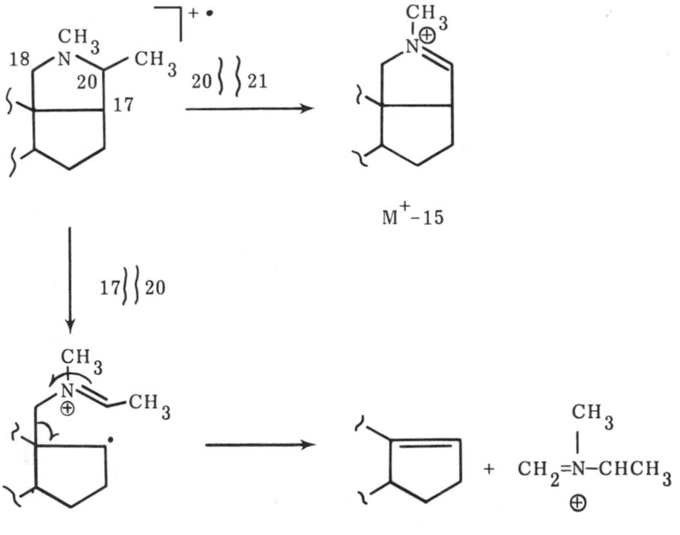

M$^+$-15

263 m/e 71

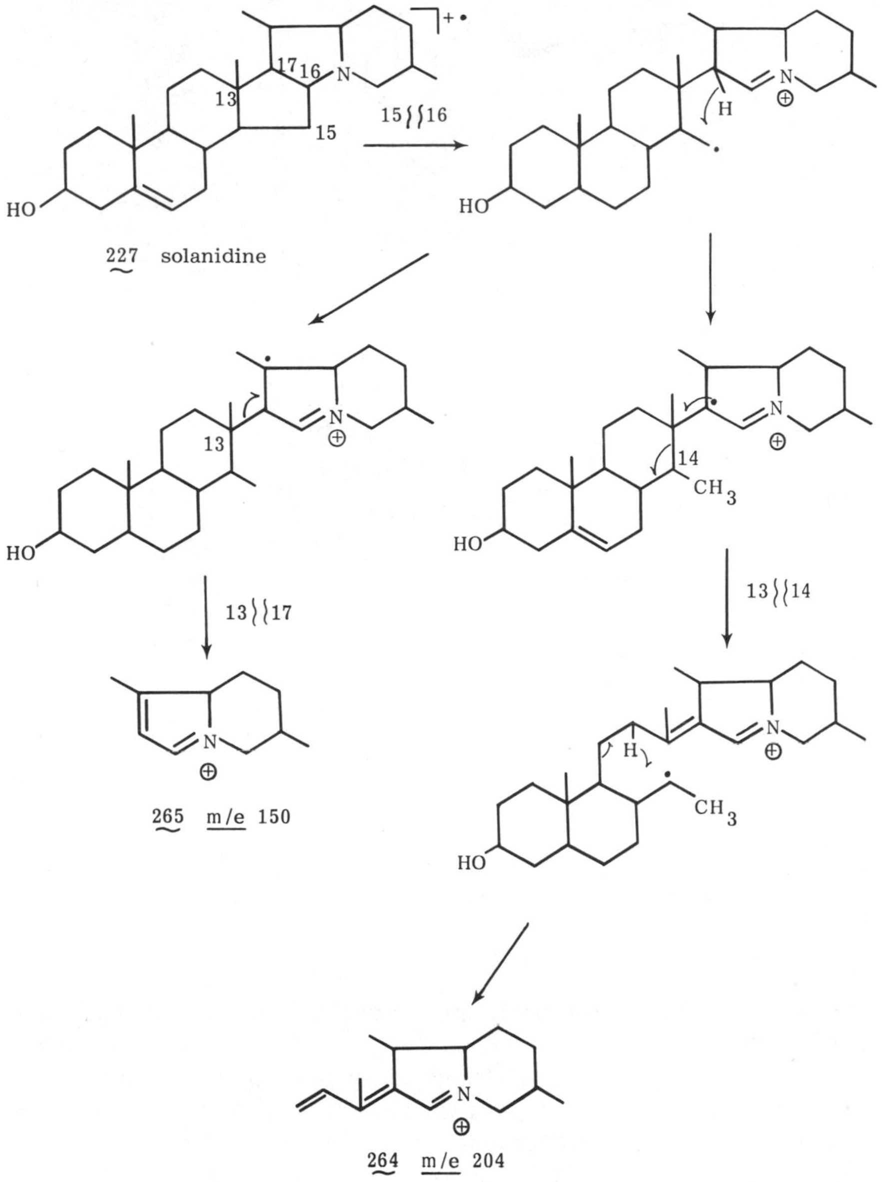

Scheme 33   Mass spectral fragmentation of solanidine (**227**).

Solasodine (**226**)-type alkaloids are also noted for giving rise to only low-molecular-weight ions, in this case ions at m/e 138 (**267**) and 114 (**268**). Again these ions contain the nitrogen atom and are apparently derived by selective fragmentation of the tetrahydrofuran ring either through the 20,22-bond or through the 16,17-bond indicated in Scheme 34.

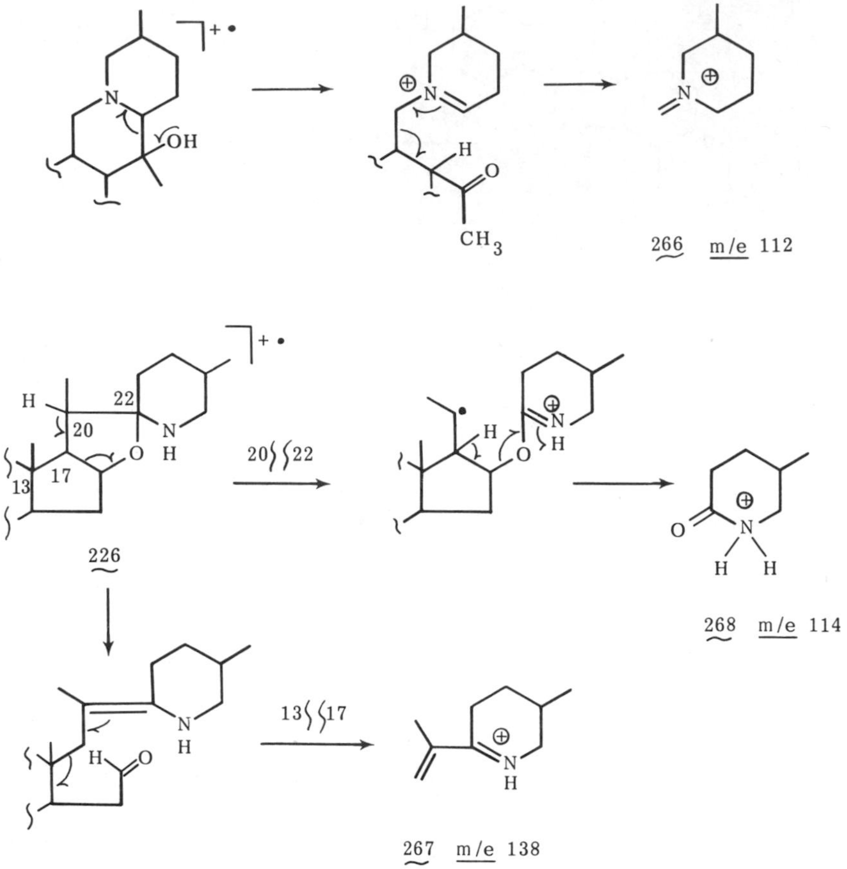

Scheme 34 Mass spectral fragmentation of solasodine (**226**).

NMR Spectra The proton NMR spectra of the *Veratrum* alkaloids were studied in the early 1960s and concentrated on the chemical shift of the methyl groups and the effect of various substituents on these shifts.

There are three methyl groups in alkaloids of the protoverine (**214**) type; two (C-19 and C-21) are tertiary and one (C-27) is secondary. As expected the C-27 signal, a doublet ($J \approx 7.1$ Hz) at 1.07 ppm is unaffected by substitution in rings A, B, C, and D. In compounds having no C-17 substituent, the C-21 methyl group appears as a singlet in the range 1.15–1.25 ppm. Acetylation of a 16-hydroxyl group shields the 21-methyl group by about 0.08 ppm and a hydroxyl group at C-17 has a similar effect. As has subsequently been demonstrated for many triterpenoid and steroid derivatives, the substituent effects are additive.

The carbon-13 NMR resonance assignments for some steroidal alkaloids are shown in Table 3.

Table 3 ^{13}C NMR Resonance Assignments for Some Steroidal Amines

	C-1	C-2	C-3	C-4	C-5	C-6	C-7	C-8	C-9	C-10	C-11	C-12	C-13
Jervine[a]	38.4	30.8	70.9	40.7	145.5	100.2	39.2	44.6	62.5	37.2	205.4	136.2	142.3
Veratramine[a]	37.6	29.8	70.1	41.9	141.9[c]	120.3	40.5	44.3	56.6	36.2	29.8	140.4	131.9
Solasodine	37.2	31.5	71.5	42.2	140.6	121.1	32.0	31.3	50.0	36.6	20.8	39.9	40.4
Tomatidine	36.8	31.3	70.7	38.0	44.7	28.5	32.1	34.9	54.2	35.4	20.9	40.0	40.7

	C-14	C-15	C-16	C-17	C-18	C-19	C-20	C-21	C-22	C-23	C-24	C-25	C-26	C-27
Jervine[a]	30.6	24.5	37.2	85.1	12.8	18.3	31.8[b]	12.1	67.1	76.6	42.3	31.4[b]	55.0	18.3
Veratramine[a]	142.9	118.6	125.4	142.9[c]	15.1	18.3	31.5[d]	18.3	67.2	69.8	34.9	31.1[d]	53.6	20.1
Solasodine	56.4	32.0	78.6	62.7	16.4	19.2[e]	41.2	15.2	98.1	34.0	30.2	31.3	47.5	19.3[e]
Tomatidine	55.6	32.5	78.3	61.8	16.8	12.2	42.8	15.7	98.7	26.5	28.5	30.8	49.9	19.2

[a] In pyridine.
[b-e] Assignments may be reversed.

214   protoverine

Biosynthesis

Biosynthetic evidence for the formation of the various steroidal alkaloid types is quite sparse, although there has been considerable biogenetic speculation as to their origin.

In Chapter 2 a considerable discussion of the intermediates in steroid biosynthesis was presented. Many of the simple alkaloids can be derived quite easily by a reductive amination of the corresponding steroid ketone, and circumstantial evidence for this hypothesis comes from the frequent co-occurrence of a closely related steroid or monoamino steroid in the plant. For example, progesterone (269) co-occurs with holaphylline (270) in *Holarrhena floribunda*. This is not to say that these compounds are biosynthetically related. In fact it has been shown that pregnenolone rather than progesterone is a precursor of holaphylline (270). [4-^{14}C]Cholesterol was a precursor of 270, but not of pregnenolone (271). If this is truly direct replacement of OH by NH$_2$, it would be analogous to the amination of sugars and purines. Cholesterol may not therefore be a true intermediate. In this case it is envisaged that pregn-5-en-3,20-dione (272) might be an intermediate. Steps from this point include amination at C-20 and functionalization at C-18. If 20-amination occurs initially, the sequence may proceed through the intermediates as shown in Scheme 35. No information is available for the sequence or veracity of these steps.

It should not be overlooked that in all the alkaloids whose stereochemistry has been vigourously determined, the 20-amino function has the α-configuration.

Not only are the steroidal alkaloids produced from steroid hormone-like precursors, but are catabolized *in vivo* to steroid hormones. For example, holaphyllamine (273) is metabolized to pregnenolone (271) in *H. floribunda*.

In spite of their novel structure there have been no studies on the biosynthesis of the *Buxus* alkaloids. A pathway was proposed in 1966 and is shown in Scheme 36. It remains to be evaluated experimentally.

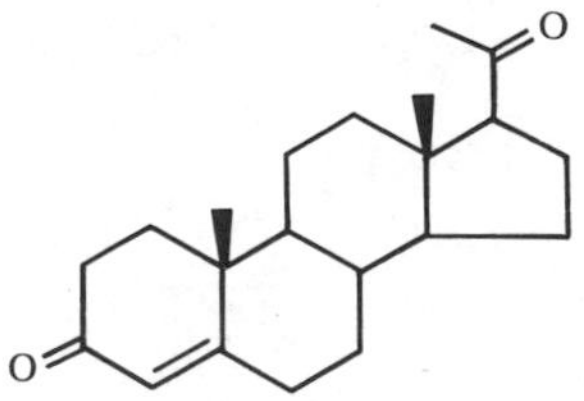

269 progesterone

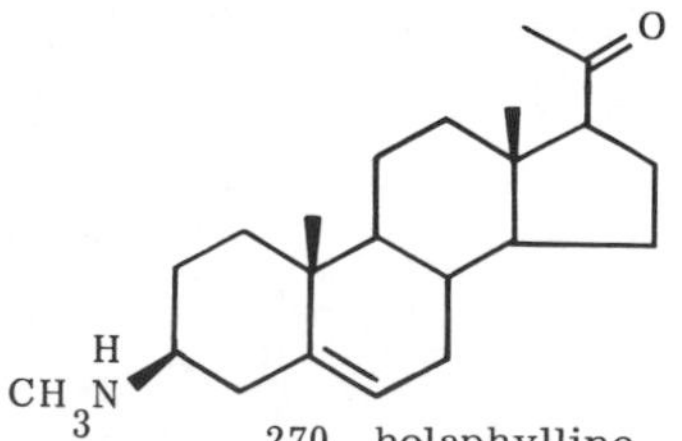

270 holaphylline

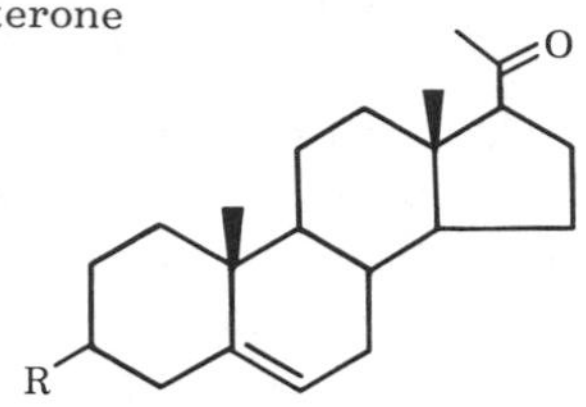

271 pregnenolone, R= OH, H
272 R= =O

Cholesterol was a precursor of tomatidine (**225**) and solasodine (**226**) in *S. lycopersicum*, but the 5,6-dihydroderivative was not evaluated.

There are clearly several alternative mechanisms for the introduction of the 25-amino group into the spirosolane alkaloids and it has been shown that the 26-oxo compound is not involved. The amination of C-26 is therefore by direct displacement of the hydroxyl group.

More interestingly, the configuration at C-25 of the spirosolanol alkaloids is determined by the enzyme that hydroxylates C-26 or C-27. An excellent illustration of this concerns the alkaloids tomatidine (**225**) and soladulcidine (**274**), which differ only in the C-22 stereochemistry and co-occur in *S. lycopersicon*. Thus 25R-[5α,6-³H₂]cholestan-3β,26-diol (**275**) was incorporated into soladulcidine (**274**), but not into tomatidine (**225**). These data suggest

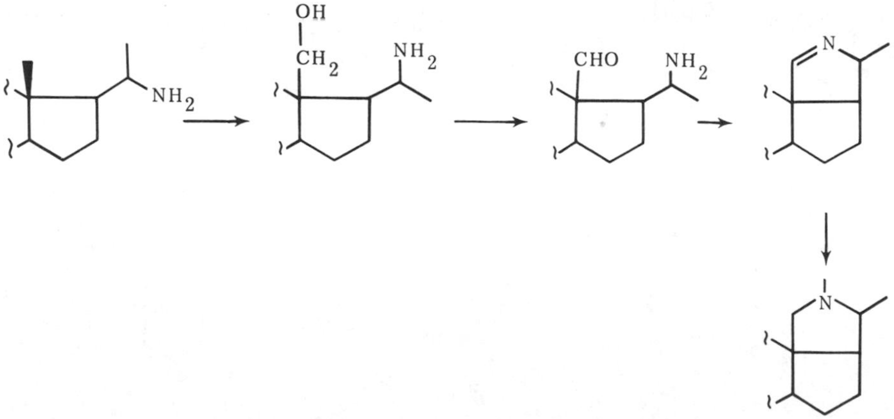

Scheme 35 Biogenesis of *Holarrhena* alkaloids.

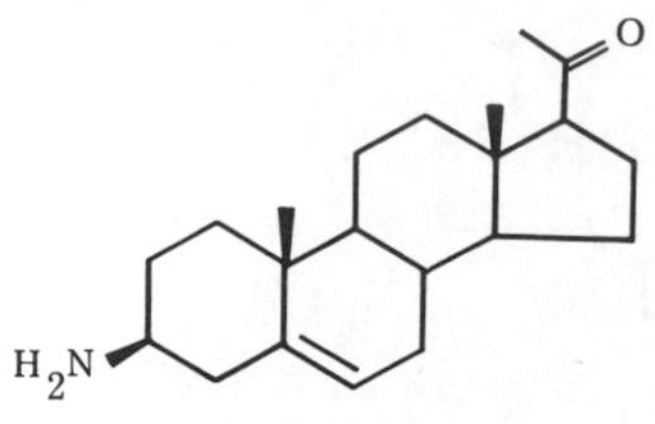

273   holaphyllamine

that hydroxylation at C-16 occurs subsequent to C-26 hydroxylation, but only one advanced intermediate has been examined to establish this.

The mechanism of closure of the E and F rings of tomatidine (**225**) and solanidine (**227**) has been studied by Canonica and associates. Using [4-^{14}C,16β-^{3}H]cholest-5-en-3β-ol (**276**) as a substrate for **225** in *Lycopersicon pimpinellifolium*, it was found that tritium was retained at C-16 in the α-position (i.e. with inversion of the configuration). This is not the usual result of biological hydroxylation of saturated carbons which normally proceeds with retention of configuration. The results exclude a C-16 keto group but do not point to a unique scheme. One mechanism that would explain the results involves an initial 16α-hydroxylation followed by a sequential nucleophilic attack at a C-22 keto group and displacement of the 16α-hydroxy group by the oxygen anion with inversion at C-16 (Scheme 37).

In the solasodine series, Tschesche and Spindler have established that **277** and its 16β-hydroxy derivative **278** are precursors of solasodine (**226**) in *S. dulcamara*.

The only group of steroidal alkaloids that have been investigated bio-

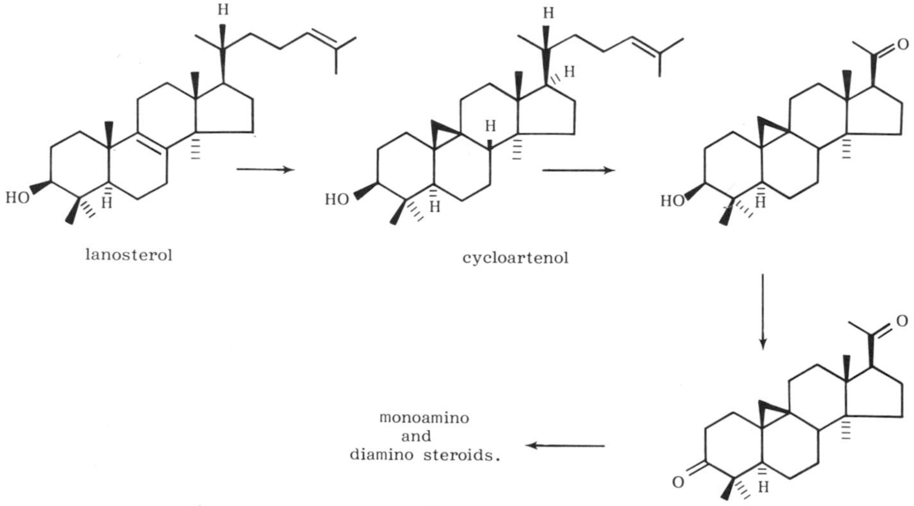

lanosterol   cycloartenol

monoamino
and
diamino steroids.

Scheme 36   Biogenesis of the *Buxus* alkaloids.

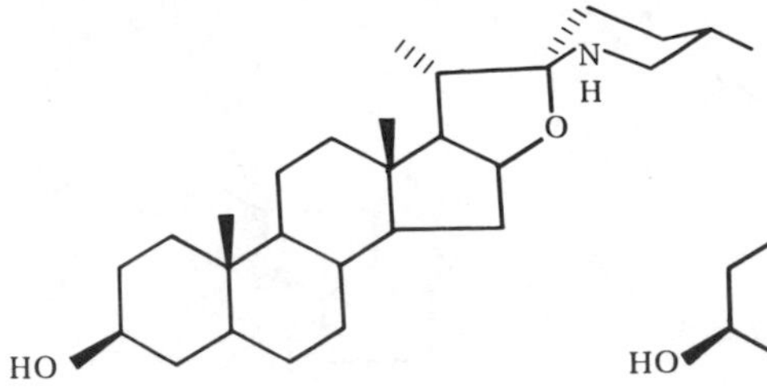

225 tomatidine

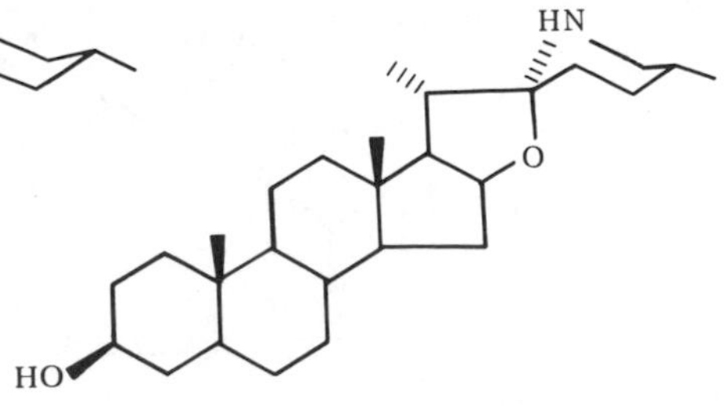

274 soladulcidine

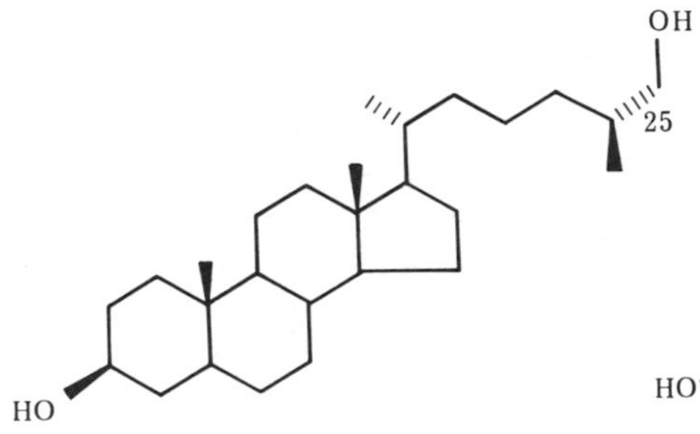

275

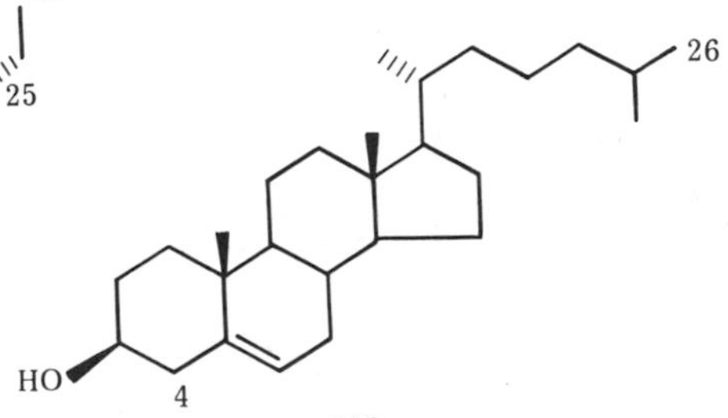

276

synthetically in any depth are those of *Veratrum grandiflorum*, and Mitsu-hashi and co-workers have been responsible for essentially all of this work. *V. grandiflorum* was chosen because it contains five types of steroidal alkaloids. Both [4-^{14}C]- and [26-^{14}C]cholesterol (276) were used as precursors in *V. grandiflorum* Loesen. and incorporated into jervine (185) and veratramine (186). But [4-^{14}C]cholesterol was not incorporated into 185 in *V. album* subsp. *lobelianum*, probably due to transportation problems. [^{14}C]11-De-oxyjervine (279) was incorporated into jervine (185) but not into veratramine (186), and the incorporation of acetate into jervine (185) was reduced in the presence of 279. This suggests that 279 is a normal precursor of 185 and that incorporation is not due to an aberrant pathway. In addition, these data

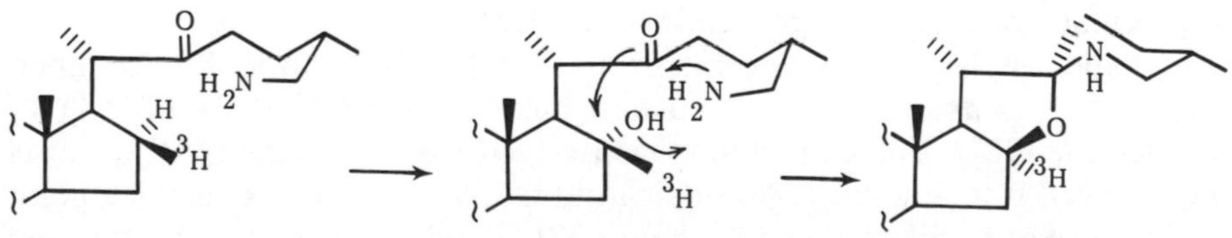

225

Scheme 37

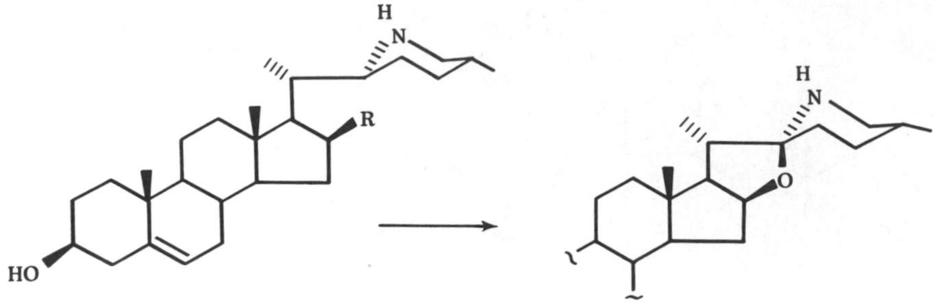

277   R=   H
278   R=   OH

226

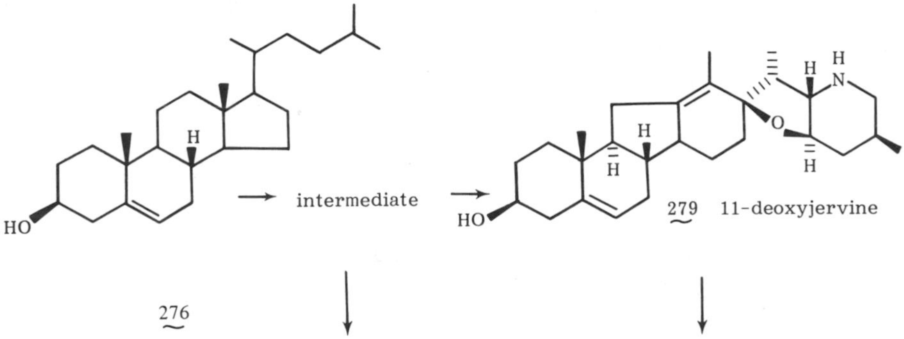

276

intermediate

279   11-deoxyjervine

186   veratramine

185   jervine

indicate an intermediate between cholesterol (276) and 11-deoxyjervine (279) and veratramine. The nature of this intermediate has recently been investigated.

Etiolated *V. grandiflorum* seedlings accumulate solanidine glycoside in the leaf, and jerveratrum alkaloids in a gradual process when the plant is illuminated. In the early stage of etiolation a different major alkaloid having the structure 280 accumulated. This alkaloid, etioline, was suggested to be a precursor of solanidine glycoside in the leaf.

Studies on the roots however indicated a different situation. For example, dormant *V. grandiflorum* accumulated hakurirodine (281), a precursor of rubijervine (282). The compound was derived from verazine (283) but was not derived from etioline (280), indicating that verazine is a branching point in the proposed biosynthesis (Scheme 38). Previous work had established that dormantinol (284) and dormantinone (285) were present in the budding plant and recently the nitrogen atom has been shown to be derived from L-arginine.

The formation of the C-nor-D-homosteroidal alkaloids such as jervine and veratramine and the protoverine-type (**286**) alkaloids has been postulated to occur from the solanidine type, specifically epirubijervine (**287**). There is no information to prove or disprove this point at present and a postulated scheme is presented in Scheme 39. Notice that to date there have been no alkaloids isolated with the skeleton **288**, nor indeed has a simple alkaloid such as **289** been obtained which would be the key intermediate for the separate formation of the jervine and veratramine types.

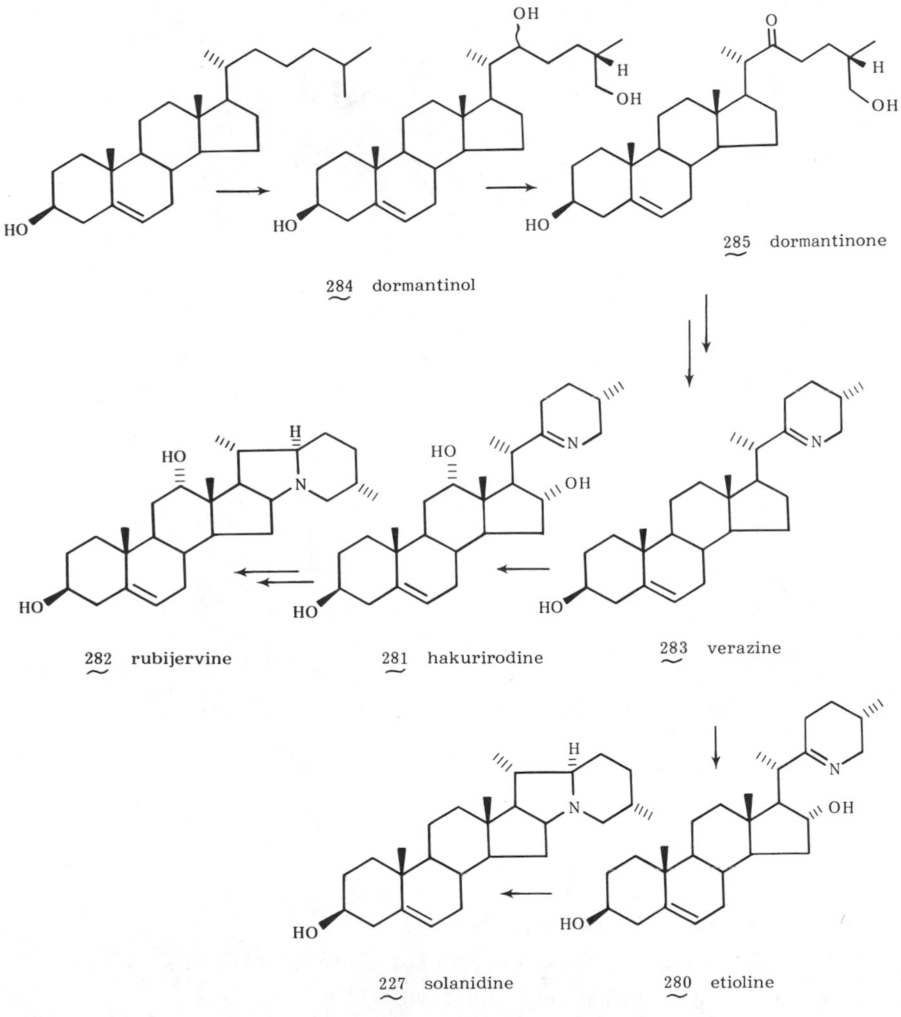

284   dormantinol

285   dormantinone

282   rubijervine

281   hakurirodine

283   verazine

227   solanidine

280   etioline

Scheme 38

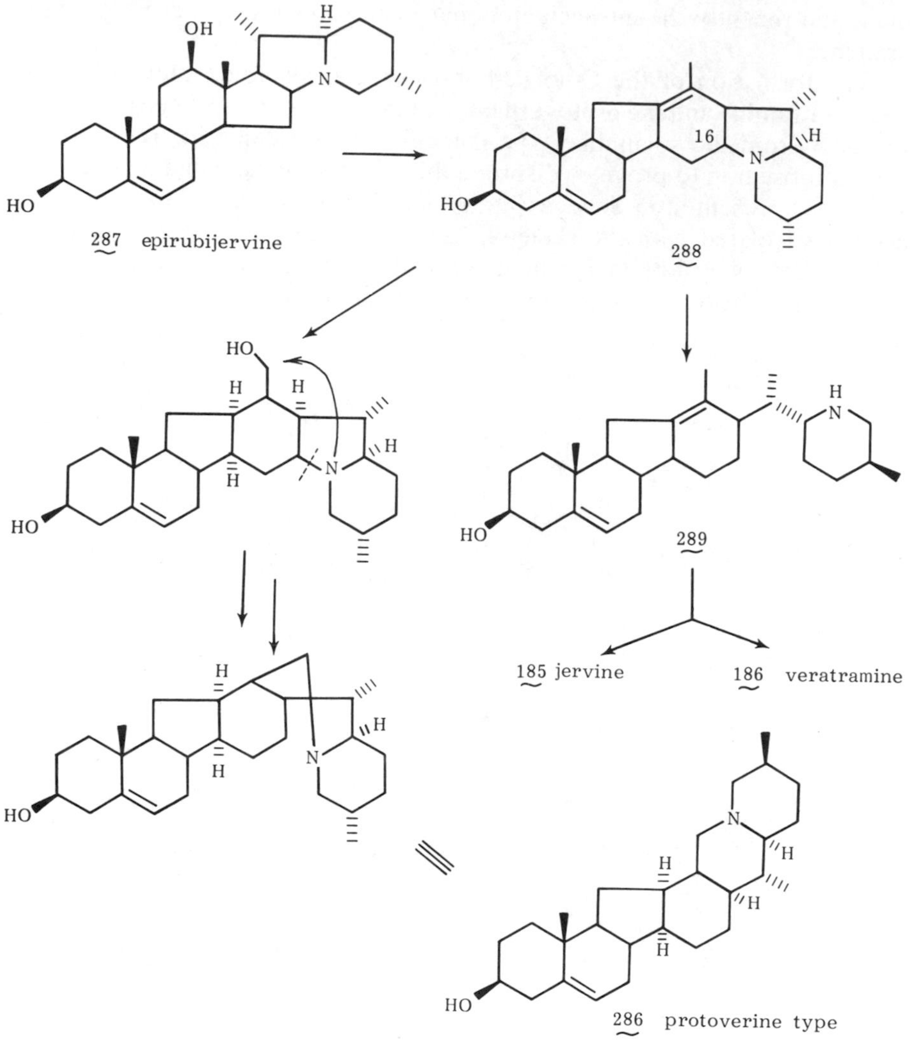

Scheme 39   Biogenesis of the *Veratrum* alkaloids.

LITERATURE

Reviews

Prelog, V., and O. Jeger, *Alkaloids NY*, **3**, 247 (1953).

Fieser, L. F., and M. Fieser, *Steroids*, Van Nostrand Reinhold, New York, 1959.

Prelog, V., and O. Jeger, *Alkaloids NY* **7**, 319 (1960).

Shoppee, C. W., in *Chemistry of the Steroids*, Butterworths, Washington, D.C., 1964, p. 433.

Kupchan, S. M., and W. E. Flacke, in E. Schlittler (Ed.), *Antihypertensive Agents,* Academic, New York, 1967, p. 429.

Schreiber, K., *Alkaloids NY* **10**, 1 (1968).

Kupchan, S. M., and A. W. By, *Alkaloids NY* **10**, 193 (1968).

Sato, Y., in S. W. Pelletier (Ed.), *Chemistry of the Alkaloids,* Van Nostrand Reinhold, New York, 1970, p. 591.

Brown, K. S., Jr., in S. W. Pelletier (Ed.), *Chemistry of the Alkaloids,* Van Nostrand Reinhold, New York, 1970, p. 631.

Tomko, J., and Z. Voticky, *Alkaloids NY* **14**, 1 (1973).

Habermehl, G. G., in K. F. Wiesner (Ed.), *MTP International Review of Science, Organic Chemistry,* Ser. 1, Vol. 9, *Alkaloids,* Butterworths, London, 1973, p. 235.

Salamandra Alkaloids

Schöpf, C., *Experientia* **17**, 285 (1961).

Habermehl, G., in W. Bucherl and E. Buckley (Eds.), *Venomous Animals and Their Venoms,* Academic, New York, 1971.

Benn, N. H., and R. Shaw, *Chem. Commun.* 288 (1973).

Habermehl, G., and A. Haaf, *Chem. Ber.* **101**, 198 (1968).

Batrachotoxinin A

Daly, J. W., B. Witkop, P. Bommer, and K. Biemann, *J. Amer. Chem. Soc.* **87**, 126 (1965).

Imhof, R., E. Gossinger, W. Graf, L. Berner-Fenz, H. Berner, R. Schaufelberger, and H. Wehrli, *Helv. Chim. Acta* **56**, 139 (1973).

Albuquerque, E. X., J. W. Daly, and B. Witkop, *Science* **172**, 995 (1971).

Albuquerque, E. X., and J. W. Daly, in P. Cuatrecasas (Ed.), *The Specificity and Action of Animal, Bacterial and Plant Toxins,* Chapman & Hall, London, 1977, p. 297.

Veratrum Alkaloids

Jeger, O., and V. Prelog, *Alkaloids NY* **7**, 363 (1960).

Kupchan, S. M., *J. Pharm. Sci.* **50**, 723 (1961).

Narayanan, C. R., *Forstschr. Chem. Org. Naturs.* **20**, 298 (1962).

Kupchan, S. M., and A. W. By, *Alkaloids NY* **10**, 193 (1968).

Synthesis

Masamune, T., M. Takasugi, A. Murai, and K. Kobayashi, *J. Amer. Chem. Soc.* **89**, 4521 (1967).

Johnson, W. S., H. A. P. de Jongh, C. E. Coverdale, J. W. Schtt, and V. Burckhardt, *J. Amer. Chem. Soc.* **89**, 4523 (1967).

Kutney, J. P., C. C. Fortes, T. Honda, Y. Murakami, A. Preston, and Y. Ueda, *J. Amer. Chem. Soc.* **99**, 964 (1977).

Solanum Alkaloids

Roddick, J., *Phytochemistry* **13**, 9 (1974).

Apocynaceae Alkaloids

Jeger, O., and V. Prelog, *Alkaloids NY* **7**, 319 (1960).

Goutarel, R., in *Les Alcaloides stéroidiques des Apocynacées*, Hermann, Paris, 1964.

Cerny, V., and F. Sorm, *Alkaloids NY* **9**, 305 (1967).

Sato, Y., in S. W. Pelletier (Ed.), *Chemistry of the Alkaloids*, Van Nostrand Reinhold, New York, p. 591 (1970).

Buxus Alkaloids

Cerny, V., and F. Sorm, *Alkaloids NY* **9**, 305 (1965).

Tomko, J., and Z. Voticky, *Alkaloids NY* **14**, 1 (1973).

Sangare, M., F. Khuong-Hun, D. Herlem, A. Milliet, B. Septe, G. Berenger, and G. Lukacs, *Tetrahedron Lett.* 1791 (1975).

Khuong-Hun-Laine, F., A. Milliet, N. E. Bisset, and R. Goutarel, *Bull. Soc. Chim. Fr.* 1216 (1966).

Spectral Data of Steroidal Alkaloids
Mass Spectra

Budzikiewicz, H., *Tetrahedron* **20**, 2267 (1964).

NMR Spectra

Ho, S., J. B. Stothers, and S. M. Kuphan, *Tetrahedron* **20**, 913 (1964).

Boll, P. M., and W. Von Philipsborn, *Acta Chem. Scand.* **19**, 1365 (1965).

Radeglia, R., G. Adam, and H. Ripperger, *Tetrahedron Lett.* 903 (1977).

Budzikiewicz, H., *Tetrahedron* **20**, 2267 (1964).

Biosynthesis of Steroidal Alkaloids

Bennett, R. D., and E. Heftmann, *Phytochemistry* **4**, 873 (1965).

Tomatidine–Solasodine

Tschesche, R., B. Goossens, and A. Töpfer, *Phytochemistry* **15**, 1387 (1976).

Canonica, L., F. Ronchetti, G. Russo, and G. Sportoletti, *Chem. Commun.* 286 (1977).

Tschesche, R., and M. Spindler, *Phytochemistry* **17**, 251 (1978).

Solanidine

Kaneko, K., H. Mitsuhashi, K. Hirayama, and S. Ohmori, *Phytochemistry* **9**, 2497 (1970).

Kaneko, K., M. Watanabe, Y. Kawakoshi, and H. Mitsuhashi, *Tetrahedron Lett.* 4251 (1971).

Kaneko, K., M. W. Tanaka, and H. Mitsuhashi, *Phytochemistry* **15**, 1391 (1976).

Jervine-Veratramine Type

Kaneko, K., H. Mitsuhashi, K. Hirayama, and N. Yoshida, *Phytochemistry* **9**, 2489 (1970).

Chapter 12

MISCELLANEOUS ALKALOIDS

There are several groups of alkaloids which cannot readily be classified in any of the preceding chapters and are therefore grouped at this point. There are no unifying structural features for these compounds and each section deals with a quite distinct and highly specialized group of compounds. Even more interesting is the range of sources and pharmacological activities exhibited by these diverse structure types.

12.1 TOXINS OF THE FROGS OF THE DENDROBATIDAE

Witkop and Daly at the National Institutes of Health have been examining the potent pharmacologic principals of frogs of the Dendrobatidae for many years. In 1969, they reported the isolation of three toxins from the skin secretions of the tropical frogs *Dendrobates pumilio* and *D. auratus*; one of these toxins, pumiliotoxin C, has the novel structure 1. Subsequently, the same group obtained histrionicotoxin (2) and related compounds from *Dendrobates histrionicus* native to Colombia, and gephyrotoxin (3) more recently from the same source. Note the *cis*-enyne units present in 2 and 3. Witkop has provided a fascinating account of the history of this work.

The potent biological activity of these compounds has stimulated intense synthetic interest, with the result that at least five total syntheses of pumiliotoxin C have been reported.

The most efficacious of these is due to Overman and Jessup. Considering the *cis* stereochemistry of the octahydroquinoline nucleus, they relied on the Diels-Alder reaction in the initial step. Indeed three chiral centers were set in the initial reaction, a cyclo addition of the dienamide 4 with *trans*-crotonaldehyde, which afforded 5 in over 60% yield. Chain extension with the Wadsworth-Emmons reagent followed by catalytic reduction afforded the keto urethane 6. This compound on deprotection spontaneously cyclized, and catalytic reduction of the intermediate imine afforded pumiliotoxin C

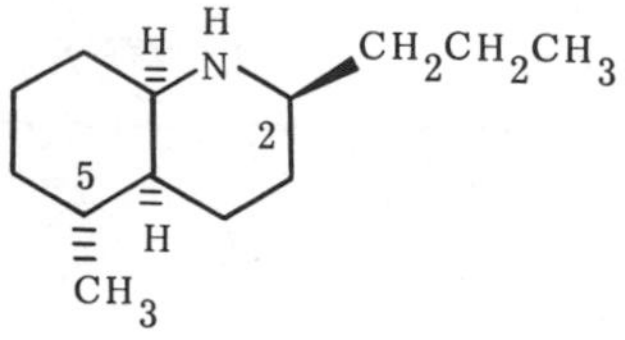

1   pumiliotoxin C

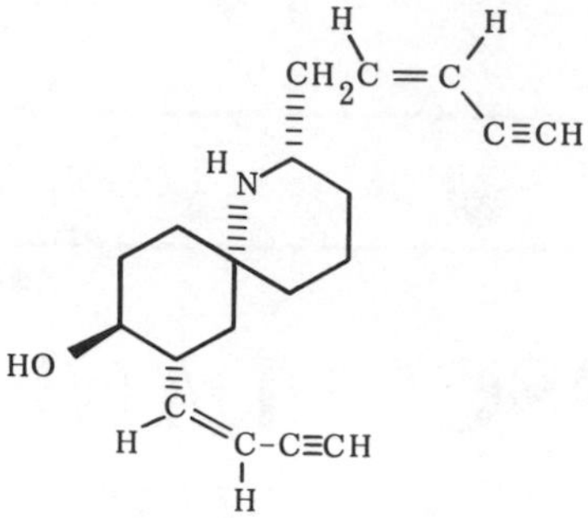

2   histrionicotoxin

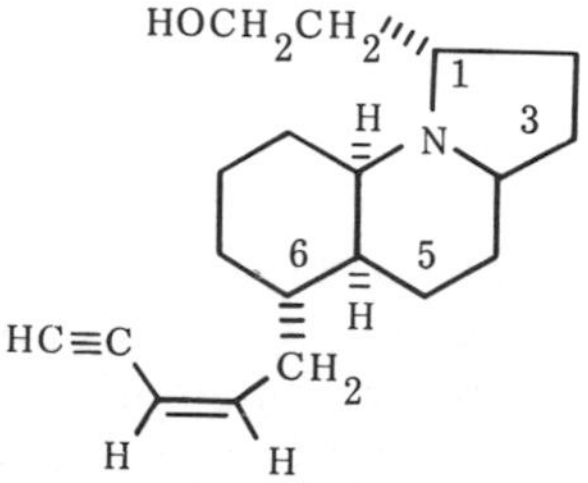

3   gephyrotoxin

(1) in high yield. The synthesis also confirmed the 2S stereochemistry of natural pumiliotoxin C.

No total syntheses of histrionicotoxin have been described, but separate routes have been discussed for the formation of the nucleus [i.e. perhydrohistrionicotoxin (7)] and for the stereoselective formation of the cis-enyne units.

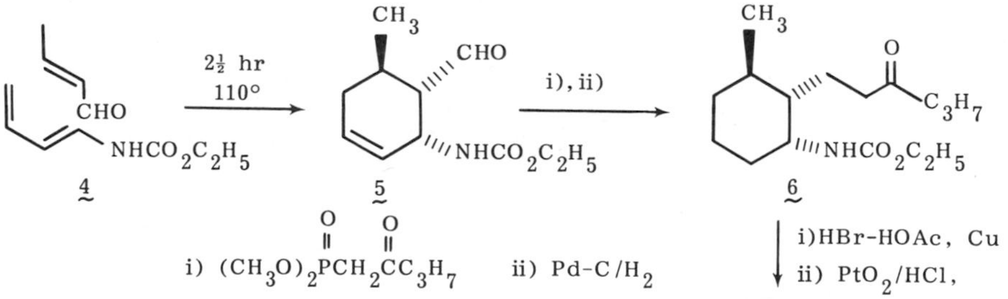

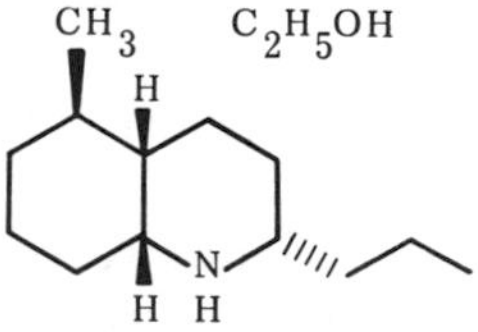

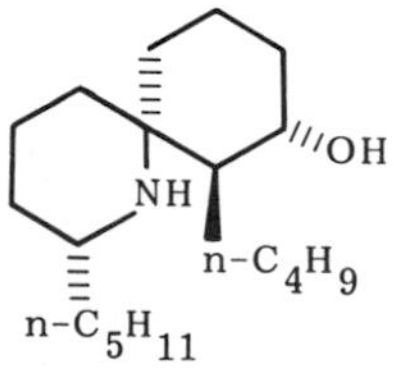

7   perhydrohistrionicotoxin

Histrionicotoxins have quite specific activity as inhibitors of neuromuscular cholinergic receptor mechanisms.

LITERATURE

Witkop, B., *Experientia* **27**, 1121 (1971).

Daly, J. W., T. Tokuyama, G. Habermehl, I. L. Karle, and B. Witkop, *Annalen* **729**, 198 (1969).

Daly, J. W., I. L. Karle, C. W. Myers, T. Tokuyama, J. A. Waters, and B. Witkop, *Proc. Natl. Acad. Sci. USA* **68**, 1870 (1971).

Daly, J. W., B. Witkop, T. Tokuyama, T. Nishikawa, and I. L. Karle, *Helv. Chim. Acta* **60**, 1128 (1977).

Overman, L. E., and P. J. Jessup, *Tetrahedron Lett.* 1253 (1977).

12.2 SAXITOXIN

The dinoflagellates, *Gonyaulax catenella* and *G. tamarensis,* produce a massive bloom, the so-called red tide, which has caused serious health problems on both the west and east coasts of North America. The red tides infest several types of shellfish in these areas, which when consumed are extremely toxic. Symptoms of toxicity include numbness of the extremities, muscular incoordination, respiratory problems, and finally death.

Saxitoxin has been isolated from the red tide and several crustaceans, including Alaska butter clams (*Saxidamus giganteus*) and mussels (*Mytilus californianus*). It is among the most toxic substances ever obtained having an LD_{50} of 5–10 µg/kg in the mouse (i.p.).

Saxitoxin proved to be a particularly perverse structure to deduce. Even the molecular formula proved difficult to obtain, and before the correct structure was deduced at least two incorrect structures, **8** and **9**, were proposed. The structure **10** was eventually elucidated by a single-crystal X-ray diffraction study of the *p*-bromobenzene sulfonate derivative.

Saxitoxin readily loses water on drying at 110°C (loss of the α-OH and

8

9

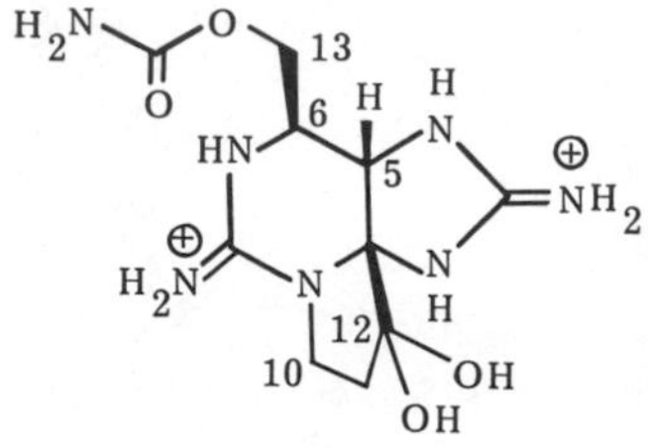

10 saxitoxin

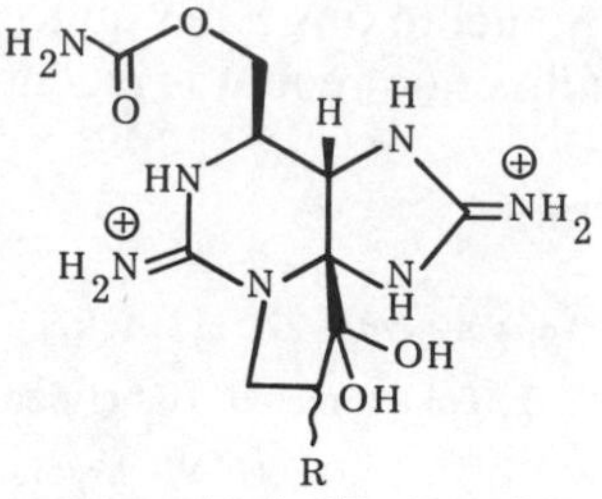

11 gonyautoxin II, R=α-OH

12 gonyautoxin III, R=β-OH

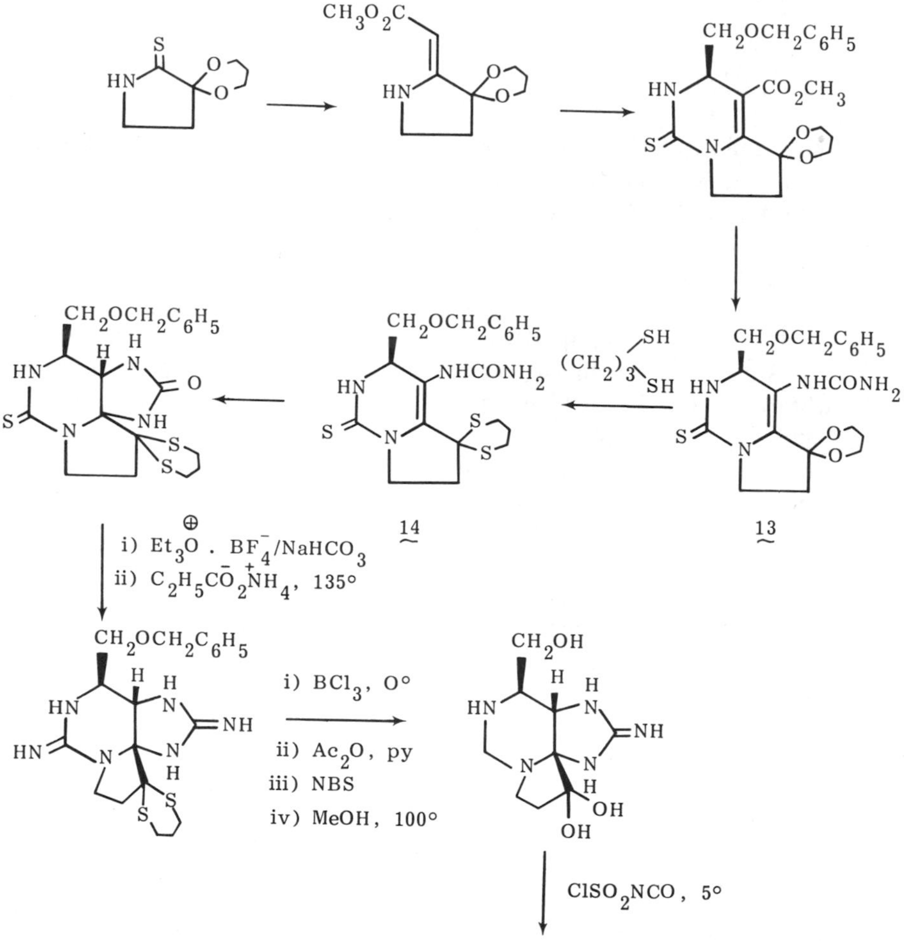

$(CH_2)_3$ SH SH
14
13
i) $Et_3\overset{\oplus}{O} \cdot \overset{-}{BF_4}$/$NaHCO_3$
ii) $C_2H_5CO_2^-\overset{+}{N}H_4$, 135°
i) BCl_3, 0°
ii) Ac_2O, py
iii) NBS
iv) MeOH, 100°
$ClSO_2NCO$, 5°

10 saxitoxin

Scheme 1 Kishi synthesis of saxitoxin (**10**).

Table 1 ^{13}C NMR Data of Saxitoxin and Gonyautoxin III (δ, ppm CDCl$_3$)

	C-2	C-4	C-5	C-6	C-8	C-10	C-11	C-12	C-13	C-14
Saxitoxin	156.1[a]	82.6	57.3	53.2	157.9[a]	43.0	33.1	98.9	63.3	159.0[a]
Gonyautoxin II	156.2[a]	81.5	57.9	53.2	158.0[a]	50.9	77.6	97.5	63.3	159.1[a]

[a] Assignments may be reversed.

adjacent H) and rehydrates easily. The NMR assignments of saxitoxin include two doublets of doublets at 4.27 and 4.05 ppm for the C-13 protons, an eight-line pattern at 3.87 ppm for the C-6 proton, and a doublet at 4.77 ppm for H-5.

Subsequently Shimizu and Nakanashi and their co-workers have reported two related toxins, gonyautoxins II (11) and III (12), from soft shell clams, *Mya arenaria* infested with *G. tamarensis*. As can be seen the compounds are 11-hydroxy derivatives of saxitoxin (10).

The ^{13}C NMR data of saxitoxin (10) and gonyautoxin II (11) are shown in Table 1.

Final proof of the correct structure of saxitoxin came from a beautiful total synthesis developed by Kishi and co-workers. The synthesis, some key steps of which are shown in Scheme 1, relies quite heavily on the subtle differences between sulfur and oxygen chemistry for its successful completion.

An example of this selective sulfur chemistry involves the thiourea derivative 13, which was acid labile. In order to proceed further it was therefore necessary to concert to the corresponding thioketal thiourea 14. The ketal group was now stable to the acidic conditions of the next step.

LITERATURE

Ghazzarossian, V. E., E. J. Schantz, H. K. Schnoes, and F. M. Strong, *Biochem. Biophys. Res. Commun.* **59**, 1219 (1974).

Schantz, E. J., V. E. Ghazarossian, H. K. Schnoes, F. M. Stong, J. P. Springer, J. O. Pezzanite, and J. Clardy, *J. Amer. Chem. Soc.* **97**, 1238 (1975).

Shimizu, Y., L. J. Buckley, M. Alam, Y. Oshima, W. E. Fallon, H. Kasai, I. Miura, V. P. Gullo, and K. Nakanishi, *J. Amer. Chem. Soc.* **98**, 5414 (1976).

Tanino, H., T. Nakata, T. Kaneko, and Y. Kishi, *J. Amer. Chem. Soc.* **99**, 2818 (1977).

12.3   SPERMIDINE AND RELATED ALKALOIDS

The two polyamines spermidine (15) and spermine (16) are analogues of the diamine putrescine (17), which is produced by reductive decarboxylation of lysine. They may be regarded therefore as being derived from a core unit of putrescine (17), which is then substituted on N either once or twice by propylamine residues.

It is considered that these compounds occur in almost all animals and microorganisms and possibly most higher plants. Detections and/or isolations from higher plants used as foodstuffs include cabbage leaves, tomato

$H_2N(CH_2)_3NH(CH_2)_4NH_2$ $H_2N(CH_2)_3NH(CH_2)_4NH(CH_2)_3NH_2$ $H_2N(CH_2)_4NH_2$

15 spermidine   16 spermine   17 putrescine

juice, apples, and spinach as well as the leaves of wheat, maize, pea, black currant, and tobacco. Spermidine and spermine also occur (12:1 ratio) in high concentrations in human semen (0.5–3.5 mg/5 ml) and crystals of spermine phosphate were first detected by van Leeuwenhoek in 1678. The name was given by Ladenburg some 210 years later, although the structure of this simple amine was not deduced until 1926.

The polyamines of human semen are formed primarily in the prostrate gland, although their function remains unclear. They may be present for their bacteriostatic effects or for stabilization of DNA (see later).

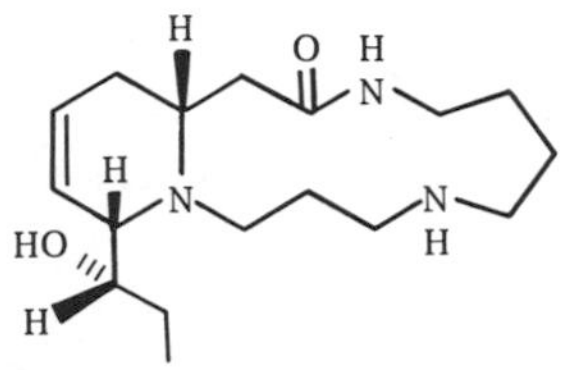

18 palustrine

Equisetum sp. (Equisetaceae)

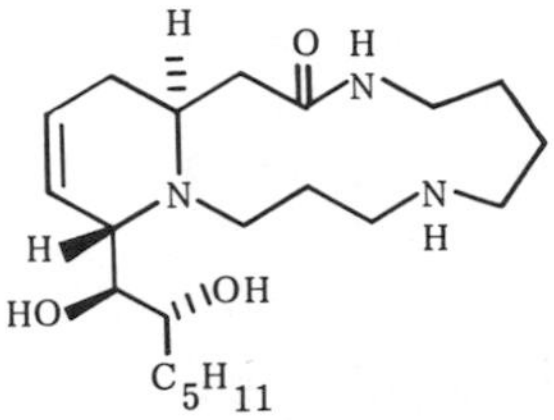

19 cannabisativine

Cannabis sativa
(Moraceae)

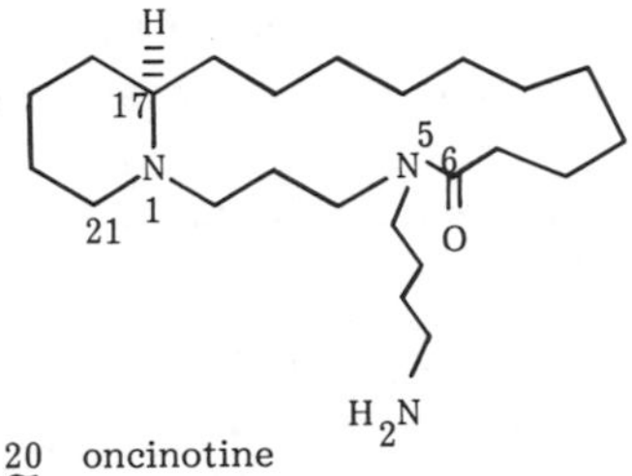

20 oncinotine

Oncinotis nitida
(Apocynaceae)

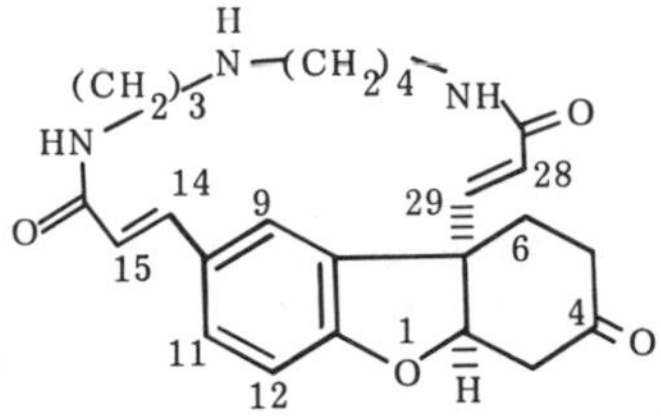

21 lunarine

Lunaria biennis (Cruciferae)

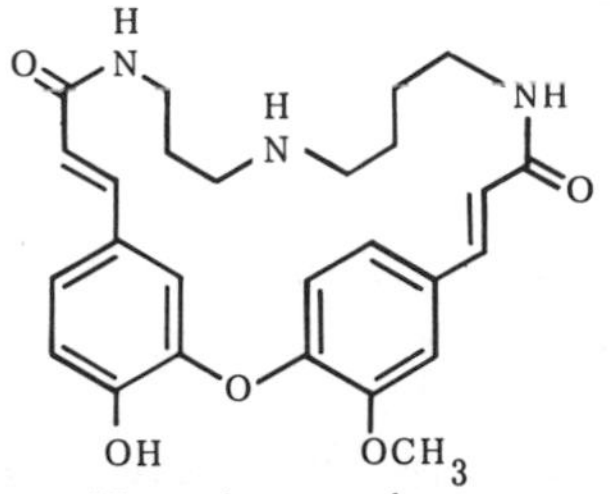

22 codonocarpine

Codonocarpus australis (Phytolaccaceae)

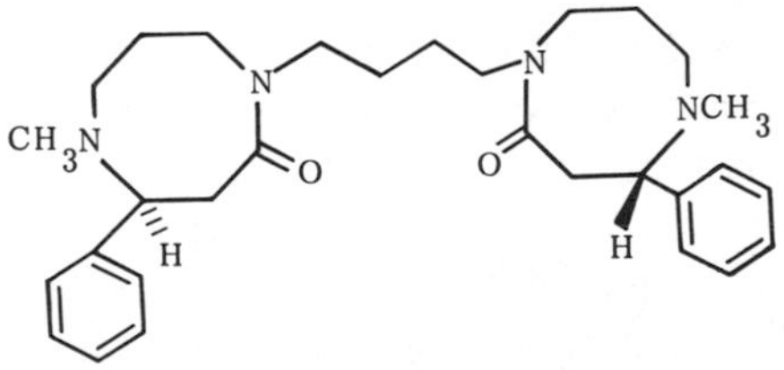

23 homaline

Homalium pronyense (Flaucourtiaceae)

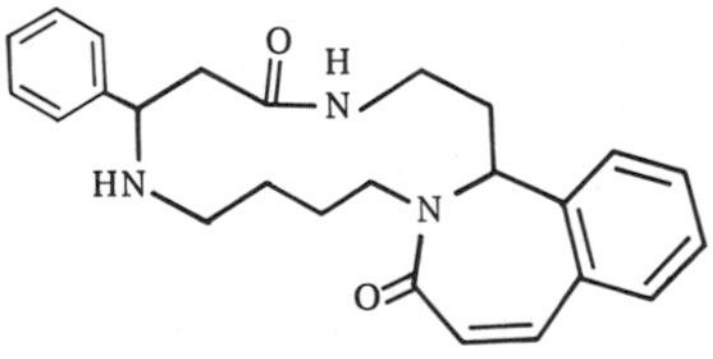

24 pleurostyline

Pleurostylia africana (Celastraceae)

Figure 1

12.3.1 Isolation and Chemistry

The structurally most interesting spermidine or spermine-containing alkaloids are those containing a macrocyclic ring. Typically these compounds are produced by the condensation of one nitrogen with a carboxylic acid, and the nucleophilic attack of a second nitrogen on an electrophilic center within the carboxylic acid–containing unit. This can clearly result in quite a substantial number of structure types, and some representative examples are shown in Figure 1. The range of plant families is worthy of note.

Acid cleavage of the lactam ring of oncinotine (20) followed by esterification, acetylation, and N-alkylation gave the quaternary species 25, which could be subjected to Hofmann degradation to yield the amine 26 and the diamide 27. CD measurement of the lithium aluminum hydride reduction product of 26 in comparison with (R)-(−)-N-methyl coniine indicated the absolute stereochemistry.

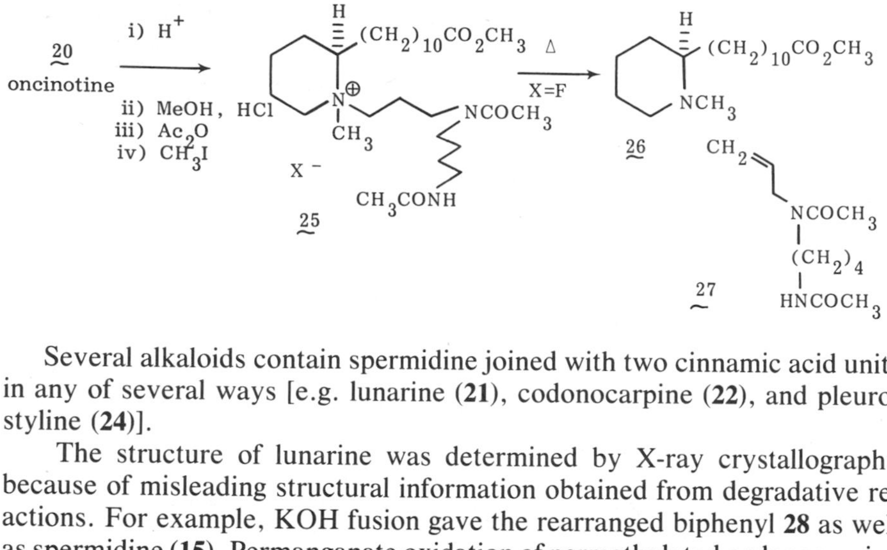

Several alkaloids contain spermidine joined with two cinnamic acid units in any of several ways [e.g. lunarine (21), codonocarpine (22), and pleurostyline (24)].

The structure of lunarine was determined by X-ray crystallography because of misleading structural information obtained from degradative reactions. For example, KOH fusion gave the rearranged biphenyl 28 as well as spermidine (15). Permanganate oxidation of permethylated codonocarpine however, gave the expected biphenyl ether 29.

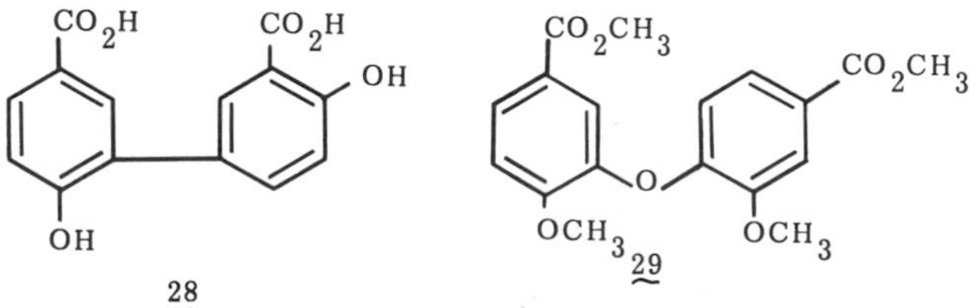

There are also some spermine alkaloids which contain two cinnamic acid residues, including the alkaloids obtained from *Homalium pronyense* such as homaline (23). Hofmann degradation of 23 eventually afforded a base 30

and catalytic benzylic cleavage of **23** followed by N-methylation gave the same product **31** as the catalytic hydrogenation product of **30**.

There has been moderate success in the synthesis of the macrocyclic spermidine–spermine alkaloids, and two of the main synthetic approaches will be described.

Condensation of the protected bromoamide **32** with the piperidine **33** afforded the amide ester **34** which could be converted to the amide amino acid **35** as shown. Transformation to the corresponding acid chloride followed by base-catalyzed cyclization afforded the lactam **36** in quite good yield. Problems were encountered in the selective hydrolysis of the acetamide group, and oncinotine (**20**) could be obtained only in quite low yield (Scheme 2).

Although lunaridine (**37**) itself has not been synthesized, the tetrahydro derivative **38** has been obtained, thereby confirming the skeleton. Phenolic oxidative coupling of dihydro-*p*-coumaric acid methyl ester gave **39**, which could be converted to the diacid ester **40** as shown. This compound was converted to the bishydroxylamine derivative **41** and thence by condensation with spermidine in THF and further modifications to **38** (Scheme 3).

12.3.2 Biosynthesis

In *E. coli*, spermidine (**15**) is formed by transfer of a propylamine residue derived from decarboxylated *S*-adenosylmethionine (**42**) to putrescine (**17**) derived from ornithine (**43**). Both putrescine and spermidine inhibit ornithine decarboxylase.

In the rat ventral prostrate gland the pathway is similar, but putrescine is required for maximal *S*-adenosylmethionine decarboxylase activity.

Richards and Coleman in 1952, in a now classic experiment, showed that potassium-deficient barley plants accumulated putrescine (**17**), and this has been confirmed in many other plants. Ornithine (**43**) was subsequently found

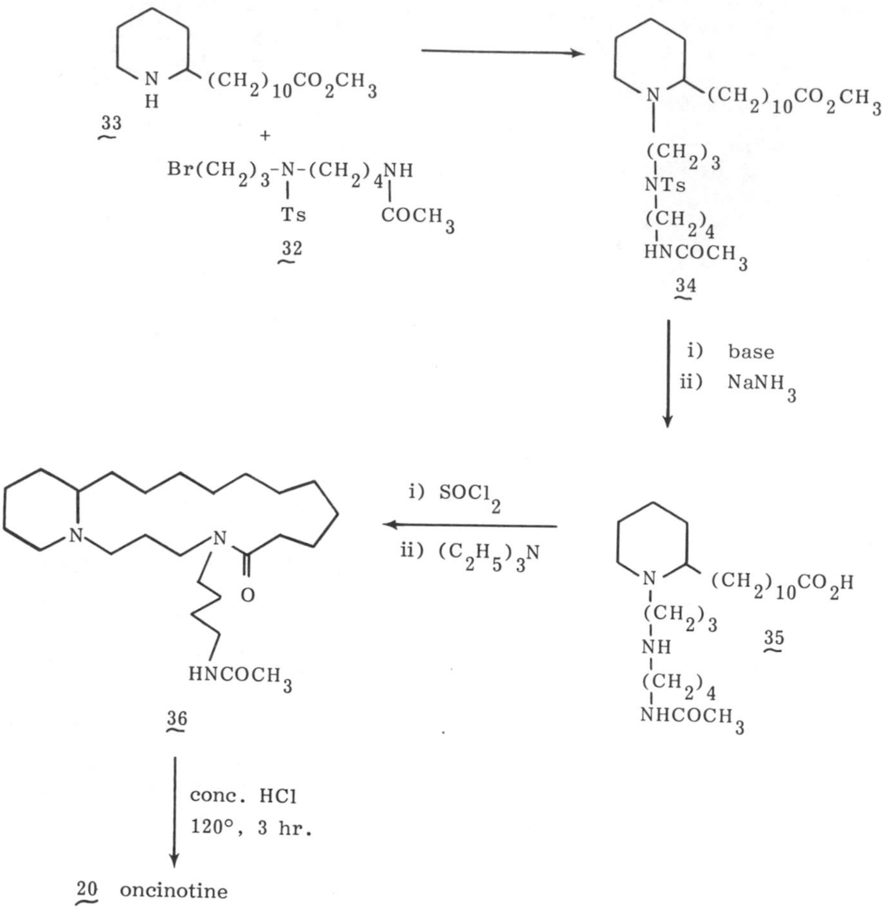

Scheme 2   Synthesis of oncinotine (**20**).

to be only poorly incorporated, and the discovery of agmatine (**44**) suggested arginine decarboxylation as an early step in putrescine (**17**) biosynthesis in barley. Indeed in potassium-deficient barley leaves arginine decarboxylase and N-carbamylputrescine amidohydrolase activities were higher than in normal plants.

Very little is known about the conversion of putrescine (**17**) to spermidine (**15**) in higher plants, but it is commonly regarded that decarboxylated S-adenoxylmethionine also provides the propylamine group in this case.

The biosynthetic pathways of putrescine and the polyamines are shown in Scheme 4. In light of the involvement of "bound" cadaverine (Chapter 4) in certain biosynthetic schemes, it would be interesting to know if the two nitrogen atoms of putrescine are biologically equivalent. The biosynthesis of the macrocyclic spermidine alkaloids does not yet appear to have been studied.

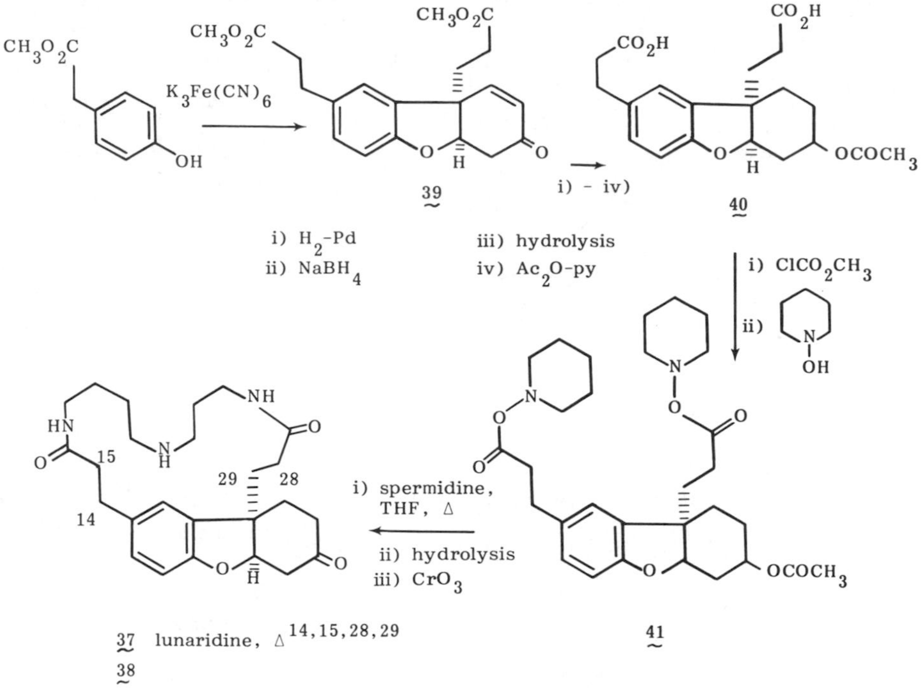

Scheme 3 Synthesis of tetrahydrolunaridine (**38**).

12.3.3 Biological Effects

In 1948, Herbst and Snell found that putrescine was an essential growth factor for the bacterium *Hemophilus parainfluenzae;* yeast extract, corn steep liquor, and pea seeds were good sources. Subsequently, spermine and spermidine were also found to possess growth-promoting activity for several types of bacteria, and for *E. coli,* spermidine has the effect of synchronizing cell division. Spermine acts to stimulate the growth of Chinese hamster cells and also tuber explants of *Helicanthus tuberosus.*

These growth-promoting properties are due to interaction with nucleic acids, particular ribosomal RNA, in which magnesium ion also plays a critical role.

Spermidine induces phenylalanine–ammonia lyase activity in excised pea pods, indicating that the polyamines possibly act as triggering devices for several other plant mechanisms. In this respect the consistently high concentrations of these amines in seeds of several plants is worth mentioning.

At higher concentrations ($\sim 5 \times 10^{-4} M$) spermine acts as an antibacterial against *S. aureus*, and several other examples of antibacterial activity of the polyamines are well established.

At 0.075 mmol/kg spermine produces renal failure within a week in

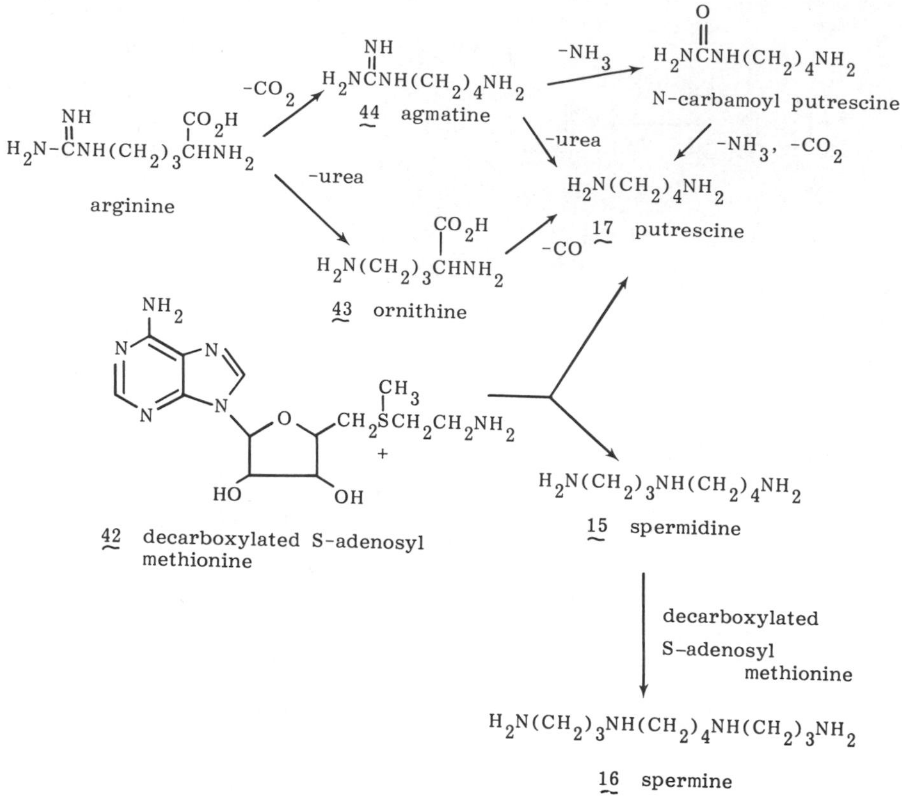

Scheme 4 Biosynthesis of putrescine (17) and the polyamines spermidine (15) and spermine (16).

several animal species, and at 0.33 mmol/kg in humans, vomiting, albuminuria hematuria, and hyperglycemia are observed.

These compounds also have the ability to antagonize the effects of antimicrobial agents as quinacrine and streptomycin.

LITERATURE

Reviews

Tabor, H., C. W. Tabor, and S. M. Rosenthal, *Ann. Rev. Biochem.* **30**, 579 (1961).

Tabor, H., and C. W. Tabor, *Pharmacol. Rev.* **16**, 245 (1964).

Smith, T. A., *Phytochemistry, 9*, 1479 (1970).

Smith, T. A., *Biol. Rev.* **46**, 201 (1971).

Smith, T. A., *Endeavor, 31*, 22 (1972).

Hesse, M., and H. Schmid, in K. F. Wiesner (Ed.), *MTP International Review of Science,* Ser. 2, Vol. 9, *Alkaloids,* Butterworths, London, 1977.

Synthesis

Guggisberg, A., P. van den Broek, M. Hesse, H. Schmid, F. Schneider, and K. Bernauer, *Helv. Chim. Acta* **59**, 3013 (1976).

Weisner, K., Z. Valenta, D. E. Orr, V. Liede, and G. Kohan, *Can. J. Chem.* **46**, 3617 (1968).

12.4 MACROCYCLIC PEPTIDE ALKALOIDS

There are two major groups of alkaloids which may be characterized as macrocyclic peptide alkaloids, that is, compounds containing large rings and derived essentially by the union of different amino acids. One of these groups occurs in higher plants, the other in fungi.

12.4.1 Macrocyclic Peptides from Higher Plants

Over 70 alkaloids containing a macrocyclic (13-, 14- or 15-membered peptide ring have been obtained from higher plants since Goutarel and Pais first obtained the adouétines from *Waltheria americana* (Sterculiaceae).

These alkaloids are particularly common in the Rhamnaceae, but have also been isolated from several other families, including the Celastraceae and Rubiaceae, where yields may vary between 0.01 and 0.1%. One problem in this area has been the variation of alkaloid distribution for a given plant species.

In general, the alkaloids are very weak bases, and being derived mainly from L-amino acids are quite strongly (−200 to −400°) levorotatory, although this is partly due to the chirality of the styrylamide group. No characteristic UV spectrum is usually observed, because the enamine moiety is not coplanar with the phenolic nucleus in solution, and this has also been observed in the solid state, where an out-of-plane twist of 73° was found for a frangulanine derivative.

The major group of alkaloids are those with a 14-membered ring, and they can be divided into three subgroups depending on the β-hydroxy amino acid participating in the ring formation: β-hydroxyleucine–frangulanine type (**45**), β-hydroxyphenylalanine–integerrine type (**46**), and *trans*-3-hydroxy-proline–amphibine type (**47**). The hydroxyleucine moiety of the frangulanine type on acid hydrolysis gives the threoisomer, but the erythroisomer is actually present in the isolate.

Ziziphine A (**48**) is a 13-membered ring alkaloid containing a biogenetically unusual β-(2-methoxy-5-hydroxy)styrylamine group. This could be established by oxidative degradation to give an amino monoaldehyde having a UV spectrum similar to 2,5-dimethoxybenzaldehyde. Hydrogenation of the parent compound shifts the UV spectrum from 268 to 290 nm.

There are 12 members of the alkaloid group containing a 15-membered isolated mainly from *Ziziphus* species; they are characterized by the direct substitution of an amino acid on the aromatic nucleus, rather than through

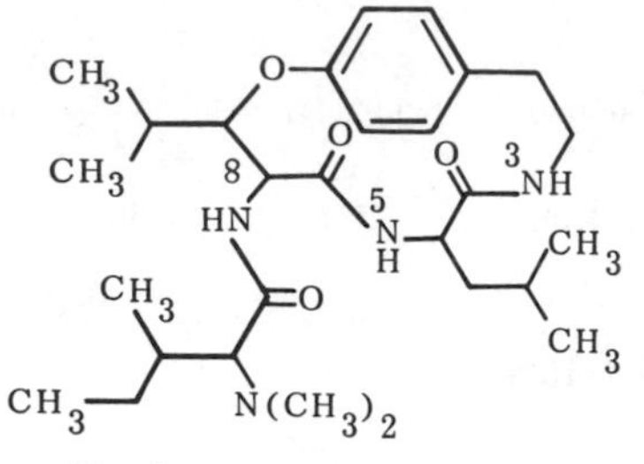

45 frangulanine

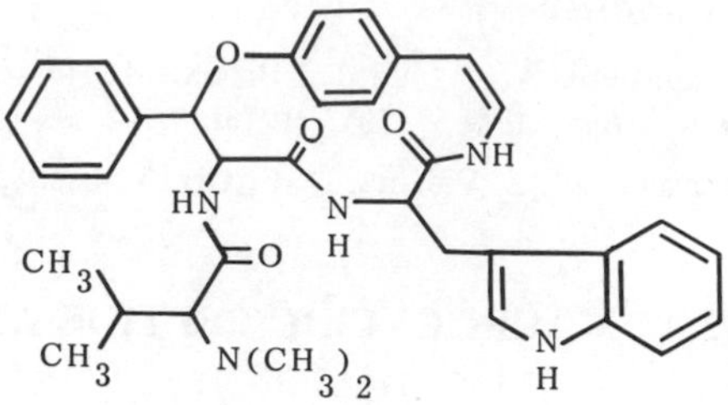

46 integerrine

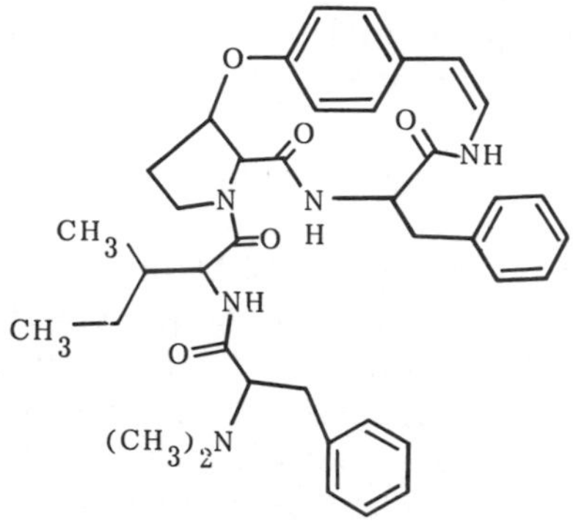

47 amphibine – B

an ether linkage. A typical representative is mucronine-A (**49**), which now does show a characteristic UV maximum at 273 nm for styrylamide moiety.

The amino acids present in these bases are relatively limited and may be obtained by hydrogenation of the double bond and complete hydrolysis with 6N hydrochloric acid at 120°C in a sealed tube.

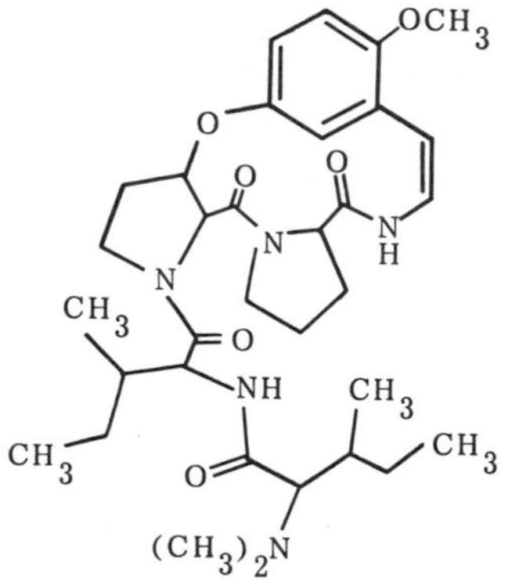

48 ziziphine-A

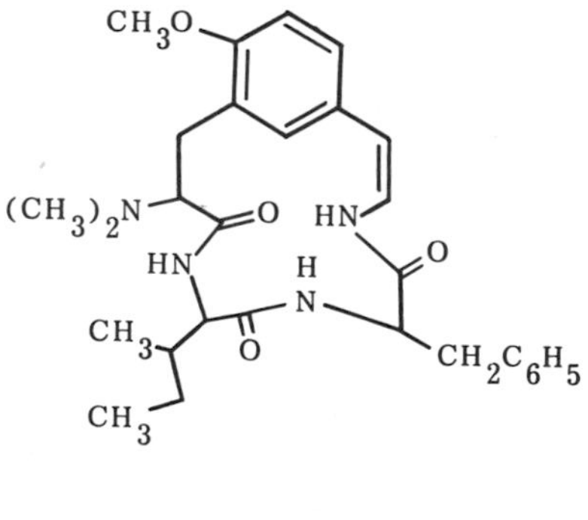

49 mucronine-A

Spectral Properties

Besides revealing the number of —NMe, —OMe, and terminal methyl groups, and also the substitution of the aromatic nucleus, the proton NMR spectrum may also reveal a number of other features.

For example, the α- and β-methine signals of hydroxyleucine occurred as a doublet at 4.40 ppm ($J = 8$ Hz), and as doublets at 4.77 ppm ($J = 8.2$ Hz), the large coupling constant being characteristic of an *erythro* configuration.

In the case of amphibine-B (**47**), the stereochemistry of the 3-hydroxy-proline moiety was deduced to be *cis* from the 5-Hz coupling constant of the 2-proton, although chemical degradation had indicated a *trans* configuration.

The amide protons also show a number of interesting features, for H-3 and H-5 are shielded by the aromatic ring and appear at 6.4 ppm, whereas H-8 appears at the more typical position of 7.7 ppm. The H-5 and H-8 protons are exchanged immediately with D_2O, whereas H-3 may not exchange even after prolonged contact.

Of greatest significance in structure determination of this group of alkaloids is the mass spectrum, as first discussed by Fehlhaber. The limitations are that the aromatic substitution cannot be determined and a distinction between a leucine or isoleucine residue is not possible. These mass spectral data have been well summarized by Tschesche and Kaussmann, and an example as it applies to the frangulanine and integerrine types as shown in Scheme 5.

In certain instances useful information may also be obtained from the mass spectra of the amino aldehydes (e.g. **50**) formed by oxidative cleavage of the macrocyclic ring (Scheme 6).

Very little can be said about the biosynthesis of these alkaloids, although Tschesche has suggested that whatever enzymes are involved cannot have very strict structural requirements since several alkaloids frequently cooccur.

Some of the alkaloids have antibiotic properties, being active against some lower fungi and gram-positive bacteria, but not against gram-negative bacteria.

The plants containing these alkaloids have a wide range of folkloric activities, but none of these can reasonably be attributed to the peptide alkaloids as yet. The problems to date have involved inadequate quantities of material available for pharmacologic testing. Hopefully these alkaloids will be subjected to more general screening in the near future.

12.4.2 Toxic Peptides of *Amanita* Species

There is a yet-more-complex group of macrocyclic peptides, the toxins of *Amanita phalloides* (Vaill. ex Fr.) Secr. Known as "death cap" or "deadly agaric," it is common mushroom in the forests of central Europe from late

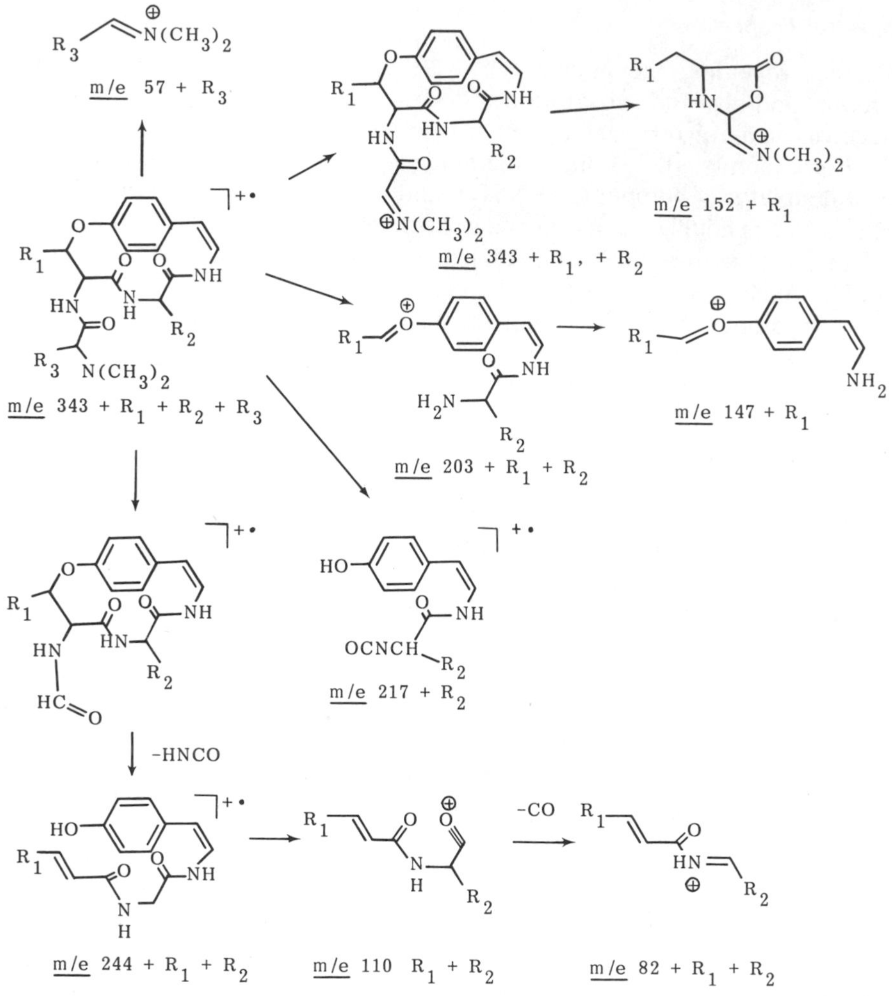

Scheme 5 Mass spectral fragmentation of the frangulanine and intergerrine-type macrocyclic peptide alkaloids.

summer through early fall, and the cap is a quite characteristic olive green color. The white *Amanita verna* (Bull. ex Fr.) Pers ex Vitt. is a corresponding North American toxic species.

Over 95% of all mushroom poisonings are thought to be due to these two species, and they are dangerous to the seeker of edible mushrooms because the symptoms of poisoning (violent emesis and diarrhea) are not apparent for several hours after ingestion. By this time the amatoxins have already begun their specific, irreversible destruction of the liver cells.

Although the number of toxins isolated is now quite moderate, only three

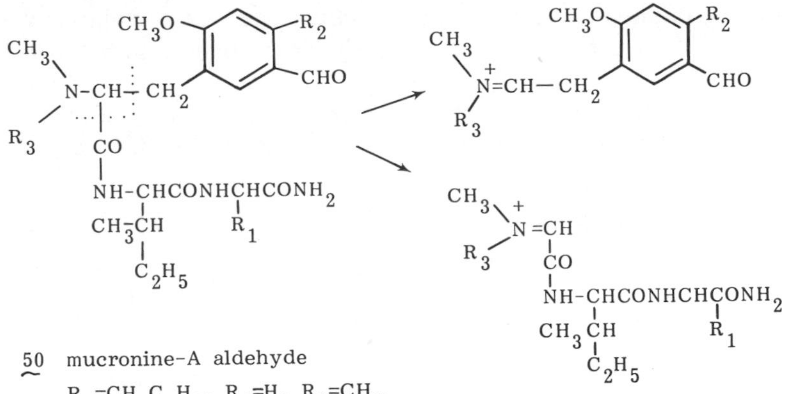

Scheme 6

R_2

CHO

CH_3O

CH_3

N–CH–CH$_2$

R_3

CO

NH–CHCONHCHCONH$_2$

CH_3CH R_1

C_2H_5

50 mucronine-A aldehyde

$R_1=CH_2C_6H_5$, $R_2=H$, $R_3=CH_3$

CH_3

$\overset{+}{N}=CH-CH_2$

R_3

CH_3O

R_2

CHO

CH_3

$\overset{+}{N}=CH$

R_3

CO

NH–CHCONHCHCONH$_2$

$CH_3 CH$ R_1

C_2H_5

Scheme 6

compounds in two groups will be considered here: the amatoxins, α-amanitin and β-amanitin, and the phallotoxin, phalloidin.

Chromatographically, the toxins are quite well separated on silica gel using butanol:acetic acid:water (4:1:1) and may be detected with cinnamic aldehyde in an atmosphere of hydrochloric acid (blue or violet color) or with diazotized sulfanilic acid (amatoxins give a red color).

As has been noted in a review, since 100 g of fresh tissue typically contains 10 mg of phalloidin, 8 mg of α-amanitin, and 5 mg of β-amanitin and the lethal dose is 0.1 mg/kg, ingestion of a moderate-size (50 g) mushroom could easily be fatal. The phallotoxins are 10–20 times less toxic than the amatoxins and normally contribute little to the overall toxic reaction. They are however much more rapid in action in lethal doses, death occurring in 1–4 hr in mice.

Phalloidin has the structure **51** and the other phallotoxins differ in the nature of the peripheral amino acid groups. Acid hydrolysis leads to L-2-hydroxytryptophan (**52**) and L-cysteine (**53**) from cleavage of the thioether bridge, and the remaining amino acids from cleavage of the peptide linkages. The thio ether bridge can be reductively removed with Raney nickel and the product is nontoxic.

The UV spectrum of the amatoxins (λ_{max} 302 nm) is quite characteristically different from that of the phallotoxins (λ_{max} 292 nm), and this was traced to the presence of sulfoxide and phenolic groups on the indole nucleus. α-Amanitin (**54**) is the amide of β-amanitin (**55**) and acid hydrolysis gives the peripheral amino acids glycine, aspartic acid, α-hydroxyproline, isoleucine, and dihydroxyisoleucine in the form of the lactone **56**. The structure of β-amanitin was recently confirmed by single-crystal X-ray analysis. This confirmed the stereochemical work of Wieland which determined the R-configuration for the β- and γ-carbon atoms of the dihydroxyisoleucine residue, and of the sulfur in the cystine residue. The γ-carbon of the hydroxy-proline unit was established to have the R-configuration by the X-ray analysis.

Perhaps the most amazing part of the cyclic peptide story of the *Amanita* mushrooms is that *A. phalloides* also contains an antitoxin, antamanide, to

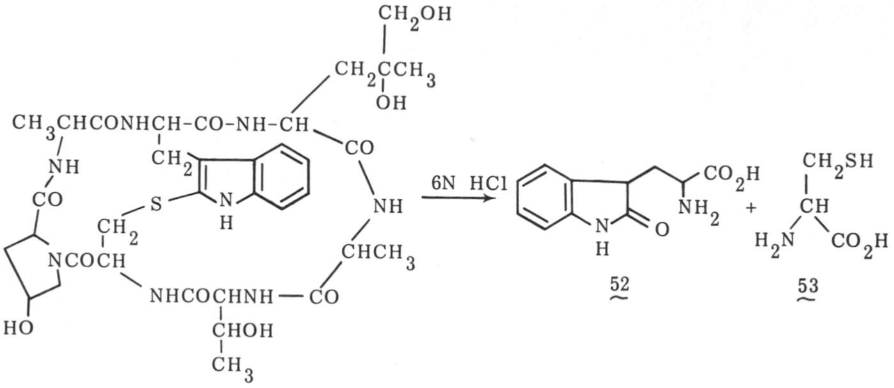

$\underline{51}$ phalloidin

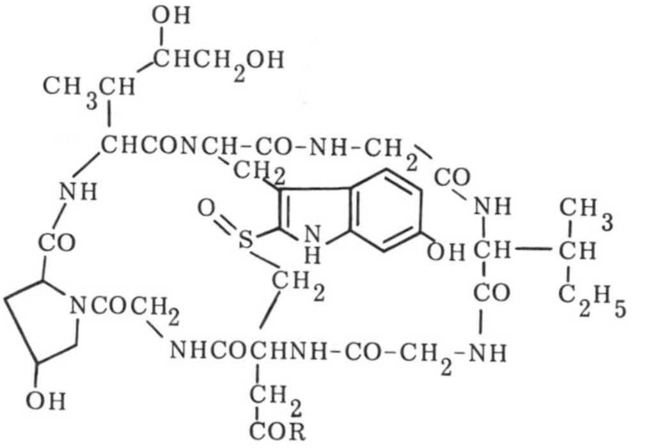

$\underline{54}$ α-amanitin, R=NH$_2$

$\underline{55}$ β-amanitin, R=OH

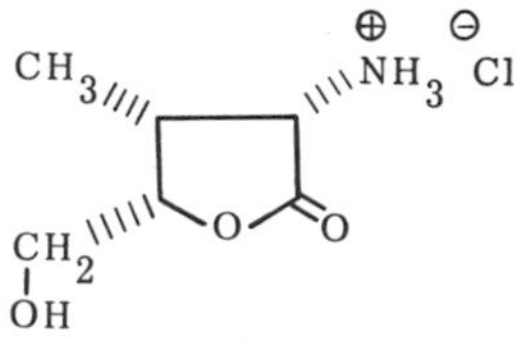

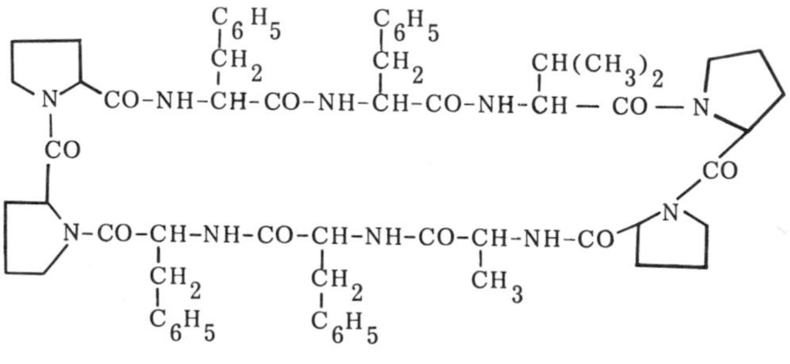

$\underline{57}$ antamanide

phalloidin! At 0.5 mg/kg it completely protects an LD_{100} dose of 5 mg/kg of phalloidin in mice. The structure **57** was determined mainly by mass spectrometry and confirmed by synthesis. The compound is a powerful complexator of sodium ions.

Biochemically it is now clear that phalloidin and amanitin differ in the primary mechanism of their cytopathogenic effects. Morphologically this is evident from observation that phalloidin affects the endoplasmic reticulum, whereas α-amanitin affects the nucleus of the cell. Indeed, α-amanitin acts by specifically inhibiting an RNA polymerase from the nucleoplasm of the liver *in vitro*. *In vivo* however the compound inhibits the synthesis of ribosomal and DNA-like RNA in the liver. The mechanism of action of phalloidin is much less well understood.

Silymarin (**58**), the antihepatotoxic principle of *Silybium marianum* (L.)

58   silymarin

Gaerta., is reported to be an antagonist of α-amanitin, although its therapeutic value in *Amanita* poisoning remains to be established.

LITERATURE

Macrocyclic Peptides from Higher Plants

Reviews

Warnhoff, E. W., *Fortschr. Chem. Org. Naturs.* **28**, 163 (1970).

Warnhoff, E. W., *Alkaloids, London* **1**, 444 (1971).

Snieckus, V. A., *Alkaloids, London* **2**, 276 (1972).

Snieckus, V. A., *Alkaloids, London* **3**, 310 (1973).

Snieckus, V. A., *Alkaloids, London* **4**, 408 (1974).

Tschesche, R., and E. U. Kaussmann, *Alkaloids NY* **15**, 165 (1975).

Snieckus, V. A., *Alkaloids, London* **5**, 274 (1975).

Tscheshe, R., *Heterocycles* **4**, 107 (1976).

Reed, J. N., and V. A. Snieckus, *Alkaloids, London* **7**, 305 (1977).

Spectral Properties

Fehlhaber, H. W., *Z. Anal. Chem.* **235**, 91 (1968).

Toxic Peptides of Amanita *Species*

Reviews

Wieland, T., and O. Wieland, *Pharmacol. Rev.* **11**, 87 (1959).

Wieland, T., *Fortschr. Chem. Org. Nature.* **25**, 214 (1967).

Wieland, T., and O. Wieland, in S. Kadis, A. Ciegler, and S. J. Ajl (Eds.), *Microbial Toxins,* Vol. 8, Academic, New York, 1972, p. 249.

Structural Studies

Gieren, A., P. Narayanan, W. Hoppe, M. Hasan, K. Michi, T. Wieland, H. O. Smith, G. Jung, and E. Breitmaier, *Ann.* 1561 (1974).

Wieland, T., B. deUrries, H. Indest, H. Faulstich, A. Gieren, M. Strum, and W. Hoppe, *Ann.* 1570 (1974).

Kostansek, E. C., W. N. Lipscomb, R. R. Yocum, and W. E. Thiessen, *J. Amer. Chem. Soc.* **99**, 1273 (1977).

12.5   MUSHROOM TOXINS OTHER THAN PEPTIDE ALKALOIDS

There are a number of interesting low-molecular-weight alkaloids derived from mushrooms, whose origin is quite different from that of the peptide alkaloids which are discussed in Section 4.2.

Accidental poisoning through ingestion of "edible" wild mushrooms is a worldwide problem and death occasionally results. The review of Tyler is an important summary of the field, and since then several other reviews have become available.

One of the simplest of these toxins is gyromitrin (**59**), obtained from several *Helvella* sp., which has a quite unusual array of functionalities. The compound was highly toxic to guinea pigs and rabbits even by inhalation. In humans, death results from nephrotoxicity and heart failure after 5–10 days. The most important of the simple mushroom toxins however, are muscarine and the simple isoxazole derivatives.

Explorers in the Kamchatka region of Siberia in the early eighteenth century observed the use of fly agaric, *Amanita muscaria* (L. ex Fr.) Hooker, as an intoxicant and narcotic, and muscarine was first recognized as the toxic principle nearly 160 years ago. However, several species of *Inocybe* and *Clitocybe* are much richer sources, and *Inocybe napipes* Lange for example may contain up to 0.7% dry weight of muscarine.

Early work was hampered by isolation problems, and it was not until 1954 that Eugster and Waser first obtained pure muscarine chloride. The structure **60** was determined some 3 years later by X-ray crystallography, and confirmed by synthesis.

Of the possible diastereoisomers and antipodes the L-(+)-isomer shown is by far the most active, exhibiting very specific muscarinic activity at postganglionic parasypathetic effector sites. At a dose of 0.01 μg/kg a hypotensive effect is observed in cats. There has been substantial interest in the synthesis not only of muscarine but of some of the other isomers, and some analogues.

Matsumoto and co-workers for example found that pyrolysis of the lactone **61** gave a tetrahydrofuran carboxylic acid **62** which can be converted

59 gyromitrin

60 L-(+)-muscarine

61   62

to (+)-**60**. The chiralities of the natural stereoisomeric muscarines were determined by chemical modification of muscarine itself. Thus (−)-allo-muscarine has the configuration 2S,3R,5R, and (+)-epi-muscarine the configuration 2S,3S,5S.

Whiting *et al.* have described a synthesis of L-(+)-allomuscarine (**63**) and L-(+)-muscarine (**60**), in which a key step is the resolution of an intermediate during N-deacetylation as shown in Scheme 7.

Potentially of course, the most effective stereoselective synthesis would be from a preformed carbohydrate precursor, and one such approach (Scheme 8) was recently reported, beginning with the furanose derivative **64** obtained from glucose.

The tetrahydrofuran nucleus is not essential for activity, and for example the cyclopentane derivative **65** shows activity comparable to muscarine.

The lack of available culture systems for producing muscarine has severely hampered biosynthetic studies, although in cultured *Clitocybe rivalosa* muscarine levels closely paralleled mycelial growth. Using several fresh stipes of *I. napipes*, [U-14C]glucose was examined as a precursor of muscarine, but incorporation was very low (<0.02%).

The principles responsible for the central nervous system stimulation and for the flyicidal effect of certain *Amanita* and *Tricholoma* species were investigated by the groups of Eugster and Takemoto, who eventually showed that the compounds involved in these activities were the same—namely muscimol (**66**), ibotenic acid (**67**), muscazone (**68**), and tricholomic acid (**69**)—and each of these compounds has been synthesized.

Ibotenic acid (**67**) is quite unstable and may decarboxylate to muscimol (**66**) during drying or on chromatography. The isoxazole compounds are detected by a characteristic reaction with the ninhydrin spray, yellow initially, followed by a gradual change over a few hours to purple.

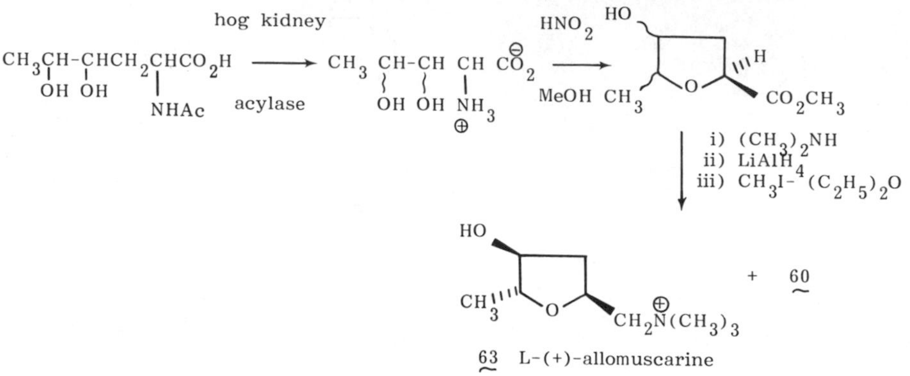

Scheme 7

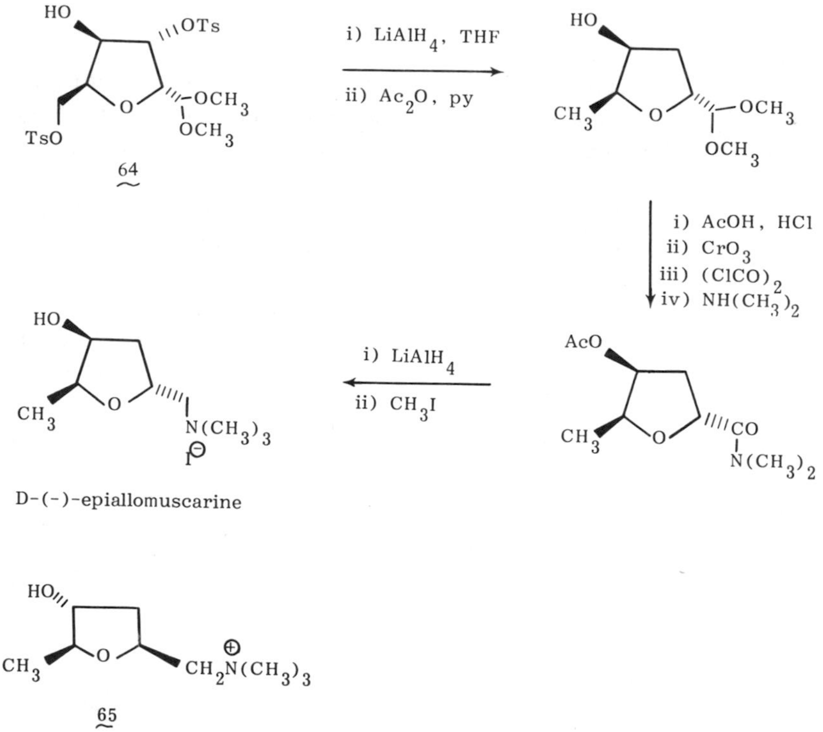

Scheme 8

66 muscimol, R=H
67 ibotenic acid, R=CO$_2$H

68 muscazone

69 tricholomic acid

Muscimol at 5 mg has a central nervous system stimulatory effect but at 15 mg orally, hallucinogenic responses were observed, including confusion, disturbance of visual preception, and space and time disorientation. Ibotenic acid at 20 mg was not psychoactive. Both compounds have sedative properties, whereas muscazone and probably tricholomic acid do not. The latter however is the most potent flyicidal of the compounds.

Perhaps the most novel property of almost any alkaloid is the commercial use of small quantities of ibotenic acid as a synergist for the flavor additive monosodium glutamate.

There appear to have been no studies on the biosynthesis of these interesting metabolites.

LITERATURE

Reviews

Tyler, V. E., Jr., *Amer. J. Pharm.* **130**, 264 (1958).

Tyler, V. E., Jr., *Progr. Chem. Toxicol.* **1**, 339 (1963).

Eugster, C. H., *Fortschr. Chem. Org. Nature.* **27**, 261 (1969).

Snieckus, V. A., *Alkaloids, London* **1**, 455 (1971).

Baker, R. W., C. H. Chothia, P. Pauling, and T. J. Petcher, *Nature* (*London*) **230**, 439 (1971).

Snieckus, V. A., *Alkaloids, London* **2**, 27 (1972).

Benedict, R. G., in S. Kadis, A. Ciegler, and S. J. Ajl (Eds.), *Microbial Toxins*, Vol. 8, Academic, New York, 1972, p. 281.

Snięckus, V. A., *Alkaloids, London* **3**, 299 (1973).

Snieckus, V. A., *Alkaloids, London* **4**, 395 (1974).

Snieckus, V. A., *Alkaloids, London* **5**, 265 (1975).

Reed, J. N., and V. A. Sniekus, *Alkaloids, London* **7**, 297 (1977).

Synthesis

Matsumoto, T., A. Ichihara, and N. Ito, *Tetrahedron* **25**, 5889 (1969).

Whiting, J., Y. K. Au-Young, and B. Belleau, *Can. J. Chem.* **50**, 3322 (1972).

Wang, P. C., Z. Lysenko, and M. M. Joullie, *Tetrahedron Lett.* 1657 (1978).

12.6 MAYTANSINOIDS

In 1972 Kupchan and co-workers reported the isolation of a quite new type of alkaloid structure, one having a macrocyclic ring of 19 atoms. The alkaloid was maytansine (**70**) and it has become one of several closely related compounds, now called maytansinoids, obtained from various plant species.

Maytansine (**70**) was first obtained (0.00002% yield) from the fruits of *Maytenus ovatus* Loes. (Celastraceae) as the principal antileukemic constituent.

The proton NMR data of maytansine are shown in **71**. Purification of

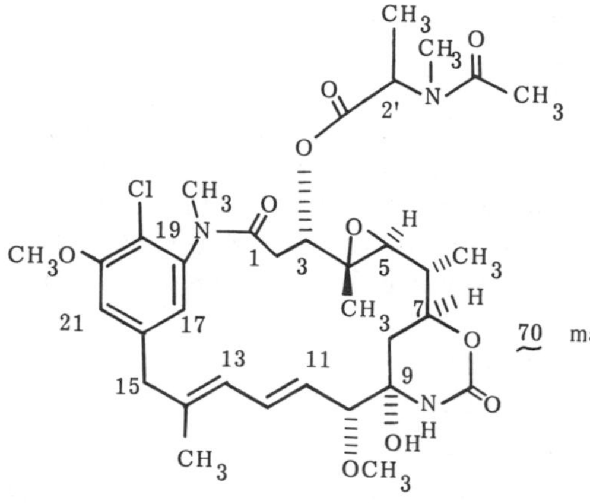

70 maytansine

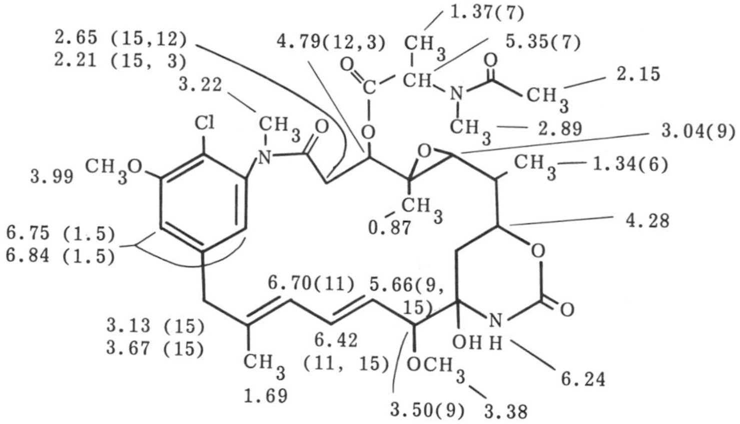

71 proton NMR data of maytansine

maytansine was facilitated by the formation of 3-bromopropyl derivative and it was this compound which was used for the X-ray structure determination. The absolute configurations are 3S, 4S, 5S, 6R, 7S, 9S, 10R, and 2'S.

From the stem bark of *Maytenus buchananii* (Loes.) R. Wilczek, the same group also obtained three homologues of maytansine, maytanprine (**72**), maytanbutine (**73**), and maytanvaline (**74**). The clue to their relationship with maytansine was established by their mass spectra. Maytansine (**70**) does not show a molecular ion, but rather an ion at *m/e* 630 corresponding to losses of both H_2O and HNCO. In **72**, **73**, and **74** these ions were observed at *m/e* 644, 658, and 672 respectively, but in all three compounds the next important ion is at *m/e* 485, which is formed by loss of the complete unit attached at C-3. These data limit the additional 14 mu in **72** and **73** to this unit and their nature was determined from the proton NMR spectrum to be an ethyl and an isopropyl group respectively.

Also obtained from *M. buchananii* were maysine (**75**) and maysenine (**76**), which are of interest because unlike the other maytansinoids, these compounds do not have antileukemic activity.

The paucity of maytansine in the *Maytenus* species so far examined led to an evaluation of other Celastraceous plants. To date the best plant source of maytansine is the stem and wood of *Putterlickia verrucosa* Szyszyl native to S. Africa, where yields as high as 12 mg/kg have been found. Also obtained from *P. verrucosa* was maytansinol (**77**), the parent alcohol of the maytansinoids. Reductive (LiAlH₄, THF, −23°C) cleavage of maytanbutine (**73**) gave **77** in 40% yield. Maytansinol has no antileukemic activity and is only marginally cytotoxic.

The Celastraceae of Kenya and South Africa are not the only source of these maytansinoids, for Wall and co-workers obtained colubrinol (**78**) and colubrinol acetate (**79**) from *Colubrina texensis* Gray (Rhamnaceae), and a fungal *Nocardia* sp. has provided related compounds.

70 maytansine R_1= COCH(CH₃)N(CH₃)COCH₃, R_2=H
72 maytanprine R_1=COCH(CH₃)N(CH₃)COC₂H₅, R_2=H
73 maytanbutine R_1=COCH(CH₃)N(CH₃)COCH(CH₃)₂, R_2=H
74 maytanvaline R_1=COCH(CH₃)N(CH₃)COCH₂CH(CH₃)₂, R_2=H
77 maytansinol R_1=H, R_2=H
78 colubrinol R_1=COCH(CH₃)N(CH₃)COCH₂CH(CH₃)₂, R_2=OH
79 colubrinol acetate R_1=COCH(CH₃)N(CH₃)COCH₂CH(CH₃)₂, R_2=OAc

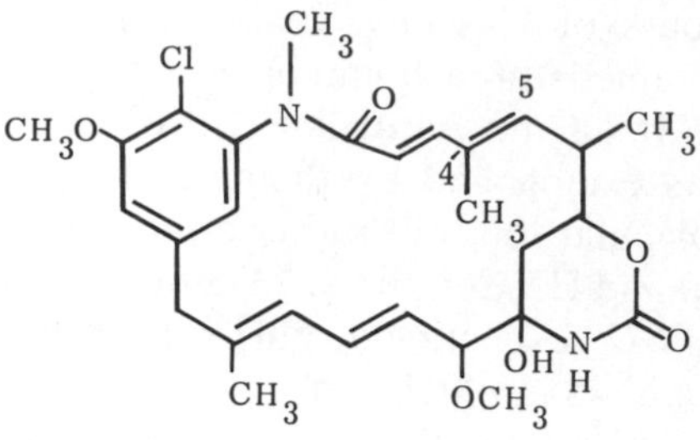

75 maysine 4,5-oxido
76 maysenine

The low yield of maytansine from natural sources has led to several interesting synthetic approaches, and a successful route is close at hand. No fewer than seven groups have reported syntheses of various parts of the macrolide nucleus.

Meyers, Corey, Bernauer, and Ganem have reported syntheses of the aromatic part of maytansine, namely the unit **80** or the less complex moieties **81** and **82**.

Corey and Ganem have reported syntheses of the carbinolamide unit **83**, Meyers a route to **84**, Vandewalle a route to **85**, and Edwards a synthesis of **86**. Fried has reported a synthesis of the unit **87**, and Corey a synthesis of **88**. The only reported synthesis of the so-called northern zone is a route developed by Meyers for **89**. This flurry of publications in the area of partial

Western zone

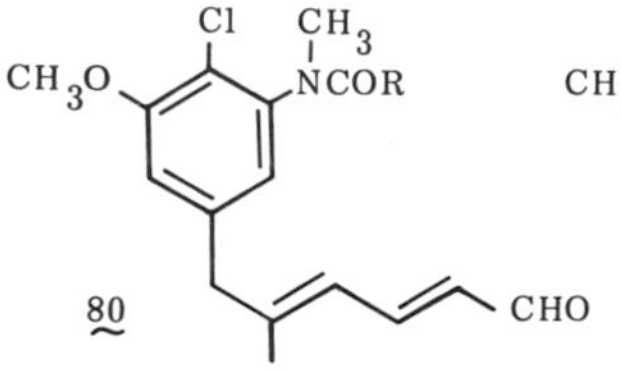

80

Corey R=CH₃
Bernauer R=OCH₂CCl₃

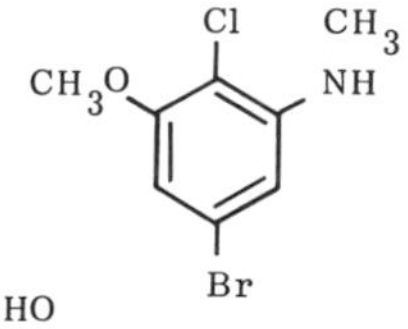

81
Meyers

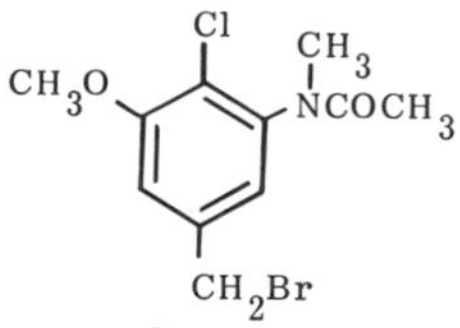

82
Ganem

Southern and eastern zone

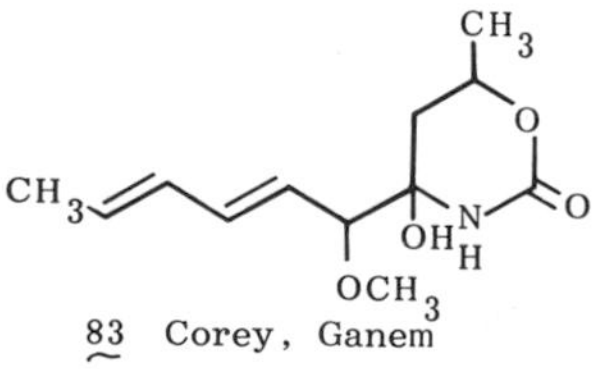

83 Corey, Ganem

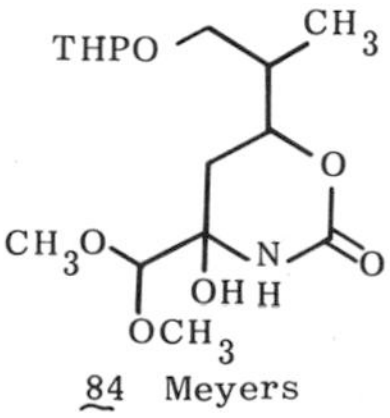

84 Meyers

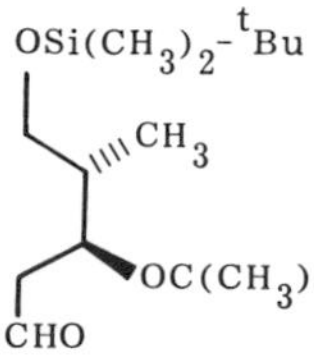

85 Vandewalle

86 Edwards

$R = -C \overset{C_6H_5}{\underset{C_6H_5}{\mid}} -p-C_6H_4OCH_3$

87 Fried

Northern zone

88 Corey

89 Meyers

synthesis is quite unprecedented, for normally only total syntheses are reported.

Maytansine is one of the most potent plant anticancer agents to have been isolated, for it displays exceptional activity in the sarcoma 180, Lewis lung carcinoma, B16 melanoma, and L-1210 and P-388 leukemias in the mouse. In addition the compound has an ED_{50} of 1×10^{-5} µg/ml in the 9KB system. Maytanbutine is even more cytotoxic (ED_{50} 8.8×10^{-7} µg/ml).

At 6×10^{-8} M maytansine irreversibly inhibits cell division in the eggs of sea urchins and clams, and also inhibits irreversibly the *in vitro* polymerization of tubulin. It is 100 times more effective than vincristine as an inhibitor of the cleavage of marine eggs.

Maytansine inhibits DNA synthesis, but not RNA or protein synthesis. *Penicillium avellaneum* UC-4376 has been suggested as a method for the quantitative microbiological assay and bioautography of the maytansinoids.

Maytansine has been evaluated in a Phase I clinical trial. Some neurotoxicity and hepatotoxicity was observed, in addition to some therepeutic benefits in the two patients with breast cancer. Typical doses were 0.5 mg/m²/day. Unexpected G. I. toxicity has curtailed further clinical evaluation.

LITERATURE

Isolation

Kupchan, S. M., Y. Komoda, W. A. Court, G. J. Thomas, R. M. Smith, A. Karim, C. J. Gilmore, R. C. Haltiwanger, and F. F. Bryan, *J. Amer. Chem. Soc.* **94**, 1356 (1972).

Kupchan, S. M., Y. Komoda, G. J. Thomas, and H. P. J. Hintz, *Chem. Commun.* 1065 (1972).

Kupchan, S. M., Y. Komoda, A. R. Braufman, R. G. Dailey, Jr., and V. A. Zimmerly, *J. Amer. Chem. Soc.* **96**, 3706 (1974).

Kupchan, S. M., A. R. Branfman, A. T. Sneden, A. K. Verma, R. G. Dailey, Jr., J. Komoda, and Y. Nagao, *J. Amer. Chem. Soc.* **97**, 5294 (1975).

Kupchan, S. M., Y. Komoda, A. R. Branfman, A. T. Sneden, W. A. Court, G. J. Thomas, H. P. H. Hintz, R. M. Smith, A. Karim, G. A. Howie, A. K. Verma, Y. Nagao, R. G. Dailey, Jr., V. A. Zimmerly, and W. C. Summer, Jr., *J. Org. Chem.* **42**, 2349 (1977) and references therein.

Wani, M. C., H. L. Taylor, and M. E. Wall, *Chem. Commun.* 390 (1973).

Synthesis

Western Zone

Kane, J. M., and A. I. Meyers, *Tetrahedron Lett.* 771 (1977).

Foy, J. E., and B. Ganem, *Tetrahedron Lett.* 775 (1977).

Corey, E. J., H. F. Wetter, A. P. Kozikowski, and A. V. Rama Rao, *Tetrahedron Lett.* 777 (1977).

Götschi, E., F. Schneider, H. Wagner, and K. Bernauer, *Helv. Chim. Acta* **60**, 1416 (1977).

Southern and Eastern Zone

Meyers, A. I., and C.-C. Shaw, *Tetrahedron Lett.* 717 (1974).

Corey, E. J., and M. G. Bock, *Tetrahedron Lett.* 2643 (1975).

Elliott, W. J., and J. Fried, *J. Org. Chem.* **41**, 2469 (1976).

Edwards, O. E., and P.-T. Ho, *Can. J. Chem.* **55**, 371 (1977).

Bonjoublian, R., and B. Ganem, *Tetrahedron Lett.* 2835 (1977).

Samson, M., P. De Clercq, H. De Wilde, and M. Vandewalle, *Tetrahedron Lett.* 3195 (1977).

Northern Zone

Meyers, A. I., C. C. Shaw, D. Horne, L. M. Trefonas, and R. J. Majeste, *Tetrahedron Lett.* 1745 (1975).

Pharmacology–Biochemistry

Wolpert-Defilippes, M. K., R. H. Adamson, R. L. Cysyk, and D. G. Johns, *Biochem. Pharmacol.* **25**, 751 (1975).

Remillard, S., L. I. Rebhun, G. A. Howie, and S. M. Kupchan, *Science* **189**, 1002 (1975).

12.7 PURINE ALKALOIDS

The most important purine alkaloids are derived from the xanthine (**90**) nucleus. Xanthine itself has not yet been found naturally, but several simple *N*-alkyl derivatives are of quite considerable significance. Prime among these is caffeine, which is 1,3,7-trimethylxanthine (**91**), together with 1,3-dimethylxanthine (**92**) (theophylline) and 3,7-dimethylxanthine (**93**) (theobromine). It is no coincidence that these alkaloids are major constituents of plants used by people throughout the world as stimulating beverages.

12.7.1 Occurrence and Chemistry

Coffee is comprised of the seeds of *Coffea arabica* L. (Rubiaceae) and other members of the *Coffea* genus. The coffee plant is indigenous to Abyssinia and parts of East Africa, but is widely cultivated in Indonesia, Sri Lanka, and South America, particularly Brazil. The seeds contain 1–2% of caffeine bound in a thermally labile complex with chlorogenic acid. Caffeine often begins to sublime during the roasting of the green bean, and this as well as several other methods have been used to more completely decaffeinate coffee.

Cola (also spelt kola) is the dried cotyledons of various species of Cola trees (Sterculiaceae), which are indigenous to the West Indies, Brazil, Java, and West Africa. *Cola nitida* (Vent.) Schott and Endl. is the principal source of kola nuts, and are important because of their caffeine content, which may be as high as 3.5%. The theobromine content is usually quite low. Guarana is the dried paste produced from the crushed seeds of *Paulinia cupana* Kunth. (Sapindaceae), which is native to Brazil and Uruguay. In Brazil it is used as a stimulating beverage because of the high (2.5–5%) caffeine content.

Mate, which is used in Paraguay as a tea and is available in many supermarkets in the United States, is comprised of the leaves of *Ilex paraguensis* St. Hill. (Aquifoliaceae), and may contain up to 2% caffeine.

Tea is the leaves and leaf buds of *Camellia sinensis* (L.) O. Kuntze (Theaceae) and is indigenous to eastern Asia. The plant is now widely cultivated in China, Japan, India, and Indonesia. Although there are a wide variety of teas available particularly in the Far East, they are fundamentally of three types, green, oolong, and black, the differences being based solely on the processing methods.

In the preparation of green tea and fresh leaves are steamed to prevent further enzymatic reaction. The leaf is then dried and rolled. In the case of black tea the enzymatic reactions are encouraged by withering the green leaves on racks and then breaking up to permit further oxidation (so-called "fermentation"). Oxidation is stopped by hot-air drying. The oolongs are prepared in a manner similar to the black teas except that the withering period of partial oxidation is quite short. The book by Quimme gives a more extensive discussion of the processes.

The caffeine content of tea is in the range 1–4% together with small quantities of theobromine and theophylline. Caffeine can be obtained commercially from tea-leaf waste and tea dust.

Cocoa, the other main world product in this series of beverages, is obtained from the seeds of *Theobroma cacao* L. (Sterculiaceae). It is produced in several countries of the world, including Colombia, Brazil, Venezuela, the West Indies, Nigeria, Ghana, Sri Lanka, and Java. The seed kernels contain 0.9–3.0% theobromine and the husks 0.2–3.0%. The annual production of theobromine is in excess of 36,000 tons.

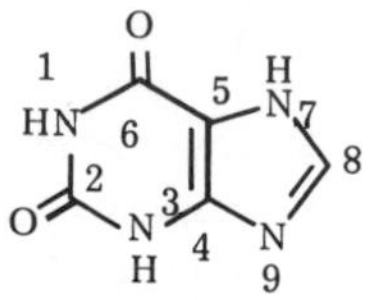

90 xanthine nucleus

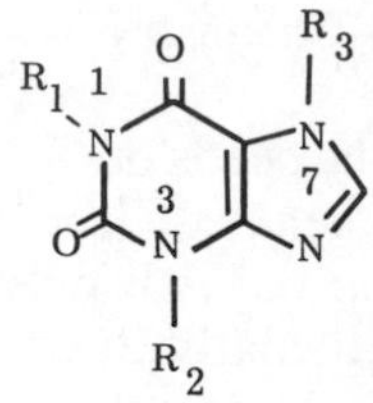

	R_1	R_2	R_3
91 caffeine	CH_3	CH_3	CH_3
92 theophylline	CH_3	CH_3	H
93 theobromine	H	CH_3	CH_3

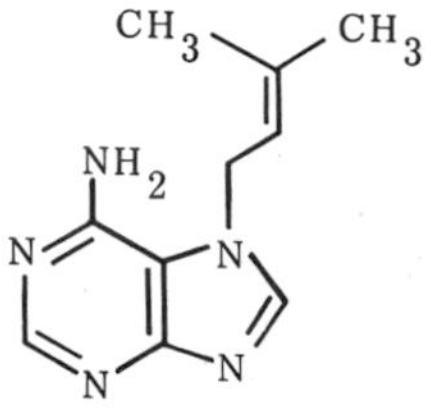

94 triacanthine

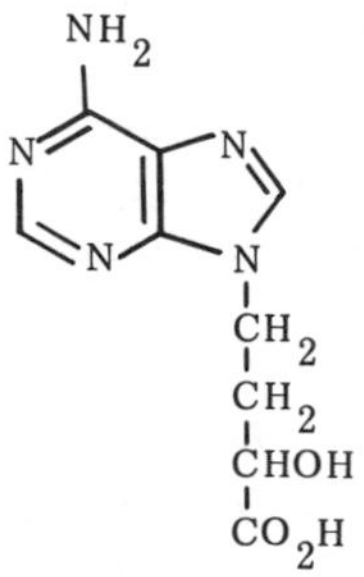

95 deoxyeritadenine

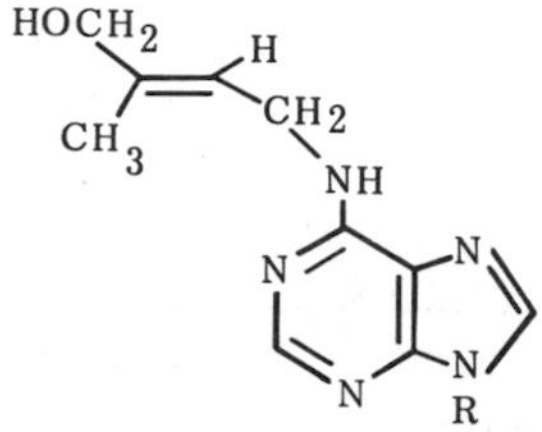

96 trans-zeatin R=H
97 lupinic acid R=CH_2CHCO_2H NH_2

Other purine alkaloids which do not contain the xanthine nucleus include triacanthine (**94**) from *Holarrhena mitis* (Apocynaceae); deoxyeritadenine (**95**) from *Leontinus edodes* (Tricholomataceae), which has hypocholestero-

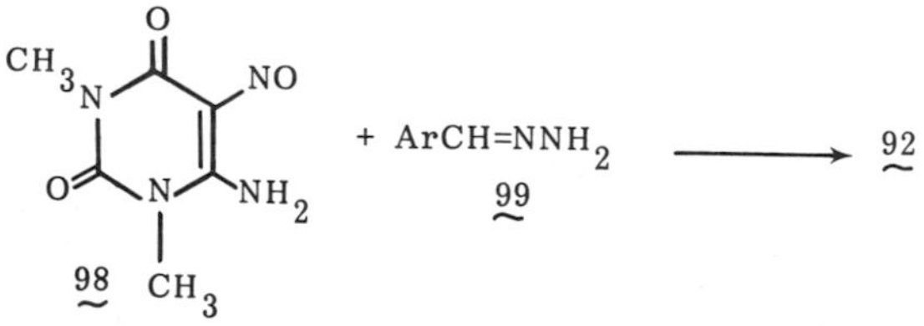

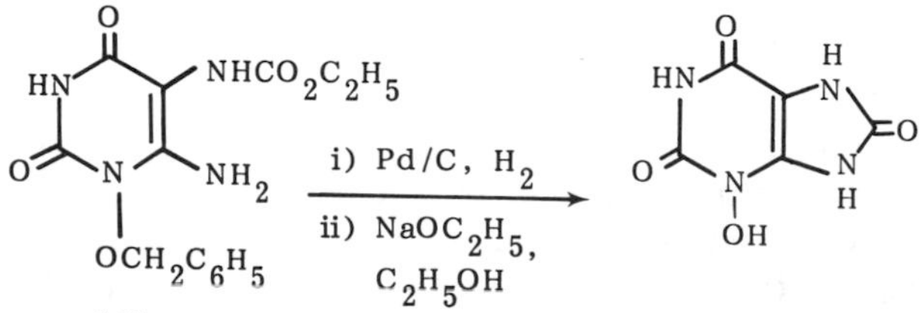

98   + ArCH=NNH$_2$ ⟶ 92
           99

101

i) Pd/C, H$_2$
ii) NaOC$_2$H$_5$, C$_2$H$_5$OH

100 3-hydroxyuric acid

lemic activity; and *trans*-zeatin (**96**) from *Zea mays*, which is a stimulator of cell division. In *Lupinus angustifolius* (Leguminosae) is metabolized to lupinic acid (**97**). The *trans* isomer of **96** is 50 times more active than *cis*-**96**.

There has been quite a flurry of activity in synthesizing the purine nucleus, mainly in an effort to prepare new pharmacologically active derivatives. Only some of the more recent synthetic efforts will be described.

Treatment of the 5-nitrosouracil **98** with the arylhydrazone **99** gave theophylline (**92**) in good yield, and 3-hydroxyuric acid (**100**), a probable carcinogenic metabolite of 3-hydroxyxanthine, was prepared by debenzylation and base-catalyzed cyclization of **101**.

Trans-zeatin (**96**) has been synthesized by Corse and Kuhnle as shown in Scheme 9. The key step is separation of the stereoisomers of **102**; an alternative route beginning with bromination of isoprene has also been reported. Two other approaches to the xanthine nucleus are shown in Scheme 10.

The ^{13}C NMR data for caffeine are shown in **103**.

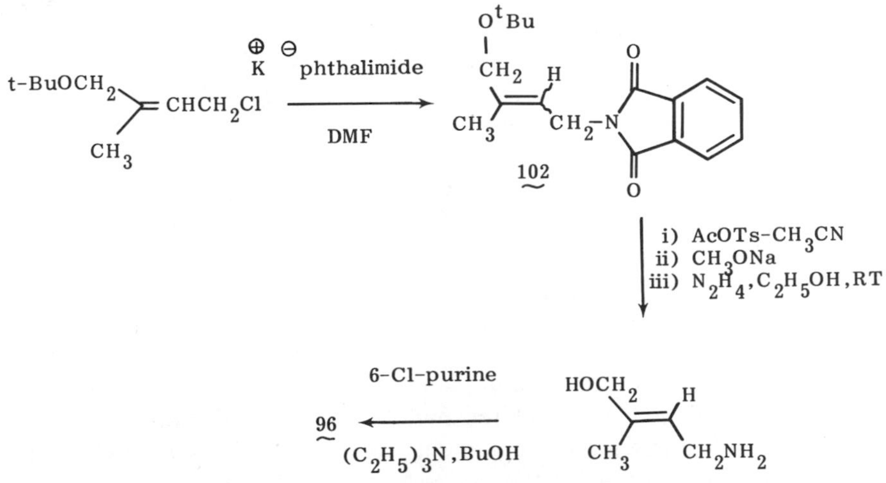

Scheme 9

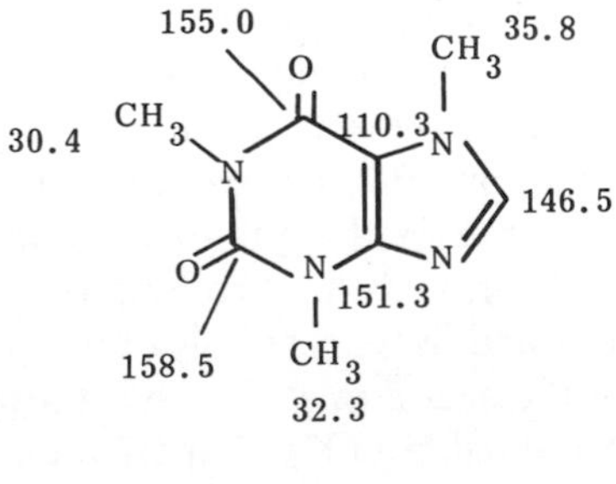

103

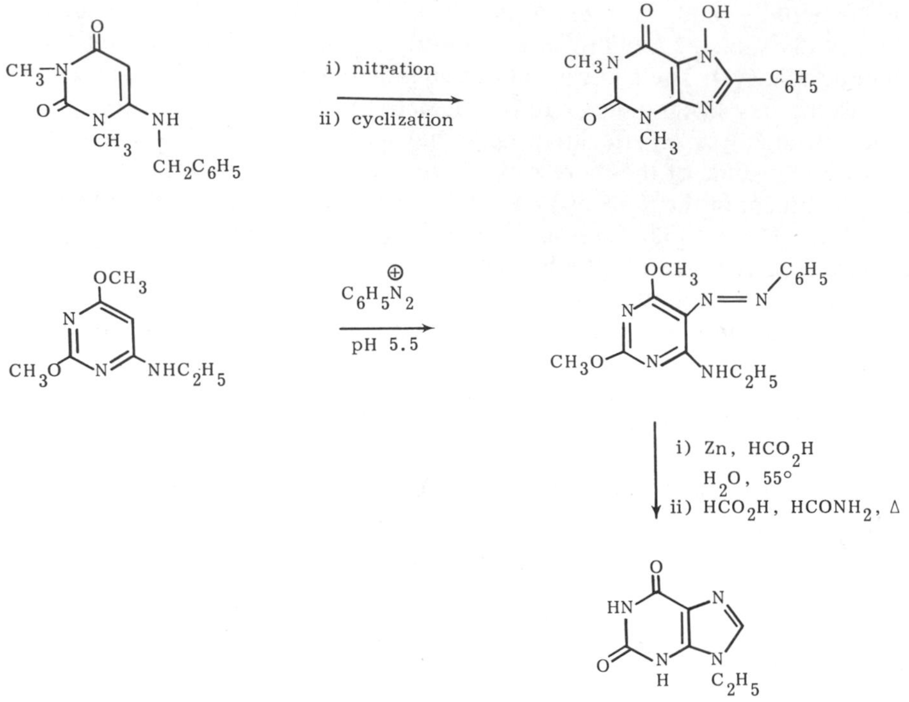

Scheme 10

12.7.2 Biosynthesis

The biosynthesis of the major xanthine metabolites has been studied in both tea and coffee plants. As might reasonably be expected, both methionine and S-adenosylmethionine are sequential precursors of the N-methyl groups. The question of the origin of the purine nucleus has been a more complex problem.

In the early 1960s two groups established that in both coffee and tea leaves the ring formation of caffeine followed the scheme of purine nucleotide biosynthesis. The question remained as to whether caffeine was produced from the nucleic acids by way of xanthine (90), 1-methylxanthine (104), and 1,3-dimethylxanthine (92) or by way of the nucleotide pool from 7-methylguanylic acid (105), 7-methylguanosine (106), 7-methylxanthosine (107), 7-methylxanthine (108), and theobromine (93) to caffeine (91).

Using both continuous and pulse feedings of tea callus tissue with $^{14}CO_2$, Looser et al. found that the second pathway was probably the correct one. In particular 7-methylxanthosine (107), 7-methylxanthine (108), and theobromine (93) stimulated caffeine biosynthesis. 7-Methylguanosine (106) however was not effective as a stimulator of caffeine biosynthesis, suggesting

that possibly the phosphate group may not be cleaved until 7-methylxan-
thylic acid (**109**) is reached. Both 7-methylxanthine (**108**) and 7-methylxan-
thosine (**107**) were also precursors of theobromine (**93**).

Suzuki and Takahashi subsequently demonstrated that hypoxanthine
(**110**), which is an established product of purine nucleotide catabolism, was

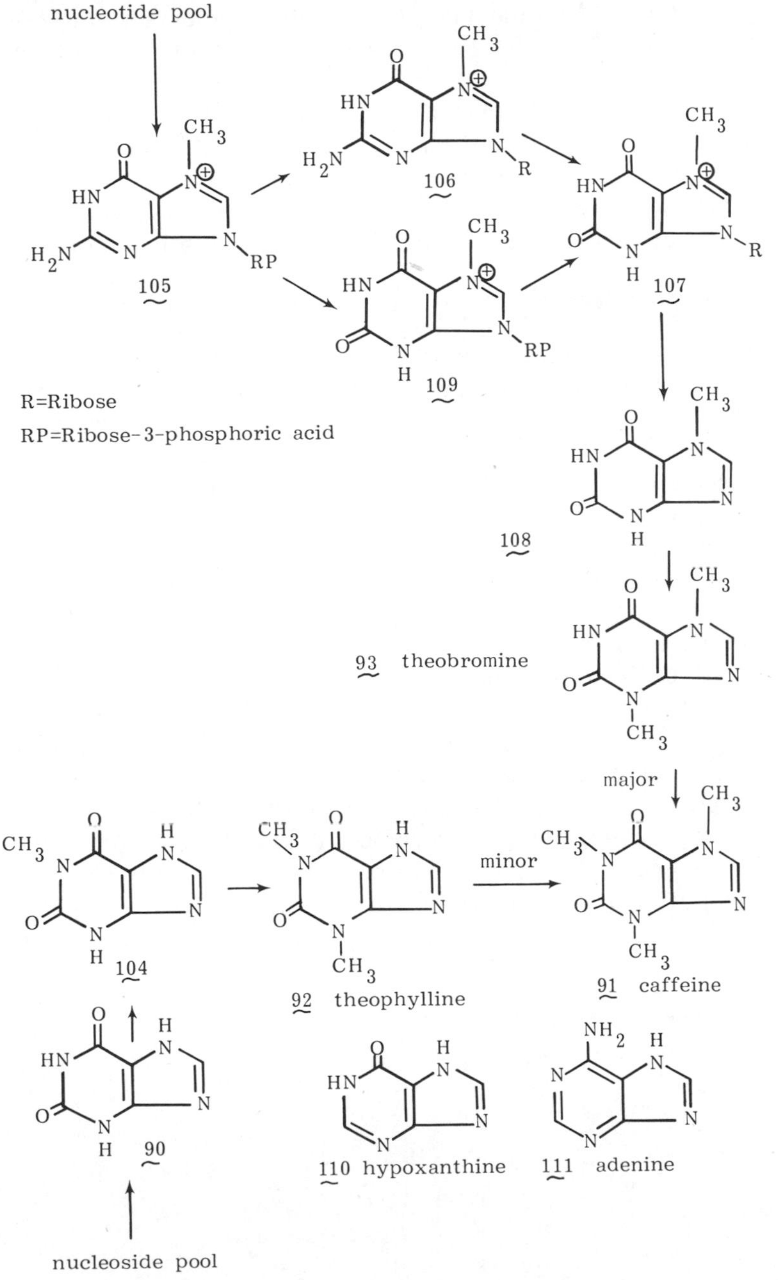

R=Ribose

RP=Ribose-3-phosphoric acid

a more effective precursor of caffeine than xanthine (**90**) in excised tea tips. Although each precursor was also catabolized by the same pathways as observed in animals to uric acid and eventually urea. It was also established that theobromine (**93**) is probably a precursor of caffeine.

Later, the same workers also showed that methionine was rapidly metabolized into an unknown compound, theobromine (**93**) and caffeine (**91**) in tea shoot tips. [8-^{14}C]Adenine (**111**) was also a precursor of caffeine, and this was suggested to be due to incorporation of radioactive purine nucleotides.

Thus theobromine (**93**) and caffeine (**91**) are derived from purine nucleotide catabolism by way of 7-methylxanthine (**108**) by an as yet unknown route. In contrast, the formation of theophylline (**92**) may involve the N-1 methylation of adenine in tRNA by way of 1-methylxanthine (**104**). Conversion of theophylline (**92**) to caffeine (**91**) is apparently a minor pathway.

12.7.3 Pharmacology

Xanthine derivatives have a number of pharmacological properties in common and five major pharmacologic actions are observed in varying degrees: (*a*) central nervous system and respiratory stimulation, (*b*) skeletal muscle stimulation, (*c*) diuresis, (*d*) cardiac stimulation, and (*e*) smooth muscle relaxation. For (*a*) and (*b*) the relative order of activity is caffeine > theobromine > theophylline, and for (*c*) and (*e*) the relative order is theophylline > theobromine > caffeine.

Caffeine increases central nervous system activity and its main effect is on the cerebral cortex, where it acts to produce clear thought and to reduce drowsiness and fatigue. The normal dose is 100–200 mg. It is a constituent of the A–P–C (aspirin–phenacetin–caffeine) products and is used in combination with ergot derivatives in some migraine preparations.

Theophylline is used as a smooth muscle, particularly a bronchial, relaxant in asthmatic preparations, where the usual dose is 200 mg. It is contraindicated in cases of cardiac edema or angina pectoris.

Theobromine as the ethylenediamine salt is used in preference to caffeine in cardiac edema and in angina pectoris. The usual dose is 300–500 mg.

Synthetic 7- and 8-substituted derivatives display a variety of interesting pharmacologic activities: for example, **112** shows spasmolytic activity, **113** shows sedative action, and **114** shows hypotensive activity in cats.

112   R_1=H, R_2=CH$_2$CH$_2$NR$_2$
113   R_1=N(CH$_2$CH$_2$OR)$_2$,R_2=H
114   R_1=CH$_2$NH$_2$, R_2=CH$_3$

LITERATURE

Reviews

Heim, F., and P. T. Hermann (Ed.), *Caffein und andere Methylxanthine*, Schattauer, Stuttgart, 1969.

Lister, J. H., in D. J. Brown (Ed.), *Heterocyclic Compounds—Fused Pyrimidines*, Vol. 2, Wiley, New York, 1971, p. 253.

Quimme, P., *Coffee and Tea*, New American Library, New York, 1976, pp. 242.

Snieckus, V. A., *Alkaloids, London* **2**, 273 (1972).

Snieckus, V. A., *Alkaloids, London* **3**, 303 (1973).

Snieckus, V. A., *Alkaloids, London* **4**, 398 (1974).

Snieckus, V. A., *Alkaloids, London* **5**, 269 (1975).

Reed, J. N., and V. A. Snieckus, *Alkaloids, London* **7**, 301 (1977).

Synthesis

Yanada, F., K. Ogiwara, M. Kanahori, and S. Nishigaki, *Chem. Commun.* 1068 (1970).

Lee, T. C., G. Stroehier, M. N. Teller, A. Myles, and G. B. Brown, *Biochemistry* **10**, 4463 (1971).

Corse, J., and J. Kuhnle, *Synthesis* 618 (1972).

Biosynthesis

Anderson, L., and M. Gibbs, *J. Biol. Chem.* **237**, 1941 (1962).

Preusser, E., and G. P. Serenkov, *Biokhimiya* **28**, 857 (1963).

Looser, E., T. W. Baumann, and H. Wanner, *Phytochemistry* **13**, 2515 (1974).

Suzuki, T., and E. Takahashi, *Biochem. J.* **146**, 79 (1975).

Suzuki, T., and E. Takahashi, *Biochem. J.* **146**, 87 (1975).

Suzuki, T., and E. Takahashi, *Phytochemistry* **15**, 1235 (1976).

Suzuki, T., and E. Takahashi, *Biochem. J.* **160**, 171 (1976).

Suzuki, T., and E. Takahashi, *Biochem. J.* **160**, 181 (1976).

ALKALOID-DETECTING REAGENTS

Precipitation

Mayer's Reagent

Mercuric chloride	1.36 g
Potassium iodide	5.00 g
Water to 100 ml	

Wagner's Reagent

Iodine	1.3 g
Potassium iodide	2.0 g
Water to 100 ml	

Dragendorff's Reagent (Kraut Modification)

Bismuth nitrate	8.0 g
Nitric acid	20.0 ml
Potassium iodide	27.2 g
Water to 100 ml	

Reineckate Salt Solution

1 g of ammonium reineckate [$NH_4(Cr(NH_3)_2(SCN)_4$], H_2O, and 0.3 g of hydroxylamine hydrochloride in 100 ml of ethanol. Solution stored in a refrigerator.

Chromatography

Antimony(III) Chloride

25 g of antimony(III) chloride in 75 g of chloroform. Heat for 10 min at 100°C after spraying.

Ceric Ammonium Sulfate

1% solution of ceric ammonium sulfate in 85% phosphoric acid.

Ceric Sulfate–Sulfuric Acid (Sonnenschein Reagent)

0.1 g of ceric sulfate suspended in 4 ml of water; add 1 g of trichloroacetic acid; heat; add concentrated sulfuric acid until turbidity disappears.

Ceric Sulfate–Sulfuric Acid

A saturated solution of ceric sulfate in 65% sulfuric acid.

Van Urk Reagent (Modified Ehrlich Reagent)

1 g of 4-dimethylaminobenzaldehyde in 50 ml of 36% hydrochloric acid and 50 ml of ethanol.

Dragendorff's Reagent

Solution a: 16 g potassium iodide dissolved in 40 ml of water; solution b: 1.7 g of bismuth nitrate and 20 g of tartaric acid dissolved in 80 ml of water. Mix 1 : 1 (v/v) of solutions a and b. Store at 0°C. To make the spray reagent, add 5 ml of the stock solution to 10 g of tartaric acid in 50 ml of water.

Ferric Chloride

1–5% solution of ferric chloride in 0.5 N hydrochloric acid.

Ferric Chloride–Perchloric Acid (Salkowski)

1 ml of 0.5 M aqueous ferric chloride solution in 50 ml of 35% perchloric acid. Warm for 5 min at 60°C after spraying.

Iodoplatinate Reagent

3 ml of 10% hexachloroplatinic(IV) acid solution in 97 ml of water mixed with 100 ml of 6% potassium iodide solution in water. The reagent should be freshly prepared.

Ninhydrin Reagent

0.3 g of ninhydrin dissolved in 100 ml of n-butanol and 3 ml of acetic acid.

Sodium Nitroprusside–Ammonia Reagent

Spray with 1% aqueous sodium nitroprusside solution and then 10% ammonium hydroxide solution.

Vanillin–Phosphoric Acid Reagent

1 g of vanillin dissolved in 100 ml of 50% aqueous phosphoric acid.

SUBJECT INDEX

occurrence, 416
partial synthesis, 417
proton NMR spectrum, 418
reduction, 417
synthesis, 417
UV spectrum, 418
Liriodenine methiodide, reduction, 417
Lobelanidine, 146
Lobelanine, 146
  in biosynthesis of lobeline, 181
  synthesis, 146
Lobelia alkaloids, biosynthesis, 181
Lobeline, 138
  biosynthesis, 181-182
  chemistry, 146
  occurrence, 146
  pharmacology, 146
  synthesis, 146
Lobinaline, 145
  biosynthesis, 181
Loganiaceae:
  alkaloids from, 4
  monoterpene indole alkaloids from,
      574, 656, 662, 684, 697, 721,
      726
  Strychnos alkaloids from, 721, 726
  yohimbinoid derivatives from, 684
Loganic acid, in biosynthesis of, gentio-
      picroside, 854
Loganin:
  in biogenesis of monoterpene alkaloids,
      847
  biosynthesis, 65, 824-825
  in biosynthesis of:
    gentiopicroside, 854
    ipecoside, 814
    monoterpene indole alkaloids, 814
    quinine, 828
    skytanthine derivatives, 853
    vindoline, 825
  reaction with ammonia, 850
Longistrobine:
  occurrence, 840
  structure determination, 841
Lophocerine:
  biosynthesis, 328
  occurrence, 327
  oxidative coupling, 328
Lophophorine:
  biosynthesis, 324
  occurrence, 319
LSD, *see* Lysergic acid diethylamide
Luciduline:
  occurrence, 177

synthesis, 177
α-Lumicolchicine, formation, 523
β-Lumicolchicine, formation, 523
γ-Lumicolchicine, formation, 523
Lunaridine, 933
Lunarine:
  occurrence, 931
  X-ray crystallographic analysis, 932
Lupanine, 156
  biogenesis, 184
  biosynthesis, 185
  in biosynthesis of:
    baptifoline, 186
    cytisine, 186
    N-methyl cytisine, 186
    thermopsine, 186
Lupinic acid, 955
Lupinine, 138, 154
  biogenesis, 184
  biosynthesis, 185
  chemistry, 154-155
  mass spectrum, 165
  occurrence, 154
  spectral properties, 155
  synthesis, 154-156
Lyadine, 661
Lycoctonine, 873
  X-ray crystallographic analysis, 875, 878
Lycoctonine skeleton:
  biogenesis, 878
  partial synthesis, 878
Lycodine, 170
  biogenesis, 189
Lycodine skeleton, in lycopodine bio-
      synthesis, 190
Lycopodine, 3, 138, 170-176
  biogenesis, 189
  biosynthesis, 190
  carbon-13 NMR spectrum, 176
  Hofmann degradation, 171
  isolation, 3
  synthesis, 173-175
  transformation to annofoline, 171-172
Lycopodium alkaloids, 170
  biosynthesis, 189-191
  chemistry, 171
  occurrence, 170
  partial synthesis, 171-173
  synthesis, 173-175
Lycoramine, 534
  synthesis, 542
Lycorenine:
  biogenesis, 549
  biosynthesis, 549-550

ORGANISM INDEX

All binomials and genera appear in *italics*.